Selected Titles in This Series

Volume

12 **Ellis Kolchin:** Selected Works of Ellis Kolchin with Commentary (Hyman Bass, Alexandru Buium, and Phyllis J. Cassidy, Editors), 1999

11 **V. S. Varadarajan:** The Selected Works of V. S. Varadarajan, 1999

10 **Maurice Auslander:** Selected Works of Maurice Auslander, Parts 1 and 2 (Idun Reiten, Sverre O. Smalø, and Øyvind Solberg, Editors), 1999

9 **Lipman Bers:** Selected Works of Lipman Bers: Papers on Complex Analysis, Parts 1 and 2 (Irwin Kra and Bernard Maskit, Editors), 1998

8 **Walter E. Thirring:** Selected Papers of Walter E. Thirring with Commentaries, 1998

7 **Robert Steinberg:** Robert Steinberg Collected Papers, 1997

6 **Julia Robinson:** The Collected Works of Julia Robinson (Solomon Feferman, Editor), 1996

5 **Freeman Dyson:** Selected Papers of Freeman Dyson with Commentary, 1996

4 **Witold Hurewicz:** Collected Works of Witold Hurewicz (Krystyna Kuperberg, Editor), 1995

3.2 **A. Adrian Albert:** A. Adrian Albert Collected Mathematical Papers: Nonassociative Algebras and Miscellany, Part 2 (Richard E. Block, Nathan Jacobson, J. Marshall Osborn, David J. Saltman, and Daniel Zelinsky, Editors), 1993

3.1 **A. Adrian Albert:** A. Adrian Albert Collected Mathematical Papers: Associative Algebras and Riemann Matrices, Part 1 (Richard E. Block, Nathan Jacobson, J. Marshall Osborn, David J. Saltman, and Daniel Zelinsky, Editors), 1993

2 **Salomon Bochner:** Collected Papers of Salomon Bochner, Parts 1–4 (Robert C. Gunning, Editor), 1992

1 **R. H. Bing:** The Collected Papers of R. H. Bing, Parts 1 and 2 (Sukhjit Singh, Steve Armentrout, and Robert J. Daverman, Editors), 1988

SELECTED WORKS OF

ELLIS KOLCHIN

WITH COMMENTARY

Ellis Kolchin

April 18, 1916–October 30, 1991

SELECTED WORKS OF

ELLIS KOLCHIN

WITH COMMENTARY

Hyman Bass
Alexandru Buium
Phyllis J. Cassidy
Editors

American Mathematical Society
Providence, Rhode Island

1991 *Mathematics Subject Classification.* Primary 00B60, 12H05;
Secondary 03C60, 11D99, 12F12, 12H20, 12Y05, 13A35, 13N10, 14E20, 14G20.

Library of Congress Cataloging-in-Publication Data

Kolchin, E. R. (Ellis Robert), 1916–
[Selections. 1999]
The selected works of Ellis Kolchin with commentary / Hyman Bass, Alexandru Buium, Phyllis J. Cassidy, editors.
p. cm.
Includes bibliographical references.
ISBN 0-8218-0542-8
1. Differential algebra. I. Bass, Hyman, 1932– . II. Buium, Alexandru, 1955– . III. Cassidy, Phyllis J. (Phyllis Joan), 1934– . IV. Title.
QA247.4.K65 1999
512′.56—dc21 98-49445
CIP

∞ The paper used in this book is acid-free and falls within the guidelines
established to ensure permanence and durability.
Visit the AMS home page at URL: `http://www.ams.org/`

10 9 8 7 6 5 4 3 2 1 04 03 02 01 00 99

Contents

Part I. The Papers of Ellis Kolchin

Books by Ellis Kolchin (not appearing in volume)

Part II. Commentary

Preface

Differential algebra, in its contemporary form, is largely the creation of Ellis Kolchin, and of his mentor, Joseph Fels Ritt. This volume assembles Kolchin's mathematical papers, which, together with his two books, archive the construction of modern differential algebra. In addition, the editors commissioned for this volume several commentaries on the history of differential algebra, and on Kolchin's work and mathematical legacy.

Ellis Robert Kolchin was born in New York City, on April 18, 1916. He studied mathematics with J. F. Ritt at Columbia University, and earned the doctorate in 1941. During World War II, he served with Naval Intelligence in Washington and the South Pacific. Kolchin accepted a position at Columbia University in 1946, and was named Adrain Professor of Mathematics in 1976. He was twice a Guggenheim Fellow, in 1954–55, at the Sorbonne, and in 1961–62, at Columbia. In 1960–1961, he was a National Science Foundation Fellow at the Université de Paris. He was also a visiting professor at the Insitute for Advanced Study in Princeton, the Tata Institute in Bombay, and the Kyoto Mathematics Institute. Fluent in Russian, he fostered contact with Soviet mathematicians, and in 1965, he was an exchange lecturer in the Soviet Union under the joint sponsorship of the U. S. and Soviet Academies of Science. In August 1975, he was Colloquium Lecturer at the American Mathematical Society Summer meeting. He was a fellow of the American Academy of Arts and Sciences and the American Association for the Advancement of Science. He retired from his position at Columbia in 1986, but continued his research in mathematics until the end of his life on October 30, 1991. Until the summer before his death, he participated actively in the weekly seminar he founded twenty-five years earlier. The all day meetings of the long-standing Kolchin seminar, interrupted only by lunch at the Moon Palace, were famous for their intensity and the free exchange of ideas and criticism. The members of the circle that formed around Kolchin were energized by his mathematical power and integrity, by his generosity in sharing his techniques and insights, by his kindness, and in general by the values he personified.

In 1930, Joseph Fels Ritt created a theory of algebraic differential equations modeled on the new algebra developed by Emmy Noether and van der Waerden, rather than on the transcendental methods of Lie. This theory, later named "differential algebra" by Kolchin, has its roots in the largely heuristic work by Lagrange, Laplace, and Poisson, and owes its origin to Ritt's desire to formulate precisely "general solution" and "singular solution." In his elimination theory for algebraic differential equations, Ritt devised powerful algorithms for deciding ideal membership and decomposition, based on the concepts of rankings and characteristic sets, analogues in differential algebra of term ordering and Gröbner bases in algebra.

Throughout Ritt's work ran the twin threads of complex function theory and constructive symbolic computation. Kolchin's first major achievement upon his return to Morningside Heights from the South Pacific was to open up Ritt theory to modern algebraic geometry, and in so doing to lead it in a new direction. Building on Ritt's foundation, but deeply influenced by Weil and Chevalley (who was then at Columbia), Kolchin wrote his fundamental papers on the Galois theory of differential fields. This is the focus of the commentaries in this volume by Armand Borel and Michael Singer. Borel's article, delivered as the First Kolchin Memorial Lecture at Columbia, discusses the differential Galois theory in the context of Kolchin's enduring interest in the theory of algebraic groups in all characteristics. This global, intensely algebraic approach stemmed from the desire he expressed in his 1948 paper on the Picard-Vessiot theory to remove the Galois groups of linear differential equations from Lie theory. Later, Kolchin's algebraic approach was to be modified by an intensification of the computer algebraic approach, by the Tannakian theory of Deligne and Katz, and by the analytic (asymptotic) approach of Martinet and Ramis. In his essay, Singer discusses these developments and surveys recent work on the calculation of differential Galois groups and the construction of differential equations with given algebraic groups as their Galois groups (the so-called inverse problem).

The remarkable intertwined development of differential algebra and model theory, a branch of logic, is the subject of Bruno Poizat's paper, reprinted from *Mathematica Japonica*. The significance of Kolchin's work for Robinson was centered on the concept of *differentially closed field*, the avatar in differential algebra of the properties that characterize algebraically closed fields in commutative algebra.

By changing the focus in differential algebra from algebraic differential equations to abstract algebraic structures equipped with a finite set of commuting derivation operators, Kolchin revealed the unifying principles of differential algebra. This is especially evident in his vast monograph, *Differential Algebra and Algebraic Groups*, where he made the important decision that, as he put it in the preface, 1 is a special case of m. Ritt's work concentrated on ordinary differential equations. By placing ordinary and partial differential equations in the same conceptual framework, and refusing to use special methods for the case $m = 1$, Kolchin at the same time blurs the distinction between finite and infinite dimensionality. The Ritt-Kolchin algorithms provide finitary methods even in the infinite-dimensional case.

Both the earliest and the latest phases of Ellis Kolchin's mathematical work are centered on the theory of groups. We alluded earlier to his interest in algebraic groups, which culminated in the strikingly original axiomatic definition in *Differential Algebra and Algebraic Groups*. In 1985, he published a second book, *Differential Algebraic Groups*, in which he deepens his discussion of algebraic structures with derivation operators. Recently, Kolchin's work on differential algebraic geometry and differential algebraic groups has been combined with the deformation theory of algebraic varieties to answer questions in Diophantine geometry. By commenting on Kolchin's papers and books, the origins of his thought in the work of Ritt, important developments in differential algebra, and recent work on differential algebraic groups and Diophantine geometry, the article by Buium and Cassidy attempts to explore the evolution of differential algebra from algebraic differential equations to problems in arithmetic.

We would like to express our thanks to Serge Lang, and to Catherine Chevalley, on behalf of Claude Chevalley, who have generously given their permission to publish here papers co-authored with Kolchin. We also, with equal gratitude, would like to express our appreciation to Kate Kolchin (Mrs. Ellis Kolchin) for granting us permission to publish the mimeographed lecture notes and the unpublished manuscript, and to the following journals and publishers for copyright permission: Advances in Mathematics (Academic Press), Annals of Mathematics (Princeton University Press), American Journal of Mathematics (Johns Hopkins University Press), Topology (Pergamon Press), The Proceedings of the International Congress of Mathematics, Moscow-1966 (Mir), Mathematica Japonica (Japanese Association of Mathematical Sciences), Pacific Journal of Mathematics (University of California), and Springer-Verlag. For her tireless and endlessly patient assistance with all the technical details involved in turning this collection of papers into a published book, our sincere thanks go to our editor, Christine M. Thivierge, at the American Mathematical Society.

In closing, we wish to acknowledge, with warmest gratitude, our debt to Kate Kolchin for her unflagging encouragement of our editorial efforts and for her warm and gracious hospitality. She opened her home to us and assisted us generously in the collection of Ellis Kolchin's papers.

Hyman Bass
Alexandru Buium
Phyllis J. Cassidy

Part I

The Papers of Ellis Kolchin

ON CERTAIN IDEALS OF DIFFERENTIAL POLYNOMIALS*

J. F. RITT AND E. R. KOLCHIN

Introduction. Let Σ be an ideal of differential polynomials in the unknowns $y_1, \cdots, y_n$. If the manifold of Σ is composed of s manifolds $\mathfrak{M}_1, \cdots, \mathfrak{M}_s$ not necessarily irreducible, none of which has a solution in common with any other, Σ has a unique representation as the product of s ideals $\Sigma_1, \cdots, \Sigma_s$ whose manifolds are, respectively, the $\mathfrak{M}_i$.†

Most of the present note is concerned with decompositions of the foregoing type and considers the case in which one of the $\mathfrak{M}_i$, say $\mathfrak{M}_1$, is composed of a single solution, that is, of a set of functions $\bar{y}_1, \cdots, \bar{y}_n$ contained in the underlying field. We shall examine, for this special case, the structure of the ideal Σ_1. Details will be given only for the case of a single unknown; the extensions to several unknowns are too obvious to require explicit mention. It will suffice, furthermore, to treat the case in which $\mathfrak{M}_1$ is composed of the solution $y=0$.

In §9, we consider a problem closely related to the theorem of decomposition stated above.

1. On the structure of Σ_1. Let Σ be an ideal of forms in the unknown y. Let $y=0$ be an essential irreducible manifold for Σ. Let Σ be the product of Σ_1 and Σ_2 where Σ_1 has $y=0$ as its manifold and Σ_2 does not admit $y=0$ as a solution. Let p be a positive integer such that y^p is contained in Σ_1.

* Presented to the Society, September 8, 1939.

† Proceedings of the National Academy of Sciences, vol. 25 (1939), p. 90. *Product* is defined in the expected way. That the intersection of the Σ_i is identical with their product follows immediately from the fact that the Σ_i, considered as *algebraic* ideals, are *paarweise teilerfremd*. See van der Waerden, *Moderne Algebra*, vol. 2, p. 46.

We shall prove that Σ_1 *is the ideal generated by* Σ *and* y^p, that is, the intersection of all ideals containing Σ and y^p.

PROOF. Obviously Σ_1 contains the ideal generated as above. What has to be proved is the converse of this fact. Let G be any form in Σ_1. Σ_2 contains a form $1-H$, where H vanishes for $y=0$. Then Σ contains $G(1-H)$ and, therefore, $G(1-H^q)$ where q is any positive integer. Now, if q is large,

$$H^q \equiv 0, \ (y^p). \tag{1}$$

Let q be fixed at a value large enough for (1) to hold, and let $M=G(1-H^q)$. Then $G\equiv M$, (y^p), and this establishes the theorem.

2. **Condition for p to be unity.** We are going to examine now the case in which Σ contains a form of the type $y+A$, where A, considered as a polynomial in the y_j, has no term of degree less than 2. Ideals of this type form a natural and interesting class; a very special example is the ideal generated by y_1^2-4y. We are going to prove that the p of §1 may be taken as unity. That is, Σ_1 *consists of all forms which vanish for* $y=0$.

What we have to prove, of course, is that Σ_1 contains y. Our procedure will be as follows. For some p, y^p is in Σ_1. Then, for any i, $y_i^{2^i p}$ is in Σ_1. Let $F=y+A$ be the form mentioned in the statement of our theorem. Then

$$y \equiv -A, \ (F). \tag{2}$$

We shall subject (2) to an iterative process and derive a relation $y\equiv K$, (F), where every term of K contains some $y_i^{2^i p}$ as a factor, i depending on the term. This will establish the theorem.

3. **Bound on degrees.** Let a form P in y be of degree g in some y_j, $j\geqq 0$. We shall show that P', the derivative of P, has a degree in y_j which does not exceed $g+1$. For let L, any term of P, be divisible by y_j^q with $q\leqq g$, and by no higher power of y. Let $L=y_j^qM$. We have, indicating first derivatives by an accent,

$$L' = qy_j^{q-1}y_{j+1}M + y_j^qM'.$$

M' consists of a set of terms, one of which will be divisible by the first power of y_j if M involves y_{j-1}. This is enough to prove our statement.

4. **The first substitution.** Let us suppose that, in addition to (2), we have a second relation $y\equiv B$, (F). In the second member of (2), let y be replaced by B and each y_j by the jth derivative of B. Then $-A$ goes over into a form C. It is easy to see that $y\equiv C$, (F).

Let r be a positive integer which is not less than the order of A in y. Let g be an integer, exceeding unity, such that each term of A is of total degree not less than g in $y, \cdots, y_r$. In A, we replace y_j by $-A^{(j)}$, $j=0, \cdots, r$, superscripts indicating differentiation.* Then $-A$ goes over into a form A_1 and $y \equiv A_1$, (F). Each term in A_1 is of total degree not less than g^2 in $y, \cdots, y_{2r}$. By §3, the $A^{(j)}, j \leqq r$, are of degree not greater than r in any one of the letters $y_{r+1}, \cdots, y_{2r}$.

Let L be a term in A_1, of total degree $d \geqq g^2$ in the y_j. The power product of degree d in L is the product of a set of power products taken from the $A^{(j)}$. If M is any of the latter power products, the total degree of M is at least g, hence at least g/r times the degree of M in any one of $y_{r+1}, \cdots, y_{2r}$. Thus, the degree of L in any one of $y_{r+1}, \cdots, y_{2r}$ is not more than $(rd)/g$.

5. **The second substitution.** Differentiating A_1, we consider the $A_1^{(j)}$ for $j=0, \cdots, r$. No $A_1^{(j)}$ is of degree exceeding r in any y_i with $2r < i \leqq 3r$. Let L, of some total degree d, be a term in some $A_1^{(j)}$. Then L, since it is derived from a term of total degree d in A_1, is of degree not exceeding $rdg^{-1}+r$ in any y_i with $r<i\leqq 2r$. As $d \geqq g^2$, we have $rdg^{-1}+r \leqq rd(g^{-1}+g^{-2})$.

In the second member of (2), we replace each y_j by $A_1^{(j)}$. We find a relation $y \equiv A_2$, (F), with each term of A_2 of total degree at least g^3. If L is a term in A_2, of some total degree d, the degree of L in any y_i with $2r<i\leqq 3r$ does not exceed rdg^{-2} and the degree of L in any y_i with $r<i\leqq 2r$ does not exceed $rd(g^{-1}+g^{-2})$.

6. **Continuation.** In the third step, we substitute the $A_2^{(j)}$ into (2). After t steps, we have $y \equiv A_t$, (F), with each term in A_t of total degree at least g^{t+1}. Let L be a term in A_t of some total degree d. Let j be any positive integer not greater than t. Then the degree of L in any y_i with $jr<i\leqq (j+1)r$ does not exceed

$$rd(g^{-j} + g^{-j-1} + \cdots + g^{-t}) < 2rdg^{-j}.$$

7. **Completion of proof.** Let t of §6 be the square of a positive integer s. The total degree of L of §6 in the y_i with $i > sr$ is no more than

$$2rd(rg^{-s} + \cdots + rg^{-t}) < 4r^2dg^{-s}.$$

Let s be so great that

$$4r^2g^{-s} < 1/2.$$

Then the total degree of L in the y_i with $i \leqq sr$ is at least $d/2$. Thus, for some particular y_i with $i \leqq sr$, the degree of L in y_i is at least

* $A^{(0)} = A$.

(3) $$\frac{d}{2(rs+1)} \geqq \frac{g^{s}+1}{2(rs+1)}.$$

We refer now to §2. If s is large, the second member of (3), if $i \leqq sr$, will exceed $2^{i}p$. This completes the proof of our theorem.

8. **Higher values of p.** It is not an unnatural conjecture that, if Σ contains a form $y^{n}+A$ with every term in A of degree greater than n, p of §1 may be taken as n. We give an example to show that the least p may exceed n.

Let Σ be the ideal generated by $F=y^3+y_1^4$. If Σ_1 contained y^3, there would exist a relation

(4) $$y^3(1-H) = MF + M_1F' + \cdots + M_rF^{(r)},$$

with H vanishing for $y=0$. For the second member of (4) to yield the term y^3 which the first member contains, it would be necessary for M to have unity as one of its terms. Then MF would have y_1^4 as a term. Equating terms of degree 4 and weight 4 for both sides of (4), we would find $y_1^4 \equiv 0$, (y^3), which is readily shown to be false.

9. **A generalization.** Let F and A be two forms in $y_1, \cdots, y_n$, both of class n and algebraically irreducible. Suppose that the general solution $\mathfrak{M}$ of A is contained in the manifold of F and is essential in that manifold. It is known how the essentiality of $\mathfrak{M}$ is reflected in the structure of F.* Suppose now that S^tF, where S^t is as in the indicated theorem of structure, has a term C_jA. It can be shown, by the method of §§2–7 above, that *there exists a relation* $AH\equiv 0$, (F), *where* H *does not hold* $\mathfrak{M}$.

Columbia University

* American Journal of Mathematics, vol. 60 (1938), p. 14.

ON THE BASIS THEOREM FOR INFINITE SYSTEMS OF DIFFERENTIAL POLYNOMIALS*

E. R. KOLCHIN

Introduction. Let $\mathcal{F}$ be a differential field of characteristic zero.† We consider an infinite system Σ of differential polynomials in the letters $y_1, \cdots, y_n$, the coefficients of the differential polynomials being in $\mathcal{F}$.‡

A finite set Φ of forms in Σ is called a *basis* of Σ if, for every form G in Σ, there is a positive integer p, dependent on G, such that G^p is in the differential ideal of Φ. If a single p will serve for every G in Σ, then we shall call the basis *strong*.

It has been shown that every system has a basis.§ Raudenbush has shown further,|| that there exist systems, not every basis of which is strong. It is now natural to ask whether or not every system of forms contains at least one strong basis.

We answer this question in the negative by showing that even a perfect differential ideal of forms may have no strong basis. The perfect differential ideal with which we work is the one generated by the form uv in the two unknowns u, v.

We employ several ideas used by Raudenbush in the second of his above mentioned papers.

1. **The assumption.** Consider a form¶ G every term of which is divisible by some u_iv_j.** Let Σ be the set of all such forms G. Then Σ is a differential ideal, and is perfect. For, if a form has a term free of, say, every u_i, then every power of the form will have such a term.

* Presented to the Society, February 25, 1939.

† For the definition of differential field, and other terms, see H. W. Raudenbush, *Ideal theory and differential equations*, Transactions of this Society, vol. 36 (1934), pp. 361–368.

‡ Throughout the rest of this paper we shall use, as is customary, the term *form* for *differential polynomial*.

§ For differential fields of meromorphic functions this was essentially shown by J. F. Ritt in his book *Differential Equations from the Algebraic Standpoint*, American Mathematical Society Colloquium Publications, vol. 14, New York, 1932. See especially §§ 7, 77. Following the work of Ritt, Raudenbush treated the case of the general differential field of characteristic zero by purely algebraic methods. See Raudenbush, loc. cit.

|| *On the analog for differential equations of the Hilbert-Netto theorem*, this Bulletin, vol. 42 (1936), pp. 371–373.

¶ For $\mathcal{F}$ we can use any differential field of characteristic zero.

** Subscripts denote derivatives.

It is easy to see that Σ is the perfect differential ideal generated by uv.

If Σ has a strong basis, then it has one consisting purely of forms u_iv_j.

Let

$$u_iv_j, \qquad i+j \leqq s, \tag{1}$$

be a strong basis for Σ, and let p be the associated positive integer. We work toward a contradiction.

We denote by α a positive integer to be fixed later.

Consider the set of all forms

$$u_{i_1}v_{j_1}\cdots u_{i_p}v_{j_p}, \qquad i_1+j_1+\cdots+i_p+j_p=\alpha. \tag{2}$$

Every such form has an expression $\sum_{\varrho=1}^{r} c_\varrho(\sum_{h=1}^{p} a_{\varrho h}u_{i_h}v_{j_h})^p$, where r is some positive integer, and the c_ϱ and the $a_{\varrho h}$ are rational numbers.* Therefore, by our assumption on the nature of the basis (1) and the integer p, every form (2) is in the differential ideal generated by the forms (1).

Hence each form (2) is a linear combination, with coefficients in F, of forms

$$\begin{gathered}(u_iv_j)_k u_{i_1}v_{j_1}\cdots u_{i_{p-1}}v_{j_{p-1}},\\ i+j\leqq s,\ i+j+k+i_1+j_1+\cdots+i_{p-1}+j_{p-1}=\alpha.\end{gathered} \tag{3}$$

Since the forms (2) are all linearly independent over $\mathcal{J}$, it follows that the number of distinct forms (2) cannot exceed the number of distinct forms (3).

We denote the number of distinct forms (2) by $R_{p,\alpha}$, and the number of distinct forms (3) by $Q_{p,\alpha}$. We thus have $R_{p,\alpha}\leqq Q_{p,\alpha}$.

In the next section we force the contradiction that $R_{p,\alpha}>Q_{p,\alpha}$ for α sufficiently large.

2. **The contradiction.** We consider those expressions (2) for which $i_1+j_1=\nu$, $(0\leqq\nu\leqq\alpha)$. The coefficient of $u_{i_1}v_{j_1}$ in (2) is then

$$u_{i_2}v_{j_2}\cdots u_{i_p}v_{j_p}, \qquad i_2+j_2+\cdots+i_p+j_p=\alpha-\nu. \tag{4}$$

The number of distinct forms (4) is $R_{p-1,\alpha-\nu}$, and therefore the number of distinct symbols† (4) is not less than $R_{p-1,\alpha-\nu}$. Since the number of expressions $u_{i_1}v_{j_1}$ with $i_1+j_1=\nu$ is $\nu+1$, the total number of sym-

* We can solve the equations $(w_1+\lambda w_2)^t=\sum_{i=0}^{t} C_{t,i}\lambda^i w_1^{t-i}w_2^i$, $(\lambda=1,\cdots,t+1)$, for $w_1w_2^{t-1}$, obtaining $w_1w_2^{t-1}=\sum_{\lambda=1}^{t+1} d_\lambda(w_1+\lambda w_2)^t$. Using this special case, we can show by induction that $w_1\cdots w_p=\sum_{\varrho=1}^{r} c_\varrho(\sum_{h=1}^{p} a_{\varrho h}w_h)^p$. Setting $w_h=u_{i_h}v_{j_h}$, we obtain the desired representation of the forms (2).

† Two distinct symbols (4) may represent the same form.

bols (2) is not less than $\sum_{\nu=0}^{\alpha}(\nu+1)R_{p-1,\alpha-\nu}$. But not more than $(p!)^2$ distinct symbols (2) can represent the same form. Hence

$$R_{p,\alpha} \geqq (p!)^{-2}\sum_{\nu=0}^{\alpha}(\nu+1)R_{p-1,\alpha-\nu}. \tag{5}$$

We now show that there exist positive numbers b_p, $(p=1, 2, \cdots)$, independent of α, such that

$$R_{p,\alpha} \geqq b_p(\alpha+1)^{2p-1}. \tag{6}$$

Obviously $R_{1,\alpha}=\alpha+1$, so that (6) holds for $p=1$. Suppose (6) holds for $p=m-1$. Then, by (5), using $[x]$ to denote the greatest integer not exceeding x, we have

$$\begin{aligned}
R_{m,\alpha} &\geqq (m!)^{-2}\sum_{\nu=0}^{\alpha}(\nu+1)b_{m-1}(\alpha-\nu+1)^{2m-3}\\
&\geqq (m!)^{-2}b_{m-1}\sum_{\nu=[\alpha/4]}^{[3\alpha/4]}(\nu+1)(\alpha-\nu+1)^{2m-3}\\
&\geqq (m!)^{-2}b_{m-1}\sum_{\nu=[\alpha/4]}^{[3\alpha/4]}([\alpha/4]+1)(\alpha/4+1)^{2m-3}\\
&\geqq (m!)^{-2}b_{m-1}(2[\alpha/4]+1)([\alpha/4]+1)(\alpha/4+1)^{2m-3}\\
&\geqq b_m(\alpha+1)^{2m-1},
\end{aligned}$$

where $b_m=(m!)^{-2}4^{-2m}b_{m-1}$. Thus (6) holds for all p.

We now consider those expressions (3) for which $i+j+k=\mu$, $(0\leqq\mu\leqq\alpha)$. The number of distinct expressions $(u_iv_j)_k$ with $i+j+k=\mu$ and with $i+j\leqq s$ does not exceed $(s+1)^2$. The coefficient of $(u_iv_j)_k$ in (3) is $u_{i_1}v_{j_1}\cdots u_{i_{p-1}}v_{j_{p-1}}$, $(i_1+j_1+\cdots+i_{p-1}+j_{p-1}=\alpha-\mu)$. Since the number of distinct forms of this kind is $R_{p-1,\alpha-\mu}$, we have for the total number of distinct forms (3):

$$Q_{p,\alpha} \leqq \sum_{\mu=0}^{\alpha}(s+1)^2R_{p-1,\alpha-\mu}. \tag{7}$$

We shall show that, for $p=1, 2, \cdots$,

$$R_{p,\alpha} \leqq (\alpha+1)^{2p-1}. \tag{8}$$

For since $R_{1,\alpha}=\alpha+1$, (8) holds for $p=1$. Suppose (8) holds for $p=m-1$. Looking at (2), it is easy to see that

$$R_{m,\alpha} \leqq \sum_{\nu=0}^{\alpha}(\nu+1)R_{m-1,\alpha-\nu}.$$

Therefore

$$R_{m,\alpha} \leqq \sum_{\nu=0}^{\alpha} (\nu + 1)(\alpha - \nu + 1)^{2m-3}$$

$$\leqq \sum_{\nu=0}^{\alpha} (\alpha + 1)(\alpha + 1)^{2m-3} = (\alpha + 1)^{2m-1}.$$

Thus (8) holds for all p.

Using (8) in (7), we find

$$Q_{p,\alpha} \leqq (s + 1)^2 \sum_{\mu=0}^{\alpha} (\alpha - \mu + 1)^{2p-3},$$

so that

$$Q_{p,\alpha} \leqq (s + 1)^2(\alpha + 1)^{2p-2}.$$

Comparing this with (6), we see that, for α sufficiently large, $R_{p,\alpha} > Q_{p,\alpha}$.

Columbia University

ANNALS OF MATHEMATICS
Vol. 42, No. 3, July, 1941

ON THE EXPONENTS OF DIFFERENTIAL IDEALS

BY ELLIS ROBERT KOLCHIN

(Received April 4, 1940)

INTRODUCTION

In the theory of polynomial ideals, in algebra, there are methods, stemming from the theorem of M. Noether, and associated with the names of E. Bertini, E. Lasker, F. S. Macauley, K. Hentzelt, H. Kapferer and P. Dubreil, for finding the exponent of an ideal, or at least a bound for the exponent.

When one seeks to create a notion of exponent for ideals of differential polynomials, one is forced, because of a situation revealed by H. W. Raudenbush,[1] to admit infinite exponents as well as finite ones. The investigation of such exponents, finite or infinite, to some extent for differential ideals of a general character, and to a deeper extent for differential ideals generated by a form in one unknown of the first order, is the object of the present paper.

If we may refer to §2 below for a definition of the *exponent of a differential ideal,* we shall proceed to enumerate our results.

Part II, which presents what is possibly the most interesting portion of our work, deals with differential ideals generated by a single form A, in one unknown, of the first order. The concept of *multiplicity* of a singular solution of A is introduced (§6) and it is shown (§7) that *if A has a singular solution of multiplicity exceeding unity, then the differential ideal generated by A has exponent infinity.* Forms A which have singular solutions, all of multiplicity unity, are discussed in §§8–10. Such singular solutions are divided into two classes, and, guided by a general theorem due to J. F. Ritt, we secure a decomposition of the differential ideal generated by A which puts these two classes into evidence (§8). In §10 it is proved, under an additional assumption (*regular type*) that *the exponent of the differential ideal generated by A is unity or two according as all the singular solutions are in the first class or at least one singular solution is in the second class.* The differential ideal generated by a form A which has *no* singular solutions, is shown, under a certain additional assumption, to have exponent *unity* (§11). Part II concludes with a discussion of a type of form which we call *hyperelliptic.* The "intermediate ideals" are found, and they are shown to fall into *chains* of a fixed length.

Part III contains a brief discussion of chains of differential ideals. A theorem is proved which gives a bound for the exponent of a so-called *principal* chain in terms of the length of the chain (§14).

Part I begins with a statement, in abstract form, of a decomposition theorem due to Ritt. After the definitions of *relative exponent* and other terms, there

[1] Bulletin of the American Mathematical Society, vol. 42 (1936), pp. 371–373.

is proved a theorem connecting the relative exponent of a differential ideal with those of the factor ideals in the decomposition just mentioned (Theorem 1). Also, a result is presented which relates the existence of a strong basis in a differential ideal to the existence of such a basis in the factor ideals (Theorem 2). The remainder of Part I deals with special questions whose treatment is important for the sequel.

In Part I, use is made of the theory of resolvents. This theory, as developed by Ritt, applies to a differential field whose elements are meromorphic functions. To permit the use of abstract fields, there is given, in an appendix, a discussion which supplements Ritt's proofs, making them valid in any differential field of characteristic zero.

Notation

Differentiation (of elements in a differential ring[2]) is indicated by subscripts; thus a_j means the j-th derivative of a. If a single letter is used with different subscripts to denote different ring elements, then differentiation is indicated by a second subscript; thus, the j-th derivative of a_i is a_{ij}.

Square brackets [ ], curled brackets { }, and parentheses ( ), when not used as symbols of aggregation, will mean, respectively, the differential ideal, the perfect differential ideal, and the (ordinary, algebraic) ideal generated by the set of elements they include. These various ideals are supposed to be formed in a fixed differential ring which underlies the discussion. When forming an (algebraic) ideal, the differential ring is considered as an (algebraic) ring.

Membership in a set is denoted in the usual way by the symbol ϵ. Set inclusion is indicated by $\subseteq$, proper inclusion by $\subset$. The symbol for the intersection of sets is $\cap$.

The notations

$$a \equiv b \qquad (m, n, \cdots),$$

$$a \equiv b \qquad [m, n, \cdots],$$

$$a \equiv b \qquad \{m, n, \cdots\}$$

will mean, respectively,

$$a - b \,\epsilon\, (m, n, \cdots), \qquad a - b \,\epsilon\, [m, n, \cdots], \qquad a - b \,\epsilon\, \{m, n, \cdots\}.$$

The product of a finite number of differential ideals is defined as their product when considered as (algebraic) ideals. This is readily seen to be a differential ideal. If $\sigma_1, \sigma_2, \cdots, \sigma_s$ are differential ideals, their product is denoted by $\sigma_1\sigma_2 \cdots \sigma_s$.

If σ is a *perfect* differential ideal in a differential ring $\mathfrak{D}$, and if $d \,\epsilon\, \mathfrak{D}$, then

[2] The definitions of differential ring, differential ideal, and other terms, are to be found in a paper by Raudenbush, Transactions of the American Mathematical Society, vol. 36 (1934), pp. 361-368.

$\sigma:d$, the quotient of σ by d, is defined as in algebra to be the totality of elements $a \in \mathfrak{D}$ such that $ad \in \sigma$. $\sigma:d$ is itself a perfect differential ideal.

If $\mathfrak{F}$ is a differential field, then $\mathfrak{F}\{u, \cdots\}$ means the differential ring obtained by the differential ring adjunction of $u, \cdots$. The result of differential field adjunction will be denoted by $\mathfrak{F}\langle u, \cdots \rangle$.

As is customary, the word *form* will be used as an abbreviation for *differential polynomial*.

By a *solution* of a set of forms in $\mathfrak{F}\{y_1, \cdots, y_n\}$, is meant a set of elements $\eta_1, \cdots, \eta_n$ of some differential extension field of $\mathfrak{F}$, which annul the forms of the set when substituted for $y_1, \cdots, y_n$, respectively.

Part I. Some General Theorems

1. The product representation

Consider a differential domain of integrity $\mathfrak{D}$ which contains the rational numbers. We suppose that every infinite system in $\mathfrak{D}$ has a basis. The set of forms in a fixed finite number of unknowns, with coefficients in an underlying differential field of characteristic zero, is such a differential domain of integrity.[3]

A set of differential ideals in $\mathfrak{D}$ will be called *separated* if $1 \equiv 0 \quad (\sigma_1, \sigma_2)$ for each pair of distinct ideals σ_1, σ_2 in the set.

A differential ideal will be said to be *connected* if it is not the intersection of two separated differential ideals. Since the two ideals are required to be separated, "intersection" may be replaced by "product."[4]

A theorem due to Ritt may be formulated abstractly as follows:

Let σ be a differential ideal in $\mathfrak{D}$. Then σ is connected if and only if $\{\sigma\}$ is. If σ is not connected, it has a representation as the intersection (= product) of a separated finite set of connected differential ideals. This representation is unique. If $\sigma_1\sigma_2 \cdots \sigma_s$ is the representation of σ, then $\{\sigma_1\}\{\sigma_2\} \cdots \{\sigma_s\}$ is the representation for $\{\sigma\}$.[5]

We shall call the representation described in this theorem the *product representation* of σ. Each σ_i will be called a *factor* of the product representation.

2. Exponents

Lemma 1: *Let σ and τ be differential ideals in $\mathfrak{D}$ such that $\sigma \subseteq \tau \subseteq \{\sigma\}$. Then the product representation of σ has the same number of factors as that of τ, and, with a suitable assignment of subscripts, we may write for these representations*

$$(1) \qquad \sigma = \sigma_1\sigma_2 \cdots \sigma_s, \qquad \tau = \tau_1\tau_2 \cdots \tau_s,$$
$$\sigma_i \subseteq \tau_i \subseteq \{\sigma_i\}, \qquad i = 1, \cdots, s.$$

Proof: The theorem of §1 shows that we may write for the product representations

[3] See the paper cited under footnote 2.

[4] See B. L. van der Waerden, *Moderne Algebra*, vol. 2, Berlin, 1931, p. 46.

[5] Ritt, Proceedings of the National Academy of Sciences, vol. 25 (1939), pp. 90–91.

$$\sigma = \sigma_1\sigma_2 \cdots \sigma_s, \qquad \tau = \tau_1\tau_2 \cdots \tau_s,$$
$$\{\sigma_i\} = \{\tau_i\}, \qquad i = 1, \cdots, s.$$

Here product is the same as intersection, and $\sigma \subseteq \tau \subseteq \tau_1$, so that

$$\sigma = (\sigma_1 \cap \tau_1) \cap \sigma_2 \cap \cdots \cap \sigma_s = (\sigma_1 \cap \tau_1)\sigma_2 \cdots \sigma_s.^{6}$$

Since the product representation is unique, this implies that $\sigma_1 = \sigma_1 \cap \tau_1$, that is $\sigma_1 \subseteq \tau_1$. Similarly, every $\sigma_i \subseteq \tau_i$.

Consider two differential ideals σ and τ in $\mathfrak{D}$. If there is a positive integer q such that $\tau^q \subseteq \sigma$, let p be the least such q; if no such q exists, let $p = \infty$. We shall call p the *exponent of σ with respect to τ*, and shall denote it by $l_\tau\sigma$. The exponent of σ with respect to $\{\sigma\}$ will be called, simply, the *exponent of σ*.

In discussing $l_\tau\sigma$ we limit ourselves to cases for which $\sigma \subseteq \tau \subseteq \{\sigma\}$. If $\tau \not\subseteq \{\sigma\}$, then $l_\tau\sigma = \infty$; if $\sigma \not\subseteq \tau$, we may let $\tau' = (\sigma, \tau)$, and then $\sigma \subseteq \tau'$, $l_\tau\sigma = l_{\tau'}\sigma$.

THEOREM 1: *Let σ and τ be differential ideals in $\mathfrak{D}$ such that $\sigma \subseteq \tau \subseteq \{\sigma\}$. Let the product representations of σ and τ be given by* (1). *Then*

$$l_\tau\sigma = \max_{i=1,\cdots,s} l_{\tau_i}\sigma_i.$$

PROOF: Let $p = \max l_{\tau_i}\sigma_i$, $q = l_\tau\sigma$. If $p = \infty$, then $q \leqq p$. Suppose $p < \infty$. Then

$$\tau^p = (\tau_1 \cdots \tau_s)^p = \tau_1^p \cdots \tau_s^p \subseteq \sigma_1 \cdots \sigma_s = \sigma,$$

so that in either case $q \leqq p$. It remains to prove that $p \leqq q$.

If $q = \infty$, then $p \leqq q$. Suppose $q < \infty$. Then

$$\sigma_1^q \cdots \sigma_s^q = (\sigma_1 \cdots \sigma_s)^q = \sigma^q \subseteq \tau = \tau_1 \cdots \tau_s.$$

But $\sigma_1^q, \cdots, \sigma_s^q$ are obviously the factors of the product representation of $\sigma_1^q \cdots \sigma_s^q$. Hence, by Lemma 1, $\sigma_i^q \subseteq \tau_i$, $i = 1, \cdots, s$, that is, $p \leqq q$.

3. Strong bases

In this section we present a theorem which will not be used in the rest of the paper, but which may be of some interest on its own account.

Let σ be a differential ideal in $\mathfrak{D}$. If σ has a strong[7] basis $b_1, \cdots, b_h$, let $q = l_\sigma[b_1, \cdots, b_h]$; of all integers so obtained, let p be the minimum. If σ has no strong basis, let $p = \infty$. We shall call p the *basis index* of σ.

---

[6] It is easy to see that $\sigma_1 \cap \tau_1, \sigma_2, \cdots, \sigma_s$ are separated. For example, because $\sigma_1, \sigma_2, \cdots, \sigma_s$ are separated, there are elements $h_1 \in \sigma_1$ and $h_2 \in \sigma_2$ such that $1 = h_1 + h_2$. Choosing, as we may, p large enough so that $h_1^p \in \tau_1$, we have $h_1^p \in \sigma_1 \cap \tau_1$, $ph_1^{p-1}h_2 + \cdots + h_2^p \in \sigma_2$ and $1 = h_1^p + ph_1^{p-1}h_2 + \cdots + h_2^p$.

[7] A basis of σ is said to be *strong* if there is a single integer such that g^t is in the differential ideal generated by the basis, for every $g \in \sigma$. There exist differential ideals which do not have a strong basis. See Theorem 5 below; also, Kolchin, Bulletin of the American Mathematical Society, vol. 45 (1939), pp. 923–926.

THEOREM 2: *Let the differential ideal* σ *in* $\mathfrak{D}$ *have the product representation* $\sigma = \sigma_1 \cdots \sigma_s$. *Then the basis index of* σ *equals the maximum of the basis indices of the* σ_i.

PROOF: We know that there exist elements m_i, with derivative $m_{i1} \in \sigma$, such that $\sigma_i = (\sigma, m_i)$, $i = 1, \cdots, s$.[8] Let $b_1, \cdots, b_h$ be a strong basis of σ, with $l_\sigma[b_1, \cdots, b_h] = p$. Then

$$\sigma_i^p = (\sigma, m_i)^p \subseteq (\sigma^p, m_i) \subseteq [b_1, \cdots, b_h, m_i].$$

Thus, for each i, $b_1, \cdots, b_h, m_i$ is a strong basis for σ_i, and $l_{\sigma_i}[b_1, \cdots, b_h, m_i] \leqq p$. Hence the basis index of σ is not less than that of each σ_i. Now let $b_1^{(i)}, \cdots, b_{h_i}^{(i)}$ be a strong basis of σ_i, and let $q \geqq l_{\sigma_i}[b_1^{(i)}, \cdots, b_{h_i}^{(i)}]$, $i = 1, \cdots, s$. Since $\sigma_i = (\sigma, m_i)$, we may write, for each i and j,

$$b_j^{(i)} = t_j^{(i)} + c_j^{(i)} m_i, \qquad t_j^{(i)} \in \sigma.$$

Then

$$\begin{aligned}\sigma^q &= (\sigma_1 \cdots \sigma_s)^q = \sigma_1^q \cdots \sigma_s^q \\ &\subseteq [b_1^{(1)}, \cdots, b_{h_1}^{(1)}] \cdots [b_1^{(s)}, \cdots, b_{h_s}^{(s)}] \\ &\subseteq [t_1^{(1)}, t_2^{(1)}, \cdots, t_{h_s}^{(s)}, m_{11}, \cdots, m_{s1}, m_1 \cdots m_s].\end{aligned}$$

Thus, σ has a strong basis whose associated exponent is less than or equal to q, so that the basis index of σ is not more than the greatest basis index of the σ_i.

4. Invariance of the exponent under differential field adjunction

In this section we specialize $\mathfrak{D}$ to a differential ring of forms.

Let $\mathfrak{F}$ be a differential field of characteristic zero. Let $\mathfrak{F}'$ be a differential extension field of $\mathfrak{F}$: $\mathfrak{F} \subseteq \mathfrak{F}'$. Throughout the present section y_i, u_i, w will denote unknowns.

If Φ is a set of forms with coefficients in $\mathfrak{F}$, we shall use (Φ), $[\Phi]$, $\{\Phi\}$ to denote the algebraic, the differential, and the perfect differential ideals, respectively, generated by Φ in the differential ring of forms with coefficients in $\mathfrak{F}$. If Φ is a set of forms with coefficients in $\mathfrak{F}'$, the algebraic, the differential, and the perfect differential ideals generated by Φ in the differential ring of forms with coefficients in $\mathfrak{F}'$ will be indicated, respectively, by $(\Phi)'$, $[\Phi]'$, and $\{\Phi\}'$. It is clear that if Σ is a differential ideal of forms with coefficients in $\mathfrak{F}$, then $(\Sigma)' = [\Sigma]'$.

THEOREM 3: *Let* Σ *and* Λ *be differential ideals in* $\mathfrak{F}\{y_1, \cdots, y_n\}$. *Then*

$$l_\Lambda \Sigma = l_{(\Lambda)'}(\Sigma)'.$$

PROOF: Let $p = l_\Lambda\Sigma$, $q = l_{(\Lambda)'}(\Sigma)'$. If $q = \infty$, then $p \leqq q$. Suppose $q < \infty$, and let $G \in \Lambda$. Then $G \in (\Lambda)'$, so that $G^q \in (\Sigma)'$. But $G^q \in \mathfrak{F}\{y_1, \cdots, y_n\}$. Hence $G^q \in \Sigma$,[9] so that $p \leqq q$.

---

[8] Ritt, loc. cit. under footnote 5.

[9] See proof in van der Waerden, loc. cit., p. 67.

If $p = \infty$, then $q \leqq p$. Suppose $p < \infty$, and let $G \in (\Lambda)'$. Then there is a relation

$$G = \sum_{\mu=1}^{m} H_\mu G_\mu, \qquad G_\mu \in \Lambda.$$

Hence

$$G^p = \sum_{i_1+\cdots+i_m=p} \frac{p!}{i_1! \cdots i_m!} H_1^{i_1} \cdots H_m^{i_m} G_1^{i_1} \cdots G_m^{i_m} \in (\Lambda^p)' \subseteq (\Sigma)',$$

so that $q \leqq p$. This completes the proof.

Theorem 3, by itself, yields no information as to how the exponent of a differential ideal is affected by a differential field adjunction, for there is no obvious reason why $\{\Sigma\}'$ should equal $(\{\Sigma\})'$. These two ideals, however, are identical. We devote the rest of this section to the proof of this fact. We use three lemmas.

LEMMA 2: *Let Ω be a prime differential ideal in $\mathfrak{F}\{u_1, \cdots, u_q, w, y_1, \cdots, y_p\}$, with basic set*

$$(2) \qquad A, A_1, \cdots, A_p$$

introducing in succession $w, y_1, \cdots, y_p$. Let A be algebraically irreducible over $\mathfrak{F}$, and let each A_i be of order zero and degree unity in y_i. Then (2) *is a basic set for $\{\Omega\}'$.*

PROOF: Let $B_1, \cdots, B_s$ be the irreducible factors of A over $\mathfrak{F}'$. Let S and I be the separant and initial, respectively, of A; let I_i be the initial ($=$ separant) of A_i. Each B_i is of the same order (call it r) in w as A is, for otherwise A would be reducible over $\mathfrak{F}$. Let T_i be the separant, J_i the initial, of B_i. We have

$$I = J_1 \cdots J_s,$$

$$S = T_1 B_2 \cdots B_s + B_1 T_2 \cdots B_s + \cdots + B_1 B_2 \cdots T_s.$$

Now, $\Omega = \{A, A_1, \cdots, A_p\} : I S I_1 \cdots I_p$. Let

$$\Omega_i = \{B_i, A_1, \cdots, A_p\} : J_i T_i I_1 \cdots I_p, \qquad i = 1, \cdots, s.$$

Then each Ω_i is a prime differential ideal in $\mathfrak{F}\{u_1, \cdots, u_q, w, y_1, \cdots, y_p\}$.[10] Moreover,

[10] PROOF: Let $CD \in \Omega_i$. For appropriate k we may write $(I_1 \cdots I_p)^k C \equiv C'$, $(I_1 \cdots I_p)^k D \equiv D'$ $[A_1, \cdots, A_p]$ where C' and D' are forms free of $y_1, \cdots, y_p$. Of course $C'D' \in \Omega_i$. We prove one of C', D' is divisible by B_i. Suppose neither is. Then there are forms M and N, free of $y_1, \cdots, y_p$, such that $E = M(C'D'T_iJ_iI_1 \cdots I_p) + NB_i$, where E is a nonzero form free of $y_1, \cdots, y_p$, of order less than r in w. Clearly $E \in \{B_i, A_1, \cdots, A_p\}$. It readily follows that $(T_iI_1 \cdots I_p)^l E \in (B_i)$ for some positive integer l. Since B_i is irreducible, this implies that $E \in (B_i)$. But the order of E in w is less than r. Hence $E = 0$. This contradiction shows that either C' or D' is divisible by B_i, say C' is. Then $C \in \Omega_i$. Thus, Ω_i is prime.

$$\begin{aligned}\{\Omega\}' &= \{\{A, A_1, \cdots, A_p\}:ISI_1 \cdots I_p\}' \\ &= \{A, A_1, \cdots, A_p\}':ISI_1 \cdots I_p \\ &= (\{B_1, A_1, \cdots, A_p\}' \cap \cdots \cap \{B_s, A_1, \cdots, A_p\}'):ISI_1 \cdots I_p \\ &= (\{B_1, A_1, \cdots, A_p\}':ISI_1 \cdots I_p) \cap \cdots \\ &\qquad\qquad \cap (\{B_s, A_1, \cdots, A_p\}':ISI_1 \cdots I_p) \\ &= (\{B_1, A_1, \cdots, A_p\}':J_1 \cdots J_sT_1B_2 \cdots B_sI_1 \cdots I_p) \\ &\qquad \cap \cdots \cap (\{B_s, A_1, \cdots, A_p\}':J_1 \cdots J_sB_1B_2 \cdots T_sI_1 \cdots I_p) \\ &= (\Omega_1:J_2 \cdots J_sB_2 \cdots B_s) \cap \cdots \cap (\Omega_s:J_1 \cdots J_{s-1}B_1 \cdots B_{s-1}) \\ &= \Omega_1 \cap \cdots \cap \Omega_s.\end{aligned}$$ [11]

$\{\Omega\}'$ does not contain a form of class $q + 1$ of order less than r in w, because Ω_i does not. Let C be of class $q + 1$ and have order r in w, and suppose that $C \in \{\Omega\}'$. $C \in \Omega_i$, and is therefore divisible by B_i, $i = 1, \cdots, s$. Hence C is divisible by A. From this it follows that (2) is a basic set for $\{\Omega\}'$.

LEMMA 3: *Let $\mathfrak{F}$ contain non-constant elements.*[12] *Let Σ be a prime differential ideal in $\mathfrak{F}\{u_1, \cdots, u_q, y_1, \cdots, y_p\}$, with arbitrary unknowns $u_1, \cdots, u_q$.*[13] *Then $(\Sigma)' = \{\Sigma\}'$.*

PROOF: It suffices to show that $\{\Sigma\}' \subseteq (\Sigma)'$. Let $A = 0$ be a resolvent for Σ, and let Ω be the associated prime differential ideal in $\mathfrak{F}\{u_1, \cdots, u_q, w, y_1, \cdots, y_p\}$, with basic set (2), where A is of order r in w, and each A_i is of order zero and degree unity in y_i.[14] We know that

$$\Sigma = \Omega \cap \mathfrak{F}\{u_1, \cdots, u_q, y_1, \cdots, y_p\}.$$

Let $G \in \{\Sigma\}'$. We may write

$$G = G_0 + G_1\omega_1 + \cdots + G_m\omega_m$$

where each $G_\mu \in \mathfrak{F}\{u_1, \cdots, u_q, y_1, \cdots, y_p\}$, and where $1, \omega_1, \cdots, \omega_m$ are elements of $\mathfrak{F}'$, linearly independent over $\mathfrak{F}$. We shall show that each $G_\mu \in \Sigma$.

Let S and I be the separant and initial of A; let I_i be the initial of A_i. Evidently there exist non-negative integers a, b, a_j such that

$$I^aS^bI_1^{a_1} \cdots I_p^{a_p}G_\mu \equiv R_\mu(\Omega), \qquad \mu = 0, 1, \cdots, m, \tag{3}$$

where each R_μ is reduced with respect to the basic set (2). Then

$$I^aS^bI_1^{a_1} \cdots I_p^{a_p}G \equiv R_0 + R_1\omega_1 + \cdots + R_m\omega_m(\Omega)'. \tag{4}$$

But, by Lemma 2, (2) is a basic set for $\{\Omega\}'$. Hence the second member of (4),

[11] The parentheses in these equations are symbols of aggregation.

[12] An element of $\mathfrak{F}$ is called *constant*, if its derivative vanishes.

[13] The u_i need not actually occur.

[14] See Appendix.

which is in $\{\Omega\}'$, yet reduced with respect to (2), must vanish. Since $1, \omega_1, \cdots, \omega_m$ are linearly independent over $\mathfrak{F}$, it follows that each R_μ vanishes. (3) then shows that each $G_\mu \in \Omega$. Since G_μ is also in $\mathfrak{F}\{u_1, \cdots, u_q, y_1, \cdots, y_p\}$, we have $G_\mu \in \Sigma$, $\mu = 0, 1, \cdots, m$.

Hence $G \in (\Sigma)'$, so that $\{\Sigma\}' \subseteq (\Sigma)'$.

LEMMA 4: *Let $\mathfrak{F}$ consist purely of constants; let $\mathfrak{F}' = \mathfrak{F}\langle x\rangle$, where x is an element whose derivative is 1. Let Σ be a prime differential ideal in $\mathfrak{F}\{y_1, \cdots, y_n\}$. Then $(\Sigma)' = \{\Sigma\}'$.*

PROOF: We must show that $\{\Sigma\}' \subseteq (\Sigma)'$. Let $G \in \{\Sigma\}'$. For a suitable non-zero element $\varphi \in \mathfrak{F}\langle x\rangle$, we may write

$$\varphi G = H_0 + H_1 x + \cdots + H_m x^m,$$

where the H_μ are in $\mathfrak{F}\{y_1, \cdots, y_n\}$. For a large integer r, $G^r \in (\Sigma)'$. Hence

$$H_0^r + rH_0^{r-1}H_1 x + \cdots + H_m^r x^{mr} \in (\Sigma)'.$$

Thus, we see that we may write

$$H_0^r + rH_0^{r-1}H_1 x + \cdots + H_m^r x^{mr} = d_1 S_1 + \cdots + d_k S_k,$$

where each $S_i \in \Sigma$, and each $d_i \in \mathfrak{F}\langle x\rangle$. Let $\omega_1, \cdots, \omega_t$ be elements of $\mathfrak{F}\langle x\rangle$ such that $1, x, \cdots, x^{mr}, \omega_1, \cdots, \omega_t$ are linearly independent over $\mathfrak{F}$, and such that each d_i is linearly dependent on $1, x, \cdots, x^{mr}, \omega_1, \cdots, \omega_t$. Then there is a relation

$$H_0^r + rH_0^{r-1}H_1 x + \cdots + H_m^r x^{mr} = T_0 + \cdots + T_{mr} x^{mr} + U_1\omega_1 + \cdots + U_t\omega_t,$$

where the T_i and U_i are in Σ. By the linear independence it follows that $H_0^r = T_0$, so that $H_0 \in \Sigma$. Hence

$$H_1 + H_2 x + \cdots + H_m x^{m-1} \in \{\Sigma\}'.$$

Continuing, we find every $H_\mu \in \Sigma$. Therefore $G \in (\Sigma)'$ and $\{\Sigma\}' \subseteq (\Sigma)'$.

We are now in a position to prove our theorem. $\mathfrak{F}$ is again any differential field of characteristic zero, and $\mathfrak{F}'$ a differential extension thereof.

THEOREM 4: *Let Σ be a perfect differential ideal in $\mathfrak{F}\{y_1, \cdots, y_n\}$. Then $(\Sigma)' = \{\Sigma\}'$.*

PROOF: Let $\Sigma = \Sigma_1 \cap \cdots \cap \Sigma_s$ be the decomposition of Σ into essential prime differential ideals. It is easy to see that

(5) $$\{\Sigma\}' = \{\Sigma_1\}' \cap \cdots \cap \{\Sigma_s\}'.$$

We shall prove that

(6) $$(\Sigma)' = (\Sigma_1)' \cap \cdots \cap (\Sigma_s)'.$$

Indeed, since $\Sigma \subseteq \Sigma_i$, we have $(\Sigma)' \subseteq (\Sigma_i)'$, $i = 1, \cdots, s$, that is

$$(\Sigma)' \subseteq (\Sigma_1)' \cap \cdots \cap (\Sigma_s)'.$$

Now let $G \in (\Sigma_i)'$, $i = 1, \cdots, s$. We may write

$$G = G_0^{(i)} + G_1^{(i)}\omega_1 + \cdots + G_m^{(i)}\omega_m ,$$

where each $G_i^{(j)} \in \Sigma_j$, and $1, \omega_1, \cdots, \omega_m$ are elements of $\mathfrak{F}'$, linearly independent over $\mathfrak{F}$. The linear independence shows that $G_i^{(1)} = \cdots = G_i^{(s)}, i = 0, 1, \cdots, m$, so that each $G_i^{(j)} \in \Sigma_1 \cap \cdots \cap \Sigma_s = \Sigma$, and $G \in (\Sigma)'$. Hence

$$(\Sigma)' \supseteq (\Sigma_1)' \cap \cdots \cap (\Sigma_s)'.$$

This proves (6).

If $\mathfrak{F}$ contains non-constants, then, by (5), (6) and Lemma 3, $\{\Sigma\}' = (\Sigma)'$.

Suppose $\mathfrak{F}$ consists purely of constants. Let $\mathfrak{F}^* = \mathfrak{F}\langle x\rangle$, where x is an element whose derivative is 1, and let $\mathfrak{F}^\ddagger = \mathfrak{F}'\langle x\rangle$. By Lemma 4 and equations analogous to (5) and (6),

$$\begin{aligned}(\Sigma)^* &= (\Sigma_1)^* \cap \cdots \cap (\Sigma_s)^* \\ &= \{\Sigma_1\}^* \cap \cdots \cap \{\Sigma_s\}^* = \{\Sigma\}^*.\end{aligned}$$

Hence, by the part of the theorem already proved,

$$\begin{aligned}(\Sigma)^\ddagger &= ((\Sigma)^*)^\ddagger = (\{\Sigma\}^*)^\ddagger \\ &= \{\{\Sigma\}^*\}^\ddagger = \{\Sigma\}^\ddagger.\end{aligned}$$

Thus, $(\Sigma)^\ddagger$ is a perfect differential ideal.

Now, $(\Sigma)^\ddagger = ((\Sigma)')^\ddagger$. Suppose that $(\Sigma)' \neq \{\Sigma\}'$. Then there is an $F \in \mathfrak{F}'\{y_1, \cdots, y_n\}$ such that $F \in \{\Sigma\}'$, $F \notin (\Sigma)'$. Clearly $F \in \{\Sigma\}^\ddagger$, that is, $F \in (\Sigma)^\ddagger = ((\Sigma)')^\ddagger$. Since $F \in \mathfrak{F}'\{y_1, \cdots, y_n\}$, this implies that $F \in (\Sigma)'$. This contradiction shows that $(\Sigma)' = \{\Sigma\}'$, and completes the proof.

Corollary 1: *Let Σ be a differential ideal in $\mathfrak{F}\{y_1, \cdots, y_n\}$. Then the exponent of Σ equals that of $(\Sigma)'$.*

Proof: By Theorems 3 and 4

$$l_{\{\Sigma\}}\Sigma = l_{(\{\Sigma\})'}(\Sigma)' = l_{\{\{\Sigma\}\}'}(\Sigma)' = l_{\{\Sigma\}'}(\Sigma)'.$$

5. A certain differential ideal

Let $\mathfrak{F}$ be any differential field of characteristic zero, and let y be an unknown. We work in the differential ring $\mathfrak{F}\{y\}$. The following theorem uses an idea due to Raudenbush.[15]

Theorem 5: *$[y]^2$ does not have a strong basis.*

Proof: Assume that $[y]^2$ has a strong basis. Then it has a strong basis consisting purely of forms $y_i y_j$. Let s be an integer so large that the set of forms

$$(7) \qquad y_i y_j, \qquad 0 \leqq i \leqq j \leqq s,$$

is a strong basis of $[y]^2$. Let p be the exponent with respect to $[y]^2$ of the differential ideal generated by the basis (7). Let α be a positive integer whose value (large) is to be given later.

---

[15] See paper cited under footnote 1.

Consider the forms

$$y_{i_1}y_{i_2}\cdots y_{i_{2p}}, \qquad i_1+\cdots+i_{2p}=\alpha. \tag{8}$$

Each form (8) is in $[y]^{2p}$. Considering degree and weight, we see that each form (8) is a linear combination, with coefficients in $\mathfrak{F}$, of forms

$$(y_iy_j)_k y_{j_1}\cdots y_{j_{2p-2}},$$
$$0 \leqq i \leqq j \leqq s, \qquad i+j+k+j_1+\cdots+j_{2p-2}=\alpha. \tag{9}$$

Hence the number of linearly independent forms (8) does not now exceed the number of forms (9). Since the totality of forms (8) are linearly independent, it follows that the number of distinct forms (8) is less than or equal to the number of forms (9).

Now, the number of expressions (8) is clearly at least

$$\frac{1}{(2p)!}\left(\frac{\alpha}{2p}+1\right)^{2p-1}$$

if we suppose that α is divisible by $2p$. For, we may assign $i_1, \cdots, i_{2p-1}$ arbitrary values from 0 to $\alpha/2p$, and then i_{2p} is uniquely determined; moreover, the number of times a given form (8) can be produced in this way is at most $(2p)!$ On the other hand, the number of expressions (9) is surely not more than

$$\tfrac{1}{2}(s+1)(s+2)(\alpha+1)^{2p-2},$$

for the number of forms in the basis (7) is $(s+1)(s+2)/2$. But if α is sufficiently large, then

$$\frac{1}{(2p)!}\left(\frac{\alpha}{2p}+1\right)^{2p-1} > \tfrac{1}{2}(s+1)(s+2)(\alpha+1)^{2p-2}.$$

Hence, the number of forms (8) exceeds the number of forms (9). This contradiction completes the proof.

COROLLARY 2: *Let $P \in \mathfrak{F}\{y\}$ be of order zero. Then the exponent of $[P]$ is ∞ or 1 according as P has or has not a multiple factor.*

PROOF: By Corollary 1 we may suppose that $\mathfrak{F}$ contains all the roots of $P = 0$. Let

$$P = b(y-\eta_1)^{k_1}\cdots(y-\eta_s)^{k_s}, \qquad \eta_i \neq \eta_j \text{ if } i \neq j,$$

be the factorization of P into linear factors. It is easy to see that

$$[P] = [(y-\eta_1)^{k_1}]\cdots[(y-\eta_s)^{k_s}] \tag{10}$$

is the product representation of $[P]$. If $k_i = 1$, then $[(y-\eta_i)^{k_i}] = [y-\eta_i]$ obviously has exponent unity. If $k_i > 1$, then $[(y-\eta_i)^{k_i}]$ has exponent ∞. For, otherwise $(y-\eta_i)^{k_i}$ would be a strong basis for $[y-\eta_i]^2$, a contradiction of Theorem 5. Since by Theorem 1 the exponent of $[P]$ equals the maximum of the exponents of the $[(y-\eta_i)^{k_i}]$, the proof is complete.

Part II. The Differential Ideal Generated by a Form in One Unknown of Order Unity

6. Preliminary remarks

Consider a form A in a single unknown y, of order unity. We seek the exponent of $[A]$. We shall work in a fixed differential field $\mathfrak{F}$ of characteristic zero, which contains the coefficients of A. By Corollary 1, we may (and do) assume, without further mention, that $\mathfrak{F}$ is extensive enough to contain certain solutions of A which we shall discuss.

The exponent of $[A]$ will be found to depend on the nature of the *singular solutions* of A, that is, the solutions of A which annul S, the separant of A.[16] A singular solution of A must annul the resultant with respect to y_1, the derivative of y, of A and S. This resultant is a nonzero form of order zero, or is 0, according as A has no multiple factor of order unity, or has such a factor, that is, according as A and S have not or have a common factor of order unity. In the former case, A has only a finite number of singular solutions, and these are easily found.

If $y = \eta$ is a singular solution of A, we consider A as a form in $z = y - \eta$, and write

$$A = \sum_{\nu=0}^{n} P_\nu z^{p_\nu} z_1^\nu, \tag{11}$$

where each P_ν either is of order zero and not divisible by z, or vanishes, and where the p_ν are definite non-negative integers.[17] We suppose that the degree of A in y_1 is n, so that $P_n \neq 0$.

Let m be the minimum of the total degrees of all terms effectively present in A considered as a polynomial in z and z_1. We shall call m the *multiplicity* of the singular solution $y = \eta$. It is obvious that if A has a multiple factor, any solution of that factor is a singular solution of multiplicity greater than unity.

7. Singular solutions of multiplicity exceeding unity

We shall prove that *if A has a singular solution of multiplicity exceeding unity, then the exponent of $[A]$ is ∞.*

(a) Let $y = \eta$ be a singular solution of A of multiplicity at least 2, and suppose that $y = \eta$ is an essential manifold of A. Let $z = y - \eta$. The product representation of $[A]$ must have a factor whose manifold is $y = \eta$, that is, a factor Σ_1 such that $\{\Sigma_1\} = \{z\}$. Now, it is known that

$$\Sigma_1 = [A, z^q], \tag{12}$$

[16] The definition here of *singular solution* is broader than that given by Ritt in his book, *Differential equations from the algebraic standpoint*, and used in the Appendix of the present paper, for A is not assumed here to be algebraically irreducible.

[17] If $P_\nu = 0$, we let p_ν be any fixed non-negative integer.

where q is any sufficiently large integer.[18] Because $y = \eta$ is of multiplicity at least 2, it follows that $A \epsilon [z^2, z_1^2]$. Hence, by (12),

$$\Sigma_1 \subseteq [z^2, z_1^2].$$

Thus, the exponent of Σ_1 is not less than that of $[z^2, z_1^2]$, which, by Theorem 5, is ∞. Therefore, using Theorem 1, we see that the exponent of $[A]$ is ∞.

(b) Suppose that $y = \eta$ is a singular solution of A of multiplicity greater than unity, which is not an essential manifold of A. Let $y - \eta = z$. We shall show first that $\{A\}$ contains a form which, when considered as a form in z, has at least one term of the first degree.

Let $C_1, \cdots, C_q$ be the distinct irreducible factors of A of order unity, and let $F = C_1 \cdots C_q$. If T denotes the separant of F, the derivative of F is a form $Tz_2 + U$, where U is a form of order not exceeding unity. Hence

$$Tz_2 \equiv -U \qquad [F].$$

It follows, by successive differentiations and eliminations, that

$$T^{a_i} z_i \equiv V_i \qquad [F], \qquad\qquad i = 2, \cdots, n+2,$$

where the a_i are appropriate integers, and the V_i are forms whose order is not greater than unity. Letting a be the maximum of the a_i, we have

$$T^a z_i \equiv W_i \qquad [F], \qquad\qquad i = 2, \cdots, n+2,$$

where $W_i = T^{a-a_i} V_i$.

Let J be the initial of F. Then J is of order zero. Clearly, there exists a non-negative integer b such that

$$J^b W_i \equiv X_i \qquad (F), \qquad\qquad i = 2, \cdots, n+2,$$

where each X_i is of order not exceeding unity, and is of degree less than n in z_1.[19] It follows that

$$J^b T^a z_i \equiv X_i \qquad [F], \qquad\qquad i = 2, \cdots, n+2.$$

The $n + 1$ forms X_i are linear combinations, with coefficients which are polynomials in z, of the n quantities $1, z_1, \cdots, z_1^{n-1}$. Hence, some linear combination of the X_i, with coefficients which are forms of order zero, not all 0, must vanish:

$$\sum_{i=2}^{n+2} K_i X_i = 0.$$

We assume, as we may, that the K_i are not all divisible by z. Referring to the last congruence, we now see that

[18] Ritt and Kolchin, Bulletin of the American Mathematical Society, vol. 45 (1939), pp. 895–898.

[19] The degree of F in z_1 is at most n.

$$J^b T^a M \equiv 0 \qquad [F],$$

where $M = \sum_{i=2}^{n+2} K_i z_i$.

Since not every K_i is divisible by z, M has at least one term of degree unity. Consider the form $y - \zeta$, where ζ is either a common solution of TJ and F, or a root of a factor of A of order zero, but is distinct from η. Let N be the product of all such forms $y - \zeta$, and let $Z = MN$.

Z, like M, has at least one term of degree unity (when considered as a form in z). Z vanishes for all solutions of F with the possible exception of $y = \eta$, and for all the solutions of A which are not solutions of F, again with the possible exception of $y = \eta$. That is, Z vanishes for all solutions of A, save possibly $y = \eta$. Since $y = \eta$ is not an essential manifold, Z must also vanish for $y = \eta$. By the analog to the Hilbert-Netto theorem,[20] then, $Z \in \{A\}$. Z is the form whose existence we were to prove.

Thus, $\{A\}$ contains a form $z_j + H$, where H, considered as a form in z, has each of its terms either of order less than j or of degree greater than unity.

Assume, now, that the exponent of $[A]$ is finite, say q. Then[21]

$$(z_{j+i_1} + H_{i_1}) \cdots (z_{j+i_q} + H_{i_q}) \equiv 0 \qquad [A]$$

for all non-negative integers $i_1, \cdots, i_q$. Since $y = \eta$ is of multiplicity at least 2, as a singular solution of A, it follows that $A \in [z^2, z_1^2]$. Hence

$$(z_{j+i_1} + H_{i_1}) \cdots (z_{j+i_q} + H_{i_q}) \equiv 0_i \qquad [z^2, z_1^2].$$

If we compare terms here of degree q and weight $qj + i_1 + \cdots + i_q$, we obtain

$$z_{j+i_1} \cdots z_{j+i_q} \equiv 0 \qquad [z^2, z_1^2], \qquad\qquad i_1, \cdots, i_q = 0, 1, 2, \cdots.$$

This is easily seen to imply, however, that $[z]^2$ has a strong basis. This contradiction of Theorem 5 completes the proof.

EXAMPLE 1. Let $A = \sum_{\nu=0}^{n} Q_\nu y_1^\nu$ $(n \geqq 1)$, where the Q_ν are forms of order zero with *constant* coefficients, and where $Q_n \not\equiv 0$.

If A has no multiple factors of order unity, the singular solutions must be constants, for they must satisfy an algebraic equation with constant coefficients;[22] since $S = \sum_{\nu=1}^{n} \nu Q_\nu y_1^{\nu-1}$, it is evident that the singular solutions of A are precisely the common zeros of Q_0 and Q_1. A singular solution is of multiplicity exceeding unity if it is a multiple zero of Q_0. Thus, *if A has a multiple factor, or if Q_0 has a multiple factor which is also a factor of Q_1, then the exponent of $[A]$ is ∞.*

8. Singular solutions of multiplicity unity; the product representation

Throughout §§8–10 we suppose that A has singular solutions, and that they are all of multiplicity unity. In the present section we shall show that all the

[20] See the paper cited under footnote 2.

[21] H_{i_k} is the i_k-th derivative of H.

[22] The equation obtained by eliminating z_1 from $A = 0$, $S = 0$.

singular solutions are essential manifolds, and shall find the product representation of $[A]$. With an additional assumption, we shall determine the exponent of $[A]$ (§10).

Since A has no singular solution of multiplicity exceeding unity, A has no multiple factors. Hence A has only a finite number of singular solutions (§6).

Let $y = \eta$ be a singular solution of A, and write $z = y - \eta$. Since S must vanish for $y = \eta$, we see from (11) that either $P_1 = 0$ or $p_1 > 0$. This means, since $y = \eta$ is of multiplicity unity, that $P_0 \not\equiv 0$ and $p_0 = 1$. (11) thus becomes

$$A = P_0 z + \sum_{\nu=1}^{n} P_\nu z^{p_\nu} z_1^\nu. \tag{13}$$

It now follows from a theorem of Ritt that $y = \eta$ is an essential manifold of A.[23]

We now turn to the product representation of $[A]$.

Let $y = \eta_i$, $i = 1, \cdots, s$, be the singular solutions of A; we denote $y - \eta_i$ by u_i. It is evident that we may write

$$[A] = \Sigma_0 \Sigma_1 \cdots \Sigma_s, \tag{14}$$

where the manifold of Σ_0 consists of the non-singular solutions of A,[24] and where the manifold of Σ_i, $i > 0$, is $y = \eta_i$, that is, $\{\Sigma_i\} = \{u_i\}$, $i = 1, \cdots, s$.

Referring to (13) and making use of a known result,[25] we see that

$$\Sigma_i = [u_i], \qquad i = 1, \cdots, s. \tag{15}$$

The determination of Σ_0 is more difficult. The first step will be to find $\{\Sigma_0\}$. If z is one of the u_i, then (13) holds. Differentiating (13) we obtain

$$A_1 = P_{01} z + \sum_{\nu=1}^{n} P_{\nu 1} z^{p_\nu} z_1^\nu + z_1 \left(P_0 + \sum_{\nu=1}^{n} p_\nu P_\nu z^{p_\nu - 1} z_1^\nu \right) + \sum_{\nu=1}^{n} \nu P_\nu z^{p_\nu} z_1^{\nu-1} \cdot z_2,$$

where $P_{\nu 1}$ is the derivative of P_ν. But by (13),

$$P_0 z \equiv -z_1 \sum_{\nu=1}^{n} P_\nu z^{p_\nu} z_1^{\nu-1} \qquad (A). \tag{16}$$

Hence

$$P_0 A_1 = P_{01} P_0 z + P_0 z_1 \sum_{\nu=1}^{n} P_{\nu 1} z^{p_\nu} z_1^{\nu-1} + P_0 z_1 \left(P_0 + \sum_{\nu=1}^{n} p_\nu P_\nu z^{p_\nu - 1} z_1^\nu \right)$$
$$+ \left(P_1 z^{p_1 - 1} \cdot P_0 z + P_0 z_1 \sum_{\nu=2}^{n} \nu P_\nu z^{p_\nu} z_1^{\nu-2} \right) z_2$$

[23] Ritt, Annals of Mathematics, vol. 37 (1936), pp. 552-617. See especially §5. Professor Ritt's proof is not applicable to abstract differential fields, as it involves function-theoretic concepts. But an abstract proof of the *sufficiency* part of the theorem (we need only the sufficiency) has recently been achieved by H. Levi, and may be expected to appear in the literature shortly.

[24] If A is algebraically irreducible, the manifold of Σ_0 is the general solution of A.

[25] See §2 of the paper cited under footnote 18.

$$\equiv z_1\Big(-P_{01}\sum_{\nu=1}^{n} P_\nu z^{p_\nu} z_1^{\nu-1} + P_0 \sum_{\nu=1}^{n} P_{\nu 1} z^{p_\nu} z_1^{\nu-1} + P_0\Big(P_0 + \sum_{\nu=1}^{n} p_\nu P_\nu z^{p_\nu - 1} z_1^{\nu}\Big)$$

$$+\Big(-P_1 z^{p_1-1}\sum_{\nu=1}^{n} P_\nu z^{p_\nu} z_1^{\nu-1} + P_0 \sum_{\nu=2}^{n} \nu P_\nu z^{p_\nu} z_1^{\nu-2}\Big) z_2\Big), \qquad (A),$$

so that

$$P_0 A_1 \equiv z_1(U z_2 + V) \qquad (A), \tag{17}$$

where

$$\begin{aligned} U &= -P_1 z^{p_1-1}\sum_{\nu=1}^{n} P_\nu z^{p_\nu} z_1^{\nu-1} + P_0 \sum_{\nu=2}^{n} \nu P_\nu z^{p_\nu} z_1^{\nu-2}, \\ V &= -P_{01}\sum_{\nu=1}^{n} P_\nu z^{p_\nu} z_1^{\nu-1} + P_0 \sum_{\nu=1}^{n} P_{\nu 1} z^{p_\nu} z_1^{\nu-1} + P_0\Big(P_0 + \sum_{\nu=1}^{n} p_\nu P_\nu z^{p_\nu-1} z_1^{\nu}\Big). \end{aligned} \tag{18}$$

From (17) and (16) we find that

$$z(P_0 U z_2 + P_0 V) \equiv 0 \qquad (A, A_1).$$

This implies that

$$\{A\} = \{z\} \cap \{A, P_0 U z_2 + P_0 V\}.$$

The two ideals in the second member here are separated, for by (18)

$$P_0 V \equiv P_0^3 \not\equiv 0 \qquad [z], \tag{19}$$

so that $P_0 U z_2 + P_0 V \not\equiv 0$ $[z]$.

Now, z represents any one of the u_i. Let

$$I_i = P_0 U, \qquad J_i = P_0 V \quad \text{when} \quad z = u_i, \qquad i = 1, \cdots, s. \tag{20}$$

Then by the above, for $i = 1, \cdots, s$,

$$\{A\} = \{u_i\} \cap \{A, I_i u_{i2} + J_i\},$$

$$u_i(I_i u_{i2} + J_i) \equiv 0, \qquad u_{i1}(I_i u_{i2} + J_i) \equiv 0 \qquad (A, A_1), \tag{21}$$

and $y = \eta_i$ is not a solution of $I_i u_{i2} + J_i$. It readily follows that

$$\{A\} = \{A, I_1 u_{12} + J_1, \cdots, I_s u_{s2} + J_s\} \cap \{u_1\} \cap \cdots \cap \{u_s\}. \tag{22}$$

The $s + 1$ differential ideals, whose intersection appears in the second member of (23), are separated, so that we may replace intersection by product.

Comparing (22) with

$$\{A\} = \{\Sigma_0\} \cap \{\Sigma_1\} \cap \cdots \cap \{\Sigma_s\}$$

(which follows from (14)) and with (15), we see that

$$\{\Sigma_0\} = \{A, I_1 u_{12} + J_1, \cdots, I_s u_{s2} + J_s\}. \tag{23}$$

We shall now obtain two forms M and N such that

(24) $M \epsilon \{u_1\} \cdots \{u_s\}, \qquad N \epsilon \{\Sigma_0\}, \qquad MN \epsilon [A], \qquad M + N = 1.$

This will enable us to determine Σ_0.

Let

$$B_i = I_i u_{i2} + J_i, \qquad i = 1, \cdots, s,$$

and let

$$Q_i = P_0 \text{ when } z = u_i, \qquad i = 1, \cdots, s.$$

Then Q_i is not divisible by u_i, and, by (19) and (20),

(25) $$B_i - Q_i^3 \equiv I_i u_{i2} \qquad (u_i, u_{i1}).$$

Since Q_i is not divisible by u_i, there is a form C_i of order zero, and an element $a_i \epsilon \mathfrak{F}$, such that

$$1 = a_i Q_i^3 + C_i u_i.$$[26]

This may be written

$$1 = [C_i u_i - a_i(B_i - Q_i^3)] + a_i B_i.$$

By (21), (25) and the definition of B_i, moreover,

$$[C_i u_i - a_i(B_i - Q_i^3)] \cdot a_i B_i \equiv 0 \qquad [A],$$

provided $I_i \equiv 0$ (u_i, u_{i1}), that is, provided $P_2 = 0$ or $p_2 > 0$ when $z = u_i$ (we refer to (20) and (18), and the fact that $p_1 > 0$). If this is not the case, then

$$[C_i u_i - a_i(B_i - Q_i^3)] \cdot (a_i B_i)^2 \equiv 0 \qquad [A],$$

because $u_{i2} B_i^2 \equiv 0$ is a consequence of $u_{i1} B_i \equiv 0$, and the latter congruence holds by (21).

Accordingly, we separate the singular solutions of A into two classes. A singular solution $y = \eta$ of A will be said to be of the *first class* if in A, considered as a form in $z = y - \eta$, the coefficient of z_1^2 is divisible by z; a singular solution which is not of the first class will be said to be of the *second class*.

We suppose the singular solutions η_i ordered so that $\eta_1, \cdots, \eta_r$ are of the first class, and $\eta_{r+1}, \cdots, \eta_s$ are of the second class. This is the same as supposing that

(26)
$$P_2 = 0 \text{ or } p_2 > 0 \text{ when } z = u_i, \qquad i = 1, \cdots, r,$$
$$P_2 \neq 0 \text{ and } p_2 = 0 \text{ when } z = u_i, \qquad i = r + 1, \cdots, s.$$

We let

(27)
$$M_i = C_i u_i - a_i(B_i - Q_i^3), \qquad i = 1, \cdots, r,$$
$$M_i = [C_i u_i - a_i(B_i - Q_i^3)]^2 + 2[C_i u_i - a_i(B_i - Q_i^3)] \cdot a_i B_i, \qquad i = r + 1, \cdots, s,$$
$$N_i = a_i B_i, \qquad i = 1, \cdots, r,$$
$$N_i = (a_i B_i)^2, \qquad i = r + 1, \cdots, s.$$

[26] It will be observed that $a_i \neq 0$.

Then $M_i \in \{u_i\}$, $N_i \in \{\Sigma_0\}$, $M_i + N_i = 1$, $M_iN_i \in [A]$, $i = 1, \cdots, s$. Letting, therefore,

$$(28)\qquad \begin{aligned} M &= M_1 \cdots M_s \\ N &= (M_1 + N_1) \cdots (M_s + N_s) - M_1 \cdots M_s = 1 - M, \end{aligned}$$

we see that (24) holds.

We refer now to the paper cited under footnote 5. The proof in that paper shows that $\Sigma_0 = [A, N]$. But, by (28), $N \equiv 0\ (N_i, \cdots, N_s)$, and $N_i = MN_i + NN_i \equiv 0\ [A, N]$. Hence $\Sigma_0 = [A, N_1, \cdots, N_s]$. Referring to (27), we see, finally, that

$$(29)\qquad \Sigma_0 = [A, B_1, \cdots, B_r, B_{r+1}^2, \cdots, B_s^2].$$

The product representation of A is given by (14), (15) *and* (29).[27]

9. Continuation

The purpose of this section is to prove that

$$(30)\qquad [A, B_1, \cdots, B_s]^2 \subseteq \Sigma_0.$$

By (29) it suffices to show that

$$(31)\qquad B_{\sigma i}B_{\tau j} \equiv 0\ (\Sigma_0), \qquad \sigma, \tau = 1, \cdots, s, \qquad i, j = 0, 1, 2, \cdots.$$

Let $Iz_2 + J$ represent any $B_\sigma = I_\sigma u_{\sigma 2} + J_\sigma$. We shall prove that

$$(32)\qquad (Iz_2 + J)_i(Iz_2 + J)_j \equiv 0 \qquad (\Sigma_0)$$

for all i, j.

From (21) we see that

$$(33)\qquad z(Iz_2 + J) \equiv 0, \qquad z_1(Iz_2 + J) \equiv 0 \qquad [A].$$

Differentiating the second congruence here $l + 1$ times, we find that

$$(34)\qquad \sum_{\lambda=0}^{l+1} \binom{l+1}{\lambda} z_{\lambda+1}(Iz_2 + J)_{l+1-\lambda} \equiv 0 \qquad [A].$$

By (29), (32) holds for $i = 0, j = 0$. We assume that (32) holds for $i = 0, 1, \cdots, l - 1$ and $j = 0$, and prove it for $i = l, j = 0$.

Multiplying (34) by $Iz_2 + J$, we obtain, using (33) and the induction assumption,

$$(l + 1)z_2(Iz_2 + J)_l(Iz_2 + J) \equiv 0 \qquad (\Sigma_0).$$

But $Iz_2(Iz_2 + J) \equiv -J(Iz_2 + J)$, (Σ_0). Hence

$$J(Iz_2 + J)_2(Iz_2 + J) \equiv 0 \qquad (\Sigma_0).$$

[27] Actually, Σ_0 may not be connected. This will occur if A has more than one factor of order unity.

However, by (20) and (19), J is congruent, modulo (z, z_1), to a nonzero element of $\mathfrak{F}$. Hence, by (33), $(Iz_2 + J)_l(Iz_2 + J) \equiv 0\ (\Sigma_0)$.

Thus (32) holds for all i when $j = 0$.

To prove (32) for all i and all j we again use induction. Assume (32) to be valid for all i and for $j = l - 1$. Differentiating (32) for $j = l - 1$ we see that

$$(Iz_2 + J)_{i+1}(Iz_2 + J)_{l-1} + (Iz_2 + J)_i(Iz_2 + J)_l \equiv 0 \qquad (\Sigma_0),$$

so that, by the induction assumption, $(Iz_2 + J)_i(Iz_2 + J)_l \equiv 0\ (\Sigma_0)$. This proves (32) for all i and j.

Thus (31) holds whenever $\sigma = \tau$. To prove (31) in the general case, consider $B_{\sigma i}B_{\tau j}$. It is sufficient, of course, to suppose that $\sigma > r$, for $B_\sigma \in \Sigma_0$ if $\sigma \leqq r$. Now, the remainder of $B_{\tau j}$ with respect to B_σ is a form in $\{\Sigma_0\}$ of order not exceeding unity. Such a form must be divisible by A. Hence, for an appropriate integer m, $I_\sigma^m B_{\tau j} \equiv 0\ [A, B_\sigma]$. Hence, since (31) holds when $\tau = \sigma$, $I_\sigma^m B_{\sigma i}B_{\tau j} \equiv 0\ (\Sigma_0)$. But $\sigma > r$, so that (see (26), (20) and (18)) I_σ is congruent modulo $(u_\sigma, u_{\sigma 1})$, to a nonzero element of $\mathfrak{F}$. Hence, by (21), $B_{\sigma i}B_{\tau j} \equiv 0\ (\Sigma_0)$.

10. The regular type

We shall say that A is of the *regular type* if

$$(35) \qquad 1 \equiv 0\ (\Sigma_0, S, I_1, \cdots, I_s).$$

In the present section *we assume that A is of the regular type.*

We shall prove that *the exponent of $[A]$ is 1 or 2 according as all of the singular solutions of A are of the first class or at least one singular solution of A is of the second class.*

(a) Let all the singular solutions of A be of the first class, and let $G \in \{\Sigma_0\}$. We shall show that $G \in \Sigma_0$.

For each i we have, for an appropriate integer b_i, $I_i^{b_i}G \equiv G_i'[B_i]$, where G_i' is of order unity, at most. Since $G_i' \in \{\Sigma_0\}$, G_i' must be divisible by A. Hence, for $i = 1, \cdots, s$, $I^{b_i}G \equiv 0\ [A, B_i]$. Similarly, for a suitable c, $S^cG \equiv 0\ [A]$. Now, by (35), there are forms X_i, and a form $T \in \Sigma_0$, such that

$$1 = T + X_0S^c + X_1I_1^{b_1} + \cdots + X_sI_s^{b_s}.$$

Therefore,

$$G = (T + X_0S^c + X_1I_1^{b_1} + \cdots + X_sI_s^{b_s})G \equiv 0 \qquad [A, B_1, \cdots, B_s].$$

But all the singular solutions of A are of the first class. Therefore, by (29), $\Sigma_0 = [A, B_1, \cdots, B_s]$. Hence, $G \in \Sigma_0$.

Thus, the exponent of Σ_0 is unity. But the exponent of $\Sigma_i = [u_i]$, $i = 1, \cdots, s$, is obviously unity. Hence, by Theorem 1, unity is also the exponent of $[A]$.

(b) Let A have at least one singular solution of the second class, and let $G \in \{\Sigma_0\}$. As in (a) we see that $G \in [A, B_1, \cdots, B_s]$. This, together with (31), proves that the exponent of Σ_0 does not exceed 2. Hence the exponent of $[A]$ is no greater than 2.

To prove that the exponent of $[A]$ is exactly 2, it suffices to exhibit a form in $\{A\}$ which is not in $[A]$. To this end, we let z be any u_i with $i > r$, and show that $\{\Sigma_0\}$ contains a form $z_h + H$, where h is a sufficiently large positive integer, and H is a form, of order less than h, contained in $[z]$.

By (35) there is a relation

$$1 = T + X_0S + X_1I_1 + \cdots + X_sI_s, \qquad T \in \Sigma_0. \tag{36}$$

Let $h(\geqq 3)$ be an integer such that $h - 1$ exceeds the orders of $T, X_0, \cdots, X_s$. If we differentiate each B_i, written as a form in z, $h - 3$ times, we obtain relations $I_iz_{h-1} + K_i \equiv 0 \ \{\Sigma_0\}$. Similarly, differentiating A $h - 2$ times we obtain $Sz_{h-1} + K_0 \equiv 0 \ \{\Sigma_0\}$. All K_i are forms of order less than $h - 1$. These relations, together with $Tz_{h-1} \equiv 0\{\Sigma_0\}$, yield, because of (36), a relation $z_{h-1} + K \equiv 0 \ \{\Sigma_0\}$, where K is a form of order less than $h - 1$.

Let α be the term of K free of z and the derivatives of z. If $\alpha = 0$, then $(z_{h-1} + K)_1$ is a form $z_h + H$ as described above; if $\alpha \neq 0$, then $\alpha(\alpha^{-1}(z_{h-1} + K))_1$ is such a form.

Now let E be the product of all the u_i other than z, and consider $E(z_h + H)$, where $z_h + H$ is the form whose existence we have just proved. We see that $\{A\}$ contains a form $Ez_h + F$, where E is a form of order zero, not divisible by z, and F is a form of order less than h.

We shall now complete the proof by showing that $Ez_h + F \not\equiv 0\ [A]$. The method will be to assume that $Ez_h + F \equiv 0\ [A]$, and to force a contradiction.

Let then

$$Ez_h + F = C_0A + C_1A_1 + \cdots + C_tA_t, \tag{37}$$

where A_j represents the j-th derivative of A. By (17), for all $i > 0$,

$$P_0A_i \equiv (z_1(Uz_2 + V))_{i-1} \qquad (A, A_1, \cdots, A_{i-1}).$$

It follows from (37), therefore, that

$$\begin{aligned} P_0^t(Ez_h + F) = D_0A &+ D_1z_1(Uz_2 + V) \\ &+ D_2(z_1(Uz_2 + V)_1 + z_2(Uz_2 + V)) \\ &+ \cdots + D_t\sum_{i=0}^{t-1}\binom{t-1}{i}z_{i+1}(Uz_2 + V)_{t-1-i}. \end{aligned} \tag{38}$$

We now consider the various forms in (38) as polynomials in the z_i. We shall modify (38) in such a way as to make evident a contradiction

Consider a power series[28] $\sum_{i=1}^{\infty} \beta_iz_1^i$ in z_1 such that A vanishes when $\Sigma\beta_iz_1^i$ is substituted for z. Such a power series exists, as by (13) $z_1 = 0, z = 0$ renders A zero and leaves $\dfrac{\partial A}{\partial z}$ nonzero.[29]

[28] All power series which we mention are formal.

[29] The "implicit function theorem" for abstract fields of characteristic zero may be proved by the method of undetermined coefficients. The proof is simpler than the one in analysis, because the series is formal and no questions of convergence arise.

Substituting $\Sigma\beta_i z_1^i$ for z in (38), we find a relation

$$(39)\quad \begin{aligned} E'z_h + F' = {} & D_1' z_1(U'z_2 + W_0) + D_2'(z_1(U'z_3 + W_1) + z_2(U'z_2 + W_0)) \\ & + \cdots + D_t' \sum_{i=0}^{t-1} \binom{t-1}{i} z_{i+1}(U'z_{t+1-i} + W_{t-1-i}). \end{aligned}$$

Here E', F', U', the D_i' and the W_i are power series in z_1 with coefficients which are polynomials in the z_i with $i > 1$. Each W_i is free of z_j with $j > i + 1$. Moreover, E' and U' involve only z_1, and have nonzero terms of degree zero. For E' this is true because $P_0^l E$ is not divisible by z; for U' because $P_2 \neq 0$ and $p_2 = 0$ (z is a u_i with $i > r$; see (26) and (18).)

Let

$$(40)\qquad w_1 = z_1, \qquad w_i = U'z_i + W_{i-2}, \qquad i = 2, 3, \cdots.$$

Since U' has a nonzero term of degree zero, the equations (40) may be inverted to obtain

$$(41)\qquad z_1 = w_1, \qquad z_i = Yw_i + Z_i, \qquad i = 2, 3, \cdots,$$

where Y is a series in w_1 with a nonzero term of degree zero, and each Z_i is a series in w_1 with coefficients which are polynomials in $w_2, \cdots, w_{i-1}$. The equations (41) define a substitution, with inverse substitution (40). Applying substitution (41) to (39), we obtain a relation

$$(42)\quad \begin{aligned} Qw_h + R = {} & L_1 w_1 w_2 + L_2 (w_1 w_3 + (Yw_2 + Z_2)w_2) \\ & + \cdots + L_t(w_1 w_{t+1} + (t-1)(Yw_2 + Z_2)w_t + \cdots + (Yw_t + Z_t)w_2), \end{aligned}$$

where Q is a series in w_1 with a nonzero term a of degree zero, and R is free of the w_i with $i \geqq h$.

For the second member of (42) to produce the term aw_h which appears in the first member, some L_i with $i \geqq h$ must have a nonzero term of degree zero, because each term in the coefficient of L_i is of at least the first degree in $w_1, w_2, \cdots, w_i$. Thus, some of the L_i have a term which is simply an element of $\mathfrak{F}$, and hence certainly terms bw_1^j ($b \in \mathfrak{F}$, $b \neq 0$, $j \geqq 0$). Let L_p be the L_i of highest subscript, with this property. Then $p \geqq h$. Of all the terms of L_p of the above mentioned type, let cw_1^k be the one of least degree. Then

$$L_p(w_1 w_{p+1} + \cdots + (Yw_p + Z_p)w_2)$$

contains the term $cw_1^{k+1}w_{p+1}$. This term does not appear in the first member of (42), and must therefore be cancelled by terms in some expression

$$(43)\qquad L_i(w_1 w_{i+1} + \cdots + (Yw_i + Z_i)w_2)$$

with $i \neq p$. But if $i <$ p, then each term of (43) is divisible by some w_j with $2 \leqq j \leqq p$. Hence $cw_1^{k+1}w_{p+1}$ must be cancelled by terms from expressions (43) with $i > p$. This implies that some L_i with $i > p$ has a term of the type bw_1^{k+1} ($b \in \mathfrak{F}$, $b \neq 0$). This contradicts the definition of L_p, and completes the proof.

The following example shows that of the forms A with *constant coefficients*, which have singular solutions, all of multiplicity unity, those which are not of the regular type are exceptional. It also shows that, for any preassigned pair of integers r, s $(0 \leqq r \leqq s,\ s > 0)$, there exist forms of the regular type having r singular solutions of the first class and $s - r$ singular solutions of the second class.

EXAMPLE 2. Let r, s, m, n be integers with $0 \leqq r \leqq s \leqq m,\ s > 0,\ n \geqq 3$, let $u_i = y - \eta_i$, $i = 1, \cdots, s$, where $\eta_1, \cdots, \eta_s$ are distinct constant elements of a differential field $\mathfrak{E}$ of characteristic zero, and let

$$Q_0 = \sum_{\mu=0}^{m-s} a_{\mu 0} y^\mu, \qquad Q_1 = \sum_{\mu=0}^{m-s} a_{\mu 1} y^\mu, \qquad Q_2 = \sum_{\mu=0}^{m-r} a_{\mu 2} y^\mu,$$

$$Q_\nu = \sum_{\mu=0}^{m} a_{\mu\nu} y^\mu, \qquad \nu = 3, \cdots, n,$$

where $a_{00}, a_{01}, \cdots, a_{mn}$ is a set of constants, algebraically independent with respect to $\mathfrak{E}$. Consider the form

$$A = Q_0 u_1 \cdots u_s + Q_1 u_1 \cdots u_s y_1 + Q_2 u_1 \cdots u_r y_1^2 + \sum_{\nu=3}^{n} Q_\nu y_1^\nu$$

with coefficients in $\mathfrak{F} = \mathfrak{E}\langle a_{00}, a_{01}, \cdots, a_{mn}\rangle$.

We shall show that the singular solutions of A are $\eta_1, \cdots, \eta_s$, where $\eta_1, \cdots, \eta_r$ are of the first class, and $\eta_{r+1}, \cdots, \eta_s$ are of the second. We shall prove, moreover, that A is of the regular type. In fact, we shall prove the stronger result that

$$(44) \qquad 1 \equiv 0 \quad (A, S, I_1, \cdots, I_s, J_1, \cdots, J_s).$$

Let R be the resultant with respect to y_1 of A and S. R is a polynomial in y and the $a_{\mu\nu}$. If we let the $a_{\mu\nu}$ take on special values, such that A becomes $u_1 \cdots u_s + y_1^n$, then S becomes $n y_1^{n-1}$, so that R becomes $n^n(u_1 \cdots u_s)^{n-1} \neq 0$. Since R is not 0 for special values of the $a_{\mu\nu}$, R does not vanish for unknown $a_{\mu\nu}$.

Now, each singular solution is a solution of R, which is of order zero and has constant coefficients. Since $R \neq 0$, this implies that all the singular solutions of A are constants. Hence, from the expression for A and the expression for S derived therefrom, we see that a singular solution of A must annul $Q_0 u_1 \cdots u_s$ and $Q_1 u_1 \cdots u_s$, that is, must be one of the η_i. That each η_i is a singular solution is obvious. Thus, the singular solutions of A are $\eta_1, \cdots, \eta_s$. It is apparent that these are of multiplicity unity, and that η_i is of the first or second class according as $i \leqq r$ or $i > r$.

We find, using (20) and (18) that

$$I_i \equiv Q_0(u_1 \cdots u_{i-1} u_{i+1} \cdots u_s)^2 I, \qquad J_i \equiv (Q_0 u_1 \cdots u_{i-1} u_{i+1} \cdots u_s)^2 J, \quad (A),$$

where

$$I = -Q_1^2u_1 \cdots u_s - Q_1Q_2u_1 \cdots u_ry_1 - Q_1 \sum_{\nu=3}^{n} Q_\nu y_1^{\nu-1} + 2Q_0Q_2u_1 \cdots u_r + Q_0 \sum_{\nu=3}^{n} \nu Q_\nu y_1^{\nu-2},$$

$$J = (Q_0u_1 \cdots u_s)' + (Q_1u_1 \cdots u_s)'y_1 + (Q_2u_1 \cdots u_r)'y_1^2 + \sum_{\nu=3}^{n} Q_\nu' y_1^\nu,$$

accents indicating differentiation with respect to y. Hence,

$$(A, S, I_1, \cdots, I_s, J_1, \cdots, J_s) = (A, S, Q_0I, Q_0^2J).$$

Suppose, now, that (44) is false. Then there is a pair of values $y = \bar{y}$, $y_1 = \bar{y}_1$ which annul the polynomials A, S, Q_0I, Q_0^2J. When $y_1 = 0$, A, S and Q_0^2J become, respectively, $Q_0u_1 \cdots u_s$, $Q_1u_1 \cdots u_s$ and $Q_0^2(Q_0u_1 \cdots u_s)'$, and these polynomials have no solution in common. Hence $\bar{y}_1 \neq 0$. Also, the resultant with respect to y_1 of S and

$$Q_1u_1 \cdots u_s + Q_2u_1 \cdots u_ry_1 + \sum_{\nu=3}^{n} Q_\nu y_1^{\nu-1} = \frac{A - Q_0u_1 \cdots u_s}{y_1}$$

is a nonzero polynomial in y, relatively prime to Q_0.[30] Hence $\bar{y}$ is not a root of $Q_0 = 0$, so that $\bar{y}$, $\bar{y}_1$ annul I and J.

Let M be the resultant with respect to y_1, N the resultant with respect to y, of J and

$$nA - y_1S = nQ_0u_1 \cdots u_s + (n-1)Q_1u_1 \cdots u_sy_1 + (n-2)Q_2u_1 \cdots u_ry_1^2 + \sum_{\nu=3}^{n-1} (n-\nu)Q_\nu y_1^\nu.$$

M is a polynomial in y, N is a polynomial in y_1, each with coefficients which are polynomials in the $a_{\mu\nu}$ *not including* a_{0n}. Moreover, M and N are not 0.[31] Now, $\bar{y}$, $\bar{y}_1$ annul M and N. Thus, $\bar{y}$ and $\bar{y}_1$ are algebraic functions of the $a_{\mu\nu}$ other than a_{0n}. Since A vanishes for $y = \bar{y}$, $y_1 = \bar{y}_1$, and since $\bar{y}_1 \neq 0$, it follows that a_{0n} is an algebraic function of the other $a_{\mu\nu}$. This contradiction completes the proof.

[30] This resultant is nonzero because it is nonzero when $A = y_1^n + u_1 \cdots u_sy_1$; it is prime to Q_0 because its coefficients are independent of $a_{00}, \cdots, a_{m-s,0}$.

[31] If we specialize some of the $a_{\mu\nu}$ so that $A = y_1^n + y_1^{n-1} + Q_0u_1 \cdots u_s$, then $J = (Q_0u_1 \cdots u_s)'$, $nA - y_1S = nQ_0u_1 \cdots u_s + y_1^n$, and $M = (Q_0u_1 \cdots u_s)'^{n-1} \neq 0$, so that M does not vanish before we specialize. If we specialize A further by setting $y_1 = 0$, then $J = (Q_0u_1 \cdots u_s)'$, $nA - y_1S = nQ_0u_1 \cdots u_s$, and N is the resultant of $nQ_0u_1 \cdots u_s$ and $(Q_0u_1 \cdots u_s)^1$, which is not 0. Thus, N is not 0 before the specialization.

EXAMPLE 3. Let $n \geqq 2$, and let $A = Qy_1^n + P$, where P and Q are relatively prime nonzero forms of order zero with constant coefficients, P being of positive degree and without multiple factors. The singular solutions of A are the roots of $P = 0$, and are all of multiplicity unity.

If $z = y - \eta$, where η is a root of $P = 0$, then $\eta_1 = 0$, and we may write $A = (P/z)z + Qz_1^n$. Thus, referring to (18), we see that

$$U = n(P/z)Qz_1^{n-2},$$
$$V = -(P/z)'Qz_1^n + (P/z)Q'z_1^n + (P/z)^2.$$

Hence, if $\eta_1, \cdots, \eta_s$ are the roots of $P = 0$, then (see (20))

$$I_i = n\left(\frac{P}{y - \eta_i}\right)^2 Qy_1^{n-2},$$
$$J_i = \frac{P}{y - \eta_i}\left(-\left(\frac{P}{y - \eta_i}\right)' Qy_1^n + \frac{P}{y - \eta_i} Q'y_1^n\right) + \left(\frac{P}{y - \eta_i}\right)^3.$$

Since the roots of $P = 0$ are distinct, it follows that $Qy_1^{n-2} \equiv 0(I_1, \cdots, I_s)$, so that $P \equiv 0(A, I_1, \cdots, I_s)$. Since P and Q are relatively prime, then, $y_1^{n-2} \equiv 0(A, I_1, \cdots, I_s)$, so that

$$\left(\frac{P}{y - \eta_i}\right)^3 \equiv 0 \;\; (A, I_1, \cdots, I_s, J_1, \cdots, J_s), \qquad i = 1, \cdots . s.$$

From this it follows that $1 \equiv 0 \;\; (A, I_1, \cdots, I_s, J_1, \cdots, J_s)$, so that A is of the regular type.

The coefficient in A of y_1^2 is divisible by $y - \eta_i$ if and only if $n > 2$ (in which case the coefficient vanishes). Thus, the singular solutions of A are all of the first class or are all of the second class according as $n > 2$ or $n = 2$. Consequently, the exponent of $Qy_1^n + P$ is 1 or 2 according as $n > 2$ or $n = 2$.

EXAMPLE 4. Let $A = xy_1^2 - 2yy_1 + y$, where x is any element whose derivative is 1. Then $S = 2(xy_1 - y)$, so that the singular solutions of A are $y = 0$ and $y = x$. We may write $A = x(y_1 - 1)^2 - 2(y - x)(y_1 - 1) - (y - x)$. Both singular solutions are of multiplicity unity, and are of the second class.

If we let $\eta_1 = 0$, $\eta_2 = x$, we have, referring to (20) and (18), $I_1 = 2(xy_1 - 2y + x)$, $I_2 = -2(xy_1 - 2y)$. Hence $1 = (I_1 + I_2)/2x$, so that A is of the regular type, and the exponent of $[A]$ is 2.

The following example shows that not every form with singular solutions, all of multiplicity unity, is of the regular type.

EXAMPLE 5. Let $A = y_1^4 + y_1^3 + P$, where P is a form of order zero, with constant coefficients, and with distinct zeros $\eta_1, \cdots, \eta_s$. The singular solutions of A are $\eta_1, \cdots, \eta_s$, all of multiplicity unity and of the first class. $S = 4y_1^3 + 3y_1^2$. By (18) and (20),

$$I_i = \left(\frac{P}{y - \eta_i}\right)^2 (4y_1^2 + 3y_1),$$

$$J_i = \frac{P}{y-\eta_i}\left(-\left(\frac{P}{y-\eta_i}\right)' y_1(y_1^3 + y_1^2) + \left(\frac{P}{y-\eta_i}\right)^2\right)$$
$$\equiv \frac{P}{y-\eta_i}\left(P\left(\frac{P}{y-\eta_i}\right)' + \left(\frac{P}{y-\eta_i}\right)^2\right) \equiv \left(\frac{P}{y-\eta_i}\right)^2 P', \qquad (A).$$

It follows by (29) that $\Sigma_0 = [A, B] = (A, [B])$, where $B = (4y_1^2 + 3y_1)y_2 + P'$. Hence

$$\begin{aligned}(\Sigma_0, S, I_1, \cdots, I_s) &= (y_1^4 + y_1^3 + P, 4y_1^2 + 3y_1, B, B_1, B_2, \cdots)\\ &= (256P^2 - 27P, 9y_1 + 64P, B, B_1, B_2, \cdots)\\ &= (256P - 27, P', 9y_1 + 64P, B_1, B_2, \cdots).\end{aligned}$$

Thus A is of the regular type if $256P - 27$ and P' are relatively prime. However, if $256P - 27$ and P' have a factor in common which is relatively prime to P'', then A is not of the regular type. This is easy to see from the congruences

$$\begin{aligned}B_1 &= (4y_1^2 + 3y_1)y_3 + (8y_1 + 3)y_2^2 + P''y_1\\ &\equiv -3y_2^2 - \tfrac{3}{4}P'' \qquad (256P - 27, 9y_1 + 64P),\\ B_2 &= (4y_1^2 + 3y_1)y_4 + 3(8y_1 + 3)y_2y_3 + \cdots\\ &\equiv -9y_2y_3 + H_2 \qquad (256P - 27, 9y_1 + 64P),\\ B_k &\equiv -3(k+1)y_2y_{k+1} + H_k \qquad (256P - 27, 9y_1 + 64P),\end{aligned}$$

where H_k is of order not exceeding k.

11. Case in which no singular solution exists

In this section we suppose that A *has no singular solutions.* By the analog of the Hilbert-Netto theorem this means that

$$1 \equiv 0 \quad (S, S_1, \cdots, S_p, A, A_1, \cdots, A_q)$$

for sufficiently large integers p and q.

We shall consider only those forms A for which p may be taken as 0, that is, for which there is a q such that

$$(45) \qquad 1 \equiv 0 \quad (S, A, A_1, \cdots, A_q).$$

We prove that *if* (45) *holds, than the exponent of* $[A]$ *is unity.*

Indeed, let $G \in \{A\}$. It is required to show that $G \in [A]$. Performing a partial reduction of G with respect to A, we see that there is a positive integer j such that $S^jG \equiv G'[A]$, where G' is of order less than or equal to unity. $G' \in \{A\}$, and hence is divisible by A. Therefore $S^jG \equiv 0$ $[A]$. Of course, $A_iG \equiv 0$ $[A]$, $i = 0, 1, \cdots, q$. By (41), however, $1 \equiv 0$ $(S^j, A, A_1, \cdots, A_q)$. Hence, $G = 1 \cdot G \equiv 0$ $[A]$, q.e.d.

The following example shows that a form A, which has no singular solution, for which (45) fails to hold, is exceptional.

EXAMPLE 6. Let $\mathfrak{F}$ be the differential field obtained by adjoining the unknowns $a_{\mu\nu}$ ($\mu = 0, \cdots, m$; $\nu = 0, \cdots, n$) to some differential field of characteristic zero. We consider in $\mathfrak{F}\{y\}$ the form

$$A = \sum_{\mu=0}^{m} \sum_{\nu=0}^{n} a_{\mu\nu} y^{\mu} y_1^{\nu}.$$

We shall show that A has no singular solutions, and that (45) holds, with $q = 1$.

A_1 is linear in y_2. Let $A_1 = Sy_2 + T$. We must prove that $1 \equiv 0(A, S, T)$. Suppose $1 \not\equiv 0(A, S, T)$. Then there is a solution $y = \bar{y}$, $y_1 = \bar{y}_1$ of the polynomials A, S, T. Now, $\bar{y}$ and $\bar{y}_1$ must annul, respectively, the resultant with respect to y_1 and the resultant with respect to y, of S and T. These resultants are nonzero polynomials in y and y_1, respectively, with coefficients which are polynomials in the $a_{\mu\nu}$ and $a_{\mu\nu 1}$, *not including* a_{00}. It follows, since $y = \bar{y}$, $y_1 = \bar{y}_1$ is a solution of A, that a_{00} is an algebraic function of the other $a_{\mu\nu}$ and the $a_{\mu\nu 1}$. This contradiction completes the proof.

There exist forms A for which (45) is satisfied for some $q > 1$, but for which (45) is not satisfied if $q = 1$. This is shown by

EXAMPLE 7. Let $A = 1 + 2y_1 + (1 - y^3)y_1^2$. Then $S = 2 + 2(1 - y^3)y_1$, and A has no singular solutions. Now, $y_1 + 1 = A - (y_1/2)S$, so that

$$(S, A) = (y^3, y_1 + 1).$$

Also, $A_1 \equiv -3y^2y_1^3 + 2(1 + (1 - y^3)y_1)y_2 \equiv 3y^2(S, A)$. Hence

$$(S, A, A_1) = (y^2, y_1 + 1).$$

Again,
$$A_2 = -6yy_1^4 - 15y^2y_1^2y_2 + 2(1 - y^3)y_2^2 + 2(1 + (1 - y^3)y_1)y_3 \equiv -6y + 2y_2^2 \ (S, A, A_1),$$

so that

$$(S, A, A_1, A_2) = (y^2, y_1 + 1, y_2^2 - 3y).$$

Finally, $A_3 = -6y_1^5 - 54yy_1^3y_2 - 36y^2y_1y_2^2 - 21y^2y_1^2y_3 + 6(1 - y^3)y_2y_3 + 2(1 + (1 - y^3)y_1)y_4 \equiv 6(1 + 9yy_2 - y_2y_3)$, (S, A, A_1, A_2), so that $1 \equiv y_2(y_3 - 9y)$, (S, A, A_1, A_2, A_3), and $1 \equiv y_2^4(y_3 - 9y)^4 \equiv 9y^2(y_3 - 9y)^4 \equiv 0 \ (S, A, A_1, A_2, A_3)$. Thus, (45) holds for $q = 3$, but for no lower value of q.

12. The intermediate ideals for hyperelliptic forms

We have seen in §10 that if A has singular solutions, all of multiplicity unity, at least one of which is of the second class, and if A is of the regular type, then $[A]$ is properly included in $\{A\}$, and has exponent 2. It is natural now to ask what differential ideals there are *between* $[A]$ and $\{A\}$, that is, what differential ideals Λ there are such that $[A] \subseteq \Lambda \subseteq \{A\}$. We shall call such a Λ an *intermediate ideal.* In the present section we answer this question completely for the forms of Example 3 with $n = 2$.

We are dealing then with a form $A = Qy_1^2 + P$, where P and Q have constant coefficients, are of order zero, and are relatively prime. P is assumed to be of positive degree, relatively prime to $P' = \frac{d}{dy}P$. Such a form A we shall call a *hyperelliptic* form.

Our first task will be to find a suitable basis for Σ_0.

Let $\eta_1, \cdots, \eta_s$ be the singular solutions of A, that is, the roots of $P = 0$. We have seen that

$$\Sigma_0 = [A, (I_1y_2 + J_1)^2, \cdots, (I_sy_2 + J_s)^2],^{32} \tag{46}$$

where

$$(y - \eta_i)(I_iy_2 + J_i) \equiv 0, \qquad y_1(I_iy_2 + J_i) \equiv 0 \qquad [A], \tag{47}$$

and

$$1 \equiv 0\ (A, I_1, \cdots, I_s). \tag{48}$$

Congruence (48) shows that there is a form K, with constant coefficients and of order less than 2, such that

$$y_2 + K \equiv 0\ (A, I_1y_2 + J_1, \cdots, I_sy_2 + J_s). \tag{49}$$

We shall show that

$$\Sigma_0 = [A, (y_2 + K)^2], \qquad \{\Sigma_0\} = [A, y_2 + K], \tag{50}$$

and that

$$(y - \eta_1) \cdots (y - \eta_s)(y_2 + K) \equiv 0, \qquad y_1(y_2 + K) \equiv 0 \qquad [A]. \tag{51}$$

Now, (51) is obvious from (47) and (49). As to (50), it is evident from (46), (49) and the fact that Σ_0 is of exponent 2, that $[A, (y_2 + K)^2] \subseteq \Sigma_0$. To prove the inclusion in the opposite direction, consider any $I_iy_2 + J_i$. We have

$$I_iy_2 + J_i = I_i(y_2 + K) + J_i - I_iK.$$

Hence $J_i - I_iK \in \{\Sigma_0\}$. Since the order of $J_i - I_iK$ does not exceed unity, this form is divisible by A. Thus,

$$I_i y_2 + J_i \equiv 0\ (A, y_2 + K),$$

so that

$$(I_iy_2 + J_i)^2 \equiv 0\ [A, (y_2 + K)^2], \qquad i = 1, \cdots, s.$$

This, in conjunction with (46), shows that $\Sigma_0 \subseteq [A, (y_2 + K)^2]$, and completes the proof of the first part of (50). The second part of (50) is now obvious.

The congruence (50) gives the basis sought for Σ_0. We have, as a result,

---

[32] Since P has constant coefficients the η_i are constants, so that $(y - \eta_i)_j = y_j$, $i = 1, \cdots, s$, $j = 1, 2, \cdots$.

(52) $$[A] = [P]\cdot[A, (y_2 + K)^2], \qquad \{A\} = [P]\cdot[A, y_2 + K],$$

for P is a constant times the product of the $y - \eta_i$.[33]

If we refer to (51), the first intermediate ideals which present themselves are of the type

(53) $$[P]\cdot[A, (y_2 + K)^2, (y - \eta_{i_1}) \cdots (y - \eta_{i_g})(y_2 + K)],$$

where $i_1, \cdots, i_g$ are distinct integers from 1 to s. If $g = s$, the ideal (53) is $[A]$, if $g = 0$, (53) is $\{A\}$.

We shall show that *the ideals* (53) *are distinct from one another*, and that *every intermediate ideal is one of the ideals* (53), that is, the ideals (53) are the only intermediate ideals.

To prove that the ideals (53) are all distinct, it suffices to show, $h = 1, \cdots, s$, that

$$[P]\cdot[A, (y_2 + K)^2, (y - \eta_{i_1}) \cdots (y - \eta_{i_{h-1}})(y_2 + K)]$$
$$\neq [P]\cdot[A, (y_2 + K)^2, (y - \eta_{i_1}) \cdots (y - \eta_{i_h})(y_2 + K)].$$

For, if

$$[P]\cdot[A, (y_2 + K)^2, (y - \eta_{i_1}) \cdots (y - \eta_{i_h})(y_2 + K)]$$

and

$$[P]\cdot[A, (y_2 + K)^2, (y - \eta_{j_1}) \cdots (y - \eta_{j_f})(y_2 + K)]$$

were equal, they would both equal

$$[P]\cdot[A, (y_2 + K)^2, (y - \eta_{k_1}) \cdots (y - \eta_{k_l})(y_2 + K)],$$

where $k_1, \cdots, k_l$ are the integers common to $i_1, \cdots, i_h$ and $j_1, \cdots, j_f$; and if

$$[P]\cdot[A, (y_2 + K)^2, (y - \eta_{i_1}) \cdots (y - \eta_{i_{h-d}})(y_2 + K)]$$
$$= [P]\cdot[A, (y_2 + K)^2, (y - \eta_{i_1}) \cdots (y - \eta_{i_h})(y_2 + K)],$$

with $d > 1$, then we would have

$$[P]\cdot[A, (y_2 + K)^2, (y - \eta_{i_1}) \cdots (y - \eta_{i_{h-1}})(y_2 + K)$$
$$= [P]\cdot[A, (y_2 + K)^2, (y - \eta_{i_1}) \cdots (y - \eta_{i_h})(y_2 + K)].$$

For $h = s$ this means we must show that

$$[A] \subset [P]\cdot[A, (y_2 + K)^2, (y - \eta_{i_1}) \cdots (y - \eta_{i_{s-1}})(y_2 + K)]$$

for all distinct $i_1, \cdots, i_{s-1}$. This will be proved, however, if we show that

(54) $$(y - \eta_{i_1}) \cdots (y - \eta_{i_{s-1}})(y_3 + K_1) \not\equiv 0 \qquad [A],$$ [34]

for the first member here is in

[33] See (10).

[34] K_j is the j-th derivative of K.

$$[P]\cdot[A, (y_2 + K)^2, (y - \eta_{i_1}) \cdots (y - \eta_{i_{s-1}})(y_2 + K)].$$[35]

The relation (54) may be established in the same way in which (37) was disproved.

From (54) it easily follows that

$$(y - \eta_{i_1}) \cdots (y - \eta_{i_{s-1}})(y_2 + K) \not\equiv 0 \qquad (\Sigma_0).$$

For $h < s$ we proceed by induction. Assume that

$$(55)\qquad \begin{aligned}&(y - \eta_{i_1}) \cdots (y - \eta_{i_{j-1}})(y - \eta_{i_{j+1}}) \cdots (y - \eta_{i_{h+1}})(y_2 + K)\\ &\qquad \not\equiv 0 \ [A, (y_2 + K)^2, (y - \eta_{i_1}) \cdots (y - \eta_{i_{h+1}})(y_2 + K)]\end{aligned}$$

for all distinct $i, \cdots, i_{h+1}$ and all $j = 1, \cdots, h + 1$. (For $h = s - 1$ this has just been proved.)

Suppose, say, that

$$\begin{aligned}&(y - \eta_1) \cdots (y - \eta_{h-1})(y_2 + K)\\ &\qquad \equiv 0 \ [A, (y_2 + K)^2, (y - \eta_1) \cdots (y - \eta_h)(y_2 + K)].\end{aligned}$$

Then, for t sufficiently large,

$$\begin{aligned}&(y - \eta_1) \cdots (y - \eta_{h-1})(y - \eta_{h+1})^t(y_2 + K)\\ &\qquad \equiv 0 \ [A, (y_2 + K)^2, (y - \eta_1) \cdots (y - \eta_{h+1})(y_2 + K)].\end{aligned}$$

Combining this congruence with

$$\begin{aligned}&(y - \eta_1) \cdots (y - \eta_{h+1})(y_2 + K)\\ &\qquad \equiv 0 \ [A, (y_2 + K)^2, (y - \eta_1) \cdots (y - \eta_{h+1})(y_2 + K)],\end{aligned}$$

we see that

$$\begin{aligned}&(y - \eta_1) \cdots (y - \eta_{h-1})(y - \eta_{h+1})(y_2 + K)\\ &\qquad \equiv 0 \ [A, (y_2 + K)^2, (y - \eta_1) \cdots (y - \eta_{h+1})(y_2 + K)].\end{aligned}$$

This contradicts (55), and completes the first part of the proof.

As for the second part, let Σ be a differential ideal between Σ_0 and $\{\Sigma_0\}$, distinct from $\{\Sigma_0\}$. We shall complete the proof by showing that Σ coincides with some ideal

$$(56)\qquad [A, (y_2 + K)^2, (y - \eta_{i_1}) \cdots (y - \eta_{i_g})(y_2 + K)].$$

[35] The first member of (54) is in $[P]$ because it is in $[y_1]$ and $y_1 \in [P]$; the first member of (54) is in $[A, (y^2 + K)^2, (y - \eta_{i_1}) \cdots (y - \eta_{i_{s-1}})(y_2 + K)]$ because $((y - \eta_{i_1}) \cdots (y - \eta_{i_{s-1}})(y_2 + K))_1 = ((y - \eta_{i_1}) \cdots (y - \eta_{i_{s-1}}))' y_1(y_2 + K) + (y - \eta_{i_1}) \cdots (y - \eta_{i_{s-1}})(y_3 + K_1) \equiv (y - \eta_{i_1}) \cdots (y - \eta_{i_{s-1}})(y_3 + K_1)$, (Σ_0). Moreover, the ideals $[P]$ and $[A, (y_2 + K)^2, (y - \eta_{i_1}) \cdots (y - \eta_{i_{s-1}})(y_2 + K)]$ are separated, because P and $y_2 + K$ have no solution in common.

Let q be the smallest integer for which some ideal (56) with $g = q$ is included in Σ. q is an integer from 1 to s. Suppose, say, that

$$\Lambda \subseteq \Sigma,$$

where

$$\Lambda = [A, (y_2 + K)^2, (y - \eta_1) \cdots (y - \eta_q)(y_2 + K)],$$

but that

$$[A, (y_2 + K)^2, (y - \eta_1) \cdots (y - \eta_{i-1})(y - \eta_{i+1}) \cdots (y - \eta_q)(y_2 + K)] \nsubseteq \Sigma,$$

$i = 1, \cdots, q$. We shall prove that $\Sigma = \Lambda$.

Assume to the contrary that $F \in \Sigma$, $F \notin \Lambda$. Since $F \in \{\Sigma_0\} = [A, y_2 + K]$, we may write

$$F = C_0 A + C_1(y_2 + K) + \cdots + C_t(y_{t+1} + K_{t-1}). \tag{57}$$

Let D_t be the remainder of C_t with respect to $y_2 + K$. Since $(y_{i+1} + K_{i-1})(y_{t+1} + K_{t-1}) \equiv 0$ (Λ), $i = 1, 2, \cdots$, we have, by (57),

$$F \equiv C_1(y_2 + K) + \cdots + C_{t-1}(y_t + K_{t-2}) + D_t(y_{t+1} + K_{t-1}), \qquad (\Lambda). \tag{58}$$

D_t is a polynomial in y, y_1. Let E_t be the sum of the terms of D_t free of y_1. Since, by (51), $(y_1(y_2 + K))_{t-1} \equiv 0$ (Λ), it follows that

$$y_1(y_{t+1} + K_{t-1}) \equiv -(t - 1)y_2(y_t + K_{t-2}) - \cdots - y_t(y_2 + K), \qquad (\Lambda). \tag{59}$$

From this and (58) we see that

$$F \equiv C_1'(y_2 + K) + \cdots + C_{t-1}'(y_t + K_{t-2}) + E_t(y_{t-1} + K_{t-1}), \qquad (\Lambda), \tag{60}$$

where the C_i' are new forms.

Let H_t be the remainder obtained on dividing E_t by $(y - \eta_1) \cdots (y - \eta_q)$. Since $((y - \eta_1) \cdots (y - \eta_q)(y_2 + K))_{t-1} \equiv 0$ (Λ),

$$\begin{aligned}(y - \eta_1) \cdots (y - \eta_q)(y_{t+1} + K_{t-1}) \\ \equiv -(t - 1)((y - \eta_1) \cdots (y - \eta_q))_1(y_t + K_{t-1}) \\ - \cdots - ((y - \eta_1) \cdots (y - \eta_q))_{t-1}(y_2 + K), \qquad (\Lambda).\end{aligned}$$

Hence, it follows from (60) that

$$F \equiv C_1''(y_2 + K) + \cdots + C_{t-1}''(y_t + K_{t-2}) + H_t(y_{t+1} + K_{t-1}), \qquad (\Lambda), \tag{61}$$

where the C_i'' are new forms.

Treating the first $t - 1$ terms in the second member of (61) as we did the second member of (57), we obtain, after $t - 1$ more steps, a relation

$$F \equiv H_1(y_2 + K) + \cdots + H_t(y_{t+1} + K_{t-1}), \qquad (\Lambda), \tag{62}$$

where each H_i is a form of order zero and degree less than q.

At least one H_i must be nonzero, else $F \in \Lambda$. We may (and do) suppose that $H_t \neq 0$.

From the congruence $y_1(y_2 + K) \equiv 0 \quad (\Lambda)$, we obtain relations

(63) $$y_1^j(y_{j+1} + K_{j-1}) \equiv 0 \qquad (\Lambda), \qquad\qquad j = 1, 2, \cdots .$$

Therefore, if we multiply (62) by y_1^{t-1} we find

$$H_t y_1^{t-1}(y_{t+1} + K_{t-1}) \equiv 0 \qquad (\Sigma).$$

Now, by (59) and (63),

$$\begin{aligned} y_1^{j-1}(y_{j+1} + K_{j-1}) &\equiv -(j-1)y_1^{j-2}y_2(y_j + K_{j-2}) \\ &\equiv (j-1)y_1^{j-2}K(y_j + K_{j-2}), \qquad (\Lambda), \end{aligned}$$

so that

$$y_1^{t-1}(y_{t+1} + K_{t-1}) \equiv (t-1)!K^{t-1}(y_2 + K), \qquad (\Lambda).$$

Hence,

$$H_t K^{t-1}(y_2 + K) \equiv 0 \qquad (\Sigma).$$

If L is the sum of the terms of K free of y_1, we find, since $y_1(y_2 + K) \equiv 0$ **(Σ)**

$$H_t L^{t-1}(y_2 + K) \equiv 0 \qquad (\Sigma).$$

But $(y - \eta_1) \cdots (y - \eta_q)(y_2 + K) \equiv 0 \ (\Sigma)$. Hence, if J is the greatest common divisor of $H_t L^{t-1}$ and $(y - \eta_1) \cdots (y - \eta_q)$,

$$J(y_2 + K) \equiv 0 \qquad (\Sigma).$$

But L is prime to $(y - \eta_1) \cdots (y - \eta_q)$, for otherwise the ideals $[y - \eta_1], \cdots, [y - \eta_q], [A, y_2 + K]$ would not be separated. Hence J is a divisor of H_t, is of degree less than q, and is a *proper* divisor of $(y - \eta_1) \cdots (y - \eta_q)$. This contradicts the definition of q as the least g for which Σ includes an ideal (56), and completes the proof.

The results just established show that the ideals between $[A]$ and $\{A\}$ can be organized into ascending *chains*, all of the same length s. That is, if $\Lambda_1, \cdots, \Lambda_{h-1}$ are intermediate ideals such that

$$[A] \subset \Lambda_1 \subset \cdots \subset \Lambda_{h-1} \subset \{A\}$$

(chain of length h), then there are $s - h$ other intermediate ideals which together with $[A], \Lambda_1, \cdots, \Lambda_{h-1}, \{A\}$ form a chain of length s. Moreover, there exist no chains of length exceeding s. This illustrates a phenomenon which we shall discuss in Part III.

Part III. Chains

13. Generalities

We return now to an abstract differential domain of integrity with basis theorem, which contains the rational numbers.

The $n + 1$ differential ideals $\sigma_0, \sigma_1, \cdots, \sigma_{n-1}, \sigma^*$ are said to form a *chain* if

(64) $$\sigma_0 \subset \sigma_1 \subset \cdots \subset \sigma_{n-1} \subset \sigma^*.$$

n is called the *length* of the chain. We shall call σ_0 and σ^* the *end ideals,* $\sigma_1, \cdots, \sigma_{n-1}$ the *intermediate ideals,* of the chain. The ideals σ_{i-1}, σ_i (also, σ_{n-1}, σ^*) will be said to be *contiguous.* A set of ideals of the chain will be called *non-contiguous* if no two ideals of the set are contiguous. A chain is called a *principal* chain if no pair of contiguous ideals have a differential ideal properly between them. A principal chain may be expressed schematically by means of dashes:

$$\sigma_0 - \sigma_1 - \cdots - \sigma_{n-1} - \sigma^*.$$

The notation $\sigma - \tau$ means that $\sigma \subset \tau$, but there is no ω with $\sigma \subset \omega \subset \tau$. A chain with end ideals σ_0, σ^* is said to be a *refinement* of the chain (64) if each intermediate ideal of (64) is an intermediate ideal of the chain in question.

Let $\sigma_0, \sigma_1, \cdots, \sigma_{n-1}, \sigma^*$ be a principal chain, and let there be a second principal chain $\sigma_0, \sigma_1, \cdots, \sigma_{i-1}, \tau_i, \sigma_{i+1}, \cdots, \sigma_{n-1}, \sigma^*$ with $\sigma_i \neq \tau_i$, that is, schematically, let

(65) $$\sigma_0 - \sigma_1 - \cdots - \sigma_{i-1} \begin{smallmatrix} \nearrow \sigma_i \searrow \\ \searrow \tau_i \nearrow \end{smallmatrix} \sigma_{i+1} - \cdots - \sigma_{n-1} - \sigma^*.$$

We shall say that the first chain is *flexible* at the intermediate ideal σ_i, and that the second chain is obtained from the first by a *flex.*

The first theorem on chains that we shall state is that of Jordan-Hölder-Schreier. A differential domain of integrity $\mathfrak{D}$ forms a group under addition. As a set of operators on this group we may consider 1. multiplication by an element of $\mathfrak{D}$, 2. differentiation. The permissible invariant subgroups of this group with operators are precisely the differential ideals of $\mathfrak{D}$. Thus, we may state

THEOREM 7 (Jordan-Hölder-Schreier): *Two chains with the same end ideals have refinements of equal length, with differential rings of remainder classes which are isomorphic (in some order).*[36]

PROOF: See van der Waerden, *Moderne Algebra,* vol. 1, 2nd edition, §46, or H. Zassenhauss, *Lehrbuch der Gruppentheorie,* vol. 1, chapter 2, §5.

As a consequence of Theorem 7, we see that all principal chains with the same end ideals have the same length.

THEOREM 8 (Ore): *Two different principal chains with the same end ideals, can be obtained from each other by a finite number of successive flexes.*

PROOF: See O. Ore, *On the foundations of abstract algebra.* I, Annals of Mathematics, vol. 36 (1935), pp. 406–434.

14. The exponent of a principal chain

By the *exponent* of the chain (64) we shall mean $l_{\sigma^*}\sigma_0$. We shall be concerned with principal chains, exclusively. Of course, if $\sigma^* \not\subseteq \{\sigma_0\}$, then the chain (64) has exponent ∞, so we may limit ourselves to chains for which $\sigma^* \subseteq \{\sigma_0\}$.

---

[36] By the differential ring of remainder classes of a chain (64) is meant the differential rings $\sigma_1/\sigma_0, \sigma_2/\sigma_1, \cdots, \sigma^*/\sigma_{n-1}$.

Our chief result may be stated as follows.

THEOREM 9: Let

$$\sigma_0 - \sigma_1 - \cdots - \sigma_{n-1} - \sigma^*, \qquad \sigma^* \subseteq \{\sigma_0\}, \tag{66}$$

be a principal chain which is flexible at r non-contiguous intermediate ideals.[37] *Then the exponent of the chain* (66) *does not exceed* $n - r + 1$.

As a consequence of Theorem 9, we see that if a differential ideal σ has finite exponent p, then there is a chain, with end ideals σ and $\{\sigma\}$, of length $p - 1$.

PROOF: We begin by showing that, if $\sigma_0 - \sigma_1$, and $\sigma_1 \subseteq \{\sigma_0\}$, then $l_{\sigma_1}\sigma_0 = 2$. Indeed, let $l_{\sigma_1}\sigma_0 > 2$. Then there is a $b \in \sigma_1$ such that $b^2 \notin \sigma_0$. Hence

$$\sigma_0 \subset [\sigma_0, b^2] \subseteq \sigma_1.$$

Since $\sigma_0 - \sigma_1$, we see that $\sigma_1 = [\sigma_0, b^2]$. In particular, $b \in [\sigma_0, b^2]$, so that

$$b \equiv A\ (\sigma_0),$$

where A is a form in b with coefficients in $\mathfrak{D}$, which has no term of degree less than 2. Using the iterative process introduced in the paper cited under footnote 19, we see that $b \equiv 0\ (\sigma_0)$. This contradiction shows that $l_{\sigma_1}\sigma_0 = 2$.

We now show that the exponent of the principal chain (66) does not exceed $n + 1$.

For chains of length 1 we have just given the proof. Let $n > 1$, and assume the result for chains of length less than n.

By the assumption, $\sigma^{*^n} \subseteq \sigma_1$. Suppose $\sigma^{*^{n+1}} \not\subseteq \sigma_0$. Then

$$\sigma_0 \subset (\sigma_0, \sigma^{*^{n+1}}) \subseteq \sigma_1.$$

Since $\sigma_0 - \sigma_1$, we see that $\sigma_1 = (\sigma_0, \sigma^{*^{n+1}})$. Hence

$$\begin{aligned} \sigma^{*^{n+1}} &= \sigma^*\sigma^{*^n} \subseteq \sigma^*\sigma_1 = \sigma^*(\sigma_0, \sigma^{*^{n+1}}) \\ &\subseteq (\sigma_0, \sigma^{*^{n+2}}) \subseteq (\sigma_0, \sigma^{*^2}\sigma_1) \\ &\subseteq (\sigma_0, (\sigma_0, \sigma^{*^{n+3}})) = (\sigma_0, \sigma^{*^{n+3}}) \\ &\subseteq \cdots \subseteq (\sigma_0, \sigma^{*^{2n}}) \subseteq (\sigma_0, \sigma_1^2) \subseteq \sigma_0. \end{aligned}$$

This contradiction proves our statement.

We are now ready to complete the proof of our theorem. What we have just shown is that the theorem is valid for $r = 0$. We proceed by induction. Let $r > 0$ and suppose the theorem holds for fewer than r non-contiguous ideals.

Let σ_i be the intermediate ideal of (66) of lowest subscript, at which the chain is flexible. Then we have, with $\sigma_i \neq \tau_i$, the situation depicted by (65). Now

$$\sigma_i - \sigma_{i+1} - \cdots - \sigma_{n-1} - \sigma^*$$

[37] Clearly, r is an integer between 0 and the greatest integer in $n/2$, inclusive.

and

$$\tau_i - \sigma_{i+1} - \cdots - \sigma_{n-1} - \sigma^*$$

are both principal chains of length $n - i$, flexible at $r - 1$ intermediate ideals. Hence, by the induction assumption, $\sigma^{*^{n-i-r+2}} \subseteq \sigma_i$, $\sigma^{*^{n-i-r+2}} \subseteq \tau_i$. But $\sigma_{i-1} \subseteq \sigma_i \cap \tau_i \subset \sigma_i$ and $\sigma_{i-1} - \sigma_i$, so that $\sigma_{i-1} = \sigma_i \cap \tau_i$. Therefore, $\sigma^{*^{n-i-r+2}} \subseteq \sigma_{i-1}$.

Suppose, now, that $\sigma^{*^{n-r+1}} \not\subseteq \sigma_0$. Let σ_k be the σ_j of least subscript, such that $\sigma^{*^{n-j-r+1}} \subseteq \sigma_j$. Then $1 \leqq k \leqq i - 1$ and

$$\sigma^{*^{n-k-r+1}} \subseteq \sigma_k, \qquad \sigma^{*^{n-k-r+2}} \not\subseteq \sigma_{k-1}, \qquad \sigma^{*^{n-k-r+3}} \subseteq \sigma_{k-1}.$$

Hence

$$\sigma_{k-1} \subset (\sigma_{k-1}, \sigma^{*^{n-k-r+2}}) \subseteq \sigma_k,$$

so that, since $\sigma_{k-1} - \sigma_k$, $\sigma_k = (\sigma_{k-1}, \sigma^{*^{n-k-r+2}})$. Thus,

$$\begin{aligned} \sigma^{*^{n-k-r+2}} &= \sigma^* \sigma^{*^{n-k-r+1}} \subseteq \sigma^* \sigma_k \\ &= \sigma^*(\sigma_{k-1}, \sigma^{*^{n-k-r+2}}) \\ &\subseteq (\sigma_{k-1}, \sigma^{*^{n-k-r+3}}) = \sigma_{k-1}. \end{aligned}$$

This contradiction completes the proof of Theorem 9.

EXAMPLE 8. Consider the chain

$$[y^{n+1}, y_1], [y^n, y_1], \cdots, [y^2, y_1], [y]$$

in any differential ring $\mathfrak{F}\{y\}$. It is easy to prove that this is a principal chain, flexible at none of its intermediate ideals. The length is n and the exponent is obviously $n + 1$. Thus, the bound on the exponent given by Theorem 9 is the exponent itself.

The necessity of the non-contiguity condition in Theorem 9 is shown by

EXAMPLE 9. Let $\mathfrak{F}$ be a differential field of characteristic zero, which contains an element x with derivative 1. Consider the chain

$$[y^3, yy_1, y_2], [y^2, y_1^2, y_2], [y^2, y_1], [y]$$

in $\mathfrak{F}\{y\}$. This is a principal chain of length 3. Since

$$[y^3, yy_1, y_2] \begin{array}{c} \nearrow [y^2, y_1^2, y_2] \searrow \\ \searrow [y^3, y_1] \nearrow \end{array} [y^2, y_1]$$

and

$$[y^2, y_1^2, y_2] \begin{array}{c} \nearrow [y^2, y_1] \searrow \\ \searrow \left[y^2, y_1 - \frac{1}{x} y\right] \nearrow \end{array} [y],$$

the chain is flexible at two intermediate ideals. Yet its exponent is 3.

APPENDIX. GENERAL SOLUTIONS AND RESOLVENTS

Liquidating an obligation incurred in Part I, we give here an abstract, purely algebraic treatment of the subject matter of Chapter 2 of Ritt's book, *Differential equations from the algebraic standpoint.*[38] We shall make free reference to this book, designating it by (*R*), and shall give proofs only when they differ from those in (*R*).

1*. The general solution of a form

We work in an abstract differential field $\mathfrak{F}$ of characteristic zero. The letters u_i, y_i and w will denote unknowns.

We establish the following fundamental lemma as a substitute for the existence theorem for differential equations.

LEMMA 1*: Let A be an algebraically irreducible form in $\mathfrak{F}\{y_1, \cdots, y_n\}$ of positive class. Let $B_1, \cdots, B_t$ be nonzero forms in $\mathfrak{F}\{y_1, \cdots, y_n\}$, reduced with respect to A. Then $B_1 \cdots B_t \notin \{A\}$.

REMARK: Lemma 1*, together with the analog to the Hilbert-Netto theorem, shows that every algebraically irreducible form of positive class has a *regular* solution, that is, a solution for which the separant S and initial I of A do not vanish. For otherwise IS would be in $\{A\}$.

PROOF: Let m be the class of A, r the order of A in y_m. Suppose $B_1 \cdots B_t \in \{A\}$, that is, $(B_1 \cdots B_t)^p \in [A]$, for some positive integer p. Now, A and $(B_1 \cdots B_t)^p$ are relatively prime. Hence there is a nonzero form C, either of class less than m, or of class m and of order less than r in y_m, such that

$$C = DA + E(B_1 \cdots B_t)^p,$$

where D and E are forms in $\mathfrak{F}\{y_1, \cdots, y_n\}$. $C \in [A]$, so that we may write, for suitable q and Q_i,

$$C = Q_0A + Q_1A_1 + \cdots + Q_qA_q. \tag{1*}$$

Now, $A_i = Sy_{m,r+i} + T_i$, where T_i is a form of order less than $r + i$ in y_m. Hence, if in (1*) we consider the forms as polynomials in the letters y_{ij}, and let

$$y_{m,r+i} = -\frac{T_i}{S}, \qquad i = 1, 2, \cdots,$$

we see that S^hC is divisible by A for a sufficiently large integer h. Since neither S nor C is so divisible, we have a contradiction, completing the proof.

We now are in a position to follow (*R*).

Let $A \in \mathfrak{F}\{y_1, \cdots, y_n\}$ be algebraically irreducible and of class n. Let S and I be the separant and initial, respectively, of A. Let Σ_1 be the set of all forms in $\mathfrak{F}\{y_1, \cdots, y_n\}$ which vanish for all regular solutions of A, or what is the same thing, let $\Sigma_1 = \{A\}:IS$.

[38] American Mathematical Society Colloquium Publications, vol. 14.

Following (R), we can show that Σ_1 *is a prime differential ideal.* The manifold of Σ_1 is called the *general solution* of A.

The following three results are proved as in (R):

$\Sigma_1 = \{A\}:S$;

Σ_1 *is an essential prime differential ideal in the decomposition of* $\{A\}$*; the other essential ideals contain* S*;*

Σ_1 *is independent of the order of the unknowns.*

2*. The resolvent of a prime differential ideal

Let Σ be a perfect differential ideal in $\mathfrak{F}\{y_1, \cdots, y_n\}$. Let

$$A_1, \cdots, A_r \tag{2*}$$

be a basic set of Σ. We denote the separant and initial of A_i, $i = 1, \cdots, r$, by S_i and I_i, respectively. Let $\Sigma' = \{A_1, \cdots, A_r\}:I_1S_1 \cdots I_rS_r$, that is, let Σ' be the set of all forms in $\mathfrak{F}\{y_1, \cdots, y_n\}$ which vanish for all regular solutions of the basic set (2*).[39] Clearly, $\Sigma \subseteq \Sigma'$.

Following (R), we see that *if* Σ *is prime, then* $\Sigma = \Sigma'$. In particular (2*) has regular solutions. Also, Σ *is the only prime differential ideal with basic set* (2*).

Let $\mathfrak{C}$ be the set of all constants (that is, elements with vanishing derivative) in $\mathfrak{F}$. $\mathfrak{C}$ itself is a differential field of characteristic zero.

We shall need the following

LEMMA 2*: $\eta_1, \cdots, \eta_n \in \mathfrak{F}$ *are linearly dependent over* $\mathfrak{C}$ *if and only if*

$$\begin{vmatrix} \eta_1 & \cdots & \eta_n \\ \eta_{11} & \cdots & \eta_{n1} \\ \cdots & \cdots & \cdots \\ \eta_{1,n-1} & \cdots & \eta_{n,n-1} \end{vmatrix} = 0. \tag{3*}$$

PROOF: If $\sum_{\nu=1}^{n} c_\nu \eta_\nu = 0$, where the c_ν are in $\mathfrak{C}$ and not all 0, then $\sum_{\nu=1}^{n} c_\nu \eta_{\nu i} = 0$, $i = 0, 1, \cdots, n-1$, so that (3*) must hold. Conversely, let (3*) hold. If $n = 1$, the lemma is obviously valid. Let $n > 1$ and suppose the lemma holds for fewer than n elements η_ν. Let

$$\mu_\nu = (-1)^{n+\nu} \begin{vmatrix} \eta_1 & \cdots & \eta_{\nu-1} & \eta_{\nu+1} & \cdots & \eta_n \\ \cdots & \cdots & \cdots & \cdots & \cdots & \cdots \\ \eta_{1,n-2} & \cdots & \eta_{\nu-1,n-2} & \eta_{\nu+1,n-2} & \cdots & \eta_{n,n-2} \end{vmatrix}, \qquad \nu = 1, \cdots, n.$$

If $\mu_n = 0$, the lemma for $n - 1$ shows that the η_ν are linearly dependent over $\mathfrak{C}$. Assume that $\mu_n \neq 0$. By (3*),

$$\sum_{\nu=1}^{n} \eta_{\nu i}\mu_\nu = 0, \qquad i = 0, 1, \cdots, n-1.\text{[40]} \tag{4*}$$

[39] A solution of (2*) is called *regular* if it does not annul $I_1S_1 \cdots I_rS_r$.

[40] For $i = n - 1$, the first member of (4*) is the development of the determinant in (3*) by the minors of the elements of the last row. For $i < n - 1$, (4*) is a statement of the vanishing of certain determinants with two identical rows.

Differentiating the first $n - 1$ equations (4*), we find

$$\sum_{\nu=1}^{n} \eta_{\nu,i+1}\mu_\nu + \sum_{\nu=1}^{n} \eta_{\nu i}\mu_{\nu 1} = 0, \qquad i = 0, 1, \cdots, n-2.$$

Taking into account the last $n - 1$ equations (4*), we see that $\sum_{\nu=1}^{n} \eta_{\nu i}\mu_{\nu 1} = 0$, $i = 0, 1, \cdots, n-2$. These equations may be written

$$\sum_{\nu=1}^{n-1} \eta_{\nu i}\mu_{\nu 1} = -\eta_{ni}\mu_{n1}, \qquad i = 0, 1, \cdots, n-2.$$

Solving these equations in $\mu_{11}, \cdots, \mu_{n-1,1}$ by Cramer's rule, we obtain

$$\mu_{\nu 1} = \frac{\mu_{n1}}{\mu_n}\mu_\nu, \qquad \nu = 1, \cdots, n-1.$$

Hence $(\mu_\nu/\mu_n)_1 = (\mu_n\mu_{\nu 1} - \mu_{n1}\mu_\nu)/\mu_n^2 = 0$, so that $\mu_\nu/\mu_n = c_\nu$, $\nu = 1, \cdots, n$, where the c_ν are constants, not all zero ($c_n = 1$). Using (4*) with $i = 0$, we see that $\sum_{\nu=1}^{n} c_\nu\eta_\nu = 0$, which completes the proof.

LEMMA 3*: *Let $\mathfrak{F}$ contain a non-constant element ξ. Let A be a nonzero form in $\mathfrak{F}\{y_1, \cdots, y_n\}$. Then there are elements $\eta_1, \cdots, \eta_n \in \mathfrak{F}$ such that $A \neq 0$ when $y_1 = \eta_1, \cdots, y_n = \eta_n$.*

PROOF: It suffices to consider the case $n = 1$, for the cases $n > 1$ follow by induction. Let $A \in \mathfrak{F}\{y\}$ be of order p, and let $Y_j = \sum_{i=0}^{p} c_i(\xi^i)_j$, $j = 0, 1, \cdots, p$, where the c_i are new unknowns.[41] The Jacobian of the $p + 1$ functions Y_j of the $p + 1$ variables c_i is

$$\omega = \begin{vmatrix} 1 & \xi & \xi^2 & \cdots & \xi^p \\ 0 & \xi_1 & (\xi^2)_1 & \cdots & (\xi^p)_1 \\ \cdots & \cdots & \cdots & \cdots & \cdots \\ 0 & \xi_p & (\xi^2)_p & \cdots & (\xi^p)_p \end{vmatrix}.$$

If ω were 0, ξ would be algebraic over $\mathfrak{C}$ (by Lemma 2*) and would therefore be a constant, so that $\omega \neq 0$. Hence $Y_0, \cdots, Y_p$, considered as polynomials in $c_0, \cdots, c_p$, are algebraically independent.[42] Therefore, if we substitute $Y_0, Y_1, \cdots, Y_p$ in A for $y, y_1, \cdots, y_p$, respectively, A will go over into a nonzero polynomial C in the c_i. Choose integers $\gamma_0, \gamma_1, \cdots, \gamma_p$ such that $C \neq 0$ when $c_i = \gamma_i$, $i = 0, 1, \cdots, p$. Then the element $\eta = \sum_{i=0}^{p} \gamma_i\xi^i \in \mathfrak{F}$ leaves A nonzero when substituted for y.

Using Lemma 3*, we can follow the proof of §25 in (R) to obtain the following result.

Let Σ be a non-trivial prime differential ideal in $\mathfrak{F}\{u_1, \cdots, u_q, y_1, \cdots, y_p\}$ with arbitrary unknowns $u_1, \cdots, u_q$ (q may be 0.) If $\mathfrak{F}$ contains a non-constant, then there are elements $\mu_1, \cdots, \mu_p \in \mathfrak{F}$, and a nonzero form $G \in \mathfrak{F}\{u_1, \cdots, u_q\}$, such that if

---

[41] $(\xi^i)_j$ is the j-th derivative of ξ^i.

[42] See O. Perron, *Algebra*, vol. 1, Berlin, 1932, p. 134.

(5*) $\bar{u}_1, \cdots, \bar{u}_q, y_1', \cdots, y_p'$ *and* $\bar{u}_1, \cdots, \bar{u}_q, y_1'', \cdots, y_p''$

are two distinct solutions of Σ, *then either* $G(\bar{u}_1, \cdots, \bar{u}_q) = 0$, *or*

$$\mu_1(y_1' - y_1'') + \cdots + \mu_p(y_p' - y_p'') \neq 0.$$

Taking over the proof in §26 of (R), we obtain:

Let $\mathfrak{F}$ *be any differential field of characteristic zero. Let* Σ *be a nontrivial prime differential ideal in* $\mathfrak{F}\{u_1, \cdots, u_q, y_1, \cdots, y_p\}$ *with arbitrary unknowns* $u_1, \cdots, u_q$. *We suppose that* $q > 0$. *Then there are forms*

$$G, M_1, \cdots, M_p \in \mathfrak{F}\{u_1, \cdots, u_p\},$$

with $G \neq 0$, *such that if* (5*) *are two distinct solutions of* Σ, *then either* $G(\bar{u}_1, \cdots, \bar{u}_q) = 0$ *or*

$$M_1(\bar{u}_1, \cdots, \bar{u}_q)\cdot(y_1' - y_1'') + \cdots + M_p(\bar{u}_1, \cdots, \bar{u}_q)\cdot(y_p' - y_p'') \neq 0.$$

Henceforth we shall deal with a non-trivial prime differential ideal Σ in $\mathfrak{F}\{u_1, \cdots, u_q, y_1, \cdots, y_p\}$, with arbitrary unknowns $u_1, \cdots, u_q$, and we shall assume that either $\mathfrak{F}$ contains non-constant elements, or the arbitrary unknowns u_i really exist. Then there exists a triad of forms G, P, $Q \in \mathfrak{F}\{u_1, \cdots, u_q, y_1, \cdots, y_p\}$, with G free of the y_i, and not 0, and with $P \notin \Sigma$, such that if (5*) are two distinct solutions of Σ for which $GP \neq 0$, then

$$\frac{Q(\bar{u}_1, \cdots, \bar{u}_q, y_1', \cdots, y_p')}{P(\bar{u}_1, \cdots, \bar{u}_q, y_1', \cdots, y_p')} \neq \frac{Q(\bar{u}_1, \cdots, \bar{u}_q, y_1'', \cdots, y_p'')}{P(\bar{u}_1, \cdots, \bar{u}_q, y_1'', \cdots, y_p'')}.$$

Following (R), we form the ideal

$$\Omega = \{\Sigma, Pw - Q\}:P \subset \mathfrak{F}\{u_1, \cdots, u_q, w, y_1, \cdots, y_p\}.$$

Ω is prime, with arbitrary unknowns $u_1, \cdots, u_q$. Further,

$$\Sigma = \Omega \cap \mathfrak{F}\{u_1, \cdots, u_q, y_1, \cdots, y_p\}.$$

Let

(6*) $$A, A_1, \cdots, A_p$$

be a basic set for Ω, introducing $w, y_1, \cdots, y_p$ in succession. We take, as we may, A algebraically irreducible.

We shall show that *each* A_i *is of order zero in* y_i *and is linear in* y_i. The equation $A = 0$ is called a *resolvent* of Σ.

Suppose that not every A_i is as described. Let A_j be the first A_i which either is of positive order or is not linear in y_i, and suppose that the order of A_j in y_j is r, with r positive.

Let U be the remainder of $ISI_1S_1 \cdots I_pS_pGP$ with respect to (6*), where the I's and J's are the initials and separants of the forms of the basic set (6*). Then U has the properties: 1. The order of U in w is not greater than that of A; 2. The order of U in y_i, $i = 1, \cdots, j$, is not greater than that of A_i; 3. U is free of $y_{j+1}, \cdots, y_p$; 4. $U \notin \Omega$.

Of all the forms in (U, Ω) which have the properties 1–4 listed for U, let V be one of least rank in y_j.

We shall prove that the order of V in y_j is less than r. Suppose it equals r. Then the degree of V in y_{jr} is no higher than that of A_j. Let J be the initial of V. Then $J \notin \Omega$.

For an appropriate integer m, we have

$$J^m A_j = EV + F,$$

where E is a nonzero form of lower degree than A_j in y_{jr}, and where F has the properties 1–3 and is of lower degree than V in y_{jr}. Hence $F \in \Omega$, so that $EV \in \Omega$, whence $E \in \Omega$.

Let $E = H_0 + H_1 y_{jr} + \cdots + H_t y_{jr} t$, where $H_t \neq 0$, and each H_i is free of $y_{j+1}, \cdots, y_p$ and is of order less than r in y_j. The initial of $J^m A_j$ is identical with that of EV. Hence $H_t \notin \Omega$.

It is easy to see that there exist positive integers $a, a_1, \cdots, a_{j-1}$ such that

$$I^a I_1^{a_1} \cdots I_{j-1}^{a_{j-1}} H_i \equiv H_i' \qquad (\Omega), \qquad\qquad i = 0, 1, \cdots, t,$$

where each H_i' is reduced with respect to the ascending set $A, A_1, \cdots, A_{j-1}$. Since $H_t \notin \Omega$, we have $H_t' \neq 0$. Thus, $H_0' + H_1' y_{jr} + \cdots + H_t' y_{jr}^t$ is a nonzero form in Ω, reduced with respect to the basic set (6*). This contradiction shows that the order of V in y_j is less than r.

Now let

$$(7^*) \qquad \bar{u}_1, \cdots, \bar{u}_q, \bar{w}, \bar{y}_1, \cdots, \bar{y}_p$$

be a solution of Ω for which V (and consequently $ISI_1S_1 \cdots I_pS_pGP$) does not vanish. For $\bar{u}_1, \cdots, \bar{u}_q, \bar{w}, \bar{y}_1, \cdots, \bar{y}_{j-1}$, the forms $A_1, \cdots, A_{j-1}$ all vanish, A_j becomes a nonzero form $\bar{A}_j$ in y_j of order r, A_i, $i = j + 1, \cdots, p$, becomes a form $\bar{A}_i$ of order zero and linear in y_i, and V becomes a form $\bar{V}$ in y_j of order less than r. It follows, by Lemma 1*, that $(y_j - \bar{y}_j)\bar{V}$ does not hold $\bar{A}_j$.

Let $\tilde{y}_j$ be a solution of $\bar{A}_j$ for which $(y_j - \bar{y}_j)\bar{V}$ does not vanish. When $y_j = \tilde{y}_j$, $\bar{A}_i$, $i = j + 1, \cdots, p$, becomes a linear form of order zero in y_i alone. The new forms $\bar{A}_i$, $i = j + 1, \cdots, p$, may be solved to obtain a solution $\tilde{y}_{j+1}, \cdots, \tilde{y}_p$.

Then $\bar{u}_1, \cdots, \bar{u}_q, \bar{w}, \bar{y}_1, \cdots, \bar{y}_{j-1}, \tilde{y}_j, \cdots, \tilde{y}_p$ is a solution of Ω for which GP does not vanish, is distinct from (7*), but has the same $w = \bar{w}$ and the same $u_i = \bar{u}_i$, $i = 1, \cdots, q$. This contradicts the fundamental property of the triad G, P, Q, and shows that A_j is of order zero in y_j.

That A_j is of degree unity in y_j may be proved as in (R). This completes the proof.

The rest of Chapter 2 in (R), except for §33, may now be taken over, word for word.

COLUMBIA UNIVERSITY,
NEW YORK, N. Y.

ON THE BASIS THEOREM FOR DIFFERENTIAL SYSTEMS

BY

E. R. KOLCHIN

One of the principal points of departure in the study of polynomials and polynomial ideals is the Hilbert basis theorem, which states that every set $\mathfrak{m}$ of polynomials in a finite number of indeterminates contains a finite subset $f_1, \cdots, f_s$ such that

$$\mathfrak{m} \subseteq (f_1 \cdots f_s).$$

As originally proved by Hilbert, this theorem applied to polynomials whose coefficients were either elements of a field, or rational integers. In keeping with the modern tendency toward abstraction, however, one now finds the theorem proved for polynomials whose coefficients are elements of a commutative ring with unit element in which every set has a finite basis.

When one turns to differential polynomials and differential ideals one finds that the exact analogue of the Hilbert theorem is lacking(1). It is not true that every system of differential polynomials Σ contains a finite subset $F_1, \cdots, F_s$ such that

$$\Sigma \subseteq [F_1 \cdots F_s](^2).$$

Instead one is forced to choose as a starting point a weakened analogue, the basis theorem of Ritt and Raudenbush. This theorem has been proved for differential polynomials in a finite number of unknowns (indeterminates) $y_1, \cdots, y_n$ with any differential field of characteristic zero as coefficient domain (3), and may be stated in either of the two following equivalent forms:

1. Every system Σ of differential polynomials has a finite subset $F_1, \cdots, F_s$ such that, for each differential polynomial $A \in \Sigma$ there is a positive integer t such that $A^t \in [F_1, \cdots, F_s]$.

2. Every system Σ of differential polynomials has a finite subset $F_1, \cdots, F_s$ such that Σ is contained in the perfect differential ideal generated by $F_1, \cdots, F_s$:

$$\Sigma \subseteq \{F_1 \cdots, F_s\} (^4).$$

Presented to the Society, February 22, 1941; received by the editors July 3, 1941.

(1) See J. F. Ritt, *Differential Equations from the Algebraic Standpoint*, American Mathematical Society Colloquium Publications, vol. 14, New York, 1932, pp. 12–13.

(2) Square brackets [ ] are used for differential ideals. Parentheses ( ) denote, as usual, (algebraic) ideals.

(3) See H. W. Raudenbush, these Transactions, vol. 36 (1934), pp. 361–368.

(4) The perfect differential ideal generated by a set is denoted by the set enclosed in braces { }.

That these two statements are equivalent (when the coefficient domain is a differential field of characteristic zero) follows from the fact that the set of all differential polynomials some powers of which are in $[F_1, \cdots, F_s]$ is a perfect differential ideal[5].

It is the object of the present paper to generalize the basis theorem of Ritt and Raudenbush, as the Hilbert basis theorem has been generalized, to permit more general coefficient domains. There is nothing in the literature, for example, which allows treatment of differential polynomials with the set of rational integers, or a differential field of nonvanishing characteristic, as the domain of coefficients.

An easy counterexample shows at the outset that there is no hope of generalizing the first statement of the theorem. In $\mathfrak{J}\{y\}$, the set of all ordinary differential polynomials in y with rational integral coefficients, the system

$$y^p, y_1^p, y_2^p, \cdots$$

where p is any integer greater than 1, is such a counterexample[6]. For, no matter what n is, no power of y_{n+1}^p is contained in $[y^p, y_1^p, \cdots, y_n^p]$. This is easy to see since y_{n+1} appears in $[y^p, y_1^p, \cdots, y_n^p]$ only in terms divisible by p or by some y_i^p $(i \leqq n)$.

On the other hand the second statement of the theorem above is susceptible of generalization, although not so wide a one as might be expected at first blush. A finite subset $b_1, \cdots, b_s$ of a subset ϕ of a differential ring $\mathfrak{R}$ is called a basis of ϕ if

$$\phi \subseteq \{b_1, \cdots, b_s\}.$$

If every subset of $\mathfrak{R}$ has a basis we say that the basis theorem holds in $\mathfrak{R}$. Our main theorem asserts that:

If $\mathfrak{R}$ is a commutative differential ring with unit element, in which the basis theorem holds, and if $\mathfrak{R}$ also satisfies a certain condition termed "regularity," then the basis theorem holds in any commutative differential ring $\mathfrak{R}'$ obtained from $\mathfrak{R}$ by a finite number of differential ring adjunctions. An example shows that the regularity condition is not superfluous.

The admittance of more general coefficient domains complicates the structure of perfect differential ideals and makes it desirable to represent, after Raudenbush, the perfect differential ideal $\{\phi\}$ generated by a set ϕ as the set-theoretic limit of a non-decreasing sequence of sets denoted by $\{\phi\}_n$. (See §1.) This permits the classification of some bases as 0-bases, 1-bases, 2-bases, and so on.

This naturally raises the question whether a set which has a basis has an

(5) Raudenbush, loc. cit., p. 363. Raudenbush neglects to state that the differential rings he considers must contain the rational number system.

(6) y_j denotes the jth derivative of y.

m-basis for some m. This question is only partially answered below and still remains for investigation. If every set in a differential ring has an m-basis for some m dependent on the set then we say that the $*$-basis theorem holds in that ring. What we show is that *if the $*$-basis theorem holds in* $\mathfrak{R}$ *then the $*$-basis theorem holds in* $\mathfrak{R}'$ ($\mathfrak{R}$, $\mathfrak{R}'$ as above). Thus we see that every set of differential polynomials in $\mathfrak{J}\{y_1, \cdots, y_n\}$ has an m-basis for some m. However, it is still unknown whether we may put a bound on m. An example shows that any such bound would depend on n.

For the sake of generality the proofs are given for partial differential rings. There is a proof for ordinary differential rings which is materially shorter and simpler, and which is not a specialization of the partial case. For its own interest we present in §11 an outline of this proof.

1. **Perfect differential ideals.** Throughout this paper $\mathfrak{R}$ will denote a commutative (partial) differential ring with r types of differentiation (or derivative operators) $\delta_1, \cdots, \delta_r$.

A differential ideal σ in $\mathfrak{R}$ is called *perfect* if σ contains an element of $\mathfrak{R}$ whenever it contains some power of that element: $a^t \in \sigma$ implies $a \in \sigma$.

Let ϕ be an arbitrary subset of $\mathfrak{R}$. There exists a perfect differential ideal in $\mathfrak{R}$ containing ϕ; for example, $\mathfrak{R}$ itself. The intersection of all perfect differential ideals containing ϕ is itself a perfect differential ideal containing ϕ, and is called the perfect differential ideal generated by ϕ; in symbols, $\{\phi\}$.

To exhibit the structure of $\{\phi\}$ we define by induction:

$\{\phi\}_0 = (\phi)$,

$\{\phi\}_n =$ set of all $a \in \mathfrak{R}$ such that $a^t \in [\{\phi\}_{n-1}]$ for some t, $n = 1, 2, \cdots$.

Each $\{\phi\}_n$ is an ideal. When $n > 0$, $\{\phi\}_n$ contains every element some power of which it contains. Moreover,

$$\{\{\phi\}_m\}_n = \{\phi\}_{m+n}$$

and

$$\phi \subseteq (\phi) = \{\phi\}_0 \subseteq \{\phi\}_1 \subseteq \{\phi\}_2 \subseteq \cdots \subseteq \{\phi\}.$$

The definitions imply that

$$\{\phi\} = \{\phi\}_0 + \{\phi\}_1 + \{\phi\}_2 + \cdots \text{ (7)}.$$

2. **Bases.** A finite subset $b_1, \cdots, b_s$ of $\phi \subseteq \mathfrak{R}$ is called a *basis* of ϕ if

$$\phi \subseteq \{b_1, \cdots, b_s\}.$$

The basis will be called an *m-basis* if

(7) If $\mathfrak{R}$ is a differential ring obtained by the differential ring adjunction of a finite number of unknowns to a differential field of characteristic 0 then $\{\phi\} = \{\phi\}_1$, as is well known. For general $\mathfrak{R}$ this is no longer true. For example, if $\mathfrak{R}$ is the totality of differential polynomials in y with rational integral coefficients, we see, because $y \in \{y^2\}$, that y_1, the derivative of y, is in $\{y^2\}$. Yet $y_1 \notin \{y^2\}_1$ because y_1 appears in $[\{y^2\}_0] = [y^2]$ only in terms which are divisible by y or by 2.

$$\phi \subseteq \{b_1, \cdots, b_s\}_m\text{(}^8\text{)}$$

One says that *the basis theorem holds in* $\mathfrak{R}$ if every subset of $\mathfrak{R}$ has a basis. If every subset of $\mathfrak{R}$ has an m-basis, with m depending on the subset, then we shall say that *the *-basis theorem holds in* $\mathfrak{R}$. If every subset has an m-basis, with a single m independent of the subset, we shall say that *the m-basis theorem holds in* $\mathfrak{R}$.

The basis theorem of Ritt and Raudenbush mentioned above is seen to be, in our terminology, a 1-basis theorem.

3. **A useful result.** Let a be an arbitrary element of $\mathfrak{R}$, ϕ an arbitrary subset of $\mathfrak{R}$. Denote the set of all elements af $(f \in \phi)$ by $a \cdot \phi$.

We shall show that

$$a \cdot \{\phi\}_m \subseteq \{a \cdot \phi\}_m.$$

Indeed, since $\{\phi\}_0 = (\phi)$, the relation in question subsists when $m = 0$. Suppose it holds for $m = k$. Let $f \in \{\phi\}_k$. We show that

$$a^t \delta_1^{i_1} \cdots \delta_r^{i_r} f \in [a \cdot \{\phi\}_k] \subseteq [\{a \cdot \phi\}_k], \qquad t = i_1 + \cdots + i_r + 1.$$

Indeed, since this relation is obvious for $i_1 + \cdots + i_r = 0$ it follows in general from the fact that

$$a^{h+1}\delta_i g = a\delta_i(a^h g) - h a^h g \delta_i a \in [a^h g].$$

Thus, $a \cdot [\{\phi\}_k] \subseteq \{a \cdot \phi\}_{k+1}$. Hence, if $g \in \{\phi\}_{k+1}$, that is, if $g^t \in [\{\phi\}_k]$, then $ag^t \in \{a \cdot \phi\}_{k+1}$, $ag \in \{a \cdot \phi\}_{k+1}$, so that $a \cdot \{\phi\}_{k+1} \subseteq \{a \cdot \phi\}_{k+1}$, q.e.d.

An easy consequence of our result is that

$$\{\phi\}_m \cdot \{\psi\}_n \subseteq \{\phi \cdot \psi\}_{m+n}.$$

4. **Maximal subsets**(9). Let $\mathfrak{M}$ be a collection of subsets of $\mathfrak{R}$ such that every transfinite sequence ϕ_ξ of subsets of $\mathfrak{R}$ in $\mathfrak{M}$ which satisfies the condition

$$\phi_\xi \subset \phi_\eta, \qquad \text{if } \xi < \eta,$$

also satisfies the condition

$$\Sigma \phi_\xi \in \mathfrak{M}.$$

We shall prove that $\mathfrak{M}$ *contains a maximal subset of* $\mathfrak{R}$, that is, a $\phi \in \mathfrak{M}$ such that $\psi \in \mathfrak{M}$ implies $\phi \not\subset \psi$.

Indeed, let ϕ_ξ be a well-ordering of $\mathfrak{M}$. Define by transfinite induction:

$\psi_1 = \phi_1$,

$\psi_\eta =$ the first ϕ_ξ such that $\psi_\nu \subset \phi_\xi$ for all $\nu < \eta$.

By the construction, no ϕ_η properly contains every ψ_η. The resulting transfinite sequence ψ_η must have a last element. For otherwise $\Sigma\psi_\eta$ would be a ϕ_ξ properly containing every ψ_η. This last element is a maximal subset.

(8) Thus, if $m \leqq n$, every m-basis is an n-basis.

(9) In this section $\mathfrak{R}$ may be an abstract set.

5. Systems of differential polynomials of bounded order. We suppose henceforth that $\mathfrak{R}$ contains a unit element 1.

Let $y_1, \cdots, y_n$ be unknowns, and let Φ be a set of (partial) differential polynomials, or forms, in $\mathfrak{R}\{y_1, \cdots, y_n\}$ (10) of bounded orders. We shall show that *if the basis (or *-basis) theorem holds in $\mathfrak{R}$ then Φ has a basis (or m-basis, for some finite m).*

Proof. Because the differential polynomials in Φ are of bounded orders, only a finite number of partial derivatives of the y_i are effectively present in the forms of Φ. Let q be the least integer such that there exists a set Φ, involving only q derivatives of the y_i which has no basis (or m-basis). By the hypothesis on $\mathfrak{R}$, $q>0$. We work toward a contradiction.

If Φ_ξ is a transfinite sequence of sets of differential polynomials in $\mathfrak{R}\{y_1, \cdots, y_n\}$ involving only q derivatives of the y_i such that

$$\Phi_\xi \subset \Phi_\eta, \qquad \text{if } \xi < \eta,$$

no Φ_ξ having a basis (or m-basis), then $\Sigma\Phi_\xi$ involves only q derivatives of the y_i and has no basis (or m-basis). For if $\Sigma\Phi_\xi$ had a basis (or m-basis) there would be a single Φ_ξ which would contain every differential polynomial of the basis, and that Φ_ξ itself would have a basis (or m-basis). Therefore, by §4, there is a maximal set of forms involving only q derivatives of the y_i which has no basis (or m-basis).

Let Φ be such a maximal set. Denote the q partial derivatives of the y_i present in Φ by $\alpha_1, \cdots, \alpha_q$.

It is clear that Φ is an ideal in $\mathfrak{R}[\alpha_1, \cdots, \alpha_q]$, for otherwise the ideal generated by Φ in $\mathfrak{R}[\alpha_1, \cdots, \alpha_q]$ would properly contain Φ, would involve only q derivatives, and would have no basis (or m-basis).

Let Φ' be the set of differential polynomials in Φ which are free of α_q.

If every element of Φ, written as a polynomial in α_q, had each coefficient in Φ', we would have $\Phi \subseteq (\Phi')$, so that Φ would have a basis (or m-basis), because Φ' does. Hence Φ contains a form in which α_q is effectively present and which, when written as a polynomial in α_q, has its leading coefficient not in Φ.

Of all such differential polynomials let

$$B = I\alpha_q^s + \cdots, \qquad I \notin \Phi,$$

be one of minimum degree s in α_q. Then, for each $G \in \Phi$, we have, for suitable t,

$$I^t G \equiv G' \ (B),$$

where $G' \in \Phi$ has its degree in α_q less than s(11). By the minimal nature of

(10) $\mathfrak{R}\{y_1, \cdots, y_n\}$ means the ring obtained by the differential ring adjunction of $y_1, \cdots, y_n$ to $\mathfrak{R}$.

(11) Here we use for the first time the fact that $\mathfrak{R}$ contains a unit element.

the degree of B it follows that $G' \in (\Phi')$. But Φ' has a basis (or m_1-basis) $D_1, \cdots, D_u$. Hence $IG \in \{B, D_1, \cdots, D_u\}$ (or $IG \in \{B, D_1, \cdots, D_u\}_{m_1}$).

Now, by the maximality of Φ, (I, Φ) has a basis (or m_2-basis) which we may write as $I, D_{u+1}, \cdots, D_v$, where each $D_i \in \Phi$. Hence, referring to §3,

$$G^2 \in G\cdot(I, \Phi) \subseteq G\cdot\{I, D_{u+1}, \cdots, D_v\}$$
$$\subseteq \{IG, D_{u+1}, \cdots, D_v\} \subseteq \{\{B, D_1, \cdots, D_u\}, D_{u+1}, \cdots, D_v\},$$
$$G \in \{B, D_1, \cdots, D_v\},$$
$$\Phi \subseteq \{B, D_1, \cdots, D_v\}$$

(or, similarly, $\Phi \subseteq \{B, D_1, \cdots, D_v\}_{m_1+m_2}$). This contradiction completes the proof.

6. **Regular differential rings.** A differential ring $\mathfrak{R}$ will be called *regular* if every prime differential ideal $\pi \subseteq \mathfrak{R}$ which contains a prime rational integer p is such that the congruence

$$a \equiv x^p \ (\pi)$$

has a solution $x \in \mathfrak{R}$ for every $a \in \mathfrak{R}$ (that is, if every element has a pth root modulo π).

If $\mathfrak{R}$ is of characteristic $p>0$ then every ideal contains p and no ideal other than $\mathfrak{R}$ itself contains a prime number different from p.

Examples of regular differential rings are:

1. every differential ring which contains the rational number system;
2. every differential ring with unit element of characteristic $q>0$ in which each element has a qth root;
3. every perfect ("vollständig") differential field;
4. the differential ring of rational integers.

7. **The basis theorem.** The theorem we shall prove is the following:

Let $\mathfrak{R}$ be a regular commutative differential ring with unit element. Let $\mathfrak{R}'$ be a commutative differential ring obtained from $\mathfrak{R}$ by the differential ring adjunction of a finite number of elements: $\mathfrak{R}' = \mathfrak{R}\{\eta_1, \cdots, \eta_n\}$ (12). *If the basis (or $*$-basis) theorem holds in $\mathfrak{R}$ then the basis (or $*$-basis) theorem holds in $\mathfrak{R}'$.*

It is necessary to prove the theorem only for the case in which the η_i are all unknowns, $\eta_i = y_i$; for if the basis theorem holds in $\mathfrak{R}\{y_1, \cdots, y_n\}$ then it is easy to see that it will continue to hold when any or all of the y_i are replaced by elements among which an algebraic differential relation subsists.

8. **The proof begun.** Assume that there exists in $\mathfrak{R}' = \mathfrak{R}\{y_1, \cdots, y_n\}$ a system which does not have a basis (or m-basis for any m).

If Σ_ξ is a transfinite sequence of such systems with $\Sigma_\xi \subset \Sigma_\eta$ whenever $\xi < \eta$ then the logical sum of the Σ_ξ is again such a system. For if the logical

(12) The η_i may be hypertranscendental over $\mathfrak{R}$ (for example, they may be unknowns) or may satisfy some algebraic differential relation with coefficients in $\mathfrak{R}$.

sum had a basis (or m-basis) then there would be a single Σ_ξ which would contain every form of the basis, and that Σ_ξ itself would have a basis (or m-basis).

By §4 it follows that there is a maximal system which has no basis (or m-basis). We let Σ be such a maximal system and seek a contradiction.

Σ is a differential ideal, for $[\Sigma]$, like Σ, has no basis (or m-basis) and therefore can not properly contain Σ. Moreover, Σ is prime. To prove this, assume to the contrary that $AB\in\Sigma$, $A\notin\Sigma$, $B\notin\Sigma$. Then (Σ, A) and (Σ, B) properly contain Σ and must have bases (or m_1- and m_2-bases, respectively), say $A, C_1, \cdots, C_u$ and $B, C_{u+1}, \cdots, C_v$, respectively, where the C_i are in Σ. Thus

$$\Sigma^2 \subseteq (\Sigma, A)(\Sigma, B) \subseteq \{A, C_1, \cdots, C_u\}\{B, C_{u+1}, \cdots, C_v\} \subseteq \{AB, C_1, \cdots, C_v\},$$

so that $\Sigma\subseteq\{AB, C_1, \cdots, C_v\}$, and Σ has a basis (or, similarly, an (m_1+m_2)-basis).

9. **The proof continued.** The object of this section is to show that Σ contains a prime rational integer p(13). To accomplish this we introduce a set of differential polynomials analogous to the "basic sets" used by Ritt.

We assume that the partial derivatives of the y_i are completely ordered by a system of marks in such a way that every partial derivative of the y_i is lower than (precedes) every other derivative of the y_i of higher order, and if α and β are two derivatives of the y_i with α lower than β then $\delta_i\alpha$ is lower than $\delta_i\beta$, $i=1, \cdots, r$. Such an ordering can always be effected(14).

Let $\sigma=\Sigma\cap\mathfrak{R}$. Clearly σ is a prime differential ideal in $\mathfrak{R}$. Since the basis (or $*$-basis) theorem holds in $\mathfrak{R}$, σ has a basis (or m-basis). Hence $\Sigma\neq(\sigma)$ so that Σ must contain forms none of whose coefficients is in σ.

Of all the forms in Σ none of whose coefficients is in σ, consider those with lowest possible leader α_1 (the leader of a form is the highest derivative of the y_i effectively present in the form). Of all those forms let A_1 be one whose degree in α_1 is as low as possible.

Of all the forms in Σ none of whose coefficients is in σ, which do not contain a proper derivative (that is, a derivative of positive order) of α_1, and which are of lower degree in α_1 than A_1, consider those with lowest possible leader α_2. Of all those let A_2 be one whose degree in α_2 is a minimum.

Continuing, at the jth step, consider, of the forms in Σ none of whose coefficients is in σ, which do not contain a proper derivative of α_i $(i=1, \cdots, j-1)$ and which are of lower degree in α_i than A_i $(i=1, \cdots, j-1)$, those forms which have the lowest leader α_j. Of all those forms let A_j be one whose degree in α_j is a minimum.

Since no α_i is a derivative of any preceding α_i, there can be only a finite

(13) If $\mathfrak{R}$ contained all the rational numbers this would suffice, for then Σ would contain $1=(1/p)\cdot p$, and would have 1 as a basis.

(14) Ritt, loc. cit., pp. 141–143.

number of the α_i[15], so that the process for defining the forms A_i must stop. Let A_s be the last A_i.

It is easy to see that if G is a form in Σ which contains no proper derivative of any α_i and whose degree in each α_i is lower than that of the corresponding A_i, then $G\in(\sigma)$.

Let I_i and S_i be the initial and separant of A_i. The coefficients of I_i are coefficients of A_i and therefore are not in σ. I_i contains no proper derivative of any α_j and is of lower degree in α_j than A_j $(j=1, \cdots, s)$. Hence $I_i \notin \Sigma$.

We shall show that at least one S_i is in Σ.

Let no S_i be in Σ. For an arbitrary form $G\in\Sigma$ there exist integers g_i, h_i such that

$$I_1^{g_1}S_1^{h_1}\cdots I_s^{g_s}S_s^{h_s}G \equiv G'[A_1, \cdots, A_s],$$

where $G'\in\Sigma$ contains no proper derivative of any α_i and is of lower degree in α_i than A_i. Thus, by the above, $G'\in(\sigma)$, so that

$$I_1^{g_1}S_1^{h_1}\cdots I_s^{g_s}S_s^{h_s}G \equiv 0[A_1, \cdots, A_s, \sigma],$$

$$I_1S_1\cdots I_sS_sG \in \{A_1, \cdots, A_s, \sigma\}_1,$$

$$I_1S_1\cdots I_sS_s\cdot\Sigma \subseteq \{A_1, \cdots, A_s, \sigma\}_1.$$

Now, Σ is prime and contains no I_i or S_i, so that $I_1S_1\cdots I_sS_s \notin \Sigma$. Hence, by the maximality of Σ, the system

$$\Sigma,\ I_1S_1\cdots I_sS_s$$

has a basis (or m_1-basis) which we may write as

$$I_1S_1\cdots I_sS_s,\ B_1, \cdots, B_t.$$

Denoting by $B_{t+1}, \cdots, B_u$ a basis (or m_2-basis) of σ, we have

$$\begin{aligned} \Sigma^2 &\subseteq \Sigma(\Sigma, I_1S_1\cdots I_sS_s) \subseteq \Sigma\{I_1S_1\cdots I_sS_s, B_1, \cdots, B_t\} \\ &\subseteq \{I_1S_1\cdots I_sS_s\cdot\Sigma, B_1, \cdots, B_t\} \subseteq \{A_1, \cdots, A_s, \sigma, B_1, \cdots, B_t\} \\ &\subseteq \{A_1, \cdots, A_s, B_1, \cdots, B_u\}, \\ \Sigma &\subseteq \{A_1, \cdots, A_s, B_1, \cdots, B_u\} \end{aligned}$$

(or, similarly, $\Sigma\subseteq\{A_1, \cdots, A_s, B_1, \cdots, B_u\}_{m_1+m_2+1}$). This contradicts the fact that Σ has no basis (or m-basis) and proves that $S_i\in\Sigma$ for at least one i.

Let S_j be the first S_i contained in Σ. S_j contains no proper derivative of any α_i and is of lower degree in α_i than A_i $(i=1, \cdots, s)$. Hence $S_j\in(\sigma)$. It follows that $n_jI_j\in\Sigma$, where n_j is the degree of A_j in α_j, so that $n_j\in\Sigma$, and one of the prime factors p of n_j must be in Σ. This completes the proof of the result at the beginning of this section.

(15) Ritt, loc. cit., pp. 135–136.

10. **The proof concluded.** Let F be any nonzero differential polynomial, γ any partial derivative of the y_i. F can be written in one and only one way as a polynomial

$$H_0 + H_1\gamma + \cdots + H_h\gamma^h, \qquad H_h \neq 0,$$

in γ of degree $h<p$, where γ does not appear in the H_i except raised to powers divisible by p. We shall call h the p-degree of F in γ. The highest derivative of the y_i in which F has a positive p-degree (if such a derivative exists) shall be called the p-leader of F. If γ is the p-leader of F and if h is the p-degree of F in γ, we shall call the coefficient of γ^h the p-initial of F, and $\partial F/\partial\gamma$ the p-separant of F.

We shall need the fact that Σ contains a form, none of whose coefficients is in σ, whose p-degree in some derivative of the y_i is positive and whose p-initial is not in Σ. To prove this assume the contrary and let G be a form of Σ, none of whose coefficients is in σ, of least possible (total) degree. Every term of G involves only powers divisible by p, else the p-degree of G in some derivative of the y_i would be positive and the p-initial of G would be a form in Σ, none of whose coefficients is in σ, of lower degree than G. Moreover, by the regularity of $\mathfrak{R}$, the coefficient of each term of G may be replaced modulo σ by the pth power of an element of $\mathfrak{R}$(16). Since $p\in\sigma$ it follows that $G\equiv H^p$ (σ), where H is the form obtained from G by replacing each term by its pth root modulo σ. H is of lower degree than G and is in Σ. This contradicts the definition of G and proves the required fact.

Of all the forms of Σ, none of whose coefficients is in σ, which involve derivatives of the y_i to a power not divisible by p and whose p-initials are not in Σ, consider those with lowest p-leader β_1. Of all those forms let B_1 be one whose p-degree in β_1 is as low as possible.

Of all the forms in Σ, none of whose coefficients is in σ, which involve derivatives of the y_i to a power not a multiple of p, whose p-initials are not in Σ, which do not contain a proper derivative of β_1 except raised to a power divisible by p, and which have a p-degree in β_1 less than that of B_1, consider those with lowest possible p-leader β_2. Of all those let B_2 be one whose p-degree in β_2 is a minimum.

Continuing, at the jth step, of all the forms in Σ, none of whose coefficients is in σ, which contain derivatives of the y_i to powers not divisible by p, whose p-initials are not in Σ, which do not contain a proper derivative of β_i except to a power divisible by p $(i=1, \cdots, j-1)$ and which have a p-degree in β_i less than that of B_i $(i=1, \cdots, j-1)$, consider those with lowest p-leader β_j. Of all those forms let B_j be one whose p-degree in β_j is as low as possible.

As with the A_i of §9, the process of defining the B_i must stop after a finite number of steps. Let B_s be the last B_i(17). Let J_i and T_i be the p-initial and p-separant, respectively, of B_i.

(16) Up to this point we have not used the regularity. Henceforth it will be important.

(17) The s here is not necessarily the same as that of §9.

If G is a form of Σ, none of whose coefficients is in σ, which contains no proper derivative of any β_i except raised to powers divisible by p and whose p-degree in each β_i is less than that of the corresponding B_i, then either G contains no derivative of the y_i that is raised to a power not divisible by p, or the p-initial of G is in Σ.

From this it can be shown that the T_i are not contained in Σ. We already know that the I_i are not in Σ.

Let α represent the highest derivative of the y_i effectively present in $B_1, \cdots, B_s$[18]. Let Σ_α denote the totality of forms in Σ which contain no derivative of the y_i which is higher than α: The forms of Σ_α are of bounded order.

We shall show that for each differential polynomial $G \in \Sigma$ there exist nonnegative integers e_i, f_i such that

$$J_1^{e_1}T_1^{f_1} \cdots J_s^{e_s}T_s^{f_s}G \equiv 0[\Sigma_\alpha].$$

Assume that this is not so. If G is a form in Σ for which such a congruence fails to hold it is easy to see that there is a relation

$$J_1^{g_1}T_1^{h_1} \cdots J_s^{g_s}T_s^{h_s}G \equiv G'[B_1, \cdots, B_s],$$

where G' is a form in Σ for which such a congruence fails to hold, which contains no proper derivative of any β_i except to a power divisible by p, and which has, in each β_i, a p-degree lower than that of the corresponding B_i. Of all forms in Σ which fail to satisfy a congruence as above, which contain no proper derivative of any β_i except to a power divisible p, and which have, in each β_i, a p-degree lower than that of the corresponding B_i, consider those with the least number of terms. Of all those forms let G be one with a minimum (total) degree. Since $\sigma \subseteq \Sigma_\alpha$, no coefficient of G is in σ. Hence, either G contains only powers divisible by p or the p-initial of G is in Σ. Suppose G contains only powers divisible by p. By the regularity of $\mathfrak{R}$, each coefficient of G may be replaced modulo σ by the pth power of an element of $\mathfrak{R}$. Hence $G \equiv H^p$ (σ), where $H \in \Sigma$, having the same number of terms as G and being of lower degree than G, satisfies a congruence as above. But this is impossible as then G would satisfy such a congruence. Thus G has a p-leader γ and the p-initial of G is in Σ:

$$G = K_0 + \cdots + K_g\gamma^g, \qquad K_g \in \Sigma.$$

Since K_g is of lower degree than G, K_g satisfies a congruence of the type in question. But $K_0 + \cdots + K_{g-1}\gamma^{g-1}$, which is in Σ and has fewer terms than G, must also satisfy such a congruence. This is impossible, however, for it implies that G itself satisfies the same kind of congruence.

Thus we have shown that

(18) α may be higher than β_i as the p-degree of each B_i in α may be 0.

$$J_1T_1 \cdots J_sT_s \cdot \Sigma \subseteq \{\Sigma_\alpha\}_1.$$

Now, by the result of §5, Σ_α has a basis (or m_1-basis), say $D_1, \cdots, D_t$. Also, by the maximality of Σ, the system

$$\Sigma,\ J_1T_1 \cdots J_sT_s$$

has a basis (or m_2-basis) which we may write as

$$J_1T_1 \cdots J_sT_s,\ D_{t+1}, \cdots, D_u \qquad D_i \in \Sigma.$$

Hence, by §3,

$$\Sigma^2 \subseteq \Sigma(\Sigma, J_1T_1 \cdots J_sT_s) \subseteq \Sigma\{J_1T_1 \cdots J_sT_s, D_{t+1}, \cdots, D_u\}$$
$$\subseteq \{J_1T_1 \cdots J_sT_s\cdot\Sigma, D_{t+1}, \cdots, D_u\} \subseteq \{D_1, \cdots, D_u\},$$
$$\Sigma \subseteq \{D_1, \cdots, D_u\}$$

(or, similarly, $\Sigma \subseteq \{D_1, \cdots, D_u\}_{m_1+m_2+1}$). This contradiction completes the proof of the theorem stated in §7.

11. **Shorter proof in the ordinary case.** We sketch in this section a shorter proof of the theorem under the assumption that we are dealing with *ordinary* differential rings, that is, differential rings with one type of differentiation.

Denote the jth derivative of any letter u by u_j.

Of the above proof we take over §§1–6.

We first show that, when the basis (or *-basis) theorem holds in $\mathfrak{R}$ and $\mathfrak{R}$ is regular, the basis (or *-basis) theorem holds in $\mathfrak{R}\{y\}$. Assuming the contrary we obtain, as in §8, a maximal system $\Sigma \subset \mathfrak{R}\{y\}$ which has no basis (or m-basis). Σ is a prime differential ideal. If F were a form in Σ whose separant S was not in Σ, $S\cdot\Sigma$ would have a basis (or m_1-basis), $B_1, \cdots, B_s$, for which we could write, for each $G \in \Sigma$, $S^qG \equiv G'[F]$, with G' of order no higher than that of F, so that we would have $S\cdot\Sigma \subseteq \{\Sigma'\}$, where Σ' is the set of forms of Σ whose orders are less than or equal to the order of F. Also, by the maximality of Σ, the system Σ, S would have a basis (or m_2-basis), say $S, B_{s+1}, \cdots, B_t$. Thus we would have

$$\Sigma^2 \subseteq \Sigma(\Sigma, S) \subseteq \Sigma\{S, B_{s+1}, \cdots, B_t\}$$
$$\subseteq \{S\cdot\Sigma, B_{s+1}, \cdots, B_t\} \subseteq \{B_1, \cdots, B_t\},$$
$$\Sigma \subseteq \{B_1, \cdots, B_t\}$$

(or, similarly, $\Sigma \subseteq \{B_1, \cdots, B_t\}_{m_1+m_2}$). This cannot be, so that every form in Σ must have its separant in Σ.

Of all forms in Σ none of whose coefficients is in Σ let A be one whose (total) degree is a minimum. Since S, the separant of A, is of lower degree than A, all the coefficients of S must be in Σ. These coefficients are coefficients of A multiplied by the exponents to which y_q appears in A. (Here q is the order of A.) Since Σ is prime and the coefficients of A are not in Σ, these exponents must be in Σ. These exponents have a common prime factor $p\in\Sigma$, and we see that y_q appears in A only to powers divisible by p. It is now easy to see that every derivative of y appears in A only to powers divisible by p; for suppose y_j is the y_i of highest subscript which appears in A to a power not a multiple of p. Then the $(q-j+1)$st derivative of A would be, terms divisible by p neglected, a form in Σ whose separant is not in Σ, an impossibility.

Now, by the regularity of $\mathfrak{R}$, we may replace modulo Σ each coefficient of A by the pth power of an element of $\mathfrak{R}$. Hence $A\equiv B^p(\Sigma)$, where $B\in\Sigma$ has no coefficient in Σ and is of lower degree than A. This completes the proof for $\mathfrak{R}\{y\}$.

Proceeding by induction, suppose the theorem has been proved for $\mathfrak{R}\{y_1, \cdots, y_{n-1}\}$ [19]. As above, we find, for a maximal system $\Sigma\subset\mathfrak{R}\{y_1, \cdots, y_n\}$, that the separant of each form of Σ must itself be in Σ. This must be true no matter how we order the unknowns. Letting A be a form in Σ, with no coefficients in Σ, of minimum degree, we see from the above that each y_{ij} appears in A only to powers divisible by a prime rational integer $p\in\Sigma$. As in the case of one unknown this leads to a contradiction and completes the proof.

12. **Examples.** From the point of view of analogy with the Hilbert basis theorem, it might be imagined that the regularity condition imposed in the basis theorem above is unnecessary. The following example shows that this is not so.

EXAMPLE 1. Let $\mathfrak{R}$ be the ordinary differential field of characteristic $p>0$ obtained from the field of rational integers modulo p by the differential field adjunction of the set of "indeterminate constants" $c_0, c_1, c_2, \cdots$, that is, each c_i is a letter whose derivative is taken to be 0, and $\mathfrak{R}=\mathfrak{J}_p\langle c_0, c_1, c_2, \cdots\rangle$. Let y be an unknown and consider, in $\mathfrak{R}\{y\}$, the system Φ:

$$y^p + c_0,\ y_1^p + c_1, \cdots,\ y_k^p + c_k, \cdots .$$

We shall show that Φ *has no basis.*

Indeed, if Φ had a basis we should have, for some k,

$$y_k^p + c_k \in \{y^p + c_0, \cdots, y_{k-1}^p + c_{k-1}\}.$$

Now, $(y_i^p+c_i)_1=py_i^{p-1}y_{i+1}+c_{i1}=0$, so that

$$[y^p + c_0, \cdots, y_{k-1}^p + c_{k-1}] = (y^p + c_0, \cdots, y_{k-1}^p + c_{k-1}).$$

But clearly $AB\in(y^p+c_0, \cdots, y_{k-1}^p+c_{k-1})$ implies that A or $B\in(y^p+c_0, \cdots, y_{k-1}^p+c_{k-1})$. Hence $(y^p+c_0, \cdots, y_{k-1}^p+c_{k-1})$ is a prime differential ideal, so that

$$\{y^p + c_0, \cdots, y_{k-1}^p + c_{k-1}\} = (y^p + c_0, \cdots, y_{k-1}^p + c_{k-1}).$$

(19) The y_i are unknowns. The jth derivative of y_i is denoted by y_{ij}.

But it is easy to verify that

$$y_k^p + c_k \notin (y^p + c_0, \cdots, y_{k-1}^p + c_{k-1}).$$

Let $\mathfrak{J}$ be the ordinary differential ring of rational integers. Let $n = 2^{m-1}$, where m is any positive integer. The following example shows that the m-basis theorem does not hold in $\mathfrak{J}\{y_1, \cdots, y_n\}$.

EXAMPLE 2. Let Φ be the system in $\mathfrak{J}\{y_1, \cdots, y_n\}$ consisting of the forms

$$y_1^2 \cdots y_n^2,\ y_{11}^2 \cdots y_{n1}^2, \cdots, y_{1k}^2 \cdots y_{nk}^2, \cdots.$$

Φ has no m-basis.

To prove this assume the contrary. Then, for some k,

$$\begin{aligned} y_{1k}^2 \cdots y_{n'k}^2 &\in \{y_1^2 \cdots y_{n'}^2, \cdots, y_{1,k-1}^2 \cdots y_{n',k-1}^2\}_m \\ &\subseteq \{2, y_1 \cdots y_{n'}, \cdots, y_{1,k-1} \cdots y_{n',k-1}\}_{m-1}, \\ y_{1k} \cdots y_{n'k} &\in \{2, y_1 \cdots y_{n'}, \cdots, y_{1,k-1} \cdots y_{n',k-1}\}_{m-1}. \end{aligned}$$

Letting $y_1 = y_2 = z_1$, $y_3 = y_4 = z_2$, $\cdots$, $y_{n-1} = y_n = z_{n'}$, where $n' = n/2 = 2^{m-2}$, we see, in the differential ring $\mathfrak{R}\{z_1, \cdots, z_{n'}\}$, that

$$\begin{aligned} z_{1k}^2 \cdots z_{n'k}^2 &\in \{2, z_1^2 \cdots z_{n'}^2, \cdots, z_{1,k-1}^2 \cdots z_{n',k-1}^2\}_{m-1} \\ &\subseteq \{2, z_1 \cdots z_{n'}, \cdots, z_{1,k-1} \cdots z_{n',k-1}\}_{m-2}, \\ z_{1k} \cdots z_{n'k} &\in \{2, z_1 \cdots z_{n'}, \cdots, z_{1,k-1} \cdots z_{n',k-1}\}_{m-2}. \end{aligned}$$

Continuing, at each step we reduce the number of unknowns by one half until we arrive, in $\mathfrak{R}\{u_1\}$, at the relation

$$u_{1k} \in \{2, u_1, \cdots, u_{1,k-1}\}_0 = (2, u_1, \cdots, u_{1,k-1}).$$

This contradiction completes the proof.

COLUMBIA UNIVERSITY,
NEW YORK, N. Y.

ANNALS OF MATHEMATICS
Vol. 43, No. 4, October, 1942

EXTENSIONS OF DIFFERENTIAL FIELDS, I

BY E. R. KOLCHIN[1]

(Received May 6, 1942)

Introduction

It is a well-known theorem of algebra that a finite algebraic extension of a field of characteristic zero K always contains a primitive element ω:

$$K(\alpha_1, \cdots, \alpha_n) = K(\omega).$$

Moreover, by means of the theory of Galois, it is possible to characterize those elements of the extension which are primitive.[2] The present paper treats the analogous problems for differential fields (ordinary or partial).

A simple example shows that the precise analog is not true without further restriction. Let $\mathfrak{F}_0$ be the ordinary differential field of rational numbers, and let α_1 and α_2 be two algebraically independent complex constants. Since α_1 and α_2 both have zero derivatives, $\mathfrak{F}_0\langle\alpha_1, \alpha_2\rangle$ is set-theoretically identical with $\mathfrak{F}_0(\alpha_1, \alpha_2)$,[3] whence it is clear that there exists no number $\beta \in \mathfrak{F}_0\langle\alpha_1, \alpha_2\rangle$ such that $\mathfrak{F}_0\langle\alpha_1, \alpha_2\rangle = \mathfrak{F}_0\langle\beta\rangle$. However, for the theorem in question to hold it suffices to place a mild condition on the differential field. In the ordinary case the condition reduces to the requirement that *the differential field contain a non-constant* (that is, an element whose derivative is different from zero), in the general (partial) case, the condition is that *the differential field contain a set of elements whose Jacobian does not vanish.*

In studying those elements of an extension $\mathfrak{G}$ of a differential field $\mathfrak{F}$ which are primitive, a theorem presents itself which bears a similarity to results from Galois' theory. However any attempt in this direction seems destined to but fragmentary results, as the concept analogous to a *normal* extension of a field is lacking, so that one must speak of isomorphisms instead of automorphisms, thereby abandoning the concept of group.

1. Generic solutions

Throughout this paper $\mathfrak{F}$ will denote a differential field of characteristic zero with m types of differentiation $\delta_1, \cdots, \delta_m$,[4] and $y_1, \cdots, y_n$ will denote unknowns (m and n are positive integers).

Let Σ be a system of differential polynomials in $\mathfrak{F}\{y_1, \cdots, y_n\}$ with mani-

[1] National Research Fellow.

[2] This is not, of course, the simplest characterization.

[3] $\mathfrak{F}\langle u, \cdots\rangle$ means the result of the differential field adjunction to $\mathfrak{F}$ of the elements $u, \cdots$. $\mathfrak{F}(u, \cdots)$ means, as usual, the result of the field adjunction to $\mathfrak{F}$ (considered as a field) of the elements $u, \cdots$. The result of differential ring adjunction is indicated by curled brackets: $\mathfrak{F}\{u, \cdots\}$.

[4] This concept has been discussed by H. W. Raudenbush, Bulletin of the American Mathematical Society, vol. 40 (1934), pp. 714–720.

fold $\mathfrak{M}$. A set $\eta_1, \cdots, \eta_n$ of elements of a differential extension field of $\mathfrak{F}$ will be called *a generic solution of* Σ *(or of* $\mathfrak{M}$*, with respect to* $\mathfrak{F}$) if a necessary and sufficient condition for a differential polynomial $F(y_1, \cdots, y_n)$ in $\mathfrak{F}\{y_1, \cdots, y_n\}$ to belong to Σ is

$$F(\eta_1, \cdots, \eta_n) = 0.$$

It is easy to see that if Σ has a generic solution, then Σ is a prime differential ideal in $\mathfrak{F}\{y_1, \cdots, y_n\}$, so that $\mathfrak{M}$ is irreducible over $\mathfrak{F}$. Conversely if Σ is a prime differential ideal other than the whole ring $\mathfrak{F}\{y_1, \cdots, y_n\}$, then Σ has a generic solution. For example, if, in the differential ring of remainder classes $\mathfrak{F}\{y_1, \cdots, y_n\}/\Sigma$, $\bar{y}_i$ is the remainder class containing y_i, then $\bar{y}_1, \cdots, \bar{y}_n$ are elements of a differential field containing $\mathfrak{F}$ (namely, the differential field of quotients of $\mathfrak{F}\{y_1, \cdots, y_n\}/\Sigma$), and $F(y_1, \cdots, y_n)$ is in Σ if and only if $F(\bar{y}_1, \cdots, \bar{y}_n) = 0$. It is not hard to see moreover, that any generic solution $\eta_1, \cdots, \eta_n$ of Σ is equivalent to $\bar{y}_1, \cdots, \bar{y}_n$, that is, $\eta_i \to \bar{y}_i$ $(i = 1, \cdots, n)$ generates an isomorphism:

$$\mathfrak{F}\langle\eta_1, \cdots, \eta_n\rangle \cong \mathfrak{F}\langle\bar{y}_1, \cdots, \bar{y}_n\rangle.$$ [5]

Now, a prime differential ideal Σ in $\mathfrak{F}\{y_1, \cdots, y_n\}$ may very well decompose, over an extension $\mathfrak{G}$ of $\mathfrak{F}$, into several essential prime differential ideals:

$$(1) \qquad \{\Sigma\} = \Lambda_1 \cap \cdots \cap \Lambda_s, \qquad \text{in } \mathfrak{G}\{y_1, \cdots, y_n\}.$$

Let $\zeta_1, \cdots, \zeta_n$ be a generic solution of some Λ_i, say of Λ_h. Then $\zeta_1, \cdots, \zeta_n$ *is a generic solution of* Σ. Indeed, it is clear that $F(y_1, \cdots, y_n) \in \Sigma$ implies $F(\zeta_1, \cdots, \zeta_n) = 0$, as $\Sigma \subseteq \Lambda_h$. Conversely, suppose that $F = F(y_1, \cdots, y_n) \in \mathfrak{F}\{y_1, \cdots, y_n\}$, and that $F(\zeta_1, \cdots, \zeta_n) = 0$. Let

$$G \in \Lambda_1 \cap \cdots \cap \Lambda_{h-1} \cap \Lambda_{h+1} \cap \cdots \cap \Lambda_s, \qquad G \notin \Lambda_h.$$

Then FG vanishes for all solutions of Σ, so that, by the Ritt analog of the *Nullstellensatz*, some power $(FG)^k$ is a linear combination, with coefficients in $\mathfrak{G}\{y_1, \cdots, y_n\}$, of differential polynomials in Σ:

$$F^kG^k = C_1S_1 + \cdots + C_lS_l \qquad (S_i \in \Sigma).$$

The coefficients of G^k, $C_1, \cdots, C_l$ are in $\mathfrak{G}$. Letting $\omega_1, \cdots, \omega_g$ be, with respect to $\mathfrak{F}$, a linearly independent linear basis of these coefficients, we find a relation

$$F^k(H_1\omega_1 + \cdots + H_g\omega_g) = T_1\omega_1 + \cdots + T_g\omega_g,$$

where each $T_i \in \Sigma$, each $H_i \in \mathfrak{F}\{y_1, \cdots, y_n\}$, and $H_1\omega_1 + \cdots + H_g\omega_g = G^k$. Equating coefficients, on both sides, of the linearly independent elements ω_i, we see that

[5] The isomorphism indicated by the symbol $\cong$ maps not only the sum and product of two elements onto the sum and product, respectively, of their images, but also the various derivatives of an element onto the corresponding derivatives of its image.

$$F^k H_i = T_i \in \Sigma \qquad (i = 1, \cdots, g).$$

But not every H_i is in Σ, for otherwise G would be in Λ_h. Hence, since Σ is a prime ideal, $F \in \Sigma$.

We use this result to prove that *if the prime differential ideal Σ in $\mathfrak{F}\{y_1, \cdots, y_n\}$ has a generic solution $\eta_1, \cdots, \eta_n$, if $\mathfrak{G}$ is a differential extension field of $\mathfrak{F}\langle\eta_1, \cdots, \eta_n\rangle$, and if no extension of $\mathfrak{G}$ contains another generic solution of Σ, then each $\eta_i \in \mathfrak{F}$.*

For, let (1) be the decomposition of $\{\Sigma\}$ into essential prime differential ideals in $\mathfrak{G}\{y_1, \cdots, y_n\}$. Any generic solution of Λ_1 is a generic solution of Σ and therefore is identical with $\eta_1, \cdots, \eta_n$. The same holds for every Λ_i, so that $s = 1$, $\{\Sigma\} = [y_1 - \eta_1, \cdots, y_n - \eta_n]$, and the only solution of Σ is $\eta_1, \cdots, \eta_n$. Assume, now, that $\eta_1 \notin \mathfrak{F}$. For some k,

$$(y_1 - \eta_1)^k = C_1 S_1 + \cdots + C_l S_l \qquad (S_i \in \Sigma).$$

We suppose that k has been chosen as low as possible, so that $1, \eta_1, \cdots, \eta_1^k$ are linearly independent over $\mathfrak{F}$. Letting $1, \eta_1, \cdots, \eta_1^k, \omega_1, \cdots, \omega_g$ be a linearly independent linear basis, with respect to $\mathfrak{F}$, of the coefficients in $(y_1 - \eta_1)^k, C_1, \cdots, C_l$, and equating coefficients of η_1^k, we arrive at the contradiction that $1 \in \Sigma$. Hence $\eta_1 \in \mathfrak{F}$, similarly, every $\eta_i \in \mathfrak{F}$.

2. Relative isomorphisms

Let $\mathfrak{G}$ be a differential extension field of $\mathfrak{F}$. By an *isomorphism of $\mathfrak{G}$ with respect to $\mathfrak{F}$* we shall mean an isomorphic mapping of $\mathfrak{G}$ onto a differential field $\mathfrak{G}'$ such that

(a) $\mathfrak{G}'$ is an extension of $\mathfrak{F}$,

(b) the isomorphic mapping leaves each element of $\mathfrak{F}$ invariant,

(c) $\mathfrak{G}$ and $\mathfrak{G}'$ have a common extension.

By means of well-ordering methods it is easy to show that an isomorphism of $\mathfrak{G}$ with respect to $\mathfrak{F}$ can be extended to an automorphism of the common extension of $\mathfrak{G}$ and its map under the isomorphism.

Concerning such relative isomorphisms we prove the following theorem:

Let $\mathfrak{G}$ be an extension of $\mathfrak{F}$, and let $\gamma \in \mathfrak{G}$. A necessary and sufficient condition that $\gamma \in \mathfrak{F}$ is that every isomorphism of $\mathfrak{G}$ with respect to $\mathfrak{F}$ leaves γ invariant. A necessary and sufficient condition that γ be a primitive element, that is, that $\mathfrak{G} = \mathfrak{F}\langle\gamma\rangle$, is that no isomorphism of $\mathfrak{G}$ with respect to $\mathfrak{F}$ other than the identity leaves γ invariant.

PROOF: A. If $\gamma \in \mathfrak{F}$, then by condition (b), every isomorphism of $\mathfrak{G}$ with respect to $\mathfrak{F}$ leaves γ invariant. Now let $\gamma \notin \mathfrak{F}$, and denote by Γ the prime differential ideal of all differential polynomials in $\mathfrak{F}\{y\}$ which vanish for $y = \gamma$. γ is a generic solution of Γ. Since $\gamma \notin \mathfrak{F}$, we know by §1 that there exists a differential field $\mathfrak{H} \supseteq \mathfrak{G}$ in which Γ has another generic solution γ'. Now, $\gamma \rightarrow \gamma'$ generates an isomorphism between $\mathfrak{F}\langle\gamma\rangle$ and $\mathfrak{F}\langle\gamma'\rangle$ which leaves invariant every element of $\mathfrak{F}$. This isomorphism can be extended to an automorphism of $\mathfrak{H}$, which automorphism in turn can be contracted to produce an isomorphism of $\mathfrak{G}$ with respect to $\mathfrak{F}$ which does not leave γ invariant.

B. If $\mathfrak{G} = \mathfrak{F}\langle\gamma\rangle$, every element of $\mathfrak{G}$ is a rational function, with coefficients in $\mathfrak{F}$, of γ and its various derivatives, so that an isomorphism of $\mathfrak{G}$ with respect to $\mathfrak{F}$ which leaves γ invariant leaves every element of $\mathfrak{G}$ invariant, that is, is the identity isomorphism. Conversely, if $\mathfrak{G} \neq \mathfrak{F}\langle\gamma\rangle$, there is an element $\alpha \in \mathfrak{G}$ such that $\alpha \notin \mathfrak{F}\langle\gamma\rangle$. By the part of the theorem already proved there is an isomorphism of $\mathfrak{G}$ with respect to $\mathfrak{F}\langle\gamma\rangle$ which does not leave α invariant. This is an isomorphism of $\mathfrak{G}$ with respect to $\mathfrak{F}$, other than the identity, which leaves γ invariant.

The existence, in certain general cases, of a primitive element will be demonstrated in §4, after the proof of a preparatory result in §3.

3. Non-vanishing of nonzero differential polynomials

The following lemma will be used in §4.

A necessary and sufficient condition that, for an arbitrary nonzero differential polynomial $A = A(y_1, \cdots, y_n) \in \mathfrak{F}\{y_1, \cdots, y_n\}$, *there exist elements* $\eta_1, \cdots, \eta_n \in \mathfrak{F}$ *such that* $A(\eta_1, \cdots, \eta_n) \neq 0$, *is that* $\mathfrak{F}$ *contain* m *elements* $\xi_1, \cdots, \xi_m$ *whose Jacobian is different from zero*:

$$J = \begin{vmatrix} \delta_1\xi_1 & \cdots & \delta_m\xi_1 \\ \cdot & \cdots & \cdot \\ \delta_1\xi_m & \cdots & \delta_m\xi_m \end{vmatrix} \neq 0.$$

PROOF: Necessity. If $\mathfrak{F}$ has the property in question, then, in particular, there are elements $\xi_1, \cdots, \xi_m$ which do not annul

$$J(y_1, \cdots, y_m) = \begin{vmatrix} \delta_1 y_1 & \cdots & \delta_m y_1 \\ \cdot & \cdots & \cdot \\ \delta_1 y_m & \cdots & \delta_m y_m \end{vmatrix}.$$

Sufficiency. It obviously suffices to consider the case $n = 1$: $A = A(y) \in \mathfrak{F}\{y\}$. Now, since $J \neq 0$ there exists an $m \times m$ matrix (α_{ij}), with elements in $\mathfrak{F}$, such that $(\alpha_{ij})(\delta_j\xi_k)$ is the unit matrix. Hence, if we introduce the operators

$$\delta_i' = \alpha_{i1}\delta_1 + \cdots + \alpha_{im}\delta_m \qquad (i = 1, \cdots, m)$$

in terms of which, in turn, the operators δ_j may be expressed

$$\delta_j = \beta_{j1}\delta_1' + \cdots + \beta_{jm}\delta_m' \qquad (j = 1, \cdots, m),$$

we shall have

$$\delta_i'\xi_k = \begin{cases} 1 & \text{if } i = k, \\ 0 & \text{if } i \neq k. \end{cases}$$

Moreover, since

$$\delta_p'\delta_q' = \sum_i \alpha_{pi}\delta_i \sum_j \alpha_{qj}\delta_j$$
$$= \sum_i \sum_j \alpha_{pi}\alpha_{qj}\delta_i\delta_j + \sum_j (\sum_i \alpha_{pi}\delta_i\alpha_{qj})\delta_j,$$

we see that

$$\delta_p'\delta_q' = \delta_q'\delta_p' + \sum_k \gamma_k^{(p,q)}\delta_k' \qquad (\gamma_k^{(p,q)} \in \mathfrak{F}).$$

Hence $A(y)$ may be expressed as a polynomial, with coefficients in $\mathfrak{F}$, in the quantities $\delta_1'^{i_1} \cdots \delta_m'^{i_m} y$:

$$A(y) = P(\cdots, \delta_1'^{i_1} \cdots \delta_m'^{i_m} y, \cdots).$$

Letting the symbols $c_{i_1 \cdots i_m}$ denote constants in $\mathfrak{F}$ such that

$$P(\cdots, c_{i_1 \cdots i_m}, \cdots) \neq 0,$$

and letting $\bar{a}_1, \cdots, \bar{a}_m$ be unknown constants (that is, indeterminates all of whose derivatives are zero), form the expression

$$\bar{\eta} = \sum \frac{c_{h_1 \cdots h_m}}{h_1! \cdots h_m!} (\xi_1 - \bar{a}_1)^{h_1} \cdots (\xi_m - \bar{a}_m)^{h_m}.$$

By the above, $\bar{\eta}$ satisfies the congruences

$$\delta_1'^{i_1} \cdots \delta_m'^{i_m} \bar{\eta} \equiv c_{i_1 \cdots i_m} \qquad (\xi_1 - \bar{a}_1, \cdots, \xi_m - \bar{a}_m).$$

Hence

$$A(\bar{\eta}) \equiv P(\cdots, c_{i_1 \cdots i_m}, \cdots) \qquad (\xi_1 - \bar{a}_1, \cdots, \xi_m - \bar{a}_m),$$

that is, $A(\bar{\eta})$ is a polynomial in the indeterminates $\bar{u}_j = \xi_j - \bar{a}_j$ $(j = 1, \cdots, m)$ with coefficients in $\mathfrak{F}$, and these coefficients are not all zero. Therefore we may choose rational values a_j for the unknown constants $\bar{a}_j$ so that, for

$$\eta = \sum \frac{c_{h_1 \cdots h_m}}{h_1! \cdots h_m!} (\xi_1 - a_1)^{h_1} \cdots (\xi_m - a_m)^{h_m},$$

we have $A(\eta) \neq 0$, q.e.d.

4. Existence of a primitive element

We are now in a position to prove our principal

THEOREM. *Let $\mathfrak{F}$ contain m elements whose Jacobian is different from zero. If $\mathfrak{F}\langle\alpha_1, \cdots, \alpha_n\rangle$ is a differential extension field of $\mathfrak{F}$ such that each α_i is a solution of a nonzero differential polynomial in $\mathfrak{F}\{y\}$, then there exists a primitive element γ:*

$$\mathfrak{F}\langle\alpha_1, \cdots, \alpha_n\rangle = \mathfrak{F}\langle\gamma\rangle.$$

By §2 we must show that there exists a $\gamma \in \mathfrak{F}\langle\alpha_1, \cdots, \alpha_n\rangle$ which is invariant under no isomorphism of $\mathfrak{F}\langle\alpha_1, \cdots, \alpha_n\rangle$ with respect to $\mathfrak{F}$. We shall prove, as a lemma, a stronger result.

Let $A_i(y_i) \in \mathfrak{F}\{y_i\}$ have the solution $y_i = \alpha_i$ $(i = 1, \cdots, n)$. We shall show that *there exist elements* $\tau_1, \cdots, \tau_n \in \mathfrak{F}$ *such that* $\tau_1 y_1 + \cdots + \tau_n y_n$ *assumes different values for different solutions of* $\{A_1(y_1), \cdots, A_n(y_n)\}$.[6] Then certainly the element $\tau_1\alpha_1 + \cdots + \tau_n\alpha_n$ will satisfy our requirements on γ.

To prove this lemma, let $z_1, \cdots, z_n, t_1, \cdots, t_n$ be new unknowns, and, in $\mathfrak{F}\{y_1, \cdots, y_n, z_1, \cdots, z_n, t_1, \cdots, t_n\}$, consider the perfect differential ideal

$$\Omega = \{A_1(y_1), \cdots, A_n(y_n), A_1(z_1), \cdots, A_n(z_n), t_1(y_1 - z_1) + \cdots + t_n(y_n - z_n)\}.$$

Let $\Omega = \Omega_1 \cap \cdots \cap \Omega_s$ be the decomposition of Ω into essential prime differential ideals, and suppose the subscripts have been assigned so that $\Omega_1, \cdots, \Omega_r$ each contains every $y_i - z_i$, whereas $\Omega_{r+1}, \cdots, \Omega_s$ each fails to contain some $y_i - z_i$. Consider an Ω_j with $j > r$. Let $\bar{\eta}_1, \cdots, \bar{\eta}_n, \bar{\zeta}_1, \cdots, \bar{\zeta}_n, \bar{\tau}_1, \cdots, \bar{\tau}_n$ be a generic solution of Ω_j. Since $\bar{\tau}_1(\bar{\eta}_1 - \bar{\zeta}_1) + \cdots + \bar{\tau}_n(\bar{\eta}_n - \bar{\zeta}_n) = 0$, and some $\bar{\eta}_i - \bar{\zeta}_i$ is different from zero, $\bar{\tau}_1, \cdots, \bar{\tau}_n$ are dependent[7] over $\mathfrak{F}\langle\bar{\eta}_1, \cdots, \bar{\eta}_n, \bar{\zeta}_1, \cdots, \bar{\zeta}_n\rangle$. But each $\bar{\eta}_i$ and each $\bar{\zeta}_i$ annul a nonzero differential polynomial with coefficients in $\mathfrak{F}$. Hence $\bar{\tau}_1, \cdots, \bar{\tau}_n$ are dependent over $\mathfrak{F}$,[8] so that Ω_j contains a nonzero differential polynomial $L_j \in \mathfrak{F}\{t_1, \cdots, t_n\}$. Now let $M(t_1, \cdots, t_n) = L_{r+1} \cdots L_s$. By the authority of §3 choose elements $\tau_1, \cdots, \tau_n$ for which $M(\tau_1, \cdots, \tau_n) \neq 0$. For any two distinct solutions $y_i = \eta_i$ $(i = 1; \cdots, n)$ and $y_i = \zeta_i$ $(i = 1, \cdots, n)$ of $\{A_1(y_1), \cdots, A_n(y_n)\}$, the $3n$ elements

$$\eta_1, \cdots, \eta_n, \zeta_1, \cdots, \zeta_n, \tau_1, \cdots, \tau_n$$

cannot be a solution of Ω. For, these elements cannot be a solution of any Ω_j with $j \leqq r$ as each such Ω_j contains every $y_i - z_i$, and they cannot be a solution of an Ω_j with $j > r$ as each such Ω_j contains $M(t_1, \cdots, t_n)$. Consequently

$$\tau_1(\eta_1 - \zeta_1) + \cdots + \tau_n(\eta_n - \zeta_n) \neq 0.$$

Since $\eta_1, \cdots, \eta_n$ and $\zeta_1, \cdots, \zeta_n$ were chosen as *any* two distinct solutions of $\{A_1(y_1), \cdots, A_n(y_n)\}$, the proof of the lemma, and therefore of the theorem, is complete.

INSTITUTE FOR ADVANCED STUDY

---

[6] We lean heavily here on the proof for the ordinary case given by J. F. Ritt, *Differential equations from the algebraic standpoint*, American Mathematical Society Colloquium Publications, vol. XIV, New York, 1932. See especially pp. 26–31.

[7] See Raudenbush, loc. cit.

[8] See Raudenbush, loc. cit.

ANNALS OF MATHEMATICS
Vol. 45, No. 2, April, 1944

EXTENSIONS OF DIFFERENTIAL FIELDS, II

BY E. R. KOLCHIN

(Received February 10, 1943)

Introduction

We consider two differential fields $\mathcal{F}$ and $\mathcal{G}$ of characteristic 0 such that $\mathcal{F} \subset \mathcal{G}$, and ask how many elements of $\mathcal{G}$ it is necessary to adjoin to $\mathcal{F}$ in order to obtain $\mathcal{G}$. The first answer to a question of this kind was furnished by J. F. Ritt,[1] who proved the following analog to an algebraic theorem of Luroth's: If $\mathcal{F}$ is an *ordinary*[2] differential field, and if $\mathcal{G} = \mathcal{F}\langle\alpha_1, \alpha_2\rangle \subseteq \mathcal{F}\langle y\rangle$, where y is an unknown, then $\mathcal{G}$ is a simple extension of $\mathcal{F}$.[3] Ritt's proof applied only to differential fields of functions meromorphic in a region of the complex plane, as it made use of function-theoretic considerations. The second result along these lines, given by the author,[4] states that if $\mathcal{F}$ is a differential field of characteristic 0 with m types of differentiation $\delta_1, \cdots, \delta_m$, if $\mathcal{F}$ contains m elements $\xi_1, \cdots, \xi_m$ whose Jacobian $|\delta_i\xi_j|$ does not vanish, if $\mathcal{G}$ is differentially algebraic over $\mathcal{F}$,[5] and if $\mathcal{G}$ is a *finite* extension of $\mathcal{F}$, then $\mathcal{G}$ is already a *simple* extension of $\mathcal{F}$.

From this result it easily follows that if the degree of differential transcendency of $\mathcal{G}$ over $\mathcal{F}$ is $d \geqq 0$, and if $\mathcal{G}$ is a finite extension of $\mathcal{F}$, then $\mathcal{G}$ is a $(d + 1)$-fold extension of $\mathcal{F}$. For example, if $\mathcal{F} \subset \mathcal{G} \subseteq \mathcal{F}\langle y\rangle$, and if $\mathcal{G}$ is a finite extension of $\mathcal{F}$, then $\mathcal{G}$ is a two-fold extension of $\mathcal{F}$. When $\mathcal{F}$ is ordinary, Ritt's theorem shows that $\mathcal{G}$ is then a simple extension of $\mathcal{F}$.

---

[1] RITT, *Differential equations from the algebraic standpoint*, American Mathematical Society Colloquium Publications, vol. XIV, New York, 1932. See chapter VIII.

[2] A differential field with m types of differentiation is called *ordinary* or *partial* according as $m = 1$, or $m > 1$.

[3] $\mathcal{F}\langle\cdots\rangle$ indicates differential field extension; $\mathcal{F}\{\cdots\}$ indicates differential ring extension.

[4] KOLCHIN, *Extensions of differential fields*, I, Annals of Mathematics, vol. 43 (1942), pp. 724–729.

[5] A differential field $\mathcal{G}$ will be called *differentially algebraic* over a differential subfield $\mathcal{F}$ if every element of $\mathcal{G}$ annuls a nonzero differential polynomial with coefficients in $\mathcal{F}$. If $\mathcal{G}$ is not differentially algebraic, it will be called *differentially transcendental*. Similarly, a single element of $\mathcal{G}$ will be spoken of as differentially algebraic or differentially transcendental over $\mathcal{F}$. We prefer these terms to "algebraically transcendental" and "hypertranscendental" (or "transcendentally transcendental") which have appeared in the literature, as more expressive. A set of elements of $\mathcal{G}$ will be called *differentially algebraically independent* over $\mathcal{F}$ if no finite subset of these elements annuls a nonzero differential polynomial with coefficients in $\mathcal{F}$. A set $\mathfrak{M}$ of differentially algebraically independent elements of $\mathcal{G}$ such that $\mathcal{G}$ is differentially algebraic over $\mathcal{F}\langle\mathfrak{M}\rangle$ will be called a *differential transcendency base* of $\mathcal{G}$ over $\mathcal{F}$. The cardinal number of a differential transcendency base of $\mathcal{G}$ over $\mathcal{F}$ will be called the *degree of differential transcendency* of $\mathcal{G}$ over $\mathcal{F}$. This number is unique. For the proof of this, and for a general discussion of these concepts, the reader is referred to two papers by H. W. RAUDENBUSH: Annals of Mathematics, vol. 34 (1933), pp. 509–517; Bulletin of the American Mathematical Society, vol. 40 (1934), pp. 714–720.

The main object of this note is to prove anew Ritt's theorem. The present proof (§1) is valid for any ordinary differential field $\mathcal{F}$ of characteristic 0, as no non-algebraic notions are employed. Also, the present statement of the theorem is slightly stronger than Ritt's, as it is not assumed here that $\mathcal{G}$ is a finite extension of $\mathcal{F}$.

It is perhaps not uninteresting that the theorem in question ceases to hold for partial differential fields. Examples are produced (§2) which show that for a partial differential field $\mathcal{F}$ one may find a differential field $\mathcal{G}$ between $\mathcal{F}$ and $\mathcal{F}\langle y\rangle$ which is a two-fold extension of $\mathcal{F}$ without being a simple extension, or which is not a finite extension at all.

1. The principal result

THEOREM. *Let $\mathcal{F}$ be an ordinary differential field of characteristic 0, let y be an unknown, and let $\mathcal{G}$ be a differential field such that $\mathcal{F} \subset \mathcal{G} \subseteq \mathcal{F}\langle y\rangle$. Then there exists an element ω in $\mathcal{G}$ such that $\mathcal{G} = \mathcal{F}\langle\omega\rangle$. If ω_1 and ω_2 are two such elements ω, then $\omega_2 = (a\omega_1 + b)/(c\omega_1 + d)$, where $a, b, c, d \in \mathcal{F}$.*

PROOF. Let Σ be the set of all differential polynomials in $\mathcal{G}\{z\}$ which vanish when $z = y$. Then Σ is a prime differential ideal the manifold of which is the general solution of some irreducible $A(z)$ in $\mathcal{G}\{z\}$. The coefficients in $A(z)$ are quotients of differential polynomials in $\mathcal{F}\{y\}$. We assume without loss of generality that one of these coefficients is unity.

Denote the least common denominator of these coefficients by $D(y)$, and let $B(y, z) = (D(y)A(z)$. Then $B(y, z) \in \mathcal{F}\{y, z\}$, and $B(y, y) = 0$. Let the orders of $B(y, z)$ in y and in z be p and q, respectively, and consider the separant $S(y, z) = \partial B(y, z)/\partial y_p$, where y_p is the p-th derivative of y.

We shall show that $S(y, y) \neq 0$. Indeed, if $S(y, y)$ vanished, then $\partial A(z)/\partial y_p$, a differential polynomial in $\mathcal{G}\{z\}$ of order not exceeding q, would vanish when $z = y$, so that $\partial A(z)/\partial y_p$, would be divisible by $A(z)$. But one of the coefficients in $A(z)$ is unity, so that the corresponding coefficient in $\partial A(z)/\partial y_p$ is zero. Thus $\partial A(z)/\partial y_p$ could not be divisible by $A(z)$ unless $\partial A(z)/\partial y_p$ were zero, which would be absurd.

We now prove that $p \leqq q$. Indeed, suppose that $p > q$, and write $B(y, z)$ as a polynomial in y_p:

$B(y, z) = I_0(y, z) + I_1(y, z)y_p + \cdots + I_n(y, z)y_p^n$. Since $B(y, y) = 0$, and each $I_i(y, y)$ is free of y_p, it follows that each $I_i(y, y) = 0$. This implies that $S(y, y) = 0$. This contradiction proves that $p \leqq q$.

Let $\omega = H(y)/K(y)$ be any coefficient in $A(z)$ which effectively involves y, with $H(y)$ and $K(y)$ free of common divisor. By the preceding, the orders of $H(y)$ and $K(y)$ are less than or equal to q. Consequently $\omega K(z) - H(z)$, which is in $\mathcal{G}\{z\}$, and which vanishes when $z = y$, is divisible by $A(z)$:

$$\omega K(z) - H(z) = \alpha A(z), \qquad \alpha \in \mathcal{G}.$$

We shall show that $\mathcal{G} = \mathcal{F}\langle\omega\rangle$. To this end let $\theta = M(y)/N(y)$ be any element of $\mathcal{G}$. Now, $\theta N(z) - M(z)$ is in $\mathcal{G}\{z\}$ and vanishes when $z = y$. Hence

$\theta N(z) - M(z) \in \Sigma$. If $\bar{y}$ is an element of some extension of $\mathcal{F}$ such that $y \to \bar{y}$ generates a relative isomorphism of $\mathcal{F}\langle y\rangle$ with respect to $\mathcal{F}\langle\omega\rangle$,[6] then $z = \bar{y}$ is a non-singular solution of $\omega K(z) - H(z)$, so that $\theta N(\bar{y}) - M(\bar{y}) = 0$. Thus, under any isomorphism of $\mathcal{F}\langle y\rangle$ relative to $\mathcal{F}\langle\omega\rangle$, θ remains invariant. Hence $\theta \in \mathcal{F}\langle\omega\rangle$, and $\mathcal{G} = \mathcal{F}\langle\omega\rangle$.[7]

Now let ω_1 and ω_2 be two elements in $\mathcal{F}\langle y\rangle$ such that $\mathcal{F}\langle\omega_1\rangle = \mathcal{F}\langle\omega_2\rangle$. We have $\omega_2 = P(\omega_1)/Q(\omega_1)$, and $\omega_1 = R(\omega_2)/S(\omega_2)$, where $P(y)$, $Q(y)$, $R(y)$, $S(y) \in \mathcal{F}\{y\}$. If $z = \omega_1'$ is any solution of $\omega_2 Q(z) - P(z)$ which does not annul $P(z)Q(z)$, then $\omega_1 \to \omega_1'$ generates an isomorphism of $\mathcal{F}\langle\omega_1\rangle$ relative to $\mathcal{F}$ under which

$$\omega_2 = P(\omega_1)/Q(\omega_1) \to P(\omega_1')/Q(\omega_1') = \omega_2\,,$$

$$\omega_1 = R(\omega_2)/S(\omega_2) \to R(\omega_2)/S(\omega_2) = \omega_1\,,$$

so that $\omega_1' = \omega_1$. Thus $\omega_2 Q(z) - P(z)$ has only one solution, and therefore is of order 0 and of degree 1, so that $\omega_2 = (a\omega_1 + b)/(c\omega_1 + d)$.

2. Examples for partial differential fields

Let $\mathcal{F}$ be a differential field of characteristic 0 with two types of differentiation δ_1 and δ_2. Let y be an unknown.

EXAMPLE 1. *Let $\mathcal{G} = \mathcal{F}\langle\delta_1 y, \delta_2 y\rangle$. Then $\mathcal{G}$ is not a simple extension of $\mathcal{F}$.*

PROOF. Suppose that ω is an element of $\mathcal{G}$ for which $\mathcal{G} = \mathcal{F}\langle\omega\rangle$. Then we can write $\omega = P(y)/Q(y)$, where $P(y)$ and $Q(y)$ are in $\mathcal{F}\{y\}$, and have no common divisor. Since $\omega \in \mathcal{F}\langle\delta_1 y, \delta_2 y\rangle$ it is clear that y can not occur undifferentiated in $P(y)$ or $Q(y)$.

Let $P(y)/Q(y)$ be of order p, and let k be the highest value of j such that $\delta_1^j\delta_2^{p-j}y$ effectively appears in $P(y)/Q(y)$.

Since $\delta_1 y \in \mathcal{F}\langle\omega\rangle$ we have $\delta_1 y = M(\omega)/N(\omega)$, where $M(\omega)$, $N(\omega) \in \mathcal{F}\{\omega\}$. Let $M(\omega)/N(\omega)$ be of order q in ω, and let h be the highest value of j such that $\delta_1^j\delta_2^{q-j}\omega$ effectively appears in $M(\omega)/N(\omega)$.

Now, $\delta_1^h\delta_2^{q-h}\omega = \delta_1^h\delta_2^{q-h}P(y)/Q(y) = [U(y)\delta_1^{h+k}\delta_2^{p+q-h-k}y + V(y)]/Q(y)^{q+1}$, where $U(y)$, $V(y)$ are of order not exceeding $p + q$ and do not contain $\delta_1^{h+k}\delta_2^{p+q-h-k}y$. Furthermore, no derivative $\delta_1^a\delta_2^b\omega$ other than $\delta_1^h\delta_2^{q-h}\omega$ appearing in $M(\omega)/N(\omega)$ effectively involves $\delta_1^{h+k}\delta_2^{p+q-h-k}y$ when expressed in terms of y. Since $\delta_1 y = M(\omega)/N(\omega)$, this implies that $h + k = 1$, $p + q - h - k = 0$. But $p \geqq 1$, so that $p = 1$, $q = 0$, $h = 0$, and $\delta_1 y$ is a rational function of ω with coefficients in $\mathcal{F}$. Similarly for $\delta_2 y$, so that $\delta_1 y$ and $\delta_2 y$ are algebraically dependent over $\mathcal{F}$, which is a contradiction.

EXAMPLE 2. *Let $\mathcal{G} = \mathcal{F}\langle\delta_1 y\cdot\delta_2 y, \delta_1^2 y\cdot\delta_2^2 y, \cdots, \delta_1^n y\cdot\delta_2^n y, \cdots\rangle$. Then $\mathcal{G}$ is not a finite extension of $\mathcal{F}$.*

---

[6] That is, an isomorphism of $\mathcal{F}\langle y\rangle$ which leaves invariant each element of $\mathcal{F}\langle\omega\rangle$. See KOLCHIN, loc. cit., p. 726.

[7] KOLCHIN, loc. cit., p. 726.

PROOF. Assume the contrary. Then for some n we may write

$$\delta_1^n y \cdot \delta_2^n y \in \mathcal{F}\langle \delta_1 y \cdot \delta_2 y, \cdots, \delta_1^{n-1} y \cdot \delta_2^{n-1} y \rangle.$$

That is, there exists a differential rational function $f(u, \cdots, u_{n-1})$ in the unknowns $u, \cdots, u_{n-1}$ with coefficients in $\mathcal{F}$ such that $\delta_1^n y \cdot \delta_2^n y = f(\delta_1 y \cdot \delta_2 y, \cdots, \delta_1^{n-1} y \cdot \delta_2^{n-1} y)$. In order that $f(\delta_1 y \cdot \delta_2 y, \cdots, \delta_1^{n-1} y \cdot \delta_2^{n-1} y)$ yield $\delta_1^n y \cdot \delta_2^n y$, $f(u, \cdots, u_{n-1})$ must effectively contain some $\delta_1^j \delta_2^j u_{n-j}$. Let k denote the greatest value of j such that $\delta_1^j \delta_2^j u_{n-j}$ is effectively contained in $f(u, \cdots, u_{n-1})$. Then $f(\delta_1 y \cdot \delta_2 y, \cdots, \delta_1^{n-1} y \cdot \delta_2^{n-1} y)$ effectively contains $\delta_1^n \delta_2^k y \cdot \delta_2^{n-k} y$, a contradiction.

OFFICE OF THE CHIEF OF ORDNANCE,
WASHINGTON, D. C.

ALGEBRAIC MATRIC GROUPS

BY E. R. KOLCHIN

DEPARTMENT OF MATHEMATICS, COLUMBIA UNIVERSITY

Communicated October 24, 1946

Let $\mathfrak{G}$ be a multiplicative group of square matrices (a_{ij}) of degree $n \geqslant 1$ with elements a_{ij} contained in an algebraically closed field $\mathcal{C}$. $\mathfrak{G}$ is called *algebraic* if there exists a set $\mathfrak{g}$ of polynomials in $\mathcal{C}[\ldots, x_{ij}, \ldots]$ such that: (*a*) for each (a_{ij}) in $\mathfrak{G}$ and each $f(\ldots, x_{ij}, \ldots)$ in $\mathfrak{g}$ we have $f(\ldots, a_{ij}, \ldots) = 0$; (*b*) every non-singular matrix (a_{ij}), with elements in $\mathcal{C}$ such that $f(\ldots, a_{ij}, \ldots) = 0$ for each $f(\ldots, x_{ij}, \ldots)$ in $\mathfrak{g}$, is in $\mathfrak{G}$. Such algebraic matric groups are encountered in the Picard-Vessiot theory of homogeneous linear ordinary differential equations, where they play a rôle similar to that played by finite permutation groups in the Galois theory of algebraic equations. Any attempt to clarify and rigorize the Picard-Vessiot theory must include an adequate treatment of algebraic matric groups, and preferably (if the Picard-Vessiot theory is to be completely algebraic) an algebraic one, independent of the theory of Lie groups. The present communication describes some results along these lines. A description of an algebraic development of the Picard-Vessiot theory is contained in the note immediately following.

If we take $\mathfrak{g}$ as large as possible, so that it is unique, then $\mathfrak{G}$ consists of the algebraic manifold of $\mathfrak{g}$ (called the *underlying manifold* of $\mathfrak{G}$) with a lower-dimensional algebraic manifold deleted. It turns out that the irreducible components of the underlying manifold of $\mathfrak{G}$ are pairwise disjoint (except for singular matrices) and all have the same dimension, and the irreducible component containing the identity matrix ϵ is the underlying manifold of a normal algebraic subgroup of $\mathfrak{G}$ of finite index (= number of irreducible components). Definitions: this subgroup is the *component of the identity* of $\mathfrak{G}$ (notation: $\mathfrak{G}^\circ$); $\mathfrak{G}$ is *connected* if $\mathfrak{G} = \mathfrak{G}^\circ$; the *dimension* of $\mathfrak{G}$ is that of its underlying manifold.

We call $\mathfrak{G}$ *anticompact* if $\mathfrak{G}$ contains no matrix $\neq \epsilon$ of finite order not divisible by the field characteristic p, and call $\mathfrak{G}$ *quasicompact* if every algebraic subgroup of $\mathfrak{G}$ of order > 1 contains such a matrix. By making joint use of the algebraic manifold properties and group properties of $\mathfrak{G}$ we can prove: (1) *$\mathfrak{G}$ is anticompact if and only if each matrix of $\mathfrak{G}$ is reducible to special triangular form* (0's below the main diagonal, 1's on it), i.e., if and only if each matrix in $\mathfrak{G}$ has all its characteristic roots equal to 1. (2) *$\mathfrak{G}$ is quasicompact if and only if each matrix in $\mathfrak{G}$ is reducible to diagonal form.*

$\mathfrak{G}$ is called *solvable* if $\mathfrak{G}$ has a normal chain in which all the factor groups are abelian (here "normal chain" signifies a normal chain in the usual sense plus the restriction that all members of the chain be algebraic). $\mathfrak{G}$ is solvable if and only if its sequence of commutator subgroups terminates with the identity group (the commutator subgroup $\mathfrak{G}'$ of $\mathfrak{G}$ is the smallest algebraic subgroup of $\mathfrak{G}$ containing $\sigma\tau\sigma^{-1}\tau^{-1}$ for all σ and τ in $\mathfrak{G}$). Using several lemmas it is possible to prove that *if $\mathfrak{G}$ is connected and solvable then $\mathfrak{G}$ is reducible to triangular form* (0's below the main diagonal). The proof employs a double induction on the matrix degree n and the normal chain length r. For $n = 1$ there is nothing to prove. Letting $n > 1$ and supposing the result verified for matrices of lower degree, we can assume that $\mathfrak{G}$ is irreducible (this requires a lemma asserting that the blocks of a reduced algebraic matric group are themselves algebraic matric groups, connected when the given reduced group is). For $r = 1$ (i.e., for abelian $\mathfrak{G}$) the result is an easy consequence of Schur's lemma. Letting $r > 1$ and supposing the result verified for matrices of degree n with lower values of r, we see that the second member $\mathfrak{G}_1$ of the normal chain for $\mathfrak{G}$ of length

r is reducible to triangular form (this requires a lemma asserting that $\mathfrak{G}'$ is connected whenever $\mathfrak{G}$ is, permitting the assumption that $\mathfrak{G}_1$ is connected). Then an argument making use of the reducibility of $\mathfrak{G}_1$ to triangular form and the abelian nature of $\mathfrak{G}/\mathfrak{G}_1$ leads to a contradiction.

Using this theorem and the results 1 and 2 above, it can be shown that *$\mathfrak{G}$ is solvable and anticompact if and only if $\mathfrak{G}$ is reducible to special triangular form,* and *if $\mathfrak{G}$ is reducible to triangular form and is quasicompact then $\mathfrak{G}$ is reducible to diagonal form.*

Finally, we investigate the extent to which the concepts "solvable," "anticompact," and "quasicompact" are broadened by the introduction of apparently more inclusive definitions by means of conditions on a normal chain

$$\mathfrak{G} = \mathfrak{G}_0 \supseteq \mathfrak{G}_1 \supseteq \ldots \supseteq \mathfrak{G}_{r-1} \supseteq \mathfrak{G}_r = \mathfrak{E}.$$

Using above-mentioned results concerning $\mathfrak{G}^\circ$ and $\mathfrak{G}'$ it is not hard to show that *if every factor group $\mathfrak{G}_{i-1}/\mathfrak{G}_i$ is abelian or finite then $\mathfrak{G}^\circ$ is solvable.*

If the definitions of anticompact and quasicompact are extended to factor groups in the obvious way, it can be proved (using a slight generalization of result 1 above) that: *if every factor group $\mathfrak{G}_{i-1}/\mathfrak{G}_1$ is anticompact or finite then $\mathfrak{G}^\circ$ is anticompact; if every $\mathfrak{G}_{i-1}/\mathfrak{G}_i$ is anticompact then $\mathfrak{G}$ is, too.* Analogous to the second part of this theorem is the result that *if every $\mathfrak{G}_{i-1}/\mathfrak{G}_i$ is quasicompact; then $\mathfrak{G}$ is, too.* I do not know whether the analog to the first part is true or false (for $p = 0$ it is obviously true, as a finite group is then quasicompact; for $p > 0$ a finite group may not be quasicompact, e.g., a group of order divisible by p).

THE PICARD-VESSIOT THEORY OF HOMOGENEOUS LINEAR ORDINARY DIFFERENTIAL EQUATIONS

By E. R. Kolchin

Department of Mathematics, Columbia University

Communicated October 24, 1946

The Galois theory of homogeneous linear ordinary differential equations as developed by Picard and Vessiot is founded on the theory of Lie groups and on the general theory of differential equations.[1] Because of the loose state of the Lie theory at the time, the weakness then of the theory of differential equations with respect to its algebraic aspects, and the over-intimate connection with the analytic point of view, the Picard-Vessiot theory suffers from a certain lack of rigor, completeness and simplicity. The present communication describes an attempt to algebraize, rigorize, round out and extend the Picard-Vessiot theory. Use is made of the Ritt theory of algebraic differential equations[2] (unavailable, of course, to Picard and Vessiot), and some results concerning algebraic groups of matrices developed for the purpose.[3]

Let $\mathfrak{F}$ be a differential field (ordinary or partial) of characteristic 0, $\mathfrak{G}$ a differential extension field thereof. A set of isomorphisms of $\mathfrak{G}$ over $\mathfrak{F}$ (i.e., isomorphisms of $\mathfrak{G}$ under which each element of $\mathfrak{F}$ is invariant) will be called *abundant* if for each differential field $\mathfrak{F}_1$ between $\mathfrak{F}$ and $\mathfrak{G}$ and each element α in $\mathfrak{G} - \mathfrak{F}_1$ there is an isomorphism in the set under which α is not invariant but every elment of $\mathfrak{F}_1$ is. By a previous paper[4] such sets of isomorphisms always exist. $\mathfrak{G}$ will be called a *normal* extension of $\mathfrak{F}$ if the group of all automorphisms of $\mathfrak{G}$ over $\mathfrak{F}$ is abundant.

Let $\mathfrak{G}$ be a normal extension of $\mathfrak{F}$, and let $\mathfrak{G}$ be an abundant group of automorphisms of $\mathfrak{G}$ over $\mathfrak{F}$ (not necessarily the group of all such automorphisms). For any differential field $\mathfrak{F}_1$ between $\mathfrak{F}$ and $\mathfrak{G}$ let $\mathfrak{G}(\mathfrak{F}_1)$ denote the group of all automorphisms in $\mathfrak{G}$ which leave invariant each element of $\mathfrak{F}_1$ (thus $\mathfrak{G}(\mathfrak{F}) = \mathfrak{G}$ and $\mathfrak{G}(\mathfrak{G}) = \mathfrak{E}$, the identity group). Then it is easy to show that *the mapping* $\mathfrak{F}_1 \rightarrow \mathfrak{G}(\mathfrak{F}_1)$ *is a one-to-one correspondence between the set of all differential fields between* $\mathfrak{F}$ *and* $\mathfrak{G}$ *and a certain set of subgroups of* $\mathfrak{G}$. Furthermore: $\mathfrak{G}(\mathfrak{F}_1)$ *is a normal subgroup of* $\mathfrak{G}$ *if and only if* $\sigma\mathfrak{F}_1 = \mathfrak{F}_1$ *for every* σ *in* $\mathfrak{G}$; *when this condition is satisfied then* $\mathfrak{F}_1$ *is a normal extension of* $\mathfrak{F}$ *and* $\mathfrak{G}/\mathfrak{G}(\mathfrak{F}_1)$ *is isomorphic with an abundant group of automorphisms of* $\mathfrak{F}_1$ *over* $\mathfrak{F}$. I do not know, in the case in which $\mathfrak{G}$ is the group of all automorphisms of $\mathfrak{G}$ over $\mathfrak{F}$, whether $\mathfrak{G}/\mathfrak{G}(\mathfrak{F}_1)$ is isomorphic with the group of all automorphisms of $\mathfrak{F}_1$ over $\mathfrak{F}$.

Henceforth, suppose that $\mathfrak{F}$ is an ordinary differential field of characteristic 0 with an algebraically closed field of constants $\mathfrak{C}$, and consider a homogeneous linear differential polynomial $L(y) = y^{(n)} + p_1y^{(n-1)} + \ldots + p_ny$ (each p_i in $\mathfrak{F}$). If $\eta_1, \ldots, \eta_n$ are n solutions of $L(y) = 0$, linearly independent over $\mathfrak{C}$, which are contained in some extension of $\mathfrak{F}$, and if the differential field $\mathfrak{G}$ obtained by adjoining $\eta_1, \ldots, \eta_n$ to $\mathfrak{F}$ contains no constants not in $\mathfrak{C}$, then $\mathfrak{G}$ will be called a *Picard-Vessiot* extension of $\mathfrak{F}$. It turns out that *every Picard-Vessiot extension* $\mathfrak{G}$ *of* $\mathfrak{F}$ *is normal, and the group* $\mathfrak{G}$ *of all automorphisms* σ *of* $\mathfrak{G}$ *over* $\mathfrak{F}$ *is (isomorphic with) an algebraic group (also denoted by* $\mathfrak{G}$*) of matrices* (k_{ij}), *with each* k_{ij} *in* $\mathfrak{C}$, *such that* $\sigma\eta_j = \sum_{i=1}^{n} k_{ij}\eta_i$ $(j = 1, \ldots, n)$. For the special case of Picard-Vessiot extensions it can be proved that the set of all groups $\mathfrak{G}(\mathfrak{F}_1)$ with $\mathfrak{F}_1$ between $\mathfrak{F}$ and $\mathfrak{G}$ is identical with the set of *all* algebraic subgroups of $\mathfrak{G}$, that when $\mathfrak{G}(\mathfrak{F}_1)$ is a normal subgroup of $\mathfrak{G}$ then $\mathfrak{G}/\mathfrak{G}(\mathfrak{F}_1)$ is isomorphic with the group of *all* automorphisms of $\mathfrak{F}_1$ over $\mathfrak{F}$, and that the *dimension* of the algebraic matric group $\mathfrak{G}$ equals the *degree of transcendency* of $\mathfrak{G}$ over $\mathfrak{F}$.

An element α of an extension of $\mathfrak{F}$ is called an *integral* of an element a of that extension if $\alpha' = a$; α is called an *exponential of an integral* of a if $\alpha' = a\alpha$. A differential extension field $\mathfrak{K}$ of $\mathfrak{F}$ will be called *liouvillian* if: (*a*) every constant in $\mathfrak{K}$ is in $\mathfrak{C}$; (*b*) $\mathfrak{C}$ is an extension of $\mathfrak{F}$ by means of integrals, exponentials of integrals, and algebraic functions, i.e., there is a monotonic sequence of differential fields $\mathfrak{F} = \mathfrak{F}_0 \subseteq \mathfrak{F}_1 \subseteq \ldots \subseteq \mathfrak{F}_r = \mathfrak{K}$ such that for each $i > 0$ $\mathfrak{F}_i$ is obtained from $\mathfrak{F}_{i-1}$ by the differential field adjunction of a single element which is either an integral of an element of $\mathfrak{F}_{i-1}$, an exponential of an integral of an element of $\mathfrak{F}_{i-1}$, or algebraic over $\mathfrak{F}_{i-1}$.

We shall have occasion to distinguish ten types of liouvillian extension, namely, extension by:

(1) integral, exponentials of integrals, and algebraic functions (i.e., any liouvillian extension),
(2) integrals and exponentials of integrals,
(3) exponentials of integrals, and algebraic functions,
(4) integrals and algebraic functions,
(5) integrals and radicals,
(6) exponentials of integrals,
(7) integrals,
(8) algebraic functions,
(9) radicals,
(10) rational functions (i.e., not a proper extension at all).

Corresponding to these ten types of liouvillian extension we consider ten types of algebraic matric groups defined by properties of the component of the identity $\mathfrak{G}^\circ$ and of $\mathfrak{G}$ itself:

(1) $\mathfrak{G}^\circ$ is solvable,
(2) $\mathfrak{G}$ is solvable,
(3) $\mathfrak{G}^\circ$ is solvable and quasicompact,
(4) $\mathfrak{G}^\circ$ is solvable and anticompact,
(5) $\mathfrak{G}$ is solvable and $\mathfrak{G}^\circ$ is anticompact,
(6) $\mathfrak{G}$ is solvable and quasicompact,
(7) $\mathfrak{G}$ is solvable and anticompact,
(8) $\mathfrak{G}$ is finite,
(9) $\mathfrak{G}$ is finite and solvable,
(10) $\mathfrak{G} = \mathfrak{E}$.

It is now possible to state the following extension of Vessiot's big theorem on "solvability by quadratures." *If the Picard-Vessiot extension* $\mathfrak{G}$ *is contained in a liouvillian extension of* $\mathfrak{F}$ *then* $\mathfrak{G}^\circ$ *is solvable. Conversely, if* $\mathfrak{G}^\circ$ *is solvable then* $\mathfrak{G}$ *is a liouvillian extension of* $\mathfrak{F}$. *In either case the liouvillian extension is of one of the types (1)–(10) if and only if the algebraic matric group* $\mathfrak{G}$ *is of the corresponding type (1)–(10).*

[1] For the literature of the Picard-Vessiot theory see: Vessiot, E., *Encyclopédie des sciences mathématiques pures et appliquées*, tome II, vol. 3, fascicule 1, 58–170, esp. pp. 152–165 (1910).

[2] A general account of this theory as of 1938 is contained in: Ritt, J. F., *American Mathematical Society Semicentennial Publications*, II, 35–55 (1938).

[3] See the immediately preceding note. Familiarity with the results of that note will be assumed in the present communication.

[4] Kolchin, E. R., *Annals of Mathematics*, **43**, 724–729 (1942).

EXTENSIONS OF DIFFERENTIAL FIELDS. III

E. R. KOLCHIN

The purpose of the present note is to show how the point of view of a preceding paper[1] can be used in developing the concepts of resolvent, dimension, and order introduced by J. F. Ritt in his theory of algebraic differential equations.[2] The present development, in addition to being simpler in some instances, has the advantage of being valid for abstract differential fields as opposed to fields of meromorphic functions of a complex variable, as used by Ritt. I shall also take the opportunity to correct mistakes in a related paper.[3] The notation and definitions used will be as in Extensions I and II.

1. **Resolvents, dimension, and order.** Let $\mathcal{J}$ be a differential field (ordinary or partial) of characteristic 0, and let $y_1, \cdots, y_n$ be unknowns. If Π is a prime differential ideal in $\mathcal{J}\{y_1, \cdots, y_n\}$ other than $\mathcal{J}\{y_1, \cdots, y_n\}$ itself then Π has a generic solution $\eta_1, \cdots, \eta_n$.

If the degree of differential transcendency of $\mathcal{J}\langle\eta_1, \cdots, \eta_n\rangle$ over $\mathcal{J}$ is q then $0 \leqq q < n$, and precisely q of the elements $\eta_1, \cdots, \eta_n$ are differentially algebraically independent over $\mathcal{J}$. Suppose, say, that $\eta_1 \cdots, \eta_q$ are independent in this way, that is, that Π does not contain a nonzero differential polynomial in $y_1, \cdots, y_q$, but does in $y_1, \cdots, y_q, y_j$ for each $j > q$. In Ritt's terminology $y_1 \cdots, y_q$ is a complete set of arbitrary unknowns for Π. It is natural to call q the *dimension* of Π (in symbols, dim Π).

Suppose henceforth that $\mathcal{J}$ is ordinary. It is easy to see that the degree of transcendency of $\mathcal{J}\langle\eta_1, \cdots, \eta_n\rangle$ over $\mathcal{J}\langle\eta_1, \cdots, \eta_q\rangle$ (both these differential fields being considered as fields) is finite. We denote the degree of transcendency of any field $\mathcal{H}$ over a subfield $\mathcal{G}$ by $\partial^0\mathcal{H}/\mathcal{G}$. It will be seen that it is natural to call the integer $\partial^0\mathcal{J}\langle\eta_1, \cdots, \eta_n\rangle/\mathcal{J}\langle\eta_1, \cdots, \eta_q\rangle$ the *order* of Π with respect to $y_1, \cdots, y_q$ (when the set of arbitrary unknowns is understood, for example when $q=0$, we use the notation: ord Π).

Presented to the Society, November 2, 1946; received by the editors October 10, 1946.

[1] Kolchin, *Extensions of differential fields*, I, Ann. of Math. vol. 43 (1942) pp. 724–729. We shall refer to this paper as *Extensions* I.

[2] The subject matter treated here, together with some of the material from *Extensions* I, is roughly parallel to the contents of §§24–31, 75 of Ritt, *Differential equations from the algebraic standpoint*, Amer. Math. Soc. Colloquium Publications, vol. 14, New York, 1932.

[3] Kolchin, *Extensions of differential fields*, II, Ann. of Math. vol. 45 (1944) pp. 358–361. We shall refer to this paper as *Extensions* II.

If $\mathcal{F}\langle\eta_1, \cdots, \eta_q\rangle$ contains a nonconstant (which is the case either when $\mathcal{F}$ does or when $q>0$) then by Extensions I there is an ω such that $\mathcal{F}\langle\eta_1, \cdots, \eta_q, \omega\rangle=\mathcal{F}\langle\eta_1, \cdots, \eta_n\rangle$. Let $A=A(\eta_1, \cdots, \eta_q, w)$ be an irreducible differential polynomial in $\mathcal{F}\langle\eta_1, \cdots, \eta_q\rangle\{w\}$, with solution $w=\omega$, of lowest possible order. Since ω and its first ord Π derivatives must be algebraically dependent over $\mathcal{F}\langle\eta_1, \cdots, \eta_q\rangle$, the order of A is not greater than ord Π. On the other hand, if the order of A is p then the pth derivative (and consequently all the derivatives) of ω is algebraically dependent over $\mathcal{F}\langle\eta_1, \cdots, \eta_q\rangle$ on ω and its first $p-1$ derivatives, so that ord $\Pi=\partial^0\mathcal{F}\langle\eta_1 \cdots, \eta_n\rangle/\mathcal{F}\langle\eta_1, \cdots, \eta_q\rangle$ $=\partial^0\mathcal{F}\langle\eta_1, \cdots, \eta_q, \omega\rangle/\mathcal{F}\langle\eta_1, \cdots, \eta_q\rangle\leqq p$. Therefore the order of A in w is $p=$ord Π. $A(y_1, \cdots, y_q, w)$ is called a *resolvent* of Π. (Actually, this is a slight generalization of Ritt's resolvent, which must be in $\mathcal{F}\{y_1, \cdots, y_q, w\}$ instead of merely in $\mathcal{F}\langle y_1, \cdots, y_q\rangle\{w\}$.)

Let $\mathcal{G}$ be a differential extension field of $\mathcal{F}$, let $\{\Pi\}=\Pi_1\cap \cdots \cap\Pi_r$ be the decomposition into prime components (that is, prime differential ideals none of which contains another) of the perfect differential ideal generated by Π in $\mathcal{G}\{y_1, \cdots, y_n\}$, and let $A_1(y_1, \cdots, y_q, w) \cdots A_s(y_1, \cdots, y_q, w)$ be the complete factorization of $A(y_1, \cdots, y_q, w)$ in $\mathcal{G}\langle y_1, \cdots, y_q\rangle\{w\}$. Each $A_i(y_1, \cdots, y_q, w)$ is of order p in w, for a factor of $A(y_1, \cdots, y_q, w)$ of order less than p would be a common factor of the coefficients in $A(y_1, \cdots, y_q, w)$ when $A(y_1, \cdots, y_q, w)$ is considered as a polynomial in w_p, the pth derivative of w. We shall now establish Ritt's result that $r=s$ and each $A_i(y_1, \cdots, y_q, w)$ is a resolvent of one Π_j. This result implies that Π decomposes if and only if $A(y_1, \cdots, y_q, w)$ factors, and that each prime component in the decomposition has the same order as Π has.

Let $\eta_1', \cdots, \eta_n'$ be a generic solution of Π_1. Then (by *Extensions* I, §1) $\eta_1', \cdots, \eta_n'$ is a generic solution of Π, so that $\eta_1'\to\eta_1, \cdots, \eta_n'\to\eta_n$ generates an isomorphism of $\mathcal{F}\langle\eta_1', \cdots, \eta_n'\rangle$ onto $\mathcal{F}\langle\eta_1, \cdots, \eta_n\rangle$. Therefore if we let ω' be the same differential rational function over $\mathcal{F}$ of $\eta_1', \cdots, \eta_n'$ that ω is of $\eta_1, \cdots, \eta_n$, we shall have

$$\mathcal{F}\langle\eta_1', \cdots, \eta_q', \omega'\rangle = \mathcal{F}\langle\eta_1', \cdots, \eta_n'\rangle.$$

Now ω' is a solution of $A'=A(\eta_1', \cdots, \eta_q', w)$, and therefore of some $A_i'=A_i(\eta_1', \cdots, \eta_q', w)$, say of A_1'. Furthermore, ω' is not a solution of two different A_i''s, for ω' does not annul the separant $\partial A'/\partial w_p=\partial(A_1' \cdots A_s')/\partial w_p$. Let ω'' be a generic solution of the prime component of $\{A_1'\}$ in $\mathcal{G}\langle\eta_1', \cdots, \eta_q'\rangle\{w\}$ not containing the separant $\partial A_1'/\partial w_p$. Then ω'' is a generic solution of the prime component of $\{A'\}$ in $\mathcal{F}\langle\eta_1', \cdots, \eta_q'\rangle\{w\}$ not containing the separant $\partial A'/\partial w_p$, so that $\omega''\rightarrow\omega'$ generates an isomorphism of $\mathcal{J}\langle\eta_1', \cdots, \eta_q', \omega''\rangle$ onto $\mathcal{J}\langle\eta_1', \cdots, \eta_q', \omega'\rangle$, and a homomorphism of $\mathcal{G}\langle\eta_1', \cdots, \eta_q'\rangle\{\omega''\}$ onto $\mathcal{G}\langle\eta_1', \cdots, \eta_q'\rangle\{\omega'\}$. Therefore, if for each $i>q$ we let η_i'' be the same differential rational function over $\mathcal{J}\langle\eta_1', \cdots, \eta_q'\rangle$ of ω'' as η_i' is of ω', then $\eta_1', \cdots, \eta_q', \eta_{q+1}'', \cdots, \eta_n''$ is a generic solution of Π and a solution of some Π_i. Since $\eta_1', \cdots, \eta_n'$ must be a solution of the same Π_i, and since one Π_i does not contain another, $\eta_1', \cdots, \eta_q', \eta_{q+1}'', \cdots, \eta_n''$ is a solution of Π_1, and indeed a generic one.

Therefore $\eta_{q+1}''\rightarrow\eta_{q+1}', \cdots, \eta_n''\rightarrow\eta_n'$ generates an isomorphism of $\mathcal{G}\langle\eta_1', \cdots, \eta_q', \eta_{q+1}'', \cdots, \eta_n''\rangle$ onto $\mathcal{G}\langle\eta_1', \cdots, \eta_n'\rangle$, A_1' is an irreducible differential polynomial in $\mathcal{G}\langle\eta_1', \cdots, \eta_q'\rangle\{w\}$, with solution $w=\omega'$, of minimal degree, and $A_1(y_1, \cdots, y_q, w)$ is a resolvent of Π_1. In the same way, every Π_i has an $A_j(y_1, \cdots, y_q, w)$ as a resolvent, so that $r\leqq s$. To show that there is no $A_j(y_1, \cdots, y_q, w)$ left over, for any j let ω_j be a generic solution of the prime component of $\{A_j'\}$ in $\mathcal{G}\langle\eta_1', \cdots, \eta_n'\rangle\{w\}$ not containing $\partial A_j'/\partial w_p$. For each $i>q$ let η_{ji} be the same differential rational function over $\mathcal{J}\langle\eta_1', \cdots, \eta_q'\rangle$ of ω_j as η_i' is of ω'. Then $\eta_1', \cdots, \eta_q', \eta_{j,q+1}, \cdots, \eta_{jn}$ is a generic solution of Π and therefore a solution of some Π_i, say Π_{i_0}. Therefore ω_j is a solution of the A_k' for which $A_k(y_1, \cdots, y_q, w)$ is a resolvent of Π_{i_0}. This implies that $A_k(y_1, \cdots, y_q, w)$ is divisible by $A_j(y_1, \cdots, y_q, w)$, so that $k=j$ and $A_j(y_1, \cdots, y_q, w)$ is a resolvent of a Π_i.

If $q=0$ and $\mathcal{J}$ consists solely of constants it is still true that each prime component of $\{\Pi\}$ has the same order as Π. To see this introduce a new unknown u and let $\mathcal{J}'=\mathcal{J}\langle u\rangle$, $\mathcal{G}'=\mathcal{G}\langle u\rangle$. The perfect differential ideal generated by Π in $\mathcal{J}'\{y_1, \cdots, y_n\}$ is clearly prime and has the same order as Π has. The prime components of the perfect differential ideal generated by Π in $\mathcal{G}'\{y_1, \cdots, y_n\}$ are the perfect differential ideals generated by $\Pi_1, \cdots, \Pi_r$, and have the same order. Therefore ord $\Pi_i=$ord Π for each i.

2. **Corrections to Extensions II.** We refer now to the proof on page 359 of *Extensions* II. The derivation of the equation $\omega K(z)-H(z)=\alpha A(z)$ is incorrect, for it rests on the unjustified assumption (see lines 18 and 17 from the bottom) that $\partial A(z)/\partial y_p\in\mathcal{G}\{z\}$. To save the proof we delete in toto lines 22–4 from the bottom ("Denote the $\cdots A(z)$:"), and replace them by the following considerations.

Let $\omega=H(y)/K(y)$ be any coefficient in $A(z)$ not merely an element of $\mathcal{J}$, with $H(y)$, $K(y)$ free of common divisor. Clearly $\omega K(z)-H(z)\in\Sigma$.

Denote the lowest common denominator of the coefficients in $A(z)$ by $D(y)$, and let $B(y, z) = D(y)A(z)$. Then $B(y, z) \in \mathcal{J}\{y, z\}$, and $B(y, y) = 0$. Since $A(z)$ is irreducible and one of the coefficients in $A(z)$ is unity, the irreducible factors of $B(y, z)$ are distinct and all have the same order in z as $A(z)$ has.

Denoting the order of $B(y, z)$ in y by p, let $B_1(y, z)$ be an irreducible factor of $B(y, z)$ of order p in y. Let Λ_1 be the prime component of $\{B_1(y, z)\}$ which contains neither of the separants of $B_1(y, z)$. No other irreducible factor of $B(y, z)$ is in Λ_1, for such a factor would have the same order in z as $B_1(y, z)$ and would be divisible by $B_1(y, z)$. Let y, ζ_1 be a generic solution of Λ_1. $B(y, z) \in \Lambda_1$ but the separant of $B(y, z)$ with respect to z is not in Λ_1 (for otherwise the separant of $B_1(y, z)$ would be in Λ_1). Therefore ζ_1 is a nonsingular solution of $A(z)$, a solution of Σ, and a solution of $\omega K(z) - H(z)$. Thus $H(y)K(z) - K(y)H(z)$ vanishes for the generic solution y, ζ_1 of Λ_1, and is in Λ_1. With order in y clearly not greater than p, $H(y)K(z) - K(y)H(z)$ must be divisible by $B_1(y, z)$.

Similarly, $H(y)K(z) - K(y)H(z)$ is divisible by all the irreducible factors $B_1(y, z), \cdots, B_s(y, z)$ of $B(y, z)$ which have order p in y. Since all these $B_i(y, z)$'s are distinct we may write

$$H(y)K(z) - K(y)H(z) = L(y, z)B_1(y, z) \cdots B_s(y, z),$$

where $L(y, z) \in \mathcal{J}\{y, z\}$. Moreover, if we denote the degree of $B(y, z)$ in y_p (the pth derivative of y) by d, we see that the degree of $H(y)K(z) - K(y)H(z)$ in y_p is not greater than d, that of $B_1(y, z) \cdots B_s(y, z)$ is d, so that $L(y, z)$ is of degree 0 in y_p, that is, of order not greater than $p-1$ in y.

Let $B_{s+1}(y, z)$ be an irreducible factor of $B(y, z)$ of order $p-1$ in y, let Λ_{s+1} be the prime component of $\{B_{s+1}(y, z)\}$ not containing the separants of $B_{s+1}(y, z)$, and let y, ζ_{s+1} be a generic solution of Λ_{s+1}. As with y, ζ_1 before, we see that y, ζ_{s+1} is a solution of $H(y)K(z) - K(y)H(z)$. But y, ζ_{s+1} is not a solution of any $B_i(y, z)$ with $i \leqq s$, for no such $B_i(y, z)$ is in Λ_{s+1}. Hence y, ζ_{s+1} is a solution of $L(y, z)$, and $L(y, z) \in \Lambda_{s+1}$. This implies, since the order of $L(y, z)$ in y is not greater than $p-1$, that $L(y, z)$ is divisible by $B_{s+1}(y, z)$.

Similarly, $L(y, z)$ is divisible by all the irreducible factors $B_{s+1}(y, z), \cdots, B_t(y, z)$ of order $p-1$ in y, so that

$$H(y)K(z) - K(y)H(z) = M(y, z)B_1(y, z) \cdots B_t(y, z),$$

where $M(y, z) \in \mathcal{J}\{y, z\}$. Moreover, if we denote the degree of $B(y, z)$ in y_p, y_{p-1} by e, we see that the degree of $H(y)K(z) - K(y)H(z)$ in y_p, y_{p-1} is not greater than e, that of $B_1(y, z) \cdots B_t(y, z)$ is e, so that $M(y, z)$ is of degree 0 in y_p, y_{p-1}, that is, of order not greater than $p-2$ in y.

Continuing in this way we finally arrive at an equation

$$H(y)K(z) - K(y)H(z) = P(z)B_1(y, z) \cdots B_w(y, z),$$

where $B_1(y, z), \cdots, B_w(y, z)$ are all the irreducible factors of $B(y, z)$. Since $H(z)$, $K(z)$ have no common divisor, $H(y)K(z)-K(y)H(z)$ has no factor free of y that is not also free of z. Therefore $P(z) \in \mathcal{J}$, and $H(y)K(z)-K(y)H(z)=aB(y, z)$, where $a \in \mathcal{J}$. The desired equation $\omega K(z)-H(z)=\alpha A(z)$ immediately follows.

The rest of the proof of the theorem as given in *Extensions* II is apparently correct.

Of the two examples given in *Extensions* II, the proof for Example 2 is incorrect, and I do not yet know whether that example is valid.

COLUMBIA UNIVERSITY

ANNALS OF MATHEMATICS
Vol. 49, No. 1, January, 1948

ALGEBRAIC MATRIC GROUPS AND THE PICARD-VESSIOT THEORY OF HOMOGENEOUS LINEAR ORDINARY DIFFERENTIAL EQUATIONS

BY E. R. KOLCHIN

(Received December 4, 1946)

TABLE OF CONTENTS

INTRODUCTION

Historical background

A central role in the Galois theory of homogeneous linear ordinary differential equations as developed by Picard and Vessiot at the end of the last century,[1]

[1] For an account of the early (and most important) papers see Vessiot [3] (numbers in square brackets refer to the References at the end of the present paper). The theory began

is played by the concept of algebraic matric group, that is, broadly speaking, multiplicative group of matrices defined by algebraic equations in the elements of the matrices. Special cases of such groups (for example: the full matric group, the unimodular group, the orthogonal group, finite matric groups) have been subject to more or less exhaustive research for many decades, but the literature seems to be devoid of any basic theory of algebraic matric groups as such. For lack of such a theory, these groups, when encountered on a large scale (as in the Picard-Vessiot theory), have been treated as special cases of Lie groups. As a result the generally brilliant theory of Picard and Vessiot suffered on the one hand from the lack of rigor of the early theory of Lie groups, and on the other hand from being too intimately bound up with the analytic point of view of the Lie theory, thereby obscuring the algebraic nature of the subject matter.

It is doubtless possible to bring the rigor of the group-theoretic aspects of the Picard-Vessiot theory up to modern standards by making reference to modern developments concerning Lie groups. But not only would this not answer the second of the above mentioned points, it would make the Picard-Vessiot theory depend on a discipline far deeper and more extensive than itself. Moreover, if it is desired to extend the theory to abstract coefficient fields, it is necessary to extend the whole machinery of Lie groups and Lie algebras to algebraic matric groups over such fields.[2]

Not only does the Picard-Vessiot theory suffer from a lack of an algebraic point of view, but some of it seems to be afflicted with a lack of clarity of ideas, or at least a lack of precise formulation of ideas. A most striking example of this affliction is afforded by Vessiot's beautiful and celebrate theorem that a homogeneous linear ordinary differential equation is "solvable by quadratures" if and only if the group is "integrable." Just what "solvable by quadratures" means is not clearly stated. A close examination of the proof reveals that solving by quadratures must permit not only the operation of integration, but also that of exponentiation (of integrals). But the confusion in terminology is sufficient to induce Picard [5] to use "quadrature" as equivalent to "integral" in the proof of the necessity, and yet to permit exponentiation in the proof of the

with Picard [1], a note amplified by Picard [2], and received great impetus from Vessiot [1] and [2]. After Picard and Vessiot, the most important contributions are those of Loewy. The presentation of the theory by Picard [5] is as full as any, but suffers from all the criticism made below. The most recent remarks published on the subject are those made in 1932 by Baer [1], who critically surveyed the general state of the theory. The theory partially outlined by Freudenthal [1] evidently is concerned with only linear phenomena, whereas the theory of Picard-Vessiot is essentially a study of certain types of differential field extensions.

[2] In the case of fields of characteristic 0, this labor could be avoided, as remarked to me orally by Professor C. Chavalley, by embedding the abstract field in the field of complex numbers, if the former field has cardinal number $\leq$ that of the continuum, or (with a slight argument) by so embedding part of the abstract field if its cardinal number is greater. Of course, if it is desired to include fields of nonzero characteristic this procedure can not be used.

sufficiency. Indeed, it seems as though the solving of any differential equation of the first order could be used, instead of integration, in the necessity arguments as given by Picard [5] and Vessiot [2]. Furthermore, the question of whether algebraic operations are permitted is not always made clear. The relationship between the necessity of using algebraic operations and the disconnectedness of the group was realized, especially by Marotte [1] and by Fano [1], but these algebraic questions were sufficiently neglected so that it was never thought necessary by anyone to ask whether the connected component of the group containing the identity matrix was actually the group of the equation over some enlarged coefficient field. Indeed, as Baer [1] points out, one of the outstanding tasks in the theory, after algebraization, is to establish an analogue to the theorem in Galois theory which asserts the one-to-one correspondence between intermediate fields and subgroups of the Galois group.

A beginning at algebraization and departure from the Lie tradition was made by Loewy [4], but he carried his work in this direction barely far enough to define in rigorous fashion the group of an equation (with respect to more general coefficient domains than thitherto) and to make clear some of its elementary properties. Beyond this there is nothing in the literature that recasts the pioneering papers of Picard and Vessiot. It seems likely that two factors contributed to this state of affairs. One is the aforementioned absence of a general theory of algebraic matric groups. The other is the inadequacy of the classical theory of differential equations for handling the algebraic aspects of differential equations. The second of these factors has been removed in recent years by the emergence of the penetrating and elegant Ritt theory of algebraic differential equations (for a general account of this theory as of 1938 see Ritt [2]). Indeed, it is difficult to imagine how this work of Picard and Vessiot could be put on a sound footing without making use of the Ritt theory.

Summary

The purposes of the present paper are, first to develop a set of theorems on algebraic matric groups, at least adequate to meet the demands of the Picard-Vessiot theory, and second, to algebraize, rigorize, round out, and augment that theory. The paper is organized into five chapters plus a list of references.

Chapter I deals with algebraic matric groups over an algebraically closed field. To emphasize the purely algebraic nature of the subject matter the proofs are carried through in a manner valid for fields of nonzero as well a zero characteristic. A matrix can be regarded as a point in n^2-dimensional space. The set of matrices in an algebraic matric group constitutes an algebraic manifold (from which is deleted an algebraic manifold of lower dimension composed of singular matrices), the underlying manifold of the group. The interplay between the group properties and the algebraic manifold properties of the algebraic matric group forms the basis for the theory. It is shown that the irreducible components of the underlying manifold are pairwise disjoint (save, perhaps, for singular matrices), and all have the same dimension; the component containing the identity matrix is the underlying manifold of a normal algebraic subgroup of finite index (the component of the identity). Following three lemmas, a Jordan-Hölder-Schreier theorem is proved in the manner of Zassenhaus [1]. Solvable algebraic matric groups are defined by means of normal chains and there is derived a necessary and sufficient condition that a set of matrices be simultaneously reducible to triangular form. This result is the same as Lie's theorem that a connected solvable Lie group of matrices is reducible to triangular form, but on the one hand is proved without recourse to infinitesimal transformations and is valid for fields of any characteristic, and on the other hand is restricted to algebraic groups. The concepts of "anticompact" and "quasicompact" algebraic matric groups are defined ($\mathfrak{G}$ is anticompact if it contains no element of finite order exceeding unity and not divisible by the field characteristic; $\mathfrak{G}$ is quasicompact if every algebraic subgroup of order greater than unity contains such an element). It is shown that an algebraic matric group is anticompact if and only if each matrix in the group has all its characteristic roots equal to unity, whereas the group is quasicompact if and only if any given matrix in it is reducible to diagonal form. Also, the influence of anticompactness and quasicompactness on triangular form is investigated. Finally, algebraic matric groups which have normal chains with certain properties different from those used in defining solvability, are studied.

Chapter II is a compilation of just those results from the Ritt theory which are needed in the sequel.

The short Chapter III contains a beginning of a Galois theory for differential fields. With the hope of some day broadening the scope of the Picard-Vessiot theory, I have carried out the first few (trivial) steps thereof in a very general setting. Without reference to linear differential equations, there is considered an extension of an arbitrary differential field (ordinary or partial) of characteristic 0. There is defined for such an extension the property of being normal, in a manner consistant with the familiar concept of normality for algebraic field extensions. The comparatively complicated nature of this definition is due to the fact that, for differential fields, the concept analogous to the splitting field of a polynomial is lacking. It is shown that an "abundant" (that is, large enough, in a precisely defined sense) group of automorphisms of the normal differential field over the ground differential field enjoys some properties similar to those of a Galois group. Namely: there is a one-to-one correspondence between intermediate differential fields and *certain* subgroups; an intermediate differential field satisfies a certain condition perhaps stronger than that of normality if and only if the corresponding subgroup is normal; and in this case the factor group is isomorphic with an abundant group of automorphisms of the intermediate differential field over the ground differential field. It is in connection with the last property that the use of abundant groups of automorphisms instead of the group of all automorphisms (which of course is abundant) is indicated. For, when the group of all automorphisms is used the factor group need not be isomorphic with the group of all automorphisms of the intermediate differential field over the ground differential field, but merely with an abundant subgroup thereof. Here, at least for the present, the general theory ends.[3] In particular, there is no characterization of those "certain" subgroups which correspond to intermediate differential fields.

Chapter IV carries this general theory forward for the special case provided by the Picard-Vessiot theory. There is defined the concept of "Picard-Vessiot extension" of a given ordinary differential field $\mathcal{F}$ of characteristic 0 with algebraically closed field of constants ($\mathcal{G}$ is a Picard-Vessiot extension of $\mathcal{F}$ if every constant in $\mathcal{G}$ is in $\mathcal{F}$ and $\mathcal{G}$ can be obtained by adjoining to $\mathcal{F}$ a fundamental system of solutions of a homogeneous linear ordinary differential equation with coefficients in $\mathcal{F}$), and such an extension is shown to be normal in the sense of Chapter III. The group of all automorphisms over the given differential field is shown to be isomorphic with an algebraic matric group. This makes immediately available the theorems of Chapters I and III. The set of all subgroups that correspond to intermediate differential fields is characterized as the set of all algebraic subgroups. The equality between the dimension of the group and the degree of transcendency of the extension is proved, the isomorphism between factor group and group of *all* automorphisms of the intermediate differential field over the ground differential field is established, and a few special topics (largely necessary for the sequel) are treated.

Chapter V is concerned with the above mentioned extension of Vessiot's big theorem. An extension of a given differential field is defined as "liouvillian"[4] if it contains no new constants and it is an extension by integrals, exponentials of integrals, and algebraic functions, that is, if the extension can be obtained by repeated adjunction of solutions of equations of the types $y' - a = 0$, $y' - ay = 0$, and $y^m + a_1y^{m-1} + \cdots + a_m = 0$. Furthermore, a hierarchy of ten different classes of liouvillian extensions is recognized, each class being defined by restricting the type of adjunctions permitted. It is then proved that a Picard-Vessiot extension is (contained in) a liouvillian extension if and only if the corresponding algebraic matric group has a solvable component of the identity. Moreover, for each of the ten classes, necessary and sufficient conditions are found for the liouvillian extension to be of that class. For example, a Picard-Vessiot extension is an extension by integrals alone if the group is solvable and anticompact, by exponentials of integrals alone if the group is solvable and quasicompact. Several consequences are pointed out. For example, if a Picard-Vessiot extension is liouvillian, it can be obtained by first making a purely algebraic adjunction and then making adjunctions solely by means of integrals and exponentials of integrals. Again, if a Picard-Vessiot

---

[3] An attempt at a general Galois theory for differential fields has been made by J. E., Eaton (*A Galois theory for differential fields*, Duke Mathematical Journal, vol. 10 (1943), pp. 751–760), but his whole paper depends on his false Theorem 1.

[4] This term was suggested by Professor Ritt, who has used the term "l-function" for elements of such fields in connection with his research in and exposition of the Liouville theory of the solvability of certain differential equations.

extension is contained in an extension by exponentials of integrals, with no new constants, then the given linear differential equation has a fundamental system of solutions such that each solution in the fundamental system is the exponential of an integral of an element of a certain extension by radicals.

It should, perhaps, be mentioned that a generalization of the Picard-Vessiot theory to allow fields of positive characteristic must await a similar generalization of at least part of the Ritt theory. As yet little has been done in this direction beyond the treatment of the Ritt-Raudenbush basis theorem in Kolchin [2].

Notation and terminology

The inclusion of an element in a set is denoted by ϵ. The set inclusion relations of being contained in and of containing are denoted by $\subseteq$ and $\supseteq$, respectively, proper inclusion by $\subset$ and $\supset$. The negation of any of these relations is indicated by a transverse line: $\notin$, $\nsubseteq$, etc.

The union (set-theoretic sum) of two sets $\mathfrak{M}$ and $\mathfrak{N}$ is denoted by $\mathfrak{M} \cup \mathfrak{N}$, their intersection by $\mathfrak{M} \cap \mathfrak{N}$. For a system of sets $\mathfrak{M}_\lambda$ the notation for union is $\bigcup \mathfrak{M}_\lambda$, for intersection is $\bigcap \mathfrak{M}_\lambda$. The set of all elements contained in $\mathfrak{M}$ but not in $\mathfrak{N}$ is denoted by $\mathfrak{M} - \mathfrak{N}$.

A subgroup $\mathfrak{H}$ of a group $\mathfrak{G}$ such that $\sigma^{-1}\mathfrak{H}\sigma = \mathfrak{H}$ for all $\sigma \in \mathfrak{G}$ is called *normal*.

Algebraic notation used is generally as in van der Waerden [1]. Ring adjunction is indicated by square brackets $\mathcal{R}[\cdots]$, field adjunction by parentheses $\mathcal{F}(\cdots)$. The degree of transcendency of a field $\mathcal{G}$ over a subfield $\mathcal{F}$ is denoted by $\partial^\circ\mathcal{G}/\mathcal{F}$.

For an algebraic manifold $\mathfrak{M}$ in m-dimensional affine space over a field $\mathcal{C}$, the set $\mathfrak{m}$ of all polynomials in $\mathcal{C}[x_1, \cdots, x_m]$ which vanish at all points of $\mathfrak{M}$ is called the *defining ideal* of $\mathfrak{M}$. A *generic zero* of $\mathfrak{m}$ or a *generic point* of $\mathfrak{M}$ is a generalized point (i.e. a point with coordinates in some extension of $\mathcal{C}$) such that $f \in \mathcal{C}[x_1, \cdots, x_m]$ vanishes at this point if and only if $f \in \mathfrak{m}$. A generic zero exists if and only if $\mathfrak{m}$ is prime and $\mathfrak{m} \subset \mathcal{C}[x_1, \cdots, x_m]$ (i.e. $\mathfrak{M}$ is irreducible and nonempty). If σ is a generalized point then $\partial^\circ\mathcal{C}(\sigma)/\mathcal{C}$ is the *dimension* of σ. If $\mathfrak{M}$ is irreducible and nonempty the dimension of a generic point of $\mathfrak{M}$ is the *dimension* of $\mathfrak{M}$ or of $\mathfrak{m}$.

A generic point of an irreducible algebraic manifold may be regarded as an ordered set of alebraic functions of a certain number (the dimension) of indeterminates or parameters. If we have several generic points (of the same or different manifolds) and if all the parameters taken together are algebraically independent over $\mathcal{C}$, then the generic points are *independent*.

For two generalized points σ and τ the substitution $\sigma \to \tau$ is a *specialization* if every polynomial in $\mathcal{C}[x_1, \cdots, x_m]$ vanishing at σ also vanishes at τ, that is, if $\sigma \to \tau$ generates a homomorphism of $\mathcal{C}[\sigma]$ onto $\mathcal{C}[\tau]$. If σ is a generic point and τ any generalized point of an algebraic manifold then $\sigma \to \tau$ is a specialization.

Every algebraic manifold is the union of a finite number of irreducible algebraic manifolds none of which contains another. These irreducible manifolds, which are unique, are the *irreducible components* of the given manifold.

Chapter I. Algebraic Matric Groups

1. Reducibility of sets of matrices

All matrices considered in this chapter are square with elements in an algebraically closed commutative field $\mathcal{C}$. The characteristic of $\mathcal{C}$ will be denoted by p. The zero matrix will be denoted by 0, the identity matrix by ι.

Two sets $\mathfrak{M}$ and $\mathfrak{N}$ of matrices of degree n are *equivalent* if there exists a nonsingular matrix ρ of degree n such that $\rho^{-1}\mathfrak{M}\rho = \mathfrak{N}$.

We shall often find it useful to think of matrices of degree n as linear operators on an n-dimensional vector space over $\mathcal{C}$. That is, if σ is a matrix (a_{ij}) we may consider an n-dimensional vector space $\mathfrak{N}$ with basis vectors $\eta_1, \cdots, \eta_n$, and think of σ as the operator defined by

$$\sigma\eta_j = \sum_i a_{ij}\eta_i, \qquad j = 1, \cdots, n.$$

If $\mathfrak{M}$ is a set of matrices, considered as linear operators on $\mathfrak{N}$, and $\mathfrak{N} = \rho^{-1}\mathfrak{M}\rho$ is an equivalent set, then $\mathfrak{N}$ can be thought of as the same set of operators, expressed with respect to the set of basis vectors $\rho\eta_1, \cdots, \rho\eta_n$ instead of $\eta_1, \cdots, \eta_n$.

A set $\mathfrak{M}$ of matrices of degree n is *reducible* if there exists an equivalent set $\mathfrak{N}$ of the form

$$\mathfrak{N} = \begin{pmatrix} \mathfrak{N}_1 & * \\ 0 & \mathfrak{N}_2 \end{pmatrix},$$

that is, for a positive integer $q < n$, every σ in $\mathfrak{N}$ has the form

$$\sigma = \begin{pmatrix} a_{11} & \cdots & a_{1q} & a_{1,q+1} & \cdots & a_{1n} \\ \cdot & \cdots & \cdot & \cdot & \cdots & \cdot \\ a_{q1} & \cdots & a_{qq} & a_{q,q+1} & \cdots & a_{qn} \\ 0 & \cdots & 0 & a_{q+1,q+1} & \cdots & a_{q+1,n} \\ \cdot & \cdots & \cdot & \cdot & \cdots & \cdot \\ 0 & \cdots & 0 & a_{n,q+1} & \cdots & a_{nn} \end{pmatrix}.$$

This is the same as saying that $\mathfrak{N}$ has a linear subspace of dimension q which is invariant under each matrix in $\mathfrak{M}$. If $\mathfrak{M}$ is not reducible then it is *irreducible*. $\mathfrak{M}$ *is completely reducible* if there is an equivalent set $\mathfrak{N}$ of the form

$$\mathfrak{N} = \begin{pmatrix} \mathfrak{N}_1 & & 0 \\ & \ddots & \\ 0 & & \mathfrak{N}_r \end{pmatrix}, \tag{1}$$

where each square block $\mathfrak{N}_i$ is irreducible, that is if $\mathfrak{N}$ can be written as the direct sum of linear subspaces invariant under every matrix in $\mathfrak{M}$, none of the linear subspaces having a subspace of lower positive dimension invariant under every matrix in $\mathfrak{M}$. It is easy to see that if a set $\mathfrak{M}$ of the form

$$\mathfrak{M} = \begin{pmatrix} \mathfrak{M}_1 & 0 \\ 0 & \mathfrak{M}_2 \end{pmatrix}$$

is completely reducible, then so are $\mathfrak{M}_1$ and $\mathfrak{M}_2$.

The following result is classical (see, e.g., Weyl [1], p. 81).

SCHUR'S LEMMA. *Let $\mathfrak{M}$ be an irreducible set of matrices. If a matrix σ commutes with every matrix in $\mathfrak{M}$, then either $\sigma = 0$ or σ is nonsingular. Thus, since $\mathcal{C}$ is algebraically closed, the only matrices commuting with every matrix in $\mathfrak{M}$ are of the form $c\iota$ (element of $\mathcal{C}$ times the identity matrix).*

A set $\mathfrak{M}$ of matrices is in *triangular form* if, for every (a_{ij}) in $\mathfrak{M}$, $a_{ij} = 0$ whenever $i > j$. We shall say that $\mathfrak{M}$ is in *special triangular form* if $\mathfrak{M}$ is in triangular form and every matrix in $\mathfrak{M}$ has all the elements on the main diagonal equal to 1. $\mathfrak{M}$ is in *diagonal form* if $a_{ij} = 0$ whenever $i \neq j$. $\mathfrak{M}$ is *reducible* to a given one of these three forms if there is an equivalent set in the given form.

By means of Schur's lemma it is easy to see that every abelian set of matrices is reducible to triangular form.[5] Indeed, the assertion is obviously true for degree $n = 1$. Let $n > 1$ and suppose the assertion true for matrices of lower degree. If the abelian set of matrices $\mathfrak{M}$ were irreducible and σ were any matrix in $\mathfrak{M}$ then σ would commute with every matrix in $\mathfrak{M}$ and therefore would be of the form $c\iota$, contradicting the irreducibility of $\mathfrak{M}$. Hence $\mathfrak{M}$ is reducible, and the result follows from the induction assumption. Actually, the form to which $\mathfrak{M}$ can be reduced may be specified more narrowly, as follows.

LEMMA 1. *An abelian set $\mathfrak{M}$ of matrices is equivalent to a set $\mathfrak{N}$ of the form* (1), *where $\mathfrak{N}_i$ is a square block of n_i rows ($n_1 + \cdots + n_r = n$, the degree of $\mathfrak{M}$) and the $\mathfrak{N}_i$-block of every σ in $\mathfrak{N}$ is in triangular form with equal elements on the main diagonal.*

For $n = 1$ this is obvious. Let $n > 1$ and suppose the result true for matrices of degree $< n$. If every matrix in $\mathfrak{M}$, assumed already in triangular form, has all its characteristic roots equal then the conclusion follows. Let $\sigma \in \mathfrak{M}$ have the characteristic equation $f(x) = (x - a_1)^{m_1} \cdots (x - a_q)^{m_q}$, $q > 1$, where $a_i \neq a_j$ when $i \neq j$. Writing $f_i(x) = f(x)\ (x - a_i)^{-m_i}$, we see that there are polynomials $p_i(x) \in \mathcal{C}[x]$ such that $1 = \sum_i p_i(x) f_i(x)$, so that $\iota = \sum_i p_i(\sigma) f_i(\sigma)$. Since a matrix satisfies its own characteristic equation, we have $f(\sigma) = 0$, and for any ζ in the vector space $\mathfrak{N}$ the element $\zeta_i = p_i(\sigma) f_i(\sigma)\zeta$ is in the linear subspace $\mathfrak{N}_i$ consisting of all η's such that $(\sigma - a_i\iota)^{m_i}\eta = 0$. It follows that $\mathfrak{N}$ is the direct sum

$$\mathfrak{N} = \mathfrak{N}_1 + \cdots + \mathfrak{N}_q$$

of the linear subspaces $\mathfrak{N}_i$. For any $\tau \in \mathfrak{M}$ we have

$$(\sigma - a_i\iota)^{m_i}\tau\mathfrak{N}_i = \tau(\sigma - a_i\iota)^{m_i}\mathfrak{N}_i = 0,$$

so that $\tau\mathfrak{N}_i \subseteq \mathfrak{N}_i$. Therefore, by choosing a set of basis vectors for $\mathfrak{N}$ consisting of bases of the individual subspaces $\mathfrak{N}_i$, we can reduce $\mathfrak{M}$ to the form

[5] This is false when $\mathcal{C}$ is not algebraically closed. E.g. the group of all matrices $\begin{pmatrix} a & b \\ -b & a \end{pmatrix}$ with real a and b not both 0, is abelian. But the characteristic roots are $a \pm b\sqrt{-1}$, so that in any reduction of this group to triangular form complex diagonal elements must appear.

(1) where each block $\mathfrak{N}_i$ is an abelian set of matrices of degree $<n$, and is, by the induction assumption, reducible to the desired form. But when each $\mathfrak{N}_i$ is so reduced then $\mathfrak{M}$ obviously is, too.

For a single matrix the form may be prescribed even more narrowly in various ways. For our purposes the following *Jordan normal form* (van der Waerden [1], vol. 2, §111) will be of importance.

LEMMA 2. *For any matrix τ there is an equivalent matrix σ of the form*

$$(2) \qquad \sigma = \begin{pmatrix} \boxed{\begin{matrix} a_1 & 1 & & 0 \\ & \ddots & \ddots & \\ & & \ddots & 1 \\ 0 & & & a_1 \end{matrix}} & & 0 \\ & \ddots & \\ 0 & & \boxed{\begin{matrix} a_t & 1 & & 0 \\ & \ddots & \ddots & \\ & & \ddots & 1 \\ 0 & & & a_t \end{matrix}} \end{pmatrix}$$

composed of square blocks along the main diagonal of the indicated type (zeros everywhere except in the main diagonal and the diagonal just above it) and zeros elsewhere.

2. Algebraic matric groups

A matrix of degree n may be regarded as a point in an n^2-dimensional affine space, the coordinates of the point being the elements of the matrix. Therefore we may speak of an algebraic manifold of matrices, and use the concepts of "defining ideal," "dimension," "generic zero" or "generic point," "specialization," etc., of modern algebraic geometry.

A (multiplicative) group $\mathfrak{G}$ of matrices of degree n is called *algebraic* if there exists an algebraic manifold in n^2-dimensional space such that: (a) every matrix σ in $\mathfrak{G}$ is a point on the algebraic manifold; and (b) every point σ on the algebraic manifold that is not a singular matrix is a matrix in $\mathfrak{G}$. If it is specified that the algebraic manifold not have an irreducible component consisting solely of singular matrices then the algebraic manifold is unique. This unique algebraic manifold will be called the *underlying manifold* of $\mathfrak{G}$.[6] The defining ideal of the underlying manifold of $\mathfrak{G}$ will be called the *defining ideal* of $\mathfrak{G}$.

EXAMPLES. 1. The *full matric group*, consisting of all nonsingular matrices of a given degree.

[6] Care must be exercised in using the words "reducible" and "irreducible." An algebraic matric group is reducible if the set of matrices in it is reducible in the sense of §1. The underlying manifold is reducible if it is the union of two algebraic manifolds neither of which contains the other.

2. The *full unimodular group*, consisting of all matrices of given degree with determinant 1.

3. The *identity group* $\mathfrak{E}$, consisting solely of the matrix ι.

4. Any *finite* matric group.

5. The *full triangular group* $\mathfrak{T}$, consisting of all matrices of given degree n which are in triangular form; and its subgroups $\mathfrak{T}_1, \cdots, \mathfrak{T}_n$ defined as follows: $\mathfrak{T}_1$ is the set of all matrices in special triangular form; for $2 \leqq i \leqq n$ $\mathfrak{T}_1$ is the set of all matrices in $\mathfrak{T}_1$ in which all the elements of the first $i - 1$ diagonals above the main diagonal are 0 (thus $\mathfrak{T}_n = \mathfrak{E}$).

6. The *full diagonal group* $\mathfrak{D}$, consisting of all nonsingular matrices of given degree which are in diagonal form.

THEOREM. *Let $\mathfrak{G}$ be an algebraic matric group. The irreducible components of the underlying manifold of $\mathfrak{G}$ are pairwise disjoint (except, possibly, for singular matrices), and all have the same dimension. The irreducible component which contains ι is the underlying manifold of a normal algebraic subgroup of $\mathfrak{G}$ of finite index equal to the number of irreducible components.*

PROOF. Let $\mathfrak{G}^*$ be the underlying manifold of $\mathfrak{G}$, and denote the irreducible components of $\mathfrak{G}^*$ by $\overset{*}{\mathfrak{G}}_0, \cdots, \overset{*}{\mathfrak{G}}_{h-1}$. Let $\overset{*}{\mathfrak{G}}_0, \cdots, \overset{*}{\mathfrak{G}}_g$ be those irreducible components which contain ι, and let $\overset{*}{\sigma}_0, \cdots, \overset{*}{\sigma}_g$ be independent generic points of $\overset{*}{\mathfrak{G}}_0, \cdots, \overset{*}{\mathfrak{G}}_g$, respectively. Now $\overset{*}{\sigma}_i\overset{*}{\sigma}_j$ is a generalized point of $\mathfrak{G}^*$, and therefore of some irreducible component $\overset{*}{\mathfrak{G}}_l$ of $\mathfrak{G}^*$. Since $\overset{*}{\sigma}_i\overset{*}{\sigma}_j \to \overset{*}{\sigma}_i \iota = \overset{*}{\sigma}_i$ is a specialization, we see that $\overset{*}{\mathfrak{G}}_l \supseteq \overset{*}{\mathfrak{G}}_i$, so that $l = i$. In the same way, $\overset{*}{\sigma}_i\overset{*}{\sigma}_j \to \iota\overset{*}{\sigma}_j = \overset{*}{\sigma}_j$ is a specialization, so that $l = j$, whence $i = j$. Since this holds for all i, j from 0 to g, we see that $g = 0$, and there is only one irreducible component $\overset{*}{\mathfrak{G}}_0$ containing ι.

Now let σ_j be any nonsingular matrix in $\overset{*}{\mathfrak{G}}_j$ ($j = 0, \cdots, h - 1$). For each j, $\sigma_j^{-1}\overset{*}{\mathfrak{G}}_j$ is an irreducible algebraic manifold contained in $\mathfrak{G}^*$, and containing ι. Therefore $\sigma_j^{-1}\overset{*}{\mathfrak{G}}_j \subseteq \overset{*}{\mathfrak{G}}_0$, $\overset{*}{\mathfrak{G}}_j \subseteq \sigma_j\overset{*}{\mathfrak{G}}_0$, and $\mathfrak{G}^* = \sigma_0\overset{*}{\mathfrak{G}}_0 \cup \cdots \cup \sigma_{h-1}\overset{*}{\mathfrak{G}}_0$. Since each $\sigma_j\overset{*}{\mathfrak{G}}_0$ is an irreducible algebraic manifold, and since the decomposition of an algebraic manifold into irreducible components is unique, it easily follows that

$$\overset{*}{\mathfrak{G}}_j = \sigma_j\overset{*}{\mathfrak{G}}_0, \qquad j = 0, \cdots, h - 1. \tag{3}$$

In the same way we find that

$$\overset{*}{\mathfrak{G}}_j = \overset{*}{\mathfrak{G}}_0\sigma_j, \qquad j = 0, \cdots, h - 1. \tag{4}$$

It follows from (3) that no element of $\mathfrak{G}$ (that is, no nonsingular matrix in $\mathfrak{G}^*$) is contained in more than one $\overset{*}{\mathfrak{G}}_j$, for if $\sigma \in \overset{*}{\mathfrak{G}}_{j_1} \cap \overset{*}{\mathfrak{G}}_{j_2}$ then $\overset{*}{\mathfrak{G}}_{j_1} = \sigma\overset{*}{\mathfrak{G}}_0 = \overset{*}{\mathfrak{G}}_{j_2}$. It also follows from (3) that the dimension of every $\overset{*}{\mathfrak{G}}_j$ is the same as that of $\overset{*}{\mathfrak{G}}_0$.

Letting $j = 0$ in (3) we see that the product of any two nonsingular matrices in $\overset{*}{\mathfrak{G}}_0$ is itself in $\overset{*}{\mathfrak{G}}_0$. Also, for any nonsingular $\sigma_0 \in \overset{*}{\mathfrak{G}}_0$ we have $\sigma_0^{-1} \in \overset{*}{\mathfrak{G}}_0$, because if $\sigma_0^{-1} \in \overset{*}{\mathfrak{G}}_j$ then $\overset{*}{\mathfrak{G}}_j = \sigma_0^{-1}\overset{*}{\mathfrak{G}}_0$, $\iota \in \overset{*}{\mathfrak{G}}_j$, and $j = 0$. Consequently the set of nonsingular matrices in $\overset{*}{\mathfrak{G}}_0$ forms an algebraic group. From (3) and (4) it is evident that this group is a normal subgroup of $\mathfrak{G}$ of index h. Thus the theorem is proved.

Henceforth, by the dimension of an algebraic matric group $\mathfrak{G}$ (notation: dim $\mathfrak{G}$) we shall mean the dimension of its underlying manifold, that is, the dimension of each of the irreducible components. The algebraic subgroup of $\mathfrak{G}$ of which the underlying manifold is the irreducible component which contains ι will be called the *component of the identity* of $\mathfrak{G}$, and will be denoted by $\mathfrak{G}^0$. $\mathfrak{G}$ will be called *connected* if its underlying manifold is irreducible, that is, if $\mathfrak{G} = \mathfrak{G}^0$.

It follows from the theorem just proved that if $\mathfrak{G}$ is not connected then $\mathfrak{G}$ contains an algebraic subgroup of finite index > 1 (for example $\mathfrak{G}^0$). Conversely, *if $\mathfrak{G}$ contains an algebraic subgroup of finite index > 1 then $\mathfrak{G}$ is not connected.* For, if the index of the algebraic subgroup $\mathfrak{H}$ is h then $\mathfrak{G} = \sigma_1\mathfrak{H} \cup \cdots \cup \sigma_h\mathfrak{H}$, so that if $\mathfrak{G}^*$ and $\mathfrak{H}^*$ are the underlying manifolds of $\mathfrak{G}$ and $\mathfrak{H}$, respectively, then $\mathfrak{G}^* = \sigma_1\mathfrak{H}^* \cup \cdots \cup \sigma_h\mathfrak{H}^*$. It is easy to see that $\sigma_i\mathfrak{H}^* \not\subseteq \sigma_j\mathfrak{H}^*$ if $i \neq j$, so that $\mathfrak{G}$ can not be connected if $h > 1$.

3. Jordan-Hölder-Schreier theorem

The following lemma will be used to show that certain subgroups of an algebraic matric group are algebraic.

LEMMA 1. *Let $\mathfrak{G}$ be an algebraic matric group, $\mathfrak{H}$ a subgroup (not assumed algebraic) such that $\mathfrak{G} - \mathfrak{H}$ is contained in an algebraic manifold of lower dimension than $\mathfrak{G}$. Then $\mathfrak{H} = \mathfrak{G}$.*

PROOF. Expressing $\mathfrak{G}$ as the union of the cosets of $\mathfrak{H}$,

$$\mathfrak{G} = \bigcup_{\lambda \geqq 1} \sigma_\lambda \mathfrak{H} \qquad (\sigma_1 = \iota),$$

we see that if there are more than one coset then $\sigma_2\mathfrak{H} \subseteq \mathfrak{G} - \mathfrak{H}$ is contained in an algebraic maifold of dimension lower than that of $\mathfrak{G}$. Hence $\mathfrak{H}$, which is a linear mapping of $\sigma_2\mathfrak{H}$, is contained in such a manifold, so that $\mathfrak{G} = \mathfrak{H} \cup (\mathfrak{G} - \mathfrak{H})$ is contained in the union of two algebraic manifolds of lower dimension, which is impossible. Therefore there is only one coset, that is, $\mathfrak{H} = \mathfrak{G}$.

By contrast with Lemma 1, the following lemma will be used to show that certain algebraic manifolds are the underlying manifolds of algebraic matric groups.

LEMMA 2. *If $\mathfrak{G}$ is a group of matrices (not assumed algebraic) and $\mathfrak{G}^*$ is the smallest algebraic manifold containing $\mathfrak{G}$, then $\mathfrak{G}^*$ is the underlying manifold of an algebraic matric group.*

PROOF. Let $\mathfrak{g}$ be the set of all polynomials in n^2 indeterminates (n = the degree of the matrices) with coefficients in $\mathcal{C}$ which vanish for all matrices in $\mathfrak{G}$, that is, let $\mathfrak{g}$ be the defining ideal of $\mathfrak{G}^*$. If $f(\xi) \in \mathfrak{g}$ then $f(\sigma\xi) \in \mathfrak{g}$ for any $\sigma \in \mathfrak{G}$, because $\sigma\tau \in \mathfrak{G}$ whenever $\sigma, \tau \in \mathfrak{G}$. Consequently, $f(\xi\tau) \in \mathfrak{g}$ for any $\tau \in \mathfrak{G}^*$, whence $\sigma\tau \in \mathfrak{G}^*$ whenever $\sigma, \tau \in \mathfrak{G}^*$. Again, since $\sigma^{-1} \in \mathfrak{G}$ whenever $\sigma \in \mathfrak{G}$, it is clear that $f(\xi^{-1})$ multiplied by a sufficiently high power of the determinant of $\mathfrak{x}$ is a polynomial in $\mathfrak{g}$. Therefore the inverse of every nonsingular matrix in $\mathfrak{G}^*$ is itself in $\mathfrak{G}^*$. Thus, the set of all nonsingular matrices in $\mathfrak{G}^*$ forms an algebraic group.

It is obvious that the intersection of two algebraic matric groups is an algebraic group. As the first application of Lemmas 1 and 2 we derive another lemma.

LEMMA 3. *Let $\mathfrak{G}$ be an algebraic matric group, $\mathfrak{H}$ an algebraic subgroup, $\mathfrak{N}$ a normal algebraic subgroup. Then $\mathfrak{H}\mathfrak{N}$ is an algebraic subgroup of $\mathfrak{G}$.*

PROOF. $\mathfrak{H}\mathfrak{N}$ is a group; we must prove it algebraic. Let $\mathfrak{H}^*$ and $\mathfrak{N}^*$ be the underlying manifolds of $\mathfrak{H}$ and $\mathfrak{N}$, respectively, and let $\mathfrak{M}^*$ be the smallest algebraic manifold containing $\mathfrak{H}\mathfrak{N}$. By Lemma 2 $\mathfrak{M}^*$ is the underlying manifold of an algebraic group $\mathfrak{M}$. Clearly $\mathfrak{H}\mathfrak{N} \subseteq \mathfrak{M}$ and $\mathfrak{M} - \mathfrak{H}\mathfrak{N} \subseteq \mathfrak{M}^* - \mathfrak{H}^*\mathfrak{N}^*$. By Lemma 1 it suffices to show that $\mathfrak{M}^* - \mathfrak{H}^*\mathfrak{N}^*$ is contained in an algebraic manifold of lower dimension than $\mathfrak{M}^*$, for then $\mathfrak{H}\mathfrak{N} = \mathfrak{M}$, so that $\mathfrak{H}\mathfrak{N}$ is algebraic. To this end let $\chi_1, \cdots, \chi_r, \nu_1, \cdots, \nu_s$ be independent generic points of the r irreducible components of $\mathfrak{H}^*$ and the s irreducible components of $\mathfrak{N}^*$. If $\mathfrak{M}_{ij}^*$ is the irreducible algebraic manifold with generic point $\chi_i\nu_j$ then clearly $\mathfrak{M}^* = \bigcup_{i,j} \mathfrak{M}_{ij}^*$. Therefore every $\sigma \in \mathfrak{M}^*$ is obtained by a specialization $\chi_i\nu_j \to \sigma$. Such a specialization can be extended to a specialization of the form $(\chi_i\nu_j, \chi_i, \nu_j) \to (\sigma, \chi_{0i}, \nu_{0j})$ provided the ideal of all polynomials in $\mathfrak{C}[\cdots, x_{ab}, \cdots; \cdots, y_{cd}, \cdots; \cdots, z_{ef}, \cdots]$ which vanish for $(x_{ab}) = \chi_i\nu_j$, $(y_{cd}) = \chi_i$, $(z_{ef}) = \nu_j$ still has a zero when (x_{ab}) is replaced by σ, that is, if for a finite basis of this ideal the polynomials obtained by replacing (x_{ab}) by σ have a common zero. By the general theory of resultants (van der Waerden [1], vol. 2, chapter 11) this will be the case provided a certain polynomial in $\mathfrak{C}[\cdots, x_{ab}, \cdots]$ which does not vanish at $\chi_i\nu_j$ still fails to vanish at σ. Thus, whenever σ does not lie on a certain algebraic submanifold of $\mathfrak{M}^*$ of lower dimension than $\mathfrak{M}^*$, there exist $\chi \in \mathfrak{H}$, $\nu \in \mathfrak{N}$ such that $\sigma = \chi\nu$.

We are now in a position to establish the following Jordan-Hölder-Schreier theorem.

THEOREM. *Any two normal chains of an algebraic matric group have isomorphic refinements.*

PROOF. Here "normal chain" means a sequence $\mathfrak{G} \supseteq \mathfrak{G}_1 \supseteq \cdots \supseteq \mathfrak{G}_{r-1} \supseteq \mathfrak{E}$ in which each member is a normal *algebraic* subgroup of its predecessor. The proof of the classical Jordan-Hölder-Schreier theorem as give by Zassenhaus [1] (pp. 50–52) consists in actually constructing groups yielding the desired refinements. Zassenhaus' construction suffices in the present case provided the constructed groups turn out to be algebraic. That this is the case follows from Lemma 3 and the fact that all the constructed groups are of the form $\mathfrak{G}_i(\mathfrak{G}_{i-1} \cap \mathfrak{H}_k)$, where $\mathfrak{G} = \mathfrak{G}_0 \supseteq \cdots \supseteq \mathfrak{G}_r = \mathfrak{E}$ and $\mathfrak{G} = \mathfrak{H}_0 \supseteq \cdots \supseteq \mathfrak{H}_s = \mathfrak{E}$ are the given normal chains.

4. Commutator groups

The *commutator* of two elements σ, τ of a group is the element $\sigma\tau\sigma^{-1}\tau^{-1}$. By the *commutator group* of an algebraic matric group we shall mean the smallest algebraic subgroup of $\mathfrak{G}$ that contains the commutator of every pair of matrices in $\mathfrak{G}$. We shall denote the commutator group of $\mathfrak{G}$ by $\mathfrak{G}'$.

Since $\mathfrak{G}'$ is an algebraic group we can form its commutator group $\mathfrak{G}''$, etc., thus obtaining the *sequence of commutator groups* $\mathfrak{G}, \mathfrak{G}', \cdots, \mathfrak{G}^{(i)}, \cdots$. Since the underlying manifolds of the groups of this sequence form a decreasing sequence, they must, beginning with some index j, all be the same. We say, in this case, that the sequence of commutator groups terminates with $\mathfrak{G}^{(j)}$.

It is easy to see that $\mathfrak{G}'$ is the smallest normal algebraic subgroup of $\mathfrak{G}$ for which the factor group is abelian.

EXAMPLE. Consider the algebraic matric groups $\mathfrak{T} = \mathfrak{T}_0, \mathfrak{T}_1, \cdots, \mathfrak{T}_n = \mathfrak{E}$ (§2 Example 5). In the product of two matrices in $\mathfrak{T}$ the elements on the main diagonal are the products of the corresponding elements in the two factor matrices. Therefore the commutator of two matrices in $\mathfrak{T}$ is always in $\mathfrak{T}_1$. In the product of two matrices in $\mathfrak{T}_i (1 \leqq i \leqq n - 1)$ the elements on the ith diagonal above the main one are the sums of the corresponding elements in the two factor matrices. Therefore the commutator of two elements in $\mathfrak{T}_i$ is always in $\mathfrak{T}_{i+1}$.

LEMMA. *If an algebraic matric group $\mathfrak{G}$ is connected then so is $\mathfrak{G}'$.*

PROOF. Let $\sigma_1, \tau_1, \sigma_2, \tau_2, \cdots$ be a sequence of independent generic points of $\mathfrak{G}^*$, the underlying manifold of $\mathfrak{G}$. Let $\mathfrak{H}_i^*$ be the irreducible manifold with generic point $\chi_i = \sigma_1\tau_1\sigma_1^{-1}\tau_1^{-1} \cdots \sigma_i\tau_i\sigma_i^{-1}\tau_i^{-1}$. Now $\chi_i = \chi_{i-1}\sigma_i\tau_i\sigma_i^{-1}\tau_i^{-1}$, so that $\chi_i \rightarrow \chi_{i-1}$ is a specialization, and $\mathfrak{H}_{i-1}^* \subseteq \mathfrak{H}_i^*$. Since each $\mathfrak{H}_i^*$ is irreducible, the dimension of $\mathfrak{H}_{i-1}^*$ is less than that of $\mathfrak{H}_i^*$ when $\mathfrak{H}_{i-1}^* \subset \mathfrak{H}_i^*$, so that there is a j such that $\mathfrak{H}_k^* = \mathfrak{H}_j^*$ for all $k \geqq j$. It is clear that $\mathfrak{H}_j^*$ is the smallest algebraic manifold containing the set $\mathfrak{K}$ of all produts of commutators of matrices in $\mathfrak{G}$. Since $\mathfrak{K}$ obviously is a group, it follows from §3 Lemma 2 that $\mathfrak{H}_j^*$ is the underlying manifold of an algebraic matric group, which is clearly $\mathfrak{G}'$. As $\mathfrak{H}_j^*$ is irreducible, $\mathfrak{G}'$ is connected.

5. Solvable algebraic matric groups

An algebraic matric group $\mathfrak{G}$ will be called *solvable* if it has a normal chain in which all the factor groups are abelian.

It is easy to see that $\mathfrak{G}$ is solvable if and only if its sequence of commutator groups terminates with $\mathfrak{E}$. Moreover, if $\mathfrak{G}$ is solvable then so is every algebraic subgroup of $\mathfrak{G}$. Consequently, if $\mathfrak{G}$ is solvable then, for a suitable j,

$$\mathfrak{G} \supseteq \mathfrak{G}^0 \supseteq \mathfrak{G}^{0'} \supseteq \cdots \supseteq \mathfrak{G}^{0(j)} = \mathfrak{E}$$

is a normal chain in which every member after the first is connected, all the factor groups after the first are abelian, and $\mathfrak{G}/\mathfrak{G}^0$ is a solvable finite group.

EXAMPLE. By the example in §4, we have $\mathfrak{T}_i' \subseteq \mathfrak{T}_{i+1}$ $(i = 0, \cdots, n - 1)$. Therefore $\mathfrak{T} = \mathfrak{T}_0 \supset \cdots \supset \mathfrak{T}_n = \mathfrak{E}$ is a normal chain of $\mathfrak{T}$ in which all the factor groups are abelian. Hence $\mathfrak{T}$ is solvable.

The following lemma will be used in §7 to permit an induction on the degree of the matrices of a group.

LEMMA. *Let $\mathfrak{G}$ be an algebraic group of matrices of degree n of the form*

$$\mathfrak{G} = \begin{pmatrix} \mathfrak{G}_1 & * \\ 0 & \mathfrak{G}_2 \end{pmatrix},$$

where the blocks $\mathfrak{G}_1$ and $\mathfrak{G}_2$ are of degree n_1 and n_2, respectively ($n_1 + n_2 = n$, $0 < n_1 < n$). Then $\mathfrak{G}_1$ and $\mathfrak{G}_2$ are algebraic matric groups homomorphic to $\mathfrak{G}$. If $\mathfrak{G}$ is connected then so are $\mathfrak{G}_1$ and $\mathfrak{G}_2$. If $\mathfrak{H}$ is a normal algebraic subgroup of $\mathfrak{G}$, and $\mathfrak{H}_k$ the corresponding subgroup of $\mathfrak{G}_k$ ($k = 1, 2$), then $\mathfrak{H}_k$ is a normal algebraic subgroup of $\mathfrak{G}_k$. If, moreover, $\mathfrak{G}/\mathfrak{H}$ is abelian then $\mathfrak{G}_k/\mathfrak{H}_k$, is, too. Consequently, if $\mathfrak{G}$ is solvable then so is $\mathfrak{G}_k$.

PROOF. It is clear that $\mathfrak{G}_k$ is a group and that if $\sigma = \begin{pmatrix} \tau_1 & * \\ 0 & \tau_2 \end{pmatrix}$ then $\sigma \to \tau_k$ is a homomorphism of $\mathfrak{G}$ onto $\mathfrak{G}_k$. To show that $\mathfrak{G}_k$ is algebraic let $\mathfrak{K}_k^*$ be the smallest algebraic manifold (in n_k^2-dimensional affine space) containing $\mathfrak{G}_k$. By §3 Lemma 2, $\mathfrak{K}_k^*$ is the underlying manifold of an algebraic group $\mathfrak{K}_k$. As in the proof of Lemma 3 in §3, we see that $\mathfrak{K}_k - \mathfrak{G}_k$ is contained in an algebraic manifold of lower dimension. Therefore, by §3 Lemma 1, $\mathfrak{G}_k = \mathfrak{K}_k$, so that $\mathfrak{G}_k$ is algebraic. The defining ideal of $\mathfrak{G}_k$ consists of all those polynomials in the defining ideal of $\mathfrak{G}$ which involve only the appropriate n_k^2 indeterminates, so that the former ideal is prime whenever the latter is, that is, $\mathfrak{G}_k$ is connected whenever $\mathfrak{G}$ is. By the part of the lemma already proved, the homomorphism $\sigma \to \tau_k$ maps $\mathfrak{H}$ onto an algebraic subgroup $\mathfrak{H}_k$ of $\mathfrak{G}_k$. It is now obvious that $\mathfrak{H}_k$ is normal in $\mathfrak{G}_k$, and that $\mathfrak{G}_k/\mathfrak{H}_k$ is abelian if $\mathfrak{G}/\mathfrak{H}$ is.

6. Anticompact and quasicompact algebraic matric groups

An algebraic matric group $\mathfrak{G}$ will be called *anticompact* if it contains no element other than ι of finite order not divisible by p (the characteristic of $\mathfrak{C}$). $\mathfrak{G}$ will be called *quasicompact* if no algebraic subgroup of $\mathfrak{G}$ of order > 1 is anticompact (equivalently: $\mathfrak{G}$ will be called quasicompact if every algebraic subgroup of $\mathfrak{G}$ of order > 1 contains an element other than ι of finite order not divisible by p, and will be called anticompact if no algebraic subgroup of $\mathfrak{G}$ of order > 1 is quasicompact). It is clear that an algebraic subgroup of $\mathfrak{G}$ is anticompact (or is quasicompact) whenever $\mathfrak{G}$ is.

The following two theorems provide characterizations of these two types of groups.

THEOREM 1. *Let u be a positive integer, $\mathfrak{G}$ be an algebraic matric group. A necessary and sufficient condition that $\mathfrak{G}$ contain no matrix of finite order not divisible by p and not dividing u, is that every matrix in $\mathfrak{G}$ have all its characteristic roots equal to u^{th} roots of unity. Thus (taking $u = 1$), $\mathfrak{G}$ is anticompact if and only if every matrix in $\mathfrak{G}$ has all its characteristic roots equal to 1, that is, each matrix in $\mathfrak{G}$ is reducible to special triangular form.*

PROOF. SUFFICIENCY. Suppose every matrix in $\mathfrak{G}$ has characteristic roots as stated. Let σ be any matrix in $\mathfrak{G}$ of finite order m not dividing u, and choose a basis for the vector space $\mathfrak{N}$ so that σ is in its Jordan normal form (2). Since m does not divide u, σ actually has a 1 in the diagonal just above the main one. Raising σ to the mth power replaces this 1 by ma_i^{m-1}. Since $\sigma^m = \iota$ and $a_i \neq 0$, m must be divisible by p.

NE ESSI Y. Suppose $\mathfrak{G}$ contains a matrix σ which has a characteristic root not a uth root of unity. We must show that $\mathfrak{G}$ then contains a matrix of finite order not divisible by p and not dividing u.

As before, choose a basis of $\mathfrak{N}$ such that σ is given by (2). It is easy to see that the element in the ith row and j^{th} column of the hth block of σ^k ($k = 0, \pm 1, \pm 2, \cdots$) is $\binom{k}{j-i} a_h^{i-j+k}$ where, as is customary,

$$\binom{x}{m} = \begin{cases} 0 & \text{for } m < 0, \\ 1 & \text{for } m = 0 \\ \dfrac{x(x-1)\cdots(x-m+1)}{m!} & \text{for } m > 0 \end{cases}$$

(this binomial coefficient $\binom{x}{m}$ is defined for any element x in an extension of $\mathcal{C}$ if $p = 0$, for any integer x if $p \neq 0$).

Let $\tau_n(a, b, c) = (t_{ij})$ be the matrix of degree n defined by

$$t_{ij} = \binom{c}{j-i} a^{i-j} b.$$

By the easily established relation

$$\sum_{i=0}^{k} \binom{c}{k-i}\binom{c'}{i} = \binom{c+c'}{k}$$

it readily follows that

$$\tau_n(a, b, c)\cdot\tau_n(a, b', c') = \tau_n(a, bb', c + c').$$

Now let $\mathfrak{H}$ be the algebraic subgroup of $\mathfrak{G}$ consisting of all matrices $\tau(b_1, \cdots, b_t, c_1, \cdots, c_t)$ in $\mathfrak{G}$ of the form

$$\tau(b_1, \cdots, b_t, c_1 \cdots, c_t) = \begin{pmatrix} \boxed{\tau_{n_1}(a_1, b_1, c_1)} & & 0 \\ & \ddots & \\ 0 & & \boxed{\tau_{n_t}(a_t, b_t, c_t)} \end{pmatrix}$$

(same number of blocks as in σ; each block of the same degree as the corresponding block in σ; $a_1, \cdots, a_t$ the same as in σ; $b_1, \cdots, b_t \in \mathcal{C}$; $c_1, \cdots, c_t \in \mathcal{C}$ if $p = 0$, $c_1, \cdots, c_t$ integers if $p \neq 0$).

If $\tau(b_1, \cdots, b_t, c_1, \cdots, c_t)$ now denotes a generic point of the underlying manifold of the component of the identity $\mathfrak{H}^0$ then each b_i either is 1 or is transcendental over $\mathcal{C}$ (because $\tau(b_1, \cdots, b_t, c_1, \cdots, c_t) \to \iota = \tau(1, \cdots, 1, 0, \cdots, 0)$ is a specialization). We first show that

$$\partial^0\mathcal{C}(b_1, \cdots, b_t, c_1, \cdots, c_t)/\mathcal{C}(b_1, \cdots, b_t) = \partial^0\mathcal{C}(c_1, \cdots, c_t)/\mathcal{C}$$

To do this it suffices to show that if some of the c_i's are algebraically dependent over $\mathcal{C}(b_1, \cdots, b_t)$ then they are over $\mathcal{C}$, too. Suppose, say, that $c_1, \cdots, c_q$ are algebraically dependent ovêr $\mathcal{C}(b_1, \cdots, b_t)$, or what is the same thing, over $\mathcal{C}(b_{i_1}, \cdots, b_{i_r})$, where $b_{i_1}, \cdots, b_{i_r}$ is a maximal set of the b_i's algebraically independent over $\mathcal{C}$. Then there is a nonzero polynomial

$$f(x_1, \cdots, x_q) \neq \cdots + \phi(b_{i_1}, \cdots, b_{i_r})x_1^{j_1} \cdots x_q^{j_q} + \cdots,$$

with coefficients which are rational functions over $\mathcal{C}$ of $b_{i_1}, \cdots, b_{i_r}$, which vanishes for $x_1 = c_1, \cdots, x_q = c_q$. We choose such a polynomial $f(x_1, \cdots, x_q)$ with as few terms as possible, and denote its degree by d. We suppose, as we may, that the coefficient in one of the terms of $f(x_1, \cdots, x_q)$ of degree d is 1.

Now $\tau(b_1, \cdots, b_t, c_1, \cdots, c_t)^k = \tau(b_1^k, \cdots, b_t^k, kc_1, \cdots, kc_t)$ is a generalized point of $\mathfrak{H}^0$ for every integral k, so that $x_1 = kc_1, \cdots, x_t = kc_t$ annuls

$$\cdots + \phi(b_{i_1}^k, \cdots, b_{i_r}^k)x_1^{j_1} \cdots x_q^{j_q} + \cdots,$$

that is, $x_1 = c_1, \cdots, x_q = c_q$ annuls the polynomials

$$\cdots + \phi(b_{i_1}^k, \cdots, b_{i_r}^k)k^{j_1+\cdots+j_q}x_1^{j_1} \cdots x_q^{j_q} \cdot + \cdots.$$

Since the coefficient in one of the terms of $f(x_1, \cdots, x_q)$ is 1, the difference between $f(x_1, \cdots, x_q)$ and one of these polynomials divided by k^d has fewer terms than $f(x_1, \cdots, x_q)$ has. Therefore this difference must be 0, and for each coefficient $\phi(b_{i_1}, \cdots, b_{i_r})$ we must have

$$\phi(b_{i_1}^k, \cdots, b_{i_r}^k) = k^{d-j_1-\cdots-j_q}\phi(b_{i_1}, \cdots, b_{i_r}).$$

Since $b_{i_1}, \cdots, b_{i_r}$ are algebraically independent over $\mathcal{C}$, it follows that each $\phi(b_{i_1}, \cdots, b_{i_r})$ is in $\mathcal{C}$, so that $c_1, \cdots, c_q$ are algebraically dependent over $\mathcal{C}$ This proves our remark concerning degrees of transcendency.

It now follows that

$$\partial^0\mathcal{C}(b_1, \cdots, b_t, c_1, \cdots, c_t)/\mathcal{C} = \partial^0\mathcal{C}(b_1, \cdots, b_t)/\mathcal{C} + \partial^0\mathcal{C}(c_1, \cdots, c_t)/\mathcal{C}.$$

Therefore, letting $\mathfrak{m}$ be the prime ideal in $\mathcal{C}[x_1, \cdots, x_t, y_1, \cdots, t_t]$ with generic zero $b_1, \cdots, b_t, c_1, \cdots, c_t$, and letting $\mathfrak{p} = \mathfrak{m} \cap \mathcal{C}[x_1, \cdots, x_t]$, $\mathfrak{q} = \mathfrak{m} \cap \mathcal{C}[y_1, \cdots, y_n]$, we see that $b_1, \cdots, b_t$ and $c_1, \cdots, c_t$ are independent generic zeros of $\mathfrak{p}$ and $\mathfrak{q}$, respectively. This implies that a specialization of $b_1, \cdots, b_t$ and a specialization of $c_1, \cdots, c_t$ can always be combined to form a specialization of $b_1, \cdots, b_t, c_1, \cdots, c_t$.

If $b_1, \cdots, b_t$ are not all 1 then $r > 0$, and we may find a positive integer v not dividing u and specialize

$$(5) \qquad b_1 \to \beta_1, \cdots, b_t \to \beta_t \qquad (\beta_1 \cdots \beta_t \neq 0)$$

in such a way that $\beta_i, \cdots, \beta_{i_r}$ are all v^{th} roots of unity other than u^{th} roots of unity. But then we may suppose that every β_i is a root of unity. To see this consider, for any $i \neq i_1, \cdots, i_r$, an irreducible nonzero polynomial $f_i(x_i, b_{i_1}, \cdots, b_{i_r}) \in \mathfrak{C}[x_1, b_{i_1}, \cdots, b_{i_r}]$ which vanishes for $x_i = b_i$. Referring to $\tau(b_1, \cdots, b_t, c_1, \cdots, c_t)^{kv+1}$ we see that

$$f_i(\beta_i^{kv+1}, \beta_{i_1}, \cdots, \beta_{i_r}) = 0, \qquad k = 0, 1, 2, \cdots.$$

We assume, as we may, that the specialization (5) is such that $f_i(x_i, \beta_{i_1}, \cdots, \beta_{i_r}) \neq 0$. Then we see that β_i^{kv+1} is the same for two different values of k, so that β_i is a root of unity.

Now, since $\tau(1, \cdots, 1, 0; \cdots, 0) = \iota \in \mathfrak{H}^0$,

$$(6) \qquad c_1 \to 0, \quad \cdots, \quad c_t \to 0$$

is a specialization. Therefore, if $b_1, \cdots, b_t$ are not all 1, we may combine the specializations (5) and (6) to obtain a matrix in $\mathfrak{H}^0$ which is in diagonal form and in which all the elements on the main diagonal are roots of unity (not all these roots of unity being u^{th} roots of unity). Such a matrix is of finite order not divisible by p and not dividing u.

It remains to consider the case in which $b_1, \cdots, b_t$ are all 1. In this case, since $\sigma^h \in \mathfrak{H}^0$ (h = index of $\mathfrak{H}^0$ in $\mathfrak{H}$), there is a smallest positive integer k such that the diagonal elements $a_1, \cdots, a_t$ af σ are all k^{th} roots of unity, and this k is not divisible by p and does not divide u (for the characteristic roots $a_1, \cdots, a_t$ are by hypothesis not all u^{th} roots of unity).

Now $\sigma = \tau(a_1, \cdots, a_t, 1, \cdots, 1)$, so that for all $j = 0, 1, 2, \cdots$

$$\tau(1, \cdots, 1, kj, \cdots, kj) = \sigma^{kj} \in \mathfrak{H}.$$

In the case $p = 0$, this means that $\mathfrak{H}$ contains an infinite number of points of the one-dimensional algebraic manifold of all matrices $\tau(1, \cdots, 1, \gamma, \cdots, \gamma)$ with $\gamma \in \mathfrak{C}$, so that $\mathfrak{H}$ contains every nonsingular such matrix, and in particular, contains $\tau(1, \cdots, 1, -1, \cdots, -1)$. In the case $p \neq 0$, $\tau(1, \cdots, 1, p^m, \cdots, p^m) = \iota$ provided $p^m \geqq n$. Since k is relatively prime to p, there is a j such that $kj \equiv -1 \pmod{p^m}$, whence $\sigma^{kj} = \tau(1, \cdots, 1, -1, \cdots, -1)$. Hence, in either case, $\mathfrak{H}$ contains $\sigma \cdot \tau(1, \cdots, 1, -1, \cdots, -1) = \tau(a_1, \cdots, a_t, 1, \cdots, 1) \cdot \tau(1, \cdots, 1, -1, \cdots, -1) = \tau(a_1, \cdots, a_t, 0, \cdots, 0)$, which is in diagonal form with diagonal elements $a_1, \cdots, a_t$. Thus $\mathfrak{H}$ contains a matrix of order k not divisible by p and not dividing u. This completes the proof.

THEOREM 2. *Let $\mathfrak{G}$ be an algebraic matric group. A necessary and sufficient condition that $\mathfrak{G}$ be quasicompact is that each matrix in $\mathfrak{G}$ be reducible to diagonal form.*

PROOF. SUFFICIENCY. Suppose every matrix in $\mathfrak{G}$ is so reducible. Let $\mathfrak{G}_1$ be an algebraic subgroup of $\mathfrak{G}$ of order > 1. We must show that $\mathfrak{G}_1$ contains an element of finite order not divisible by p. Let σ be an element of $\mathfrak{G}_1$ other than ι, and choose a basis for the vector space $\mathfrak{R}$ in such a way that σ is in diagonal form. Let $\mathfrak{G}_2$ be the set of all matrices in $\mathfrak{G}_1$ which are in diagonal form. If $\mathfrak{G}_2^0 = \mathfrak{E}$ then each diagonal element of σ is a root of unity and σ itself is of finite order not divisible by p. If $\mathfrak{G}_2^0 \neq \mathfrak{E}$ then a generic point of the underlying manifold of $\mathfrak{G}_2^0$ can be specialized to a matrix τ_0 in which all the diagonal elements are roots of unity not all 1. Then τ_0 is of finite order not divisible by p.

NECESSITY. Suppose $\mathfrak{G}$ is quasicompact, and let σ be any matrix in $\mathfrak{G}$ other than ι. We must show that σ is reducible to diagonal form. As in the proof of Theorem 1, let σ be give by (2), and introduce the groups $\mathfrak{H}$, $\mathfrak{H}^0$ and the generic point $\tau(b_1, \cdots, b_t, c_1, \cdots, c_t)$ of the underlying manifold of $\mathfrak{H}^0$.

We first show that $\tau(b_1, \cdots, b_t, c_1, \cdots, c_t)$ is in diagonal form. Suppose that this is false. Then there is a specialization

$$c_1 \to \gamma_1, \cdots, c_t \to \gamma_t \quad (\gamma_1, \cdots, \gamma_t \in \mathfrak{C}, \text{ not every } \gamma_i = 0).$$

Again, since $\iota \in \mathfrak{H}^0$,

$$b_1 \to 1, \cdots, b_t \to 1$$

is a specialization. As we saw in the proof of Theorem 1, we can combine these two specializations to obtain a specialization of $\tau(b_1, \cdots, b_t, c_1, \cdots, c_t)$. Therefore the algebraic group $\mathfrak{K}$, consisting of all matrices in $\mathfrak{H}$ in which all elements on the main diagonal are 1, properly contains $\mathfrak{E}$. But by Theorem 1, $\mathfrak{K}$ is anticompact, so that $\mathfrak{K} = \mathfrak{E}$. This contradiction shows that every matrix in $\mathfrak{H}^0$ is in diagonal form.

Now suppose that σ is not in diagonal form, that is, that not every block of σ in (2) is of degree 1. Then σ is contained in an irreducible component $\mathfrak{H}_1^*$ of the underlying manifold of $\mathfrak{H}$ for which

$$\sigma \cdot \tau(b_1, \cdots, b_t, 0, \cdots, 0) = \tau(a_1, \cdots, a_t, 1, \cdots, 1) \cdot \tau(b_1, \cdots, b_t, 0, \cdots, 0)$$
$$= \tau(a_1 b_1, \cdots, a_t b_t, 1, \cdots, 1)$$

is a generic point (see §2). As in the proof of Theorem 1, let $b_{i_1}, \cdots, b_{i_r}$ be a maximal set of b_i's algebraically independent over $\mathfrak{C}$. For each $i \neq i_1, \cdots, i_r$ there is a nonzero polynomial $g_i(x_i, a_{i_1} b_{i_1}, \cdots, a_{i_r} b_{i_r}) \in \mathfrak{C}[x_i, a_{i_1} b_{i_1}, \cdots, a_{i_r} b_{i_r}]$ such that $g_i(a_i b_i, a_{i_1} b_{i_1}, \cdots, a_{i_r} b_{i_r}) = 0$. We now specialize

$$b_1 \to \beta_1, \cdots, b_t \to \beta_t \qquad (\beta_1 \cdots \beta_t \neq 0)$$

in such a way that $a_{i_1}\beta_{i_1}, \cdots, a_{i_r}\beta_{i_r}$ are all roots of unity and every $g_i(x_i, a_{i_1}\beta_{i_1}, \cdots, a_{i_r}\beta_{i_r}) \neq 0$. If v is a multiple of the order of $\mathfrak{H}/\mathfrak{H}^0$ such that $(a_{i_1}\beta_{i_1})^v = \cdots = (a_{i_r}\beta_{i_r})^v = 1$, then $\tau(a_1\beta_1, \cdots, a_t\beta_t, 1, \cdots, 1)^{kv+1} \in \mathfrak{H}_1^*$ for $k = 0, 1, \cdots$. Therefore

$$g((a_i\beta_i)^{kv+1}, a_{i_1}\beta_{i_1}, \cdots, a_{i_r}\beta_{i_r}) = 0, k = 0, 1, \cdots,$$

so that every $a_i\beta_i$ is a root of unity.

Letting s be the smallest positive integer such that every $(a_i\beta_i)^s = 1$, we see that s is not divisible by p, and $\tau(a_1\beta_1, \cdots, a_t\beta_t, 1, \cdots, 1)^s = \tau(1, \cdots, 1, s, \cdots, s)$ is a matrix in $\mathfrak{H}$ which is not in diagonal form and in which all the elements on the main diagonal are 1. As above, this leads to a contradiction. Therefore σ is in diagonal form and the proof is complete.

7. Reducibility to triangular form

In this section a necessary and sufficient condition is derived for a set of matrices to be reducible to triangular form, and the bearing of anticompactness and quasicompactness on reducibility to triangular form is studied.

THEOREM 1. *A necessary and sufficient condition that a set of matrices of degree n be reducible to triangular form is that the set be contained in the underlying manifold of a connected solvable algebraic matric group.*

PROOF. SUFFICIENCY. It is enough to show that a connected solvable algebraic matric group $\mathfrak{G}$ is reducible to triangular form. For $n = 1$ this is obvious. Let $n > 1$ and suppose the theorem verified for matrices of lower degree. If $\mathfrak{G}$ is reducible then by the induction assumption and the lemma of §5, $\mathfrak{G}$ is clearly reducible to triangular form. We suppose, then, that $\mathfrak{G}$ is irreducible and seek a contradiction.

By §5, $\mathfrak{G}$ has a normal chain

$$\mathfrak{G} = \mathfrak{G}_0 \supset \mathfrak{G}_1 \supset \cdots \supset \mathfrak{G}_h = \mathfrak{E}$$

in which each $\mathfrak{G}_i$ is connected and each $\mathfrak{G}_{i-1}/\mathfrak{G}_i$ is abelian. If the chain length h equals 1 then $\mathfrak{G}$ is abelian and (by §1 Lemma 1) reducible to triangular form, and therefore is reducible. Suppose that $h > 1$, and that every connected solvable algebraic matric group of degree n which has a normal chain as above of length $< h$ is reducible. Then $\mathfrak{G}_1$ is reducible, and therefore reducible to triangular form. Consequently the vector space $\mathfrak{N}$ contains a vector η which is a characteristic vector for every $\sigma_1 \in \mathfrak{G}_1$. Let $\mathfrak{N}_1$ be the linear subspace of $\mathfrak{N}$ spanned by all such vectors η. For any $\sigma \in \mathfrak{G}$, $\sigma_1 \in \mathfrak{G}_1$ we have $\sigma^{-1}\sigma_1\sigma \in \mathfrak{G}_1$, so that, for any given vector η as above, there exists a $c \in \mathfrak{C}$ such that $\sigma^{-1}\sigma_1\sigma\eta = c\eta$, $\sigma_1\sigma\eta = c\sigma\eta$. That is, $\sigma\eta$ is a characteistic vector for every $\sigma_1 \in \mathfrak{G}_1$, whence $\sigma\eta \in \mathfrak{N}_1$, and $\mathfrak{G}\mathfrak{N}_1 = \mathfrak{N}_1$. Since $\mathfrak{G}$ is irreducible this means that $\mathfrak{N}_1 = \mathfrak{N}$, so that $\mathfrak{G}_1$ is reducible to diagonal form.

We now write $\mathfrak{N}$ as a direct sum

$$\mathfrak{N} = \mathfrak{N}' + \cdots + \mathfrak{N}^{(k)}$$

of linear subspaces $\mathfrak{N}^{(i)}$ such that:

(a) for each $\sigma_1 \in \mathfrak{G}_1$ there is a $c_i \in \mathfrak{C}$ such that $\sigma_1\zeta = c_i\zeta$ for all $\zeta \in \mathfrak{N}^{(i)}$;

(b) if $i \neq j$ there is a $\sigma_1 \in \mathfrak{G}_1$ for which $c_i \neq c_j$. For any $\sigma \in \mathfrak{G}$, $\sigma_1 \in \mathfrak{G}_1$ we have $\sigma^{-1}\sigma_1\sigma \in \mathfrak{G}_1$, so that, for each i, there is a $c_i \in \mathfrak{C}$ such that for every $\zeta \in \mathfrak{N}^{(i)}$ we have $\sigma^{-1}\sigma_1\sigma\zeta = c_i\zeta$, $\sigma_1\sigma\zeta = c_i\sigma\zeta$. This shows that all the vectors in $\sigma\mathfrak{N}^{(i)}$ are in the same $\mathfrak{N}^{(j)}$, whence $\sigma\mathfrak{N}^{(i)} = \mathfrak{N}^{(j)}$. Thus, the matrices of $\mathfrak{G}$ permute the linear subspaces $\mathfrak{N}', \cdots, \mathfrak{N}^{(k)}$ among themselves, so that there is a homomorphism of $\mathfrak{G}$ onto a group of permutations of k objects. The kernel of this homomorphism is the group $\mathfrak{H}$ consisting of all matrices in $\mathfrak{G}$ which leave each $\mathfrak{N}^{(i)}$ invariant. $\mathfrak{H}$ is clearly an algebraic subgroup of $\mathfrak{G}$ of finite index (equal to the order of the permutation group). Since $\mathfrak{G}$ is connected, it follows from the remark at the end of §2 that $\mathfrak{G} = \mathfrak{H}$. Since $\mathfrak{G}$ is irreducible we must have $\mathfrak{N}' = \mathfrak{N}$, that is, every matrix in $\mathfrak{G}_1$ is of the form $c\iota$.

The commutator of any two matrices σ, τ in $\mathfrak{G}$ is in $\mathfrak{G}_1$, so that $\sigma\tau\sigma^{-1} = c\tau$ for some c in $\mathfrak{C}$. Taking determinants we see that $c^n = 1$, so that the underlying manifold of $\mathfrak{G}$ falls apart into a finite number of algebraic submanifolds, each one characterized by a different n^{th} root of unity c in the equation $\sigma\tau\sigma^{-1} = c\tau$ for τ. Since the underlying manifold of $\mathfrak{G}$ is irreducible, and since no matrix $\tau \neq 0$ can satisfy two such equations, c can assume only one value, which must be 1. Therefore $\mathfrak{G}$ is abelian. This contradicts the assumption $h > 1$, and completes the proof of the sufficiency.

NECESSITY. It suffices to show that the full triangular group $\mathfrak{T}$ (§2 Example 5) is connected and solvable. $\mathfrak{T}$ is obviously connected, for its underlying manifold is linear. $\mathfrak{T}$ is solvable by the example of §5.

THEOREM 2. *An algebraic matric group is solvable and anticompact if and only if it is reducible to special triangular form.*

PROOF. By the example of §5 and Theorem 1 of §6, $\mathfrak{G}$ is solvable and anticompact if it is reducible to special triangular form. Conversely, let $\mathfrak{G}$ be solvable and anticompact. By Theorem 1 $\mathfrak{G}^0$ is reducible to triangular form, and by §6 Theorem 1 every matrix in $\mathfrak{G}$ has all its characteristic roots equal to 1. Now $\mathfrak{G}/\mathfrak{G}^0$ is solvable, so that there is a segment of a normal chain

$$\mathfrak{G} \supset \mathfrak{G}_1 \supset \cdots \supset \mathfrak{G}_{h-1} \supset \mathfrak{G}^0$$

in which each factor group is abelian. Therefore, to prove the theorem it suffices to prove the following

LEMMA. *If $\mathfrak{G}$ is an algebraic matric group, and $\mathfrak{N}$ is a normal algebraic subgroup such that $\mathfrak{N}$ is in special triangular form and $\mathfrak{G}/\mathfrak{N}$ is abelian, then $\mathfrak{G}$ is reducible to triangular form.*

PROOF. For degree $n = 1$ this is obvious. Let $n > 1$ and assume the lemma established for matrices of lower degree. Let $\mathfrak{N}_1$ be the linear subspace of the vector space $\mathfrak{N}$ consisting of all vectors η such that $\nu\eta = \eta$ for all $\nu \in \mathfrak{N}$. Since $\mathfrak{N} \subseteq \mathfrak{T}_1$, the dimension n_1 of $\mathfrak{N}_1$ is at least 1. If $n_1 = n$ then $\mathfrak{N} = \mathfrak{E}$, $\mathfrak{G}$ is abelian and (by §1 Lemma 1) reducible to triangular form. Suppose $n_1 < n$. For any $\sigma \in \mathfrak{G}$, $\nu \in \mathfrak{N}$ we have $\sigma^{-1}\nu\sigma \in \mathfrak{N}$, so that for any $\eta \in \mathfrak{N}_1$ we have $\sigma^{-1}\nu\sigma\eta = \eta$, $\nu\sigma\eta = \sigma\eta$. It follows that $\mathfrak{N}_1$ is invariant under every matrix in $\mathfrak{G}$ so that there is a matrix ρ for which

$$\rho^{-1}\mathfrak{G}\rho = \begin{pmatrix} \mathfrak{G}_1 & * \\ 0 & \mathfrak{G}_2 \end{pmatrix}.$$

Since $\mathfrak{N} \subseteq \mathfrak{G}$ we have, correspondigly,

$$\rho^{-1}\mathfrak{N}\rho = \begin{pmatrix} \mathfrak{N}_1 & * \\ 0 & \mathfrak{N}_2 \end{pmatrix},$$

and (by the lemma of §5) $\mathfrak{N}_i$ is a normal algebraic subgroup of the algebraic matric group $\mathfrak{G}_i$ such that $\mathfrak{G}_i/\mathfrak{N}_i$ is abelian ($i = 1, 2$). But $\mathfrak{N}_1$ consists solely of the identity matrix, so that on the one hand $\mathfrak{G}_1$ is abelian and therefore reducible to triangular form, and on the other there is an obvious homomorphism between $\mathfrak{N}$ and $\mathfrak{N}_2$. Since $\mathfrak{N} \subseteq \mathfrak{T}_1$ we see (by Theorem 1 and §6 Theorem 1) that $\mathfrak{N}$ is solvable and anticompact. By virtue of the homomorphism and §6 Theorem 1 the same may be said of $\mathfrak{N}_2$. Consequently there is a normal chain

$$\mathfrak{G}_2 \supseteq \mathfrak{N}_2 \supset \mathfrak{N}_{21} \supset \cdots \supset \mathfrak{N}_{2u} = \mathfrak{E}$$

in which each factor group is abelian. By the induction assumption we may start with $\mathfrak{N}_{2u} = \mathfrak{E}$ and prove successively that $\mathfrak{N}_{2,u-1}, \cdots, \mathfrak{N}_{21}, \mathfrak{N}_2, \mathfrak{G}_2$ are reducible to triangular form. Since $\mathfrak{G}_1$ and $\mathfrak{G}_2$ are both reducible to triangular form, so is $\mathfrak{G}$.

THEOREM 3. *If a quasicompact algebraic matric group is reducible to triangular form then it is reducible to diagonal form.*

PROOF. Let $\mathfrak{G}$ be quasicompact, and assume, without loss of generality, that $\mathfrak{G}$ is already in triangular form: $\mathfrak{G} \subseteq \mathfrak{T}$. Since $\mathfrak{G}$ is quasicompact it contains no anticompact algebraic subgroup of order > 1. Therefore $\mathfrak{G} \cap \mathfrak{T}_1$ (which by §6 Theorem 1 is anticompact) must be $\mathfrak{E}$. But $\mathfrak{G} = \mathfrak{G}/\mathfrak{E} = \mathfrak{G}/\mathfrak{G} \cap \mathfrak{T}_1$ is isomorphic with $\mathfrak{G}\mathfrak{T}_1/\mathfrak{T}_1 \subseteq \mathfrak{T}/\mathfrak{T}_1$.[7] By the example of §5, $\mathfrak{T}/\mathfrak{T}_1$ is abelian; hence $\mathfrak{G}$ is abelian, too. By §1, Lemma 1, $\mathfrak{G}$ is therefore equivalent to a group $\mathfrak{N}$ of the form (1), where for each i the $\mathfrak{N}_i$-block of every matrix in $\mathfrak{N}$ is in triangular form with equal elements on the main diagonal. Since every matrix in $\mathfrak{G}$ is reducible to diagonal form (§6 Theorem 2), the same is true of the $\mathfrak{N}_i$-block of every matrix in $\mathfrak{N}$ (see the remark on complete reducibility made in §1); and since all the elements on the main diagonal of the $\mathfrak{N}_i$-block are equal, the $\mathfrak{N}_i$-block is already in diagonal form. This holds for every i, so that $\mathfrak{N}$ is in diagonal form.

8. Algebraic matric groups with certain types of normal chains

An algebraic matric group is solvable if it has a normal chain in which each factor group is abelian. We now investigate groups which contain normal chains with certain other types of factor groups.

THEOREM 1. *If an algebraic matric group $\mathfrak{G}$ has a normal chain*

$$\mathfrak{G} = \mathfrak{G}_0 \supset \mathfrak{G}_1 \supset \cdots \supset \mathfrak{G}_r = \mathfrak{E} \tag{7}$$

in which each factor group is either abelian or finite, then $\mathfrak{G}^0$ is solvable.

PROOF. For any algebraic subgroup $\mathfrak{H}$ of $\mathfrak{G}$, $\mathfrak{H} \cap \mathfrak{G}_{i-1}/\mathfrak{H} \cap \mathfrak{G}_i$ is isomorphic with a subgroup of $\mathfrak{G}_{i-1}/\mathfrak{G}_i$. Therefore every algebraic subgroup $\mathfrak{H} \neq \mathfrak{E}$ has a normal chain of positive length with the same properties as (7). Now consider the sequence of commutator groups for $\mathfrak{G}^0$: $\mathfrak{G}^0, \mathfrak{G}^{0\prime}, \cdots, \mathfrak{G}^{0(j)}, \cdots$. By §4 this sequence terminates with some $\mathfrak{G}^{0(j)}$. This $\mathfrak{G}^{0(j)}$ contains no normal

[7] VAN DER WAERDEN [1] vol. 1, p. 148, or ZASSENHAUS [1] p. 32.

algebraic proper subgroup with abelian factor group. Also, by the lemma of §3, this $\mathfrak{G}^{0(j)}$ is connected, and therefore, by the final remark of §2, contains no algebraic proper subgroup of finite index. Consequently $\mathfrak{G}^{0(j)}$ does not have a normal chain of positive length like (7), so that $\mathfrak{G}^{0(j)} = \mathfrak{E}$. This means that the sequence of commutator groups of $\mathfrak{G}^0$ terminates with $\mathfrak{E}$, whence $\mathfrak{G}^0$ is solvable.

We now extend the concepts of "anticompact" and "quasicompact" to factor groups, in the following natural manner. Let $\mathfrak{G}$ be an algebraic matric group, $\mathfrak{N}$ a normal algebraic subgroup thereof. The factor group $\mathfrak{G}/\mathfrak{N}$ is anticompact if it contains no element, other than the identity, of finite order not divisible by p; $\mathfrak{G}/\mathfrak{N}$ is quasicompact if there is no algebraic group $\mathfrak{H}$, with $\mathfrak{N} \subset \mathfrak{H} \subseteq \mathfrak{G}$, such that $\mathfrak{H}/\mathfrak{N}$ is anticompact.

THEOREM 2. *If an algebraic matric group $\mathfrak{G}$ has a normal chain* (7) *in which each factor group is either anticompact or finite, then $\mathfrak{G}^0$ is anticompact. If all the factor groups are anticompact then $\mathfrak{G}$ is anticompact.*

PROOF. Every algebraic subgroup $\mathfrak{H} \neq \mathfrak{E}$ of $\mathfrak{G}$ has a normal chain of positive length with the same properties as (7). In particular, if $\mathfrak{G}^0 \neq \mathfrak{E}$ then $\mathfrak{G}^0$ has such a normal chain:

$$\mathfrak{G}^0 \supset \mathfrak{H}_1 \supset \cdots \supset \mathfrak{H}_s = \mathfrak{E}.$$

Since $\mathfrak{G}^0$ is connected, $\mathfrak{G}^0/\mathfrak{H}_1$ is not finite, so that $\mathfrak{G}^0/\mathfrak{H}_1$ is anticompact and $\dim \mathfrak{G}^0 > \dim \mathfrak{H}_1$. We now use induction on $\dim \mathfrak{G}^0$. If $\dim \mathfrak{G}^0 = 0$ (that is, if $\mathfrak{G}^0 = \mathfrak{E}$) then $\mathfrak{G}^0$ is anticompact. Let $\dim \mathfrak{G}^0 > 0$ and suppose the theorem established for groups of lower dimension. Then $\mathfrak{H}_1^0$ is anticompact, and, by §6 Theorem 1, all the charateristic roots of each matrix in $\mathfrak{H}_1^0$ are 1. If $\mathfrak{G}^0$ contains a matrix σ of finite order j not divisible by p then $\sigma \in \mathfrak{H}_1$, for otherwise $\mathfrak{G}^0/\mathfrak{H}_1$ would not be anticompact. On the other hand $\sigma \notin \mathfrak{H}_1^0$, for $\mathfrak{H}_1^0$ is anticompact. Therefore j divides the order of $\mathfrak{H}_1/\mathfrak{H}_1^0$. If this order is u then it follows from §6 Theorem 1 that all the characteristic roots of each matrix in $\mathfrak{G}^0$ are u^{th} roots of unity. But $\mathfrak{G}^0$ is connected, so it easily follows that all these u^{th} roots of unity are actually 1. Therefore, by §6 Theorem 1, $\mathfrak{G}^0$ is anticompact.

If all the factor groups are anticompact and σ is any element of $\mathfrak{G}$ other than ι, consider the $\mathfrak{G}_j$ of highest subscript such that $\sigma \in \mathfrak{G}_j$. Since $\mathfrak{G}_j/\mathfrak{G}_{j+1}$ is anticompact, no finite power σ^k with k not divisible by p is contained in $\mathfrak{G}_{j+1}$, so that no $\sigma^k = \iota$. Thus $\mathfrak{G}$ is anticompact.

THEOREM 3. *If an algebraic matric group $\mathfrak{G}$ has a normal chain* (7) *in which each factor group is quasicompact, then $\mathfrak{G}$ is quasicompact.*[8]

PROOF. Let $\mathfrak{H} \neq \mathfrak{E}$ be any algebraic subgroup of $\mathfrak{G}$, and consider the $\mathfrak{G}_j$ of highest subscript such that $\mathfrak{G}_j \cap \mathfrak{H} \neq \mathfrak{E}$. Since $\mathfrak{G}_j/\mathfrak{G}_{j+1}$ is quasicompact,

[8] It would be of interest to know whether or not the following is true: *If each factor group is either quasicompact or finite then $\mathfrak{G}^0$ is quasicompact.* When $p = 0$ this is implied by Theorem 3, for then every finite group is quasicompact. But when $p \neq 0$ there are finite groups (groups of order divisible by p) which are not quasicompact.

$\mathfrak{G}_j \cap \mathfrak{H}$ contains a matrix $\sigma \neq \iota$ such that $\sigma^k \in \mathfrak{G}_{j+1}$ for some positive k not divisible by p. But $\mathfrak{G}_{j+1} \cap \mathfrak{H} = \mathfrak{E}$, so that $\sigma^k = \iota$, $\mathfrak{H}$ contains an element other than ι of finite order not divisible by p, and $\mathfrak{G}$ is quasicompact.

CHAPTER II. SOME RESULTS FROM THE THEORY OF ALGEBRAIC DIFFERENTIAL EQUATIONS

9. Differential rings, fields, and ideals

Let m be a positive integer. A *differential ring with m types of differentiation* is a (commutative) ring together with m operators $\delta_1, \cdots, \delta_m$ (called *derivatives*) such that:

(a) $\delta_i a$ is defined and is an element of the ring for every element a in the ring, $i = 1, \cdots, m$;

(b) $\delta_i \delta_j = \delta_j \delta_i$, $i, j = 1, \cdots, m$;

(c) $\delta_i(a + b) = \delta_i a + \delta_i b$ for all a, b in the ring, $i = 1, \cdots, m$;

(d) $\delta_i(ab) = \delta_i a \cdot b + a \cdot \delta_i b$ for all a, b in the ring, $i = 1, \cdots, m$.

A *differential field* is defined similarly. A differential ring or field is *ordinary* or *partial* according as $m = 1$ or $m > 1$. In the ordinary case we shall write a' for $\delta_1 a$.

If $\mathcal{R}_1$ and $\mathcal{R}_2$ are two differential rings with m types of differentiation, $\mathcal{R}_1$ is a *differential subring* of $\mathcal{R}_2$, and $\mathcal{R}_2$ is a *differential extension ring* of $\mathcal{R}_1$, provided $\mathcal{R}_1$ is a subring of $\mathcal{R}_2$ and $\delta_i a$ for $\mathcal{R}_2$ coincides with $\delta_i a$ for $\mathcal{R}_1$ whenever $a \in \mathcal{R}_1$ $(i = 1, \cdots, m)$. *Differential subfields* and *differential extension fields* are defined similarly.

Let $\mathcal{R}_1$ be a differential subring of the differential ring $\mathcal{R}_2$ and consider a subset Φ of $\mathcal{R}_2$. The set of all polynomials with coefficients in $\mathcal{R}_1$ in a finite number of elements of the form $\delta_1^{i_1} \cdots \delta_m^{i_m} a$ ($a \in \Phi$; each i_ν a nonnegative integer), forms a differential extension ring of $\mathcal{R}_1$, denoted by $\mathcal{R}_1\{\Phi\}$. If Φ is a finite set $a_1, \cdots, a_n$ then we write $\mathcal{R}_1\{a_1, \cdots, a_n\}$.

Let $\mathcal{F}_1$ be a differential subfield of the differential field $\mathcal{F}_2$ and consider a subset Φ of $\mathcal{F}_2$. The set of all rational functions, in a finite number of elements of the form $\delta_1^{i_1} \cdots \delta_m^{i_m} a$($a \in \Phi$; each i_ν a nonnegative integer), with coefficients in $\mathcal{F}_1$ and with nonvanishing denominator, forms a differential extension field of $\mathcal{F}_1$ denoted by $\mathcal{F}_1\langle\Phi\rangle$ (in the finite case: $\mathcal{F}_1\langle a_1, \cdots, a_n\rangle$).

An element c of a differential ring or field with m types of differentiation is a *constant* if $\delta_i c = 0$, $i = 1, \cdots, m$. The set of all constants is a differential subring or subfield, respectively, of the original differential ring or field, and is called the *ring* or *field of constants* thereof.

Let $\mathcal{R}$ be a differential ring. A *differential ideal* in $\mathcal{R}$ is a subset of $\mathcal{R}$ which is an ideal in $\mathcal{R}$ when $\mathcal{R}$ is considered as a ring, and which is closed with respect to differentiation. A differential ideal is *perfect* if it contains an element of $\mathcal{R}$ whenever it contains some power of the element. A differential ideal is *prime* if, whenever it contains the product of two elements in $\mathcal{R}$, it contains one of those elements. Clearly, a prime differential ideal is perfect.

If Φ is a subset of $\mathcal{R}$, the intersection of all differential ideals containing Φ

is itself a differential ideal, the differential ideal generated by Φ, denoted by $[\Phi]$. If $\mathcal{R}$ contains a unit element then $[\Phi]$ consists of all linear combinations, with coefficients in $\mathcal{R}$, of a finite number of elements of the form $\delta_1^{i_1} \cdots \delta_m^{i_m} a$ ($a \in \Phi$; each i_ν a nonnegative integer). The intersection of all perfect differential ideals containing Φ is itself a perfect differential ideal, the perfect differential ideal generated by Φ, denoted by $\{\Phi\}$. If $\mathcal{R}$ contains all the rational numbers, then $\{\Phi\}$ consists of all elements a in R for which there is a positive integer k such that $a^k \in [\Phi]$.[9]

10. Differential polynomials

Let $\mathcal{F}$ be a differential field of characteristic 0, and introduce a symbol y, called an *unknown* (unknowns play the same rôle in the theory of differential fields as indeterminates do in the theory of fields). A *differential polynomial* in y with coefficients in $\mathcal{F}$, is a polynomial, with coefficients in $\mathcal{F}$, in a finite number of the new symbols $\delta_1^{i_1} \cdots \delta_m^{i_m} y$ ($i_1, \cdots, i_m$ nonnegative integers). When the derivatives of differential polynomials are defined in the obvious way, the set of all differential polynomials in y with coefficients in $\mathcal{F}$ becomes a differential ring, denoted by $\mathcal{F}\{y\}$. Similarly, it is possible to introduce any finite number n of unknowns $y_1, \cdots, y_n$ and to form the differential ring $\mathcal{F}\{y_1, \cdots, y_n\}$ of differential polynomials in $y_1, \cdots, y_n$ with coefficients in $\mathcal{F}$.

THEOREM. *If Σ is a perfect differential ideal in $\mathcal{F}\{y_1, \cdots, y_n\}$, then there exist a finite number of prime differential ideals $\Pi_1, \cdots, \Pi_r$ in $\mathcal{F}\{y_1, \cdots, y_n\}$ such that*

$$\Sigma = \Pi_1 \cap \cdots \cap \Pi_r, \qquad \Pi_i \not\subset \Pi_j \text{ for } i \neq j.$$

The set $\Pi_1, \cdots, \Pi_r$ is unique.[10]

We shall call the prime differential ideals $\Pi_1, \cdots, \Pi_r$ the *prime components of* Σ.

11. Solutions

Let $\mathcal{F}$ be a differential field of characteristic 0, and let $y_1, \cdots, y_n$ be unknowns. If Φ is a set of differential polynomials in $\mathcal{F}\{y_1, \cdots, y_n\}$, a *solution* of Φ is an ordered set of n elements $\eta_1, \cdots, \eta_n$ of a differential extension field of $\mathcal{F}$ such that every member of Φ becomes 0 when each $\delta_1^{i_1} \cdots \delta_m^{i_m} y_j$ is replaced by the corresponding $\delta_1^{i_1} \cdots \delta_m^{i_m} \eta_j$. Clearly the solutions of Φ are identical with those of $\{\Phi\}$.

A solution of Φ is a *generic solution* of Φ if it is a solution of no differential polynomial in $\mathcal{F}\{y_1, \cdots, y_n\}$ other than those in Φ. Φ has a generic solution if and only if Φ is a prime differential ideal $\subset \mathcal{F}\{y_1, \cdots, y_n\}$. If $\eta_1, \cdots, \eta_n$

---

[9] For proof in the ordinary case see RAUDENBUSH [1], p. 363.

[10] This theorem, first proved by Ritt for differential fields of meromorphic functions, is an easy consequence of the Ritt-Raudenbush basis theorem. See RAUDENBUSH [1] where the basis theorem and the present theorem are proved in the ordinary case for abstract differential fields. Since the abstract basis theorem also holds in the partial case (see KOLCHIN [2]), the same is true of the present theorem.

is a generic solution and $\bar{\eta}_1, \cdots, \bar{\eta}_n$ a solution of the same prime differential ideal, then the substitution $\eta_1 \to \bar{\eta}_1, \cdots, \eta_n \to \bar{\eta}_n$ generates a homomorphism[11] of $\mathcal{F}\{\eta_1, \cdots, \eta_n\}$ onto $\mathcal{F}\{\bar{\eta}_1, \cdots, \bar{\eta}_n\}$.

THEOREM. *Let $F \in \mathcal{F}\{y_1, \cdots, y_n\}$, $\Phi \subseteq \mathcal{F}\{y_1, \cdots, y_n\}$. If every solution of Φ is a solution of F, then $\mathcal{F} \in \{\Phi\}$.*[12]

12. Relative isomorphisms

Let $\mathcal{F}$ be a differential field of characteristic 0, $\mathcal{G}$ a differential extension field of $\mathcal{F}$. An *isomorphism of $\mathcal{G}$ with respect to* or *over* $\mathcal{F}$ is an isomorphism[13] of $\mathcal{G}$ onto a differential field $\mathcal{G}'$ such that:

(a) $\mathcal{G}'$ is a differential extension field of $\mathcal{F}$;

(b) $\mathcal{G}$ and $\mathcal{G}'$ have a common differential extension field;

(c) the isomorphism leaves each element of $\mathcal{F}$ invariant. An isomorphism σ of $\mathcal{G}$ over $\mathcal{F}$ can be extended to an isomorphism over $\mathcal{F}$ of any common differential extension field of $\mathcal{G}$ and $\sigma\mathcal{G}$.[14]

THEOREM. *Let $\gamma \in \mathcal{G}$. A necessary and sufficient condition that $\gamma \in \mathcal{F}$ is that every isomorphism of $\mathcal{G}$ over $\mathcal{F}$ leave γ invariant.*[15]

13. Order

Let $\mathcal{F}$ be an ordinary differential field of characteristic 0. If $\eta_1, \cdots, \eta_n$ are elements of a differential extension field, then we may think of $\mathcal{F}$ and $\mathcal{F}<\eta_1, \cdots, \eta_n>$ as fields (not differential fields), and consider the degree of transcendency $\partial^0\mathcal{F}<\eta_1, \cdots, \eta_n>/\mathcal{F}$. We shall call this number the *order of* $\eta_1, \cdots, \eta_n$ over $\mathcal{F}$. When each η_i is a solution of a nonzero differential polynomial in $\mathcal{F}\{y\}$ then the order of $\eta_1, \cdots, \eta_n$ over $\mathcal{F}$ is finite.

The *order* of a prime differential ideal Π in $\mathcal{F}\{y_1, \cdots, y_n\}$ is defined as the order over $\mathcal{F}$ of a generic solution $\eta_1, \cdots, \eta_n$ of Π, and will be denoted by ord Π. If, for each i, Π contains a nonzero differential polynomial in y_i alone, then ord Π is finite.

THEOREM 1. *If Π is a prime differential ideal in $\mathcal{F}\{y_1, \cdots, y_n\}$ of finite order,*

---

[11] That is, a single valued mapping which preserves addition, multiplication, and the m types of differentiation.

[12] This analog of the Hilbert-Netto *Nullstellensatz* is due to Ritt ([1] ch. 7) who proved it for the case in which the elements of $\mathcal{F}$ are meromorphic functions of a complex variable. A simple proof for abstract ordinary differential fields of characteristic 0 appears in RAUDENBUSH [1]. Raudenbush's proof is obviously applicable in the partial case, too.

[13] That is, a one-to-one mapping which preserves addition, multiplication, and the m types of differentiation.

[14] See KOLCHIN [3], where this is incorrectly stated. To correct the error there: on p. 726 15th line from the bottom for "automorphism of the" read "isomorphism of any"; on p. 726 1st line from the bottom for "automorphism" read "isomorphism"; on p. 727 1st line for "automorphism" read "isomorphism."

[15] For proof see KOLCHIN [3].

and if $\eta_1, \cdots, \eta_n$ *is a solution of* Π *with order equal to* ord Π, *then* $\eta_1, \cdots, \eta_n$ *is a generic solution of* Π.[16]

THEOREM 2. *Let* $\mathcal{G}$ *be a differential extension field of* $\mathcal{F}$. *If* Π *is a prime differential ideal in* $\mathcal{F}\{y_1, \cdots, y_n\}$ *and if, in* $\mathcal{G}\{y_1, \cdots, y_n\}$, Π_1 *is a prime component of* $\{\Pi\}$, *then* ord Π_1 = ord Π.[17]

14. Dependence

Let $\mathcal{F}$ be an ordinary[18] differential field, and denote the field of constants of $\mathcal{F}$ by $\mathcal{C}$.

THEOREM 1. *A finite set of elements* $\eta_1, \cdots, \eta_n$ *in* $\mathcal{F}$ *is linearly dependent over* $\mathcal{C}$ *if and only if the wronskian determinant*

$$W(\eta_1, \cdots, \eta_n) = \begin{vmatrix} \eta_1 & \cdots & \eta_n \\ \eta_1' & \cdots & \eta_n' \\ \cdot & \cdots & \cdot \\ \eta_1^{(n-1)} & \cdots & \eta_n^{(n-1)} \end{vmatrix}$$

vanishes.[19]

Since the form of $W(\eta_1, \cdots, \eta_n)$ is independent of $\mathcal{F}$ the vanishing of $W(\eta_1, \cdots, \eta_n)$ is a necessary and sufficient condition for the linear dependence of $\eta_1, \cdots, \eta_n$ over the field of constants of *any* differential field containing $\eta_1, \cdots, \eta_n$, so that we may speak simply of linear dependence (or independence) over constants.

THEOREM 2. *Let* $\mathfrak{m}'$ *be a set of polynomials in* $\mathcal{F}[x_1, \cdots, x_n]$. *Then there is a set* $\mathfrak{m}$ *of polynomials in* $\mathcal{C}[x_1, \cdots, x_n]$ *such that* n *constants in a differential extension field of* $\mathcal{F}$ *form a zero of* $\mathfrak{m}'$ *if and only if they form a zero of* $\mathfrak{m}$. *In particular, if* $\gamma_1, \cdots, \gamma_n$ *are constants in a differential extension field of* $\mathcal{F}$, *then* $\partial^0\mathcal{F}(\gamma_1, \cdots, \gamma_n)/\mathcal{F} = \partial^0\mathcal{C}(\gamma_1, \cdots, \gamma_n)/\mathcal{C}$.

PROOF. By the Hilbert basis theorem $\mathfrak{m}'$ may be replaced by a finite subset $\mathfrak{m}''$. Of the coefficients appearing in the various polynomials of $\mathfrak{m}''$, let $\alpha_1, \cdots, \alpha_h$ be a maximal set that is linearly independent over $\mathcal{C}$ (and therefore over any set of constants). Then every P in $\mathfrak{m}''$ can be written in the form $P = Q_1\alpha_1 + \cdots + Q_h\alpha_a$, where each $Q_i \in \mathcal{C}[x_1, \cdots, x_n]$. By the linear independence

[16] In case $\mathcal{F}$ contains nonconstants this is a special case of a theorem due to Gourin [1]. The case in which $\mathcal{F}$ consists solely of constants can, as indicated by Gourin, be reverted to the former case by the differential field adjunction of an unknown to $\mathcal{F}$. Gourin's proof made use of Ritt's concept of resolvent, which at the time had been developed only for differential fields of meromorphic functions of a complex variable (see RITT [1] ch. 2). Subsequently, Ritt's results have been obtained for abstract differential fields of characteristic 0 (see KOLCHIN [4], or [1] p. 776), so that Gourin's theorem may be regarded as proved in the abstract case.

[17] This is part of a result due to RITT [1], §75. Ritt's proof is function-theoretic. For an abstract proof see KOLCHIN [4].

[18] The results of this section can be extended to partial differential fields, but as this is not necessary for the purposes of the present paper, we avoid the longer discussion this would require.

[19] The proof in the abstract case is straightforward; e.g., see KOLCHIN [1].

of $\alpha_1, \cdots, \alpha_h$ we see that n constants form a zero of P if and only if they do so for $Q_1, \cdots, Q_h$. The theorem now quickly follows.

15. Homogeneous linear ordinary differential equations

Let $\mathcal{F}$ be an ordinary differential field of characteristic 0 with field of constants $\mathcal{C}$, let y be an unknown, n a positive integer, $p_1, \cdots, p_n \in \mathcal{F}$, and consider the homogeneous linear differential polynomial

$$L(y) = y^{(n)} + p_1 y^{(n-1)} + \cdots + p_n y_n.$$

THEOREM. *There exists a differential extension field of $\mathcal{F}$ which contains n solutions of $L(y)$ which are linearly independent over constants.*[20] *No such extension of $\mathcal{F}$ contains more than n solutions of $L(y)$ linearly independent over constants.*

PROOF. Let $y_1, \cdots, y_n$ be new unknowns. Since the order of the wronskian determinant $W(y_1, \cdots, y_n)$ is $< n$ it is easy to see that $W(y_1, \cdots, y_n) \notin \{L(y_1), \cdots, L(y_n)\}$. By the theorem of §11 there therefore exists a solution $\eta_1, \cdots, \eta_n$ of $\{L(y_1), \cdots, L(y_n)\}$ for which $W(\eta_1, \cdots, \eta_n) \neq 0$. By §14 Theorem 1, $\eta_1, \cdots, \eta_n$ are solutions of $L(y)$, linearly independent over constants. Now suppose we have $n + 1$ solutions $\eta_1, \cdots, \eta_{n+1}$ of $L(y)$. Then $L(\eta_i) = 0$ $(i = 1, \cdots, n + 1)$ is a system of $n + 1$ homogeneous linear equations in the $n + 1$ quantities $1, p_1, \cdots, p_n$. Since not all these quantities are 0, the determinant of the system vanishes. But the determinant is the wronskian $W(\eta_1, \cdots, \eta_{n+1})$, so that $\eta_1, \cdots, \eta_{n+1}$ are linearly dependent over constants.

A set of n linearly independent solutions as above is a *fundamental system of solutions of* $L(y)$.

CHAPTER III. NORMAL DIFFERENTIAL EXTENSION FIELDS

16. Normal differential extension fields

Let $\mathcal{F}$ be a differential subfield of a differential field $\mathcal{G}$ of characteristic 0. Denote by $\mathbf{F}$ the set of all differential subfields of $\mathcal{G}$ which contain $\mathcal{F}$.

A set of isomorphisms of $\mathcal{G}$ over $\mathcal{F}$ will be called *abundant* if, for each $\mathcal{F}_1 \in \mathbf{F}$ and each $\alpha \in \mathcal{G} - \mathcal{F}_1$, there is an isomorphism in the set which leaves invariant each element of $\mathcal{F}_1$ but which does not leave α invariant. It follows from Chapter II §12 that an abundant set of such isomorphisms always exists.

$\mathcal{G}$ will be called a *normal* extension of $\mathcal{F}$ if the set of all *auto*morphisms of $\mathcal{G}$ over $\mathcal{F}$ is abundant. E.g., if $\mathcal{G}$ is a finite algebraic extension of $\mathcal{F}$ and if $\mathcal{G}$ is normal in the sense of Galois theory, then $\mathcal{G}$ is normal in the sense just defined. In general, if $\mathcal{G}$ is a normal extension of $\mathcal{F}$ then obviously $\mathcal{G}$ is a normal extension of any $\mathcal{F}_1 \in \mathbf{F}$.

We define the product $\sigma\tau$ of two automorphisms σ, τ as the automorphism obtained by first operating with τ and then with σ. Then the set of all auto-

[20] As remarked by BAER [1], it would be of interest to know whether there exist n such solutions $\eta_1, \ldots, \eta_n$ such that $\mathcal{F}\langle\eta_1, \ldots, \eta_n\rangle$ contains no constant transcendental over $\mathcal{C}$.

morphisms of $\mathcal{G}$ over $\mathcal{F}$ is a group. In the above mentioned case in which $\mathcal{G}$ is a normal finite algebraic extension of $\mathcal{F}$, this group coincides with the Galois group of $\mathcal{G}$ over $\mathcal{F}$.

Let $\mathcal{G}$ be a normal extension of $\mathcal{F}$ and let $\mathfrak{G}$ be an abundant group of automorphisms of $\mathcal{G}$ over $\mathcal{F}$ (not necessarily the group of all such automorphisms). For each $\mathcal{F}_1 \in \mathbf{F}$ let $\mathfrak{G}(\mathcal{F}_1)$ be the subgroup of $\mathfrak{G}$ consisting of all automorphisms which leave invariant every element of $\mathcal{F}_1$. In particular, $\mathfrak{G}(\mathcal{F}) = \mathfrak{G}$, $\mathfrak{G}(\mathcal{G}) = \mathfrak{E}$ (the group consisting solely of the identity automorphism ι).

Let $\mathbf{G}$ denote the set of all subgroups of $\mathfrak{G}$ of the form $\mathfrak{G}(\mathcal{F}_1)$, where $\mathcal{F}_1 \in \mathbf{F}$.

THEOREM 1. *For any $\mathcal{F}_1$ in $\mathbf{F}$ the set of all elements in $\mathcal{G}$ invariant under every automorphism in $\mathfrak{G}(\mathcal{F}_1)$ is $\mathcal{F}_1$. Therefore the correspondence $\mathcal{F}_1 \to \mathfrak{G}(\mathcal{F}_1)$ is a one-to-one mapping of $\mathbf{F}$ onto $\mathbf{G}$.*

PROOF. $\mathcal{F}_1$ is clearly contained in the set mentioned in the theorem. To show that this set has no other elements consider any $\alpha \in \mathcal{G} - \mathcal{F}_1$. Since $\mathcal{G}$ is a normal extension of $\mathcal{F}$, there exists a $\sigma \in \mathfrak{G}$ which leaves invariant each element of $\mathcal{F}_1$ (so that $\sigma \in \mathfrak{G}(\mathcal{F}_1)$, but which does not leave α invariant. Therefore α is not contained in the set in question, and the theorem follows.

It is easy to see that for any set of differential fields $\mathcal{F}_\lambda$ in $\mathbf{F}$ we have

$$\mathfrak{G}(\mathcal{F}\langle \bigcup \mathcal{F}_\lambda \rangle) = \bigcap \mathfrak{G}(\mathcal{F}_\lambda),$$

$$\mathfrak{G}(\bigcap \mathcal{F}_\lambda) = \text{smallest group in } \mathbf{G} \text{ containing } \bigcup \mathfrak{G}(\mathcal{F}_\lambda).$$

THEOREM 2. *Let $\mathcal{F}_1 \in \mathbf{F}$. A necessary and sufficient condition that $\mathfrak{G}(\mathcal{F}_1)$ be a normal subgroup of $\mathfrak{G}$ is that $\sigma\mathcal{F}_1 = \mathcal{F}_1$ for every σ in $\mathfrak{G}$. When this condition is fulfilled the factor group $\mathfrak{G}/\mathfrak{G}(\mathcal{F}_1)$ is isomorphic with an abundant group of automorphisms of $\mathcal{F}_1$ over $\mathcal{F}$ (in particular, $\mathcal{F}_1$ is a normal extension of $\mathcal{F}$).*

PROOF. If $\sigma\mathcal{F}_1 = \mathcal{F}_1$ for every $\sigma \in \mathfrak{G}$ then, for each $\tau \in \mathfrak{G}(\mathcal{F}_1)$, $\sigma^{-1}\tau\sigma$ leaves each element of $\mathcal{F}_1$ invariant; that is, $\sigma^{-1}\tau\sigma \in \mathfrak{G}(\mathcal{F}_1)$, and $\mathfrak{G}(\mathcal{F}_1)$ is a normal subgroup of $\mathfrak{G}$.

Conversely, let $\mathfrak{G}(\mathcal{F}_1)$ be a normal subgroup of $\mathfrak{G}$. Then, for all $\sigma \in \mathfrak{G}$ and $\tau \in \mathfrak{G}(\mathcal{F}_1)$ we have $\sigma^{-1}\tau\sigma \in \mathfrak{G}(\mathcal{F}_1)$, so that $\tau\sigma\alpha = \sigma\alpha$ for all $\alpha \in \mathcal{F}_1$. Thus $\sigma\alpha$ is invariant under every $\tau \in \mathfrak{G}(\mathcal{F}_1)$ and is, by Theorem 1, in $\mathcal{F}_1$. It follows that $\sigma\mathcal{F}_1 = \mathcal{F}_1$. This implies that every $\sigma \in \mathfrak{G}$ induces an automorphism of $\mathcal{F}_1$ over $\mathcal{F}$ (the contraction of σ to the domain $\mathcal{F}_1$). Two elements σ, σ_1 of $\mathfrak{G}$ induce the same automorphism of $\mathcal{F}_1$ if and only if $\sigma^{-1}\sigma_1 \in \mathfrak{G}(\mathcal{F}_1)$, that is, if and only if they are in the same coset of $\mathfrak{G}(\mathcal{F}_1)$ in $\mathfrak{G}$. For any differential field $\mathcal{F}_2$ between $\mathcal{F}$ and $\mathcal{F}_1$, and any $\alpha \in \mathcal{F}_1 - \mathcal{F}_2$, there is (since $\mathcal{G}$ is a normal extension of $\mathcal{F}$) a σ in $\mathfrak{G}$ which leaves invariant every element of $\mathcal{F}_2$ but which does not leave α invariant. The automorphism of $\mathcal{F}_1$ induced by σ has the same properties. Therefore the set of all automorphisms of $\mathcal{F}_1$ over $\mathcal{F}$ induced by elements of $\mathfrak{G}$ is an abundant group, isomorphic with $\mathfrak{G}/\mathfrak{G}(\mathcal{F}_1)$.

REMARK 1. Theorem 2 does not assert that the normality of $\mathcal{F}_1$ over $\mathcal{F}$ implies the normality of $\mathfrak{G}(\mathcal{F}_1)$ in $\mathfrak{G}$. Nor does it assert, in the case in which $\mathfrak{G}$ is the group of *all* automorphisms of $\mathcal{G}$ over $\mathcal{F}$, that $\mathfrak{G}/\mathfrak{G}(\mathcal{F}_1)$ is isomorphic with the group of all automorphisms of $\mathcal{F}_1$ over $\mathcal{F}$, but merely with an abundant subgroup thereof.[21]

REMARK 2. If $\mathcal{F}_1$ is a normal *algebraic* extension of F of *finite* degree, then $\mathcal{F}_1$ contains with any element α all the roots of the defining equation of α over $\mathcal{F}$, so that $\sigma\alpha \in \mathcal{F}_1$ for all α in $\mathcal{F}_1$ and every σ in $\mathfrak{G}$. Therefore $\sigma\mathcal{F}_1 = \mathcal{F}_1$ and $\mathfrak{G}(\mathcal{F}_1)$ is a normal subgroup of $\mathfrak{G}$. Furthermore, $\mathfrak{G}/\mathfrak{G}(\mathcal{F}_1)$ is isomorphic with the group of all automorphisms of $\mathcal{F}_1$ over $\mathcal{F}$, that is, with the Galois group of $\mathcal{F}_1$ over $\mathcal{F}$; for a Galois group can contain no abundant proper subgroup.

REMARK 3. A non-trivial characterization of **G** remains an open problem. If $\mathcal{G}$ is a normal infinite algebraic extension then the group $\mathfrak{G}$ of all automorphisms of $\mathcal{G}$ over $\mathcal{F}$ is the Galois group of $\mathcal{G}$ over $\mathcal{F}$ (for the Galois theory of infinite algebraic extensions see Krull [1]), and **G** is the set of all subgroups of $\mathfrak{G}$ which are closed in a certain natural topology (see Krull [1]). A subgroup of $\mathfrak{G}$ is abundant it if is everywhere dense in $\mathfrak{G}$ in the abovementioned topology. When $\mathfrak{G}$ is not the whole group of automorphisms, but merely an abundant group, then **G** is the set of all subgroups of $\mathfrak{G}$ closed in $\mathfrak{G}$.

CHAPTER IV. PICARD-VESSIOT EXTENSIONS

17. Picard-Vessiot extensions and their isomorphisms

Throughout this chapter $\mathcal{F}$ is an ordinary differential field of characteristic 0. The field of constants of $\mathcal{F}$, denoted by $\mathcal{C}$, is assumed to be algebraically closed. The letters $y, y_1, \cdots, y_n$ are used for unknowns.

A differential extension field $\mathcal{G}$ of $\mathcal{F}$ will be called a *Picard-Vessiot extension* of $\mathcal{F}$ if:

(a) there exists a homogeneous linear ordinary differential polynomial

$$L(y) = y^{(n)} + p_1 y^{(n-1)} + \cdots + p_n y \quad (n \geqq 1, \text{each } p_i \in \mathcal{F})$$

with a fundamental system of solutions $\eta_1, \cdots, \eta_n$ such that

$$\mathcal{F} < \eta_1, \cdots, \eta_n > = \mathcal{G};$$

(b) the field of constants of $\mathcal{G}$ is $\mathcal{C}$.

We consider a Picard-Vessiot extension $\mathcal{G} = \mathcal{F}<\eta_1, \cdots, \eta_n>$ as above, and investigate the nature of the isomorphisms of $\mathcal{G}$ over $\mathcal{F}$. To do this, let Γ be the prime differential ideal in $\mathcal{F}\{y_1, \cdots, y_n\}$ consising of all differential polynomials which vanish for $y_1 = \eta_1, \cdots, y_n = \eta_n$ (so that $\eta_1, \cdots, \eta_n$ is a generic solution of Γ).

If κ_{ij} $(i, j = 1, \cdots, n)$ are n^2 indeterminate constants then the substitution

$$y_j \to \sum_{i=1}^{n} \kappa_{ij}\eta_i \qquad (j = 1, \cdots, n)$$

[21] Whether such assertions are possible is an open question. In the following chapter it will be shown that, for a special type of differential field extension, the second assertion may be made.

defines a mapping of $\mathcal{G}\{y_1, \cdots, y_n\}$ into $\mathcal{G}[\cdots, \kappa_{ij}, \cdots]$. Let $\mathfrak{g}'$ be the image of Γ under this mapping. By Chapter II §14 Theorem 2 there is an ideal $\mathfrak{g}$ in $\mathcal{C}[\cdots, \kappa_{ij}, \cdots]$ such that n^2 constants k_{ij} $(i, j = 1, \cdots, n)$ in an extension of $\mathcal{G}$ form a zero of $\mathfrak{g}'$ if and only if they do so for $\mathfrak{g}$. We take $\mathfrak{g}$ as large as possible, so that any polynomial some power of which is in $\mathfrak{g}$ is itself in $\mathfrak{g}$.

If σ is any isomorphism of $\mathcal{G}$ over $\mathcal{F}$ then, for $j = 1, \cdots, n, \sigma\eta_j, \eta_1, \cdots, \eta_n$ are $n + 1$ solutions of $L(y)$, so that (Chapter II §15) there are n^2 constants k_{ij} in $\mathcal{F}<\mathcal{G}, \sigma\mathcal{G}>$ such that

$$(8) \qquad \sigma\eta_j = \sum_{i=1}^{n} k_{ij}\eta_i, \qquad j = 1, \cdots, n.$$

Since $\sigma\eta_1, \cdots, \sigma\eta_n$ is a solution of Γ, the k_{ij}'s must form a zero of $\mathfrak{g}$. Since $\sigma\eta_1, \cdots, \sigma\eta_n$ (like $\eta_1, \cdots, \eta_n$) are linearly independent over constants, the determinant $| k_{ij} | \neq 0$.

Conversely, let k_{ij} $(i, j = 1, \cdots, n)$ be constants in an extension of $\mathcal{G}$ which form a zero of $\mathfrak{g}$ for which $| k_{ij} | \neq 0$. The equations (8) define a mapping σ of $\mathcal{F}\{\eta_1, \cdots, \eta_n\}$ for which $\sigma\eta_1, \cdots, \sigma\eta_n$ is a solution of Γ. Therefore σ is a homomorphism of $\mathcal{F}\{\eta_1, \cdots, \eta_n\}$ onto $\mathcal{F}\{\sigma\eta_1, \cdots, \sigma\eta_n\}$. Now

$$\partial^0\mathcal{F}<\eta_1, \cdots, \eta_n, \sigma\eta_1, \cdots, \sigma\eta_n>/\mathcal{F} = \partial^0\mathcal{F}<\eta_1, \cdots, \eta_n>/\mathcal{F} + \partial^0\mathcal{F}<\eta_1, \cdots, \eta_n, \sigma\eta_1, \cdots, \sigma\eta_n>/\mathcal{F}<\eta_1, \cdots, \eta_n>$$

$$\partial^0\mathcal{F}<\eta_1, \cdots, \eta_n, \sigma\eta_1, \cdots, \sigma\eta_n>/\mathcal{F} = \partial^0\mathcal{F}<\sigma\eta_1, \cdots, \sigma\eta_n>/\mathcal{F} + \partial^0\mathcal{F}<\eta_1, \cdots, \eta_n, \sigma\eta_1, \cdots, \sigma\eta_n>/\mathcal{F}<\sigma\eta_1, \cdots, \sigma\eta_n>$$

so that

$$\partial^0\mathcal{F}<\eta_1, \cdots, \eta_n>/\mathcal{F} - \partial^0\mathcal{F}<\sigma\eta_1, \cdots, \sigma\eta_n>\mathcal{F} = \partial^0\mathcal{F}<\eta_1, \cdots, \eta_n, \sigma\eta_1, \cdots, \sigma\eta_n>/\mathcal{F}<\sigma\eta_1, \cdots, \sigma\eta_n> - \partial^0\mathcal{F}<\eta_1, \cdots, \eta_n, \sigma\eta_1, \cdots, \sigma\eta_n>/\mathcal{F}<\eta_1, \cdots, \eta_n>.$$

But clearly

$$\mathcal{F}<\eta_1, \cdots, \eta_n, \sigma\eta_1, \cdots, \sigma\eta_n> = \mathcal{F}<\eta_1, \cdots, \eta_n, k_{11}, k_{12}, \cdots, k_{nn}> = \mathcal{F}<\sigma\eta_1, \cdots, \sigma\eta_n, k_{11}, k_{12}, \cdots, k_{nn}>.$$

Therefore, if we let $\mathcal{C}'$ be the field of constants of $\mathcal{F}<\sigma\eta_1, \cdots, \sigma\eta_n>$ (so that $\mathcal{C}' \supseteq \mathcal{C}$), we find (using Chapter II §15 Theorem 2):

$$\partial^0\mathcal{F}<\eta_1, \cdots, \eta_n>/\mathcal{F} - \partial^0\mathcal{F}<\sigma\eta_1, \cdots, \sigma\eta_n>/\mathcal{F} = \partial^0\mathcal{C}'(\cdots, k_{ij}, \cdots)/\mathcal{C}' - \partial^0\mathcal{C}(\cdots, k_{ij}, \cdots)/\mathcal{C} \leqq 0.$$

But σ is a homomorphism, so that we obviously have

$$\partial^0\mathcal{F}<\eta_1, \cdots, \eta_n>/\mathcal{F} \geqq \partial^0\mathcal{F}<\sigma\eta_1, \cdots, \sigma\eta_n>/\mathcal{F}.$$

Therefore

$$\partial^0\mathcal{F}<\eta_1, \cdots, \eta_n>/\mathcal{F} = \partial^0\mathcal{F}<\sigma\eta_1, \cdots, \sigma\eta_n>/\mathcal{F}.$$

It follows, by Gourin's theorem (Chapter II §13 Theorem 1), that $\sigma\eta_1, \cdots, \sigma\eta_n$ is a generic solution of Γ, so that the homomorphism σ is actually an isomorphism of $\mathcal{F}\{\eta_1, \cdots, \eta_n\}$, and can clearly be extended in a unique way to an isomorphism of $\mathcal{G}$ over $\mathcal{F}$. Thus we have proved:

The equations (8) define a one-to-one correspondence between isomorphisms σ of $\mathcal{G}$ over $\mathcal{F}$ and nonsingular matrices (k_{ij}) of degree n in which the elements k_{ij} are constants contained in some differential extention field of $\mathcal{G}$ which form a solution of $\mathfrak{g}$.

Since $\mathcal{F}<\mathcal{G}, \sigma\mathcal{G}> = \mathcal{G}<\cdots, k_{ij}, \cdots>$, it is easy to see that the isomorphism σ is an automorphism if and only if $k_{ij} \in \mathcal{C}(i, j = 1, \cdots, n)$. It is also easy to see that the product of two automorphisms of $\mathcal{G}$ over $\mathcal{F}$ corresponds through (8) to the product of the respective matrices of the automorphisms. Therefore we have proved:

THEOREM. *The equations* (8) *establish an isomorphism between the group of all automorphisms of $\mathcal{G}$ over $\mathcal{F}$ and an algebraic matric group over $\mathcal{C}$ of degree n.*

We shall, when danger of confusion does not arise, use the same letter to denote both the automorphism and its matrix, and shall regard the group of automorphisms as an algebraic matric group. We denote this group by $\mathfrak{G}$. The defining ideal of $\mathfrak{G}$ is $\mathfrak{g}$.

EXAMPLE 1. The *general* homogeneous linear ordinary differential equation of order n is $y^{(n)} + u_1 y^{(n-1)} + \cdots + u_n y = 0$, where $u_1, \cdots, u_n$ are new unknowns. If $\mathcal{F}_0$ is any ordinary differential field of characteristic 0 with respect to which $u_1, \cdots, u_n$ are still unknowns, if $\mathcal{C}$ is the field of constants of $\mathcal{F}_0$ (and therefore of $\mathcal{F}_0<u_1, \cdots, u_n>$), and if $\eta_1, \cdots, \eta_n$ are any n solutions of the general equation, linearly independent over constants, contained in some extension of $\mathcal{F}_0<u_1, \cdots, u_n>$, then $\eta_1, \cdots, \eta_n$ annul no nonzero differential polynomial in $\mathcal{F}_0\{y_1, \cdots, y_n\}$. $\mathcal{G} = \mathcal{F}_0<u_1, \cdots, u_n, \eta_1, \cdots, \eta_n>$ is a Picard-Vessiot extension of $\mathcal{F} = \mathcal{F}_0<u_1, \cdots, u_r>$ for which the group of automorphisms is the full matric group over $\mathcal{C}$ of degree n.

EXAMPLE 2. If the coefficients $p_1, \cdots, p_n$ in $L(y)$ are all constants then it is easy to see that $\mathfrak{G}$ is reducible to triangular form. If $\mathcal{F}$ contains an element x such that $x' = 1$ then $\mathfrak{G}$ is reducible to diagonal form.

18. Normality

Let $\alpha = P(\eta_1, \cdots, \eta_n)/Q(\eta_1, \cdots, \eta_n)$ and $\beta = R(\eta_1, \cdots, \eta_n)/S(\eta_1, \cdots, \eta_n)$ be any two elements of $\mathcal{G}$. An isomorphism τ of $\mathcal{G}$ over $\mathcal{F}$ will satisfy $\tau\alpha = \beta$ if and only if the elements of the matrix of τ satisfy the equation

$$S(\eta_1, \cdots, \eta_n)P(\textstyle\sum \kappa_{i1}\eta_i, \cdots, \sum \kappa_{in}\eta_i)$$
$$= R(\eta_1, \cdots, \eta_n)Q(\textstyle\sum \kappa_{i1}\eta_i, \cdots, \sum \kappa_{in}\eta_i)$$

in the indeterminates κ_{ij}. By Chapter II §14 Theorem 2 we see that the conditon $\tau\alpha = \beta$ can be expressed algebraically over $\mathcal{C}$.

Now, since $\mathcal{C}$ is algebraically closed, a set of polynomials over $\mathcal{C}$ which has a zero, has a zero with coordinates in $\mathcal{C}$. Therefore, if there exists an isomorphism of $\mathcal{G}$ over $\mathcal{F}$ the matrix of which has elements satisfying a given set of algebraic conditions over $\mathcal{C}$, then there exists an automorphism of $\mathcal{G}$ over $\mathcal{F}$ (that is, an element of $\mathfrak{G}$) satisfying the same conditions.

In particular, if $\mathcal{F}_1$ is any differential field such that $\mathcal{F} \subseteq \mathcal{F}_1 \subseteq \mathcal{G}$, if $\alpha \in \mathcal{G} - \mathcal{F}_1$, and if there exists an isomorphism σ' of $\mathcal{G}$ over $\mathcal{F}$ such that $\sigma'\alpha_1 = \alpha_1$ for all $\alpha_1 \in \mathcal{F}_1$ and $\sigma'\alpha \neq \alpha$, then there exists a $\sigma \in \mathfrak{G}$ with the same properties. But by Chapter II §12 such an isomorphism σ' always exists. Thus, we have the following result.

THEOREM 1. *Every Picard-Vessiot extension of $\mathcal{F}$ is a normal extension of $\mathcal{F}$.*

In conformity with Chapter III we shall denote the set of all differential fields $\mathcal{F}_1$ such that $\mathcal{F} \subseteq \mathcal{F}_1 \subseteq \mathcal{G}$ by $\mathbf{F}$, the subgroup of $\mathfrak{G}$ consisting of all automorphisms of $\mathcal{G}$ over $\mathcal{F}_1$ ($\mathcal{F}_1 \in \mathbf{F}$) by $\mathfrak{G}(\mathcal{F}_1)$, the set of all groups $\mathfrak{G}(\mathcal{F}_1)$ with $\mathcal{F}_1 \in \mathbf{F}$ by $\mathbf{G}$.

Since $\mathcal{G}$ is a normal extension of $\mathcal{F}$ we have at our disposal the theorems of Chapter III. In the special case of Picard-Vessiot extensions, however, Theorem 2 of Chapter III can be strengthened (see Remarks 1 and 2 at the end of that chapter) by the following result.

THEOREM 2. *Let $\mathcal{F}_1 \in \mathbf{F}$. If $\mathfrak{G}(\mathcal{F}_1)$ is a normal subgroup of $\mathfrak{G}$ then $\mathfrak{G}/\mathfrak{G}(\mathcal{F}_1)$ is isomorphic with the group of all automorphisms of $\mathcal{F}_1$ over $\mathcal{F}$.*

PROOF. We know from Chapter III that $\sigma\mathcal{F}_1 = \mathcal{F}_1$, so that σ induces an automorphism σ_0 of $\mathcal{F}_1$ over $\mathcal{F}$, and the set of automorphisms σ_0 so induced is a group isomorphic with $\mathfrak{G}/\mathfrak{G}(\mathcal{F}_1)$. It remains to show that every automorphism σ_0 of $\mathcal{F}_1$ over $\mathcal{F}$ can be so induced, that is, that every such σ_0 can be extended to an automorphism of $\mathcal{G}$ over $\mathcal{F}$. That this can be done is shown by the following more general result.

LEMMA. *Let $\mathcal{F}_1, \mathcal{F}_2 \in \mathbf{F}$. If σ_0 is an isomorphism over $\mathcal{F}$ of $\mathcal{F}_1$ onto $\mathcal{F}_2$ then σ_0 can be extended to an automorphism $\sigma \in \mathfrak{G}$.*

PROOF. By Chapter II, §12, σ_0 can be extended to an isomorphism σ' of $\mathcal{G}$ over $\mathcal{F}$. By the first two paragraphs of the present section there is an automorphism $\sigma \in \mathfrak{G}$ coinciding with σ' for all α in $\mathcal{G}$ for which $\sigma'\alpha \in \mathcal{G}$.

19. Characterization of G

For any $\mathcal{F}_1 \in \mathbf{F}$ we have $\mathcal{F}_1\langle\eta_1, \cdots, \eta_n\rangle = \mathcal{G}$, so that $\mathcal{G}$ is a Picard-Vessiot extension of F_1. Therefore $\mathfrak{G}(\mathcal{F}_1)$ is an algebraic group, that is, every element of $\mathbf{G}$ is an algebraic subgroup of $\mathfrak{G}$. We shall now prove the converse thereby obtaining the following result.

THEOREM. *$\mathbf{G}$ is the set of all algebraic subgroups of $\mathfrak{G}$.*

PROOF. It remains to show that if $\mathfrak{G}_1$ is an algebraic subgroup of $\mathfrak{G}$ then $\mathfrak{G}_1 \in \mathbf{G}$. Let $\mathcal{F}_1$ be the set of all elements of $\mathcal{G}$ invariant under every $\tau \in \mathfrak{G}_1$. Clearly $\mathfrak{G}_1 \subseteq \mathfrak{G}(\mathcal{F}_1)$. If $\mathfrak{G}_1 = \mathfrak{G}(\mathcal{F}_1)$ then $\mathfrak{G}_1 \in \mathbf{G}$. Suppose $\mathfrak{G}_1 \subset \mathfrak{G}(\mathcal{F}_1)$.

There is a polynomial $f(\cdots, \kappa_{ij}, \cdots) \in \mathcal{C}[\cdots, \kappa_{ij}, \cdots]$ which vanishes for every $\sigma = (k_{ij}) \in \mathfrak{G}_1$ but not for every $\sigma = (k_{ij}) \in \mathfrak{G}(\mathcal{F}_1)$. If we denote the inverse of the wronskian matrix $(\eta_j^{(i)})$ by (H_{ij}), then

$$f\left(\cdots, \sum_{l=0}^{n-1} H_{li} y_j^{(l)}, \cdots\right)$$

is a differential polynomial in $\mathcal{G}\{y_1, \cdots, y_n\}$ which has the solution

$$y_1 = \sigma\eta_1, \cdots, y_n = \sigma\eta_n \tag{9}$$

for all $\sigma \in \mathfrak{G}_1$ but not for all $\sigma \in \mathfrak{G}(\mathcal{F}_1)$. Of all differential polynomials in $\mathcal{G}\{y_1, \cdots, y_n\}$ which have this property let $F = F(y_1, \cdots, y_n)$ be one with fewest possible terms. We assume without loss of generality that one of the coefficients in F is 1.

For any $\tau \in \mathfrak{G}_1$ denote by $F_\tau = F_\tau(y_1, \cdots, y_n)$ the differential polynomial obtained by replacing each coefficient in F by its image under τ. Then

$$F_\tau(\sigma\eta_1, \cdots, \sigma\eta_n) = \tau F(\tau^{-1}\sigma\eta_1, \cdots, \tau^{-1}\sigma\eta_n) = 0$$

whenever $\sigma \in \mathfrak{G}_1$, so that (9) is a solution of $F - F_\tau$ whenever $\sigma \in \mathfrak{G}_1$. Since $F - F_\tau$ has fewer terms than F has, $F - F_\tau$ must admit the solution (9) for all $\sigma \in \mathfrak{G}(\mathcal{F}_1)$. Therefore, for any $\gamma \in \mathcal{G}$, $F - \gamma(F - F_\tau)$ has (9) as a solution for every $\sigma \in \mathfrak{G}_1$ but not for every $\sigma \in \mathfrak{G}(\mathcal{F}_1)$. If $F - F_\tau$ were not 0 there would be a $\gamma \in \mathcal{G}$ for which $F - \gamma(F - F_\tau)$ had fewer terms than F, which would contradict the definition of F. Hence $F - F_\tau = 0$, that is, each coefficient in F is invariant under every $\tau \in \mathfrak{G}_1$ and is therefore in $\mathcal{F}_1$. It follows that $F(\sigma\eta_1, \cdots, \sigma\eta_n) = \sigma F(\eta_1, \cdots, \eta_n) = 0$ for all $\sigma \in \mathfrak{G}(\mathcal{F}_1)$. This contradiction completes the proof.

20. Dimension

For each j, η_j is a solution of the differential polynomial $L(y)$ so that every $\eta_j^{(i)}$ with $i \geqq n$ can be expressed linearly (and therefore rationally) in terms of $\eta_j, \eta_j', \cdots, \eta_j^{(n-1)}$. It follows that the degree of transcendency of $\mathcal{G}$ over $\mathcal{F}$ is at most n^2: $\partial^0\mathcal{G}/\mathcal{F} \leqq n^2$.

Again, $\mathfrak{G}$ is contained in the full matric group, which is an algebraic matric group of dimension n^2. Therefore the dimension of $\mathfrak{G}$ is at most n^2: $\dim \mathfrak{G} \leqq n^2$.

THEOREM. $\dim \mathfrak{G} = \partial^0\mathcal{G}/\mathcal{F}$.

As in §17, let Γ be the prime differential ideal in $\mathcal{F}\{y_1, \cdots, y_n\}$ consisting of all differential polynomials vanishing for $y_1 = \eta_1, \cdots, y_n = \eta_n$. Let Γ' be the perfect differential ideal generated by Γ in $\mathcal{G}\{y_1, \cdots, y_n\}$, and let

$$\Gamma' = \Gamma_1 \cap \cdots \cap \Gamma_s$$

be the representation of Γ' as the intersection of its prime components Γ_i in $\mathcal{G}\{y_1, \cdots, y_n\}$ (see Chapter II §10).

The substitution $y_j \to \sum_i \kappa_{ij}\eta_i$ $(j = 1, \cdots, n)$ defines a mapping which sends every element of $\mathcal{G}\{y_1, \cdots, y_n\}$ into an element of $\mathcal{G}[\cdots, \kappa_{ij}, \cdots]$. By considering the matrix (H_{ij}) inverse to $(\eta_j^{(i)})$ we see as in §4 that this is a mapping of $\mathcal{G}\{y_1, \cdots, y_n\}$ *onto* $\mathcal{G}[\cdots, \kappa_{ij}, \cdots]$. By considering generic solutions of the Γ_i's, it is not difficult to see that if an element of $\mathcal{G}\{y_1, \cdots, y_n\}$ is mapped thus onto 0, then the element is contained in Γ'.

For each h let $\mathfrak{g}_h'$ be the image under this mapping of Γ_h. Since Γ_h is a prime ideal, and every element of $\mathcal{G}[\cdots, \kappa_{ij}, \cdots]$ is the image of some element of $\mathcal{G}\{y_1, \cdots, y_n\}$, it is easy to see that $\mathfrak{g}_h'$ is a prime ideal in $\mathcal{G}[\cdots, \kappa_{ij}, \cdots]$.

Let $\overset{*}{\eta}_1, \cdots, \overset{*}{\eta}_n$ be a generic solution of Γ_h. Then each $\overset{*}{\eta}_j$ is a solution of $L(y)$, so that

$$\overset{*}{\eta}_j = \sum_{i=1}^{n} \overset{*}{k}_{ij}\eta_i \qquad (j = 1, \cdots, n)$$

where the $\overset{*}{k}_{ij}$'s are constants in some extension of $\mathcal{G}$. Clearly every element of $\mathfrak{g}_h'$ vanishes for $(\kappa_{ij}) = (\overset{*}{k}_{ij})$. Conversely, if $f(\cdots, \kappa_{ij}, \cdots) \in \mathcal{G}[\cdots, \kappa_{ij}, \cdots]$ vanishes for $(\kappa_{ij}) = (\overset{*}{k}_{ij})$ then $f(\cdots, \sum_l H_{li}y_j^{(l)}, \cdots) \in \mathcal{G}\{y_1, \cdots, y_n\}$ vanishes for $y_1 = \overset{*}{\eta}_1, \cdots, y_n = \overset{*}{\eta}_n$, so that $f(\cdots, \sum_l H_{li}y_j^{(l)}, \cdots) \in \Gamma_h$, and $f(\cdots, \kappa_{ij}, \cdots) \in \mathfrak{g}_h'$. Thus $(\overset{*}{k}_{ij})$ is a generic zero of $\mathfrak{g}_h'$.

By Chapter I §14 Theorem 2, we see that $\mathfrak{g}_h'$ is equivalent to a prime ideal $\mathfrak{g}_h$ in $\mathcal{C}[\cdots, \kappa_{ij}, \cdots]$ of which $(\overset{*}{k}_{ij})$ is again a generic zero. Now, $(\kappa_{ij}) = (k_{ij})$ will be a solution of $\mathfrak{g}_1 \cap \cdots \cap \mathfrak{g}_s$ if and only if $y_1 = \sum k_{i1}\eta_i, \cdots, y_n = \sum k_{in}\eta_i$ is a solution of Γ, that is, if and only if $(\kappa_{ij}) = (k_{ij})$ is a solution of $\mathfrak{g}$, the defining ideal of $\mathfrak{G}$. Therefore $\mathfrak{g} = \mathfrak{g}_1 \cap \cdots \cap \mathfrak{g}_s$.

For each h we have

$$\dim \mathfrak{g}_h = \partial^0\mathcal{C}(\cdots, \overset{*}{k}_{ij}, \cdots)/\mathcal{C} = \partial^0\mathcal{G}(\cdots, \overset{*}{k}_{ij}, \cdots)/\mathcal{G} = \partial^0\mathcal{G}\langle\overset{*}{\eta}_1, \cdots, \overset{*}{\eta}_n\rangle/\mathcal{G} = \text{ord } \Gamma_h .$$

But by Ritt's theorem (Chapter II §13 Theorem 2), ord Γ_h = ord Γ, and, since $\eta_1, \cdots, \eta_n$ is a generic solution of Γ, ord $\Gamma = \partial^0\mathcal{F}\langle\eta_1, \cdots, \eta_n\rangle/\mathcal{F} \leqq \partial^0\mathcal{G}/\mathcal{F}$. Therefore $\dim \mathfrak{g}_h = \partial^0\mathcal{G}/\mathcal{F}$. Since this holds for $h = 1, \cdots, s$ this proves anew that the underlying manifold of $\mathfrak{G}$ breaks up into irreducible manifolds of equal dimension (Chapter I §2), and also shows that $\dim \mathfrak{G} = \partial^0\mathcal{G}/\mathcal{F}$.

21. Adjunction of new elements

Let $\mathfrak{M}$ be a set of elements in some differential extension field of $\mathcal{G}$ in which every constant is an element of $\mathcal{C}$, and write $\mathcal{F}^\ddagger = \mathcal{F}\langle\mathfrak{M}\rangle$, $\mathcal{G}^\ddagger = \mathcal{G}\langle\mathfrak{M}\rangle$. Then $\mathcal{G}^\ddagger = \mathcal{F}^\ddagger\langle\eta_1, \cdots, \eta_n\rangle$, so that $\mathcal{G}^\ddagger$ is a Picard-Vessiot extension of $\mathcal{F}^\ddagger$. Denote the group of all automorphisms of $\mathcal{G}^\ddagger$ over $\mathcal{F}^\ddagger$ by $\mathfrak{G}^\ddagger$.

THEOREM. *$\mathfrak{G}^\ddagger$ is isomorphic with $\mathfrak{G}(\mathcal{F}^\ddagger \cap \mathcal{G})$.*

PROOF. If $\sigma^\ddagger \in \mathfrak{G}^\ddagger$ then $\sigma^\ddagger$ obviously leaves invariant each element of $\mathcal{F}$. Also, if we denote the matrix of $\sigma^\ddagger$ by $(k_{ij}^\ddagger)$, then $\sigma^\ddagger\eta_j = \sum_i k_{ij}^\ddagger\eta_i$, and each $k_{ij}^\ddagger \in \mathcal{C}$, so that $\sigma^\ddagger\mathcal{G} = \mathcal{G}$. Therefore $\sigma^\ddagger$ induces an automorphism σ of $\mathcal{G}$ over $\mathcal{F}$: $\sigma\alpha = \sigma^\ddagger\alpha(\alpha \in \mathcal{G})$. The matrix of σ is identical with that of $\sigma^\ddagger$. Therefore $\mathfrak{G}^\ddagger$ is isomorphic with a subgroup $\mathfrak{G}(\mathcal{F}_1)$ of $\mathfrak{G}$, where $\mathcal{F}_1 \in \mathbf{F}$. Now, every element of $\mathcal{F}^\ddagger \cap \mathcal{G}$ is invariant under every $\sigma^\ddagger \in \mathfrak{G}^\ddagger$ and therefore under every $\sigma \in \mathfrak{G}(\mathcal{F}_1)$.

On the other hand, an element of $\mathcal{G}$ which is invariant under every $\sigma \in \mathfrak{G}(\mathcal{F}_1)$, that is, under every $\sigma^\ddagger \in \mathfrak{G}^\ddagger$, must be in $\mathcal{F}^\ddagger$ and therefore in $\mathcal{F}^\ddagger \cap \mathcal{G}$. Therefore the set of all elements of $\mathcal{G}$ invariant under every $\sigma \in \mathfrak{G}(\mathcal{F}_1)$ is $\mathcal{F}^\ddagger \cap \mathcal{G}$, so that $\mathcal{F}_1 = \mathcal{F}^\ddagger \cap \mathcal{G}$.

22. Linear reducibility of $L(y)$

The homogeneous linear ordinary differential polynomial $L(y)$ is called *linearly reducible* over $\mathcal{F}$ if there exist two homogeneous linear differential polynomials $J(y), K(y) \in \mathcal{F}\{y\}$ of positive order such that $L(y) = K(J(y))$.[22]

THEOREM 1. *$L(y)$ is linearly reducible if and only if the algebraic matric group $\mathfrak{G}$ of degree n is reducible.*[23]

PROOF. Let $L(y) = K(J(y))$, with $J(y)$ of order $m (0 < m < n)$. A solution of $J(y)$ is a solution of $L(y)$, so that every solution of $J(y)$ lying in an extension of $\mathcal{G}$ is a linear combination with constant coefficients of $\eta_1, \cdots, \eta_n$. Since such linear combinations exist, and since $\mathcal{C}$ is algebraically closed, it is easy to see that such linear combinations exist with coefficients in $\mathcal{C}$, that is, there exists a $\zeta = c_1\eta_1 + \cdots + c_n\eta_n \in \mathcal{G}$ such that $J(\zeta) = 0$. Therefore the set of all solutions of $J(y)$ in $\mathcal{G}$, which is a linear subspace of the n-dimensional linear space spaned by $\eta_1, \cdots, \eta_n$, has dimension > 0 and $< n$. But this subspace is clearly an invariant one for each $\sigma \in \mathfrak{G}$, so that $\mathfrak{G}$ is reducible.

Conversely, let $\mathfrak{G}$ be reducible. Then, after a proper choice of the fundamental system of solutions $\eta_1, \cdots, \eta_n$, we may suppose that there is an m $(0 < m < n)$ such that every $\sigma \in \mathfrak{G}$ is of the form

$$\sigma = \begin{pmatrix} \sigma_1 & * \\ 0 & \sigma_2 \end{pmatrix},$$

where σ_1 is a square matrix of degree m and σ_2 is a square matrix of degree $n - m$. From this it follows that the coefficients in the homogeneous linear differential polynomial $J(y) = W(y, \eta_1, \cdots, \eta_m)/W(\eta_1, \cdots, \eta_m)$ in y (quotient of two wronskian determinants) are all invariant under every $\sigma \in \mathfrak{G}$, so that $J(y)$ has all its coefficients in $\mathcal{F}$. Since every solution of $J(y)$ is a solution of $L(y)$ it is easy to see that there exists a homogeneous linear $K(y) \in \mathcal{F}\{y\}$ such that $L(y) = K(J(y))$.

Using the same methods it is easy to prove

THEOREM 2. *$\mathfrak{G}$ is reducible to the form*

$$\begin{pmatrix} \mathfrak{G}_1 & & * \\ & \ddots & \\ 0 & & \mathfrak{G}_k \end{pmatrix},$$

[22] This definition is equivalent to one first given by FROBENIUS [1] for more special coefficient domains.

[23] First proved by BEKE [1].

where $\mathfrak{G}_i$ *is a group of matrices of degree* n_i $(n_1 + \cdots + n_k = n)$, *if and only if there exist* k *homogeneous linear differential polynomials* $L_1(y), \cdots, L_k(y)$, *with coefficients in* $\mathcal{F}$, *of orders* $n_1, \cdots, n_k$, *respectively, such that* $L(y) = L_k(\cdots L_1(y) \cdots)$. *When this is the case then* $\mathfrak{G}_i$ *is isomorphic with the group of all automorphisms of* $\mathcal{F}<\zeta_{i1}, \cdots, \zeta_{in_i}>$ *over* $\mathcal{F}$, *where* $\zeta_{i1}, \cdots, \zeta_{in_i}$ *are* n_i *solutions of* $L_i(y)$ *in* $\mathcal{G}$, *linearly independent over constants.*

This theorem was first stated by Loewy [1]. Loewy's proof seems to show merely that $\mathfrak{G}_i$ is an abundant subgroup of the group of all automorphisms. That $\mathfrak{G}_i$ is an algebraic group, and therefore the full group of automorphisms, follows from the lemma in Chater I §5.

Chapter V. Liouvillian Extensions

23. Integrals and exponentials of integrals

By an *integral* of an element a of a differential field, we shall mean any solution of the differential equation $y' = a$.

Two integrals of a which are contained in the same extension of the given differential field containing a clearly differ by a constant.

As in Chapter IV, let $\mathcal{F}$ be a differential field of characteristic 0 with an algebraically closed field of constants $\mathcal{C}$.

Let α be an integral of $a \in \mathcal{F}$ such that $\alpha \notin \mathcal{F}$ and every constant in $\mathcal{F}<\alpha>$ is in $\mathcal{C}$. Then $a \neq 0$, and $1, \alpha$ are linearly independent over constants. Since $1, \alpha$ are solutions of the homogeneous linear ordinary differential polynomial $y'' - (a'/a)y'$, we see that $\mathcal{F}<\alpha>$ is a Picard-Vessiot extension of $\mathcal{F}$. For any σ in the group $\mathfrak{G}_I$ of all automorphisms of $\mathcal{F}<\alpha>$ over $\mathcal{F}$ we have $\sigma 1 = 1$, $\sigma\alpha = c + \alpha$ (because $\sigma\alpha$, like α, is a solution of $y' = a$), so that the matrix of σ is $\begin{pmatrix} 1 & c \\ 0 & 1 \end{pmatrix}$. For every $\sigma \neq \iota$ we have $c \neq 0$. Because $\mathfrak{G}_I$ is algebraic and contains the matrix $\begin{pmatrix} 1 & hc \\ 0 & 1 \end{pmatrix}$ of σ^h for $h = 0, 1, \cdots$, it follows that $\mathfrak{G}_I$ contains the matrix $\begin{pmatrix} 1 & k \\ 0 & 1 \end{pmatrix}$ for every $k \in \mathcal{C}$. Thus $\mathfrak{G}_I$ is isomorphic with the additive group of $\mathcal{C}$, and contains no proper algebraic subgroup other than $\mathfrak{E}$ (in particular, $\mathfrak{G}_I$ is abelian and anticompact. It follows that α is transcendental over $\mathcal{F}$, and that there is no differential field properly between $\mathcal{F}$ and $\mathcal{F}<\alpha>$.

By an *exponential of an integral* of an element a of a differential field we shall mean any solution of the differential polynomial $y' - ay$.[24] If α and $\beta \neq 0$ are both exponentials of an integral of a contained in a common differential field then $(\alpha/\beta)' = (\beta\alpha' - \alpha\beta')/\beta^2 = (\beta a\alpha - \alpha a\beta)/\beta^2 = 0$, so that $\alpha = k\beta$, with k a constant.

[24] Thus, the concept "exponential of a" has not been defined. It could be defined as an exponential of an integral of a', but for the purposes of the present paper nothing would be gained thereby.

Let α be an exponential of an integral of $a \in \mathcal{F}$ such that every constant in $\mathcal{F}\langle\alpha\rangle$ is in $\mathcal{C}$. Then $\mathcal{F}\langle\alpha\rangle$ is a Picard-Vessiot extension of $\mathcal{F}$. For each automorphism σ of $\mathcal{F}\langle\alpha\rangle$ over $\mathcal{F}$ we have $\sigma\alpha = k\alpha$ $(k \in \mathcal{C})$, so that σ has the matrix (k) of degree 1. If α is algebraic over $\mathcal{F}$ then the group of all such automorphisms is of finite order h, $\sigma^h = \iota$, and $k^h = 1$. Therefore $\sigma(\alpha^h) = (\sigma\alpha)^h = (k\alpha)^h = \alpha^h$ for each σ, so that $\alpha^h \in \mathcal{F}$. If α is transcendental over $\mathcal{F}$ then the group of all automorphisms of $\mathcal{F}\langle\alpha\rangle$ over $\mathcal{F}$, which group we now call $\mathfrak{G}_E$, is of dimension 1 and is isomorphic with the group of all matrices (k) of degree 1 with nonzero $k \in \mathcal{C}$ (i.e., isomorphic with the multiplicative group of $\mathcal{C}$). The only algebraic subgroups of $\mathfrak{G}_E$ are cyclic groups of finite order, and precisely one such subgroup of each order exists (thus, $\mathfrak{G}_E$ is abelian and quasicompact). If $\mathfrak{G}_E^h$ is the one of order h and $\mathcal{F}^h$ is the differential subfield of $\mathcal{F}\langle\alpha\rangle$ corresponding to $\mathfrak{G}_E^h$, then $\mathcal{F}^h = \mathcal{F}\langle\alpha^h\rangle$. Now, α^h is an exponential of an integral of ha, so that $\mathfrak{G}_E/\mathfrak{G}_E^h$, which is isomorphic with the group of automorphisms of $\mathcal{F}\langle\alpha^h\rangle$ over $\mathcal{F}$, is isomorphic with $\mathfrak{G}_E$.

24. Liouvillian extensions

A differential extension field $\mathcal{H} \supseteq \mathcal{F}$, all the constants of which are in $\mathcal{C}$, will be called a *liouvillian* extension of $\mathcal{F}$ if $\mathcal{H}$ is an extension of $\mathcal{F}$ by integrals, exponentials of integrals, and algebraic functions, i.e., if there exists a finite sequence of elements $\alpha_1, \cdots, \alpha_r \in \mathcal{H}$ such that:

(a) for each i, either α_i is an integral of an element of $\mathcal{F}\langle\alpha_1, \cdots, \alpha_{i-1}\rangle$, or α_i is an exponential of an integral of such an element, or α_i is algebraic over $\mathcal{F}\langle\alpha_1, \cdots, \alpha_{i-1}\rangle$;

(b) $\mathcal{H} = \mathcal{F}\langle\alpha_1, \cdots, \alpha_r\rangle$;

(c) every constant in $\mathcal{H}$ is in $\mathcal{F}$ (i.e., te field of constants of $\mathcal{H}$ is $\mathcal{C}$).

For a given liouvillian extension of $\mathcal{F}$ it may not be necessary to employ all three types of elements α_i. We shall distinguish ten types of liouvillian extensions, namely, extensions by

(1) integrals, exponentials of integrals, and algebraic functions,
(2) integrals and exponentials of integrals,
(3) exponentials of integrals, and algebraic functions,
(4) integrals and algebraic functions,
(5) integrals and radicals,
(6) exponentials of integrals,
(7) integrals,
(8) algebraic functions,
(9) radicals,
(10) rational functions.

Thus, every liouvillian extension is of the first type, and the only one of the last type is $\mathcal{F}$ itself. There is no need to discuss separately extensions by exponentials of integrals and radicals, for such an extension is an extension by exponentials of integrals (if $\alpha^h = a \in \mathcal{F}$, then $\alpha' - h^{-1}a^{-1}a'\alpha = 0$).

We remark that if $\mathcal{H} = \mathcal{F}\langle\alpha_1, \cdots, \alpha_r\rangle$ is a liouvillian extension of $\mathcal{F}$, with $\alpha_1, \cdots, \alpha_r$ as above, then $\mathcal{H}$ is contained in a liouvillian extension $\mathcal{F}\langle\beta_1, \cdots, \beta_r\rangle$, where $\beta_1, \cdots, \beta_r$ have the same properties as $\alpha_1, \cdots, \alpha_r$ above, and for each i $\mathcal{F}\langle\beta_1, \cdots, \beta_i\rangle$ is a normal extension of $\mathcal{F}\langle\beta_1, \cdots, \beta_{i-1}\rangle$. Indeed, when α_i is an integral or an exponential of an integral of an element of $\mathcal{F}\langle\alpha_1, \cdots, \alpha_{i-1}\rangle$ then by §23 $\mathcal{F}\langle\alpha_1, \cdots, \alpha_i\rangle$ is a Picard-Vessiot extension of $\mathcal{F}\langle\alpha_1, \cdots, \alpha_{i-1}\rangle$ and *a fortiori* normal. When α_i is algebraic over $\mathcal{F}\langle\alpha_1, \cdots, \alpha_{i-1}\rangle$ we may replace α_i by a primitive element of the splitting field over $\mathcal{F}\langle\alpha_1, \cdots, \alpha_{i-1}\rangle$ of the irreducible equation satisfied by α_i.

25. The principal theorem and some consequences

We return now to a Picard-Vessiot extension $\mathcal{G} = \mathcal{F}\langle\eta_1, \cdots, \eta_n\rangle$ as in Chapter IV, and list certain properties which the group $\mathfrak{G}$ of all automorphisms of $\mathcal{G}$ over $\mathcal{F}$ and its component of the identity $\mathfrak{G}^0$ may possess:

(1) $\mathfrak{G}^0$ is solvable:
(2) $\mathfrak{G}$ is solvable;
(3) $\mathfrak{G}^0$ is solvable and quasicompact;
(4) $\mathfrak{G}^0$ is solvable and anticompact;
(5) $\mathfrak{G}$ is solvable and $\mathfrak{G}^0$ is anticompact;
(6) $\mathfrak{G}$ is solvable and quasicompact;
(7) $\mathfrak{G}$ is solvable and anticompact;
(8) $\mathfrak{G}$ is finite;
(9) $\mathfrak{G}$ is solvable and finite;
(10) $\mathfrak{G} = \mathfrak{E}$.

It will be observed that we have here distinguished as many types of algebraic matric groups as we have of liouvillian extensions in §24.

THEOREM. *Let i be a positive integer $\leqq 10$. If $\mathcal{G}$ is contained in a liouvillian extension of $\mathcal{F}$ of type (i) then $\mathfrak{G}$ is an algebraic matric group of type (i). Conversely, if $\mathfrak{G}$ is an algebraic matric group of type (i) then $\mathcal{G}$ is a liouvillian extension of $\mathcal{F}$ of type (i).*

Several remarks can be made on the basis of this theorem:

1. If $\mathcal{G}$ is a liouvillian extension of $\mathcal{F}$, then there is an $\mathcal{F}^0$ between $\mathcal{F}$ and $\mathcal{G}$ such that $\mathcal{F}^0$ is a normal algebraic extension of $\mathcal{F}$, and $\mathcal{G}$ is an extension of $\mathcal{F}^0$ by integrals and exponentials of integrals (indeed, $\mathcal{F}^0$ is the differential field such that $\mathfrak{G}(\mathcal{F}^0) = \mathfrak{G}^0$). If $\mathcal{G}$ is a liouvillian extension of $\mathcal{F}$ of one of the types (3), (4), (5) then $\mathcal{G}$ is a liouvillian extension of $\mathcal{F}^0$ of, respectively, the types (6), (7), (7) (and in the last case $\mathcal{F}^0$ is an extension of $\mathcal{F}$ by radicals).

2. If $\mathcal{G}$ is a liouvillian extension of $\mathcal{F}$ and $\mathcal{F}^0$ is as above, then the homogeneous linear differential polynomial $L(y)$ splits completely over $\mathcal{F}^0$ into "factors" of order 1, i.e., there are n differential polynomials $L_i(y) = y' - a_iy$ $(a_i \in \mathcal{F}^0)$ such that $L(y) = L_n(\cdots L_1(y) \cdots)$ (see Chapter I §7 Theorem 1 and Chapter IV §22 Theorem 2). If $\mathcal{G}$ is an extension of $\mathcal{F}$ by integrals then $L(y)$ splits completely over $\mathcal{F}$ (see Chapter I §7 Theorem 2).

3. If $\mathcal{G}$ is an extension by exponentials of integrals and algebraic functions then $L(y)$ has a fundamental system of solutions in $\mathcal{G}$ composed of exponentials of integrals of elements of $\mathcal{F}^0$ ($\mathcal{F}^0$ as above) (see Chapter I §7 Theorems 1 and 2).

4. If $\mathcal{G}$ is an extension of $\mathcal{F}$ by integrals and also an extension of $\mathcal{F}$ by exponentials of integrals, then $\mathcal{G} = \mathcal{F}$. If $\mathcal{G}$ is an extension by integrals and algebraic functions, and also by exponentials of integrals and algebraic functions, then $\mathcal{G}$ is algebraic over $\mathcal{F}$.

5. If $\dim \mathfrak{G} > n(n - 1)/2$ (the dimension of the full triangular group $\mathfrak{T}$) then $\mathcal{G}$ is not contained in a liouvillian extension of $\mathcal{F}$. In particular, the general homogeneous linear ordinary differential equation of order $n > 1$ does not have a fundamental system of solutions in any liouvillian extension of $\mathcal{F}_0\langle u, \cdots, u_n\rangle$, where $\mathcal{F}_0$ is any differential field of characteristic zero over which $u_1, \cdots, u_n$ are unknowns (see Chapter I §7 Theorem 1 and Chapter IV §17 Example 1).

26. The proof, first half

Let $\mathcal{G}$ be contained in a liouvillian extension $\mathcal{F}\langle\alpha_1, \cdots, \alpha_r\rangle$ where $\alpha_1, \cdots, \alpha_r$ are as in §24 (see final remark in §24).

We shall show that $\mathfrak{G}$ has a normal chain in which each factor group is abelian or finite. This is obvious for $r = 0$. Assume, inductively, that $r > 0$ and that this result is verified for lower values of r.

By Chapter IV §21, $\mathcal{G}\langle\alpha_1\rangle$ is a Picard-Vessiot extension of $\mathcal{F}\langle\alpha_1\rangle$, and the group of all automorphisms of $\mathcal{G}\langle\alpha_1\rangle$ over $\mathcal{F}\langle\alpha_1\rangle$ is isomorphic with $\mathfrak{G}(\mathcal{G} \cap \mathcal{F}\langle\alpha_1\rangle)$. By the induction assumption, then, $\mathfrak{G}(\mathcal{G} \cap \mathcal{F}\langle\alpha_1\rangle)$ has a normal chain in which each factor group is abelian or finite. Now, $\mathcal{F} \subseteq \mathcal{G} \cap \mathcal{F}\langle\alpha_1\rangle \subseteq \mathcal{F}\langle\alpha_1\rangle$.

If α_1 is an integral of an element of $\mathcal{F}$ then (§23) either $\mathcal{G} \cap \mathcal{F}\langle\alpha_1\rangle = \mathcal{F}$, in which case $\mathfrak{G}(\mathcal{G} \cap \mathcal{F}\langle\alpha_1\rangle) = \mathfrak{G}$, or $\mathcal{G} \cap \mathcal{F}\langle\alpha_1\rangle = \mathcal{F}\langle\alpha_1\rangle$. In the latter case, by §23, $\sigma(\mathcal{G} \cap \mathcal{F}\langle\alpha_1\rangle) = \sigma\mathcal{F}\langle\alpha_1\rangle = \mathcal{F}\langle\sigma\alpha_1\rangle = \mathcal{F}\langle\alpha_1 + c\rangle = \mathcal{F}\langle\alpha_1\rangle = \mathcal{G} \cap \mathcal{F}\langle\alpha_1\rangle$ for every $\sigma \in \mathfrak{G}$, so that (Chapter III Theorem 2) $\mathfrak{G}(\mathcal{G} \cap \mathcal{F}\langle\alpha_1\rangle)$ is a normal subgroup of $\mathfrak{G}$ with (Chapter IV §18 Therem 2) factor group isomorphic with the group of all automorphisms of $\mathcal{F}\langle\alpha_1\rangle$ over $\mathcal{F}$, i.e., with the abelian group $\mathfrak{G}_I$ (see §23).

If α_1 is an exponential of an integral of an element of $\mathcal{G}$ then (§23) either $\mathcal{G} \cap \mathcal{F}\langle\alpha_1\rangle = \mathcal{F}\langle\beta\rangle$, where β is an hth root of an element of $\mathcal{F}$, or $\mathcal{G} \cap \mathcal{F}\langle\alpha_1\rangle = \mathcal{F}\langle\alpha_1^h\rangle$, for some positive integer h. In the former case $\sigma(\mathcal{G} \cap \mathcal{F}\langle\alpha_1\rangle) = \mathcal{G} \cap \mathcal{F}\langle\alpha_1\rangle$ is obviously true for all $\sigma \in \mathfrak{G}$ so that $\mathfrak{G}(\mathcal{G} \cap \mathcal{F}\langle\alpha_1\rangle)$ is a normal subgroup of $\mathfrak{G}$ with factor group isomorphic with the group of all automorphisms of $\mathcal{F}\langle\beta\rangle$ over $\mathcal{F}$, i.e., with a cyclic group of finite order. In the latter case, by §23, $\sigma(\mathcal{G} \cap \mathcal{F}\langle\alpha_1\rangle) = \sigma\mathcal{F}\langle\alpha_1^h\rangle = \mathcal{F}\langle\sigma\alpha_1^h\rangle = \mathcal{F}\langle c^h\alpha_1^h\rangle = \mathcal{F}\langle\alpha_1^h\rangle = \mathcal{G} \cap \mathcal{F}\langle\alpha_1\rangle$ for all $\sigma \in \mathfrak{G}$, so that $\mathfrak{G}(\mathcal{G} \cap \mathcal{F}\langle\alpha_1\rangle)$ is a normal subgroup of $\mathfrak{G}$ with factor group isomorphic with the group of all automorphisms of $\mathcal{F}\langle\alpha_i^h\rangle$ over $\mathcal{F}$, i.e. with the abelian group $\mathfrak{G}_E$ (see §23).

If $\mathcal{F}\langle\alpha_1\rangle$ is a normal algebraic extension of $\mathcal{F}$, then all the roots of the defining equation of α_1 over $\mathcal{F}$ are contained in $\mathcal{F}\langle\alpha_1\rangle$, so that $\sigma(\mathcal{G} \cap \mathcal{F}\langle\alpha_1\rangle) = \mathcal{G} \cap \mathcal{F}\langle\alpha_1\rangle$ for all $\sigma \in \mathfrak{G}$, and $\mathfrak{G}(\mathcal{G} \cap \mathcal{F}\langle\alpha_1\rangle)$ is a normal subgroup of $\mathfrak{G}$ with factor group isomorphic with the group of all automorphisms of $\mathcal{G} \cap \mathcal{F}\langle\alpha_1\rangle$ over $\mathcal{F}$, i.e., with a finite group.

Thus, in any case, $\mathfrak{G}$ has a normal chain $\mathfrak{G} \supseteq \mathfrak{G}(\mathcal{G} \cap \mathcal{F}\langle\alpha_1\rangle) \supseteq \cdots \supseteq \mathfrak{E}$ in which every factor group is abelian or finite.

A glance at the proof reveals, furthermore, that when α_1 is an integral of an element of $\mathcal{F}$ then $\mathfrak{G}/\mathfrak{G}(\mathcal{G} \cap \mathcal{F}\langle\alpha_1\rangle)$ is not only abelian but also anticompact, when α_1 is an exponential of an integral of an element of $\mathcal{F}$ then $\mathfrak{G}/\mathfrak{G}(\mathcal{G} \cap \mathcal{F}\langle\alpha_1\rangle)$ is not only abelian but also quasicompact, when α_1 is expressible over $\mathcal{F}$ by means of radicals then $\mathfrak{G}/\mathfrak{G}(\mathcal{G} \cap \mathcal{F}\langle\alpha_1\rangle)$ is not only finite but also solvable.

It follows, using Theorems 1, 2, and 3 of Chapter I §8, that $\mathfrak{G}^0$ is solvable, and, when the liouvillian extension $\mathcal{F}\langle\alpha_1, \cdots, \alpha_r\rangle$ is of one of the types (1), $\cdots$, (10), then $\mathfrak{G}$ is an algebraic matric group of the corresponding type.

27. The proof, second half

Now let $\mathfrak{G}$ be an algebraic matric group of one of the ten types introduced above. In the case of type (10) ($\mathfrak{G} = \mathfrak{E}$) it is obvious that $\mathcal{G} = \mathcal{F}$, so that $\mathcal{G}$ is a liouvillian extension of $\mathcal{F}$ of type (10). In the case of type (8) ($\mathfrak{G}$ is finite) then $\mathcal{G}$ is a normal finite algebraic extension of $\mathcal{F}$ with $\mathfrak{G}$ as its Galois group (so that $\mathcal{G}$ is a liouvillian extension of $\mathcal{F}$ of type (8)), and if $\mathfrak{G}$ is moreover solvable (type (9)) then $\mathcal{G}$ is an extension of $\mathcal{F}$ by radicals (a liouvillian extension of $\mathcal{F}$ of type (9)).

Types (1), (2), (3), (4), (5), (6) can be reduced to the cases in which $\mathfrak{G}$ is connected and of type (2), (2), (6), (7), (7), (6), respectively, by observing that if $\mathfrak{G}^0 = \mathfrak{G}(\mathcal{F}^0)$ then $\mathcal{F}^0$ is a normal finite algebraic extension of $\mathcal{F}$ (an extension by radicals when $\mathfrak{G}$ is of type (2), (5), or (6)), and $\mathcal{G}$ is a Picard-Vessiot extension of $\mathcal{F}^0$ such that the group of all automorphisms of $\mathcal{G}$ over $\mathcal{F}^0$ is an algebraic matric group of type (2), (2), (6), (7), (7), (6), in the respective instances.

We consider, then, three cases. Suppose, firstly, that $\mathfrak{G}$ is solvable and anticompact (type (7)). By Chapter I, §7, Theorem 2, $\mathfrak{G}$ is reducible to special triangular form, that is, $L(y)$ has a fundamental system of solutions $\eta_1, \cdots, \eta_n$ in $\mathcal{G} = \mathcal{F}\langle\eta_1, \cdots, \eta_n\rangle$ such that, for every $\sigma \in \mathfrak{G}$,

$$\sigma\eta_j = \sum_{i=1}^{j} k_{ij}\eta_i, \qquad j = 1, \cdots, n, \tag{10}$$

where each $k_{jj} = 1$. If we divide through all these equations by $\sigma\eta_1 = \eta_1$, and differentiate once, we find

$$\sigma\left(\frac{\eta_j}{\eta_1}\right)' = \sum_{i=2}^{j} k_{ij}\left(\frac{\eta_i}{\eta_1}\right)', \qquad j = 2, \cdots, n.$$

This is a system of relations in the $n - 1$ elements $(\eta_2/\eta_1)', \cdots, (\eta_n/\eta_1)'$ of the same form as the relations (10) in the n elements $\eta_1, \cdots, \eta_n$. Therefore, by making the appropriate induction assumption, we may suppose that $\mathcal{F}\langle(\eta_2/\eta_1)',$

$\cdots, (\eta_n/\eta_1)'>$ is an extension of $\mathcal{F}$ by integrals. Since, for $j = 2, \cdots, n$, η_j is η_1 (an element of $\mathcal{F}$) multiplied by an integral of $(\eta_j/\eta_1)'$ it follows that $\mathcal{G}$ is an extension of $\mathcal{F}$ by integrals (and therefore a liouvillian extension of $\mathcal{F}$ of type (7)).

Suppose, secondly, that $\mathfrak{G}$ is connected, solvable and quasicompact (type (6)). By Chapter I, §7, Theorems 1 and 3, $\mathfrak{G}$ is reducible to diagonal form. Thus, we may assume that $\eta_1, \cdots, \eta_n$ is a fundamental system of solutions of $L(y)$ such that

$$\sigma\eta_j = k_j\eta_j, \qquad j = 1, \cdots, n,$$

for every $\sigma \in \mathfrak{G}$. It follows that

$$\sigma(\eta_j'/\eta_j) = \eta_j'/\eta_j, \qquad j = 1, \cdots, n,$$

for every $\sigma \in \mathfrak{G}$, so that $\eta_j'/\eta_j \in \mathcal{F}$ $(j = 1, \cdots, n)$, and each η_j is an exponential of an integral of an element of $\mathcal{F}$. In particular, $\mathcal{G}$ is a liouvillian extension of $\mathcal{F}$ of type (6).

Suppose, finally, that $\mathfrak{G}$ is connected and solvable (type (2)). By Chapter I, §7, Theorem 1, we may assume that every $\sigma \in \mathfrak{G}$ satisfies equations of the form (10), where we no longer assume that $k_{jj} = 1$. Dividing through all the equations (10) by $\sigma\eta_1 = k_{11}\eta_1$ and differentiating once, we find that

$$\sigma\left(\frac{\eta_j}{\eta_1}\right)' = \sum_{i=2}^{j} \frac{k_{ij}}{k_{11}}\left(\frac{\eta_i}{\eta_1}\right)', \qquad j = 2, \cdots, n.$$

These are relations in the $n - 1$ elements $(\eta_2/\eta_1)', \cdots, (\eta_n/\eta_1)'$ of the same form as the relations (10) in the n elements $\eta_1, \cdots, \eta_n$. Therefore, by making the appropriate induction assumption, we may suppose that $\mathcal{F}<(\eta_2/\eta_1)', \cdots, (\eta_n/\eta_1)'>$ is an extension of $\mathcal{F}$ by integrals and exponentials of integrals. Since, for $j = 2, \cdots, n$, η_j is η_1 (an exponential of an integral of an element of $\mathcal{F}$) multiplied by an integral of $(\eta_j/\eta_1)'$, it follows that $\mathcal{G}$ is an extension of $\mathcal{F}$ by integrals and exponentials of integrals, and therefore a liouvillian extension of $\mathcal{F}$ of type (2).

References

R. Baer. 1. A note on the status of the Picard-Vessiot theory, included among comments by O. Haupt in Felix Klein's *Vorlesungen über hypergeometrische Funktionen*, Berlin, 1933.

E. Beke. 1. *Die Irreducibilität der homogenen linearen Differentialgleichungen*, Math. Ann., vol. 45 (1894), pp. 278–294.

G. Fano. 1. *Ueber lineare homogene Differentialgleichungen mit algebraischen Relationen zwischen den Fundamentallösungen*, Math. Ann., vol. 53 (1900), pp. 493–590.

H. Freudenthal. 1. *Zur "Galoisschen" Theorie der linearen Differentialgleichungen*, K. Akad. van Wetensch. Amsterdam, Proceedings of the section of Sciences, vol. 34 (1931), pp. 1124–1126.

G. Frobenius. 1. *Ueber den Begriff der Irreducibilität in den Theorie der linearen Differentialgleichungen*, Journal f. d. r. u. a. Math., vol. 76 (1873), pp. 236–270.

E. Gourin. 1. *On irreducible systems of algebraic differential equations*, Bull. Amer. Math. Soc., vol. 39 (1933), pp. 593–595.

E. R. KOLCHIN. 1. *On the exponents of differential ideals*, Ann. of Math., (2) vol. 42 (1941), pp. 740–777.

2. *On the basis theorem for differential systems*, Trans. Amer. Math. Soc., vol. 52 (1942), pp. 115–127.

3. *Extensions of differential fields, I*, Ann. of Math., (2) vol. 43 (1942), pp. 724–729.

4. *Extensions of differential fields, III*, Bull. Amer. Math. Soc., vol. 53 (1947), pp. 397–401.

W. KRULL. 1. *Galoissche Theorie der unendlichen algebraischen Erweiterungen*, Math. Ann., vol. 100 (1928), pp. 687–698.

A. LOEWY. 1. *Ueber die irreduciblen Factoren eines linearen homogenen Differentialausdrückes*, Ber. über die Verh. der Königlich Sächsischen Gesellschaft der Wissenschaften zu Leipzig, Math.-phys. Kl., vol. 51 (1902), pp. 1–13.

2. *Über reduzible lineare homogene Differentialgleichungen*, Math. Ann., vol. 56 (1902), pp. 549–584.

3. *Über die Adjunktion von Integralen linearer homogener Differentialgleichungen*, Math. Ann., vol. 59 (1904), pp. 435–448.

4. *Die Rationalitätsgruppe einer linearen homogenen Differentialgleichungen*, Math. Ann., vol. 65 (1908), pp. 129–160.

5. *Über die Irreduzibilität der linearen homogenen Substitutsionsgruppen und Differentialgleichungen*, Math. Ann., vol. 70 (1911), pp. 94–109.

F. MAROTTE. 1. *Les équations différentielles linéaires et la théorie des groupes*, Ann. de la Fac. des Sci. de Toulouse, (1) vol. 12 (1898), pp. H1–H92.

É. PICARD. 1. *Sur les groupes de transformation des équations différentielles linéaires*, Comptes rendus, Paris, vol. 96 (1883), pp. 1131–1134.

2. *Sur les équations différentielles linéaires et les groupes algébriques de transformations*, Ann. de la Fac. des Sci. de Toulouse, (1) vol. 1 (1887), pp. A1–A15.

3. *Sur les groupes de transformations des équations différentielles linéaires*, Comptes rendus, Paris, vol. 119 (1894), pp. 584–589 or Math. Ann., vol. 46 (1895), pp. 161–166.

4. *Sur l'extension des idées de Galois à la théorie des équations différentielles*, Comptes rendus, Paris, vol. 121 (1895), pp. 789–792 or Math. Ann., vol. 47 (1896), pp. 155–156.

5. *Traité d'Analyse*, vol. 3, chapter 17, Paris, 1898 or 1908 or 1928 (reprinted as *Analogies entre la théorie des équations différentielles linéaires et la théorie des équations algébriques*, Paris, 1936).

H. W. RAUDENBUSH. 1. *Ideal theory and algebraic differential equations*, Trans. Amer. Math. Soc., vol. 36 (1934), pp. 361–368.

J. F. RITT. 1. *Differential equations from the algebraic standpoint*, Amer. Math. Soc. Colloq. Publications, vol. 14, New York, 1932.

2. *Algebraic aspects of the theory of differential equations*, Amer. Math. Soc. Semicentennial Publications, vol. 2 (1938), pp. 35–55.

E. VESSIOT. 1. *Sur les équations différentielles linéaires*, Comptes rendus, Paris, vol. 112 (1891), pp. 778–780.

2. *Sur les intégrations des équations différentielles linéaires*, Ann. Sci. de l'École Norm. Sup., (3) vol. 9 (1892), pp. 192–280.

3. *Methodes d'intégration elémentaires*, Encyclopédie des sci. math. pures et appliquées, tome II, vol. 3, fascicule 1 (1910), pp. 58–170 (esp. pp. 152–165).

B. L. VAN DER WAERDEN. 1. *Moderne Algebra*; vol. I, Berlin, 1937; vol. II, Berlin, 1940.

HERMANN WEYL. 1. *The classical groups*, Princeton, 1939.

HANS ZASSENHAUS. 1. *Lehrbuch der Gruppentheorie*, vol. 1, Leipzig and Berlin, 1937.

COLUMBIA UNIVERSITY

Annals of Mathematics
Vol. 49, No. 4, October, 1948

ON CERTAIN CONCEPTS IN THE THEORY OF ALGEBRAIC MATRIC GROUPS

By E. R. Kolchin

(Received October 15, 1947)

Introduction.

The present paper is a continuation of Chapter I of my previous paper *Algebraic matric groups and the Picard-Vessiot theory of homogeneous linear ordinary differential equations*, Ann. of Math., (2) vol. 49 (1948), pp. 1–42. That paper is referred to below as "PV".

As in PV, we deal here with algebraic matric groups over an algebraically closed field $\mathcal{C}$ of arbitrary characteristic p. Several concepts which were introduced in PV in connection with an algebraic matric group $\mathfrak{G}$, which were discussed there, and which played an important role in the Picard-Vessiot theory as there developed, are here further illuminated. These are the concept of "component of the identity" of $\mathfrak{G}$, and the concepts of "connectedness", "solvability", "anticompactness", and "quasicompactness", properties which $\mathfrak{G}$ may possess.

In PV (§6 Theorem 1) it was shown that $\mathfrak{G}$ is anticompact (no element of finite order > 1 not divisible by p) if and only if each matrix in $\mathfrak{G}$ is reducible to special triangular form (0's below the main diagonal, 1's on it). It is shown in §1 below that this already implies that the whole group $\mathfrak{G}$ is reducible to special triangular form (and therefore is solvable). This is not unexpected in the light of classical theorems concerning Lie groups and Lie algebras,[1] but the present proof has the merit of being purely algebraic (not employing Lie algebras, and being valid for nonzero as well as zero characteristic p).

Similarly, in PV (§6 Theorem 2) it was shown that $\mathfrak{G}$ is quasicompact (no anticompact algebraic subgroup of order > 1) if and only if each matrix in $\mathfrak{G}$ is reducible to diagonal form. It is shown in §2 below that when $\mathfrak{G}$ is connected this already implies that the whole group $\mathfrak{G}$ is reducible to diagonal form (and therefore is abelian).

A partial converse of these two results is given in §3, where it is shown that if $\mathfrak{G}$ is abelian then $\mathfrak{G} = \mathfrak{G}_q \times \mathfrak{G}_a$ (direct product), where $\mathfrak{G}_q$ and $\mathfrak{G}_a$ are simultaneously reducible to diagonal form and to special triangular form, respectively. $\mathfrak{G}_q$ and $\mathfrak{G}_a$ are unique.

The result of §4 is mainly for application in §5, but is perhaps not without interest in itself. It is shown that if $\mathfrak{G}$ is connected, if k is an integer not divisible by p, and if τ is a generic point of the underlying manifold of $\mathfrak{G}$, then so is τ^k such a generic point. Thus, the set of matrices in $\mathfrak{G}$ which are not k^{th} powers

[1] A nilpotent Lie algebra is solvable. See, for example, C. Chevalley, *Algebraic Lie algebras*, Ann. of Math., (2) vol. 48 (1947), pp. 91–100 (esp. p. 94).

of matrices in $\mathfrak{G}$ lie in a lower-dimensional algebraic submanifold of the underlying manifold of $\mathfrak{G}$.

Of the five concepts listed above, only anticompactness is defined in abstract group-theoretic terms. The definitions of the others are algebraic-group-theoretic, that is, bring in properties of the underlying manifold of $\mathfrak{G}$. From the point of view of the Picard-Vessiot theory this is a blemish, because the criteria for a Picard-Vessiot extension $\mathcal{G}$ of a differential field $\mathcal{F}$ to be a liouvillian extension of any one of several types are thus made to depend on the particular representation of the group of automorphisms of $\mathcal{G}$ over $\mathcal{F}$ as an algebraic matric group, or more precisely, on the particular choice of homogeneous linear ordinary differential equation $L(y) = 0$ with fundamental system of solutions $\eta_1, \cdots, \eta_n$ such that $\mathcal{G} = \mathcal{F}\langle \eta_1, \cdots, \eta_n \rangle$.

The object of §5 below is to give, as far as is possible, abstract group criteria which permit definitions of component of the identity, connectedness, solvability, and quasicompactness of an algebraic matric group $\mathfrak{G}$ without reference to the underlying manifold of $\mathfrak{G}$. The first result in this direction is to the effect that if $\mathfrak{G}$ is connected then $\mathfrak{G}$ contains no normal (not necessarily algebraic) proper subgroup of finite index not divisible by p, and no subgroup of finite index k such that $k!$ is not divisible by p. In the case $p = 0$ (the case of interest for the Picard-Vessiot theory) this permits us to say that $\mathfrak{G}$ is connected if and only if $\mathfrak{G}$ has no subgroup of finite index, and more generally, that the component of the identity of $\mathfrak{G}$ is the (uniquely defined) smallest subgroup of finite index. For $p \neq 0$ an example is given of two group-isomorphic algebraic matric groups $\mathfrak{G}_1$ and $\mathfrak{G}_2$ such that $\mathfrak{G}_1$ is connected and $\mathfrak{G}_2$ is not.

Using a lemma from PV, it is shown that $\mathfrak{G}$ is solvable as an algebraic matric group if (and only if, of course) $\mathfrak{G}$ is solvable as an abstract group.

Thus, of the five concepts mentioned above, all except "quasicompactness" are definable group theoretically. It is therefore mildly surprising to learn that there exist two group-isomorphic connected algebraic matric groups $\mathfrak{G}_1$ and $\mathfrak{G}_2$ such that $\mathfrak{G}_1$ is quasicompact and $\mathfrak{G}_2$ is not. The situation is at least partly redeemed by the fact that there exists a group criterion for a connected algebraic matric group *of given dimension* to be quasicompact, and such a criterion is given. Since, in the Picard-Vessiot theory, the dimension of $\mathfrak{G}$ is tied up with the extension $\mathcal{G}$ over $\mathcal{F}$ in an invariant manner (dim $\mathfrak{G}$ = degree of transcendency of $\mathcal{G}$ over $\mathcal{F}$), this criterion is satisfactory from the point of view of that theory.

1. Anticompactness.

We prove the following result.

THEOREM. *If $\mathfrak{G}$ is an anticompact algebraic matric group then $\mathfrak{G}$ is reducible to special triangular form.*

Since $\mathfrak{G}$ is anticompact, all the characteristic roots of every matrix in $\mathfrak{G}$ are 1 (PV §6 Theorem 1). Hence, it suffices to prove the following lemma.

LEMMA. *Let $\mathfrak{G}$ be a multiplicative system of matrices such that every matrix in*

$\mathfrak{G}$ has all its characteristic roots equal to 1. *Then $\mathfrak{G}$ is reducible to special triangular form.*

For matrices of degree $n = 1$ the lemma holds trivially. Let $n > 1$ and suppose the lemma verified for lower values of n. Suppose, as we may, that $\mathfrak{G}$ contains a matrix $\sigma \neq \iota$ (the identity matrix). Without loss of generality we assume that σ is in Jordan normal form:

$$\sigma = \begin{pmatrix} \sigma_1 & & \\ & \ddots & \\ & & \sigma_t \end{pmatrix},$$

where each σ_i is of the form

$$\sigma_i = \begin{pmatrix} 1 & 1 & & \\ & \ddots & \ddots & \\ & & \cdot & 1 \\ & & & 1 \end{pmatrix} \tag{1}$$

and is of degree n_i $(n_1 + \cdots + n_t = n)$.

Let τ be any matrix in $\mathfrak{G}$. We write τ in the form

$$\tau = \begin{pmatrix} \tau_{11} & \cdots & \tau_{1t} \\ \cdot & \cdots & \cdot \\ \tau_{t1} & \cdots & \tau_{tt} \end{pmatrix}$$

where τ_{ij} is a rectangular matrix of n_i rows and n_j columns. Then

$$\sigma\tau = \begin{pmatrix} \sigma_1\tau_{11} & \sigma_1\tau_{12} & \cdots & \sigma_1\tau_{1t} \\ \sigma_2\tau_{21} & \sigma_2\tau_{22} & \cdots & \sigma_2\tau_{2t} \\ \cdot & \cdot & \cdots & \cdot \\ \sigma_t\tau_{t1} & \sigma_t\tau_{t2} & \cdots & \sigma_t\tau_{tt} \end{pmatrix}.$$

For any matrix $\rho = (r_{ij})$ we let Tr (ρ) denote the trace of ρ and let $D(\rho)$ denote the sum of the elements in the first diagonal below the main diagonal, that is, $D(\rho) = \sum r_{i+1,i}$. Then, referring to (1), we see that $\mathrm{Tr}\,(\sigma_i\tau_{ii}) = \mathrm{Tr}\,(\tau_{ii}) + D(\tau_{ii})$ Moreover, the trace of every matrix in $\mathfrak{G}$ is clearly n. Therefore $n = \mathrm{Tr}\,(\sigma\tau) = \sum \mathrm{Tr}\,(\sigma_i\tau_{ii}) = \sum \mathrm{Tr}\,(\tau_{ii}) + \sum D(\tau_{ii}) = \mathrm{Tr}\,(\tau) + \sum D(\tau_{ii}) = n + \sum D(\tau_{ii})$, so that $\sum D(\tau_{ii}) = 0$. Thus, every τ in $\mathfrak{G}$ lies on a certain $(n^2 - 1)$-dimensional linear manifold in n^2-dimensional space, so that the maximum number of linearly independent matrices in $\mathfrak{G}$ is $< n^2$. Now, Burnside's theorem (see, e.g., van der Waerden, *Moderne Algebra*, Vol. 2, p. 183 (1st ed.), p. 197 (2nd ed.)) asserts that an irreducible multiplicative system of $n \times n$ matrices over an algebraically closed field contains n^2 linearly independent matrices. Therefore we conclude that $\mathfrak{G}$ is reducible. The matrices in each of the principal blocks into which $\mathfrak{G}$ reduces form a multiplicative system, have all their characteristic roots equal to 1, and have degree $< n$. Hence each of these blocks is reducible to special triangular form, and $\mathfrak{G}$ is, too.

2. Quasicompactness.

Before we prove our theorem we need some lemmas. The first one, for which the proof is not given, is a very special case of a theorem proved by Eli Gourin (*On irreducible polynomials in several variables which become reducible when the variables are replaced by powers of themselves*, Trans. Amer. Math. Soc., Vol. 32 (1930), pp. 485–501) in completing certain earlier results of J. F. Ritt (*A factorization theory for functions* $\sum_{i=1}^{n} a_i e^{\alpha_i x}$, Trans. Amer. Math. Soc., vol. 29 (1927), pp. 584–596). Gourin's statement and proof of the theorem were for the field of all complex numbers, but both remain valid, with more or less obvious modifications, for any algebraically closed field $\mathcal{C}$ of arbitrary characteristic p.

LEMMA 1. *Let $f(x_1, \cdots, x_n)$ be a polynomial in the indeterminates $x_1, \cdots, x_n$ with coefficients in $\mathcal{C}$, let $f(x_1, \cdots, x_n)$ be irreducible and have more than two terms. Then there exists a finite set of positive integers $t_1, \cdots, t_s$ such that, if t is a positive integer for which $f(x_1^t, \cdots, x_n^t)$ is reducible, then t is divisible by some t_i.*

LEMMA 2. *Let $\mathfrak{G}$ be a connected algebraic matric group in diagonal form, of dimension g and degree n. Let $(\delta_{ij}\alpha_j)$ be a generic point of the underlying manifold of $\mathfrak{G}$, with $\alpha_1, \cdots, \alpha_g$ algebraically independent over $\mathcal{C}$. Then, for each $l = g + 1, \cdots, n$, there exist integers $m_l \neq 0, m_{l1}, \cdots, m_{lg}$ such that*

$$\alpha_1^{m_{l1}} \cdots \alpha_g^{m_{lg}} \alpha_l^{m_l} = 1 \qquad (l = g + 1, \cdots, n).$$

PROOF. Let $f_l(x_1, \cdots, x_g, x_l)$ be an irreducible polynomial with coefficients in $\mathcal{C}$ of lowest degree in x_l such that $f_l(\alpha_1, \cdots, \alpha_g, \alpha_l) = 0$. Since, for every integer k, $(\delta_{ij}\alpha_j^k) = (\delta_{ij}\alpha_j)^k$ is a specialization of $(\delta_{ij}\alpha_j)$, we have

$$f_l(\alpha_1^k, \cdots, \alpha_g^k, \alpha_l^k) = 0,$$

so that $f_l(\alpha_1^k, \cdots, \alpha_g^k, x_l^k)$ is divisible by $f_l(\alpha, \cdots, \alpha_g, x_l)$, whence $f_l(x_1^k, \cdots, x_g^k, x_l^k)$ is divisible by $f_l(x_1, \cdots, x_g, x_l)$. Thus $f_l(x_1^k, \cdots, x_g^k, x_l^k)$ is reducible for $k = 2, 3, \cdots$. It follows from Lemma 1 that $f_l(x_1, \cdots, x_g, x_l)$ has at most two terms. Since no α_i is 0, and $\alpha_1, \cdots, \alpha_g$ are algebraically independent, we have $\alpha_1^{m_{l1}} \cdots \alpha_g^{m_{lg}} \alpha_l^{m_l} = a_l \in \mathcal{C}(m_{l1}, \cdots, m_{lg}, m_l$ integers; $m_l \neq 0)$. Since ι is a specialization of the generic point $(\delta_{ij}\alpha_j)$, we see that $a_l = 1$.

LEMMA 3. *Let $a_1, \cdots, a_n \in \mathcal{C}(a_1 \cdots a_n \neq 0)$ have the property that there do not exist integers $m_1, \cdots, m_n$ not all 0 such that $a_1^{m_1} \cdots a_m^{n^n} = 1$. Then the smallest algebraic matric group containing $(\delta_{ij}a_j)$ is the full diagonal group $\mathfrak{D}$.*

PROOF. Let $\mathfrak{H}$ be the smallest algebraic matric group containing $(\delta_{ij}a_j)$. Then $\mathfrak{H} \subseteq \mathfrak{D}$. Let $(\delta_{ij}\alpha_j)$ be a generic point of the underlying manifold of the component of the identity $\mathfrak{H}^0$. Let $\dim \mathfrak{H} = g$, and suppose that $\alpha_1, \cdots, \alpha_g$ are algebraically independent over $\mathcal{C}$. If g were $<n$ we should have, by Lemma 2, a relation $\alpha_1^{m_1} \cdots \alpha_g^{m_g} \alpha_{g+1}^{m_{g+1}} = 1$. If we denote the index of $\mathfrak{H}^0$ in $\mathfrak{H}$ by k then $(\delta_{ij}a_j)^k \in \mathfrak{H}^0$, so that we should have $a_1^{m_1 k} \cdots a_{g+1}^{m_{g+1} k} = 1$, contradicting our hypothesis. Therefore $\dim \mathfrak{H} = n$, $\mathfrak{H}^0 = \mathfrak{D}$, $\mathfrak{H} = \mathfrak{D}$.

LEMMA 4. *Let $\alpha_1, \cdots, \alpha_n$ be elements of some extension field of $\mathcal{C}$, let $\alpha_1, \cdots, \alpha_g$ be algebraically independent over $\mathcal{C}$, and let $f(x_1, \cdots, x_n)$ be a polynomial over $\mathcal{C}$*

such that $f(\alpha_1, \cdots, \alpha_n) \neq 0$. *If* $\mathcal{C}$ *is either of characteristic* 0 *or transcendental over its prime field, then there exists a specialization* $(\alpha_1, \cdots, \alpha_n) \to (a_1, \cdots, a_n)$ *such that* $a_1, \cdots, a_n \in \mathcal{C}$, $f(a_1, \cdots, a_n) \neq 0$, *and there do not exist integers* $m_1, \cdots, m_g$ *not all* 0 *for which* $a_1^{m_1} \cdots a_g^{m_g} = 1$.

PROOF. Since $\alpha_1, \cdots, \alpha_g$ are algebraically independent over $\mathcal{C}$, any g elements $a_1, \cdots, a_g$ have the property that $(\alpha_1, \cdots, \alpha_g) \to (a_1, \cdots, a_g)$ is a specialization. This specialization can be extended to one of $(\alpha_1, \cdots, \alpha_n)$ such that $f(\alpha_1, \cdots, \alpha_n) \nrightarrow 0$ provided that $a_1, \cdots, a_g$ fail to annul a certain set $\mathfrak{m}$ of nonzero polynomials. If $\mathcal{C}$ is of characteristic 0 then we may always find g distinct primes $p_1, \cdots, p_g$ which fail to annul $\mathfrak{m}$, that is, for which there exists a specialization $(\alpha_1, \cdots, \alpha_n) \to (a_1, \cdots, a_n)$ with $a_1 = p_1, \cdots, a_g = p_g$ and $f(a_1, \cdots, a_n) \neq 0$. If $\mathcal{C}$ is transcendental over its prime field then $\mathcal{C}$ contains an element u transcendental over the prime field, and g distinct elements $b_1, \cdots, b_g$ algebraic over the prime field, such that $u - b_1, \cdots, u - b_g$ fail to annul $\mathfrak{m}$, that is, such that there exists a specialization $(\alpha_1, \cdots, \alpha_n) \to (a_1, \cdots, a_n)$ with $a_1 = u - b_1, \cdots, a_g = u - b_g$ and $f(a_1, \cdots, a_n) \neq 0$. In either case it is clear that $a_1^{m_1} \cdots a_g^{m_g}$ can not equal 1 if $m_1, \cdots, m_g$ are integers not all 0.

We are now in a position to prove the following result.

THEOREM. *If* $\mathfrak{G}$ *is a connected quasicompact algebraic matric group, then* $\mathfrak{G}$ *is reducible to diagonal form.*

PROOF. Let σ be a generic point of the underlying manifold of $\mathfrak{G}$, and let $\alpha_1, \cdots, \alpha_n$ be the characteristic roots of σ. If the degree of transcendency $\partial^0\mathcal{C}(\alpha_1, \cdots, \alpha_n)/\mathcal{C}$ is g, assume that these characteristic roots are so ordered that $\alpha_1, \cdots, \alpha_g$ are algebraically independent over $\mathcal{C}$.

We suppose temporarily that $\mathcal{C}$ is either of characteristic 0 or transcendental over its prime field. We specialize $(\sigma, \alpha_1, \cdots, \alpha_n) \to (\rho, a_1, \cdots, a_n)$, where $\rho \in \mathfrak{G}$, $a_1, \cdots, a_n \in \mathcal{C}$, $a_1 \cdots a_n \neq 0$, and there do not exist integers $m_1, \cdots, m_g$ not all 0 such that $a_1^{m_1} \cdots a_g^{m_g} = 1$ (Lemma 4). By PV §6 Theorem 2, each matrix in the connected quasicompact group $\mathfrak{G}$ is reducible to diagonal form. Without loss of generality, we assume that ρ is already in diagonal form and $\rho = (\delta_{ij}a_j)$. It follows from Lemma 3 that $\mathfrak{G}$ contains a connected algebraic subgroup $\mathfrak{D}_0$ in diagonal form such that, for a generic point $(\delta_{ij}\beta_j)$ of the underlying manifold of $\mathfrak{D}_0$, the elements $\beta_1, \cdots, \beta_g$ are algebraically independent over $\mathcal{C}$. Clearly $\partial^0\mathcal{C}(\beta_1, \cdots, \beta_n)/\mathcal{C} = g$.

We now suppose that $\mathfrak{G}$ is not in diagonal form, and seek a contradiction. $\mathfrak{G}$ contains a matrix $\tau = (b_{ij})$ with the two properties:

(a) τ is not in diagonal form;

(b) all the principal minors (i.e., minors along the main diagonal) of τ are nonzero. This follows immediately from the obvious fact that a generic point of the underlying manifold of $\mathfrak{G}$ has these properties.

Now consider the n forms

$$A_1(x_1, \cdots, x_n) = \sum_{1 \leq i \leq n} b_{ii} x_i,$$

$$A_2(x_1, \cdots, x_n) = \sum_{1 \leqq i < j \leqq n} \begin{vmatrix} b_{ii} & b_{ij} \\ b_{ji} & b_{jj} \end{vmatrix} x_i x_j,$$

$$A_3(x_1, \cdots, x_n) = \sum_{1 \leqq i < j < k \leqq n} \begin{vmatrix} b_{ii} & b_{ij} & b_{ik} \\ b_{ji} & b_{jj} & b_{jk} \\ b_{ki} & b_{kj} & b_{kk} \end{vmatrix} x_i x_j x_k,$$

. . . .

$$A_n(x_1, \cdots, x_n) = \begin{vmatrix} b_{11} & \cdots & b_{1n} \\ \cdot & \cdots & \cdot \\ b_{n1} & \cdots & b_{nn} \end{vmatrix} x_1 x_2 \cdots x_n.$$

The system of equations

$$(2) \qquad A_i(x_1, \cdots, x_n) = 0 \qquad (i = 1, \cdots, n)$$

has no solution other than $(0, \cdots, 0)$; for if $(\xi_1, \cdots, \xi_n)$ is a solution then $A_n(\xi_1, \cdots, \xi_n) = 0$ implies that some $\xi_i = 0$, say $\xi_n = 0$; whence $A_{n-1}(\xi_1, \cdots, \xi_{n-1}, 0) = 0$ implies that some ξ_i with $i < n$ is 0, say $\xi_{n-1} = 0$; etc. It follows that for any $\lambda_1, \cdots, \lambda_n$ the algebraic manifold in n-dimensional projective space defined by

$$A_i(x_1, \cdots, x_n) = \lambda_i x_0^i \qquad (i = 1, \cdots, n),$$

which is obviously of dimension $\geqq 0$, is actually of dimension $= 0$. For otherwise the intersection of this manifold with the hyperplane $x_0 = 0$ would be of dimension $\geqq 0$, i.e., the equations (2) would have a solution other than $(0, \cdots, 0)$.

Now let

$$(3) \qquad \bar{A}_i = A_i(\beta_1, \cdots, \beta_n) \qquad (i = 1, \cdots, n).$$

By the result just proved, for any solution of

$$A_i(x_1, \cdots, x_n) = \bar{A}_i x_0^i \qquad (i = 1, \cdots, n),$$

each x_i/x_0 is algebraic over $\mathcal{C}(\bar{A}_1, \cdots, \bar{A}_n)$. Since $(1, \beta_1, \cdots, \beta_n)$ is such a solution, each β_i is algebraic over $\mathcal{C}(\bar{A}_1, \cdots, \bar{A}_n)$. On the other hand, each $\bar{A}_i$ is obviously algebraic over $\mathcal{C}(\beta_1, \cdots, \beta_n)$. Hence

$$\partial^0 \mathcal{C}(\bar{A}_1, \cdots, \bar{A}_n)/\mathcal{C} = \partial^0 \mathcal{C}(\beta_1, \cdots, \beta_n)/\mathcal{C} = g.$$

Now $\sigma \to (\delta_{ij}\beta_j)\tau$ is a specialization over $\mathcal{C}$. This can be extended to a specialization of $(\sigma, \alpha_1^{\delta_1}, \cdots, \alpha_n^{\delta_n})$, where each δ_i is $+1$. (For the elementary facts concerning specializations used in the present paper see André Weil, *Foundations of algebraic geometry*, Amer. Math. Soc. Colloq. Publ. vol. 21, pp. 26–31). Since $\alpha_1, \cdots, \alpha_n$ are the characteristic roots of σ, and since the characteristic equation of $(\delta_{ij}\beta_j)\tau$ is of degree n and has no vanishing root, it is easy to see that no $\alpha_i^{\delta_i}$ specializes to 0, so that we may take every $\delta_i = 1$. That is, the specialization $\sigma \to (\delta_{ij}\beta_j)\tau$ can be extended to a specialization $(\sigma, \alpha_1, \cdots, \alpha_n) \to ((\delta_{ij}\beta_j)\tau, \gamma_1, \cdots, \gamma_n)$. Of course $\gamma_1, \cdots, \gamma_n$ are the characteristic roots of $(\delta_{ij}\beta_j)\tau$. The characteristic equation of $(\delta_{ij}\beta_j)\tau$ is readily seen to be $x^n - \bar{A}_1 x^{n-1} + \cdots + (-1)^n \bar{A}_n = 0$. Therefore $\partial^0\mathcal{C}(\gamma_1, \cdots, \gamma_n)/\mathcal{C} = \partial^0\mathcal{C}(\bar{A}_1, \cdots, \bar{A}_n)/\mathcal{C} = g$. Since $\partial^0\mathcal{C}(\alpha_1, \cdots, \alpha_n)/\mathcal{C} = g$, the specialization $(\alpha_1, \cdots, \alpha_n) \to (\gamma_1, \cdots, \gamma_n)$ is generic, i.e., generates an isomorphism of $\mathcal{C}(\alpha_1, \cdots, \alpha_n)$ onto $\mathcal{C}(\gamma_1, \cdots, \gamma_n)$. Since $\iota \in \mathfrak{G}$, $(\alpha_1, \cdots, \alpha_n) \to (1, \cdots, 1)$ is a specialization, whence $(\gamma_1, \cdots, \gamma_n) \to (1, \cdots, 1)$ is, too. Now $\bar{A}_1, \cdots, \bar{A}_n$ are the elementary symmetric functions of $\gamma_1, \cdots, \gamma_n$, so the last specialization implies that $(\gamma_1, \cdots, \gamma_n, \bar{A}_1, \cdots, \bar{A}_n) \to (1, \cdots, 1, \binom{n}{1}, \cdots, \binom{n}{n})$ is a specialization, too. This in turn can be extended to a specialization of

$$(\gamma_1, \cdots, \gamma_n, \bar{A}_1, \cdots, \bar{A}_n, \beta_1^{\delta_1}, \cdots, \beta_n^{\delta_n}),$$

where each $\delta_i = \pm 1$ and where $\beta_i^{-1} \to 0$ if $\delta_i = -1$. If some exponents δ_i were -1 then one of them, say δ_j, would have the property that the last specialization could be extended to include $(\beta_j^{-1}\beta_1, \cdots, \beta_j^{-1}\beta_n)$. Under such an extended specialization the relations

$$A_i(\beta_j^{-1}\beta_1, \cdots, \beta_j^{-1}\beta_n) = \beta_j^{-i}\bar{A}_i \qquad (i = 1, \cdots, n),$$

which follow from (3), would yield a solution of (2) other than $(0, \cdots, 0)$. It follows that every $\delta_i = 1$, i.e., there is a specialization

$$(\gamma_1, \cdots, \gamma_n, \bar{A}_1, \cdots, \bar{A}_n, \beta_1, \cdots, \beta_n) \to (1, \cdots, 1, \tbinom{n}{1}, \cdots, \tbinom{n}{n}, b_1, \cdots, b_n),$$

where $b_1, \cdots, b_n \in \mathcal{C}$. In particular, the matrix $(\delta_{ij}\beta_j)\tau$ with characteristic roots $\gamma_1, \cdots, \gamma_n$ specializes to the matrix $(\delta_{ij}b_j)\tau \in \mathfrak{G}$ with characteristic roots $1, \cdots, 1$. Since every matrix in $\mathfrak{G}$ is reducible to diagonal form, we have $(\delta_{ij}b_j)\tau = \iota$, or $\tau = (\delta_{ij}b_j^{-1})$, contradicting the property (a) of τ, above.

Thus, the theorem is proved in the case that $\mathcal{C}$ is of characteristic 0 or is transcendental over its prime field. Now suppose that $\mathcal{C}$ is of characteristic $p \neq 0$ and algebraic over its prime field. Let u be an indeterminate and let $\mathcal{C}^\dagger$ be the algebraic closure of $\mathcal{C}(u)$. Let $\mathfrak{G}^*$ denote the underlying manifold of $\mathfrak{G}$, $\mathfrak{g}$ the defining ideal of $\mathfrak{G}^*$; let $\mathfrak{G}^{\dagger *}$ be the algebraic manifold of $\mathfrak{g}$ over $\mathcal{C}^\dagger$. Then $\mathfrak{G}^{\dagger *}$ is an irreducible algebraic manifold, a point of $\mathfrak{G}^{\dagger *}$ is a generalized point of $\mathfrak{G}^*$, and a generalized point of $\mathfrak{G}^*$ with coordinates in $\mathcal{C}^\dagger$ is a point of $\mathfrak{G}^{\dagger *}$. Therefore, if $\sigma^\dagger, \tau^\dagger \in \mathfrak{G}^{\dagger *}$ then $\sigma^\dagger\tau^\dagger \in \mathfrak{G}^{\dagger *}$, and if, moreover, $\sigma^\dagger$ is nonsingular then $\sigma^{\dagger -1} \in \mathfrak{G}^{\dagger *}$. It follows that $\mathfrak{G}^{\dagger *}$ is the underlying manifold of an algebraic matric group $\mathfrak{G}^\dagger$. Since $\mathfrak{G}^{\dagger *}$ is irreducible, $\mathfrak{G}^\dagger$ is connected. Now, every $\sigma^\dagger \in \mathfrak{G}^\dagger$ is reducible to diagonal form, for otherwise $\sigma^\dagger$, which is a generalized point of $\mathfrak{G}^*$, could be specialized to a matrix in $\mathfrak{G}$ not reducible to diagonal form, and this is impossible. Therefore $\mathfrak{G}^\dagger$ is quasicompact. Since $\mathcal{C}^\dagger$ is transcendental over its prime field, it follows from the part of the theorem already proved that $\mathfrak{G}^\dagger$ is reducible to diagonal form. Therefore $\mathfrak{G}$ is, too, and the proof of the theorem is complete.

REMARK 1. In contrast to the lemma of §1, it is not true that any multiplicative system of matrices, such that each matrix in the system is reducible to diagonal form, is reducible to diagonal form. Indeed, if $\mathcal{C}$ is the field of all complex numbers, the group of all real proper orthogonal matrices serves as a counterexample.

REMARK 2. In contrast to the theorem of §1, the hypothesis of connectedness in the present theorem is not superfluous. For example, the algebraic group of all nonsingular matrices of either of the forms $\begin{pmatrix} a & 0 \\ 0 & b \end{pmatrix}$, $\begin{pmatrix} 0 & a \\ b & 0 \end{pmatrix}$ with $a, b \in \mathcal{C}$ (of characteristic $\neq 2$) is quasicompact (for every matrix is reducible to diagonal form). But this group is not reducible to diagonal form.

3. Abelian algebraic groups.

Let σ be a matrix of degree n in Jordan normal form:

$$\sigma = \begin{bmatrix} \boxed{\begin{matrix} a_1 & 1 & & \\ & \ddots & \ddots & \\ & & \cdot & 1 \\ & & & a_1 \end{matrix}} & & \\ & \ddots & \\ & & \boxed{\begin{matrix} a_t & 1 & & \\ & \ddots & \ddots & \\ & & \cdot & 1 \\ & & & a_t \end{matrix}} \end{bmatrix}$$

(square blocks of the indicated types along the main diagonal of degrees $n_1, \cdots, n_t (n_1 + \cdots + n_t = n)$, 0's elsewhere). As in PV §6, we define

$$t_{ij} = \binom{c}{j-i} a^{i-j} b,$$

$\tau_n(a, b, c) = (t_{ij})$, a square matrix of degree n,

$$\tau(b_1, \cdots, b_t, c_1, \cdots, c_t) = \begin{bmatrix} \boxed{\tau_{n_1}(a_1, b_1, c_1)} & & \\ & \ddots & \\ & & \boxed{\tau_{n_t}(a_t, b_t, c_t)} \end{bmatrix}$$

(square blocks along the main diagonal of degrees $n_1, \cdots, n_t$, 0's elsewhere; $a_1, \cdots, a_t$ the same as in σ). We consider the matrix $\tau(b_1, \cdots, b_t, c_1, \cdots, c_t)$ to be defined for $b_1, \cdots, b_t, c_1, \cdots, c_t$ in any extension of $\mathcal{C}$ if the field characteristic $p = 0$, for $b_1, \cdots, b_t$ in any extension of $\mathcal{C}$ and $c_1, \cdots, c_t$ integers if $p \neq 0$. The following properties are easy to establish:

$$(4)\quad \begin{aligned} &\tau(a_1, \cdots, a_t, 1, \cdots, 1) = \sigma, \\ &\tau(b_1, \cdots, b_t, c_1, \cdots, c_t)\tau(b_1', \cdots, b_t', c_1', \cdots, c_t') \\ &\qquad = \tau(b_1b_1', \cdots, b_tb_t', c_1 + c_1', \cdots, c_t + c_t'), \\ &\tau(1, \cdots, 1, p^m, \cdots, p^m) = \iota \text{ for } m \text{ sufficiently large} \end{aligned}$$

(if $p = 0$ then $m \geqq 1$ suffices, if $p \neq 0$ then $p^m \geqq n$ suffices). It is easy to see that the set of all nonsingular matrices of the form $\tau(b_1, \cdots, b_t, c_1, \cdots, c_t)$ with $b_1, \cdots, b_t, c_1, \cdots, c_t \in \mathcal{C}$ if $p = 0$, or with $b_1, \cdots, b_t \in \mathcal{C}$ and $c_1, \cdots, c_t$ integers if $p \neq 0$, forms an algebraic matric group. Similarly, the set of all nonsingular matrices $\tau(b_1, \cdots, b_t, c, \cdots, c)$ does, too.

LEMMA 1. *Let $\mathfrak{H}$ be the smallest algebraic matric group containing*

$$\sigma = \tau(a_1, \cdots, a_t, 1, \cdots, 1).$$

If $\tau(b_1, \cdots, b_t, c, \cdots, c) \in \mathfrak{H}$ then both $\tau(b_1, \cdots, b_t, 0, \cdots, 0)$ and $\tau(1, \cdots, 1, c, \cdots, c)$ are in $\mathfrak{H}$.

PROOF. Let $\tau(\beta_1, \cdots, \beta_t, \gamma, \cdots, \gamma)$ be a generic point of the underlying manifold of the component of the identity $\mathfrak{H}^0$. Then $\sigma\tau(\beta_1, \cdots, \beta_t, \gamma, \cdots, \gamma) = \tau(a_1\beta_1, \cdots, a_t\beta_t, \gamma + 1, \cdots, \gamma + 1)$ is a generic point of the irreducible component of the underlying manifold of $\mathfrak{H}$ which contains σ.

By PV §6, any specialization of $(\beta_1, \cdots, \beta_t)$ and any specialization of γ can be combined to yield a specialization of $(\beta_1, \cdots, \beta_t, \gamma)$, and therefore of $\tau(\beta_1, \cdots, \beta_t, \gamma, \cdots, \gamma)$. Since $\tau(1, \cdots, 1, 0, \cdots, 0) = \iota \in \mathfrak{H}^0$, it follows that both $\tau(\beta_1, \cdots, \beta_t, 0, \cdots, 0)$ and $\tau(1, \cdots, 1, \gamma, \cdots, \gamma)$ are specializations of $\tau(\beta_1, \cdots, \beta_t, \gamma, \cdots, \gamma)$.

If $p = 0$ then γ is transcendental over $\mathcal{C}$ (because $\mathfrak{H}^0$ contains infinitely many powers $\sigma^j = \tau(a_1^j, \cdots, a_t^j, j, \cdots, j)$), so that $\mathfrak{H}$ contains every matrix $\tau(1, \cdots, 1, c, \cdots, c)$ with $c \in \mathcal{C}$, and the lemma quickly follows.

Suppose $p \neq 0$, and let $\beta_{i_1}, \cdots, \beta_{i_g}$ be a maximal subset of $\beta_1, \cdots, \beta_t$ algebraically independent over $\mathcal{C}$. Then for each $i \neq i_1, \cdots, i_g$ there is a nonzero polynomial $f_i(x_{i_1}, \cdots, x_{i_g}, x_i) \in \mathcal{C}[x_{i_1}, \cdots, x_{i_g}, x_i]$ such that

$$f_i(a_{i_1}\beta_{i_1}, \cdots, a_{i_g}\beta_{i_g}, a_i\beta_i) = 0.$$

There clearly exists a specialization $(\beta_1, \cdots, \beta_t) \rightarrow (\bar{\beta}_1, \cdots, \bar{\beta}_t)$ such that $\bar{\beta}_1, \cdots, \bar{\beta}_t \in \mathcal{C}, \bar{\beta}_1 \cdots \bar{\beta}_t \neq 0, f_i(a_{i_1}\beta_{i_1}, \cdots, a_{i_g}\beta_{i_g}, x_i) \neq 0$ $(i \neq i_1, \cdots, i_g)$, and $a_{i_1}\bar{\beta}_{i_1}, \cdots, a_{i_g}\beta_{i_g}$ are all roots of unity. If l is a multiple of the order of $\mathfrak{H}/\mathfrak{H}^0$ such that $(a_{i_1}\bar{\beta}_{i_1})^l = \cdots = (a_{i_g}\beta_{i_g})^l = 1$, then

$$\tau(a_1\beta_1, \cdots, a_t\beta_t, \gamma + 1, \cdots, \gamma + 1)^{jl+1}$$

is a specialization of $\tau(a_1\beta_1, \cdots, a_t\beta_t, \gamma + 1, \cdots, \gamma + 1)$, so that

$$f_i(a_{i_1}\bar{\beta}_{i_1}, \cdots, a_{i_g}\bar{\beta}_{i_g}, (a_i\bar{\beta}_i)^{jl+1}) = 0$$

for every integer j. It follows that $(a_i\bar{\beta}_i)^{jl+1}$ assumes equal values for two distinct integers j, so that $a_i\bar{\beta}_i$ is a root of unity, too. Thus, there is a smallest positive integer k such that $(a_1\bar{\beta}_1)^k = \cdots = (a_t\bar{\beta}_t)^k = 1$. Of course, k is not divisible by p, so that there is an integer j such that $kj \equiv 1 \pmod{p^m}$ (see (4)). For this j, $\tau(1, \cdots, 1, 1, \cdots, 1) = \tau((a_1\bar{\beta}_1)^{kj}, \cdots, (a_t\bar{\beta}_t)^{kj}, kj, \cdots, kj) = \tau(a_1\bar{\beta}_1, \cdots, a_t\bar{\beta}_t, 1, \cdots, 1)^{kj} \in \mathfrak{H}$. Therefore $\mathfrak{H}$ contains $\tau(1, \cdots, 1, c, \cdots, c) = \tau(1, \cdots, 1, 1, \cdots, 1)^c$ for every integer c, and the lemma quickly follows.

An almost immediate consequence of this lemma is the following result.

THEOREM. *Every abelian algebraic matric group* $\mathfrak{A}$ *is a direct product* $\mathfrak{A} = \mathfrak{A}_q \times \mathfrak{A}_a$, *where* $\mathfrak{A}_q$ *and* $\mathfrak{A}_a$ *are algebraic subgroups of* $\mathfrak{A}$ *simultaneously reducible to diagonal form and to special triangular form, respectively. The direct factors* $\mathfrak{A}_q$, $\mathfrak{A}_a$ *are unique.*

PROOF. We assume, without loss of generality, that $\mathfrak{A}$ is of the form

$$\mathfrak{A} = \begin{bmatrix} \mathfrak{A}_1 & & & 0 \\ & \cdot & & \\ & & \cdot & \\ & & & \cdot \\ 0 & & & \mathfrak{A}_t \end{bmatrix}, \tag{5}$$

where each block $\mathfrak{A}_i$ is in triangular form with all the elements on the main diagonal the same (PV §1 Lemma 1). Let $\mathfrak{A}_q = \mathfrak{A} \cap \mathfrak{D}$, $\mathfrak{A}_a = \mathfrak{A} \cap \mathrm{T}_1$ ($\mathfrak{D}$ is the full diagonal group, $\mathfrak{T}_1$ is the full special triangular group). Clearly, $\mathfrak{A}_q \cap \mathfrak{A}_a = \mathfrak{E}$ (the identity group). By the lemma, if $\sigma \in \mathfrak{A}$ then $\sigma = \sigma_q\sigma_a$, where $\sigma_q, \sigma_a \in \mathfrak{A}$, σ_q is reducible to diagonal form, and σ_a is reducible to special triangular form. According to the reduction (5) we have

$$\sigma_q = \begin{bmatrix} \sigma_{q1} & & & 0 \\ & \cdot & & \\ & & \cdot & \\ & & & \cdot \\ 0 & & & \sigma_{qt} \end{bmatrix},$$

where each σ_{qi} is in triangular form with equal characteristic roots. Since σ_q is reducible to diagonal form, each σ_{qi} is, too. Since all the characteristic roots of σ_{qi} are the same, this means that σ_{qi} is already in diagonal form, so that σ_q is, too, that is, $\sigma_q \in \mathfrak{A}_q$. Again, since σ_a is reducible to special triangular form, all its characteristic roots are 1, so that σ_a is already in special triangular form, $\sigma_a \in \mathfrak{A}_a$. Thus, $\mathfrak{A} = \mathfrak{A}_q \times \mathfrak{A}_a$. If we also have $\mathfrak{A} = \mathfrak{A}'_q \times \mathfrak{A}'_a$, where $\mathfrak{A}'_q$ is reducible to diagonal form and $\mathfrak{A}'_a$ is reducible to special triangular form, then for any $\sigma'_q \in \mathfrak{A}'_q$, $\sigma'_a \in \mathfrak{A}'_a$ we see as above that $\sigma'_q \in \mathfrak{A}_q$, $\sigma'_a \in \mathfrak{A}_a$, whence $\mathfrak{A}'_q \subseteq \mathfrak{A}_q$, $\mathfrak{A}'_a \subseteq \mathfrak{A}_a$. From this it follows that the representation $\mathfrak{A} = \mathfrak{A}_q \times \mathfrak{A}_a$ is unique.

We shall call $\mathfrak{A}_q$ and $\mathfrak{A}_a$ the *quasicompact* and *anticompact factors*, respectively, of the abelian algebraic matric group $\mathfrak{A}$.

It is easy to see that the components of the identity of $\mathfrak{A}$, $\mathfrak{A}_q$, and $\mathfrak{A}_a$ enjoy the properties

$$\mathfrak{A}^0 = (\mathfrak{A}_q)^0 \times (\mathfrak{A}_a)^0,$$
$$(\mathfrak{A}_q)^0 = (\mathfrak{A}^0)_q, \qquad (\mathfrak{A}_a)^0 = (\mathfrak{A}^0)_a.$$

If τ_q and τ_a are independent generic points of the underlying manifolds of $\mathfrak{A}_q^0$ and $\mathfrak{A}_a^0$, respectively, then $\tau_q\tau_a$ is clearly a generic point of the underlying manifold of $\mathfrak{A}^0$. It follows that

$$\dim \mathfrak{A} = \dim \mathfrak{A}_q + \dim \mathfrak{A}_a. \tag{6}$$

4. k^{th} powers.

We require three lemmas in this section.

LEMMA 1. *Let $\mathfrak{G}$ be a connected algebraic matric group in diagonal form, let k be a nonzero integer. Then for every $\sigma \in \mathfrak{G}$ there exists a $\rho \in \mathfrak{G}$ such that $\rho^k = \sigma$.*

PROOF. Let $\tau = (\delta_{ij}\beta_j)$ be a generic point if the underlying manifold of $\mathfrak{G}$. Precisely dim $\mathfrak{G}$ of the elements $\beta_1, \cdots, \beta_n$ are algebraically independent over $\mathfrak{C}$. Therefore precisely dim $\mathfrak{G}$ of the elements $\beta_1^k, \cdots, \beta_n^k$ are algebraically independent over $\mathfrak{C}$, so that τ^k is a generic point of the underlying manifold of $\mathfrak{G}$. Therefore, if $\sigma = (\delta_{ij}a_j) \in \mathfrak{G}$, then $(\beta_1^k, \cdots, \beta_n^k) \to (a_1, \cdots, a_n)$ is a specialization. This can be extended to a specialization $(\beta_1^k, \cdots, \beta_n^k, \beta_1^{\delta_1}, \cdots, \beta_n^{\delta_n}) \to (a_1, \cdots, a_n, \alpha_1, \cdots, \alpha_n)$, where each $\delta_i = \pm 1$ and if $\delta_i = -1$ then $\alpha_i = 0$. If some δ_i were actually -1 the relation $1 - \beta_i^k(\beta_i^{\delta_i})^k = 0$ would specialize to $1 - a_i \cdot 0 = 0$, which is impossible. Therefore every $\delta_i = 1$, and $\rho = (\delta_{ij}\alpha_j)$ has the property that $(\tau^k, \tau) \to (\sigma, \rho)$ is a specialization. It follows that $\rho \in \mathfrak{G}$ and $\rho^k = \sigma$.

LEMMA 2. *Let $\mathfrak{G}$ be an algebraic matric group in special triangular form, let k be an integer not divisible by the field characteristic p. Then, for every $\sigma \in \mathfrak{G}$ there is a $\rho \in \mathfrak{G}$ such that $\rho^k = \sigma$.*

PROOF. Let $\sigma \in \mathfrak{G}$, and suppose that σ is in Jordan normal form. Then $\sigma = \tau(1, \cdots, 1, 1, \cdots, 1)$, and $\mathfrak{G}$ contains $\sigma^j = \tau(1, \cdots, 1, j, \cdots, j)$ for every integer j. If $p = 0$ then $\mathfrak{G}$ contains $\tau(1, \cdots, 1, c, \cdots, c)$ for every $c \in \mathfrak{C}$, in particular for $c = k^{-1}$, and $\rho = \tau(1, \cdots, 1, k^{-1}, \cdots, k^{-1})$ has the desired property $\rho^k = \sigma$. If $p \neq 0$ then there is an integer j such that $kj = 1 \pmod{p^m}$, whence (see (4)) $\rho = \tau(1, \cdots, 1, j, \cdots, j)$ has the desired property $\rho^k = \sigma$.

LEMMA 3. *Let $\mathfrak{C}$ be either of characteristic $p = 0$ or transcendental over its prime field. Let $\mathfrak{G}$ be a connected algebraic matric group. For each $\sigma \in \mathfrak{G}$ let $\mathfrak{H}_\sigma$ be the smallest algebraic matric group containing σ. Then*[2] *the set of all matrices $\sigma \in \mathfrak{G}$ such that $\mathfrak{H}_{\sigma q}$ is connected is not contained in an algebraic manifold of dimension $<$ dim $\mathfrak{G}$.*

PROOF. Let τ be a generic point of the underlying manifold of $\mathfrak{G}$ and denote the characteristic roots of τ by $\beta_1, \cdots, \beta_n$. If σ is any matrix in $\mathfrak{G}$, then the specialization $\tau \to \sigma$ can always be extended to a specialization $(\tau, \beta_1, \cdots, \beta_n) \to (\sigma, \bar{\beta}_1, \cdots, \bar{\beta}_n)$. The $\bar{\beta}_i$'s are, of course, the characteristic roots of σ. Letting

[2] It is easy to see, using *PV* §3 Lemma 1, that $\mathfrak{H}_\sigma$ is always abelian. Therefore we may consider, as in §3, its quasicompact factor $\mathfrak{H}_{\sigma q}$.

$\beta_{i_1}, \cdots, \beta_{i_g}$ be a maximal subset of $\beta_1, \cdots, \beta_n$ algebraically independent over $\mathcal{C}$, we suppose that σ has the property that there exists a specialization as above such that no power product $\bar{\beta}_{i_1}^{m_1} \cdots \bar{\beta}_{i_g}^{m_g}$ with integral exponents not all 0 is equal to 1. By §2 Lemma 4 above, the set of all such matrices $\sigma \in \mathfrak{G}$ is not contained in a lower-dimensional algebraic submanifold of the underlying manifold of $\mathfrak{G}$. By §2 Lemma 3, $\mathfrak{H}_{\sigma q}$ is of dimension g. Assuming without loss of generality that σ is in Jordan normal form, we see that $\mathfrak{H}_{\sigma q}^0$ is in diagonal form, so that the underlying manifold of $\mathfrak{H}_{\sigma q}^0$ has a generic point of the form $\rho = (\delta_{ij}\gamma_j)$. The specialization $\tau \to \rho$ can be extended to a specialization of $(\tau, \beta_1, \cdots, \beta_n)$, so that, for a suitable permutation $j_1, \cdots, j_n$ of $1, \cdots, n$, $(\beta_{j_1}, \cdots, \beta_{j_n}) \to (\gamma_1, \cdots, \gamma_n)$ is a specialization. Since $(\beta_{j_1}, \cdots, \beta_{j_n})$ and $(\gamma_1, \cdots, \gamma_n)$ both have the same dimension g, the inverse mapping is a specialization, too. Since $\rho^k \in \mathfrak{H}_{\sigma q}^0$ for every integer k, $(\gamma_1, \cdots, \gamma_n) \to (\gamma_1^k, \cdots, \gamma_n^k)$ is a specialization, whence $(\beta_1, \cdots, \beta_n) \to (\beta_1^k, \cdots, \beta_n^k)$ is, too, and $(\beta_1, \cdots, \beta_n) \to (\bar{\beta}_1^k, \cdots, \bar{\beta}_n^k)$ is a specialization for every integer k. Denoting the element in the ith row and i^{th} column of σ by a_i (so that $a_1, \cdots, a_n$ is a permutation of $\bar{\beta}_1, \cdots, \bar{\beta}_n$), we see, for a suitable permutation $l_1, \cdots, l_n$ of $1, \cdots, n$, that $(\beta_{l_1}, \cdots, \beta_{l_n}) \to (a_1^k, \cdots, a_n^k)$ is a specialization for every integer k. It easily follows that $(\beta_{l_1}, \cdots, \beta_{l_n}) \to (b_1, \cdots, b_n)$ is a specialization for every matrix $(\delta_{ij}b_j)$ in $\mathfrak{H}_{\sigma q}$, and therefore, for the matrix ρ. Thus, we may take $(j_1, \cdots, j_n) = (l_1, \cdots, l_n)$, and we see that $(\delta_{ij}\beta_{l_j})$ is a generic point of the underlying manifold of $\mathfrak{H}_{\sigma q}^0$. By the above, then, every $(\delta_{ij}b_j)$ in $\mathfrak{H}_{\sigma q}$ is in $\mathfrak{H}_{\sigma q}^0$, so that $\mathfrak{H}_{\sigma q}$ is connected, and the proof is complete.

THEOREM. *Let $\mathfrak{G}$ be a connected algebraic matric group, k an integer not divisible by the field characteristic p. If τ is a generic point of the underlying manifold of $\mathfrak{G}$, then so is τ^k.*

PROOF. We suppose, temporarily, that either $p = 0$ or $\mathcal{C}$ is transcendental over its prime field. Let σ be a matrix in $\mathfrak{G}$ such that $\mathfrak{H}_{\sigma q}$ is connected. By the theorem of §3, $\mathfrak{H}_\sigma = \mathfrak{H}_{\sigma q} \times \mathfrak{H}_{\sigma a}$. By Lemmas 1 and 2 every matrix in $\mathfrak{H}_{\sigma q}$ has a kth root in $\mathfrak{H}_{\sigma q}$, and every matrix in $\mathfrak{H}_{\sigma a}$ has a kth root in $\mathfrak{H}_{\sigma a}$. Therefore every matrix in $\mathfrak{H}_\sigma$ has a kth root in $\mathfrak{H}_\sigma$, and in particular, there is a $\rho \in \mathfrak{H}_\sigma$ such that $\rho^k = \sigma$. It follows that $\tau^k \to \sigma$ is a specialization. By Lemma 3, the set of all such σ's is not contained in an algebraic manifold of dimension $<$ dim $\mathfrak{G}$, so that the dimension of the point τ^k equals that of the generic point τ, and τ^k is also a generic point of the underlying manifold of $\mathfrak{G}$.

To remove the temporary restriction on $\mathcal{C}$, now suppose that $p \neq 0$ and $\mathcal{C}$ is algebraic over its prime field. As in the proof of the theorem of §2, we introduce an indeterminate u, the algebraic closure $\mathcal{C}^\dagger$ of the field $\mathcal{C}(u)$, and the connected algebraic matric group $\mathfrak{G}^\dagger$ which is the extension of $\mathfrak{G}$ to the field $\mathcal{C}^\dagger$. By the part of the theorem already proved, a generic point $\tau^\dagger$ of the underlying manifold of $\mathfrak{G}^\dagger$ has the property that $\tau^{\dagger k}$ is also a generic point of the underlying manifold of $\mathfrak{G}^\dagger$. But it is easy to see that a generic point of the underlying manifold of $\mathfrak{G}^\dagger$ is also a generic point of the underlying manifold of $\mathfrak{G}$. Since the kth power $\tau^{\dagger k}$ of the generic point $\tau^\dagger$ of the underlying manifold of $\mathfrak{G}$ is itself a generic point, the same is true of every generic point τ, and the proof of the theorem is complete.

REMARK. The hypothesis in the theorem that k be not divisible by p can not be dispensed with. To see this, consider the connected algebraic group of all matrices $\begin{pmatrix} 1 & c \\ 0 & 1 \end{pmatrix}$ with $c \in \mathfrak{C}$. The matrix $\tau = \begin{pmatrix} 1 & u \\ 0 & 1 \end{pmatrix}$, with u an indeterminate, is a generic point of the underlying manifold, and $\tau^p = \begin{pmatrix} 1 & 0 \\ 0 & 1 \end{pmatrix} = \iota$ is not such a generic point.

5. Group-theoretic characterizations.

By definition (PV §6), an algebraic matric group $\mathfrak{G}$ over an (algebraically closed) field $\mathfrak{C}$ of characteristic p is anticompact when $\mathfrak{G}$ contains no element of finite order >1 not divisible by p. Thus, the criterion for anticompactness is purely group-theoretic; two faithful representations of an abstract group as algebraic matric groups over fields of the same characteristic are either both anticompact or both not anticompact. On the other hand the definitions of connectedness and of component of the identity (PV §2), of solvability (PV §5), and of quasicompactness (PV §6), all make reference to the underlying manifold of $\mathfrak{G}$. Thus it is conceivable, for each of these properties, that two group-theoretically ismorphic algebraic matric groups exist, one enjoying the property, the other failing to do so. The purpose of the present section is to ascertain to what extent this may occur, and to what extent group-theoretic criteria (as opposed to algebraic-group-theoretic criteria) can be given.

THEOREM 1. *If an algebraic matric group $\mathfrak{G}$ over a field $\mathfrak{C}$ of characteristic p is connected then $\mathfrak{G}$ contains no normal proper subgroup of finite index not divisible by p, and no proper subgroup of finite index k such that $k!$ is not divisible by p.*

PROOF. Let $\mathfrak{H}$ be a subgroup (not necessarily algebraic) of $\mathfrak{G}$ of finite index k not divisible by p. If $\mathfrak{H}$ is a normal subgroup then $\sigma^k \in \mathfrak{H}$ for every $\sigma \in \mathfrak{G}$. By the theorem of §4, the set of all matrices of $\mathfrak{G}$ which are not k^{th} powers of matrices of $\mathfrak{G}$ is contained in an algebraic manifold of dimension $<$ dim $\mathfrak{G}$, so that $\mathfrak{G} - \mathfrak{H}$ is contained in such a manifold. By PV §3 Lemma 1, this implies that $\mathfrak{G} = \mathfrak{H}$. If $\mathfrak{H}$ is not normal then it no longer follows that $\sigma^k \in \mathfrak{H}$ for every $\sigma \in \mathfrak{G}$. But it does follow that $\sigma^{k!} \in \mathfrak{H}$ for every $\sigma \in \mathfrak{G}$, as is easy to see. If $k!$ is not divisible by p, then $\mathfrak{G} - \mathfrak{H}$ is contained in an algebraic manifold of dimension $<$ dim $\mathfrak{G}$, and $\mathfrak{G} = \mathfrak{H}$.

REMARK 1. As a corollary we see that if $\mathfrak{G}$ is an algebraic matric group over a field of *characteristic* 0 then $\mathfrak{G}$ is connected if and only if $\mathfrak{G}$ has no proper subgroup of finite index and, more generally, that the component of the identity $\mathfrak{G}^0$ can be characterized as the smallest subgroup of $\mathfrak{G}$ of finite index. The first statement is immediate from Theorem 1. The second follows from the observations that $\mathfrak{G}^0$ is a subgroup of $\mathfrak{G}$ of finite index, and if $\mathfrak{H}$ is any subgroup of $\mathfrak{G}$ of finite index then $\mathfrak{G}^0 \cap \mathfrak{H}$ is a subgroup of $\mathfrak{G}^0$ of finite index, so that (since $\mathfrak{G}^0$ is connected) $\mathfrak{G}^0 \cap \mathfrak{H} = \mathfrak{G}^0$, or $\mathfrak{G}^0 \subseteq \mathfrak{H}$.

REMARK 2. When $p \neq 0$ no purely group-theoretic criterion exists. To see this, let $\mathfrak{G}_1$ be the set of all matrices of the form

$$\begin{pmatrix} 1 & c & 0 & 0 \\ 0 & 1 & 0 & 0 \\ 0 & 0 & 1 & d \\ 0 & 0 & 0 & 1 \end{pmatrix}$$

in which $c, d \in \mathcal{C}$, and let $\mathfrak{G}_a$ be the set of all such matrices in which $c \in \mathcal{C}$ and d has one of the values $0, 1, \cdots, p - 1$. $\mathfrak{G}_1$ is isomorphic with the additive group of the vector space over $\mathcal{C}$ of dimension 2. If we regard this vector space as a linear space over the prime field $\mathcal{C}_p$ of integers (mod p), we see that $\mathfrak{G}_1$ is isomorphic with the additive group of the vector space over $\mathcal{C}_p$ of dimension equal to the linear dimension of $\mathcal{C}$ over $\mathcal{C}_p$. $\mathfrak{G}_2$ is isomorphic with the direct product of the additive group of $\mathcal{C}$ and the additive group of $\mathcal{C}_p$, which again is seen to be isomorphic to a vector space over $\mathcal{C}_p$ of dimension equal to the linear dimension of $\mathcal{C}$ over $\mathcal{C}_p$. Thus, $\mathfrak{G}_1$ and $\mathfrak{G}_2$ are group-theoretically isomorphic. But $\mathfrak{G}_1$ is connected and $\mathfrak{G}_2$ is not.

THEOREM 2. *An algebraic matric group $\mathfrak{G}$ is solvable if and only if it is solvable as an abstract group.*

PROOF. If $\mathfrak{G}$ is solvable as an algebraic matric group there is a chain $\mathfrak{G} = \mathfrak{G}_0 \supset \cdots \supset \mathfrak{G}_r = \mathfrak{E}$ in which each $\mathfrak{G}_i$ is a normal algebraic subgroup of $\mathfrak{G}_{i-1}$ such that $\mathfrak{G}_{i-1}/\mathfrak{G}_i$ is abelian. Therefore $\mathfrak{G}$ is solvable as an abstract group. Conversely, if $\mathfrak{G}$ is solvable as an abstract group, then there is a chain $\mathfrak{G} = \mathfrak{H}_0 \supset \cdots \supset \mathfrak{H}_r = \mathfrak{E}$ in which each $\mathfrak{H}_i$ is a normal subgroup of $\mathfrak{H}_{i-1}$ such that $\mathfrak{H}_{i-1}/\mathfrak{H}_i$ is abelian. By PV §3 Lemma 2, the smallest algebraic manifold containing $\mathfrak{H}_i$ is the underlying manifold of an algebraic matric group, call it $\mathfrak{G}_i$. Clearly $\mathfrak{G} = \mathfrak{G}_0 \supseteq \cdots \supseteq \mathfrak{G}_r = \mathfrak{E}$. We must show that each $\mathfrak{G}_i$ is a normal subgroup of $\mathfrak{G}_{i-1}$ with abelian factor group.

Let $f(\xi) = f(\cdots, x_{ij}, \cdots)$ be any polynomial in $\mathcal{C}[\cdots, x_{ij}, \cdots]$ vanishing whenever ξ is replaced by a matrix in $\mathfrak{H}_i$, that is, any polynomial in the defining ideal of $\mathfrak{G}_i$. For every $\sigma, \tau \in \mathfrak{H}_{i-1}$ we have $\sigma\tau\sigma^{-1}\tau^{-1} \in \mathfrak{H}_i$, so that $f(\sigma\tau\sigma^{-1}\tau^{-1}) = 0$. For a sufficiently large positive integer k, $|\xi|^k f(\sigma\xi\sigma^{-1}\xi^{-1})$ is a polynomial $g(\xi) = g(\cdots, x_{ij}, \cdots) \in \mathcal{C}[\cdots, x_{ij}, \cdots]$. Since $g(\tau)$ vanishes for every $\tau \in \mathfrak{H}_{i-1}$, $g(\tau)$ must also vanish for every $\tau \in \mathfrak{G}_{i-1}$, so that $f(\sigma\tau\sigma^{-1}\tau^{-1}) = 0$ for every $\sigma \in \mathfrak{H}_{i-1}$ and every $\tau \in \mathfrak{G}_{i-1}$. In a similar fashion, we see that this implies that $f(\sigma\tau\sigma^{-1}\tau^{-1}) = 0$ for every $\sigma, \tau \in \mathfrak{G}_{i-1}$. Since this is true for every $f(\xi)$ in the defining ideal of $\mathfrak{G}_i$, this means that $\sigma\tau\sigma^{-1}\tau^{-1} \in \mathfrak{G}_i$ for every $\sigma, \tau \in \mathfrak{G}_{i-1}$. Thus $\mathfrak{G}_i$ is a normal subgroup of $\mathfrak{G}_{i-1}$ with abelian factor group $\mathfrak{G}_{i-1}/\mathfrak{G}_i$.

REMARK 3. Thus, for an algebraic matric group over a field of characteristic p, the concepts "anticompact" and "solvable", and, in the case $p = 0$ (but not otherwise), "connected" and "component of the identity", are all definable group-theoretically. The same may not be said, however, of the concept "quasicompact". This is shown by the following example. Let $\mathfrak{G}_1$ be the set of all nonsingular matrices of the form $\begin{pmatrix} a & 0 \\ 0 & a \end{pmatrix}$ with $a \in \mathcal{C}$, and let $\mathfrak{G}_2$ be the set of all nonsingular matrices of the form $\begin{pmatrix} b & c \\ 0 & b \end{pmatrix}$ with $b, c \in \mathcal{C}$. $\mathfrak{G}_1$ is a connected algebraic matric group, isomorphic with the multiplicative group of the field $\mathcal{C}$. Taking $\mathcal{C}$ to be the field of all complex numbers, we see that $\mathfrak{G}_1$ is group-theoretically isomorphic with the direct product $R \times R_1$, where R is the additive group of all real numbers, and R_1 is the additive group of all real numbers (mod 1). $\mathfrak{G}_2$ is a connected algebraic matric group, and may be written as the direct product $\mathfrak{G}_3 \times \mathfrak{G}_1$, where $\mathfrak{G}_3$ is the group of all matrices of the form $\begin{pmatrix} 1 & c \\ 0 & 1 \end{pmatrix}$ with c complex. $\mathfrak{G}_3$ is clearly isomorphic with the additive group of all complex numbers, that is, with $R \times R$. Thus, $\mathfrak{G}_2$ is isomorphic with $R \times R \times R \times R_1$. But all finite-dimensional vector spaces over an infinite field are isomorphic with each other, or more precisely, their additive groups are group-isomorphic. Therefore $R \times R \times R$ is isomorphic with R, so that $\mathfrak{G}_1$ is isomorphic with $\mathfrak{G}_2$. But $\mathfrak{G}_1$ is quasicompact and $\mathfrak{G}_2$ is not.

Since quasicompactness can not be defined group-theoretically, it is natural to ask what criteria, in addition to group-theoretic ones, suffice to characterize quasicompactness. To this end we prove the following result.

THEOREM 3. *A connected algebraic matric group $\mathfrak{G}$ of dimension g is quasicompact if and only if:*

1. *$\mathfrak{G}$ is abelian;*
2. *there exists a positive integer h such that, if $r(k)$ denotes the number of elements $\sigma \in \mathfrak{G}$ for which $\sigma^k = \iota$, then*

$$k^g \leqq r(k) \leqq hk^g \tag{7}$$

for all positive integers k relatively prime to h.

PROOF. If the connected algebraic matric group $\mathfrak{G}$ is quasicompact then, by the theorem of §2, $\mathfrak{G}$ is reducible to diagonal form, and therefore is abelian. Hence it is easy to see that the theorem follows from the results of §3 and the following lemma.

LEMMA. *Let $\mathfrak{A}$ be a connected abelian algebraic matric group, let $r(k)$ denote the number of matrices $\sigma \in \mathfrak{A}$ for which $\sigma^k = \iota$, and let $g = \dim \mathfrak{A}_q$. Then there exists a positive integer h such that (7) holds for all positive integers k relatively prime to h.*

PROOF. By §3 we may write $\mathfrak{A} = \mathfrak{A}_q \times \mathfrak{A}_a$, where $\mathfrak{A}_q$ and $\mathfrak{A}_a$ are the quasicompact and anticompact factors, respectively, of $\mathfrak{A}$, and are connected. If k is a positive integer not divisible by p then it is obvious that $r(k)$ equals the number of matrices $\sigma \in \mathfrak{A}_q$ for which $\sigma^k = \iota$. By §2, $\mathfrak{A}_q$ is reducible to diagonal form, so that we may assume that the underlying manifold of $\mathfrak{A}_q$ has a generic point $(\delta_{ij}\alpha_j)$ with $\alpha_1, \cdots, \alpha_g$ algebraically independent over $\mathcal{C}$. If a matrix $\sigma = (\delta_{ij}a_j) \in \mathfrak{A}_q$ satisfies $\sigma^k = \iota$ (positive integer k not divisible by p), then each a_j is a kth root of unity. Referring to §2 Lemma 2, above, we see that for each choice of $a_1, \cdots, a_g$ there are fewer than $m_{g+1}m_{g+2} \cdots m_n$ matrices $(\delta_{ij}a_j)$ which are in $\mathfrak{A}_q$. Therefore, the number of matrices $\sigma \in \mathfrak{A}_q$ for which $\sigma^k = \iota$ is $r(k) \leqq k^g m_{g+1} \cdots m_n$.

On the other hand, let $a_1, \cdots, a_g$ be any k^{th} roots of unity, where k is a positive integer not divisible by p and relatively prime to $s = m_{g+1} \cdots m_n$. By the lemma of PV §5, the set of all matrices $(\delta_{ij}b_j)_{i,j=1,\cdots,g}$ which can be enlarged to a matrix $(\delta_{ij}b_j)_{i,j=1,\cdots,n}$ in $\mathfrak{A}_q$ is a connected algebraic group. The dimension of this group is obviously g, that is, it is the full diagonal group of degree g. Therefor $\mathfrak{A}_q$ contains at least one enlargement $(\delta_{ij}a_j)_{i,j=1,\cdots,n}$ of the matrix $(\delta_{ij}a_j)_{i,j=1,\cdots,g}$ By §2 Lemma 2, each a_j with $j > g$ is an $(sk)^{\text{th}}$ root of unity. Therefore $\tau = (\delta_{ij}a_j^s)$ is a matrix in $\mathfrak{A}_q$ with the property that $\tau^k = \iota$. Moreover, since k and s are relatively prime, two different choices of kth roots of unity $a_1, \cdots, a_g$ lead to two different matrices $\tau = (\delta_{ij}a_j^s)$. As there are k^g different choices possible for $a_1, \cdots, a_g$, we see that $r(k) \geqq k^g$, and the proof is complete.

REMARK 4. Theorem 3 gives a group-theoretic criterion for the quasicompactness of a *connected* algebraic matric group of *given dimension*. As shown above (Remark 3), we can not find such a criterion if we omit specification of the dimension. We can, however, *in the case* $p = 0$, omit the assumption of connectedness. For in that case $\mathfrak{G}$ is quasicompact if and only if G^0 is (PV §8 Theorem 3), and we already have a group-theoretic characterization of $\mathfrak{G}^0$ (Remark 1, above).

COLUMBIA UNIVERSITY

EXISTENCE THEOREMS CONNECTED WITH THE PICARD-VESSIOT THEORY OF HOMOGENEOUS LINEAR ORDINARY DIFFERENTIAL EQUATIONS

E. R. KOLCHIN

1. Introduction. The Picard-Vessiot theory, as recently reformulated by the author,[1] deals with an abstract ordinary differential field $\mathcal{F}$ of characteristic 0 having an algebraically closed field of constants $\mathcal{C}$, and a differential extension field $\mathcal{G}$ over $\mathcal{F}$ with the two properties:

(a) There exists a homogeneous linear differential polynomial $L(y)=y^{(n)}+p_1y^{(n-1)}+\cdots+p_ny$ (each p_i in $\mathcal{F}$) which has a fundamental system of solutions $\eta_1, \cdots, \eta_n$ such that $\mathcal{G}=\mathcal{F}\langle\eta_1, \cdots, \eta_n\rangle$;[2]

(b) The field of constants of $\mathcal{G}$ is $\mathcal{C}$.

Such a $\mathcal{G}$ is called a *Picard-Vessiot* extension of $\mathcal{F}$. It is to be noted that the extension $\mathcal{G}$ is given, and the existence of the differential polynomial $L(y)$ with the properties (a) and (b) is postulated. It is not immediately apparent, and it would be of interest to know, whether a given $L(y)$, with coefficients p_i in $\mathcal{F}$, always has a fundamental system of solutions $\eta_1, \cdots, \eta_n$ such that $\mathcal{F}\langle\eta_1, \cdots, \eta_n\rangle$ is a Picard-Vessiot extension of $\mathcal{F}$ (that is, contains no constant not in $\mathcal{C}$). This question was posed by R. Baer (in his critical note on the then current status of the Picard-Vessiot theory, included among comments by O. Haupt in F. Klein's *Vorlesungen über hypergeometrische Funktionen*, Berlin, 1933), who remarked that the difficulty lay not in proving the existence of a fundamental system of solutions (see PV, §15), but in proving the existence of one which brings in no new constants.

A differential extension field $\mathcal{K}$ of $\mathcal{F}$ may be an extension of $\mathcal{F}$ by integrals, exponentials of integrals, and algebraic functions. If it is, and if the field of constants of $\mathcal{K}$ is still $\mathcal{C}$, then $\mathcal{K}$ is called a *liouvillian* extension of $\mathcal{F}$. The Picard-Vessiot theory provides a group-theoretic answer to the question of when a Picard-Vessiot extension $\mathcal{G}$ of $\mathcal{F}$ is

Presented to the Society, April 17, 1948; received by the editors November 7, 1947.

[1] *Algebraic matric groups and the Picard-Vessiot theory of homogeneous linear ordinary differential equations*, Ann. of Math. (2) vol. 49 (1948) pp. 1–42. This paper, referred to below as "PV", contains the necessary background for the present note.

[2] The notation $\mathcal{F}\langle\cdots\rangle$ indicates, as usual, differential field adjunction. Thus $\mathcal{F}\langle\eta_1, \cdots, \eta_n\rangle$ is the differential field consisting of all differential rational functions of $\eta_1, \cdots, \eta_n$ with coefficients in $\mathcal{F}$.

(contained in) a liouvillian extension $\mathcal{K}$ (PV, §25). It is natural to broaden this question in two respects: first, by relinquishing the requirement that the field of constants of $\mathcal{K}$ be $\mathcal{C}$ (that is, by demanding only that $\mathcal{K}$ be an extension of $\mathcal{F}$ by integrals, exponentials of integrals, and algebraic functions, without demanding that $\mathcal{K}$ be liouvillian); second, by inquiring whether or not $L(y)$ has at least one solution contained in such an extension $\mathcal{K}$.

It is the purpose of the present note to show how answers to the points raised in the two preceding paragraphs can be obtained (see Theorems 2 and 4, below) as corollaries to a general theorem on algebraic differential equations due to J. F. Ritt. This theorem can be formulated in the following way ($\mathcal{F}$ is an ordinary differential field of characteristic 0, $y_1, \cdots, y_n$ are unknowns, and m is a nonnegative integer less than n).

RITT'S THEOREM. *Let* Π *be a prime differential ideal in* $\mathcal{F}\{y_1, \cdots, y_n\}$,[3] *let* J *be a differential polynomial in* $\mathcal{F}\{y_1, \cdots, y_n\}$ *but not in* Π, *let* $\Pi_0 = \Pi \cap \mathcal{F}\{y_1, \cdots, y_m\}$. *Then there exists a differential polynomial* J_0 *in* $\mathcal{F}\{y_1, \cdots, y_m\}$ *but not in* Π_0 *such that every solution of* Π_0 *which is not a solution of* J_0 *can be completed*[4] *into a solution of* Π *which is not a solution of* J.

In Ritt's proof (Trans. Amer. Math. Soc. vol. 48 (1940) pp. 542–552; see especially pp. 543–545) it is assumed that $\mathcal{F}$ consists of functions of a complex variable meromorphic in a given region, but it is not very difficult to modify his proof to obtain a purely algebraic one, valid for abstract $\mathcal{F}$.

2. **The existence theorems.** We work with an ordinary differential field $\mathcal{F}$ of characteristic 0 with an algebraically closed field of constants $\mathcal{C}$.

THEOREM 1. *Let* Σ *be a proper subset of* $\mathcal{F}\{y_1, \cdots, y_n\}$, *and let* J *be a differential polynomial in* $\mathcal{F}\{y_1, \cdots, y_n\}$ *but not in the perfect differential ideal* $\{\Sigma\}$. *Then* Σ *has a solution* $\eta_1, \cdots, \eta_n$ *for which* $J \neq 0$ *and the field of constants of* $\mathcal{F}\langle\eta_1, \cdots, \eta_n\rangle$ *is* $\mathcal{C}$.

PROOF. Since $J \notin \{\Sigma\}$, we have $J \notin \Pi$, where Π is one of the prime components of $\{\Sigma\}$ (in the representation of $\{\Sigma\}$ as an intersection

[3] The notation $\mathcal{F}\{\cdots\}$ indicates, as usual, differential ring adjunction. Thus $\mathcal{F}\{y_1, \cdots, y_n\}$ is the differential ring consisting of all differential polynomials in $y_1, \cdots, y_n$ with coefficients in $\mathcal{F}$.

[4] Following Ritt, we say that a solution $\eta_1, \cdots, \eta_m$ of Π_0 can be *completed* into a solution of Π provided there exist elements $\eta_{m+1}, \cdots, \eta_n$ in some extension of $\mathcal{F}\langle\eta_1, \cdots, \eta_m\rangle$ such that $\eta_1, \cdots, \eta_n$ is a solution of Π.

of prime differential ideals none of which contains any other). It follows that Π has a solution $\zeta_1, \cdots, \zeta_n$, which is not a solution of J, such that each ζ_i is differentially algebraic over $\mathcal{F}$. (To see this let $y_1, \cdots, y_m$ be a complete set of arbitrary unknowns for Π; by Ritt's theorem any solution of Π_m which is not a solution of a certain differential polynomial $J_0 \notin \Pi_m$ can be completed to a solution of Π which is not a solution of J; but $\Pi_m = (0)$, so that any set $\zeta_1, \cdots, \zeta_m$ of differentially algebraic elements not annulling J_0 can be completed to a solution $\zeta_1, \cdots, \zeta_n$ of Π not annulling J; clearly this solution has the desired property.) We suppose that of all solutions $\zeta_1, \cdots, \zeta_n$ of Π, not annulling J and having the property that each ζ_i is differentially algebraic over $\mathcal{F}$, ours leads to the smallest value of $p = $ degree of transcendency of $\mathcal{F}\langle \zeta_1, \cdots, \zeta_n \rangle$ over $\mathcal{F}$.

Let γ be a constant in $\mathcal{F}\langle \zeta_1, \cdots, \zeta_n \rangle$, and suppose that γ is transcendental over $\mathcal{C}$. Then (PV, §14, Theorem 2) γ is transcendental over $\mathcal{F}$, too. Introducing a new unknown w, let Γ be the prime differential ideal in $\mathcal{F}\{w, y_1, \cdots, y_n\}$ with generic solution (PV, §11) $\gamma, \zeta_1, \cdots, \zeta_n$. Then $\Gamma_0 = \Gamma \cap \mathcal{F}\{w\}$, which has the generic solution γ, equals $[w']$. By Ritt's theorem there is a $J_0 \in \mathcal{F}\{w\}$, with $J_0 \notin [w']$, such that every solution of $[w']$ (that is, any constant) not annulling J_0 can be completed into a solution of Γ not annulling J. Since there obviously exists a constant $c \in \mathcal{C}$ for which $J_0 \neq 0$, Γ has a solution $c, \eta_1, \cdots, \eta_n$ for which $J \neq 0$. By Gourin's theorem (PV, §13) we have: degree of transcendency of $\mathcal{F}\langle c, \eta_1, \cdots, \eta_n \rangle$ over $\mathcal{F}$ is less than degree of transcendency of $\mathcal{F}\langle \gamma, \zeta_1, \cdots, \zeta_n \rangle$ over $\mathcal{F}$. Therefore $\eta_1, \cdots, \eta_n$ is a solution of Π not annulling J with the property that each η_i is differentially algebraic over $\mathcal{F}$, and we have: degree of transcendency of $\mathcal{F}\langle \eta_1, \cdots, \eta_n \rangle$ over $\mathcal{F}$ is less than p. This contradicts the definition of p, shows that γ is algebraic over $\mathcal{C}$, and proves the theorem.

THEOREM 2. *If $L(y) = y^{(n)} + p_1 y^{(n-1)} + \cdots + p_n y$, where each $p_i \in \mathcal{F}$, then $L(y)$ has a fundamental system of solutions $\eta_1, \cdots, \eta_n$ such that the field of constants of $\mathcal{F}\langle \eta_1, \cdots, \eta_n \rangle$ is $\mathcal{C}$.*

PROOF. The wronskian determinant $W(y_1, \cdots, y_n)$ is not contained in $\{L(y_1), \cdots, L(y_n)\}$, as $W(y_1, \cdots, y_n)$ is of order $n-1$. By Theorem 1, $\{L(y_1), \cdots, L(y_n)\}$ has a solution $\eta_1, \cdots, \eta_n$ such that $W(\eta_1, \cdots, \eta_n) \neq 0$ and the field of constants of $\mathcal{F}\langle \eta_1, \cdots, \eta_n \rangle$ is $\mathcal{C}$. Since $W(\eta_1, \cdots, \eta_n) \neq 0$, the elements $\eta_1, \cdots, \eta_n$ are linearly independent over constants, and constitute a fundamental system of solutions of $L(y)$.

REMARK. As observed by the referee, Theorem 2 extends to systems $L_i(y_1, \cdots, y_n) = 0$, $i = 1, \cdots, n$, where $L_i(y_1, \cdots, y_n) = y_i'$

$-\sum_{j=1}^{n} a_{ij}y_j$ (each $a_{ij}\in\mathcal{F}$). It is merely necessary to introduce n^2 unknowns y_{ij} and to note that $D(\cdots, y_{ij}, \cdots)=\det |y_{ij}|$ is of order 0 and therefore not contained in $\{\cdots, L_i(y_{j1}, \cdots, y_{jn}), \cdots\}$. The proof then proceeds as above.

We now consider differential fields which are extensions of $\mathcal{F}$ by integrals, exponentials of integrals, and algebraic functions, that is, extensions of the form $\mathcal{F}\langle\alpha_1, \cdots, \alpha_m\rangle$ where either $\alpha_i' \in\mathcal{F}\langle\alpha_1, \cdots, \alpha_{i-1}\rangle$ or $\alpha_i'/\alpha_i\in\mathcal{F}\langle\alpha_1, \cdots, \alpha_{i-1}\rangle$ or α_i is algebraic over $\mathcal{F}\langle\alpha_1, \cdots, \alpha_{i-1}\rangle$, $i=1, \cdots, m$.

If $\mathcal{K}$ is such an extension of $\mathcal{F}$, and if in addition the field of constants of $\mathcal{K}$ is again $\mathcal{C}$, then $\mathcal{K}$ is said to be a *liouvillian* extension of $\mathcal{F}$.

In any case, we introduce (corresponding to PV, §24) ten types of extensions by integrals, exponentials of integrals and algebraic functions, namely, extensions by

(1) integrals, exponentials of integrals, and algebraic functions,
(2) integrals and exponentials of integrals,
(3) exponentials of integrals,
(4) integrals and algebraic functions,
(5) integrals and radicals,
(6) exponentials of integrals,
(7) integrals,
(8) algebraic functions,
(9) radicals,
(10) rational functions.

THEOREM 3. *Let $\Sigma\subset\mathcal{F}\{y_1, \cdots, y_n\}$, $J\in\mathcal{F}\{y_1, \cdots, y_n\}$, $J\notin\{\Sigma\}$. If Σ has a solution $\zeta_1, \cdots, \zeta_n$ for which $J\neq 0$ such that $\mathcal{F}\langle\zeta_1, \cdots, \zeta_n\rangle$ is contained in an extension of $\mathcal{F}$ of one of the types* (1)–(10), *then Σ has a solution $\eta_1, \cdots, \eta_n$ for which $J\neq 0$ such that $\mathcal{F}\langle\eta_1, \cdots, \eta_n\rangle$ is contained in a liouvillian extension of $\mathcal{F}$ of the same type.*

PROOF. Suppose $\mathcal{F}\langle\zeta_1, \cdots, \zeta_n\rangle\subseteq\mathcal{F}\langle\alpha_1, \cdots, \alpha_m\rangle$, where each α_i is, appropriately, either an integral of an element of $\mathcal{F}\langle\alpha_1, \cdots, \alpha_{i-1}\rangle$, an exponential of an integral of such an element, or algebraic over $\mathcal{F}\langle\alpha_1, \cdots, \alpha_{i-1}\rangle$. Then there exist differential polynomials $P_i(u_1, \cdots, u_m)$, $Q(u_1, \cdots, u_m)\in\mathcal{F}\{u_1, \cdots, u_m\}$, where $u_1, \cdots, u_m$ are new unknowns, such that $\zeta_i=P_i(\alpha_1, \cdots, \alpha_m)/Q_i(\alpha_1, \cdots, \alpha_m)$, $i=1, \cdots, m$. There also exist, for $i=1, \cdots, m$, differential polynomials $M_i(u_1, \cdots, u_{i-1})$, $N_i(u_1, \cdots, u_{i-1})\in\mathcal{F}\{u_1, \cdots, u_{i-1}\}$ such that either $\alpha_i' =M_i(\alpha_1, \cdots, \alpha_{i-1})/N_i(\alpha_1, \cdots, \alpha_{i-1})$, or $\alpha_i'/\alpha_i =M_i(\alpha_1, \cdots, \alpha_{i-1})/N_i(\alpha_1, \cdots, \alpha_{i-1})$, or $\alpha_iN_i(\alpha_1, \cdots, \alpha_{i-1})$ is integral and algebraic over $\mathcal{F}\{\alpha_1, \cdots, \alpha_{i-1}\}$. Let Λ be the prime differential ideal in $\mathcal{F}\{y_1, \cdots, y_n, u_1, \cdots, u_m\}$ with generic solution $\zeta_1, \cdots, \zeta_n, \alpha_1, \cdots, \alpha_m$. By Theorem 1, Λ has a solution $\eta_1, \cdots, \eta_n, \beta_1, \cdots, \beta_m$ for which $J(\eta_1, \cdots, \eta_n)Q_1(\beta_1, \cdots, \beta_m) \cdots Q_n(\beta_1, \cdots, \beta_m)N_1 \cdots N_m(\beta_1, \cdots, \beta_{m-1})\beta_1 \cdots \beta_m \neq 0$, such that the field of constants of $\mathcal{F}\langle\eta_1, \cdots, \eta_n, \beta_1, \cdots, \beta_m\rangle$ is $\mathcal{C}$. It is now obvious that $\eta_1, \cdots, \eta_n$ is a solution of Σ not annulling J, that $\mathcal{F}\langle\eta_1, \cdots, \eta_n\rangle \subseteq \mathcal{F}\langle\beta_1, \cdots, \beta_m\rangle$, and that $\mathcal{F}\langle\beta_1, \cdots, \beta_m\rangle$ is a liouvillian extension of $\mathcal{F}$ of the same type as $\mathcal{F}\langle\alpha_1, \cdots, \alpha_m\rangle$.

A homogeneous linear differential polynomial $L(y) = y^{(n)} + p_1 y^{(n-1)} + \cdots + p_n y$, with each p_i in $\mathcal{F}$, is *linearly reducible* over $\mathcal{F}$ if there exist two homogeneous linear differential polynomials $M(y)$ and $N(y)$ with coefficients in $\mathcal{F}$ and of positive order, such that $L(y) = M(N(y))$. If $L(y)$ is not linearly reducible over $\mathcal{F}$ it is *linearly irreducible* over $\mathcal{F}$.

THEOREM 4. *Let $L(y) = y^{(n)} + p_1 y^{(n-1)} + \cdots + p_n y$, with each p_i in $\mathcal{F}$, be linearly irreducible over $\mathcal{F}$. If $L(y)$ has one solution contained in an extension of $\mathcal{F}$ of one of the types* (1)–(10), *then $L(y)$ has a fundamental system of solutions $\eta_1, \cdots, \eta_n$ such that $\mathcal{F}\langle\eta_1, \cdots, \eta_n\rangle$ is a liouvillian extension of $\mathcal{F}$ of the same type.*

PROOF. By Theorem 3, $L(y)$ has a solution η contained in a liouvillian extension $\mathcal{H}$ of $\mathcal{F}$ of the required type. By Theorem 2, $L(y)$ has a fundamental system of solutions $\zeta_1, \cdots, \zeta_n$ such that the field of constants of $\mathcal{H}\langle\zeta_1, \cdots, \zeta_n\rangle$ is $\mathcal{C}$. Therefore η is a linear combination over $\mathcal{C}$ of $\zeta_1, \cdots, \zeta_n$, so that $\mathcal{G} = \mathcal{F}\langle\zeta_1, \cdots, \zeta_n\rangle$ is a Picard-Vessiot extension of $\mathcal{F}$ containing η. Letting $\mathfrak{G}$ be the group of all automorphisms of $\mathcal{G}$ over $\mathcal{F}$, we see that, in the linear space over $\mathcal{C}$ spanned by $\zeta_1, \cdots, \zeta_n$, the linear subspace spanned by the set of all elements $\sigma\eta$ ($\sigma \in \mathfrak{G}$) is invariant under $\mathfrak{G}$. Since $L(y)$ is linearly irreducible, the only invariant linear subspaces are the zero space and the whole space (PV, §22, Theorem 1). Therefore there exist n automorphisms $\sigma_1, \cdots, \sigma_n$ in $\mathfrak{G}$ such that $\sigma_1\eta, \cdots, \sigma_n\eta$ are linearly independent over $\mathcal{C}$.

Now, σ_1 can be extended to an isomorphism τ_1 of $\mathcal{H}_1 = \mathcal{H}\langle\zeta_1, \cdots, \zeta_n\rangle$ over $\mathcal{F}$; σ_2 can be extended to an isomorphism τ_2 of $\mathcal{H}_2 = \mathcal{F}\langle\mathcal{H}_1, \tau_1\mathcal{H}_1\rangle$ over $\mathcal{F}$; $\cdots$; σ_n can be extended to an isomorphism τ_n of $\mathcal{H}_n = \mathcal{F}\langle\mathcal{H}_{n-1}, \tau_{n-1}\mathcal{H}_{n-1}\rangle$ over $\mathcal{F}$. Therefore $\sigma_1\eta, \cdots, \sigma_n\eta$ is a fundamental system of solutions of $L(y)$ contained in $\mathcal{H}_0 = \mathcal{F}\langle\tau_1\mathcal{H}, \cdots, \tau_n\mathcal{H}\rangle$. That is, $\sigma_1\eta, \cdots, \sigma_n\eta$ is a solution of $\{L(y_1), \cdots, L(y_n)\}$ for which $W(\sigma_1\eta, \cdots, \sigma_n\eta) \neq 0$, such that $\mathcal{F}\langle\sigma_1\eta, \cdots, \sigma_n\eta\rangle \subseteq \mathcal{H}_0$. Now, $\mathcal{H}_0$ is clearly an extension of $\mathcal{F}$ of the same type that $\mathcal{H}$ is. Therefore, by Theorem 3, $\{L(y_1), \cdots, L(y_n)\}$ has a solution $\eta_1, \cdots, \eta_n$ for which $W(\eta_1, \cdots, \eta_n) \neq 0$, such that

$\mathcal{J}\langle\eta_1, \cdots, \eta_n\rangle$ is contained in a liouvillian extension of $\mathcal{J}$ of the required type. It follows that $\eta_1, \cdots, \eta_n$ is a fundamental system of solutions of $L(y)$, and (see PV, §25) $\mathcal{J}\langle\eta_1, \cdots, \eta_n\rangle$ is itself a liouvillian extension of $\mathcal{J}$ of the required type.

COLUMBIA UNIVERSITY

Algebraic groups and differential equations

E. R. Kolchin

Introduction

My point of departure is the Picard-Vessiot theory. That theory deals with an ordinary differential field F of characteristic C with algebraically closed field of constants C, and with a differential extension field G of F such that (a) there exists a homogeneous linear differential polynomial $y^{(n)} - p_1 y^{(n-1)} - \dots - p_n y$ with coefficients p_i in F which has a fundamental system of solutions $\eta_1, \dots, \eta_n$ for which $F\langle \eta_1, \dots, \eta_n \rangle = G$, and (b) the field of constants of G is C (such a G being called a Picard-Vessiot extension of F). The main results of the Picard-Vessiot theory are : 1) The group of automorphisms of G over F is isomorphic with an algebraic matric group $\mathfrak{G}$ over C; 2) G is normal over F in the sense that for every differential field F_1 intermediate to F and G and for every element α in G but not in F_1, there exists an automorphism of G over F_1 under which α is not invariant (this implies that there is a 1 - to-1 galois correspondence between the intermediate differential fields and certain subgroups of $\mathfrak{G}$; these certain sub-groups are precisely the algebraic ones, the degree of transcendence of G over F equals the dimension of $\mathfrak{G}$, and if $\mathfrak{G}_1$ is a normal algebraic subgroup of $\mathfrak{G}$ and F_1 the corresponding intermediate differential field then F_1 is normal over F and the group of automorphisms of F_1 over F is isomorphic with $\mathcal{G}/\mathcal{G}_1$); 3) G is (contained in) an extension of F by integrals, exponentials of integrals, and algebraic functions if and only if the component of the identity of $\mathcal{G}$ is solvable.

There are three directions in which it is natural to try to extend the results of the Picard-Vessiot theory: 1) within the frame-work of that theory to investigate other types of "solvability", that is, to seek types of extensions other than those of 3) above and to seek conditions under which a Picard-Vessiot extension G is contained in extensions of these types; 2) to investigate normal extensions other than Picard-Vessiot ones (if such exist); 3) to extend the concept of algebraic group, that is, a group defined on an algebraic variety (or a portion thereof) in a suitable way.

The purpose of the present talk is to describe some initial gropings in all three directions.

1. Algebraic groups.

Let $\mathcal{G}$ be a group, the elements of which are points in a finite cartesian product $\prod P_{n_i}(C)$ of projective spaces $P_{n_i}(C)$ over an algebraically closed field C of characteristic 0. Let $\mathcal{T}$ be the graph of the mapping $(\sigma, \tau) \longrightarrow \sigma\tau$ of $\mathcal{G} \times \mathcal{G}$ onto $\mathcal{G}$ (so that $\mathcal{T} \subseteq \prod P_{n_i}(C) \times \prod P_{n_i}(C) \times \prod P_{n_i}(C)$). For any set $\mathcal{M}$ in a cartesian product of projective spaces over C we let $\mathcal{M}^*$ denote the smallest algebraic variety containing $\mathcal{M}$. We shall say that $\mathcal{G}$ is an <u>algebraic group</u> provided:

1) $\mathfrak{G}^* - \mathfrak{G}$ is an algebraic variety;

2) $\mathfrak{T}^* - \mathfrak{T}$ is an algebraic variety;

3) $(\mathfrak{G} \times \mathfrak{G} \times \mathfrak{G}^*) \cap \mathfrak{T}^* = (\mathfrak{G} \times \mathfrak{G}^* \times \mathfrak{G}) \cap \mathfrak{T}^*$

$$= (\mathfrak{G}^* \times \mathfrak{G} \times \mathfrak{G}) \cap \mathfrak{T}^* = \mathfrak{T}.$$

We call $\mathfrak{G}^*$ the <u>underlying variety</u> of $\mathfrak{G}$.[1]

---

[1] Note. It was brought out in the discussion following the talk that a definition of algebraic group given by A. Weil in a forthcoming monograph, applicable to any abstract algebraic variety over any field, coincides with the definition given here for fields of characteristic 0. In Weil's definition 2), 3) are replaced by the requirement that the mappings $(\sigma, \tau) \longrightarrow \sigma\tau$, $\sigma \longrightarrow \sigma^{-1}$ be rational and regular at every point of $\mathfrak{G} \times \mathfrak{G}$, $\mathfrak{G}$ respectively.

It is easy to show that: if $\mathfrak{G}_1^*$, $\mathfrak{G}_2^*$ are two distinct irreducible components of $\mathfrak{G}^*$ then $\mathfrak{G}_1^* \cap \mathfrak{G}_2^* \cap \mathfrak{G}$ is empty; all the irreducible components have the same dimension (called the dimension of the group); the unique irreducible component containing ι (the identity element of $\mathfrak{G}$) is the underlying variety of a normal algebraic subgroup $\mathfrak{G}^0$ of $\mathfrak{G}$; the index of $\mathfrak{G}^0$ in $\mathfrak{G}$ is the number of irreducible components of $\mathfrak{G}^*$. We call $\mathfrak{G}^0$ the <u>component of the identity</u> of $\mathfrak{G}$, and say that $\mathfrak{G}$ is <u>connected</u> if $\mathfrak{G} = \mathfrak{G}^0$.

Let $\mathfrak{H}$ be a subgroup (not assumed algebraic) of the algebraic group $\mathfrak{G}$. It is not hard to see that $\mathfrak{H}^* \cap \mathfrak{G}$

is an algebraic group; furthermore, if $\mathfrak{G} - \mathfrak{H}$ is contained in an algebraic variety of lower dimension than that of $\mathfrak{G}$ then $\mathfrak{H} = \mathfrak{G}$.

As a consequence of these two remarks it follows that if $\mathfrak{N}$ is a normal algebraic subgroup of the algebraic group $\mathfrak{G}$ and if $\mathfrak{H}$ is an algebraic subgroup of $\mathfrak{G}$ then $\mathfrak{H}\mathfrak{N}$ is an algebraic group.

This fact makes it possible to take over Zassenhaus' proof of the Jordan-Holder-Schreier theorem to obtain: any two algebraic normal chains of an algebraic group $\mathfrak{G}$ have isomorphic refinements. (We call a normal chain algebraic if all the groups in the chain are algebraic.) A particular algebraic normal chain is the chain

$$\mathfrak{G} \supseteq \mathfrak{G}^{0} \supseteq \mathfrak{G}^{0'} \supseteq \mathfrak{G}^{0''} \supseteq \cdots \mathfrak{G}^{0(j)} \supsetneqq \{1\}$$

made up of $\mathfrak{G}^0$ and the sequence of commutator groups of $\mathfrak{G}^0$. This normal algebraic chain is algebraic because, in general, the commutator group of a connected algebraic group is itself a connected algebraic group.

As examples of algebraic groups we have:

1) any finite group (whose elements are points in a cartesian product of projective spaces);

2) any algebraic matric group;

3)the irredicible curve in $P_2(C)$ defined by the form

$$x_0x_2^{2} - 4x_1^{3} + g_2x_0^{2}x_1 + g_3x_0^{3} \quad (g_2, g_3 \in C,\ g_2^{3} - 27g_3^{2} \neq 0),$$

with the law of composition for two independent generic points

$\xi = (1:\xi_1:\xi_2)$, $\eta = (1:\eta_1:\eta_2)$ defined by $\xi\eta = (1:\zeta_1:\zeta_2)$ where

$$\zeta_1 = -\xi_1 - \eta_1 + \frac{1}{4}\left(\frac{\xi_2-\eta_2}{\xi_1-\eta_1}\right)^2,$$

$$\zeta_2 = \frac{\xi_2\eta_1 - \xi_1\eta_2}{\xi_1-\eta_1} + (\xi_1+\eta_1)\frac{\xi_2-\eta_2}{\xi_1-\eta_1} - \frac{1}{4}\left(\frac{\xi_2-\eta_2}{\xi_1-\eta_1}\right)^3$$

(addition theorem for Weierstrass $\wp$-function!), and with the law of composition for algebraic points defined by specialization (the identity element of this group is (1:0:0)).

This last algebraic group is not isomorphic with any algebraic matric group of dimension 1 over any algebraically closed field.

2. $\wp$-functions of integrals.

Let F by an ordinary differential field of characteristic 0 with algebraically closed field of constants C. Let $a \in F$, $g_2, g_3 \in C$ ($a \neq 0$, $g_2^3 - 27g_3^2 \neq 0$). An element α of a differential extension field of F will be called a <u>$\wp$-function of an integral of a</u> (with respect to g_2, g_3) provided α is in the general solution of the differential polynomial $y'^2 - a^2(4y^3 - g_2y - g_3)$, that is, provided $\alpha'^2 = a^2(4\alpha^3 - g_3\alpha - g_3) \neq 0$.

If α is a $\wp$ - function of an integral of a, then the point $(1:\alpha:\alpha'/a)$ is an element of the 1-dimensional algebraic group mentioned at the end of § 1 (which group we now call $\mathfrak{G}_\wp$). It is a simple matter to verify that if β is another $\wp$-function of an integral of a (with respect to the same g_2, g_3) contained together with α in an extension of F then the coordinate ratios of the point $(1:\alpha:\alpha'/a)^{-1}$ $(1:\beta:\beta'/a)$ are constants (i.e. have derivative 0). Using this fact we can show the following:

Let α be a $\wp$-function of an element of F such that α is transcendental over F and the field of constants of F $\langle\alpha\rangle$ is C. Then: F $\langle\alpha\rangle$ is a normal extension of F; the mapping

$$\sigma \longrightarrow (1:\alpha:\alpha'/a)^{-1}\ (1:\sigma\alpha:\sigma\alpha'/a)$$

is an isomorphism of the group of all automorphisms σ of F $\langle\alpha\rangle$ over F onto the algebraic group $\mathfrak{G}_\wp$; the subgroups of $\mathfrak{G}_\wp$ corresponding to the differential fields intermediate to F and F $\langle\alpha\rangle$ are precisely the algebraic subgroups.

As a consequence we see that the only differential fields intermediate to F and F $\langle\alpha\rangle$ are of finite degree under F $\langle\alpha\rangle$.

3. Solvability of homogeneous linear ordinary differential equations.

Let F by an ordinary differential field of characteristic 0, with algebraically closed field of constants C; let G be a Picard-Vessiot extension of F.

Suppose there exists a differential field $H \supseteq G$ such that $H = F\langle\alpha_1,\dots,\alpha_r\rangle$, where for each i = 1,2,... .,r α_i is either algebraic over $F\langle\alpha_1,\dots,\alpha_{i-1}\rangle$, or an integral of an element of $F\langle\alpha_1,\dots,\alpha_{i-1}\rangle$, or an exponential of an integral of an element of $F\langle\alpha_1,\dots,\alpha_{i-1}\rangle$, or a $\wp$ - function of an integral of an element of $F\langle\alpha_1,\dots,\alpha_{i-1}\rangle$. Then it can be shown there exists such an H for which the field of constants is C and for which $F\langle\alpha_1,\dots,\alpha_i\rangle$ is a normal extension of

$F\langle\alpha_1,\ldots,\alpha_{i-1}\rangle$, $i = 1,2,\ldots,r$; we assume H has these two properties.

It can then be proved that $\mathfrak{G}^s$ has an algebraic normal chain $\mathfrak{G}^s = \mathfrak{G}^s_0 \supseteq \mathfrak{G}^s_1 \supseteq \cdots \supseteq \mathfrak{G}^s_r = \{\iota\}$ such that for each $i = 1,2,\ldots r$ $\mathfrak{G}^s_{i-1}/\mathfrak{G}^s_i$ is a homomorphic image of the group of all automorphisms of $F\langle\alpha_1,\ldots,\alpha_i\rangle$ over $F\langle\alpha_1,\ldots,\alpha_{i-1}\rangle$. Since these groups are all finite orabelian it follows that the component of the identity $\mathfrak{G}^{s\,0}$ is solvable. Thus we have:

If a Picard-Vessiot extension is contained in an extension by algebraic functions, integrals, exponentials of integrals, and $\wp$-functions of integrals, then the Picard-Vessiot extension is an extension by algebraic functions, integrals, and exponentials of integrals. In other words, if a homogeneous linear ordinary differential equation with coefficients in F is solvable by algebraic functions, integrals, exponentials of integrals, and $\wp$-functions of integrals, then it is solvable with algebraic functions, integrals, and exponentials of integrals.

TWO PROOFS OF A THEOREM ON ALGEBRAIC GROUPS

C. CHEVALLEY AND E. KOLCHIN

1. Introduction. We shall be considering $(n \times n)$-matrices with coefficients in some field K, which will be assumed to contain infinitely many elements. Such matrices may also be considered to represent endomorphisms of a vector space V over K, in which a base has been selected once and for all. We shall also be considering functions $F(x_1, \cdots, x_m)$ of m arguments in V, with values in K; such a function is called a *polynomial function* if its value may be expressed as a polynomial (with coefficients in K) in the mn coordinates of its arguments with respect to our base in V. The polynomial functions of m arguments form a ring, which will be denoted by $\mathfrak{o}_m$; this ring is isomorphic with the ring of polynomials in mn variables with coefficients in K (because K is infinite). The ring $\mathfrak{o}_m$ may therefore be imbedded in its field of quotients, which will be denoted by $\mathfrak{R}_m$; the elements of $\mathfrak{R}_m$ will be called the *rational expressions* in m elements of V.

Let s be any inversible matrix. Then we may associate to s an automorphism $\eta(s)$ of the ring $\mathfrak{o}_m$ which maps any polynomial function F upon the function $\eta(s)F$ defined by

$$(\eta(s)F)(x_1, \cdots, x_m) = F(sx_1, \cdots, sx_m).$$

The automorphism $\eta(s)$ may be extended to an automorphism of the field $\mathfrak{R}_m$ which we shall still denote by $\eta(s)$.

Let G be a group of $(n \times n)$-matrices (it will always be assumed, when we speak of a group of matrices, that the neutral element of the group is the unit matrix, which implies that every element of the group is an inversible matrix). An element R of $\mathfrak{R}_m$ is called an *invariant* of the group G if we have $\eta(s)R = R$ for every $s \in G$. It is clear that, if P and Q are invariants of G in $\mathfrak{o}_m$, and $Q \neq 0$, then PQ^{-1} is an invariant. However, not every invariant in $\mathfrak{R}_m$ may be represented as a quotient of two polynomial invariants. But, let R be any invariant of G in $\mathfrak{R}_m$, and let R be represented in the form PQ^{-1}, where P and Q are relatively prime to each other in the ring $\mathfrak{o}_m$. Then, if $s \in G$, $\eta(s)P = MP$ and $\eta(s)Q = MQ$, where M is an element of $\mathfrak{o}_m$. On the other hand, it is easily seen that $\eta(s)$ cannot raise the degree of a polynomial function; it follows that M is a scalar. This leads to the following definition: an element P of $\mathfrak{o}_m$ is called a *semi-invariant* of G if there exists a function M on G with values in K such that $\eta(s)P$

Received by the editors November 16, 1949.

$=M(s)P$ for all $s\in G$. The function $M(s)$ is uniquely determined if $P\neq 0$, and is called the *weight* of P. Thus we see that a necessary and sufficient condition for an element R of $\mathfrak{R}_m$ to be an invariant of G is that R be representable as the quotient of two semi-invariants belonging to the same weight; and, if this is the case, then any representation of R in the form of an irreducible fraction with terms in $\mathfrak{o}_m$ is a representation of R as a quotient of semi-invariants.

We may identify the set of all $(n\times n)$-matrices with coefficients in K with the set of systems of n elements of V; the elements of $\mathfrak{o}_n$ may therefore be considered as polynomial functions of matrices. A function of matrices is a polynomial function when its value for any matrix s may be expressed as a polynomial in the elements of s. A group G of matrices is called an *algebraic group* when the condition for an inversible matrix s to belong to G may be expressed by a system of equations of the form $F(s)=0$, where each F is a polynomial function. Given any $m>0$ and a set E of elements of $\mathfrak{o}_m$, it is not difficult to see that the group G of all inversible matrices s such that $\eta(s)R=R$ for every $R\in E$ is an algebraic group. Similarly, given any set E' of elements of $\mathfrak{o}_m$, the set of inversible matrices s such that $\eta(s)P$ is a scalar multiple of P whenever $P\in E'$ is an algebraic group. Our purpose is to establish partial converses of these statements, namely, Theorems 1 and 2 below.

THEOREM 1. *Let G be an algebraic group of $(n\times n)$-matrices. Then there exists a finite subset E of $\mathfrak{R}_n$ such that G consists of all inversible matrices s such that $\eta(s)R=R$ for all $R\in E$.*

THEOREM 2. *Let G be an algebraic group of $(n\times n)$-matrices. Then there exists a finite subset E' of $\mathfrak{o}_n$ such that G consists of all inversible matrices s such that $\eta(s)P$ is a scalar multiple of P whenever $P\in E'$.*

We shall give two proofs of these two theorems. The first one (§II) is pretty straightforward, and goes through without any restriction on the characteristic of the basic field. The second proof (§III) applies only to the case where the basic field is of characteristic 0, but it has the interest of connecting our problem with a seemingly altogether different question, namely, the algebraic theory of differential equations. The first proof gives at the same time proofs for both theorems; the second one establishes Theorem 1. We shall see now that Theorem 1 implies trivially Theorem 2. Assume that Theorem 1 is true; let G be an algebraic group, and let E be a set of elements of $\mathfrak{R}_n$ which has the property stated in Theorem 1. We may obviously assume that this set does not contain any element of the basic field K. Let R_i $(1\leqq i\leqq h)$ be the elements of E; write $R_i=P_iQ_i^{-1}$, where P_i and Q_i are in $\mathfrak{o}_n$ and relatively prime to each other. Then P_i and Q_i are semi-invariants of G of the same weight, which shows that P_i+Q_i is a semi-invariant. Let E' be the set composed of the elements P_i, Q_i, and P_i+Q_i $(1\leqq i\leqq h)$, and let s be an inversible matrix such that $\eta(s)P_i=a_iP_i$, $\eta(s)Q_i=b_iQ_i$, and $\eta(s)(P_i+Q_i)=c_i(P_i+Q_i)$ $(1\leqq i\leqq h)$, where the a_i's ,the b_i's, and the c_i's are scalars. Since R_i is not in K, P_i and Q_i are linearly independent over K, and it follows immediately that $a_i=b_i$, whence $\eta(s)R_i=R_i$ $(1\leqq i\leqq h)$, and therefore $s\in G$. This shows that Theorem 2 is true.

REMARK. In Theorems 1 and 2, we characterized an algebraic group either by its rational invariants or by its polynomial semi-invariants. It is not true that every algebraic group can be characterized by its polynomial invariants, even if we do not place any restriction on the number m of arguments. This is easily seen by taking for G the group of all inversible diagonal matrices, for then the only polynomial invariants of G (in any number m of arguments) are the constant functions. On the other hand, it is not true that every algebraic group can be characterized by its rational invariants in $\mathfrak{R}_{n-1}$, as one can see without difficulty by taking G to consist of all matrices in special triangular form (0's above the main diagonal, 1's along the main diagonal).

In a final section (§IV) we apply Theorem 2 to show that a factor group of two algebraic groups always has a faithful representation. More precisely, we prove the following result.

THEOREM 3. *Let G be an algebraic group of matrices, let H be a distinguished algebraic subgroup of G. Then there exists a representation ρ of G such that the kernel of ρ is H.*

It can be shown, although we do not do so here, that when the coefficient field is algebraically closed, then the group $\rho(G)$ is itself algebraic.

2. **The first proof.** Let G be an algebraic group of $(n\times n)$-matrices. We have said already that the polynomial functions of n arguments in V are the same thing as the polynomial functions of matrices. Let $\mathfrak{a}$ be the ideal of polynomial functions of n arguments which are 0 at every point of G. If s is an inversible matrix, then a necessary and sufficient condition for s to belong to G is that $\eta(s)$ (operating on $\mathfrak{o}_n$) should map $\mathfrak{a}$ into itself. For, if $s\in G$ and $P\in\mathfrak{a}$, then we have, for any matrix t, $(\eta(s)P)(t)=P(st)$; if $t\in G$, then so is st, which proves that $\eta(s)P\in\mathfrak{a}$. Conversely, let s be an inversible matrix such that $\eta(s)$ maps $\mathfrak{a}$ into itself. If $P\in\mathfrak{a}$, then $\eta(s)P$ is 0 at every point of G, and, in particular, at the neutral element I, whence

$P(s)=(\eta(s)P)\ (I)=0$; since G is algebraic, it follows that $s\in G$.

If r is any integer, denote by L_r the set of polynomial functions of matrices which are represented by polynomials of degrees less than or equal to r in the elements of the matrix. Then it is clear that, for any matrix s, $\eta(s)$ maps L_r into itself. Set $A_r=L_r\cap\mathfrak{a}$; then, if $s\in G$, $\eta(s)$ also maps A_r into itself. Now, the ideal $\mathfrak{a}$ has a finite set of generators; we select r in such a way that A_r contains a set of generators of $\mathfrak{a}$. Then, if an inversible matrix s is such that $\eta(s)$ maps A_r into itself, $\eta(s)$ maps $\mathfrak{a}$ into itself, whence $s\in G$.

The space L_r is finite-dimensional, for it has a base composed of the functions which may be expressed as monomials of degrees less than or equal to r in the coefficients of a matrix. Let $\{u_1, \cdots, u_a\}$ be a base of L_r which contains a base $\{u_1, \cdots, u_b\}$ of A_r. We construct the exterior algebra E over L_r and the subspace E_b of E which is spanned by the exterior products of b elements of L_r. This space has a base $\{v_1, \cdots, v_c\}$ which is composed of the exterior products $u_{i_1}\cdots u_{i_b}$, where $i_1<\cdots<i_b$; we assume that $v_1=u_1\cdots u_b$. Let s be any inversible matrix; since $\eta(s^{-1})=(\eta(s))^{-1}$ (as follows immediately from the definition), the restriction $\eta_r(s)$ of $\eta(s)$ to L_r is an automorphism of this vector space. This automorphism may be extended to an automorphism $\zeta(s)$ of the exterior algebra E, whose restriction $\zeta_b(s)$ to E_b is the bth skew-symmetric power of $\eta_r(s)$ (cf. N. Bourbaki, *Éléments de mathématique*, *Algèbre*, III, §5, no. 7). If $\eta_r(s)$ maps A_r into itself, then it is clear that $\zeta_b(s)$ maps v_1 upon a scalar multiple of itself. Conversely, assume that $\zeta_b(s)v_1=ev_1$, where e is a scalar. Since $\zeta_b(s)$ is an automorphism, we have $e\neq 0$. Now, the elements of A_r are the elements u of L_r such that $uv_1=0$ (cf. N. Bourbaki, loc. cit., Corollary to Proposition 3, §7, no. 3). Thus, if $u\in A_r$, we have $(\zeta(s))(uv_1)=(\eta_r(s)u)ev_1=0$, whence $\eta_r(s)u\in A_r$, which shows that $\eta_r(s)$ maps A_r into itself.

If P is any given polynomial function of matrices, the coefficients of the expression of $\eta(s)P$ as a polynomial in the coefficients of the matrix argument are obviously polynomial functions of s. It follows immediately that the elements of the matrix which represents $\eta_r(s)$ with respect to the base $\{u_1, \cdots, u_a\}$ are polynomial functions of s. The coefficients of the matrix which represents $\zeta_b(s)$ with respect to the base $\{v_1, \cdots, v_c\}$ are certain minors of the preceding matrix, which proves that they are polynomial functions of s. Set $\zeta_b(s)v_i = \sum_{j=1}^{c}P_{ij}(s)v_j$ $(1\leqq i\leqq c)$, and denote by $\mathfrak{P}(s)$ the matrix $(P_{ij}(s))$. It follows immediately from the definition that, for any matrices s and s', we have $\eta(ss')=\eta(s')\eta(s)$, whence $\eta_r(ss')=\eta_r(s')\eta_r(s)$ and $\zeta_b(ss')=\zeta_b(s')\zeta_b(s)$, which proves that the matrix $\mathfrak{P}(ss')$ is $\mathfrak{P}(s)\mathfrak{P}(s')$.

If $s \in G$, then we have $P_{1j}(s)=0$ $(2 \leqq j \leqq c)$, whence, for any matrix t, $P_{1j}(st)=P_{11}(s)P_{1j}(t)$ for $1 \leqq j \leqq c$. This means that each P_{1j} is a semi-invariant of G. Now, let s be any inversible matrix such that $P_{1j}(st) = a_j P_{1j}(t)$ identically in t, each a_j being a scalar. If I is the unit matrix, we have $P_{1j}(I)=0$ $(2 \leqq j \leqq c)$; it follows that $P_{1j}(s)=0$ $(2 \leqq j \leqq c)$, which means that $\zeta_b(s)$ maps v_1 upon a scalar multiple of itself, therefore the $\eta_r(s)$ maps A_r into itself and that $s \in G$. Thus we see that the set E' composed of the elements P_{1j} $(1 \leqq j \leqq c)$ has the property stated in Theorem 2. The function P_{11} is not equal to 0 (for $P_{11}(I)=1$), and it follows from the formulas $P_{1j}(st)=P_{11}(s)P_{1j}(t)$ $(1 \leqq j \leqq c)$ that the rational expressions $R_j = P_{1j}P_{11}^{-1}$ are invariants of G. Let s be an inversible matrix such that $\eta(s)R_j = R_j$ $(1 \leqq j \leqq c)$; then, since $R_j(I)=0$ $(2 \leqq j \leqq c)$, we see that $P_{1j}(s)=R_j(I)P_{11}(s)=0$ for $j>1$, and we conclude as above that $s \in G$. Thus the set E composed of $R_1, \cdots, R_c$ has the property stated in Theorem 1.

3. **The second proof.** Let G be an algebraic group of $(n \times n)$-matrices with coefficients in a field K of characteristic 0. We first show that in proving Theorem 1 it is permissible to assume that K is algebraically closed. Suppose then that K is not algebraically closed and that Theorem 1 is known to be true for algebraically closed fields. Let $\mathfrak{a}$ be the ideal of all polynomial functions over K which vanish for every $s \in G$, and let $\bar{G}$ be the set of all inversible matrices $\bar{s}$ with coefficients in the algebraic closure $\bar{K}$ of K such that $P(\bar{s})=0$ for every $P \in \mathfrak{a}$. The first step is to show that $\bar{G}$ is an algebraic group of matrices with coefficients in $\bar{K}$. Let $\bar{P}$ be a polynomial function over $\bar{K}$ such that $\bar{P}(s)=0$ for all $s \in G$. If we write $\bar{P} = \sum \omega_i P_i$, where $\omega_1, \omega_2, \cdots$ are elements of $\bar{K}$ linearly independent over K and each P_i is a polynomial function over K, we find that $P_i(s)=0$ for all $s \in G$, so that each $P_i \in \mathfrak{a}$. It follows that if we let $\mathfrak{G}$ denote the smallest algebraic variety over $\bar{K}$ which contains G, then $\bar{G}$ consists of the inversible matrices in $\mathfrak{G}$. Therefore[1] $\bar{G}$ is an algebraic group over $\bar{K}$. This being so, there exists a finite set $\bar{E}$ of rational invariants of $\bar{G}$ with the property stated in Theorem 1. For each $\bar{R} \in \bar{E}$ we may write $\bar{R} = \sum \omega_i R_i$ where $\omega_1, \omega_2, \cdots$ are elements of $\bar{K}$ linearly independent over K and each R_i is a rational expression over K; let E be the finite set of all rational expressions R_i obtained as $\bar{R}$ runs over $\bar{E}$. If $s \in G$, then for each $\bar{R} \in \bar{E}$ we have $\eta(s)\bar{R} = \bar{R}$, that is, $\sum \omega_i \eta(s) R_i = \sum \omega_i R_i$. It follows that E is a set of rational invariants of G. Conversely, if s is an inversible matrix with coefficients in K such that $\eta(s)R = R$ for all $R \in E$, then for every $\bar{R} \in \bar{E}$ we have $\eta(s)\bar{R} = \sum \omega_i \eta(s) R_i = \sum \omega_i R_i = \bar{R}$,

[1] E. Kolchin, Ann. of Math. vol. 49 (1948) pp. 1–42; see §3, Lemma 2.

so that $s \in \bar{G}$, whence $s \in G$. Therefore E has the property stated in Theorem 1. This shows that it suffices to prove Theorem 1 when the coefficient field is algebraically closed.

Assume then that K is algebraically closed. By defining a derivation $c \to c'$ in K by the formula $c' = 0$ for all $c \in K$ we make K an ordinary differential field. Let

$$y_1, \cdots, y_n, y_1', \cdots, y_n', \cdots, y_1^{(i)}, \cdots, y_n^{(i)}, \cdots$$

be an infinite sequence of elements of an extension field of K which are algebraically independent over K, and let $\mathcal{G}$ denote the field obtained by adjoining all these elements to K. It is well known that the differential field structure of K can be extended to $\mathcal{G}$ in one and only one way so that the derivative of $y_j^{(i)}$ is $y_j^{(i+1)}$ $(i \geqq 0,\ 1 \leqq j \leqq n)$; in the language of differential field extensions we then have $\mathcal{G} = K\langle y_1, \cdots, y_n \rangle$. It is easy to see that the field of constants of $\mathcal{G}$ (that is, the field of all elements $R \in \mathcal{G}$ such that $R' = 0$) is K.

If $s = (a_{ij})$ is any matrix in the group $GL(n, K)$ of all inversible $(n \times n)$-matrices with coefficients in K, then the substitution $y_j \to \sum_{h=1}^{n} a_{hj} y_h$ $(1 \leqq j \leqq n)$ defines an automorphism over K of the differential field $\mathcal{G}$; under this automorphism $y_j^{(i)} \to \sum_{h=1}^{n} a_{hj} y_h^{(i)}$. The mapping which in this way assigns to each $s \in GL(n, K)$ an automorphism of $\mathcal{G}$ is an isomorphism of the group $GL(n, K)$ into the group of all automorphisms of $\mathcal{G}$ over K; therefore we may and henceforth do identify each matrix with the corresponding automorphism of $\mathcal{G}$.

The set of all elements $R \in \mathcal{G}$ such that $sR = R$ for all $s \in GL(n, K)$ is a differential subfield $\mathcal{E}$ of $\mathcal{G}$. $W(z, y_1, \cdots, y_n)/W(y_1, \cdots, y_n) = L(z)$ (quotient of two Wronskian determinants) is a homogeneous linear differential polynomial in z with coefficients in $\mathcal{G}$; since these coefficients obviously are invariant under every $s \in GL(n, K)$ they must belong to $\mathcal{E}$. It is obvious that $y_1, \cdots, y_n$ are n solutions of the differential equation $L(z) = 0$ and are linearly independent over K; also $\mathcal{E}\langle y_1, \cdots, y_n \rangle = \mathcal{G}$. Therefore $\mathcal{G}$ is a Picard-Vessiot extension of $\mathcal{E}$ and $GL(n, K)$ is the group of all automorphisms of $\mathcal{G}$ over $\mathcal{E}$.[2]

Since G is an algebraic subgroup of $GL(n, K)$ we know by the Picard-Vessiot theory that the set of all elements of $\mathcal{G}$ which are invariant under every $s \in G$ is a differential field $\mathcal{J}$ such that $\mathcal{E} \subseteq \mathcal{J} \subseteq \mathcal{G}$ and such that $s \in GL(n, K)$ belongs to G if $sR = R$ for all $R \in \mathcal{J}$.

---

[2] The relevant facts concerning the Picard-Vessiot theory can be found in the paper cited in footnote 1.

Now it is easy to see that the differential field $\mathcal{J}$ is finitely generated over $\mathcal{E}$; therefore if m is large and if we let $\mathcal{J}_m$ denote the set of all elements of $\mathcal{J}$ which are of order[3] less than or equal to m, we shall have the following criterion: a matrix $s\in GL(n, K)$ belongs to G if and only if $sR=R$ for all $R\in\mathcal{J}_m$. In the language of §1 this means that G is characterized by its rational invariants in $m+1$ arguments.

To complete the proof of Theorem 1 it remains to show that G is characterized by its rational invariants in n arguments, that is, that in this criterion we may replace $\mathcal{J}_m$ by $\mathcal{J}_{n-1}$; we do this by proving that if the criterion holds for a given $m\geqq n$ then it continues to hold when m is replaced by $m-1$. To this end we observe that each y_j is a zero of the homogeneous linear differential polynomial in z

$$W(y_1, \cdots, y_n)^{-1}\begin{vmatrix} z, & y_1, & \cdots & y_n, \\ z' & y_1' & \cdots & y_n' \\ \cdot & \cdot & \cdots & \cdot \\ z^{(n-1)} & y_1^{(n-1)} & \cdots & y_n^{(n-1)} \\ z^{(m)} & y_1^{(m)} & \cdots & y_n^{(m)} \end{vmatrix}$$

which we write as $M(z)=z^{(m)}+q_1z^{(n-1)}+\cdots+q_nz$; the coefficients $q_1, \cdots, q_n$ are clearly invariant under every $s\in GL(n, K)$ and therefore belong to $\mathcal{E}$. We observe further that the elements $y_j^{(i)}$ $(0\leqq i\leqq m-1,\ 1\leqq j\leqq n)$ and q_k $(1\leqq k\leqq n)$ are algebraically independent over K. Now let $R\in\mathcal{J}_m$. By means of the equation $M(y_j)=0$ $(1\leqq j\leqq n)$ we write $R=AB^{-1}$, where A and B are polynomials in $q_1, \cdots, q_n$ without common divisor and with coefficients which are rational expressions over K in the elements $y_j^{(i)}$ $(0\leqq i\leqq m-1,\ 1\leqq j\leqq n)$; we assume without loss of generality that one of the coefficients in A is 1. Because of this assumption, and the lack of common divisor of A and B, and the algebraic independence mentioned above, every $s\in GL(n, K)$ which leaves R invariant also leaves invariant all the coefficients in A and B; therefore all these coefficients are invariants of G and consequently belong to $\mathcal{J}_{m-1}$. It follows that s leaves every $R\in\mathcal{J}_{m-1}$ invariant if and only if s leaves every $R\in\mathcal{J}_m$ invariant. This establishes the criterion with m replaced by $m-1$ and, as we have seen, completes the proof of Theorem 1.

4. Factor groups. Let G again be an algebraic group of $(n\times n)$-matrices with coefficients in an infinite field K, and let H be a distinguished algebraic subgroup of G. Denoting by E_r' the set of all

[3] An element of $\mathcal{G}$ is of order less than or equal to m if the element belongs to $K(y_1, \cdots, y_n, y_1', \cdots, y_n', \cdots, y_1^{(m)}, \cdots, y_n^{(m)})$.

semi-invariants of H in $\mathfrak{o}_n$ which are of degree less than or equal to r, we see by Theorem 2 applied to H that for large r an inversible matrix s belongs to H if and only if $\eta(s)P$ is a scalar multiple of P for all $P\in E_r'$. It is obvious that the elements of E_r' of a given weight form a vector space. We now show that a sum $\sum_{i=1}^{h} P_i$ of nonzero elements of E_r' of distinct weights $M_1, \cdots, M_h$ is never 0. For $h=1$ this is trivial; let $h>1$ and suppose the statement already verified for sums of fewer than h terms. If $\sum_{i=1}^{h} P_i$ were zero, we would have $\sum_{i=1}^{h} P_iM_i(t)-\eta(t)\sum_{i=1}^{h} P_i=0$ for every $t\in H$ and therefore $\sum_{i=1}^{h-1}P_i\cdot(M_i(t)-M_h(t))=0$, whence by the induction assumption $P_i\cdot(M_i(t)-M_h(t))=0$, $1\leqq i\leqq h=1$. Since M_i and M_h are distinct, there exists a $t_i\in H$ such that $M_i(t_i)\neq M_h(t_i)$, so that we would have $P_i=0$ for $1\leqq i\leqq h-1$ and therefore also for $1\leqq i\leqq h$.

As a consequence we conclude that there exist a finite number of weights $M_1, \cdots, M_h$ such that the vector space W spanned by E_r' can be written as a direct sum $W=W_1+\cdots+W_h$, where W_i consists of all the elements of E_r' which have weight M_i. Now if P is a semi-invariant of H of some weight M and if $s\in G$, then we have, for all $t\in H$, $\eta(t)(\eta(s)P)=\eta(s)\ (\eta(sts^{-1})P)=\eta(s)(M(sts^{-1})\cdot P)=M(sts^{-1})\cdot(\eta(s)P)$, so that $\eta(s)P$ is a semi-invariant of H of weight N where N is the function on H defined by $N(t)=M(s^{-1}st)$. Moreover, if the degree of P is less than or equal to r, then so is the degree of $\eta(s)P$. It follows that the restriction $\xi(s)$ of $\eta(s)$ to W is a vector space automorphism of W which maps each W_i onto some W_j. Therefore if we introduce a base in each space W_i the restriction of $\xi(s)$ to W_i may be identified with an inversible matrix $\xi_i(s)$, and $\xi(t)$ itself then identified with a matrix which is composed of h^2 blocks, h of these blocks being $\xi_1(s), \cdots, \xi_h(s)$ and the others being 0. The element $s\in G$ belongs to H if and only if all these nonzero blocks $\xi_i(s)$ are on the main diagonal and are scalar matrices.

Let d_i be the dimension of W_i, so that $\xi_i(s)$ is a $(d_i\times d_i)$-matrix. The mapping $\alpha\rightarrow\xi_i(s)^{-1}\alpha\xi_i(s)$ is an automorphism of the vector space consisting of all $(d_i\times d_i)$-matrices α with coefficients in K. Choosing a basis for this vector space, we denote the matrix of this automorphism by $\rho_i(s)$; then $\rho_i(s)$ is an inversible $(d_i^2\times d_i^2)$-matrix. Finally, let $\rho(s)$ be the matrix composed of h^2 blocks, h of these blocks being $\rho_1(s), \cdots, \rho_h(s)$ and the others being 0, the arrangement of these nonzero blocks in $\rho(s)$ being the same as the arrangement of the blocks $\xi_1(s), \cdots, \xi_h(s)$ in $\xi(s)$. Thus $\rho(s)$ is an inversible $(d\times d)$-matrix, where $d=\sum_{i=1}^{h} d_i^2$.

It can be seen without great difficulty that the mapping $\rho:s\rightarrow\rho(s)$ is a representation of G. If $s\in G$, then $\rho(s)$ is the unit matrix if and only if each nonzero block $\rho_i(s)$ is on the main diagonal of $\rho(s)$ and is itself a unit matrix, that is, if and only if each nonzero block $\xi_i(s)$ is on the main diagonal of $\xi(s)$ and is a scalar matrix, that is, if and only if $s \in H$. Thus the kernel of the representation ρ is H. This proves Theorem 3.

COLUMBIA UNIVERSITY

PICARD-VESSIOT THEORY OF PARTIAL DIFFERENTIAL FIELDS

E. R. KOLCHIN

Introduction.[1] The purpose of this note is to extend to partial differential fields the Picard-Vessiot theory of homogeneous linear ordinary differential equations.[2] Although this can be done by extending the procedure of PV to partial differential fields, it is perhaps not without interest to see how the theory in the partial case can be deduced from the ordinary theory, and the latter is what is done below.

If $\mathcal{J}$ is an ordinary differential field of characteristic 0 with algebraically closed field of constants and $\mathcal{G}$ is an extension of $\mathcal{J}$ with the same field of constants, then (PV, §17) $\mathcal{G}$ is a Picard-Vessiot extension of $\mathcal{J}$ if there exists a homogeneous linear differential equation $y^{(n)}+p_1y^{(n-1)}+ \cdots +p_ny=0$ (each $p_i \in \mathcal{J}$) with a fundamental system of solutions $\eta_1, \cdots, \eta_n$ such that $\mathcal{G}=\mathcal{J}\langle\eta_1, \cdots, \eta_n\rangle$. Since p_i is then plus or minus the quotient of two determinants W_i/W_0, where $W_l=\det (\eta_k^{(j)})_{0\leqq j\leqq n, j\neq n-l; \leqq k\leqq n}$, it is the same thing to require that $\mathcal{G}$ contain elements $\eta_1, \cdots, \eta_n$ linearly independent over constants such that $\mathcal{G}=\mathcal{J}\langle\eta_1, \cdots, \eta_n\rangle$ and such that each W_i/W_0 belongs to $\mathcal{J}$. It is this property which is generalized below to define "Picard-Vessiot extension" in the case of partial differential fields.

1. **Linear dependence over constants.** Let $\mathcal{G}$ be a partial differential field of characteristic 0 with m derivations $\delta_1, \cdots, \delta_m$; denote the field of constants of $\mathcal{G}$ by $\mathcal{C}$. Let $u_1, \cdots, u_m$ be elements of some extension of $\mathcal{G}$ which are differentially algebraically independent over $\mathcal{G}$, and consider the partial differential field $\mathcal{G}\langle u_1, \cdots, u_m\rangle$. The operator $D=\sum u_i\delta_i$ is then a derivation of $\mathcal{G}\langle u_1, \cdots, u_m\rangle$ and therefore induces in $\mathcal{G}\langle u_1, \cdots, u_m\rangle$ a structure of an *ordinary* differential field; we denote the ordinary differential field so defined by $\mathcal{G}_D$. The field of constants of $\mathcal{G}_D$ is $\mathcal{C}$; indeed if we order the expressions $\delta_1^{t_1} \cdots \delta_m^{t_m}u_j$ lexicographically and if $\delta_1^{a_1} \cdots \delta_m^{a_m}u_b$ is the highest such expression effectively present in some element $\phi \in \mathcal{G}_D$ ($\phi \notin \mathcal{G}$), then $\delta_1^{a_1+1}\delta_2^{a_2} \cdots \delta_m^{a_m}u_b$ is effectively present in $D\phi$, so that $D\phi \neq 0$.

Let $y_1, \cdots, y_n$ be indeterminates, and let $\theta_1, \cdots, \theta_n$ denote

Received by the editors November 20, 1951.

[1] The research on which this paper is based was done in connection with a contract with the Office of Naval Research.

[2] It is assumed that the reader is familiar with my paper on this subject in Ann. of Math. vol. 49 (1948) pp. 1–42, which will be referred to as PV.

differential operators of the form $\delta_1^{i_1} \cdots \delta_m^{i_m}$, where each i_ν is an integer not less than 0. Then we may form the determinant $W_{\theta_1 \cdots \theta_n} = W_{\theta_1 \cdots \theta_n}(y_1, \cdots, y_n) = \det(\theta_i y_j)_{1 \leq i \leq n, 1 \leq j \leq n}$. The wronskian $\det(D^i y_j)_{0 \leq i \leq n-1, 1 \leq j \leq n}$ obviously can be expressed as a linear combination of determinants $W_{\theta_1 \cdots \theta_n}$ in which each θ_i is of order not greater than $n-1$.

LEMMA 1. *Let $\eta_1, \cdots, \eta_n \in \mathcal{G}$. If $\eta_1, \cdots, \eta_n$ are linearly dependent over $\mathcal{C}$, then $W_{\theta_1 \cdots \theta_n}(\eta_1, \cdots, \eta_n) = 0$ for every choice of $\theta_1, \cdots, \theta_n$. Conversely if $W_{\theta_1 \cdots \theta_n}(\eta_1, \cdots, \eta_n) = 0$ for every choice of $\theta_1, \cdots, \theta_n$ of order not greater than $n-1$, then $\eta_1, \cdots, \eta_n$ are linearly dependent over $\mathcal{C}$.*

PROOF. If $\sum c_j \eta_j = 0$ ($c_1, \cdots, c_n \in \mathcal{C}$, not all 0), then $\sum c_j \theta_i \eta_j = 0$ ($1 \leq i \leq n$) for all choices of $\theta_1, \cdots, \theta_n$, whence $W_{\theta_1 \cdots \theta_n}(\eta_1, \cdots, \eta_n) = 0$. Conversely if $W_{\theta_1 \cdots \theta_n}(\eta_1, \cdots, \eta_n) = 0$ for all choices of $\theta_1, \cdots, \theta_n$ of order not greater than $n-1$ then, by a remark made above, the wronskian $\det(D^i \eta_j) = 0$; since $\mathcal{G}_D$ is an ordinary differential field it is known (PV, §14) that this implies that $\eta_1, \cdots, \eta_n$ are linearly dependent over the field of constants of $\mathcal{G}_D$, that is over $\mathcal{C}$.

Since the coefficients in $W_{\theta_1 \cdots \theta_n}$ are integers, the lemma shows that $\eta_1, \cdots, \eta_n$ are linearly dependent (or independent) over $\mathcal{C}$ if and only if they are so over the field of constants of any partial differential field containing $\eta_1, \cdots, \eta_n$. Because of this we may speak simply of linear dependence (or independence) over constants.

2. **Two lemmas.** We continue with $\mathcal{G}$, $\mathcal{C}$, $\mathcal{G}_D$ as in §1.

LEMMA 2. *Let $\eta_1, \cdots, \eta_n \in \mathcal{G}$ be linearly independent over $\mathcal{C}$, so that $W_{\theta_{01} \cdots \theta_{0n}}(\eta_1, \cdots, \eta_n) \neq 0$ for some choice of operators $\theta_{01}, \cdots, \theta_{0n}$ of order not greater than $n-1$; let $\mathcal{J}$ be a partial differential subfield of $\mathcal{G}$ which contains*

(1) $$W_{\theta_1 \cdots \theta_n}(\eta_1, \cdots, \eta_n)/W_{\theta_{01} \cdots \theta_{0n}}(\eta_1, \cdots, \eta_n)$$

for every choice of $\theta_1, \cdots, \theta_n$ of order not greater than n. Then $\mathcal{J}$ contains (1) *for every choice of $\theta_1, \cdots, \theta_n$ regardless of order.*

PROOF. Each η_j is a zero of the partial differential polynomial

(2) $$W_{\theta_0 \theta_1 \cdots \theta_n}(y, \eta_1, \cdots, \eta_n)/W_{\theta_{01} \cdots \theta_{0n}}(\eta_1, \cdots, \eta_n)$$

for every choice of $\theta_0, \theta_1, \cdots, \theta_n$. When $\theta_0, \theta_1, \cdots, \theta_n$ are all of order not greater than n, then, by the hypothesis of the lemma, each coefficient in (2) is in $\mathcal{J}$. Therefore if σ is any isomorphism of $\mathcal{G}$ over $\mathcal{J}$ (PV, §12), then $\sigma\eta_j$ is a zero of (2) whenever $\theta_0, \theta_1, \cdots, \theta_n$ are of order not greater than n. It follows from Lemma 1 that $\sigma\eta_j, \eta_1, \cdots, \eta_n$

are linearly dependent over constants, that is, $\sigma\eta_j = \sum k_{ij}\eta_i$ $(1 \leqq j \leqq n)$, where the elements k_{ij} are constants in an extension of $\mathcal{G}$. Therefore $\sigma W_{\theta_1 \cdots \theta_n}(\eta_1, \cdots, \eta_n) = \det (k_{ij}) \cdot W_{\theta_1 \cdots \theta_n}(\eta_1, \cdots, \eta_n)$, so that every expression (1) is invariant under every isomorphism σ of $\mathcal{G}$ over $\mathcal{J}$, whence (PV, §12) every expression (1) belongs to $\mathcal{J}$.

If $\mathcal{J}$ is any partial differential subfield of $\mathcal{G}$, then $\mathcal{J}\langle u_1, \cdots, u_m\rangle$ is a partial differential subfield of $\mathcal{G}\langle u_1, \cdots, u_m\rangle$; it is clear that we may also regard $\mathcal{J}\langle u_1, \cdots, u_m\rangle$ as an *ordinary* differential subfield of $\mathcal{G}_D$, which we shall denote by $\mathcal{J}_D$. If $\eta_1, \cdots, \eta_n$ are any elements of $\mathcal{G}$, then we naturally denote by $\mathcal{J}_D\langle \eta_1, \cdots, \eta_n\rangle$ the ordinary differential field extension, which set-theoretically coincides with $\mathcal{J}_D((D^i\eta_j)_{0 \leqq i < \infty, 1 \leqq j \leqq n})$.

LEMMA 3. *Under the hypotheses of Lemma 2 and the assumption that* $\mathcal{G} = \mathcal{J}\langle \eta_1, \cdots, \eta_n\rangle$ *we have* $\mathcal{G}_D = \mathcal{J}_D\langle \eta_1, \cdots, \eta_n\rangle$.

PROOF. We must show for every j and for every operator $\theta = \delta_1^{i_1} \cdots \delta_m^{i_m}$ that $\theta\eta_j \in \mathcal{J}_D\langle \eta_1, \cdots, \eta_n\rangle$. Now $(\eta_1, \cdots, \eta_n)$ is a fundamental system of solutions of the homogeneous linear ordinary differential equation

$$W_{\theta_{01} \cdots \theta_{0n}}(\eta_1, \cdots, \eta_n)^{-1} W(y, \eta_1, \cdots, \eta_n) = 0, \tag{3}$$

where $W(y, \eta_1, \cdots, \eta_n)$ is the wronskian determinant (with respect to the derivation D) of $y, \eta_1, \cdots, \eta_n$; (3) is of order n and obviously has its coefficients in $\mathcal{J}_D$. Let

$$D^k y + v_1 D^{k-1} y + \cdots + v_k y = 0 \tag{4}$$

be the homogeneous linear ordinary differential equation of lowest order with coefficients in $\mathcal{J}_D$ and with solution η_1. Every solution of (4) is a solution of (3), so there exist k linear combinations (with constant coefficients) $\zeta_1 = \eta_1, \zeta_2, \cdots, \zeta_k$ of $\eta_1, \cdots, \eta_n$ such that $(\zeta_1, \cdots, \zeta_k)$ is a fundamental system of solutions of (4). Since $\zeta_1, \cdots, \zeta_k$ are linearly independent over constants, there exist operators $\lambda_1, \cdots, \lambda_k$ of order not greater than $k-1$ such that $W_{\lambda_1 \cdots \lambda_k}(\zeta_1, \cdots, \zeta_k) \neq 0$.

Every isomorphism of $\mathcal{G}$ over $\mathcal{J}$ obviously maps $\zeta_1, \cdots, \zeta_k$ into k solutions of (4) linearly independent over constants, and therefore performs a linear substitution (with constant coefficients and nonzero determinant) on $\zeta_1, \cdots, \zeta_k$; it follows that every such isomorphism leaves invariant the element

$$W_{\theta_1 \cdots \theta_k}(\zeta_1, \cdots, \zeta_k)/W_{\lambda_1 \cdots \lambda_k}(\zeta_1, \cdots, \zeta_k)$$

for every choice of $\theta_1, \cdots, \theta_k$, so that every such element belongs to

$\mathcal{F}$. Thus for each θ the equation

$$W_{\theta\lambda_1\cdots\lambda_k}(y, \zeta_1, \cdots, \zeta_k)/W_{\lambda_1\cdots\lambda_k}(\zeta_1, \cdots, \zeta_k) = 0,$$

which is satisfied by $\zeta_1=\eta_1$, has its coefficients in $\mathcal{F}$. It follows that for each operator θ there exist elements $a_{\theta 1}, \cdots, a_{\theta k}\in\mathcal{F}$ such that

$$\theta\eta_1 = \sum_{j=1}^{k} a_{\theta j}\lambda_j\eta_1. \tag{5}$$

Consequently for each integer $i\geqq 0$ there exist elements $v_{i1}, \cdots, v_{ik} \in\mathcal{F}_D$ such that

$$D^i\eta_1 = \sum_{j=1}^{k} v_{ij}\lambda_j\eta_1.$$

If the determinant of $(v_{ij})_{0\leqq i\leqq k-1, 1\leqq j\leqq k}$ were 0, then $\eta_1, D\eta_1, \cdots, D^{k-1}\eta_1$ would be linearly dependent over $\mathcal{F}_D$, that is, η_1 would satisfy a homogeneous linear differential equation over $\mathcal{F}_D$ of order less than k. Therefore this determinant is nonzero and we can express each $\lambda_j\eta_1$ linearly over $\mathcal{F}_D$ in terms of $\eta_1, D\eta_1, \cdots, D^{k-1}\eta_1$, so that $\lambda_j\eta_1 \in\mathcal{F}_D\langle\eta_1\rangle$, whence, by (5), $\theta\eta_1\in\mathcal{F}_D\langle\eta_1\rangle$ for every differential operator θ. Similarly $\theta\eta_j\in\mathcal{F}_D\langle\eta_j\rangle$ for each j.

3. **Picard-Vessiot theory.** Let $\mathcal{F}$ be a partial differential field of characteristic 0 with m derivations $\delta_1, \cdots, \delta_m$ and with algebraically closed field of constants $\mathcal{C}$, and let $\mathcal{G}$ be an extension of $\mathcal{F}$. We shall call $\mathcal{G}$ a *Picard-Vessiot* extension of $\mathcal{F}$ if: (1) the field of constants of $\mathcal{G}$ is $\mathcal{C}$; (2) $\mathcal{G}$ contains a finite family of elements $(\eta_1, \cdots, \eta_n)$ which is linearly independent over constants, has the property that $\mathcal{F}\langle\eta_1, \cdots, \eta_n\rangle=\mathcal{G}$, and has the property that

$$W_{\theta_1\cdots\theta_n}(\eta_1, \cdots, \eta_n)/W_{\theta_{01}\cdots\theta_{0n}}(\eta_1, \cdots, \eta_n) \in \mathcal{F}$$

for all choices of $\theta_1, \cdots, \theta_n$ of order not greater than n and some fixed $\theta_{01}, \cdots, \theta_{0n}$ such that $W_{\theta_{01}\cdots\theta_{0n}}(\eta_1, \cdots, \eta_n)\neq 0$. As seen in the introduction, for "partial" differential fields with one derivation (case $m=1$), that is, for ordinary differential fields, this coincides with the concept of Picard-Vessiot extension already known.

Let $\mathcal{G}$ be such a Picard-Vessiot extension of $\mathcal{F}$, and let $\mathcal{F}_D$ and $\mathcal{G}_D$ be as before. By Lemma 3, $\mathcal{G}_D$ is an ordinary Picard-Vessiot extension of $\mathcal{F}_D$. It follows that the equations

$$\sigma\eta_j = \sum_{i=1}^{n} s_{ij}\eta_i \qquad (1 \leqq j \leqq n), \tag{6}$$

establish an isomorphism between the group of all automorphisms σ of $\mathcal{G}_D$ over $\mathcal{F}_D$ and a certain algebraic group $\mathfrak{G}$ of matrices (s_{ij}) over $\mathcal{C}$. By (6) we see that $\sigma\mathcal{G}=\mathcal{G}$; also, for each $\alpha\in\mathcal{G}$ we have

$$\sum u_i(\sigma\delta_i\alpha - \delta_i\sigma\alpha) = \sigma\sum u_i\delta_i\alpha - \sum u_i\delta_i\sigma\alpha = \sigma D\alpha - D\sigma\alpha = 0$$

whence, since $u_1, \cdots, u_m$ are differentially algebraically independent over $\mathcal{G}$, $\sigma\delta_i\alpha=\delta_i\sigma\alpha$ $(1\leqq i\leqq m)$. It follows that the restriction to $\mathcal{G}$ of every automorphism of $\mathcal{G}_D$ over $\mathcal{F}_D$ is an automorphism of $\mathcal{G}$ over $\mathcal{F}$.

Conversely if we start with an automorphism of $\mathcal{G}$ over $\mathcal{F}$, it can be extended to a unique automorphism of $\mathcal{G}\langle u_1, \cdots, u_m\rangle$ over $\mathcal{F}\langle u_1, \cdots, u_m\rangle$, which is a fortiori an automorphism of $\mathcal{G}_D$ over $\mathcal{F}_D$. Summarizing these remarks we have the following theorem.

THEOREM 1. *If $\mathcal{G}$ is a Picard-Vessiot extension of $\mathcal{F}$, then $\mathcal{G}_D$ is a Picard-Vessiot extension of $\mathcal{F}_D$, the mapping which to each automorphism of $\mathcal{G}_D$ over $\mathcal{F}_D$ assigns its restriction to $\mathcal{G}$ is an isomorphism of the group of all automorphisms of $\mathcal{G}_D$ over $\mathcal{F}_D$ onto the group of all automorphisms of $\mathcal{G}$ over $\mathcal{F}$, and the equations* (6) *establish an isomorphism of each of these groups with an algebraic matric group over $\mathcal{C}$.*

By virtue of this theorem we identify the three groups mentioned therein and denote them all by $\mathfrak{G}$. If $\mathcal{F}_1$ is a partial differential field between $\mathcal{F}$ and $\mathcal{G}$ and if we denote the group of all automorphisms of $\mathcal{G}$ over $\mathcal{F}_1$ by $\mathfrak{G}(\mathcal{F}_1)$, then $\mathfrak{G}(\mathcal{F}_1)$ is also (identified with) the group of all automorphisms of $\mathcal{G}_D$ over $\mathcal{F}_{1D}$ (where $\mathcal{F}_{1D}$ is $\mathcal{F}_1\langle u_1, \cdots, u_m\rangle$ considered as an ordinary differential field with derivation D); an element in $\mathcal{G}$ but not in $\mathcal{F}_1$ fails to belong to $\mathcal{F}_{1D}$ and therefore (because $\mathcal{G}_D$ is normal over $\mathcal{F}_D$) is not an invariant of $\mathfrak{G}(\mathcal{F}_1)$. Consequently $\mathcal{G}$ is normal over $\mathcal{F}$ (PV, §16). Of course $\mathfrak{G}(\mathcal{F}_1)$ is an algebraic subgroup of $\mathfrak{G}$; the fact that every algebraic subgroup of $\mathfrak{G}$ is some $\mathfrak{G}(\mathcal{F}_1)$ with $\mathcal{F}_1$ a partial differential field between $\mathcal{F}$ and $\mathcal{G}$ can be proved exactly as in the ordinary case (PV, §19). Thus $\mathcal{F}_1\rightarrow\mathfrak{G}(\mathcal{F}_1)$ is a one-to-one mapping of the set of all partial differential fields between $\mathcal{F}$ and $\mathcal{G}$ onto the set of all algebraic subgroups of $\mathfrak{G}$. Since there is a similar one-to-one mapping of the set of all ordinary differential fields between $\mathcal{F}_D$ and $\mathcal{G}_D$ onto the set of all algebraic subgroups of $\mathfrak{G}$, we see that every differential field between $\mathcal{F}_D$ and $\mathcal{G}_D$ is the form $\mathcal{F}_{1D}$, where $\mathcal{F}_1$ is a partial differential field between $\mathcal{F}$ and $\mathcal{G}$. The degree of transcendence of $\mathcal{G}$ over $\mathcal{F}_1$ obviously equals that of $\mathcal{G}_D$ over $\mathcal{F}_{1D}$, which by the ordinary Picard-Vessiot theory (PV, §20) equals the dimension of the algebraic group $\mathfrak{G}(\mathcal{F}_1)$. Thus we have the following theorem.

THEOREM 2. *Let $\mathcal{G}$ be a Picard-Vessiot extension of $\mathcal{J}$. Then $\mathcal{J}_1 \to \mathfrak{G}(\mathcal{J}_1)$ is a one-to-one mapping of the set of all partial differential fields between $\mathcal{J}$ and $\mathcal{G}$ onto the set of all algebraic subgroups of $\mathfrak{G}$, such that the transcendence degree of $\mathcal{G}$ over $\mathcal{J}_1$ equals the dimension of $\mathfrak{G}(\mathcal{J}_1)$. Also, $\mathcal{J}_1 \to \mathcal{J}_{1D}$ is a one-to-one mapping of the set of all partial differential fields between $\mathcal{J}$ and $\mathcal{G}$ onto the set of all ordinary differential fields between $\mathcal{J}_D$ and $\mathcal{G}_D$.*

Continuing, let $\mathcal{J}_1$ be a partial differential field between $\mathcal{J}$ and $\mathcal{G}$. If σ_1 is an isomorphism over $\mathcal{J}$ of $\mathcal{J}_1$ into $\mathcal{G}$, then σ_1 can obviously be extended to an isomorphism over $\mathcal{J}\langle u_1, \cdots, u_m\rangle$ of $\mathcal{J}_1\langle u_1, \cdots, u_m\rangle$ into $\mathcal{G}\langle u_1, \cdots, u_m\rangle$, which is of course also an isomorphism over $\mathcal{J}_D$ of $\mathcal{J}_{1D}$ into $\mathcal{G}_D$. By the ordinary Picard-Vessiot theory (PV, §19, lemma) this can in turn be extended to an automorphism of $\mathcal{G}_D$, and by Theorem 1 the restriction of this to $\mathcal{G}$ is an automorphism of $\mathcal{G}$. This proves the following theorem.

THEOREM 3. *If $\mathcal{G}$ is a Picard-Vessiot extension of $\mathcal{J}$, and $\mathcal{J}_1$ is a partial differential field between $\mathcal{J}$ and $\mathcal{G}$, then every isomorphism over $\mathcal{J}$ of $\mathcal{J}_1$ into $\mathcal{G}$ can be extended to an automorphism of $\mathcal{G}$.*

Exactly as in the ordinary case (PV, §16, Theorem 2, §18, Theorem 2) we have the following

COROLLARY. *$\mathfrak{G}(\mathcal{J}_1)$ is a normal subgroup if and only if $\sigma\mathcal{J}_1 = \mathcal{J}_1$ for every automorphism σ of $\mathcal{G}$ over $\mathcal{J}$; when this is the case the mapping which to each such σ assigns its restriction to $\mathcal{J}_1$ is a homomorphism with kernel $\mathfrak{G}(\mathcal{J}_1)$ of $\mathfrak{G}$ onto the group of all automorphisms of $\mathcal{J}_1$ over $\mathcal{J}$.*

4. **Liouvillian extensions.** Suppose as before that the field of constants $\mathcal{C}$ of $\mathcal{J}$ is algebraically closed.

Let α be an element of an extension of $\mathcal{J}$ such that the field of constants of $\mathcal{J}\langle\alpha\rangle$ is $\mathcal{C}$, $\alpha \notin \mathcal{J}$, and $\delta_i\alpha \in \mathcal{J}$ $(1 \leqq i \leqq m)$. Now $W_{1\delta_i}(1, \alpha) = \delta_i\alpha \in \mathcal{J}$, and these are not all zero (for otherwise α would be a constant and belong to $\mathcal{J}$); also, $W_{1,\delta_i\delta_j}(1, \alpha) = \delta_i\delta_j\alpha \in \mathcal{J}$, and $W_{\theta_1\theta_2}(1, \alpha) = 0$ if both θ_1, θ_2 have order not less than 1. Therefore $\mathcal{J}\langle\alpha\rangle = \mathcal{J}\langle 1, \alpha\rangle$ is a Picard-Vessiot extension of $\mathcal{J}$. Since $\mathcal{J}\langle\alpha\rangle_D = \mathcal{J}_D\langle\alpha\rangle$ and $D\alpha = \sum u_i\delta_i\alpha \in \mathcal{J}_D$ we see (PV, §23) that we may identify the group of all automorphisms of $\mathcal{J}\langle\alpha\rangle$ over $\mathcal{J}$ with the abelian algebraic group of all matrices of degree two of the form

$$\begin{pmatrix} 1 & c \\ 0 & 1 \end{pmatrix}, \qquad c \in \mathcal{C}.$$

Starting afresh, let α be an element of an extension of $\mathcal{J}$ such that the field of constants of $\mathcal{F}\langle\alpha\rangle$ is $\mathcal{C}$, $\alpha\notin\mathcal{F}$, and $\alpha^{-1}\delta_i\alpha\in\mathcal{F}$ $(1\leqq i\leqq m)$. We have $W_1(\alpha)=\alpha$, $W_{\delta_i}(\alpha)=\delta_i\alpha$ so that $W_{\delta_i}(\alpha)/W_1(\alpha)=\alpha^{-1}\delta_i\alpha\in\mathcal{F}$. Therefore $\mathcal{F}\langle\alpha\rangle$ is a Picard-Vessiot extension of $\mathcal{F}$. Since $\alpha^{-1}D\alpha=\alpha^{-1}\sum u_i\delta_i\alpha\in\mathcal{F}_D$, we see (PV, §23) that we may identify the group of all automorphisms of $\mathcal{F}\langle\alpha\rangle$ over $\mathcal{F}$ either (if α is transcendental over $\mathcal{F}$) with the group of all matrices of degree one of the form (c), with $c\in C$, $c\neq 0$, or (if α is algebraic of degree h over $\mathcal{F}$) with the group of all such matrices with c an hth root of unity (in which case we have $\alpha^h\in\mathcal{F}$).

Let $\mathcal{K}$ be an extension of $\mathcal{F}$. We shall call $\mathcal{K}$ a *liouvillian* extension of $\mathcal{F}$ if the field of constants of $\mathcal{K}$ is $\mathcal{C}$ and if $\mathcal{K}$ contains a finite sequence of elements $\alpha_1, \cdots, \alpha_r$ such that $\mathcal{F}\langle\alpha_1, \cdots, \alpha_r\rangle=\mathcal{K}$ and for each j $(1\leqq j\leqq r)$ either α_j is algebraic over $\mathcal{F}\langle\alpha_1, \cdots, \alpha_{j-1}\rangle$, or $\delta_i\alpha_j\in\mathcal{F}\langle\alpha_1, \cdots, \alpha_{j-1}\rangle$ $(1\leqq i\leqq m)$, or $\alpha_j^{-1}\delta_i\alpha_j\in\mathcal{F}\langle\alpha_1, \cdots, \alpha_{j-1}\rangle$ $(1\leqq i\leqq m)$. The concept of liouvillian extension, already familiar for ordinary differential fields, is hereby extended to partial differential fields.

Suppose that, as above, $\mathcal{K}$ is a liouvillian extension of $\mathcal{F}$ with $\mathcal{K}=\mathcal{F}\langle\alpha_1, \cdots, \alpha_r\rangle$. It follows that $\mathcal{K}_D$ is a liouvillian extension of $\mathcal{F}_D$ with $\mathcal{K}_D=\mathcal{F}_D\langle\alpha_1, \cdots, \alpha_r\rangle$.

If a Picard-Vessiot extension $\mathcal{G}$ of $\mathcal{F}$ is contained in a liouvillian extension $\mathcal{K}$ of $\mathcal{F}$, it follows from the above that $\mathcal{G}_D$ is an ordinary Picard-Vessiot extension of $\mathcal{F}_D$ contained in the (ordinary) liouvillian extension $\mathcal{K}_D$ of $\mathcal{F}_D$. By the ordinary theory (PV, §25) this implies that the group $\mathfrak{G}$ of all automorphisms of $\mathcal{G}$ over $\mathcal{F}$ has a solvable component of the identity $\mathfrak{G}^0$.

Conversely, suppose the component of the identity $\mathfrak{G}^0$ is solvable. If $\mathcal{F}^0$ is the partial differential field between $\mathcal{F}$ and $\mathcal{G}$ such that $\mathfrak{G}^0=\mathfrak{G}(\mathcal{F}^0)$, then $\mathcal{G}$ is a Picard-Vessiot extension of $\mathcal{F}^0$, $\mathcal{F}^0$ is an algebraic extension of $\mathcal{F}$, and the group of all automorphisms of $\mathcal{G}$ over $\mathcal{F}^0$ is $\mathfrak{G}^0$. Since $\mathfrak{G}^0$ is solvable we may assume (PV, §7, Theorem 1) that $\mathfrak{G}^0$ (identified with an algebraic matric group by means of the equations (6)) is in triangular form, that is, for each $\sigma\in\mathfrak{G}^0$ we have:

$$\sigma\eta_j = \sum_{i=1}^{j} s_{ij}\eta_i \qquad (1 \leqq j \leqq n),$$

each s_{ij} being an element of $\mathcal{C}$. It follows that

$$\sigma\delta_k(\eta_j/\eta_1) = \sum_{i=2}^{j} s_{11}^{-1} s_{ij}\delta_k(\eta_i/\eta_1) \qquad (2 \leqq j \leqq n)$$

for each k, from which it easily follows that $\mathcal{F}^0\langle\delta_k(\eta_2/\eta_1), \cdots,$

$\delta_k(\eta_n/\eta_1)\rangle$ is a Picard-Vessiot extension of $\mathcal{F}^0$ for which the corresponding group of automorphisms is in triangular form and therefore solvable. Now if $n=1$ it is obvious that $\mathcal{G}$ is a liouvillian extension of $\mathcal{F}^0$ and therefore of $\mathcal{F}$. Let $n>1$ and suppose that for all lower values of n we know that the group's having a solvable component of the identity implies that the Picard-Vessiot extension is liouvillian. Then, for each k, $\mathcal{F}^0\langle\delta_k(\eta_2/\eta_1), \cdots, \delta_k(\eta_n/\eta_1)\rangle$ is a liouvillian extension of $\mathcal{F}^0$, so that $\mathcal{F}^0\langle(\delta_k(\eta_j/\eta_1))_{1\leqq k\leqq m, 2\leqq j\leqq n}\rangle$ is a liouvillian extension of $\mathcal{F}^0$. Now $\sigma\eta_1=s_{11}\eta_1$, so that $\sigma(\eta_1^{-1}\delta_k\eta_1)=\eta_1^{-1}\delta_k\eta_1$ for all $\sigma\in\mathfrak{G}^0$, whence $\eta_1^{-1}\delta_k\eta_1\in\mathcal{F}^0$ $(1\leqq k\leqq n)$. It follows that $\mathcal{G}=\mathcal{F}^0\langle\eta_1, \cdots, \eta_n\rangle$ is a liouvillian extension of $\mathcal{F}^0$, and therefore of $\mathcal{F}$. Thus we have proved the following result.

THEOREM 4. *Let $\mathcal{G}$ be a Picard-Vessiot extension of $\mathcal{F}$, and let $\mathfrak{G}$ be the group of all automorphisms of $\mathcal{G}$ over $\mathcal{F}$ (suitably identified with an algebraic matric group over $\mathcal{C}$). If $\mathcal{G}$ is contained in a liouvillian extension of $\mathcal{F}$, then the component of the identity $\mathfrak{G}^0$ is solvable. Conversely, if $\mathfrak{G}^0$ is solvable, then $\mathcal{G}$ is itself a liouvillian extension of $\mathcal{F}$.*

It is easy to refine this theorem by introducing, as in the ordinary case (PV, §§24, 25), ten types of liouvillian extensions and ten types of algebraic matric groups.

COLUMBIA UNIVERSITY

GALOIS THEORY OF DIFFERENTIAL FIELDS.*

By E. R. Kolchin.[1]

Table of Contents.

* Received April 21, 1953. This paper was an address delivered before the New York meeting of the American Mathematical Society on April 26, 1952, by the invitation of the Committee to Select Hour Speakers for Eastern Sectional Meetings; it was originally submitted to the Bulletin of the American Mathematical Society on October 29, 1952, but it could not be published among the invited addresses in the Bulletin because of its length.

[1] Part of the research on which this paper is based was done in connection with a contract with the Office of Naval Research.

Introduction.

1. Differential fields and meaning of normality. The study of algebraic equations has led to the concept of field, and thence to the beginnings of algebraic geometry and to Galois theory. In much the same way, the study of algebraic differential equations has, in modern times, led to the concept of *differential field,* and thence, in the work of the late J. F. Ritt, to the extensive theory of differential algebra, which in its elementary parts bears considerable analogy to the elementary parts of algebraic geometry (see Ritt [8]). A differential field is a commutative field, in the usual sense, together with a finite family of operators $\delta_1, \cdots, \delta_m$ each of which maps the field into itself as a derivation and which commute in pairs; a differential field is said to be *ordinary* or *partial* according as the number m equals or exceeds 1.[2] *In the present paper the expression "differential field" always stands for "differential field of characteristic* 0." The purpose of the present paper is to develop a Galois theory for such differential fields.

A main problem in initiating such a theory is to find a suitable definition of normal extension of a differential field. Now, two special cases of a Galois theory already exist, and it is natural to look to these examples for hints, and to require that any general theory developed generalize these two cases. One of these cases is the Galois theory of differential field extensions of finite degree, which is the classical Galois theory, since a relative field isomorphism of such an extension is automatically a differential field isomorphism. The other case is the Picard-Vessiot theory (see Kolchin [3] and [6]; a certain

[2] By permitting m to be 0 it is possible to subsume the concept of field under that of differential field; we shall not pursue this possibility further here, and when we refer to a differential field it will always be understood that $m \geq 1$.

familiarity of the reader with the contents of these two papers will be assumed).

In the classical Galois theory an algebraic field extension of characteristic 0 is normal if the field of invariants of the group of all automorphisms of the extension over the ground-field is the ground-field itself. If an extension has this property it follows that it has the same property when considered as an extension of any intermediate field; indeed, the fundamental theorem of Galois theory could not hold were this not the case. When we turn to differential fields, however, the state of affairs is different. If $\mathcal{F}$ is a differential field and $\mathcal{G}$ is an extension of $\mathcal{F}$ such that every invariant of the group of all automorphisms of $\mathcal{G}$ over $\mathcal{F}$ belongs to $\mathcal{F}$, and if $\mathcal{F}_1$ is a differential field between $\mathcal{F}$ and $\mathcal{G}$, it does not follow, even if $\mathcal{G}$ is finitely generated and differentially algebraic over $\mathcal{F}$, that every invariant of the group of all automorphisms of $\mathcal{G}$ over $\mathcal{F}_1$ belongs to $\mathcal{F}_1$ (see the example in footnote 7). Accordingly, we define $\mathcal{G}$ to be *weakly normal* over $\mathcal{F}$ if the invariants of the group of all automorphisms of $\mathcal{G}$ over $\mathcal{F}$ all belong to $\mathcal{F}$, and $\mathcal{G}$ to be *normal* over $\mathcal{F}$ if $\mathcal{G}$ is weakly normal over every differential field between $\mathcal{F}$ and $\mathcal{G}$. The latter is the same definition as given in Kolchin [3], § 16; in that paper it was shown, and indeed it is obvious, that when $\mathcal{G}$ is normal over $\mathcal{F}$ in this sense then there is a one-to-one Galois correspondence between the set of all differential fields intermediate to $\mathcal{F}$ and $\mathcal{G}$ and a *certain* set of subgroups of the group $\mathfrak{G}$ of all automorphisms of $\mathcal{G}$ over $\mathcal{F}$. If the subgroup corresponding to an intermediate differential field is normal then the intermediate differential field is a normal extension of $\mathcal{F}$ (but not conversely!). Aside from the fact that in this definition of normality we demand what is essentially the conclusion of the theorem we wish to prove, there remain two blemishes, one of which we can remove, the other of which we can not. The first blemish is that, when the subgroup $\mathfrak{G}(\mathcal{F}_1)$ of $\mathfrak{G}$ corresponding to $\mathcal{F}_1$ is normal, so that $\mathcal{F}_1$ is a normal extension of $\mathcal{F}$, the factor group $\mathfrak{G}/\mathfrak{G}(\mathcal{F}_1)$ need not be isomorphic with the group of all automorphisms of $\mathcal{F}_1$ over $\mathcal{F}$. This situation is remedied by defining a set of isomorphisms of $\mathcal{G}$ over $\mathcal{F}$ to be *abundant* if, for every intermediate differential field $\mathcal{F}_1$ and every element α of $\mathcal{G}$ not in $\mathcal{F}_1$, there exists an isomorphism σ in the set which leaves every element of $\mathcal{F}_1$ invariant but which does not leave α invariant; clearly $\mathcal{G}$ is normal if and only if the group of all automorphisms of $\mathcal{G}$ over $\mathcal{F}$ is abundant. Furthermore, if $\mathcal{G}$ is normal over $\mathcal{F}$, and $\mathfrak{G}$ is *any* abundant group of automorphisms of $\mathcal{G}$ over $\mathcal{F}$ (not necessarily the full automorphism group), the above mentioned results continue to hold and, when $\mathfrak{G}(\mathcal{F}_1)$ is a normal subgroup of $\mathfrak{G}$, so that $\mathcal{F}_1$ is a normal extension of $\mathcal{F}$, then $\mathfrak{G}/\mathfrak{G}(\mathcal{F}_1)$ is isomorphic to an abundant group of automorphisms of $\mathcal{F}_1$ over $\mathcal{F}$. The second and more serious blemish is that we have no characterization of those "certain" subgroups which correspond to the intermediate differential fields. To avoid this defect we seek a more stringent definition.

In the classical Galois theory a normal extension is characterized also by the property that every relative isomorphism of the extension into any overfield of the extension is actually an automorphism. It would be unreasonable to demand the analogous property for differential fields, as this would exclude even the Picard-Vessiot extensions; indeed it can be shown that the extension would then be a normal algebraic extension in the classical sense. However, a hint of how to proceed is contained in the Picard-Vessiot theory. Let $\mathcal{G}$ be a Picard-Vessiot extension of $\mathcal{F}$, presupposing thereby that $\mathcal{F}$ and $\mathcal{G}$ are subject to the restrictions that $\mathcal{F}$ and $\mathcal{G}$ have the same field of constants $\mathcal{C}$, that $\mathcal{C}$ is algebraically closed, and that $\mathcal{G}$ is finitely generated and of finite transcendence degree over $\mathcal{F}$; it is easy to verify that $\mathcal{F}$ and $\mathcal{G}$ have the property that if σ is any isomorphism of $\mathcal{G}$ over $\mathcal{F}$ into an extension of $\mathcal{G}$ and if $\mathcal{C}_\sigma$ denotes the field of constants of the compositum $\mathcal{G}\langle\sigma\mathcal{G}\rangle$ then

$$(1) \qquad \mathcal{G}\langle\sigma\mathcal{G}\rangle = \mathcal{G}\langle\mathcal{C}_\sigma\rangle = (\sigma\mathcal{G})\langle\mathcal{C}_\sigma\rangle.$$

This is the property which we use, in the general case subject to the above restrictions, for our definition, which may be formulated in the following manner. We define an isomorphism σ of $\mathcal{G}$ into an extension of $\mathcal{G}$ to be *strong* if (1) holds; obviously every automorphism is a strong isomorphism. We then say, when $\mathcal{F}$ and $\mathcal{G}$ are subject to the above restrictions, that $\mathcal{G}$ is *strongly normal* over $\mathcal{F}$ if every isomorphism of $\mathcal{G}$ over $\mathcal{F}$ is strong.[3] As indicated in the following summary, it is this type of normality which appears to be the fruitful one.

2. Summary. Chapter I contains various results from elementary differential algebra which are used in the succeeding chapters. Several of these extend to partial differential fields results which are already known in the ordinary case. One theorem proved asserts the existence, for any differential field, of a suitably defined *universal extension*; roughly speaking, a given extension is universal if it is so big that all elements of all extensions we ever have occasion to introduce may be taken in the given extension. The use of a universal extension, which follows the now well-known procedure of

[3] By pursuing further the possibility mentioned in footnote [2] and suitably formulating the definition of constant, we could make the classical concept of normal algebraic extension of finite degree of a field of characteristic 0 a special case of concept of strongly normal extension of a differential field.

modern algebraic geometry (see Weil [9]), makes it possible to avoid certain logical difficulties connected with phrases like "the set of all extensions" (of a given differential field).

Chapter II contains a detailed study of strong isomorphisms of an extension $\mathcal{G}$ of a differential field $\mathcal{F}$, subject to the restrictions above. It is shown that it is possible, in a natural way, to introduce a multiplication in the set $\mathfrak{G}^*$ of all strong isomorphisms of $\mathcal{G}$ over $\mathcal{F}$, with respect to which $\mathfrak{G}^*$ becomes a group; the group $\mathfrak{G}$ of all automorphisms of $\mathcal{G}$ over $\mathcal{F}$ is then a subgroup of $\mathfrak{G}^*$. The concept of specialization is defined for an isomorphims of $\mathcal{G}$ into an extension of $\mathcal{G}$ (or more generally, for a family of such isomorphisms): σ' is called a *specialization* of σ if the family of elements $(\sigma'\alpha)_{\alpha\varepsilon\mathcal{G}}$ is a specialization over $\mathcal{G}$ of the family $(\sigma\alpha)_{\alpha\varepsilon\mathcal{G}}$. If σ is a strong isomorphism of $\mathcal{G}$ over $\mathcal{F}$ so is every specialization of σ. Most of the important facts concerning specializations of strong isomorphisms follow from Proposition 9 of Chapter II, which asserts that if $\sigma_1, \cdots, \sigma_p$ are strong isomorphisms of $\mathcal{G}$ over $\mathcal{F}$ and if $\gamma_{ik} \varepsilon \mathcal{E}_{\sigma_i} (1 \leq i \leq p, 1 \leq k \leq q_i)$ then, roughly speaking, a specialization of $(\gamma_{ik})_{1\leq i\leq p,\, 1\leq k\leq q_i}$ over $\mathcal{E}$ can, under certain general conditions, be extended to a specialization of $(\sigma_1, \cdots, \sigma_p)$ in such a way that various inequalities are preserved. An algebraico-geometric structure is introduced into $\mathfrak{G}^*$ in the following way. A subset $\mathfrak{M}^*$ of $\mathfrak{G}^*$ is called an *irreducible set* in $\mathfrak{G}^*$ if $\mathfrak{M}^*$ contains an element σ^* such that $\mathfrak{M}^*$ is the set of all specializations of σ^*; σ^* is then called a *generic element* of $\mathfrak{M}^*$, and the transcendence degree of $\mathcal{G}\langle\sigma^*\mathcal{G}\rangle$ over $\mathcal{G}$, which is the same as the transcendence degree of $\mathcal{E}_{\sigma^*}$ over $\mathcal{E}$ and does not depend on the choice of generic element σ^*, is called the *dimension* of $\mathfrak{M}^*$. A subset $\mathfrak{M}^*$ of $\mathfrak{G}^*$ is called an *algebraic set* in $\mathfrak{G}^*$ if $\mathfrak{M}^*$ is the union of a finite set of irreducible sets in $\mathfrak{G}^*$; the definition and elementary properties of the components of an algebraic set quickly follow. This algebraico-geometric structure in $\mathfrak{G}^*$ induces a similar structure in $\mathfrak{G}$. Some propositions are proved about algebraic sets in $\mathfrak{G}$ which are analogous to some elementary results in algebraic geometry. Finally, by combining the group structure and algebraico-geometric structure of $\mathfrak{G}$ we arrive at the concept of *algebraic group* in $\mathfrak{G}$. Several simple results about such algebraic groups are proved which are like certain known results on algebraic matric groups (Kolchin [3]) and, more generally, group varities in the sense of Weil [10].

In Chapter III, after a brief discussion of normal extensions of differential fields, the results of Chapter II are applied to develop a Galois theory of strongly normal extensions. It is shown that strong normality implies normality, but not conversely. Let $\mathcal{G}$ be a strongly normal extension of $\mathcal{F}$.

The group $\mathfrak{G}$ of all automorphisms of $\mathcal{G}$ over $\mathcal{F}$ is itself algebraic, and there is a one-to-one Galois correspondence between the set of all intermediate differential fields and a certain set of subgroups of $\mathfrak{G}$; this certain set is characterized as the set of all algebraic groups in $\mathfrak{G}$. The transcendence degree of $\mathcal{G}$ over any intermediate differential field $\mathcal{F}_1$ is proved to be equal to the dimension of the corresponding group $\mathfrak{G}(\mathcal{F}_1)$; the component of $\mathfrak{G}$ containing the identity (which component is unique and is a normal algebraic subgroup of $\mathfrak{G}$ of finite index) corresponds to the relative algebraic closure of $\mathcal{F}$ in $\mathcal{G}$. It is shown that if $\mathcal{F}_1$ is an intermediate differential field then the following conditions are equivalent: 1) $\mathcal{F}_1$ is strongly normal over $\mathcal{F}$; 2) $\mathcal{F}_1$ is normal over $\mathcal{F}$; 3) $\mathcal{F}_1$ is weakly normal over $\mathcal{F}$; 4) $\sigma\mathcal{F}_1 \subseteq \mathcal{F}_1$ for every $\sigma \varepsilon \mathfrak{G}$; 5) $\mathfrak{G}(\mathcal{F}_1)$ is a normal subgroup of $\mathfrak{G}$. And when these conditions are satisfied, the factor group $\mathfrak{G}/\mathfrak{G}(\mathcal{F}_1)$ is isomorphic with the group of all automorphisms of $\mathcal{F}_1$ over $\mathcal{F}$.

The remainder of Chapter III is devoted to three special types of extension. An element α is defined to be *primitive* over a differential field $\mathcal{F}$ if $\delta_i\alpha \varepsilon \mathcal{F}$ $(1 \leq i \leq m)$, to be *exponential* over $\mathcal{F}$ if $\alpha \neq 0$ and $\alpha^{-1}\delta_i\alpha \varepsilon \mathcal{F}$ $(1 \leq i \leq m)$, and to be *weierstrassian* over $\mathcal{F}$ if α is not a constant, and there exist two elements $g_2, g_3 \varepsilon \mathcal{C}$ with the polynomial $4y^3 - g_2y - g_3$ having simple roots only and m elements $a_1, \cdots, a_m \varepsilon \mathcal{F}$ such that $(\delta_i\alpha)^2 = a_i^2(4\alpha^3 - g_2\alpha - g_3)$ $(1 \leq i \leq m)$. In all three cases, if the field of constants of $\mathcal{F}\langle\alpha\rangle$ is $\mathcal{C}$, $\mathcal{F}\langle\alpha\rangle$ is strongly normal over $\mathcal{F}$. In the first two cases $\mathcal{F}\langle\alpha\rangle$ is a Picard-Vessiot extension of $\mathcal{F}$, but in the third case it is not, unless it is algebraic; indeed, it can be shown that if α is weierstrassian over $\mathcal{F}$ and if it is possible to find a family $\mathcal{F}_0, \mathcal{F}_1, \cdots, \mathcal{F}_r$ of differential fields such that $\mathcal{F}_0 = \mathcal{F}$, $\mathcal{F}_i$ is a Picard-Vessiot extension of $\mathcal{F}_{i-1}$ $(1 \leq i \leq r)$, and $\alpha \varepsilon \mathcal{F}_r$, then α is algebraic over $\mathcal{F}$. Reciprocally, it is shown that if a Picard-Vessiot extension can be obtained by a sequence of adjunctions of algebraic, primitive, exponential, and weierstrassian elements, then it can be obtained by adjunction of algebraic, primitive, and exponential elements alone. When α is transcendental over $\mathcal{F}$ then, in all three cases, $\mathcal{F}\langle\alpha\rangle$ is of transcendence degree 1 over $\mathcal{F}$. Conversely, it is proved that every strongly normal (and indeed every weakly normal) extension of $\mathcal{F}$ of transcendence degree 1 can be obtained from $\mathcal{F}$ by combining with algebraic adjunctions an adjunction of one of these three types. The proof of this converse, which is long and in places involves complicated computations, makes use of the well-known theorem that the group of automorphisms of an algebraic function field of one variable over an algebraically closed field of characteristic 0 is finite if the genus exceeds 1.

3. Problems. Various problems remain for investigation; we mention three, which are related.

First, there is the connection between algebraic groups of automorphisms as defined herein, and group varieties as defined by A. Weil. Is it always possible to identify the component of the identity of an algebraic group with a group variety? Conversely, is every group variety identifiable with an algebraic group?

Second, there is the task of characterizing, if possible, by algebraic-group properties, those strongly normal extensions which are Picard-Vessiot extensions.

Third, there is the significance of solvability, or even of commutativity, of the group of automorphisms of a strongly normal extension. For what sort of strongly normal extension is the group abelian? It is conceivable that investigation of this question will lead into the theory of abelian functions.

4. Notation. The notation used is more or less the same as in Kolchin [3], and is reasonably standard. We mention only that the degree of transcendence and the degree of differential transcendence of $\mathfrak{G}$ over $\mathfrak{F}$ are denoted by $\partial^0\mathfrak{G}/\mathfrak{F}$ and $\nabla^0\mathfrak{G}/\mathfrak{F}$ respectively.

Chapter I. Differential-algebraic preliminaries.

1. A lemma on polynomial ideals. Let K be a field of characteristic 0,[4] and let $y_1, \cdots, y_n$ be indeterminates. We shall prove the following lemma, which collects certain known facts in a form convenient for future use.

LEMMA. *Let $\mathfrak{p}$ be a prime ideal of $K[y_1, \cdots, y_n]$ of dimension d. For every extension L of K the ideal $L \cdot \mathfrak{p}$ generated by $\mathfrak{p}$ in $L[y_1, \cdots, y_n]$ is equal to its own radical; the minimal prime ideal divisors $\mathfrak{p}_1, \cdots, \mathfrak{p}_r$ of $L \cdot \mathfrak{p}$ all have dimension d; every generic zero of every $\mathfrak{p}_i$ is a generic zero of $\mathfrak{p}$, and every generic zero of $\mathfrak{p}$ is a zero of precisely one $\mathfrak{p}_i$. There exists, independent of L, an irreducible polynomial R with coefficients in K such that for every extension L of K the number of minimal prime ideal divisors of $L \cdot \mathfrak{p}$ equals the number of irreducible factors into which R splits over L.*

Proof. Let $\mathfrak{P}$ be the radical of $L \cdot \mathfrak{p}$, so that $\mathfrak{P} = \mathfrak{p}_1 \cap \cdots \cap \mathfrak{p}_r$ where $\mathfrak{p}_1, \cdots, \mathfrak{p}_r$ are the minimal prime ideal divisors of $L \cdot \mathfrak{p}$. Let $(\eta_{i1}, \cdots, \eta_{in})$ be a generic zero of $\mathfrak{p}_i$; $(\eta_{i1}, \cdots, \eta_{in})$ is obviously a zero of $\mathfrak{p}$. If

[4] This condition, which suffices for the present purposes, can be relaxed.

$F \varepsilon K[y_1, \cdots, y_n]$ vanishes at $(\eta_{i1}, \cdots, \eta_{in})$ then $F \varepsilon \mathfrak{p}_i$, so that if we let G be a polynomial in $L[y_1, \cdots, y_n]$ such that $G \varepsilon \bigcap_{j \neq i} \mathfrak{p}_j$, $G \not\varepsilon \mathfrak{p}_i$ then $FG \varepsilon \mathfrak{P}$, whence for some exponent $e > 0$ we have $F^e G^e \varepsilon L \cdot \mathfrak{p}$. Therefore there exist elements $\lambda_k \varepsilon L$ linearly independent over K such that we may write $F^e G^e = \Sigma \lambda_k P_k$, where each $P_k \varepsilon \mathfrak{p}$, and $G^e = \Sigma \lambda_k G_k$, where each $G_k \varepsilon K[y_1, \cdots, y_n]$. From this we see that $\Sigma \lambda_k F^e G_k = \Sigma \lambda_k P_k$, so that $F^e G_k = P_k \varepsilon \mathfrak{p}$ for every k; since not every G_k belongs to $\mathfrak{p}$ (for otherwise G would belong to $\mathfrak{p}_i$) and since $\mathfrak{p}$ is prime we conclude that $F \varepsilon \mathfrak{p}$. This shows that $(\eta_{i1}, \cdots, \eta_{in})$ is a generic zero of $\mathfrak{p}$.

Let $(\eta_1, \cdots, \eta_n)$ be any generic zero of $\mathfrak{p}$. For the sake of definiteness we suppose that $\eta_1, \cdots, \eta_d$ are algebraically independent over K; then there exists an element ω such that $K(\eta_1, \cdots, \eta_n) = K(\eta_1, \cdots, \eta_d, \omega)$. Let w be a new indeterminate and let R be a polynomial in $K[y_1, \cdots, y_d, w]$ of as low degree as possible which vanishes at $(\eta_1, \cdots, \eta_d, \omega)$, so that R is irreducible over K. Because $(\eta_{i1}, \cdots, \eta_{in})$ is also a generic zero of $\mathfrak{p}$, $(\eta_{i1}, \cdots, \eta_{in})$ is a generic specialization [5] of $(\eta_1, \cdots, \eta_n)$ over K, and therefore can be extended to a generic specialization $(\eta_{i1}, \cdots, \eta_{in}, \omega_i)$ of $(\eta_1, \cdots, \eta_n, \omega)$ over K. Now $(\eta_{i1}, \cdots, \eta_{id}, \omega_i)$ is a zero of R and therefore of some irreducible factor of R over L; moreover $(\eta_{i1}, \cdots, \eta_{id}, \omega_i)$ is a zero of only one of these irreducible factors, for otherwise $(\eta_{i1}, \cdots, \eta_{id}, \omega_i)$ and therefore $(\eta_1, \cdots, \eta_d, \omega)$ would be a zero of $\partial R/\partial w$ which is of lower degree than R. We denote the irreducible factor of R over L which vanishes at $(\eta_{i1}, \cdots, \eta_{id}, \omega_i)$ by R_i.

Let $(\eta'_{i1}, \cdots, \eta'_{id}, \omega'_i)$ be a generic zero of the prime ideal $\mathfrak{r}_i$ generated by R_i in $L[y_1, \cdots, y_d, w]$. Then $(\eta_{i1}, \cdots, \eta_{id}, \omega_i)$ is a specialization of $(\eta'_{i1}, \cdots, \eta'_{id}, \omega'_i)$ over L and a generic specialization of $(\eta'_{i1}, \cdots, \eta'_{id}, \omega'_i)$ over K; therefore there exist elements $\eta'_{i,d+1}, \cdots, \eta'_{in}$ such that $(\eta_{i1}, \cdots, \eta_{in}, \omega_i)$ is a generic specialization of $(\eta'_{i1}, \cdots, \eta'_{in}, \omega'_i)$ over K. Now $(\eta'_{i1}, \cdots, \eta'_{in})$ is a zero of $\mathfrak{p}$ and therefore of $\mathfrak{p}_j$ for some j; since $(\eta_{i1}, \cdots, \eta_{in})$ must be a zero of this $\mathfrak{p}_j$ and since $\mathfrak{p}_j \not\subset \mathfrak{p}_i$ if $i \neq j$ it follows that $i = j$. As $\eta'_{i1}, \cdots, \eta'_{id}$ are obviously algebraically independent over L we have

$$\partial^0 L(\eta'_{i1}, \cdots, \eta'_{in})/L \geq d = \partial^0 K(\eta_{i1}, \cdots, \eta_{in})/K$$
$$\geq \partial^0 L(\eta_{i1}, \cdots, \eta_{in})/L = \dim \mathfrak{p}_i,$$

so that $(\eta'_{i1}, \cdots, \eta'_{in})$ is a generic zero of $\mathfrak{p}_i$ and $\dim \mathfrak{p}_i = d$. It follows that $(\eta'_{i1}, \cdots, \eta'_{in})$ is a generic specialization of $(\eta_{i1}, \cdots, \eta_{in})$ over L, so that

[5] The very elementary facts concerning specializations over a field used in this paper can be found in Weil [9], chapter II.

$(\eta'_{i1},\cdots,\eta'_{in},\omega'_i)$ is a generic specialization of $(\eta_{i1},\cdots,\eta_{in},\omega_i)$ over L, and $(\eta_{i1},\cdots,\eta_{id},\omega_i)$ is a generic zero of $\mathfrak{r}_i$.

If $R_i=R_j$ then $\mathfrak{r}_i=\mathfrak{r}_j$ and $(\eta_{i1},\cdots,\eta_{id},\omega_i)$ is a generic specialization of $(\eta_{j1},\cdots,\eta_{jd},\omega_j)$ over L, so that $(\eta_{i1},\cdots,\eta_{in})$ is a generic specialization of $(\eta_{j1},\cdots,\eta_{jn})$ over L, whence $\mathfrak{p}_i=\mathfrak{p}_j$ and $i=j$. Thus $R_1,\cdots,R_r$ are distinct irreducible factors of R over L. Let S be any irreducible factor of R over L, let $\mathfrak{s}$ denote the prime ideal generated by S in $L[y_1,\cdots,y_d,w]$, and let $(\zeta_1,\cdots,\zeta_d,\theta)$ be a generic zero of $\mathfrak{s}$; $(\zeta_1,\cdots,\zeta_d,\theta)$ is clearly a generic specialization of $(\eta_1,\cdots,\eta_d,\omega)$ over K and therefore can be extended to a generic specialization $(\zeta_1,\cdots,\zeta_n,\theta)$ of $(\eta_1,\cdots,\eta_n,\omega)$. $(\zeta_1,\cdots,\zeta_n)$ is a zero of $\mathfrak{p}$ and therefore of $\mathfrak{p}_i$ for some i; therefore $(\zeta_1,\cdots,\zeta_d,\theta)$ is a zero of R_i, so that R_i is divisible by S. It follows that $R_1,\cdots,R_r$ are all the irreducible factors of R over L, so that the number of minimal prime ideal divisors of $L\cdot\mathfrak{p}$ equals the number of irreducible factors of R over L. Also $(\eta_1,\cdots,\eta_n)$, which obviously is a zero of some $\mathfrak{p}_i$, is a zero of only one $\mathfrak{p}_i$; for if $(\eta_1,\cdots,\eta_n)$ were a zero of $\mathfrak{p}_i$ and $\mathfrak{p}_j$ $(i\neq j)$ then $(\eta_1,\cdots,\eta_d,\omega)$ would be a zero of R_i and R_j, and therefore of $\partial R/\partial w$, which is of lower degree than R.

It remains to prove that $\mathfrak{P}=L\cdot\mathfrak{p}$, and to do this it suffices to show that $\mathfrak{P}\subseteq L\cdot\mathfrak{p}$. If $F\in\mathfrak{P}$ then we may write $F=\Sigma\lambda_jF_j$, where each F_j belongs to $K[y_1,\cdots,y_n]$ and (λ_j) is a family of elements of L linearly independent over K. Let $A_{d+1},B_{d+1},\cdots,A_n,B_n$ be polynomials in $K[y_1,\cdots,y_d,w]$ such that

$$\eta_k=A_k(\eta_1,\cdots,\eta_d,\omega)/B_k(\eta_1,\cdots,\eta_d,\omega),\ d+1\leq k\leq n.$$

Then there exists a single exponent $e\geq 0$ such that for each j

$$(B_{d+1}\cdots B_n)^eF_j\equiv G_j(B_{d+1}y_{d+1}-A_{d+1},\cdots,B_ny_n-A_n),$$

where $G_j\in K[y_1,\cdots,y_d,w]$. It is easy to see that $\Sigma\lambda_jG_j$ vanishes at $(\eta_{i1},\cdots,\eta_{id},\omega_i)$ for each i and therefore is divisible by R. Because the λ_j's are linearly independent over K it easily follows that each G_j is divisible by R, so that each G_j vanishes at $(\eta_1,\cdots,\eta_d,\omega)$, each $(B_{d+1}\cdots B_n)^eF_j$ vanishes at $(\eta_1,\cdots,\eta_n,\omega)$, each F_j vanishes at $(\eta_1,\cdots,\eta_n)$, each $F_j\in\mathfrak{p}$, and $F\in L\cdot\mathfrak{p}$.

2. Prime differential ideals and differential field extension. Let $\mathcal{F}$ be a differential field and let $y_1,\cdots,y_n$ denote indeterminates. If Π is a prime differential ideal of the differential ring $\mathcal{F}\{y_1,\cdots,y_n\}$ and $(\eta_1,\cdots,\eta_n)$ is a generic zero of Π then the degree of differential transcendence

7

$\nabla^0 \mathcal{F}\langle\eta_1, \cdots, \eta_n\rangle/\mathcal{F}$ is called the *dimension* of Π (notation: $\dim \Pi$); $\dim \Pi$ does not depend on the particular generic zero used, and $0 \leq \dim \Pi \leq n$. The degree of transcendence $\partial^0 \mathcal{F}\langle\eta_1, \cdots, \eta_n\rangle/\mathcal{F}$ is called the *order* of Π (notation: $\operatorname{ord} \Pi$); $\operatorname{ord} \Pi$, which does not depend on the particular generic zero used, either is an integer ≥ 0 (finite order) or is ∞ (infinite order). The following result is well-known for the case in which $\mathcal{F}$ is an ordinary differential field (see e. g. Ritt [8], pp. 50-51).

PROPOSITION 1. *Let Π be a prime differential ideal of $\mathcal{F}\{y_1, \cdots, y_n\}$. For every extension $\mathcal{G}$ of $\mathcal{F}$ the ideal $\mathcal{G} \cdot \Pi$ generated by Π in $\mathcal{G}\{y_1, \cdots, y_n\}$ is a perfect differential ideal; the minimal prime differential ideal divisors $\Pi_1, \cdots, \Pi_r$ of $\mathcal{G} \cdot \Pi$ all have the same dimension as Π, and all have the same order as Π; every generic zero of every Π_i is a generic zero of Π, and every generic zero of Π is a zero of precisely one Π_i. There exists, independent of $\mathcal{G}$, an irreducible polynomial R with coefficients in $\mathcal{F}$ such that for every extension $\mathcal{G}$ of $\mathcal{F}$ the number of minimal prime differential ideal divisors of $\mathcal{G} \cdot \Pi$ equals the number of irreducible factors into which R splits over $\mathcal{G}$.*

Proof. Let $\mathcal{R}$, $\mathcal{R}'$ denote $\mathcal{F}\{y_1, \cdots, y_n\}$, $\mathcal{G}\{y_1, \cdots, y_n\}$ respectively. For each integer $k \geq 0$ let $\mathcal{R}_k$, $\mathcal{R}'_k$ denote the set of all elements of $\mathcal{R}$, $\mathcal{R}'$ respectively which do not have order $> k$. We shall consider $\mathcal{R}_k$ and $\mathcal{R}'_k$ as polynomial rings over $\mathcal{F}$ and $\mathcal{G}$ respectively (each element of $\mathcal{R}_k$ and of $\mathcal{R}'_k$ is a polynomial in the expressions $\delta_1^{i_1} \cdots \delta_m^{i_m} y_j$ with $0 \leq i_1 + \cdots + i_m \leq k$, $1 \leq j \leq n$).

For every $k \geq 0$, $\Pi \cap \mathcal{R}_k$ is a prime ideal of $\mathcal{R}_k$, so that (§ 1) $\mathcal{G} \cdot (\Pi \cap \mathcal{R}_k)$ is an ideal of $\mathcal{R}'_k$ which is equal to its own radical. Clearly $\mathcal{G} \cdot \Pi$ is a differential ideal of $\mathcal{R}'$; if $\mathcal{G} \cdot \Pi$ were not perfect there would exist an $F \varepsilon \mathcal{R}'$ and an integer $e > 0$ with $F \not\varepsilon \mathcal{G} \cdot \Pi$, $F^e \varepsilon \mathcal{G} \cdot \Pi$, so that for large k we would have $F \not\varepsilon \mathcal{G} \cdot (\Pi \cap \mathcal{R}_k)$, $F^e \varepsilon \mathcal{G} \cdot (\Pi \cap \mathcal{R}_k)$, contradicting the fact that $\mathcal{G} \cdot (\Pi \cap \mathcal{R}_k)$ is its own radical. Thus $\mathcal{G} \cdot \Pi$ is a perfect differential ideal of $\mathcal{R}'$.

We now assert that $\mathcal{G} \cdot (\Pi \cap \mathcal{R}_k) = (\mathcal{G} \cdot \Pi) \cap \mathcal{R}'_k$. Indeed, it is obvious that $\mathcal{G} \cdot (\Pi \cap \mathcal{R}_k) \subseteq (\mathcal{G} \cdot \Pi) \cap \mathcal{R}'_k$. Suppose then that $F \varepsilon (\mathcal{G} \cdot \Pi) \cap \mathcal{R}'_k$. Then we may write

$$F = \sum P_k \phi_k \tag{1}$$

where the elements $\phi_k \varepsilon \mathcal{G}$ are linearly independent over $\mathcal{F}$ and each $P_k \varepsilon \Pi$. Fixing our attention on any derivative $\delta_1^{i_1} \cdots \delta_m^{i_m} y_j$ of order $i_1 + \cdots + i_m > k$, let C_k denote the coefficient in P_k of any fixed positive power of this derivative; since F is not of order $> k$, (1) yields the relation $\sum C_k \phi_k = 0$, so that each

$C_k = 0$. It follows that no P_k has order $> k$, so that each $P_k \varepsilon \mathcal{R}_k$, and $F \varepsilon \mathfrak{G} \cdot (\Pi \cap \mathcal{R}_k)$. This proves our assertion.

Since the perfect differential ideal $\mathfrak{G} \cdot \Pi$ is the intersection of its minimal prime differential ideal divisors $\Pi_1, \cdots, \Pi_r$ it is a consequence of the above assertion that

$$\mathfrak{G} \cdot (\Pi \cap \mathcal{R}_k) = (\Pi_1 \cap \mathcal{R}'_k) \cap \cdots \cap (\Pi_r \cap \mathcal{R}'_k).$$

Now $\Pi_i \not\subset \Pi_j$ if $i \neq j$, so that if k is sufficiently great $\Pi_i \cap \mathcal{R}'_k \not\subset \Pi_j \cap \mathcal{R}'_k$ whenever $i \neq j$. Taking k large enough for this to be the case, we see that $\Pi_1 \cap \mathcal{R}'_k, \cdots, \Pi_r \cap \mathcal{R}'_k$ are the minimal prime ideal divisors of $\mathfrak{G} \cdot (\Pi \cap \mathcal{R}_k)$ in $\mathcal{R}'_k$.

It is obvious that Π_i is of dimension no higher than Π. If Π_i had lower dimension than Π then there would exist a subset $z_1, \cdots, z_t$ of $y_1, \cdots, y_n$ such that Π contains no nonzero differential polynomial in $z_1, \cdots, z_t$ alone but Π_i does, that is (for k large) $\Pi \cap \mathcal{R}_k$ contains no nonzero polynomial in the expressions $\delta_1^{i_1} \cdots \delta_m^{i_m} z_l$ with $0 \leq i_1 + \cdots + i_m \leq k$, $1 \leq l \leq t$, but $\Pi_i \cap \mathcal{R}'_k$ does; this would imply that the prime polynomial ideal $\Pi_i \cap \mathcal{R}'_k$ of $\mathcal{R}'_k$ has lower dimension than the prime polynomial ideal $\Pi \cap \mathcal{R}_k$ of $\mathcal{R}_k$, contradicting the lemma of § 1. Therefore each Π_i is of the same dimension as Π.

Again, the order of Π_i obviously equals the limit (including the possibility ∞) as k becomes infinite of the dimension of $\Pi_i \cap \mathcal{R}'_k$, which by the lemma equals the limit of the dimension of $\Pi \cap \mathcal{R}_k$, which equals the order of Π.

If $(\eta_{i1}, \cdots, \eta_{in})$ is a generic zero of Π_i then obviously

$$(\delta_1^{i_1} \cdots \delta_m^{i_m} \eta_{ij})_{0 \leq i_1 + \ldots + i_m \leq k,\ 1 \leq j \leq n}$$

is a generic zero of $\Pi_i \cap \mathcal{R}'_k$ and therefore (by the lemma) a generic zero of $\Pi \cap \mathcal{R}_k$; since k can be arbitrarily large this means that $(\eta_{i1}, \cdots, \eta_{in})$ is a generic zero of Π.

If $(\eta_1, \cdots, \eta_n)$ is a generic zero of Π then $(\eta_1, \cdots, \eta_n)$ is a zero of at least one Π_i; if it were a zero of Π_i for two distinct values of i then for k large $(\delta_1^{i_1} \cdots \delta_m^{i_m} \eta_j)_{0 \leq i_1 + \ldots + i_m \leq k,\ 1 \leq j \leq n}$ would be a generic zero of $\Pi \cap \mathcal{R}_k$ and a zero of $\Pi_i \cap \mathcal{R}'_k$ for two distinct values of i, contradicting the lemma.

It remains to prove the existence of a polynomial R as described in the statement of the proposition. To this end let $k(\mathfrak{G})$ be the smallest integer such that, for all $k \geq k(\mathfrak{G})$, $\Pi_1 \cap \mathcal{R}'_k, \cdots, \Pi_r \cap \mathcal{R}'_k$ are the minimal prime ideal divisors of $\mathfrak{G} \cdot (\Pi \cap \mathcal{R}_k)$, that is such that, for all $k \geq k(\mathfrak{G})$, $\Pi_i \cap \mathcal{R}'_k \not\subset \Pi_j \cap \mathcal{R}'_k$ whenever $i \neq j$. We shall show below that $k(\mathfrak{G})$ is an increasing function of $\mathcal{G}$, that is, if $\mathcal{H}$ is an extension of $\mathcal{G}$ then $k(\mathcal{G}) \leq k(\mathcal{H})$. Assuming this result, let us see how we can complete the proof of the proposition. By the lemma for each $k \geq 0$ there exists, independent of $\mathcal{G}$, an irreducible polynomial R_k with coefficients in $\mathcal{F}$ such that the number of minimal prime ideal divisors of $\mathcal{G} \cdot (\Pi \cap \mathcal{R}_k)$ in $\mathcal{R}'_k$ equals the number of irreducible factors of R_k over $\mathcal{G}$. Now the number of irreducible factors of R_k over $\mathcal{G}$ equals the number of irreducible factors of R_k over the relative algebraic closure $\mathcal{F}^0$ of $\mathcal{F}$ in $\mathcal{G}$. Therefore

number of minimal prime differential ideal divisors of $\mathcal{G} \cdot \Pi =$
number of minimal prime ideal divisors of $\mathcal{G} \cdot (\Pi \cap \mathcal{R}_k)$ (all $k \geq k(\mathcal{G})$) $=$
number of irreducible factors of R_k over $\mathcal{G}$ (all $k \geq k(\mathcal{G})$) $=$
number of irreducible factors of R_k over $\mathcal{F}^0$ (all $k \geq k(\mathcal{G})$) $=$
number of irreducible factors of R_k over $\mathcal{F}^0$ (all $k \geq k(\mathcal{F}^0)$) $=$
number of irreducible factors of R_k over $\mathcal{G}$ (all $k \geq k(\mathcal{F}^0)$).

Thus if we let $\mathcal{F}'$ be an algebraic closure[6] of $\mathcal{F}$ and set $R = R_l$, where $l = k(\mathcal{F}')$, then $l \geq k(\mathcal{F}^0)$ and R does not depend on $\mathcal{G}$, and the number of minimal prime differential ideal divisors of $\mathcal{G} \cdot \Pi$ equals the number of irreducible factors of R over $\mathcal{G}$.

We now show that $k(\mathcal{G})$ is an increasing function of $\mathcal{G}$. Let $\mathcal{H}$ be an extension of $\mathcal{G}$, let $\mathcal{R}'' = \mathcal{H}\{y_1, \cdots, y_n\}$, and let $\mathcal{R}''_k$ denote the ring of all elements of $\mathcal{R}''$ which do not have order $> k$. $\mathcal{H} \cdot \Pi$ is a perfect differential ideal of $\mathcal{R}''$ and $\mathcal{H} \cdot \Pi = \mathcal{H} \cdot (\mathcal{G} \cdot \Pi) = \mathcal{H} \cdot (\Pi_1 \cap \cdots \cap \Pi_r)$. Now if Π' is any ideal of R' and if for some $P'' \varepsilon \mathcal{H} \cdot \Pi'$ we write $P'' = \sum \phi_l P'_l$, where each $P'_l \varepsilon \mathcal{R}'$ and the elements ϕ_l of $\mathcal{H}$ are linearly independent over $\mathcal{G}$, then it is easy to see that each $P'_l \varepsilon \Pi'$. It follows that

$$(\mathcal{H} \cdot \Pi_1) \cap \cdots \cap (\mathcal{H} \cdot \Pi_r) = \mathcal{H} \cdot (\Pi_1 \cap \cdots \cap \Pi_r),$$

so that $\mathcal{H} \cdot \Pi = (\mathcal{H} \cdot \Pi_1) \cap \cdots \cap (\mathcal{H} \cdot \Pi_r)$. Now if $i_1 \neq i_2$ then a minimal prime differential ideal divisor Λ_1 of $\mathcal{H} \cdot \Pi_{i_1}$ can not be contained in a minimal prime differential ideal divisor Λ_2 of $\mathcal{H} \cdot \Pi_{i_2}$, because otherwise for every k we would have $\Lambda_2 \cap \mathcal{R}''_k \supseteq \mathcal{H} \cdot (\Pi_{i_1} \cap \mathcal{R}'_k + \Pi_{i_2} \cap \mathcal{R}'_k)$ and $\Lambda_2 \cap \mathcal{R}''_k$

[6] If K is any field of characteristic 0 and K' is an algebraic closure of K then every derivation of K has a unique extension which is a derivation of K' (see, for example, Bourbaki [1], chapter V, § 9, proposition 5, p. 139); moreover, it is easy to verify that if two derivations of K commute then there extended derivations of K' commute. It follows that every differential field $\mathcal{F}$ has an algebraically closed algebraic differential field extension; we call any such extension an *algebraic closure* of the differential field $\mathcal{F}$. Any two algebraic closures of $\mathcal{F}$ are isomorphic over $\mathcal{F}$.

would have lower dimension than $\Pi_{i_2} \cap \mathcal{R}'_k$ in contradiction to the lemma of § 1 (since for k large $\Lambda_2 \cap \mathcal{R}''_k$ is a minimal prime ideal divisor of $\mathcal{H} \cdot (\Pi_{i_2} \cap \mathcal{R}'_k)$). Therefore if we denote the minimal prime differential ideal divisors of $\mathcal{H} \cdot \Pi_i$ by $\Pi_{i1}, \cdots, \Pi_{i,s(i)}$ $(1 \leq i \leq r)$ then the minimal prime differential ideal divisors of $\mathcal{H} \cdot \Pi$ are the ideals Π_{ij} $(1 \leq i \leq r,$ $1 \leq j \leq s(i))$. If $k < k(\mathcal{G})$ then there exist i, i' with $i \neq i'$ such that $\Pi_i \cap \mathcal{R}'_k \subseteq \Pi_{i'} \cap \mathcal{R}'_k$; for these i, i' we have

$$(\Pi_{i1} \cap \mathcal{R}''_k) \cap \cdots \cap (\Pi_{i,s(i)} \cap \mathcal{R}''_k) = (\mathcal{H} \cdot \Pi_i) \cap \mathcal{R}''_k$$
$$\subseteq (\mathcal{H} \cdot \Pi_{i'}) \cap \mathcal{R}''_k \subseteq \Pi_{i'1} \cap \mathcal{R}''_k,$$

so that for some j we have $\Pi_{ij} \cap \mathcal{R}''_k \subseteq \Pi_{i'1} \cap \mathcal{R}''_k$, whence $k < k(\mathcal{H})$. It follows that $k(\mathcal{G}) \leq k(\mathcal{H})$. As we have seen, this completes the proof of Proposition 1.

3. Specializations over differental fields. For purposes of convenience we extend the language of specializations, as used in algebraic geometry, to differential fields. Let $\mathcal{F}$ be a differential field and let $(\eta_j)_{j \varepsilon J}$ be an indexed family of elements of some extension of $\mathcal{F}$. A family $(\zeta_j)_{j \varepsilon J}$, with the same set of indices J, of elements of some extension of $\mathcal{F}$ will be called a *specialization of* $(\eta_j)_{j \varepsilon J}$ *over* $\mathcal{F}$ if, for every finite subset $j_1, \cdots, j_n$ of J, every differential polynomial in $\mathcal{F}\{y_1, \cdots, y_n\}$ which vanishes at $(\eta_{j_1}, \cdots, \eta_{j_n})$ also vanishes at $(\zeta_{j_1}, \cdots, \zeta_{j_n})$. If $(\zeta_j)_{j \varepsilon J}$ is a specialization of $(\eta_j)_{j \varepsilon J}$ over $\mathcal{F}$ such that $(\eta_j)_{j \varepsilon J}$ is a specialization of $(\zeta_j)_{j \varepsilon J}$ over $\mathcal{F}$ then we say that $(\zeta_j)_{j \varepsilon J}$ is a *generic* specialization of $(\eta_j)_{j \varepsilon J}$ over $\mathcal{F}$. If I is a subset of J and $(\zeta_j)_{j \varepsilon J}$ is a specialization of $(\eta_j)_{j \varepsilon J}$ over $\mathcal{F}$ then $(\zeta_i)_{i \varepsilon I}$ is a specialization of $(\eta_i)_{i \varepsilon I}$ over $\mathcal{F}$; we say in this case that the specialization $(\zeta_j)_{j \varepsilon J}$ of $(\eta_j)_{j \varepsilon J}$ over $\mathcal{F}$ is an *extension* of the specialization $(\zeta_i)_{i \varepsilon I}$ of $(\eta_i)_{i \varepsilon I}$ over $\mathcal{F}$. If $(\zeta_j)_{j \varepsilon J}$ is a generic specialization of $(\eta_j)_{j \varepsilon J}$ over $\mathcal{F}$ then there exists a unique isomorphism of $\mathcal{F}\langle(\eta_j)_{j \varepsilon J}\rangle$ onto $\mathcal{F}\langle(\zeta_j)_{j \varepsilon J}\rangle$ over $\mathcal{F}$ which maps η_j onto ζ_j for every $j \varepsilon J$. If $(\zeta_j)_{j \varepsilon J}$ is a generic specialization of $(\eta_j)_{j \varepsilon J}$ over $\mathcal{F}$ and if $(\eta'_{j'})_{j' \varepsilon J'}$ is any family of elements of some extension of $\mathcal{F}\langle(\eta_j)_{j \varepsilon J}\rangle$, then the specialization can be extended to a generic specialization

$$((\zeta_j)_{j \varepsilon J}, (\zeta'_{j'})_{j' \varepsilon J'}) \text{ of } ((\eta_j)_{j \varepsilon J}, (\eta'_{j'})_{j' \varepsilon J'})$$

over $\mathcal{F}$. The following proposition is well-known in the case of ordinary differential fields (Ritt [8], p. 49).

PROPOSITION 2. *If* $(\zeta_j)_{j \varepsilon J}$ *is a specialization of* $(\eta_j)_{j \varepsilon J}$ *over* $\mathcal{F}$ *then*

$$\nabla^0 \mathcal{F}\langle(\zeta_j)_{j \varepsilon J}\rangle/\mathcal{F} \leq \nabla^0 \mathcal{F}\langle(\eta_j)_{j \varepsilon J}\rangle/\mathcal{F}$$

and

$$\partial^0\mathfrak{F}\langle(\zeta_j)_{j\varepsilon J}\rangle/\mathfrak{F} \leq \partial^0\mathfrak{F}\langle(\eta_j)_{j\varepsilon J}\rangle/\mathfrak{F};$$

if in addition $\partial^0\mathfrak{F}\langle(\eta_j)_{j\varepsilon J}\rangle/\mathfrak{F}$ *is finite and equal to* $\partial^0\mathfrak{F}\langle(\zeta_j)_{j\varepsilon J}\rangle/\mathfrak{F}$ *then the specialization is generic.*

Proof. The first part of the proposition is obvious. We prove the second part. Since $\partial^0\mathfrak{F}\langle(\eta_j)_{j\varepsilon J}\rangle/\mathfrak{F}$ and $\partial^0\mathfrak{F}\langle(\zeta_j)_{j\varepsilon J}\rangle/\mathfrak{F}$ are finite and equal there exist a finite subset K of J such that

$$\partial^0\mathfrak{F}\langle(\zeta_j)_{j\varepsilon J'}\rangle/\mathfrak{F} = \partial^0\mathfrak{F}\langle(\zeta_j)_{j\varepsilon J}\rangle/\mathfrak{F} = \partial^0\mathfrak{F}\langle(\eta_j)_{j\varepsilon J}\rangle/\mathfrak{F} = \partial^0\mathfrak{F}\langle(\eta_j)_{j\varepsilon J'}\rangle/\mathfrak{F}$$

for every finite subset J' of J which contains K. It is clear that $(\zeta_j)_{j\varepsilon J}$ is a generic specialization of $(\eta_j)_{j\varepsilon J}$ if $(\zeta_j)_{j\varepsilon J'}$ is a generic specialization of $(\eta_j)_{j\varepsilon J'}$ for every finite subset J' of J which contains K. It follows that we may assume that J is finite. Making this assumption, it is easy to see that there exists an integer $k_0 \geq 0$ such that

$$\mathfrak{F}\langle(\eta_j)_{j\varepsilon J}\rangle = \mathfrak{F}\left((\delta_1^{i_1}\cdots\delta_m^{i_m}\eta_j)_{0\leq i_1+\ldots+i_m\leq k,\, j\varepsilon J}\right),$$

$$\mathfrak{F}\langle(\zeta_j)_{j\varepsilon J}\rangle = \mathfrak{F}\left((\delta_1^{i_1}\cdots\delta_m^{i_m}\zeta_j)_{0\leq i_1+\ldots+i_m\leq k,\, j\varepsilon J}\right)$$

for every integer $k \geq k_0$. By a well-known result concerning specializations over a field (for example, see Weil [9], p. 28, Theorem 3) it follows that if we regard $\mathfrak{F}$ as a field then $(\delta_1^{i_1}\cdots\delta_m^{i_m}\zeta_j)_{0\leq i_1+\ldots+i_m\leq k,\, j\varepsilon J}$ is a generic specialization of $(\delta_1^{i_1}\cdots\delta_m^{i_m}\eta_j)_{0\leq i_1+\ldots+i_m\leq k,\, j\varepsilon J}$ over $\mathfrak{F}$ for every $k \geq k_0$. This implies that $(\zeta_j)_{j\varepsilon J}$ is a generic specialization of $(\eta_j)_{j\varepsilon J}$ over the differential field $\mathfrak{F}$.

COROLLARY. *A zero* $(\zeta_1,\cdots,\zeta_n)$ *of a prime differential ideal* Π *of* $F\{y_1,\cdots,y_n\}$ *of finite order is generic if and only if*

$$\partial^0\mathfrak{F}\langle\zeta_1,\cdots,\zeta_n\rangle/\mathfrak{F} = \operatorname{ord}\Pi.$$

4. Constants. Let $\mathfrak{F}$ be a differential field, and denote the field of constants of $\mathfrak{F}$ by $\mathfrak{C}$. By the *order* of a differential operator $\delta_1^{i_1}\cdots\delta_m^{i_m}$ we mean the inteer $i_1+\cdots+i_m$. If $\theta_1,\cdots,\theta_n$ are differential operators of the form $\delta_1^{i_1}\cdots\delta_m^{i_m}$ $(0\leq i_1<\infty,\cdots,0\leq i_m<\infty)$ then we use $W_{\theta_1\ldots\theta_n}$ to denote the differential polynomial defined by $W_{\theta_1\ldots\theta_n} = \det(\theta_i y_j)$.

PROPOSITION 3. *The elements* $\eta_1,\cdots,\eta_n$ *of* $\mathfrak{F}$ *are linearly dependent over* $\mathfrak{C}$ *if and only if* $W_{\theta_1\ldots\theta_n}(\eta_1,\cdots,\eta_n)=0$ *for all choices of* $\theta_1,\cdots,\theta_n$ *of order* $< n$.

This has been proved in Kolchin [6].

A consequence of this proposition is that if $\eta_1, \cdots, \eta_n$ are linearly dependent (or independent) over the field of constants of some differential field containing them then they are linearly dependent (or independent) over the field of constants of any differential field containing them; therefore we may speak simply of linear dependence or independence *over constants.*

COROLLARY 1. *Let L be a homogeneous linear polynomial in $\mathfrak{F}[u_1, \cdots, u_q]$. There exist a finite number of homogeneous linear polynomials $L_1, \cdots, L_r$ in $\mathfrak{C}[u_1, \cdots, u_q]$ such that q constants in an extension of $\mathfrak{F}$ form a zero of L if and only if they form a zero of $L_1, \cdots, L_r$.*

Proof. Write $L = \sum_{i=1}^{r} L_i\alpha_i$, where $\alpha_1, \cdots, \alpha_r$ are elements of $\mathfrak{F}$ linearly independent over $\mathfrak{C}$ (and therefore over constants) and each L_i is a homogeneous linear polynomial in $\mathfrak{C}[u_1, \cdots, u_q]$. If $\gamma_1, \cdots, \gamma_q$ are constants then so is $L_i(\gamma_1, \cdots, \gamma_q)$, $1 \leq i \leq r$, so that $L(\gamma_1, \cdots, \gamma_q) = 0$ if and only if $L_i(\gamma_1, \cdots, \gamma_q) = 0$, $1 \leq i \leq r$.

COROLLARY 2. *Let $\mathfrak{m}'$ be a set of polynomials in $\mathfrak{F}[u_1, \cdots, u_q]$. There exists a set $\mathfrak{m}$ of polynomials in $\mathfrak{C}[u_1, \cdots, u_q]$ such that q constants in an extension of $\mathfrak{F}$ form a zero of $\mathfrak{m}'$ if and only if they form a zero of $\mathfrak{m}$.*

Proof. This follows from Corollary 1 since each polynomial in $u_1, \cdots, u_q$ is a linear combination of power products in $u_1, \cdots, u_q$.

COROLLARY 3. *Let $\gamma_1, \cdots, \gamma_q$ be constants in an extension of $\mathfrak{F}$. Then $\partial^0\mathfrak{F}\langle\gamma_1, \cdots, \gamma_q\rangle/\mathfrak{F} = \partial^0\mathfrak{C}\langle\gamma_1, \cdots, \gamma_q\rangle/\mathfrak{C}$.*

Proof. By Corollary 2, $\gamma_{i_1}, \cdots, \gamma_{i_d}$ are algebraically dependent over $\mathfrak{F}$ if and only if they are over $\mathfrak{C}$.

COROLLARY 4. *If $\mathfrak{G}$ is an extension of $\mathfrak{F}$ with field of constants $\mathfrak{D}$ and $\mathfrak{E}$ is a differential subfield of $\mathfrak{F}$ then $\mathfrak{E}\langle\mathfrak{D}\rangle \cap \mathfrak{F} = \mathfrak{E}\langle\mathfrak{C}\rangle$.*

Proof. If $\alpha \in \mathfrak{E}\langle\mathfrak{C}\rangle$ then obviously $\alpha \in \mathfrak{E}\langle\mathfrak{D}\rangle \cap \mathfrak{F}$. Conversely, let $\alpha \in \mathfrak{E}\langle\mathfrak{D}\rangle \cap \mathfrak{F}$. Then there exist elements $e_1, \cdots, e_r \in \mathfrak{E}$ linearly independent over constants and elements $d_1, \cdots, d_r, d'_1, \cdots, d'_r \in \mathfrak{D}$ with $d'_1, \cdots, d'_r$ not all 0 (so that $\sum d'_ie_i \neq 0$) such that $\alpha = \sum d_ie_i/\sum d'_ie_i$, that is $\sum d'_ie_i\alpha - \sum d_ie_i = 0$. This means that $e_1\alpha, \cdots, e_r\alpha, e_1, \cdots, e_r$ are linearly dependent over constants, and therefore over $\mathfrak{C}$; thus there exist elements $c_1, \cdots, c_r, c'_1, \cdots, c'_r \in \mathfrak{C}$ not all 0 such that $\sum c'_ie_i\alpha - \sum c_ie_i = 0$. Since $e_1, \cdots, e_r$ are linearly independent over constants it follows that $c'_1, \cdots, c'_r$ are not all 0, so that $\sum c'_ie_i \neq 0$, whence $\alpha = \sum c_ie_i/\sum c'_ie_i \in \mathfrak{E}\langle\mathfrak{C}\rangle$.

COROLLARY 5. *Let $\mathcal{D}$ be a field of constants containing $\mathcal{C}$ and contained in some extension of $\mathcal{F}$. Then the field of constants of $\mathcal{F}\langle\mathcal{D}\rangle$ is $\mathcal{D}$.*

Proof. Let e be a nonzero constant in $\mathcal{F}\langle\mathcal{D}\rangle$. Then we may write $e\sum_{j=1}^{s}\alpha''_j d''_j = \sum_{i=1}^{r}\alpha'_i d'_i$, where $d'_i, d''_j \varepsilon \mathcal{D}$, $\alpha'_i, \alpha''_j \varepsilon \mathcal{F}$, $\sum_{j=1}^{s}\alpha''_j d''_j \neq 0$. We suppose that of all such equations ours is one for which s is minimal, and also, without loss of generality, that $\alpha''_s = 1$. For each $k (1 \leq k \leq m)$, $e\sum_{j=1}^{s-1}(\delta_k\alpha''_j)d''_j = \sum_{i=1}^{r}(\delta_k\alpha'_i)d'_i$, and this would contradict the minimal nature of s unless $\sum_{j=1}^{s}(\delta_k\alpha''_j)d''_j = 0$; thus $\sum_{j=1}^{s}\alpha''_j d''_j$ is a constant. It follows that our corollary will be proved if we show that every nonzero constant $a \varepsilon \mathcal{F}\langle\mathcal{D}\rangle$ of the form $a = \sum_{i=1}^{q}\alpha_i d_i$ $(\alpha_i \varepsilon \mathcal{F}, d_i \varepsilon \mathcal{D})$ belongs to $\mathcal{D}$. To this end we may suppose that $\alpha_1, \cdots, \alpha_q$ are linearly independent over constants and that every $d_i \neq 0$. Since $\sum_{i=1}^{q}(\delta_k\alpha_i)d_i = \delta_k a = 0$ $(1 \leq k \leq m)$ it follows from Corollary 1 that there exist elements $c_1, \cdots, c_q$ of $\mathcal{C}$ not all 0 such that $\sum_{i=1}^{q}(\delta_k\alpha_i)c_i = 0$ $(1 \leq k \leq m)$, so that the element $c = \sum_{i=1}^{q}\alpha_i c_i$ belongs to $\mathcal{C}$. Now, $\alpha_1, \cdots, \alpha_q$ are linearly independent and

$$\sum_{i=1}^{q}\alpha_i(ac_i - cd_i) = a\sum_{i=1}^{q}\alpha_i c_i - c\sum_{i=1}^{q}\alpha_i d_i = 0,$$

so that each $ac_i - cd_i = 0$, whence $a \varepsilon \mathcal{D}$.

5. Universal extensions. Let $\mathcal{F}^*$ be a differential field and let $\mathcal{F}$ be a differential subfield of $\mathcal{F}^*$. We shall call $\mathcal{F}^*$ a *universal extension* of $\mathcal{F}$ if, for every finitely generated differential field extension $\mathcal{F}_1$ of $\mathcal{F}$ with $\mathcal{F}_1 \subseteq \mathcal{F}^*$ and every integer $n > 0$ and every prime differential ideal Π of $\mathcal{F}_1\{y_1, \cdots, y_n\}$ not containing 1, there exists a generic zero $(\eta_1, \cdots, \eta_n)$ of Π with $\eta_1, \cdots, \eta_n \varepsilon \mathcal{F}^*$. A necessary and sufficient condition for an extension $\mathcal{F}^*$ of $\mathcal{F}$ to be universal is that for every finitely generated extension $\mathcal{F}_1$ of $\mathcal{F}$ with $\mathcal{F}_1 \subseteq \mathcal{F}^*$ and every finitely generated extension $\mathcal{G}$ of $\mathcal{F}_1$ there exist an isomorphim of $\mathcal{G}$ over $\mathcal{F}_1$ into $\mathcal{F}^*$ (that is, an isomorphism σ of $\mathcal{G}$ into $\mathcal{F}^*$ such that $\sigma a = a$ for every $a \varepsilon \mathcal{F}_1$). If $\mathcal{F}^*$ is a universal extension of $\mathcal{F}$ then $\mathcal{F}^*$ is a universal extension of every finitely generated extension of $\mathcal{F}$ contained in $\mathcal{F}^*$, and $\mathcal{F}^*$ is also a universal extension of every differential subfield of $\mathcal{F}$. If $\mathcal{F}^*$ is a universal extension of $\mathcal{F}$ then the degree of differential transcendence of $\mathcal{F}^*$ over $\mathcal{F}$ is infinite, and (because of the Ritt basis theorem) for every integer $n > 0$ every differential ideal in $\mathcal{F}^*\{y_1, \cdots, y_n\}$ not containing 1 has a zero $(\eta_1, \cdots, \eta_n)$ with $\eta_1, \cdots, \eta_n \in \mathcal{F}^*$; in particular $\mathcal{F}^*$ is algebraically closed, and the field of constants of $\mathcal{F}^*$ is an alegbraically closed extension of the field of constants of $\mathcal{F}$ of infinite degree of transcendence.

We shall prove below that every differential field $\mathcal{F}$ has a universal extension $\mathcal{F}^*$. Once this fact is known it is possible to define the *manifold* of a set Φ of differential polynomials in $\mathcal{F}\{y_1, \cdots, y_n\}$ as the set of all zeros $(\eta_1, \cdots, \eta_n)$ of Φ with $\eta_1, \cdots, \eta_n \in \mathcal{F}^*$; this use of universal extensions extends the well-known procedure of modern algebraic geometry, and gives a workable definition of manifold free of the logical difficulty involved in using "the set of all extension of $\mathcal{F}$" (see Ritt [8], footnote 2 on p. 21).

Let Π be a prime differential ideal of $\mathcal{F}\{y_1, \cdots, y_n\}$. If $\mathcal{G}$ is an extension of $\mathcal{F}$ the ideal $\mathcal{G} \cdot \Pi$ of $\mathcal{G}\{y_1, \cdots, y_n\}$ is a perfect differential ideal (Proposition 1); we shall say that Π is *absolutely prime* if $\mathcal{G} \cdot \Pi$ is prime for every extension $\mathcal{G}$. If $\mathcal{G} \cdot \Pi$ is prime when we take for $\mathcal{G}$ some algebraically closed extension of $\mathcal{F}$ then Π is absolutely prime (because of Proposition 1 and the fact that every polynomial R over $\mathcal{F}$ which is irreducible over an algebraically closed extension of $\mathcal{F}$ is absolutely irreducible). In particular, if $\mathcal{F}$ is algebraically closed then every prime differential ideal of $\mathcal{F}\{y_1, \cdots, y_n\}$ is absolutely prime.

PROPOSITION 4. *Let I be a nonempty set of indices; for each $i \in I$ let n_i be an integer > 0; let $(y_{ij})_{i \in I, 1 \leq j \leq n_i}$ be a family of indeterminates; for each $i \in I$ let Π_i be an absolutely prime differential ideal of $\mathcal{F}\{y_{i1}, \cdots, y_{in_i}\}$ not containing 1. Then the ideal Π generated by $\bigcup_{i \in I} \Pi_i$ in $\mathcal{F}\{(y_{ij})_{i \in I, 1 \leq j \leq n_i}\}$ is a prime differential ideal not containing 1. If I is finite then Π is absolutely prime and* $\operatorname{ord} \Pi = \sum_{i \in I} \operatorname{ord} \Pi_i$.

Proof. That Π is a differential ideal is obvious. To prove that Π is prime and $1 \notin \Pi$ it suffices to consider the case in which I is finite; by induction then the entire proposition can be reduced to the case in which I consists of two elements. Accordingly, let I consist of the numbers 1 and 2. Then Π consists of all differential polynomials P which can be writen in the form

$$(2) \qquad P = \sum_{k_1} C_{2k_1} P_{1k_1} + \sum_{k_2} C_{1k_2} P_{2k_2} (P_{ik_i} \in \Pi_i, C_{ik} \in \mathcal{F}\{y_{i1}, \cdots, y_{in}\}).$$

Let $(\eta_{11}, \cdots, \eta_{1n_1})$ be a generic zero of Π_1. Since Π_2 is absolutely prime the ideal Λ_2 generated by Π_2 in $\mathcal{F}\langle\eta_{11}, \cdots, \eta_{1n_1}\rangle\{y_{21}, \cdots, y_{2n_2}\}$ is a prime differential ideal, and obviously $1 \notin \Lambda_2$; let $(\eta_{21}, \cdots, \eta_{2n_2})$ be a generic zero of Λ_2. We shall show that $(\eta_{11}, \cdots, \eta_{1n_1}, \eta_{21}, \cdots, \eta_{2n_2})$ is a generic zero of Π, thereby proving that Π is prime, that $1 \not\in \Pi$, and that

$$\begin{aligned}\operatorname{ord} \Pi &= \partial^0 \mathcal{F}\langle \eta_{11}, \cdots, \eta_{1n}, \eta_{21}, \cdots, \eta_{2n_2}\rangle / \mathcal{F} \\ &= \partial^0 \mathcal{F}\langle \eta_{11}, \cdots, \eta_{1n_1}, \eta_{21}, \cdots, \eta_{2n_2}\rangle / \mathcal{F}\langle \eta_{11}, \cdots, \eta_{1n_1}\rangle \\ &\quad + \partial^0 \mathcal{F}\langle \eta_{11}, \cdots, \eta_{1n_1}\rangle / \mathcal{F} = \operatorname{ord} \Lambda_2 + \operatorname{ord} \Pi_1 = \operatorname{ord} \Pi_2 + \operatorname{ord} \Pi_1.\end{aligned}$$

It is clear from (2) that $(\eta_{11}, \cdots, \eta_{1n_1}, \eta_{21}, \cdots, \eta_{2n_2})$ is a zero of Π. Let P be any differential polynomial in $\mathcal{F}\{y_{11}, \cdots, y_{1n_1}, y_{21}, \cdots, y_{2n_2}\}$ which vanishes at $(\eta_{11}, \cdots, \eta_{1n_1}, \eta_{21}, \cdots, \eta_{2n_2})$. Then

$$P(\eta_{11}, \cdots, \eta_{1n_1}, y_{21}, \cdots, y_{2n_2}) \in \Lambda_2.$$

We now write

$$\begin{aligned}P(\eta_{11}, &\cdots, \eta_{1n_1}, y_{21}, \cdots, y_{2n_2}) \\ &= \sum_{k_2} C_{1k_2}(\eta_{11}, \cdots, \eta_{1n_1}) P_{2k_2}(y_{21}, \cdots, y_{2n_2}),\end{aligned}$$

where

$$C_{1k_2}(y_{11}, \cdots, y_{1n_1}) \in \mathcal{F}\{y_{11}, \cdots, y_{1n_1}\},\ P_{2k_2}(y_{21}, \cdots, y_{2n_2}) \in \mathcal{F}\{y_{21}, \cdots, y_{2n_2}\},$$

and the elements $C_{1k_2}(\eta_{11}, \cdots, \eta_{1n_1})$ of $\mathcal{F}\langle \eta_{11}, \cdots, \eta_{1n_1}\rangle$ are linearly independent over $\mathcal{F}$; it is easy to see, since $\Lambda_2 = \mathcal{F}\langle \eta_{11}, \cdots, \eta_{1n_1}\rangle \cdot \Pi_2$, that each $P_{2k_2} \in \Pi_2$. Let

$$Q = P - \sum_{k_2} C_{1k_2} P_{2k_2},$$

so that $Q(\eta_{11}, \cdots, \eta_{1n_1}, y_{21}, \cdots, y_{2n_2}) = 0$; if we write

$$Q = \sum_{k_1} C_{2k_1} P_{1k_1},$$

where the C_{2k_1} are distinct power products in $y_{21}, \cdots, y_{2n_2}$ and their derivatives of various orders, and each $P_{1k_1} \in \mathcal{F}\{y_{11}, \cdots, y_{1n_1}\}$, then each $P_{1k_1}(\eta_{11}, \cdots, \eta_{1n_1}) = 0$, so that each $P_{1k_1} \in \Pi_1$. It follows that P can be written in the form (2) and therefore belongs to Π; therefore $(\eta_{11}, \cdots, \eta_{1n_1}, \eta_{21}, \cdots, \eta_{2n_2})$ is a generic zero of Π.

To complete the proof it remains to show that Π is absolutely prime. To this end let $\mathcal{G}$ be any extension of $\mathcal{F}$. Clearly $\mathcal{G} \cdot \Pi$ is the ideal generated by $(\mathcal{G} \cdot \Pi_1) \cup (\mathcal{G} \cdot \Pi_2)$ in $\mathcal{G}\{y_{11}, \cdots, y_{1n_1}, y_{21}, \cdots, y_{2n_2}\}$, and $\mathcal{G} \cdot \Pi$, $\mathcal{G} \cdot \Pi_2$ are prime (because Π_1, Π_2 are absolutely prime); therefore by what we have already proved $\mathcal{G} \cdot \Pi$ is prime. Thus Π is absolutely prime.

REMARK. We observe from the proof that the hypothesis in Proposition 4 that each Π_i be absolutely prime may be weakened. It is enough to assume that, for each $i \in I$, Π_i is prime and $\mathcal{F}_i \cdot \Pi_i$ is prime whenever $\mathcal{F}_i$ is an extension of $\mathfrak{F}$ obtained by the adjunction of generic zeros of a finite number of Π_j's with $j \neq i$. Except for the statement that Π is *absolutely* prime, the conclusion of Proposition 4 is then valid (Π still being prime).

THEOREM. *Every differential field has a universal extension.*

Proof. We lose no generality in assuming that the given differential field $\mathfrak{F}$ is algebraically closed, for a universal extension of an algebraic closure of $\mathfrak{F}$ is a universal extension of $\mathfrak{F}$. We shall show that for every algebraically closed differential field $\mathfrak{G}$ there exists an extension $\mathfrak{G}^\dagger$ of $\mathfrak{G}$ with the following two properties: 1) $\mathfrak{G}^\dagger$ is algebraically closed; 2) for every integer $n > 0$ and for every prime differential ideal Π of $\mathfrak{G}\{y_1, \cdots, y_n\}$ not containing 1 there exists a generic zero $(\eta_1, \cdots, \eta_n)$ of Π with $\eta_1, \cdots, \eta_n \in \mathfrak{G}^\dagger$. Once this is done we can define inductively a sequence of differential fields $\mathfrak{F}^{(k)}$ such that $\mathfrak{F}^{(0)} = \mathfrak{F}$ and $\mathfrak{F}^{(k+1)} = \mathfrak{F}^{(k)\dagger}$ for every integer $k \geq 0$; the union $\mathfrak{F}^* = \cup \mathfrak{F}^{(k)}$ will then be a differential field which, as is easy to see, is a universal extension of $\mathfrak{F}$.

Let $\mathfrak{P}_n$ be the set of all prime differential ideals in $\mathfrak{G}\{y_1, \cdots, y_n\}$ which do not contain 1; since $\mathfrak{G}$ is algebraically closed, every element of $\mathfrak{P}_n$ is absolutely prime. Let $(y_{n\Pi j})_{1 \leq n < \infty, \Pi \in \mathfrak{P}_n, 1 \leq j \leq n}$ be a family of indeterminates. For each $\Pi \in \mathfrak{P}_n$ let $\Lambda(n, \Pi)$ denote the set which is obtained when in all the differential polynomials in Π we replace y_j by $y_{n\Pi j}$ $(1 \leq j \leq n)$; $\Lambda(n, \Pi)$ is obviously an absolutely prime differential ideal of $\mathfrak{G}\{y_{n\Pi 1}, \cdots, y_{n\Pi n}\}$ which does not contain 1. It follows from Proposition 4 that the ideal Λ generated by $\bigcup_{1 \leq n < \infty, \Pi \in \mathfrak{P}_n} \Lambda(n, \Pi)$ in the differential ring $\mathfrak{R} = \mathfrak{G}\{(y_{n\Pi j})_{1 \leq n < \infty, \Pi \in \mathfrak{P}_n, 1 \leq j \leq n}\}$ is a prime differential ideal not containing 1. The differential ring of residue classes $\mathfrak{R}/\Lambda$ is therefore a differential domain of integrity, which can be embedded in its differential field of quotients $\mathfrak{G}'$. Since $1 \notin \Lambda$, the canonical homomorphism h of $\mathfrak{R}$ onto $\mathfrak{R}/\Lambda$ maps $\mathfrak{G}$ isomorphically; therefore we may identify each element $a \in \mathfrak{G}$ with its image $h(a) \in \mathfrak{G}'$. With this identification $\mathfrak{G}$ becomes a differential subfield of $\mathfrak{G}'$. It is now easy to see that if we set $\eta_{n\Pi j} = h(y_{n\Pi j})$ for all n, Π, j then, for each n and each $\Pi \in \mathfrak{P}_n$, $(\eta_{n\Pi 1}, \cdots, \eta_{n\Pi n})$ is a generic zero of $\Lambda(n, \Pi)$, and consequently a generic zero of Π. Therefore if we let $\mathfrak{G}^\dagger$ be an algebraic closure of $\mathfrak{G}'$ then $\mathfrak{G}^\dagger$ will have the required properties 1), 2) above. As we have seen, this suffices to prove the theorem.

Chapter II. Algebraic groups of automorphisms.

Throughout the rest of this paper $\mathcal{G}$ *will denote a differential field with algebraically closed field of constants* $\mathcal{C}$, *and* $\mathcal{F}$ *will denote a differential subfield of* $\mathcal{G}$, *with the same field of constants* $\mathcal{C}$, *such that* $\mathcal{G}$ *is finitely generated and of finite transcendence degree over* $\mathcal{F}$. *The relative algebraic closure of* $\mathcal{F}$ *in* $\mathcal{G}$ *will be denoted by* $\mathcal{F}^0$. *All differential fields mentioned will tacitly be assumed to lie in a universal extension of* $\mathcal{G}$ *fixed once and for all; in particular, every isomorphism of* $\mathcal{G}$ *will be an isomorphism into this universal extension. The identity isomorphism of* $\mathcal{G}$ *will be denoted by* ι. *The field of constants of the universal extension will be denoted by* $\mathcal{C}^*$.

1. Specializations of isomorphisms. Let $\sigma_1, \cdots, \sigma_p, \tau_1, \cdots, \tau_p$ be isomorphisms of $\mathcal{G}$; we shall say that $(\tau_1, \cdots, \tau_p)$ *is a specialization* of $\sigma_1, \cdots, \sigma_p)$ if $(\tau_i\alpha)_{1\leq i\leq p,\,\alpha\varepsilon\mathcal{G}}$ is a specialization of $(\sigma_i\alpha)_{1\leq i\leq p,\,\alpha\varepsilon\mathcal{G}}$ over $\mathcal{G}$. If $(\tau_1, \cdots, \tau_p)$ is a specialization of $(\sigma_1, \cdots, \sigma_p)$ such that $(\sigma_1, \cdots, \sigma_p)$ is a specialization of $(\tau_1, \cdots, \tau_p)$ the we shall say that $(\tau_1, \cdots, \tau_p)$ is a *generic* specialization of $(\sigma_1, \cdots, \sigma_p)$. A specialization which is not generic will be called *nongeneric*. The relation "τ is a generic specialization of σ" is an equivalence on the set of all isomorphisms of $\mathcal{G}$, and two isomorphisms of $\mathcal{G}$ which are in this relation will accordingly be called *equivalent*.

2. Isolated isomorphisms. We shall say that σ is an *isolated* isomorphism of $\mathcal{G}$ over $\mathcal{F}$ if σ is an isomorphism of $\mathcal{G}$ over $\mathcal{F}$ such that there does not exist an isomorphism of $\mathcal{G}$ over $\mathcal{F}$ of which σ is a nongeneric specialization.

Let $\eta_1, \cdots, \eta_n$ be elements such that $\mathcal{G} = \mathcal{F}\langle\eta_1, \cdots, \eta_n\rangle$. If $\sigma_1, \cdots, \sigma_p, \tau_1, \cdots, \tau_p$ are isomorphisms of $\mathcal{G}$ over $\mathcal{F}$ then $(\tau_1, \cdots, \tau_p)$ is a specialization of $(\sigma_1, \cdots, \sigma_p)$ if and only if $(\tau_i\eta_j)_{1\leq i\leq p,\,1\leq j\leq n}$ is a specialization of $(\tau_i\eta_j)_{1\leq i\leq p,\,1\leq j\leq n}$ over $\mathcal{G}$.

Let Π be the prime differential ideal of $\mathcal{F}\{y_1, \cdots, y_n\}$ with generic zero $(\eta_1, \cdots, \eta_n)$, so that (Chapter I, Proposition 1) $\mathcal{G}\cdot\Pi$ is a perfect differential ideal of $\mathcal{G}\{y_1, \cdots, y_n\}$, and let $\Pi_1, \cdots, \Pi_h$ be the minimal prime differential ideal divisors of $\mathcal{G}\cdot\Pi$. Let $(\eta_{i1}, \cdots, \eta_{in})$ be a generic zero of Π_i; then $(\eta_{i1}, \cdots, \eta_{in})$ is also a generic zero of Π, so that there exists a unique isomorphism χ_i of $\mathcal{G}$ over $\mathcal{F}$ such that $\chi_i\eta_j = \eta_{ij}$ $(1 \leq j \leq n)$. It is obvious that if $i \neq i'$ then χ_i is not equivalent to $\chi_{i'}$. If σ is any isomorphism of $\mathcal{G}$ over $\mathcal{F}$ then $(\sigma\eta_1, \cdots, \sigma\eta_n)$ is a generic zero of Π, so that (Chapter I,

Proposition 1) $(\sigma\eta_1, \cdots, \sigma\eta_n)$ is a zero of precisely one Π_i; therefore σ is a specialization of precisely one χ_i. We have thus proved the following result.

PROPOSITION 1. *$\chi_1, \cdots, \chi_n$ are inequivalent isolated isomorphisms of $\mathfrak{G}$ over $\mathfrak{F}$, and every isomorphism of $\mathfrak{G}$ over $\mathfrak{F}$ is a specialization of precisely one of these.*

By Proposition 1 an isomorphism σ of $\mathfrak{G}$ over $\mathfrak{F}$ is isolated if and only if σ is equivalent to χ_i for some i, that is (Chapter I, Proposition 2) if and only if

$$\begin{aligned}\partial^0\mathfrak{G}\langle\sigma\mathfrak{G}\rangle/\mathfrak{G} &= \partial^0\mathfrak{G}\langle\sigma\eta_1, \cdots, \sigma\eta_n\rangle/\mathfrak{G} \\ &= \partial^0\mathfrak{G}\langle\chi_i\eta_1, \cdots, \chi_i\eta_n\rangle/\mathfrak{G} = \operatorname{ord}\Pi_i = \operatorname{ord}\Pi = \partial^0\mathfrak{F}\langle\eta_1, \cdots, \eta_n\rangle/\mathfrak{F} \\ &= \partial^0\mathfrak{G}/\mathfrak{F}.\end{aligned}$$

Since $\mathfrak{G}\langle\sigma\mathfrak{G}\rangle = (\sigma\mathfrak{G})\langle\mathfrak{G}\rangle$ and $\partial^0\mathfrak{G}/\mathfrak{F} = \partial^0(\sigma\mathfrak{G})/\mathfrak{F}$ it follows that σ is an isolated isomorphism of $\mathfrak{G}$ over $\mathfrak{F}$ if and only if σ^{-1} is an isolated isomorphism of $\sigma\mathfrak{G}$ over $\mathfrak{F}$. Thus we have the following result.

PROPOSITION 2. *Let σ be an isomorphism of $\mathfrak{G}$ over $\mathfrak{F}$. The following three statements are equivalent:* 1) *σ is an isolated isomorphism of $\mathfrak{G}$ over $\mathfrak{F}$;* 2) *σ^{-1} is an isolated isomorphism of $\sigma\mathfrak{G}$ over $\mathfrak{F}$;* 3) $\partial^0\mathfrak{G}\langle\sigma\mathfrak{G}\rangle/\mathfrak{G} = \partial^0\mathfrak{G}/\mathfrak{F}$.

Now let σ be any isomorphism of $\mathfrak{G}$ over $\mathfrak{F}$, and suppose that σ leaves invariant some element $\zeta \in \mathfrak{G}$ which is transcendental over $\mathfrak{F}$. Let Λ be the prime differential ideal of $\mathfrak{F}\{z, y_1, \cdots, y_n\}$ with generic zero $(\zeta, \eta_1, \cdots, \eta_n)$ and let $\Lambda_1, \cdots, \Lambda_k$ be the minimal prime differential ideal divisors of the perfect differential ideal $\mathfrak{G}\cdot\Lambda$ of $\mathfrak{G}\{z, y_1, \cdots, y_n\}$. $(\zeta, \sigma\eta_1, \cdots, \sigma\eta_n)$ is a generic zero of Λ and therefore a zero of Λ_i for some i; let $(\zeta', \eta'_1, \cdots, \eta'_n)$ be a generic zero of Λ_i. Then $(\zeta', \eta'_1, \cdots, \eta'_n)$ is a generic zero of Λ, so that there exists a unique isomorphism σ' of $\mathfrak{G}$ over $\mathfrak{F}$ such that $\sigma'\zeta = \zeta'$, $\sigma'\eta_1 = \eta'_1, \cdots, \sigma'\eta_n = \eta'_n$; it is clear that σ is a specialization of σ'. If ζ were invariant under σ', that is if we had $\zeta' = \zeta$, then we would have (because ζ is transcendental over $\mathfrak{F}$)

$$\begin{aligned}&\partial^0\mathfrak{G}\langle\zeta', \eta'_1, \cdots, \eta'_n\rangle/\mathfrak{G} \\ &\quad \leq \partial^0\mathfrak{F}\langle\zeta', \eta'_1, \cdots, \eta'_n\rangle/\mathfrak{F}\langle\zeta'\rangle < \partial^0\mathfrak{F}\langle\zeta', \eta'_1, \cdots, \eta'_n\rangle/\mathfrak{F}\langle\zeta'\rangle \\ &\qquad + \partial^0\mathfrak{F}\langle\zeta'\rangle/\mathfrak{F} = \partial^0\mathfrak{F}\langle\zeta', \eta'_1, \cdots, \eta'_n\rangle/\mathfrak{F},\end{aligned}$$

or in other words $\operatorname{ord}\Lambda_i < \operatorname{ord}\Lambda$, contradicting Chapter I, Proposition 1. Therefore $\sigma'\zeta \neq \zeta$; since $\sigma\zeta = \zeta$ this means that σ is a nongeneric specialization of σ', so that σ can not be an isolated isomorphism of $\mathfrak{G}$ over $\mathfrak{F}$.

This shows that the field of invariants of an isolated isomorphism of $\mathcal{G}$ over $\mathcal{F}$ must be contained in $\mathcal{F}^0$.

If an element of $\mathcal{G}$ is invariant under every isolated isomorphism of $\mathcal{G}$ over $\mathcal{F}$ (or, equivalently, under the isolated isomorphisms $\chi_1, \cdots, \chi_h$) then the element is invariant under every isomorphism of $\mathcal{G}$ over $\mathcal{F}$, and therefore (Kolchin [3], § 12) belongs to $\mathcal{F}$.

Consider again the prime differential ideal Π of $\mathcal{F}\{y_1, \cdots, y_n\}$ with generic zero $(\eta_1, \cdots, \eta_n)$. It is a consequence of Chapter I, Proposition 1 that the minimal prime differential ideal divisors of the perfect differential ideal $\mathcal{F}^0 \cdot \Pi$ of $\mathcal{F}^0\{y_1, \cdots, y_n\}$ are h in number, one being contained in and generating each of the minimal prime differential ideal divisors $\Pi_1, \cdots, \Pi_h$ of $\mathcal{G} \cdot \Pi$; we denote the minimal prime differential ideal divisor of $\mathcal{F}^0 \cdot \Pi$ which is contained in Π_i by $\Pi_i{}^0$, so that $\mathcal{G} \cdot \Pi_i{}^0 = \Pi_i$. Now $(\eta_1, \cdots, \eta_n)$ is a zero of precisely one $\Pi_i{}^0$, say of $\Pi_1{}^0$; because

$$\partial^0 \mathcal{F}^0\langle\eta_1, \cdots, \eta_n\rangle/\mathcal{F}^0 = \partial^0 \mathcal{F}\langle\eta_1, \cdots, \eta_n\rangle/\mathcal{F} = \operatorname{ord} \Pi = \operatorname{ord} \Pi_1$$

we see (Chapter I, Proposition 2) that $(\eta_1, \cdots, \eta_n)$ is a generic zero of $\Pi_1{}^0$. Also, the identity isomorphism ι is a specialization of precisely one χ_i. Because $(\eta_1, \cdots, \eta_n) = (\iota\eta_1, \cdots, \iota\eta_n)$ is a zero of $\Pi_1{}^0$ and $\Pi_1 = \mathcal{G} \cdot \Pi_1{}^0$, ι must be a specialization of χ_1. It follows from this that the restriction of χ_1 to $\mathcal{F}^0$ is the identity isomorphism of $\mathcal{F}^0$. If σ is any isomorphism of $\mathcal{G}$ such that the restriction of σ to $\mathcal{F}^0$ is the identity then $(\sigma\eta_1, \cdots, \sigma\eta_n)$ is a zero of $\Pi_1{}^0$ (because $(\eta_1, \cdots, \eta_n)$ is) and consequently a zero of Π_1, so that σ is a specialization of χ_1.

Collecting these remarks we have the following result.

Proposition 3. *The field of invariants of an isolated isomorphism of $\mathcal{G}$ over $\mathcal{F}$ is contained in $\mathcal{F}^0$; the field of invariants of a complete set of representatives of the h equivalence classes of isolated isomorphisms of $\mathcal{G}$ over $\mathcal{F}$ is $\mathcal{F}$. If σ_0 is an isolated isomorphism of $\mathcal{G}$ over $\mathcal{F}$ of which ι is a specialization then the field of invariants of σ_0 is $\mathcal{F}^0$; every isomorphism of $\mathcal{G}$ which leaves all the elements of $\mathcal{F}^0$ invariant is a specialization of σ_0.*

3. Strong isomorphisms. If σ is any isomorphism of $\mathcal{G}$ we denote the field of constants of $\mathcal{G}\langle\sigma\mathcal{G}\rangle$ by $\mathcal{C}_\sigma$.

We shall say that an isomorphism σ of $\mathcal{G}$ is *strong* if

$$\sigma\mathcal{G} \subset \mathcal{G}\langle\mathcal{C}^*\rangle, \; \mathcal{G} \subset (\sigma\mathcal{G})\langle\mathcal{C}^*\rangle.$$

By Chapter I, Corollary 4 to Proposition 3 (with $\mathcal{G}$, $\mathcal{G}\langle\sigma\mathcal{G}\rangle$, $\mathcal{G}\langle\mathcal{C}^*\rangle$ playing the role of $\mathcal{E}$, $\mathcal{F}$, $\mathcal{G}$ in that Corollary) the first of these inclusions is equivalent to $\sigma\mathfrak{G} \subset \mathfrak{G}\langle\mathcal{C}_\sigma\rangle$; similarly, the second of these inclusions is equivalent to $\mathfrak{G} \subset (\sigma\mathfrak{G})\langle\mathcal{C}_\sigma\rangle$. Consequently σ is strong if and only if

$$\mathfrak{G}\langle\mathcal{C}_\sigma\rangle = \mathfrak{G}\langle\sigma\mathfrak{G}\rangle = (\sigma\mathfrak{G})\langle\mathcal{C}_\sigma\rangle. \tag{1}$$

Obviously every automorphism of $\mathfrak{G}$ is strong.

Let $\Pi, \eta_1, \cdots, \eta_n$ have the same significance as in § 2. Because the field of constants $\mathcal{C}$ of $\mathfrak{F}$ is algebraically closed it is easy to see that every polynomial irreducible over $\mathfrak{F}$ remains irreducible over $\mathfrak{F}\langle\mathcal{C}^*\rangle$. It follows (Chapter I, Proposition 1) that the differential ideal $\Pi^* = \mathfrak{F}\langle\mathcal{C}^*\rangle \cdot \Pi$ of $\mathfrak{F}\langle\mathcal{C}^*\rangle\{y_1, \cdots, y_n\}$ is prime, and $\operatorname{ord}\Pi^* = \operatorname{ord}\Pi$.

Let σ be any strong isomorphism of $\mathfrak{G}$ over $\mathfrak{F}$; $(\sigma\eta_1, \cdots, \sigma\eta_n)$ is a generic zero of Π, and therefore a zero of Π^*. For every finite subset $\mathfrak{c}$ of $\mathcal{C}^*$ we have (Chapter I, Corollary 3 to Proposition 3)

$$\begin{aligned}\partial^0\mathfrak{F}\langle\mathfrak{c}\rangle\langle\sigma\eta_1, \cdots, \sigma\eta_n\rangle/\mathfrak{F}\langle\mathfrak{c}\rangle &= \partial^0(\sigma\mathfrak{G})\langle\mathfrak{c}\rangle/\mathfrak{F}\langle\mathfrak{c}\rangle \\ &= \partial^0(\sigma\mathfrak{G})\langle\mathfrak{c}\rangle/\mathfrak{F} - \partial^0\mathfrak{F}\langle\mathfrak{c}\rangle/\mathfrak{F} = \partial^0(\sigma\mathfrak{G}\langle\mathfrak{c}\rangle/\sigma\mathfrak{G} + \partial^0(\sigma\mathfrak{G})/\mathfrak{F} \\ &\quad - \partial^0\mathfrak{F}\langle\mathfrak{c}\rangle/\mathfrak{F} = \partial^0\mathcal{C}\langle\mathfrak{c}\rangle/\mathcal{C} + \operatorname{ord}\Pi - \partial^0\mathcal{C}\langle\mathfrak{c}\rangle/\mathcal{C} = \operatorname{ord}\Pi^*;\end{aligned}$$

since this holds for every finite subset $\mathfrak{c}$ of $\mathcal{C}^*$ we infer that

$$\partial^0\mathfrak{F}\langle\mathcal{C}^*\rangle\langle\sigma\eta_1, \cdots, \sigma\eta_n\rangle/\mathfrak{F}\langle\mathcal{C}^*\rangle = \operatorname{ord}\Pi^*,$$

so that (Chapter I, Corollary to Proposition 2) $(\sigma\eta_1, \cdots, \sigma\eta_n)$ is a generic zero of Π^*. In the same way we also show that $(\eta_1, \cdots, \eta_n)$ is a general zero of Π^*, so that $(\sigma\eta_1, \cdots, \sigma\eta_n)$ is a generic specialization of $(\eta_1, \cdots, \eta_n)$ over $\mathfrak{F}\langle\mathcal{C}^*\rangle$. Therefore there exists a unique isomorphism σ^* of $\mathfrak{F}\langle\mathcal{C}^*\rangle\langle\eta_1, \cdots, \eta_n\rangle$ over $\mathfrak{F}\langle\mathcal{C}^*\rangle$ onto $\mathfrak{F}\langle\mathcal{C}^*\rangle\langle\sigma\eta_1, \cdots, \sigma\eta_n\rangle$ such that $\sigma^*\eta_1 = \sigma\eta_1, \cdots, \sigma^*\eta_n = \sigma\eta_n$, that is there exists a unique isomorphism σ^* of $\mathfrak{G}\langle\mathcal{C}^*\rangle$ over $\mathfrak{F}\langle\mathcal{C}^*\rangle$ which extends σ. Since by (1)

$$\sigma^*(\mathfrak{G}\langle\mathcal{C}^*\rangle) = (\sigma\mathfrak{G})\langle\mathcal{C}^*\rangle = (\sigma\mathfrak{G})\langle\mathcal{C}_\sigma\rangle\langle\mathcal{C}^*\rangle = \mathfrak{G}\langle\mathcal{C}_\sigma\rangle\langle\mathcal{C}^*\rangle = \mathfrak{G}\langle\mathcal{C}^*\rangle,$$

we see that σ^* is an automorphism of $\mathfrak{G}\langle\mathcal{C}^*\rangle$.

Now let us start at the other end with any automorphism σ^* of $\mathfrak{G}\langle\mathcal{C}^*\rangle$ over $\mathfrak{F}\langle\mathcal{C}^*\rangle$. The restriction σ of σ^* to $\mathfrak{G}$ is then an isomorphism of $\mathfrak{G}$ over $\mathfrak{F}$. Obviously $\sigma\mathfrak{G} \subset \mathfrak{G}\langle\mathcal{C}^*\rangle$ and $\mathfrak{G} \subset (\sigma\mathfrak{G})\langle\mathcal{C}^*\rangle$, so that σ is strong.

We have thus proved the following result.

PROPOSITION 4. *The mapping which to each automorphism of $\mathfrak{G}\langle\mathcal{C}^*\rangle$ over $\mathfrak{F}\langle\mathcal{C}^*\rangle$ assigns its restriction to $\mathfrak{G}$ is one-to-one onto the set of all strong isomorphisms of $\mathfrak{G}$ over $\mathfrak{F}$.*

In virtue of Proposition 4 we may identify each strong isomorphism of $\mathcal{G}$ over $\mathcal{F}$ with the automorphism of $\mathcal{G}\langle\mathcal{C}^*\rangle$ over $\mathcal{F}\langle\mathcal{C}^*\rangle$ of which it is the restriction. This identification permits us to multiply any two strong isomorphisms of $\mathcal{G}$ over $\mathcal{F}$, and to consider the set of all of them as a group. We shall denote this group of all strong isomorphisms of $\mathcal{G}$ over $\mathcal{F}$ by $\mathfrak{G}^*$, and the subgroup of $\mathfrak{G}^*$ consisting of all automorphisms of $\mathcal{G}$ over $\mathcal{F}$ by $\mathfrak{G}$.

If σ is a strong isomorphism of $\mathcal{G}$ over $\mathcal{F}$, application of σ^{-1} to (1) shows that $\mathcal{C}_\sigma \subseteq \mathcal{C}_{\sigma^{-1}}$; interchanging σ and σ^{-1} reverses the inclusion; we conclude that

$$\mathcal{C}_\sigma = \mathcal{C}_{\sigma^{-1}}. \tag{2}$$

4. Specializations of strong isomorphisms.

PROPOSITION 5. *A specialization of a strong isomorphism of $\mathcal{G}$ is always strong.*

Proof. Let σ be a strong isomorphism of $\mathcal{G}$. By (1), for each $\alpha \in \mathcal{G}$ we may write a relation $\sigma\alpha = \sum_{i=1}^{r} a_i\beta_i / \sum_{i=1}^{r} b_i\beta_i$, where a_i, b_i are constants, $\beta_1, \cdots, \beta_r$ are elements of $\mathcal{G}$ linearly independent over constants, and $\sum b_i\beta_i \neq 0$. Therefore $\beta_1, \cdots, \beta_r, \beta_1\sigma\alpha, \cdots, \beta_r\sigma\alpha$ are linearly dependent over constants so that (Chapter I, Proposition 3) the differential polynomial $W_{\theta_1\ldots\theta_{2r}}(\beta_1, \cdots, \beta_r, \beta_1 y, \cdots, \beta_r y) \in \mathcal{G}\{y\}$ vanishes at $\sigma\alpha$ for all choices of the differential operators $\theta_1, \cdots, \theta_{2r}$ of order $< 2r$. If σ' is a specialization of σ then this differential polynomial also vanishes at $\sigma'\alpha$ so that (Chapter I, Proposition 3) there exist elements $a'_1, \cdots, a'_r, b'_1, \cdots, b'_r \in \mathcal{C}_{\sigma'}$ not all 0 such that $\sum a'_i\beta_i - \sum b'_i\beta_i\sigma'\alpha = 0$. Since $\beta_1, \cdots, \beta_r$ are linearly independent over constants not every b'_i is 0 so that $\sum b'_i\beta_i \neq 0$ and

$$\sigma'\alpha = \sum a'_i\beta_i / \sum b'_i\beta_i \in \mathcal{G}\langle\mathcal{C}_{\sigma'}\rangle.$$

Thus $\mathcal{G}\langle\sigma'\mathcal{G}\rangle = \mathcal{G}\langle\mathcal{C}_{\sigma'}\rangle$.

Again by (1), for each $\alpha \in \mathcal{G}$ we may write $\alpha = \sum_{i=1}^{r} a_i\sigma\beta_i / \sum_{i=1}^{r} b_i\sigma\beta_i$, where a_i, b_i are constants, $\sigma\beta_1, \cdots, \sigma\beta_r$ are elements of $\sigma\mathcal{G}$ linearly independent over constants, and $\sum b_i\sigma\beta_i \neq 0$ (so that $\beta_1, \cdots, \beta_r$ are elements of $\mathcal{G}$ linearly independent over constants and $\sum b_i\beta_i \neq 0$). Therefore the differential polynomial $W_{\theta_1\ldots\theta_{2r}}(y_1, \cdots, y_r, \alpha y_1, \cdots, \alpha y_r) \in \mathcal{G}\{y_1, \cdots, y_r\}$ vanishes at $(\sigma\beta_1, \cdots, \sigma\beta_r)$ for every choice of $\theta_1, \cdots, \theta_{2r}$ of order $< 2r$, so that this differential polynomial also vanishes at $(\sigma'\beta_1, \cdots, \sigma'\beta_r)$. This implies that there exist constants $a'_i, b'_i \in \mathcal{C}_{\sigma'}$ not all 0 such that $\sum a'_i\sigma'\beta_i - \sum \alpha b'_i\sigma'\beta_i = 0$;

because $\beta_1, \cdots, \beta_r$ are linearly independent over constants $\sigma'\beta_1, \cdots, \sigma'\beta_r$ are too, so that $\sum b'_i\sigma'\beta_i \neq 0$. Therefore

$$\alpha = \sum a'_i\sigma'\beta_i / \sum b'_i\sigma'\beta_i \in (\sigma'\mathcal{G})\langle\mathcal{C}_{\sigma'}\rangle,$$

so that $\mathcal{G}\langle\sigma'\mathcal{G}\rangle = (\sigma'\mathcal{G})\langle\mathcal{C}_{\sigma'}\rangle$. It follows that σ' satisfies (1) and σ' is strong.

REMARK. We observe from the proof of Proposition 5 that if σ is an isomorphism of $\mathcal{G}$ which satisfies the first (second) equation (1) then every specialization of σ also satisfies the first (second) equation (1).

PROPOSITION 6. *If $\sigma_1, \cdots, \sigma_p$ are strong isomorphisms of $\mathcal{G}$ over $\mathcal{F}$ and if $(\tau_1, \cdots, \tau_p)$ is a specialization of $(\sigma_1, \cdots, \sigma_p)$ then $(\tau_1^{-1}, \tau_1^{-1}\tau_2, \cdots, \tau_1^{-1}\tau_p)$ is a specialization of $(\sigma_1^{-1}, \sigma_1^{-1}\sigma_2, \cdots, \sigma_1^{-1}\sigma_p)$.*

Proof. Let $F \in \mathcal{G}\{(z_{1j})_{1\leq j\leq q}, (z_{ij})_{2\leq i\leq p,\, 1\leq j\leq q}\}$; if we denote the coefficients in F by $\beta_1, \cdots, \beta_r$ then we may write

$$F((z_{1j})_{1\leq j\leq q}, (z_{ij})_{2\leq i\leq p,\, 1\leq j\leq q})$$
$$= G((z_{1j})_{1\leq j\leq q}, (z_{ij})_{2\leq i\leq p,\, 1\leq j\leq q}, (\beta_k)_{1\leq k\leq r}),$$

where $G \in \mathcal{F}\{(z_{1j})_{1\leq j\leq q}, (z_{ij})_{2\leq i\leq p,\, 1\leq j\leq q}, (y_k)_{1\leq k\leq r}\}$. If F vanishes at $((\sigma_1^{-1}\alpha_j)_{1\leq j\leq q}, (\sigma_1^{-1}\sigma_i\alpha_j)_{2\leq i\leq p,\, 1\leq j\leq q})$, that is if G vanishes at

$$((\sigma_1^{-1}\alpha_j)_{1\leq j\leq q}, (\sigma_1^{-1}\sigma_i\alpha_j)_{2\leq i\leq p,\, 1\leq j\leq q}, (\beta_k)_{1\leq k\leq r}),$$

then application of σ_1 shows that G vanishes at

$$((\alpha_j)_{1\leq j\leq q}, (\sigma_i\alpha_j)_{2\leq i\leq p,\, 1\leq j\leq q}, (\sigma_1\beta_k)_{1\leq k\leq r});$$

since $(\tau_1, \cdots, \tau_p)$ is a specialization of $(\sigma_1, \cdots, \sigma_p)$ this implies that G vanishes at

$$((\alpha_j)_{1\leq j\leq q}, (\tau_i\alpha_j)_{2\leq i\leq p,\, 1\leq j\leq q}, (\tau_1\beta_k)_{1\leq k\leq r});$$

application of τ_1^{-1} shows that G vanishes at

$$((\tau_1^{-1}\alpha_j)_{1\leq j\leq q}, (\tau_1^{-1}\tau_i\alpha_j)_{2\leq i\leq p,\, 1\leq j\leq q}, (\beta_k)_{1\leq k\leq r}),$$

that is that F vanishes at $((\tau_1^{-1}\alpha_j)_{1\leq j\leq q}, (\tau_1^{-1}\tau_i\alpha_j)_{2\leq i\leq p,\, 1\leq j\leq q})$. It follows that $\tau_1^{-1}, \tau_1^{-1}\tau_2, \cdots, \tau_1^{-1}\tau_p)$ is a specialization of $(\sigma_1^{-1}, \sigma_1^{-1}\sigma_2, \cdots, \sigma_1^{-1}\sigma_p)$.

Let $\sigma_1, \cdots, \sigma_p$ be isomorphisms of $\mathcal{G}$ which have the following property: whenever $\tau_1, \cdots, \tau_p$ are isomorphisms of $\mathcal{G}$ such that τ_i is a specialization of σ_i $(1 \leq i \leq p)$ then $(\tau_1, \cdots, \tau_p)$ is a specialization of $(\sigma_1, \cdots, \sigma_p)$; we shall say under these circumstances that $\sigma_1, \cdots, \sigma_p$ are *independent.*

8

PROPOSITION 7. *The strong isomorphisms $\sigma_1, \cdots, \sigma_p$ of $\mathcal{G}$ over $\mathcal{F}$ are independent if and only if $\partial^0 \mathcal{G}\langle\sigma_1\mathcal{G}, \cdots, \sigma_p\mathcal{G}\rangle/\mathcal{G} = \sum_{i=1}^{p} \partial^0 \mathcal{G}\langle\sigma_i\mathcal{G}\rangle/\mathcal{G}$.*

Proof. Let Λ_i denote the prime differential ideal in $\mathcal{G}\{y_{i1}, \cdots, y_{im}\}$ with generic zero $(\sigma_i\eta_i, \cdots, \sigma_i\eta_n)$, where as before $\eta_1, \cdots, \eta_n$ generate $\mathcal{G} : \mathcal{G} = \mathcal{F}\langle\eta_1, \cdots, \eta_n\rangle$. If $(\eta'_{i1}, \cdots, \eta'_{in})$ is a generic zero of Λ_i then $(\eta'_{i1}, \cdots, \eta'_{in})$ is a generic specialization of $(\sigma_i\eta_1, \cdots, \sigma_i\eta_n)$ over $\mathcal{G}$, so that there exists an isomorphism of $\mathcal{G}\langle\sigma_i\mathcal{G}\rangle = \mathcal{G}\langle\sigma_i\eta_1, \cdots, \sigma_i\eta_n\rangle$ onto $\mathcal{G}\langle\eta'_{i1}, \cdots, \eta'_{in}\rangle$ over $\mathcal{G}$; since $\mathcal{G}\langle\sigma_i\mathcal{G}\rangle = \mathcal{G}\langle\mathcal{C}_{\sigma_i}\rangle$ we have $\mathcal{G}\langle\eta'_{i1}, \cdots, \eta'_{in}\rangle = \mathcal{G}\langle\mathcal{C}'_i\rangle$, where $\mathcal{C}'_i$ is the field of constants of $\mathcal{G}\langle\eta'_{i1}, \cdots, \eta'_{in}\rangle$. The field of constants of $\mathcal{G}$ is $\mathcal{C}$, which is algebraically closed; it easily follows that every polynomial which is irreducible over $\mathcal{G}$ remains irreducible over

$$\mathcal{G}_{i_0} = \mathcal{G}\langle\mathcal{C}'_i, \cdots, \mathcal{C}'_{i_0-1}, \mathcal{C}'_{i_0+1}, \cdots, \mathcal{C}'_p\rangle = \mathcal{G}\langle(\eta'_{ij})_{1\leq i\leq p,\, i\neq i_0,\, 1\leq j\leq n}\rangle,$$

so that (Chapter I, Proposition 1) $\mathcal{G}_{i_0}\cdot\Lambda_{i_0}$ is prime $(1 \leq i_0 \leq p)$. It follows (see remark following the proof of Proposition 4 of Chapter I) that the ideal Λ generated by $\bigcup_{i=1}^{p} \Lambda_i$ in $\mathcal{G}\{(y_{ij})_{1\leq i\leq p, 1\leq j\leq n}\}$ is a prime differential ideal. Since the order of each Λ_i is finite so is the order of Λ, and $\operatorname{ord}\Lambda = \sum \operatorname{ord}\Lambda_i$. Therefore (Chapter I, Corollary to Proposition 2) $(\sigma_i\eta_j)_{1\leq i\leq p,\, 1\leq j\leq n}$ is a generic zero of Λ if and only if

$$\begin{aligned}\partial^0\mathcal{G}\langle\sigma_1\mathcal{G}, \cdots, \sigma_p\mathcal{G}\rangle/\mathcal{G} &= \partial^0\mathcal{G}\langle(\sigma_i\eta_j)_{1\leq i\leq p,\, 1\leq j\leq n}\rangle/\mathcal{G} = \operatorname{ord}\Lambda \\ &= \sum\operatorname{ord}\Lambda_i = \sum\partial^0\mathcal{G}\langle\sigma_i\eta_1, \cdots, \sigma_i\eta_n\rangle/\mathcal{G} = \partial^0\mathcal{G}\langle\sigma_i\mathcal{G}\rangle/\mathcal{G}.\end{aligned}$$

It is easy to see, however, that $\sigma_1, \cdots, \sigma_p$ are independent if and only if $(\sigma_i\eta_j)_{1\leq i\leq p, 1\leq j\leq n}$ is a generic zero of Λ. This completes the proof.

PROPOSITION 8. *If $\sigma_1, \cdots, \sigma_p$ are independent strong isomorphisms of $\mathcal{G}$ over $\mathcal{F}$ and if $\tau_1, \cdots, \tau_p$ are isomorphisms of $\mathcal{G}$ such that τ_i is a specialization of σ_i $(1 \leq i \leq p)$ then $(\tau_1, \tau_1\tau_2, \cdots, \tau_1\tau_p)$ is a specialization of $(\sigma_1, \sigma_1\sigma_2, \cdots, \sigma_1\sigma_p)$.*

Proof. By (1) and (2) we have

$$\begin{aligned}\partial^0\mathcal{G}\langle\sigma_1^{-1}\mathcal{G}, \sigma_2\mathcal{G}, \cdots, \sigma_p\mathcal{G}\rangle/\mathcal{G} &= \partial^0\mathcal{G}\langle\mathcal{C}_{\sigma_1}, \mathcal{C}_{\sigma_2}, \cdots, \mathcal{C}_{\sigma_p}\rangle\mathcal{G} \\ &= \partial^0\mathcal{G}\langle\sigma_1\mathcal{G}, \sigma_2\mathcal{G}, \cdots, \sigma_p\mathcal{G}\rangle/\mathcal{G},\end{aligned}$$

which (Proposition 7) equals

$$\sum_{1\leq i\leq p}\partial^0\mathcal{G}\langle\sigma_i\mathcal{G}\rangle/\mathcal{G} = \partial^0\mathcal{G}\langle\sigma_1^{-1}\mathcal{G}\rangle/\mathcal{G} + \sum_{2\leq i\leq p}\partial^0\mathcal{G}\langle\sigma_i\mathcal{G}\rangle/\mathcal{G}.$$

Therefore (Proposition 7) $\sigma_1^{-1}, \sigma_2, \cdots, \sigma_p$ are independent. If τ_i is a specialization of σ_i $(1 \leq i \leq p)$ then (Proposition 6) τ_1^{-1} is a specialization of σ_1^{-1}, whence $(\tau_1^{-1}, \tau_2, \cdots, \tau_p)$ is a specialization of $(\sigma_1^{-1}, \sigma_2, \cdots, \sigma_p)$. By Proposition 6 it follows that $(\tau_1, \tau_1\tau_2, \cdots, \tau_1\tau_p)$ is a specialization of $(\sigma_1, \sigma_1\sigma_2, \cdots, \sigma_1\sigma_p)$.

If $W(X_1, \cdots, X_p)$ is a word in $X_1, \cdots, X_p$, that is if it is an element of the free group generated by $X_1, \cdots, X_p$, and if $\sigma_1, \cdots, \sigma_p$ are strong isomorphisms of $\mathcal{G}$ over $\mathcal{F}$, then $W(\sigma_1, \cdots, \sigma_p)$, the meaning of which is obvious, is itself a strong isomorphism of $\mathcal{G}$ over $\mathcal{F}$.

If σ is a strong isomorphism of $\mathcal{G}$ over $\mathcal{F}$ then, since $\mathcal{G}$ is finitely generated over $\mathcal{F}$, $\sigma\mathcal{G}$ is too, so that $\mathcal{G}\langle\sigma\mathcal{G}\rangle = \mathcal{G}\langle\mathcal{C}_\sigma\rangle$ is finitely generated over $\mathcal{G}$; using Chapter I, § 4, it is not difficult to see that then $\mathcal{C}_\sigma$ is finitely generated over $\mathcal{C}$.

PROPOSITION 9. *Let $\sigma_1, \cdots, \sigma_p$ be strong isomorphisms of $\mathcal{G}$ over $\mathcal{F}$, let $\gamma_{i1}, \cdots, \gamma_{iq_i}$ be constants such that $\mathcal{C}_{\sigma_i} = \mathcal{C}\langle\gamma_{i1}, \cdots, \gamma_{iq_i}\rangle$ $(1 \leq i \leq p)$, let $W_1(X_1, \cdots, X_p), \cdots, W_r(X_1, \cdots, X_p)$ be words, let $\xi_1, \cdots, \xi_r$ be elements of $\mathcal{G}$, and let N be a differential polynomial in $\mathcal{G}\{w_1, \cdots, w_r\}$ which does not vanish at $(W_1(\sigma_1, \cdots, \sigma_p)\xi_1, \cdots, W_r(\sigma_1, \cdots, \sigma_p)\xi_r)$. Then there exists a polynomial $M \in \mathcal{C}[(u_{ik})_{1\leq i\leq p, 1\leq k\leq q_i}]$ which does not vanish at $(\gamma_{ik})_{1\leq i\leq p, 1\leq k\leq q_i}$ and which has the following property: if $(c_{ik})_{1\leq i\leq p, 1\leq k\leq q_i}$ is a family of constants which is a specialization of $(\gamma_{ik})_{1\leq i\leq p, 1\leq k\leq q_i}$ over $\mathcal{C}$ which does not annul M then for each i $(1 \leq i \leq p)$ there exists a unique isomorphism τ_i of $\mathcal{G}$ such that $((\tau_i\alpha)_{\alpha\in\mathcal{G}}, (c_{ik})_{1\leq k\leq q_i})$ is a specialization of $((\sigma_i\alpha)_{\alpha\in\mathcal{G}}, (\gamma_{ik})_{1\leq k\leq q_i})$ over $\mathcal{G}$, $\mathcal{C}_{\tau_i} = \mathcal{C}\langle c_{i1}, \cdots, c_{iq_i}\rangle$, N does not vanish at*

$$(W_1(\tau_1, \cdots, \tau_p)\xi_1, \cdots, W_r(\tau_1, \cdots, \tau_p)\xi_r),$$

and, for every finite family

$$(W'_1(X_1, \cdots, X_p), \cdots, W'_s(X_1, \cdots, X_p))$$

of words,

$$(W'_1(\tau_1, \cdots, \tau_p), \cdots, W'_s(\tau_1, \cdots, \tau_p))$$

is a specialization of

$$(W'_1(\sigma_1, \cdots, \sigma_p), \cdots, W'_s(\sigma_1, \cdots, \sigma_p)).$$

Proof. By (1) there exist

$$A_{ij}, B_{ij} \in \mathcal{F}\{y_1, \cdots, y_n\}[u_{i1}, \cdots, u_{iq_i}] \qquad (1 \leq i \leq p, 1 \leq j \leq n),$$

$$C_{ik}, D_{ik} \in \mathcal{F}\{y_1, \cdots, y_n, z_{i1}, \cdots, z_{in}\} \qquad (1 \leq i \leq p, 1 \leq k \leq q_i),$$

$$E_{ij}, F_{ij} \in \mathcal{F}\{z_{i1}, \cdots, z_{in}\}[u_{i1}, \cdots, u_{iq_i}] \qquad (1 \leq i \leq p, 1 \leq j \leq n)$$

such that, for all i, j, k,

$$B_{ij}(\eta_1, \cdots, \eta_n, \gamma_{i1}, \cdots, \gamma_{iq_i}) \neq 0,$$

$$D_{ik}(\eta_1, \cdots, \eta_n, \sigma_i\eta_1, \cdots, \sigma_i\eta_n) \neq 0,$$

$$F_{ij}(\sigma_i\eta_1, \cdots, \sigma_i\eta_n, \gamma_{i1}, \cdots, \gamma_{iq_i}) \neq 0,$$

(3) $$\sigma_i\eta_j = A_{ij}(\eta_1, \cdots, \eta_n, \gamma_{i1}, \cdots, \gamma_{iq_i})/B_{ij}(\eta_1, \cdots, \eta_n, \gamma_{i1}, \cdots, \gamma_{iq_i}),$$

$$\gamma_{ik} = C_{ik}(\eta_1, \cdots, \eta_n, \sigma_i\eta_1, \cdots, \sigma_i\eta_n)/D_{ik}(\eta_1, \cdots, \eta_n, \sigma_i\eta_1, \cdots, \sigma_i\eta_n),$$

(4) $$\eta_j = E_{ij}(\sigma_i\eta_1, \cdots, \sigma_i\eta_n, \gamma_{i1}, \cdots, \gamma_{iq_i})/F_{ij}(\sigma_i\eta_1, \cdots, \sigma_i\eta_n, \gamma_{i1}, \cdots, \gamma_{iq_i}).$$

Let $(c_{ik})_{1\leq i\leq p,\, 1\leq k\leq q_i}$ be a family of constants which is a specialization of $(\gamma_{ik})_{1\leq i\leq p,\, 1\leq k\leq q_i}$ over $\mathcal{C}$, and therefore (Chapter I, Corollary 2 to Proposition 3) over $\mathcal{G}$. If

(5) $$\prod_{i,j} B_{ij}(\eta_1, \cdots, \eta_n, c_{i1}, \cdots, c_{iq_i}) \neq 0$$

and if we set

(6) $$\zeta_{ij} = A_{ij}(\eta_1, \cdots, \eta_n, c_{i1}, \cdots, c_{iq_i})/B_{ij}(\eta_1, \cdots, \eta_n, c_{i1}, \cdots, c_{iq_i})$$

then $((\zeta_{ij})_{1\leq i\leq p,\, 1\leq j\leq n}, (c_{ik})_{1\leq i\leq p,\, 1\leq k\leq q_i})$ is a specialization of

$$((\sigma_i\eta_j)_{1\leq i\leq p,\, 1\leq j\leq n}, (\gamma_{ik})_{1\leq i\leq p,\, 1\leq k\leq q_i})$$

over $\mathcal{G}$. If moreover

(7) $$\prod_{i,k} D_{ik}(\eta_1, \cdots, \eta_n, \zeta_{i1}, \cdots, \zeta_{in}) \neq 0$$

then

(8) $$c_{ik} = C_{ik}(\eta_1, \cdots, \eta_n, \zeta_{i1}, \cdots, \zeta_{in})/D_{ik}(\eta_1, \cdots, \eta_n, \zeta_{i1}, \cdots, \zeta_{in}).$$

If in addition

(9) $$\prod_{i,j} F_{ij}(\zeta_{i1}, \cdots, \zeta_{in}, c_{i1}, \cdots, c_{iq_i}) \neq 0$$

then

(10) $$\eta_j = E_{ij}(\zeta_{i1}, \cdots, \zeta_{in}, c_{i1}, \cdots, c_{iq_i})/F_{ij}(\zeta_{i1}, \cdots, \zeta_{in}, c_{i1}, \cdots, c_{iq_i}).$$

Assuming that (5), (7), (9) hold we see that $(\zeta_{i1}, \cdots, \zeta_{in})$ is a specialization of $(\sigma_i\eta_1, \cdots, \sigma_i\eta_n)$ over $\mathcal{G}$, and therefore of $(\eta_1, \cdots, \eta_n)$ over $\mathcal{F}$, such that

$$\partial^0\mathcal{F}\langle\zeta_{i1}, \cdots, \zeta_{in}\rangle/\mathcal{F} - \partial^0\mathcal{F}\langle\eta_1, \cdots, \eta_n\rangle/\mathcal{F}$$

$$= \partial^0\mathcal{F}\langle\eta_1, \cdots, \eta_n, \zeta_{i1}, \cdots, \zeta_{in}\rangle/\mathcal{F} - \partial^0\mathcal{F}\langle\eta_1, \cdots, \eta_n, \zeta_{i1}, \cdots, \zeta_{in}\rangle/\mathcal{F}\langle\zeta_{i1}, \cdots, \zeta_{i}$$

$$- \partial^0\mathcal{F}\langle\eta_1, \cdots, \eta_n, \zeta_{i1}, \cdots, \zeta_{in}\rangle/\mathcal{F}$$

$$+ \partial^0\mathcal{F}\langle\eta_1, \cdots, \eta_n, \zeta_{i1}, \cdots, \zeta_{in}\rangle/\mathcal{F}\langle\eta_1, \cdots, \eta_n\rangle,$$

which by (6), (8), (10) equals

$$-\partial^0\mathcal{F}\langle\zeta_{i1},\cdots,\zeta_{in},c_{i1},\cdots,c_{iq_i}\rangle/\mathcal{F}\langle\zeta_{i1},\cdots,\zeta_{in}\rangle$$
$$+\partial^0\mathcal{F}\langle\eta_1,\cdots,\eta_n,c_{i1},\cdots,c_{iq_i}\rangle/\mathcal{F}\langle\eta_1,\cdots,\eta_n\rangle,$$

which by Chapter I, Corollary 3 to Proposition 3, is ≥ 0. It follows (Chapter I, Proposition 2) that $(\zeta_{i1},\cdots,\zeta_{in})$ is a generic specialization of $(\eta_1,\cdots,\eta_n)$ over $\mathcal{F}$, so that there exists a unique isomorphism τ_i of $\mathcal{G}=\mathcal{F}\langle\eta_1,\cdots,\eta_n\rangle$ over $\mathcal{F}$ such that $\tau_i\eta_j=\zeta_{ij}$ $(1\leq j\leq n)$. By (6), (8), (10)

$$\mathcal{G}\langle\tau_i\mathcal{G}\rangle=\mathcal{G}\langle c_{i1},\cdots,c_{iq_i}\rangle=(\tau_i\mathcal{G})\langle c_{i1},\cdots,c_{iq_i}\rangle,$$

so that $\mathcal{C}_{\tau_i}=\mathcal{C}\langle c_{i1},\cdots,c_{iq_i}\rangle$. It is apparent that τ_i is the unique isomorphism of $\mathcal{G}$ such that $((\tau_i\alpha)_{\alpha\varepsilon G},(c_{ik})_{1\leq k\leq q_i})$ is a specialization of $((\sigma_i\alpha)_{\alpha\varepsilon G},(\gamma_{ik})_{1\leq k\leq q_i})$ over $\mathcal{G}$.

By (4) and (10)

$$(11)\quad \sigma_i^{-1}\eta_j=E_{ij}(\eta_1,\cdots,\eta_n,\gamma_{i1},\cdots,\gamma_{iq_i})/F_{ij}(\eta_1,\cdots,\eta_n,\gamma_{i1},\cdots,\gamma_{iq_i}),$$

$$(12)\quad \tau_i^{-1}\eta_j=E_{ij}(\eta_1,\cdots,\eta_n,c_{i1},\cdots,c_{iq_i})/F_{ij}(\eta_1,\cdots,\eta_n,c_{i1},\cdots,c_{iq_i}).$$

From (3), (6), (11), (12) it is not difficult to see that for every word $W(X_1,\cdots,X_p)$ there exist

$$G_j^W,H_j^W\,\varepsilon\,\mathcal{F}\{y_1,\cdots,y_n\}[(u_{ik})_{1\leq i\leq p,\,1\leq k\leq q_i}]$$

such that

$$H_j^W(\eta_1,\cdots,\eta_n,(\gamma_{ik}))\neq 0,\qquad H_j^W(\eta_1,\cdots,\eta_n,(c_{ik}))\neq 0$$

and

$$W(\sigma_1,\cdots,\sigma_p)\eta_j=G_j^W(\eta_1,\cdots,\eta_n,(\gamma_{ik}))/H_j^W(\eta_1,\cdots,\eta_n,(\gamma_{ik})),$$

$$W(\tau_1,\cdots,\tau_p)\eta_j=G_j^W(\eta_1,\cdots,\eta_n,(c_{ik}))/H_j^W(\eta_1,\cdots,\eta_n,(c_{ik})).$$

It follows, for any finite family of words

$$W'_1(X_1,\cdots,X_p),\cdots,W'_s(X_1,\cdots,X_p),$$

that $(W'_l(\tau_1,\cdots,\tau_p)\eta_j)_{1\leq l\leq s,\,1\leq j\leq n}$ is a specialization of

$$W'_l(\sigma_1,\cdots,\sigma_p)\eta_j)_{1\leq l\leq s,\,1\leq j\leq n}$$

over $\mathcal{G}$, so that

$$(W'_1(\tau_1,\cdots,\tau_p),\cdots,W'_s(\tau_1,\cdots,\tau_p))$$

is a specialization of

$$(W'_1(\sigma_1,\cdots,\sigma_p),\cdots,W'_s(\sigma_1,\cdots,\sigma_p)).$$

It also follows that for each h $(1\leq h\leq r)$ there exist

$$I_h,J_h\,\varepsilon\,\mathcal{F}\{y_1,\cdots,y_n\}[(u_{ik})_{1\leq i\leq p,\,1\leq k\leq q_i}]$$

with

$$J_h(\eta_1, \cdots, \eta_n, (\gamma_{ik})) \neq 0, \qquad J_h(\eta_1, \cdots, \eta_n, (c_{ik})) \neq 0$$

such that

$$W_h(\sigma_1, \cdots, \sigma_p)\xi_h = I_h(\eta_1, \cdots, \eta_n, (\gamma_{ik}))/J_h(\eta_1, \cdots, \eta_n, (\gamma_{ik})),$$

$$W_h(\tau_1, \cdots, \tau_p)\xi_h = I_h(\eta_1, \cdots, \eta_n, (c_{ik}))/J_h(\eta_1, \cdots, \eta_n, (c_{ik})),$$

and therefore that there exist

$$U, V \in \mathfrak{F}\{y_1, \cdots, y_n\}[(u_{ik})_{1\leq i\leq p, 1\leq k\leq q_i}]$$

such that

$$V(\eta_1, \cdots, \eta_n), (\gamma_{ik})) \neq 0, \qquad V(\eta_1, \cdots, \eta_n(c_{ik})) \neq 0$$

and

$$N(W_1(\sigma_1, \cdots, \sigma_p)\xi_1, \cdots, W_r(\sigma_1, \cdots, \sigma_p)\xi_r)$$
$$= U(\eta_1, \cdots, \eta_n, (\gamma_{ik}))/V(\eta_1, \cdots, \eta_n, (\gamma_{ik})),$$

$$N(W_1(\tau_1, \cdots, \tau_p)\xi_1, \cdots, W_r(\tau_1, \cdots, \tau_p)\xi_r)$$
$$= U(\eta_1, \cdots, \eta_n, (c_{ik}))/V(\eta_1, \cdots, \eta_n, (c_{ik})),$$

Now there exists a polynomial $M' \in \mathfrak{G}[(u_{ik})_{1\leq i\leq p, 1\leq k\leq q_i}]$ which has the two properties that it does not vanish at $(\gamma_{ik})_{1\leq i\leq p, 1\leq k\leq q_i}$ and that if it does not vanish at $(c_{ik})_{1\leq i\leq p, 1\leq k\leq q_i}$ then (5), (7), (9) hold and $U(\eta_1, \cdots, \eta_n, (c_{ik})) \neq 0$. Therefore (Chapter I, Corollary 2 to Proposition 3) there exists a polynomial $M \in \mathcal{C}[(u_{ik})_{1\leq i\leq p, 1\leq k\leq q_i}]$ with the same two properties. This M has the property described in the statement of the proposition.

COROLLARY 1. *Let $\eta_1, \cdots, \eta_n$ be elements of $\mathfrak{G}$ such that $\mathfrak{G} = \mathfrak{F}\langle\eta_1, \cdots, \eta_n\rangle$, let q, r be integers such that $1 \leq q \leq r$, let $W_1(X_1, \cdots, X_p), \cdots, W_r(X_1, \cdots, X_p)$ be words, let $\sigma_1, \cdots, \sigma_p$ be strong isomorphisms of $\mathfrak{G}$ over $\mathfrak{F}$, and let $Q \in \mathfrak{G}\{(y_{ij})_{1\leq i\leq r, 1\leq j\leq n}\}$ not vanish at $(W_i(\sigma_1, \cdots, \sigma_p)\eta_j)_{1\leq i\leq r, 1\leq j\leq n}$. Then there exists a $P \in \mathfrak{G}\{(y_{ij})_{1\leq i\leq q, 1\leq j\leq n}\}$ which does not vanish at $(W_i(\sigma_1, \cdots, \sigma_p)\eta_j)_{1\leq i\leq q, 1\leq j\leq n}$ such that for every specialization $(\rho'_1, \cdots, \rho'_q)$ of $(W_1(\sigma_1, \cdots, \sigma_p), \cdots, W_q(\sigma_1, \cdots, \sigma_p))$ for which $P((\rho'_i\eta_j)_{1\leq i\leq q, 1\leq j\leq n}) \neq 0$ there exists strong isomorphisms $\sigma'_1, \cdots, \sigma'_p$ of G such that $\rho'_i = W_i(\sigma'_1, \cdots, \sigma'_p)$ $(1 \leq i \leq q)$,*

$$(\sigma'_1, \cdots, \sigma'_p, W_1(\sigma'_1, \cdots, \sigma'_p), \cdots, W_r(\sigma'_1, \cdots, \sigma'_p))$$

is a specialization of

$$(\sigma_1, \cdots, \sigma_p, W_1(\sigma_1, \cdots, \sigma_p), \cdots, W_r(\sigma_1, \cdots, \sigma_p)),$$

and $Q((W_i(\sigma'_1, \cdots, \sigma'_p)\eta_j)_{1\leq i\leq r, 1\leq j\leq n}) \neq 0$.

COROLLARY 2. *Let σ be a strong isomorphism of $\mathcal{G}$ over $\mathcal{F}$, let $\xi_1, \cdots, \xi_s \varepsilon \mathcal{G}$, let $N \varepsilon \mathcal{G}\{w_1, \cdots, w_s\}$ not vanish at $(\sigma\xi_1, \cdots, \xi\sigma_s)$. Then there exists an automorphism τ of $\mathcal{G}$ over $\mathcal{F}$ such that τ is a specialization of σ and $N(\tau\xi_1, \cdots, \tau\xi_s) \neq 0$.*

COROLLARY 3. *Let σ be a strong isomorphism of $\mathcal{G}$ over $\mathcal{F}$, let $\mathcal{F}_1$ be a differential field between $\mathcal{F}$ and $\mathcal{G}$, and suppose that the restriction σ' of σ to $\mathcal{F}_1$ is a strong isomorphism of $\mathcal{F}_1$. Then there exist a finite number of nongeneric specializations $\sigma'_1, \cdots, \sigma'_s$ of σ' such that every specialization τ' of σ' which is not a specialization of σ'_j $(1 \leq j \leq s)$ is the restriction to $\mathcal{F}_1$ of some specialization τ of σ.*

Proof. Let $\gamma_1, \cdots, \gamma_p, \gamma_{p+1}, \cdots, \gamma_q$ be constants such that $\mathcal{F}_1\langle\sigma'\mathcal{F}_1\rangle = \mathcal{F}_1\langle\gamma_1, \cdots, \gamma_p\rangle$, $\mathcal{G}\langle\sigma\mathcal{G}\rangle = \mathcal{G}\langle\gamma_1, \cdots, \gamma_q\rangle$. By Proposition 9 there exists a polynomial $M \varepsilon \mathcal{C}[u_1, \cdots, u_q]$ with $M(\gamma_1, \cdots, \gamma_q) \neq 0$ such that whenever $c_1, \cdots, c_q$ are constants such that $(c_1, \cdots, c_q)$ is a specialization of $(\gamma_1, \cdots, \gamma_q)$ over $\mathcal{C}$ with $M(c_1, \cdots, c_n) \neq 0$ then there exist an isomorphism τ of $\mathcal{G}$ such that $((\tau\alpha)_{\alpha \varepsilon \mathcal{G}}, c_1, \cdots, c_q)$ is a specialization of $((\sigma\alpha)_{\alpha \varepsilon \mathcal{G}}, \gamma_1, \cdots, \gamma_q)$ over $\mathcal{G}$, and a unique isomorphism τ' of $\mathcal{F}_1$ such that $((\tau'\alpha)_{\alpha \varepsilon \mathcal{F}_1}, c_1, \cdots, c_p)$ is a specialization of $((\sigma'\alpha)_{\alpha \varepsilon \mathcal{F}_1}, \gamma_1, \cdots, \gamma_p)$ over $\mathcal{F}_1$. There also exists a polynomial $K \varepsilon \mathcal{C}[u_1, \cdots, u_p]$ with $K(\gamma_1, \cdots, \gamma_p) \neq 0$ such that every specialization $(c_1, \cdots, c_p)$ of $(\gamma_1, \cdots, \gamma_p)$ with $K(c_1, \cdots, c_p) \neq 0$ can be extended to a specialization $(c_1, \cdots, c_q)$ of $(\gamma_1, \cdots, \gamma_q)$ over $\mathcal{C}$ with $M(c_1, \cdots, c_q) \neq 0$. We now write, for $1 \leq i \leq p$,

$$\gamma_i = R_i(\sigma'\theta_1, \cdots, \sigma'\theta_l)/S(\sigma'\theta_1, \cdots, \sigma'\theta_l),$$

where $\theta_1, \cdots, \theta_l$ are elements of $\mathcal{F}_1$, such that $\mathcal{F}_1 = \mathcal{F}\langle\theta_1, \cdots, \theta_l\rangle$, R_i and $S \varepsilon \mathcal{F}_1\{y_1, \cdots, y_l\}$, and $S(\sigma'\theta_1, \cdots, \sigma'\theta_l) \neq 0$. If τ' is any specialization of σ' such that $S(\tau'\theta_1, \cdots, \tau'\theta_l) \neq 0$ and if we set

$$c_i = R_i(\tau'\theta_1, \cdots, \tau'\theta_l)/S(\tau'\theta_1, \cdots, \tau'\theta_l),$$

then $c_1, \cdots, c_p$ are constants such that $(c_1, \cdots, c_p)$ is a specialization of $(\gamma_1, \cdots, \gamma_p)$ over $\mathcal{C}$. If in addition

$$K\left(\frac{R_1(\tau'\theta_1, \cdots, \tau'\theta_l)}{S(\tau'\theta_1, \cdots, \tau'\theta_l)}, \cdots, \frac{R_p(\tau'\theta_1, \cdots, \tau'\theta_l)}{S(\tau'\theta_1, \cdots, \tau'\theta_l)}\right) \neq 0,$$

that is $K(c_1, \cdots, c_p) \neq 0$, then we may extend $(c_1, \cdots, c_p)$ to a specialization $(c_1, \cdots, c_q)$ of $(\gamma_1, \cdots, \gamma_q)$ over $\mathcal{C}$ such that $M(c_1, \cdots, c_q) \neq 0$, so that there exists an isomorphism τ of $\mathcal{G}$ such that $((\tau\alpha)_{\alpha \varepsilon \mathcal{G}}, c_1, \cdots, c_q)$ is a specialization of $((\sigma\alpha)_{\alpha \varepsilon \mathcal{G}}, \gamma_1, \cdots, \gamma_q)$ over $\mathcal{G}$. Under these circumstances τ will be a specialization of σ, and τ' will be the restriction of τ to $\mathcal{F}_1$. Thus, the specializations τ' of σ' which can *not* be extended to a specialization of σ have the property that $(\tau'\theta_1, \cdots, \tau'\theta_l)$ is a zero of SL, where L is a differential polynomial in $\mathcal{F}_1\{y_1, \cdots, y_l\}$ obtained by multiplying $K(S^{-1}R_1, \cdots, S^{-1}R_l)$ by a power of S; of course $(\sigma'\theta_1, \cdots, \sigma'\theta_l)$ is not a zero of SL. Now let Σ be the prime differential ideal of $\mathcal{F}_1\{y_1, \cdots, y_l\}$ with generic zero $(\sigma'\theta_1, \cdots, \sigma'\theta_l)$, and let $\Sigma_1, \cdots, \Sigma_s$ be the minimal prime differential ideal divisors of $\{\Sigma, SL\}$ in $\mathcal{F}_1\{y_1, \cdots, y_l\}$. If τ' is a specialization of σ' which can not be extended to a specialization of σ then $(\tau'\theta_1, \cdots, \tau'\theta_l)$ is a zero of Σ_j for some j. Let $(\psi_{j1}, \cdots, \psi_{jl})$ be a generic zero of this Σ_j; then $(\psi_{j1}, \cdots, \psi_{jl})$ is a zero of Σ. hence a specialization of $(\sigma'\theta_1, \cdots, \sigma'\theta_l)$ over $\mathcal{F}_1$ and a fortiori over $\mathcal{F}$, and therefore a specialization of $(\theta_1, \cdots, \theta_l)$ over $\mathcal{F}$. But $(\psi_{j1}, \cdots, \psi_{jl})$ admits $(\tau'\theta_1, \cdots, \tau'\theta_l)$ as a specialization over $\mathcal{F}_1$ and therefore over $\mathcal{F}$, so that $(\theta_1, \cdots, \theta_l)$ is a specialization of $(\psi_{j1}, \cdots, \psi_{jl})$ over $\mathcal{F}$. Thus $(\psi_{j1}, \cdots, \psi_{jl})$ is a generic specialization of $(\theta_1, \cdots, \theta_l)$ over $\mathcal{F}$, so that there is a unique isomorphism σ'_j of $\mathcal{F}_1$ over $\mathcal{F}$ such that $\sigma'_j\theta_i = \psi_{ji}$ $(1 \leq i \leq l)$. Clearly τ' is a specialization of σ'_j, σ'_j is a specialization of σ', and because

$$\partial^0\sigma'_j\mathcal{F}_1/\mathcal{F}_1 = \operatorname{ord} \Sigma_j < \operatorname{ord} \Sigma = \partial^0\sigma'\mathcal{F}_1/\mathcal{F}_1$$

the latter specialization is nongeneric. This completes the proof of Corollary 3.

5. Algebraic sets. A subset $\mathfrak{M}^*$ of the group $\mathfrak{G}^*$ of all strong isomorphisms of $\mathcal{G}$ over $\mathcal{F}$ will be called an *irreducible set in* $\mathfrak{G}^*$ if $\mathfrak{M}^*$ contains an element σ^* such that $\mathfrak{M}^*$ is the set of all specializations of σ^*; any such σ^* will then be called a *generic element* of $\mathfrak{M}^*$, and the transcendence degree $\partial^0\mathcal{G}\langle\sigma^*\mathcal{G}\rangle/\mathcal{G}$, which does not depend on the particular generic element σ^* employed, will be called the *dimension* of $\mathfrak{M}^*$ (notation: $\dim \mathfrak{M}^*$). It follows from Chapter I, Projosition 2, that if σ^* and τ^* are any two isomorphisms of $\mathcal{G}$ over $\mathcal{F}$ such that τ^* is a specialization of σ^* then $\partial^0\mathcal{G}\langle\tau^*\mathcal{G}\rangle/\mathcal{G} \leq \partial^0\mathcal{G}\langle\sigma^*\mathcal{G}\rangle/\mathcal{G}$, and we have equality here if and only if the specialization is generic. Consequently if $\mathfrak{M}^*$ and $\mathfrak{N}^*$ are irreducible sets in $\mathfrak{G}^*$ such that $\mathfrak{M}^* \supseteq \mathfrak{N}^*$ then $\dim \mathfrak{M}^* \geq \dim \mathfrak{N}^*$, and $\dim \mathfrak{M}^* = \dim \mathfrak{N}^*$ if and only if $\mathfrak{M}^* = \mathfrak{N}^*$. A subset $\mathfrak{M}^*$ of $\mathfrak{G}^*$ will be called an *algebraic set in* $\mathfrak{G}^*$ if $\mathfrak{M}^*$ is the union of a finite number of irreducible sets in $\mathfrak{G}^*$. It is easy to see that an algebraic set in $\mathfrak{G}^*$ can be written in one and only one way as the union of a finite set of irreducible sets in $\mathfrak{G}^*$ none of which contains another; these unique irreducible sets, which are the maximal irreducible sets contained in the algebraic set, will be called the *components* of the algebraic set.

It is a consequence of Corollary 2 of Proposition 9 that every nonempty algebraic set in $\mathfrak{G}^*$ has a nonempty intersection with the group $\mathfrak{G}$ of all automorphisms of $\mathcal{G}$ over $\mathcal{F}$, and that if $\mathfrak{M}^*$, $\mathfrak{N}^*$ are algebraic sets in $\mathfrak{G}^*$ with $\mathfrak{M}^* \neq \mathfrak{N}^*$ then $\mathfrak{M}^* \cap \mathfrak{G} \neq \mathfrak{N}^* \cap \mathfrak{G}$. Accordingly we shall call a subset $\mathfrak{M}$ of $\mathfrak{G}$ an *irreducible set in* $\mathfrak{G}$ or an *algebraic set in* $\mathfrak{G}$ if $\mathfrak{M}$ is the intersection with $\mathfrak{G}$ of, respectively, an irreducible set in $\mathfrak{G}^*$ or an algebraic set in $\mathfrak{G}^*$; this set in $\mathfrak{G}^*$, which is unique, we shall denote by $\mathfrak{M}^*$. If $\mathfrak{M}$ is an irreducible set in $\mathfrak{G}$ we define the *dimension* of $\mathfrak{M}$ (notation: $\dim \mathfrak{M}$) by the formula $\dim \mathfrak{M} = \dim \mathfrak{M}^*$, and define *generic element* of $\mathfrak{M}$ as a generic element of $\mathfrak{M}^*$ (so that a generic element of $\mathfrak{M}$ need not be an element of $\mathfrak{M}$); if σ^* is a generic element of $\mathfrak{M}$ we thus have $\dim \mathfrak{M} = \partial^0 \mathcal{G}\langle \sigma^* \mathcal{G} \rangle / \mathcal{G}$. If $\mathfrak{M}$ and $\mathfrak{N}$ are two irreducible sets in $\mathfrak{G}$ with $\mathfrak{M} \supseteq \mathfrak{N}$ then $\dim \mathfrak{M} \geq \dim \mathfrak{N}$, and equality of dimension implies that $\mathfrak{M} = \mathfrak{N}$. An algebraic set in $\mathfrak{G}$ can be written in one and only one way as the union of a finite set of irreducible sets in $\mathfrak{G}$ none of which contains another; these unique irreducible sets, which are the maximal irreducible sets in $\mathfrak{G}$ contained in the algebraic set, will be called the *components* of the algebraic set. If $\mathfrak{M}$ is an algebraic set in $\mathfrak{G}$ and $\mathfrak{M}_1, \cdots, \mathfrak{M}_p$ are its components then $\mathfrak{M}^*_1, \cdots, \mathfrak{M}^*_p$ are the components of $\mathfrak{M}^*$.

PROPOSITION 10. *Every nonempty set of algebraic sets in* $\mathfrak{G}$ *has a minimal element.*

Proof. To each algebraic set $\mathfrak{M}$ we associate the sequence

$$k(\mathfrak{M}) = (k_d(\mathfrak{M}))_{0 \leq d < \infty},$$

where $k_d(\mathfrak{M})$ is the number of components of $\mathfrak{M}$ of dimension d; since the dimension of every irreducible set is $\leq \partial^0 \mathcal{G}/\mathcal{F}$ we have $k_d(\mathfrak{M}) = 0$ for all $d > \partial^0 \mathcal{G}/\mathcal{F}$. We introduce an order into the set of sequences $k(\mathfrak{M})$ by writing $k(\mathfrak{M}) \leq k(\mathfrak{N})$ whenever either $k(\mathfrak{M}) = k(\mathfrak{N})$ or $k(\mathfrak{M}) \neq k(\mathfrak{N})$ and the last nonzero difference $k_d(\mathfrak{N}) - k_d(\mathfrak{M})$ is positive. It is easy to see that if $\mathfrak{M} \subseteq \mathfrak{N}$ then $k(\mathfrak{M}) \leq k(\mathfrak{N})$; since it is obvious that in any nonempty set of algebraic sets in G there exists one for which the associated sequence is minimal, the proof is complete.

PROPOSITION 11. *If* $\mathfrak{M}_1, \cdots, \mathfrak{M}_p$ *are irreducible sets in* $\mathfrak{G}$ *then there exists an independent family of generic elements of* $\mathfrak{M}_1, \cdots, \mathfrak{M}_p$; *if* $\rho_i \in \mathfrak{M}^*_i$ $(1 \leq i \leq p)$ *then* $\rho_1, \cdots, \rho_p$ *are independent generic elements of* $\mathfrak{M}_1, \cdots, \mathfrak{M}_p$ *if and only if* $\partial^0 \mathcal{G}\langle \rho_1 \mathcal{G}, \cdots, \rho_p \mathcal{G} \rangle / \mathcal{G} = \sum_{i=1}^{p} \dim \mathfrak{M}_i$.

Proof. Let σ_i be a generic element of $\mathfrak{M}_i$, and let $\gamma_{i1}, \cdots, \gamma_{iq_i}$ be constants such that $\mathcal{B}_{\sigma_i} = \mathcal{B}\langle\gamma_{i1},\cdots,\gamma_{iq_i}\rangle$, $1 \leq i \leq p$. It is obvious that there exist generic specializations $(\delta_{i1},\cdots,\delta_{iq_i})$ of $(\gamma_{i1},\cdots,\gamma_{iq_i})$ over $\mathcal{B}$, $1 \leq i \leq p$, such that

$$\partial^0 \mathcal{B}\langle(\delta_{ij})_{1\leq i\leq p,\, 1\leq j\leq q_i}\rangle/\mathcal{B} = \sum_{i=1}^{p} \partial^0 \mathcal{B}\langle\delta_{i1},\cdots,\delta_{iq_i}\rangle/\mathcal{B}.$$

By Proposition 9 there exists a strong isomorphism τ_i of $\mathcal{G}$ over $\mathcal{F}$ such that τ_i is a specialization of σ_i and $\mathcal{B}_{\tau_i} = \mathcal{B}\langle\delta_{i1},\cdots,\delta_{iq_i}\rangle$. Because

$$\partial^0 \mathcal{G}\langle\tau_i\mathcal{G}\rangle/\mathcal{G} = \partial^0 \mathcal{B}\langle\delta_{i1},\cdots,\delta_{iq_i}\rangle/\mathcal{B}$$
$$= \partial^0 \mathcal{B}\langle\gamma_{i1},\cdots,\gamma_{iq_i}\rangle/\mathcal{B} = \partial^0 \mathcal{G}\langle\sigma_i\mathcal{G}\rangle/\mathcal{G}$$

it follows that

$$\partial^0 \mathcal{G}\langle\tau_i\eta_1,\cdots,\tau_i\eta_n\rangle/\mathcal{G} = \partial^0 \mathcal{G}\langle\sigma_i\eta_1,\cdots,\sigma_i\eta_n\rangle/\mathcal{G},$$

where $\eta_1,\cdots,\eta_n$ are elements of $\mathcal{G}$ such that $\mathcal{G} = \mathcal{F}\langle\eta_1,\cdots,\eta_n\rangle$; therefore (Chapter I, Proposition 2) $(\tau_i\eta_1,\cdots,\tau_i\eta_n)$ is a generic specialization of $(\sigma_i\eta_1,\cdots,\sigma_i\eta_n)$ over $\mathcal{G}$, so that τ_i is a generic specialization of σ_i and therefore a generic element of $\mathfrak{M}_i$. Because

$$\partial^0 \mathcal{G}\langle\tau_1\mathcal{G},\cdots,\tau_p\mathcal{G}\rangle/\mathcal{G} = \partial^0 \mathcal{G}\langle(\delta_{ik})_{1\leq i\leq p,\, 1\leq k\leq q_i}\rangle/\mathcal{G}$$
$$= \partial^0 \mathcal{B}\langle(\delta_{ik})_{1\leq i\leq p,\, 1\leq k\leq q_i}\rangle/\mathcal{B} = \sum \partial^0 \mathcal{B}\langle\delta_{i1},\cdots,\delta_{iq_i}\rangle/\mathcal{B}$$
$$= \sum \partial^0 \mathcal{B}_{\tau_i}/\mathcal{B} = \sum \partial^0 \mathcal{G}\langle\tau_i\mathcal{G}\rangle/\mathcal{G},$$

we see from Proposition 7 that $\tau_1,\cdots,\tau_p$ are independent. If ρ_i is an element of $\mathfrak{M}^*_i$ $(1 \leq i \leq p)$ then $\rho_1,\cdots,\rho_p$ are independent (Proposition 7) if and only if $\partial^0 \mathcal{G}\langle\rho_1\mathcal{G},\cdots,\rho_p\mathcal{G}\rangle/\mathcal{G} = \sum \partial^0 \mathcal{G}\langle\rho_i\mathcal{G}\rangle/\mathcal{G}$, and are generic if and only if $\partial^0 \mathcal{G}\langle\rho_i\mathcal{G}\rangle/\mathcal{G} = \dim \mathfrak{M}_i$ $(1 \leq i \leq p)$.

PROPOSITION 12. *If $\mathfrak{M}$ is an irreducible set in $\mathfrak{G}$ with generic element σ and if $\tau \in \mathfrak{G}$ then $\tau\mathfrak{M}$ and $\mathfrak{M}\tau$ and $\mathfrak{M}^{-1}$ are irreducible sets in $\mathfrak{G}$ with generic elements $\tau\sigma$ and $\sigma\tau$ and σ^{-1} respectively, and* $\dim \tau\mathfrak{M} = \dim \mathfrak{M}\tau = \dim \mathfrak{M}^{-1} = \dim \mathfrak{M}$.

Proof. The set consisting of the single element τ is obviously an irreducible set of dimension 0, and

$$\partial^0 \mathcal{G}\langle\sigma\mathcal{G},\tau\mathcal{G}\rangle/\mathcal{G} = \partial^0 \mathcal{G}\langle\sigma\mathcal{G}\rangle/\mathcal{G} = \dim \mathfrak{M} + 0;$$

therefore by Proposition 11 σ,τ are independent, so that (Proposition 8) every element of $\tau\mathfrak{M}$ is a specialization of $\tau\sigma$. Therefore if $\mathfrak{N}$ is the irreducible set in $\mathfrak{G}$ with generic element $\tau\sigma$ then $\tau\mathfrak{M} \subseteq \mathfrak{N}$. Similarly, $\tau^{-1}\mathfrak{N}$ is contained in the irreducible set in $\mathfrak{G}$ with generic element $\tau^{-1}(\tau\sigma) = \sigma$, that is in $\mathfrak{M}$, so that $\mathfrak{N} \subseteq \tau\mathfrak{M}$. Therefore $\tau\mathfrak{M} = \mathfrak{N}$, so that $\tau\mathfrak{M}$ is irreducible and has generic element $\tau\sigma$. The proof for $\mathfrak{M}\tau$ is similar and for $\mathfrak{M}^{-1}$ is even simpler. Finally

$$\dim \tau\mathfrak{M} = \partial^0 \mathcal{G}\langle\tau\sigma\mathcal{G}\rangle/\mathcal{G} = \partial^0(\tau^{-1}\mathcal{G})\langle\sigma\mathcal{G}\rangle/\tau^{-1}\mathcal{G}$$
$$= \partial^0 \mathcal{G}\langle\sigma\mathcal{G}\rangle/\mathcal{G} = \dim \mathfrak{M},$$

$$\dim \mathfrak{M}\tau = \partial^0 \mathcal{G}\langle\sigma\tau\mathcal{G}\rangle/\mathcal{G} = \partial^0 \mathcal{G}\langle\sigma\mathcal{G}\rangle/\mathcal{G} = \dim \mathfrak{M},$$

and $\dim \mathfrak{M}^{-1} = \dim \mathfrak{M}$ by (2).

PROPOSITION 13. *If Φ is a nonempty set of algebraic sets in $\mathfrak{G}$ then $\bigcap_{\mathfrak{M}\in\Phi} \mathfrak{M}$ is an algebraic set in $\mathfrak{G}$.*

Proof. If Φ contains only one element the conclusion is obvious. Suppose next that Φ contains precisely two elements, say $\mathfrak{M}_1$ and $\mathfrak{M}_2$, and assume first that $\mathfrak{M}_1$ and $\mathfrak{M}_2$ are irreducible; let σ_1 and σ_2 be generic elements of $\mathfrak{M}_1$ and $\mathfrak{M}_2$ respectively. Letting $\eta_1, \cdots, \eta_n$ be elements of $\mathcal{G}$ such that $\mathcal{G} = \mathcal{F}\langle\eta_1, \cdots, \eta_n\rangle$, we let M_i denote the prime differential ideal in $\mathcal{G}\{y_1, \cdots, y_n\}$ with generic zero $(\sigma_i\eta_1, \cdots, \sigma_i\eta_n)$, $i = 1, 2$. Let $\mathrm{N}_1, \cdots, \mathrm{N}_p$ be the minimal prime differential ideal divisors of the perfect differential ideal $\{\mathrm{M}_1, \mathrm{M}_2\}$ in $\mathcal{G}\{y_1, \cdots, y_n\}$, and let $(\zeta_{i1}, \cdots, \zeta_{in})$ be a generic zero of N_i, $1 \leq i \leq p$. It is clear that $(\zeta_{i1}, \cdots, \zeta_{in})$ is a specialization of $(\sigma_1\eta_1, \cdots, \sigma_1\eta_n)$ and of $(\sigma_2\eta_1, \cdots, \sigma_2\eta_n)$ over $\mathcal{G}$ and a fortiori over $\mathcal{F}$, so that $(\zeta_{i1}, \cdots, \zeta_{in})$ is a specialization of $(\eta_1, \cdots, \eta_n)$ over $\mathcal{F}$. For every $\sigma \in \mathfrak{M}_1 \cap \mathfrak{M}_2$, $(\sigma\eta_1, \cdots, \sigma\eta_n)$ is a zero of $\{\mathrm{M}_1, \mathrm{M}_2\}$ and hence of some N_i and therefore is a specialization of $(\zeta_{i1}, \cdots, \zeta_{in})$ over $\mathcal{G}$ for some i. Let I be the set of all integers i with $1 \leq i \leq p$ such that $\mathfrak{M}_1 \cap \mathfrak{M}_2$ contains an element σ for which $(\sigma\eta_1, \cdots, \sigma\eta_n)$ is a specialization of $(\zeta_{i1}, \cdots, \zeta_{in})$. For each $i \in I$ there exists a $\sigma \in \mathfrak{M}_1 \cap \mathfrak{M}_2$ such that $(\sigma\eta_1, \cdots, \sigma\eta_n)$ is a specialization of $(\zeta_{i1}, \cdots, \zeta_{in})$ over $\mathcal{G}$ and a fortiori over $\mathcal{F}$, so that $(\eta_1, \cdots, \eta_n)$ is a specialization of $(\zeta_{i1}, \cdots, \zeta_{in})$ over $\mathcal{F}$. By the above, $(\zeta_{i1}, \cdots, \zeta_{in})$ is then a generic specialization of $(\eta_1, \cdots, \eta_n)$ over $\mathcal{F}$, so that there exists a unique isomorphism τ_i of $\mathcal{G}$ over $\mathcal{F}$ such that $\tau_i\eta_1 = \zeta_{i1}, \cdots, \tau_i\eta_n = \zeta_{in}$. Let $\mathfrak{N}_i$ be the irreducible set in $\mathfrak{G}$ with generic element τ_i. Each element of $\mathfrak{M}_1 \cap \mathfrak{M}_2$ is a specialization of some τ_i and therefore belongs to $\bigcup_{i\in I} \mathfrak{N}_i$. Conversely, each element of $\bigcup_{i\in I} \mathfrak{N}_i$ is a specialization of some τ_i and hence also of σ_1 and σ_2 and therefore belongs to $\mathfrak{M}_1 \cap \mathfrak{M}_2$. Thus $\mathfrak{M}_1 \cap \mathfrak{M}_2 = \bigcup_{i\in I} \mathfrak{N}_i$, so that $\mathfrak{M}_1 \cap \mathfrak{M}_2$ is algebraic.

Continuing with the case in which Φ consists of $\mathfrak{M}_1$ and $\mathfrak{M}_2$, now abandon the assumption that $\mathfrak{M}_1$ and $\mathfrak{M}_2$ are irreducible. Denoting the components of $\mathfrak{M}_i$ by $\mathfrak{M}_{i1}, \cdots, \mathfrak{M}_{ip_i}$ $(i = 1, 2)$, we have $\mathfrak{M}_1 \cap \mathfrak{M}_2 = \bigcup_{j,k} \mathfrak{M}_{1j} \cap \mathfrak{M}_{2k}$. By what has already been proved, each $\mathfrak{M}_{1j} \cap \mathfrak{M}_{2k}$ is algebraic, so that $\mathfrak{M}_1 \cap \mathfrak{M}_2$ is too. The proposition thus holds when Φ consists of two algebraic sets. The extension to the case in which Φ consists of any finite number of algebraic sets is immediate by induction.

Now let Φ be arbitrary. Let Ψ be the set of all intersections of finite nonempty subsets of Φ. By the case already known, each element of Ψ is an algebraic set in $\mathfrak{G}$. By Proposition 10, Ψ has a minimal element, say $\mathfrak{N}$. It is obvious that $\mathfrak{M} \cap \mathfrak{N} \in \Psi$ for each $\mathfrak{M} \in \Phi$; since $\mathfrak{M} \cap \mathfrak{N} \subseteq \mathfrak{N}$ and $\mathfrak{N}$ is minimal in Ψ this means that $\mathfrak{M} \cap \mathfrak{N} = \mathfrak{N}$, that is $\mathfrak{N} \subseteq \mathfrak{M}$. Therefore $\mathfrak{N} = \bigcap_{\mathfrak{M} \in \Phi} \mathfrak{M}$, and the latter is algebraic.

6. Algebraic groups. By an *algebraic group* in $\mathfrak{G}$ we shall mean a subset of $\mathfrak{G}$ which is a subgroup of $\mathfrak{G}$ and at the same time an algebraic set in $\mathfrak{G}$. If $\mathfrak{H}$ is an algebraic group in $\mathfrak{G}$ and $\mathfrak{H}^*$ is the algebraic set in $\mathfrak{G}^*$ such that $\mathfrak{H} = \mathfrak{G} \cap \mathfrak{H}^*$ then $\mathfrak{H}^*$ is a subgroup of $\mathfrak{G}^*$. To prove this we observe that for two independent generic points σ^*, τ^* of two not necessarily distinct components of $\mathfrak{H}^*$ (see Proposition 11) we have $\sigma^{*-1}\tau^* \in \mathfrak{H}^*$, for otherwise by Proposition 9 we could find elements $\sigma, \tau \in \mathfrak{H}$ such that $\sigma^{-1}\tau \notin \mathfrak{H}$; it follows from Proposition 6 that $\sigma^{-1}\tau \in \mathfrak{H}^*$ whenever $\sigma, \tau \in \mathfrak{H}^*$, so that $\mathfrak{H}^*$ is a group.

THEOREM 1. *Let $\mathfrak{H}$ be an algebraic group in $\mathfrak{G}$. The components of $\mathfrak{H}$ are pairwise disjoint; the component $\mathfrak{H}^0$ of $\mathfrak{H}$ which contains the identity automorphism ι is a normal algebraic subgroup of $\mathfrak{H}$ of finite index; the components of $\mathfrak{H}$ are the cosets of $\mathfrak{H}^0$ in $\mathfrak{H}$.*

Proof. Let $\mathfrak{H}_1, \mathfrak{H}_2$ be components of $\mathfrak{H}$ which contain ι and let σ^*_1, σ^*_2 be independent generic elements of $\mathfrak{H}_1, \mathfrak{H}_2$. Since $\sigma^*_1\sigma^*_2$ belongs to $\mathfrak{H}^*$ it belongs to some component $\mathfrak{H}^*_0$ of $\mathfrak{H}^*$. By Proposition 8, $\sigma^*_1 = \sigma^*_1\iota$ is a specialization of $\sigma^*_1\sigma^*_2$ so that $\mathfrak{H}^*_1 \subseteq \mathfrak{H}^*_0$ and therefore $\mathfrak{H}^*_1 = \mathfrak{H}^*_0$, and similarly $\sigma^*_2 = \iota\sigma^*_2$ is a specialization of $\sigma^*_1\sigma^*_2$ so that $\mathfrak{H}^*_2 \subseteq \mathfrak{H}^*_0$ and $\mathfrak{H}^*_2 = \mathfrak{H}^*_0$; therefore $\mathfrak{H}^*_1 = \mathfrak{H}^*_2$ so that $\mathfrak{H}_1 = \mathfrak{H}_2$. Thus precisely one of the components of $\mathfrak{H}$ contains ι; we denote this component by $\mathfrak{H}^0$. Now let $\mathfrak{H}'$ be any component of $\mathfrak{H}$; if σ' is any element of $\mathfrak{H}'$ then (Proposition 12) $\sigma'^{-1}\mathfrak{H}'$ is an irreducible subset of $\mathfrak{H}$ containing ι so that $\sigma'^{-1}\mathfrak{H}' \subseteq \mathfrak{H}^0$ and $\mathfrak{H}' \subseteq \sigma'\mathfrak{H}^0$. Since (Proposition 12) $\sigma'\mathfrak{H}^0$ is an irreducible subset of $\mathfrak{H}$ and $\mathfrak{H}'$ is a component of $\mathfrak{H}$ this implies that $\mathfrak{H}' = \sigma'\mathfrak{H}^0$. Similarly we find that $\mathfrak{H}' = \mathfrak{H}^0\sigma'$, so that $\sigma'\mathfrak{H}^0 = \mathfrak{H}^0\sigma'$. If we choose $\mathfrak{H}'$ as $\mathfrak{H}^0$ we see that $\sigma\mathfrak{H}^0 = \mathfrak{H}^0$

for every $\sigma \varepsilon \mathfrak{H}^0$ so that $\mathfrak{H}^0$ is a subgroup of $\mathfrak{H}$; since $\sigma'\mathfrak{H}^0 = \mathfrak{H}^0\sigma'$ for every $\sigma' \varepsilon \mathfrak{H}$, $\mathfrak{H}^0$ is a normal subgroup of $\mathfrak{H}$. By the above, the components of $\mathfrak{H}$ are the cosets of $\mathfrak{H}^0$ in $\mathfrak{H}$, and are therefore pairwise disjoint.

COROLLARY 1. *The components of an algebraic group in* $\mathfrak{G}$ *all have the same dimension.*

Proof. This follows from the Theorem and Proposition 12.

By the *dimension* of an algebraic group $\mathfrak{H}$ in $\mathfrak{G}$ we shall mean the dimension of any one of its components. The component containing ι we shall call the *component of the identity* of $\mathfrak{H}$, and we shall always denote it by $\mathfrak{H}^0$.

COROLLARY 2. *An algebraic group in* $\mathfrak{G}$ *is irreducible if and only if it has no algebraic subgroup of finite index* > 1.

Proof. If $\mathfrak{H}$ is not irreducible then $\mathfrak{H}^0$ is an algebraic subgroup of $\mathfrak{H}$ of finite index > 1. Conversely, if $\mathfrak{K}$ is an algebraic subgroup of $\mathfrak{H}$ of finite index > 1 then so is the component of the identity $\mathfrak{K}^0$ of $\mathfrak{K}$; as $\mathfrak{H}$ is the union of the left cosets of $\mathfrak{K}^0$ in $\mathfrak{H}$ and these are irreducible sets (Proposition 12) none of which contains another, $\mathfrak{H}$ is not irreducible.

PROPOSITION 14. *Let* $\mathfrak{h}$ *be a subgroup of* $\mathfrak{G}$ *which is contained in at least one algebraic set in* $\mathfrak{G}$. *Then the intersection* $\mathfrak{H}$ *of all the algebraic sets in* $\mathfrak{G}$ *which contain* $\mathfrak{h}$ *is an algebraic group in* $\mathfrak{G}$.

Proof. By Proposition 13, $\mathfrak{H}$ is an algebraic set in $\mathfrak{G}$. Obviously $\mathfrak{H}$ can be characterized as the smallest algebraic set in $\mathfrak{G}$ which contains $\mathfrak{h}$. If σ is any element of $\mathfrak{h}$ then $\mathfrak{H}\sigma$, which by Proposition 12 is an algebraic set, is obviously the smallest algebraic set containing $\mathfrak{h}\sigma$. Since $\mathfrak{h}\sigma = \mathfrak{h}$ this means that $\mathfrak{H}\sigma = \mathfrak{H}$, so that $\mathfrak{H}\mathfrak{h} = \mathfrak{H}$. Now let σ be any element of $\mathfrak{H}$. Obviously $\sigma\mathfrak{H}$ is the smallest algebraic set containing $\sigma\mathfrak{h}$; but $\sigma\mathfrak{h} \subseteq \mathfrak{H}\mathfrak{h} = \mathfrak{H}$, so that $\sigma\mathfrak{H} \subseteq \mathfrak{H}$. By Proposition 12, $\mathfrak{H}^{-1}$ is algebraic and therefore is the smallest algebraic set containing $\mathfrak{h}^{-1}$; since $\mathfrak{h}^{-1} = \mathfrak{h}$ this means that $\mathfrak{H}^{-1} = \mathfrak{H}$, so that $\mathfrak{H}$ is a group.

PROPOSITION 15. *Let* $\mathfrak{H}$ *be an algebraic group in* $\mathfrak{G}$ *and let* $\mathfrak{h}$ *be a subgroup of* $\mathfrak{H}$ *such that* $\mathfrak{H} - \mathfrak{h}$ *is contained in the union of a finite family of irreducible sets in* $\mathfrak{G}$ *each of dimension* $< \dim \mathfrak{H}$. *Then* $\mathfrak{h} = \mathfrak{H}$.

Proof. Expressing $\mathfrak{H}$ as the union of cosets of $\mathfrak{h}$, $\mathfrak{H} = \bigcup_{i \varepsilon I} \sigma_i \mathfrak{h}$ $(\sigma_{i_0} = \iota)$, we see that if there were more than one coset there would exist an $i_1 \varepsilon I$

with $i_0 \neq i_1$; for this i_1 we would have $\sigma_{i_1}\mathfrak{h} \subseteq \mathfrak{H} - \mathfrak{h}$, so that $\sigma_{i_1}\mathfrak{h}$ would be contained in a finite union of irreducible sets each of dimension $< \dim \mathfrak{H}$, whence (Proposition 12) so would $\mathfrak{h}$. This would imply that the same is true of $\mathfrak{H} = \mathfrak{h} \cup (\mathfrak{H} - \mathfrak{h})$, which is impossible.

PROPOSITION 16. *Let $\mathfrak{H}$ be an algebraic group in $\mathfrak{G}$, $\mathfrak{K}$ an algebraic subgroup of $\mathfrak{H}$, $\mathfrak{N}$ a normal algebraic subgroup of $\mathfrak{H}$. Then $\mathfrak{K}\mathfrak{N}$ is an algebraic group in $\mathfrak{G}$.*

Proof. $\mathfrak{K}\mathfrak{N}$ is a group; we must prove it is algebraic. Let (Proposition 11) $\kappa_1, \cdots, \kappa_r, \nu_1, \cdots, \nu_s$ be independent generic elements of the r components of $\mathfrak{K}$ and the s components of $\mathfrak{N}$, let $\mathfrak{M}_{ij}$ be the irreducible set in $\mathfrak{G}$ with generic element $\kappa_i\nu_j$, and let $\mathfrak{M}$ be the intersection of all the algebraic sets in $\mathfrak{G}$ which contain $\mathfrak{K}\mathfrak{N}$. By Proposition 8, $\mathfrak{K}\mathfrak{N} \subseteq \bigcup \mathfrak{M}_{ij}$, so that (Proposition 14) $\mathfrak{M}$ is an algebraic group and $\mathfrak{M} \subseteq \bigcup \mathfrak{M}_{ij}$. If we had $\kappa_i\nu_j \not\varepsilon \mathfrak{M}^*$ then (Proposition 9) there would exist $\kappa' \varepsilon \mathfrak{K}$, $\nu'_j \varepsilon \mathfrak{N}$ such that $\kappa'_i\nu'_j \not\varepsilon \mathfrak{M}$, which is impossible; it follows that $\mathfrak{M} = \bigcup \mathfrak{M}_{ij}$, so that $\dim \mathfrak{M} \geq \dim \mathfrak{M}_{ij}$. By Corollary 1 to Proposition 9, the set of elements of $\mathfrak{M}_{ij}$ which do not belong to $\mathfrak{K}\mathfrak{N}$ lies in an algebraic set properly contained in $\mathfrak{M}_{ij}$, that is, in a finite union of irreducible sets all of dimension $< \dim \mathfrak{M}_{ij}$. It follows that $\mathfrak{M} - \mathfrak{K}\mathfrak{N}$ is contained in a finite union of irreducible sets each of dimension $< \dim \mathfrak{M}$, so that (Proposition 15) $\mathfrak{K}\mathfrak{N} = \mathfrak{M}$, whence $\mathfrak{K}\mathfrak{N}$ is algebraic.

PROPOSITION 17. *Let $\mathfrak{K}$ be an algebraic subgroup of an algebraic group $\mathfrak{H}$ in $\mathfrak{G}$, and let $N(\mathfrak{K})$ be the normalisor of $\mathfrak{K}$ in $\mathfrak{H}$. Then $N(\mathfrak{K})$ is an algebraic group in $\mathfrak{G}$.*

Proof. The intersection $\mathfrak{N}$ of all the algebraic sets containing $N(\mathfrak{K})$ is an algebraic group (Proposition 14); we shall show that $N(\mathfrak{K}) = \mathfrak{N}$, thereby proving that $N(\mathfrak{K})$ is algebraic. Let $\mathfrak{N}_0$ be any component of $\mathfrak{N}$ and let σ_0 be a generic element of $\mathfrak{N}_0$. Suppose $\sigma_0\mathfrak{K}^*\sigma_0^{-1} \not\subseteq \mathfrak{K}^*$. Using Proposition 9 we see that there exists an element $\tau \varepsilon \mathfrak{K}$ such that $\sigma_0\tau\sigma_0^{-1} \not\varepsilon \mathfrak{K}^*$. Letting $\eta_1, \cdots, \eta_n$ be elements of $\mathfrak{g}$ such that $\mathfrak{g} = \mathfrak{F}\langle\eta_1, \cdots, \eta_n\rangle$, we see that therefore there exists a $Q \varepsilon \mathfrak{g}\{w_1, \cdots, w_n\}$ which vanishes at $(\sigma\eta_1, \cdots, \sigma\eta_n)$ for every $\sigma \varepsilon \mathfrak{K}$ but which does not vanish at $(\sigma_0\tau\sigma_0^{-1}\eta_1, \cdots, \sigma_0\tau\sigma_0^{-1}\eta_n)$. By Corollary 1 to Proposition 9 there exists a $P \varepsilon \mathfrak{g}\{y_1, \cdots, y_n\}$ which does not vanish at $(\sigma_0\eta_1, \cdots, \sigma_0\eta_n)$ such that whenever σ'_0 is a specialization of σ_0 for which $P(\sigma'_0\eta_1, \cdots, \sigma'_0\eta_n) \neq 0$ then $Q(\sigma'_0\tau\sigma'^{-1}_0\eta_1, \cdots, \sigma'_0\tau\sigma'^{-1}_0\eta_n) \neq 0$, that is then $\sigma'_0\tau\sigma'^{-1}_0 \not\varepsilon \mathfrak{K}$, so that $\sigma'_0 \not\varepsilon N(\mathfrak{K})$. Thus $P(\sigma\eta_1, \cdots, \sigma\eta_n) = 0$ for every $\sigma \varepsilon N(\mathfrak{K}) \cap \mathfrak{N}_0$. It easily follows that $N(\mathfrak{K}) \cap \mathfrak{N}_0$ is contained in an algebraic set properly contained in $\mathfrak{N}_0$, so that $N(\mathfrak{K})$ is contained in an algebraic set properly contained in $\mathfrak{N}$. This contradicts the definition of $\mathfrak{N}$, and proves that $\sigma_0\mathfrak{K}^*\sigma_0^{-1} \subseteq \mathfrak{K}^*$. From this it easily follows, using Corollary 1 to Proposition 9, that $\sigma\mathfrak{K}\sigma^{-1} \subseteq \mathfrak{K}$ for all $\sigma \in \mathfrak{N}$ save possibly those in a finite union of irreducible sets all of dimension $< \dim \mathfrak{N}$, that is $\mathfrak{N} - \mathfrak{N}(K)$ is contained in such a union. From Proposition 15 it now follows that $N(\mathfrak{K}) = \mathfrak{N}$.

Chapter III. Galois theory of strongly normal extensions.

We recall that the conditions and conventions set forth at the beginning of Chapter II remain in force in the present chapter.

1. Normal extensions. We recall (Kolchin [3], Chapter III) two definitions: 1) a set of isomorphisms of $\mathcal{G}$ over $\mathcal{F}$ is said to be *abundant* if for every differential field $\mathcal{F}_1$ between $\mathcal{F}$ and $\mathcal{G}$ and every element α in $\mathcal{G}$ but not in $\mathcal{F}_1$ there exists an isomorphism, in the set, which leaves invariant each element of $\mathcal{F}_1$ but which does not leave α invariant; 2) $\mathcal{G}$ is said to be a *normal* extension of $\mathcal{F}$ if the set of all automorphisms of $\mathcal{G}$ over $\mathcal{F}$ is abundant. We now introduce the following definition: $\mathcal{G}$ is said to be a *weakly normal* extension of $\mathcal{F}$ if for every element α in $\mathcal{G}$ but not in $\mathcal{F}$ there exists an automorphism of $\mathcal{G}$ over $\mathcal{F}$ which does not leave α invariant. It is clear that if $\mathcal{G}$ is normal over $\mathcal{F}$ then $\mathcal{G}$ is weakly normal over $\mathcal{F}$, and that $\mathcal{G}$ is normal over $\mathcal{F}$ if and only if $\mathcal{G}$ is weakly normal over every differential field between $\mathcal{F}$ and $\mathcal{G}$. Whether $\mathcal{G}$ can be weakly normal over $\mathcal{F}$ without being normal over $\mathcal{F}$ is an open question.[7]

The following result was proved in Kolchin [3], § 16.

THEOREM 1. *Let $\mathcal{G}$ be a normal extension of $\mathcal{F}$, let $\mathfrak{G}$ be an abundant group of automorphisms of $\mathcal{G}$ over $\mathcal{F}$* (not necessarily the group of all such automorphisms), *and for each differential field $\mathcal{F}_1$ between $\mathcal{F}$ and $\mathcal{G}$ let*

[7] If we relax the conditions on $\mathcal{G}$ and $\mathcal{F}$ by dropping the requirement that every constant in $\mathcal{G}$ belong to $\mathcal{F}$ then the answer to this question is affirmative, as is shown by the following example. Let $\mathcal{F}$ be the field of all algebraic numbers, let θ be a transcendental number, let $\mathcal{G} = \mathcal{F}(\theta)$, and make $\mathcal{F}$ and $\mathcal{G}$ into ordinary differential fields by defining $\delta_1 a = 0$ for every $a \in \mathcal{G}$. The automorphisms of $\mathcal{G}$ over $\mathcal{F}$ may be identified with the fractional linear substitutions $\theta \to (a\theta + b)(c\theta + d)^{-1}$ with $a, b, c\, d \in \mathcal{F}$ and $ad - bc \neq 0$. The only elements of $\mathcal{G}$ invariant under $\theta \to \theta + 1$ are those of $\mathcal{F}$, so that $\mathcal{G}$ is weakly normal over $\mathcal{F}$ (in the relaxed sense). But $\theta \notin \mathcal{F}(\theta^3 + \theta)$, and the only fractional linear substitution leaving $\theta^3 + \theta$ invariant is $\theta \to \theta$, which leaves θ invariant, too; therefore $\mathcal{G}$ is not normal over $\mathcal{F}$ (in the relaxed sense).

$\mathfrak{G}(\mathcal{F}_1)$ *denote the group of all elements of* $\mathfrak{G}$ *which leave invariant each element of* $\mathcal{F}_1$. *Then for each* $\mathcal{F}_1$ *the field of all elements of* $\mathcal{G}$ *invariant under every element of* $\mathfrak{G}(\mathcal{F}_1)$ *is* $\mathcal{F}_1$, *so that* $\mathcal{F}_1 \to \mathfrak{G}(\mathcal{F}_1)$ *is a one-to-one mapping of the set of all differential fields between* $\mathcal{F}$ *and* $\mathcal{G}$ *onto a certain set of subgroups of* $\mathfrak{G}$. *A necessary and sufficient condition that* $\mathfrak{G}(\mathcal{F}_1)$ *be a normal subgroup of* $\mathcal{G}$ *is that* $\sigma\mathcal{F}_1 \subseteq \mathcal{F}_1$ *for every* $\sigma \in \mathfrak{G}$, *and when this condition is satisfied then the mapping which to each element of* $\mathfrak{G}$ *assigns its restriction to* $\mathcal{F}_1$ *is a homomorphism with kernel* $\mathfrak{G}(\mathcal{F}_1)$ *of* $\mathfrak{G}$ *onto an abundant group of automorphisms of* $\mathcal{F}_1$ *over* $\mathcal{F}$.

We shall give an example which will show that even when $\mathfrak{G}$ is taken as the group of *all* automorphisms of $\mathcal{G}$ over $\mathcal{F}$ and $\mathfrak{G}(\mathcal{F}_1)$ is a normal subgroup of $\mathfrak{G}$ (so that $\mathcal{F}_1$ is a normal extension of $\mathcal{F}$) the factor group $\mathfrak{G}/\mathfrak{G}(\mathcal{F}_1)$ need not be isomorphic to the group of all automorphisms of $\mathcal{F}_1$ over $\mathcal{F}$, but merely to an abundant subgroup thereof; the example will also show that it is possible for an intermediate differential field $\mathcal{F}_1$ that $\mathcal{F}_1$ be a normal extension of $\mathcal{F}$ and $\mathfrak{G}(\mathcal{F}_1)$ fail to be a normal subgroup of $\mathfrak{G}$. To discuss this example we shall need the following lemma.

LEMMA 1. *Let* p, q *be integers not both zero, let* a, b, c, d *be nonzero elements of* $\mathcal{C}$, *let* X_1, X_2, X_3, X_4 *be indeterminates, and let* $f \in \mathcal{C}(X_1, X_2, X_3, X_4)$. *If* f *is invariant under the substitution*

$$(X_1, X_2, X_3, X_4) \to (X_1 + d, aX_2, bX_3, cX_2^pX_3^qX_4) \tag{1}$$

then $f \in \mathcal{C}(X_2, X_3)$.

Proof. We suppose as we may that $f \neq 0$; then f is uniquely expressible in the form

$$f = \sum_{i=0}^{m} g_iX_4^i / \sum_{j=0}^{m} h_jX_4^j \,(g_i, h_j \in \mathcal{C}(X_1, X_2, X_3), g_m \neq 0, h_n = 1),$$

where $\sum g_iX_4^i$ and $\sum h_jX_4^j$ have no common factor as polynomials in X_4. Because of the invariance of f under the indicated substitution we find that

$$g_i(X_1 + d, aX_2, bX_3) = g_i(X_1, X_2, X_3)(cX_2^pX_3^q)^{n-i},$$

$$h_j(X_1 + d, aX_2, bX_3) = h_j(X_1, X_2, X_3)(cX_2^pX_3^q)^{n-j}.$$

But the degree in X_2 of the numerator of any nonzero element $\phi \in \mathcal{C}(X_1, X_2, X_3)$ minus the degree in X_2 of the denominator of ϕ is obviously invariant under the substitution

$$(X_1, X_2, X_3) \to (X_1 + d, aX_2, bX_3); \tag{2}$$

therefore $p(n-i)=0$ for every i such that $g_i \neq 0$. Similarly, regarding degrees in X_3 instead of X_2, we see that $q(n-i)=0$ whenever $g_i \neq 0$. Since p and q are not both 0 this implies that $g_i = 0$ whenever $i \neq n$. In the same way we find that $h_j = 0$ whenever $j \neq n$. Therefore $m = n = 0$, and $f \in \mathcal{C}(X_1, X_2, X_3)$; thus we may write $f = f(X_1, X_2, X_3)$, and f is invariant under the substitution (2).

Now the set $\mathfrak{H}$ of all matrices

$$\tau(\alpha, \beta, \delta) = \begin{bmatrix} 1 & \delta & 0 & 0 \\ 0 & 1 & 0 & 0 \\ 0 & 0 & \alpha & 0 \\ 0 & 0 & 0 & \beta \end{bmatrix} \quad (\alpha, \beta, \delta \in \mathcal{C}, \alpha\beta \neq 0)$$

such that $f(X_1 + \delta, \alpha X_2, \beta X_3) = f(X_1, X_2, X_3)$, is an algebraic matic group, and $\tau(a, b, d) \in \mathfrak{H}$. It follows (Kolchin [5], § 3, Lemma 1) that $\tau(1, 1, d) \in \mathfrak{H}$, that is, $f(X_1 + d, X_2, X_3) = f(X_1, X_2, X_3)$. Since $d \neq 0$ it is easy to conclude that f is free of X_1, that is $f \in \mathcal{C}(X_2, X_3)$.

Example. Let $\mathcal{F}$ be the field of all complex numbers and let $\mathcal{G}$ be the field $\mathcal{F}(x, e^x, e^{ix}, e^{x^2})$ obtained by the field adjunction to $\mathcal{F}$ of the functions x, e^x, e^{ix}, e^{x^2} of the complex variable x ($i = \sqrt{-1}$); with respect to the operator d/dx, $\mathcal{G}$ is an ordinary differential field with field of constants $\mathcal{F}$.

We shall first show that if $\mathcal{F}_1$ is a differential field between $\mathcal{F}$ and $\mathcal{G}$ then either $\mathcal{F}_1$ is between $\mathcal{F}\langle x\rangle$ and $\mathcal{G}$ or else $\mathcal{F}_1$ is between $\mathcal{F}$ and $\mathcal{F}\langle e^x, e^{ix}\rangle$. Indeed, suppose $x \notin \mathcal{F}_1$, and let $\theta \in \mathcal{F}_1$; we must prove that $\theta \in \mathcal{F}\langle e^x, e^{ix}\rangle$. Since $x \notin \mathcal{F}_1$ there exists an isomorphism of $\mathcal{G}$ over $\mathcal{F}_1$ under which x is not invariant; it easily follows that there exist nonzero complex numbers a_0, b_0, c_0, d such that $(x + d, a_0 e^x, b_0 e^{ix}, c_0 e^{(x+d)^2})$ is a generic specialization of $(x, e^x, e^{ix}, e^{x^2})$ over $\mathcal{F}_1$. Letting $c = c_0 e^{d^2}$ and letting f be an element of $\mathcal{F}(X_1, X_2, X_3, X_4)$ such that $\theta = f(x, e^x, e^{ix}, e^{x^2})$, we see that

$$f(x + d, a_0 e^x, b_0 e^{ix}, c e^{2dx} e^{x^2}) = f(x, e^x, {}^{ix}, e^{x^2}). \tag{3}$$

If f were not free of X_4 this would mean that $e^x, e^{ix}, e^{2dx}, e^{x^2}$ are algebraically dependent over $\mathcal{F}(x)$, and this would imply that $x, ix, 2dx, x^2$ are linearly dependent over the ring of integers, that is, there would exist integers p, q, r with $r \neq 0$ and p, q not both 0 such that $2d = (p + iq)/r$. Choosing complex numbers a, b such that $a_0 = a^r$, $b_0 = b^r$ we would then have

$$f(x + d, a^r e^x, b^r e^{ix}, c e^{[(p+iq)/r]x} e^{x^2}) = f(x, e^x, e^{ix}, e^{x^2}).$$

Since $e^{x/r}, e^{ix/r}, e^{x^2}$ are algebraically independent over $\mathcal{F}(x)$, this would imply

9

that $f(X_1, X_2{}^r, X_3{}^r, X_4)$ is invariant under the substitution (1), so that by Lemma 1 we would have $f \in \mathfrak{F}(X_2, X_3)$. It follows that f is free of X_4. Therefore we may write (3) in the form

$$f(x+d, a_0e^x, b_0{}^{ix}, c_0e^xe^{x^2}) = f(x, e^x, e^{ix}, e^{x^2}),$$

whence f is invariant under the substitution (1) (with $(a_0, b_0, c_0, d, 1, 0)$ instead of (a, b, c, d, p, q)). It follows from Lemma 1 that $f \in \mathfrak{F}(X_2, X_3)$, so that $\theta \in \mathfrak{F}\langle e^x, e^{ix}\rangle$. This completes the proof that $\mathfrak{F}_1$ is either between $\mathfrak{F}\langle x\rangle$ and $\mathfrak{G}$ or between $\mathfrak{F}$ and $\mathfrak{F}\langle e^x, e^{ix}\rangle$.

It is easy to see that for each choice of nonzero complex numbers a, b, c and integers p, q there exists a unique automorphism $\sigma = \sigma(p, q, a, b, c)$ of $\mathfrak{G}$ over $\mathfrak{F}$ such that $\sigma x = x + \frac{1}{2}(p + iq)$, $\sigma e^x = ae^x$, $\sigma e^{ix} = be^{ix}$, $\sigma e^{x^2} = ce^{px}e^{qix}e^{x^2}$, and every automorphism of $\mathfrak{G}$ over $\mathfrak{F}$ is of this form; let us denote the group of all these automorphisms by $\boldsymbol{\mathfrak{G}}$.

Let $\mathfrak{F}_1$ be between $\mathfrak{F}$ and $\mathfrak{G}$. If $\mathfrak{F}_1$ is between $\mathfrak{F}\langle x\rangle$ and $\mathfrak{G}$ then, since $\mathfrak{G} = \mathfrak{F}\langle x\rangle\langle e^x, e^{ix}, e^{x^2}\rangle$ is obviously a Picard-Vessiot extension of $\mathfrak{F}\langle x\rangle$, $\mathfrak{G}$ is normal over $\mathfrak{F}\langle x\rangle$ and therefore weakly normal over $\mathfrak{F}_1$. Suppose, then, that $\mathfrak{F}_1$ is between $\mathfrak{F}$ and $\mathfrak{F}\langle e^x, e^{ix}\rangle$, and let α be an element of $\mathfrak{G}$ but not of $\mathfrak{F}_1$. If $\alpha \notin \mathfrak{F}_1\langle e^x, e^{ix}\rangle$ then, by Lemma 1 and the algebraic independence of e^x, e^{ix}, e^{x^2} over $\mathfrak{F}(x)$, the automorphism $\sigma(1, 0, 1, 1, 1)$ does not leave α invariant but obviously leaves each element of $\mathfrak{F}_1$ invariant; suppose, then, that $\alpha \in \mathfrak{F}\langle e^x, e^{ix}\rangle$. Since $\mathfrak{F}\langle e^x, e^{ix}\rangle$ is obviously a Picard-Vessiot extension of $\mathfrak{F}$, there exists an automorphism σ_0 of $\mathfrak{F}\langle e^x, e^{ix}\rangle$ over $\mathfrak{F}_1$ such that $\sigma_0\alpha \neq \alpha$; now there exist nonzero complex numbers a, b such that $\sigma_0 e^x = ae^x$, $\sigma_0 e^{ix} = be^{ix}$. It is clear that the element $\sigma(0, 0, a, b, 1)$ of $\boldsymbol{\mathfrak{G}}$ is an extension of σ_0, and therefore does not leave α invariant but leaves each element of $\mathfrak{F}_1$ invariant. Thus again $\mathfrak{G}$ is a weakly normal extension of $\mathfrak{F}_1$. Since $\mathfrak{F}_1$ is arbitrary between $\mathfrak{F}$ and $\mathfrak{G}$, $\mathfrak{G}$ is a normal extension of $\mathfrak{F}$.

Now let $\mathfrak{F}_2 = \mathfrak{F}\langle x, e^x, e^{x^2}\rangle$, $\mathfrak{F}_3 = \mathfrak{F}\langle x\rangle$. It is clear that $\sigma\mathfrak{F}_3 = \mathfrak{F}_3$ for every $\sigma = \sigma(p, q, a, b, c) \in \boldsymbol{\mathfrak{G}}$, so that $\boldsymbol{\mathfrak{G}}(\mathfrak{F}_3)$ is a normal subgroup of $\boldsymbol{\mathfrak{G}}$; the restrictions to $\mathfrak{F}_3$ of the elements of $\boldsymbol{\mathfrak{G}}$ are the automorphisms $\sigma(d)$ of $\mathfrak{F}_3$ over $\mathfrak{F}$ defined by

$$\sigma(d)x = x + d,$$

with d an arbitrary number of the form $\frac{1}{2}(p + iq)$, p and q being integers, and the group of all these restrictions is not the group of all automorphisms of $\mathfrak{F}_3$ over $\mathfrak{F}$, that is, is not the group of all automorphisms $\sigma(d)$ with d an arbitrary complex number, but is merely an abundant subgroup thereof.

Finally, there exists an automorphism $\sigma \in \boldsymbol{\mathfrak{G}}$ such that $\sigma F_2 \not\subseteq F_2$ (for example $\sigma = \sigma(0, 1, 1, 1, 1))$, so that $\mathfrak{G}(\mathcal{F}_2)$ is not a normal subgroup of $\mathfrak{G}$; nevertheless $\mathcal{F}_2$ is a normal extension of $\mathcal{F}$, as can be shown by the method used to prove that $\mathcal{G}$ is a normal extension of $\mathcal{F}$.

2. Strongly normal extensions. We make the following definition: $\mathcal{G}$ is said to be a *strongly normal* extension of $\mathcal{F}$ if every isomorphism of $\mathcal{G}$ over $\mathcal{F}$ is strong. It is obvious that if $\mathcal{G}$ is strongly normal over $\mathcal{F}$ then $\mathcal{G}$ is strongly normal over every differential field between $\mathcal{F}$ and $\mathcal{G}$.

PROPOSITION 1. *A necessary and sufficient condition that $\mathcal{G}$ be a strongly normal extension of $\mathcal{F}$ is that $\chi\mathcal{G} \subseteq \mathcal{G}\langle\mathcal{C}^*\rangle$ for every isolated isomorphism χ of $\mathcal{G}$ over $\mathcal{F}$.*

Proof. That the condition is necessary is obvious; suppose then that the condition is satisfied. The differential field $\chi\mathcal{G}$, isomorphic with $\mathcal{G}$, also must have this property, so that $\mathcal{G} = \chi^{-1}(\chi\mathcal{G}) \subset (\chi\mathcal{G})\langle\mathcal{C}^*\rangle$, and χ is strong. It follows (Chapter II, Propositions 2 and 5) that every isomorphism of $\mathcal{G}$ over $\mathcal{F}$ is strong, so that $\mathcal{G}$ is strongly normal over $\mathcal{F}$.

PROPOSITION 2. *If $\mathcal{G}$ is strongly normal over $\mathcal{F}$ then $\mathcal{G}$ is normal over $\mathcal{F}$.*

Proof. Let $\mathcal{F}_1$ be a differential field between $\mathcal{F}$ and $\mathcal{G}$, and let α be an element of $\mathcal{G}$ not in $\mathcal{F}_1$. There exists an isomorphism σ of $\mathcal{G}$ over $\mathcal{F}_1$ such that $\sigma\alpha - \alpha \neq 0$, and by hypothesis σ is strong. By Corollary 2 to Proposition 9 of Chapter II there exists an automorphism τ of $\mathcal{G}$ over $\mathcal{F}_1$ which is a specialization of σ such that $\tau\alpha - \alpha \neq 0$. Therefore $\mathcal{G}$ is normal over $\mathcal{F}$.

That $\mathcal{G}$ can be normal over $\mathcal{F}$ without being strongly normal over $\mathcal{F}$ is shown by the example of § 1; using the notation of that example we easily see that for each complex number d there exists a unique isomorphism σ of $\mathcal{G}$ over $\mathcal{F}$ such that $\sigma x = x + d$, $\sigma e^x = e^d e^x$, $\sigma e^{ix} = e^{id} e^{ix}$, $\sigma e^{x^2} = e^{d^2} e^{2dx} e^{x^2}$, and it is obvious that if $2d$ is not a gaussian integer then $e^{2dx} \notin \mathcal{G} = \mathcal{F}\langle x, e^x, e^{ix}, e^{x^2}\rangle$ so that $\mathcal{G}\langle\sigma\mathcal{G}\rangle \not\subseteq \mathcal{G}\langle\mathcal{C}_\sigma\rangle = \mathcal{G}$, whence $\mathcal{G}$ is not strongly normal over $\mathcal{F}$.

PROPOSITION 3. *If $\mathcal{G}$ is a strongly normal extension of $\mathcal{F}$ then the group of all automorphisms of $\mathcal{G}$ over $\mathcal{F}$ is algebraic.*

Proof. By Proposition 1 of Chapter II there exists a finite number of isomorphisms $\chi_1, \cdots, \chi_h$ of $\mathcal{G}$ over $\mathcal{F}$ such that every isomorphism of $\mathcal{G}$ over $\mathcal{F}$ is a specialization of one of these; by hypothesis χ_i is strong and is therefore a generic element of an irreducible set $\mathfrak{M}_i$ of automorphisms of $\mathcal{G}$ over $\mathcal{F}$. The group of all automorphisms of $\mathcal{G}$ over $\mathcal{F}$ is $\bigcup \mathfrak{M}_i$ and therefore is algebraic.

3. The fundamental theorems. Whenever $\mathcal{G}$ is strongly normal over $\mathcal{F}$ we shall denote the group of all automorphisms of $\mathcal{G}$ over $\mathcal{F}$ by $\mathfrak{G}$, and for each intermediate differential field $\mathcal{F}_1$ we shall denote the group of all automorphisms of $\mathcal{G}$ over $\mathcal{F}_1$ by $\mathfrak{G}(\mathcal{F}_1)$.

THEOREM 2. *If $\mathcal{G}$ is strongly normal over $\mathcal{F}$, then the mapping $\mathcal{F}_1 \to \mathfrak{G}(\mathcal{F}_1)$ is one-to-one from the set of all differential fields between $\mathcal{F}$ and $\mathcal{G}$ onto the set of all algebraic groups in $\mathfrak{G}$, and has the property that* $\dim \mathfrak{G}(\mathcal{F}_1) = \partial^0 \mathcal{G}/\mathcal{F}_1$.

Proof. By Proposition 2 and Theorem 1 the mapping is one-to-one, and by Proposition 3 $\mathfrak{G}(\mathcal{F}_1)$ is algebraic; a generic element of a component of $\mathfrak{G}(\mathcal{F}_1)$ is an isolated isomorphism of $\mathcal{G}$ over $\mathcal{F}_1$ whence (Chapter II, Proposition 2) $\dim \mathfrak{G}(\mathcal{F}_1) = \partial^0 \mathcal{G}/\mathcal{F}_1$. It remains to prove the mapping is onto. To this end let $\mathfrak{G}_1$ be any algebraic group in $\mathfrak{G}$, and let $\mathcal{F}_1$ be the differential field of invariants of $\mathfrak{G}_1$; we shall show that $\mathfrak{G}_1 = \mathfrak{G}(\mathcal{F}_1)$, thereby proving that $\mathfrak{G}_1$ is algebraic and completing the proof of the theorem. Now, it is obvious that $\mathfrak{G}_1 \subseteq \mathfrak{G}(\mathcal{F}_1)$. Suppose that $\mathfrak{G}_1 \neq \mathfrak{G}(\mathcal{F}_1)$. Then, if $\eta_1, \cdots, \eta_n$ are elements such that $\mathcal{G} = \mathcal{F}\langle \eta_1, \cdots, \eta_n \rangle$, there exists a differential polynomial in $\mathcal{G}\{y_1, \cdots, y_n\}$ which vanishes at

$$(\sigma\eta_1, \cdots, \sigma\eta_n) \tag{4}$$

for every $\sigma \varepsilon \mathfrak{G}_1$ but not for every $\sigma \varepsilon \mathfrak{G}(\mathcal{F}_1)$; of all such differential polynomials let F be one with a minimal number of terms, and assume without loss of generality that one of the coefficients in F is 1. Let τ be any element of $\mathfrak{G}_1$ and let F_τ be the differential polynomial obtained when each coefficient ϕ in F is replaced by $\tau\phi$. Then

$$F_\tau(\sigma\eta_1, \cdots, \sigma\eta_n) = \tau F(\tau^{-1}\sigma\eta_1, \cdots, \tau^{-1}\sigma\eta_n) = 0$$

for every $\sigma \varepsilon \mathfrak{G}_1$, so that $F - F_\tau$ vanishes at (4) for every $\sigma \varepsilon \mathfrak{G}_1$. Now $F - F_\tau$ has fewer terms than F has, so that $F - F_\tau$ must vanish at (4) for every $\sigma \varepsilon \mathfrak{G}(\mathcal{F}_1)$. If $F - F_\tau$ were not 0 there would exist an element $\gamma \varepsilon \mathcal{G}$ such that $F - \gamma(F - F_\tau)$ has fewer terms than F has; but $F - \gamma(F - F_\tau)$ obviously vanishes at (4) for every $\sigma \varepsilon \mathfrak{G}_1$ but not for every $\sigma \varepsilon \mathfrak{G}(\mathcal{F}_1)$, and therefore can not have fewer terms that F has. It follows that $F - F_\tau = 0$. Since this is true for every $\tau \varepsilon \mathfrak{G}_1$, each coefficient in F belongs to F_1. From this it follows that F vanishes at (4) for every $\sigma \in \mathfrak{G}(\mathcal{F}_1)$. This contradiction proves that $\mathfrak{G}_1 = \mathfrak{G}(\mathcal{F}_1)$, and completes the proof of the theorem.

THEOREM 3. *If $\mathcal{G}$ is strongly normal over $\mathcal{F}$, $\mathcal{F}_1$ is a differential field between $\mathcal{F}$ and $\mathcal{G}$, and σ_1 is an isomorphism of $\mathcal{F}_1$ over $\mathcal{F}$ into $\mathcal{G}$, then there exists an automorphism $\sigma \in \mathfrak{G}$ which is an extension of σ_1.*

Proof. σ_1 can be extended to an isomorphism σ' of $\mathcal{G}$ over $\mathcal{F}$. By Corollary 2 to Proposition 9 of Chapter II there exist an automorphism σ of $\mathcal{G}$ which is a specialization of σ' and which therefore coincides with σ_1 on $\mathcal{F}_1$.

THEOREM 4. *Let $\mathcal{G}$ be a strongly normal extension of $\mathcal{F}$ and let $\mathcal{F}_1$ be a differential field between $\mathcal{F}$ and $\mathcal{G}$. Then the following five conditions are equivalent.* 1) *$\mathcal{F}_1$ is strongly normal over $\mathcal{F}$*; 2) *$\mathcal{F}_1$ is normal over $\mathcal{F}$*; 3) *$\mathcal{F}_1$ is weakly normal over $\mathcal{F}$*; 4) *$\sigma\mathcal{F}_1 \subseteq \mathcal{F}_1$ for every $\sigma \in \mathfrak{G}$*; 5) *$\mathfrak{G}(\mathcal{F}_1)$ is a normal subgroup of $\mathfrak{G}$. When these conditions are satisfied then the mapping, which to each $\sigma \in \mathfrak{G}$ assigns the restriction of σ to $\mathcal{F}_1$, is a homomorphism with kernel $\mathfrak{G}(\mathcal{F}_1)$ of $\mathfrak{G}$ onto the group of all automorphisms of $\mathcal{F}_1$ over $\mathcal{F}$.*

Proof. We already know that 1) implies 2) and that 2) implies 3); also, by Theorem 1, 4) is equivalent to 5). To prove the equivalence of the five conditions it suffices to show that 3) implies 5) and that 4) implies 1). To settle the first point suppose that $\mathfrak{G}(\mathcal{F}_1)$ is not a normal subgroup of $\mathfrak{G}$, so that the normalisor $N(\mathfrak{G}(\mathcal{F}_1)) \neq \mathfrak{G}$; by Proposition 17 of Chapter II, $N(\mathfrak{G}(\mathcal{F}_1))$ is an algebraic group in $\mathfrak{G}$, so that by Theorem 2 there exists a differential field $\mathcal{F}_2$ between $\mathcal{F}$ and $\mathcal{F}_1$ with $\mathcal{F}_2 \neq \mathcal{F}$, such that $N(\mathfrak{G}(\mathcal{F}_1)) = \mathfrak{G}(\mathcal{F}_2)$. Let α be any element of $\mathcal{F}_2$ not in $\mathcal{F}$. If σ_1 is any automorphism of $\mathcal{F}_1$ over $\mathcal{F}$ then (Theorem 3) σ_1 can be extended to an automorphism $\sigma \in \mathfrak{G}$; since $\sigma\mathcal{F}_1 = \sigma_1\mathcal{F}_1 = \mathcal{F}_1$, we see that $\tau\sigma\beta = \sigma\beta$ for every $\beta \in \mathcal{F}_1$ and every $\tau \in \mathfrak{G}(\mathcal{F}_1)$, that is $\sigma^{-1}\tau\sigma\beta = \beta$, so that $\sigma^{-1}\tau\sigma \in \mathfrak{G}(\mathcal{F}_1)$, whence $\sigma \in N(\mathfrak{G}(\mathcal{F}_1)) = \mathfrak{G}(\mathcal{F}_2)$. Therefore $\sigma_1\alpha = \sigma\alpha = \alpha$, that is, every automorphism of $\mathcal{F}_1$ over $\mathcal{F}$ leaves α invariant, so that $\mathcal{F}_1$ is not weakly normal over $\mathcal{F}$. Thus, if $\mathcal{F}_1$ is weakly normal over $\mathcal{F}$ then $\mathfrak{G}(\mathcal{F}_1)$ is a normal subgroup of $\mathfrak{G}$.

To settle the second point, suppose that $\mathcal{F}_1$ is not strongly normal over $\mathcal{F}$. Then (Proposition 1) there exists an isomorphism σ_1 of $\mathcal{F}_1$ over $\mathcal{F}$ such that $\sigma_1\mathcal{F}_1 \not\subseteq \mathcal{F}_1\langle\mathcal{C}^*\rangle$, and σ_1 can be extended to an isomorphism σ of $\mathcal{G}$ over $\mathcal{F}$. Let θ be an element of $\sigma_1\mathcal{F}_1 = \sigma\mathcal{F}_1$ which does not belong to

$\mathcal{F}_1\langle\mathcal{C}^*\rangle$. We claim there exists a $\tau \in \mathfrak{G}(\mathcal{F}_1)$ such that $\tau\theta \neq \theta$. Indeed, since $\theta \in \sigma\mathcal{F}_1 \subseteq \mathcal{G}\langle\sigma\mathcal{G}\rangle = \mathcal{G}\langle\mathcal{C}\sigma\rangle$, we may write

$$\theta B(c_1, \cdots, c_r) = \sum_{i=0}^{s-1} A_i(c_1, \cdots, c_r) d^i,$$

where $B, A_0, \cdots, A_{s-1}$ are polynomials in $\mathcal{G}[u_1, \cdots, u_r]$ without common divisor, one of the coefficients in B is 1, $c_1, \cdots, c_r$ are elements of $\mathcal{C}\sigma$ algebraically independent over $\mathcal{C}$ (and hence over $\mathcal{G}$), and d is an element of $\mathcal{C}\sigma$ which is algebraic of some degree s over $\mathcal{G}\langle c_1, \cdots, c_r\rangle$. If $\tau \in \mathfrak{G}(\mathcal{F}_1)$ has the property that $\tau\theta = \theta$ then,

$$\sum_{i=0}^{s-1} (B_\tau(c_1, \cdots, c_r) A_i(c_1, \cdots, c_r) - B(c_1, \cdots, c_r) A_{i\tau}(c_1, \cdots, c_r)) d^i = 0$$

(where in general for any polynomial C we denote by C_τ the polynomial obtained upon replacing each coefficient in C by its image under τ), whence $B_\tau A_i = B A_{i\tau}$ $(0 \leq i \leq s-1)$. Because $B, A_0, \cdots, A_{s-1}$ have no common divisor and one of the coefficients in B is 1, it follows that $B_\tau = B$, $A_{i\tau} = A_i$ $(0 \leq i \leq s-1)$. Therefore if θ were invariant under every $\tau \in \mathfrak{G}(\mathcal{F}_1)$ then so would every coefficient in B and each A_i, and these coefficients would all belong to $\mathcal{F}_1$, contradicting the fact that $\theta \notin \mathcal{F}_1\langle\mathcal{C}^*\rangle$. This establishes our claim that for some $\tau \in \mathfrak{G}(\mathcal{F}_1)$ we have $\tau\theta \neq \theta$.

Now $\theta \in \sigma\mathcal{F}_1$, so that there exists an element $\zeta \in \mathcal{F}_1$ such that $\theta = \sigma\zeta$. For this ζ and the above τ we have $\tau\sigma\zeta \neq \sigma\zeta$. By Chapter II, Proposition 9, there exists an automorphism σ_0 of $\mathcal{G}$ which is a specialization of σ for which $\tau\sigma_0\zeta \neq \sigma_0\zeta$. Since $\tau \in \mathfrak{G}(\mathcal{F}_1)$ this means that $\sigma_0\zeta \notin \mathcal{F}_1$ and since $\zeta \in \mathcal{F}_1$ this shows that $\sigma_0\mathcal{F}_1 \not\subseteq \mathcal{F}_1$. We have thus shown that if $\mathcal{F}_1$ is not strongly normal over $\mathcal{F}$ then $\sigma_0\mathcal{F}_1 \not\subseteq \mathcal{F}_1$ for some $\sigma_0 \in \mathfrak{G}$, so that 4) implies 1). This completes the proof that the five conditions are equivalent.

When these conditions are satisfied then (Theorem 1) the mapping which to each automorphism in $\mathfrak{G}$ assigns its restriction to $\mathcal{F}_1$ is a homomorphism with kernel $\mathfrak{G}(\mathcal{F}_1)$ of $\mathfrak{G}$ *into* the group of all automorphisms of $\mathcal{F}_1$ over $\mathcal{F}$. That this homomorphism is *onto* follows from Theorem 3.

COROLLARY. *If $\mathcal{G}$ and $\mathcal{H}$ are strongly normal extension of $\mathcal{F}$ such that the field of constants of $\mathcal{F}\langle\mathcal{G}, \mathcal{H}\rangle$ is $\mathcal{C}$, then $\mathcal{F}\langle\mathcal{G}, \mathcal{H}\rangle$ and $\mathcal{G} \cap \mathcal{H}$ are strongly normal over $\mathcal{F}$.*

Proof. Let σ be any isomorphism of $\mathcal{F}\langle\mathcal{G}, \mathcal{H}\rangle$ over $\mathcal{F}$; the restrictions of σ to $\mathcal{G}$ and to $\mathcal{H}$ are isomorphisms of $\mathcal{G}$ and of $\mathcal{H}$ over $\mathcal{F}$ and, since $\mathcal{G}$ and $\mathcal{H}$ are strongly normal over $\mathcal{F}$, we have

$$\begin{aligned}\sigma(\mathcal{F}\langle\mathcal{G}, \mathcal{H}\rangle) &= \mathcal{F}\langle\sigma\mathcal{G}, \sigma\mathcal{H}\rangle \\ &\subseteq \mathcal{F}\langle\mathcal{G}\langle\mathcal{C}^*\rangle, \mathcal{H}\langle\mathcal{C}^*\rangle\rangle = \mathcal{F}\langle\mathcal{G}, \mathcal{H}\rangle\langle\mathcal{C}^*\rangle.\end{aligned}$$

Therefore (Proposition 1) $\mathcal{F}\langle\mathcal{G},\mathcal{H}\rangle$ is strongly normal over $\mathcal{F}$. Now let τ be any automorphism of $\mathcal{F}\langle\mathcal{G},\mathcal{H}\rangle$ over $\mathcal{F}$. Since $\mathcal{G}$ and $\mathcal{H}$ are both strongly normal over $\mathcal{F}$ we see by Theorem 4 that $\tau\mathcal{G}\subseteq\mathcal{G}$ and $\tau\mathcal{H}\subseteq\mathcal{H}$, whence $\tau(\mathcal{G}\cap\mathcal{H})\subseteq\mathcal{G}\cap\mathcal{H}$, so that (Theorem 4) $\mathcal{G}\cap\mathcal{H}$ is strongly normal over $\mathcal{F}$.

THEOREM 5. *Let $\mathfrak{m}$ be a set of elements such that the field of constants of $\mathcal{G}\langle\mathfrak{m}\rangle$ is $\mathcal{C}$, let $\mathcal{F}^\dagger=\mathcal{F}\langle\mathfrak{m}\rangle$ and $\mathcal{G}^\dagger=\mathcal{G}\langle\mathfrak{m}\rangle$, and denote the group of all automorphisms of $\mathcal{G}^\dagger$ over $\mathcal{F}^\dagger$ by $\mathfrak{G}^\dagger$. If $\mathcal{G}$ is strongly normal over $\mathcal{F}$ then $\mathfrak{G}^\dagger$ is strongly normal over $\mathcal{F}^\dagger$, and the mapping which to each $\sigma^\dagger\in\mathfrak{G}^\dagger$ assigns the restriction σ of $\sigma^\dagger$ to $\mathcal{G}$ is an isomorphism of $\mathfrak{G}^\dagger$ onto $\mathfrak{G}(\mathcal{F}^\dagger\cap\mathcal{G})$ which maps every algebraic group $\mathfrak{M}^\dagger$ in $\mathfrak{G}^\dagger$ onto an algebraic group $\mathfrak{M}$ in $\mathfrak{G}$ of the same dimension; if $\mathfrak{M}^\dagger$ is irreducible then so is $\mathfrak{M}$.*

Proof. Let $\rho^\dagger$ be any isomorphism of $\mathcal{G}^\dagger$ over $\mathcal{F}^\dagger$ and let ρ be the restriction of $\rho^\dagger$ to $\mathcal{G}$. Then ρ is an isomorphism of $\mathcal{G}$ over $\mathcal{F}$, and

$$\mathcal{G}^\dagger\langle\rho^\dagger\mathcal{G}^\dagger\rangle=\mathcal{G}\langle\mathfrak{m}\rangle\langle(\rho\mathcal{G})\langle\mathfrak{m}\rangle\rangle=\mathcal{G}\langle\mathcal{C}_\rho\rangle\langle\mathfrak{m}\rangle=\mathcal{G}^\dagger\langle\mathcal{C}_\rho\rangle;$$

it follows (Proposition 1) that $\mathcal{G}^\dagger$ is strongly normal over $\mathcal{F}^\dagger$, and also (Corollary 5 to Proposition 3 of Chapter 1) that the field of constants of $\mathcal{G}^\dagger\langle\rho^\dagger\mathcal{G}^\dagger\rangle$ is $\mathcal{C}_\rho$.

The mapping $\sigma^\dagger\to\sigma$ is obviously a homomorphism of $\mathfrak{G}^\dagger$ into $\mathfrak{G}$. If σ is the identity automorphism of $\mathcal{G}$ then $\sigma^\dagger$ is obviously the identity automorphism of $\mathfrak{G}^\dagger$; therefore this homomorphism is an isomorphism of $\mathfrak{G}^\dagger$ onto some subgroup $\mathfrak{G}_1$ of $\mathfrak{G}$. An element $\alpha\in\mathcal{G}$ is invariant under every $\sigma\in\mathfrak{G}_1$, that is under every $\sigma^\dagger\in\mathfrak{G}^\dagger$, if and only if $\alpha\in\mathcal{F}^\dagger$; it follows that $\mathfrak{G}(\mathcal{F}^\dagger\cap\mathcal{G})$ is the smallest algebraic group in $\mathfrak{G}$ containing $\mathfrak{G}_1$, and therefore (Chapter II, Proposition 14) also the smallest algebraic set containg $\mathfrak{G}_1$.

Now let $\mathfrak{M}^\dagger$ be any irreducible set in $\mathfrak{G}^\dagger$, let $\rho^\dagger$ be a generic element of $\mathfrak{M}^\dagger$, and let ρ be the restriction of $\rho^\dagger$ to $\mathcal{G}$. Let $\mathfrak{M}$ be the set of all restrictions to $\mathcal{G}$ of elements of $\mathfrak{M}^\dagger$, and let $\mathfrak{M}_0$ be the irreducible set in $\mathfrak{G}$ with generic element ρ. Because every element of $\mathfrak{M}^\dagger$ is a specialization of $\rho^\dagger$, every element of $\mathfrak{M}$ is a specialization of ρ, so that $\mathfrak{M}\subseteq\mathfrak{M}_0$. If $\gamma_1,\cdots,\gamma_q$ are constants such that $\mathcal{C}_\rho=\mathcal{C}\langle\gamma_1,\cdots,\gamma_q\rangle$ then, by Proposition 9 of Chapter II and the fact that the field of constants of $\mathcal{G}^\dagger\langle\rho^\dagger\mathcal{G}^\dagger\rangle$ is $\mathcal{C}_\rho$, we know that there exists a polynomial $M\in\mathcal{C}[u_1,\cdots,u_q]$ with $M(\gamma_1,\cdots,\gamma_q)\neq0$ such that whenever $c_1,\cdots,c_q$ are constants with $M(c_1,\cdots,c_q)\neq0$ then there exists a unique isomorphism $\tau^\dagger$ of $\mathcal{G}^\dagger$ over $\mathcal{F}^\dagger$ for which $((\tau^\dagger\alpha)_{\alpha\in\mathcal{G}^\dagger},c_1,\cdots,c_q)$ is a specialization of $((\rho^\dagger\alpha)_{\alpha\in\mathcal{G}^\dagger},\gamma_1,\cdots,\gamma_q)$ over $\mathcal{G}^\dagger$. But it is easy to see that if $\eta_1,\cdots,\eta_n$ are elements of $\mathcal{G}$ such that $\mathcal{G}=\mathcal{F}\langle\eta_1,\cdots,\eta_n\rangle$, then there exists a differential polynomial

$K \in \mathcal{G}\{y_1, \cdots, y_n\}$ with $K(\rho\eta_1, \cdots, \rho\eta_n) \neq 0$ which has the following property: whenever τ is an isomorphism of $\mathcal{G}$ over $\mathcal{F}$ which is a specialization of ρ with $K(\tau\eta_1, \cdots, \tau\eta_n) \neq 0$ then there exist unique constants $c_1, \cdots, c_q$ with $M(c_1, \cdots, c_q) \neq 0$ such that $((\tau\alpha)_{\alpha \in G}, c_1, \cdots, c_q)$ is a specialization of $((\rho\alpha)_{\alpha \in G}, \gamma_1, \cdots, \gamma_q)$ over $\mathcal{G}$. It follows that every specialization τ of ρ such that $K(\tau\eta_1, \cdots, \tau\eta_n) \neq 0$ is the restriction to $\mathcal{G}$ of a specialization $\tau^\dagger$ of $\rho^\dagger$, and that $\tau^\dagger$ is an automorphism of $\mathcal{G}^\dagger$ if τ is an automorphism of $\mathcal{G}$. From this it is not difficult to conclude that $\mathfrak{M}_0 - \mathfrak{M}$ is contained in a finite union of irreducible sets in $\mathfrak{G}$ of lower dimension that $\mathfrak{M}_0$.

Suppose now that, in addition to being an irreducible set in $\mathfrak{G}^\dagger$, $\mathfrak{M}^\dagger$ is also a group; then $\mathfrak{M}$ is a group in $\mathfrak{G}$. If we let $\mathfrak{N}$ denote the intersection of all the algebraic sets in $\mathfrak{G}$ which contain $\mathfrak{M}$ then $\mathfrak{M} \subseteq \mathfrak{N} \subseteq \mathfrak{M}_0$. By Chapter I, Proposition 14, $\mathfrak{N}$ is an algebraic group. Also, the irreducible set $\mathfrak{M}_0$ is the union of $\mathfrak{N}$ and a finite union of irreducible sets of lower dimension than $\mathfrak{M}_0$, so that $\mathfrak{M}_0 = \mathfrak{N}$, $\mathfrak{M}_0$ is an algebraic group in $\mathfrak{G}$, and (Chapter II, Proposition 15) $\mathfrak{M}_0 = \mathfrak{M}$. Therefore $\mathfrak{M}$ is an irreducible algebraic group in $\mathfrak{G}$ of dimension

$$= \partial^0 \mathcal{G}\langle\rho\mathcal{G}\rangle/\mathcal{G} = \partial^0 \mathcal{E}\rho/\mathcal{E} = \partial^0 \mathcal{G}^\dagger\langle\rho^\dagger\mathcal{G}^\dagger\rangle/\mathcal{G}^\dagger = \dim \mathfrak{M}^\dagger.$$

Thus every irreducible algebraic group in $\mathfrak{G}^\dagger$ is mapped onto an irreducible algebraic group in $\mathfrak{G}$ of the same dimension. Since every algebraic group is the union of the finite number of cosets of its component of the identity, a similar remark holds for not necessarily irreducible algebraic groups. In particular, the image of $\mathfrak{G}^\dagger$ is $\mathfrak{G}_1 = \mathfrak{G}(\mathcal{F}^\dagger \cap \mathcal{G})$.

COROLLARY. *Let $\mathcal{G}$ be strongly normal over $\mathcal{F}$, and let $\mathcal{G}_0, \mathcal{G}_1, \cdots, \mathcal{G}_r$ be differential fields such that*

$$\mathcal{F} = \mathcal{G}_0 \subseteq \mathcal{G}_1 \subseteq \cdots \subseteq \mathcal{G}_r, \mathcal{G} \subseteq \mathcal{G}_r,$$

and $\mathcal{G}_i$ is a strongly normal extension of $\mathcal{G}_{i-1}$ $(1 \leq i \leq r)$; denote the group of all automorphisms of $\mathcal{G}_i$ over $\mathcal{G}_{i-1}$ by $\mathfrak{G}_i$ $(1 \leq i \leq r)$. Then

$$\mathfrak{G} = \mathfrak{G}(\mathcal{G}_0 \cap \mathcal{G}) \supseteq \mathfrak{G}(\mathcal{G}_1 \cap \mathcal{G}) \supseteq \cdots \supseteq \mathfrak{G}(\mathcal{G}_r \cap \mathcal{G}) = \{\iota\} \tag{5}$$

is a normal chain of subgroups of $\mathfrak{G}$, $\mathfrak{G}_i(\mathcal{G}_i \cap \mathcal{G}\langle\mathcal{G}_{i-1}\rangle)$ is a normal subgroup of $\mathfrak{G}_i$, and for $i = 1, 2, \cdots, r$

$$\mathfrak{G}(\mathcal{G}_{i-1} \cap \mathcal{G})/\mathfrak{G}(\mathcal{G}_i \cap \mathcal{G}) \approx \mathfrak{G}_i/\mathfrak{G}_i(\mathcal{G}_i \cap \mathcal{G}\langle\mathcal{G}_{i-1}\rangle), \tag{6}$$

$$\begin{aligned} \dim \mathfrak{G}(\mathcal{G}_{i-1} \cap \mathcal{G}) &- \dim \mathfrak{G}(\mathcal{G}_i \cap \mathcal{G}) \\ &= \dim \mathfrak{G}_i - \dim \mathfrak{G}_i(\mathcal{G}_i \cap \mathcal{G}\langle\mathcal{G}_{i-1}\rangle). \end{aligned} \tag{7}$$

Proof. If $r=1$ the assertions follow immediately from Theorem 4. Let $r>1$ and suppose the corollary proved for lower values of r.

By the corollary to Theorem 4, $\mathcal{G}_1 \cap \mathcal{G}$ is strongly normal over $\mathcal{F}$ so that (Theorem 4) $\mathfrak{G}(\mathcal{G}_1 \cap \mathcal{G})$ is a normal subgroup $\mathfrak{G} = \mathfrak{G}(\mathcal{G}_0 \cap \mathcal{G})$ and $\mathfrak{G}_1(\mathcal{G}_1 \cap \mathcal{G})$ is a normal subgroup of $\mathfrak{G}_1$. The two factor groups $\mathfrak{G}/\mathfrak{G}(\mathcal{G}_1 \cap \mathcal{G})$ and $\mathfrak{G}_1/\mathfrak{G}_1(\mathcal{G}_1 \cap \mathcal{G})$ are isomorphic to the group of all automorphisms of $\mathcal{G}_1 \cap \mathcal{G}$ over $\mathcal{F}$, and therefore to each other, so that (6) holds for $i=1$. By Theorem 2, also (7) holds for $i=1$, as in that case both members equal $\partial^0(\mathcal{G}_1 \cap \mathcal{G})/\mathcal{F}$.

To complete the proof we consider the group $\mathfrak{G}^\dagger$ of all automorphisms of $\mathcal{G}\langle\mathcal{G}_1\rangle$ over $\mathcal{F}\langle\mathcal{G}_1\rangle = \mathcal{G}_1$. By the theorem $\mathcal{G}\langle\mathcal{G}_1\rangle$ is strongly normal over $\mathcal{G}_1$. By the corollary (case $r-1$)

$$(8)\qquad \begin{aligned}\mathfrak{G}^\dagger &= \mathfrak{G}^\dagger(\mathcal{G}_1 \cap \mathcal{G}\langle\mathcal{G}_1\rangle)\\ &\supseteq \mathfrak{G}^\dagger(\mathcal{G}_2 \cap \mathcal{G}\langle\mathcal{G}_1\rangle) \supseteq \cdots \supseteq \mathfrak{G}^\dagger(\mathcal{G}_r \cap \mathcal{G}\langle\mathcal{G}_1\rangle) = \{\iota\}\end{aligned}$$

is a normal chain of subgroups of $\mathfrak{G}^\dagger$, $\mathfrak{G}_i(\mathcal{G}_i \cap \mathcal{G}\langle\mathcal{G}_{i-1}\rangle)$ is a normal subgroup of $\mathfrak{G}_i$ $(2 \leq i \leq r)$, and

$$\mathfrak{G}^\dagger(\mathcal{G}_{i-1} \cap \mathcal{G}\langle\mathcal{G}_1\rangle)/\mathfrak{G}^\dagger(\mathcal{G}_i \cap \mathcal{G}\langle\mathcal{G}_1\rangle) \approx \mathfrak{G}_i/\mathfrak{G}_i(\mathcal{G}_i \cap \mathcal{G}\langle\mathcal{G}_{i-1}\rangle),$$

$$\begin{aligned}\dim \mathfrak{G}^\dagger(\mathcal{G}_{i-1} \cap \mathcal{G}\langle\mathcal{G}_1\rangle) &- \dim \mathfrak{G}^\dagger(\mathcal{G}_i \cap \mathcal{G}\langle\mathcal{G}_1\rangle)\\ &= \dim \mathfrak{G}_i - \dim \mathfrak{G}_i(\mathcal{G}_i \cap \mathcal{G}\langle\mathcal{G}_{i-1}\rangle)\end{aligned}$$

for $i = 2, \cdots, r$. By the theorem, the mapping $\sigma^\dagger \to \sigma$ which assigns to each $\sigma^\dagger \in \mathfrak{G}^\dagger$ the restriction σ of $\sigma^\dagger$ to $\mathcal{G}$ is a dimension-preserving isomorphism which maps the normal chain (8) onto the normal chain

$$\mathfrak{G}(\mathcal{G}_1 \cap \mathcal{G}) \supseteq \mathfrak{G}(\mathcal{G}_2 \cap \mathcal{G}) \supseteq \cdots \supseteq \mathfrak{G}(\mathcal{G}_r \cap \mathcal{G}) = \{\iota\}.$$

It now quickly follows that (6) and (7) hold for $2 \leq i \leq r$, and the proof of the corollary is complete.

4. Primitive elements. An element α will be called *primitive* over $\mathcal{F}$ if $\delta_i\alpha \in \mathcal{F}$ $(1 \leq i \leq m)$. It is obvious that if α is primitive over $\mathcal{F}$ with $\delta_i\alpha = a_i$ $(1 \leq i \leq m)$, and if β is primitive over $\mathcal{F}$ with $\delta_i\beta = b_i$ $(1 \leq i \leq m)$, and if we set $\eta = \alpha + \beta$, then η is primitive over $\mathcal{F}$ with $\delta_i\eta = a_i + b_i$ $(1 \leq i \leq m)$.

Let α be primitive over $\mathcal{F}$ and suppose that the field of constants of $\mathcal{F}\langle\alpha\rangle$ is $\mathcal{C}$. For every isomorphism σ of $\mathcal{F}\langle\alpha\rangle$ over $\mathcal{F}$

$$\delta_i(\sigma\alpha - \alpha) = \sigma\delta_i\alpha - \delta_i\alpha = \delta_i\alpha - \delta_i\alpha = 0 \qquad (1 \leq i \leq m),$$

so that $c(\sigma) = \sigma\alpha - \alpha$ is a constant. Because $\sigma\alpha = \alpha + c(\sigma)$ $\mathcal{F}\langle\alpha\rangle$ is strongly normal over $\mathcal{F}$ and every isomorphism of $\mathcal{F}\langle\alpha\rangle$ over $\mathcal{F}$ is strong. If σ_1, σ_2 are two isomorphisms of $\mathcal{F}\langle\alpha\rangle$ over $\mathcal{F}$ then

$$\sigma_1\sigma_2\alpha = \sigma_1(\alpha + c(\sigma_2)) = \alpha + c(\sigma_1) + c(\sigma_2),$$

so that $c(\sigma_1\sigma_2) = c(\sigma_1) + c(\sigma_2)$. Since $c(\sigma) = 0$ only when σ is the identity automorphism ι of $\mathcal{F}\langle\alpha\rangle$, it follows that $\sigma \to c(\sigma)$ is an isomorphism of the group of all automorphisms of $\mathcal{F}\langle\alpha\rangle$ over $\mathcal{F}$ into $\mathcal{C}^+$ (the additive group of $\mathcal{C}$). If $\alpha \in \mathcal{F}$ the automorphism group consists solely of ι, and the corresponding group in $\mathcal{C}^+$ is the zero group. Suppose $\alpha \notin \mathcal{F}$. Then there exists an automorphism σ of $\mathcal{F}\langle\alpha\rangle$ over $\mathcal{F}$ different from ι, and therefore with $c(\sigma) \neq 0$. Now if c is a constant then $\alpha + c$ is a generic specialization of α over $\mathcal{F}$ if and only if $\alpha + c$ is a specialization of α over $\mathcal{F}$, that is if and only if $\alpha + c$ is a zero of every differential polynomial in $\mathcal{F}\{y\}$ which vanishes at α, and this takes place if and only if c is a zero of a certain ideal of polynomials in $\mathcal{C}[u]$. Since this ideal of polynomials has infinitely many zeros, namely $c(\sigma^n) = nc(\sigma)$ for every integer n, it is the zero ideal, so that $\alpha + c$ is a generic specialization of α over $\mathcal{F}$ and $c = c(\sigma)$ for a suitable isomorphism σ of $\mathcal{F}\langle\alpha\rangle$ over $\mathcal{F}$ for every constant c. Summarizing: *Let α be primitive over $\mathcal{F}$ and the field of constants of $\mathcal{F}\langle\alpha\rangle$ be $\mathcal{C}$. Then $\mathcal{F}\langle\alpha\rangle$ is strongly normal over $\mathcal{F}$; either $\alpha \in \mathcal{F}$, or else α is transcendental over $\mathcal{F}$ and the mapping $\sigma \to \sigma\alpha - \alpha$ is an isomorphism of the group of all automorphisms of $\mathcal{F}\langle\alpha\rangle$ over $\mathcal{F}$ onto $\mathcal{C}^+$ and there exists no differential field between $\mathcal{F}$ and $\mathcal{F}\langle\alpha\rangle$ other than $\mathcal{F}$ and $\mathcal{F}\langle\alpha\rangle$.*

We note the trivial fact that, with the law of composition $(c_1, c_2) \to c_1 + c_2$, one-dimensional affine space becomes a group variety $\boldsymbol{D}$ in the sense of Weil [10], and that $\mathcal{C}^+$ is the subgroup of $\boldsymbol{D}$ consisting of all points of $\boldsymbol{D}$ which are rational over $\mathcal{C}$.

5. Exponential elements. An element α will be called *exponential* over $\mathcal{F}$ if $\alpha \neq 0$ and $\alpha^{-1}\delta_i\alpha \in \mathcal{F}$ $(1 \leq i \leq m)$. It is obvious that if α is exponential over $\mathcal{F}$ with $\alpha^{-1}\delta_i\alpha = a_i$ $(1 \leq i \leq m)$, and if β is exponential over $\mathcal{F}$ with $\beta^{-1}\delta_i\beta = b_i$ $(1 \leq i \leq m)$, and if we set $\eta = \alpha\beta$, the η is exponential over $\mathcal{F}$ with $\eta^{-1}\delta_i\eta = a_i + b_i$ $(1 \leq i \leq m)$.

Let α be exponential over $\mathcal{F}$ and suppose that the field of constants of $\mathcal{F}\langle\alpha\rangle$ is $\mathcal{C}$. For every isomorphism σ of $\mathcal{F}\langle\alpha\rangle$ over $\mathcal{F}$

$$\begin{aligned}\delta_i(\alpha^{-1}\sigma\alpha) &= \alpha^{-2}(\alpha\cdot\delta_i\sigma\alpha - \sigma\alpha\cdot\delta_i\alpha)\\ &= \alpha^{-2}(\alpha\cdot\sigma\alpha\cdot\sigma(\alpha^{-1}\delta_i\alpha) - \sigma\alpha\cdot\alpha\cdot\alpha^{-1}\delta_i\alpha) = 0 \qquad (1 \leq i \leq m),\end{aligned}$$

so that $c(\sigma) = \alpha^{-1}\sigma\alpha$ is a constant. Because $\sigma\alpha = c(\sigma)\alpha$, $\mathcal{F}\langle\alpha\rangle$ is strongly normal over $\mathcal{F}$ and every isomorphism of $\mathcal{F}\langle\alpha\rangle$ over $\mathcal{F}$ is strong. If σ_1, σ_2 are two isomorphism of $\mathcal{F}\langle\alpha\rangle$ over $\mathcal{F}$ then

$$\sigma_1\sigma_2\alpha = \sigma_1(c(\sigma_2)\alpha) = c(\sigma_1)c(\sigma_2)\alpha,$$

so that $c(\sigma_1\sigma_2) = c(\sigma_1)c(\sigma_2)$. Since $c(\sigma) = 1$ only when $\sigma = \iota$, it follows that $\sigma \rightarrow c(\sigma)$ is an isomorphism of the group of all automorphisms of $\mathcal{F}\langle\alpha\rangle$ over $\mathcal{F}$ into $\mathcal{C}^\times$ (the multiplicative group of nonzero elements of $\mathcal{C}$). If α is algebraic over $\mathcal{F}$ then the automorphism group is finite, say of order d; since every finite subgroup of $\mathcal{C}^\times$ is cyclic the automorphism group is cyclic, being generated by a single automorphism, say σ. It follows that $c(\sigma)$ is a primitive d-th root of unity, so that $\sigma(\alpha^d) = (c(\sigma)\alpha)^d = \alpha^d$, whence $\alpha^d \in \mathcal{F}$. If α is transcendental over $\mathcal{F}$ then the automorphism group is infinite, and it follows, much as in the case of primitive elements, that $c\alpha$ is a generic specialization of α over $\mathcal{F}$ for every nonzero constant c, so that $\sigma \rightarrow c(\sigma)$ is an isomorphism of the automorphism group of $\mathcal{F}\langle\alpha\rangle$ over $\mathcal{F}$ onto $\mathcal{C}^\times$; furthermore, if $\mathcal{F}_1$ is a differential field between $\mathcal{F}$ and $\mathcal{F}\langle\alpha\rangle$ then, since α is exponential over $\mathcal{F}_1$, the mapping $\sigma \rightarrow c(\sigma)$ maps the group of all automorphisms of $\mathcal{F}\langle\alpha\rangle$ over $\mathcal{F}_1$ either onto a cyclic group of some finite order d (in which case $\alpha^d \in \mathcal{F}_1$, whence $\mathcal{F}_1 = \mathcal{F}\langle\alpha^d\rangle$) or else onto the whole group $\mathcal{C}^\times$ (in which case $\mathcal{F}_1 = \mathcal{F}$). Summarizing: *Let α be exponential over $\mathcal{F}$ and the field of constants of $\mathcal{F}\langle\alpha\rangle$ be $\mathcal{C}$. Then $\mathcal{F}\langle\alpha\rangle$ is strongly normal over $\mathcal{F}$; either there exists an integer $d > 0$ such that $\alpha^d \in \mathcal{F}$, or else α is transcendental over $\mathcal{F}$ and the mapping $\sigma \rightarrow \alpha^{-1}\sigma\alpha$ is an isomorphism of the group of all automorphisms of $\mathcal{F}\langle\alpha\rangle$ over $\mathcal{F}$ onto $\mathcal{C}^\times$ and the only differential fields between $\mathcal{F}$ and $\mathcal{F}\langle\alpha\rangle$ are those of the form $\mathcal{F}\langle\alpha^d\rangle$, where d is an integer ≥ 0.*

We note the trivial fact that, with the law of composition $(c_1, c_2) \rightarrow c_1c_2$, one-dimensional affine space with the origin deleted becomes a group variety $\boldsymbol{E}$, and that $\mathcal{C}^\times$ is the subgroup of $\boldsymbol{E}$ consisting of all points of $\boldsymbol{E}$ which are rational over $\mathcal{C}$.

6. Weierstrassian elements. An element α will be called *weierstrassian* over $\mathcal{F}$ if α is not a constant and there exist two elements $g_2, g_3 \in \mathcal{C}$ with $27g_3^2 - g_2^3 \neq 0$ and m elements $a_1, \cdots, a_m \in \mathcal{F}$ such that

$$(\delta_i\alpha)^2 = a_i^2(4\alpha^3 - g_2\alpha - g_3) \qquad (1 \leq i \leq m).$$

The condition $27g_3^2 - g_2^3 \neq 0$ is equivalent to the condition that the polynomial $4y^3 - g_2y - g_3$ have no multiple root; the constants g_2, g_3, which are uniquely determined if α is transcendental over $\mathcal{F}$, will be called the *invariants* of α. Since any a_i may obviously be replaced by $-a_i$ *we may suppose (and in the future we always shall suppose, without expressly mentioning the fact) that* $(\delta_1\alpha:\cdots:\delta_m\alpha)=(a_1:\cdots:a_m)$.

In order to study weierstrassian elements with invariants g_2, g_3 we consider the irreducible algebraic curve in the projective plane defined by the equation

$$X_0X_2^2-4X_1^3+g_2X_0^2X_1+g_3X_0^3=0.$$

This curve, which is of genus 1, has precisely one point on the line at infinity $X_0=0$, namely the point $(0:0:1)$. On this curve there exists a law of composition, which we shall write multiplicatively, with respect to which the points of the curve form a group, the unity element being the point of infinity; the curve is, in the language of Weil [10], a group variety (actually an abelian variety) of dimension 1. If $(1:\xi_1:\xi_2)$ and $(1:\eta_1:\eta_2)$ are two points of this curve and if $\eta_1\neq\xi_1$, then their product is given by the formulae

$$(9)\quad\begin{cases}(1:\xi_1:\xi_2)(1:\eta_1:\eta_2)=(1:\zeta_1:\zeta_2),\\ \zeta_1=-\xi_1-\eta_1+\frac{1}{4}\left(\dfrac{\xi_2-\eta_2}{\xi_1-\eta_1}\right)^2,\\ \zeta_2=-\frac{1}{2}(\xi_2+\eta_2)+\frac{3}{2}(\xi_1+\eta_1)\dfrac{\xi_2-\eta_2}{\xi_1-\eta_1}-\frac{1}{4}\left(\dfrac{\xi_2-\eta_2}{\xi_1-\eta_1}\right)^3;\end{cases}$$

the inverse of any point $(1:\xi_1:\xi_2)$ of the curve is the point $(1:\xi:-\xi_2)$, so that there are precisely three points of order 2, namely the points $(1:e_1:0)$, $(1:e_2:0)$, $(1:e_3:0)$, where e_1, e_2, e_3 are the roots of $4y^3-g_2y-g_3$; the square of any point $(1:\xi_1:\xi_2)$ with $\xi_2\neq 0$ is given by the formulae

$$(10)\quad\begin{cases}(1:\xi_1:\xi_2)^2=(1:\zeta_1:\zeta_2),\\ \zeta_1=-2\xi_1+4^{-1}(6\xi_1^2-\frac{1}{2}g_2)^2\xi_2^{-2},\\ \zeta_2=-\xi_2+3\xi_1(6\xi_1^2-\frac{1}{2}g_2)\xi_2^{-1}-4^{-1}(6\xi_1^2-\frac{1}{2}g_2)^3\xi_2^{-3}.\end{cases}$$

These facts are well-known and are not difficult to verify directly. We shall denote this group variety by $\boldsymbol{W}(g_2, g_3)$.

If α is weierstrassian over $\mathcal{F}$ with $(\delta_i\alpha)^2=a_i^2(4\alpha^3-g_2\alpha-g_3)$ $(1\leq i\leq m)$ then, since α is not a constant, $\delta_i\alpha\neq 0$ for at least one value of i and $a_i\neq 0$ whenever $\delta_i\alpha\neq 0$. By the convention made above, the point $(1:\alpha:a_i^{-1}\delta_i\alpha)$ does not depend on the choice of i from among those values for which $a_i\neq 0$; this point obviously belongs to $\boldsymbol{W}(g_2, g_3)$.

LEMMA 2. *Let* α *be weierstrassian over* $\mathfrak{F}$ *with*

$$(\delta_i\alpha)^2 = a_i^2(4\alpha^3 - g_2\alpha - g_3) \qquad (1 \leq i \leq m)$$

and $a_{i_0} \neq 0$, *let* β *be weierstrassian over* $\mathfrak{F}$ *with*

$$(\delta_i\beta)^2 = b_i^2(4\beta^3 - g_2\beta - g_3) \qquad (1 \leq i \leq m)$$

and $b_{j_0} \neq 0$, *suppose that*

$$(1:\alpha:a_{i_0}^{-1}\delta_{i_0}\alpha)(1:\beta:b_{j_0}^{-1}\delta_{j_0}\beta) \neq (0:0:1),$$

and let

$$(1:\alpha:a_{i_0}^{-1}\delta_{i_0}\alpha)(1:\beta:b_{j_0}^{-1}\delta_{j_0}\beta) = (1:\eta:\zeta).$$

Then $\delta_i\eta = (a_i + b_i)\zeta$ $(1 \leq i \leq m)$, *so that either* η *and* ζ *are both constants or else* η *is weierstrassian over* $\mathfrak{F}$ *with*

$$(\delta_i\eta)^2 = (a_i + b_i)^2(4\eta^3 - g_2\eta - g_3) \qquad (1 \leq i \leq m).$$

Proof. Suppose first that $\alpha \neq \beta$. Then by (9) and a simple computation we find that

$$(11) \quad \zeta = (\alpha - \beta)^{-3}(-\beta^2(3\alpha + \beta) + \tfrac{1}{4}g_2(\alpha + 3\beta) + g_3)a_{i_0}^{-1}\delta_{i_0}\alpha + (\alpha - \beta)^{-3}(\alpha^2(\alpha + 3\beta) - \tfrac{1}{4}g_2(3\alpha + \beta) - g_3)b_{j_0}^{-1}\delta_{j_0}\beta.$$

On the other hand

$$\delta_i\alpha = a_i a_{i_0}^{-1}\delta_{i_0}\alpha, \qquad \delta_i\beta = b_i b_{j_0}^{-1}\delta_{j_0}\beta,$$

$$\delta_i(a_{i_0}^{-1}\delta_{i_0}\alpha) = (6\alpha^2 - \tfrac{1}{2}g_2)a_i, \qquad \delta_i(b_{j_0}^{-1}\delta_{j_0}\beta) = (6\beta^2 - \tfrac{1}{2}g_2)b_i,$$

so that from (9) we find that

$$\begin{aligned}\delta_i\eta &= -\delta_i\alpha - \delta_i\beta + \tfrac{1}{2}\frac{a_{i_0}^{-1}\delta_{i_0}\alpha - b_{j_0}^{-1}\delta_{j_0}\beta}{\alpha - \beta}\left(\frac{\delta_i(a_{i_0}^{-1}\delta_{i_0}\alpha - b_{j_0}^{-1}\delta_{j_0}\beta)}{\alpha - \beta} - \frac{(a_{i_0}^{-1}\delta_{i_0}\alpha - b_{j_0}\delta_{j_0}\beta)(\delta_i\alpha - \delta_i\beta)}{(\alpha - \beta)^2}\right) \\ &= -a_i a_{i_0}^{-1}\delta_{i_0}\alpha - b_i b_{j_0}^{-1}\delta_{j_0}\beta \\ &\quad + \tfrac{1}{2}\frac{a_{i_0}^{-1}\delta_{i_0}\alpha - b_{j_0}^{-1}\delta_{j_0}\beta}{\alpha - \beta}\ \frac{(6\alpha^2 - \tfrac{1}{2}g_2)a_i - (6\beta^2 - \tfrac{1}{2}g_2)b_i}{\alpha - \beta} \\ &\quad - \tfrac{1}{2}\frac{(a_{i_0}^{-1}\delta_{i_0}\alpha - b_{j_0}^{-1}\delta_{j_0}\beta)^2(a_i a_{i_0}^{-1}\delta_{i_0}\alpha - b_i b_{j_0}^{-1}\delta_{j_0}\beta)}{(\alpha - \beta)^3}.\end{aligned}$$

The coefficient of a_i here is easily seen to be the second member of (11), and likewise for the coefficient of b_i here. It follows that $\delta_i\eta = (a_i + b_i)\zeta$.

Now suppose that $\alpha = \beta$. Because by hypothesis

$$(1:\alpha:a_{i_0}^{-1}\delta_{i_0}\alpha)(1:\beta:b_{j_0}^{-1}\delta_{j_0}\beta) \neq (0:0:1),$$

we have $a_i = b_i$ $(1 \leq i \leq m)$. Therefore (10) is applicable in computing $(1:\eta:\zeta)$, and we find on the one hand

$$\begin{aligned}\zeta &= -a_{i_0}^{-1}\delta_{i_0}\alpha + 3\alpha(6\alpha^2 - \tfrac{1}{2}g_2)(a_{i_0}^{-1}\delta_{i_0}\alpha)^{-1} \\ &\quad - \tfrac{1}{4}(6\alpha^2 - \tfrac{1}{2}g_2)^3(a_{i_0}^{-1}\delta_{i_0}\alpha)^{-3} \\ &= (-1 + 3\alpha(6\alpha^2 - \tfrac{1}{2}g_2)(4\alpha^3 - g_2\alpha - g_3)^{-1} \\ &\quad - \tfrac{1}{4}(6\alpha^2 - \tfrac{1}{2}g_2)^3(4\alpha^3 - g_2\alpha - g_3)^{-2})a_{i_0}^{-1}\delta_{i_0}\alpha,\end{aligned}$$

and on the other hand

$$\eta = -2\alpha + \tfrac{1}{4}(6\alpha^2 - \tfrac{1}{2}g_2)^2(4\alpha^3 - g_2\alpha - g_3)^{-1},$$

so that

$$\begin{aligned}\delta_i\eta &= (-2 + 6\alpha(6\alpha^2 - \tfrac{1}{2}g_2)(4\alpha^3 - g_2\alpha - g_3)^{-1} \\ &\quad - \tfrac{1}{2}(6\alpha^2 - \tfrac{1}{2}g_2)^3(4\alpha^3 - g_2\alpha - g_3)^{-2})\delta_i\alpha,\end{aligned}$$

whence $\delta_i\eta = 2a_i\zeta = (a_i + b_i)\zeta$, q. e. d.

Now let α be an element which is weierstrassian over $\mathcal{F}$ with

$$(\delta_i\alpha)^2 = a_i^2(4\alpha^3 - g_2\alpha - g_3) \qquad (1 \leq i \leq m),$$

and suppose that the field of constants of $\mathcal{F}\langle\alpha\rangle$ is $\mathcal{C}$. We let i_0 denote any subscript such that $a_{i_0} \neq 0$. For any isomorphism σ of $\mathcal{F}\langle\alpha\rangle$ over $\mathcal{F}$,

$$\begin{aligned}&(1:\sigma\alpha:a_{i_0}^{-1}\delta_{i_0}\sigma\alpha)(1:\alpha:-a_{i_0}^{-1}\delta_{i_0}\alpha) \\ &\qquad = (1:\sigma\alpha:a_{i_0}^{-1}\delta_{i_0}\sigma\alpha)(1:\alpha:a_{i_0}^{-1}\delta_{i_0}\alpha)^{-1}\end{aligned}$$

is a point of the group variety $\boldsymbol{W}(g_2, g_3)$; we denote this point by $P(\sigma)$. If $\sigma = \iota$ then obviously $P(\sigma) = (0:0:1)$. Suppose $\sigma \neq \iota$; then $P(\sigma) \neq (0:0:1)$, and we may write $P(\sigma) = (1:c_1(\sigma):c_2(\sigma))$. It follows from Lemma 2 that $\delta_i c_1(\sigma) = (a_i - a_i)c_2(\sigma) = 0$ $(1 \leq i \leq m)$, so that $c_1(\sigma)$ and, therefore, $c_2(\sigma)$ are constants. Therefore for every isomorphism σ of $\mathcal{F}\langle\alpha\rangle$ over $\mathcal{F}$ we have

$$(1:\sigma\alpha:a_{i_0}^{-1}\delta_{i_0}\sigma\alpha) = P(\sigma)(1:\alpha:a_{i_0}^{-1}\delta_{i_0}\alpha)$$

and $\sigma\alpha \in \mathcal{F}\langle\alpha\rangle\langle\mathcal{C}^*\rangle$, so that $\mathcal{F}\langle\alpha\rangle$ is strongly normal over $\mathcal{F}$ and σ is strong. If σ_1, σ_2 are two isomorphisms of $\mathcal{F}\langle\alpha\rangle$ over $\mathcal{F}$ then

$$\begin{aligned}&(1:\sigma_1\sigma_2\alpha:a_{i_0}^{-1}\delta_{i_0}\sigma_1\sigma_2\alpha) \\ &\qquad = P(\sigma_2)(1:\sigma_1\alpha:c_{i_0}^{-1}\delta_{i_0}\sigma_1\alpha) = P(\sigma_2)P(\sigma_1)(1:\alpha:a_{i_0}^{-1}\delta_{i_0}\alpha),\end{aligned}$$

so that $P(\sigma_1\sigma_2) = P(\sigma_1)P(\sigma_2)$. Since $P(\sigma) = (0:0:1)$ only when $\sigma = \iota$, it follows that $\sigma \to P(\sigma)$ is an isomorphism of the group of all automorphisms of $\mathcal{F}\langle\alpha\rangle$ over $\mathcal{F}$ into the group $W(g_2, g_3; \mathcal{C})$ consisting of all points of $W(g_2, g_3)$ which are rational over $\mathcal{C}$.

Let c_1, c_2 be any two constants such that $(1:c_1:c_2) \in W(g_2, g_3)$ and set $(1:\beta:\beta') = (1:c_1:c_2)(1:\alpha:a_{i_0}^{-1}\delta_{i_0}\alpha)$. Now if (β, β') is a specialization of $(\alpha, a_{i_0}^{-1}\delta_{i_0}\alpha)$ over $\mathcal{F}$ then the specialization is generic, and this will be the case if and only if $\beta = \sigma\alpha$, $\beta' = a_{i_0}^{-1}\delta_{i_0}\sigma\alpha$ for some isomorphism σ of $\mathcal{F}\langle\alpha\rangle$ over $\mathcal{F}$, that is, if and only if $(1:c_1:c_2) = P(\sigma)$ for some such σ. On the other hand (β, β') is a specialization of $(\alpha, a_{i_0}^{-1}\delta_{i_0}\alpha)$ over $\mathcal{F}$ if and only if (c_1, c_2) is a zero of a certain set of polynomials with coefficients in $\mathcal{F}\langle\alpha\rangle$, and therefore if and only if (c_1, c_2) is a zero of a certain set of polynomials with coefficients in $\mathcal{C}$. It follows that $\sigma \to P(\sigma)$ maps the group of all automorphisms of $\mathcal{F}\langle\alpha\rangle$ over $\mathcal{F}$ onto the intersection with $W(g_2, g_3; \mathcal{C})$ of a subgroup of $W(g_2, g_3)$ which is a subvariety (not necessarily irreducible) of $W(g_2, g_3)$. Of course the subvarieties of an irreducible curve other than the curve itself are finite.

We may summarize these facts as follows: *Let α be weierstrassian over $\mathcal{F}$ and the field of constants of $\mathcal{F}\langle\alpha\rangle$ be $\mathcal{C}$. Then $\mathcal{F}\langle\alpha\rangle$ is strongly normal over $\mathcal{F}$; either α is algebraic over $\mathcal{F}$ and the mapping*

$$\sigma \to (1:\alpha:a_{i_0}^{-1}\delta_{i_0}\alpha)^{-1}(1:\sigma\alpha:a_{i_0}^{-1}\delta_{i_0}\sigma\alpha)$$

is an isomorphism of the group of all automorphisms of $\mathcal{F}\langle\alpha\rangle$ over $\mathcal{F}$ onto a finite subgroup of $W(g_2, g_3; \mathcal{C})$, or else α is transcendental over $\mathcal{F}$ and this mapping is an isomorphism onto $W(g_2, g_3; \mathcal{C})$. It can be shown, although we do not do so here, that if $\mathcal{F}_1$ is a differential field between $\mathcal{F}$ and $\mathcal{F}\langle\alpha\rangle$ other than $\mathcal{F}$ then $\mathcal{F}_1$ contains an element α_1 weierstrassian over $\mathcal{F}$ such that $\mathcal{F}_1 = \mathcal{F}\langle\alpha_1\rangle$.

7. Picard-Vessiot extensions. Let $\mathcal{G}$ be a Picard-Vessiot extension of $\mathcal{F}$. Then (Kolchin [3] and [6]), for suitable generators $\eta_1, \cdots, \eta_n$ of $\mathcal{G}$ over $\mathcal{F}$, every isomorphism σ of $\mathcal{G}$ over $\mathcal{F}$ satisfies equations

$$\sigma\eta_j = \sum_{i=1}^{n} c_{ij}(\sigma)\eta_i \qquad (1 \leq j \leq n), \tag{12}$$

where each $c_{ij}(\sigma)$ is a constant, and these equations establish a one-to-one correspondence between the set of all isomorphisms σ of $\mathcal{G}$ over $\mathcal{F}$ and a certain set of invertible matrices $(c_{ij}(\sigma))$ of degree n with constant coordinates. It follows that $\mathcal{G}$ is strongly normal over $\mathcal{F}$, and each isomorphism of $\mathcal{G}$ over $\mathcal{F}$ is strong. The mapping $\sigma \to (c_{ij}(\sigma))$ is an isomorphism of the algebraic group $\mathfrak{G}$ of all automorphisms of $\mathcal{G}$ over $\mathcal{F}$ onto a certain algebraic matric group $\mathfrak{G}_M$ over $\mathcal{C}$; algebraic subgroups of $\mathfrak{G}$ are mapped thereby onto algebraic subgroups of $\mathfrak{G}_M$ of the same dimension. If σ, τ are isomorphisms of $\mathcal{G}$ over $\mathcal{F}$ then τ is a specialization of σ if and only if $(c_{ij}(\tau))$ is a specialization of $(c_{ij}(\sigma))$ over $\mathcal{C}$.

We shall say that a differential field $\mathcal{H}$ is an extension of $\mathcal{F}$ by algebraic, primitive, exponential and weierstrassian elements if $\mathcal{H}$ contains a finite family of elements $\alpha_1, \cdots, \alpha_r$ such that $\mathcal{H} = \mathcal{F}\langle\alpha_1, \cdots, \alpha_r\rangle$ and for each i $(1 \leq i \leq r)$ α_i is either algebraic, or primitive, or exponential, or weierstrassian over $\mathcal{F}\langle\alpha_1, \cdots, \alpha_{i-1}\rangle$. If $\mathcal{H} = \mathcal{F}\langle\alpha_1, \cdots, \alpha_r\rangle$ and for each i $(1 \leq i \leq r)$ α_i is either algebraic, or primitive, or exponential over $\mathcal{F}\langle\alpha_1, \cdots, \alpha_{i-1}\rangle$, and if the field of constants of $\mathcal{H}$ is $\mathcal{C}$, then $\mathcal{H}$ is called a *liouvillian* extension of $\mathcal{F}$.

THEOREM 6. *If a Picard-Vessiot extension of $\mathcal{F}$ is contained in an extension of $\mathcal{F}$ by algebraic, primitive, exponential, and weierstrassian elements with field of constants $\mathcal{C}$,*[8] *then the Picard-Vessiot extension is a liouvillian extension of $\mathcal{F}$.*

Proof. By the hypothesis, the results of the preceding three sections, and the corollary to Theorem 5, the group of all automorphisms of the Picard-Vessiot extension of $\mathcal{F}$ has a normal chain of algebraic subgroups in which each factor group is either finite or abelian. Therefore (Kolchin [3], § 8, Theorem 1) the component of the identity of this group of automorphisms is solvable. It follows that the Picard-Vessiot extension is a liouvillian extension.

8. Extensions of transcendence degree 1; formulation of the theorem. In §§ 4-6 we saw that if α is transcendental and either primitive or exponential or weierstrassian over $\mathcal{F}$, and if the field of constants of $\mathcal{F}\langle\alpha\rangle$ is $\mathcal{C}$, then $\mathcal{F}\langle\alpha\rangle$ is a strongly normal extension of $\mathcal{F}$ of transcendence degree 1. We shall now state a theorem which implies that every strongly normal (indeed, every weakly normal) extension of $\mathcal{F}$ of transcendence degree 1 can be obtained by combining the adjunction of an element of one of these three types with algebraic adjunctions.

Let $\mathcal{G}$ be a weakly normal extension of $\mathcal{F}$. It is a simple matter to see that the relative algebraic closure $\mathcal{F}^0$ of $\mathcal{F}$ in $\mathcal{G}$ is a normal algebraic extension of $\mathcal{F}$ (in the classical sense) of finite degree. If the group $\mathfrak{G}$ of all automorphisms of $\mathcal{G}$ over $\mathcal{F}$ is finite then $\mathcal{G} = \mathcal{F}^0$; therefore if $\mathcal{G}$ is trans-

[8] By considerations similar to those of Kolchin [4] it can be shown that this restriction on the field of constants of the extension may be omitted.

cendental over $\mathcal{F}$ then $\mathfrak{G}$ is infinite, and it easily follows that the group $\mathfrak{G}^0$ of all automorphisms of $\mathcal{G}$ over $\mathcal{F}^0$ is infinite. We therefore may state our theorem in the following form.

THEOREM 7. *Let $\mathcal{G}$ be of transcendence degree* 1 *over $\mathcal{F}$, let $\mathcal{F}$ be relatively algebraically closed in $\mathcal{G}$, and let the group $\mathfrak{G}$ of all automorphisms of $\mathcal{G}$ over $\mathcal{F}$ be infinite. Then there exists an element $\alpha \in \mathcal{G}$ such that either α is primitive over $\mathcal{F}$ and $\mathcal{G} = \mathcal{F}\langle\alpha\rangle$, or α is exponential over $\mathcal{F}$ and $\mathcal{G} = \mathcal{F}\langle\alpha\rangle$, or α is weierstrassian over $\mathcal{F}$ and $\mathcal{G}$ is an algebraic*[9] *extension of $\mathcal{F}\langle\alpha\rangle$. In the last case, if $\mathcal{F}$ is algebraically closed then the weierstrassian element α may be chosen so that $\mathcal{G} = \mathcal{F}\langle\alpha\rangle$.*

This theorem will be proved in §§ 9-11.

An immediate consequence of Theorem 7, the results of §§ 4-6, and the Corollary of Theorem 5, is the following.

COROLLARY. *Let $\mathcal{G}$ be any strongly normal extension of $\mathcal{F}$. If $\mathcal{G}$ is contained in an extension of $\mathcal{F}$ by algebraic, primitive, exponential, and weierstrassian elements then the group $\mathfrak{G}$ of all automorphisms of $\mathcal{G}$ over $\mathcal{F}$ has a normal chain $\mathfrak{G} = \mathfrak{G}_0 \supseteq \cdots \supseteq \mathfrak{G}_s = \{\iota\}$ of algebraic groups such that* $\dim \mathfrak{G}_{i-1} - \dim \mathfrak{G}_i \leq 1$ $(1 \leq i \leq s)$. *Conversely, if $\mathfrak{G}$ has such a normal chain then $\mathcal{G}$ is itself an extension of $\mathcal{F}$ by algebraic, primitive, exponential, and weierstrassian elements.*

9. The proof begun: reduction to the case of algebraically closed ground field. We shall show in this section that if Theorem 7 holds when $\mathcal{F}$ is algebraically closed then it holds in general.

Let $\mathcal{F}^\dagger$ be the algebraic closure of $\mathcal{F}$ and let $\mathcal{G}^\dagger = \mathcal{G}\langle\mathcal{F}^\dagger\rangle$. Since $\mathcal{F}$ is relatively algebraically closed in $\mathcal{G}$, the degree of each element of $\mathcal{F}^\dagger$ over $\mathcal{F}$ equals its degree over $\mathcal{G}$. Now every $P \in \mathcal{F}^\dagger\{y_1, \cdots, y_n\}$ can be written in the form $P = \sum_{i=0}^{d-1} P_i\lambda^i$, where each $P_i \in \mathcal{F}\{y_1, \cdots, y_n\}$ and λ is an element of $\mathcal{F}^\dagger$ of some degree d over $\mathcal{F}$, and therefore over $\mathcal{G}$. Consequently a family $(\eta_1, \cdots, \eta_n)$ of elements of $\mathcal{G}$ is a zero of P if and only if it is a zero of $P_0, P_1, \cdots, P_{d-1}$. From this it follows that every automorphism of $\mathcal{G}$ over $\mathcal{F}$ can be extended to an automorphism of $\mathcal{G}^\dagger$ over $\mathcal{F}^\dagger$. From the hypothesis of Theorem 7 it therefore follows that there are infinitely many automorphisms of $\mathcal{G}^\dagger$ over $\mathcal{F}^\dagger$. By the assumption that the theorem holds when the ground

[9] Abelian.

field is algebraically closed we conclude that there exists an element η such that $\mathcal{G}^\dagger = \mathcal{F}^\dagger\langle\eta\rangle$ and η is either primitive or exponential or weierstrassian over $\mathcal{F}^\dagger$. Thus there exist elements $a_1, \cdots, a_m \in \mathcal{F}^\dagger$ such that either $\delta_i\eta = a_i$ $(1 \leq i \leq m)$ or $\eta^{-1}\delta_i\eta = a_i$ $(1 \leq i \leq m)$ or $(4\eta^3 - g_2\eta - g_3)^{-1}(\delta_i\eta)^2 = a_i^2$ $(1 \leq i \leq m)$, where in the last case $g_2, g_3 \in \mathcal{C}$, $27g_3^2 - g_2^3 \neq 0$, and $(\delta_1\eta : \cdots : \delta_m\eta) = (a_1 : \cdots : a_m)$.

Let σ be any automorphism of $\mathcal{G}^\dagger$ over $\mathcal{F}^\dagger$ other than the identity. We may write, in the respective cases, $\sigma\eta = \eta + c$ or $\sigma\eta = c\eta$ or $(1 : \sigma\eta : a_i^{-1}\delta_i\sigma\eta) = (1 : c_1 : c_2)(1 : \eta : a_i^{-1}\delta_i\eta)$, where in the first case $c \in \mathcal{C}^+$, in the second case $c \in \mathcal{C}^\times$, and in the third case $(1 : c_1 : c_2) \in \boldsymbol{W}(g_2, g_3; \mathcal{C})$ and $a_i \neq 0$. Now let τ be any automorphism of $\mathcal{G}^\dagger$ over $\mathcal{G}$. It is clear that $\tau\mathcal{F}^\dagger = \mathcal{F}^\dagger$; therefore $\sigma\tau\theta = \tau\theta = \tau\sigma\theta$ for all $\theta \in \mathcal{F}^\dagger$. If σ happens to be one of the infinitely many automorphisms of $\mathcal{G}^\dagger$ over $\mathcal{F}^\dagger$ which are extensions of automorphisms of $\mathcal{G}$ over $\mathcal{F}$ then $\sigma\tau\theta = \sigma\theta = \tau\sigma\theta$ for every $\theta \in \mathcal{G}$ whence, since $\mathcal{G}^\dagger = \mathcal{G}\langle\mathcal{F}^\dagger\rangle$, $\sigma\tau = \tau\sigma$. Because $\mathcal{G}^\dagger = \mathcal{F}^\dagger\langle\eta\rangle$ we may write $\tau\eta = f(\eta)$, where $f \in \mathcal{F}^\dagger\langle y\rangle$, and for arbitrary σ we shall have $\sigma\tau = \tau\sigma$ if and only if $\sigma\tau\eta = \tau\sigma\eta$, that is $f(\eta + c) = f(\eta) + c$ in the first case, $f(c\eta) = cf(\eta)$ in the second case, and

$$f\left(-\eta - c_1 + \tfrac{1}{4}\left(\frac{a_i^{-1}\delta_i\eta - c_2}{\eta - c_1}\right)^2\right) = -f(\eta) - c_1 + \tfrac{1}{4}\left(\frac{a_i^{-1}\delta_i f(\eta) - c_2}{f(\eta) - c_1}\right)^2$$

in the third case. Since this condition is satisfied for infinitely many choices of $c \in \mathcal{C}^+$ in the first case, of $c \in \mathcal{C}^\times$ in the second case, and of $(1 : c_1 : c_2) \in \boldsymbol{W}(g_2, g_3; \mathcal{C})$ in the third case, it must be satisfied identically. Thus σ commutes with every automorphism of $\mathcal{G}^\dagger$ over $\mathcal{G}$.

Let $\tau_1, \cdots, \tau_n$ be automorphisms of $\mathcal{G}^\dagger$ over $\mathcal{G}$ such that the restrictions of $\tau_1, \cdots, \tau_n$ to $\mathcal{G}\langle\eta\rangle$ are distinct and constitute the set of all isomorphisms of $\mathcal{G}\langle\eta\rangle$ over $\mathcal{G}$ (so that n equals the degree of $\mathcal{G}\langle\eta\rangle$ over $\mathcal{G}$). Since $\sigma\tau_j = \tau_j\sigma$, we have, in the respective cases, $\sigma\tau_j\eta = \tau_j\eta + c$, or $\sigma\tau_j\eta = c\tau_j\eta$, or

$$(1 : \sigma\tau_j\eta : \sigma\tau_j(a_i^{-1}\delta_i\eta)) = (1 : c_1 : c_2)(1 : \tau_j\eta : \tau_j(a_i^{-1}\delta_i\eta)).$$

We now consider the first case. Letting $\alpha = \sum_j \tau_j\eta$, we see that $\alpha \in \mathcal{G}$; also, $\delta_i\alpha = \sum_j \tau_j a_i \in \mathcal{F}$, so that α is primitive over $\mathcal{F}$. Furthermore, $\sigma\alpha = \alpha + nc$, so that $\alpha \notin \mathcal{F}^\dagger$; it follows (§ 4) that $\mathcal{G}^\dagger = \mathcal{F}^\dagger\langle\eta\rangle = \mathcal{F}^\dagger\langle\alpha\rangle$. If θ is any element of $\mathcal{G}$ then we may write $\theta = \sum \phi_i\alpha^i / \sum \psi_i\alpha^i$, where $\sum \phi_i y^i$ and $\sum \psi_i y^i$ are relatively prime polynomials in $\mathcal{F}^\dagger[y]$ such that the leading coefficient in $\sum \psi_i y^i$ is 1. For any automorphism τ of $\mathcal{G}^\dagger$ over $\mathcal{G}$ we have $\tau\alpha = \alpha$ and $\tau\theta = \theta$, so that $\sum \tau\phi_i\alpha^i / \sum \tau\psi_i\alpha^i = \sum \phi_i\alpha^i / \sum \psi_i\alpha^i$, whence $(\sum \tau\phi_i\alpha^i)(\sum \psi_i\alpha^i) = (\sum \phi_i\alpha^i)(\sum \tau\psi_i\alpha^i)$. Because of the relative primeness mentioned above and the fact that $\sum \psi_i y^i$ has leading coefficient 1 we conclude that $\sum \tau\phi_i\alpha^i = \sum \phi_i\alpha^i$ and $\sum \tau\psi_i\alpha^i = \sum \psi_i\alpha^i$, so that $\tau\phi_i = \phi_i$ and $\tau\psi_i = \psi_i$ for all i. Since τ is any automorphism of $\mathcal{G}^\dagger$ over $\mathcal{G}$ it follows that ϕ_i and ψ_i belong to $\mathcal{G} \cap \mathcal{F}^\dagger = \mathcal{F}$, so that $\theta \in \mathcal{F}\langle\alpha\rangle$. Thus $\mathcal{G} = \mathcal{F}\langle\alpha\rangle$, and the reduction is complete in the first case.

We turn to the second case. We assert that there exist an integer $r > 0$ and a nonzero element $\phi \in \mathcal{F}^\dagger$ such that $(\phi\eta)^r \in \mathcal{G}$. Indeed, $\sigma \prod \tau_j\eta = c^n \prod \tau_j\eta$, so that if we set $\chi = \eta^{-n} \prod \tau_j\eta$ then $\sigma\chi = \chi$, whence $\chi \in \mathcal{F}^\dagger$; letting ψ be an element of $\mathcal{F}^\dagger$ such that $\psi^n = \chi$, we find that $(\psi\eta)^n = \prod \tau_j\eta \in \mathcal{G}$, which proves our assertion. Of all pairs r, ϕ as above let us suppose we have chosen one for which r is as small as possible. Let $\alpha = (\phi\eta)^r$, so that $\alpha \in \mathcal{G}$; α is exponential over $\mathcal{F}^\dagger$, so that $\alpha^{-1}\delta_i\alpha \in \mathcal{F}^\dagger \cap \mathcal{G} = \mathcal{F}$ $(1 \leq i \leq m)$, whence α is exponential over $\mathcal{F}$. We shall show that $\mathcal{G} = \mathcal{F}\langle\alpha\rangle$, thereby completing the reduction in the second case. Indeed, if θ is any element of $\mathcal{G}$ then $\theta \in \mathcal{G}^\dagger = \mathcal{F}^\dagger\langle\eta\rangle = \mathcal{F}^\dagger\langle\phi\eta\rangle$, so that we may write $\theta = \sum \phi_i(\phi\eta)^i / \sum \psi_i(\phi\eta)^i$, where $\sum \phi_i y^i$ and $\sum \psi_i y^i$ are relatively prime polynomials in $\mathcal{F}^\dagger[y]$ and one of the coefficients ϕ_0, ψ_0 is 1. Since $(\phi\eta)^r \in \mathcal{G}$, for any automorphism τ of $\mathcal{G}^\dagger$ over $\mathcal{G}$ we may write $\tau(\phi\eta) = e\phi\eta$ where e is some r-th root of 1. As $\tau\theta = \theta$, we have $\sum \tau\phi_i \cdot e^i(\phi\eta)^i / \sum \tau\psi_i \cdot e^i(\phi\eta)^i = \sum \phi_i(\phi\eta)^i / \sum \psi_i(\phi\eta)^i$, so that $(\sum \tau\phi_i \cdot e^i(\phi\eta)^i)(\sum \psi_i(\phi\eta)^i) = (\sum \phi_i(\phi\eta)^i)(\sum \tau\psi_i \cdot e^i(\phi\eta)^i)$. Because of the relative primeness mentioned above and the fact that ϕ_0 or ψ_0 is 1, we conclude that $\sum \tau\phi_i \cdot e^i(\phi\eta)^i = \sum \phi_i(\phi\eta)^i$, $\sum \tau\psi_i \cdot e^i(\phi\eta)^i = \sum \psi_i(\phi\eta)^i$, so that $\tau\phi_i = e^{-i}\phi_i$, $\tau\psi_i = e^{-i}\psi_i$. Consider any value of i which is not divisible by r; writing $i = qr + r'$, where $0 < r' < r$, we find that $\tau(\phi_i(\phi\eta)^{r'}) = e^{-i}\phi_i e^{r'}(\phi\eta)^{r'} = \phi_i(\phi\eta)^{r'}$. Since τ is any automorphism of $\mathcal{G}^\dagger$ over $\mathcal{G}$ this implies that $\phi_i(\phi\eta)^{r'} \in \mathcal{G}$. Letting ϕ' be an element of $\mathcal{F}^\dagger$ such that $\phi'^{r'} = \phi_i\phi^{r'}$, we see that $(\phi'\eta)^{r'} \in \mathcal{G}$; because of the minimal nature of r and the relation $0 < r' < r$, we conclude that $\phi' = 0$, whence $\phi_i = 0$. Similarly $\psi_i = 0$ for all i not divisible by r. On the other hand, if i is divisible by r then $\tau\phi_i = e^{-i}\phi_i = \phi_i$, so that $\phi_i \in \mathcal{G} \cap \mathcal{F}^\dagger = \mathcal{F}$, and similarly, $\psi_i \in \mathcal{F}$. It follows that $\theta \in \mathcal{F}\langle(\phi\eta)^r\rangle = \mathcal{F}\langle\alpha\rangle$, so that $\mathcal{G} = \mathcal{F}\langle\alpha\rangle$.

Finally we consider the third case. It is apparent that the point $(1 : \eta : a_i^{-1}\delta_i\eta)^{-n} \prod (1 : \tau_j\eta : \tau_j(a_i^{-1}\delta_i\eta))$ of $\boldsymbol{W}(g_2, g_3)$ is invariant under σ. Since σ is any automorhpism of $\mathcal{G}^\dagger$ over $\mathcal{F}^\dagger$ other than the identity, this point is rational over $\mathcal{F}^\dagger$. Because $\mathcal{F}^\dagger$ is algebraically closed and $\boldsymbol{W}(g_2, g_3)$ is a complete curve in the projective plane, this point can be written in the form P^n, where P is a point of $\boldsymbol{W}(g_2, g_3)$ which is rational over $\mathcal{F}^\dagger$. For this P we have $(1 : \eta : a_i^{-1}\delta_i\eta)^n P^n = \prod (1 : \tau_j\eta : \tau_j(a_i^{-1}\delta_i\eta))$, which is clearly rational over $\mathcal{G}$. Now η is transcendental over $\mathcal{F}^\dagger$, so that $(1 : \eta : a_i^{-1}\delta_i\eta)$ is a generic point of the curve $\mathbf{W}(g_2, g_3)$ over $\mathcal{F}^\dagger$; since P is rational over $\mathcal{F}^\dagger$ $(1:\eta:a_i^{-1}\delta_i\eta)P$ is also a generic point of $\mathbf{W}(g_2, g_3)$ over $\mathcal{F}^\dagger$, whence $(1:\eta:a_i^{-1}\delta_i\eta)^n P^n$ is, too. Since the field of constants of $\mathcal{G}^\dagger$ is clearly $\mathcal{C}$, which is contained in $\mathcal{F}^\dagger$, it follows from Lemma 2 (§ 6) that $(1:\eta:a_i^{-1}\delta_i\eta)^n P^n = (1:\alpha:\beta)$, where α is weierstrassian over $\mathcal{F}^\dagger$ with invariants g_2, g_3. But by the above, $\alpha, \beta \in \mathcal{G}$, so that α is weierstrassian over $\mathcal{G} \cap \mathcal{F}^\dagger = \mathcal{F}$. Also, α is transcendental over $\mathcal{F}$, so that $\mathcal{G}$ is algebraic over $\mathcal{F}\langle\alpha\rangle$. This completes the reduction in the third and final case.

10. The proof continued: case of genus 0. We assume now, in addition to the hypothesis of Theorem 7, that $\mathcal{F}$ is algebraically closed. Since $\mathcal{G}$ is of transcendence degree 1 over $\mathcal{F}$, we may regard $\mathcal{G}$ as a field of algebraic functions of one variable over $\mathcal{F}$; furthermore, every automorphism of the differential field $\mathcal{G}$ over $\mathcal{F}$ is obviously an automorphism of the algebraic function field $\mathcal{G}$. It is a well-known theorem that if the genus of such an algebraic function field is greater than 1 then the group of its automorphisms is finite (for a proof in the general (abstract) case see Iwasawa and Tamagawa [7]). It follows that the algebraic function field $\mathcal{G}$ has genus 0 to 1. In the present section we treat the case of genus 0 and show that in this case there exists an element α, which is either primitive or exponential over $\mathcal{F}$, such that $\mathcal{G} = \mathcal{F}\langle\alpha\rangle$. In the next section we shall treat the case of genus 1.

Assuming then that $\mathcal{G}$ has genus 0, we see that $\mathcal{G}$ is a purely transcendental extension of $\mathcal{F}$ (see for example Chevalley [2], Chapter II, § 2), that is, that there exists a single element θ transcendental over $\mathcal{F}$ such that $\mathcal{G} = \mathcal{F}(\theta)$. Since $\delta_i\theta \in \mathcal{G}$ for each i, there exist polynomials $P_1, \cdots, P_m$, $Q \in \mathcal{F}[y]$ with $Q \neq 0$ such that

$$(13) \qquad \delta_i\theta = Q(\theta)^{-1}P_i(\theta), \qquad 1 \leq i \leq m.$$

It is obvious that θ is not a constant, so that $P_i \neq 0$ for at least one value of i. If $c \in \mathcal{C}$ is not a zero of Q nor of any nonzero P_i, and if $k = \max(\deg P_1, \cdots, \deg P_m, \deg Q)$, then $\bar{Q}(y) = Q(y^{-1} + c)y^k$ and the nonzero expression $\bar{P}_i(y) = - P_i(y^{-1} + c)y^k$ are polynomials of degree k; but if we let $\bar{\theta} = (\theta - c)^{-1}$, so that $\mathcal{G} = \mathcal{F}(\bar{\theta})$, then $\delta_i\bar{\theta} = \bar{Q}(\bar{\theta})^{-1}\bar{P}_i(\bar{\theta})\bar{\theta}^2$ $(1 \leq i \leq m)$. Therefore we lose no generality in assuming that $\deg P_i = 2 + \deg Q$ for all values of i such that $P_i \neq 0$; we assume too, as we obviously may, that $P_1, \cdots, P_m, Q$ have no common divisor and that the leading coefficient in Q is 1. We denote the degree of Q by d.

Every automorphism of a simple transcendental extension is given by a fractional linear substitution. Therefore if σ is any automorphism of the differential field $\mathcal{G}$ over $\mathcal{F}$ then there exist elements $a_{11}, a_{12}, a_{21}, a_{22}$ in $\mathcal{F}$ such that

$$\sigma\theta = (a_{21}\theta + a_{22})^{-1}(a_{11}\theta + a_{12}), \qquad | \sigma | \neq 0,$$

where $|\sigma| = a_{11}a_{22} - a_{12}a_{21}$. Applying σ to each side of (13) we find

$$\begin{aligned}(a_{11}\delta_i\theta &+ \delta_i a_{11}\cdot\theta + \delta_i a_{12})(a_{21}\theta + a_{22})^{-1} \\ &- (a_{11}\theta + a_{12})(a_{21}\theta + a_{22})^{-2}(a_{21}\delta_i\theta + \delta_i a_{21}\cdot\theta + \delta_i a_{22}) \\ &= Q((a_{21}\theta + a_{22})^{-1}(a_{11}\theta + a_{12}))^{-1}P_i((a_{21}\theta + a_{22})^{-1}(a_{11}\theta + a_{12})),\end{aligned}$$

which we rewrite (using (13)) in the form

$$\begin{aligned}(14)\quad Q((a_{21}\theta &+ a_{22})^{-1}(a_{11}\theta + a_{12}))(a_{21}\theta + a_{22})^{d}\cdot(|\sigma|\, P_i(\theta) + A_i(\theta)Q(\theta)) \\ &= Q(\theta)P_i((a_{21}\theta + a_{22})^{-1}(a_{11}\theta + a_{12}))(a_{21}\theta + a_{22})^{d+2},\end{aligned}$$

where

$$\begin{aligned}(15)\quad A_i(y) = (a_{21}\delta_i a_{11} &- a_{11}\delta_i a_{21})y^2 + (a_{22}\delta_i a_{11} - a_{11}\delta_i a_{22} \\ &+ a_{21}\delta_i a_{12} - a_{12}\delta_i a_{21})y + a_{22}\delta_i a_{12} - a_{12}\delta_i a_{22}.\end{aligned}$$

It follows from (14) that for each i

$$Q((a_{21}y + a_{22})^{-1}(a_{11}y + a_{12}))(a_{21}y + a_{22})^{d}\cdot |\sigma|\, P_i(y)$$

is divisible by $Q(y)$, whence the polynomial

$$Q((a_{21}y + a_{22})^{-1}(a_{11}y + a_{12}))(a_{21}y + a_{22})^{d}$$

is so divisible. It follows that the fractional linear transformation

$$(16)\qquad x \to (a_{21}x + a_{22})^{-1}(a_{11}x + a_{12})$$

permutes the zeros of Q. Since this must happen for each of the infinitely many automorphisms σ of $\mathcal{G}$ over $\mathcal{F}$, and since a fractional linear transformation is uniquely determined by its values at three points, Q can have no more than two distinct zeros.

Supose Q has two distinct zeros ξ_1, ξ_2, so that $Q = (y - \xi_1)^{h_1}(y - \xi_2)^{h_2}$, where $h_1 + h_2 = d$; we suppose for the sake of definiteness that $h_1 \leq h_2$. For each σ in the group of all automorphisms of $\mathcal{G}$ over $\mathcal{F}$ the transformation (16) permutes ξ_1, ξ_2. The subgroup of all automorphisms σ for which (16) leaves ξ_1 and ξ_2 invariant is obviously of index 2, and is therefore infinite. For every σ of this infinite subgroup we have $(a_{21}\xi_j + a_{22})^{-1}(a_{11}\xi_j + a_{12}) = \xi_j$, that is $a_{21}\xi_j^2 + (a_{22} - a_{11})\xi_j - a_{12} = 0$ $(j = 1, 2)$, so that

$$(17)\qquad (a_{21} : a_{11} - a_{22} : -a_{12}) = (1 : \xi_1 + \xi_2 : \xi_1\xi_2).$$

Now because $Q((a_{21}y+a_{22})^{-1}(a_{11}y+a_{12}))(a_{21}y+a_{22})^d$ is divisible by $Q(y)$ and obviously has the same degree as $Q(y)$, the quotient of these two polynomials is in $\mathfrak{F}$; since $Q(y)=(y-\xi_1)^{h_1}(y-\xi_2)^{h_2}$, an easy computation shows that this quotient is $(a_{11}-a_{21}\xi_1)^{h_1}(a_{11}-a_{21}\xi_2)^{h_2}$. But

$$(a_{11}-a_{21}\xi_1)(a_{11}-a_{21}\xi_2)=|\sigma|,$$

so that

$$Q((a_{21}y+a_{22})^{-1}(a_{11}y+a_{12}))(a_{21}y+a_{22})^d=|\sigma|^{h_1}(-a_{21}\xi_2+a_{11})^{h_1-h_2}Q(y).$$

It follows from (14) that

$$\begin{aligned}|\sigma|^{h_1}(-a_{21}\xi_2+a_{11})^{h_2-h_1}(|\sigma|P_i(y)+A_i(y)Q(y))&\\ =P_i((a_{21}y+a_{22})^{-1}(a_{11}y+a_{12}))(a_{21}y+a_{22})^{d+2}&.\end{aligned}$$

Replacing y by ξ_j we find that

$$|\sigma|^{h_1+1}(-a_{21}\xi_2+a_{11})^{h_2-h_1}P_i(\xi_j)=P_i(\xi_j)(a_{21}\xi_j+a_{22})^{d+2}.$$

Now, for at least one value of i, $P_i(\xi_j)\neq 0$, for otherwise $P_1,\cdots,P_m,Q$ would have the common factor $y-\xi_j$. Therefore

$$|\sigma|^{h_1+1}(-a_{21}\xi_2+a_{11})^{h_2-h_1}=(a_{21}\xi_j+a_{22})^{d+2}.$$

Since the left member here is the same for both values of j, the same must be true for the right member, that is $(a_{21}\xi_1+a_{22})^{d+2}=(a_{21}\xi_2+a_{22})^{d+2}$, so that $a_{21}\xi_1+a_{22}=\mu(a_{21}\xi_2+a_{22})$, where μ is one of the $(d+2)$-th roots of unity. But this equation and (17) together admit only a finite number of solutions $(a_{11}:a_{12}:a_{21}:a_{22})$. This contradicts the infinite number of possibilities for σ, and proves that Q can not have two distinct zeros.

Suppose now that Q has precisely one zero ξ. If we set $\bar\theta=\theta-\xi$, so that $\mathfrak{G}=\mathfrak{F}(\bar\theta)$, then $\delta_i\bar\theta=\bar\theta^{-d}(P_i(\bar\theta+\xi)-\delta_i\xi\cdot\bar\theta^d)=\bar\theta^{-d}\bar P_i(\bar\theta)$, where $\bar P_i(y)=P_i(y+\xi)-\delta_i\xi\cdot y^d$; whenever $P_i\neq 0$ then $\bar P_i\neq 0$ and is of degree $d+2$; if $P_i=0$ then, choosing some j such that P_j is not divisible by $y-\xi$ and letting P_{ji} denote the polynomial obtained by replacing each coefficient ϕ in P_j by $\delta_i\phi$, we find that

$$\begin{aligned}0&=\delta_j(Q(\theta)^{-1}P_i(\theta))=\delta_j\delta_i\theta=\delta_i\delta_j\theta\\&=\delta_i(Q(\theta)^{-1}P_j(\theta))=\delta_i((\theta-\xi)^{-d}P_j(\theta))\\&=-d(\theta-\xi)^{-d-1}((\theta-\xi)^{-d}P_i(\theta)-\delta_i\xi)P_j(\theta)\\&\quad+(\theta-\xi)^{-d}(P_{ji}(\theta)+P'_j(\theta)(\theta-\xi)^{-d}P_i(\theta))\\&=d(\theta-\xi)^{-d-1}\delta_i\xi\cdot P_j(\theta)+(\theta-\xi)^{-d}P_{ji}(\theta),\end{aligned}$$

so that $d\delta_i\xi \cdot P_j$ is divisible by $y - \xi$, whence $\delta_i\xi = 0$, and $\bar{R}_i = 0$. Therefore we lose no generality in assuming that $Q = y^d$. For every automorphism σ of $\mathfrak{G}$ over $\mathfrak{F}$ the transformation (16) then leaves 0 invariant, so that $a_{12} = 0$, and

$$Q((a_{21}y + a_{22})^{-1}(a_{11}y + a_{12}))(a_{21}y + a_{22})^d = a_{11}{}^d y^d = a_{11}{}^d Q(y).$$

It follows from (14) that

$$(18) \quad a_{11}{}^d(|\sigma| P_i(y) + A_i(y)y^d) = P_i((a_{21}y + a_{22}))^{-1}a_{11}y)(a_{21}y + a_{22})^{d+2}.$$

Replacing y by 0 here we find that

$$a_{11}{}^d \mid \sigma \mid P_i(0) = P_i(0)a_{22}{}^{d+2}.$$

But $|\sigma| = a_{11}a_{22} \neq 0$, and $P_i(0) \neq 0$ for some i; therefore $a_{11}{}^{d+1} = a_{22}{}^{d+1}$, whence $a_{22} = \mu a_{11}$, where now μ is one of the $(d+1)$-th roots of unity. By (15) we thus find that

$$A_i(y) = (a_{21}\delta_i a_{11} - a_{11}\delta_i a_{21})y^2 = - a_{11}{}^2\delta_i(a_{11}{}^{-1}a_{21}) \cdot y^2,$$

so that (18) becomes

$$\mu P_i(y) - \delta_i(a_{11}{}^{-1}a_{21}) \cdot y^{d+2} = P_i((a_{11}{}^{-1}a_{21}y + \mu)^{-1}y)(a_{11}{}^{-1}a_{21}y + \mu)^{d+2}.$$

Equating coefficients of y here we obtain

$$(d+2)P_i(0)a_{11}{}^{-1}a_{21} = P'_i(0)(\mu - 1).$$

Since $P_i(0) \neq 0$ for some i, and since we already know that $a_{12} = 0$ and $a_{22} = \mu a_{11}$, this contradicts the fact that the number of automorphisms σ is infinite. Therefore Q does not have a zero, so that $Q = 1$, and each P_i which is different from 0 has degree 2.

Thus we may write

$$P_i = p_{i0} + p_{i1}y + p_{i2}y^2 \qquad (p_{ij} \in \mathfrak{F},\ p_{i2} \neq 0 \text{ if } P_i \neq 0).$$

From this an easy computation shows that

$$\begin{aligned}\delta_j\delta_i\theta = \delta_j p_{i0} + p_{j0}p_{i1} + (\delta_j p_{i1} + 2p_{j0}p_{i2} + p_{j1}p_{i1})\theta \\ + (\delta_j p_{i2} + p_{j2}p_{i1} + 2p_{j1}p_{i2})\theta^2 + 2p_{j2}p_{i2}\theta^3.\end{aligned}$$

Since $\delta_j\delta_i = \delta_i\delta_j$ this implies that

$$(19) \qquad \begin{cases} \delta_j p_{i0} + p_{j0}p_{i1} = \delta_i p_{j0} + p_{i0}p_{j1}, \\ \delta_j p_{i1} + 2p_{j0}p_{i2} = \delta_i p_{j1} + 2p_{i0}p_{j2}, \\ \delta_j p_{i2} + p_{j1}p_{i2} = \delta_i p_{j2} + p_{i1}p_{j2}. \end{cases}$$

Let σ_0 be any automorphism of $\mathcal{G}$ over $\mathcal{F}$ such that $\sigma_0\theta \neq \theta$. It is easy to verify that the conditions

$$\delta_i(\tfrac{1}{2}p_{j1} + p_{j2}\theta) = \delta_j(\tfrac{1}{2}p_{i1} + p_{i2}\theta)$$

hold for all i and j. These conditions imply that the differential ideal $[\delta_1 z + (\frac{1}{2}p_{11} + p_{12}\theta)z, \cdots, \delta_m z + (\frac{1}{2}p_{m1} + p_{m2}\theta)z]$ in $\mathcal{G}\{z\}$ does not contain z and thus has a zero $\zeta_1 \neq 0$; it is not difficult to see that we may take ζ_1 so that the field of constants of $\mathcal{G}\langle\zeta_1\rangle$ is $\mathcal{C}$. Similarly, there exists a zero $\zeta_2 \neq 0$ of the differential ideal

$$[\delta_1 z + (\tfrac{1}{2}p_{11} + p_{12}\sigma_0\theta)z, \cdots, \delta_m z + (\tfrac{1}{2}p_{m1} + p_{m2}\sigma_0\theta)z]$$

in $\mathcal{G}\langle\zeta_1\rangle\{z\}$ such that the field of constants of $\mathcal{G}\langle\zeta_1, \zeta_2\rangle$ is $\mathcal{C}$. For any pair θ_1, θ_2 of operators of the form $\delta_1^{i_1} \cdots \delta_m^{i_m}$ we write

$$W_{\theta_1,\theta_2}(z_1, z_2) = \theta_1 z_1 \cdot \theta_2 z_2 - \theta_2 z_1 \cdot \theta_1 z_1.$$

It is easy to verify that

$$W_{1,\delta_i}(\zeta_1, \zeta_2) = p_{i2}\zeta_1\zeta_2(\theta - \sigma_0\theta), \tag{20}$$

which is different from 0 for at least one value of i, and that

$$W_{\theta_1,\theta_2}(\zeta_1, \zeta_2)\zeta_1^{-1}\zeta_2^{-1}(\theta - \sigma_0\theta)^{-1} \varepsilon \mathcal{F}$$

for every pair θ_1, θ_2 of operators of order ≤ 2. Therefore (Kolchin [6], § 3) $\mathcal{H} = \mathcal{F}\langle\zeta_1, \zeta_2\rangle$ is a Picard-Vessiot extension of $\mathcal{F}$. Of course $\mathcal{G} \subseteq \mathcal{H}$.

We denote the group of all automorphisms of $\mathcal{H}$ over $\mathcal{F}$ by $\mathfrak{H}$. By the Picard-Vessiot theory (Kolchin [6]) each element $\tau \varepsilon \mathfrak{H}$ may be identified with an element $(b_{ij}) = (b_{ij})_{1 \leq i \leq 2,\ 1 \leq j \leq 2}$ of an algebraic matric group over $\mathcal{C}$ by means of equations

$$\tau\zeta_j = b_{1j}\zeta_1 + b_{2j}\zeta_2 \qquad (j = 1, 2).$$

Let σ be any automorphism of $\mathcal{G}$ over $\mathcal{F}$ distinct from the identity and σ_0; σ can (Kolchin [6], § 3, Theorem 3) be extended to an element $\tau \varepsilon \mathfrak{H}$. Writing $\tau = (b_{ij})$ we have

$$\begin{aligned} 0 = \tau 0 &= \tau(\delta_i\zeta_1 + (\tfrac{1}{2}p_{i1} + p_{i2}\theta)\zeta_1) \\ &= b_{11}\delta_i\zeta_1 + b_{21}\delta_i\zeta_2 + (\tfrac{1}{2}p_{i1} + p_{i2}\sigma\theta)(b_{11}\zeta_1 + b_{21}\zeta_2) \\ &= - b_{11}(\tfrac{1}{2}p_{i1} + p_{i2}\theta)\zeta_1 - b_{21}(\tfrac{1}{2}p_{i1} + p_{i2}\sigma_0\theta)\zeta_2 \\ &\qquad\qquad + (\tfrac{1}{2}p_{i1} + p_{i2}\sigma\theta)(b_{11}\zeta_1 + b_{21}\zeta_2) \\ &= b_{11}p_{i2}(\sigma\theta - \theta)\zeta_1 + b_{21}p_{i2}(\sigma\theta - \sigma_0\theta)\zeta_2; \end{aligned}$$

since $\det(b_{ij}) \neq 0$ this implies that $b_{11}b_{21} \neq 0$ and $\zeta_1^{-1}\zeta_2 \varepsilon \mathcal{G}$.

A straightforward computation shows that $\delta_i(\zeta_1\zeta_2(\theta-\sigma_0\theta))=0$ $(1\leq i\leq m)$, so that $\zeta_1\zeta_2(\theta-\sigma_0\theta)=c\,\varepsilon\,\mathcal{C}$. From (20) we see at once that $W_{1,\delta_i}(\zeta_1,\zeta_2)=cp_{i2}\,\varepsilon\,\mathfrak{F}$, so that

$$\det(b_{ij})\cdot W_{1,\delta_i}(\zeta_1,\zeta_2)=W_{1,\delta_i}(\tau\zeta_1,\tau\zeta_2)=\tau W_{1,\delta_i}(\zeta_1,\zeta_2)=W_{1,\delta_i}(\zeta_1,\zeta_2)$$

whence $\det(b_{ij})=1$ for all $\tau\,\varepsilon\,\mathfrak{H}$. Also, $\zeta_1^{-2}W_{1,\delta_i}(\zeta_1,\zeta_2)=\delta_i(\zeta_1^{-1}\zeta_2)\,\varepsilon\,\mathfrak{G}$, so that $\zeta_1^2\,\varepsilon\,\mathfrak{G}$, whence $\zeta_1^2,\zeta_1\zeta_2,\zeta_2^2\,\varepsilon\,\mathfrak{G}$. Since $\delta_i\zeta_1^2=-2\zeta_1^2(\frac{1}{2}p_{i1}+p_{i2}\theta)$, it follows that $\theta\,\varepsilon\,\mathfrak{F}\langle\zeta_1^2,\zeta_1\zeta_2,\zeta_2^2\rangle$, so that $\mathfrak{G}=\mathfrak{F}\langle\zeta_1^2,\zeta_1\zeta_2,\zeta_2^2\rangle$. Therefore $\partial^0\mathfrak{H}/\mathfrak{F}=\partial^0\mathfrak{G}/\mathfrak{F}=1$, whence $\dim\mathfrak{H}=1$.

It follows that the component of the identity $\mathfrak{H}^0$ is reducible either to diagonal form or else to special triangular form, that is, there exist two linear combinations ω_1, ω_2 of ζ_1, ζ_2 over $\mathcal{C}$, which are linearly independent over constants, such that either for every $\tau\,\varepsilon\,\mathfrak{H}^0$ there exists a nonzero $b\,\varepsilon\,\mathcal{C}$ for which $\tau\omega_1=b\omega_1$, $\tau\omega_2=b^{-1}\omega_2$, or else for every $\tau\,\varepsilon\,\mathfrak{H}^0$ there exists a $b\,\varepsilon\,\mathcal{C}$ for which $\tau\omega_1=\omega_1$, $\tau\omega_2=b\omega_1+\omega_2$.

In the former case $\tau(\omega_j^{-1}\delta_i\omega_j)=\omega_j^{-1}\delta_i\omega_j$ for every $\tau\,\varepsilon\,\mathfrak{H}^0$ so that $\omega_j^{-1}\delta_i\omega_j$ is algebraic over $\mathfrak{F}$; but

$$\omega_j^{-1}\delta_i\omega_j=\tfrac{1}{2}\omega_j^{-2}\delta_i(\omega_j^2)\,\varepsilon\,\mathfrak{F}\langle\zeta_1^2,\zeta_1\zeta_2,\zeta_2^2\rangle=\mathfrak{G}$$

and $\mathfrak{F}$ is relatively algebraically closed in $\mathfrak{G}$, so that $\omega_j^{-1}\delta_i\omega_j\,\varepsilon\,\mathfrak{F}$, that is, ω_j is exponential over $\mathfrak{F}$. Therefore for every τ in $\mathfrak{H}$ there exists a nonzero $b\,\varepsilon\,\mathcal{C}$ such that $\tau\omega_1=b\omega_1$, $\tau\omega_2=b^{-1}\omega_2$. Consequently $\omega_1\omega_2$ is invariant under every $\tau\,\varepsilon\,\mathfrak{H}$ and belongs to $\mathfrak{F}$, so that

$$\mathfrak{G}=\mathfrak{F}\langle\zeta_1^2,\zeta_1\zeta_2,\zeta_2^2\rangle=\mathfrak{F}\langle\omega_1^2,\omega_1\omega_2,\omega_2^2\rangle=\mathfrak{F}\langle\omega_1^2\rangle.$$

Setting $\alpha=\omega_1^2$ we see that $\alpha^{-1}\delta_i\alpha=2\omega_1^{-1}\delta_i\omega_1\,\varepsilon\,\mathfrak{F}$, so that α is exponential over $\mathfrak{F}$, and also that $\mathfrak{G}=\mathfrak{F}\langle\alpha\rangle$.

In the latter case $\tau\omega_1=\omega_1$ for every $\tau\,\varepsilon\,\mathfrak{H}^0$, so that ω_1 is algebraic over $\mathfrak{F}$ whence, since $\omega_1^2\,\varepsilon\,\mathfrak{F}\langle\zeta_1^2,\zeta_1\zeta_2,\zeta_2^2\rangle=\mathfrak{G}$, we have $\omega_1^2\,\varepsilon\,\mathfrak{F}$. Since $W_{1,\delta_i}(\zeta_1,\zeta_2)=cp_{i2}\,\varepsilon\,\mathfrak{F}$, we also have $W_{1,\delta_i}(\omega_1,\omega_2)\,\varepsilon\,\mathfrak{F}$, so that $\delta_i(\omega_1^{-1}\omega_2)$ $=\omega_1^{-2}W_{1,\delta_i}(\omega_1,\omega_2)\,\varepsilon\,\mathfrak{F}$. Therefore if we set $\alpha=\omega_1^{-1}\omega_2$ then α is primitive over $\mathfrak{F}$ and $\mathfrak{G}=\mathfrak{F}\langle\omega_1^2,\omega_1\omega_2,\omega_2^2\rangle=\mathfrak{F}\langle\omega_1^2,\omega_1^2\alpha,\omega_1^2\alpha^2\rangle=\mathfrak{F}\langle\alpha\rangle$.

This completes the treatment of the case of genus 0.

11. The proof concluded. Case of genus 1. We consider now the remaining case in which $\mathfrak{F}$ is algebraically closed and $\mathfrak{G}$ is of genus 1. It is known (for example see Chevalley [2], Chapter II, § 3) that in this case there exist two elements α,β in $\mathfrak{G}$ such that $\mathfrak{G}=\mathfrak{F}(\alpha,\beta)$ and $\beta^2=P(\alpha)$, where P is a cubic polynomial in $\mathfrak{F}[y]$ which does not have a multiple root. Replacing α, β by suitable elements $a\alpha + b$, $c\beta (a, b, c \in \mathfrak{F})$, we lose no generality in supposing that

$$(21) \qquad \beta^2 = 4\alpha^3 - g_2\alpha - g_3,$$

where $g_2 \in \mathfrak{C}$, $g_3 \in \mathfrak{F}$, and $27g_3{}^2 - g_2{}^3 \neq 0$.

We shall prove that then $g_3 \in \mathfrak{C}$, $\mathfrak{G} = \mathfrak{F}\langle\alpha\rangle$, and there exist elements $a_1, \cdots, a_m \in \mathfrak{F}$ (not all 0) such that

$$(22) \qquad (\delta_i\alpha)^2 = a_i{}^2(4\alpha - g_2\alpha - g_3), \qquad (1 \leq i \leq m$$

(so that α is weierstrassian over $\mathfrak{F}$). This will complete the proof of Theorem 7.

We begin by observing that α is transcendental over $\mathfrak{F}$; since the field of constants of $\mathfrak{G}$ is $\mathfrak{C}$, α is not a constant. If $\delta_i\alpha = 0$ then (22) holds with $a_i = 0$. Let i be any index such that $\delta_i\alpha \neq 0$; in what follows we shall keep i fixed, and for every element ξ of $\mathfrak{G}$ we shall denote $\delta_i\xi$ by ξ'.

Clearly there exist polynomials $A, B, C \in \mathfrak{F}[y]$, without common divisor and with the leading coefficient in C equal to 1, such that

$$(23) \qquad \alpha' = \frac{A(\alpha) + B(\alpha)\beta}{C(\alpha)}.$$

Applying δ_i to both members of (21) we find that $2\beta\beta' = (12\alpha^2 - g_2)\alpha' - g'_3$; from this, (21), and (23) we obtain

$$(24) \quad \beta' = \frac{(12\alpha^2 - g_2)(4\alpha^3 - g_2\alpha - g_3)B(\alpha) + ((12\alpha^2 - g_2)A(\alpha) - g'_3C(\alpha))\beta}{2C(\alpha)(4\alpha^3 - g_2\alpha - g_3)}.$$

Now $(1:\alpha:\beta)$ is a point of the group variety $\boldsymbol{W}(g_2, g_3)$ defined in § 6. Since there exist infinitely many automorphisms over $\mathfrak{F}$ of the differential field $\mathfrak{G}$, there exist infinitely many such automorphisms σ such that

$$(25) \qquad (1:\sigma\alpha:\sigma\beta) = (1:a:b)(1:\alpha:\beta),$$

where $a, b \in \mathfrak{F}$ and $(1:a:b) \in \boldsymbol{W}(g_2, g_3)$,[10] that is, such that

[10] This follows from the known fact that, in the group of all automorphisms of the algebraic function field G, the subgroup of those automorphisms for which an equation of the form (25) holds is of finite index. To see this observe that this subgroup acts transitively on the set of all places of the function field G, and that there exists a place $\mathfrak{p}$ at which α has a pole of order 2 and β has a pole of order 3; the above fact is then a consequence of a second fact, namely that the group of all automorphisms of the function field G which leave $\mathfrak{p}$ invariant is finite. This second fact is in turn an easy

$$(26)\qquad \begin{cases} \sigma\alpha = -\alpha - a + \frac{1}{4}\left(\dfrac{\beta - b}{\alpha - a}\right)^2, \\ \sigma\beta = -\frac{1}{2}(\beta + b) + \frac{3}{2}\,\dfrac{\beta - b}{\alpha - a} - \frac{1}{4}\left(\dfrac{\beta - b}{\alpha - a}\right)^3. \end{cases}$$

These equations may, with the help of (21), be rewritten in the form

$$(27)\qquad \begin{cases} \sigma\alpha = \dfrac{4a\alpha^2 + (4a^2 - g_2)\alpha - (g_2 a + 2g_3) - 2b\beta}{4(\alpha - a)^2}, \\ \sigma\beta = \dfrac{(4\alpha^3 + 12a\alpha^2 - 3g_2\alpha - g_2 a - 4g_3)b + ((-12a^2 + g_2)\alpha - 4a^3 + 3g_2 a + 4g_3)\beta}{4(\alpha - a)^3}. \end{cases}$$

Applying δ_i to the first equation (26), and making use of (21), (23), and (24), we obtain

$$(\sigma\alpha)' = (U + V\beta)4^{-1}(\alpha - a)^{-1}(4\alpha^3 - g_2\alpha - g_3)^{-1}C(\alpha)^{-1},$$

where

$$\begin{aligned}(28)\quad U = &-4a'(\alpha - a)^3(4\alpha^3 - g_2\alpha - g_3)C(\alpha) - 4(\alpha - a)^3(4\alpha^3 - g_2\alpha - g_3)A(\alpha) \\ &+ (\alpha - a)(12\alpha^2 - g_2)(4\alpha^3 - g_2\alpha - g_3)A(\alpha) - g'_3(\alpha - a)(4\alpha^3 - g_2\alpha - g_3)C(\alpha) \\ &- b(\alpha - a)(12\alpha^2 - g_2)(4\alpha^3 - g_2\alpha - g_3)B(\alpha) \\ &+ ((12a^2 - g_2)a' - g'_3)(\alpha - a)(4\alpha^3 - g_2\alpha - g_3)C(\alpha) \\ &+ 2b(4\alpha^3 - g_2\alpha - g_3)(A(\alpha) - a'C(\alpha)) - 2(4\alpha^3 - g_2\alpha - g_3)^2B(\alpha),\end{aligned}$$

and

$$\begin{aligned}(29)\quad V = &-4(\alpha - a)^3(4\alpha^3 - g_2\alpha - g_3)B(\alpha) + (\alpha - a)(12\alpha^2 - g_2)(4\alpha^3 - g_2\alpha - g_3)B(\alpha) \\ &- 2b'(\alpha - a)(4\alpha^3 - g_2\alpha - g_3)C(\alpha) - b(\alpha - a)((12\alpha^2 - g_2)A(\alpha) - g'_3C(\alpha)) \\ &- 2(4\alpha^3 - g_2\alpha - g_3)(A(\alpha) - a'C(\alpha)) + 2b(4\alpha^3 - g_2\alpha - g_3)B(\alpha).\end{aligned}$$

On the other hand, by (23)

$$\sigma(\alpha') = \frac{A(\sigma\alpha) + B(\sigma\alpha)\sigma\beta}{C(\sigma\alpha)}.$$

Since $(\sigma\alpha)' = \sigma(\alpha')$ we therefore find, with the help of (27), that

---

consequence of the Riemann-Roch theorem; indeed if σ_0 is any automorphism which leaves $\mathfrak{p}$ invariant then $\sigma_0\alpha$ has a pole of order 2 at $\mathfrak{p}$ and $\sigma_0\beta$ has a pole of order 3 there, whence (for example see Chevalley [2], chapter II, corollary to theorem 6) $\sigma_0\alpha = c_1\alpha + c_2$, $\sigma_0\beta = c_3\beta + c_4\alpha + c_5$, where $c_1, \cdots, c_5 \in \mathcal{F}$ and $c_1c_3 \neq 0$; since σ_0 must preserve equation (21), an easy computation shows that $c_2 = c_4 = c_5 = 0$, $c_1^3 - c_3^2 = 0$, $g_2(c_1 - c_3^2) = 0$, $g_3(c_3^2 - 1) = 0$, so that there are only a finite number of possibilities for σ_0. For this short proof I am indebted to M. Rosenlicht.

$$(30)\qquad 4(\alpha-a)^3(4\alpha^3-g_2\alpha-g_3)C(\alpha)\Big[A\Big(W-\frac{b\beta}{2(\alpha-a)^2}\Big)$$
$$+B\Big(W-\frac{b\beta}{2(\alpha-a)^2}\Big)$$
$$\times\frac{(4\alpha^3+12a\alpha^2-3g_2\alpha-g_2a-4g_3)b+((-12a^2+g_2)\alpha-4a^3+3g_2a+4g_3)\beta}{4(\alpha-a)^3}\Big]$$
$$=C\Big(W-\frac{b\beta}{2(\alpha-a)^2}\Big)(U+V\beta),$$

where

$$W=\frac{4a\alpha^2+(4a^2-g_2)\alpha-(g_2a+2g_3)}{4(\alpha-a)^2}.$$

Because of (21) the left hand member here is a linear combination of 1 and β with coefficients which belong to $\mathcal{F}(\alpha)$, have denominators that are powers of $\alpha-a$, and have numerators that are divisible by $(4\alpha^3-g_2\alpha-g_3)C(\alpha)$. The right hand member can also be expressed as a linear combination of 1 and β, the coefficient of 1 being

$$(31)\qquad \sum_{j\geq 0}\frac{1}{(2j)!}C^{(2j)}(W)\frac{(4a^3-g_2a-g_3)^j(4\alpha^3-g_2\alpha-g_3)^j}{2^{2j}(\alpha-a)^{4j}}U$$
$$-\sum_{j\geq 0}\frac{1}{(2j+1)!}C^{(2j+1)}(W)\frac{(4a^3-g_2a-g_3)^j(4\alpha^3-g_2\alpha-g_3)^{j+1}}{2^{2j+1}(\alpha-a)^{4j+2}}bV,$$

and the coefficient of β being

$$(32)\qquad \sum_{j\geq 0}\frac{1}{(2j)!}C^{(2j)}(W)\frac{(4a^3-g_2a-g_3)^j(4\alpha^3-g_2\alpha-g_3)^j}{2^{2j}(\alpha-a)^{4j}}V$$
$$-\sum_{j\geq 0}\frac{1}{(2j+1)!}C^{(2j+1)}(W)\frac{(4a^3-g_2a-g_3)^j(4\alpha^3-g_2\alpha-g_3)^j}{2^{2j+1}(\alpha-a)^{4j+2}}bU.$$

Therefore (31) and (32) are both expressible as quotients in which the denominator is a power of $\alpha-a$ and the numerator is a polynomial in $\mathcal{F}[\alpha]$ divisible by $(4\alpha^3-g_2\alpha-g_3)C(\alpha)$.

Observing from (28) that U is divisible by $4\alpha^3-g_2\alpha-g_3$, and from (29) that each term of V is so divisible except for

$$-b(\alpha-a)((12\alpha^2-g_2)A(\alpha)-g'_3C(\alpha)),$$

and recalling that α is transcendental over $\mathcal{F}$, we conclude, on substituting for α in (32) any root e of $4y^3-g_2y-g_3$, that

$$C\left(\frac{4ae^2+(4a^2-g_2)e-(g_2a+2g_3)}{4(e-a)^2}\right)$$
$$\times b(e-a)((12e^2-g_2)A(e)-g'_3C(e))=0.$$

Since this is true for infinitely many points $(1:a:b)$ of the curve $\mathbf{W}(g_2, g_3)$, this implies that $(12e^2 - g_2)A(e) - g'_3C(e) = 0$. Because this equation holds for each of the three roots e of $4y^3 - g_2y - g_3$ we conclude that

$$(33) \qquad (12y^2 - g_2)A(y) - g'_3C(y) \equiv 0 \qquad (\mathrm{mod}\ 4y^3 - g_2y - g_3).$$

Returning now to (31) we see that if r denotes the degree of $C(y)$ and if we multiply (31) by $(\alpha - a)^{2r}$ then we obtain a polynomial in $\mathcal{F}[\alpha]$ divisible by $C(\alpha)$, that is, we have a congruence in $\mathcal{F}[\alpha]$ of the form

$$LU + MbV \equiv 0 \qquad (\mathrm{mod}\ C(\alpha)).$$

Since every subgroup of $\mathbf{W}(g_2, g_3)$ which contains $(1:a:b)$ also contains $1:a:-b) = (1:a:b)^{-1}$, this congruence continues to hold if we replace b by $-b$. Using the two congruences (one with b and one with $-b$), and observing from (28) and (29) that

$$\begin{aligned} U \equiv {} & (-4(\alpha - a)^3 + (\alpha - a)(12\alpha^2 - g_2) + 2b)(4\alpha^3 - 2g_2\alpha - g_3)A(\alpha) \\ & - (b(\alpha - a)(12\alpha^2 - g_2) + 2(4\alpha^3 - g_2\alpha - g_3))(4\alpha^3 - g_2\alpha - g_3)B(\alpha) \\ & \qquad (\mathrm{mod}\ C(\alpha)) \end{aligned}$$

and

$$\begin{aligned} V \equiv {} & - (b(\alpha - a)(12\alpha^2 - g_2) + 2(4\alpha^3 - g_2\alpha - g_3))A(\alpha) \\ & + (-4(\alpha - a)^3 + (\alpha - a)(12\alpha^2 - g_2) + 2b)(4\alpha^3 - g_2\alpha - g_3)B(\alpha) \\ & \qquad (\mathrm{mod}\ C(\alpha)), \end{aligned}$$

we obtain the following two congruences in $\mathcal{F}[\alpha]$, in which b no longer appears:

$$\begin{aligned} (34) \quad & [\sum_{j \geqq 0} \frac{1}{(2j)!2^{2j}} C^{(2j)}(W)(\alpha - a)^{2r-4j} \\ & \quad \times (4a^3 - g_2a - g_3)^j(4\alpha^3 - g_2\alpha - g_3)^j(12\alpha^2 - g_2 - 4(\alpha - a)^2) \\ & + \sum_{j \geqq 0} \frac{1}{(2j+1)!2^{2j+1}} C^{(2j+1)}(W)(\alpha - a)^{2r-4j-2} \\ & \quad \times (4a^3 - g_2a - g_3)^{j+1}(4\alpha^3 - g_2\alpha - g_3)^j(12\alpha^2 - g_2)](\alpha - a)A(\alpha) \\ & - [\sum_{j \geqq 0} \frac{1}{(2j)!2^{2j}} C^{(2j)}(W)(\alpha - a)^{2r-4j} \\ & \quad \times (4a^3 - g_2a - g_3)^j(4\alpha^3 - g_2\alpha - g_3)^{j+1} \\ & + \sum_{j \geqq 0} \frac{1}{(2j+1)!2^{2j+1}} C^{(2j+1)}(W)(\alpha - a)^{2r-4j-2} \\ & \quad \times (4a^3 - g_2a - g_3)^{j+1}(4\alpha^3 - g_2\alpha - g_3)^{j+1}]2B(\alpha) \\ & \qquad \equiv 0 \ (\mathrm{mod}\ C(\alpha)), \end{aligned}$$

and

$$(35)\quad [\sum_{j\geqq 0}\frac{1}{(2j)!2^{2j}}C^{(2j)}(W)(\alpha-a)^{2r-4j}$$
$$\times(4a^3-g_2a-g_3)^j(4\alpha^3-g_2\alpha-g_3)^j$$
$$+\sum_{j\geqq 0}\frac{1}{(2j+1)!2^{2j+1}}C^{(2j+1)}(W)(\alpha-a)^{2r-4j-2}$$
$$\times(4a^3-g_2a-g_3)^j(4\alpha^3-g_2\alpha-g_3)^{j+1}]2A(\alpha)$$
$$-[\sum_{j\geqq 0}\frac{1}{(2j)!2^{2j}}C^{(2j)}(W)(\alpha-a)^{2r-4j}$$
$$\times(4a^3-g_2a-g_3)^j(4\alpha^3-g_2\alpha-g_3)^j(12\alpha^2-g_2)$$
$$+\sum_{j\geqq 0}\frac{1}{(2j+1)!2^{2j+1}}C^{(2j+1)}(W)(\alpha-a)^{2r-4j-2}$$
$$\times(4a^3-g_2a-g_3)^j(4\alpha^3-g_2\alpha-g_3)^{j+1}(12\alpha^2-g_2-4(\alpha-a)^2)](\alpha-a)B(\alpha)$$
$$\equiv 0 \pmod{C(\alpha)}.$$

Since the congruences (34) and (35) hold for infinitely many points $(1:a:b)\,\varepsilon\,\boldsymbol{W}(g_2,g_3)$ with $a,b\,\varepsilon\,\mathfrak{F}$, that is for infinitely many elements $a\,\varepsilon\,\mathfrak{F}$, they must hold for all elements a. In particular, they hold for a equal to α. Now the leading coefficient in $C(y)$ is 1; therefore, when we replace a by α, $C^{(k)}(W)(\alpha-a)^{2r-2k}$ becomes $(r!(r-k)!)2^{r-k}(4\alpha^3-g_2\alpha-g_3)^{r-k}$. Consequently, when a is replaced by α, (34) becomes

$$-[\sum_{j\geqq 0}\binom{r}{2j}2^{r-4j}(4\alpha^2-g_2\alpha-g_3)^{r+1}$$
$$+\sum_{j\geqq 0}\binom{r}{2j+1}2^{r-4j-2}(4\alpha^3-g_2\alpha-g_3)^{r+1}]2B(\alpha)\equiv 0 \quad \pmod{C(\alpha)},$$

so that

$$B(\alpha)(4\alpha^3-g_2\alpha-g_3)^{r+1}\equiv 0 \qquad \pmod{C(\alpha)}.$$

In the same way, on replacing a by α in (35), we find that

$$A(\alpha)(4\alpha^3-g_2\alpha-g_3)^{r}\equiv 0 \qquad \pmod{C(\alpha)}.$$

Since A, B, C have no common divisor, it follows that

$$(4y^3-g_2y-g_3)^{r+1}\equiv 0 \qquad \pmod{C(y)}.$$

Therefore if C were of degree >0 then C would have a root e in common with $4y^3-g_2y-g_3$, and we would obtain, on replacing α by e in (35)

$$C\left(\frac{4ae^2+4(a^2-g_2)e-(g_2a+2g_3)}{4(e-a)^2}\right)(e-a)^{2r}$$
$$\times\,(2A(e)-(e-a)(12e^2-g_2)B(e))=0,$$

so that we would have $A(e)=0$, $B(e)=0$, contradicting the fact that A, B, C are without common divisor. Consequently

$$C(y)=1. \tag{36}$$

Now A and B are not both 0, for otherwise by (23) we would have $\delta_i\alpha=\alpha'=0$, contrary to assumption. Let p denote the maximum of the degrees of A and B, and denote the coefficients of y^p in A and B by c_A and c_B respectively. Suppose p were >0. Multiplying both sides of (30) by $(\alpha-a)^{2p}$, then expressing each side as a linear combination, over $\mathfrak{F}(\alpha)$, of 1 and β, and then equating coefficients of 1, we would obtain an equation in $\mathfrak{F}[\alpha]$; if in this equation we equated the coefficients, right and left of α^{3p+6} we would obtain

$$32c_A-(48b+32)c_B=0;$$

since this would hold for infinitely many points $(1:a:b)\in\mathfrak{W}(g_2,g_3)$, that is, for infinitely many values of b, we would have $c_A=c_B=0$, which is impossible. Therefore $p=0$, so that $A=c_A\in\mathfrak{F}$, $B=c_B\in\mathfrak{F}$. By (36) and (33) we further conclude that $c_A=0$ and $g'_3=0$, whence from (23) $\alpha'=c_B\beta$. Writing $c_B=a_i$ we see, by (21), that (22) holds whence $\mathfrak{G}=\mathfrak{F}(\alpha,\beta)=\mathfrak{F}\langle\alpha\rangle$. To complete the proof of the theorem it remains to show that $g_3\in\mathfrak{C}$. We have just seen that $\delta_ig_3=0$ for those values of i for which $\delta_i\alpha\neq0$; we must prove that $\delta_kg_3=0$ for all values of k such that $\delta_k\alpha=0$. For such i and k we have

$$\delta_k((\delta_i\alpha)^2)=2\delta_i\cdot\alpha\cdot\delta_k\delta_i\alpha=2\delta_i\alpha\cdot\delta_i\delta_k\alpha=0$$

so that, because of (22),

$$0=\delta_k(a_i^2(4\alpha^3-g_2\alpha-g_3))=2a_i\cdot\delta_ka_i\cdot(4\alpha^3-g_2\alpha-g_3)+a_i^2(-\delta_kg_3);$$

since α is transcendental over $\mathfrak{F}$, this implies that $\delta_ka_i=0$ and $\delta_kg_3=0$. Thus $g_3\in\mathfrak{C}$, and the proof of Theorem 7 is complete.

COLUMBIA UNIVERSITY.

REFERENCES.

[1] N. Bourbaki, "*Algèbre*," Chapters IV-V (Actualités scientifiques et industrielles 1102), Hermann et Cie., Paris, 1950.

[2] C. Chevalley, "*Introduction to the theory of algebraic functions of one variable*," (Mathematical Surveys VI), American Mathematical Society, New York, 1951.

[3] E. R. Kolchin, "Algebraic matric groups and the Picard-Vessiot theory of homogeneous linear ordinary differential equations," *Annals of Mathematics*, vol. 49 (1948), pp. 1-42.

[4] ———, "Existence theorems connected with the Picard-Vessiot theory of homogeneous linear ordinary differential equations," *Bulletin of the American Mathematical Society*, vol. 54 (1948), pp. 927-932.

[5] ———, "On certain concepts in the theory of algebraic matric groups," *Annals of Mathematics*, vol. 49 (1948), pp. 774-789.

[6] ———, "Picard-Vessiot theory of partial differential fields," *Proceedings of the American Mathematical Society*, vol. 3 (1952), pp. 596-603.

[7] K. Iwasawa and T. Tamagawa, "On the group of automorphisms of a function field," *Journal of the Mathematical Society of Japan*, vol. 3 (1951), pp. 137-147.

[8] J. F. Ritt, "*Differential algebra*," (American Mathematical Society Colloquium Publications, vol. 33), New York, 1950.

[9] A. Weil, "*Foundations of algebraic geometry*," (ibid., vol. 29), New York, 1946.

[10] ———, "*Variétés abeliennes et courbes algébriques* (Actualités scientifiques et industrielles 1064), Hermann et Cie., Paris, 1948.

COLLOQUE HENRI POINCARÉ
Institut Henri Poincaré
11 rue Pierre Curie
P A R I S 5°

Octobre 1954

Differential fields and group varieties,

by E. R. KOLCHIN.

First Lecture.

1.- Introduction.

As every one knows, the study of algebraic equations with numerical coefficients has led, step by step, to the concept of abstract field, the modern theory of fields (including the Galois theory), and modern algebraic geometry. In much the same way, the study of algebraic differential equations has, in more recent times, led to the concept of abstract differential field and thence, in the work of Ritt (see especially [7]), to the theory of differential algebra, which in its elementary parts bears considerable analogy to the elementary parts of algebraic geometry. Still more recently, I have been interested (see [2]) in developing defferential algebra in the direction of a Galois theory of differential fields. This work gives, I hope, a proper approach to a theory toward which the pioneering researches of Picard and Vessiot groped.

In addition to facilitating the study of certain questions concerning differential field extensions and algebraic differential equations, this theory derives some interest from the fact that it has a point of contact with a seemingly different subject, namely the theory of algebraic group varieties, developed in its full generality by Weil [9] . My purpose here is to elaborate on this point of contact, and I shall do this in the second lecture. The first lecture will be an outline of those parts of the Galois theory of differential fields just mentioned which are needed for an understanding of the second lecture ; although these are not new they are, I believe, not widely known.

2.- Differential fields.

A differential field is a commutative field together with a commutative finite family of operators each of which maps the field into itself as a derivation. For example, if m is an integer $\geqslant 1$, V is a connected open set in the space

ELLIS KOLCHIN 249

of m complex variables $x_1, \ldots, x_m$, $\mathcal{F}$ is a set of meromorphic functions on V closed with respect to the rational operations and the m partial differentiations $\frac{\partial}{\partial x_1}, \ldots, \frac{\partial}{\partial x_m}$, and $\delta_1, \ldots, \delta_m$ are operators on $\mathcal{F}$ such that $\delta_i u = \partial u/\partial x_i$ $(1 \leq i \leq m)$ for every $u \in \mathcal{F}$, then $\mathcal{F}$ (together with the family of operators $\delta_1, \ldots, \delta_m$) is a differential field. In general, a differential field is said to be ordinary or partial according as the number of operators equals or exceeds 1 .

Let $\mathcal{F}$ and $\mathcal{G}$ be two differential fields with the same family of operators, $\delta_1, \ldots, \delta_m$.

The three statements " $\mathcal{F}$ is a differential subfield of $\mathcal{G}$ " , " $\mathcal{G}$ is a differential overfield of $\mathcal{F}$ " , and " $\mathcal{G}$ is an extension of $\mathcal{F}$ " have obvious meanings, and are equivalent.

Let $\mathcal{G}$ be a differential overfield of $\mathcal{F}$. If Φ is a subset of $\mathcal{G}$, the smallest differential subfield of $\mathcal{G}$ containing both $\mathcal{F}$ and Φ is called the extension of $\mathcal{F}$ generated by Φ, and is denoted by $\mathcal{F}\langle \Phi \rangle$; if in addition Φ is finite then the extension is said to be finitely generated, or of finite type.

An element c such that $\delta_i c = 0$ $(1 \leq i \leq m)$ is called a constant. The set of all constants in $\mathcal{F}$ is a differential subfield of $\mathcal{F}$, called the field of constants of $\mathcal{F}$.

A mapping σ of $\mathcal{G}$ into a differential field with the same family operators $\delta_1, \ldots, \delta_m$ is called an isomorphism if it preserves addition and multiplication (i.e. is a field isomorphism) and in addition $\sigma \delta_i u = \delta_i \sigma u$ $(1 \leq i \leq m)$ for every $u \in \mathcal{G}$; if in addition σ leaves invariant every element of $\mathcal{F}$ then σ is said to be an isomorphism of $\mathcal{G}$ over $\mathcal{F}$ or an $\mathcal{F}$-isomorphism of $\mathcal{G}$.

Let $(\alpha_j)_{j \in J}$ and $(\alpha'_j)_{j \in J}$ be two families of elements of $\mathcal{G}$, having the same set of indices J . If every differential polynomial with coefficients in $\mathcal{F}$ vanishing at $(\alpha_j)_{j \in J}$ (i.e. every polynomial with coefficients in $\mathcal{F}$ vanishing at $(\delta_1^{i_1} \ldots \delta_m^{i_m} \alpha_j)_{0 \leq i_1 < \infty, \ldots, 0 \leq i_m < \infty, j \in J}$) also vanishes at $(\alpha'_j)_{j \in J}$, then $(\alpha'_j)_{j \in J}$ is said to be a specialization of $(\alpha_j)_{j \in J}$ over $\mathcal{F}$. If in addition $(\alpha_j)_{j \in J}$ is a specialization of $(\alpha'_j)_{j \in J}$ over $\mathcal{F}$ then $(\alpha'_j)_{j \in J}$ is said to be a generic specialization of $(\alpha_j)_{j \in J}$ over $\mathcal{F}$; in that case there exists a unique $\mathcal{F}$-isomorphism of $\mathcal{F}\langle(\alpha_j)_{j \in J}\rangle$ onto $\mathcal{F}\langle(\alpha'_j)_{j \in J}\rangle$ which maps each α_j onto the α'_j of the same index.

An extension $\mathcal{F}^*$ of $\mathcal{F}$ is said to be universal if, for every finitely generated extension $\mathcal{F}_1$ of $\mathcal{F}$ contained in $\mathcal{F}^*$, and every finitely generated extension $\mathcal{F}_2$ of $\mathcal{F}_1$ (not necessarily contained in $\mathcal{F}^*$) there exists an $\mathcal{F}_1$-isomorphism of $\mathcal{F}_2$ into $\mathcal{F}^*$. A universal extension $\mathcal{F}^*$ of $\mathcal{F}$ is also a universal extension of each differention subfield of $\mathcal{F}$ and of each finitely generated extension of $\mathcal{F}$ contained in $\mathcal{F}^*$. It is not difficult to prove that every differential field of characteristic 0 has a universal extension, and that the field of constants of such an extension is an extension, algebraically closed and of infinite transcendence degree, of the field of constants of the given differential field.

3.- Strongly normal extensions, and algebraic groups.

Let $\mathcal{F}$ be a differential field of characteristic 0 , and let $\mathcal{G}$ be an extension of $\mathcal{F}$. We assume that :

1/ $\mathcal{G}$ is a finitely generated extension of $\mathcal{F}$ of finite transcendence degree, and
2/ $\mathcal{F}$ and $\mathcal{G}$ have the same field of constants $\mathcal{C}$, which is algebraically closed.
Moreover, we suppose given once and for all a universal extension of $\mathcal{G}$. Every differential field mentioned will be contained in this universal extension ; in particular, every isomorphism of $\mathcal{G}$ will be into this universal extension. We denote the field of constants of the universal extension by $\mathcal{C}^*$.

If σ is any isomorphism of $\mathcal{G}$, we let $\mathcal{C}_\sigma$ denote the field of constants of the compositum $\mathcal{G}\langle\sigma\mathcal{G}\rangle$.

An isomorphism σ of $\mathcal{G}$ is called strong if $\sigma\mathcal{G}\subset\mathcal{G}\langle\mathcal{C}^*\rangle$ and $\mathcal{G}\subset(\sigma\mathcal{G})\langle\mathcal{C}^*\rangle$. Obviously every automorphism of $\mathcal{G}$ is a strong isomorphism of $\mathcal{G}$.

$\mathcal{G}$ is said to be a strongly normal extension of $\mathcal{F}$ if every $\mathcal{F}$-isomorphism of $\mathcal{G}$ is strong.

Let $\mathcal{G}$ be a strongly normal extension of $\mathcal{F}$. Let $\mathfrak{G}$ be the group of $\mathcal{F}$-automorphism of $\mathcal{G}$, and let $\mathfrak{G}^*$ be the set of $\mathcal{F}$-isomorphisms of $\mathcal{G}$; then $\mathfrak{G}\subset\mathfrak{G}^*$. If $\mathcal{F}_1$ is an intermediate differential field (that is an extension of $\mathcal{F}$ contained in $\mathcal{G}$), we denote the set of $\mathcal{F}_1$-automorphisms and the set of $\mathcal{F}_1$-isomorphisms by $\mathfrak{G}(\mathcal{F}_1)$ and $\mathfrak{G}^*(\mathcal{F}_1)$ respectively.

The principal theorem of the Galois theory of differential fields asserts that

$$\mathcal{F}_1 \dashrightarrow \mathcal{G}(\mathcal{F}_1)$$

is a one-to-one mapping of the set of all intermediate differential fields into a certain set of subgroups of $\mathcal{G}$. But in order to describe this certain set of subgroups we need a few definitions.

If σ, $\sigma' \in \mathcal{G}^*$, σ' is said to be a specialization (or a generic specialization) of σ if $(\sigma'\alpha)_{\alpha \in \mathcal{G}}$ is a specialization (or a generic specialization) of $(\sigma\alpha)_{\alpha \in \mathcal{G}}$ over $\mathcal{G}$.

A subset $\mathcal{M}^*$ of $\mathcal{G}^*$ is called an irreducible set in $\mathcal{G}^*$ if there exists an element $\sigma^* \in \mathcal{G}^*$ such that an element $\sigma \in \mathcal{G}^*$ belongs to $\mathcal{M}^*$ if and only if σ is a specialization of σ^*; the element σ^*, which obviously belongs to $\mathcal{M}^*$, is unique up to generic specialization, and is called a generic element of $\mathcal{M}^*$. The degree of transcendence of $\mathcal{G}\langle\sigma^*\mathcal{G}\rangle$ over $\mathcal{G}$, which is the same as that of $\mathcal{C}_{\sigma^*}$ over $\mathcal{C}$, is called the dimension of $\mathcal{M}^*$.

A subset $\mathcal{M}^*$ of $\mathcal{G}^*$ is called an algebraic set in $\mathcal{G}^*$ if $\mathcal{M}^*$ is the union of a finite number of irreductible sets in $\mathcal{G}^*$.

Two distinct algebraic sets in $\mathcal{G}^*$ have distinct intersections with $\mathcal{G}$. Therefore we may define an algebraic or irreducible set in $\mathcal{G}$ as the intersection with $\mathcal{G}$ of an algebraic or irreducible set in $\mathcal{G}^*$. If $\mathcal{M}$ is an algebraic set in $\mathcal{G}$, the algebraic set $\mathcal{M}^*$ in $\mathcal{G}^*$ such that $\mathcal{M} = \mathcal{M}^* \cap \mathcal{G}$ is unique; $\mathcal{M}^*$ is irreducible if and only if $\mathcal{M}$ is. If $\mathcal{M}$ is irreducible, its dimension is defined by the formula $\dim \mathcal{M} = \dim \mathcal{M}^*$.

Each algebraic set is the union of a unique finite set of irreducible sets none of which contains another. These irreducible sets are called the components of the given algebraic set.

$\mathcal{G}$ has a group structure, whereas $\mathcal{G}^*$ does not (because the product of two isomorphisms is in general not defined). But it can be shown that each isomorphism of $\mathcal{G}$ over $\mathcal{F}$ can be extended to a unique automorphism of $\mathcal{G}\langle \mathcal{C}^*\rangle$ over $\mathcal{F}\langle \mathcal{C}^*\rangle$; on the other hand, it is obvious that each automorphism of $\mathcal{G}\langle \mathcal{C}^*\rangle$ over $\mathcal{F}\langle \mathcal{C}^*\rangle$ is the extension of a unique automorphism of $\mathcal{G}$. Therefore we may identify each isomorphism of $\mathcal{G}$ over $\mathcal{F}$ with the automorphism of $\mathcal{G}\langle \mathcal{C}^*\rangle$ which extends it. After this identification, $\mathcal{G}^*$ is a group (of which $\mathcal{G}$ is a subgroup).

By an algebraic group (in $\mathcal{G}$ or in $\mathcal{G}^*$) we mean an algebraic set (in $\mathcal{G}$ or in $\mathcal{G}^*$) which is also a subgroup (of $\mathcal{G}$ or of $\mathcal{G}^*$). The components of an algebraic group are always disjoint. The component containing the identity is a normal subgroup ; the components are the cosets relative to the component of the identity, and all have the same dimension, which is called the dimension of the given algebraic group.

Now we are in a position to complete the statement of the principal theorem, by asserting that <u>a subgroup $\mathcal{G}_1$ of $\mathcal{G}$ corresponds to an intermediate differential field if and only if $\mathcal{G}_1$ is algebraic</u> ; also, if $\mathcal{G}_1 = \mathcal{G}(\mathcal{F}_1)$, then $\dim \mathcal{G}_1$ = the degree of transcendence of $\mathcal{G}$ over $\mathcal{F}_1$. Moreover, <u>$\mathcal{F}_1$ is itself strongly normal over $\mathcal{F}$ if and only if $\mathcal{G}_1$ is a normal subgroup of $\mathcal{G}$, and in this case the mapping which to each automorphism of $\mathcal{G}$ corresponds its restriction to $\mathcal{F}_1$, is a homomorphism, with kernel $\mathcal{G}_1$, of $\mathcal{G}$ onto the group of $\mathcal{F}$-automorphisms of $\mathcal{F}_1$</u> .

$\mathcal{G}$ itself obviously corresponds to $\mathcal{F}$; therefore $\mathcal{G}$ is algebraic. The component of the identity of $\mathcal{G}$ corresponds to the relative algebraic closure of $\mathcal{F}$ in $\mathcal{G}$; thus, $\mathcal{G}$ is irreducible if and only if $\mathcal{F}$ is relatively algebraically closed in $\mathcal{G}$.

For example, suppose $\mathcal{G}$ is an ordinary differential field (we write $\alpha' = \delta_1 \alpha$, $\alpha'' = \delta_1^2$, etc.), and is the extension of $\mathcal{F}$ generated by a fundamental system of solutions $\eta_1 , \dots , \eta_n$ of a homogeneous linear differential equation of order n ,

$$(1) \qquad y^{(n)} + p_1 y^{(n-1)} + \dots + p_{n-1} y' + p_n y = 0 ,$$

with coefficients $p_1 , \dots , p_n \in \mathcal{F}$; $\mathcal{G}$ is then called a <u>Picard-Vessiot</u> extension of $\mathcal{F}$ (see [3]). If σ is an $\mathcal{F}$-isomorphism of $\mathcal{G}$, $\sigma \eta_j$ is a solution of (1) and therefore is a linear combination with constant coefficients (i.e. with coefficients in C^*) of $\eta_1 , \dots , \eta_n$:

$$(2) \qquad \sigma \eta_j = \sum_{i=1}^{n} c_{ij} \eta_i \qquad (1 \leq i \leq n)$$

It follows from (2) that $\sigma \mathcal{G} \subset \mathcal{G}\langle C^* \rangle$. Also, it is almost immediate that the matrix (c_{ij}) is invertible ; therefore $\mathcal{G} \subset (\sigma \mathcal{G})\langle C^* \rangle$. Thus, every $\mathcal{F}$-isomorphism σ of $\mathcal{G}$ is strong, so that $\mathcal{G}$ is a strongly normal extension of $\mathcal{F}$.

Let us return to the general case. An algebraic group resembles, in many respects, a group variety, that is, an abstract algebraic variety (as defined by

Weil) which is a multiplicative group such that the mappings (s,t) ---> st and s ---> s^{-1} are rational (see [9]). In the special case of a Picard-Vessiot extension, the mapping σ ---> (c_{ij}) defined by (2) is an isomorphism of the algebraic group $\mathcal{G}$ into an algebraic linear group (as studied by several authors, especially by Chevalley [1]), a very special case of a group variety. One is thus led to ask if, in the general case, an algebraic group can always be realized as a group variety.

In the second lecture it will be shown that each irreducible algebraic group is isomorphic to a group variety over a field of characteristic 0 , and also the converse will be proved. Thus, the irreducible algebraic groups and the group varieties over fields of characteristic 0 are essentially the same things.

For bibliography see end of second Lecture.

COLLOQUE HENRI POINCARÉ
Institut Henri Poincaré
11 rue Pierre Curie
PARIS 5°

Octobre 1954

Differential fields and group varieties,

by E. R. KOLCHIN.

Second Lecture.

1.- Formulation of the theorem.

In this lecture I shall sketch the proof of a theorem which shows the connection between algebraic groups and group varieties. The proof, which contains little new, consists largely in combining in proper sequence certain results and ideas of Chevalley, S. Nakano, Rosenlicht and Weil, together with certain results of my own.

Let $\mathcal{C}$ be a differential field of constants with family of operators $\delta_1, \dots, \delta_m$; we assume that $\mathcal{C}$ is algebraically closed and of characteristic 0. We choose once and for all a universal extension $\mathcal{F}^*$ of $\mathcal{C}$ and denote its fields of constants by $\mathcal{C}^*$. $\mathcal{C}^*$ is an algebraically closed extension of $\mathcal{C}$ of infinite transcendence degree, and therefore may be used as a universal domain for algebraic geometry.

If $\mathcal{G}$ is a strongly normal extension of a differential field $\mathcal{F}$ with field of constants $\mathcal{C}$, admitting $\mathcal{F}^*$ as a universal extension, and $\mathcal{G}^*$ resp. $\mathcal{G}$ is the algebraic group of $\mathcal{F}$-isomorphisms resp. $\mathcal{F}$-automorphisms of $\mathcal{G}$, and if on the other hand V^* is a group variety, in the algebraic geometry with universal domain $\mathcal{C}^*$, having $\mathcal{C}$ as a field of definition, we shall call birational isomorphism of $\mathcal{G}^*$ onto V^* a mapping f of $\mathcal{G}^*$ onto V^* such that :

1) f is a group isomorphism of $\mathcal{G}^*$ onto V^*,
2) $\mathcal{C}_\sigma = \mathcal{C}(f(\sigma))$ for every $\sigma \in \mathcal{G}^*$, and
3) (for $\sigma, \sigma' \in \mathcal{G}^*$) σ' is a specialization of σ if and only if $f(\sigma')$ is a specialization of $f(\sigma)$ over $\mathcal{C}$ (in the usual sense of the word in algebraic geometry).

We also say, under the same circumstances, that the inverse mapping f^{-1} is a birational isomorphism of V^* onto $\mathcal{G}^*$. If f is such a birational isomorphism of $\mathcal{G}^*$ onto V^* then f maps $\mathcal{G}$ onto the group V consisting of all points of V^* which are rational over $\mathcal{C}$.

We can now state our theorem as follows : Every algebraic group $\mathfrak{G}^{\times}$ as above which is irreducible is birationally isomorphic to some group variety defined over $\mathcal{C}$, and conversely.

The first part of the theorem has been proved independently by H. Matsumura [5] . His proof is longer but, as opposed to the present one, has the merit of not requiring any special theorem on group varieties. The special theorem we use is one proved by Weil [10] and later refined by Rose licht [9] . To state this theorem we need the following definition due to Weil : If W is any algebraic variety, a normal law of composition on W is a rational mapping of $W \times W$ into W (not necessarily everywhere defined) which 1) is associative, and 2) has the property that if we denote the law of composition multiplicatively and if s , t are independent generic points of W over K then $K(s,st) = K(t,st) = K(s,t)$, where K is a field of definition of W and the law of composition. Weil's theorem is the following : If W is an algebraic variety with a normal law of composition then there exists a birational correspondence T between W and a group variety V which transforms the normal law of composition on W into the group composition on V . Rosenlicht's contribution is to add : If K is a field of definition of W and its normal law of composition as above, and if the points of W which are rational over K are dense in W (for example, if K is algebraically closed), then we may take V and T so that K is also a field of definition of V , the group composition on V , and T .

2.- Proof of theorem, first part.

Consider an irreducible algebraic grpup $\mathfrak{G}^{\times}$ as above. There exist generic elements $\sigma^{\times}$, $\tau^{\times}$ of $\mathfrak{G}^{\times}$ which are 'independent" (see [3]), that is, which have the property that if σ is a specialization of $\sigma^{\times}$ and τ is a specialization of $\tau^{\times}$ then (σ, τ) is a specializatiion of $(\sigma^{\times}, \tau^{\times})$; it follows that then $\sigma^{\times}\tau^{\times}$ is a generic element of $\mathfrak{G}^{\times}$.

Let $c_1(\sigma^{\times}), \ldots , c_r(\sigma^{\times})$ be constants such that $\mathcal{C}_{\sigma^{\times}} = \mathcal{C}(c_1(\sigma^{\times}), \ldots , c_r(\sigma^{\times}))$, and write $c(\sigma^{\times}) = (c_1(\sigma^{\times}), \ldots , c_r(\sigma^{\times}))$; define $c(\tau^{\times})$ and $c(\sigma^{\times}\tau^{\times})$ in the obvious way. Then $c(\sigma^{\times})$, $c(\tau^{\times})$, and $c(\sigma^{\times}\tau^{\times})$ are generic points over $\mathcal{C}$ of a certain variety $W^{\times}$ in r-dimensional affine space, and $(c(\sigma^{\times}) c(\tau^{\times})) \dashrightarrow c(\sigma^{\times}\tau^{\times})$ defines a normal law of composition on $W^{\times}$ over $\mathcal{C}$. Therefore, by the Weil-Rosenlicht result just mentioned, there exists a group variety $V^{\times}$ defined over $\mathcal{C}$, and for each generic element $\rho^{\times}$ of $\mathfrak{G}^{\times}$ a generic point $g(\rho^{\times}$

of V^* over $\mathcal{C}$, such that $\mathcal{C}_{\rho^*} = \mathcal{C}(g(\rho^*))$ and $g(\sigma^* \tau^*) = g(\sigma^*)\, g(\tau^*)$. Thus, $g(\sigma^* \tau^*)\, g(\tau^*)^{-1} = g(\sigma^*)$ is independent of the choice of τ^*. It easily follows that if σ is <u>any</u> element of $\mathcal{G}^*$, and if we choose a generic element τ^* of $\mathcal{G}^*$ such that σ, τ^* are independent, then the point $g(\sigma\tau^*)\, g(\tau^*)^{-1}$ of V^* does not depend on the choice of τ^*. Therefore we may define a mapping f of $\mathcal{G}^*$ into V^* by the formula $f(\sigma) = g(\sigma\tau^*)\, g(\tau^*)^{-1}$. It is not difficult to show now that f is a birational isomorphism of $\mathcal{G}^*$ onto V^*.

The device used here in defining f is like the one used by S. Nakano [7] in another connection.

3.- <u>Proof of theorem, second part.</u>

Let V^* be any group variety defined over $\mathcal{C}$ in the algebraic geometry with $\mathcal{C}^*$ as universal domain, let $\mathcal{G}_0^*$ denote the field of rational functions on V^*, and let $\mathcal{G}_0$ denote the field of those rational functions on V^* for which $\mathcal{C}$ is a field of definition. We identify each constant funtion with its value, so that $\mathcal{C} \subset \mathcal{G}_0$ and $\mathcal{C}^* \subset \mathcal{G}_0^*$; both these extensions are of transcendence degree equal to the dimension of V^*, which we denote by r, and we have $\mathcal{G}_0^* = \mathcal{G}_0(\mathcal{C}^*)$.

The elements of V^* operate on $\mathcal{G}_0^*$ in a natural way ; namely, for each $s \in V^*$ we consider the mapping ρ_s of $\mathcal{G}_0^*$ into itself defined by the formula $(\rho_s f)(t) = f(ts)$. Each ρ_s is a $\mathcal{C}^*$-automorphism of $\mathcal{G}_0^*$, and the mapping $s \dashrightarrow \rho_s$ is an isomorphism of the group V^*. Furthermore,

$$(1) \qquad \mathcal{G}_0(\rho_s \mathcal{G}_0) = \mathcal{G}_0(\mathcal{C}(s)) = (\rho_s \mathcal{G}_0)(\mathcal{C}(s))$$

for every $s \in V^*$.

It is known (e.g. see [6]) that there exist r derivations $D_1, \dots, D_r$ of $\mathcal{G}_0^*$ over $\mathcal{C}^*$ which are linearly independent over $\mathcal{G}_0^*$, which are <u>invariant</u> in the sense that $\rho_s D_i = D_i \rho_s$ ($s \in V^*$, $1 \leq i \leq r$), and which map $\mathcal{G}_0$ into itself. This fact, which is central in the proof, was pointed out to me in this connection by Chevalley.

Let $(u_{ij})_{1 \leq i \leq r, 0 \leq j < \infty}$ be a family of indeterminates, and write $\mathcal{F} = \mathcal{C}((u_{ij}))$, $\mathcal{G} = \mathcal{G}_0((u_{ij}))$. The fields $\mathcal{C}^*((u_{ij}))$ and $\mathcal{G}_0^*((u_{ij}))$ may then be denoted by $\mathcal{F}(\mathcal{C}^*)$ and $\mathcal{G}(\mathcal{C}^*)$ respectively. For each $s \in V^*$ the automorphism ρ_s of $\mathcal{G}_0^*$ over $\mathcal{C}^*$ can be extended to a unique automorphism (which we also denote by ρ_s) of $\mathcal{G}(\mathcal{C}^*)$ over $\mathcal{F}(\mathcal{C}^*)_s$. Similarly, each derivation D_i of $\mathcal{G}_0^*$ over $\mathcal{C}^*$ can be extended to a unique derivation (which we also denote by D_i)

of $\mathcal{G}(\mathcal{C}^*)$ over $\mathcal{C}^*$ such that $D_i u_{ij} = u_{i0}^{-1} u_{i,j+1}$ and $D_i u_{i'j} = 0$ $(i' \neq i)$. The extended D_i and ρ_s continue to satisfy the equations $\rho_s D_i = D_i \rho_s$, and the only elements of $\mathcal{G}(\mathcal{C}^*)$ annihilated by every D_i are those belonging to $\mathcal{C}^*$.

We now let $\delta_1, \ldots, \delta_m$ operate on $\mathcal{G}(\mathcal{C}^*)$ by defining.

$$\delta_1 \alpha = \sum_{i=1}^{r} u_{i0} D_i \alpha , \quad \delta_2 \alpha = 0 , \ldots , \quad \delta_m \alpha = 0$$

for every $\alpha \in \mathcal{G}(\mathcal{C}^*)$. It is clear that $\mathcal{G}(\mathcal{C}^*)$ (together with the family of operators $\delta_1, \ldots, \delta_m$) is a differential field of which $\mathcal{C}, \mathcal{F}, \mathcal{G}, \mathcal{C}^*$ are differential subfields. $\mathcal{G}(\mathcal{C}^*)$ itself may now be written $\mathcal{G}\langle \mathcal{C}^* \rangle$. Furthermore, if we denote u_{i0} by u_i, then $\delta_1^j u_i = u_{ij}$, so that $\mathcal{F} = \mathcal{C}\langle u_1, \ldots, u_r \rangle$ and $\mathcal{G} = \mathcal{F}\langle \mathcal{G}_0 \rangle$. It is not difficult to see, now, that $\mathcal{C}^*$ is the field of constants of $\mathcal{G}\langle \mathcal{C}^* \rangle$, and that $\mathcal{C}$ is the field of constants of both $\mathcal{F}$ and $\mathcal{G}$.

$\mathcal{G}$ is a finitely generated extension of the differential field $\mathcal{C}$; therefore there exists an isomorphism of $\mathcal{G}$ over $\mathcal{C}$ into the universal extension $\mathcal{F}^*$. It is easy to see that $\mathcal{C}^*$ and the image of $\mathcal{G}$, just like $\mathcal{C}^*$ and $\mathcal{G}$ itself, are linearly disjoint over $\mathcal{C}$. It follows that the isomorphism can be extended to a unique isomorphism of $\mathcal{G}\langle \mathcal{C}^* \rangle$ over $\mathcal{C}^*$ into $\mathcal{F}^*$. By virtue of this isomorphism we identify $\mathcal{G}\langle \mathcal{C}^* \rangle$ with a differential subfield of $\mathcal{F}^*$, so that henceforth all out differential fields are to be thought of as lying in $\mathcal{F}^*$.

We now observe that $\rho_s \delta_i = \delta_i \rho_s$ $(1 \leq i \leq m)$ for each $s \in V^*$, so that every ρ_s is an automorphism of $\mathcal{G}\langle \mathcal{C}^* \rangle$ over $\mathcal{F}\langle \mathcal{C}^* \rangle$ as differential fields (not merely as fields). This proves that each ρ_s, identified with its restriction to $\mathcal{G}$, is a strong isomorphism of $\mathcal{G}$ over $\mathcal{F}$. This can be seen also from the equations.

$$(2) \qquad \mathcal{G}\langle \rho_s \mathcal{G} \rangle = \mathcal{G}\langle \mathcal{C}(s) \rangle = (\rho_s \mathcal{G})\langle \mathcal{C}(s) \rangle$$

which follow from (1). From (2) we see that ρ_s is an automorphism of $\mathcal{G}$ if and only if $s \in V$, that is, s is rational over $\mathcal{C}$. It is not difficult to see, moreover, that (for $s, s' \in V^*$) s' is a specialization of s over $\mathcal{C}$ if and only if $\rho_{s'}$ is a specialization of ρ_s when both are considered as isomorphisms of $\mathcal{G}$.

Now, $\mathcal{F}$ is relatively algebraically closed in $\mathcal{G}$. Therefore there exists an $\mathcal{F}$-isomorphism σ^* of $\mathcal{G}$ such that every $\mathcal{F}$-isomorphism of $\mathcal{G}$ is a specialization of σ^* (see [3], ch. II, paragraph 2), and the transcendence degree of $\mathcal{G}\langle \sigma^* \mathcal{G} \rangle$ over $\mathcal{G}$ equals r. If s^* is a generic point of V^* over $\mathcal{C}$ then by (2) the transcendence degree of $\mathcal{G}\langle \rho_{s^*}\mathcal{G}\rangle$ over $\mathcal{G}$ is also r, so that ([2], ch. II, paragraph 2) ρ_{s^*} is a generic specialization of σ^*. Thus, every $\mathcal{F}$-isomorphism of $\mathcal{G}$ is a specialization of ρ_{s^*}, and therefore ([2], ch. II, paragraph 4) is strong; in other words, $\mathcal{G}$ is a strongly normal extension of $\mathcal{F}$. It is now a relatively simple matter to show that the mapping $s \dashrightarrow \rho_s$ ($s \in V^*$) is a birational isomorphism of V^* onto the algebraic group $\mathfrak{G}^*$ of all strong isomorphisms of $\mathcal{G}$ over $\mathcal{F}$.

4.- Two applications.

Let V^* be a group variety defined over an algebraically closed field $\mathcal{C}$ of characteristic 0.

An immediate corollary of the theorem just proved and the (completed) principal theorem stated in the first lecture is the following : *If N^* is a normal subgroup of V^* and also a bunch of varieties on V^* defined over $\mathcal{C}$, then there exists a homomorphism of V^* with kernel N^* onto a group variety defined over $\mathcal{C}$.* This result is not new (e.g. see [7]), and moreover remains valid without restriction on field characteristic, but the fact that it is a consequence of theorems on differential equations is perhaps not without interest.

Similarly, it is easy to derive a second corollary : *Let W^* be a subset of V^*, a necessary and sufficient condition that W^* be a subgroup of V^* and a bunch (of varieties on V^* defined over $\mathcal{C}$, is that) there exist a finite subset E of the field $\mathcal{G}_0$ of rational functions on V^* defined over $\mathcal{C}$ such that W^* is the set of all elements $s \in V^*$ for which $\rho_s\varphi = \varphi$ for every $\varphi \in E$.* This generalizes, in the case of field characteristic 0, a theorem on algebraic linear groups [2].

Bibliography.

[1] C. CHEVALLEY - Théorie des groupes de Lie, Tome II : Groupes algébriques, Paris, 1951.

[2] C. CHEVALLEY - E. KOLCHIN - Two proofs of a theorem on algebraic groups, Proc. Amer. math. Soc., 2 (1951), pp. 126-134.

[3] E.R. KOLCHIN - Galois theory of differential fields, Amer. J. Math., 75 (1953 pp. 753-824.

[4] E.R. KOLCHIN - Algebraic matric groups and the Picard-Vessiot theory of homogeneous linear ordinary differential equations, Ann. of Math 49 (1948), pp. 1-42.

[5] H. MATSUMURA - A paper to appear in Kyoto Math. Memoirs.

[6] S. NAKANO - On invariant differential forms on a group variety, J. Math. Soc. Jap., 2 (1951), pp. 216-227.

[7] S. NAKANO - Note on group varieties, Kyoto Math. Memoirs, 27 (1952), pp. 55-66.

[8] J.F. RITT - Differential algebra, Amer. Math. Soc. Colloq. Publ. 33, New York, 1950.

[9] M. ROSENLICHT - Generalized Jacobian varieties, Ann. of Math., 59 (1954), pp. 505-530.

[10] A. WEIL - Variétés abéliennes et courbes algébriques, Paris 1948.

[Les numéros de renvois bibliographiques indiqués dans le tirage provisoire de la 1e conférence : [2], [3], [7] et [9] sont à remplacer respectivement par : [3], [4], [8] et [10] pour correspondre aux références indiquées dans la bibliographie ci-dessus.]

ON THE GALOIS THEORY OF DIFFERENTIAL FIELDS.*

By E. R. Kolchin.[1]

Introduction.

1. Summary. In a preceding paper [7] there was presented a Galois theory, for a certain kind of differential field extension called strongly normal. The Galois group of a strongly normal extension is endowed with a structure very much like that of a group variety, as studied by Weil [14]. In the present paper the study of such Galois groups is renewed for the purpose of clarifying their connection with group varieties on the one hand and with strongly normal extensions on the other.

Consider a differential field $\mathcal{F}$ of characteristic 0 with algebraically closed field of constants $\mathcal{C}$, and a strongly normal extension $\mathcal{G}$ of $\mathcal{F}$; suppose given a universal extension $\mathcal{F}^*$ of $\mathcal{G}$, and denote the field of constants of $\mathcal{F}^*$ by $\mathcal{C}^*$. For reasons of convenience we use the term "Galois group of $\mathcal{G}$ over $\mathcal{F}$" for the group of all (ipso facto strong) isomorphisms of $\mathcal{G}$ over $\mathcal{F}$, and not for its subgroup consisting of all automorphisms of $\mathcal{G}$ over $\mathcal{F}$. The field $\mathcal{C}^*$, being algebraically closed and of finite transcendence degree over $\mathcal{C}$, may be used as a universal domain for algebraic geometry (Weil [13], ch. I, § 1); it is the group varieties of this algebraic geometry, defined over $\mathcal{C}$ and slightly generalized to permit group varieties which are reducible (i. e. which have more than one component), which we consider. By an "algebraic group" we mean either a Galois group or a group variety as above.

Chapter I is primarily a study of "rational" homomorphisms of algebraic groups into algebraic groups; these seem to be the homomorphisms appropriate to the consideration of algebraic groups. Rational homomorphisms of group varities into group varieties, without restriction to fields of characteristic 0, were considered by Weil [14] (who omitted the adjective "rational").

Specializing the concept of rational homomorphism we obtain that of birational isomorphism. In Chapter II it is shown that every Galois group is birationally isomorphic to a group variety, and conversely that every *irreducible* group variety is birationally isomorphic to a Galois group; this

* Received June 29, 1955.

[1] Some of the results in this paper were obtained in connection with a contract with the National Science Foundation, and some while the author was a Guggenheim fellow.

answers a question raised in [7], p. 759. The first of these results has already been proved by Matsumura [8] for irreducible Galois groups (the extension to reducible Galois groups is fairly trivial); the present proof, although shorter, makes use of a special theorem due to Weil and refined by Rosenlicht (see Chapter II, § 1 below), whereas Matsumura's proof has the merit of being more self-contained. The proof of the second of these results makes use of the fact that the vector space of derivations of the field of rational functions on an irreducible group variety has a base consisting of invariant derivations (see Chapter II, § 2 below); this fact, which is central in the proof, was pointed out to me in this connection by Chevalley. Whether or not the result can be extended to reducible group varieties remains open. Chapter II also contains certain consequences of these two results.

In Chapter III the concept of a rational homomorphism of the Galois group $\mathfrak{G}$ of $\mathcal{G}$ over $\mathcal{F}$ into a group variety G is generalized to obtain that of a rational crossed homomorphism, or a cocycle of dimension 1, or a 1-cocycle, of $\mathfrak{G}$ into G. This leads, in much the usual way, to the set $H^1(\mathfrak{G}, G)$ of cohomology classes of dimension 1 of $\mathfrak{G}$ into G. Since G is not assumed to be commutative, $H^1(\mathfrak{G}, G)$ is not in general a group. It is proved that $H^1(\mathfrak{G}, G)$ is trivial provided *either* G is a full matric group *or* $\mathcal{F}$ is algebraically closed; also, if G is commutative (so that $H^1(\mathfrak{G}, G)$ is a group) then every element of $H^1(\mathfrak{G}, G)$ has finite order. Using the first of these results, it is shown that the strongly normal extension $\mathcal{G}$ of $\mathcal{F}$ is a Picard-Vessiot extension if and only if its Galois group $\mathfrak{G}$ is birationally isomorphic to an algebraic group of matrices; this answers another question raised in [7], p. 759. Similarly, the last of these results is used to give a short proof of the characterization given in [7] of all strongly normal extensions of transcendence degree 1.

2. Assumptions and notation. Throughout this paper $\mathcal{F}$ denotes a differential field of characteristic 0 with derivation operators $\delta_1, \cdots, \delta_m$ and algebraically closed field of constants $\mathcal{C}$, and $\mathcal{G}$ denotes a strongly normal extension of $\mathcal{F}$. We fix a universal extension $\mathcal{F}^*$ of $\mathcal{G}$ and denote the field of constants of $\mathcal{F}^*$ by $\mathcal{C}^*$; except when the contrary is explicitly stated, all differential fields considered are differential subfields of $\mathcal{F}^*$, so that every isomorphism of a differential field has image in $\mathcal{F}^*$.

Because $\mathcal{G}$ is strongly normal over $\mathcal{F}$, every isomorphism σ of $\mathcal{G}$ over $\mathcal{F}$ is strong; denoting the field of constants of $\mathcal{G}\langle\sigma\mathcal{G}\rangle$ by $\mathcal{C}(\sigma)$ (instead of by $\mathcal{C}_\sigma$ as in [7]), we can express this fact by writing

$$\mathcal{G}\langle\sigma\mathcal{G}\rangle = \mathcal{G}\langle\mathcal{C}(\sigma)\rangle = (\sigma\mathcal{G})\langle\mathcal{C}(\sigma)\rangle.$$

As in [7], we identify each such σ with the unique automorphism of $\mathcal{G}\langle\mathcal{C}^*\rangle$ over $\mathcal{F}\langle\mathcal{C}^*\rangle$ which extends σ; the set of all isomorphisms of $\mathcal{G}$ over $\mathcal{F}$ is thereby identified with the group of all automorphisms of $\mathcal{G}\langle\mathcal{C}^*\rangle$ over $\mathcal{F}\langle\mathcal{C}^*\rangle$. We call this group the *Galois group* of $\mathcal{G}$ over $\mathcal{F}$ and denote it by $\mathfrak{G}$ (instead of by $\mathfrak{G}^*$ as in [7]); we recall that $\mathfrak{G}$ is algebraic in the sense defined in [7]. The group of all automorphisms of $\mathcal{G}$ over $\mathcal{F}$ is a subgroup of $\mathfrak{G}$. If $\sigma_1, \cdots, \sigma_p \in \mathfrak{G}$ we denote by $\mathcal{C}(\sigma_1, \cdots, \sigma_p)$ the field of constants of $\mathcal{G}\langle\sigma_1\mathcal{G}, \cdots, \sigma_p\mathcal{G}\rangle$, that is, the compositum of the p fields $\mathcal{C}(\sigma_1), \cdots, \mathcal{C}(\sigma_p)$. It is easy to see that $\mathcal{C}(\sigma^{-1}) = \mathcal{C}(\sigma)$ and $\mathcal{C}(\sigma\tau, \sigma) = \mathcal{C}(\sigma, \tau) = \mathcal{C}(\sigma\tau, \tau)$ for all $\sigma, \tau \in \mathfrak{G}$.

Chapter I. Rational homomorphisms.

1. Varieties. The field $\mathcal{C}^*$ is an algebraically closed extension of $\mathcal{C}$ of infinite transcendence degree, and therefore may be used as a universal domain for algebraic geometry. We find it convenient to deviate from Weil's terminology, in order to permit varieties which are not irreducible. Specifically, we adopt the following terminology, which is sufficient for our purpose: [2] we call *irreducible variety* what Weil calls abstract variety; we call *variety* any finite union of irreducible subvarieties of some irreducible variety (in Weil's terminology, a bunch of subvarieties of an abstract variety), or the union of a disjoint finite set of such. Every irreducible variety of dimension > 0 which we consider, except those for which the contrary is explicitly stated, is assumed to have $\mathcal{C}$ as a field of definition. Similarly, when we speak of a specialization of a point it is always to be understood, except when the contrary is stated, that the specialization is over $\mathcal{C}$. We find it convenient sometimes to call a subset of a variety *algebraic* to indicate that the subset is a subvariety. What we mean by the *components* of a variety is clear.

Let V and W be varieties, and denote the components of V by $V_1, \cdots, V_h$.

[2] An intrinsic definition of abstract algebraic variety, encompassing the case of reducible varieties, has been given by Serre [12]. The present definition, which involves an imbedding in some irreducible variety, is adopted here in the interest of brevity. This is, of course, merely a difference of procedure and not of substance. However, in the discussion below of rational mappings, the definition adopted for what it means for a rational mapping to be defined at a particular point is weaker than the definition that naturally arises in developing farther the ideas of Serre. Specifically, the present definition leads to a notion of isomorphism of varieties such that two components of a variety can have different higher intersection properties from the corresponding components of an isomorphic variety; on the other hand, because tthe definition here is weaker, Proposition 1 below is stronger. Of course, when dealing with varieties for which the components are disjoint (which by Proposition 1 is the case for group varieties) this difference in the two definitions disappears.

By a *rational mapping* of V into W we mean any sequence $f=(f_1,\cdots,f_h)$ where, for each i, f_i is a rational mapping of V_i into some component of W (that is, f_i is what is called by Weil [14] a function defined on V_i with values in a component of W); by a *field of definition* of f we mean a common field of definition of $f_1,\cdots,f_h$. All rational mappings except those for which the contrary is stated, are assumed to have $\mathcal{E}$ as a field of definition. We say the rational mapping f is *defined* at a point $s\in V$ if, for every point $t\in V$ of which s is a specialization, there exists an index i such that $t\in V_i$ and f_i is defined at t, and for all such indices i the value $f_i(t)$ is the same. If f is defined at s then there exists an i such that f_i is defined at s, and the point $f_i(s)$ is independent of the choice of i; we call this point the *value* of f at s and denote it by $f(s)$. This definition insures that if s' is a specialization of $s\in V$ and f is defined at s', then f is defined at s and $(s',f(s'))$ is a specialization of $(s,f(s))$. If ϕ is a mapping (in the set-theoretic sense) of a subset of V into W, then we permit ourselves to say that ϕ is *rational* if there exists a rational mapping f of V into W such that f is defined at precisely those points $s\in V$ at which ϕ is defined and $f(s)=\phi(s)$ at all such points s; in other words, we identify each rational mapping of V into W with a mapping, which we call rational.

By a *biregular birational mapping of V onto W* we mean a one-to-one mapping f of V onto W such that f and f^{-1} are both rational and everywhere defined.

By a *group variety* we mean any variety G with a group structure such that (when the group composition is written multiplicatively) the mapping $(s,s')\rightarrow ss'$ of $G\times G$ into G and the mapping $s\rightarrow s^{-1}$ of G into G are rational. Using an argument which is familiar in several special cases and analogous situations ([5], § 2, Th., p. 10; Weil [14], § 4, Prop. 9, pp. 39-40; Chevalley [3], Ch. I, § 3, Th. 2, p. 86; [7], Ch. II, Th. 1, p. 788), it is easy to prove the following result.

PROPOSITION 1. *Precisely one component of any group variety G contains the unity element; this component G^0 is a normal subgroup of finite index, and the cosets of G relative to G^0 are the components of G.*

Thus, the components of G are disjoint and all have the same dimension, which we call the dimension of G and denote by $\dim G$.

PROPOSITION 2. *If $\mathfrak{h}$ is a subgroup of the group variety G consisting solely of points rational over $\mathcal{E}$, and H is the smallest subvariety of G containing $\mathfrak{h}$, then H is a subgroup of G.*

Proof. For any $s \varepsilon \mathfrak{h}$, sH is the smallest subvariety of G containing $s\mathfrak{h} = \mathfrak{h}$, so that $sH = H$; therefore $\mathfrak{h}H = H$. For every $t \varepsilon H$ which is rational over $\mathcal{K}$, Ht is the smallest subvariety of G containing $\mathfrak{h}t$, and $\mathfrak{h}t \subset \mathfrak{h}H = H$, so that $Ht \subset H$; this inclusion obviously continues to hold without the restriction that t be rational over $\mathcal{K}$, so that $HH \subset H$. Finally, H^{-1} is the smallest subvariety of G containing $\mathfrak{h}^{-1} = \mathfrak{h}$, so that $H^{-1} = H$. Therefore H is a group.

PROPOSITION 3. *If $\mathfrak{g}$ is a subgroup of the group variety G such that $G - \mathfrak{g}$ is contained in a subvariety of G all the components of which are of lower dimension than G, then $\mathfrak{g} = G$.*

Proof. If $\mathfrak{g}$ were different from G, a coset $s\mathfrak{g}$ of G relative to $\mathfrak{g}$ would be contained in $G - \mathfrak{g}$ and therefore in a subvariety of G all the components of which have dimension $< \dim G$, and we obviously may suppose s rational over $\mathcal{K}$; therefore $\mathfrak{g}$ would be contained in such a variety, so that $G = \mathfrak{g} \cup (G - \mathfrak{g})$ would, too, contradiction.

The proofs of Propositions 2 and 3 are the same as those of the analogous propositions on algebraic groups of strong isomorphisms of a differential field (see [7], Ch. II, Props. 14, 15, p. 789, for the case of algebraic groups of automorphisms).

2. Rational homomorphisms of group varieties. Consider two group varieties G and H, and a rational mapping f of G into H such that $f(st) = f(s)f(t)$ whenever s and t are points of G such that f is defined at s, t, and st (or what amounts to the same thing, whenever s and t are independent generic points of components of G). We call such a rational mapping a *rational homomorphism* of G into H.

PROPOSITION 4. *A necessary and sufficient condition that f be a rational homomorphism of the group variety G into the group variety H is that f be a homomorphism of the group G into the group H having the two properties: 1) for all $s \varepsilon G$, $\mathcal{K}(f(s)) \subset \mathcal{K}(s)$; 2) for all $s, s' \varepsilon G$ with s' a specialization of s, $f(s')$ is a specialization of $f(s)$.*

Proof. Let f be a rational homomorphism of G into H, and let s' be any point of G. Choose independent generic points s, t of components of G such that s' is a specialization of s; there exists a specialization t' of t such that t' is rational over $\mathcal{K}$ and f is defined at t' and $s't'$. Then f is defined at st', and $f(s) = f(st')f(t')^{-1}$. This makes it clear that f is defined at s', that is, f is defined at every point of G. Thus, f is a homomorphism (in the group theoretic sense) of the group G into the group H, and f obviously has the two required properties.

Conversely, let f be any homomorphism of the group G into the group H having these two properties. Letting $s_1, \cdots, s_n$ be independent generic points of the n components of G, we see from 1) that there exists a rational mapping g of G into H such that $g(s_i) = f(s_i)$ $(1 \leqq i \leqq h)$; we then see from 2) that $g(s) = f(s)$ for every generic point of every component of G, so that $g(st) = g(s)g(t)$ whenever s and t are independent generic points of components of G; in other words, g is a rational homomorphism of G into H. Therefore g is everywhere defined and, by 2), $g(s) = f(s)$ for every $s \varepsilon G$.

PROPOSITION 5. *Let f be a rational homomorphism of a group variety G into a group variety H. If N is any algebraic set in H then $f^{-1}(N)$ is an algebraic set in G. If M is an algebraic subgroup of G then $f(M)$ is an algebraic subgroup of H.*

Proof. Let A be the smallest algebraic set in G containing $f^{-1}(N)$, let A_0 be any component of A, and let s be a generic point of A_0. If $f(s)$ were not a point of N there would exist a proper algebraic subset A_1 of A_0 such that every specialization s' of s with $s' \not\varepsilon A_1$ has the property that $f(s') \not\varepsilon N$, that is $f^{-1}(N)$ would be contained in the algebraic set obtained from A on replacing the component A_0 by A_1, so that A would not be the smallest alegbraic set in G containing $f^{-1}(N)$. Therefore $f(s) \varepsilon N$, so that $f(A_0) \subset N$, whence $f(A) \subset N$. Thus $f^{-1}(N) = A$, which is algebraic.

Let B be the smallest algebraic set in H containing $f(M)$; obviously B is also the smallest algebraic set in H containing $f(M')$, where M' is the group of all points of M which are rational over $\mathcal{C}$, so that by Proposition 2 B is an algebraic group. Let $s_1, \cdots, s_h$ be generic points of all the components of M, and let B_i be the irreducible algebraic set in H with generic point $f(s_i)$. Clearly $B = B_1 \cup \cdots \cup B_h$. For each i there exists a proper algebraic subset C_i of B_i such that every specialization t_i' of $f(s_i)$ with $t_i' \not\varepsilon C_i$ has the property that $t_i' = f(s_i')$ for some specialization s_i' of s_i. Thus, $B - f(M)$ is contained in an algebraic set every component of which is a proper subset of a component of B. It follows from Proposition 3 that $f(M) = B$.

COROLLARY. *The kernel and the image of a rational homomorphism are algebraic.*

We call a rational homomorphism of a group variety G into a group variety H a *birational isomorphism* of G onto H provided the kernel is trivial, the image is H, and the inverse is a rational homomorphism of H into G. It follows from Proposition 4 that f is a birational isomorphism of G onto H if and only if f is an isomorphism of the group G onto the group H such that: 1) $\mathcal{C}(s) = \mathcal{C}(f(s))$ for every $s \varepsilon G$; 2) for all s, s' in G, s' is a specialization of s if and only if $f(s')$ is a specialization of $f(s)$.

3. Rational homomorphisms in general. Henceforth we use the term "algebraic group" to mean *either* a Galois group (of a strongly normal extension of a differential field with field of constants $\mathcal{C}$), *or else* a group variety. We generalize to arbitrary algebraic groups the notion of rational homomorphism previously defined for group varieties.

By a *rational homomorphism* of an algebraic group G into an algebraic group H we mean any homomorphism f of the group G into the group H having the two properties that whenever $s \varepsilon G$ then $\mathcal{C}(f(s)) \subset \mathcal{C}(s)$, and whenever $s, s' \varepsilon G$ and s' is a specialization of s then $f(s')$ is a specialization of $f(s)$. If in addition the kernel of f is trivial and the image of f is H, so that f is an isomorphism of the group G onto the group H, and if f^{-1} is a rational homomorphism of H into G, then we say that f is a *birational isomorphism* of G onto H. In other words, a birational isomorphism of G onto H is an isomorphism f of the group G onto the group H such that $\mathcal{C}(f(s)) = \mathcal{C}(s)$ $(s \varepsilon G)$ and s' is a specialization of s if and only if $f(s')$ is a specialization $f(s)$ $(s, s' \varepsilon G)$.

Example 1. Let $\mathcal{H}$ be a differential field between $\mathcal{F}$ and $\mathcal{G}$ which is strongly normal over $\mathcal{F}$, and denote the Galois group of $\mathcal{H}$ over $\mathcal{F}$ by $\mathfrak{H}$. We know ([7], Ch. III, Th. 4, p. 797) that the algebraic subgroup $\mathfrak{G}(\mathcal{H})$ of $\mathfrak{G}$ corresponding to the intermediate differential field $\mathcal{H}$ is normal, and that the mapping r which to each $\sigma \varepsilon \mathfrak{G}$ corresponds the restriction of σ to $\mathcal{H}$ is a homomorphism, with kernel $\mathfrak{G}(\mathcal{H})$ and image $\mathfrak{H}$, of the group $\mathfrak{G}$ into the group $\mathfrak{H}$. For every $\sigma \varepsilon \mathfrak{G}$, the field of constants of $\mathcal{H}\langle r(\sigma)\mathcal{H}\rangle = \mathcal{H}\langle \sigma\mathcal{H}\rangle$ is obviously contained in the field of constants of $\mathcal{G}\langle \sigma\mathcal{G}\rangle$, that is, $\mathcal{C}(r(\sigma)) \subset \mathcal{C}(\sigma)$. Also, if $(\sigma'\alpha)_{\alpha \varepsilon \mathcal{G}}$ is a specialization of $(\sigma\alpha)_{\alpha \varepsilon \mathcal{G}}$ over $\mathcal{G}$ then $(r(\sigma')\alpha)_{\alpha \varepsilon \mathcal{H}} = (\sigma'\alpha)_{\alpha \varepsilon \mathcal{H}}$ is a specialization of $(r(\sigma)\alpha)_{\alpha \varepsilon \mathcal{H}} = (\sigma\alpha)_{\alpha \varepsilon \mathcal{H}}$ over $\mathcal{H}$, that is, if σ' is a specialization of σ then $r(\sigma')$ is a specialization of $r(\sigma)$. Therefore r is a rational homomorphism of $\mathfrak{G}$ into $\mathfrak{H}$.

Example 2. If $\mathcal{G}$ is a Picard-Vessiot extension of $\mathcal{F}$ then ([5] and [6]) $\mathcal{G}$ has a fundamental system of generators, that is, a finite number of elements

$\eta_1, \cdots, \eta_n$ linearly independent over constants such that $\mathcal{G} = \mathcal{F}\langle \eta_1, \cdots, \eta_n \rangle$ and such that for every $\sigma \varepsilon \mathfrak{G}$ there exists an invertible matrix of constants

$$c(\sigma) = (c_{ij}(\sigma))_{1 \leqq i \leqq n,\, 1 \leqq j \leqq n}$$

for which

$$\sigma\eta_j = \sum_{i=1}^{n} c_{ij}(\sigma)\eta_i \qquad (1 \leqq j \leqq n).$$

The mapping $c: \sigma \to c(\sigma)$ $(\sigma \varepsilon \mathfrak{G})$ is an isomorphism of $\mathfrak{G}$ onto an algebraic subgroup of the group variety $GL(n)$ of all invertible square matrices of degree n with coordinates in $\mathcal{C}^*$. It is not difficult to see that c is a birational isomorphism.

It is obvious that if f is a rational homomorphism of an algebraic group G into an algebraic group H and g is a rational homomorphism of H into an algebraic group I, then $g \circ f$ is a rational homomorphism of G into I.

PROPOSITION 6. *Let f and f' be rational homomorphisms of an algebraic group G into group varieties H and H' respectively. If f and f' have the same kernel and have respective images H and H', then the isomorphism ϕ of H onto H' such that $\phi \circ f = f'$ is birational.*

Proof. Letting s be any element of G we see that for a generic specialization s' of s, $f(s) = f(s')$ if and only if $f'(s) = f'(s')$; this means that an isomorphism of $\mathcal{C}(s)$ leaves invariant each element of $\mathcal{C}(f(s))$ if and only if it leaves invariant each element of $\mathcal{C}(f'(s))$, so that $\mathcal{C}(f(s)) = \mathcal{C}(f'(s))$; thus $\mathcal{C}(t) = \mathcal{C}(\phi(t))$ for every $t \varepsilon H$. Now, if s is a generic point of a component of G then $f(s)$ is a generic point of a component of H, and every generic point of a component of H is of this form; similarly for H'. If t' is a generic specialization of $t = f(s)$, then there exists a generic specialization s' of s such that $t' = f(s')$, and obviously $f'(s')$ is a generic specialization of $f'(s)$. Therefore there exists a rational mapping g of H into H' such that $g(t) = \phi(t)$ for every generic point t of a component of H. If t, t' are independent generic points of components of H then we may write $t = f(s)$, $t' = f(s')$, where s, s' are independent generic elements of components of G, so that $g(tt') = g(f(s)f(s')) = g(f(ss'))$ $= \phi(f(ss')) = f'(ss') = f'(s)f'(s') = \phi(t)\phi(t') = g(t)g(t')$. Therefore g is a rational homomorphism of H into H'. It follows that $g \circ f$ is a rational homomorphism of G into H' coinciding with f' at every generic point of every component of G and therefore at every point whatever, that is $g \circ f = f'$. Therefore $\phi = g$, so that ϕ is a rational homomorphism of H into H'.

Reversing the roles of f and f' we see that ϕ^{-1} is rational too, so that ϕ is birational.

COROLLARY. *If f is a rational homomorphism of a group variety G into a group variety H such that the kernel of f is trivial and the image of f is H, then f is a birational isomorphism of G onto H.*

Proof. In the proposition set $H' = G$, $f' =$ the identity mapping of G.

PROPOSITION 7. *Let G be an algebraic group, G_1 a normal algebraic subgroup of G of finite index, and K any normal algebraic subgroup of G. Then every rational homomorphism f_1 of G_1 with kernel $G_1 \cap K$ and image a group variety can be extended to a rational homomorphism f of G with kernel K and image a group variety.*

Proof. We extend f_1 in two stages; first to a rational homomorphism of G_1K with kernel K and the same image as f_1, and then to a rational homomorphism of G. G_1K is the union of its cosets relative to G_1; these are finite in number and each is an algebraic set, so that G_1K is a normal algebraic subgroup of G. Every element s of G_1K is of the form $s = s_0a$, where $s_0 \in G_1$ and $a \in K$, and $f_1(s_0)$ clearly depends only on s and not on the particular choice of s_0 and a. Therefore we may define a mapping g of G_1K into the image H_1 of f_1 by the formula $g(s_0a) = f_1(s_0)$. The computation $g(s_0'a's_0a) = g(s_0's_0 \cdot s_0^{-1}a's_0a) = f_1(s_0's_0) = f_1(s_0')f_1(s_0) = g(s_0'a')g(s_0a)$ shows that g is a homomorphism of G_1K with image H_1, and clearly g has kernel K. Now every coset of G_1K relative to G_1 contains an element of K, and therefore an element $a \in K$ such that $\mathscr{k}(a) = \mathscr{k}$. Therefore every element $s \in G_1K$ can be written in the form $s = s_0a$ with $s_0 \in G_1$, $a \in K$, and $\mathscr{k}(a) = \mathscr{k}$, so that $\mathscr{k}(g(s)) = \mathscr{k}(f(s_0)) \subset \mathscr{k}(s_0) = \mathscr{k}(s_0a) = \mathscr{k}(s)$; furthermore, if s' is a specialization of s then s' belongs to the same coset as s, so that $s' = s_0'a$, where $s_0' \in G_1$; obviously s_0' is a specialization of s_0, so that $f_1(s_0')$ is a specialization of $f_1(s_0)$, whence $g(s')$ is a specialization of $g(s)$. Thus, the homomorphism g is rational.

Let $s_1, \cdots, s_r$ be a complete set of representatives of the cosets of G relative to G_1K; we assume without loss of generality that $\mathscr{k}(s_i) = \mathscr{k}$ $(1 \leqq i \leqq r)$ and s_1 is the unity element of G. For each index i the mapping ϕ_i of G_1K defined by $\phi_i(s) = s_iss_i^{-1}$ is clearly a birational isomorphism of G_1K onto G_1K. For each pair of indices i, j there is an index $k(i,j)$ and an element $s(i,j) \in G_1K$ such that $s_is_j = s(i,j)s_{k(i,j)}$. Let $H_2, \cdots, H_r$ be varieties biregularly birationally equivalent to H_1 such that $H_1, H_2, \cdots, H_r$ are disjoint, and let h_i be a biregular birational mapping of H_1 into H_i $(1 \leqq i \leqq r)$, where h_1 is the identity mapping of H_1.

Since g and $g \circ \phi_i$ are rational homomorphisms of G_1K with the same kernel K and image H_1, there exists by Proposition 6 a birational isomorphism ψ_i of H_1 onto H_1 such that $\psi_i \circ g = g \circ \phi_i$. Furthermore, the mapping ρ_{ij} of H_1 defined by $\rho_{ij}(t) = t \cdot g(s(i,j))$ is a biregular birational mapping of H_1 onto H_1. Also, the law of composition, which we denote by μ_1, of the group variety H_1 is an everywhere defined rational mapping of $H_1 \times H_1$ into H_1. Consequently the mapping

$$\mu_{ij} = h_{k(i,j)} \circ \rho_{ij} \circ \mu_1 \circ (h_i^{-1} \times (\psi_i \circ h_j^{-1}))$$

is an everywhere defined rational mapping of $H_i \times H_j$ into the variety $H = H_1 \cup \cdots \cup H_r$, and it is easy to see that $\mu_{11} = \mu_1$. There is a unique everywhere defined rational law of composition μ on H which extends every μ_{ij}; since μ extends μ_1 we may write $\mu(t,t') = tt'$ for $t, t' \varepsilon H$.

Let f be the mapping of G into H defined by $f(ss_i) = h_i(g(s))$ $(s \varepsilon G_1K, 1 \leq i \leq r)$. Clearly f has image H, extends f_1, and because

$$\begin{aligned} f(ss_i \cdot s's_j) &= f(s \cdot s_i s' s_i^{-1} \cdot s_i s_j) = f(s\phi_i(s')s(i,j)s_{k(i,j)}) \\ &= h_{k(i,j)}(g(s)g(\phi_i(s'))g(s(i,j))) \\ &= (h_{k(i,j)} \circ \rho_{ij} \circ \mu_1 \circ (h_i^{-1} \times (\psi_i \circ h_j)))(h_i(g(s)), h_j(g(s'))) \\ &= h_i(g(s))h_j(g(s')) = f(ss_i)f(s's_j), \end{aligned}$$

f is a homomorphism of the group G. Therefore H, with the law of composition μ, is a group. To prove that H is a group variety it remains to show that $t \to t^{-1}$ $(t \varepsilon H)$ is an everywhere defined rational mapping of H into H. Now, the mappings λ_i and ρ_i of H defined by $\lambda_i(t) = f(s_i)^{-1}t$ and $\rho_i(t) = tf(s_i)^{-1}$ are everywhere defined rational mappings of H into H mapping H_i into H_1, and the mapping ν of H_1 defined by $\nu(t) = t^{-1}$ is an everywhere defined rational mapping of H_1 into H; but the mapping $t \to t$ $(t \varepsilon H)$ obviously coincides with $\lambda_i \circ \nu \circ \rho_i$ on H_i and therefore is rational and everywhere defined.

Finally, if $s \varepsilon G_1K$ then

$$\mathcal{E}(f(ss_i)) = \mathcal{E}(h_i(g(s))) = \mathcal{E}(g(s)) \subset \mathcal{E}(s) = \mathcal{E}(ss_i),$$

and if $s's_i$ is a specialization of ss_i then s' is a specialization of s, $g(s')$ is a specialization of $g(s)$, and $f(s's_i) = h_i(g(s'))$ is a specialization of $f(ss_i) = h_i(g(s))$, so that f is rational.

COROLLARY. *A birational isomorphism of the component of the identity $\mathfrak{G}^0$ of the Galois group $\mathfrak{G}$ onto a group variety can always be extended to a birational isomorphism of $\mathfrak{G}$ onto a group variety.*

Proof. In the proposition set $G = \mathfrak{G}$, $G_1 = \mathfrak{G}^0$, and $K =$ the group consisting of the identity automorphism.

17

Chapter II. Birational isomorphisms between Galois groups and group varieties.

1. Galois groups onto group varieties. In this section we shall prove the following theorem.

THEOREM 1. *The Galois group* $\mathfrak{G}$ *of the strongly normal extension* $\mathscr{G}$ *of the differential field* $\mathscr{F}$ *of characteristic* 0 *with algebraically closed field of constants* $\mathscr{C}$ *is always birationally isomorphic to a group variety defined over* $\mathscr{C}$.

Proof. By the corollary to Proposition 7 of Chapter I, we may assume that $\mathfrak{G}$ is irreducible. To prove the theorem under this added hypothesis, we use a theorem due to Weil ([14], Th. 15, p. 54), and later refined by Rosenlicht ([11], Th. 4, p. 511). To state this theorem we need the following definition due to Weil ([14], p. 51): If W is any irreducible variety, a *normal law of composition* on W is a rational mapping of $W \times W$ into W which, if we denote the law of composition multiplicatively, has the two properties that 1) $\mathscr{C}(s, st) = \mathscr{C}(t, st) = \mathscr{C}(t, s)$ for independent generic points s and t of W, and 2) $(st)u$ $s(tu)$ for independent generic points s, t, and u of W. Weil's theorem is: If W is an irreducible variety with a normal law of composition, then there exists a birational correspondence T between W and a group variety G which transforms the normal law of composition on W into the group composition on G (T, G, and the group composition on G not necessarily defined over $\mathscr{C}$). Rosenlicht's contribution is to add: If the points of W which are rational over $\mathscr{C}$ form a dense (Zariski topology) subset of W (which is true in the present case because $\mathscr{C}$ is algebraically closed), then we may take G and T so that T, G, and the group composition on G are defined over $\mathscr{C}$.

Let σ, τ be generic elements of $\mathfrak{G}$ which are independent ([7], pp. 777-778); then $\sigma\tau$ is also a generic element of $\mathfrak{G}$. Let $\gamma_1, \cdots, \gamma_r$ be constants such that $\mathscr{C}(\sigma) = \mathscr{C}(\gamma_1, \cdots, \gamma_r)$. Each γ_i can be expressed as the quotient of two polynomials over $\mathscr{G}$ in a finite number of elements $\sigma\eta_1, \cdots, \sigma\eta_n$, where each $\eta_j \in \mathscr{G}$; fixing two such polynomials, for each generic element ρ of $\mathfrak{G}$ let $c_i(\rho)$ be the element obtained on replacing each $\sigma\eta_j$ by $\rho\eta_j$ in the quotient of the two polynomials. Then $c_i(\sigma) = \gamma_i$, and for each generic element ρ of $\mathfrak{G}$, $\mathscr{C}(\rho) = \mathscr{C}(c_1(\rho), \cdots, c_r(\rho))$. Also, writing $c(\rho) = (c_1(\rho), \cdots, c_r(\rho))$, we see that $c(\sigma)$ is a generic point of a certain variety W in r-dimensional affine space, that $(c(\sigma), c(\tau))$ is a generic point of $W \times W$, and that $c(\sigma\tau)$ is a generic point of W with $\mathscr{C}(c(\sigma\tau)) = \mathscr{C}(\sigma\tau) \subset \mathscr{C}(\sigma, \tau) = \mathscr{C}(c(\sigma), c(\tau))$.

Therefore there exists a rational mapping of $W \times W$ into W which maps $(c(\sigma), c(\tau))$ onto $c(\sigma\tau)$. It is now a simple matter to see that this rational mapping is a normal law of composition on W. Therefore, by the Weil-Rosenlicht result just mentioned, there exists a group variety G, and a birational mapping g' between W and G, such that $g'(c(\sigma\tau)) = g'(c(\sigma))g'(c(\tau))$ whenever σ, τ are independent generic elements of $\mathfrak{G}$. If for each generic element σ of $\mathfrak{G}$ we define $g(\sigma) = g'(c(\sigma))$ then g is a one-to-one mapping of the set of generic elements of $\mathfrak{G}$ onto the set of generic points of G, such that $\mathcal{K}(\sigma) = \mathcal{K}(g(\sigma))$ for every generic element σ of $\mathfrak{G}$, and such that $g(\sigma\tau) = g(\sigma)g(\tau)$ for all pairs σ, τ of independent generic elements of $\mathfrak{G}$; it is easy to see, moreover, that if σ, τ, σ', τ' are generic elements of $\mathfrak{G}$ then (σ', τ') is a specialization of (σ, τ) if and only if $(g(\sigma'), g(\tau'))$ is a specialization of $(g(\sigma), g(\tau))$. Every generic element σ of $\mathfrak{G}$ thus has the property that $g(\sigma\tau)g(\tau)^{-1}$ has the same value for all choices of generic element τ of $\mathfrak{G}$ such that σ, τ are independent. It is now a simple matter to prove that every element σ of $\mathfrak{G}$ has this property. Indeed, if $\sigma \in \mathfrak{G}$ and if τ, τ' are generic elements of $\mathfrak{G}$ such that σ, τ are independent and σ, τ' are independent, we may find a generic element τ'' of $\mathfrak{G}$ such that σ, τ, τ'' are independent and σ, τ', τ'' are independent; obviously then $\sigma\tau''$, $\tau''^{-1}\tau$ are independent generic points of $\mathfrak{G}$ as are $\tau'', \tau''^{-1}\tau$, so that

$$g(\sigma\tau)g(\tau)^{-1} = g(\sigma\tau'' \cdot \tau''^{-1}\tau)g(\tau'' \cdot \tau''^{-1}\tau)^{-1}$$
$$= g(\sigma\tau'')g(\tau''^{-1}\tau)(g(\tau'')g(\tau''^{-1}\tau))^{-1} = g(\sigma\tau'')g(\tau'')^{-1},$$

and similarly $g(\sigma\tau')g(\tau')^{-1} = g(\sigma\tau'')g(\tau'')^{-1}$, whence $g(\sigma\tau)g(\tau)^{-1} = g(\sigma\tau')g(\tau')^{-1}$.

This being established, we may define a mapping f of $\mathfrak{G}$ into G by the formula $f(\sigma) = g(\sigma\tau)g(\tau)^{-1}$ $(\sigma \in \mathfrak{G})$, where τ represents any generic element of $\mathfrak{G}$ such that σ, τ are independent. If $\sigma, \sigma' \in \mathfrak{G}$ we may choose a generic element τ such that each of the three pairs σ, τ and σ', τ and $\sigma, \sigma'\tau$ is independent, so that

$$f(\sigma\sigma') = g(\sigma\sigma'\tau)g(\tau)^{-1} = g(\sigma \cdot \sigma'\tau)g(\sigma'\tau)^{-1} \cdot g(\sigma'\tau)g(\tau)^{-1} = f(\sigma)f(\sigma')$$

and f is a homomorphism of the group $\mathfrak{G}$ into the group G. If ν is any element of the kernel of f and τ is a generic element of $\mathfrak{G}$ such that ν, τ are independent, so that $\nu\tau$ is generic on $\mathfrak{G}$, then $g(\nu\tau) = f(\nu\tau) = f(\tau) = g(\tau)$, whence $\nu\tau = \tau$ and ν is the identity; thus, the kernel of f is trivial. For any $\sigma \in \mathfrak{G}$ we may find a generic element τ of $\mathfrak{G}$ such that $\mathcal{K}(\mathcal{K}(\sigma), \mathcal{K}(f(\sigma)))$ and $\mathcal{K}(\tau)$ are linearly disjoint over $\mathcal{K}$, whence in particular σ, τ are independent; for such τ we have

$$\mathcal{K}(f(\sigma), g(\tau)) = \mathcal{K}(g(\sigma\tau)g(\tau)^{-1}, g(\tau)) = \mathcal{K}(g(\sigma\tau), g(\tau)) = \mathcal{K}(\sigma\tau, \tau) = \mathcal{K}(\sigma, \tau),$$

whence it easily follows that $\mathcal{C}(f(\sigma)) = \mathcal{C}(\sigma)$. Again, if σ and σ' are any element of $\mathfrak{G}$ and we choose generic elements τ and τ' of $\mathfrak{G}$ such that each pair σ, τ and σ', τ' is independent then ([7], Ch. II, Props. 6, 8, pp. 777, 778) σ' is a specialization of σ if and only if $(\tau'\sigma', \tau')$ is a specialization of $(\tau\sigma, \tau)$; this takes place if and only if $(g(\tau'\sigma'), g(\tau'))$ is a specialization of $(g(\tau\sigma), g(\tau))$, that is, if and only if $(f(\tau')f(\sigma'), f(\tau'))$ is a specialization of $(f(\tau)f(\sigma), f(\tau))$, which in turn occurs if and only if $f(\sigma')$ is a specialization of $f(\sigma)$. To prove that f is a birational isomorphism of $\mathfrak{G}$ onto G it remains to show that G is the image of f. Now, the image contains $f(\tau)$, where τ is a generic element of $\mathfrak{G}$, and $f(\tau)$ is a generic element of G; therefore the smallest subvariety of G conaining the image is G itself. It follows (with the help of [7], Ch. II, Prop. 9, p. 779) that the complement of the image in G is contained in a proper subvariety of G, so that (Ch. I, Prop. 3) the image of f is G.

2. Irreducible group variety onto Galois group. We now prove the following partial converse to Theorem 1.

THEOREM 2. *Every irreducible group variety defined over the algebraically closed field $\mathcal{C}$ of characteristic 0 is birationally isomorphic to the Galois group of a strongly normal extension of a differential field with field of constants $\mathcal{C}$.*

Proof. Let G be an irreducible group variety defined over $\mathcal{C}$, and set $r = \dim G$. The field $\mathcal{H}_0{}^*$ of all rational functions on G (not necessarily defined over $\mathcal{C}$) is a finitely generated field extension of transcendence degree r of its subfield of all constant functions on G, which subfield we identify with $\mathcal{C}^*$ by identifying each constant function with its value. The subfield $\mathcal{H}_0$ of $\mathcal{H}_0{}^*$ consisting of all rational functions on G for which $\mathcal{C}$ is a field of definition, is a finitely generated field extension of transcendence degree r of $\mathcal{C}$. $\mathcal{H}_0$ and $\mathcal{C}^*$ are linearly disjoint over $\mathcal{C}$ (so that their intersection is $\mathcal{C}$), and their compositum is $\mathcal{H}_0{}^*$.

For each point $s \in G$ there exists a unique automorphism ρ_s of the field $\mathcal{H}_0{}^*$ with the following property: if $f \in \mathcal{H}_0{}^*$ and $t \in G$ then $\rho_s f$ is defined at t if and only if f is defined at ts and, when this is the case, $(\rho_s f)(t) = f(ts)$. The computation

$$((\rho_s\rho_{s'})f)(t) = (\rho_s(\rho_{s'}f))(t) = (\rho_{s'}f)(ts) = f(tss') = (\rho_{ss'}f)(t)$$

shows that

(1) $$\rho_{ss'} = \rho_s\rho_{s'} \quad (s, s' \in G).$$

Also, if $\rho_s f = f$ for every $f \in \mathcal{H}_0{}^*$, that is if $f(ts) = f(t)$ for all $f \in \mathcal{H}_0{}^*$ and a generic point t of G such that t, s are independent (over $\mathcal{E}$), then obviously s is the unity element of G, that is

(2) "$\rho_s =$ identity automorphism" implies "$s =$ unity element of G."

If $s \varepsilon G$, and if t denotes any point of G (in particular, any generic point of G such that s, t are independent over $\mathcal{E}$), then $\mathcal{E}(t, ts) = \mathcal{E}(t, s) = \mathcal{E}(ts, s)$. It follows that

$$\mathcal{H}_0(\rho_s\mathcal{H}_0) = \mathcal{H}_0(\mathcal{E}(s)) = (\rho_s\mathcal{H}_0)(\mathcal{E}(s)). \tag{3}$$

Also if $s' \varepsilon G$ and we take t generic with s, t independent and s', t independent then s' is a specialization of s over $\mathcal{E}$ if and only if s' is a specialization of s over $\mathcal{E}(t)$, which occurs if and only if $(f(ts'))_{f \varepsilon \mathcal{H}_0}$ is a specialization of of $(f(ts))_{f \varepsilon \mathcal{H}_0}$ over $\mathcal{E}(t)$, that is, $((\rho_{s'}f)(t))_{f \varepsilon \mathcal{H}_0}$ is a specialization of $((\rho_s f)(t))_{f \varepsilon \mathcal{H}_0}$ over $\mathcal{E}(t)$; it follows that

$$(4) \qquad \begin{cases} s' \text{ is a specialization of } s \text{ over } \mathcal{E} \text{ if and only if} \\ (\rho_{s'}f)_{f \varepsilon \mathcal{H}_0} \text{ is a specialization of } (\rho_s f)_{f \varepsilon \mathcal{H}_0} \text{ over } \mathcal{H}_0. \end{cases}$$

The set of all derivations of $\mathcal{H}_0{}^*$ over $\mathcal{E}^*$ is a vector space over $\mathcal{H}_0{}^*$ of dimension r. It is known (for example, see Nakano [9]) that this space has a base $D_1, \cdots, D_r$ composed of derivations each of which maps $\mathcal{H}_0$ into itself and is "right-invariant" in the sense that

$$\rho_s \circ D_i = D_i \circ \rho_s \qquad (s \varepsilon G, 1 \leqq i \leqq r). \tag{5}$$

An element of $\mathcal{H}_0{}^*$ annihilated by $D_1, \cdots, D_r$ is annihilated by every derivation of $\mathcal{H}_0{}^*$ over $\mathcal{E}^*$, and therefore belongs to $\mathcal{E}^*$.

Now let $(u_{ij})_{1 \leqq i \leqq r, 0 \leqq j < \infty}$ be a family of indeterminates, and let

$$\mathcal{E} = \mathcal{E}((u_{ij})_{1 \leqq i \leqq r, 0 \leqq j < \infty}), \ \mathcal{H} = \mathcal{H}_0((u_{ij})_{1 \leqq i \leqq r, 0 \leqq j < \infty});$$

the fields $\mathcal{E}^*((u_{ij})_{1 \leqq i \leqq r, 0 \leqq j < \infty})$ and $\mathcal{H}_0{}^*((u_{ij})_{1 \leqq i \leqq r, 0 \leqq j < \infty})$ may then be denoted by $\mathcal{E}(\mathcal{E}^*)$ and $\mathcal{H}(\mathcal{E}^*)$, respectively. For each $s \varepsilon G$ the automorphism ρ_s of $\mathcal{H}_0{}^*$ over $\mathcal{E}^*$, can be extended to a unique automorphism (which we also denote by ρ_s) of $\mathcal{H}(\mathcal{E}^*)$ over $\mathcal{E}(\mathcal{E}^*)$. For each i the invariant derivation D_i of $\mathcal{H}_0{}^*$ over $\mathcal{E}^*$ can be extended to a unique derivation (which we also denote by D_i) of $\mathcal{H}(\mathcal{E}^*)$ over $\mathcal{E}^*$ with the property that $D_i u_{ij} = u_{i0}{}^{-1} u_{i,j+1}$ and $D_i u_{i'j} = 0$ whenever $i' \neq i$ $(0 \leqq j < \infty)$. For the extended automorphisms and derivations (1), (2), (5) continue to hold, and (3), (4) yield

$$\mathcal{H}(\rho_s\mathcal{H}) = \mathcal{H}(\mathcal{E}(s)) = (\rho_s\mathcal{H})(\mathcal{E}(s)) \qquad (s \varepsilon G) \tag{6}$$

$$(7) \qquad \begin{cases} s' \text{ is a specialization of } s \text{ over } \mathcal{E} \text{ if and only if} \\ (\rho_{s'}\alpha)_{\alpha \varepsilon \mathcal{H}} \text{ is a specialization of } (\rho_s \alpha)_{\alpha \varepsilon \mathcal{H}} \text{ over } \mathcal{H}. \end{cases}$$

Furthermore, if α is any element of $\mathcal{H}(\mathcal{C}^*)$, then α may be written as a rational fraction with coefficients in $\mathcal{H}_0^*$ in the indeterminate u_{ij}, and if α contains u_{kl} effectively but for every $l' > l$ fails to contain $u_{kl'}$ effective then it is clear that $u_{k,l+1}$ is effectively present in $D_k\alpha$, so that $D_k\alpha \neq 0$; thus the only elements of $\mathcal{H}(\mathcal{C}^*)$ annihilated by every D_i are the elements of $\mathcal{C}^*$:

$$(8) \qquad D_i\alpha = 0 \, (1 \leqq i \leqq r) \text{ if and only if } \alpha \in \mathcal{C}^*.$$

We now make $\mathcal{H}(\mathcal{C}^*)$ into a differential field (not necessarily contained in the universal extension $\mathcal{F}^*$ of $\mathcal{G}$ introduced in Section 1 of Chapter I) with derivation operators $\delta_1, \cdots, \delta_m$ by defining

$$(9) \qquad \delta_1\alpha = \sum u_{i0} D_i\alpha, \qquad \delta_k\alpha = 0 \, (2 \leqq k \leqq m)$$

for every $\alpha \in \mathcal{H}(\mathcal{C}^*)$. It is clear that $\mathcal{C}$, $\mathcal{E}$, $\mathcal{H}$, $\mathcal{C}^*$ all are differential subfields of $\mathcal{H}(\mathcal{C}^*)$, so that $\mathcal{H}(\mathcal{C}^*)$ may now be written as $\mathcal{H}\langle\mathcal{C}^*\rangle$; furthermore, if we denote u_{i0} by u_i then $u_{ij} = \delta_1{}^j u$ $(1 \leqq i \leqq r, 0 \leqq j < \infty)$, so that we may write $\mathcal{E} = \mathcal{C}\langle u_1, \cdots, u_r\rangle$ and $\mathcal{H} = \mathcal{E}\langle\mathcal{H}_0\rangle$.

If $\alpha \in \mathcal{H}\langle\mathcal{C}^*\rangle$ contains u_{kl} effectively but for every $l' > l$ fails to contain $u_{kl'}$ effectively, then $\delta_1\alpha = \sum u_{i0}D_i\alpha$ contains $u_{k,l+1}$ effectively, so that $\delta_1\alpha \neq 0$. Therefore if $\delta_i\alpha = 0$ then $\alpha \in \mathcal{H}_0^*$, and from the equation $\sum u_{i0}D_i\alpha = 0$ and the fact that $u_{10}, \cdots, u_{r0}$ are linearly independent over $\mathcal{H}_0^*$ we conclude that $D_i\alpha = 0$ $(1 \leqq i \leqq r)$. It follows from (8) that the field of constants of the differential field $\mathcal{H}\langle\mathcal{C}^*\rangle$ is $\mathcal{C}^*$; $\mathcal{C}^*$ is therefore also the field of constants of $\mathcal{E}\langle\mathcal{C}^*\rangle$, and $\mathcal{C}$ is the field of constants of $\mathcal{H}$ and of $\mathcal{E}$.

Before proceeding farther, we rid ourselves of the awkward circumstance that $\mathcal{H}\langle\mathcal{C}^*\rangle$ is not in general a differential subfield of $\mathcal{F}^*$. In the first place, $\mathcal{H} = \mathcal{E}\langle\mathcal{H}_0\rangle = \mathcal{C}\langle u_1, \cdots, u_r, \mathcal{H}_0\rangle$, so that $\mathcal{H}$ is a finitely generated differential field extension of $\mathcal{C}$; since $\mathcal{F}^*$ is a universal extension of $\mathcal{C}$ (see [7], Ch. I, § 5, p. 768), there exists an isomorphism over $\mathcal{C}$ of $\mathcal{H}$ onto a differential subfield $\mathcal{H}'$ of $\mathcal{F}^*$. Secondly, a family of elements of $\mathcal{H}$ linearly independent over $\mathcal{C}$ is also linearly independent over $\mathcal{C}^*$ (see [7], Ch. I, § 4, Prop. 3 and remark following, pp. 766-767), so that $\mathcal{H}$ and $\mathcal{C}^*$ are linearly disjoint over $\mathcal{C}$; it follows that the isomorphism over $\mathcal{C}$ of $\mathcal{H}$ onto $\mathcal{H}'$ can be extended to a unique isomorphism over $\mathcal{C}^*$ of $\mathcal{H}\langle\mathcal{C}^*\rangle$ onto $\mathcal{H}'\langle\mathcal{C}^*\rangle$. By virtue of this isomorphism we identify $\mathcal{H}\langle\mathcal{C}^*\rangle$ with $\mathcal{H}'\langle\mathcal{C}^*\rangle$, so that henceforth all our differential fields are differential subfields of $\mathcal{F}^*$.

By (5) and (9) we know that $\rho_s\delta_i\alpha = \delta_i\rho_s\alpha$ $(s \in G, 1 \leqq i \leqq m, \alpha \in \mathcal{H}\langle\mathcal{C}^*\rangle)$, so that each ρ_s is an automorphism of $\mathcal{H}\langle\mathcal{C}^*\rangle$ as a differential field (not merely as a field). This proves ([7], Ch. II, § 3, Prop. 4, p. 775) that each ρ_s, identified with its restriction to $\mathcal{H}$, is a strong isomorphism of $\mathcal{H}$ over $\mathcal{E}$.

It follows from (6) that if t is a generic point of G then the transcendence degree of $\mathcal{H}\langle\rho_t\mathcal{H}\rangle$ over $\mathcal{H}$ equals r, which in turn equals the transcendence degree of $\mathcal{H}$ over $\mathcal{E}$. Consequently ([7], Ch. II, § 2, Prop. 2, p, 773) ρ_t is an isolated isomorphism of $\mathcal{H}$ over $\mathcal{E}$. Since $\mathcal{E}$ is obviously relatively algebraically closed in $\mathcal{H}$, it follows ([7], Ch. II, § 2, Prop. 3, p. 774) that every isomorphism of $\mathcal{H}$ over $\mathcal{E}$ is a specialization of ρ_t, and therefore ([7], Ch. II, § 4, Prop. 5, p. 776) is strong. Thus $\mathcal{H}$ is a strongly normal extension of $\mathcal{E}$; we denote its Galois group by $\mathfrak{H}$.

By (1), the mapping $f: s \rightarrow \rho_s$ ($s \in G$) is a homomorphism of G into $\mathfrak{H}$. By (2) the kernel of f is trivial. By (6) and [7], Ch. I, § 4, Cor. 5 to Prop. 3, p. 768,

$$\mathcal{B}(\rho_s) = \mathcal{B}(s) \qquad (s \in G),$$

and by (7), $\rho_{s'}$ is a specialization of ρ_s if and only if s' is a specialization of s (over $\mathcal{B}$). Therefore f is a rational homomorphism of G into $\mathfrak{H}$. Since the Galois group $\mathfrak{H}$ is by Theorem 1 birationally isomorphic with a group variety, it follows from the corollary to Proposition 5 in Chapter I that the image of f is an algebraic subgroup of $\mathfrak{H}$. Since the image contains a generic point $f(t)$ of $\mathfrak{H}$, it follows that the image of f is $\mathfrak{H}$ itself. Again making use of the fact that $\mathfrak{H}$ is birationally isomorphic with a group variety, we conclude from the corollary to Proposition 6 in Chapter I that f is a birational isomorphism of G onto $\mathfrak{H}$.

3. Applications. It follows from Theorem 1 that many propositions about group varieties can be generalized by permitting each group variety in the statement of a given proposition to be any algebraic group (that is, either a Galois group or a group variety). For example (as we have already had occasion to observe in Section 2):

PROPOSITION 1. *Proposition 5 of Chapter I and its corollary, and Proposition 6 of Chapter I and its corollary can be generalized to arbitrary algebraic groups.*

Another example:

PROPOSITION 2. *If* $\sigma_1, \cdots, \sigma_n \in \mathfrak{G}$ *and if* $(\sigma_1', \cdots, \sigma_n')$ *is a specialization of* $(\sigma_1, \cdots, \sigma_n)$, *then* $(W_i(\sigma'_1, \cdots, \sigma_n'))_{i \in I}$ *is a specialization of* $(W_i(\sigma_1, \cdots, \sigma_n))_{i \in I}$, *where* $(W_i(X_1, \cdots, X_n))_{i \in I}$ *is any family of words in* n *indeterminates* $X_1, \cdots, X_n$.

This is an immediated consequence of the fact that the same proposition holds for group varieties.

On the other hand some propositions on group varieties can not be indiscriminately generalized in this way because such a generalization may depend on Theorem 2, which we have proved only for *irreducible* group varieties. For example, we have not the right to assert the generalization of Proposition 7 of Chapter I, or of its corollary.

It is perhaps of interest to observe that not only can we prove results on Galois groups by appealing to previously proved analogous results on group varieties, but also we can work in the opposite direction. For example, we can prove that, roughly speaking, a factor group of a group variety is itself a group variety. More precisely:

PROPOSITION 3. *If G is a group variety and K is a normal algebraic subgroup of G then, there exists a rational homomorphism of G with kernel K and image a group variety.*

This result is contained (except for the rather trivial extension here to group varieties which are not necessarily irreducible) in much stronger results of Nakano [10]; the novelty here lies in the fact that the result on group varieties is obtained as a consequence of theorems on algebraic differential equations. The proof runs as follows. By Proposition 1 of Chapter I, the result holds for a group variety G and kernel K provided it holds for the group variety G^0 (= the component of the unity of G) and kernel $G^0 \cap K$, that is, we may assume that G is irreducible. Making this assumption, we know from Theorem 2 that G is birationally isomorphic with a Galois group, say f is a birational isomorphism of G onto $\mathfrak{G}$; $f(K)$ is a *normal* algebraic subgroup of $\mathfrak{G}$ and therefore is the Galois group $\mathfrak{G}(\mathcal{H})$ of $\mathcal{G}$ over a differential field $\mathcal{H}$ intermediate to $\mathcal{F}$ and $\mathcal{G}$ and *strongly normal* over $\mathcal{F}$. Therefore (see Example 1 of Chapter I, § 4) the mapping r, which to each $\sigma \varepsilon \mathfrak{G}$ corresponds the restriction of σ to $\mathcal{H}$, is a rational homomorphism of $\mathfrak{G}$ with kernel $\mathfrak{G}(\mathcal{H})$ and image the Galois group $\mathfrak{H}$ of $\mathcal{H}$ over $\mathcal{F}$. By Theorem 1, there exists a birational isomorhpism h of $\mathfrak{H}$ onto a group variety H. It is now immediate that the composite mapping $h \circ r \circ f$ is a rational homomorphism of G with kernel K and image H.

Similarly, we can prove the following result.

PROPOSITION 4. *If a rational homomorphism of an algebraic group G has kernel K and image H, then* $\dim G = \dim K + \dim H$.

This too is contained in results of Nakano [10]. Our proof uses the fact that the dimension of the Galois group of a strongly normal differential field extension equals the transcendence degree of the extension. We omit the details, which are rather obvious.

Finally, we generalize to arbitrary connected group varieties (but only in the case of field characteristic 0, of course) a theorem on algebraic matric groups (Chevalley [3], Ch. II, § 4, Th. 1, part 2), p. 84, or Chevalley and Kolchin [4], § 1, Th. 2).

PROPOSITION 5. *The notation being as in § 2, if W is a subset of an irreducible group variety G then a necessary and sufficient condition for W to be an algebraic subgroup of G is that there exist a finite subset S, of the field $\mathcal{H}_0$ of rational functions on G which are defined over $\mathcal{C}$, such that W is the set of all points $s \in G$ for which $\rho_s \alpha = \alpha$ for every $\alpha \in S$.*

Indeed, we saw in the proof of Theorem 2 that $\mathcal{H} = \mathcal{E}\langle \mathcal{H}_0 \rangle$ is a strongly normal extension of $\mathcal{E} = \mathcal{C}\langle u_1, \cdots, u_r \rangle$, and $s \rightarrow \rho_s$ $(s \in G)$ is a birational isomorphism f of G onto the Galois group $\mathfrak{H}$ of $\mathcal{H}$ over $\mathcal{E}$. W is an algebraic subgroup of G if and only if $f(W)$ is an algebraic subgroup of $\mathfrak{H}$, that is, is the Galois group of $\mathcal{H}$ over a differential field $\mathcal{E}_1$ intermediate to $\mathcal{E}$ and $\mathcal{H}$; now every intermediate differential field is a finitely generated differential field extension, and it is easy to see that $\mathcal{E}_1 = \mathcal{E}\langle \alpha_1, \cdots, \alpha_p \rangle$ for some finite set S of elements $\alpha_1, \cdots, \alpha_p$ which belong to $\mathcal{H}_0$. Therefore $f(W)$ is the Galois group of $\mathcal{H}$ over $\mathcal{E}_1$ if and only if W is the set of all elements $s \in G$ such that $\rho_s \alpha_i = \alpha_i$ $(1 \leqq i \leqq p)$.

Chapter III. One-dimensional cohomology.

1. **Definition of $H^1(\mathfrak{G}, G)$.** The universal extension $\mathcal{F}^*$ of $\mathcal{G}$ is, like its field of constants $\mathcal{C}^*$, an algebraically closed field extension of $\mathcal{C}$ of infinite transcendence degree and therefore may be used, like $\mathcal{C}^*$, as a universal domain for algebraic geometry. Furthermore, since $\mathcal{C}^* \subset \mathcal{F}^*$, every variety V in the algebraic geometry based on $\mathcal{C}^*$ as universal domain determines in an obvious way a variety $V^\dagger$ in the algebraic geometry based on $\mathcal{F}^*$ as universal domain, such that a point of $V^\dagger$ belongs to V if and only if the point is rational over $\mathcal{C}^*$. It is obvious that if V is a group variety in the algebraic geometry based on $\mathcal{C}^*$ then $V^\dagger$ is a group variety in the algebraic geometry based on $\mathcal{F}^*$.

Now, a mapping f of the Galois group $\mathfrak{G}$ of $\mathcal{G}$ over $\mathcal{F}$ into a group variety G may also be regarded as a mapping of $\mathfrak{G}$ into $G^\dagger$; since $\mathcal{C}(f(\sigma))$ consists of constants for any $\sigma \in \mathfrak{G}$, the condition "$\mathcal{C}(f(\sigma)) \subset \mathcal{C}(\sigma)$" is equivalent to "$\mathcal{G}(f(\sigma)) \subset \mathcal{G}(\sigma\mathcal{G})$," for any $\sigma, \sigma' \in \mathfrak{G}$ the condition "$f(\sigma')$ is a specialization of $f(\sigma)$ over $\mathcal{C}$" is equivalent to the condition "$f(\sigma')$ is a specialization of $f(\sigma)$ over $\mathcal{G}$" and the condition "$f(\sigma\sigma') = f(\sigma)f(\sigma')$" may, if we wish, be written in the form "$f(\sigma\sigma') = f(\sigma) \cdot \sigma f(\sigma')$." Therefore the following definition generalizes the notion of rational homomorphism of $\mathfrak{G}$ into G:

By a *rational crossed homomorphism*, or a *cocycle of dimension* 1, or a 1-*cocycle*, of the Galois group $\mathfrak{G}$ into the group variety G we mean any mapping f of $\mathfrak{G}$ *into* $G^\dagger$ such that $\mathcal{G}(f(\sigma)) \subset \mathcal{G}(\sigma\mathcal{G})$ for all $\sigma \in \mathfrak{G}$, such that "σ' is a specialization of σ" implies "$f(\sigma')$ is a specialization of $f(\sigma)$ over $\mathcal{G}$" for all $\sigma, \sigma' \in \mathfrak{G}$, and such that $f(\sigma\sigma') = f(\sigma)\cdot\sigma f(\sigma')$ for all $\sigma, \sigma' \in \mathfrak{G}$.

As remarked above, every rational homomorphism of $\mathfrak{G}$ into G is a rational crossed homomorphism of $\mathfrak{G}$ into G. Also, if α is any point of $G^\dagger$ *which is rational over* $\mathcal{G}$, then it is evident that the mapping $\sigma \to \alpha^{-1}\sigma\alpha$ $(\sigma \in \mathfrak{G})$ is a rational crossed homomorphism of $\mathfrak{G}$ into G; such a rational crossed homomorphism is said to be *trivial*, or to be a *coboundary of dimension* 1 or a 1-*coboundary* of $\mathfrak{G}$ into G.

We denote the set of all 1-cocyles of $\mathfrak{G}$ into G by $Z^1(\mathfrak{G}, G)$, and the set of all 1-coboundaries by $B^1(\mathfrak{G}, G)$, so that $B^1(\mathfrak{G}, G) \subset Z^1(\mathfrak{G}, G)$. The mapping which to each element of $\mathfrak{G}$ corresponds the unity element 1 of G is obviously a 1-coboundary (because $1^{-1}\sigma 1 = 1$ for every $\sigma \in \mathfrak{G}$); we denote it by 1. In general, as we have made no assumption of commutativity for G, $Z^1(\mathfrak{G}, G)$ and $B^1(\mathfrak{G}, G)$ are not groups under the law of composition $(f, g) \to fg$, where $(fg)(\sigma) = f(\sigma)g(\sigma)$, because fg need not be an element of $Z^1(\mathfrak{G}, G)$.

If $f \in Z^1(\mathfrak{G}, G)$, and if $\alpha \in G^\dagger$ is rational over $\mathcal{G}$, then the mapping g of $\mathfrak{G}$ defined by $g(\sigma) = \alpha^{-1}f(\sigma)\sigma\alpha$ is obviously an element of $Z^1(\mathfrak{G}, G)$, which we call *cohomologous* to f. The relation "g is cohomologous to f" is an equivalence on $Z^1(\mathfrak{G}, G)$. An equivalence class with respect to this equivalence we call a *cohomology class of dimension* 1 or a 1-*cohomology class* of $\mathfrak{G}$ into G; we denote the set of all these by $H^1(\mathfrak{G}, G)$. $B^1(\mathfrak{G}, G)$ itself is a 1-cohomology class; when we think of this class as an element of $H^1(\mathfrak{G}, G)$ we frequently permit ourselves to denote it by 1, and we also permit ourselves to write $H^1(\mathfrak{G}, G) = 1$ when we wish to assert that $B^1(\mathfrak{G}, G)$ is the only 1-cohomology class of $\mathfrak{G}$ into G.

Of course, if G is commutative then $Z^1(\mathfrak{G}, G)$ and $B^1(\mathfrak{G}, G)$ are commutative groups, and $H^1(\mathfrak{G}, G)$ is the factor group $Z^1(\mathfrak{G}, G)/B^1(\mathfrak{G}, G)$; we then call it the *first cohomology group* of $\mathfrak{G}$ into G.

2. Determination of $H^1(\mathfrak{G}, G)$ in some general cases. The computation of $H^1(\mathfrak{G}, G)$ in various cases below is facilitated by the observation that two 1-cocycles f and g of $\mathfrak{G}$ into G are equal provided $f(\sigma) = g(\sigma)$ for all $\sigma \in \mathfrak{G}$ which are automorphisms of $\mathcal{G}$ over $\mathcal{F}$); this follows from [7], Ch. II, Cor. 2 to Prop. 9, p. 783.

Our results on $H^1(\mathfrak{G}, G)$ are summarized in the following theorem. We recall that $\mathfrak{G}$ is the Galois group of the strongly normal extension $\mathcal{G}$

of the differential field $\mathfrak{F}$ of characteristic 0 with algebraically closed field of constants $\mathfrak{C}$. G denotes any group variety defined over $\mathfrak{C}$. We denote the group of all invertible square matrices of degree n with constant coordinates by $GL(n)$; $GL(n)$ is an irreducible group variety of dimension n^2.

THEOREM 1. 1) $H^1(\mathfrak{G}, GL(n)) = 1$.

2) *If* $\mathfrak{F}$ *is algebraically closed then* $H^1(\mathfrak{G}, G) = 1$.

3) *If* G *is commutative then every element of the group* $H^1(\mathfrak{G}, G)$ *is of finite order.*

Proof of 1). Let $f \varepsilon Z^1(\mathfrak{G}, GL(n))$, and let τ be a generic element of the component of the identity $\mathfrak{G}^0$ of $\mathfrak{G}$. Since $\mathfrak{G}(f(\tau)) \subset \mathfrak{G}(\tau\mathfrak{G})$, there must exist a finite number of elements $\tau\xi_1, \cdots, \tau\xi_p \varepsilon \tau\mathfrak{G}$ which are algebraically independent over $\mathfrak{G}$, and a finite number of elements $\tau\eta_1, \cdots, \tau\eta_q \varepsilon \tau\mathfrak{G}$ which are linearly independent over $\mathfrak{G}(\tau\xi_1, \cdots, \tau\xi_p)$, such that

$$\mathfrak{G}(f(\tau)) \subset \sum_{j=1}^{q} \mathfrak{G}(\tau\xi_1, \cdots, \tau\xi_p) \cdot \tau\eta_j.$$

Therefore there exist matrices $A_j = A_j(X_1, \cdots, X_p)$, $1 \leqq j \leqq q$, square and of degree n, with coordinates which are polynomials over $\mathfrak{G}$ in indeterminates $X_1, \cdots, X_p$, and there exists a nonzero polynomial

$$B = B(X_1, \cdots, X_p) \varepsilon \mathfrak{G}[X_1, \cdots, X_p],$$

such that

$$f(\tau) = \sum_{j=1}^{q} (\tau\eta_j) B(\tau\xi_1, \cdots, \tau\xi_p)^{-1} A_j(\tau\xi_1, \cdots, \tau\xi_p). \tag{1}$$

This equation continues to hold, of course, if τ is replaced by any generic element of $\mathfrak{G}^0$. We assume without loss of generality that B does not have a factor of degree > 0 which divides every coordinate of every A_j, and we assume, too, that one of the coefficients in B is 1. For any $\sigma \varepsilon \mathfrak{G}^0$ which is an *auto*morphism of $\mathfrak{G}$, $\sigma^{-1}\tau$ is a generic element of $\mathfrak{G}^0$. On the other hand, $f(\tau) = f(\sigma \cdot \sigma^{-1}\tau) = f(\sigma) \cdot \sigma f(\sigma^{-1}\tau)$. Therefore by (1)

$$\sum_{j=1}^{q} (\tau\eta_j) B(\tau\xi_1, \cdots, \tau\xi_p)^{-1} A_j(\tau\xi_1, \cdots, \tau\xi_p)$$
$$= f(\sigma) \cdot \sum_{j=1}^{q} (\tau\eta_j) B_\sigma(\tau\xi_1, \cdots, \tau\xi_p)^{-1} A_{j\sigma}(\tau\xi_1, \cdots, \tau\xi_p),$$

where B_σ is obtained from B by replacing each coefficient by its image under σ, and similarly for $A_{j\sigma}$. Because $\tau\eta_1, \cdots, \tau\eta_q$ are linearly independent over $\mathfrak{G}(\tau\xi_1, \cdots, \tau\xi_p)$, it follows that

$$B(\tau\xi_1, \cdots, \tau\xi_p)^{-1} A_j(\tau\xi_1, \cdots, \tau\xi_p)$$
$$= f(\sigma) B_\sigma(\tau\xi_1, \cdots, \tau\xi_p)^{-1} A_{j\sigma}(\tau\xi_1, \cdots, \tau\xi_p), \qquad 1 \leqq j \leqq p.$$

Because $\tau\xi_1, \cdots, \tau\xi_p$ are algebraically independent over $\mathfrak{G}$, it follows that

$$B_\sigma A_j = f(\sigma) \cdot BA_{j\sigma}, \qquad 1 \leqq j \leqq p.$$

Because no factor of B of degree > 0 divides every coordinate of every A_j, it follows that B_σ is a multiple of B by an element of $\mathfrak{G}$, and because a coefficient in B is 1 we conclude that $B_\sigma = B$, so that

$$(2) \qquad A_j = f(\sigma) A_{j\sigma}, \qquad 1 \leqq j \leqq p.$$

Now, it is a consequence of (1) that if $Y_1, \cdots, Y_q$ are additional indeterminates then

$$\det\left(\sum_{j=1}^{q} Y_j A_j(X_1, \cdots, X_p)\right) \neq 0.$$

Therefore there exist element $a_1, \cdots, a_p, b_1, \cdots, b_q \varepsilon \mathfrak{F}$ such that if we form the matrix $\alpha = \sum b_j A_j(a_1, \cdots, a_p)$ then $\det \alpha \neq 0$; obviously the coordinates of α are elements of $\mathfrak{G}$. However,

$$\sigma A_j(a_1, \cdots, a_p) = A_{j\sigma}(\sigma a_1, \cdots, \sigma a_p) = A_{j\sigma}(a_1, \cdots, a_p).$$

Therefore by (2) we have $\alpha = f(\sigma)\sigma\alpha$. Thus, if we let g be the 1-cocycle cohomologous to f defined by the equation $g(\sigma') = \alpha^{-1} f(\sigma')\sigma'\alpha$ for all $\sigma' \varepsilon \mathfrak{G}$, then $g(\sigma) = 1$ for every $\sigma \varepsilon \mathfrak{G}^0$.

If σ' and σ'' belong to the same coset of $\mathfrak{G}$ relative to $\mathfrak{G}^0$, then $g(\sigma'') = g(\sigma' \cdot \sigma'^{-1}\sigma'') = g(\sigma')\sigma' g(\sigma'^{-1}\sigma'') = g(\sigma')$, so that g is constant on each coset. Since every coset contains an automorphism of $\mathfrak{G}$, $g(\sigma')$ is always rational over $\mathfrak{G}$; but, for any $\sigma \varepsilon \mathfrak{G}^0$, we have $g(\sigma') = g(\sigma\sigma') = g(\sigma)\sigma g(\sigma')$ $= \sigma g(\sigma')$, so that $g(\sigma')$, is always rational over the relative algebraic closure $\mathfrak{F}^0$ of $\mathfrak{F}$ in $\mathfrak{G}$. Of course, if $\sigma_1, \cdots, \sigma_h$ form a complete set of representatives of the cosets of $\mathfrak{G}$ relative to $\mathfrak{G}^0$, then the restrictions of $\sigma_1, \cdots, \sigma_h$ to $\mathfrak{F}^0$ are precisely the elements of the Galois group of $\mathfrak{F}^0$ over $\mathfrak{F}$. Now, if $T_1, \cdots, T_h$ are indeterminates, then $\det(\sum T_i g(\sigma_i))$ is a polynomial in $\mathfrak{F}^0[T_1, \cdots, T_h]$, and is different from 0 (because, for example, at $(1, 0, \cdots, 0)$ it has the value $\det(g(\sigma_1))$). Therefore (Bourbaki [1], Ch. V, § 10, Th. 4, p. 157) there exists an element $a \varepsilon \mathfrak{F}^0$ such that $\det(\sum(\sigma_i a) g(\sigma_i)) \neq 0$. Writing $\beta = \sum(\sigma_i a) g(\sigma_i)$, we have for any $\sigma' \varepsilon \mathfrak{G}$,

$$\sigma'\beta = \sum(\sigma'\sigma_i a)\sigma' g(\sigma_i) = \sum(\sigma'\sigma_i a) g(\sigma')^{-1} g(\sigma'\sigma_i) = g(\sigma')^{-1}\beta.$$

Letting $\gamma = \beta^{-1}\alpha^{-1}$ we therefore obtain $\gamma^{-1}\sigma'\gamma = \alpha\beta\sigma'\beta^{-1} \cdot \sigma'\alpha^{-1} = \alpha g(\sigma')\sigma'\alpha^{-1}$ $= f(\sigma')$ for every $\sigma' \varepsilon \mathfrak{G}$, so that $f \varepsilon B^1(\mathfrak{G}, GL(n))$.

Proof of 2). Since $\mathfrak{F}$ is algebraically closed $\mathfrak{G}$ is irreducible. Let $f \varepsilon Z^1(\mathfrak{G}, G)$, and let τ be a generic element of $\mathfrak{G}$. Letting $\eta_1, \cdots, \eta_n$ be elements of $\mathfrak{G}$ such that $\mathfrak{G} = \mathfrak{F}(\eta_1, \cdots, \eta_n)$, we see that

$$\mathfrak{G}(\tau\mathfrak{G}) = \mathfrak{F}(\eta_1, \cdots, \eta_n, \tau\eta_1, \cdots, \tau\eta_n);$$

also, $(\eta_1,\cdots,\eta_n)$ and $(\tau\eta_1,\cdots,\tau\eta_n)$ are independent over $\mathfrak{F}$ (because the transcendence degree of $\mathfrak{F}(\eta_1,\cdots,\eta_n,\tau\eta_1,\cdots,\tau\eta_n)$ over $\mathfrak{F}$ equals that of $\mathfrak{F}(\eta_1,\cdots,\eta_n)$ over $\mathfrak{F}$ plus that of $\mathfrak{F}(\tau\eta_1,\cdots,\tau\eta_n)$ over $\mathfrak{F}$).

Let $f_i(\tau)$ denote a representative in affine space of the abstract point $f(\tau)$. Because $f(\tau)$ is rational over $\mathfrak{G}(\tau\mathfrak{G})$ there exist rational fractions $R_{i1},\cdots,R_{id_i}$ in $2n$ indeterminates, with coefficients in $\mathfrak{F}$ and denominators different from 0 at $(\eta_1,\cdots,\eta_n,\tau\eta_1,\cdots,\tau\eta_n)$, such that

$$f_i(\tau)=(R_{i1}(\eta_1,\cdots,\eta_n,\tau\eta_1,\cdots,\tau\eta_n),\cdots,R_{id_i}(\eta_1,\cdots,\eta_n,\tau\eta_1,\cdots,\tau\eta_n))$$

For any $\sigma\,\varepsilon\,\mathfrak{G}$ which is an *auto*morphism of $\mathfrak{G}$, the corresponding representative of $\sigma f(\sigma^{-1}\tau)$ is obviously given by

$$\sigma f_i(\sigma^{-1}\tau)=(R_{i1}(\sigma\eta_1,\cdots,\sigma\eta_n,\tau\eta_1,\cdots,\tau\eta_n),\cdots,R_{id_i}(\sigma\eta_1,\cdots,\sigma\eta_n,\tau\eta_1,\cdots,\tau\eta_n)).$$

Since $\mathfrak{F}$ is algebraically closed, there exists an affine point $(b_1,\cdots,b_n)$ which is a specialization in the sense of algebraic geometry of $(\tau\eta_1,\cdots,\tau\eta_n)$ over $\mathfrak{F}$, with each $b_i\,\varepsilon\,\mathfrak{F}$ and with the denominator of each R_{ij} different from 0 at $(\eta_1,\cdots,\eta_n,b_1,\cdots,b_n)$; application of the automorphism σ to $R_{ij}(\eta_1,\cdots,\eta_n,b_1,\cdots,b_n)$ shows that the denominator of each R_{ij} is different from 0 at $(\sigma\eta_1,\cdots,\sigma\eta_n,b_1,\cdots,b_n)$. Since $(\eta_1,\cdots,\eta_n)$ and $(\tau\eta_1,\cdots,\tau\eta_n)$ are independent over $\mathfrak{F}$, $(\eta_1,\cdots,\eta_n,b_1,\cdots,b_n)$ is a specialization of $(\eta_1,\cdots,\eta_n,\tau\eta_1,\cdots,\tau\eta_n)$ over $\mathfrak{F}$, that is, $(b_1,\cdots,b_n)$ is a specialization of $(\tau\eta_1,\cdots,\tau\eta_n)$ *over* $\mathfrak{G}$. Therefore if we set

$$g_i=(R_{i1}(\eta_1,\cdots,\eta_n,b_1,\cdots,b_n),\cdots,R_{id_i}(\eta_1,\cdots,\eta_n,b_1,\cdots,b_n))$$

then $(g_i,\sigma g_i)$ is a specialization of $(f_i(\tau),\sigma f(\sigma^{-1}\tau))$ over $\mathfrak{G}$ for every i. Accordingly, the affine points g_i are representatives of an abstract point g which is rational over $\mathfrak{G}$, and $(g,\sigma g)$ is a specialization of $(f(\tau),\sigma f(\sigma^{-1}\tau))$ over $\mathfrak{G}$. But $\tau=\sigma\cdot\sigma^{-1}\tau$, so that $f(\tau)=f(\sigma)\sigma f(\sigma^{-1}\tau)$, and $f(\sigma)$ is rational over $\mathfrak{G}$. Therefore $g=f(\sigma)\sigma g$ for all $\sigma\,\varepsilon\,\mathfrak{G}$ which are automorphisms of $\mathfrak{G}$, hence for all $\sigma\,\varepsilon\,\mathfrak{G}$; that is, $f\,\varepsilon\,B^1(\mathfrak{G},G)$.

Proof of 3). Let $\mathfrak{F}'$ be the algebraic closure of $\mathfrak{F}$. $\mathfrak{F}'$ is a normal algebraic extension of $\mathfrak{F}$, and $\mathfrak{F}'\cap\mathfrak{G}$ is the relative algebraic closure $\mathfrak{F}^0$ of $\mathfrak{F}$ in $\mathfrak{G}$. Therefore every automorphism τ of $\mathfrak{F}'$ over $\mathfrak{F}^0$ can be extended to a unique automorphism (which we also denote by τ) of the compositum $\mathfrak{F}'\langle\mathfrak{G}\rangle=\mathfrak{G}\langle\mathfrak{F}'\rangle$ over $\mathfrak{G}$ (the group of automorphisms of $\mathfrak{F}'$ over $\mathfrak{F}^0$ thereby becoming identified with the group of automorphisms of $\mathfrak{G}\langle\mathfrak{F}'\rangle$ over $\mathfrak{G}$); also ([7], Ch. III, § 3, Th. 5, p. 799) $\mathfrak{G}\langle\mathfrak{F}'\rangle$ is a strongly normal extension of $\mathfrak{F}'$, and every isomorphism σ of $\mathfrak{G}$ over $\mathfrak{F}^0$ can be extended to a unique isomorphism (which we also denote by σ) of $\mathfrak{G}\langle\mathfrak{F}'\rangle$ over $\mathfrak{F}'$ (the component of identity $\mathfrak{G}^0$ of $\mathfrak{G}$ thereby becoming identified with the Galois group of $\mathfrak{G}\langle\mathfrak{F}'\rangle$ over $\mathfrak{F}'$). Since every element of $\mathfrak{G}\langle\mathfrak{F}'\rangle$ can be expressed as the quotient of two elements of the form $\sum \psi_i\phi_i$ ($\psi_i \varepsilon \mathcal{G}, \phi_i \varepsilon \mathcal{F}'$), and since when σ is an *auto*morphism

$$\sigma\tau(\sum \psi_i\phi_i) = \sigma \sum \psi_i\tau\phi_i = \sum (\sigma\psi_i)\tau\phi_i = \tau \sum (\sigma\psi_i)\phi_i = \tau\sigma(\sum \psi_i\phi_i),$$

we see that $\sigma\tau = \tau\sigma$. It follows that if $\alpha_1, \cdots, \alpha_h$ are the conjugates of $\alpha \varepsilon \mathcal{G}\langle\mathcal{F}'\rangle$ over $\mathcal{G}$ then $\sigma\alpha_1, \cdots, \sigma\alpha_h$ are the conjugates of $\sigma\alpha$ over $\mathcal{G}$ (where σ is any element of $\mathfrak{G}^0$ which is an automorphism of $\mathcal{G}$).

Now let $f \varepsilon Z^1(\mathfrak{G}, G)$. Since $\mathfrak{G}^0$ can be identified with the Galois group of $\mathcal{G}\langle\mathcal{F}'\rangle$ over $\mathcal{F}'$, since (for $\sigma \varepsilon \mathfrak{G}^0$) $\mathcal{G}\langle\mathcal{F}'\rangle(f(\sigma)) \subset \mathcal{G}\langle\mathcal{F}'\rangle(\sigma(\mathcal{G}\langle\mathcal{F}'\rangle))$, and since (for $\sigma, \sigma' \varepsilon \mathfrak{G}^0$) "$\sigma'$ is a specialization of σ (as isomorphism of $\mathcal{G}\langle\mathcal{F}'\rangle$)" implies "$\sigma'$ is a specialization of σ (as isomorphisms of $\mathcal{G}$)" which implies "$f(\sigma')$ is a specialization of $f(\sigma)$ over $\mathcal{G}$" which (because $\mathcal{G}\langle\mathcal{C}^*\rangle$ and $\mathcal{G}\langle\mathcal{F}'\rangle$ are linearly disjoint over $\mathcal{G}$) in turn implies "$f(\sigma')$ is a specialization of $f(\sigma)$ over $\mathcal{G}\langle\mathcal{F}'\rangle$," we see that the restriction of f to $\mathfrak{G}^0$ may be considered as a rational crossed homomorphism of the Galois group of $\mathcal{G}\langle\mathcal{F}'\rangle$ over $\mathcal{F}'$ into G. By 2) above, therefore, there exists an $\alpha \varepsilon \mathfrak{G}^\dagger$ rational over $\mathcal{G}\langle\mathcal{F}'\rangle$, such that $f(\sigma) = \alpha^{-1}\sigma\alpha$ for every $\sigma \varepsilon \mathfrak{G}^0$ which is an automorphism of $\mathcal{G}$. Denoting the conjugates of α over $\mathcal{G}$ by $\alpha_1, \cdots, \alpha_h$ we see by the above (since $f(\sigma)$ is rational over $\mathcal{G}$) that $f(\sigma) = \alpha_i^{-1}\sigma\alpha_i$ ($1 \leqq i \leqq h$), Since G is by assumption commutative, we theerfore have $(f(\sigma))^h = \beta^{-1}\sigma\beta$, where $\beta = \alpha_1 \cdots \alpha_h$ is rational over $\mathcal{G}$. Therefore the 1-cocycle g cohomologous to f^h, defined by $g(\sigma) = \beta(f(\sigma))^h\sigma\beta^{-1}$ ($\sigma \varepsilon \mathfrak{G}$), has the value 1 at every element of $\mathfrak{G}^0$. It follows, as in the proof of 1) above, that g is constant on each coset and that $g(\sigma)$ is rational over $\mathcal{F}^0$ for every $\sigma \varepsilon \mathfrak{G}$. Consequently, letting $\sigma_1, \cdots, \sigma_k$ denote a complete set of representatives of the cosets of $\mathfrak{G}$ relative to $\mathfrak{G}^0$, and setting $\gamma = g(\sigma_1) \cdots g(\sigma_k)$, we find that

$$\sigma\gamma = \prod \sigma g(\sigma_j) = \prod g(\sigma)^{-1} g(\sigma\sigma_j) = (g(\sigma))^{-k}\gamma$$

for every $\sigma \varepsilon \mathfrak{G}$, so that $(f(\sigma))^{hk} = \beta^{-k}\gamma\sigma(\gamma^{-1}\beta^k)$ for every $\sigma \varepsilon \mathfrak{G}$ and $f^{hk} \varepsilon B^1(\mathfrak{G}, G)$.

3. Applications. Theorem 1 of Chapter II asserts that there exists a birational isomorphism of the Galois group $\mathfrak{G}$ onto a group variety G. Now such a birational isomorphism, or more generally any rational homomorphism f of $\mathfrak{G}$ into G, is a 1-cocycle of $\mathfrak{G}$ into G. If we can prove that f is a 1-coboundary then there exists a point α of G (or, more precisely, of $G^\dagger$) which is rational over $\mathcal{G}$ such that $f(\sigma) = \alpha^{-1}\sigma\alpha$, that is, such that $\sigma\alpha = \alpha f(\sigma)$ for every $\sigma \varepsilon \mathfrak{G}$. In the case in which f is actually an isomorphism, we see that α is invariant under σ only when σ is the identity isomorphism, so that $\mathcal{G} = \mathcal{F}\langle\alpha\rangle$. More generally, if f is a homomorphism with *finite* kernel then $\mathcal{G}$ is a normal algebraic extension of $\mathcal{F}\langle\alpha\rangle$ with Galois group isomorphic to the kernel of f. These remarks are the basis of the use of Theorem 1 in the proofs of the following two theorems.

THEOREM 2. *A necessary and sufficient condition that the strongly normal extension* $\mathcal{G}$ *of* $\mathcal{F}$ *be a Picard-Vessiot extension is that its Galois group* $\mathfrak{G}$ *be birationally isomorphic to an algebraic group of matrices.*

Proof. If $\mathcal{G}$ is a Picard-Vessiot extension of $\mathcal{F}$ then, as we have already observed (Chapter I, § 3, Example 2), $\mathfrak{G}$ is birationally isomorphic to an algebraic matric group. Conversely, let f be a birational isomorphism of $\mathfrak{G}$ onto an algebraic subgroup of $GL(n)$. By Theorem 1, there exists an invertible square matrix $\alpha = (\alpha_{ij})_{1 \leqq i \leqq n,\, 1 \leqq j \leqq n}$ with coordinates $\alpha_{ij} \varepsilon \mathcal{G}$ such that $\sigma\alpha = \alpha f(\sigma)$ for every $\sigma \varepsilon \mathfrak{G}$, and $\mathcal{G} = \mathcal{F}\langle\alpha\rangle$. Therefore if $\eta_1, \cdots, \eta_p$ form a maximal subset of the set of elements α_{ij} linearly independent over $\mathcal{C}$ (and therefore over every field of constants) then for each $\sigma \varepsilon \mathfrak{G}$ there exists an invertible matrix $(c_{ij}(\sigma))_{1 \leqq i \leqq p,\, 1 \leqq j \leqq p}$ with constant coordinates $c_{ij}(\sigma)$, such that $\sigma\eta_j = \sum_{1 \leqq i \leqq p} c_{ij}(\sigma)\eta_i$ $(1 \leqq j \leqq p)$. It is now clear that if $\theta_1, \cdots, \theta_p$ are arbitrary operators of the form $\delta_1^{i_1} \cdots \delta_m^{i_m}$ then the determinant

$$W_{\theta_1 \cdots \theta_p}(\eta_1, \cdots, \eta_p) = \det(\theta_i \eta_j)_{1 \leqq i \leqq p,\, 1 \leqq j \leqq p}$$

has the property that

$$\sigma W_{\theta_1 \cdots \theta_p}(\eta_1, \cdots, \eta_p) = \det(c_{ij}(\sigma)) \cdot W_{\theta_1 \cdots \theta_p}(\eta_1, \cdots, \eta_p),$$

so that the ratio of any two nonzero determinants of this type is invariant under $\mathfrak{G}$ and hence is an element of $\mathcal{F}$. Thus ([6], § 3, p. 599) $\mathcal{G}$ is a Picard-Vessiot extension of $\mathcal{F}$.

COROLLARY. *If* $\mathcal{G}$ *is a Picard-Vessiot extension of* $\mathcal{F}$ *and* $\mathcal{H}$ *is any differential field between* $\mathcal{F}$ *and* $\mathcal{G}$ *which is a strongly normal extension*[3] *of* $\mathcal{F}$ *then* $\mathcal{H}$ *is a Picard-Vessiot extension of* $\mathcal{F}$.

Proof. Since $\mathcal{G}$ is Picard-Vessiot over $\mathcal{F}$ there exists a birational isomorphism f of $\mathfrak{G}$ onto an algebraic matric group G. The Galois group $\mathfrak{G}(\mathcal{H})$ of $\mathcal{G}$ over $\mathcal{H}$ is a normal algebraic subgroup of $\mathfrak{G}$, therefore $K = f(\mathfrak{G}(\mathcal{H}))$ is a normal algebraic subgroup of G. Now it is known (Chevalley and Kolchin [4], § 1, Th. 3) that there exists a matric representation g of G with kernel K; furthermore, although it is not specifically stated in [4], the proof there (§ 4) provides such a g which is a rational homomorphism of G. If we denote the image of g by H then (Chapter I, corollary to Proposition 5) H is an algebraic matric group. Clearly $g \circ f$ is a rational homomorphism of $\mathfrak{G}$ with kernel $\mathfrak{G}(\mathcal{H})$ and image H. Now (Chapter I, § 3, Example 1) the mapping r of $\mathfrak{G}$, which to each $\sigma \varepsilon \mathfrak{G}$ corresponds the restriction of σ

[3] Or even ([7], Ch. III, § 3, Th. 4, p. 797) a weakly normal extension.

to $\mathcal{H}$, is a rational homomorphism of $\mathfrak{G}$ with kernel $\mathfrak{G}(\mathcal{H})$ and image the Galois group $\mathfrak{H}$ of $\mathcal{H}$ over $\mathcal{F}$. Therefore (Chapter II, § 3, Proposition 1) there exists a birational isomorphism of $\mathfrak{H}$ onto H so that, by the present theorem, $\mathcal{H}$ is a Picard-Vessiot extension of $\mathcal{F}$.

THEOREM 3. *If the strongly normal extension $\mathcal{G}$ of $\mathcal{F}$ is of transcendence degree* 1 *and if $\mathcal{F}$ is relatively algebraically closed in $\mathcal{G}$ then there exists an element $\alpha \in \mathcal{G}$ such that either α is primitive over $\mathcal{F}$ and $\mathcal{G} = \mathcal{F}\langle\alpha\rangle$, or α is exponential over $\mathcal{F}$ and $\mathcal{G} = \mathcal{F}\langle\alpha\rangle$, or α is weierstrassian over $\mathcal{F}$ and $\mathcal{G}$ is an abelian algebraic extension of $\mathcal{F}\langle\alpha\rangle$ of finite degree. In the last case, if $\mathcal{F}$ is algebraically closed then the weierstrassian element α may be chosen so that $\mathcal{G} = \mathcal{F}\langle\alpha\rangle$.*

This theorem is not new, being the main substance of Th. 7 of [7], Ch. III, p. 809. We include it here to show how it can be derived as a rather easy consequence of Theorem 1, and Theorem 1 of Chapter II; this makes it possible to avoid the long, tedious, and intricate computations of the proof in [7], pp. 809-823.

Proof. It follows from the hypothesis that the Galois group $\mathfrak{G}$ has dimension 1 and is irreducible; by Theorem 1 of Chapter II there exists a birational isomorphism c of $\mathfrak{G}$ onto a group variety G defined over $\mathcal{C}$, which must also be of dimension 1 and irreducible. It is known (see Weil [14], § V, No. 38, pp. 68-69) that every such group variety is birationally isomorphic to (and therefore we may assume equal to) either the additive group of the field $\mathcal{C}^*$ of all constants or the multiplicative group of $\mathcal{C}^*$, or else a group variety which is a complete curve of genus 1 in projective space. In the last case we may always assume (for example, see Chevalley [2], Ch. II, §3) that the curve has an equation of the form $X_0X_2{}^2 = P(X_0, X_1)$, where P is a homogenous polynomial in $\mathcal{C}[X_0, X_1]$ of degree 3 without multiple factor; applying a suitable transformation $(X_0, X_1, X_2) \rightarrow (X_0, aX_0 + bX_1, cX_2)$ if necessary, we lose no generality in supposing that

$$P(X_0, X_1) = 4X_1{}^3 - g_2X_0{}^2X_1 - g_3X_0{}^3,$$

where $g_2, g_3 \in \mathcal{C}$. We may take the law of composition so that the group unity is $(0:0:1)$, which is the sole point at infinity on the curve, and such that the product $(1:\eta:\zeta)$ of two points $(1:\alpha:\beta)$ and $(1:a:b)$ with $\alpha \neq a$ is given by

$$(3) \qquad \eta = -\alpha - a + 4^{-1}(\alpha - a)^{-2}(\beta - b)^2,$$

$$(4) \qquad \zeta = -2^{-1}(\beta + b) + 2^{-1}3(\alpha - a)^{-1}(\beta - b) - 4^{-1}(\alpha - a)^{-3}(\beta - b)^3.$$

Consider the first case. By Theorem 1, there exists an element $\alpha \varepsilon \mathcal{G}$ such that $c(\sigma) = \sigma\alpha - \alpha$ for every $\sigma \varepsilon \mathfrak{G}$. Since $c(\sigma) = 0$ only when σ is the identity, only the identity in $\mathfrak{G}$ leaves α invariant, so that $\mathcal{G} = \mathcal{F}\langle\alpha\rangle$. Applying the derivation operator δ_i to both sides of the equation $\sigma\alpha = \alpha + c(\sigma)$ we obtain $\sigma\delta_i\alpha = \delta_i\alpha (\sigma \varepsilon \mathfrak{G})$, so that $\delta_i\alpha \varepsilon \mathcal{F}$ $(1 \leqq i \leqq m)$, and α is primitive over $\mathcal{F}$.

Next consider the second case. By Theorem 1, there exists a nonzero element $\alpha \varepsilon \mathcal{G}$ such that $c(\sigma) = \alpha^{-1}\sigma\alpha$ for every $\sigma \varepsilon \mathfrak{G}$. Since $c(\sigma) = 1$ only when σ is the identity, only the identity in $\mathfrak{G}$ leaves α invariant, so that $\mathcal{G} = \mathcal{F}\langle\alpha\rangle$. From the equation $\sigma\alpha = \alpha c(\sigma)$ we find that $\sigma\delta_i\alpha = (\delta_i\alpha)c(\sigma)$, whence $\sigma(\alpha^{-1}\delta_i\alpha) = \alpha^{-1}\delta_i\alpha$ $(\sigma \varepsilon \mathfrak{G})$, so that $\alpha^{-1}\delta_i\alpha \varepsilon \mathcal{F}$ $(1 \leqq i \leqq m)$, and α is exponential over $\mathcal{F}$.

Consider finally the third case. By Theorem 1, there exists a point $(1:\alpha:\beta)$ on the curve $G^\dagger$, with $\alpha, \beta \varepsilon \mathcal{G}$, and there exists an integer $n > 0$ ($n = 1$ if $\mathcal{F}$ is algebraically closed), such that $c(\sigma)^n = (1:\alpha:\beta)^{-1}$ $(1:\sigma\alpha:\sigma\beta)$ for every $\sigma \varepsilon \mathfrak{G}$. Not every element of $\mathfrak{G}$ has order n, so that $c(\sigma)^n$ fails to be the identity for some $\sigma \varepsilon \mathfrak{G}$; therefore $(1:\alpha:\beta)$ is not an invariant of the whole group $\mathfrak{G}$, hence is not rational over $\mathcal{F}$. Since $\mathcal{F}$ is relatively algebraically closed in $\mathcal{G}$ and $\mathcal{G}$ is of transcendence degree 1 over $\mathcal{F}$, $\mathcal{G}$ is an algebraic extension of $\mathcal{F}\langle\alpha, \beta\rangle$ which clearly must be of finite degree, normal, and indeed abelian; if $\mathcal{F}$ is algebraically closed, so that $n = 1$, then $\mathcal{G} = \mathcal{F}\langle\alpha, \beta\rangle$. For any $\sigma \varepsilon \mathfrak{G}$ which is not of order n (in particular, for a generic element of $\mathfrak{G}$) we may write $c(\sigma)^n = (1:a:b)$, where a, b are constants. Then $\sigma\alpha$ and $\sigma\beta$ are given by the second members of (3) and (4), respectively. Making use of the fact that

$$(5) \qquad \beta^2 = 4\alpha^3 - g_2\alpha - g_3,$$

these may be rewritten to obtain

$$(6) \qquad \sigma\alpha = 4^{-1}(\alpha - a)^{-2}[4a\alpha^2 + (4a^2 - g_2)\alpha - (g_2a + 2g_3) - 2b\beta],$$

$$(7) \qquad \sigma\beta = 4^{-1}(\alpha - a)^{-3}[(4\alpha^3 + 12a\alpha^2 - 3g_2\alpha - g_2a - 4g_3)b + ((-12a^2 + g_2)\alpha - 4a^3 + 3g_2a + 4g_3)\beta].$$

From (5) we find $2\beta\delta_i\beta = (12\alpha^2 - g_2)\delta_i\alpha$. Therefore application of δ_i to (6) yields

$$\begin{aligned}\sigma\delta_i\alpha &= 4^{-1}(\alpha - a)^{-2}((8a\alpha + 4a^2 - g_2)\delta_i\alpha - 2b)\delta_i\beta \\ &\quad - 2^{-1}(\alpha - a)^{-3}(4a\alpha^2 + (4a^2 - g_2)\alpha - g_2a - 2g_3 - 2b\beta)\delta_i\alpha \\ &= 4^{-1}(\alpha - a)^{-2}((8a\alpha + 4a^2 - g_2)\beta - b(12\alpha^2 - g_2))\beta^{-1}\delta_i\alpha \\ &\quad - 2^{-1}(\alpha - a)^{-3}((4a\alpha^2 + (4a^2 - g_2)\alpha - g_2a - 2g_3)\beta \\ &\qquad - 2b(4\alpha^3 - g_2\alpha - g_3))\beta^{-1}\delta_i\alpha.\end{aligned}$$

The last member equals $\beta^{-1}\delta_i\alpha$ multiplied by the second member of (7). Therefore $\sigma\delta_i\alpha = (\sigma\beta)\beta^{-1}\delta_i\alpha$, that is $\sigma(\beta^{-1}\delta_i\alpha) = \beta^{-1}\delta_i\alpha$ $(\sigma \varepsilon \mathfrak{G})$, so that $\beta^{-1}\delta_i\alpha \,\varepsilon\, \mathfrak{F}$ $(1 \leqq i \leqq m)$. It follows that $\mathfrak{F}\langle\alpha,\beta\rangle = \mathfrak{F}\langle\alpha\rangle$, and also (with the help of (5) that $(4\alpha^3 - g_2\alpha - g_3)^{-1}$ $(\delta_i\alpha)^2$ is the square of an element of $\mathfrak{F}$ $(1 \leqq i \leqq m)$, that is, that α is weierstrassian over $\mathfrak{F}$.

COLUMBIA UNIVERSITY
AND UNIVERSITY OF PARIS.

REFERENCES.

[1] N. Bourbaki, "*Algèbre*," Chapters IV-V (Actualités scientifiques et industrielles 1102), Hermann et Cie., Paris, 1950.

[2] C. Chevalley, "*Introduction to the theory of algebraic functions of one variable*" (Mathematical Surveys VI), American Mathematical Society, New York, 1951.

[3] ———, "*Théorie des groupes de Lie, Tome II: Groupes algébriques*" (Actualités scientifiques et industrielles 1152), Hermann et Cie., Paris, 1951.

[4] ———, and E. Kolchin, "Two proofs of a theorem on algebraic groups," *Proceedings of the American Mathematical Society*, vol. 2 (1951), pp. 126-134.

[5] E. R. Kolchin, "Algebraic matric groups and the Picard-Vessiot theory of homogeneous linear ordinary differential equations," *Annals of Mathematics*, vol. 49 (1948), pp. 1-42.

[6] ———, "Picard-Vessiot theory of partial differential fields," *Proceedings of the American Mathematical Society*, vol. 3 (1952), pp. 596-603.

[7] ———, "Galois theory of differential fields," *American Journal of Mathematics*, vol. 75 (1953), pp. 753-824.

[8] H. Matsumura, "Automorphism-groups of differential fields and group varieties," *Memoirs of the College of Science, University of Kyoto, Series A*, vol. 28 (1954), pp. 283-292.

[9] S. Nakano, "On invariant differential forms on a group variety," *Journal of the Mathematical Society of Japan*, vol. 2 (1951), pp. 216-227.

[10] ———, "Note on group varieties," *Memoirs of the College of Science, University of Kyoto, Series A*, vol. 27 (1952), pp. 55-66.

[11] M. Rosenlicht, "Generalized Jacobian varieties," *Annals of Mathematics*, vol. 59 (1954), pp. 505-530.

[12] J.-P. Serre, "Faisceaux algébriques cohérents," *Annals of Mathematics*, vol. 61 (1955), pp. 197-278.

[13] A. Weil, "*Foundations of Algebraic geometry*" (American Mathematical Society Colloquium Publications, vol. 29), New York, 1946.

[14] ———, "Variétés abéliennes et courbes algébrique (Actualités scientifiques et industrielles 1064), Hermann et Cie., Paris, 1948.

ALGEBRAIC GROUPS AND THE GALOIS THEORY OF DIFFERENTIAL FIELDS.*

By Ellis Kolchin[1] and Serge Lang.

The purpose of this note is to draw together morely closely the Galois theory of differential fields and the theory of algebraic groups and homogeneous spaces.

Given a strongly normal extension of a differential field of characteristic 0, it is known (see [2] or Matsumura [3]) that the Galois group is isomorphic to the group of points of an algebraic group which are rational over the field of constants. We show below that the extension itself has a model which is a principal homogeneous space for the algebraic group, such that the operation of the elements of the Galois group on the extension is induced by the operation of the corresponding elements of the algebraic group on the space. The Galois theory can then be deduced from known facts concerning principal homogeneous spaces (Rosenlicht [4]). This provides an analysis of the Galois theory into two parts, one depending on the foundations of differential algebra and the other on the foundations of the theory of algebraic groups.

There remains almost untouched a converse problem which, for the sake of simplicity, we state for connected algebraic groups, or as we shall also say, group varieties. Given a differential field F of characteristic 0 with algebraically closed field of constants C, a group variety G defined over C, and a principal homogeneous space V for G defined over F, to determine all extensions of the differential field structure on F to the function field $F(V)$ which makes $F(V)$ a strongly normal extension of F with automorphism group induced by the group G_C of points of G rational over C. Put another way, the problem is to describe all possible ways of extending the derivation operators of F to derivation operators on $F(V)$ which are invariant relative to G_C, which commute with each other, and which have precisely C as field of constants. We observe here merely that the first condition alone

* Received August 3, 1957.

[1] The work of this author was assisted by a grant from the National Science Foundation.

presents no difficulty, i.e. that it is easily shown that the derivations of F can be extended to invariant derivations of $F(V)$.

1. Algebraic groups. We recall here briefly the notions of algebraic group and principal homogeneous space. For the general theory, see Weil [5] and [6], and Rosenlicht [4].

An *algebraic set* V is a finite union of varieties, $V = \bigcup V_i$, which for the purposes of this paper we shall always assume disjoints; the V_i are the components of V. Let V, W be two algebraic sets. A rational map $f\colon V \to W$ is an algebraic set, having one component f_i for each component V_i of V such that f_i is a rational map in the usual sense of V_i into a component of W. We say that f is defined at a point $u \in U_i$ if f_i is defined at that point, and we write $f(u) = f_i(u)$.

An *algebraic group* is an algebraic set G together with an everywhere defined rational map of $G \times G$ into G, relative to which G is a group, having the further property that the inverse mapping $x \to x^{-1}$ is an everywhere defined rational map of G into G.

If G is an algebraic group, a *transformation space* for G is an algebraic set V together with an everywhere defined rational map of $V \times G$ into V (the operation of G on V) which, if we denote the image of (v, x) by vx, satisfies the conditions

$$v(xy) = (vx)y \text{ and } ve = v$$

for all $v \in V$, $x \in G$, $y \in G$ (e denoting the unity element of G.) A transformation space is a *homogeneous space* if, given any $v, w \in V$, there exists an $x \in G$ such that $vx = w$. The homogeneous space is *principal* if this point x is determined uniquely and rationally. By the latter we mean that there exists an everywhere defined rational map $\lambda\colon V \times V \to G$ such that $x = \lambda(v, w)$. If V is a principal homogeneous space for G then the number of components of V equals that of G and each component of V is a principal homogeneous space for the component G_0 of G containing e, G_0 being a normal subgroup of G the cosets of which are precisely the components of G.

We shall say that the algebraic group G is defined over a field k if all the components of G, of its group law, and of its inverse mappings are defined over k. Similarly, we shall say that the principal homogeneous space V for G is defined over k if G is defined over k, and if all the components of V, of the operation of G on V, and of λ are also defined over k. Although this notion of field of definition may present some drawbacks, it is adequate for our present purpose.

2. Differential-algebraic preliminaries. Let F be a differential field of characteristic 0. Denote the field of constants of F by C. We work in a fixed universal extension F^* of F in the sense of [1], and denote the field of constants of F^* by C^*. The facts stated in the following propositions were proved in [1].

PROPOSITION 1. *The fields F and C^* are linearly disjoint over C.*

By an extension of F, we shall always mean a differential field extension contained in the universal domain. Similarly, an isomorphism will always mean a differential field isomorphism. When we wish to consider only the field structure, we shall say field extension, and field isomorphism.

Let E be an extension of F, and suppose that as a field extension E is finitely generated. Let F_0 be the algebraic closure of F in E. We recall if σ and σ' are isomorphisms of E, σ' is called a *specialization* of σ if the family $(\sigma'\alpha)_{\alpha \in E}$ is a specialization of $(\sigma\alpha)_{\alpha \in E}$ over E in the usual sense of algebraic geometry. The specialization is called *generic* if also σ is a specialization of σ'. An isomorphism of E over F is called *isolated* if it is not a nongeneric specialization of any isomorphism of E over F.

PROPOSITION 2. (a) *There exists a finite set of isolated isomorphisms of E over F such that every isomorphism of E over F is a specialization of precisely one of these.*

(b) *An isomorphism σ of E over F is isolated if and only if the transcendence degree of the compositum $E \cdot \sigma E$ over E equals that of E over F.*

(c) *An alement of E invariant under an isolated isomorphism of E over F belongs to F_0.*

(d) *If σ is isolated then σ' is a specialization of σ if and only if their restriction to F_0 coincide. Every isomorphism of F_0 over F is the restriction to F_0 of an isolated isomorphism of E over F.*

Suppose further that E has the same field of constants C as F has, and that C is algebraically closed. We recall that an isomorphism σ of E is called *strong* if $\sigma E \subset E \cdot C^*$ and $E \subset \sigma E \cdot C^*$. Letting $C(\sigma)$ denote the field of constants of $E \cdot \sigma E$, we can see with the help of Proposition 1 that σ is strong if and only if

$$E \cdot C(\sigma) = E \cdot \sigma E = \sigma E \cdot C(\sigma).$$

Every automorphism of E is obviously a strong isomorphism of E.

PROPOSITION 3. (a) *Every strong isomorphism of E over F has a specialization which is an automorphism of E.*

(b) *An isomorphism of E over F is strong if and only if it is the restriction to E of an automorphism of $E\cdot C^*$ over $F\cdot C^*$.*

As the automorphism mentioned in (b) is evidently unique, we may identify the strong isomorphism with it. The set of strong isomorphisms of E over F is then identified with the group of automorphisms of $E\cdot C^*$ over $F\cdot C^*$. We denote this group by $\mathfrak{G}$.

Let $\sigma,\tau\in\mathfrak{G}$. As $C(\sigma^{-1})$ is the field of constants of $E\cdot\sigma^{-1}E$, $\sigma C(\sigma^{-1}) = C(\sigma^{-1})$ must be the field of constants of $\sigma(E\cdot\sigma^{-1}E)=E\cdot\sigma E$. Thus

$$C(\sigma^{-1})=C(\sigma).$$

Writing $C(\sigma,\tau)$ to denote the compositum $C(\sigma)\cdot C(\tau)$, we see that

$$\begin{aligned}E\cdot C(\sigma,\sigma\tau)&=E\cdot C(\sigma)\cdot C(\sigma\tau)=E\cdot\sigma E\cdot\sigma\tau E=E\cdot\sigma(E\cdot\tau E)=E\cdot\sigma(E\cdot C(\tau))\\&=E\cdot\sigma E\cdot C(\tau)=E\cdot C(\sigma)\cdot C(\tau)=E\cdot C(\sigma,\tau).\end{aligned}$$

It follows with the help of Proposition 1 (applied to E instead of F) that $C(\sigma,\tau)=C(\sigma,\sigma\tau)$. Therefore $C(\tau^{-1},\sigma^{-1})=C(\tau^{-1},\tau^{-1}\sigma^{-1})=C(\tau,\sigma\tau)$. Thus

$$C(\sigma,\sigma\tau)=C(\sigma,\tau)=C(\sigma\tau,\tau). \tag{1}$$

We observe also, using Propositions 2(b) and Proposition 1, that $\sigma\in\mathfrak{G}$ is isolated if and only if the transcendence degree of $C(\sigma)$ over C equals that of E over F.

We suppose, finally, that the extension E of F is *strongly normal,* that is, that in addition to all the hypotheses made above, every isomorphism of E over F is strong. $\mathfrak{G}$ is then called the *Galois group* of E over F.

3. The theorem and its proof. We temporarily impose the restriction that $F=F_0$. Then by Proposition 2(a) and (d), all isolated elements of $\mathfrak{G}$ are generic specializations of each other.

Let $\sigma_0\in\mathfrak{G}$ be isolated, and let W be any model of the field $C(\sigma_0)$ over C. This means that W is a variety defined over C and there exists a generic point x of W over C such that $C(\sigma_0)=C(x)$. If σ is any isolated element of $\mathfrak{G}$ then σ is a generic specialization of σ_0 so that there exists an isomorphism $E\cdot\sigma_0 E\approx E\cdot\sigma E$ over E mapping $\sigma_0\alpha$ onto $\sigma\alpha$ $(\alpha\in E)$. This isomorphism maps $C(\sigma_0)$ onto $C(\sigma)$ and therefore maps x onto a generic point of W over C, which we call x_σ (so that in particular $x_{\sigma_0}=x$) and which has the property that $C(\sigma)=C(x_\sigma)$. Because $E\cdot\sigma_0E=E\cdot C(\sigma_0)$ the isomorphism $E\cdot\sigma_0 E \approx E\cdot\sigma E$ over E is determined by its restriction to $C(\sigma_0)$, so that σ is determined by x_σ.

Furthermore if y is any point of W which is rational over C^* and generic over C, there exists a unique field isomorphism $C(x)\approx C(y)$ over C mapping x onto y, and because E and C^* are linearly disjoint over C this field isomorphism can be extended to a field isomorphism ϕ over E of $E\cdot C(x) = E\cdot C(\sigma_0) = E\cdot\sigma_0 E$ onto $E\cdot C(y)$. As ϕ maps $C(\sigma)$ into $C(y)$, ϕ commutes on $C(\sigma)$ with the derivation operators of the differential field structure, and as ϕ is the identity on E, ϕ commutes on E with these derivation operators, so that ϕ commutes on $E\cdot C(\sigma)$ with these operators. In other words, ϕ is a differential field isomorphism. Letting $\tau=\phi\sigma_0$, we see that $y=x_\tau$. We have thus proved that $\sigma\rightarrow x_\sigma$ is a bijective mapping of the set of isolated elements of G onto the set of points of W which are rational over C^* and generic over C.

Let $\sigma,\tau\in\mathfrak{G}$. If in addition to being isolated, σ and τ are independent, in the sense that $C(\sigma)$ and $C(\tau)$ are linearly disjoint over C, then (1) shows that $\sigma\tau$ is isolated. Furthermore, by Proposition 1, $E\cdot\sigma E = E\cdot C(\sigma)$ and $E\cdot\tau E = E\cdot C(\tau)$ are linearly disjoint over E. If σ' and τ' are also independent isolated elements of $\mathfrak{G}$, then the isomorphism $E\cdot\sigma E\approx E\cdot\sigma' E$ over E mapping $(\sigma\alpha)_{\alpha\in E}$ onto $(\sigma'\alpha)_{\alpha\in E}$ and the analogous isomorphism $E\cdot\tau E \approx E\cdot\tau' E$ have a unique common extension.

$$E\cdot\sigma E\cdot\tau E\approx E\cdot\sigma' E\cdot\tau' E. \tag{2}$$

For any element $\alpha\in E$, $\tau\alpha$ can be expressed in the form $\tau\alpha = f(x_\tau)$, where f is a rational function on W defined over E. Hence we have $\tau'\alpha = f(x_{\tau'})$, $\sigma\tau\alpha = f^\sigma(x_\tau)$, and $\sigma'\tau'\alpha = f^{\sigma'}(x_{\tau'})$. From this it is clear that the isomorphism (2) maps $(\sigma\tau\alpha)_{\alpha\in E}$ onto $(\sigma'\tau'\alpha)_{\alpha\in E}$ and therefore maps $x_{\sigma\tau}$ onto $x_{\sigma'\tau'}$. Thus $(x_{\sigma'}, x_{\tau'}, x_{\sigma'\tau'})$ is a specialization of $(x_\sigma, x_\tau, x_{\sigma\tau})$ over C (and even over E).

It follows that we may define a law of composition on W by the formula $x_\sigma x_\tau = x_{\sigma\tau}$, this law of composition being defined over C. From (1) we have $C(x_\sigma, x_\sigma x_\tau) = C(x_\sigma, x_\tau) = C(x_\sigma x_\tau, x_\tau)$. Furthermore, if ω is an isolated element of $\mathfrak{G}$ such that ω,σ,τ are independent, i.e. such that $x_\omega, x_\sigma, x_\tau$ are independent generic points of W, then it is clear that $x_\omega(x_\sigma x_\tau) = (x_\omega x_\sigma)x_\tau$. Thus we have a normal law of composition on W, so that by the fundamental theorem of Wiel [5], W is birationally equivalent over C to a group variety defined over C. Thus we may suppose that W is itself a group variety, and we henceforth denote it by G.

Since we have assumed that $F=F_0$, E is a regular extension of F.

Therefore there exists a model V for E over F, i.e. a variety V defined over F such that E may be written $E = F(v)$ where v is a generic point of V over F. Letting σ be an isolated element of $\mathfrak{G}$, so that x_σ is a point of G which is rational over C^* and generic over C (and therefore over E), we define a rational map of $V \times G$ into V by the formula

$$(v, x_\sigma) \rightarrow v x_\sigma = \sigma v.$$

It follows from the above considerations that this rational map, which is defined over F, does not depend on which isolated element σ is used. It is also easy to see from what precedes that if σ and τ are independent isolated elements of $\mathfrak{G}$, then

$$F(vx_\sigma, x_\sigma) = F(v, x_\sigma) = F(v, vx_\sigma)$$

and

$$v(x_\sigma x_\tau) = (vx_\sigma)x_\tau.$$

It follows from another result of Weil ([6] Prop. 3) that V is birationally equivalent over F to a principal homogeneous space for G defined over F, and therefore we may assume that V itself is such a space. This principal homogeneous space is, moreover, uniquely determined up to an everywhere defined birational correspondence compatible with the operation of G.

This being the case, we see that every point $a \in G$ (generic or not) such that v is generic on V over $F(a)$ induces a field isomorphism σ_a of $E = F(v)$ over F mapping v onto va. If a is any point of the group G_{C^*} of points of G rational over C^*, then v is automatically generic on V over $F(a)$, because E and C^* are linearly disjoint over C, and hence E and $F \cdot C^*$ are linearly disjoint over F. In this case, σ_a *is* an isomorphism of the differential field structure. Indeed, for elements of G_{C^*} which are generic over C this has been proved above, and for an arbitrary $a \in G_{C^*}$, there exists an element $x \in G_{C^*}$ such that x^{-1} and xa are generic over C, and $\sigma_a = \sigma_{x^{-1}}\sigma_{xa}$.

The mapping $a \rightarrow \sigma_a$ $(a \in G_{C^*})$ is obviously an isomorphism of G_{C^*} into $\mathfrak{G}$. It is actually onto $\mathfrak{G}$, for given any $\tau \in \mathfrak{G}$ we can select an isolated $\tau' \in \mathfrak{G}$ so that τ and τ' are independent. Then $\tau\tau'$ and τ'^{-1} are isolated and $\tau = (\tau\tau')\tau'^{-1}$.

We have now proved our main result in the special case that $F = F_0$. We shall indicate below how to extend it to the general case, and we state here the general theorem which for the case $F = F_0$ summarizes the above discussion.

THEOREM. *Let F be a differential field of characteristic* 0 *with algebraically closed field of constants C, and let E be a strongly normal extension* *of F, all lying in a universal extension F^* with field of constants C^*. Let $\mathfrak{G}$ be the Galois group of E over F. Then there exists an algebraic group G defined over C, and a principal homogeneous spave V for G, defined over the algebraic closure F_0 of F in E, having the following properties.*

(a) *V is a model of E over F, that is, any component of V has a point v generic over F_0 such that $E = F_0(v)$, and these components are all conjugates to each other over F.*

(b) *For each $\sigma \in \mathfrak{G}$ there exists a unique point $x_\sigma \in G_{C^*}$ (the group of points of G which are rational over C^*) such that $\sigma v = v x_\sigma$, and the mapping $\sigma \rightarrow x_\sigma$ $(\sigma \in \mathfrak{G})$ is a group isomorphism of $\mathfrak{G}$ onto G_{C^*}.*

(c) *This isomorphism is rational in the sense that $C(\sigma) = C(x_\sigma)$ and that σ' is a specialization of σ if and only if $x_{\sigma'}$ is a specialization of x_σ over C.*

In order to extend the result proved for $F = F_0$ to the general case, we apply the proved result to the extension E of F_0. We thus obtain a connected algebraic group G_0 defined over C and a principal homogeneous space V_0 defined over F_0 such that V_0 has a point v generic over F_0 with $E = F_0(v)$, and the Galois group $\mathfrak{G}_0$ of E over F_0 is isomorphic with G_{0C^*} in the manner described in the statement of the theorem. Every $\sigma \in \mathfrak{G}$ maps F_0 onto the set of elements of σE which are algebraic over F, so that if σ is an automorphism of E then $\sigma F_0 = F_0$. Since by Proposition 3, σ always has a specialization which is an automorphism of E, which by Proposition 2(d) coincides with σ on F_0, we see that $\sigma F = F_0$ for every $\sigma \in \mathfrak{G}$. It now follows by Proposition 2(d) and (a) that the mapping

$$\sigma \rightarrow \text{restriction of } \sigma \text{ to } F_0$$

is a homomorphism of $\mathfrak{G}$ with kernel $\mathfrak{G}_0$ onto a finite group of automorphism of F_0 with fixed field F. If σ_i $(i = 1, \cdots, r)$ are the elements of the finite set of itsolated isomorphisms mentioned in Proposition 2(a) (with say $\sigma_0 \in \mathfrak{G}_0$) and if $\mathfrak{G}_i$ is the set of specializations of σ_i, then the $\mathfrak{G}_i$ are cosets of $\mathfrak{G}_0$ in $\mathfrak{G}$.

Using the fact (Proposition 3(a)) that each $\mathfrak{G}_i$ contains an element σ_i' which is an automorphism of E, we follow [2] to extend the isomorphism $\mathfrak{G}_0 \approx G_0$ to an isomorphism $\mathfrak{G} \approx G$ (rational in the sense of (c)), where G is an algebraic group of which the component of the unity element is G_0. The varieties $V_i = V_0^{\sigma_i'}$ determined by the restrictions of the σ_i' to F_0 are precisely the conjugates of V_0 over F, and for each i the point $v_i = \sigma_i' v_0$ is a generic point of V_i over F_0. We now define the operation of G on the algebraic set $V = \bigcup V_i$ in the obvious manner, namely by setting $v_i x_{\sigma_j} = \sigma_j v_i$ for all i and j. We leave to the reader the verification that this makes V into a principal homogeneous space for G with the stated properties.

COLUMBIA UNIVERSITY.

REFERENCES.

[1] E. R. Kolchin, "Galois theory of differential fields," *American Journal of Mathematics*, vol. 75 (1953), pp. 753-824.

[2] ———, "On the Galois theory of differential fields," *American Journal of Mathematics*, vol. 77 (1955), pp. 868-894.

[3] H. Matsumura, "Automorphism-groups of differential fields and group varieties," *Memoirs of the College of Science, University of Kyoto, Series A*, vol. 28 (1954), pp. 283-292.

[4] M. Rosenlicht, "Some basic theorems on algebraic groups," *American Journal of Mathematics*, vol. 78 (1956), pp. 401-443.

[5] A. Weil, "On algebraic groups of transformation," *American Journal of Mathematics*, vol. 77 (1955), pp. 355-391.

[6] ———, "On algebraic groups and homogeneous spaces," *American Journal of Mathematics*, vol. 77 (1955), pp. 493-512.

RATIONAL APPROXIMATION TO SOLUTIONS OF ALGEBRAIC DIFFERENTIAL EQUATIONS

E. R. KOLCHIN[1]

Introduction. It was observed by Liouville (C. R. Acad. Sci. Paris, vol. 18 (1844) pp. 910–911; J. Math. Pures Appl. vol. 16 (1851) pp. 133–142) that an element α of the field $\boldsymbol{C}$ of complex numbers which is algebraic of degree n (over the ring $\boldsymbol{Z}$ of rational integers) can not be approximated very well by rational numbers, in the following sense: there exists a real number $\gamma>0$ such that $|\alpha-p/q| \geqq \gamma/|q|^n$ for all $p, q\in\boldsymbol{Z}$ with $q\neq 0$ and $p/q\neq\alpha$. Using this theorem Liouville gave the first examples of transcendental numbers. The proof depends only on the circumstance that every nonzero element of $\boldsymbol{Z}$ has absolute value $\geqq 1$ and the following two obvious facts (in the statement of which f denotes the polynomial of degree n vanishing at α): (i) α is an isolated point of the set of zeros of f; (ii) $f(y/z)$ can be written as a fraction in which the numerator is a polynomial in y and z and the denominator is z^n. It follows that Liouville's theorem has an abstract version in which $\boldsymbol{C}$ and $\boldsymbol{Z}$ are replaced by an arbitrary nontrivially valued field and a nonzero subring thereof in which each nonzero element has value $\geqq 1$. Since the field $K((X^{-1}))$ of power series in the reciprocal of an indeterminate X over a given commutative field K admits a valuation for which the series $u=c_mX^{-m}+c_{m+1}X^{-(m+1)}+\cdots$ (with $c_m\neq 0$) has the value $|u|=e^{-m}$, and the polynomial ring $K[X]$ is a subring of $K((X^{-1}))$ in which every nonzero element has value $\geqq 1$, Liouville's theorem applies in this situa-

Received by the editors July 25, 1958.

[1] This paper was prepared in connection with a grant from the National Science Foundation.

tion, that is, his theorem is also a result on the approximation of algebraic functions by rational ones.

An algebraic equation may be considered as an algebraic differential equation of order zero, so that algebraic functions are special cases of differentially algebraic ones. The purpose of this note is to extend Liouville's theorem (abstract version) to an ordinary differential field with valuation which is subject to a certain not unnatural condition. A weaker and slightly less general result for the special case of power series has been obtained by E. Maillet (*Nombres transcendents*, Paris, 1906). The present treatment is, I believe, more transparent.

The condition on the valuation, which can not be omitted, and which in the case of $K((X^{-1}))$ is equivalent to the condition that the field characteristic be 0, is used in §1 to define "valued differential field." The fact (i) above is suitably generalized in §3. Since a differential polynomial in general has infinitely many solutions, the generalization is not quite obvious; it is proved using a lemma on wronskian determinants established in §2. The generalization of fact (ii) presents no difficulty provided the exponent n ($=\deg f$) is replaced by a generally bigger number called the "denomination" of f. When the differential polynomial f is of the order 0, its denomination coincides with its degree. This (and a slight generalization thereof) is described in §4. The proof of the approximation theorem itself, which is then trivial, is found in §5. An almost immediate consequence of the theorem is that a power series $\sum_{k=0}^{\infty} c_k X^{s_k}$, with nonzero coefficients c_k in a field K of characteristic 0 and with strictly increasing integral exponents $s_k>0$ such that the sequence of ratios s_{k+1}/s_k is unbounded, is differentially transcendental over $K(X)$.

It remains to make the obvious remark, in view of the deep Thue-Siegel-Roth improvement of Liouville's theorem (see K. F. Roth, Mathematika vol. 2 (1955) pp. 1–20), that it would be desirable to obtain a similar improvement in the present theorem.

Terminology and notation. In what follows, all differential rings and fields are ordinary and commutative; the derivative of an element a of a differential ring or field is denoted by a', and for each natural number k the kth derivative of a is denoted by $a^{(k)}$. All valuations are nonarchimedean; the value group of any valuation is written multiplicatively, and the value of an element a is denoted by $|a|$.

1. Valued differential fields.

DEFINITION. A *valued differential field* is a differential field endowed with a valuation satisfying the following condition: there exist elements α and β of the value group such that

$$(1)\qquad \alpha|a| \leqq |a'| \leqq \beta|a|$$

for every element a of the differential field with $|a|<1$. The elements α and β are then called, respectively, *lower* and *upper bounds* of the valued differential field. The valued differential field is said to be *trivial* or *nontrivial* according as the value group is trivial or not.

LEMMA 1. *If $\mathfrak{F}$ is a nontrivial valued differential field with lower bound α and upper bound β, then the second inequality in* (1) *holds for every $a\in\mathfrak{F}$, and the first inequality in* (1) *holds for every $a\in F$ with $|a|\neq 1$.*

PROOF. When $|a|<1$ then (1) holds by hypothesis. When $|a|>1$ then $a\neq 0$ and $|a^{-1}|<1$; since $|a'/a|=|(a^{-1})'/a^{-1}|$, again (1) holds. It remains to prove the second part of (1) when $|a|=1$. By hypothesis there exists a nonzero element $b\in\mathfrak{F}$ with $|b|\neq 1$, whence $|ab|\neq 1$, and therefore $|b'/b|\leqq\beta$ and $|a'/a+b'/b|=|(ab)'/ab|\leqq\beta$; if $|a'/a|\leqq|b'/b|$ then obviously $|a'/a|\leqq\beta$, whereas if $|a'/a|>|b'/b|$ then $|a'/a|=|a'/a+b'/b|\leqq\beta$.

COROLLARY 1. *Hypothesis as in Lemma* 1, *let n be a natural number. Then $|a^{(n)}|\leqq\beta^n|a|$ for every $a\in\mathfrak{F}$, and $|a^{(n)}|\geqq\alpha^n|a|$ for every $a\in\mathfrak{F}$ with $|a|>1$ and $|a|>1/\alpha^{n-1}$.*

PROOF. Induction on n.

COROLLARY 2. *Let $\mathfrak{F}$ be a nontrivial valued differential field. Every nonzero constant c in $\mathfrak{F}$ has value $|c|=1$. The characteristic of $\mathfrak{F}$ is* 0.

PROOF. If there existed a nonzero constant c with $|c|\neq 1$ we would have $0=|0|=|c'|\geqq\alpha|c|>0$. If the characteristic were $p\neq 0$ then, for every nonzero $a\in\mathfrak{F}$, a^p would be a nonzero constant, we would have $|a|^p=|a^p|=1$, and the valuation would be trivial.

2. **Wronskian determinants.** If $(a_1,\cdots,a_n)$ is a finite sequence of elements of a differential ring with unity element, we denote the wronskian determinant $\det(a_j^{(i-1)})_{1\leq i\leq n,\ 1\leq j\leq n}$ by $W(a_1,\cdots,a_n)$. The wronskian determinant equals 1 when $n=0$ and equals a_1 when $n=1$. If $n>0$ and we expand the determinant by the minors of the last row, we obtain the formula

$$(2)\qquad \sum_{j=1}^{n}(-1)^{n+j}a_j^{(n-1)}W(a_1,\cdots,a_{j-1},a_{j+1},\cdots,a_n)=W(a_1,\cdots,a_n).$$

Applying the same device to the determinant obtained on replacing the last row by one of the other rows, we obtain the formula

$$\sum_{j=1}^{n} (-1)^{n+j} a_j^{(i)} W(a_1, \cdots, a_{j-1}, a_{j+1}, \cdots, a_n) = 0 \tag{3}$$

$$(0 \leqq i \leqq n-2).$$

LEMMA 2. *If $b, a_1, \cdots, a_n$ are elements of a differential ring with unity element then $W(ba_1, \cdots, ba_n) = b^n W(a_1, \cdots, a_n)$.*

PROOF. If $n=0$ or 1 this is obvious. Let $n>1$ and suppose the result known for lower values of n. Then, by (2) and (3),

$$\begin{aligned} W(ba_1, \cdots, ba_n) &\\ = \sum_{j=1}^{n} (-1)^{n+j} (ba_j)^{(n-1)} W(ba_1, \cdots, ba_{j-1}, ba_{j+1}, \cdots, ba_n) &\\ = \sum_{j=1}^{n} (-1)^{n+j} \sum_{i=0}^{n-1} \binom{n-1}{i} b^{(n-1-i)} a_j^{(i)} b^{n-1} W(a_1, \cdots, a_{j-1}, a_{j+1}, \cdots, a_n) &\\ = b^{n-1} \sum_{i=0}^{n-1} \binom{n-1}{i} b^{(n-1-i)} \sum_{j=1}^{n} (-1)^{n+j} a_j^{(i)} W(a_1, \cdots, a_{j-1}, a_{j+1}, \cdots, a_n) &\\ = b^{n-1} b W(a_1, \cdots, a_n) = b^n W(a_1, \cdots, a_n). & \end{aligned}$$

LEMMA 3. *Let $a_1, \cdots, a_n$ be elements of a nontrivial valued field with lower bound α and upper bound β.*

(a) $|W(a_1, \cdots, a_n)| \leqq \beta^{n(n-1)/2} |a_1 \cdots a_n|$.

(b) *If* $|a_j| < (\alpha/\beta)^{j-2} |a_{j-1}|$ $(2 \leqq j \leqq n)$ *then* $|W(a_1, \cdots, a_n)| \geqq \alpha^{n(n-1)/2} |a_1 \cdots a_n|$.

REMARK. The condition in part (b) can be weakened.

PROOF. The lemma is obvious if $n=0$ or 1, and also if $a_j=0$ for some j. Let $n>1$ and $a_j \neq 0$ for every j, and suppose the lemma proved for lower values of n. By Lemma 2,

$$\begin{aligned} W(a_1, \cdots, a_n) &= a_1^n W(1, a_2/a_1, \cdots, a_n/a_1) \\ &= a_1^n W((a_2/a_1)', \cdots, (a_n/a_1)'). \end{aligned}$$

Then

$$\begin{aligned} |W(a_1, \cdots, a_n)| &\leqq |a_1|^n \beta^{(n-1)(n-2)/2} |(a_2/a_1)' \cdots (a_n/a_1)'| \\ &\leqq |a_1|^n \beta^{(n-1)(n-2)/2} \beta |a_2/a_1| \cdots \beta |a_n/a_1| \\ &= \beta^{n(n-1)/2} |a_1 \cdots a_n|. \end{aligned}$$

Also, under the condition of part (b),

$$|(a_{j+1}/a_1)'| \leqq \beta |a_{j+1}/a_1| < \beta(\alpha/\beta)^{j-1}|a_j/a_1| = (\alpha/\beta)^{j-2}\alpha|a_j/a_1|$$
$$\leqq (\alpha/\beta)^{j-2}|(a_j/a_1)'| \quad (2 \leqq j \leqq n-1),$$

so that

$$|W(a_1, \cdots, a_n)| \geqq |a_1|^n \alpha^{(n-1)(n-2)/2}|(a_2/a_1)'| \cdots |(a_n/a_1)'|$$
$$\geqq |a_1|^n \alpha^{(n-1)(n-2)/2}\alpha|a_2/a_1| \cdots \alpha|a_n/a_1|$$
$$= \alpha^{n(n-1)/2}|a_1 \cdots a_n|.$$

3. **Simple zeros.** Let P be an element of the differential polynomial ring in a differential indeterminate y over a differential field. A zero of P which fails to be a zero of $\partial P/\partial y^{(i)}$ for at least one natural number i is said to be *simple.*

LEMMA 4. *Let $\mathfrak{F}$ be a nontrivial valued differential field, and let $u \in \mathfrak{F}$ be a simple zero of a differential polynimial $P \in \mathfrak{F}\{y\}$. Then there exists an element γ_0 of the value group such that $|v-u| \geqq \gamma_0$ for every $v \in \mathfrak{F}$ which is a zero of P different from u.*

PROOF. Translating by u, we may suppose that $u=0$. Then $P(0)=0$, and there is a natural number i such that $(\partial P/\partial y^{(i)})(0) \neq 0$. Let n denote the biggest such i. Then we may write $P=L+R$, where

$$L = \sum_{i=0}^{n} b_i y^{(i)}, \qquad b_i \in \mathfrak{F} \ (0 \leqq i \leqq n), \qquad b_n \neq 0$$

and

$$R = \sum_k c_k y^{(i_{k1})} \cdots y^{(i_{k,h(k)})}, \qquad c_k \in \mathfrak{F}, \qquad h(k) \geqq 2 \ (\text{all } k).$$

For any zero $v \in \mathfrak{F}$ of P with $|v| \leqq 1$,

$$|L(v)| = \left|\sum_k c_k v^{(i_{k1})} \cdots v^{(i_{k,h(k)})}\right|$$
$$\leqq \sup_k |c_k|\,|v^{(i_{k1})}| \cdots |v^{(i_{k,h(k)})}|$$
$$\leqq \sup_k |c_k|\,\beta^{i_{k1}+\cdots+i_{k,h(k)}}|v|^{h(k)}$$
$$\leqq \gamma_1 |v|^2,$$

where $\gamma_1 = \sup_k |c_k|\beta^{i_{k1}+\cdots+i_{k,h(k)}}$ is independent of v.

Now assume the lemma false. Then $\mathfrak{F}$ contains zeros $v_1, \cdots, v_{n+1}$ of P, distinct from 0, such that

$$|v_1| < 1, \qquad |v_1| < \gamma_1^{-1}\alpha^n(\alpha/\beta)^{n(n-1)/2}|b_n|,$$
$$|v_j| < (\beta/\alpha)^{j-2}|v_{j-1}| \quad (2 \leqq j \leqq n+1).$$

By the above, $|L(v_j)| \leqq \gamma_1 |v_j|^2$ $(1 \leqq j \leqq n+1)$. By the definition of L,

$$\sum_{i=0}^{n} b_i v_j^{(i)} = L(v_j) \qquad (1 \leqq j \leqq n+1).$$

Considering this as a system of linear equations for $b_0, b_1, \cdots, b_n$, and solving for b_n, we find that

$$W(v_1, \cdots, v_{n+1})b_n = \sum_{j=1}^{n+1} (-1)^{n+1+j} W(v_1, \cdots, v_{j-1}, v_j, \cdots, v_{n+1})L(v_j).$$

It follows by Lemma 3 that

$$\begin{aligned} \alpha^{n(n+1)/2}|v_1 \cdots v_{n+1}|\,|b_n| &\leqq |W(v_1, \cdots, v_{n+1})b_n| \\ &\leqq \sup_j |W(v_1, \cdots, v_{j-1}, v_{j+1}, \cdots, v_{n+1})L(v_j)| \\ &\leqq \sup_j \beta^{n(n-1)/2}|v_1 \cdots v_{j-1}v_{j+1} \cdots v_{n+1}|\,|\gamma_1|\,|v_j|^2 \\ &= \sup_j \beta^{n(n-1)/2}\gamma_1|v_1 \cdots v_{n+1}|\,|v_j| \\ &= \beta^{n(n-1)/2}\gamma_1|v_1 \cdots v_{n+1}|\,|v_1|, \end{aligned}$$

so that $|v_1| \geqq \gamma_1^{-1}\alpha^n(\alpha/\beta)^{n(n-1)/2}|b_n|$. This contradicts the definition of v_1 and completes the proof.

4. **Denomination.** A nonzero differential polynomial P in a differential indeterminate y is a linear combination with nonzero coefficients of certain monomials $y^{e_0}y'^{e_1} \cdots y^{(k)e_k} \cdots$. The biggest of the natural numbers $\sum_k (k+1)e_k$ will be called the *denomination* of P. If z is another differential indeterminate, and we denote the denomination of P by d, then $P(y/z)$ can be written with denominator z^d, that is, $z^dP(y/z)$ is a differential polynomial in y and z.

More generally, for each natural number $s \geqq 1$, the biggest of the natural numbers $\sum_k (k+s)e_k$ will be called the *s-denomination* of P. (The 1-denomination of P is thus its denomination.) If we denote the s-denomination of P by d_s then $P(y/z^s)$ can be written with denominator z^{d_s}.

Now let $\mathfrak{F}$ be a differential field, let $\mathfrak{Z}$ be a differential subring of $\mathfrak{F}$, and let u be an element of $\mathfrak{F}$ which is differentially algebraic over $\mathfrak{Z}$. The smallest natural number d such that there exists a nonzero differential polynomial in $\mathfrak{Z}\{y\}$ having denomination d and vanishing at u

will be called the *denomination of u over $\mathfrak{Z}$*. The definition of the *s-denomination of u over $\mathfrak{Z}$* is similar.

Suppose now that the characteristic of $\mathfrak{F}$ is zero.

If $P \in \mathfrak{Z}\{y\}$ vanishes at u and has s-denomination equal to that of u over $\mathfrak{Z}$, then u is a simple zero of P. More precisely, if the order of P is n then $\partial P/\partial y^{(n)}$ does not vanish at u. For $\partial P/\partial y^{(n)}$ is obviously nonzero and of lower s-denomination than P.

5. The approximation theorem.

THEOREM. *Let $\mathfrak{F}$ be a nontrivial valued differential field, let $\mathfrak{Z}$ be a nonzero differential subring of $\mathfrak{F}$ such that $|a| \geqq 1$ for every nonzero element $a \in \mathfrak{Z}$, let $u \in \mathfrak{F}$ be differentially algebraic over $\mathfrak{Z}$, let s be any natural number $\geqq 1$, and denote the s-denomination of u over $\mathfrak{Z}$ by d_s. Then there exists an element γ of the value group such that*

$$|u - a/b^s| \geqq \gamma/|b|^{d_s}$$

for all elements $a, b \in \mathfrak{Z}$ with $b \neq 0$ and $u \neq a/b^s$.

PROOF. Let $P \in \mathfrak{Z}\{y\}$ vanish at u and have s-denomination d_s. By Lemma 4 there is an element γ_0 of the value group such that $|u - v| \geqq \gamma_0$ for every zero v of P with $v \in \mathfrak{F}$ and $v \neq u$; we obviously may suppose that $\gamma_0 < 1$. Since $P(u) = 0$ we may write

$$P(y) = \sum_k c_k(y - u)^{(i_{k1})} \cdots (y - u)^{(i_{k,h(k)})},$$

where $c_k \in \mathfrak{F}$ and $h(k) \geqq 1$ for each k. Setting

$$\gamma_1 = \sup_k |c_k| \beta^{i_{k1} + \cdots + i_{k,h(k)}},$$

we therefore see (using the first part of Lemma 1) that if $|u - a/b^s| < \gamma_0$ then $|P(a/b^s)| \leqq \gamma_1 |a/b^s - u|$ and $P(a/b^s) \neq 0$; as $b^{d_s}P(a/b^s) \in \mathfrak{Z}$ we have $|b^{d_s}P(a/b^s)| \geqq 1$, so that $|a/b^s - u| \geqq \gamma_1^{-1}/|b|^{d_s}$. On the other hand, if $|u - a/b^s| \geqq \gamma_0$ then obviously $|u - a/b^s| \geqq \gamma_0/|b|^{d_s}$. Setting $\gamma = \inf(\gamma_0, \gamma_1^{-1})$, we find the theorem proved.

COLUMBIA UNIVERSITY

EXISTENCE OF INVARIANT BASES

ELLIS KOLCHIN AND SERGE LANG[1]

Let K be a field, G a group of automorphisms of K, and M a vector space over K on which G acts in such a way that $\sigma(aD)=\sigma a\cdot\sigma D$ for $\sigma\in G$, $a\in K$, and $D\in M$. The problem arises to find whether M has a basis consisting of invariant elements under G. In other words, letting K_0 be the fixed field under G, and M_0 the set of fixed elements of M under G so that M_0 is a vector space over K_0, to find out whether M is isomorphic to the tensor product

$$M \approx K \otimes_{K_0} M_0$$

under the natural map. We shall see that this is so if and only if a certain cocycle of G in the full linear group is trivial.

In some applications, a rational structure is added to K and G, namely K is the function field of a principal homogeneous space over a group variety G. We shall show that the cocycle involved is then determined rationally. This leads us into a discussion of rational cocycles in §3, and of their comparison with the ordinary cocycles of Galois theory, i.e. where G is a finite Galois group. All cocycles involved with coefficients in the full linear group split, and in fact the Galois cohomology (in dimension 1) of the group variety of units in an algebra is trivial (Propositions 2 and 5).

1. **The invariant subspace.** Let K be a field, and G a group of automorphisms of K. By a (G, K)-*space* M we shall mean a vector space over K which is also a unitary G-module, such that

$$\sigma(aD) = \sigma a\cdot\sigma D$$

for $a\in K$, $D\in M$ and $\sigma\in G$. An element D of M is said to be invariant under G if $\sigma D=D$ for all $\sigma\in G$. A basis $(D)=(D_i)$ of M over K will be called *invariant* if $\sigma D_i=D_i$ for every $\sigma\in G$ and every i.

PROPOSITION 1. *Let M be a finite dimensional (G, K)-space. If $(D)=(D_1, \cdots, D_m)$ is any basis of M, and if*

$$A(\sigma) = (a_{ij}(\sigma)) \qquad (i, j = 1, \cdots, m)$$

is the matrix defined by

$$\sigma D_j = \sum_\nu a_{\nu j}(\sigma) D_\nu \qquad (1 \leqq j \leqq m),$$

Received by the editors April 1, 1959.

[1] This paper was prepared in connection with contract NONR 266(57).

then $A(\sigma)\cdot\sigma A(\tau)=A(\sigma\tau)$. *A necessary and sufficient condition that* M *have an invariant basis is that there exist an invertible matrix* B *with coordinates in* K *such that* $A(\sigma)=B^{-1}\sigma B$.

PROOF. That $A(\sigma)$ satisfies the cocycle relation is easy to see. Suppose $A(\sigma)=B^{-1}\sigma B$. Using matrix notation, we may write $\sigma(D)={}^tA(\sigma)(D)$, where for any matrix X, we denote by ${}^tX$ the transpose of X. Let a new basis (D') be defined by $(D')={}^tB^{-1}(D)$. Then

$$\sigma(D') = \sigma{}^tB^{-1}\sigma(D) = {}^t\sigma(B)^{-1}\cdot{}^tA(\sigma)\cdot(D) = {}^tB^{-1}(D) = (D')$$

and thus (D') is invariant. Conversely, if (D') is an invariant basis, define the matrix B by the relation $(D)={}^tB(D')$. Then

$${}^tA(\sigma)(D) = \sigma(D) = {}^t\sigma B\cdot\sigma(D') = {}^t\sigma B\cdot(D') = {}^t\sigma B\cdot{}^tB^{-1}(D)$$

and $A(\sigma)=B^{-1}\sigma B$. This proves our proposition.

If we denote by M_0 the set of G-invariant elements of M and by K_0 the fixed field of K under G, then M_0 is a vector space over K_0. If M admits an invariant basis, then one sees immediately that M is isomorphic to the tensor product

$$K\otimes_{K_0}M_0 \approx M$$

under the natural map $a\otimes D\to aD$, $a\in K$, $D\in M_0$.

We observe that the set of K_0-linear transformations of K, denoted by $\mathrm{End}_{K_0}(K)$, forms a (G, K)-space in a natural fashion: If $D\in\mathrm{End}_{K_0}(K)$, and $\sigma\in G$, then one defines

$$(\sigma D)(a) = \sigma(D(\sigma^{-1}a)),$$

and verifies trivially that $\sigma(aD)=\sigma a\cdot\sigma D$.

In the applications, one frequently takes the subspace of $\mathrm{End}_{K_0}(K)$ consisting of the derivations of K over K_0, or a finite dimensional space of linear transformations over a subfield of K.

REMARK. Proposition 1 and its proof generalizes so that K can be a ring (with unity, not necessarily commutative) and M can be a unitary K-module with finite basis. In this situation, a matrix B over K with m rows and n columns is *invertible* if there exists a matrix C over K with n rows and m columns such that BC is the unity matrix of degree m and CB is the unity matrix of degree n. C is then unique and is denoted by B^{-1}. The generalized proposition states that if (D) is a basis of m elements, and $A(\sigma)$ is defined as above, then $A(\sigma)\cdot\sigma A(\tau)=A(\sigma\tau)$, and a necessary and sufficient condition that there exist an invariant basis of n elements is that there exist an invertible matrix B over K with n rows and m columns such that $A(\sigma)=B^{-1}\sigma B$.

2. **Galois cohomology in dimension 1.** As we have seen in the last section, it is useful to have a criterion to split a 1-cocycle. We shall give one in this section.

Let G be a group variety (i.e. a connected algebraic group) defined over a field k. Let K be a finite Galois extension of k with Galois group $\mathfrak{g}$. Then $\mathfrak{g}$ operates as a group of automorphisms of the subgroup of G consisting of the points of G which are rational over K. We denote this subgroup by G_K. Suppose we are given a family $(x_\sigma)_{\sigma\in\mathfrak{g}}$ of points of G_K satisfying

$$x_\sigma \cdot \sigma x_\tau = x_{\sigma\tau}, \qquad \sigma, \tau \in \mathfrak{g}.$$

Such a family is called a cocycle of $\mathfrak{g}$ in G_K. The set of these cocycles is denoted by $Z^1(\mathfrak{g}, G_K)$. We say that (x_σ) is *cohomologous* to (y_σ) and write $(x_\sigma)\sim(y_\sigma)$ if there exists an element $z\in G_K$ such that

$$y_\sigma = z^{-1}x_\sigma\sigma z$$

for each $\sigma\in\mathfrak{g}$. This is obviously an equivalence relation between cocycles. The set of equivalence classes is called the *first cohomology set* of $\mathfrak{g}$ in G_K and is denoted by $H^1(\mathfrak{g}, G_K)$. If $z\in G_K$ then $(z^{-1}\sigma z)$ is a cocycle, and such a cocycle is called a coboundary of $\mathfrak{g}$ in G_K. The set of all such coboundaries is denoted by $B^1(\mathfrak{g}, G_K)$ and is itself an equivalence class, i.e. an element of $H^1(\mathfrak{g}, G_K)$. (Of course, if G is commutative, these sets are groups, and $H^1(\mathfrak{g}, G_K)$ is a commutative group.)

If $L\supset K$ is another Galois extension of k, then there is a natural map of $H^1(\mathfrak{g}_{K/k}, G_K)$ into $H^1(\mathfrak{g}_{L/k}, G_L)$ obtained by inflation: A cocycle for $\mathfrak{g}_{K/k}$ determines one for $\mathfrak{g}_{L/k}$ simply by extending the function to cosets of the subgroup of $\mathfrak{g}_{L/k}$ of which $\mathfrak{g}_{K/k}$ is a factor group. It is trivially seen that this natural map is actually *injective* and we may take the injective limit of these cohomology sets as L becomes larger and larger. The limiting set, union of all $H^1(\mathfrak{g}_{L/k}, G_L)$, will be denoted by $H^1(k, G)$.

We are interested in a noncommutative group, namely the full linear group. More generally, let A_0 be a finite dimensional associative algebra with unity element over the field k, let Ω be a universal domain containing k, and let A be the algebra over Ω which is the tensor product of A_0 with Ω. Let $e_1, \cdots, e_m$ be a linear basis of A_0 over k, and therefore of A over Ω. Expressing elements of A in terms of this basis, $x=\sum\xi_ie_i$, one sees that there exists a polynomial $P(X_1, \cdots, X_m)\in k[X_1, \cdots, X_m]$ such that x is invertible if and only if $P(\xi_1, \cdots, \xi_m)\neq 0$, or as we shall abbreviate, $P(x)\neq 0$. These invertible elements therefore form a group variety defined over k (it is a k-open subset of affine m-space in the Zariski topology), and we shall denote it by $\Gamma(A)$ or simply Γ. The general linear group $GL(m)$ is an example of such a group variety defined over the prime field.

PROPOSITION 2. *Let* $\Gamma=\Gamma(A)$ *be as above the group variety of units in an algebra defined over* k, *and let* K *be a Galois extension of* k *of finite degree, with Galois group* $\mathfrak{g}$. *Then* $H^1(\mathfrak{g}, \Gamma_K)$ *is trivial, that is, every cocycle is a coboundary.*

PROOF. The case in which k is finite is a special case of the fact that $H^1(\mathfrak{g}, G_K)$ is trivial for any group variety defined over a finite field k (see [2]). In the infinite case, the theorem is proved in [1]. We reproduce the proof here for the convenience of the reader. Let (x_τ) be in $Z^1(\mathfrak{g}, \Gamma_K)$. If $(t_\tau)_{\tau\in\mathfrak{g}}$ is a family of elements of Ω, algebraically independent over K, then $P(\sum_\tau t_\tau x_\tau)\neq 0$ because this polynomial does not vanish when one t_τ is replaced by 1 and all the others by 0. It is well known (see, for instance, Bourbaki, *Algèbre*, Chapter V, §10, Theorem 4, p. 57) that this implies the existence of an element $\alpha\in K$ such that $P(\sum(\tau\alpha)x_\tau)\neq 0$. Writing $y=\sum(\tau\alpha)x_\tau$ we see that $y\in\Gamma_K$ and $\sigma y=\sum(\sigma\tau\alpha)\sigma x_\tau=\sum(\sigma\tau\alpha)x_\sigma^{-1}x_{\sigma\tau}=x_\sigma^{-1}y$, so that $(x_\sigma)=(y\sigma y^{-1})$ is in $B^1(\mathfrak{g}, \Gamma_K)$. This concludes the proof.

3. **Rational cohomology.** We recall the notion of a homogeneous space and use the terminology of Weil [4]. Let V be a variety and G a group variety. Suppose we are given a rational map

$$f\colon V\times G\to V$$

which is everywhere defined. Given $v\in V$ and $x\in G$, we write vx instead of $f(v, x)$. We say that V is a *transformation space* for G if

$$v(xy) = (vx)y \qquad \text{and} \qquad ve = v$$

for $x, y\in G$ and $v\in V$. Here, as usual, e denotes the unity element of G. We say that the transformation space V is defined over k if G, V and the rational map f are defined over k.

We observe that if Ω is the universal domain, then G can be viewed as a group of automorphisms of the function field $\Omega(V)$. Indeed, for each $x\in G$, we have the automorphism σ_x such that

$$(\sigma_x f)(v) = f(vx)$$

whenever f is a function in $\Omega(V)$, and v is a generic point of V over a field of definition for f over which x is rational. In the same manner, if the transformation space V is defined over k, then G_k is a group of automorphisms of $k(V)$.

A transformation space is said to be a *homogeneous space* if given two points $v, w\in V$, there exists $x\in G$ such that $vx=w$. We say that

V is a *principal* homogeneous space if the element x is uniquely and rationally determined. By this we mean that there exists an everywhere defined rational map μ: $V \times V \to G$ such that $x = \mu(v, w)$. One may write symbolically $x = v^{-1}w$. The principal space V is said to be defined over k if, as a transformation space it is defined over k, and the rational map μ is defined over k.

Let now G, G' be group varieties, let V be a transformation space of G, all these being defined over k. A rational map

$$f: V \times G \to G'$$

of $V \times G$ into G', defined over k, is said to be a 1-*cocycle* if it satisfies the relation

$$f(v, x)f(vx, y) = f(v, xy)$$

whenever v, x, y are independent generic points of V, G, and G' over k. *It then follows that $f(v, x)$ is defined whenever x is any point and v is a generic point of V over $k(x)$*, because we can write

$$f(v, x) = f(v, xy)f(vx, y)^{-1}$$

with y generic over $k(v, x)$. The 1-cocycles form a set denoted by $Z_k^1(G, V, G')$. We say that two cocycles are *cohomologous* and write $f \sim g$, if there exists a rational map ϕ: $V \to G'$ defined over k such that $f(v, x) = \phi(v)^{-1}g(v, x)\phi(vx)$. This establishes an equivalence relation among the cocycles, and the equivalence classes form the first cohomology set $H_k^1(G, V, G')$. The cocycles in the identity class, i.e. those of type $\phi(v)^{-1}\phi(vx)$ are called *coboundaries*, and form a set $B_k^1(G, V, G')$.

(If G' is commutative, and written additively, we can define cocycles in higher dimension, an r-cocycle being by definition a rational map

$$f: V \times G \times \cdots \times G \to G'$$

defined over k satisfying the coboundary formula

$$\begin{aligned} 0 = (\delta f)(v, x_1, \cdots, x_{r+1}) &= f(vx_1, x_2, \cdots, x_{r+1}) \\ &+ \sum (-1)^\nu f(v, x_1, \cdots, x_\nu x_{\nu+1}, \cdots, x_{r+1}) \\ &+ (-1)^{r+1} f(v, x_1, \cdots, x_r). \end{aligned}$$

One then has an rth cohomology group for $r \geqq 0$.)

We return to the noncommutative case, and assume that V is a principal homogeneous space for the group variety G, all defined over k. Let $f \in Z_k^1(G, V, G')$. If $w_0 \in V$ has the property that f is defined at $(u, u^{-1}w_0)$ for u generic on V over $k(w_0)$ and if we define

$\phi(v)=f(v, v^{-1}w_0)^{-1}$, so that ϕ is a rational map of V into G' defined over $k(w_0)$, then $f(v, x)=\phi(v)^{-1}\phi(vx)$. Thus f is trivial as an element of $Z^1_{k(w_0)}(G, V, G')$. It follows that if the points of V which are rational over k are dense in the Zariski topology, then every element of $Z^1_k(G, V, G')$ is a coboundary. (This is the case for instance if k is separably closed.)

Now we transform the above cocycles into a homogeneous form. Let λ: $V\times V\to V\times G$ be the canonical map such that $\lambda(u, v)=(u, u^{-1}v)$. Then the inverse of λ is a rational map sending (v, x) onto (v, vx). Given a cocycle $f\in Z^1_k(G, V, G')$, there exists therefore a rational map F: $V\times V\to G'$ which makes the following diagram commutative:

$$\begin{array}{ccc} V\times G & \xrightarrow{f} & G' \\ & \nwarrow^{\lambda} \quad \nearrow_{F} & \\ & V\times V & \end{array}$$

and the mapping F satisfies the relation

$$F(u, v)F(v, w) = F(u, w)$$

whenever u, v, w are independent generic points of V over k. We may of course start with an arbitrary variety V defined over k and a group variety G' with such a mapping. We may thus define homogeneous cocycles $Z^1_k(V, G')$, and coboundaries $B^1_k(V, G')$, these being rational maps of type $F(u, v)=\phi(u)^{-1}\phi(v)$ where ϕ: $V\to G'$ is a rational map defined over k. This allows us to define $H^1_k(V, G')$ for any variety V defined over k.

PROPOSITION 3. *Let V be a principal homogeneous space for G, and let G' be another group variety defined over k. Then there is a bijective mapping between $H^1_k(G, V, G')$ and $H^1_k(V, G')$ given by*

$$f(v, x) = F(v, vx).$$

The proof is trivial.

As Serre has pointed out to us, we can inject $H^1_k(V, G')$ into the Galois cohomology set $H^1(k, G')$ as defined in §2. The way this is done is described in the following proposition.

PROPOSITION 4. *Let V be a principal homogeneous space for G, and let G' be another group variety, defined over k. For each $\overline{F}\in H^1_k(V, G')$ choose a representative cocycle F in $Z^1_k(V, G')$, and choose a finite Galois extension K of k with group denoted by $\mathfrak{g}$ such that V has a point v_0 rational over K for which $F(u, v_0)$ is defined when u is generic over K.*

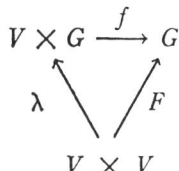

For each $\sigma \in \mathfrak{g}$, *let* $x_\sigma = F(u, v_0)^{-1}F(u, \sigma v_0)$, *where* u, v_0 *are chosen as above. Then* $(x_\sigma)_{\sigma\in\mathfrak{g}}$ *is a cocycle in* $Z^1(\mathfrak{g}, G'_K)$, *the corresponding element* $\bar{x}$ *of* $H^1(k, G')$ *is independent of the choice of* F, K, v_0, u *and the mapping* $\bar{F} \to \bar{x}$ *is an injection*

$$H_k^1(V, G') \to H^1(k, G').$$

PROOF. From the coboundary relation one verifies immediately that for independent generic points u, v of V over K, we have

$$F(u, v_0)^{-1}F(u, \sigma v_0) = F(v, v_0)^{-1}F(v, \sigma v_0),$$

so that each x_σ is rational over K, and (x_σ) is a cocycle. The rest of the proof is straightforward and is left to the reader.

In particular, we get

COROLLARY. *Let* V *be a principal homogeneous space for* G *and let* G' *be another group variety, defined over* k. *If* $H^1(k, G')$ *is trivial, then so is* $H_k^1(V, G')$.

Examples of group varieties G' for which $H^1(k, G')$ is trivial are:

The group variety of units in a finite dimensional algebra as we showed in Proposition 2, and in particular the full linear groups $GL(m)$;

The additive and multiplicative groups of the universal domain, this being Hilbert's Theorem 90 in its multiplicative and additive forms;

The group varieties G' having a normal sequence $G' = G_0 \supset G_1 \supset \cdots \supset G_r = 1$ defined over k, with each G_{i-1}/G_i either the additive or multiplicative group as above.

One may ask how it is possible to characterize those elements of $H^1(k, G')$ which come from an element of $H_k^1(V, G')$. It is known [3, Proposition 4] that $H^1(k, G')$ is in bijective correspondence with the set of isomorphism classes of principal homogeneous spaces of G' over k, and the reader may easily verify that the image of $H_k^1(V, G')$ in $H^1(k, G')$ corresponds to those principal homogeneous spaces of G', defined over k, which have a rational point in a field $k(v)$ where v is a generic point of V over k. The reader will also note that a cocycle $F: V \times V \to G'$ determines a principal homogeneous space of G' through which it can be factored [4, Proposition 4] and that this space corresponds precisely to the one determined by the cocycle described in Proposition 4 (Serre).

4. Rational determination of the invariant subspace. Let V be a principal homogeneous space for the group variety G. For each point

x of G, there is a unique automorphism σ_x of the function field $\Omega(V)$ over the universal domain Ω, such that, if $f \in \Omega(V)$ then $\sigma_x f$ is defined at a point v of V whenever f is defined at vx, and $(\sigma_x f)(v) = f(vx)$. The mapping $x \rightarrow \sigma_x$ is a group isomorphism permitting us to identify G with a group of automorphisms of $\Omega(V)$ over Ω. If k is a field of definition of V, and if we make the assumption that x is rational over k, then σ_x maps $k(V)$ onto itself. Therefore without this assumption, σ_x maps $k(V)$ into $k(x)(V)$.

Let f: $V \times G \rightarrow W$ be a rational map of $V \times G$ into some variety W, and let x be a point of G. Assume that the following condition is satisfied. For some field of definition k for f (i.e. for V, G, W and the graph of f), and for some generic point v of V over $k(x)$, f is defined at (v, x). Then there exists a unique rational map f_x: $V \rightarrow W$ defined over $k(x)$, such that $f_x(v) = f(v, x)$. When the above condition is satisfied, we shall say that f_x is *meaningful*. Of course, if x is generic on G over k, then f_x must be meaningful.

Let M be a vector space over $\Omega(V)$, of finite dimension m, which is a $(G, \Omega(V))$-space. If $(D) = (D_1, \cdots, D_m)$ is a basis of M, then for each point x of G, every $\sigma_x D_j$ is a linear combination of the basis vectors D_i with coefficients in $\Omega(V)$. We shall call the basis *rational* if there exist rational functions f_{ij} on $V \times G$ such that, for some common field of definition k of V and the f_{ij}, and for every generic point x of G over k, we have

$$(1) \qquad \sigma_x D_j = \sum_i f_{ijx} D_i, \qquad (1 \leqq j \leqq m).$$

A simple computation shows that when this is the case then, for independent generic points x, y of G over k, we have

$$f(v, xy) = f(v, x) f(vx, y)$$

where we denote by f the matrix (f_{ij}). In other words, f is a 1-cocycle in $Z^1_k(G, V, GL(m))$. It follows (§3) that f_x (i.e. each f_{ijx}) is meaningful for every point x of G, and also, that Equation (1) holds for every point x of G since each point of G can be expressed as a product of generic points ($x = xy \cdot y^{-1}$). In particular, any common field of definition of V and the f_{ij} must enjoy the same properties that have been attributed to k above. Such a common field of definition will be called a *field of rationality* of the rational basis (D). It is obvious that every invariant basis of M is rational, and admits as field of rationality any field of definition of the principal homogeneous space V.

It is almost immediate that if one basis of a $(G, \Omega(V))$-space is rational, then so are all its bases. By a *rational* $(G, \Omega(V))$-*space* we shall mean a $(G, \Omega(V))$-space with rational bases. The nature of such spaces is completely described by the following theorem, the proof of which follows that of Proposition 1, making use of the results of §2 and §3.

THEOREM. *Let M be a finite dimensional $(G, \Omega(V))$-space. A necessary and sufficient condition that M be rational is that M have an invariant basis. If (D) is any rational basis of M, and k is a field of rationality of (D), then there exists an invertible matrix over $k(V)$ transforming (D) into an invariant basis of M.*

PROOF. Since invariant bases are rational, the sufficiency is clear. To prove the necessity and the final part of the theorem, let (D) be a rational basis of M with field of definition k. Denoting the corresponding cocycle in $Z_k^1(G, V, GL(m))$ by $f=(f_{ij})$, we conclude from §§2, 3 that there exist rational functions $\phi_{ij}\in k(V)$ such that the matrix $\phi=(\phi_{ij})$ is invertible and $f(v, x)=\phi(v)^{-1}\phi(vx)$ for $x\in G$ and v generic on V over $k(x)$. Setting $E_j=\sum_i\psi_{ij}D_i$ $(1\leqq j\leqq m)$, where $(\psi_{ij})=(\phi)^{-1}$, we conclude that (E) is an invariant basis of M.

In the applications of the above theorem, one is usually given a subset $M_{k(V)}$ of M which is a vector space over $k(V)$ such that we have an isomorphism

$$M \approx \Omega(V) \otimes_{k(V)} M_{k(V)}$$

under the natural map $f\otimes D\rightarrow fD$, and such that a basis of $M_{k(V)}$ over $k(V)$ is a rational basis of M, having k as field of rationality. One may then say that M is *rationally $k(V)$-extended.* In that case one may say that an element D of M is *defined* over a field $k'\supset k$ if D lies in $k'(V)M_{k(V)}$. If M_k^0 is the set of elements of M which are invariant and defined over k, then M_k^0 is a vector space over k, and our theorem shows that we have an isomorphism

$$k(V) \otimes_k M_k^0 \approx M_{k(V)}$$

again under the map $f\otimes D\rightarrow fD$.

BIBLIOGRAPHY

1. E. Kolchin, *On the Galois theory of differential fields*, Amer. J. Math. vol. 77 (1955) pp. 868–894.

2. S. Lang, *Algebraic groups over finite fields*, Amer. J. Math. vol. 78 (1956) pp. 555–563.

3. S. Lang and J. Tate, *Principal homogeneous spaces over abelian varieties*, ibid. vol. 80 (1958) pp. 659–684.

4. A. Weil, *Algebraic groups and homogeneous spaces*, ibid. vol. 77 (1955) pp. 493–512.

COLUMBIA UNIVERSITY

ABELIAN EXTENSIONS OF DIFFERENTIAL FIELDS.*

By E. R. Kolchin.[1]

Introduction. The Galois theory of differential fields ([2], [3]) is concerned with strongly normal extensions of a differential field $\mathcal{F}$ with algebraically closed field of constants $\mathcal{C}$ of characteristic 0. The Galois group $\mathfrak{G}$ of such an extension $\mathcal{G}$ has a sort of algebraico-geometric structure and is shown to be birationally isomorphic with an algebraic group variety (not necessarily connected); more precisely, there is an algebraic group G in the algebraic geometry having for universal domain the fixed universal differential field extension $\mathcal{U}$ of $\mathcal{G}$, G being defined over $\mathcal{C}$, such that $\mathfrak{G}$ is birationally isomorphic with the group $G_{\mathcal{K}}$ consisting of the points of G which are rational over the field of constants $\mathcal{K}$ of $\mathcal{U}$ ($G_{\mathcal{K}}$ is an algebraic group in the algebraic geometry with universal domain $\mathcal{K}$). This permits the classification of strongly normal extensions by means of algebraic groups: a strongly normal extension of $\mathcal{F}$ with Galois group birationally isomorphic to the subgroup $G_{\mathcal{K}}$ of an algebraic group G defined over $\mathcal{C}$, or to an algebraic subgroup of $G_{\mathcal{K}}$, will be called a *G-extension* of $\mathcal{F}$.

If $\mathcal{G}$ is a G-extension of $\mathcal{F}$ and if

$$G = G_0 \supset G_1 \supset \cdots \supset G_r = 1$$

is a normal chain of algebraic subgroups of G, all defined over $\mathcal{C}$, then there is a corresponding tower

$$\mathcal{F} = \mathcal{F}_0 \subset \mathcal{F}_1 \subset \cdots \subset \mathcal{F}_r = \mathcal{G}$$

of differential fields such that $\mathcal{F}_i$ is a G_{i-1}/G_i-extension of $\mathcal{F}_{i-1}$ $(1 \leqq i \leqq r)$. Now, by Chevalley's structure theorem (see Rosenlicht [4], p. 439, or Barsotti [1], p. 47) G has a canonical normal chain

$$G \supset G^0 \supset H^0 \supset 1 \tag{1}$$

defined over $\mathcal{C}$, G^0 being the component of the unity element of G and H^0 being the greatest connected linear algebraic subgroup of G^0, such that the

* Received October 26, 1959.

[1] This work was done in connection with a contract with the Office of Naval Research.

factor groups G/G^0 and G^0/H^0 are, respectively, finite and abelian (i.e. an abelian variety). Therefore $\mathcal{F}$ and $\mathcal{G}$ appear in a corresponding tower

$$(2) \qquad \mathcal{F} \subset \mathcal{F}^0 \subset \mathcal{E}^0 \subset \mathcal{G}.$$

But a finite group corresponds to an extension of finite degree, so that $\mathcal{F}^0$ is finite algebraic over $\mathcal{F}$ (indeed, $\mathcal{F}^0$ is the algebraic closure of $\mathcal{F}$ in $\mathcal{G}$), and ([3] p. 891) a linear group corresponds to a Picard-Vessiot extension, so that $\mathcal{G}$ is Picard-Vessiot over $\mathcal{E}^0$. Thus, the gap in our knowledge of the structure of $\mathcal{G}$ over $\mathcal{F}$ is in the extension from $\mathcal{F}^0$ to $\mathcal{E}^0$. The purpose of the present paper is to close this gap, that is, to characterize the strongly normal extensions which are A-extensions for some abelian variety A. We shall call any such extension an *abelian* extension. The characterization which we find is in terms of solutions of certain kinds of differential equations, and generalizes the result previously obtained ([2] p. 809, [3] p. 892) in the case in which the extension is of transcendence degree 1, i.e. A is of dimension 1. The remainder of the present introduction is intended to describe this characterization more fully, in its natural more general setting which includes, among others, the above mentioned characterization of strongly normal extensions with linear groups.

Let G be any *connected* algebraic group defined over $\mathcal{k}$, of dimension say n. If ω is a differential on G and α is a point of G at which ω is finite, there is induced a differential $\omega(\alpha)$ of $\mathcal{U}$ over $\mathcal{K}$ (if $\omega = df$ and the rational function f on G is defined over $\mathcal{K}$ then $\omega(\alpha)$ maps each derivation $\mathfrak{d}$ of $\mathcal{U}$ over $\mathcal{K}$ onto $\mathfrak{d}f(\alpha)$). Every birational automorphism ϕ of G induces a contravariant automorphism ϕ^* of the vector space $\mathfrak{J}(G)$ over $\mathcal{U}$ of *invariant* (i.e. left invariant) differentials on G; in particular, the inner automorphism $\tau(\beta)$: $s \to \beta s \beta^{-1}$ determined by an element β of G determines an automorphism $\tau(\beta)^*$ of $\mathfrak{J}(G)$. It is "well-known" that if ω is invariant, and therefore finite at every point of G, then

$$(3) \qquad \omega(\alpha\beta) = (\tau(\beta^{-1})^*\omega)(\alpha) + \omega(\beta)$$

for all points α and β of G. This formula (which when G is the additive or multiplicative group of $\mathcal{U}$ becomes, respectively, $\mathfrak{d}(\alpha+\beta) = \mathfrak{d}\alpha + \mathfrak{d}\beta$ or $(\alpha\beta)^{-1}\mathfrak{d}(\alpha\beta) = \alpha^{-1}\mathfrak{d}\alpha + \beta^{-1}\mathfrak{d}\beta$, $\mathfrak{d}$ representing an arbitrary derivation of $\mathcal{U}$ over $\mathcal{K}$, and when G is the general linear group $GL(r)$ becomes $(\alpha\beta)^{-1}\mathfrak{d}(\alpha\beta) = \beta^{-1}(\alpha^{-1}\mathfrak{d}\alpha)\beta + \beta^{-1}\mathfrak{d}\beta$) has been given a rigorous exposition by Rosenlicht [5], but unfortunately only in the case of commutative groups G; therefore we give in § 1 a discussion supplementing that of Rosenlicht, establishing the desired formula in full generality.

We define a point α of G to be a *G-primitive* over $\mathcal{F}$ if the following two conditions are satisfied: 1° the field of constants of $\mathcal{F}\langle\alpha\rangle$ is $\mathcal{C}$; 2° $\langle\delta, \omega(\alpha)\rangle \in \mathcal{F}$ for every derivation operator δ of the differential field and every invariant differential ω on G defined over $\mathcal{C}$. If the derivation operators are denoted by $\delta_1, \cdots, \delta_m$ and if n linearly independent differentials $\omega_1, \cdots, \omega_n$ defined over $\mathcal{C}$ are expressed in the form

$$\omega_j = \sum_{1 \leqq \nu \leqq n} f_{j\nu}\, dx_\nu \qquad (1 \leqq j \leqq n),$$

where the $f_{j\nu}$ and the x_ν are rational functions on G defined over $\mathcal{C}$ and defined at α, and $x_1, \cdots, x_n$ are algebraically independent over $\mathcal{C}$, then the condition 2° above is equivalent to the condition that α satisfy a system of differential equations

$$\sum_{1 \leqq \nu \leqq n} f_{j\nu}(\alpha)\,\delta_i x_\nu(\alpha) = a_{ij} \qquad (1 \leqq i \leqq m, 1 \leqq j \leqq n),$$

where the a_{ij} are elements of $\mathcal{F}$.

By a *G-primitive extension* of $\mathcal{F}$, we mean an extension of $\mathcal{F}$ of the form $\mathcal{F}\langle\alpha\rangle$, where α is a G-primitive over $\mathcal{F}$.

It is not difficult to prove, making use of (3), that a necessary and sufficient condition for α to be a G-primitive over $\mathcal{F}$ is that the field of constants of $\mathcal{F}\langle\alpha\rangle$ be $\mathcal{C}$ and $\sigma\alpha \cdot \alpha^{-1} \in G_{\mathcal{K}}$ for every $\mathcal{F}$-isomorphism σ of $\mathcal{F}\langle\alpha\rangle$ onto an extension of $\mathcal{F}$ in $\mathcal{U}$. It follows almost immediately that every G-primitive extension of $\mathcal{F}$ is a G-extension of $\mathcal{F}$.

Suppose now that $\mathcal{G}$ is a G-extension of $\mathcal{F}$, so that there exists an injective rational homomorphism $\gamma\colon \mathfrak{G} \to G_{\mathcal{K}}$, $\mathfrak{G}$ denoting the Galois group of $\mathcal{G}$ over $\mathcal{F}$. Regarding γ as a rational crossed homomorphism of $\mathfrak{G}$ into G, we find by the above that *if γ splits in $G_{\mathcal{G}}$ (in particular if the cohomology set $H^1(\mathfrak{G}, G)$ is trivial) then $\mathcal{G}$ is a G-primitive extension of $\mathcal{F}$.* It is known ([3] p. 887) that $H^1(\mathfrak{G}, G)$ is trivial whenever G is the additive group of $\mathcal{U}$, or the multiplicative group of $\mathcal{U}$, or the general linear group $GL(r)$ for any r; this recaptures the characterizations ([2] p. 809 and [3] p. 891) of the strongly normal extensions corresponding to these groups as extensions of the form $\mathcal{F}\langle\alpha\rangle$, where in the first case α is an element such that $\delta\alpha \in \mathcal{F}$ for each derivation operator δ (extension by a *primitive* element), in the second case α is an element such that $\alpha^{-1}\delta\alpha \in \mathcal{F}$ for each δ (extension by an *exponential* element), and in the third case α is an invertible $r \times r$ matrix such that $\alpha^{-1}\delta\alpha$ has its coordinates in $\mathcal{F}$ for each δ (*Picard-Vessiot* extension).

If G is commutative, in particular if G is an abelian variety A, then we do not know in general that $H^1(\mathfrak{G}, G)$, which is now a group, is trivial; but we do know ([3] p. 887) that every element of $H^1(\mathfrak{G}, G)$ is of finite order,

9

so that for a suitable exponent $h > 0$ the cocycle γ^h splits in $G_{\mathcal{G}}$. It follows in this case that there is a G-primitive α such that $\mathcal{F} \subset \mathcal{F}\langle\alpha\rangle \subset \mathcal{G}$ and the Galois group of $\mathcal{G}$ over $\mathcal{F}\langle\alpha\rangle$ equals the finite group $\mathfrak{G}(h)$ of all elements $\sigma \in \mathfrak{G}$ with $\sigma^h = 1$.

Returning to the general case of a G-extension $\mathcal{G}$ with arbitrary algebraic group G defined over $\mathcal{C}$, we see that thse normal chain (1) has a refinement

$$G \supset G^0 \supset H \supset H^0 \supset 1,$$

where H is the set of points $s \in G^0$ such that $s^h \in H$, and therefore the group variety $A = G^0/H$ is abelian and H/H^0 is finite abelian, and that the tower (2) has a corresponding refinement

$$\mathcal{F} \subset \mathcal{F}^0 \subset \mathcal{E} \subset \mathcal{E}^0 \subset \mathcal{G},$$

where $\mathcal{E}$ is A-primitive over $\mathcal{F}^0$ and $\mathcal{E}^0$ is an abelian extension of $\mathcal{E}$ of finite degree. Since H/H^0 is finite and H^0 is linear, it easily follows that H is birationally isomorphic with a linear algebraic group, and therefore that $\mathcal{G}$ is a Picard-Vessiot extension of $\mathcal{E}$. Thus *every strongly normal extension has a three-storey tower*

$$\mathcal{F} \subset \mathcal{F}^0 \subset \mathcal{E} \subset \mathcal{G}$$

in which $\mathcal{F}^0$ (the algebraic closure of $\mathcal{F}$ in $\mathcal{G}$) is a normal extension of $\mathcal{F}$ of finite degree, $\mathcal{E}$ is an A-primitive extension of $\mathcal{F}^0$ for some abelian variety A defined over $\mathcal{C}$, and $\mathcal{G}$ is a Picard-Vessiot extension of $\mathcal{E}$.

In the final § 3 the A-primitive extensions are given a function-theoretic characterization in the classical case in which $\mathcal{F}$ consists of functions of m complex variables meromorphic in some region of $\boldsymbol{C}^m$. It is shown that if A is an abelian variety of dimension n defined over the field $\boldsymbol{C}$ of complex numbers then *every A-primitive extension of $\mathcal{F}$ is of the form $\mathcal{F}(f_1(\zeta), \cdots, f_r(\zeta))$, where $f_1(z), \cdots, f_r(z)$ are abelian functions of $z = (z_1, \cdots, z_n)$ which generate a nondegenerate abelian function field, and $\zeta = (\zeta_1, \cdots, \zeta_n)$ is a family of n functions of m complex variables, each ζ_i being primitive over $\mathcal{F}$.*

1. Induced differentials. This section supplements the discussion of Rosenlicht [5]; a familiarity with the terminology and notation as well as with the methods and results of that paper is assumed. We find it convenient, for any element u of a vector space and any element v of the dual space, to denote the value of v at u by $\langle u, v\rangle$.

Let V be an algebraic variety defined over a field k, let K be a subfield of the universal domain $\mathcal{U}$ (not necessarily distinct from $\mathcal{U}$) with $k \subset K$, and let $\mathfrak{D}(K/k)$ denote the vector space over K formed by the derivations of K over k. We recall (Rosenlicht [5], p. 54) that if α is a simple point on V which is rational over K then there is a unique K-linear mapping $\mathfrak{d} \to \mathfrak{d}_\alpha$, from $\mathfrak{D}(K/k)$ to the space of tangent vectors to V at α which are rational over K, defined by the condition that $\mathfrak{d}_\alpha f = \mathfrak{d} f(\alpha)$ $(f \in \mathfrak{o}_\alpha \cap k(V))$; also, if ω is any differential on V which is finite at α and defined over K, then the local component ω_α of ω at α is a K-linear mapping of the tangent space to V at α into $\mathcal{U}$ which maps the tangent vectors rational over K into K. The composite of the two mappings $\mathfrak{d} \to \mathfrak{d}_\alpha$ and ω_α then gives a K-linear mapping of $\mathfrak{D}(K/k)$ into K, that is, a differential of K over k; we call it the differential of K over k *induced by ω at α*, and following Rosenlicht denote it by $\omega(\alpha)$, or by $\omega_{K/k}(\alpha)$ when the greater detail is desirable.

If $f_i, g_i \in \mathfrak{o}_\alpha \cap k(V)$ $(1 \leqq i \leqq r)$ and $\omega = \sum f_i dg_i$ then

$$\langle \mathfrak{d}, \omega(\alpha) \rangle = \sum f_i(\alpha) \mathfrak{d} g_i(\alpha).$$

It is easy to see that if a subfield k_0 of k is also a field of definition of V and ω is defined over k_0 then $\langle \mathfrak{d}, \omega(\alpha) \rangle \in k_0(\alpha) \cdot k_0(\mathfrak{d} k_0(\alpha))$. Furthermore, if V' is also a variety defined over k, and $\phi\colon V \to V'$ is a rational mapping defined over k such that ϕ is defined at α and $\phi(\alpha)$ is a simple point on V', and ω' is a differential on V' finite at $\phi(\alpha)$ and defined over K, then $\omega'(\phi(\alpha)) = (\phi^* \omega')(\alpha)$.

We are interested in the case in which V is a group variety G and ω is an invariant (meaning left invariant) differential on G. Every point of G is simple, and ω is finite everywhere. In order to prove the result we require, we use the notion of a rational crossed homomorphism.

Let G and G' be connected algebraic groups. We say that *G operates on G' by g* if g is a group homomorphism of G into the group of automorphisms of G' such that the mapping $G \times G' \to G'$ defined by $(s, s') \to g(s)s'$ is rational. When such is the case $g(s)$ is a birational automorphism of G', defined over $k_0(s)$, whenever k_0 is a common field of definition of G, of G', and of the mapping $(s, s') \to g(s)s'$.

Suppose G operates on G' by g. A *rational crossed homomorphism* (or 1-*cocycle*) of G into G' (relative to g) is a rational mapping $\zeta\colon G \to G'$ such that, if k_0 is a common field of definition of G, of G', of the mapping $(s, s') \to g(s)s'$, and of ζ, and if (s, t) is a generic point of $G \times G$ over k_0, then $\zeta(st) = \zeta(s) \cdot g(s)\zeta(t)$. It is easy to see that this implies that ζ is defined at every point of G, $\zeta(st) = \zeta(s) \cdot g(s)\zeta(t)$ for all points s, t of G, and $\zeta(1) = 1$. (We denote the unity element of every multiplicative group by 1.) The notion of rational crossed homomorphism includes that of rational homomorphism, the latter corresponding to the case in which the operation g is trivial, that is $g(s)$ is the identity automorphism of G' for every $s \in G$.

Consider a rational crossed homomorphism $\zeta\colon G \to G'$. ζ induces a homomorphism $X \to \zeta X$ of the tangent space to G at 1 into the tangent space to G' at $1 = \zeta(1)$; since these tangent spaces are canonically isomorphic (as vector spaces) to the respective Lie algebras $\mathfrak{L}(G)$ and $\mathfrak{L}(G')$ of G and G', we obtain a linear mapping $D \to \zeta D$ of $\mathfrak{L}(G)$ into $\mathfrak{L}(G')$. Thus, by definition, $(\zeta D)_1 = \zeta D_1$ $(D \in L(G))$. (We use a point of a variety as a subscript to a derivation or differential on the variety to indicate the local component at that point.)

This being so for any rational crossed homomorphism, in particular it is so for any rational homomorphism, including the automorphism $g(s)\colon G' \to G'$.

Returning to ζ, we note that the crossed homomorphism property $\zeta(st) = \zeta(s)\cdot g(s)\zeta(t)$ can be written in the form $\zeta T_s = T_{\zeta(s)} g(s)\zeta$, where for any element v of any group we use T_v to denote the left multiplication mapping $x \to vx$ of that group into itself. It follows that for any $D \in \mathfrak{L}(G)$ we may write $\zeta D_s = \zeta T_s D_1 = T_{\zeta(s)} g(s)\zeta D_1 = T_{\zeta(s)}(g(s)\zeta D)_1 = (g(s)\zeta D)_{\zeta(s)}$, so that

$$(g(s)\zeta D)_{\zeta(s)} = \zeta D_s \qquad (D \in \mathfrak{L}(G), s \in G).$$

Therefore the homomorphism $\omega' \to \zeta^*\omega'$ which ζ induces from the space of differentials on G' into the space of differentials on G has the property that $\langle D, \zeta^*\omega'\rangle(s) = \langle \zeta D_s, \omega'_{\zeta(s)}\rangle = \langle (g(s)\zeta D)_{\zeta(s)}, \omega'_{\zeta(s)}\rangle = \langle g(s)\zeta D, \omega'\rangle(\zeta(s))$. But if ω' is an element of the space $\mathfrak{J}(G')$ of *invariant* differentials on G' then $\langle g(s)\zeta D, \omega'\rangle(t')$ is independent of the element $t' \in G'$, so that we may write

$$(4) \qquad \langle D, \zeta^*\omega'\rangle(s) = \langle g(s)\zeta D, \omega'\rangle(s) \quad (D \in \mathfrak{L}(G), \omega' \in \mathfrak{J}(G'), s \in G).$$

In particular, a rational homomorphism $G \to G'$ induces vector space homomorphisms $\mathfrak{L}(G) \to \mathfrak{L}(G')$ and $\mathfrak{J}(G') \to \mathfrak{J}(G)$ each of which is the transpose of the other.

If $\nu\colon G \to G'$ is a rational homomorphism, and if $\zeta'\colon G' \to G''$ is a rational crossed homomorphism (G' operating on the connected algebraic group G'' by g'), then G operates on G'' by $g'\nu$, and $\zeta'\nu$ is a rational crossed homomorphism of G into G''; furthermore, $(\zeta'\nu)D = \zeta'(\nu D)$ $(D \in \mathfrak{L}(G))$.

We now apply these considerations to the differentials $\omega(\alpha) = \omega_{K/k}(\alpha)$ of K over k induced by an invariant differential ω on a connected algebraic group G; as before, G is defined over k, ω is defined over K, and $\alpha \in G_K$ (the group of points of G which are rational over K).

As examples of rational homomorphisms we have the two projections

$$\pi_j\colon G \times G \to G, \qquad \pi_j(s_1, s_2) = s_j,$$

the three injections

$$\iota_1\colon G\to G\times G,\quad \iota_1 s=(s,1),$$
$$\iota_2\colon G\to G\times G,\quad \iota_2 s=(1,s),$$
$$\Delta\colon G\to G\times G,\quad \Delta s=(s,s),$$

the identity mapping $\iota\colon G\to G$, and the trivial endomorphism

$$\epsilon\colon G\to G,\qquad \epsilon s=1.$$

Obviously $\pi_1\iota_1=\pi_2\iota_2=\pi_1\Delta=\pi_2\Delta=\iota$, and $\pi_1\iota_2=\pi_2\iota_1=\epsilon$; also $\epsilon D=0$ $(D\in\mathfrak{L}(G))$. Furthermore, $\pi_1{}^*\mathcal{U}(G)$ is the subfield of $\mathcal{U}(G\times G)$ consisting of the rational functions f on $G\times G$ such that $f(s_1,s_2)$ is independent of s_2, and similarly for $\pi_2{}^*\mathcal{U}(G)$, so that $\mathcal{U}(G\times G)$ is precisely the compositum

$$\mathcal{U}(G\times G)=\pi_1{}^*\mathcal{U}(G)\cdot\pi_2{}^*\mathcal{U}(G). \tag{5}$$

For any $D\in\mathfrak{L}(G)$,

$$(\Delta D-\iota_1 D-\iota_2 D)_1\pi_1{}^*f=(\pi_1(\Delta D-\iota_1 D-\iota_2 D)_1)f$$
$$=(\pi_1\Delta D-\pi_1\iota_1 D-\pi_1\iota_2 D)_1 f=(\iota D-\iota D-\epsilon D)_1 f=0$$

for all $f\in\mathfrak{o}_1$, so that

$$((\Delta D-\iota_1 D-\iota_2 D)\pi_1{}^*f)(s,t)=(\Delta D-\iota_1 D-\iota_2 D)_{(s,t)}\pi_1{}^*f=0$$

provided $f\in\mathfrak{o}_s$; choosing (s,t) generic on $G\times G$ over a common field of definition of G, of D, and of f, we see that then $f\in\mathfrak{o}_s$ and

$$((\Delta D-\iota_1 D-\iota_2 D)\pi_1{}^*f)(s,t)=0;$$

thus $(\Delta D-\iota_1 D-\iota_2 D)\pi_1{}^*f=0$ for every $f\in\mathcal{U}(G)$. Similarly

$$(\Delta D-\iota_1 D-\iota_2 D)\pi_2{}^*f=0$$

for every $f\in\mathcal{U}(G)$, so that by (5) $\Delta D=\iota_1 D+\iota_2 D$ $(D\in\mathfrak{L}(G))$.

Now, for each $t\in G$ let $\tau(t)$ denote the inner automorphism of G defined by t: $\tau(t)s=tst^{-1}$. Then G operates on itself by τ, and the reciprocation mapping

$$\rho\colon G\to G,\qquad \rho(s)=s^{-1}$$

is a rational crossed homomorphism. Also, $G\times G$ operates on G by $\tau\pi_2$, and the mapping

$$\chi\colon G\times G\to G,\qquad \chi(s,t)=st^{-1}$$

is a rational crossed homomorphism. Obviously $\chi\Delta=\epsilon$, $\chi\iota_1=\iota$, and $\chi\iota_2=\rho$. Therefore $0=\epsilon D=\chi\Delta D=\chi\iota_1 D+\chi\iota_2 D=\iota D+\rho D$, so that

$$\rho D=-D\qquad (D\in\mathfrak{L}(G)).$$

It is easy to see that the two mappings $D \to \iota_j D$ of $\mathfrak{L}(G)$ into $\mathfrak{L}(G \times G)$ are injective. Also, if $D, E \in \mathfrak{L}(G)$ and $\iota_1 D + \iota_2 E = 0$ then $D = \iota D + \epsilon E = \pi_1 \iota_1 D + \pi_1 \iota_2 E = \pi_1(\iota_1 D + \iota_2 E) = 0$ and similarly $E = 0$. Therefore $\iota_1 \mathfrak{L}(G) \cap \iota_2 \mathfrak{L}(G) = \{0\}$, so that

$$\dim(\iota_1\mathfrak{L}(G) + \iota_2\mathfrak{L}(G)) = \dim \iota_1\mathfrak{L}(G) + \dim \iota_2\mathfrak{L}(G)$$
$$= 2 \dim \mathfrak{L}(G) = \dim \mathfrak{L}(G \times G),$$

whence

$$\mathfrak{L}(G \times G) = \iota_1\mathfrak{L}(G) + \iota_2\mathfrak{L}(G) \quad \text{(direct sum)}.$$

Since $\chi\iota_1 D - \pi_1\iota_1 D + \pi_2\iota_1 D = \iota D - \iota D + \epsilon D = 0$ and $\chi\iota_2 D - \pi_1\iota_2 D + \pi_2\iota_2 D = \rho D - \epsilon D + \iota D = 0$, we conclude that

$$\chi D = \pi_1 \hat{D} - \pi_2 \hat{D} \qquad (\hat{D} \in \mathfrak{L}(G \times G)). \tag{6}$$

For any $\alpha, \beta \in G_K$ (α, β) is a point of $(G \times G)_K$ so that for any $\mathfrak{d} \in \mathfrak{D}(K/k)$ we have the tangent vector $\mathfrak{d}_{(\alpha,\beta)}$ to $G \times G$ at (α, β) and $\mathfrak{d}_{(\alpha,\beta)}$ is rational over K; also, there is a unique $D \in \mathfrak{L}(G \times G)$ with $\hat{D}_{(\alpha,\beta)} = \mathfrak{d}_{(\alpha,\beta)}$. Since χ is obviously defined over k it follows that $\langle \mathfrak{d}, \omega(\alpha\beta^{-1})\rangle = \langle \mathfrak{d}, \omega(\chi(\alpha,\beta))\rangle = \langle \mathfrak{d}, (\chi^*\omega)(\alpha,\beta)\rangle = \langle \mathfrak{d}_{(\alpha,\beta)}, \chi^*\omega_{(\alpha,\beta)}\rangle = \langle D, \chi^*\omega\rangle(\alpha,\beta)$; by the general formula (4) this equals $\langle \tau(\beta)\chi\hat{D}, \omega\rangle(\alpha,\beta)$, and by (6) this in turn equals

$$\begin{aligned}
\langle \tau(\beta)(\pi_1\hat{D} - \pi_2\hat{D}), \omega\rangle(\alpha,\beta) &= \langle D, \pi_1{}^*\tau(\beta)^*\omega - \pi_2{}^*\tau(\beta)^*\omega\rangle(\alpha,\beta) \\
&= \langle \mathfrak{d}_{(\alpha,\beta)}, (\pi_1{}^*\tau(\beta)^*\omega)_{(\alpha,\beta)} - (\pi_2{}^*\tau(\beta)^*\omega)_{(\alpha,\beta)}\rangle \\
&= \langle \mathfrak{d}, (\pi_1{}^*\tau(\beta)^*\omega)(\alpha,\beta) - \pi_2{}^*\tau(\beta)^*\omega)(\alpha,\beta)\rangle \\
&= \langle \mathfrak{d}, (\tau(\beta)^*\omega)(\alpha) - (\tau(\beta)^*\omega)(\beta).
\end{aligned}$$

Thus, $\omega(\alpha\beta^{-1}) = (\tau(\beta)^*\omega)(\alpha) - (\tau(\beta)^*\omega)(\beta)$. Applying this formula to $\tau(\beta^{-1})^*\omega$ instead of ω, and then replacing α by $\alpha\beta$, we obtain the desired formula (3) of the introduction.

2. G-extensions and G-primitives.

In this section $\mathcal{F}$ denotes a commutative differential field of characteristic 0 with algebraically closed field of constants $\mathcal{C}$; the derivation operators of $\mathcal{F}$ are denoted by $\delta_1, \cdots, \delta_m$. $\mathcal{U}$ denotes a fixed universal extension of $\mathcal{F}$ (in the sense of [2] p. 768) and $\mathcal{K}$ denotes the field of constants of $\mathcal{U}$. We identify the derivation operators $\delta_1, \cdots, \delta_m$ in an obvious way with derivations of $\mathcal{U}$ over $\mathcal{K}$.

$\mathcal{U}$ is an algebraically closed extension of $\mathcal{F}$ of infinite transcendence degree, and therefore can be used as a universal domain for algebraic geometry. Whenever we introduce algebraico-geometric notions it always is in the algebraic geometry based on $\mathcal{U}$ as universal domain. The algebraic varieties we have occasion to introduce, other than points, are actually defined over $\mathcal{C}$; if ω is a differential on such a variety, the symbol $\omega(\alpha)$ always denotes the differential $\omega_{\mathcal{U}/\mathcal{K}}(\alpha)$ of $\mathcal{U}$ over $\mathcal{K}$ induced by ω at α, so that α can be any simple point of the variety at which ω is finite (and therefore, in the case of an invariant differential on a group variety, can be any point whatsoever).

Let G be a connected algebraic group defined over $\mathcal{C}$; denote by $G_{\mathcal{K}}$ the group consisting of the points of G which are rational over $\mathcal{K}$.

It is apparent from the definition of induced differentials in § 1 that *if* $\gamma \in G$, *then a necessary and sufficient condition that* $\gamma \in G_{\mathcal{K}}$ *is that* $\langle \delta_i, \omega(\gamma)\rangle = 0$ *for every index i and every invariant differential* ω *on* G.

It follows from this that *if* $\alpha, \beta \in G$ *then a necessary and sufficient condition that* $\beta\alpha^{-1} \in G_{\mathcal{K}}$ *is that* $\langle \delta_i, \omega(\alpha)\rangle = \langle \delta_i, \omega(\beta)\rangle$ *for every index i and every invariant differential* ω *on* G. Indeed, setting $\gamma = \beta\alpha^{-1}$ we have, by formula (3),

$$\begin{aligned}\langle \delta_i, \omega(\beta)\rangle &= \langle \delta_i, \omega(\gamma\alpha)\rangle \\ &= \langle \delta_i, (\tau(\alpha^{-1})^*\omega)(\gamma)\rangle + \langle \delta_i, \omega(\alpha)\rangle,\end{aligned}$$

and as ω runs through the set of invariant differentials on G so does $\tau(\alpha^{-1})^*\omega$.

It is also obvious from the definition of induced differentials that *if* ω *is an invariant differential on* G *which is defined over* $\mathcal{C}$ *and* $\alpha \in G$, *then* $\langle \delta_i, \omega(\alpha)\rangle \in \mathcal{C}\langle\alpha\rangle$ $(1 \leqq i \leqq m)$. *If in addition* σ *is an isomorphism of the differential field* $\mathcal{C}\langle\alpha\rangle$ *over* $\mathcal{C}$ *onto an extension of* $\mathcal{C}$ *in* $\mathcal{U}$, *then* $\sigma\langle \delta_i, \omega(\alpha)\rangle = \langle \delta_i, \omega(\sigma\alpha)\rangle$ $(1 \leqq i \leqq m)$. To prove the second point we observe that there exist uniformizing coordinates $x_1, \cdots, x_n$ on G at α with each $x_j \in \mathcal{C}(G)$, and we may write $\omega = \sum f_j\, dx_j$ with each $f_j \in \mathfrak{o}_\alpha \cap \mathcal{C}(G)$; then

$$\langle \delta_i, \omega(\alpha)\rangle = \sum_{1 \leqq j \leqq n} f_j(\alpha)\delta_i x_j(\alpha),$$

and the result follows.

Let α be a G-primitive over $\mathcal{F}$. We recall from the introduction that this means: 1° the field of constants of $\mathcal{F}\langle\alpha\rangle$ is $\mathcal{C}$; 2° $\langle \delta_i, \omega(\alpha)\rangle \in \mathcal{F}$ for each index i and every invariant differential ω on G defined over $\mathcal{C}$. Then, for each isomorphism σ of $\mathcal{F}\langle\alpha\rangle$ over $\mathcal{F}$ onto an extension of $\mathcal{F}$ in $\mathcal{U}$, $\langle \delta_i, \omega(\sigma\alpha)\rangle = \sigma\langle \delta_i, \omega(\alpha)\rangle = \langle \delta_i, \omega(\alpha)\rangle$, so that the point $\gamma(\sigma) = \alpha^{-1}\cdot\sigma\alpha$ is in $G_{\mathcal{K}}$. Therefore $\mathcal{F}\langle\alpha\rangle\cdot\sigma(\mathcal{F}\langle\alpha\rangle) = \mathcal{F}\langle\alpha, \sigma\alpha\rangle = \mathcal{F}\langle\alpha, \gamma(\sigma)\rangle = \mathcal{F}\langle\sigma\alpha, \gamma(\sigma)\rangle$, so that $\mathcal{F}\langle\alpha\rangle$ is strongly normal over $\mathcal{F}$, and ([2] p. 768, cor. 5) the field of constants of $\mathcal{F}\langle\alpha\rangle\cdot\sigma(\mathcal{F}\langle\alpha\rangle)$ is $\mathcal{C}(\gamma(\sigma))$. Identifying the isomorphisms of $\mathcal{F}\langle\alpha\rangle$ over $\mathcal{F}$ with the unique automorphisms of $\mathcal{F}\langle\alpha\rangle\cdot\mathcal{K}$ over $\mathcal{F}\cdot\mathcal{K}$ to which they extend ([2], Ch. II, § 3), we find that $\alpha\cdot\gamma(\sigma\tau) = \sigma\tau\alpha = \sigma(\alpha\cdot\gamma(\tau)) = \alpha\cdot\gamma(\sigma)\cdot\gamma(\tau)$, so that $\gamma(\sigma\tau) = \gamma(\sigma)\gamma(\tau)$. Since $\gamma(\sigma) = 1$ only if $\sigma = 1$, the mapping $\sigma \to \gamma(\sigma)$ is a birational isomorphism of the Galois group of $\mathcal{F}\langle\alpha\rangle$ over $\mathcal{F}$ onto an algebraic subgroup of $G_{\mathcal{K}}$. In particular, *every G-primitive extension of $\mathcal{F}$ is a G-extension of $\mathcal{F}$.*

Conversely, suppose that we are given a G-extension of $\mathcal{F}$, that is, a strongly normal extension $\mathcal{G}$ of $\mathcal{F}$ such that there exists an injective rational homomorphism $\gamma\colon \mathfrak{G} \to G_{\mathcal{K}}$, $\mathfrak{G}$ denoting the Galois group of $\mathcal{G}$ over $\mathcal{F}$.

If γ splits over $\mathcal{G}$, i.e. if there exists a point $\alpha \in G_{\mathcal{G}}$ such that $\gamma(\sigma) = \alpha \cdot \sigma\alpha^{-1}$ $(\sigma \in \mathfrak{G})$, then

$$\sigma\langle\delta_i, \omega(\alpha)\rangle = \langle\delta_i, \omega(\sigma\alpha)\rangle = \langle\delta_i, \omega(\gamma(\sigma^{-1})\alpha)\rangle$$
$$= \langle\delta_i, (\tau(\alpha^{-1})^*\omega)(\gamma(\sigma^{-1}))\rangle + \langle\delta_i, \omega(\alpha)\rangle = \langle\delta_i, \omega(\alpha)\rangle \quad (\sigma \in \mathfrak{G}),$$

so that $\langle\delta_i, \omega(\alpha)\rangle \in \mathcal{F}$ for each index i and each invariant differential ω on G defined over $\mathcal{C}$, i.e. α is a G-primitive over $\mathcal{F}$. Since γ is injective, $\sigma\alpha = \alpha$ only if $\sigma = 1$, so that $\mathcal{F}\langle\alpha\rangle = \mathcal{G}$. Thus we have proved, in particular, that *if $\mathcal{G}$ is a G-extension of $\mathcal{F}$ with Galois group $\mathfrak{G}$ and if $H^1(\mathfrak{G}, G) = 1$ then $\mathcal{G}$ is a G-primitive extension of $\mathcal{F}$.*

If we do not assume that γ splits over $\mathcal{G}$, but suppose instead that G is commutative, then ([3] p. 887), for some integer $h > 0$, γ^h splits over $\mathcal{G}$: there exist a point $\alpha \in G_{\mathcal{G}}$ such that $\gamma(\sigma)^h = \alpha \cdot \sigma\alpha^{-1}$ $(\sigma \in \mathfrak{G})$. The same reasoning as above then proves: *If G is commutative and $\mathcal{G}$ is a G-extension of $\mathcal{F}$ with Galois group $\mathfrak{G}$, then there exist a G-primitive extension $\mathcal{H}$ of $\mathcal{F}$ and an integer $h > 0$ such that $\mathcal{F} \subset \mathcal{H} \subset \mathcal{G}$ and the Galois group of $\mathcal{G}$ over $\mathcal{H}$ is the finite group $\mathfrak{G}(h)$ of all elements σ of $\mathfrak{G}$ with $\sigma^h = 1$.*

3. Abelian extensions and abelian functions. In this section we consider the classical case in which $\mathcal{F}$ is a differential field consisting of functions meromorphic in some region of complex m-space $\boldsymbol{C}^m$, the derivation operators being the partial derivations $\partial/\partial x_1, \cdots, \partial/\partial x_m$ with respect to the m coordinate functions $x_1, \cdots, x_m$, and the field of constants of $\mathcal{F}$ being the complex number field $\boldsymbol{C}$. We seek to characterize the A-primitive extensions of $\mathcal{F}$ in function-theoretic terms, A denoting an arbitrary abelian variety defined over $\boldsymbol{C}$.

To this end we recall certain facts about abelian varieties defined over $\boldsymbol{C}$ and abelian functions. If Γ is a free abelian subgroup of $\boldsymbol{C}^n$ with $2n$ generators, the field $\mathcal{A}$ of all meromorphic functions on $\boldsymbol{C}^n$ admitting as periods the elements of Γ is called the abelian function field with period group Γ. $\mathcal{A}$ is said to be degenerate if by means of some invertible linear transformation on $\boldsymbol{C}^n$ the elements of $\mathcal{A}$ can all be expressed as meromorphic functions of fewer than n variables, that is, if there exist complex numbers

$k_1, \cdots, k_n$ not all 0 such that $\sum k_j \partial f/\partial z_j = 0$ $(f \in \mathcal{A})$, $z = (z_1, \cdots, z_n)$ denoting the usual coordinate functions on $\boldsymbol{C}^n$, and $\mathcal{A}$ is said to be nondegenerate otherwise. There are well known necessary and sufficient conditions on Γ for $\mathcal{A}$ to be nondegenerate (the period relations). When $\mathcal{A}$ is nondegenerate then Γ is of rank $2n$ over the field of real numbers, and there exist an abelian variety A of dimension n defined over $\boldsymbol{C}$ and a biholomorphic group isomorphism $\boldsymbol{C}^n/\Gamma \approx A_{\boldsymbol{C}}$ (these groups being endowed with their usual complex analytic structures). The composite of the canonical projection $\boldsymbol{C}^n \to \boldsymbol{C}^n/\Gamma$ and this isomorphism is a surjective holomorphic homomorphism $p\colon \boldsymbol{C}^n \to A_{\boldsymbol{C}}$ with kernel Γ such that $\mathcal{A} = \boldsymbol{C}(p(z))$. Conversely, if A is any abelian variety of dimension n defined over $\boldsymbol{C}$, there exist a nondegenerate abelian function field $\mathcal{A}$ and a surjective holomorphic homomorphism $p\colon \boldsymbol{C}^n \to A_{\boldsymbol{C}}$ exactly as above. These facts, or a reasonable facsimile of them, can be found in Chapter VI of Weil's book [6] (see especially his Theorem 5, page 130).

Now, $\mathcal{A}$ is obviously a differential field relative to the derivation operators $\partial/\partial z_1, \cdots, \partial/\partial z_n$, the field of constants of $\mathcal{A}$ being $\boldsymbol{C}$, and for each $k \in \boldsymbol{C}^n$ the translation mapping $f(z) \to f(z+k)$ $(f(z) \in \mathcal{A})$ is an automorphism of this differential field over $\boldsymbol{C}$. It follows by § 2 that if $(\omega_1, \cdots, \omega_n)$ denotes a basis of the space of invariant differentials on A defined over $\boldsymbol{C}$ then this automorphism maps $\langle \partial/\partial z_j, \omega_{j'}(p(z))\rangle$ onto $\langle \partial/\partial z_j, \omega_{j'}(p(z+k))\rangle$, which by § 2 equals $\langle \partial/\partial z_j, \omega_{j'}(p(z))\rangle + \langle \partial/\partial z_j, \omega_{j'}(p(k))\rangle = \langle \partial/\partial z_j, \omega_{j'}(p(z))\rangle$. Thus $\langle \partial/\partial z_j, \omega_{j'}(p(z))\rangle$ is invariant under every translation, that is, is a complex constant, which we denote by $a_{jj'}$ $(1 \leqq j \leqq n, 1 \leqq j' \leqq n)$. In other words, *$p(z)$ is an A-primitive over $\boldsymbol{C}$.* Because $\mathcal{A}$ is nondegenerate, the matrix $(a_{jj'})$ is invertible.

Shifting our attention to the differential field $\mathcal{F}$ of functions of m complex variables $x_1, \cdots, x_m$ meromorphic in some region D of $\boldsymbol{C}^m$ (m possibly different from n), let $\zeta(x) = (\zeta_1(x_1, \cdots, x_m), \cdots, \zeta_n(x_1, \cdots, x_m))$ be a sequence of n functions of $x = (x_1, \cdots, x_m)$ meromorphic in some subregion of D. Then

$$\langle \partial/\partial x_i, \omega_{j'}(p(\zeta(x)))\rangle$$
$$= \sum_j \partial\zeta_j(x)/\partial x_i \cdot \langle \partial/\partial z_j, \omega_{j'}(p(z))\rangle = \sum_j \partial\zeta_j(x)/\partial x_i \cdot a_{jj'},$$

so that we may write the matrix equation

$$(7) \qquad (\langle \partial/\partial x_i, \omega_{j'}(p(\zeta(x)))\rangle) = (\partial\zeta_j(x)/\partial x_i)\,(a_{jj'}).$$

Therefore *if each $\zeta_j(x)$ is primitive over $\mathcal{F}$, i.e. $\partial\zeta_j(x)/\partial x_i \in \mathcal{F}$ $(1 \leqq i \leqq m, 1 \leqq j \leqq n)$, then $p(\zeta(x))$ is an A-primitive over $\mathcal{F}$.*

Conversely, let q be any A-primitive over $\mathcal{F}$, so that we may write

$$\langle \partial/\partial x_i, \omega_{j'}(q)\rangle = f_{ij'}(x) \in \mathfrak{F} \qquad (1 \leqq i \leqq m, 1 \leqq j \leqq n).$$

It is easy to see that for each index j' the m functions $f_{ij'}(x)$ satisfy the integrability conditions $\partial f_{ij'}(x)/\partial x_{i'} = \partial f_{i'j'}(x)/\partial x_i$ $(1 \leqq i \leqq m, 1 \leqq l' \leqq m)$, and therefore for each j the m functions $g_{ij}(x)$ defined by the matrix equation $(g_{ij}(x)) = (f_{ij'}(x))(a_{jj'})^{-1}$ satisfy similar integrability conditions. Consequently there exist functions $\zeta_j(x)$ meromorphic in some subregion of D such that $\partial\zeta_j(x)/\partial x_i = g_{ij}(x)$ $(1 \leqq i \leqq m, 1 \leqq j \leqq n)$. Therefore by (7)

$$(\langle \partial/\partial x_i, \omega_{j'}(p(\zeta(x)))\rangle)$$
$$= (g_{ij}(x))(a_{jj'}) = (f_{ij'}(x)) = (\langle \partial/\partial x_i, \omega_{j'}(q)\rangle).$$

It follows by § 2 that $q \cdot p(\zeta(x))^{-1} \in A_{\boldsymbol{C}}$, i. e. $q \cdot p(\zeta(x))^{-1} = p(k)$ for some $k = (k_1, \cdots, k_n) \in \boldsymbol{C}^n$, whence $q = p(\zeta(x) + k)$. Since each $\zeta_j(x) + k_j$ is primitive over $\mathfrak{F}$, we have the following result: *In the classical case of a differential field $\mathfrak{F}$ of functions meromorphic in a region of complex m-space $\boldsymbol{C}^m$ with field of constants $\boldsymbol{C}$, a necessary and sufficient condition that a point $q \in A$ be an A-primitive over $\mathfrak{F}$ is that $q = p(\zeta)$, where $\zeta = (\zeta_1, \cdots, \zeta_n)$ and each ζ_j is primitive over $\mathfrak{F}$.* This means that the A-primitive extensions of $\mathfrak{F}$ are precisely the extensions of the form $\mathfrak{F}\langle p(\zeta)\rangle = \mathfrak{F}(p(\zeta))$, and therefore: *Every A-primitive extension of $\mathfrak{F}$ is obtained by adjoining to $\mathfrak{F}$ finitely many functions $f_1(\zeta), \cdots, f_r(\zeta)$,* where *$f_1(z), \cdots, f_r(z)$ are abelian functions which generate the nondegenerate abelian function field $\mathfrak{A}$ and $\zeta = (\zeta_1, \cdots, \zeta_n)$ is a sequence of n functions which are primitive over $\mathfrak{F}$.*

COLUMBIA UNIVERSITY.

REFERENCES.

[1] Iacopo Barsotti, "Un teorema di struttura per le varieta gruppali," *Rendiconti dell'Academia Nazionale dei Lincei*, vol. 18 (1955), pp. 43-50.

[2] E. R. Kolchin, "Galois theory of differential fields," *American Journal of Mathematics*, vol. 75 (1953), pp. 753-824.

[3] ———, "On the Galois theory of differential fields," *ibid.*, vol. 77 (1955), pp. 868-894.

[4] Maxwell Rosenlicht, "Some basic theorems on algebraic groups," *ibid.*, vol. 78 (1956), pp. 401-443.

[5] ———, "A note on derivations and differentials on algebraic varieties," *Portugaliae mathematica*, vol. 16 (1957), pp. 43-55.

[6] André Weil, *Introduction à l'étude des variétés kaehlériennes* (Actualités scientifiques et industrielles 1267), 1958.

Séminaire DUBREIL-PISOT
(Algèbre et Théorie des nombres)
14e année, 1960/61, n° 7

9 janvier 1961

LE THÉORÈME DE LA BASE FINIE POUR LES POLYNÔMES DIFFÉRENTIELS

par Ellis R. KOLCHIN

Introduction. - Le théorème de la base finie de Hilbert est un outil précieux dans l'étude des équations algébriques, et il serait bien agréable d'avoir un théorème semblable dans la théorie des équations différentielles algébriques. Malheureusement, l'énoncé tout à fait analogue dans cette dernière théorie est inexact. Mais RITT a démontré (voir [2]) l'analogue du théorème plus faible dans lequel on affirme le principe des "chaînes ascendantes" pour l'ensemble des idéaux parfaits au lieu de l'ensemble de tous les idéaux (un idéal est dit "parfait" lorsqu'il contient un élément toutes les fois qu'il contient une puissance de cet élément) ; on sait que, pour beaucoup d'applications, le théorème affaibli suffit. Or, le théorème de Ritt a été démontré dans le cas où les coefficients forment un "corps différentiel" de caractéristique 0 . J'ai démontré [1] le théorème sous une hypothèse plus générale mais encore assez restrictive, donnant en même temps un contre-exemple pour quelques corps différentiels de caractéristique $\neq 0$. SEIDENBERG a repris la question et a démontré [3] que le théorème reste vrai pour un corps différentiel quelconque si on se borne à une classe d'idéaux plus étroite.

Dans ce qui suit, j'aborde le théorème dans une formulation plus large, et j'obtiens un résultat qui contient tous les résultats connus, et même un peu plus.

Avant de pouvoir le formuler, il faut préciser certaines définitions et développer certains résultats préliminaires.

1. Anneaux différentiels.

Un anneau (resp. corps) différentiel est un anneau (resp. corps) $\mathcal{R}$ muni d'un ensemble fini d'opérateurs Δ tel que

$$\delta(a+b) = \delta a + \delta b \quad ,$$

$$\delta(ab) = (\delta a) b + a(\delta b) \quad ,$$

$$(\delta\delta') a = \delta(\delta' a)$$

pour tous $a \in \mathcal{R}$, $b \in \mathcal{R}$, $\delta \in \Delta$, $\delta' \in \Delta$; l'ensemble Δ s'appelle l'ensemble d'opérateurs de dérivation de $\mathcal{R}$. Nous ne considérerons que des anneaux et des corps commutatifs.

Soient $\mathcal{R}$ un anneau différentiel, Δ son ensemble d'opérateurs de dérivations. Notons Θ le monoïde commutatif libre engendré par Δ ; si les éléments de Δ sont notés $\delta_1, \ldots, \delta_m$, alors les éléments de Θ sont les produits $\delta_1^{i_1} \ldots \delta_m^{i_m}$, chaque i_μ étant un entier naturel. On peut définir d'une façon unique une opération de Θ sur $\mathcal{R}$ prolongeant l'opération de Δ sur $\mathcal{R}$ telle que $(\theta\theta')a = \theta(\theta' a)$ pour $\theta \in \Theta$, $\theta' \in \Theta$, $a \in \mathcal{R}$ et que $1a = a$ pour $a \in \mathcal{R}$ (1 désignant l'élément-unité de Θ) ; Θ s'appelle l'ensemble d'opérateurs dérivés de $\mathcal{R}$. Si

$$\theta = \delta_1^{i_1} \ldots \delta_m^{i_m} \in \Theta ,$$

l'entier $r = \sum i_\mu$ s'appelle l'ordre de θ , et se note ord θ ; et, pour chaque $a \in \mathcal{R}$, θa s'appelle un dérivé de a d'ordre r .

Un élément $c \in \mathcal{R}$, tel que $\delta c = 0$ pour chaque $\delta \in \Delta$, s'appelle une constante. L'ensemble des constantes de $\mathcal{R}$ est un sous-anneau différentiel de $\mathcal{R}$ (et dans le cas où $\mathcal{R}$ est un corps différentiel, un sous-corps différentiel de $\mathcal{R}$) ; on l'appelle l'anneau (ou, selon le cas, le corps) des constantes de $\mathcal{R}$. Si $\mathcal{R}$ est un corps de caractéristique $p \neq 0$, alors $\mathcal{R}^p$ est un sous-corps du corps des constantes de $\mathcal{R}$.

Si $\mathcal{R}'$ est aussi un anneau différentiel, ayant le même ensemble d'opérateurs de dérivation Δ , une application $f : \mathcal{R} \to \mathcal{R}'$ s'appelle homomorphisme de $\mathcal{R}$ dans $\mathcal{R}'$, si elle est un homomorphisme d'anneaux et si, de plus, $f(\delta a) = \delta f(a)$ pour chaque $a \in \mathcal{R}$ et chaque $\delta \in \Delta$. L'image $f(\mathcal{R})$ d'un homomorphisme $f : \mathcal{R} \to \mathcal{R}'$ est un sous-anneau différentiel de $\mathcal{R}'$. Le noyau k de f est un idéal de $\mathcal{R}$ ayant la propriété $\delta k \subset k$ pour chaque $\delta \in \Delta$. Un idéal de $\mathcal{R}$ ayant cette propriété s'appelle un idéal différentiel.

Si k est un idéal différentiel quelconque de $\mathcal{R}$, et si $\delta \in \Delta$, deux éléments a , b d'une même classe de restes de $\mathcal{R}$ suivant k ont toujours leurs dérivés δa , δb dans une même classe de restes, de sorte que l'on peut définir sur l'anneau de restes $\mathcal{R}/k$ une structure d'anneau différentiel unique telle que l'homomorphisme canonique d'anneaux $g : \mathcal{R} \to \mathcal{R}/k$ est un homomorphisme d'anneaux différentiels. On appelle $\mathcal{R}/k$ l'anneau différentiel de restes de $\mathcal{R}$ suivant k. L'application bijective $\alpha \to g(\alpha)$ de l'ensemble d'idéaux α de $\mathcal{R}$ avec $\alpha \supset k$ sur l'ensemble de tous les idéaux de $\mathcal{R}/k$ (dont l'application réciproque est donnée par $\alpha' \to g^{-1}(\alpha')$) a la propriété :

α est différentiel si et seulement si $g(\alpha)$ l'est.

Pour chaque partie $\Sigma \subset \mathcal{R}$, l'intersection de tous les idéaux différentiels de $\mathcal{R}$ contenant Σ est un idéal différentiel contenant Σ ; il s'appelle l'idéal différentiel de $\mathcal{R}$ engendré par Σ et se note $[\Sigma]$. Evidemment, $[\Sigma]$ est égal à l'idéal $(\Theta\Sigma)$ engendré par l'ensemble des dérivés θs $(\theta \in \Theta\ ,\ s \in \Sigma)$.

Si Σ est une partie non vide multiplicative de $\mathcal{R}$, et si $\delta \in \Delta$, deux éléments a/s , b/t de l'anneau de fractions $\Sigma^{-1}\mathcal{R}$ avec $a/s = b/t$ ont toujours la propriété :

$$(s\ \delta a - a\ \delta s)/s^2 = (t\ \delta b - b\ \delta t)/t^2 \qquad ;$$

on peut donc définir sur $\Sigma^{-1}\mathcal{R}$, de façon évidente, une structure d'anneau différentiel, et alors l'homomorphisme canonique d'anneaux $h : \mathcal{R} \to \Sigma^{-1}\mathcal{R}$ devient un homomorphisme d'anneaux différentiels. On appelle $\Sigma^{-1}\mathcal{R}$ l'anneau différentiel de fractions de $\mathcal{R}$ sur Σ . En appelant Σ- premier chaque idéal $\mathfrak{k}$ de $\mathcal{R}$ tel que $\mathfrak{k}:s = \mathfrak{k}$ pour chaque $s \in \Sigma$, on a une application biunivoque $\mathfrak{k} \to (\Sigma^{-1}\mathcal{R})\ h(\mathfrak{k}) = \Sigma^{-1}\mathfrak{k}$ de l'ensemble des idéaux Σ-premiers de $\mathcal{R}$ sur l'ensemble des idéaux de $\Sigma^{-1}\mathcal{R}$, dont l'application réciproque est donnée par $\mathfrak{a}' \to h^{-1}(\mathfrak{a}')$; $\mathfrak{a}$ est différentiel si et seulement si $\Sigma^{-1}\mathfrak{a}'$ l'est.

2. Polynômes différentiels.

Soit $\mathcal{R}$ un anneau différentiel admettant l'ensemble d'opérateurs de dérivation Δ et l'ensemble d'opérateurs dérivés Θ ; désignons par δ_1 , ... , δ_m les éléments de Δ .

Considérons une famille $(u_i)_{i\in I}$ d'éléments d'un sur-anneau différentiel $\mathcal{S}$ de $\mathcal{R}$. Le sous-anneau de $\mathcal{S}$ engendré par les éléments de $\mathcal{R}$ et tous les dérivés θu_i $(\theta \in \Theta\ ,\ i \in I)$ est un sous-anneau différentiel de $\mathcal{S}$, que l'on note $\mathcal{R}\{(u_i)_{i\in I}\}$. Si la famille $(\theta u_i)_{\theta\in\Theta, i\in I}$ est algébriquement liée sur $\mathcal{R}$, on dit que la famille $(u_i)_{i\in I}$ est différentiellement algébriquement liée sur $\mathcal{R}$; sinon, on dit que $(u_i)_{i\in I}$ est différentiellement algébriquement libre sur $\mathcal{R}$, ou bien que $(u_i)_{i\in I}$ est une famille d'indéterminées différentielles (sur $\mathcal{R}$). On voit sans difficulté que, pour chaque ensemble I d'indices, il existe une famille $(y_i)_{i\in I}$ d'indéterminées différentielles.

Soit $(y_i)_{i\in I}$ une famille d'indéterminées différentielles sur $\mathcal{R}$. Les éléments de $\mathcal{R}\{(y_i)_{i\in I}\}$ s'appellent polynômes différentiels en $(y_i)_{i\in I}$, à coefficients dans $\mathcal{R}$ (ou sur $\mathcal{R}$). Si $(u_i)_{i\in I}$ est une famille quelconque d'éléments d'un sur-anneau différentiel de $\mathcal{R}$, ayant le même ensemble d'indices I , il existe un homomorphisme unique $\mathcal{R}\{(y_i)_{i\in I}\} \to \mathcal{R}\{(u_i)_{i\in I}\}$ laissant fixe chaque élément de $\mathcal{R}$ et envoyant y_i sur u_i pour chaque $i \in I$.

On l'appelle l'homomorphisme de substitution de (u_i) à (y_i) ; si $A \in \mathcal{R}\{(y_i)_{i \in I}\}$, on désigne l'image de A par cet homomorphisme par $A((u_i)_{i \in I})$.

3. Rangements. Réduction.

Avec la même notation que dans le § 2, soit $\mathcal{S} = \mathcal{R}\{y_1 , \ldots , y_n\}$ l'anneau différentiel des polynômes différentiels sur $\mathcal{R}$ en une famille finie $(y_1 , \ldots , y_n)$ d'indéterminées différentielles.

Pour faciliter l'étude systématique de $\mathcal{S}$, il est commode d'introduire un type d'ordre sur l'ensemble des dérivés θy_j $(\theta \in \Theta , \ 1 \leq j \leq n)$. Un ordre sur cet ensemble s'appelle un rangement de $(y_1 , \ldots , y_n)$ si cet ordre est total et si les deux conditions suivantes sont remplies :

$$u < \delta u \quad ,$$

$$u < v \implies \delta u < \delta v \quad ,$$

pour tous les dérivés u et v et chaque $\delta \in \Delta$. On voit sans peine qu'un rangement est toujours un bon ordre, c'est-à-dire un ordre par rapport auquel l'ensemble des dérivés θy_j est bien ordonné. Si $u < v$ on dit alors que u est de rang inférieur à v , etc. Un rangement s'appelle séquentiel si son ensemble ordonné est isomorphe à l'ensemble ordonné $\mathbb{N}$ des entiers naturels, c'est-à-dire si, pour chaque dérivé v , le nombre de dérivés u avec $u < v$ est fini. Par exemple, on obtient un rangement séquentiel en ordonnant l'ensemble des dérivés $\delta_1^{i_1} , \ldots , \delta_m^{i_m} y_j$ lexiccographiquement par rapport à $(\Sigma i_\mu , i_1 , \ldots , i_m , j)$.

Soit donné un rangement de $(y_1 , \ldots , y_n)$. Si $A \in \mathcal{S}$ et $A \notin \mathcal{R}$, on appelle leader de A et on note u_A le dérivé du plus haut rang figurant dans A . On peut alors écrire

$$A = A_0 + A_1 u_A + \ldots + A_d u_A^d \quad , \qquad A_d \neq 0 \quad ,$$

où les A_i sont dans $\mathcal{S}$ et ne contiennent que des dérivés de rang inférieur à u_A ; on appelle initial de A et on note I_A le polynôme différentiel A_d . Le polynôme différentiel $\partial A/\partial u_A = A_1 + \ldots + d.A_d u^{d-1}$ s'appelle le séparant de A et se note S_A .

On étend la notion de rang comparatif à l'ensemble $\mathcal{S}$ tout entier en convenant que $A \leq B$ si :

a. $A = 0$; ou

b. $A \in \mathcal{R}$, $A \neq 0$, $B \neq 0$; ou

c. $A \notin \mathcal{R}$, $B \notin \mathcal{R}$, et ou bien $u_A < u_B$, ou bien $u_A = u_B$ et $\deg_{u_A} A \leq \deg_{u_B} B$.

On n'obtient pas ainsi une relation d'ordre sur $\mathcal{S}$, mais seulement une relation de pré-ordre, puisqu'on peut avoir $A \leq B$ et $B \leq A$ sans $A = B$. Mais ce pré-ordre est bon.

Si $A \in \mathcal{S}$, $A \notin \mathcal{R}$, alors $\delta A - S_A\, \delta u_A < \delta u_A$ pour chaque $\delta \in \Delta$. On voit par récurrence que, pour chaque $\theta \in \Theta$ avec $\operatorname{ord}\theta > 0$, il existe un $T_\theta \in \mathcal{S}$ avec $T_\theta < \theta u_A$ tel que $\theta A = S_A\, \theta u_A - T_\theta$. Ce fait nous permettra de démontrer un lemme pour les polynômes différentiels un peu analogue (bien que beaucoup plus compliqué) au théorème de division euclidienne.

Si $A \in \mathcal{S}$, $A \notin \mathcal{R}$, un élément $B \in \mathcal{S}$ s'appelle réduit par rapport à A si B ne contient aucun dérivé de u_A d'ordre > 0 et, de plus, $\deg_{u_A} B < \deg_{u_A} A$.

Un ensemble $\mathcal{A} \subset \mathcal{S}$ s'appelle autoréduit si $\mathcal{A}$ ne contient aucun élément de $\mathcal{R}$ et chaque élément de $\mathcal{A}$ est réduit par rapport à chaque autre élément de $\mathcal{A}$. Deux éléments distincts d'un ensemble autoréduit ont leurs leaders distincts. On en conclut que chaque ensemble autoréduit est fini. On dit que B est réduit par rapport à $\mathcal{A}$, si B est réduit par rapport à chaque élément de $\mathcal{A}$.

LEMME de réduction. - Soit $\mathcal{A}$ un ensemble autoréduit dans $\mathcal{S}$, et soit $B_i \in \mathcal{S}$ $(1 \leq i \leq r)$. Il existe des entiers naturels $s(A)$, $t(A)$ $(A \in \mathcal{A})$ et des éléments $B_i^* \in \mathcal{S}$ $(1 \leq i \leq r)$ tels que

$$\left.\begin{array}{l} B_i^* \text{ est réduit par rapport à } \mathcal{A}, \\ \prod_{A\in\mathcal{A}} S_A^{s(A)} I_A^{t(A)} . B_i \equiv B_i^* \pmod{[\mathcal{A}]} \end{array}\right\} \quad (1 \leq i \leq r) .$$

Si chaque B_i est réduit par rapport à $\mathcal{A}$, le resultat est évident ; sinon, on procède (en utilisant le fait signalé ci-dessus) par récurrence sur le plus grand rang des dérivés v tels que

ou bien v est un dérivé d'ordre > 0 d'un leader u_A avec $A \in \mathcal{A}$ et v se trouve dans un B_i au moins,

ou bien $v = u_A$ pour un $A \in \mathcal{A}$ et $\deg_v B_i \geq \deg_{u_A} A$ pour un i au moins.

Remarquons que l'on peut définir la notion de rang comparatif sur l'ensemble des parties autoréduites de $\mathcal{S}$. A savoir, si $\mathcal{A}$ et $\mathcal{B}$ sont deux ensembles autoréduits dont les éléments (dans l'ordre des rangs croissants) sont $A_1, \ldots, A_r$ et $B_1, \ldots, B_s$, respectivement ; on pose $\mathcal{A} \leq \mathcal{B}$ si ou bien, il existe un entier k avec $1 \leq k \leq r$ et $1 \leq k \leq s$ tel que A_i et B_i soient du même rang $(1 \leq i < k)$ et A_k soit de rang inférieur à B_k,

ou bien $r \geq s$ et A_i et B_i sont du même rang $(1 \leq i \leq s)$.

On définit ainsi une relation de bon pré-ordre sur l'ensemble de toutes les parties autoréduites de $\mathfrak{S}$.

4. Extensions séparables et extensions quasi-séparables.

Soit L une extension d'un corps K de caractéristique p. Rappelons que L est dite séparable sur K si :

ou bien $p = 0$,

ou bien $p \neq 0$ et les corps L^p et K sont linéairement disjoints sur K^p.

Nous allons introduire, pour une extension de corps, une propriété moins forte que la propriété d'être séparable.

Appelons $x = (x_i)_{i \in I}$ une famille d'éléments de L séparablement liée sur K s'il existe un polynôme $f \in K[X] = K[(X_i)_{i \in I}]$ tel que $f(x) = 0$ et $\frac{\partial f}{\partial X_i}(x) \neq 0$ pour un indice i au moins, et séparablement libre sur K dans le cas contraire. Dire que x est séparablement liée sur K équivaut à dire qu'une coordonnée x_i de x est séparablement algébrique sur l'extension du corps K engendrée par les autres coordonnées.

D'autre part, si J est une partie de I telle que la sous-famille $(x_i)_{i \in J}$ soit une base de transcendance de l'extension $K(x)$ de K, et si $I - J$ est fini, alors $\operatorname{card}(I - J)$ est independant du choix de J ; nous disons alors que la famille x est de codimension algébrique finie sur K et que l'entier naturel $\operatorname{card}(I - J)$ est sa codimension algébrique sur K.

Cela dit, on a le critère suivant de séparabilité d'une extension L du corps K :

pour que L soit séparable sur K, il faut et il suffit que chaque famille d'éléments de L qui est séparablement libre sur K soit de codimension algébrique 0 sur K (i. e. soit algébriquement libre sur K).

Nous dirons que L est quasi-séparable sur K si chaque famille d'éléments de L qui est séparablement libre sur K est de codimension algébrique finie sur K.

Evidemment, si L est séparable sur K, L est quasi-séparable. On peut démontrer que, si L est de type fini (en qualité d'extension du corps K), alors L est quasi-séparable sur K.

Ces notions s'appliquent, naturellement, aux corps différentiels.

Soit $\mathfrak{F}$ un corps différentiel de caractéristique p ; désignons par $\mathcal{C}$ le corps de constantes de $\mathfrak{F}$. Nous dirons que $\mathfrak{F}$ est différentiellement parfait (resp. différentiellement quasi-parfait) si chaque sur-corps différentiel de $\mathfrak{F}$ est séparable (resp. quasi-séparable) sur $\mathfrak{F}$.

Si $p = 0$, évidemment $\mathfrak{F}$ est différentiellement parfait, donc différentiellement quasi-parfait.

Si $p \neq 0$, on voit aisément que $\mathfrak{F}$ est différentiellement parfait si et seulement si $\mathcal{C} = \mathfrak{F}^p$.

Plus difficile à démontrer, mais également vrai, est le critère de quasi-perfection différentielle. $\mathfrak{F}$ (de caractéristique $p \neq 0$) est différentiellement quasi-parfait si et seulement si le degré $[\mathcal{C} : \mathfrak{F}^p]$ est fini. Nous omettrons la démonstration.

5. Idéaux séparables et idéaux quasi-séparables.

Soit R un anneau commutatif et soit R_0 un sous-anneau de R .

Si p est un idéal premier de R et si $f : R \to R/p$ désigne l'homomorphisme canonique, le corps de fractions de $f(R)$ est une extension du corps de fractions de $f(R_0)$; nous dirons que p est séparable (resp. quasi-séparable) sur R_0 si cette extension est séparable (resp. quasi-séparable).

Cette notion de séparabilité peut être généralisée. Remarquons que p est séparable sur R_0 si et seulement si

ou bien la caractéristique p de $f(R_0)$ est 0 ,

ou bien $p \neq 0$, et $f(R)^p$ et $f(R_0)$ sont linéairement disjoints sur $f(R_0)^p$.

Considérons un idéal $\mathfrak{k}$ quelconque de R , et notons encore $f : R \to R/\mathfrak{k}$ l'homomorphisme canonique. Nous dirons que $\mathfrak{k}$ est séparable sur R_0 si $\mathfrak{k} = R$, ou si $\mathfrak{k} \neq R$ et les deux conditions suivantes sont remplies.

(i) $a \in R_0$, $a \notin \mathfrak{k}$, $b \in R$, $b \notin \mathfrak{k}$ $\Longrightarrow$ $ab \notin \mathfrak{k}$ (de sorte que $\mathfrak{k} \cap R_0$ est un idéal premier de R_0 et $f(R_0)$ est intègre).

(ii) Ou bien la caractéristique p de $f(R_0)$ est 0 ou bien $p \neq 0$ et $f(R)^p$ et $f(R_0)$ sont linéairement disjoints sur $f(R_0)^p$.

Il est évident que, si $g : R \to R'$ est un épimorphisme d'anneaux avec $\ker g \subset \mathfrak{k}$, alors $\mathfrak{k}$ est un idéal de R séparable sur R_0 (resp. un idéal premier de R quasi-séparable sur R_0) si et seulement si $g(\mathfrak{k})$ est un idéal de R' séparable sur $g(R_0)$ (resp. un idéal premier de R' quasi-séparable sur $g(R_0)$).

6. Systèmes conservatifs.

Soit R un anneau.

Considérons un ensemble $\mathfrak{C}$ d'idéaux de R tel que les trois conditions suivantes soient satisfaites.

(i) L'intersection d'un ensemble quelconque d'éléments de $\mathfrak{C}$ est elle-même un élément de $\mathfrak{C}$.

(ii) La réunion d'un ensemble non vide d'éléments de $\mathfrak{C}$ qui est totalement ordonné (par inclusion) est toujours un élément de $\mathfrak{C}$.

(iii) Si $c \in \mathfrak{C}$ et $s \in R$, alors $c:s^\infty \in \mathfrak{C}$ (où $c:s^\infty = \bigcup (c : s^n)$).

Nous appellerons un tel ensemble $\mathfrak{C}$ un système conservatif de R.

Exemples.

1° L'ensemble de tous les idéaux de R.

2° L'ensemble réduit au seul élément R.

3° L'ensemble formé par R et tous les idéaux de R séparables sur R_0 et ayant avec R_0 une intersection donnée.

4° L'ensemble de tous les idéaux parfaits de R. Un idéal $\mathfrak{a}$ de R s'appelle parfait si

$$(x \in R \text{ et } x^2 \in \mathfrak{a}) \Longrightarrow (x \in \mathfrak{a}) \quad .$$

5° L'ensemble de tous les idéaux différentiels d'un anneau différentiel.

6° L'intersection d'un ensemble non vide quelconque de systèmes conservatifs de R.

Un système conservatif dont tous les éléments sont parfaits (resp. séparables sur R_0, resp. différentiels) s'appellera parfait (resp. séparable sur R_0, resp. différentiel).

Soit $F : \mathfrak{C} \to \mathfrak{C}'$ une application de $\mathfrak{C}$ dans un système conservatif d'un anneau R', et supposons que $F(\bigcap_{c \in M} c) = \bigcap_{c \in M} F(c)$ pour tous les ensembles $M \subset \mathfrak{C}$. Alors F est une application croissante :

$$a \subset b \Longrightarrow F(a) \subset F(b) \quad .$$

Si, de plus,

$$F(\bigcup_{c \in T} c) = \bigcup_{c \in T} F(c)$$

pour chaque ensemble $T \subset \mathfrak{C}$ totalement ordonné, et si, pour chaque $c \in \mathfrak{C}$ et chaque $s' \in R'$, il existe un $c_0 \in \mathfrak{C}$ tel que $F(c) : s'^{\infty} = F(c_0)$, alors on dit que F est une application conservative, ou un homomorphisme, de $\mathfrak{C}$ dans $\mathfrak{C}'$. Si F est un homomorphisme bijectif, alors F^{-1} est un homomorphisme de $\mathfrak{C}'$ dans $\mathfrak{C}$; on dit alors que F est un isomorphisme. On peut voir, dans le cas général, que l'image $F(\mathfrak{C})$ est un système conservatif.

Exemples.

1° Soit R_0 un sous-anneau de R. L'application $c \to c \cap R_0$ $(c \in \mathfrak{C})$ est conservative. Désignons l'image (système conservatif de R_0) par $\mathfrak{C}|R_0$.

2° Soit $f : R \to R'$ un épimorphisme. L'ensemble de tous les $c \in \mathfrak{C}$ avec $c \supset \ker f$ est un système conservatif de R, et $c \to f(c)$ en est un isomorphisme sur un système conservatif de R', que nous notons $f(\mathfrak{C})$. Quand f est l'homomorphisme canonique $R \to R/\mathfrak{t}$ nous désignons $f(\mathfrak{C})$ par $\mathfrak{C}/\mathfrak{t}$.

3° Soit Σ une partie non vide multiplicative de R. L'ensemble de tous les $c \in \mathfrak{C}$ qui sont Σ-premiers est un système conservatif de R, et $c \to \Sigma^{-1} c$ en est un isomorphisme sur un système conservatif de $\Sigma^{-1} R$ que nous notons $\Sigma^{-1} \mathfrak{C}$.

Une application conservative $F : \mathfrak{C} \to \mathfrak{C}'$ s'appelle parfaite si $F(c)$ est parfait chaque fois que c l'est. Les applications conservatives des exemples ci-dessus sont parfaites, et (dans les deux derniers exemples) leurs inverses également.

Soit $\mathfrak{C}$ un système conservatif de R. Si Σ est une partie de R, on note $(\Sigma)_{\mathfrak{C}}$ l'intersection de tous les éléments de $\mathfrak{C}$ contenant Σ, c'est-à-dire le plus petit élément de $\mathfrak{C}$ qui contient Σ. Si $c \in \mathfrak{C}$, $c = (\Sigma)_{\mathfrak{C}}$, et Σ est finie, on dit que Σ est une $\mathfrak{C}$-base de c.

Pour un Σ quelconque, on appelle $\mathfrak{C}$-composant de Σ chaque élément minimal de l'ensemble des éléments de $\mathfrak{C}$ qui sont premiers et qui contiennent Σ. Chaque élément premier de $\mathfrak{C}$ qui contient Σ contient un $\mathfrak{C}$-composant de Σ.

LEMME 1. - *Si* $a \in (\Sigma)_{\mathfrak{C}}$, *il existe alors une partie finie* Φ *de* Σ *avec* $a \in (\Phi)_{\mathfrak{C}}$.

Démonstration par récurrence sur card(Σ) .

LEMME 2. - Soit $\mathfrak{C}$ un système conservatif parfait de R ; soient Σ et T des parties de R . Alors

$$(\Sigma T)_{\mathfrak{C}} = (\Sigma)_{\mathfrak{C}} \cap (T)_{\mathfrak{C}} \quad .$$

$$(\Sigma T)_{\mathfrak{C}} : \Sigma = \bigcap_{s\in\Sigma} (\Sigma T)_{\mathfrak{C}} : s = \bigcap_{s\in\Sigma} (\Sigma T)_{\mathfrak{C}} : s^{\infty}$$

est un élément de $\mathfrak{C}$ contenant T , donc contenant $(T)_{\mathfrak{C}}$; il en résulte que $(\Sigma T)_{\mathfrak{C}} : (T)_{\mathfrak{C}}$ est un élément de $\mathfrak{C}$ contenant Σ , donc contenant $(\Sigma)_{\mathfrak{C}}$; ainsi $(\Sigma)_{\mathfrak{C}} (T)_{\mathfrak{C}} \subset (\Sigma T)_{\mathfrak{C}}$. Puisque l'idéal $(\Sigma T)_{\mathfrak{C}}$ est parfait, le résultat en découle.

COROLLAIRE. - Si $\mathfrak{C}$ est un système conservatif parfait, alors chaque élément de $\mathfrak{C}$ est l'intersection de ses $\mathfrak{C}$-composants.

Soit $c \in \mathfrak{C}$. Si $x \notin c$, il existe un élément de $\mathfrak{C}$ contenant c mais ne contenant pas x , donc il existe un tel élément maximal m_x ; en utilisant le lemme 2, on voit que m_x est premier, donc que m_x contient un $\mathfrak{C}$-composant p_x de c . Evidemment $\bigcap_{x\notin c} p_x = p$.

7. Systèmes conservatifs rittiens.

Pour un système conservatif $\mathfrak{C}$ les deux conditions suivantes sont équivalentes.

(i) Chaque élément possède une $\mathfrak{C}$-base.

(ii) Chaque partie non vide de $\mathfrak{C}$ possède un élément maximal. Si $\mathfrak{C}$ satisfait ces conditions, et si, de plus, $\mathfrak{C}$ est parfait, on dit que $\mathfrak{C}$ est rittien.

Pour les systèmes conservatifs rittiens, le corollaire ci-dessus peut être beaucoup précisé.

THÉORÈME 1. - Si $\mathfrak{C}$ est un système conservatif rittien, chaque $c \in \mathfrak{C}$ est l'intersection d'un ensemble fini d'idéaux premiers dans $\mathfrak{C}$ dont aucun ne contient un autre ; cet ensemble fini est unique, étant l'ensemble de tous les composants de c .

L'existence se démontre par récurrence sur c (en munissant $\mathfrak{C}$ de l'ordre opposé à l'inclusion). L'unicité ne présente aucune difficulté.

PROPOSITION 1. - Soit $\mathfrak{C}$ un système conservatif rittien.

a. Chaque système conservatif contenu dans $\mathfrak{C}$ est rittien.

b. Chaque image d'un homomorphisme parfait de $\mathfrak{C}$ est rittien.

(a) est évident. Quant à (b), remarquons que, si (c'_i) ... est une suite strictement croissante dans $F(\mathfrak{C})$, et si l'on note c_i l'intersection de tous les $c \in \mathfrak{C}$ avec $F(c) = c'_i$, alors (c_i) est une suite strictement croissante dans $\mathfrak{C}$.

LEMME 3. - Si le système conservatif parfait $\mathfrak{C}$ de R n'est pas rittien, l'ensemble des éléments de $\mathfrak{C}$ qui ne possèdent pas une $\mathfrak{C}$-base a un élément maximal. Un tel élément maximal est toujours premier.

L'existence d'un élément maximal résulte du lemme de Zorn ; que cet élément soit premier est une conséquence du lemme 2.

PROPOSITION 2. - Soient R_0 et R_1 deux sous-anneaux de R, R_1 étant un sur-anneau de R_0 de type fini, et soit $\mathfrak{C}$ un système conservatif parfait de R. Si $\mathfrak{C}|R_0$ est rittien alors $\mathfrak{C}|R_1$ est rittien.

On peut supposer que $R_1 = R_0[v]$, v étant un élément de R. Supposons la proposition fausse. Il existe alors (lemme 3) dans l'ensemble des éléments de $\mathfrak{C}|R_1$ sans $\mathfrak{C}|R_1$-base un élément maximal $\mathfrak{m}$, et $\mathfrak{m}$ est premier. $\mathfrak{C}|R_0$ étant rittien, $\mathfrak{m} \cap R_0$ a une $\mathfrak{C}|R_0$-base Ψ_1 ; évidemment

$$(\mathfrak{m} \cap R_0)_{\mathfrak{C}|R_1} = (\Psi_1)_{\mathfrak{C}|R_1} ,$$

d'où

$$\mathfrak{m} \neq (\mathfrak{m} \cap R_0)_{\mathfrak{C}|R_1} ,$$

et il existe donc un polynôme $f = a_0 X^n + \ldots + a_n \in R_0[X]$ avec $f(v) \in \mathfrak{m}$ et $f(v) \notin (\mathfrak{m} \cap R_0)_{\mathfrak{C}|R_1}$. En choisissant n aussi petit que possible, on a $n > 0$ et $a_0 \notin \mathfrak{m}$. A cause de la maximalité de $\mathfrak{m}$, $(a_0, \mathfrak{m})_{\mathfrak{C}|R_1}$ a une $\mathfrak{C}|R_1$-base et (lemme 1) même une $\mathfrak{C}|R_1$-base de la forme $\{a_0\} \cup \Psi_2$ avec $\Psi_2 \subset \mathfrak{m}$.

Or, pour chaque $w \in \mathfrak{m}$, on peut écrire $w = g(v)$, où $g \in R_0[X]$; en divisant g par f, on a $a_0^k g = qf + r$ avec $\deg r < n$, donc

$$r(v) \in (\mathfrak{m} \cap R_0)_{\mathfrak{C}|R_1} = (\Psi_1)_{\mathfrak{C}|R_1} ,$$

d'où

$$a_0^k w \in (f(v), \Psi_1)_{\mathfrak{C}|R_1} .$$

Ainsi,

$$a_0 \mathfrak{m} \subset (f(v), \Psi_1)_{\mathfrak{C}|R_1} ,$$

d'où (voir le lemme 2)

$$\mathfrak{m} = \mathfrak{m} \cap (a_0 , \mathfrak{m})_{\mathfrak{C}|R_1} = \mathfrak{m} \cap (a_0 , \Psi_2)_{\mathfrak{C}|R_1} = (a_0 \mathfrak{m} , \Psi_2)_{\mathfrak{C}|R_1} = (f(v) , \Psi_1 , \Psi_2)_{\mathfrak{C}|R_1} ,$$

et ceci contredit le fait que $\mathfrak{m}$ n'a pas une $\mathfrak{C}|R_1$-base.

8. Un lemme.

Soit $\mathcal{S} = \mathcal{R}\{y_1 , \ldots , y_n\}$ l'anneau différentiel des polynômes différentiels en une famille finie d'indéterminées différentielles sur l'anneau différentiel $\mathcal{R}$.

LEMME 4. - Soit $\mathfrak{p}$ un idéal différentiel premier de $\mathcal{S}$, quasi-séparable sur $\mathcal{R}$; soit donné un rangement séquentiel de $(y_1 , \ldots , y_n)$; soit $\mathfrak{A}$ une partie auto-réduite de $\mathfrak{p}$ telle que $S_\Lambda \notin \mathfrak{p}$ pour chaque $\Lambda \in \mathfrak{A}$, de rang minimal. Notons V l'ensemble des dérivés des y_i qui ne sont dérivés d'ordre > 0 d'aucun leader u_Λ avec $\Lambda \in \mathfrak{A}$. Il existe alors une partie finie Y de V telle que chaque élément de $\mathfrak{p}$ qui est réduit par rapport à $\mathfrak{A}$ est dans l'idéal $(\mathfrak{p} \cap \mathcal{R}[Y])$ de $\mathcal{S}$.

Soit W l'ensemble des éléments $w \in V$ tels que seulement un nombre fini des dérivés de w soient dans V . On voit que, si $v \in V - W$, alors il existe un opérateur de dérivation δ tel que $\delta v \in V - W$. On peut démontrer que l'ensemble W est fini.

Remarquons ensuite que, si un élément $P \in \mathfrak{p}$ est réduit par rapport à $\mathfrak{A}$ et $P \notin \mathcal{R}$, alors $S_P \in \mathfrak{p}$; c'est une conséquence de la minimalité du rang de $\mathfrak{A}$. Or, si $P \in \mathfrak{p} \cap \mathcal{R}[V - W]$, alors P est réduit par rapport à $\mathfrak{A}$. En utilisant le fait que le rangement est séquentiel, et le lemme de réduction (§ 3), on peut démontrer alors que, pour chaque $P \in \mathfrak{p} \cap \mathcal{R}[V - W]$,

$$\partial P/\partial v \in \mathfrak{p} \qquad (v \in V - W) \qquad .$$

Notons $f : \mathcal{S} \to \mathcal{S}/\mathfrak{p}$ l'homomorphisme canonique. Ce que nous venons de dire montre que, si T est une partie de $V - W$, alors la famille $(f(v))_{v \in T}$ est séparablement libre sur $f(\mathcal{R})$ (c'est-à-dire, sur le corps de fractions de $f(\mathcal{R})$).

Supposons le lemme faux. Notons V_i l'ensemble des éléments de V d'ordre $< i$, et posons $q_i = \mathrm{card}(V_i)$; on voit alors que, pour chaque $i \in \mathbb{N}$, il existe un entier $i' > i$ tel que $\mathfrak{p} \cap \mathcal{R}[V_{i'}]$ contienne un élément qui n'est pas dans $(\mathfrak{p} \cap \mathcal{R}[V_i])$. Cela entraîne que le degré de transcendance de $f(\mathcal{R}[V_{i'}])$ sur $f(\mathcal{R}[V_i])$ est $< q_{i'} - q_i$. En posant

$$i^{(0)} = i \quad , \quad i^{(\nu+1)} = i^{(\nu)'} \quad (\nu \in \underset{\sim}{N}) \qquad ,$$

on voit que le degré de transcendance de $f(\mathcal{R}[V_i(h)])$ sur $f(\mathcal{R})$ est

$$\leqslant q_i + \sum_{0\leqslant\nu<h} (q_{i^{(\nu+1)}} - q_{i^{(\nu)}} - 1) = q_{i^{(h)}} - h \qquad .$$

Donc, pour chaque i , il existe un $i^* > i$ tel que le degré de transcendance de $f(\mathcal{R}[V_{i^*}])$ sur $f(\mathcal{R})$ soit $< q_{i^*} - q_i$.

Choisissons $i(0) \in \underset{\sim}{N}$ assez grand pour que $W \subset V_{i(0)}$, et posons

$$i(\nu + 1) = i(\nu)^* \quad (\nu \in \underset{\sim}{N}) \qquad .$$

Le degré de transcendance de $f(\mathcal{R}[V_{i(\nu+1)}])$ sur $f(\mathcal{R})$ est $< q_{i(\nu+1)} - q_{i(\nu)}$, donc la famille $(f(v))_{v\in V_{i(\nu+1)} - V_{i(\nu)}}$ est algébriquement liée sur $f(\mathcal{R})$.
Cela étant ainsi pour chaque ν , on voit, en posant

$$V' = \bigcup_{\nu\in \underset{\sim}{N}} (V_{i(\nu+1)} - V_{i(\nu)}) = V - V_{i(0)} \subset V - W \qquad ,$$

que la famille $(f(v))_{v\in V'}$ est de codimension algébrique infinie sur $f(\mathcal{R})$. Mais cette famille est séparablement libre sur $f(\mathcal{R})$. Donc p n'est pas quasi-séparable sur $\mathcal{R}$.

9. Le théorème de la base.

THÉORÈME 2. - Soient $\mathcal{R}$ un anneau différentiel, $\mathcal{S}$ un sur-anneau différentiel de $\mathcal{R}$ de type fini, et $\mathfrak{C}$ un système conservatif parfait différentiel de $\mathcal{S}$. Si $\mathfrak{C}|\mathcal{R}$ est rittien et si chaque élément premier de $\mathfrak{C}$ est quasi-séparable sur $\mathcal{R}$, alors $\mathfrak{C}$ est rittien.

Il existe un $\mathcal{R}$-épimorphisme d'un anneau différentiel de polynômes différentiels $\mathcal{R}\{y_1 , \dots , y_n\}$ sur $\mathcal{S}$; on peut donc supposer que même $\mathcal{S} = \mathcal{R}\{y_1 , \dots , y_n\}$. Supposons le théorème faux. Alors (lemme 3) il existe dans l'ensemble des éléments de $\mathfrak{C}$ sans $\mathfrak{C}$-base, un élément maximal p , et p est premier. En adoptant la notation du lemme 4 de la proposition 2, on voit qu'il existe une partie finie Ψ de $p \cap \mathcal{R}[Y]$ telle que

$$p \cap \mathcal{R}[Y] = (\Psi)_{\mathfrak{C}|\mathcal{R}[Y]} \qquad ,$$

d'où

$$(p \cap \mathcal{R}[Y]) \subset (\Psi)_{\mathfrak{C}} \qquad .$$

Si $B \in \mathfrak{p}$, alors (lemme de réduction) on peut écrire

$$\prod_{\Lambda \in \mathfrak{A}} S_\Lambda^{s(\Lambda)} I_\Lambda^{t(\Lambda)} . B \equiv B^* \pmod{[\mathfrak{A}]} \quad ,$$

où B^* est réduit par rapport à $\mathfrak{A}$. En posant $H = \prod_{\Lambda \in \mathfrak{A}} S_\Lambda I_\Lambda$, on en conclut que

$$H\mathfrak{p} \subset (\mathfrak{A} , \Psi)_{\mathfrak{C}} \quad .$$

D'autre part, on voit sans peine que $I_\Lambda \notin \mathfrak{p}$ $(\Lambda \in \mathfrak{A})$, de sorte que $H \notin \mathfrak{p}$; donc $(H , \mathfrak{p})_{\mathfrak{C}}$ a une $\mathfrak{C}$-base, et même (lemme 1) une $\mathfrak{C}$-base de la forme $\{H\} \cup \Phi$, où $\Phi \subset \mathfrak{p}$. Donc, grâce au lemme 2,

$$\mathfrak{p} = \mathfrak{p} \cap (H , \mathfrak{p})_{\mathfrak{C}} = \mathfrak{p} \cap (H , \Phi)_{\mathfrak{C}} = (H\mathfrak{p} , \Phi)_{\mathfrak{C}} = (\mathfrak{A} , \Psi , \Phi)_{\mathfrak{C}} \quad ,$$

de sorte que $\mathfrak{A} \cup \Psi \cup \Phi$ est une $\mathfrak{C}$-base de $\mathfrak{p}$.

Un anneau différentiel, dans lequel l'ensemble de tous les idéaux différentiels parfaits est un système conservatif rittien, s'appelle <u>rittien</u>.

COROLLAIRE 1. - <u>L'anneau de polynômes différentiels $\mathcal{R}\{y_1 , \dots , y_n\}$ est rittien si et seulement si $\mathcal{R}$ est rittien et, pour chaque idéal différentiel premier $\mathfrak{p}_0$ de $\mathcal{R}$, le corps différentiel de fractions de $\mathcal{R}/\mathfrak{p}_0$ est différentiellement quasi-parfait.</u>

Le seul point qui n'est pas conséquence du théorème est le fait que si $\mathcal{R}\{y_1 , \dots , y_n\}$ est rittien et si $\mathfrak{p}_0$ est un idéal différentiel premier de $\mathcal{R}$, alors le corps différentiel de fractions de $\mathcal{R}/\mathfrak{p}_0$ (corps que nous notons $\mathfrak{F}$) est différentiellement quasi-parfait. Or, on voit aisément que si $\mathcal{R}\{y_1 , \dots , y_n\}$ est rittien, alors $(\mathcal{R}/\mathfrak{p}_0)\{y_1 , \dots , y_n\}$ est rittien, donc $\mathfrak{F}\{y_1 , \dots , y_n\}$ aussi. Si $\mathfrak{F}$ n'était pas différentiellement quasi-parfait, il y aurait (voir critère de quasi-perfection différentielle) une suite infinie (γ_i) de constantes dans $\mathfrak{F}$ telle que $\gamma_i \notin \mathfrak{F}^p(\gamma_0 , \gamma_1 , \dots , \gamma_{i-1})$ pour chaque i ; les idéaux $\mathfrak{p}_i = (y_1^p - \gamma_0 , (\delta y_1)^p - \gamma_1 , \dots , (\delta^i y_1)^p - \gamma_i)$ de $\mathfrak{F}\{y_1 , \dots , y_n\}$ seraient différentiels et premiers (a fortiori parfaits), et formeraient une suite infinie strictement croissante ; donc $\mathfrak{F}\{y_1 , \dots , y_n\}$ ne serait pas rittien.

COROLLAIRE 2. - <u>Si $\mathfrak{F}$ est un corps différentiel, l'anneau de polynômes différentiels $\mathfrak{F}\{y_1 , \dots , y_n\}$ est rittien si et seulement si $\mathfrak{F}$ est différentiellement quasi-parfait.</u>

COROLLAIRE 3. - <u>Si $\mathfrak{F}$ est un corps différentiel, le système conservatif de tous les idéaux différentiels parfaits de $\mathfrak{F}\{y_1 , \dots , y_n\}$ qui sont séparables sur</u> $\mathfrak{F}$

est rittien.

Ce dernier corollaire est le théorème de Seidenberg.

BIBLIOGRAPHIE

[1] KOLCHIN (Ellis R.). - On the basis theorem for differential systems, Trans. Amer. math. Soc., t. 52, 1942, p. 115-127.

[2] RITT (Joseph Fels). - Differential algebra. - New York, American mathematical Society, 1950 (Amer. math. Soc. Coll. Publ., 33).

[3] SEIDENBERG (A.). - Some basis theorems in differential algebra (caracteristic b , arbitrary), Trans. Amer. math. Soc., t. 73, 1952, p. 174-190.

THE NOTION OF DIMENSION IN THE THEORY OF ALGEBRAIC DIFFERENTIAL EQUATIONS[1]

BY E. R. KOLCHIN

Communicated by L. Bers, March 18, 1964

Consider a system of algebraic differential equations

$$P(y_1, \cdots, y_n) = 0 \qquad (P \in \Sigma)$$

with coefficients in a differential field $\mathfrak{F}$ (ordinary or partial); here Σ is any subset of the differential polynomial algebra $\mathfrak{A} = \mathfrak{F}\{y_1, \cdots, y_n\}$ over $\mathfrak{F}$. Denote the set of all solutions of this system by $\mathfrak{Z}(\Sigma)$. We seek a measure of the size of $\mathfrak{Z}(\Sigma)$. The analogous question for systems of algebraic equations (i.e. for affine algebraic geometry) has a satisfactory answer in the notion of dimension.

In the classical literature, where $\mathfrak{F}$ consists of meromorphic functions on some region of complex m-space, the solution is said to depend on a certain number d of arbitrary functions of m variables; if $d=0$ then the solution is said to depend on a certain number of arbitrary functions of $m-1$ variables; and so on. Of course, except in certain special cases, what this means (how these numbers are defined) is not made precise, and general results are therefore wanting.

The Ritt theory (see [1]) contains the beginning of a general answer to the question (when $\mathfrak{F}$ is of characteristic 0). First Σ is replaced by the perfect differential ideal $\mathfrak{a}$ generated by Σ; this is harmless since $\mathfrak{Z}(\Sigma) = \mathfrak{Z}(\mathfrak{a})$. Then $\mathfrak{a}$ is expressed as the intersection of its components, $\mathfrak{a} = \mathfrak{p}_1 \cap \cdots \cap \mathfrak{p}_r$; since $\mathfrak{Z}(\mathfrak{a}) = \mathfrak{Z}(\mathfrak{p}_1) \cup \cdots \cup \mathfrak{Z}(\mathfrak{p}_r)$, the question is reduced to the case in which Σ is a prime differential ideal $\mathfrak{p}$ of $\mathfrak{A}$. Finally, one takes a generic zero $\eta = (\eta_1, \cdots, \eta_n)$ of $\mathfrak{p}$, and computes the differential transcendence degree $d(\mathfrak{p})$ of the differential field extension $\mathfrak{F}\langle\eta\rangle$ of $\mathfrak{F}$; $d(\mathfrak{p})$ is called the *differential dimension* of $\mathfrak{p}$, or of $\mathfrak{Z}(\mathfrak{p})$, and is the "correct" definition for what is classically called the number of arbitrary functions of m variables in the solution of the system $P=0$ $(P \in \mathfrak{p})$. Moreover, if $\mathfrak{p}'$ is another prime differential ideal of $\mathfrak{A}$ subject to the inclusion $\mathfrak{p} \subset \mathfrak{p}'$ (or, equivalently, to the inclusion $\mathfrak{Z}(\mathfrak{p}) \supset \mathfrak{Z}(\mathfrak{p}')$) then $d(\mathfrak{p}) \geqq d(\mathfrak{p}')$; however, when the inclusions are strict the inequality need not be so. This shows that $d(\mathfrak{p})$ is not a sufficiently fine measure of the size of $\mathfrak{Z}(\mathfrak{p})$.

In what follows we present another measure, which is sufficiently fine, and describe its relation to $d(\mathfrak{p})$ and some of its other properties; it is vaguely reminiscent of Hilbert's "characteristic function" for

[1] This research was supported by the National Science Foundation.

homogeneous polynomial ideals.[2] For the sake of simplicity we continue to suppose that the differential field $\mathfrak{F}$ is of characteristic 0; we denote the derivation operators of $\mathfrak{F}$ by $\delta_1, \cdots, \delta_m$. We omit the proofs. All details, as well as generalization to nonzero characteristic, will appear in a book now in preparation.

We recall that a polynomial $f \in \mathbf{R}[X]$ in one indeterminate is said to be *numerical* if $f(s) \in \mathbf{Z}$ for all sufficiently big $s \in \mathbf{N}$. Any f can be written in the form

$$f = \sum_k a_k \binom{X+k}{k},$$

where $a_k \in \mathbf{R}$ and

$$\binom{X+k}{k} = (X+1)(X+2)\cdots(X+k)/k!;$$

f is numerical if and only if $a_k \in \mathbf{Z}$ for every k.[3] We define $f \leqq g$ to mean that $f(s) \leqq g(s)$ for all sufficiently big $s \in \mathbf{N}$; this totally orders $\mathbf{R}[X]$, and well orders the set of all numerical polynomials which are $\geqq 0$.

For any $s \in \mathbf{N}$ there are

$$n\binom{s+m}{m}$$

derivatives $\delta_1^{i_1} \cdots \delta_m^{i_m} y_j$ with $i_1 + \cdots + i_m \leqq s$ and $1 \leqq j \leqq n$; they may be regarded as indeterminates over the field $\mathfrak{F}$, and therefore the ring

$$\mathfrak{A}_s = \mathfrak{F}\left[(\delta_1^{i_1} \cdots \delta_m^{i_m} y_j)_{i_1+\cdots+i_m \leqq s, 1 \leqq j}\right]_n$$

is a polynomial algebra over $\mathfrak{F}$ in the usual sense. For any prime differential ideal $\mathfrak{p}$ of $\mathfrak{A}$, $\mathfrak{p} \cap \mathfrak{A}_s$ is a prime ideal of $\mathfrak{A}_s$ and hence has a dimension.

THEOREM 1. *Let $\mathfrak{p}$ be a prime differential ideal of $\mathfrak{A}$. There exists a unique numerical polynomial $\omega_\mathfrak{p}$ such that* $\dim(\mathfrak{p} \cap \mathfrak{A}_s) = \omega_\mathfrak{p}(s)$ *for all sufficiently big $s \in \mathbf{N}$.*

[2] For a general discussion of the characteristic function see e.g., [2, pp. 230–237]. By using a filtration instead of a grading one can equally well introduce the characteristic function in the nonhomogeneous theory.

[3] See e.g., [2, p. 233].

We call $\omega_\mathfrak{p}$ the *differential dimension polynomial* of $\mathfrak{p}$. Some of its properties are given by the following result.

THEOREM 2. *Let $\mathfrak{p}$ be a prime differential ideal of $\mathfrak{A}$.*

(a)
$$0 \leqq \omega_\mathfrak{p} \leqq n\binom{X+m}{m},$$

so that $\deg \omega_\mathfrak{p} \leqq m$.

(b) $\omega_\mathfrak{p}=0$ *if and only if* $Z(\mathfrak{p})$ *is a finite set.*

(c) $\omega_\mathfrak{p}=n\binom{X+m}{m}$ *if and only if* $\mathfrak{p}=(0)$.

(d) *If we write* $\omega_\mathfrak{p}=\sum_{0\leq k\leq n} a_k(\mathfrak{p})\binom{X+m}{m}$ *then* $a_m(\mathfrak{p})=d(\mathfrak{p})$.

Thus, $\omega_\mathfrak{p}$ contains at least as much information about $\mathfrak{p}$ as $d(\mathfrak{p})$ does. The following observation shows that $\omega_\mathfrak{p}$ gives an adequate measure of the size of $Z(\mathfrak{p})$.

THEOREM 3. *Let $\mathfrak{p}$, $\mathfrak{p}'$ be prime differential ideals of $\mathfrak{A}$ with $\mathfrak{p}\subset\mathfrak{p}'$, $\mathfrak{p}\neq\mathfrak{p}'$. Then $\omega_\mathfrak{p}>\omega_{\mathfrak{p}'}$.*

We recall (see [1, pp. 30–31 and 166–167]) that for an irreducible differential polynomial $P\in\mathfrak{A}$ Ritt defined the notions of *general* component and *singular* component of P; precisely one of the components of P is general, the rest all are singular.

THEOREM 4. *Let $\mathfrak{p}$ be a prime differential ideal of $\mathfrak{A}$. A necessary and sufficient condition that $\mathfrak{p}$ be the general component of an irreducible differential polynomial in $\mathfrak{A}$ of order e is that*

$$\omega_\mathfrak{p} = n\binom{X+m}{m} - \binom{X+m-e}{m}.$$

It follows that when this is the case then

$$\omega_\mathfrak{p} = (n-1)\binom{X+m}{m} + e\binom{X+m-1}{m-1} + \cdots,$$

that is, $a_m(\mathfrak{p})=n-1$ and $a_{m-1}(\mathfrak{p})=e$.

We remark that $\omega_\mathfrak{p}$ is a birational invariant but not a differential birational invariant. By this we mean that if $\mathfrak{p}$, $\mathfrak{q}$ are prime differential ideals of $\mathfrak{A}$ with respective generic zeros η, ζ then the condition

$\mathfrak{F}(\eta)=\mathfrak{F}(\zeta)$ implies that $\omega_{\mathfrak{p}}=\omega_{\mathfrak{q}}$ but the weaker condition $\mathfrak{F}\langle\eta\rangle=\mathfrak{F}\langle\zeta\rangle$ does not. Nevertheless, $\omega_{\mathfrak{p}}$ carries with it certain differential birational invariants. One example is $a_m(\mathfrak{p})$. The following result provides others.

THEOREM 5. *For each prime differential ideal* $\mathfrak{p}$ *of* $\mathfrak{A}$ *with* $\omega_{\mathfrak{p}}\neq 0$ *set* $m'(\mathfrak{p})=\deg\omega_{\mathfrak{p}}$ *and* $d'(\mathfrak{p})=a_{m'(\mathfrak{p})}(\mathfrak{p})$. *Then* $m'(\mathfrak{p})$ *and* $d'(\mathfrak{p})$ *are differential birational invariants.*

We call $m'(\mathfrak{p})$ the *differential type* of $\mathfrak{p}$ and call $d'(\mathfrak{p})$ the *typical differential dimension* of $\mathfrak{p}$. The following result interprets these invariants and justifies the terminology somewhat.

THEOREM 6. *Let* $\mathfrak{p}$ *be a prime differential ideal of* $\mathfrak{A}$ *having differential type* m' *and typical differential dimension* d'; *let* η *be a generic zero of* $\mathfrak{p}$ *and set* $\mathfrak{G}=\mathfrak{F}\langle\eta\rangle$. *Then there exists an* $m'\times m$ *matrix* $(c_{i'i})$ *over the field of constants* $\mathfrak{C}$ *of* $\mathfrak{F}$ *of rank* m' *such that if we set* $\delta'_{i'}=\sum_{1\leq i\leq m} c_{i'i}\delta_i$ $(1\leq i'\leq m')$ *and regard* $\mathfrak{F}$ *and* $\mathfrak{G}$ *as differential fields with the* m' *derivation operators* $\delta'_1, \cdots, \delta'_{m'}$ *then* $\mathfrak{G}$ *is a finitely generated differential field extension of* $\mathfrak{F}$ *of differential transcendence degree* d'. *The matrices* $(c_{i'i})$ *having this property form an open subset of* $\mathfrak{C}^{m'm}$ *in the Zariski* $\mathfrak{C}$-*topology.*

In the classical terminology we could say that the solution of the system $P=0$ $(P\in\mathfrak{p})$ depends on d' arbitrary functions of m' variables (but not on any arbitrary functions of more than m' variables).

It would be interesting to find other differential birational invariants.

BIBLIOGRAPHY

1. J. F. Ritt, *Differential algebra*, Amer. Math. Soc. Colloq. Publ. Vol. 33, Amer. Math. Soc., Providence, R. I., 1950.

2. Oscar Zariski and Pierre Samuel, *Commutative algebra*, Vol. II, Van Nostrand, Princeton, N. J., 1960.

COLUMBIA UNIVERSITY

Topology Vol. 3, Suppl. 2,pp. 309–318. Pergamon Press, 1965. Printed in Great Britain

SINGULAR SOLUTIONS OF ALGEBRAIC DIFFERENTIAL EQUATIONS AND A LEMMA OF ARNOLD SHAPIRO

E. R. KOLCHIN†

(*Received* 10 *April* 1963)

INTRODUCTION

WE CONSIDER differential equations of the form $A = 0$ where A is a differential polynomial in finitely many differential indeterminates $y_1, \ldots, y_n$ with coefficients in a differential field $\mathcal{F}$ of characteristic zero; that is, A is an element of the differential polynomial algebra $\mathcal{A} = \mathcal{F}\{y_1 \ldots, y_n\}$. We denote the derivation operators of $\mathcal{F}$ by $\delta_1, \ldots, \delta_m$. We suppose fixed, once for all, a universal extension $\mathcal{U}$ of $\mathcal{F}$ (see [3], pp. 768–771).

Ritt showed (see [6], p. 13 and pp. 165–166) that if $\mathfrak{a}$ is any perfect differential ideal of $\mathcal{A}$ then $\mathfrak{a}$ is the intersection of finitely many prime differential ideals of $\mathcal{A}$ none of which contains any other; these primes, which are unique, are the *prime components* of $\mathfrak{a}$. The prime components are especially interesting when $\mathfrak{a}$ is the perfect differential ideal $\{A\}$ generated by an irreducible $A \in \mathcal{A}$.

In order to describe the situation in that case we consider a total ordering of the set of all derivatives $\delta_1^{i_1} \ldots \delta_m^{i_m} y_j$ $(0 \leqslant i_1 < \infty, \ldots, 0 \leqslant i_m < \infty, 1 \leqslant j \leqslant n)$ such that for all such derivatives u, v and all δ_i

$$u < \delta_i u,$$

$$u < v \Rightarrow \delta_i u < \delta_i v.$$

We call such an ordering a *ranking* of $y_1, \ldots, y_n$; rankings exist (e.g. we may order the derivatives $\delta_1^{i_1} \ldots \delta_m^{i_m} y_j$ lexicographically with respect to $(\Sigma i_\mu, j, i_1, \ldots, i_m))$ but are in general not unique. Given a ranking, the highest derivative u present in A is called the *leader* of A, and the partial derivative $\partial A/\partial u$ is called the *separant* of A; of course, a different choice of ranking may give to A a different leader and separant.

Ritt called a zero of A (i.e., a solution of the differential equation $A = 0$) *singular* if it is a zero of every separant of A. He showed (see [6], p. 31 and p. 167) that among the prime components of A there is one, which we shall denote by $\mathfrak{P}(A)$, with the following property: $\mathfrak{P}(A)$ contains *no* separant of A whereas each other prime component of A contains *every* separant of A. $\mathfrak{P}(A)$ is called the *general* component of A, the others are called the *singular*

† This work was supported by the National Science Foundation.

components of A. Thus, every zero of a singular component of A is a singular zero of A, and every nonsingular zero of A is a zero of $\mathfrak{P}(A)$, but a singular zero of A may be a zero of $\mathfrak{P}(A)$.

It is a remarkable result of Ritt (see [6], pp. 57–62 and pp. 167–170, and also Hillman [1], p. 163) that every singular component of A is the general component of another irreducible differential polynomial in $\mathscr{A}$. Furthermore, he gave ([6], p. 109 and pp. 175–176) an algorithm (modulo the possibility of factorization of polynomials over $\mathscr{F}$) for finding a finite set of irreducible polynomials the general components of which include among them the singular components of A, and then established a criterion (the famous low-power theorem (see [6], pp. 64–70 and pp. 170–172)) for determining, given an irreducible $B \in \mathscr{A}$, whether $\mathfrak{P}(B)$ is a singular component of A.

There remains the problem, posed by Ritt, of determining, for a given zero of A, the components of A which admit that zero. In the light of the above this reduces to a number of problems of the following type: given a zero of A, to determine whether or not it is a zero of $\mathfrak{P}(A)$. It is not difficult to see, moreover, that it suffices to be able to solve this problem when the zero is $(0, \ldots, 0)$. Thus, we are led to the following problem:

Given an irreducible differential polynomial $A \in \mathscr{F}\{y_1, \ldots, y_n\}$ *which vanishes at* $(0, \ldots, 0)$, *to determine whether* $(0, \ldots, 0)$ *is a zero of* $\mathfrak{P}(A)$.

This problem is wide open. As yet, only very special cases have been solved. Two principal tools have been used in these special cases, as follows:

To prove that $(0, \ldots, 0)$ *is a zero of* $\mathfrak{P}(A)$. Let $P_j, Q_{jj'}$ $(1 \leqslant j \leqslant n,\ 1 \leqslant j' \leqslant n)$ be power series over $\mathscr{U}$ in an indeterminate constant c, which vanish at 0, such that $\det(Q_{jj'} \neq 0$; for each nonzero $F \in \mathscr{F}\{y_1, \ldots, y_n\}$ let $F*$ denote the leading coefficient of

$$F(P_1 + \sum_{j_1} Q_{1j_1}y_{j_1}, \ldots, P_n + \sum_{j_n} Q_{nj_n}y_{j_n})$$

(i.e., the lowest nonzero coefficient when considered as a power series in c over $\mathscr{U}\{y_1, \ldots, y_n\}$). *If the leader of* $A(P_1 + \sum_{j_1} Q_{1j_1}y_{j_1}, \ldots, P_n + \sum_{j_n} Q_{nj_n}y_{j_n})$ *is present in* $F*$, *or if* $S* \notin \{A*\}$ *for some separant* S *of* A, *then* $(0, \ldots, 0)$ *is a zero of* $\mathfrak{P}(A)$. This result, generalizing results of Hillman and of Ritt, is an almost immediate consequence of Hillman's leading coefficient theorem. (For an efficient proof of the leading coefficient theorem see Hillman and Mead [2]; for an indication of how this theorem leads to the above result see Hillman [1], §§ 7–8.).

To prove that $(0, \ldots, 0)$ *is not a zero of* $\mathfrak{P}(A)$. Suppose that $m = 1$ (ordinary differential equations) and that A has more than one term. It is a consequence of two results of Levi ([4], §§ 38–41 and §§ 44–52) that if A has a term $y_1^{e_1} \ldots y_n^{e_n}$ of order 0 such that for every other term T and each y_k either the degree of T in $y_k, y_k', y_k'', \ldots$ is $> e_k$ or T is divisible by $y_k^{e_k}$, or if $n = 1$ and A has a term $y_1^{f_0}y_1'^{f_1} \ldots y_1^{(r)f_r}$ such that for every other term T and $y_1^{(k)}(0 \leqslant k \leqslant r)$ the degree of T in $y_1^{(k)}, y_1^{(k+1)}, y_1^{(k+2)}, \ldots$ is $> f_k + f_{k+1} + \ldots + f_r$, then $(0, \ldots, 0)$ is not a zero of $\mathfrak{P}(A)$.

We come at last to the point of the present paper. This is to present a result which broadens considerably the class of differential polynomials A for which it is known that

(0, ... , 0) is not a zero of $\mathfrak{P}(A)$. To do this we introduce the notion of 'domination' of one differential monomial over another (§ 3 below), and establish a key lemma (§ 5) about this notion which generalizes both of Levi's results mentioned above. The proof depends on a lemma of Levi (§ 2 below). The domination lemma yields our proposition (§ 6), a special case of which can be stated as follows (m and n now being arbitrary).

Let A have more than one term. If A has a term which is dominated by every other term of A then (0, ... , 0) *is not a zero of* $\mathfrak{P}(A)$.

A crucial role in the proof of the domination lemma is played (§ 4) by a combinatorial lemma proved by Arnold Shapiro (unpublished). His lemma is presented in § 1.

NOTATION

We consistently use the following notation:

N, **Q**, **R** denote respectively the set of natural numbers, of rational numbers, of real numbers.

For any set K, $\mathfrak{P}(K)$ denotes the set of all subsets of K, and card K denotes the cardinal number of K. The empty set is denoted by ϕ.

Θ denotes the set of all derivative operators $\delta_1^{i_1} \dots \delta_m^{i_m}$ ($i_1 \in \mathbf{N}, \dots, i_m \in \mathbf{N}$) of the differential field $\mathscr{F}$ (and of all the various differential rings and fields considered).

$[A_1, \dots, A_r]$ denotes the differential ideal generated by the elements $A_1, \dots, A_r$ of a given differential ring. $\{A_1, \dots, A_r\}$ denotes the perfect differential ideal generated by these elements; since in the cases considered the differential ring contains **Q**, $\{A_1, \dots A_r\}$ is the set of all elements B such that $B^s \in [A_1, \dots, A_r]$ for some $s \in \mathbf{N}$.

§ 1. SHAPIRO'S LEMMA

SHAPIRO'S LEMMA. *Let K be a finite set, let $(a_k)_{k \in K}$ be a family with $a_k \in \mathbf{R}$* and $a_k \geqslant 0$ *($k \in K$), let $(x_J)_{J \in \mathfrak{P}(K) - \mathfrak{P}(\phi)}$ be a family with $x_J \in \mathbf{R}$ and $x_J \geqslant 0$ ($J \in \mathfrak{P}(K) - \mathfrak{P}(\phi)$), and suppose that*

$$\sum_{J \in \mathfrak{P}(K) - \mathfrak{P}(K - I)} x_J > \sum_{i \in I} a_i \qquad (I \in \mathfrak{P}(K) - \mathfrak{P}(\phi)). \tag{1}$$

Then there exist numbers $x_{J,j} \in \mathbf{R}$ with $x_{J,j} \geqslant 0$ ($J \in \mathfrak{P}(K) - \mathfrak{P}(\phi)$, $j \in J$) such that

$$\sum_{j \in J} x_{J,j} = x_J \qquad (J \in \mathfrak{P}(K) - \mathfrak{P}(\phi))$$

and

$$\sum_{J \ni j} x_{J,j} > a_j \qquad (j \in K).$$

Remarks. (1) We may think of the elements of K as representing the vertices of a simplex, the nonempty subsets of K as representing the faces of that simplex, the numbers a_j as forming a system of masses located at the vertices, and the numbers x_J as forming a system of masses located on the faces. The lemma then asserts that if, for each face I, the sum of the masses of the second system located on the faces touching I exceeds the sum of the masses of the first system located at the vertices of I, then the mass on each face can be redistributed among the vertices of that face in such a way that, for each vertex, the redistributed mass of the second system at the vertex exceeds the mass of the first system there.

(2) The proof shows that $x_{J,j}$ may be taken in the field $\mathbf{Q}((a_k)_{k\in K}, (x_J)_{J\in\mathfrak{P}(K)-\mathfrak{P}(\phi)})$.

Proof. Let $s = \text{card } K$; we may suppose that $s > 0$ as otherwise the result is trivial. Then there exists a $J \in \mathfrak{P}(K) - \mathfrak{P}(\phi)$ with $x_J > 0$, and therefore there exists a unique $r \in \mathbf{N}$ such that $x_J \neq 0$ for some J with card $J = r$ but $x_J = 0$ for all J with card $J > r$; of course $1 \leqslant r \leqslant s$. Let t denote the number of elements $J \in \mathfrak{P}(K)$ with card $J = r$ and $x_J \neq 0$; then $t > 0$. If $r = 1$ then the nonzero masses of the second system are already all at the vertices, so that the result is trivial. Therefore we may assume that $r > 1$ We assume, too, that the result has been proved for lower values of (s, r, t) in the lexicographically well-ordered set $\mathbf{N}^3$.

Fix some $I_0 \in \mathfrak{P}(K)$ with card $I_0 = r$ and $x_{I_0} \neq 0$, and fix some $k \in I_0$; let I_1 denote the set of elements of I_0 other than k. Then $K - I_0 \subset K - I_1 \subset K$, so that the system of inequalities (1) can be written as three subsystems:

(1a) corresponding to $I \in \mathfrak{P}(K - I_0) - \mathfrak{P}(\phi)$;
(1b) corresponding to $I \in \mathfrak{P}(K - I_1) - \mathfrak{P}(K - I_0)$;
(1c) corresponding to $I \in \mathfrak{P}(K) - \mathfrak{P}(K - I_1)$.

The left members in (1a) contain neither of the terms x_{I_0}, x_{I_1} and the left members in (1c) contain both these terms; the left members in (1b) contain x_{I_0} but not x_{I_1}. It follows that if $\xi \in \mathbf{R}$, $\xi > 0$, and if we replace x_{I_0} by $x_{I_0} - \xi$ and x_{I_1} by $x_{I_1} + \xi$, then (1a) and (1c) remain valid. The system (1b) remains valid provided ξ is sufficiently small. If (1b) remains valid for $\xi = x_{I_0}$ then the replacement transforms the original system (1) into a similar system with a lower value for (s, r, t). We may therefore suppose that at least one of the inequalities (1b) fails after the replacement using $\xi = x_{I_0}$. Then there is a smallest value for ξ, and we denote it simply by ξ, such that after the replacement (1b) fails to hold; using this ξ we see that (1b) becomes a system (1b′) of *weak* inequalities in the same direction. Obviously $0 < \xi \leqslant x_{I_0}$, and at least one of the weak inequalities (1b′) is an equality.

From among all the $I \in \mathfrak{P}(K) - I_1) - \mathfrak{P}(K - I_0)$ for which the corresponding inequality (1b′) is an equality, choose a maximal one, say K', and set $K'' = K - K'$. Then

$$(2) \qquad \sum_{J\in\mathfrak{P}(K)-\mathfrak{P}(K'')} x_J = \sum_{i\in K'} a_i.$$

Consider any $I'' \in \mathfrak{P}(K'') - \mathfrak{P}(\phi)$; writing $I = K' \cup I''$, we see that either $I \in \mathfrak{P}(K - I_1)$ and I corresponds to an equality (1c) or else $I \in \mathfrak{P}(K - I_1) - \mathfrak{P}(K - I_0)$ and I corresponds to an inequality (1b′). In either case the inequality is strict. Subtracting from it the equation (2) we obtain

$$(1'') \qquad \sum_{J\in\mathfrak{P}(K'')-\mathfrak{P}(K''-I'')} x_J > \sum_{i\in I''} a_i \qquad (I'' \in \mathfrak{P}(K'') - \mathfrak{P}(\phi)).$$

On the other hand, if we start with some $I \in \mathfrak{P}(K') - \mathfrak{P}(\phi)$ then either $I \in \mathfrak{P}(K - I_0) - \mathfrak{P}(\phi)$ and we have a strict inequality (1a) or else $I \in \mathfrak{P}(K - I_1) - \mathfrak{P}(K - I_0)$ and we have a weak inequality (1b′). If we now reduce ξ slightly, still keeping it positive, then the inequalities (1a) remain valid, and the inequalities (1b′) all become strict; that is, we regain (1b); furthermore, if the amount by which we reduce ξ is sufficiently small than (1″) remains valid, too. Then, in addition to (1″) we obtain (denoting I by I')

$$(3) \qquad \sum_{J \in \mathfrak{P}(K) - \mathfrak{P}(K-I')} x_J > \sum_{i \in I'} a_i \qquad (I' \in \mathfrak{P}(K') - \mathfrak{P}(\phi)).$$

If $J \in \mathfrak{P}(K) - \mathfrak{P}(K - I')$ then $J \in \mathfrak{P}(K) - \mathfrak{P}(K - K')$ and therefore this J does not occur in the left side of (1″). For each $J \in \mathfrak{P}(K) - \mathfrak{P}(K - K')$ we now decrease x_J and increase $x_{J \cap K'}$ by the same amount x_J (that is, we shift the entire mass x_J from the face J to the face $J \cap K'$). This does not affect the inequalities (1″), and replaces the inequalities (3) by

$$(1') \qquad \sum_{J \in \mathfrak{P}(K') - \mathfrak{P}(K'-I')} x_J > \sum_{i \in I'} a_i \qquad (I' \in \mathfrak{P}(K') - \mathfrak{P}(\phi)).$$

Since card $K' < s$ and card $K'' < s$, the lemma holds for each of the two systems (1′) and (1″). It is now a simple matter to see that the lemma holds for the original system (1).

COROLLARY. *Let K be a finite set, let $a_k \in \mathbf{N}(k \in K)$, let $x_J \in \mathbf{N}(J \in \mathfrak{P}(K) - \mathfrak{P}(\phi))$, and suppose that*

$$\sum_{J \in \mathfrak{P}(K) - \mathfrak{P}(K-I)} x_J > \sum_{i \in I} a_i \qquad (I \in \mathfrak{P}(K) - \mathfrak{P}(\phi)).$$

Then, for each sufficiently big $h \in \mathbf{N}$, there exist $y_{J,j} \in \mathbf{N}(J \in \mathfrak{P}(K) - \mathfrak{P}(\phi), j \in J)$ such that

$$\sum_{j \in J} y_{J,j} = hx_J \qquad (J \in \mathfrak{P}(K) - \mathfrak{P}(\phi))$$

and

$$\sum_{J \ni j} y_{J,j} > ha_j \qquad (j \in K).$$

Proof. There exist (see second remark after Shapiro's lemma) *rational* numbers $x_{J,j}$ satisfying the conclusion of that lemma. There obviously exists a $\xi > 0$, smaller than every nonzero $x_{J,j}$, such that if we set

$$x'_{J,j} = \begin{cases} x_{J,j} - \xi & (x_{J,j} \neq 0) \\ 0 & (x_{J,j} = 0) \end{cases}$$

then $\sum_{J \ni j} x'_{J,j} > a_j$ $(j \in K)$; of course $\sum_{j\ J} x'_{J,j} \leqslant x_J$. For any $h \in \mathbf{N}$ with $h > \xi^{-1}$ there exist $x''_{J,j} \in h^{-1}\mathbf{N}$ such that $x'_{J,j} \leqslant x''_{J,j} \leqslant x_{J,j}$ and for such $x''_{J,j}$ obviously $\sum_{J \in j} x''_{J,j} \leqslant x_J (J \in \mathfrak{P}(K) - \mathfrak{P}(\phi))$ and $\sum_{J \ni j} x''_{J,j} > a_j (j \in K)$. For each $J \in \mathfrak{P}(K) - \mathfrak{P}(\phi)$ fix an element $i(J) \in J$ and set $y_{J,j} = hx''_{J,j}(j \in J, j \neq i(J))$, $y_{J,i(J)} = hx''_{J,i(J)} + hx_J - \sum_{j \in J} hx''_{J,j}$. It is easy to see that the numbers $y_{J,j}$ have the required properties.

§ 2. LEVI'S LEMMA

Our point of departure is the following result concerning differential polynomials in $y_1, \ldots, y_n$ and a number of other differential indeterminates $u_{ij}(1 \leqslant i \leqslant n, 0 \leqslant j \leqslant r_i)$.

LEVI'S LEMMA. *Let $G_1, \ldots, G_n$ be differential polynomials in $\mathbf{Q}\{y_1, \ldots, y_n, (u_{ij})_{1 \leqslant i \leqslant n,\, 0 \leqslant j \leqslant r_i}\}$ given by $G_i = u_{i0} y_i^{q_i} + \sum_{1 \leqslant j \leqslant r_i} u_{ij} M_{ij}$ $(1 \leqslant i \leqslant n)$ where, for each i, $q_i \in \mathbf{N}$ and $M_{i1}, \ldots, M_{ir_i}$ are* *differential monomials in* $y_1, \ldots, y_n$ *of degree* $> q_i$. Then there exist a monomial

$$U = u_{10}^{d_1} \ldots u_{n0}^{d_n} \text{ in } u_{10}, \ldots, u_{n0}$$

and a differential polynomial $Y \in \mathbf{Q}\{y_1, \ldots, y_n, (u_{ij})_{1 \leqslant i \leqslant n, 0 \leqslant j \leqslant r_i}\}$ *with the following properties*:

Y *is homogeneous in* $(\theta u_{ij})_{\theta \in \Theta, 0 \leqslant j \leqslant r_i}$ *of degree* d_i $(1 \leqslant i \leqslant n)$;
the degree of Y *in* $(\theta u_{i0})_{\theta \in \Theta, 1 \leqslant i \leqslant n}$ *is* $< d_1 + \ldots + d_n$;
$Y \in [y_1, \ldots, y_n]$;
$y_i(U + Y) \in \{G_1, \ldots, G_r\}$ $(1 \leqslant i \leqslant n)$.

Levi proved this result for ordinary differential polynomials ([4] §§ 32–36, § 47), and for partial differential polynomials in the case $r = 1$ ([5], p. 118). The general lemma can be established in the same way with little extra difficulty.

§ 3. DOMINATION

We deal with differential monomials in $y_1, \ldots, y_n$, that is, with products of powers of derivatives θy_i $(\theta \in \Theta, 1 \leqslant i \leqslant n)$.

By a *prime factor* of such a differential monomial M we mean a derivative θy_i which divides M. If V is *any* set of derivatives θy_i we let ΘV denote the set of all derivatives of the y_i which can be written in the form θv $(\theta \in \Theta, v \in V)$. The product of all the prime factors w of M with $w \in \Theta V$, each w taken the same number of times as it occurs in M, is a differential monomial which we denote by M_V.

Let M and N be differential monomials. We shall say that N *dominates* M if, for every set V, the following condition is satisfied:

$$\text{either } \deg M_V < \deg N_V \text{ or } M_V = N_V.$$

Since $M_V = M_{\Theta V} = M_{(\Theta V) \cap V(M)}$, where $V(M)$ denotes the set of all prime factors of M, it suffices to verify this condition for every nonempty set V with $V \subset V(M)$. If, for every nonempty V with $V \subset V(M)$, N satisfies the stronger condition

$$\deg M_V < \deg N_V$$

then we shall say that N *strongly* dominates M.

It is easy to see that there exists a biggest set W of prime factors of M such that $M_W = N_W$. If N dominates M then a necessary and sufficient condition that N strongly dominate M is that W be empty. We shall call W the *weakness* of N over M.

If N_k dominates M and W_k denotes the weakness of N_k over M $(1 \leqslant k \leqslant r)$ then the weakness of $\prod_{1 \leqslant k \leqslant r} N_k$ over M^r is $\bigcap_{1 \leqslant k \leqslant r} W_k$.

§ 4. FACTORIAL DOMINATION

If N_k dominates (resp. strongly dominates) M_k $(1 \leqslant k \leqslant r)$ then $\prod_{1 \leqslant k \leqslant r} N_k$ dominates (resp. strongly dominates) $\prod_{1 \leqslant k \leqslant r} M_k$. If N_k dominates M $(1 \leqslant k \leqslant r)$ and, for at least one k,

N_k strongly dominates M then $\prod_{1 \leqslant k \leqslant r} N_k$ strongly dominates M^r. It follows from the former statement that if $M = \prod_{1 \leqslant k \leqslant r} v_k^{a_k}$, where $v_1, \ldots, v_r$ are the distinct prime factors of M, and if $N = \prod_{1 \leqslant k \leqslant r} N_k$ where N_k dominates (resp. strongly dominates) $v_k^{a_k}$ $(1 \leqslant k \leqslant r)$, then N dominates (resp. strongly dominates) M. We shall say in such a case that N dominates (resp. strongly dominates) M *factorially*.

If N_l dominates (resp. strongly dominates) M_l factorially $(1 \leqslant l \leqslant s)$ then $\prod_{1 \leqslant l \leqslant s} N_l$ dominates (resp. strongly dominates) $\prod_{1 \leqslant l \leqslant s} M_l$ factorially. If N_l dominates M factorially $(1 \leqslant l \leqslant s)$ and, for at least one l, N_l strongly dominates M factorially then $\prod_{1 \leqslant l \leqslant r} N_l$ strongly dominates M^s factorially.

Shapiro's lemma enters at this point.

FIRST PRELIMINARY LEMMA. *If N_l strongly dominates M $(1 \leqslant l \leqslant s)$ then, for all $(i_1, \ldots, i_s) \in \mathbf{N}^s$ for which the sum $h = i_1 + \ldots + i_s$ is sufficiently big, $\prod_{1 \leqslant l \leqslant s} N_l^{i_l}$ strongly dominates M^h factorially.*

Proof. Write $M = \prod_{k \in K} v_k^{a_k}$ with K a finite set and the elements v_k $(k \in K)$ the distinct prime factors of M. For each nonempty set $J \subset K$ let x_{lJ} denote the number of prime factors v of N_l such that v is a derivative of v_k for every $h \in J$ and is not a derivative of v_k for any $k \in K - J$ (each v being counted as many times as it occurs in N_l). Because N_l strongly dominates M we have for each l

$$\sum_{J \in \mathfrak{P}(K) - \mathfrak{P}(K - I)} x_{lJ} > \sum_{i \in I} a_i \qquad (I \in \mathfrak{P}(K) - \mathfrak{P}(\phi)).$$

By the corollary to Shapiro's lemma there exist an $h_0 \in \mathbf{N}$ and, for each (J, j) with $J \in \mathfrak{P}(K) - \mathfrak{P}(\phi)$ and $j \in J$, a $y_{lJj} \in \mathbf{N}$ such that

$$\sum_{j \in J} y_{lJj} = h_0 x_J \quad \text{and} \quad \sum_{J \in j} y_{lJj} > h_0 a_j.$$

It follows that we may write $N_l^{h_0} = \prod_{j \in K} N_{lj}$ where, for each $j \in K$, N_{lj} is a differential monomial of which the degree in $(\theta v_j)_{\theta \in \Theta}$ is $\geqslant h_0 a_j + 1$. Let $(i_1, \ldots, i_s) \in \mathbf{N}^s$ and write $i_l = q_l h_0 + r_l$ with $q_l, r_l \in \mathbf{N}$ and $r_l < h_0$. Then $\prod_{1 \leqslant l \leqslant s} N_l^{i_l} = \prod_{j \in K} \prod_{1 \leqslant l \leqslant s} N_{lj}^{q_l} \cdot \prod_{1 \leqslant l \leqslant s} N_l^{r_l}$. For each $j \in K$ the degree of $\prod_{1 \leqslant l \leqslant s} N_{lj}^{q_l}$ in $(\theta v_j)_{\theta \in \Theta}$ is

$$\geqslant \sum_{1 \leqslant l \leqslant s} q_l(h_0 a_j + 1) = \sum_{1 \leqslant l \leqslant s} (i_l - r_l) h_0^{-1}(h_0 a_j + 1) \geqslant (h - s(h_0 - 1)) h_0^{-1}(h_0 a_j + 1),$$

where $h = \sum_{1 \leqslant l \leqslant s} i_l$, so that this degree is $> a_j h$ provided $h > s(h_0 - 1)(h_0 a_j + 1)$. Whenever this is the case then $\prod_{1 \leqslant l \leqslant s} N_l^{i_l}$ strongly dominates M^h factorially.

SECOND PRELIMINARY LEMMA. *Let $F = \sum_{0 \leqslant l \leqslant s} u_l M_l \in \mathbf{Q}\{y_1, \ldots, y_n, u_0, \ldots, u_s\}$, where $M_0, M_1, \ldots M_s$ are differential monomials in $y_1, \ldots, y_n$ such that $M_0 \neq 1$ and $M_l \neq M_0$ $(1 \leqslant l \leqslant s)$. If each M_l with $l \neq 0$ dominates (resp. strongly dominates) M_0 then the ideal (F)*

contains a differential polynomial $G = u_0^a M_0^a + \sum_{1 \leq l \leq t} U_l N_l$, *where* $a \in \mathbf{N}$, $a \neq 0$, *each* U_l *is a monomial in* $u_0, u_1, \ldots, u_s$ *different from* u_0^a *of degree* a, *and each* N_l *is a differential monomial in* $y_1, \ldots, y_n$ *different from* M_0^a *which dominates (resp. strongly dominates)* M_0^a *factorially.*

Proof. Suppose that each M_l with $l \neq 0$ strongly dominates M_0. Raising both sides of the congruence $u_0 M_0 \equiv -\sum_{1 \leq l \leq s} u_l M_l \pmod{F}$ to an odd power a, we obtain a congruence $u_0^a M_0^a \equiv -\sum u_{l_1} \ldots u_{l_a} M_{l_1} \ldots M_{l_a} \pmod{F}$; by the first preliminary lemma we may choose a so big that each $M_{l_1} \ldots M_{l_a}$ strongly dominates M_0^a factorially, and therefore the differential polynomial $G = u_0^a M_0^a + \sum u_{l_1} \ldots u_{l_a} M_{l_1} \ldots M_{l_a}$ satisfies the 'resp.' part of the conclusion.

Now suppose merely that M_l dominates M_0 $(1 \leq l \leq s)$. Let Λ_0 denote the set of indices l with $l \neq 0$ such that M_l dominates M_0 factorially. For each l with $l \neq 0$ and $l \notin \Lambda_0$ the weakness of M_l over M_0 is a subset of the set of prime factors of M_0. Denote the distinct weaknesses of the various M_l with $l \neq 0$ and $l \notin \Lambda_0$ by $W_1, \ldots, W_k$ and for each W_j let Λ_j denote the set of indices l with $l \neq 0$ and $l \notin \Lambda_0$ such that the weakness of M_l over M_0 is W_j; we choose the notation so that $W_k \not\subset W_j$ $(1 \leq j \leq k-1)$. Then $F = u_0 M_0 + \sum_{0 \leq j \leq k} \sum_{l \in \Lambda_j} u_l M_l$. Set $\pi = \operatorname{card} \bigcup_{1 \leq j \leq k} \mathfrak{P}(W_j)$.

If $\pi = 0$ we may take $G = F$. Let $\pi > 0$ and suppose the result proved for lower values of π. Then $k > 0$. Raising both sides of the congruence $u_0 M_0 + \sum_{l \in \Lambda_0} u_l M_l \equiv -\sum_{1 \leq j \leq k} \sum_{l \in \Lambda_j} u_l M_l \pmod{F}$ to an odd power h, we obtain on the left $u_0^h M_0^h$ plus a number of terms UN with U a monomial in u_0, u_l $(l \in \Lambda_0)$ different from u_0^h of degree h and N a differential monomial in $y_1, \ldots, y_n$ different from M_0^h which dominates M_0^h factorially. On the right we obtain a sum of terms $-UN = -u_{l_1} \ldots u_{l_h} M_{l_1} \ldots M_{l_h}$; for any such term either some index l_i is in a Λ_j with $1 \leq j \leq k-1$ or $l_i \in \Lambda_k$ $(1 \leq i \leq h)$. In the former case the weakness of N over M_0^h is a subset of W_j for some j with $1 \leq j \leq k-1$. In the latter case we may write $M_{l_i} = M'_{l_i} M_{l_i W_k} = M'_{l_i} M_{0 W_k}$ $(1 \leq i \leq h)$ and $M_0 = M'_0 M_{0 W_k}$, and each M'_{l_i} strongly dominates M'_0; by the first preliminary lemma we may choose h so that $M'_{l_1} \ldots M'_{l_h}$ strongly dominates $M_0'^h$ factorially, and then $N = M_{l_1} \ldots M_{l_h} = M'_{l_1} \ldots M'_{l_h} M_{0 W_k}^h$ dominates $M_0^h = M_0'^h M_{0 W_k}^h$ factorially. Transposing to the left side all the terms on the right we obtain on the left a differential polynomial.

$$F^* = U_0 M_0^* + \sum_{0 \leq j \leq k^*} \sum_{l \in \Lambda_j^*} U_j M_j^* \in (F)$$

where $\Lambda_0^*, \ldots, \Lambda_{k^*}^*$ are disjoint finite sets not containing 0, $U_0 = u_0^h$, each U_l with $l \neq 0$ is a monomial in $u_0, u_1, \ldots, u_s$ different from u_0^h of degree h, $M_0^* = M_0^h$, every M_l^* with $l \in \Lambda_0^*$ is a differential monomial in $y_1, \ldots, y_n$ different from M_0^h which dominates M_0^h factorially, for each index j with $1 \leq j \leq k^*$ all the M_l^* with $l \in \Lambda_j^*$ are differential monomials in $y_1, \ldots, y_n$ different from M_0^h which dominate M_0^h and have over M_0^h one and the'same weakness W_j^*, and each of these weaknesses $W_1^*, \ldots, W_{k^*}^*$ is a subset of some W_j with $1 \leq j \leq k-1$. It follows from the last remark that the number $\pi^* = \operatorname{card} \bigcup_{1 \leq j \leq k^*} \mathfrak{P}(W_j^*)$ has the property that $\pi^* < \pi$. Therefore we may apply the present lemma to F^*, and the existence of a differential polynomial $G \in (F)$ with the required property quickly follows.

§ 5. THE DOMINATION LEMMA

We now come to the main lemma from which our results on singular solution will quickly follow.

DOMINATION LEMMA. *Let* $F = \sum_{0 \leq l \leq s} u_l M_l \in \mathbf{Q}\{y_1, \ldots, y_n, u_0, \ldots, u_s\}$ *where* $M_0, \ldots, M_s$ *are differential monomials in* $y_1, \ldots, y_n$ *such that* $M_0 \neq 1$ *and* $M_l \neq M_0$ $(1 \leq l \leq s)$. *If each* M_l *with* $l \neq 0$ *dominates (resp. strongly dominates)* M_0 *then there exist a nonzero* $e \in \mathbf{N}$ *and a* $Y \in \mathbf{Q}\{y_1, \ldots, y_n, u_0, \ldots, u_s\}$ *with the following properties*:

Y is homogeneous in $(\theta u_l)_{\theta \in \Theta,\ 0 \leq l \leq s}$ *of degree* e;

the degree of Y in $(\theta u_0)_{\theta \in \Theta}$ *is* $< e$;

$Y \in [y_1, \ldots, y_n]$ *(resp.* $Y \in \{M_0\}$*)*;

$M_0 (u_0^e + Y) \in \{F\}$.

Proof. Write $M_0 = v_1^{q_1} \ldots v_t^{q_t}$, where $v_1, \ldots v_t$ are the distinct prime factors of M_0. Suppose first that $t = 1$. For each l either M_l is divisible by $v_1^{q_1}$ or the degree of M_l in Θv_1 is $> q_1$. Therefore we may write

$$F = (u_0 + \sum_{l \in \Lambda'} u_l L_l) v_1^{q_1} + \sum_{l \in \Lambda''} u_l L_l N_l(v_1)$$

where Λ', Λ'' are disjoint sets the union of which is the set of indices $1, 2, \ldots, s$ (with $\Lambda' = \phi$ if each M_l with $l \neq 0$ strongly dominates M_0), each L_l is a differential monomial in $y_1, \ldots, y_n$, deg $L_l > 0$ $(l \in \Lambda')$, and each N_l with $l \in \Lambda''$ is a differential monomial (in some new differential indeterminate z) of degree $> q_1$. We may apply Levi's lemma (case $n = 1$) to the differential polynomial $G = \bar{u}_0 z^{q_1} + \sum_{l \in \Lambda''} \bar{u}_l N_l \in \mathbf{Q}\{z, \bar{u}_0, (\bar{u}_l)_{l \in \Lambda''}\}$ to show the existence of a differential polynomial $z(\bar{u}_0^e + Z) \in \{G\}$ with Z homogeneous in $((\theta \bar{u}_0)_{\theta \in \Theta}, (\theta \bar{u}_l)_{\theta \in \Theta,\ l \in \Lambda''})$ of degree e, the degree of Z in $(\theta \bar{u}_0)_{\theta \in \Theta}$ strictly smaller than e, and $Z \in [z]$. Since the substitution of $(v_1, u_0 + \sum_{l \in \Lambda'} u_l L_l, (u_l L_l)_{l \in \Lambda''})$ for $(z, \bar{u}_0, (u_l)_{l \in \Lambda''})$ maps G onto F, the desired result follows.

Now suppose that $t > 1$ and that the lemma has been proved for lower values of t. By the second preliminary lemma, (F) contains a differential polynomial $G = u_0^a v_1^{q_1 a} \ldots v_t^{q_t a} + \sum_{1 \leq j \leq r} U_j N_{1j} M_j'$, where each U_j is a monomial in $u_0, \ldots, u_s$ other than u_0^a of degree a, each N_{1j} is a differential monomial in $y_1, \ldots, y_n$ which dominates (resp. strongly dominates) $v_1^{q_1 a}$, each M_j' is a differential monomial in $y_1, \ldots, y_n$ which dominates (resp. strongly dominates) $v_2^{q_2 a} \ldots v_t^{q_t a}$, and $N_{1j} M_j' \neq M_0^a$. Replacing G by $(u_0^a M_0^a)^h + (\sum_{1 \leq j \leq r} U_j N_{1j} M_j')^h$ if necessary, h denoting an odd natural number > 1, we may even suppose that $M_j' \neq v_2^{q_2 a} \ldots v_t^{p_t a}$ $(1 \leq j \leq r)$. Then we may apply the present lemma (case $t - 1$) to the differential polynomial $F' = u_0' v_2^{q_2 a} \ldots v_t^{q_t a} + \sum_{1 \leq j \leq r} u_j' M_j' \in \mathbf{Q}\{y_1, \ldots, y_n, u_0', \ldots, u_r'\}$ to prove the existence of a nonzero $e' \in \mathbf{N}$ and a $Y' \in \mathbf{Q}\{y_1, \ldots, y_n, u_0', \ldots, u_r'\}$ with Y' homogeneous in $(\theta u_j')_{\theta \in \Theta,\ 1 \leq j \leq r}$ of degree e', the degree of Y' in $(\theta u_0')_{\theta \in \Theta}$ strictly smaller than e', $Y' \in [y_1, \ldots, y_n]$ (resp. $Y' \in \{v_2^{q_2 a} \ldots v_t^{q_t a}\} = \{v_2 \ldots v_t\}$), and $v_2 \ldots v_t(u_0'^e + Y') \in \{F'\}$. Substituting $(u_0^a v_1^{q_1 a}, U_1 N_{11}, \ldots, U_r N_{1r}$ for $(u_0', u_1', \ldots, u_r')$ we see that $\{F\}$ contains a differential polynomial $v_2 \ldots v_t (u_0^{a_1} v_1^{c_1} + \sum_{l \in \Lambda_1} U_{1l} M_{1l})$, where a_1 and c_1 are nonzero natural numbers, each U_{1l} is the product if a rational number with a differential monomial in

$u_0, \dots, u_s$ of degree a_1 and of degree in $(\theta u_0)_{\theta\in\Theta}$ strictly smaller than a_1, and each M_{1l} is a differential monomial in $y_1, \dots, y_n$ different from $v_1^{c_1}$ which dominates (resp. strongly dominates) $v_1^{c_1}$. Let Λ_1'' denote the set of indices $l \in \Lambda_1$ such that M_{1l} strongly dominates $v_1^{c_1}$, and set $\Lambda_1' = \Lambda_1 - \Lambda_1''$ (so that under the 'resp.' hypothesis $\Lambda_1' = \phi$). For each $l \in \Lambda_1'$ we may write $M_{1l} = L_{1l}v_1^{c_1}$ with L_{1l} a differential monomial in $y_1, \dots, y_n$ of degree > 0. Thus, the perfect differential ideal $\{F\}: M_0$ contains $(u_0^{a_1} + \sum_{l\in\Lambda_1'} U_{1l}L_{1l})v_1^{c_1} + \sum_{l\in\Lambda_1''} U_{1l}M_{1l}$. Similarly, for each $k \in \mathbf{N}$ with $1 \leqslant k \leqslant t$, $\{F\}: M_0$ contains a differential polynomial $(u_0^{a_k} + \sum_{l\in\Lambda_k'} U_{kl}L_{kl})v_k^{c_k} + \sum_{l\in\Lambda_k''} U_{kl}M_{kl}$ with entirely analogous properties. An easy application of Levi's lemma (case $n = t$) now completes the proof.

§ 6. SINGULAR SOLUTIONS

Let A be an irreducible differential polynomial in $\mathscr{F}\{y_1, \dots, y_n\}$, and suppose that $(0, \dots, 0)$ is a singular solution of the differential equation $A = 0$. A sufficient condition that $(0, \dots, 0)$ not be a zero of the general component $\mathfrak{P}(A)$ is provided by the domination lemma. We formulate the result in the following more general setting.

PROPOSITION. *Let* $\mathfrak{P}$ *be a prime differential ideal of* $\mathscr{F}\{y_1, \dots, y_n\}$, *and suppose that* $\mathfrak{P}$ *contains a differential polynomial* $\sum_{0\leqslant l\leqslant s} C_l M_l(B_1, \dots, B_r)$, *where:* $C_l \in \mathscr{F}\{y_1, \dots, y_n\}$ $(0 \leqslant l \leqslant s)$ *and* $C_0(0, \dots 0) \neq 0$; $B_k \in \mathscr{F}\{y_1, \dots, y_n\}$ *and* $B_k(0, \dots, 0) = 0$ $(1 \leqslant k \leqslant r)$; $M_0, M_1, \dots, M_s$ *are differential monomials in differential indeterminates* $z_1, \dots, z_r$ *with* $M_k \neq M_0$ $(1 \leqslant k \leqslant r)$ *such that* M_k *dominates* $M_0 (1 \leqslant k \leqslant r)$; and $M_0(B_1, \dots, B_r) \notin \mathfrak{P}$. *Then* $(0, \dots, 0)$ *is not a zero of* $\mathfrak{P}$.

Proof. By the domination lemma $\mathfrak{P}$ contains a differential polynomial $M_0(B_r, \dots, B_r)$ $(C_0^e + Y(B_1, \dots, B_r))$ where $e \in \mathbf{N}$ and $Y \in [z_1, \dots, z_r]$ in $\mathbf{Q}\{z_1, \dots, z_r, C_0, \dots, C_s\}$, so that $\mathfrak{P}$ contains the differential polynomial $C_0^e + Y(B_1, \dots, B_r)$ which does not vanish at $(0, \dots, 0)$.

REFERENCES

1. ABRAHAM HILLMAN: On the differential algebra of a single differential polynomial, *Ann. Math, Princeton* **56** (1952), 157–168.
2. ABRAHAM HILLMAN and DAVID MEAD: *Amer. J. Math.*, to be published.
3. E. R. KOLCHIN: Galois theory of differential fields, *Amer. J. Math.* **75** (1953), 753–824.
4. HOWARD LEVI: On the structure of differential polynomials and on their theory of ideals, *Trans. Amer. Math. Soc.* **51** (1942), 532–568.
5. HOWARD LEVI: The low power theorem for partial differential polynomials, *Ann. Math., Princeton* **46** (1945), 113–119.
6. J. F. RITT: Differential algebra, *Colloq. Lect. Amer. Math. Soc.* **33** (1950).

Columbia University., N.Y., U.S.A.

SOME PROBLEMS IN DIFFERENTIAL ALGEBRA

E. R. KOLCHIN

Introduction

It is 36 years since the publication of Ritt's first paper in differential algebra [1], and almost 16 years since his death. The fact is that during the twenty years preceding his death the subject progressed at a much faster rate than afterward. This is in no way due to a lack of worthwhile problems, but rather is a reflection of the scanty attention these problems have received. The experience and techniques of algebrists and especially algebraic geometers are particularly suited to differential algebra, and it is surprising that so few of them have turned their talents in this direction. I hope that my description of a few of the more challenging unsolved problems will lure some into accepting the challenge.

Consider a differential field $\mathfrak{F}$, i.e., a field in the usual sense (which we shall suppose to have characteristic 0) together with a finite number of mutually commuting derivation operators $\delta_1, \ldots, \delta_m$; when $m = 1$ the differential field is *ordinary* (and we usually write a' instead of $\delta_1 a$), and when $m > 1$ it is *partial*. Our problems have to do with algebraic differential equations of the form $A = 0$, where A is a differential polynomial over $\mathfrak{F}$ in a finite number n of differential indeterminates (i.e. A is a polynomial, with coefficients in $\mathfrak{F}$, in the derivatives $\delta_1^{i_1} \ldots \delta_m^{i_m} Y_j$ $(0 \leqslant i_1 \leqslant \infty, \ldots, 0 \leqslant i_m < \infty, 1 \leqslant j \leqslant n)$. The set $\mathfrak{R} = \mathfrak{F}\{Y_1, \ldots Y_n\}$ of all differential polynomials over $\mathfrak{F}$ in $Y_1, \ldots, Y_n$ is a differential ring. A subset Σ of $\mathfrak{R}$ defines a system of differential equations: $A = 0$ $(A \in \Sigma)$. A solution $(\eta_1, \ldots, \eta_n)$ of this system, having coordinates in an extension (i.e. differential field extension) of $\mathfrak{F}$, is also called a *zero* of Σ. It is convenient (albeit not essential) to work in a fixed *universal* extension $\mathfrak{U}$ of $\mathfrak{F}$. Such a $\mathfrak{U}$, which always exists, is, roughly speaking, so large that it contains enough elements for every purpose we can have.

An ideal $\mathfrak{A}$ of $\mathfrak{R}$ is *differential* if $\delta_i \mathfrak{A} \subset \mathfrak{A}$ $(1 \leqslant i \leqslant m)$, and is *perfect* if the condition $A^2 \in \mathfrak{A}$ implies that $A \in \mathfrak{A}$. The smallest differential ideal containing the set Σ is denoted by $[\Sigma]$, and the smallest perfect differential ideal containing Σ is denoted by $\{\Sigma\}$; because the field characteristic is 0, $\{\Sigma\}$ is the radical of $[\Sigma]$. The zeros of Σ are the same as the zeros of $\{\Sigma\}$. A fundamental fact is that Σ always has a finite subset Φ such that $\{\Phi\} = \{\Sigma\}$; this is the Ritt-Raudenbush basis theorem. As a consequence, every perfect differential ideal $\mathfrak{A}$ is the intersection of finitely many prime differential ideals none of which contains another; these are the minimal prime differential ideals containing $\mathfrak{A}$, and are called the *components* of $\mathfrak{A}$. These general facts from the Ritt theory, as well as others referred to below, can be found in his book [2].

1. Singular solutions

Ritt's theory of components is especially interesting when $\mathfrak{A}$ is the perfect differential ideal $\{A\}$ generated by an irreducible differential polynomial $A \in \mathfrak{R}$. Consider any total ordering of the set of all derivatives $\delta_1^{i_1} \ldots \delta_m^{i_m} Y_j$, such that for all such derivatives u and v and all δ_i

$$\text{(i) } u < \delta_i u, \qquad \text{(ii) } u < v \Rightarrow \delta_i u < \delta_i v;$$

such orderings exist (e.g. the lexicographic ordering relative to $(i_1 + \ldots + i_m, j, i_1, \ldots, i_m)$) but in general are not unique. The highest derivative u present in A is called the *leader* of A, and the partial derivative $\partial A/\partial u$ is called the *separant* of A; of course, a different choice of ordering may give to A a different leader and separant. A zero of A is called *singular* if it is a zero of every separant of A.

Among the components of $\{A\}$ there is one, which we denote by $\mathfrak{p}(A)$, that contains *no* separant, whereas each other component contains *every* separant: $\mathfrak{p}(A)$ is the *general* component, and the others are the *singular* components, of A. Every zero of a singular component is singular, but a singular zero may be a zero of the general component. It is remarkable that all the singular components of A are the general components of other irreducible differential polynomials in $\mathfrak{R}$; furthermore, there is an effective procedure for finding these others. Yet, given a zero $\eta = (\eta_1, \ldots, \eta_n)$ and a component, no general method is known for deciding whether η is a zero of that component. Because it is possible to translate by η (extending $\mathfrak{F}$ if necessary), this question reduces to the following problem posed by Ritt:

Given an irreducible differential polynomial $A \in \mathfrak{R}$; to determine whether $(0, \ldots, 0)$ *is a zero of* $\mathfrak{p}(A)$.

Only very special cases have been solved. When $m = 1$, $n = 1$ (*ordinary* differential equation in *one* unknown), and the order of A is $\leqslant 2$ the solution was given by Ritt [3]. Beyond this there are only certain sufficient conditions and other necessary conditions, which are far apart; these conditions, successively developed by Ritt,.

H. Levi, A. Hillman, and me, are described in [4] where references are given.

The condition that $(0, \ldots, 0)$ be a zero of $\mathfrak{p}(A)$ is equivalent to the condition that $\mathfrak{p}(A) \subset [Y_1, \ldots, Y_n]$. Thus, the above problem about A is a special case of the following: given a prime differential ideal $\mathfrak{p}$ of $\mathfrak{R}$, to determine whether or not $\mathfrak{p}(A) \subset \mathfrak{p}$. A second interesting special case is that in which $\mathfrak{p}$ is the general component of another irreducible differential polynomial B: *Given A and B, to determine whether $\mathfrak{p}(A) \subset \mathfrak{p}(B)$.* Here, too, the results are meager.

2. Extensions of differential specializations

Let $(\eta_1, \ldots, \eta_n)$ and $(\zeta_1, \ldots, \zeta_n)$ be points of $\mathfrak{U}^n$. We say that $(\zeta_1, \ldots, \zeta_n)$ is a *differential specialization of* $(\eta_1, \ldots, \eta_n)$ over $\mathfrak{F}$, and write $(\eta_1, \ldots, \eta_n) \underset{\mathfrak{F}}{\longrightarrow} (\zeta_1, \ldots, \zeta_n)$, if every differential polynomial in $\mathfrak{F}\{Y_1, \ldots, Y_n\}$ that vanishes at $(\eta_1, \ldots, \eta_n)$ vanishes also at $(\zeta_1, \ldots, \zeta_n)$. When this is the case, and when k is an integer with $1 \leqslant k < n$, then obviously $(\eta_1, \ldots, \eta_k) \underset{\mathfrak{F}}{\longrightarrow} (\zeta_1, \ldots, \zeta_k)$; we say that the former differential specialization is an *extension* of this one.

Let $(\eta_1, \ldots, \eta_n)$ and k be given, and let $B \in \mathfrak{F}\{Y_1, \ldots, Y_n\}$, $B(\eta_1, \ldots, \eta_n) \neq 0$. It is known ([5], [6], [7]) that there exists a nonzero $B_0 \in \mathfrak{F}\{Y_1, \ldots, Y_k\}$ with $B_0(\eta_1, \ldots, \eta_k) \neq 0$ such that every differential specialization $(\eta_1, \ldots, \eta_k) \underset{\mathfrak{F}}{\longrightarrow} (\zeta_1, \ldots \zeta_k)$ with $B_0(\zeta_1, \ldots, \zeta_k) \neq 0$ can be extended to a differential specialization $(\eta_1, \ldots, \eta_n) \underset{\mathfrak{F}}{\longrightarrow} (\zeta_1, \ldots, \zeta_n)$ with $B(\zeta_1, \ldots, \zeta_n) \neq 0$.

It might be expected, in analogy with the usual notion of specialization as used in algebraic geometry, that every differential specialization $(\eta_1, \ldots, \eta_k) \underset{\mathfrak{F}}{\longrightarrow} (\zeta_1, \ldots, \zeta_k)$ can be extended to some differential specialization $(\eta_1, \ldots, \eta_n) \underset{\mathfrak{F}}{\longrightarrow} (\zeta_1, \ldots, \zeta_n)$, provided that some of the elements $\zeta_{k+1}, \ldots, \zeta_n$ are allowed to be ∞ in an appropriate sense. It is noteworthy that this is not so, even in the simple case in which $m = 1$, $n = 2$, $k = 1$, and $\mathfrak{F}$ is the ordinary differential field $\mathbf{C}(x)$ of rational functions of a complex variable x. For example, if we take $\eta_1 = \Gamma(x)$, where $\Gamma(x)$ is the usual gamma function, and $\eta_2 = \wp\left(\int \Gamma(x)^{-1}dx\right)$, where $\wp$ is a doubly periodic function of Weierstrass, then $\eta_1 \underset{\mathfrak{F}}{\longrightarrow} 0$ but there does not exist any ζ_2 such that $(\eta_1, \eta_2) \underset{\mathfrak{F}}{\longrightarrow} (0, \zeta_2)$ or $(\eta_1, \eta_2^{-1}) \underset{\mathfrak{F}}{\longrightarrow} (0, \zeta_2)$.

The problem, then, is *to find a criterion that a differential specialization* $(\eta_1, \ldots, \eta_k) \underset{\mathfrak{F}}{\longrightarrow} (\zeta_1, \ldots, \zeta_k)$ *can be extended to a differential specialization* $(\eta_1, \ldots, \eta_n) \underset{\mathfrak{F}}{\longrightarrow} (\zeta_1, \ldots, \zeta_n)$.

A special case is met in the problem of indeterminate forms. Let $F, G \in \mathfrak{F}\{Y_1, \ldots, Y_n\}$ be relatively prime, with $G \neq 0$, and suppose that F and G vanish at $(0, \ldots, 0)$. The problem is to assign a value to the quotient F/G at $(0, \ldots, 0)$. If elements $t_1, \ldots, t_n \in \mathfrak{U}$ are differentially algebraically independent over $\mathfrak{F}$ and we set $u = F(t_1, \ldots, t_n)/G(t_1, \ldots, t_n)$, it is natural to say that F/G admits the value α at $(0, \ldots, 0)$ when $(t_1, \ldots, t_n, u) \underset{\mathfrak{F}}{\longrightarrow} (0, \ldots, 0, \alpha)$. Thus, the problem becomes that of finding the extensions of $(t_1, \ldots, t_n) \underset{\mathfrak{F}}{\longrightarrow} (0, \ldots, 0)$ to $(t_1, \ldots, t_n, u)$. This is equivalent to determining the elements $\alpha \in \mathfrak{U}$ such that $(0, \ldots, 0, \alpha)$ is a zero of the general component of the differential polynomial $Y_{n+1}G - F \in \mathfrak{F}\{Y_1, \ldots, Y_{n+1}\}$. Ritt conjectured [8] that either α is unique (possibly ∞) or else α is completely arbitrary: he proved his conjecture in the case $m = 1$, $n = 1$, $\operatorname{ord}(FG) = 1$.

3. Differential dimension polynomials

A prime differential ideal $\mathfrak{p}$ of $\mathfrak{F}\{Y_1, \ldots, Y_n\}$ has a *generic zero*: this is a zero $(\eta_1, \ldots, \eta_n)$ such that every element of $\mathfrak{F}\{Y_1, \ldots, Y_n\}$ that vanishes at $(\eta_1, \ldots, \eta_n)$ is an element of $\mathfrak{p}$. The extension $\mathfrak{G} = \mathfrak{F}\langle\eta_1, \ldots, \eta_n\rangle$ has a finite differential transcendence degree that we denote by $d(\mathfrak{p})$. This number $d(\mathfrak{p})$ is called the *differential dimension* of $\mathfrak{p}$, and is the "correct" definition of what in the classical literature is called the number of arbitrary functions of m complex variables on which the solution of the system $P = 0$ $(P \in \mathfrak{p})$ depends; it measures the size of the set of zeros of $\mathfrak{p}$. If $\mathfrak{q}$ is another prime differential ideal, and if $\mathfrak{p} \subset \mathfrak{q}$, then $d(\mathfrak{p}) \geqslant d(\mathfrak{q})$, but if the inclusion is strict, the inequality need not be. For this reason a finer measure of the size is desirable. This is provided by the *differential dimension polynomial* $\omega_{\mathfrak{p}}$ of $\mathfrak{p}$; we recall its definition [9]. For any natural number s the ring $\mathfrak{R}_s = \mathfrak{F}[(\delta_1^{i_1} \ldots \delta_m^{i_m} Y_j)_{i_1 + \ldots + i_m \leq s,\, 1 \leq j \leq n}]$ is a polynomial algebra over $\mathfrak{F}$ in $\binom{s+m}{m} n$ indeterminates; the intersection $\mathfrak{p} \cap \mathfrak{R}_s$ is a prime ideal of $\mathfrak{R}_s$ and therefore has a dimension in the usual sense; there exists a unique polynomial $\omega_{\mathfrak{p}}$ in one variable over $\mathbf{Q}$ such that $\dim(\mathfrak{p} \cap \mathfrak{R}_s) = \omega_{\mathfrak{p}}(s)$ for all sufficiently big natural numbers s. This polynomial has degree $\leqslant m$, and therefore can be written in the form $\omega_{\mathfrak{p}}(s) = \sum_{0\leq i\leq m} \alpha_i(\mathfrak{p}) \binom{s-i}{i}$. The coefficients $\alpha_i(\mathfrak{p})$ are then integers. We order these differential dimension polynomials by defining $\omega_{\mathfrak{p}} \geqslant \omega_{\mathfrak{q}}$ to mean that $\omega_{\mathfrak{p}}(s) \geqslant \omega_{\mathfrak{q}}(s)$ for all sufficiently big integers s, i.e. we order them lexicographically with respect to $(a_m(\mathfrak{p}), \ldots, a_0(\mathfrak{p}))$. Then the inclusion $\mathfrak{p} \subset \mathfrak{q}$ implies the inequality $\omega_{\mathfrak{p}} \geqslant \omega_{\mathfrak{q}}$, and when the inclusion is strict so is the inequality. The connection with the differential dimension is given by the equation $a_m(\mathfrak{p}) = d(\mathfrak{p})$.

It may happen that $a_m(\mathfrak{p}) = 0$. The biggest natural number $\tau = \tau(\mathfrak{p})$ such that $\alpha_\tau(\mathfrak{p}) \neq 0$ is called the *differential type* of $\mathfrak{p}$, and $\alpha_\tau(\mathfrak{p})$ is called the *typical differential dimension of* $\mathfrak{p}$. This terminology is justified by the fact that if we regard $\mathfrak{F}$ and $\mathfrak{G}$ as differential fields with respect to τ new derivation operators $\delta'_{i'} = \sum_{1\leq i\leq m} c_{ii'}\delta_i$ $(1 \leqslant i' \leqslant \tau)$, where the $c_{ii'}$ are constants in $\mathfrak{F}$ subject to a certain inequation, then $\mathfrak{G}$ becomes a finitely generated extension of $\mathfrak{F}$ of differential transcendence degree $\alpha_\tau(\mathfrak{p})$.

The differential dimension polynomial $\omega_{\mathfrak{p}}$ is a birational invariant but not a differential birational invariant; this means that if η and ζ are generic zeros of $\mathfrak{p}$ and $\mathfrak{q}$, respectively, then the condition $\mathfrak{F}(\eta) = \mathfrak{F}(\zeta)$ implies that $\omega_{\mathfrak{p}} = \omega_{\mathfrak{q}}$ but the weaker condition $\mathfrak{F}\langle\eta\rangle = \mathfrak{F}\langle\zeta\rangle$ does not. Nevertheless, $\omega_{\mathfrak{p}}$ carries certain differential birational invariants with it. An obvious example is the differential dimension $a_m(\mathfrak{p})$; two others are $\tau(\mathfrak{p})$ and $\alpha_{\tau(\mathfrak{p})}(\mathfrak{p})$. It would be interesting to discover other differential birational invariants, for any such invariant should have great significance for $\mathfrak{p}$. I should mention that an interesting connection between $\omega_{\mathfrak{p}}$ and the characteristic polynomials of Hilbert-Serre has recently been discovered by Joseph L. Johnson; his work will appear in his Columbia University dissertation.

Consider a subset Σ of $\mathfrak{F}\{Y_1, \ldots, Y_n\}$. If the elements of Σ have bounded orders then the differential dimension polynomials of the components of $\{\Sigma\}$ are subject to certain restriction. Specifically, *if for each Y_j the order in Y_j of every element of Σ is $\leqslant e_j$, then for any component $\mathfrak{p}$ of $\{\Sigma\}$, the condition $a_m(\mathfrak{p}) = 0$ implies the condition* $a_{m-1}(\mathfrak{p}) \leqslant \sum e_j$. (This generalizes to any m the result proved by Ritt [2, p. 135] for $m = 1$.) There are grounds for conjecturing that in general

$$\alpha_{\tau(\mathfrak{p})}(\mathfrak{p}) \leqslant \sum_{1\leq j\leq n} \binom{e_j + m - \tau(\mathfrak{p}) - 1}{m - \tau(\mathfrak{p})}.$$

18—1220

When $\tau(\mathfrak{p}) = m$ this states that $a_m(\mathfrak{p}) \leqslant m$, which is obvious, and when $\tau(\mathfrak{p}) = m - 1$ this reduces to the result just mentioned.

When the set $\Sigma \subset \mathfrak{F}\{Y_1, \ldots, Y_n\}$ consists of precisely n differential polynomials $F_1, \ldots, F_n$ then there are two further conjectures. Set $e_{ij} = \mathrm{ord}_{Y_j} F_i$ $(1 \leqslant i \leqslant n,\ i \leqslant j \leqslant n)$ and $h = \max_{\pi} (e_{1\pi(1)} + \ldots + e_{n\pi(n)})$, where π runs through the symmetric group $\mathfrak{S}_n$. This number h, heuristically arrived at by Jacobi, is in general smaller than the number $\sum e_j$. The first of the conjectures then is: *For any component* $\mathfrak{p}$ *of* $\{F_1, \ldots, F_n\}$, *if* $a_m(\mathfrak{p}) = 0$ *then* $a_{m-1}(\mathfrak{p}) \leqslant h$. Ritt [10] proved this conjecture in two special cases: 1° $m = 1$, $n = 2$; 2° $m = 1$, each F_j linear. Aside from these cases, nothing seems to be known. The second of the conjectures is: *For any component* $\mathfrak{p}$ *of* $\{F_1, \ldots, F_n\}$, *if* $a_m(\mathfrak{p}) = a_{m-1}(\mathfrak{p}) = 0$ *then* $\omega_{\mathfrak{p}} = 0$. When the F_j are linear this may be regarded as a precise formulation of a conjecture made by Janet [11]. Even in this case the problem is open.

4. Picard-Vessiot theory

Now let $\mathfrak{F}$ be an ordinary differential field and denote its field of constants by $\mathfrak{C}$; denote the field of constants of $\mathfrak{U}$ by $\mathfrak{K}$. Consider a homogeneous linear differential polynomial $L = Y^{(n)} + a_1 Y^{(n-1)} + \ldots + a_n Y$ with coefficients in $\mathfrak{F}$. If $\mathfrak{C}$ is algebraically closed then L has a fundamental system of zeros $(\eta_1, \ldots, \eta_n)$ such that the extension $\mathfrak{G} = \mathfrak{F}\langle\eta_1, \ldots, \eta_n\rangle$ of $\mathfrak{F}$ has the same field of constants $\mathfrak{C}$, and then the set of all isomorphisms over $\mathfrak{F}$ of $\mathfrak{G}$ onto an extension of $\mathfrak{F}$ in $\mathfrak{U}$ has a natural group structure; by means of the fundamental system of zeros $(\eta_1, \ldots, \eta_n)$ this group can be identified with an algebraic subgroup G of $GL_k(n)$ defined over $\mathfrak{C}$. This G can be called the *Galois group of* L over $\mathfrak{F}$. Of course, a different choice of fundamental systems of zeros in general gives a different Galois group; G is determined only up to conjugation in $GL(n)$ by a matrix that is rational over $\mathfrak{C}$. (This ambiguity can be removed by taking for G an algebraic subgroup of the group $GL(V)$ of all automorphisms of the vector space V of zeros of L.)

As in ordinary Galois theory, there are two general problems:

1° *Given* L, *to determine* G.

2° *Given* G, *to find an* L *of which* G *is the Galois group*.

Of course, the nature of these problems depends on the differential field $\mathfrak{F}$ which is given. For example, when $\mathfrak{F} = \mathfrak{C}$ then it is easy to see that, for every L, G is reducible to triangular form (and hence is solvable). A. Białynicki-Birula [12] has shown that if $\mathfrak{F}$ has finite transcendence degree over $\mathfrak{C}$ and G is a connected and nilpotent algebraic subgroup of $GL(n)$ defined over $\mathfrak{C}$, then there exists a homogeneous linear differential polynomial L with coefficients in $\mathfrak{F}$ such that the Galois group of L over $\mathfrak{F}$ is isomorphic to G. Aside from this and some work in a slightly different direction by Lawrence Goldman [13], even when $\mathfrak{F} = \mathbf{C}(x)$ very little of a general nature is known. For the second problem this is not unexpected, in view of the history of the analogous problem in classical Galois theory. The obstinacy of the first problem is somewhat of a surprise, and it now appears that we should be grateful for almost any kind of information about G; the difficulty stems in part from the fact that a zero of L that is not a zero of any *linear* differential polynomial over $\mathbf{C}(x)$ of lower order may very well be a zero of a nonlinear one.

5. Rational approximation

Let me close with my favorite problem. This is to prove for algebraic differential equations an analog of the Thue-Siegel-Roth theorem [14] on approximation to algebraic numbers by rational ones.

Let K be any field and x be an indeterminate over K. The field $K((x^{-1}))$ of formal power series in x^{-1} over K contains the field $K(x^{-1}) = K(x)$ of rational fractions in x over K. There is a discrete nonarchimedean valuation of $K((x^{-1}))$ such that $|\sum_{k \geqslant r} c_k (x^{-1})^k| = e^{-r}$ when $c_r \neq 0$; $K((x^{-1}))$ is the completion of $K(x)$ with respect to this valuation.

Relative to the derivation operator d/dx, $K((x^{-1}))$ is an ordinary differential field; $K(x)$ is a differential subfield of $K((x^{-1}))$, and $K[x]$ is a differential subring of $K(x)$ such that $|b| \geqslant 1$ for every nonzero $b \in K[x]$. Suppose henceforth that the characteristic of K is 0. Then the field of constants of $K((x^{-1}))$ is K, and $|b'| = |b| e$ for every $b \in K((x^{-1}))$ such that $|b| \neq 1$.

Consider any differential polynomial P in one differential indeterminate y, and let z be another differential indeterminate. There is a smallest natural number d such that $P(y/z) z^d$ is a differential polynomial in y, z; we call this smallest d the *denomination* of P. (When P is a polynomial, i. e. a differential polynomial of order 0, then the denomination of P is its degree.) An *element* $u \in K((x^{-1}))$ that is differentially algebraic over $K(x)$ is said to have denomination d, if u is a zero of a differential polynomial in $K(x)\{y\}$ of denomination d but is not a zero of one of denomination $< d$.

It is known ([15], [16]) that *if $u \in K((x^{-1}))$ is differentially algebraic over $K(x)$ and of denomination d then there exists a real number $\alpha > 0$ such that*

$$\left| u - \frac{a}{b} \right| > \frac{\alpha}{|b|^d}$$

18*

for all $a, b \in K[x]$ *with* $b \neq 0$ *and* $u \neq a/b$. This generalizes the analog for algebraic functions of Liouville's simple precurser of the Thue-Siegel-Roth theorem.

The problem, then, is *to show that the exponent d here can be replaced by any number greater than* 2. This seems to be extremely difficult, and at present any improvement in the exponent would be of interest. One of the difficulties arises from the fact that in Ritt's process of reducing a differential polynomial A, by a differential polynomial B (analog of Euclidean division of polynomials), it is necessary to multiply A by a power of the separant of B; this spoils all estimates.

Dept. of Mathematics,
Columbia University,
New York, USA

REFERENCES

[1] Ritt J. F., Manifolds of functions defined by systems of algebraic differential equations, *Trans. Amer. Math. Soc.*, **32** (1930), 369-398.
[2] Ritt J. F., Differential algebra, Amer. Math. Soc. Colloquium Publications, **33** (1950), viii + 184 pp.
[3] Ritt J. F., On the singular solutions of algebraic differential equations, *Annals of Math.*, **37** (1936), 552-617.
[4] Kolchin E. R., Singular solutions of algebraic differential equations and a lemma of Arnold Shapiro, *Topology*, **3**, Suppl. 2 (1965), 309-318.
[5] Ritt J. F., On a type of algebraic differential manifold, *Trans. Amer. Math. Soc.*, **48** (1940), 542-552.
[6] Seidenberg A., An elimination theory for differential algebra, Univ. California Publ. Math. (N.S.), 3 (1956), 31-65.
[7] Rosenfeld Azriel, Specializations in differential algebra, *Trans. Amer. Math. Soc.*, **90** (1959), 394-407.
[8] Ritt J. F., Indeterminate expressions involving an analytic function and its derivatives, *Monatshefte für Math.*, **43** (1936), 97-104.
[9] Kolchin E. R., The notion of dimension in the theory of algebraic differential equations, *Bul. Amer. Math. Soc.*, **70** (1964), 570-573.
[10] Ritt J. F., Jacobi's problem on the order of a system of differential equations, *Annals of Math.*, **36** (1935), 303-312.
[11] Janet M., Sur les systèmes aux derivées partielles comprenant autant d'équations que de fonctions inconnues, *Comptes Rendues Acad. Sci. Paris*, **172** (1921), 1637-1639.
[12] Białynicki-Birula A., On the inverse problem of Galois Theory of differential fields, *Bul. Amer. Math. Soc.*, **69** (1963), 960-964.
[13] Goldman Lawrence, Specialization and Picard-Vessiot theory, *Trans. Amer. Math. Soc.*, **85** (1957), 327-356.
[14] Roth K. F., Rational approximation to algebraic numbers, *Mathematika*, **2** (1955), 1-20.
[15] Maillet E., Nombres transcendents, Paris, 1906.
[16] Kolchin E. R., Rational approximation to solutions of algebraic differential equations, *Proc. Amer. Math. Soc.*, **10** (1959), 238-244.

ALGEBRAIC GROUPS AND ALGEBRAIC DEPENDENCE.

By E. R. Kolchin.[1]

Introduction. The purpose of this note is to show how a certain rather trifling theorem about algebraic groups can be used, with the help of the Galois theory of differential fields, to prove that if certain functions are algebraically dependent then they satisfy an algebraic equation of a very special kind.

Typical of the results that can be obtained in this way is one due to Ostrowski [1]. He dealt with a differential field $\mathcal{F}$ of meromorphic functions that contains the field $\boldsymbol{C}$ of complex numbers, and with functions $\eta_1, \cdots, \eta_n$ that are primitive over $\mathcal{F}$ (i.e. that have the property that their derivatives are in $\mathcal{F}$). He showed that if $\eta_1, \cdots, \eta_n$ are algebraically dependent over $\mathcal{F}$ then there exist $c_1, \cdots, c_n \in \boldsymbol{C}$ not all 0 and an $\alpha \in \mathcal{F}$ such that $\sum c_j\eta_j = \alpha$.

A seemingly different sort of result, that follows in the same way albeit with a little more effort, is due to Siegel [2] (Satz 9). He considered nonzero complex numbers $a_1, \cdots, a_k$ having distinct squares, and complex numbers $\nu_1, \cdots, \nu_l$ enjoying the property that none of the numbers $\nu_j + \frac{1}{2}$, and none of the numbers $\nu_j \pm \nu_{j'}$ with $j \neq j'$, are in the ring $\boldsymbol{Z}$ of integers. Letting $J_\nu(x)$, $Y_\nu(x)$ denote (for each $\nu = \nu_1, \cdots, \nu_l$) a fundamental system of solutions of the Bessel differential equation

$$y'' + x^{-1}y' + (1 - \nu^2x^{-2})y = 0,$$

he showed that the $3kl$ functions

$$J_{\nu_j}(a_ix), Y_{\nu_j}(a_ix), J'_{\nu_j}(a_ix) \qquad (1 \leqq i \leqq k, 1 \leqq j \leqq l)$$

are algebracally independent over $\boldsymbol{C}(x)$.

These and other results are derived below in an abstract setting, following the formulation and proof of the theorem on algebraic groups.

1. *K*-simple algebraic groups. Throughout this section K denotes a field that is contained in a fixed field U (the universal field) that is algebraically closed and of infinite transcendence degree over K. *We assume throughout that the field characteristic is* 0. This guarantees that if G is

Received June 5, 1967.

[1] This work was partially supported by the National Science Foundation (grants GP-5303 and GP-6535).

an algebraic group defined over K and H is an algebraic subgroup of G then the condition that H be defined over K is equivalent to the condition that H be K-closed in G; also, if G' and G'' are algebraic groups defined over K and $f'\colon G \to G'$ and $f''\colon G \to G''$ are K-homomorphisms (i.e. rational homomorphisms defined over K) such that f' is surjective and has kernel contained in that of f'', then there exists a unique K-homomorphism $g\colon G' \to G''$ such that $g \circ f' = f''$. For any subfield L of U containing a field of definition of G, we denote by G_L the group consisting of all points of G that are rational over L.

We say that an algebraic group G is *K-simple* if G satisfies the following three condition: G is defined over K; $\dim G > 0$; every normal K-closed subgroup of G other than G has dimension 0.

For example, every one dimensional connected algebraic group defined over K is K-simple; in particular, the additive group U and the multiplicative group U^* are K-simple. It is well known (L. E. Dickson, 1901) that the unimodular groups $SL(n)$ with $n \geqq 2$ are K-simple (for a modern proof see [3], or Dieudonné [4] p. 38.)

In general, since the component of the identity of an algebraic group defined over K is itself defined over K, a K-simple algebraic group must be connected. It follows that every normal K-closed proper subgroup of a K-simple algebraic group G is central. Indeed, if the subgroup $F \neq G$ is normal and K-closed then, for any $u \in F$, the formula $v \mapsto vuv^{-1}$ defines a continuous mapping of G into F; since G is connected and F is finite the image reduces to a single point, which evidently is u, so that u is central in G. A similar argument shows that if Z is the center of G then Z/F is the center of G/F (for any $u \in G$ such that uF is in the center of Z/F, $v \mapsto uvu^{-1}v^{-1}$ is a continuous mapping of G into F). Therefore G is commutative if and only if G/F is, and when G is not commutative then G/Z contains no normal K-closed proper subgroup other than 1.

THEOREM. *Let $G_1, \cdots, G_n$ be K-simple algebraic groups; for each index j let Z_j denote the center of G_j and $\pi_j\colon G_j \to G_j/Z_j$ denote the canonical homomorphism. Let G be a K-closed proper subgroup of the direct product $\times_{1 \leqq i \leqq n} G_i$. Then one of the following three conditions is satisfied.*

(i) *There exists an index j such that the image of G under the canonical homomorphism $\times_{1 \leqq i \leqq n} G_i \to G_j$ is not G_j.*

(ii) *There exist two distinct indices j and k with G_j and G_k each noncommutative, and a K-isomorphism $f\colon G_j/Z_j \to G_k/Z_k$, such that $f(\pi_j(x_j)) = \pi_k(x_k)$ for every $(x_1, \cdots, x_n) \in G$.*

(iii) *There exist* l *distinct indices* $j(1),\cdots,j(l)$ *with* $l\geqq 2$ *and* $G_j(1),\cdots,G_j(l)$ *each commutative, and* l *surjective* K*-homomorphisms* $f_\lambda\colon G_{j(\lambda)}\to G_{j(l)}$ $(1\leqq\lambda\leqq l)$, *such that* $\prod_{1\leqq\lambda\leqq l} f_\lambda(x_{j(\lambda)})=1$ *for every* $(x_1,\cdots,x_n)\in G$.

Proof. If $n=1$ the conclusion is obvious. Let $n>1$ and assume the theorem proved for lower values of n. Denote the restrictions to G of the canonical projections $\times_{1\leqq i\leqq n} G_i\to G_j$ and $\times_{1\leqq i\leqq n} G_i\to\times_{1\leqq i\leqq n-1} G_i$ by $p_j\colon G\to G_j$ and $p^n\colon G\to\times_{1\leqq i\leqq n-1} G_i$, respectively; because of the inductive assumption, we may suppose that p_n and p^n are surjective. The kernel of p^n may be written as $1\times\cdots\times 1\times F_n$, where F_n is a K-closed subgroup of G_n. Since this kernel is normal in G, and p_n is surjective, F_n must be normal in G_n; since G_n is K-simple and

$$\begin{aligned}\dim F_n&=\dim\operatorname{Ker}(p^n)=\dim G-\dim\operatorname{Im}(p^n)\\&<\dim\times_{1\leqq i\leqq n} G_i-\dim\times_{1\leqq i\leqq n-1} G_i=\dim G_n,\end{aligned}$$

F_n must be finite with $F_n\subset Z_n$.

Let π'_n denote the canonical homomorphism $G_n\to G_n/F_n$. Since

$$\operatorname{Ker}(p^n)=1\times\cdots\times 1\times F_n\subset\operatorname{Ker}(\pi'_n\circ p_n),$$

there exists a K-homomorphism $g\colon\times_{1\leqq i\leqq n-1} G_i\to G_n/F_n$ such that $g\circ p^n=\pi'_n\circ p_n$. For each index j with $j<n$ let h_j denote the canonical injection $G_j\to\times_{1\leqq i\leqq n-1} G_i$, and set $g_j=g\circ h_j$. Then $g(x_1,\cdots,x_{n-1})=\prod_{1\leqq j\leqq n-1} g_j(x_j)$ for every $(x_i,\cdots,x_{n-1})\in\times_{1\leqq i\leqq n-1} G_i$. If $j,j'<n$ and $j\neq j'$ then every element of $h_j(G_j)$ commutes with every element of $h_{j'}(G_{j'})$, and hence every element of $g_j(G_j)$ commutes with every element of $g_{j'}(G_{j'})$. Since

$$\prod_{1\leqq i\leqq n-1} g_i(G_i)=g(\times_{1\leqq i\leqq n-1} G_i)=g(p^n(G))=\pi'_n(p_n(G))=G_n/F_n,$$

this implies that each $g_j(G_j)$ is normal in G_n/F_n, so that $g_j(G_j)$ (which is connected) is either 1 or G_n/F_n; correspondingly, $\operatorname{Ker}(g_j)$ is either G_j or a finite central subgroup of G_j. Furthermore, $g_j(G_j)=G_n/F_n$ for at least one index $j<n$, and if there is another such index, say j', then (because every element of $g_j(G_j)$ commutes with every element of $g_{j'}(G_{j'})$) G_n/F_n is commutative, and hence G_n is, too.

Suppose that G_n is not commutative. Then there exists a unique $j<n$ such that $g_j(G_j)=G_n/F_n$, and g_j followed by the canonical homomorphism

$G_n/F_n \to G_n/Z_n$ is a surjective K-homomorphism $g'\colon G_j \to G_n/Z_n$ the kernel of which is a finite central subgroup of G_j; the quotient of G_j by this kernel is K-isomorphic to G_n/Z_n and hence has no center, so that the kernel must be Z_j. Therefore there exists a K-isomorphism $f\colon G_j/Z_j \to G_n/Z_n$ such that $f \circ \pi_j = g'$, and evidently $f(\pi_j(x_j)) = \pi_n(x_n)$ for every $(x_1, \cdots, x_n) \in G$.

Suppose, finally, that G_n is commutative, and let $j(1), \cdots, j(l-1)$ denote the distinct indices $j < n$ such that $g_j(G_j) = G_n/F_n$; clearly, $l \geqq 2$. For each $\lambda = 1, \cdots, l-1$ the kernel $F_{j(\lambda)}$ of $g_{j(\lambda)}$ is finite and $G_{j(\lambda)}/F_{j(\lambda)}$ is commutative, too. Letting $g_n\colon G_n \to G_n/F_n$ denote the surjective K-homomorphism defined by the formula $x_n \mapsto \pi'_n(x_n)^{-1}$, and setting $j(l) = n$, we find that $\prod_{1 \leqq \lambda \leqq l} g_{j(\lambda)}(x_{j(\lambda)}) = 1$ for every $(x_1, \cdots, x_n) \in G$. If m denotes the lowest common multiple of the orders of the elements of F_n, then the mapping $w\colon G_n \to G_n$ defined by the formula $w(x_n) = x_n{}^m$ is a surjective K-homomorphism with finite kernel containing F_n, and therefore there exists a K-homomorphism $e\colon G_n/F_n \to G_n$ such that $e \circ \pi'_n = w$. Setting $f_\lambda = e \circ g_{j(\lambda)}$ $(1 \leqq \lambda \leqq l)$, we find that $\prod_{1 \leqq \lambda \leqq l} f_\lambda(x_{j(\lambda)}) = 1$ for every $(x_1, \cdots, x_n) \in G$. This completes the proof of the theorem.

We remark for later use that in certain cases the theorem can be improved. If the groups G_j and G_k appearing in the statement of condition (ii) are both equal to $SL(r)$, where $r \geqq 2$, then the K-isomorphim $f\colon G_j/Z_j \to G_k/Z_k$ is induced by a K-isomorphism $\bar{f}\colon G_j \to G_k$, i.e. by a K-automorphism of $SL(r)$. (This can be deduced, with only a little difficulty, from the known facts about the automorphisms of $SL(r)$ and $PSL(r)$ as given in Dieudonné [4]; a complete proof is given in [3].) Evidently $x_k^{-1}\bar{f}(x_j) \in Z_k$ for every $(x_1, \cdots, x_n) \in G$. The center Z_k of $SL(r)$ consists of the scalar matrices $\rho 1_r$ with ρ an element of the group $\boldsymbol{P}_r$ of r-th roots of unity (1_r here denoting the $r \times r$ unity matrix). Hence we can write $x_k^{-1}\bar{f}(x_j) = \gamma(x_1, \cdots, x_n)1_r$, where $\gamma(x_1, \cdots, x_n) \in \boldsymbol{P}_r$, and an easy computation shows that $\gamma\colon G \to \boldsymbol{P}_r$ is a K-homomorphism. But a K-automorphism of $SL(r)$ is of one of the following two types

$$x \mapsto bxb^{-1} \quad (x \in SL(r)),$$
$$x \mapsto b\check{x}b^{-1} \quad (x \in SL(r)),$$

where $b \in GL_k(r)$ and $\check{x}$ denotes the transpose of the inverse of x. (This can be deduced from the results given in Dieudonné [4]; a complete proof appears in [3].) Therefore either

$$x_k = \gamma(x_1, \cdots, x_n) b x_j b^{-1} \qquad ((x_1, \cdots, x_n) \in G)$$

or else

$$x_k = \gamma(x_1, \cdots, x_n) b \check{x}_j b^{-1} \qquad ((x_1, \cdots, x_n) \in G).$$

When $r = 2$ this improved conclusion can be simplified, because then

$$\check{x} = \begin{pmatrix} 0 & -1 \\ 1 & 0 \end{pmatrix} x \begin{pmatrix} 0 & -1 \\ 1 & 0 \end{pmatrix}^{-1}$$

as is easy to verify. Therefore if the groups G_j and G_k appearing in the statement of condition (ii) are known to be $SL(2)$ then that condition implies the following: *there exist a matrix $b \in GL_K(2)$ and a K-homomorphism $\gamma\colon G \to \boldsymbol{P}_2$ such that*

$$x_k = \gamma(x_1, \cdots, x_n) b x_j b^{-1} \qquad ((x_1, \cdots, x_n) \in G).$$

2. Ostrowski's theorem and its generalizations. Let $\mathcal{F}$ be a differential field of characteristic 0 with derivation operators $\delta_1, \cdots, \delta_m$ (the differential field being ordinary or partial according as $m = 1$ or $m > 1$). Denote the field of constants of $\mathcal{F}$ by $\mathcal{C}$. By "extension" we shall henceforth mean "differential field extension." For convenience we embed $\mathcal{F}$ in a universal extension $\mathcal{U}$ and denote the field of constants of $\mathcal{U}$ by $\mathcal{K}$; then $\mathcal{K}$ is an algebraically closed field extension of $\mathcal{C}$ of infinite transcendence degree. Consider n elements $\eta_1, \cdots, \eta_n \in \mathcal{U}$ such that the field of constants of the differential field $\mathcal{G} = \mathcal{F}\langle \eta_1, \cdots, \eta_n \rangle$ is $\mathcal{C}$.

Ostrowski's theorem asserts that *if each η_j is primitive over $\mathcal{F}$* (i.e. has the property that $\delta_i \eta_j \in \mathcal{F}$ $(1 \leq i \leq m)$) *and $\eta_1, \cdots, \eta_n$ are algebraically dependent over $\mathcal{F}$ then there exist constants $c_1, \cdots, c_n \in \mathcal{C}$ not all 0 such that $\sum c_j \eta_j \in \mathcal{F}$.* To start the proof, observe that each $\mathcal{F}\langle \eta_j \rangle$ is strongly normal over $\mathcal{F}$, and therefore $\mathcal{G}$ is, too. For every element σ of the Galois group $\mathfrak{G}(\mathcal{G}/\mathcal{F})$, $\sigma \eta_j = \eta_j + b_j(\sigma)$, where $b_j(\sigma) \in \mathcal{K}$, and the formula $\sigma \mapsto (b_1(\sigma), \cdots, b_n(\sigma))$ defines an isomorphism between $\mathfrak{G}(\mathcal{G}/\mathcal{F})$ and a $\mathcal{C}$-closed subgroup G of $\mathcal{K}^n$. Because $\eta_1, \cdots, \eta_n$ are algebraically dependent over $\mathcal{F}$, the transcendence degree of $\mathcal{G}$ over $\mathcal{F}$ is $< n$, and therefore $\dim G < n$. By the theorem of § 1, either there exists an index j such that the j-th projection of G is trivial, or else there exist distinct indices $j(1), \cdots, j(l)$ with $l \geq 2$ and surjective $\mathcal{C}$-homomorphisms $f_\lambda\colon \mathcal{K} \to \mathcal{K}$ $(1 \leq \lambda \leq l)$ such that $\sum_{1 \leq \lambda \leq l} f_\lambda(b_{j(\lambda)}(\sigma)) = 0$ for every $\sigma \in \mathfrak{G}(\mathcal{G}/\mathcal{F})$. In the former case $\sigma \eta_j = \eta_j$ for every $\sigma \in \mathfrak{G}(\mathcal{G}/\mathcal{F})$ whence $\eta_j \in \mathcal{F}$. In the latter case, since every $\mathcal{C}$-homomorphism $\mathcal{K} \to \mathcal{K}$ is of the form $x \mapsto cx$ with $c \in \mathcal{C}$, there exist non zero constants $c_{j(1)}, \cdots, c_{j(l)} \in \mathcal{C}$ such that $\sum_{1 \leq \lambda \leq l} c_{j(\lambda)} b_{j(\lambda)}(\sigma) = 0$ for every σ; but then $\sigma(\sum c_{j(\lambda)} \eta_{j(\lambda)}) = \sum c_{j(\lambda)} \eta_{j(\lambda)}$ for every σ, so that $\sum c_{j(\lambda)} \eta_{j(\lambda)} \in \mathcal{F}$.

The same method of proof gives a multiplicative analog of this theorem. It uses the fact that every $\mathcal{C}$-homomorphism $\mathcal{K}^* \to \mathcal{K}^*$ is of the form $x \to x^e$ with $e \in \mathbf{Z}$. The result is: *if each* η_j *is exponential over* $\mathcal{F}$ (i. e. has the property that $\eta_j \neq 0$ and $\eta_j^{-1}\delta_i\eta_j \in \mathcal{F}$ $(1 \leqq i \leqq m)$) *and* $\eta_1, \cdots, \eta_n$ *are algebraically dependent over* $\mathcal{F}$ *then there exist exponents* $e_1, \cdots, e_n \in \mathbf{Z}$ *not all* 0 *such that* $\prod \eta_j^{e_j} \in \mathcal{F}$.

The additive and multiplicative results can even be combined as follows: *Let* $\eta_1, \cdots, \eta_n, \zeta_1, \cdots, \zeta_r$ *be elements of an extension of* $\mathcal{F}$ *having field of constants* $\mathcal{C}$, *with each* η_j *primitive over* $\mathcal{F}$ *and each* ζ_k *exponential over* $\mathcal{F}$. *If* $\eta_1, \cdots, \eta_n, \zeta_1, \cdots, \zeta_r$ *are algebraically dependent over* $\mathcal{F}$ *then there exists either a nontrivial relation of the form* $\sum c_j\eta_j \in \mathcal{F}$ *or else one of the form* $\prod \zeta_k^{e_k} \in \mathcal{F}$. The proof is the same, using, in addition to the facts already mentioned, the fact that the only $\mathcal{C}$-homomorphisms $\mathcal{K} \to \mathcal{K}^*$ or $\mathcal{K}^* \to \mathcal{K}$ are trivial.

There is also an elliptic analog of the theorem. Let elements $g_2, g_3 \in \mathcal{C}$ have the property that $g_2^3 - 27g_3^2 \neq 0$. The elliptic curve

$$X_0X_2^2 = 4X_1^3 - g_2X_0^2X_1 - g_3X_0^3$$

in the projective plane, which curve we shall denote by $W(g_2, g_3)$, has a well known structure of algebraic group with $(0:0:1)$ as the unity element; this algebraic group is commutative and defined over $\mathbf{Q}(g_2, g_3) \subset \mathcal{C}$. For any $e \in \mathbf{Z}$ the mapping $w \mapsto w^e$ of $W(g_2, g_3)$ into itself is a $\mathcal{C}$-endomorphism. Under certain exceptional circumstances $W(g_2, g_3)$ has other $\mathcal{C}$-endomorphisms (*complex multiplications*), namely, when (and only when) the so called modulus $j = 64 \cdot 27g_2^3/(g_2^3 - 27g_3^2)$ is a root of one of a certain sequence of polynomials $F_h(X, X) \in \mathbf{Z}[X]$. (For details, see Weber [5].)

If ξ is any nonconstant element of an extension of $\mathcal{F}$, there exists an element η such that $(1:\xi:\eta) \in W(g_2, g_3)$; clearly, η is nonconstant and unique up to a factor ± 1. The nonconstant element ξ is said to be *Weierstrassian* over $\mathcal{F}$ (for the coefficients g_2, g_3) if $\eta^{-1}\delta_i\xi \in \mathcal{F}$ $(1 \leqq i \leqq m)$, i. e. if there exist elements $a_1, \cdots, a_m \in \mathcal{F}$ such that

$$(\delta_i\xi)^2 = a_i^2(4\xi^3 - g_2\xi - g_3) \qquad (1 \leqq i \leqq m).$$

The elliptic analog can now be formulated as follows. *Let* $\xi_1, \cdots, \xi_n$ *be elements of an extension of* $\mathcal{F}$ *having field of constants* $\mathcal{C}$, *with each* ξ_j *Weierstrassian over* $\mathcal{F}$ *for the coefficients* g_2, g_3, *and suppose there is no complex multiplication; for each index* j, *fix* η_j *so that* $(1:\xi_j:\eta_j) \in W(g_2, g_3)$. *If* $\xi_1, \cdots, \xi_n$ *are algebraically dependent over* $\mathcal{F}$ *then there exist exponents* $e_1, \cdots, e_n \in \mathbf{Z}$ *not all* 0 *such that* $\prod_{1 \leqq j \leqq n} (1:\xi_j:\eta_j)^{e_j}$ *is rational over* $\mathcal{F}$. The proof is essentially the same as in the other two cases, making use of the following facts: the differential field $\mathcal{G} = \mathcal{F}\langle\xi_1, \cdots, \xi_n\rangle$ is strongly normal over $\mathcal{F}$; the formula

$$\sigma \mapsto ((1:\xi_j:\eta_j)^{-1}(1:\sigma\xi_j:\sigma\eta_j))_{1\leq j\leq n}$$

defines an isomorphism of the Galois group $\mathfrak{G}(\mathcal{G}/\mathcal{F})$ onto a $\mathcal{C}$-closed subgroup G of $W_{\mathcal{K}}(g_2, g_3)^n$ having dimension equal to the transcendence degree of $\mathcal{G}$ over $\mathcal{F}$; the only $\mathcal{C}$-endomorphisms of $W(g_2, g_3)$ are of the form $w \mapsto w^e$ with $e \in \mathbf{Z}$ (i.e. $W(g_2, g_3)$ is without complex multiplication).

It is easy to see that the elliptic theorem can be combined with the additive and multiplicative ones. We can even include, in such a combined theorem, Weierstrassian elements corresponding to different elliptic curves $W(g'_2, g'_3), \cdots, W(g_2^{(r)}, g_3^{(r)})$, each without complex multiplication, subject only to the condition that when $k \neq l$ then there does not exist a nontrivial $\mathcal{C}$-homomorphism $W(g_2^{(k)}, g_3^{(k)}) \mapsto W(g_2^{(l)}, g_3^{(l)})$. This condition means (see Weber [5]) that the corresponding moduli $j^{(k)}, j^{(l)}$ must not satisfy any of a certain sequence of polynomial equations $F_h(j^{(k)}, j^{(l)}) = 0$.

3. Siegel's theorem. In this section we suppose, for the sake of simplicity, that the differential field $\mathcal{F}$ is *ordinary*, and for any element $\alpha \in \mathcal{U}$ we write α' instead of $\delta_1\alpha$.

For each integer j with $1 \leqq j \leqq n$, let $(\eta_{j1}, \cdots, \eta_{jr})$ be a fundamental system of solutions of a homogeneous linear differential equation

$$y^{(r)} + p_{j1}y^{(r-1)} + \cdots + p_{jr}y = 0$$

of order $r \geqq 2$ with coefficients in $\mathcal{F}$, and suppose that the field of constants of the differential field $\mathcal{G} = \mathcal{F}\langle(\eta_{jk})_{1\leq j\leq n, 1\leq k\leq r}\rangle$ is $\mathcal{C}$. For brevity, denote the Wronskian matrix $(\eta_{jk}{}^{(i-1)})_{1\leq i\leq r, 1\leq k\leq r}$ by η_j. Then $\mathcal{G}$ is a Picard-Vessiot extension of $\mathcal{F}$, as is the differential field $\mathcal{G}_j = \mathcal{F}\langle\eta_{j1}, \cdots, \eta_{jr}\rangle = \mathcal{F}(\eta_j)$ for each j. The canonical homomorphism $\mathfrak{G}(\mathcal{G}/\mathcal{F}) \to \mathfrak{G}(\mathcal{G}_j/\mathcal{F})$ of Galois groups, that to each element $\sigma \in \mathfrak{G}(\mathcal{G}/\mathcal{F})$ associates the restriction of σ to $\mathcal{G}_j$, is surjective. Now, the mapping $\sigma_j \mapsto \eta_j^{-1}\sigma_j\eta_j(\sigma_j \in \mathcal{G}_j/\mathcal{F}))$ identifies $\mathfrak{G}(\mathcal{G}_j/\mathcal{F})$ with a $\mathcal{C}$-closed subgroup G_j of $GL_{\mathcal{K}}(r)$. It follows that if, for each $\sigma \in \mathfrak{G}(\mathcal{G}/\mathcal{F})$, we set $c_j(\sigma) = \eta_j^{-1}\sigma\eta_j$, then the formula $\sigma \mapsto (c_1(\sigma), \cdots, c_n(\sigma))$ defines an isomorphism between the Galois group $\mathfrak{G}(\mathcal{G}/\mathcal{F})$ and a $\mathcal{C}$-closed subgroup G of $\times_{1\leqq j\leqq n} G_j$. The trancendence degree of $\mathcal{G}$ (resp. $\mathcal{G}_j$) over $\mathcal{F}$ equals the dimension of G (resp. G_j). Furthermore, $G_j \subset SL_{\mathcal{K}}(r)$ if and only if the differential equation $y' + p_{j1}y = 0$ has a nontrivial solution in $\mathcal{F}$, i.e. the Wronskian determinant $\det \eta_j$ is an element of $\mathcal{F}$.

We now prove that *if for each j the equation $y' + p_{j1}y = 0$ has a non-trivial solution in $\mathfrak{F}$ and the transcendence degree of $\mathfrak{G}$ over $\mathfrak{F}$ is $< n(r^2 - 1)$, then either there exists an index j such that the transcendence degree of $\mathfrak{G}_j$ over $\mathfrak{F}$ is $< r^2 - 1$, or else there exist two distinct indices j, j', and a non zero element $\alpha \in \mathfrak{G}$ with $\alpha^r \in \mathfrak{F}$, and a matrix $a \in GL_{\mathfrak{F}}(r)$, and a matrix $b \in GL_{\mathcal{C}}(r)$, such that one of the two conditions*

$$\eta_{j'} = \alpha a \eta_j b, \qquad \eta_{j'} = \alpha a \check{\eta}_j b$$

is satisfied.

Indeed, if the transcendence degree of each $\mathfrak{G}_j$ over $\mathfrak{F}$ is $r^2 - 1$ then $G_j = SL_{\mathcal{K}}(r)$ for every j. By the theorem in § 1 and the subsequent remarks, there exist distinct indices j and j', a homomorphism $\gamma\colon \mathfrak{S}(\mathfrak{G}/\mathfrak{F}) \to \boldsymbol{P}_r$, and a matrix $b' \in GL_{\mathcal{C}}(r)$, such that either $c_{j'}(\sigma) = \gamma(\sigma) b' c_j(\sigma) b'^{-1}$ for every $\sigma \in \mathfrak{S}(\mathfrak{G}/\mathfrak{F})$ or else $c_{j'}(\sigma) = \gamma(\sigma) b' \check{c}_j(\sigma) b'^{-1}$ for every $\sigma \in \mathfrak{S}(\mathfrak{G}/\mathfrak{F})$. Setting $u = \eta_{j'} b' \eta_j^{-1}$ in the former case and $u = \eta_{j'} b'{}^t\eta_j^{-1}$ in the latter, we find by straightforward computations that $\sigma u = \gamma(\sigma) u$ for every $\sigma \in \mathfrak{S}(\mathfrak{G}/\mathfrak{F})$. Hence if we let α be a non zero coordinate of u and set $a = \alpha^{-1} u$, then $\sigma(\alpha^r) = (\gamma(\sigma)\alpha)^r = \alpha^r$ and $\sigma a = a$ for every $\sigma \in \mathfrak{S}(\mathfrak{G}/\mathfrak{F})$, so that $\alpha^r \in \mathfrak{F}$ and $a \in GL_{\mathfrak{F}}(r)$. Setting $b = b'^{-1}$ we therefore find that $\eta_{j'} = u \eta_j b'^{-1} = \alpha a \eta_j b$ or $\eta_{j'} = u \eta_j b'^{-1} = \alpha a \eta_j b$. This proves the desired result.

We observe that if we replace the fundamental system of solutions $(\eta_{j1}, \cdots, \eta_{jr})$ by another, via a linear substitution with matrix $b \in GL_{\mathcal{C}}(r)$ then the matrix $\eta_j = (\eta_{jk}{}^{(i-1)})$ is replaced by $\eta_j b$. Coupling this observation with the remark, made in § 1, that

$$\check{x} = \begin{pmatrix} 0 & -1 \\ 1 & 0 \end{pmatrix} x \begin{pmatrix} 0 & -1 \\ 1 & 0 \end{pmatrix}^{-1}$$

for every $x \in GL(2)$, we find the following result.

For each index j with $1 \leqq j \leqq n$ let the differential equation

$$y'' + p_j y' + q_j y = 0$$

have coefficients in $\mathfrak{F}$ and have Galois group over $\mathfrak{F}$ identified with $SL_{\mathcal{K}}(2)$; suppose that the extension of $\mathfrak{F}$ obtained by adjoining to $\mathfrak{F}$ a fundamental system of solutions of each of these equations has transcendence degree $< 3n$ and has field of constants $\mathcal{C}$. Then there exist two distinct indices j, k, and fundamental systems of solutions (η_j, ζ_j), (η_k, ζ_k) of the j-th and k-th equations, respectively, and a non zero element α of the extension with $\alpha^2 \in \mathfrak{F}$, and a matrix $\begin{pmatrix} a_{11} & a_{12} \\ a_{21} & a_{22} \end{pmatrix} \in GL_{\mathfrak{F}}(2)$, such that

$$\begin{pmatrix}\eta_k & \zeta_k \\ \eta_k' & \zeta_k'\end{pmatrix} = \alpha\begin{pmatrix}a_{11} & a_{12} \\ a_{21} & a_{22}\end{pmatrix}\begin{pmatrix}\eta_j & \zeta_j \\ \eta_j' & \zeta_j'\end{pmatrix}.$$

Now, it is known (and is proved in the appendix, below) that a necessary and sufficient condition, that the Galois group of the Bessel equation

$$y'' + x^{-1}y' + (1 - \nu^2 x^{-2})y = 0$$

over $\boldsymbol{C}(x)$ be $SL_{\mathcal{K}}(2)$, is that $\frac{1}{2} + \nu \notin \boldsymbol{Z}$. Therefore the proof of Siegel's result described in the introduction is reduced to the proof of the following technical lemma which is in effect a long exercise.

Let a, b, ν, $\rho \in \boldsymbol{C}$, and let (η, ζ) and (ϕ, ψ) be, respectively, fundamental systems of solutions of the differential equations

$$y'' + x^{-1}y' + (a^2 - \nu^2 x^{-2})y = 0 \quad \textit{and} \quad y'' + x^{-1}y' + (b^2 - \rho^2 x^{-2})y = 0.$$

Suppose that $\frac{1}{2} + \nu$, $\frac{1}{2} + \rho \notin \boldsymbol{Z}$ and that

$$\begin{pmatrix}\phi & \psi \\ \phi' & \psi'\end{pmatrix} = \alpha\begin{pmatrix}a_{11} & a_{12} \\ a_{21} & a_{22}\end{pmatrix}\begin{pmatrix}\eta & \zeta \\ \eta' & \zeta'\end{pmatrix}$$

where $\begin{pmatrix}a_{11} & a_{12} \\ a_{21} & a_{22}\end{pmatrix} \in GL_{\boldsymbol{C}(x)}(2)$ and $\alpha^2 \in \boldsymbol{C}(x)$. Then $a^2 = b^2$ and either $\nu + \rho \in \boldsymbol{Z}$ or $\nu - \rho \in \boldsymbol{Z}$.

For the sake of completeness we perform the exercise. Modifying α and (a_{ij}) by suitable nonzero factors in $\boldsymbol{C}(x)$, we may suppose that $\alpha^2 = A$ and $a_{ij} = A_{ij}/B$, where A, B, A_{11}, A_{12}, A_{21}, $A_{22} \in C[x]$, A and B are nonzero and have highest coefficient 1, A and A' are relatively prime, and B, A_{11}, A_{12}, A_{21}, A_{22} have greatest common divisor 1. Evidently $\alpha' = \alpha A'/2A$. Now,

$$\begin{pmatrix}\eta & \zeta \\ \eta' & \zeta'\end{pmatrix}' = \begin{pmatrix}\eta' & \zeta' \\ \eta'' & \zeta''\end{pmatrix} = \begin{pmatrix}\eta' & \zeta' \\ -x^{-1}\eta' - (a^2 - \nu^2 x^{-2})\eta & -x^{-1}\zeta' - (a^2 - \nu^2 x^{-2})\zeta\end{pmatrix}$$
$$= \begin{pmatrix}0 & 1 \\ -(a^2 - \nu^2 x^{-2}) & -x^{-1}\end{pmatrix}\begin{pmatrix}\eta & \zeta \\ \eta' & \zeta'\end{pmatrix}$$

and there is an analogous equation for $\begin{pmatrix}\phi & \psi \\ \phi' & \psi'\end{pmatrix}$. Therefore

$$\alpha\begin{pmatrix}0 & 1 \\ -(b^2 - \rho^2 x^{-2}) & -x^{-1}\end{pmatrix}\begin{pmatrix}a_{11} & a_{12} \\ a_{21} & a_{22}\end{pmatrix}\begin{pmatrix}\eta & \zeta \\ \eta' & \zeta'\end{pmatrix} = \begin{pmatrix}\phi & \psi \\ \phi' & \psi'\end{pmatrix}'$$
$$= (\alpha'\begin{pmatrix}a_{11} & a_{12} \\ a_{21} & a_{22}\end{pmatrix} + \alpha\begin{pmatrix}a_{11}' & a_{12}' \\ a_{21}' & a_{22}'\end{pmatrix})\begin{pmatrix}\eta & \zeta \\ \eta' & \zeta'\end{pmatrix} + \alpha\begin{pmatrix}a_{11} & a_{12} \\ a_{21} & a_{22}\end{pmatrix}\begin{pmatrix}\eta & \zeta \\ \eta' & \zeta'\end{pmatrix}'$$
$$= (\alpha'\begin{pmatrix}a_{11} & a_{12} \\ a_{21} & a_{22}\end{pmatrix} + \alpha\begin{pmatrix}a_{11}' & a_{12}' \\ a_{11}' & a_{12}'\end{pmatrix}$$
$$+ \alpha\begin{pmatrix}a_{11} & a_{12} \\ a_{21} & a_{22}\end{pmatrix}\begin{pmatrix}0 & 1 \\ -(a^2 - \nu^2 x^{-2}) & -x^{-1}\end{pmatrix})\begin{pmatrix}\eta & \zeta \\ \eta' & \zeta'\end{pmatrix},$$

whence

$$\begin{pmatrix} a_{21}' & a_{22}' \\ a_{21}' & a_{22}' \end{pmatrix} = \begin{pmatrix} 0 & 1 \\ -(b^2 - \rho^2 x^{-1}) & -x^{-1} \end{pmatrix} \begin{pmatrix} a_{11} & a_{12} \\ a_{21} & a_{22} \end{pmatrix} - \frac{\alpha'}{\alpha} \begin{pmatrix} a_{11} & a_{12} \\ a_{21} & a_{22} \end{pmatrix}$$
$$- \begin{pmatrix} a_{11} & a_{12} \\ a_{21} & a_{22} \end{pmatrix} \begin{pmatrix} 0 & 1 \\ -(a^2 - \nu^2 x^{-2}) & -x^{-1} \end{pmatrix}.$$

Since $a_{ij} = A_{ij}/B$ and $\alpha'/\alpha = A'/2A$ this yields the equations

$$(\#)\begin{cases} x^2A(A_{11}'B - A_{11}B') = (-\tfrac{1}{2}x^2A'A_{11} + (a^2x^2 - \nu^2)AA_{12} + x^2AA_{21})B, \\ x^2A(A_{12}'B - A_{12}B') = (-x^2AA_{11} + (xA - \tfrac{1}{2}x^2A')A_{12} + x^2AA_{22})B, \\ x^2A(A_{21}'B - A_{21}B') = (-(b^2x^2 - \rho^2)AA_{11} \\ \qquad\qquad - (xA + \tfrac{1}{2}x^2A')A_{21} + (a^2x^2 - \nu^2(AA_{22})B, \\ x^2A(A_{22}'B - A_{22}B') = (-(b^2x^2 - \rho^2)AA_{12} - x^2AA_{21} - \tfrac{1}{2}x^2A'A_{22})B. \end{cases}$$

Let $c \neq 0$ be a root of B of multiplicity k. By (#), c is a root of each $x^2AA_{ij}B'$ of multiplicity $\geqq k$, and hence is a root (of necessity simple) of A. By (#), for each (i, j) the polynomial

$$x^2AA_{ij}B' - \tfrac{1}{2}x^2A'A_{ij}B = \tfrac{1}{2}x^2A_{ij}(2AB' - A'B)$$

is divisible by AB, and hence c is a root of $2AB' - A'B$ of multiplicity $\geqq 1 + k$; but when this last polynomial is expanded in powers of $x - c$ the coefficient of $(x - c)^k$ is seen to be different from 0. This contradiction shows that B has no nonzero root, so that $B = x^k$ for some $k \in \boldsymbol{N}$.

Let $c \neq 0$ be a root of A; by the above, c is not a root of $A'B$, and therefore by (#), for each (i, j) c is a root of A_{ij} of some multiplicity k_{ij}. If $\min k_{ij} = k_{11}$ then the first equation (#) shows that the polynomial $x^2AA_{11}'B + \tfrac{1}{2}x^2A'A_{11}B = \tfrac{1}{2}x^2B(2AA_{11}' + A'A_{11})$ has c as a root of multiplicity $\geqq 1 + k_{11}$, contradicting the evident fact that the coefficient of $(x - c)^{k_{11}}$ in this polynomial is not 0. In the same way we arrive at a contradiction if $\min k_{ij}$ equals k_{12} or k_{21} or k_{22}. Therefore A can have no nonzero root, so that $A = x^h$ where $h = 0$ or $h = 1$.

Let $p = \max \deg A_{ij}$, and denote the coefficient of x^p in A_{ij} by u_{ij}. Examining the terms in (#) or degree $2 + h + p + k$, we find that

$$\begin{cases} \qquad a^2u_{12} + u_{21} \qquad = 0, \\ -u_{11} \qquad\qquad + u_{22} = 0, \\ -b^2u_{11} \qquad\qquad + a^2u_{22} = 0, \\ \qquad -b^2u_{12} - u_{21} \qquad = 0. \end{cases}$$

As u_{11}, u_{12} can not both be 0 (otherwise every u_{ij} would be 0), we infer that $a^2 = b^2$.

The first and last equations (#) show that $\nu^2AA_{12}B$ and $\rho^2AA_{12}B$ are divisible by x^{k+1+h}, so that $(\nu^2-\rho^2)A_{12}$ is divisible by x; if A_{12} is not divisible by x it follows that $\nu^2-\rho^2=0$, whence $\nu+\rho=0\in\mathbf{Z}$ or $\nu-\rho\in\mathbf{Z}$. Therefore we may suppose that A_{12} is divisible by x, say $A_{12}=x\bar{A}_{12}$. Then the equations (#) reduce to

$$(\#\#)\begin{cases} xA_{11}'=(k-\frac{h}{2})A_{11}+(a^2x^2-\nu^2)\bar{A}_{12}+xA_{21},\\ x\bar{A}_{12}'=-A_{11}+(k-\frac{h}{2})\bar{A}_{12}+A_{22},\\ x^2A_{21}'=-(a^2x^2-\rho^2)A_{11}+(k-1+\frac{h}{2})xA_{21}+(a^2x^2-\nu^2)A_{22},\\ xA_{22}'=-(a^2x^2-\rho^2)\bar{A}_{12}-xA_{21}+(k-\frac{h}{2})A_{22}. \end{cases}$$

The first equation (##) plus $k-\frac{h}{2}$ times the second minus the fourth yields the congruence

$$((k-\frac{h}{2})^2-\nu^2-\rho^2)\bar{A}_{12}\equiv 0 \pmod{x},$$

and $-\rho^2$ times the first plus $k-\frac{h}{2}$ times the third plus ν^2 times the fourth yields the congruence

$$2\nu^2\rho^2\bar{A}_{12}\equiv 0 \pmod{x}.$$

If x does not divide $\bar{A}_{12}$ then either $\nu=0$, $(k-\frac{h}{2})^2-\rho^2=0$, $\rho=\pm(k-\frac{h}{2})$, $\rho=\pm k$ (because $\frac{1}{2}+\rho\notin\mathbf{Z}$), $\nu+\rho\in\mathbf{Z}$, or else $\rho=0$ and again $\nu+\rho\in\mathbf{Z}$. Therefore we may suppose that x divides $\bar{A}_{12}$. But then the second and third equations (##) yield the congruences

$$-A_{11}+A_{22}\equiv 0,\quad \rho^2A_{11}-\nu^2A_{22}\equiv 0 \pmod{x},$$

so that $(\nu^2-\rho^2)A_{11}\equiv 0 \pmod{x}$. Supposing as we may that $\nu^2-\rho^2\neq 0$, we infer that x divides A_{11} and hence A_{22}. Setting $A_{11}=xB_{11}$, $\bar{A}_{12}=xB_{12}$, $A_{21}=B_{21}$, $B_{22}=xB_{22}$, we find that the equations (##) reduce to

$$(\#\#\#)\begin{cases} xB_{11}'=(k-1-\frac{h}{2})B_{11}+(a^2x^2-\nu^2)B_{12}+B_{21},\\ xB_{12}'=-B_{11}+(k-1-\frac{h}{2})B_{12}+B_{22},\\ xB_{21}'=-(a^2x^2-\rho^2)B_{11}+(k-1-\frac{h}{2})B_{21}+(a^2x^2-\nu^2)B_{22},\\ xB_{22}'=-(a^2x^2-\rho^2)B_{12}-B_{21}+(k-1-\frac{h}{2})B_{22}. \end{cases}$$

We now show that $h=0$ (so that $A=1$ and $\alpha=\pm 1$). Assume $h=1$ (so that $k-1-\frac{h}{2}\neq 0$), and let $q=\max\deg B_{ij}$. The first equation (###) shows that $\deg B_{12}\leqq q-2$; then the second equation (###) shows that $\deg(B_{22}-B_{11})\leqq q-2$, and the sum of the first and fourth shows that $\deg(B_{11}+B_{22})\leqq q-2$, whence $\deg B_{11}\leqq q-2$, $\deg B_{22}\leqq q-2$, $q=B_{21}\geqq 2$. Therefore we may set

$$B_{11}=v_{11}x^{q-2}+\cdots,B_{12}=v_{12}x^{q-2}+\cdots,$$
$$B_{21}=v_{21}x^{q}+\cdots,B_{22}=v_{22}x^{q-2}+\cdots,$$

and of course $v_{21}\neq 0$. Then the first three equations (###) show that

$$\begin{cases} a^2v_{12}+v_{21} &=0,\\ -v_{11}+(k-1-\frac{h}{2}-q+2)v_{12}+v_{22} &=0,\\ -a^2v_{11}+(k-1-\frac{h}{2}-q)v_{21}+a^2v_{22} &=0; \end{cases}$$

the first of these equations shows that $v_{21}=-a^2v_{12}$ (so that $v_{12}\neq 0$ and $a\neq 0$) and therefore the third implies that

$$-v_{11}-(k-1-\frac{h}{2}-q)v_{12}+v_{22}=0,$$

which together with the second shows that $(2k-h-2q)v_{12}=0$. Since $v_{12}\neq 0$, we find the contradiction $1=h=2k-2q$. This shows that $h=0$.

This being the case, the equations (###) yield the system of congruences

$$\begin{cases} (k-1)B_{11}-\nu^2B_{12}+B_{21} &\equiv 0 \pmod{x},\\ -B_{11}+(k-1)B_{12}+B_{22} &\equiv 0 \pmod{x},\\ \rho^2B_{11}+(k-1)B_{21}-\nu^2B_{22} &\equiv 0 \pmod{x},\\ \rho^2B_{12}-B_{21}+(k-1)B_{22} &\equiv 0 \pmod{x}. \end{cases}$$

But the Wronskian determinants $\det\begin{pmatrix}\eta & \zeta\\ \eta' & \zeta'\end{pmatrix}$ and $\det\begin{pmatrix}\phi & \psi\\ \phi' & \psi'\end{pmatrix}$ both are solutions of the differential equation $y'+x^{-1}y=0$, and therefore

$$\det\begin{pmatrix}a_{11} & a_{12}\\ a_{21} & a_{22}\end{pmatrix}=\det\begin{pmatrix}\phi & \psi\\ \phi' & \psi'\end{pmatrix}\cdot\det\begin{pmatrix}\eta & \zeta\\ \eta' & \zeta'\end{pmatrix}^{-1}\in\boldsymbol{C};$$

since $B^2\det\begin{pmatrix}a_{11} & a_{12}\\ a_{21} & a_{22}\end{pmatrix}=A_{11}A_{22}-A_{12}A_{21}=x^2(B_{11}B_{22}-B_{12}B_{21})$, we infer that x divides B. If the determinant of the system of congruences were not 0 we should have $B_{21}\equiv 0 \pmod{x}$, i.e. $A_{21}\equiv 0 \pmod{x}$, and x would be a common factor of A_{11}, A_{12}, A_{21}, A_{22}, B. Hence the determinant is 0, i.e.

$$(\nu-\rho+k-1)(\nu+\rho+k-1)(\nu-\rho-k+1)(\nu+\rho-k+1)=0.$$

It follows, finally, that $\nu+\rho$ or $\nu-\rho$ is in $\boldsymbol{Z}$.

Appendix. The Galois group of the Bessel equation.

Consider the Bessel equation $y''+x^{-1}y'+(1-\nu^2x^{-2})y=0$, where $\nu\in\boldsymbol{C}$. Denote the Galois group of this equation over $\boldsymbol{C}(x)$ by G. Since the equation $y'+x^{-1}y=0$ has the nontrivial solution $x^{-1}\in\boldsymbol{C}(x)$, $G\subset SL_{\mathcal{K}}(2)$.

If $\nu-\frac{1}{2}\in\boldsymbol{Z}$ we may (replacing ν by $-\nu$ if necessary) suppose that the number $s=\nu-\frac{1}{2}$ is $\geqq 0$, and then set

$$\eta_{\pm}=e^{\pm ix}\sum_{0\leqq k\leqq s}\left(\frac{(s+k)!}{(s-k)!k!}\right)(\pm i)^k2^{-k}x^{-k-\frac{1}{2}};$$

it is easy to see that η_+, η_- form a fundamental system of solutions of the Bessel equation, that they are exponential over $\boldsymbol{C}(x)$, and that their product is in $\boldsymbol{C}(x)$. It follows that *if* $\nu-\frac{1}{2}\in\boldsymbol{Z}$ *then* G (relative to this fundamental system of solutions) *consists of all the matrices* $\begin{pmatrix}c & 0\\ 0 & c^{-1}\end{pmatrix}$ *with* $c\in\mathcal{K}^*$. We shall show that *if* $\nu-\frac{1}{2}\notin\boldsymbol{Z}$ *then* $G=SL_{\mathcal{K}}(2)$.

Suppose $G\neq SL_{\mathcal{K}}(2)$, so that $\dim G\leqq 2$; then the component of the identity G^0 is solvable, so that, relative to a suitable fundamental system of solutions (η,ζ), G^0 is triangular, i.e. $\sigma\eta$ is a constant times η for every $\sigma\in G^0$; hence the function $\phi=\eta'/\eta$ is invariant under G^0, so that ϕ is algebraic over $\boldsymbol{C}(x)$. Now, ϕ has an expansion in (possibly fractional) powers of x^{-1}; because ϕ satisfies the Riccati equation

$$y'+y^2+x^{-1}y+1-\nu^2x^{-2}=0,$$

in any such expansion we find that no nonintegral power occurs, that no negative power occurs, and that the constant term is $\pm i$. Similarly, for each $c\in\boldsymbol{C}$ with $c\neq 0$, in any expansion of ϕ in powers of $x-c$ nonintegral powers do not occur, negative powers other than $(x-c)^{-1}$ do not occur, and if $(x-c)^{-1}$ occurs it has coefficient 1. Finally, any expansion of ϕ in powers of x has no nonintegral term and is of the form $bx^{-1}+\cdots$ where $b^2=\nu^2$. This shows that $\phi=a+bx^{-1}+\sum_{1\leqq k\leqq s}(x-c_k)^{-1}$, where $a^2=-1$ and $c_1,\cdots,c_s$ are the poles of ϕ other than 0. Substituting this expression for ϕ in the left side of the Riccati equation and then multiplying by $x^2\prod_{1\leqq k\leqq s}(x-c_k)^2$, we obtain a polynomial in x. The coefficient of x^{2s+1} in this polynomial is

$2a(b+\frac{1}{2}+s)$. Since ϕ is a solution of the Riccati equation, this coefficient vanishes; since $b=\pm\nu$ we conclude that $\nu-\frac{1}{2}\in\boldsymbol{Z}$.

COLUMBIA UNIVERSITY.

REFERENCES.

[1] A. Ostrowski, "Sur les relations algébriques entre les intégrales indéfinies," *Acta Mathematica*, vol. 78 (1946), pp. 315-318.

[2] C. L. Siegel, "Über einige Anwendungen diophantischer Approximationen," *Abh. der preussischen Akad. der Wissenschaften, phys.-math. Klasse*, Jahrgang 1929, Nr. 1 (1930), pp. 1-70.

[3] E. R. Kolchin, *The birational automorphisms of $SL(n)$ and its quotients*, Columbia University, Department of Mathematics Report, 1967.

[4] J. Dieudonné, "La géométrie des groupes classiques," *Ergebnisse der Math. und ihrer Grenzgebiete*, neue Folge 5 (1955).

[5] H. Weber, *Lehrbuch der Algebra*, vol. 3, Braunschweig (1908).

DIFFERENTIAL POLYNOMIALS AND STRONGLY NORMAL EXTENSIONS.

By E. R. KOLCHIN[1] and T. SOUNDARARAJAN.

Let $\mathcal{F}$ be an ordinary differential field of characteristic zero, fix a universal extension $\mathcal{U}$ of $\mathcal{F}$ and an element $t \in \mathcal{U}$ that is differentially transcendental over $\mathcal{F}$, and denote the field of constants of $\mathcal{F}$ resp. $\mathcal{U}$ by $\mathcal{C}$ resp. $\mathcal{K}$. Let y be a differential indeterminate, and let $\mathcal{F}\{y\}$ denote the differential polynomial algebra in y over $\mathcal{F}$, that is $\mathcal{F}\{y\} = \mathcal{F}[y, y', y'', \cdots]$.

The field of constants of the differential field $\mathcal{F}\langle t\rangle$ is $\mathcal{C}$. If $\mathcal{E}$ is a differential field with $\mathcal{F} \subset \mathcal{E} \subset \mathcal{F}\langle t\rangle$ then, by Ritt's analog of Lüroth's theorem (see [3], page 52), there exists an element $u \in \mathcal{F}\langle t\rangle$ such that $\mathcal{E} = \mathcal{F}\langle u\rangle$. It is natural to ask for conditions under which $\mathcal{F}\langle t\rangle$ is a strongly normal extension of $\mathcal{E}$.[2] The analogous question for fields was considered in [4] and [5].

The purpose of this note is to give a complete answer to our question in the case in which u is given as a differential polynomial in t. Let $A \in \mathcal{F}\{y\}$, $A \notin \mathcal{F}$, and set $m = \operatorname{ord} A$, $n = \deg A$. We shall prove the following result.

THEOREM 1. *A necessary and sufficient condition that $\mathcal{F}\langle t\rangle$ be strongly normal over $\mathcal{F}\langle A(t)\rangle$ is that A can be written in the form $A = \alpha(L+\beta)^n + \gamma$ with $\alpha, \beta, \gamma \in \mathcal{F}$, $\alpha \neq 0$, $L = y^{(m)} + \alpha_1 y^{(m-1)} + \cdots + \alpha_m y$ and $\alpha_1, \cdots, \alpha_m \in \mathcal{F}$, that $\mathcal{F}$ contain a fundamental system of zeros of L, and in the case $n > 1$ that $\mathcal{F}$ contain a zero of $L + \beta$. When this condition is satisfied then $\mathcal{F}\langle t\rangle$ is a Picard-Vessiot extension of $\mathcal{F}\langle A(t)\rangle$ and its Galois group is $\mathcal{C}$-isomorphic to the group $G(m,n)$ of all matrices*

$$g(c_1, \cdots, c_m; a) = \left(\begin{array}{c|c} & c_1 \\ & \cdot \\ 1_m & \cdot \\ & \cdot \\ & c_m \\ \hline 0 \cdots 0 & a \end{array}\right)$$

with $c_1, \cdots, c_m \in \mathcal{K}$ and a an n-th root of unity.

Received April 27, 1971.

[1] The work of the first listed author was supported by a grant from the National Science Foundation.

[2] For an exposition of the Galois theory of differential fields see [1] and [2]. In those two papers the unnecessary restriction was imposed that $\mathcal{C}$ be algebraically closed. In the present paper this restriction is not made.

We shall also prove the following result.

THEOREM 2. *A necessary and sufficient condition that $\mathcal{F}\langle A(t)\rangle$ be the set of fixed elements of a group of automorphisms of the differential field $\mathcal{F}\langle t\rangle$ is that the condition in Theorem 1 be satisfied and that $\mathcal{F}$ contain all the n-th roots of unity.*

Remark. When an extension of a differential field has a group of automorphisms for which the fixed set is the given differential field the extension is said to be *weakly normal.* It follows from the two theorems that weak normality of $\mathcal{F}\langle t\rangle$ over $\mathcal{F}\langle A(t)\rangle$ implies strong normality, but not conversely unless $\mathcal{C}$ contains the n-th roots of unity.

The proof of these theorems makes use of the fact that if $P, Q \in \mathcal{F}\{y\}$, $PQ \notin \mathcal{F}$, and $\text{g.c.d}(P, Q) = 1$ then the element $v = P(t)/Q(t)$ of $F\langle t\rangle$ is differentially transcendental over $\mathcal{F}$, the differential polynomial

$$vQ - P \in \mathcal{F}\langle v\rangle\{y\}$$

is irreducible over $\mathcal{F}\langle v\rangle$, and t is a generic zero of its general component; this is well known and easy to prove. A consequence of this fact is that $\operatorname{tr\,deg} \mathcal{F}\langle t\rangle/\mathcal{F}\langle v\rangle = \max(\operatorname{ord} P, \operatorname{ord} Q)$. A second consequence is that if $v \in \mathcal{F}\langle A(t)\rangle$ then either $\max(\operatorname{ord} P, \operatorname{ord} Q) > m$ or else

$$\max(\deg_{y^{(m)}} P, \deg_{y^{(m)}} Q) \geqq \deg_{y^{(m)}} A.$$

A third consequence is that every automorphism of $\mathcal{F}\langle t\rangle$ over $\mathcal{F}$ is given by a fractional linear substitution $t \mapsto (ct + d)^{-1}(at + b)$ with $a, b, c, d \in \mathcal{F}$ and $ad - bc \neq 0$; a simple induction argument shows, for any $k > 0$, that then

$$t^{(k)} \mapsto (ad - bc)(ct + d)^{-2} t^{(k)} + \cdots$$

where the dots here stand for an element of $\mathcal{F}(t)[t', \cdots, t^{(k-1)}]$.

Proof of sufficiency. Starting with Theorem 1, let $A = \alpha(L + \beta)^n + \gamma$ as above, let $\eta_1, \cdots, \eta_m \in \mathcal{F}$ form a fundamental system of zeros of L, and let $\xi \in \mathcal{F}$ be a zero of $L + \beta$ or be 0 according as $n > 1$ or $n = 1$. Then

$$L' - \frac{(L(t+\xi)^n)'}{nL(t+\xi)^n} L$$

is a homogeneous linear differential polynomial of order $m + 1$ having coefficients in $\mathcal{F}\langle A(t)\rangle$ and admitting $\eta_1, \cdots, \eta_m$, $t + \xi$ as a fundamental system of zeros. Therefore $\mathcal{F}\langle t\rangle$ is a strongly normal (indeed, a Picard-Vessiot) extension of $\mathcal{F}\langle A(t)\rangle$, and for any element σ of its Galois group G, $\sigma\eta_i = \eta_i$ $(1 \leqq i \leqq m)$ and

$$\sigma(t+\xi) = c_1(\sigma)\eta_1 + \cdots + c_m(\sigma)\eta_m + a(\sigma)(t+\xi)$$

where $c_1(\sigma), \cdots, c_m(\sigma)$, $a(\sigma) \in \mathcal{K}$; the condition $\sigma L(t+\xi)^n = L(t+\xi)^n$ shows that $a(\sigma)^n = 1$, and hence the formula $\sigma \mapsto g(c_1(\sigma), \cdots, c_m(\sigma); a(\sigma))$ defines an injective $\mathcal{C}$-homomorphism $f\colon G \to G(m,n)$. Because $L(t+\xi)^e$ can not be in $\mathfrak{F}\langle L(t+\xi)^n\rangle = \mathfrak{F}\langle A(t)\rangle$ for any integer e with $1 \leqq e < n$, $a(\sigma)$ runs over the group of all n-th roots of unity as σ runs over G; because $\dim G = \operatorname{tr\,deg} \mathfrak{F}\langle t\rangle / \mathfrak{F}\langle A(t)\rangle = m$, $(c_1(\sigma), \cdots, c_m(\sigma))$ runs over $\mathcal{K}^m$ as σ runs over G. It follows that f is a $\mathcal{C}$-isomorphism. When $\mathcal{C}$ contains the n-th roots of unity then evidently the group $G_{\mathcal{C}}(m,n)$, consisting of the elements of $G(m,n)$ that are rational over $\mathcal{C}$, is dense in $G(m,n)$ in the Zariski topology, and hence also $G_{\mathcal{C}}$ is dense in G, so that the fixed field of $G_{\mathcal{C}}$ is $\mathfrak{F}\langle A(t)\rangle$. Because $G_{\mathcal{C}}$ is the group of automorphisms of $\mathfrak{F}\langle t\rangle$ over $\mathfrak{F}\langle A(t)\rangle$, this proves the sufficiency in Theorem 2, too.

Proof of necessity in Theorem 2. Let $\mathfrak{F}\langle A(t)\rangle$ be the set of fixed elements of some group G of automorphisms of $\mathfrak{F}\langle t\rangle$. When A is regarded as a polynomial in $(y^{(i)})_{1\leqq i\leqq m}$ with coefficients in $\mathfrak{F}[y]$ its terms are of the form $P \prod_{1\leqq i\leqq m} y^{(i)k_i}$ with $P \in \mathfrak{F}[y]$ and $k_i \in \boldsymbol{N}$ $(1 \leqq i \leqq m)$. Let

$$(\alpha y^{e_0} + \cdots + \epsilon) \prod_{1\leqq i\leqq m} y^{(i)e_i}$$

denote the highest nonzero term in A when the terms are ordered lexicographically with respect to $(k_m, \cdots, k_1)$; here $e_0 \in \boldsymbol{N}$ and $\alpha, \cdots, \epsilon \in \mathfrak{F}$ and $\alpha \neq 0$. Consider any $\sigma \in G$, and write $\sigma t = (ct+d)^{-1}(at+b)$ with $a, b, c, d \in \mathfrak{F}$ and $ad - bc \neq 0$. The condition $\sigma A(t) = A(t)$ shows that

$$\begin{aligned}(\alpha((ct+d)^{-1}(at+b))^{e_0} + \cdots + \epsilon) \prod_{1\leqq i\leqq m} ((ad-bc)(ct+d)^{-2}t^{(i)} + \cdots)^{e_i} + \cdots \\ = (\alpha t^{e_0} + \cdots + \epsilon) \prod_{1\leqq i\leqq m} t^{(i)e_i} + \cdots.\end{aligned}$$

Equating coefficients of the monomial $\prod_{1\leqq i\leqq m} t^{(i)e_i}$, and setting $e = \sum_{0\leqq i\leqq m} e_i$, we find that

$$\begin{aligned}(\alpha((ct+d)^{-1}(at+b))^{e_0} + \cdots + \epsilon)(ad-bc)^{e-e_0}(ct+d)^{-2e+2e_0} \\ = \alpha t^{e_0} + \cdots + \epsilon,\end{aligned}$$

whence

$$\begin{aligned}(ad-bc)^{e-e_0}(\alpha(at+b)^{e_0} + \cdots + \epsilon(ct+d)^{e_0} \\ = (\alpha t^{e_0} + \cdots + \epsilon)(ct+d)^{2e-e_0}.\end{aligned}$$

It follows that $at+b$ is divisible by $ct+d$, and therefore $c = 0$, we can take

$d=1$, and then $a^e=1$. In particular, a is a constant, so that $\sigma t^{(i)}=(at+b)^{(i)}$ $=at^{(i)}+b^{(i)}$ for every i.

Now write $A=A_0 y^{(m)e_m}+\cdots+A_{e_m}$, where each A_j is a differential polynomial of order $<m$ and $A_0\neq 0$. Then

$$\sigma A(t)=(\sigma A_0(t))(at^{(m)}+b^{(m)})^{e_m}+\cdots+\sigma A_{e_m}(t).$$

Comparing coefficients of $t^{(m)e_m}$ in the equation $\sigma A(t)=A(t)$, and also coefficients of $t^{(m)e_m-1}$, we find the two equations

$$(*)\qquad \sigma A_0(t)a^{e_m}=A_0(t),$$

$$(**)\qquad \sigma A_0(t)e_m a^{e_m-1}b^{(m)}+\sigma A_1(t)a^{e_m-1}=A_1(t).$$

From (*) and the equation $a^e=1$ it follows that $\sigma A_0(t)^e=A_0(t)^e$, whence $A_0(t)^e\in\mathcal{F}\langle A(t)\rangle$; since $\operatorname{ord} A_0<m$, this implies that $A_0{}^e\in\mathcal{F}$, whence also $A_0\in\mathcal{F}$. Therefore $A_0=\alpha$, $e_i=0$ $(0\leqq i<m)$, and $e_m=e$. From (**) it therefore follows that

$$(***)\qquad \sigma(t^{(m)}+e^{-1}\alpha^{-1}A_1(t))=a(t^{(m)}+e^{-1}\alpha^{-1}A_1(t)),$$

so that $(t^{(m)}+e^{-1}\alpha^{-1}A_1(t))^e$ is fixed by G and hence is in $\mathcal{F}\langle A(t)\rangle$. As the difference $A-\alpha(y^{(m)}+e^{-1}\alpha^{-1}A_1)^e$ has order $\leqq m$, and as its degree in $y^{(m)}$ is $<e$, this difference is in $\mathcal{F}$; thus, $A=\alpha(y^{(m)}+e^{-1}\alpha^{-1}A_1)^e+\gamma$ where $\gamma\in\mathcal{F}$. By (**), $A_1(at+b)=aA_1(t)-\alpha eb^{(m)}$; taking the partial derivative with respect to $t^{(i)}$ of each side, we find that

$$\frac{\partial A_1}{\partial y^{(i)}}(at+b)a=a\frac{\partial A_1}{\partial y^{(i)}}(t).$$

Thus, $\dfrac{\partial A_1}{\partial y^{(i)}}(t)$ is fixed by G and hence is in $\mathcal{F}\langle A(t)\rangle$; since $\operatorname{ord}\dfrac{\partial A_1}{\partial y^{(i)}}<m$, this implies that $\dfrac{\partial A_1}{\partial y^{(i)}}\in\mathcal{F}$ $(0\leqq i<m)$, so that

$$e^{-1}\alpha^{-1}A_1=\alpha_1 y^{(m-1)}+\cdots+\alpha_m y+\beta$$

where $\alpha_1,\cdots,\alpha_m,\beta\in\mathcal{F}$. Also, $n=e$. Setting

$$L=y^{(m)}+\alpha_1 y^{(m-1)}+\cdots+\alpha_m y,$$

we therefore find that $A=\alpha(L+\beta)^n+\gamma$.

By (***), $\sigma(L(t)+\beta)=a(L(t)+\beta)$. Hence the formula $\sigma\mapsto a$ defines a homomorphism of G into the group of n-th roots of unity; it is surjective, because otherwise there would exist a positive integer $k<n$ such that $(L(t)+\beta)^k$ is fixed for G, i.e. such that $(L(t)+\beta)^k\in\mathcal{F}\langle A(t)\rangle$. Therefore $\mathcal{F}$ contains all the n-th roots of unity. Also,

$$aL(t)+L(b)+\beta=L(at+b)+\beta=\sigma(L(t)+\beta)=aL(t)+a\beta,$$

so that $L(b)+(1-a)\beta=0$. This shows that when $n>1$ then $\mathcal{F}$ contains a zero of $L+\beta$ (for any σ with $a\neq 1$, $(1-a)^{-1}b$ is such a zero).

Finally, let $\eta_1,\cdots,\eta_k$ be a maximal set of elements of F that are zeros of L and are linearly independent over constants, and let H denote the Wronskian determinant.

$$H=\det\begin{bmatrix} y & \eta_1 & \cdots\eta_k \\ y' & \eta_1' & \cdots\eta_k' \\ \cdot & \cdot & \cdot \\ \cdot & \cdot & \cdot \\ \cdot & \cdot & \cdot \\ y^{(k)} & \eta_1^{(k)} & \cdots\eta_k^{(k)} \end{bmatrix}.$$

Then $0\leqq k\leqq m$, H is a nonzero homogeneous linear differential polynomial in $\mathcal{F}\{y\}$ of order k, and every zero of L in $\mathcal{F}$ is a zero of H. Let ξ denote a zero of $L+\beta$ in $\mathcal{F}$ if $n>1$ and denote 0 otherwise. For any $\sigma\in G$, $(1-a)\xi$ and b are zeros of $L+(1-a)\beta$; therefore $(1-a)\xi-b$ is a zero of L in $\mathcal{F}$, and hence is a zero of H. Therefore

$$\begin{aligned}\sigma H(t-\xi)=H(at+b-\xi)&=H(at+b-\xi+(1-a)\xi-b)\\ &=aH(t-\xi),\end{aligned}$$

so that $H(t-\xi)^n$ is fixed by G and hence is in $\mathcal{F}\langle A(t)\rangle$. Since $H(t-\xi)^n \notin F$, $H(t-\xi)$ must have order $\geqq m$, so that $k=m$.

Proof of necessity in Theorem 1. Let $\mathcal{F}\langle t\rangle$ be strongly normal over $\mathcal{F}\langle A(t)\rangle$, and let $\mathcal{C}_a$ denote the algebraic closure of $\mathcal{C}$. Then $\mathcal{C}_a\mathcal{F}\langle t\rangle$ is a strongly normal extension of $\mathcal{C}_a\mathcal{F}\langle A(t)\rangle$, which has field of constants $\mathcal{C}_a$. But a strongly normal extension of a differential field with algebraically closed field of constants is weakly normal (because its group of automorphisms is dense in its Galois group). It follows by Theorem 2 that $A=\alpha(L+\beta)^n+\gamma$, where $L=y^{(m)}+\alpha_1 y^{(m-1)}+\cdots+\alpha_m y$ and $\alpha,\beta,\gamma,\ \alpha_1,\cdots,\alpha_m\in\mathcal{C}_a\mathcal{F}$; when $n=1$ we may replace β,γ by $0,\ \beta+\gamma$, that is, we may suppose then that $\beta=0$. Since $A\in\mathcal{F}\{y\}$, it is easy to see that $\alpha,\beta,\gamma,\ \alpha_1,\cdots,\alpha_m\in\mathcal{F}$.

Also by Theorem 2, there exist elements $\zeta_1,\cdots,\zeta_m\in\mathcal{C}_a\mathcal{F}$ that are linearly independent over constants and are zeros of L. Because $\mathcal{C}_a\mathcal{F}=F[C_a]$, we can write $\zeta_i=\sum_{1\leqq k\leqq r}\eta_{ik}\gamma_k$ $(1\leqq i\leqq m)$ with $\eta_{ik}\in\mathcal{F}$ and $\gamma\in\mathcal{C}_a$, and with $\gamma_1,\cdots,\gamma_r$ linearly independent over $\mathcal{C}$; because $\mathcal{F}$ and $\mathcal{C}_a$ are linearly disjoint over $\mathcal{C}$, $\gamma_1,\cdots,\gamma_r$ are linearly independent over $\mathcal{F}$. Since the element

η_{ik} $(1 \leqq i \leqq m, 1 \leqq k \leqq r)$ obviously generate the vector space of zeros of L, some set of these elements is a basis of this vector space. Thus $\mathfrak{F}$ contains a fundamental system of zeros of L.

Finally, if $n > 1$ then, again by Theorem 2, some element $\xi \in \mathcal{E}_a\mathfrak{F}$ is a zero of $L + \beta$. Writing $\xi = \sum_{1 \leqq k \leqq r} \xi_k \gamma_k$, with $\xi_1, \cdots, \xi_r \in \mathfrak{F}$ and $\gamma_1, \cdots, \gamma_r \in \mathcal{E}_a$ and $\gamma_1, \cdots, \gamma_r$ linearly independent over $\mathcal{E}$ (and hence over $\mathfrak{F}$, too) and $\gamma_1 = 1$, we see that $(L(\xi_1) + \beta)\gamma_1 + \sum_{1 \leqq k \leqq r} L(\xi_k)\gamma_k = L(\xi) + \beta = 0$, so that $L(\xi_1) + \beta = 0$.

COLUMBIA UNIVERSITY
AND
MADURAI UNIVERSITY.

REFERENCES.

[1] E. R. Kolchin, "Galois theory of differential fields," *American Journal of Mathematics*, vol. 75 (1953), pp. 753-824.

[2] ———, "On the Galois theory of differential fields," *American Journal of Mathematics*, vol. 77 (1955). pp. 868-894.

[3] J. F. Ritt, "Differential algebra" (*Amer. Math. Soc. Colloq. Pub.* No. 33), New York, 1951.

[4] T. Soundararajan, "Normal subfields of simple extension fields," *Monatsh. Math.*, vol. 69 (1965), pp. 256-268.

[5] ———, "Normal polynomials in simple extension fields II," *Monatsh. Math.*, vol. 72 (1968), pp. 432-444.

Reprinted from ADVANCES IN MATHEMATICS
All Rights Reserved by Academic Press, New York and London

Vol. 12, No. 2, February 1974
Printed in Belgium

Constrained Extensions of Differential Fields

E. R. KOLCHIN*

Department of Mathematics, Columbia University, New York, New York 10027

TO SAMUEL EILENBERG ON HIS 60TH BIRTHDAY, 30 SEPTEMBER 1973

INTRODUCTION

In 1959, A. Robinson [5] defined, in the context and language of his theory of model-complete theories, the notion of what he called "differentially closed" ordinary differential field of characteristic 0. This notion in the theory of differential fields is analogous in some but not all ways to that of algebraically closed fields in the theory of fields. Indeed, the crucial property of such a differential field $\mathscr{H}$ is that every finite system of algebraic differential equations and inequations with coefficients in $\mathscr{H}$ that has a solution rational over an extension of $\mathscr{H}$ has a solution rational over $\mathscr{H}$; moreover, every ordinary differential field of characteristic 0 has a differentially algebraic extension that is "differentially closed". On the other hand, every $\mathscr{H}$ as above has a nontrivial differentially algebraic extension, and, therefore, the term "differentially closed" is not entirely appropriate.

In recent years the subject has seen a small renewal of activity, all by logicians working in the broad framework of model theory. Two papers are particularly relevant here. In the first, L. Blum [1] somewhat simplified Robinson's definition and, making use of a result of M. D. Morley [4], proved that every ordinary differential field of characteristic 0 has what she called a "differential closure" (or "prime differentially closed" extension), that is, a "differentially closed" extension that can be embedded in every "differentially closed" extension. In the second, S. Shelah [9] derived a criterion for an extension to be such a "differential closure" and proved it unique up to isomorphism. His contribution, which dealt with any "complete first-order totally transcendental theory", was perhaps the most difficult step of all. Further accounts of some or all of these results are contained in Robinson [6], Sacks [7], and Sacks [8].

* Research supported by a grant from the National Science Foundation.

The time now seems ripe for a unified exposition of the whole theme, in the setting and language of differential algebra, with the aim of making as explicit as possible the several ideas involved and the relations among them. What follows is an attempt at such an exposition. The differential fields considered here can be partial as well as ordinary ones. The bibliographic references in differential algebra are almost all to my recent book [2], which is cited as DAAG. The differential algebraic terminology used here follows DAAG and is more or less standard. The terminology adopted in connection with the particular subject matter of this paper is not that of Robinson *et al.*, but rather is one that develops naturally from that of differential algebra.

The basic notion is that of a constrained family of elements of an extension, as defined in DAAG, Chap. III, Section 10. This leads in Section 2, below, to the notion of constrained extension (called "atomic" by the logicians), thence in Section 3 to the notion of constrainedly closed (= Robinson's "differentially closed") differential fields, and finally in Section 4 to the notion of constrained closure (= Blum's "differential closure") of a differential field. Shelah's main results are proved in Section 7 after some preparatory material due to him in Sections 5 and 6. In Section 8 it is shown that a constrained closure of an ordinary differential field of constants $\mathscr{C}$ is isomorphic to a proper differential subfield of itself and hence contains an infinite strictly descending sequence of constrained closures of $\mathscr{C}$. This result, which refutes a conjecture of Sacks [8], is obtained as a consequence of the fact that the nonconstant solutions of the differential equation $dy/dx = y^3 - y^2$ are algebraically independent over $\mathscr{C}(x)$; this fact follows from a result about polynomials established in the appendix. (The same differential equation, among others, has previously been considered by M. Rosenlicht, who showed (unpublished) that if a solution η is transcendental over a differential field $\mathscr{F}$, then the constants of $\mathscr{F}\langle\eta\rangle$ are in $\mathscr{F}$.)[1] In the ultimate Section 9, a theorem is proved that implies that every strongly normal extension is a constrained extension. This fact has evidently been discovered independently, in one form or another, by several people (including Blum, Kovacic, and myself), but to my knowledge has not appeared in print.

[1] After this paper was written, I received from Rosenlicht a preprint of a paper [6a] in which he proves the result in Section 8. On the way, he considers the same differential equation in a much broader context, and obtains the fact mentioned above and much more. In two other papers, of which I have just become aware, C. Wood [10, 11] deals with ordinary differential fields of nonzero characteristic.

I am greatly indebted to Blum for conducting me on a private guided tour of the relevant work of Robinson, herself, and Shelah. Without that tour this paper would not have been written.

1. Preliminaries

Throughout this paper, $\mathscr{F}$ denotes a differential field *of characteristic* 0 relative to the derivation operators $\delta_1, \ldots, \delta_m$. The term "extension" of a differential field always means "differential field extension."

The letters y and z, with or without various subscripts, always denote differential indeterminates.

If, for some nonzero $n \in \mathbf{N}$, there exists a nonzero differential polynomial in $\mathscr{F}\{y_1, \ldots, y_n\}$ that vanishes identically on $\mathscr{F}^n$, then (see DAAG, Chap. II, Section 5) for every nonzero $n \in \mathbf{N}$ there exists such a differential polynomial. According as this is or is not the case, we shall say that the differential field $\mathscr{F}$ is *degenerate* or *nondegenerate*. It is known (see DAAG, Chap. II, Section 6, Corollary to Theorem 3) that $\mathscr{F}$ is nondegenerate if and only if $\mathscr{F}$ contains m elements $x_1, \ldots, x_m$ with nonvanishing Jacobian:

$$\det(\delta_i x_{i'})_{1 \leqslant i \leqslant m, 1 \leqslant i' \leqslant m} \neq 0.$$

Consider an extension $\mathscr{G}$ of $\mathscr{F}$ and a finite family $\eta = (\eta_1, \ldots, \eta_n)$ of elements of an extension of $\mathscr{G}$. Then $\mathscr{G}$ contains a finitely generated extension $\mathscr{G}_0$ of $\mathscr{F}$ with the properties that $\mathscr{G}_0\langle\eta\rangle$ and $\mathscr{G}$ are linearly disjoint over $\mathscr{G}_0$, that $\omega_{\eta/\mathscr{G}_0} = \omega_{\eta/\mathscr{G}}$, and that every differential specialization of η over $\mathscr{G}_0$ is a differential specialization of η over $\mathscr{G}$. This follows from the fact that the defining differential ideal of η in $\mathscr{G}\{y_1, \ldots, y_n\}$ has a differential field of definition that is a finitely generated extension of $\mathscr{F}$ (see DAAG, Chap. III, Section 3, Proposition 1) and from the definition of the differential transcendence polynomial $\omega_{\eta/\mathscr{G}}$ (see DAAG, Chap. II, Section 12). Of course, $\mathscr{G}_0$ can be replaced by any larger finitely generated extension of $\mathscr{F}$ in $\mathscr{G}$.

2. Constrained Extensions

Recall (see DAAG, Chap. III, Section 10) that a family $\eta = (\eta_j)_{j \in J}$ of elements of an extension of $\mathscr{F}$ is said to be *constrained* over $\mathscr{F}$ if there exists a differential polynomial $C \in \mathscr{F}\{(y_j)_{j \in J}\}$ with $C(\eta) \neq 0$ such that $C(\eta') = 0$ for every nongeneric differential specialization η' of η over $\mathscr{F}$. Any such C is called a *constraint* of η over $\mathscr{F}$.

By a *constrained extension* of $\mathcal{F}$, we shall mean an extension such that every finite family of elements of it is constrained over $\mathcal{F}$. Every constrained extension is differentially algebraic, but not conversely. Every algebraic extension is constrained. The field of constants of a constrained extension of $\mathcal{F}$ is algebraic over the field of constants of $\mathcal{F}$ (see DAAG, Chap. III, Section 10, Proposition 7).

When $\mathcal{F}$ is nondegenerate, a necessary and sufficient condition that an extension $\mathcal{G}$ of $\mathcal{F}$ be constrained is that each element of $\mathcal{G}$ be constrained over $\mathcal{F}$. Indeed, for any $\eta_1, \ldots, \eta_n \in \mathcal{G}$, there exists an element $\zeta \in \mathcal{G}$ such that $\mathcal{F}\langle\eta_1, \ldots, \eta_n\rangle = \mathcal{F}\langle\zeta\rangle$ (see DAAG, Chap. II, Section 8, Proposition 9); if the condition is satisfied, then ζ is constrained over $\mathcal{F}$ and, hence, so is $(\eta_1, \ldots, \eta_n)$ (see DAAG, Chap. III, Section 10, Proposition 7).

PROPOSITION 1. *Let η be a finite family of elements of an extension of $\mathcal{F}$. If η is constrained over $\mathcal{F}$, then $\mathcal{F}\langle\eta\rangle$ is a constrained extension of $\mathcal{F}$.*

Proof. Let ζ be any finite family of elements of $\mathcal{F}\langle\eta\rangle$. Then $\mathcal{F}\langle\eta\rangle = \mathcal{F}\langle\eta, \zeta\rangle$, so that (η, ζ) is constrained over $\mathcal{F}$ and hence ζ is also (see DAAG, Chap. III, Section 10, Proposition 7).

PROPOSITION 2. *Let $\mathcal{G}$ be an extension of $\mathcal{F}$, and $\mathcal{H}$ be an extension of $\mathcal{G}$.*

(a) *If $\mathcal{G}$ is constrained over $\mathcal{F}$ and $\mathcal{H}$ is constrained over $\mathcal{G}$, then $\mathcal{H}$ is constrained over $\mathcal{F}$.*

(b) *If $\mathcal{H}$ is constrained over $\mathcal{F}$, then $\mathcal{G}$ is constrained over $\mathcal{F}$ and, provided $\mathcal{G}$ is finitely generated over $\mathcal{F}$, $\mathcal{H}$ is constrained over $\mathcal{G}$.*

Proof. (a) Consider any finite family ζ of elements of $\mathcal{H}$, and fix a constraint D of ζ over $\mathcal{G}$. As noted in Section 1, $\mathcal{G}$ contains a finitely generated extension $\mathcal{G}_0$ of $\mathcal{F}$ that contains all the coefficients in D, such that every differential specialization ζ' of ζ over $\mathcal{G}_0$ is a differential specialization of ζ over $\mathcal{G}$. When $D(\zeta') \neq 0$, then ζ' is a generic differential specialization of ζ over $\mathcal{G}$ and hence over $\mathcal{G}_0$ too. Therefore, ζ is constrained over $\mathcal{G}_0$. Fix a finite family η of elements of $\mathcal{G}_0$ such that $\mathcal{F}\langle\eta\rangle = \mathcal{G}_0$. By hypothesis, η is constrained over $\mathcal{F}$. Therefore, (η, ζ) is constrained over $\mathcal{F}$ and hence ζ is also (see DAAG, Chap. III, Section 10, Proposition 7).

(b) It is obvious that $\mathcal{G}$ is constrained over $\mathcal{F}$. Supposing that $\mathcal{G}$ is a finitely generated extension of $\mathcal{F}$, say $\mathcal{G} = \mathcal{F}\langle\eta\rangle$, where η is a finite family of generators, let ζ be any finite family of elements of $\mathscr{H}$. Then (η, ζ), a finite family of elements of $\mathscr{H}$, is constrained over $\mathscr{F}$, and, hence, (see DAAG, Chap, III, Section 10, Proposition 7) ζ is constrained over the differential field $\mathscr{F}\langle\eta\rangle = \mathscr{G}$.

3. Constrainedly Closed Differential Fields

It is obvious that every differential field is a constrained extension of itself. A differential field that has no other constrained extension will be called *constrainedly closed.* It is clear that a constrainedly closed differential field is algebraically closed.

Proposition 3. *Let $n \in \mathbf{N}$, $n \neq 0$, and let $\mathscr{U}$ be a universal differential field that is an extension of $\mathscr{F}$. The following three conditions are equivalent:*

(i°) *$\mathscr{F}$ is constrainedly closed.*

(ii°) *Every element of $\mathscr{U}^n$ that is constrained over $\mathscr{F}$ is in $\mathscr{F}^n$.*

(iii°) *For every prime differential ideal $\mathfrak{p}$ of $\mathscr{F}\{y_1, \dots, y_n\}$ and every $C \in \mathscr{F}\{y_1, \dots, y_n\}$ with $C \notin \mathfrak{p}$, there exists an $\eta \in \mathscr{F}^n$ such that*

$$P(\eta) = 0 \ (P \in \mathfrak{p}), \qquad C(\eta) \neq 0.$$

Remark 1. Condition (i°) is independent of the choice of n and hence so are conditions (ii°) and (iii°). Similarly, condition (ii°) is independent of the choice of $\mathscr{U}$.

Remark 2. When the differential fields are ordinary and we choose $n = 1$, a prime differential ideal $\mathfrak{p}$ is the same thing as the general component $\mathfrak{p}_{\mathscr{F}}(A)$ of an irreducible differential polynomial $A \in \mathscr{F}\{y\}$. Since the separant S_A of A is not in $\mathfrak{p}_{\mathscr{F}}(A)$, an element $C \in \mathscr{F}\{y\}$ has the property that $C \notin \mathfrak{p}_{\mathscr{F}}(A)$ if and only if the remainder R of $S_A C$ with respect to A is not 0, and when this is the case, the ideal (A, R) contains a nonzero differential polynomial B of lower order than A. Therefore the proposition shows that when the differential fields are ordinary, a necessary and sufficient condition that $\mathscr{F}$ be constrainedly closed is that, for every pair (A, B) of elements of $\mathscr{F}\{y\}$ with A irreducible and $B \neq 0$ and $\operatorname{ord} B < \operatorname{ord} A$, there exist an element $\eta \in \mathscr{F}$ such that $A(\eta) = 0$ and $B(\eta) \neq 0$.

Remark 3. Condition (iii°) can be expressed in terms of the differential Zariski topology relative to $\mathscr{F}$ (see DAAG, Chap. IV, Section 4) in the following way: For every $\mathscr{F}$-closed subset $\mathscr{M}$ of $\mathscr{U}^n$, the set of points of $\mathscr{M}$ that are rational over $\mathscr{F}$ is $\mathscr{F}$-dense in $\mathscr{M}$.

Proof. We first show that (ii°) and (iii°) are equivalent. Let (ii°) be satisfied, and let $\mathfrak{p}$, C be as in (iii°). There exists a family $\zeta = (\zeta_1, \ldots, \zeta_n)$ of elements of some extension of $\mathscr{F}$ such that $P(\zeta) = 0$ ($P \in \mathfrak{p}$), $C(\zeta) \neq 0$, and (by DAAG, Chap. III, Section 10, Proposition 6) there exists a differential specialization η of ζ over $\mathscr{F}$ such that η is constrained over $\mathscr{F}$ with constraint C. As noted in Section 1, $\mathscr{F}$ has a finitely generated differential subfield $\mathscr{F}_0$ with $C \in \mathscr{F}_0\{y_1, \ldots, y_n\}$ such that every differential specialization of η over $\mathscr{F}_0$ is a differential specialization of η over $\mathscr{F}$. Because $\mathscr{F}_0$ is finitely generated, $\mathscr{U}$ is a universal extension of $\mathscr{F}_0$ (see DAAG, Chap. III, Section 7, Proposition 4); therefore, $\mathscr{U}^n$ contains a generic differential specialization η' of η over $\mathscr{F}_0$. By the above, η' is a differential specialization of η over $\mathscr{F}$, and since evidently $C(\eta) \neq 0$, η' is a generic differential specialization of η over $\mathscr{F}$, and hence η' is constrained over $\mathscr{F}$. By (ii°) then $\eta' \in \mathscr{F}^n$, and hence also $\eta \in \mathscr{F}^n$. Therefore (iii°) is satisfied.

Conversely, let (iii°) be satisfied, and let $\eta \in \mathscr{U}^n$ be constrained over $\mathscr{F}$, say with constraint C. Let $\mathfrak{p}$ denote the defining differential ideal of η in $\mathscr{F}\{y_1, \ldots, y_n\}$. Then $\mathfrak{p}$ is prime and $C \notin \mathfrak{p}$ whence, by (iii°), there exists an $\eta' \in \mathscr{F}^n$ such that $P(\eta') = 0$ ($P \in \mathfrak{p}$) and $C(\eta') \neq 0$. Because of the equations here, η' is a differential specialization of η over $\mathscr{F}$, and because of the inequation, the differential specialization is generic. Since $\eta' \in \mathscr{F}^n$, it follows that $\eta \in \mathscr{F}^n$. Therefore (ii°) is satisfied. Thus, (ii°) and (iii°) are equivalent.

It is obvious that (i°) implies (ii°). It remains to show that (ii°) implies (i°). Let (ii°) be satisfied (and hence also (iii°), and let $\mathscr{G}$ be any constrained extension of $\mathscr{F}$, not necessarily contained in $\mathscr{U}$. If an element $\eta \in \mathscr{U}$ is constrained over $\mathscr{F}$, then so is the family $(\eta, \ldots, \eta) \in \mathscr{U}^n$, whence by (ii°) $(\eta, \ldots, \eta) \in \mathscr{F}^n$, so that $\eta \in \mathscr{F}$. Thus, (ii°) remains satisfied when n is replaced by 1, and therefore (iii°) does too. Now, (iii°) is independent of the choice of $\mathscr{U}$, and hence (ii°) is also. Since $\mathscr{G}$ can be embedded in a universal differential field (see DAAG, Chap. III, Section 7, Theorem 2), this implies that every element of $\mathscr{G}$ is in $\mathscr{F}$, that is, $\mathscr{G} = \mathscr{F}$.

COROLLARY 1. *Every universal differential field of characteristic* 0 *is constrainedly closed.*

COROLLARY 2. *Let* $n \in \mathbf{N}$, $n \neq 0$, *and let* $\mathcal{G}$ *be a constrainedly closed extension of* $\mathcal{F}$. *If every element of* $\mathcal{G}^n$ *that is constrained over* $\mathcal{F}$ *is in* $\mathcal{F}^n$, *then* $\mathcal{F}$ *is constrainedly closed.*

Proof. If an element $\alpha \in \mathcal{G}$ is algebraic over $\mathcal{F}$, then the element $(\alpha,..., \alpha) \in \mathcal{G}^n$ is obviously constrained over $\mathcal{F}$, whence $\alpha \in \mathcal{F}$; since $\mathcal{G}$ is algebraically closed, this shows that $\mathcal{F}$ is algebraically closed. Now let η be any element of any constrained extension of $\mathcal{F}$, let $\mathfrak{p}$ be the defining differential ideal of η in $\mathcal{F}\{y\}$, and fix a constraint C of η over $\mathcal{F}$. Since $\mathcal{F}$ is algebraically closed, the differential ideal $\mathcal{G}\mathfrak{p}$ of $\mathcal{G}\{y\}$ is prime, and of course $C \notin \mathcal{G}\mathfrak{p}$. By the proposition (with $\mathcal{F}$ and n now $\mathcal{G}$ and 1), $\mathcal{G}$ has an element ζ that is a zero of $\mathcal{G}\mathfrak{p}$ but not of the constraint C. This ζ is a generic differential specialization of η over $\mathcal{F}$, and hence, like η, is constrained over $\mathcal{F}$. By the hypothesis, then $\zeta \in \mathcal{F}$, whence also $\eta \in \mathcal{F}$.

COROLLARY 3. *Let* $\mathcal{G}$ *be a constrained finitely generated extension of* $\mathcal{F}$, *and let* $\varphi: \mathcal{F} \rightarrow \mathcal{H}$ *be a homomorphism of* $\mathcal{F}$ *into a constrainedly closed differential field* $\mathcal{H}$. *Then* φ *can be extended to a homomorphism* $\mathcal{G} \rightarrow \mathcal{H}$.

Proof. Fix finitely many elements $\eta_1,..., \eta_n \in \mathcal{G}$ with $\mathcal{G} = \mathcal{F}\langle\eta_n,..., \eta_n\rangle$, and let $\mathfrak{p}$ denote the defining differential ideal of $(\eta_1,..., \eta_n)$ in $\mathcal{F}\{y_1,..., y_n\}$. Then $(\eta_1,..., \eta_n)$ is constrained over $\mathcal{F}$. Fix a constraint C of $(\eta_1,..., \eta_n)$ over $\mathcal{F}$. Every prime differential ideal of $\mathcal{F}\{y_1,..., y_n\}$ properly containing $\mathfrak{p}$ contains C. Therefore $\mathfrak{p}^\varphi$ is a prime differential ideal of $\varphi(\mathcal{F})\{y_1,..., y_n\}$, $C^\varphi \notin \mathfrak{p}^\varphi$, and every prime differential ideal of $\varphi(\mathcal{F})\{y_1,..., y_n\}$ properly containing $\mathfrak{p}^\varphi$ contains C^φ. Also, $C^\varphi \notin \mathfrak{q}$ for some component $\mathfrak{q}$ of the perfect differential ideal $\mathcal{H}\mathfrak{p}^\varphi$ of $\mathcal{H}\{y_1,..., y_n\}$. By the proposition, $\mathcal{H}^n$ contains a zero $(\zeta_1,..., \zeta_n)$ of $\mathfrak{q}$ with $C^\varphi(\zeta_1,..., \zeta_n) \neq 0$. By the above, $(\zeta_1,..., \zeta_n)$ is a generic zero of $\mathfrak{p}^\varphi$. Therefore, φ extends to a homomorphism $\mathcal{G} \rightarrow \mathcal{H}$ with $\eta_j \mapsto \zeta_j$ $(1 \leqslant j \leqslant n)$.

4. CONSTRAINED CLOSURES: EXISTENCE

Since every differential field has a universal extension, Corollary 1 to Proposition 3 shows that every differential field of characteristic 0 has a constrainedly closed extension. A constrainedly closed extension of $\mathcal{F}$ that can be embedded in every constrainedly closed extension of $\mathcal{F}$ will be called a *constrained closure* of $\mathcal{F}$.

It is obvious that any constrainedly closed extension of $\mathscr{F}$ contained in a constrained closure of $\mathscr{F}$ is itself a constrained closure of $\mathscr{F}$.

THEOREM 1. *There exists a constrained closure of $\mathscr{F}$. It is a constrained extension of $\mathscr{F}$, and its cardinal number equals that of $\mathscr{F}$.*

Proof. Fix a constrainedly closed extension $\mathscr{G}$ of $\mathscr{F}$, and let $(\mathscr{G}_i)_{i\in I}$ be a family of constrainedly closed extensions of $\mathscr{F}$ in $\mathscr{G}$ such that every constrainedly closed extension of $\mathscr{F}$ in $\mathscr{G}$ is $\mathscr{G}_i$ for some i. Then $I \neq \varnothing$. Let $\mathfrak{M}$ denote the set of all pairs $((\mathscr{F}_i)_{i\in I}, (f_{ij})_{i\in I, j\in I})$ with the following three properties:

$$\text{for each } i,\ \mathscr{F}_i \text{ is a constrained extension of } \mathscr{F} \text{ in } \mathscr{G}_i;$$
$$\text{for each } (i,j),\ f_{ij} \text{ is an isomorphism } \mathscr{F}_j \approx \mathscr{F}_i \text{ over } \mathscr{F};$$
$$\text{for each } (i,j,k),\ f_{ij}\circ f_{jk} = f_{ik}.$$

Then $\mathfrak{M} \neq \varnothing$, because we can take $\mathscr{F}_i = \mathscr{F} (i \in I)$ and $f_{ij} = \mathrm{id}_{\mathscr{F}}$ $((i,j) \in I^2)$. For any two elements $((\mathscr{F}_i), (f_{ij}))$ and $((\mathscr{F}_i'), (f_{ij}'))$ of $\mathfrak{M}$, define

$$(\mathscr{F}_i), (f_{ij})) \leqslant ((\mathscr{F}_i'), (f_{ij}'))$$

to mean that $\mathscr{F}_i \subset \mathscr{F}_i'$ for every i and f_{ij}' extends f_{ij} for every (i,j). This evidently makes $\mathfrak{M}$ an inductive ordered set. By Zorn's lemma, $\mathfrak{M}$ has a maximal element. Fix such a maximal element $((\mathscr{F}_i), (f_{ij}))$.

We claim that each $\mathscr{F}_i$ is constrainedly closed. To establish this, fix any index $j \in I$, and consider any element $\eta_j \in \mathscr{G}_j$ that is constrained over $\mathscr{F}_j$, say with constraint B. Let $\mathfrak{p}$ denote the defining differential ideal of η_j in $\mathscr{F}_j\{y\}$. For each $i \in I$ with $i \neq j$, $\mathfrak{p}^{f_{ij}}$ is a prime differential ideal of $\mathscr{F}_i\{y\}$ not containing $B^{f_{ij}}$; therefore, $\mathfrak{p}^{f_{ij}}$ has a generic zero η_i' in some extension of $\mathscr{F}_i$, and $B^{f_{ij}}(\eta_i') \neq 0$. It is easy to see that η_i' is constrained over $\mathscr{F}_i$ with constraint $B^{f_{ij}}$. Fix a component $\mathfrak{p}_i'$ of the perfect differential ideal $\mathscr{G}_i\mathfrak{p}^{f_{ij}}$ of $\mathscr{G}_i\{y\}$; it is known (see DAAG, Chap. III, Section 6, Proposition 3) that $\mathfrak{p}_i' \cap \mathscr{F}_i\{y\} = \mathfrak{p}^{f_{ij}}$, and hence $B^{f_{ij}} \notin \mathfrak{p}_i'$. It follows by Proposition 3 that there exists an element $\eta_i \in \mathscr{G}_i$ that is a zero of $\mathfrak{p}_i'$ (and hence of $\mathfrak{p}^{f_{ij}}$) but not of $B^{f_{ij}}$. This η_i is a differential specialization of η_i' over $\mathscr{F}_i$ with $B^{f_{ij}}(\eta_i) \neq 0$, and hence is a generic one; therefore η_i is a generic zero of $\mathfrak{p}^{f_{ij}}$. It follows that the isomorphism $f_{ij}\colon \mathscr{F}_j \approx \mathscr{F}_i$ extends to an isomorphism $f_{ij}'\colon \mathscr{F}_j\langle\eta_j\rangle \approx \mathscr{F}_i\langle\eta_i\rangle$ with $f_{ij}'(\eta_j) = \eta_i$. By Propositions 1 and 2, $\mathscr{F}_i\langle\eta_i\rangle$ is a constrained extension of $\mathscr{F}$, as is $\mathscr{F}_j\langle\eta_j\rangle$. All this is for fixed j and every $i \neq j$. Define $f_{jj}' = \mathrm{id}_{\mathscr{F}_j\langle\eta_j\rangle}$; for all $i \neq j$, define $f_{ji}' = f_{ij}'^{-1}$; for all $i, i' \neq j$, define $f_{ii'}' = f_{ij}' \circ f_{ji'}'$. Then $((\mathscr{F}_i\langle\eta_i\rangle)_{i\in I}, (f_{ii'}')_{i\in I, i'\in I})$ is an element of $\mathfrak{M}$ that is greater than or equal to the element $((\mathscr{F}_i), (f_{ii'}))$. By the maximality of the latter, then $\mathscr{F}_i\langle\eta_i\rangle = \mathscr{F}_i$ $(i \in I)$; in particular, $\eta_j \in \mathscr{F}_j$. Because of Corollary 2 to Proposition 3, this establishes our claim.

Thus, there exists a constrainedly closed constrained extension $\mathscr{F}^\dagger$ of $\mathscr{F}$ in $\mathscr{G}$ such that $\mathscr{F}^\dagger$ can be embedded in every constrainedly closed extension of $\mathscr{F}$ in $\mathscr{G}$ (for example, each $\mathscr{F}_i$ is such an $\mathscr{F}^\dagger$). Now, let $\mathscr{G}'$ be any constrainedly closed extension of $\mathscr{F}$. We shall show that $\mathscr{F}^\dagger$ can be embedded in $\mathscr{G}'$. This will prove that $\mathscr{F}^\dagger$ is a constrained closure of $\mathscr{F}$.

Replacing $\mathscr{G}'$ by an isomorphic copy, we may suppose that $\mathscr{G}$ and $\mathscr{G}'$ have a common extension $\mathscr{H}$, and since $\mathscr{H}$ can be replaced by any larger differential field, for example a universal one, we may even suppose that $\mathscr{H}$ is constrainedly closed. Concluding for $\mathscr{H}$ as we did for $\mathscr{G}$, above, we see that there exists a constrainedly closed constrained extension $\mathscr{F}^\ddagger$ of $\mathscr{F}$ in $\mathscr{H}$ such that $\mathscr{F}^\ddagger$ can be embedded in every constrainedly closed extension of $\mathscr{F}$ in $\mathscr{H}$. In particular, $\mathscr{F}^\ddagger$ can be embedded (a) in $\mathscr{G}$ and (b) in $\mathscr{G}'$. Because of (a), $\mathscr{F}^\dagger$ can be embedded in $\mathscr{F}^\ddagger$ and hence, because of (b), $\mathscr{F}^\dagger$ can be embedded in $\mathscr{G}'$.

It remains to prove that card $\mathscr{F}^\dagger =$ card $\mathscr{F}$. This is an immediate consequence of Corollary 1 to Proposition 3 and the following result inadvertently omitted from DAAG.

Proposition 4. *$\mathscr{F}$ has a universal extension of the same cardinal number as $\mathscr{F}$.*

Proof. It suffices to prove that $\mathscr{F}$ has a semiuniversal extension $\mathscr{F}'$ with card $\mathscr{F}' =$ card $\mathscr{F}$, because then there exists an infinite sequence

$$\mathscr{F} \subset \mathscr{F}' \subset \mathscr{F}'' \subset \cdots$$

in which each term beyond the first is a semiuniversal extension of the preceding term having the same cardinal number, and the union $\mathscr{F} \cup \mathscr{F}' \cup \mathscr{F}'' \cup \cdots$ is a universal extension of $\mathscr{F}$. Now, by the Ritt–Raudenbush basis theorem, for each n, the set $\Lambda(n)$ of all prime differential ideals of $\mathscr{F}\{y_1, \ldots, y_n\}$ has cardinal number $\leqslant$ the cardinal number of the set of finite subsets of $\mathscr{F}\{y_1, \ldots, y_n\}$, whence card $\Lambda(n) =$ card $\mathscr{F}$. Setting $\Lambda = \bigcup_{n \in \mathbf{N}} \Lambda(n)$, we see that card $\Lambda =$ card $\mathscr{F}$. Fix a universal extension $\mathscr{U}$ of $\mathscr{F}$. Each $\mathfrak{p} \in \Lambda$ has a generic zero $\eta_\mathfrak{p}$ every coordinate of which is in $\mathscr{U}$. The differential field $\mathscr{F}' = \mathscr{F}\langle(\eta_\mathfrak{p})_{\mathfrak{p}\in\Lambda}\rangle$ is a semiuniversal extension of $\mathscr{F}$ and, by the above, card $\mathscr{F}' =$ card $\mathscr{F}$.

PROPOSITION 5. *Let $\mathscr{F}^\dagger$ be a constrained closure of $\mathscr{F}$, and let $\mathscr{E}$ be a finitely generated extension of $\mathscr{F}$ in $\mathscr{F}^\dagger$. Then $\mathscr{F}^\dagger$ is a constrained closure of $\mathscr{E}$.*

Proof. We first show that it suffices to prove that *some* constrained closure of $\mathscr{F}$ has the property alleged for $\mathscr{F}^\dagger$. Indeed, suppose that $\mathscr{F}^\ddagger$ is a constrained closure of $\mathscr{F}$ with that property. Then there exists an embedding $f: \mathscr{F}^\dagger \to F^\ddagger$ over $\mathscr{F}$. By hypothesis, $\mathscr{F}^\ddagger$ is a constrained closure of $f(\mathscr{E})$. Since $f(\mathscr{E}) \subset f(\mathscr{F}^\dagger) \subset \mathscr{F}^\ddagger$ and $f(\mathscr{F}^\dagger)$ is constrainedly closed, this implies that $f(\mathscr{F}^\dagger)$ is a constrained closure of $f(\mathscr{E})$ and, hence, that $\mathscr{F}^\dagger$ is a constrained closure of $\mathscr{E}$.

This being the case, fix any constrainedly closed extension $\mathscr{H}$ of $\mathscr{F}$. Let $\mathfrak{M}$ denote the set of all well ordered subsets Ω of $\mathscr{H}$ such that, for each $\alpha \in \Omega$, α is constrained over $\mathscr{F}\langle\Omega_\alpha\rangle$ (where Ω_α denotes the initial segment of Ω determined by α, that is, Ω_α is the set of elements of Ω that are $< \alpha$). For any two elements Ω_1, Ω_2 of $\mathfrak{M}$, define $\Omega_1 \leqslant \Omega_2$ to mean that Ω_1 is an initial segment of Ω_2. This makes Ω an inductive ordered set, which then must have a maximal element. Let Ω be such a maximal element. Then $\mathscr{F}\langle\Omega\rangle$ is constrainedly closed, for otherwise (by Corollary 2 to Proposition 3) there would exist an element $\beta \in \mathscr{H} - \mathscr{F}\langle\Omega\rangle$ constrained over $\mathscr{F}\langle\Omega\rangle$, and the set consisting of β and the elements of Ω would, with the obvious well ordering, be an element of $\mathfrak{M}$ strictly greater than Ω. Let $\mathscr{E}'$ be any finitely generated extension of $\mathscr{F}$ in $\mathscr{F}\langle\Omega\rangle$, and let $\mathscr{H}'$ be any constrainedly closed extension of $\mathscr{E}'$. We shall show that $\mathscr{F}\langle\Omega\rangle$ can be embedded in $\mathscr{H}'$ over $\mathscr{E}'$. This will show that $\mathscr{F}\langle\Omega\rangle$ is a constrained closure of $\mathscr{E}'$ and, by the above, will complete the proof of the proposition.

Let $\mathfrak{N}$ denote the set of all pairs (Λ, g) with Λ an initial segment of Ω and g an embedding over $\mathscr{E}'$ of $\mathscr{E}'(\Lambda)$ in $\mathscr{H}'$. For any two elements (Λ_1, g_i), (Λ_2, g_2) of $\mathfrak{N}$, define $(\Lambda_1, g_1) \leqslant (\Lambda_2, g_2)$ to mean $\Lambda_1 \subset \Lambda_2$ and g_1 is the restriction of g_2 to $E'(\Lambda_1)$. This makes $\mathfrak{N}$ an inductive ordered set, which then must have a maximal element. Let (Λ, g) be such a maximal element. If Λ were not Ω, there would be an $\alpha \in \Omega$ with $\Omega_\alpha = \Lambda$; α would be constrained over $\mathscr{F}\langle\Omega_\alpha\rangle$ and hence, by Proposition 2(b), over the differential field $\mathscr{F}\langle\Omega_\alpha\rangle\,\mathscr{E}' = \mathscr{E}'\langle\Lambda\rangle$; therefore, by Corollary 3 to Proposition 3, the embedding $g: \mathscr{E}'\langle\Lambda\rangle \to \mathscr{H}'$ could be extended to an embedding $g': \mathscr{E}'\langle\Lambda'\rangle \to \mathscr{H}'$ (Λ' denoting the initial segment of Ω consisting of α and the elements of Λ), and (Λ', g') would be an element of $\mathfrak{N}$ strictly greater than (Λ, g). This shows that g is an embedding of $\mathscr{E}'\langle\Omega\rangle = \mathscr{F}\langle\Omega\rangle$ in $\mathscr{H}'$ over $\mathscr{E}'$.

5. Small Extensions

Given any family $(K_i)_{i\in I}$ of field extensions of a field K, all contained in some large field, we say that the *family is* (or the *fields* K_i $(i \in I)$ *are*) *linearly disjoint* over K if, for each index $i_0 \in I$, K_{i_0} and $K(\bigcup_{i\in I-\{i_0\}} K_i)$ are linearly disjoint over K in the usual sense. When this is the case, if $(I_j)_{j\in J}$ is a partition of I and we set $L_j = K(\bigcup_{i\in I_j} K_i)$ $(j \in J)$, then $(L_j)_{j\in J}$ is linearly disjoint over K.

Lemma 1. *Let $\mathscr{F}_i$ $(i \in I)$ and $\mathscr{E}$ be extensions of $\mathscr{F}$, all contained in an extension of $\mathscr{F}$. Suppose that $(\mathscr{F}_i)_{i\in I}$ is linearly disjoint over $\mathscr{F}$ and $\mathscr{E}$ is finitely generated. Then there exists a finite subset I_0 of I such that the differential fields $\mathscr{F}_i$ $(i \in I - I_0)$ and $\mathscr{E}$ are linearly disjoint over $\mathscr{F}$.*

Proof. For each subset J of I, write $\mathscr{F}_J = \mathscr{F}(\bigcup_{i\in J} \mathscr{F}_i)$. As noted in Section 1, $\mathscr{F}_I$ contains a finitely generated extension $\mathscr{G}_0$ of $\mathscr{F}$ such that $\mathscr{G}_0\mathscr{E}$ and $\mathscr{F}_I$ are linearly disjoint over $\mathscr{G}_0$. For some finite $I_0 \subset I$, $\mathscr{G}_0 \subset \mathscr{F}_{I_0}$ and hence $\mathscr{F}_{I_0}\mathscr{E}$ and $\mathscr{F}_I$ are linearly disjoint over $\mathscr{F}_{I_0}$. Because $\mathscr{F}_{I_0}$ and $\mathscr{F}_{I-I_0}$ are linearly disjoint over $\mathscr{F}$, and $\mathscr{F}_{I-I_0} \subset \mathscr{F}_I$, it follows that $\mathscr{F}_{I_0}\mathscr{E}$ and $\mathscr{F}_{I-I_0}$ are linearly disjoint over $\mathscr{F}$, so that $\mathscr{E}$ and $\mathscr{F}_{I-I_0}$ are also. This implies that the $\mathscr{F}_i$ $(i \in I - I_0)$ and $\mathscr{E}$ are linearly disjoint over $\mathscr{F}$.

Consider an extension $\mathscr{H}$ of $\mathscr{F}$ and a subset Σ of $\mathscr{H}$. For each element $\alpha \in \Sigma$, let Σ_α denote the set of elements of Σ other than α. Call Σ a *set of conjugates* over $\mathscr{F}$ if all the elements have the same defining differential ideal $\mathfrak{p}$ in $\mathscr{F}\{y\}$; call Σ *independent* over $\mathscr{F}$ if the differential fields $\mathscr{F}\langle\alpha\rangle$ $(\alpha \in \Sigma)$ are linearly disjoint over $\mathscr{F}$; call Σ *symmetric* over $\mathscr{F}$ if, for every permutation π of Σ, $(\pi\alpha)_{\alpha\in\Sigma}$ is a generic differential specialization of $(\alpha)_{\alpha\in\Sigma}$ over $\mathscr{F}$. When Σ is a set of conjugates over $\mathscr{F}$, as above, with defining differential ideal $\mathfrak{p}$ in $\mathscr{F}\{y\}$, a necessary and sufficient condition that Σ be independent over $\mathscr{F}$ is that, for each $\alpha \in \Sigma$, the defining differential ideal of α in $\mathscr{F}\langle\Sigma_\alpha\rangle\{y\}$ be $\mathscr{F}\langle\Sigma_\alpha\rangle\mathfrak{p}$. When Σ is symmetric over $\mathscr{F}$, then Σ is a set of conjugates over $\mathscr{F}$; when Σ is an independent set of conjugates over $\mathscr{F}$, then Σ is symmetric over $\mathscr{F}$. When Σ is symmetric over $\mathscr{F}$, then Σ is symmetric over every differential subfield of $\mathscr{F}$ and, for any subset Σ' of Σ, Σ' is symmetric over $\mathscr{F}$ and $\Sigma - \Sigma'$ is symmetric over $\mathscr{F}\langle\Sigma'\rangle$; also, if every finite subset of Σ is symmetric over $\mathscr{F}$, then so is Σ.

Lemma 2. *Let $\mathscr{H}$ be an extension of $\mathscr{F}$ and Σ be a subset of $\mathscr{H}$ that is symmetric over $\mathscr{F}$. Then Σ has a finite subset Φ such that $\Sigma - \Phi$ is an independent set of conjugates over $\mathscr{F}\langle\Phi\rangle$.*

Proof. We may evidently suppose that card $\Sigma > 1$. As before, for each $\alpha \in \Sigma$, let Σ_α denote the set of elements of Σ different from α; also, let $\mathfrak{p}_\alpha$ denote the defining differential ideal of α in $\mathscr{F}\langle\Sigma_\alpha\rangle\{y\}$. There is (see DAAG, Chap. III, Section 3) a smallest extension $\mathscr{F}_\alpha$ of $\mathscr{F}$ that is a differential field of definition of $\mathfrak{p}_\alpha$, and $\mathscr{F}_\alpha$ is finitely generated over $\mathscr{F}$; therefore, there is a finite subset $\Phi(\alpha)$ of Σ_α such that $\mathscr{F}_\alpha \subset \mathscr{F}\langle\Phi(\alpha)\rangle$. Of course, $\mathscr{F}_\alpha$ is the smallest differential field of definition also of the perfect differential ideal $\mathscr{F}\langle\Sigma\rangle\,\mathfrak{p}_\alpha$ of $\mathscr{F}\langle\Sigma\rangle\{y\}$. Fix $\beta \in \Sigma$ so that $\mathscr{F}\langle\Sigma\rangle\,\mathfrak{p}_\beta$ is maximal in the set of all the perfect differential ideals $\mathscr{F}\langle\Sigma\rangle\,\mathfrak{p}_\alpha$ $(\alpha \in \Sigma)$, and set $\mathscr{F}' = \mathscr{F}_\beta$, $\mathfrak{p}' = \mathfrak{p}_\beta \cap \mathscr{F}'\{y\}$. Consider any $\alpha \in \Sigma - \Phi(\beta)$. Because $\Sigma - \Phi(\beta)$ is symmetric over $\mathscr{F}\langle\Phi(\beta)\rangle$ and hence over $\mathscr{F}'$, α is, like β, a zero of $\mathfrak{p}'$. Since $\mathfrak{p}' \subset \mathscr{F}'\{y\} \subseteq \mathscr{F}\langle\Phi(\beta)\rangle\{y\} \subset \mathscr{F}\langle\Sigma_\alpha\rangle\{y\}$ and α is a generic zero of $\mathfrak{p}_\alpha$, this means that $\mathfrak{p}' \subset \mathfrak{p}_\alpha$, so that $\mathscr{F}\langle\Sigma\rangle\,\mathfrak{p}_\beta = \mathscr{F}\langle\Sigma\rangle\,\mathfrak{p}' \subset \mathscr{F}\langle\Sigma\rangle\,\mathfrak{p}_\alpha$, and hence, by the maximality of $\mathscr{F}\langle\Sigma\rangle\,\mathfrak{p}_\beta$, $\mathscr{F}\langle\Sigma\rangle\,\mathfrak{p}_\beta = \mathscr{F}\langle\Sigma\rangle\,\mathfrak{p}_\alpha$. It follows, for every $\alpha \in \Sigma - \Phi(\beta)$, that $\mathscr{F}_\alpha = \mathscr{F}'$, that $\mathfrak{p}'$ is the defining differential ideal of α in $\mathscr{F}'\{y\}$, and that $\mathfrak{p}_\alpha = \mathscr{F}\langle\Sigma_\alpha\rangle\,\mathfrak{p}' = \mathscr{F}'\langle\Sigma_\alpha - \Phi(\beta)\rangle\,\mathfrak{p}'$. Therefore, $\Sigma - \Phi(\beta)$ is an independent set of conjugates over the differential field $\mathscr{F}' = \mathscr{F}\langle\Phi(\beta)\rangle$.

We shall say that $\mathscr{H}$ is a *small extension* of $\mathscr{F}$ (or that $\mathscr{H}$ is *small over* $\mathscr{F}$) if every subset of $\mathscr{H}$ that is symmetric over $\mathscr{F}$ is denumerable.

Consider a differential field $\mathscr{G}$ with $\mathscr{F} \subset \mathscr{G} \subset \mathscr{H}$. It follows from the definition that *if $\mathscr{H}$ is a small extension of $\mathscr{F}$, then $\mathscr{G}$ is small over $\mathscr{F}$ and $\mathscr{H}$ is small over $\mathscr{G}$.*

PROPOSITION 6. *Every constrained closure of $\mathscr{F}$ is a small extension of $\mathscr{F}$.*

Proof. Let $\mathscr{F}^\dagger$ be a constrained closure of $\mathscr{F}$, and let Σ be a subset of $\mathscr{F}^\dagger$ that is symmetric over $\mathscr{F}$. By Lemma 2, Σ has a finite subset Φ such that $\Sigma - \Phi$ is an independent set of conjugates over $\mathscr{F}\langle\Phi\rangle$, and by Proposition 5, $\mathscr{F}^\dagger$ is a constrained closure of $\mathscr{F}\langle\Phi\rangle$. To prove that Σ is denumerable, it suffices to show that $\Sigma - \Phi$ is denumerable. Therefore, we may replace $\mathscr{F}$, $\mathscr{F}^\dagger$, Σ by $\mathscr{F}\langle\Phi\rangle$, $\mathscr{F}^\dagger$, $\Sigma - \Phi$, that is, we may suppose that Σ is an independent set of conjugates over $\mathscr{F}$. The elements of Σ all are generic zeros of the same prime differential ideal $\mathfrak{p}$ of $\mathscr{F}\{y\}$.

Assume that Σ is not denumerable. Then Σ has an infinite denumerable subset Σ', say consisting of the elements α_n $(n \in \mathbf{N})$. Since $\mathscr{F}^\dagger$ is a constrainedly closed extension of $\mathscr{F}\langle\Sigma'\rangle$, Theorem 1 shows that $\mathscr{F}^\dagger$ contains a constrained closure $\mathscr{H}$ of $\mathscr{F}\langle\Sigma'\rangle$. Since $\mathscr{H}$ is a constrainedly closed extension of $\mathscr{F}$, there exists an embedding $f: \mathscr{F}^{\dagger} \to \mathscr{H}$ over $\mathscr{F}$. The differential fields $\mathscr{F}\langle f(\alpha)\rangle$ $(\alpha \in \Sigma)$ are, like the $\mathscr{F}\langle\alpha\rangle$ $(\alpha \in \Sigma)$, linearly disjoint over $\mathscr{F}$. Hence, by Lemma 1, for each $n \in \mathbf{N}$ there exists a finite subset Σ_n of Σ such that the differential fields $\mathscr{F}\langle f(\alpha)\rangle$ $(\alpha \in \Sigma - \Sigma_n)$ and $\mathscr{F}\langle\alpha_0, \ldots, \alpha_{n-1}\rangle$ are linearly disjoint over $\mathscr{F}$. The set $T = \Sigma - \bigcup_{n \in \mathbf{N}} \Sigma_n$ is nondenumerable and the differential fields $\mathscr{F}\langle f(\alpha)\rangle$ $(\alpha \in T)$ and $\mathscr{F}\langle\Sigma'\rangle$ are linearly disjoint over $\mathscr{F}$.

Fix $\beta \in T$. By the linear disjointness, $f(\beta) \notin \Sigma'$ and the defining differential ideal of $f(\beta)$ in $\mathscr{F}\langle\Sigma'\rangle\{y\}$ is $\mathscr{F}\langle\Sigma'\rangle\mathfrak{p}$. Because $f(\beta) \in \mathscr{H}$, $f(\beta)$ is constrained over $\mathscr{F}\langle\Sigma'\rangle$, say with constraint $C \in \mathscr{F}\langle\Sigma'\rangle\{y\}$, and, hence, we can fix $n \in \mathbf{N}$ so that $C \in \mathscr{F}\langle\alpha_0, \ldots, \alpha_{n-1}\rangle\{y\}$. Because of the linear disjointness, the defining differential ideal of α_n in $\mathscr{F}\langle\alpha_0, \ldots, \alpha_{n-1}\{y\}$ is $\mathscr{F}\langle\alpha_0, \ldots, \alpha_{n-1}\rangle\mathfrak{p}$, which does not contain C. Therefore, α_n is a zero of $\mathscr{F}\langle\Sigma'\rangle\mathfrak{p}$ with $C(\alpha_n) \neq 0$, so that α_n is a generic differential specialization of $f(\beta)$ over $\mathscr{F}\langle\Sigma'\rangle$. Since $\alpha_n \in \Sigma'$, this means that $f(\beta) \in \Sigma'$, in contradiction to the above.

6. Normality in an Extension

Let $\mathscr{H}$ be an extension of $\mathscr{F}$. Call a subset Σ of $\mathscr{H}$ *normal over $\mathscr{F}$ in $\mathscr{H}$* if every set of conjugates over $\mathscr{F}$ in $\mathscr{H}$ that intersects Σ is a subset of Σ. It is clear that Σ is normal over $\mathscr{F}$ in $\mathscr{H}$ if and only if $\Sigma = \bigcup_{i \in I} \Sigma_i$, where, for each index $i \in I$, Σ_i is a maximal set of conjugates over $\mathscr{F}$ in $\mathscr{H}$, that is, Σ_i is the set of all elements of $\mathscr{H}$ that are generic zeros of a prime differential ideal $\mathfrak{p}_i$ of $\mathscr{F}\{y\}$.

When Σ is normal over $\mathscr{F}$ in $\mathscr{H}$ and $\mathscr{F}_1$ is a differential field with $\mathscr{F} \subset \mathscr{F}_1 \subset \mathscr{H}$, then Σ is normal over $\mathscr{F}_1$ in $\mathscr{H}$.

If a differential field $\mathscr{G}$ with $\mathscr{F} \subset \mathscr{G} \subset \mathscr{H}$ is normal over $\mathscr{F}$ in $\mathscr{H}$ (as a subset of $\mathscr{H}$, in the sense defined above), then $\mathscr{G} = \mathscr{F}\langle\Sigma\rangle$ for some set $\Sigma \subset \mathscr{H}$ that is normal over $\mathscr{F}$ in $\mathscr{H}$. (For example, take $\Sigma = \mathscr{G}$.) Conversely, if Σ is a subset of $\mathscr{H}$ that is normal over $\mathscr{F}$ in $\mathscr{H}$, *and if $\mathscr{H}$ is a constrainedly closed constrained extension of $\mathscr{F}$*, then the differential field $\mathscr{G} = \mathscr{F}\langle\Sigma\rangle$ is normal over $\mathscr{F}$ in $\mathscr{H}$. Indeed, let T be any set of conjugates over $\mathscr{F}$ in $\mathscr{H}$ containing an element $\beta \in \mathscr{G}$; then $\beta = P(\alpha_1, \ldots, \alpha_n)/Q(\alpha_1, \ldots, \alpha_n)$, where $P, Q \in \mathscr{F}\{y_1, \ldots, y_n\}$ and $\alpha_1, \ldots, \alpha_n \in \Sigma$ and $Q(\alpha_1, \ldots, \alpha_n) \neq 0$. For any $\beta' \in T$, there is a homomorphism $\mathscr{F}\langle\beta\rangle \to \mathscr{H}$ over $\mathscr{F}$ with $\beta \mapsto \beta'$, and by Corollary 3 to Proposition 3, this can be extended to a homomorphism $\mathscr{F}\langle\beta, \alpha_1, \ldots, \alpha_n\rangle \to \mathscr{H}$. Denoting the image of α_j by α_j', we see that $\alpha_j' \in \Sigma$ because Σ is normal over $\mathcal{F}$ in $\mathcal{H}$, and hence that $\beta' = P(\alpha_1', \ldots, \alpha_n')/Q(\alpha_1', \ldots, \alpha_n') \in \mathcal{G}$. Thus, $T \subset \mathcal{G}$.

PROPOSITION 7. *Let $\mathcal{H}$ be a constrainedly closed constrained extension of $\mathcal{F}$, and let $\mathcal{G}$ be a differential field with $\mathcal{F} \subset \mathcal{G} \subset \mathcal{H}$. If $\mathcal{G}$ is either normal over $\mathcal{F}$ in $\mathcal{H}$ or algebraic over $\mathcal{F}$, then $\mathcal{H}$ is a constrained extension of $\mathcal{G}$.*

Proof. First, let $\mathcal{G}$ be normal over $\mathcal{F}$ in $\mathcal{H}$, and let $\eta = (\eta_1, \ldots, \eta_n)$ be any finite family of elements of $\mathcal{H}$. Then (see Section 1) $\mathcal{G}$ contains a finitely generated extension $\mathcal{G}_0$ of $\mathcal{F}$ such that $\omega_{\eta/\mathcal{G}_0} = \omega_{\eta/\mathcal{G}}$ and every differential specialization of η over $\mathcal{G}_0$ is a differential specialization of η over $\mathcal{G}$. By Proposition 2, η is constrained over $\mathcal{G}_0$. Fix a constraint $C \in \mathcal{G}_0\{y_1, \ldots, y_n\}$ of η over $\mathcal{G}_0$. It suffices to show that C is a constraint of η over $\mathcal{G}$.

Consider any differential specialization $\eta' = (\eta_1', \ldots, \eta_n')$ of η over $\mathcal{G}$ with $C(\eta') \neq 0$. Then $\mathcal{G}$ contains a finitely generated extension $\mathcal{G}_1'$ of $\mathcal{G}_0$ such that $\omega_{\eta'/\mathcal{G}_1'} = \omega_{\eta'/\mathcal{G}}$ and every differential specialization of η' over $\mathcal{G}_1'$ is a differential specialization of η' over $\mathcal{G}$. Because $C(\eta') \neq 0$, η' is a generic differential specialization of η over $\mathcal{G}_0$, that is, there exists an isomorphism $\mathcal{G}_0\langle\eta'\rangle \approx \mathcal{G}_0\langle\eta\rangle$ with $\eta_j' \mapsto \eta_j$ $(1 \leqslant j \leqslant n)$. By Corollary 3 to Proposition 3, this can be extended to an isomorphism $\mathcal{G}_1'\langle\eta'\rangle \approx \mathcal{G}_1\langle\eta\rangle$ mapping $\mathcal{G}_1'$ onto a finitely generated extension $\mathcal{G}_1$ of $\mathcal{G}_0$ with $\mathcal{G}_1 \subset \mathcal{H}$. Because $\mathcal{G}_1' \subset \mathcal{G}$ and $\mathcal{G}$ is normal over $\mathcal{F}$ in $\mathcal{H}$, $\mathcal{G}_1 \subset \mathcal{G}$. It follows that

$$\omega_{\eta'/\mathcal{G}} = \omega_{\eta'/\mathcal{G}_1'} = \omega_{\eta/\mathcal{G}_1} \geqslant \omega_{\eta/\mathcal{G}},$$

so that (see DAAG, Chap. III, Section 5, Proposition 2) η' is a generic differential specialization of η over $\mathcal{G}$. This shows that C is a constraint of η over $\mathcal{G}$.

Now let $\mathcal{G}$ be algebraic over $\mathcal{F}$. The algebraic closure $\mathcal{G}^0$ of $\mathcal{G}$ in $\mathcal{H}$ is evidently normal over $\mathcal{F}$ in $\mathcal{H}$, so that, by the above, $\mathcal{H}$ is constrained over $\mathcal{G}^0$. But $\mathcal{G}^0$ is constrained over $\mathcal{G}$, and hence, by Proposition 2(a), $\mathcal{H}$ is constrained over $\mathcal{G}$.

LEMMA 3. *Let $\varphi: \mathcal{F} \approx \mathcal{F}'$ be an isomorphism of differential fields, and let $\mathcal{H}$ and $\mathcal{H}'$ be constrainedly closed constrained extensions of $\mathcal{F}$ and $\mathcal{F}'$, respectively. Let $(\mathfrak{p}_i)_{i \in I}$ be a family of prime differential ideals of $\mathcal{F}\{y\}$ and, for each index $i \in I$, let $C_i \in \mathcal{F}\{y\}$, $C_i \notin \mathfrak{p}_i$, let Σ_i denote the set of all* *elements of $\mathscr{H}$ that are zeros of $\mathfrak{p}_i$ but not of C_i, and let Σ_i' denote the set of all elements of $\mathscr{H}'$ that are zeros of $\mathfrak{p}_i^{\varphi}$ but not of C_i^{φ}. Suppose, for each differential field isomorphism ψ: $\mathscr{G} \approx \mathscr{G}'$ extending φ with $\mathscr{F} \subset \mathscr{G} \subset \mathscr{H}$ and $\mathscr{G}$ normal over $\mathscr{F}$ in $\mathscr{H}$ and with $\mathscr{F}' \subset \mathscr{G}' \subset \mathscr{H}'$ and $\mathscr{G}'$ normal over $\mathscr{F}'$ in $\mathscr{H}'$, and for each index $i \in I$, that ψ can be extended to an isomorphism $\mathscr{G}\langle\Sigma_i\rangle \approx \mathscr{G}'\langle\Sigma_i'\rangle$. Then φ can be extended to an isomorphism $\mathscr{F}\langle\bigcup_{i\in I}\Sigma_i\rangle \approx \mathscr{F}'\langle\bigcup_{i\in I}\Sigma_i'\rangle$.*

Proof. Let $\mathfrak{M}$ denote the set of all pairs (J, ψ) with $J \subset I$ and ψ an isomorphism $\mathscr{F}\langle\bigcup_{i\in J}\Sigma_i\rangle \approx \mathscr{F}'\langle\bigcup_{i\in J}\Sigma_i'\rangle$ extending φ. For any (J_1, ψ_1), $(J_2, \psi_2) \in \mathfrak{M}$, define $(J_1, \psi_1) \leqslant (J_2, \psi_2)$ to mean that $J_1 \subset J_2$ and ψ_2 extends ψ_1. Then $\mathfrak{M}$ is an inductive ordered set and, by Zorn's lemma, $\mathfrak{M}$ has a maximal element. Let (J, ψ) be such a maximal element, and set $\mathscr{G} = \mathscr{F}\langle\bigcup_{i\in J}\Sigma_i\rangle$, $\mathscr{G}' = \mathscr{F}'\langle\bigcup_{i\in J}\Sigma_i'\rangle$. The set $\bigcup_{i\in J}\Sigma_i$ is normal over $\mathscr{F}$ in $\mathscr{H}$ and therefore $\mathscr{G}$ is also; similarly, $\mathscr{G}'$ is normal over $\mathscr{F}'$ in $\mathscr{H}'$. If J were not I, we could fix $i_0 \in I - J$ and, by hypothesis, could then extend ψ to an isomorphism $\mathscr{G}\langle\Sigma_{i_0}\rangle \approx \mathscr{G}'\langle\Sigma_{i_0}'\rangle$, contradicting the maximality of (J, ψ). Therefore, $J = I$ and the lemma is proved.

If $\mathscr{E}$ is a differential subfield of a differential field $\mathscr{H}$ and $\mathfrak{a}$ is a perfect differential ideal of $\mathscr{E}\{y_1, \ldots, y_n\}$, then $\mathscr{H}\mathfrak{a}$ is a perfect differential ideal of $\mathscr{H}\{y_1, \ldots, y_n\}$. It follows by the Ritt–Raudenbush basis theorem (see DAAG, Chap. III, Section 4, Corollary 5 to Theorem 1) that, for fixed $\mathscr{H}$ and variable $(\mathscr{E}, \mathfrak{a})$, we can argue by induction on $\mathscr{H}\mathfrak{a}$. More precisely, in proving a proposition about $(\mathscr{E}, \mathfrak{a})$, we may argue under the inductive hypothesis that the proposition is known for all pairs $(\mathscr{E}_1, \mathfrak{a}_1)$ like $(\mathscr{E}, \mathfrak{a})$ such that $\mathscr{H}\mathfrak{a}_1$ properly contains $\mathscr{H}\mathfrak{a}$.

LEMMA 4. *Let φ: $\mathscr{F} \approx \mathscr{F}'$ be an isomorphism of differential fields, and let $\mathscr{H}$ and $\mathscr{H}'$ be constrainedly closed small constrained extensions of $\mathscr{F}$ and $\mathscr{F}'$, respectively. Let $\mathfrak{a}$ be a perfect differential ideal of $\mathscr{F}\{y\}$, and let $C \in \mathscr{F}\{y\}$, $C \notin \mathfrak{a}$. Let Σ denote the set of all elements of $\mathscr{H}$ that are zeros of $\mathfrak{a}$ but not of C, and let Σ' denote the set of all elements of $\mathscr{H}'$ that are zeros of $\mathfrak{a}^{\varphi}$ but not of C^{φ}. Then φ can be extended to an isomorphism $\mathscr{F}\langle\Sigma\rangle \approx \mathscr{F}'\langle\Sigma'\rangle$.*

Proof. We argue by induction on $\mathscr{H}\mathfrak{a}$. An easy application of Zorn's lemma shows that $\mathscr{H}$ has a maximal subset Λ that is independent over $\mathscr{F}$ and consists of generic zeros of $\mathfrak{a}$; of course, $\Lambda \subset \Sigma$, Λ is symmetric over $\mathscr{F}$, and if $\mathfrak{a}$ is not prime (or if $\mathfrak{a}$ is prime but a generic zero of $\mathfrak{a}$ is not constrained over $\mathscr{F}$) then $\Lambda = \varnothing$. If Λ is finite, then by Corollary 3 to Proposition 3, φ can be extended to an isomorphism φ_1: $\mathscr{F}\langle\Lambda\rangle \approx \mathscr{F}'\langle\Lambda'\rangle$ mapping Λ onto a subset Λ' of Σ', and evidently Λ' is finite and independent over $\mathscr{F}'$ and consists of generic zeros of $\mathfrak{a}^\varphi$; moreover, Λ' is a maximal such subset of $\mathscr{H}'$, because if Λ_1' were a strictly larger one, then φ_1^{-1} could be extended to an isomorphism $\mathscr{F}'\langle\Lambda_1'\rangle \approx \mathscr{F}\langle\Lambda_1\rangle$ mapping Λ_1' onto a subset Λ_1 of $\mathscr{H}$ that violated the maximality of Λ.

First consider the case in which there exists a maximal Λ as above that is finite. Fix φ_1 and Λ' as above, and set $\mathscr{F}_1 = \mathscr{F}\langle\Lambda\rangle$, $\mathscr{F}_1' = \mathscr{F}'\langle\Lambda'\rangle$. The extensions $\mathscr{H}$ and $\mathscr{H}'$ of $\mathscr{F}_1$ and $\mathscr{F}_1'$, respectively, are small and (by Proposition 2(b)) constrained. Let $(\mathfrak{p}_i)_{i\in I}$ be a family such that, as i runs through I, $\mathfrak{p}_i$ runs through the set of all prime differential ideals of $\mathscr{F}_1\{y\}$ that properly contain $\mathscr{F}_1\mathfrak{a}$ but do not contain C. Because of the maximality of Λ, the defining differential ideal in $\mathscr{F}_1\{y\}$ of any element of $\Sigma - \Lambda$ is $\mathfrak{p}_i$ for some i. Therefore, if we let Σ_i denote the set of all elements of $\mathscr{H}$ that are zeros of $\mathfrak{p}_i$ but not of C, then $\Sigma - \Lambda = \bigcup_{i\in I}\Sigma_i$; similarly, if we let Σ_i' denote the set of all elements of $\mathscr{H}'$ that are zeros of $\mathfrak{p}_i^\varphi$ but not of C^φ, then $\Sigma' - \Lambda' = \bigcup_{i\in I}\Sigma_i'$. Consider any differential field isomorphism ψ: $\mathscr{G} \approx \mathscr{G}'$ extending φ_1 with $\mathscr{F}_1 \subset \mathscr{G} \subset \mathscr{H}$ and $\mathscr{G}$ normal over $\mathscr{F}_1$ in $\mathscr{H}$ and with $\mathscr{F}_1' \subset \mathscr{G}' \subset \mathscr{H}'$ and $\mathscr{G}'$ normal over $\mathscr{F}_1'$ in $\mathscr{H}'$; the extensions $\mathscr{H}$ and $\mathscr{H}'$ of $\mathscr{G}$ and $\mathscr{G}'$, respectively, are small and (by Proposition 7) constrained, and by the above, each $\mathscr{H}\mathfrak{p}_i$ properly contains $\mathscr{H}\mathfrak{a}$; therefore, we may suppose that, for each $i\in I$, ψ can be extended to an isomorphism $\mathscr{G}\langle\Sigma_i\rangle \approx \mathscr{G}'\langle\Sigma_i'\rangle$. It follows by Lemma 3 that φ_1 can be extended to an isomorphism $\mathscr{F}_1\langle\Sigma - \Lambda\rangle \approx \mathscr{F}_1'\langle\Sigma' - \Lambda'\rangle$, so that φ can be extended to an isomorphism $\mathscr{F}\langle\Sigma\rangle \approx \mathscr{F}'\langle\Sigma'\rangle$.

Now consider the case in which every maximal set Λ as above is infinite. Then $\mathfrak{a}$ is prime and hence $\mathfrak{a}^\varphi$ is also, and every maximal subset Λ' of $\mathscr{H}'$ that is independent over $\mathscr{F}'$ and consists of generic zeros of $\mathfrak{a}^\varphi$ is infinite. Fix such sets Λ and Λ'. Since Λ is symmetric over $\mathscr{F}$ and $\mathscr{H}$ is small over $\mathscr{F}$, the elements of Λ can be written in an infinite sequence

$$\Lambda = \{\alpha_0, \alpha_1, \alpha_2, \ldots\},$$

and, similarly, so can the elements of Λ',

$$\Lambda' = \{\alpha_0', \alpha_1', \alpha_2', \ldots\}.$$

We are going to define by induction an infinite sequence of differential field isomorphisms

$$\varphi_n\colon \mathscr{F}_n \approx \mathscr{F}_n' \qquad (n \in \mathbf{N})$$

with $(\mathscr{F}_0, \mathscr{F}_0', \varphi_0) = (\mathscr{F}, \mathscr{F}', \varphi)$ such that, for every $n > 0$:

$$\mathscr{F}_{n-1} \subset \mathscr{F}_n \subset \mathscr{F}\langle\Sigma\rangle, \qquad \mathscr{F}'_{n-1} \subset \mathscr{F}_n' \subset \mathscr{F}'\langle\Sigma'\rangle;$$

$$\alpha_j \in \mathscr{F}_n, \qquad \alpha_j' \in \mathscr{F}_n' \qquad (0 \leqslant j < n);$$

$\mathscr{F}_n$ contains every $\alpha \in \Sigma$ such that the defining differential ideal of α in $\mathscr{F}_{n-1}\{y\}$ properly contains $\mathscr{F}_{n-1}\mathfrak{a}$;

$\mathscr{F}_n'$ contains every $\alpha' \in \Sigma'$ such that the defining differential ideal of α' in $\mathscr{F}'_{n-1}\{y\}$ properly contains $\mathscr{F}_{n-1}\mathfrak{a}^\varphi$;

$\mathscr{H}$ and $\mathscr{H}'$ are constrained extensions of $\mathscr{F}_n$ and $\mathscr{F}_n'$, respectively;

$$\varphi_n \text{ extends } \varphi_{n-1}.$$

Indeed, let $n > 0$, and suppose that isomorphisms $\varphi_j\colon \mathscr{F}_j \approx \mathscr{F}_j'$ $(0 \leqslant j < n)$ have been defined with the desired properties. By Corollary 3 to Proposition 3, φ_{n-1} can be extended to an isomorphism $\mathscr{F}_{n-1}\langle\alpha_{n-1}\rangle \approx \mathscr{F}'_{n-1}\langle\beta'_{n-1}\rangle$ mapping α_{n-1} onto some element $\beta'_{n-1} \in \Sigma'$, and the inverse of this isomorphism can be extended to an isomorphism

$$\mathscr{F}'_{n-1}\langle\beta'_{n-1}, \alpha'_{n-1}\rangle \approx \mathscr{F}_{n-1}\langle\alpha_{n-1}, \beta_{n-1}\rangle$$

mapping α'_{n-1} onto some element $\beta_{n-1} \in \Sigma$. Denote the inverse of this last isomorphism by φ'_{n-1}. By Proposition 2(b), $\mathscr{H}$ and $\mathscr{H}'$ are constrained extensions of $\mathscr{F}_{n-1}\langle\alpha_{n-1}, \beta_{n-1}\rangle$ and $\mathscr{F}'_{n-1}\langle\beta'_{n-1}, \alpha'_{n-1}\rangle$, respectively. Let $(\mathfrak{p}_i)_{i\in I}$ be a family such that, as i runs through I, $\mathfrak{p}_i$ runs through the set of all prime differential ideals of $\mathscr{F}_{n-1}\langle\alpha_{n-1}, \beta_{n-1}\rangle\{y\}$ that properly contain $\mathscr{F}_{n-1}\langle\alpha_{n-1}, \beta_{n-1}\rangle\mathfrak{a}$ but do not contain C; let Σ_i denote the set of all elements of $\mathscr{H}$ that are zeros of $\mathfrak{p}_i$ but not of C, and let Σ_i' denote the set of all elements of $\mathscr{H}'$ that are zeros of $\mathfrak{p}_i^{\varphi'}$ but not of C^φ. For any $\alpha \in \Sigma$, if the defining differential ideal of α in $\mathscr{F}_{n-1}\{y\}$ properly contains $\mathscr{F}_{n-1}\mathfrak{a}$, then the defining differential ideal of α in $\mathscr{F}_{n-1}\langle\alpha_{n-1}, \beta_{n-1}\rangle\{y\}$ properly contains $\mathscr{F}_{n-1}\langle\alpha_{n-1}, \beta_{n-1}\rangle\mathfrak{a}$ and hence is $\mathfrak{p}_i$ for some i, whence $\alpha \in \Sigma_i$; similarly, for any $\alpha' \in \Sigma'$, if the defining differential ideal of α' in $\mathscr{F}'_{n-1}\{y\}$ properly contains $\mathscr{F}'_{n-1}\mathfrak{a}$, then $\alpha' \in \Sigma_i'$ for some i. Now consider any differential field isomorphism $\psi\colon \mathscr{G} \approx \mathscr{G}'$ extending φ'_{n-1} with $\mathscr{F}_{n-1}\langle\alpha_{n-1}, \beta_{n-1}\rangle \subset \mathscr{G} \subset \mathscr{H}$ and $\mathscr{G}$ normal over $\mathscr{F}_{n-1}\langle\alpha_{n-1}, \beta_{n-1}\rangle$ in $\mathscr{H}$ and with $\mathscr{F}'_{n-1}\langle\beta'_{n-1}, \alpha'_{n-1}\rangle \subset \mathscr{G}' \subset \mathscr{H}'$ and $\mathscr{G}'$ normal over $\mathscr{F}'_{n-1}\langle\beta'_{n-1}, \alpha'_{n-1}\rangle$ in $\mathscr{H}'$; $\mathscr{H}$ and $\mathscr{H}'$ are constrainedly closed small constrained extensions of $\mathscr{G}$ and $\mathscr{G}'$, respectively (see Proposition 7); because each $\mathscr{H}\mathfrak{p}_i$ properly contains $\mathscr{H}\mathfrak{a}$, we may suppose that, for each $i \in I$, ψ can be extended to an isomorphism $\mathcal{G}\langle\Sigma_i\rangle \approx \mathcal{G}'\langle\Sigma_i'\rangle$. It follows by Lemma 3 that φ_{n-1}' can be extended to an isomorphism

$$\varphi_n\colon \mathcal{F}_{n-1}\langle\alpha_{n-1}, \beta_{n-1}\rangle\Big\langle\bigcup_{i\in I}\Sigma_i\Big\rangle \approx \mathcal{F}_{n-1}'\langle\beta_{n-1}', \alpha_{n-1}'\rangle\Big\langle\bigcup_{i\in I}\Sigma_i'\Big\rangle.$$

Setting

$$\mathcal{F}_n = \mathcal{F}_{n-1}\langle\alpha_{n-1}, \beta_{n-1}\rangle\Big\langle\bigcup_{i\in I}\Sigma_i\Big\rangle \quad\text{and}\quad \mathcal{F}_n' = \mathcal{F}_{n-1}'\langle\beta_{n-1}', \alpha_{n-1}'\rangle\Big\langle\bigcup_{i\in I}\Sigma_i'\Big\rangle,$$

we see that $\mathcal{F}_n$ is normal over $\mathcal{F}_{n-1}\langle\alpha_{n-1}, \beta_{n-1}\rangle$ in $\mathcal{H}$ and, hence, by Proposition 7, that $\mathcal{H}$ is constrained over $\mathcal{F}_n$ and, similarly, that $\mathcal{H}'$ is constrained over $\mathcal{F}_n'$. This completes the definition of the infinite sequence of isomorphisms $\varphi_n\colon \mathcal{F}_n \approx \mathcal{F}_n'$ with the stated properties.

Finally, consider any $\alpha \in \Sigma$. By the maximality of Λ, the defining differential ideal of α in $\mathcal{F}\langle\Lambda\rangle\{y\}$ properly contains $\mathcal{F}\langle\Lambda\rangle\mathfrak{a}$. Hence, for some $n \in \mathbf{N}$, the defining differential ideal of α in $\mathcal{F}\langle\alpha_0, \alpha_1, \ldots, \alpha_{n-2}\rangle\{y\}$ properly contains $\mathcal{F}\langle\alpha_0, \alpha_1, \ldots, \alpha_{n-2}\rangle\mathfrak{a}$. Therefore, the defining differential ideal of α in $\mathcal{F}_{n-1}\{y\}$ properly contains $\mathcal{F}_{n-1}\mathfrak{a}$ and $\alpha \in \mathcal{F}_n$. Thus, $\mathcal{F}\langle\Sigma\rangle = \bigcup \mathcal{F}_n$ and, similarly, $\mathcal{F}'\langle\Sigma'\rangle = \bigcup \mathcal{F}_n'$. The unique mapping $\bigcup \mathcal{F}_n \to \bigcup \mathcal{F}_n'$ that extends every φ_n is therefore an isomorphism $\mathcal{F}\langle\Sigma\rangle \approx \mathcal{F}'\langle\Sigma'\rangle$.

7. Constrained Closures: Uniqueness up to Isomorphism

We are now in a position to prove the following theorem, which characterizes constrained closures and establishes their uniqueness up to isomorphism.

Theorem 2. (a) *A necessary and sufficient condition that an extension $\mathcal{H}$ of $\mathcal{F}$ be a constrained closure of $\mathcal{F}$ is that $\mathcal{H}$ be a constrainedly closed small constrained extension of $\mathcal{F}$.*

(b) *Let $\varphi\colon \mathcal{F} \approx \mathcal{F}'$ be an isomorphism of differential fields, and let $\mathcal{F}^\dagger$ and $\mathcal{F}'^\dagger$ be constrained closures of $\mathcal{F}$ and $\mathcal{F}'$, respectively. Then φ can be extended to an isomorphism $\mathcal{F}^\dagger \approx \mathcal{F}'^\dagger$.*

Proof. (a) The necessity follows from the definition of constrained closure, Theorem 1, and Proposition 6. To prove the sufficiency, let the condition be satisfied, and observe that by Theorem 1 $\mathcal{F}$ has a constrained closure $\mathscr{F}^\dagger$, that by the necessity $\mathscr{F}^\dagger$ satisfies the condition, and that by Lemma 4 (with $\mathscr{F}' = \mathscr{F}$, $\varphi = \mathrm{id}_{\mathscr{F}}$, $\mathscr{H}' = \mathscr{F}^\dagger$, $\mathfrak{a} = (0)$, $C = 1$) there exists an isomorphism $\mathscr{H} \approx \mathscr{F}^\dagger$ over $\mathscr{F}$; therefore $\mathscr{H}$ is a constrained closure of $\mathscr{F}$.

(b) Again use Lemma 4 (with $\mathscr{H} = \mathscr{F}^\dagger$, $\mathscr{H}' = \mathscr{F}'^\dagger$, $\mathfrak{a} = (0)$, $C = 1$).

COROLLARY 1. *Let $\mathscr{F}^\dagger$ be a constrained closure of $\mathscr{F}$ and $\mathscr{G}$ be an extension of $\mathscr{F}$ in $\mathscr{F}^\dagger$. If $\mathscr{G}$ is finitely generated over $\mathscr{F}$, or algebraic over $\mathscr{F}$, or normal over $\mathscr{F}$ in $\mathscr{F}^\dagger$, then $\mathscr{F}^\dagger$ is a constrained closure of $\mathscr{G}$.*

Proof. $\mathscr{F}^\dagger$ is constrainedly closed, is small over $\mathscr{G}$, and is constrained over $\mathscr{G}$ (by Proposition 2(b) in the finitely generated case, by Proposition 7 in the other two cases). Therefore, part (a) of the theorem applies.

Remark. The result in the finitely generated case is precisely Proposition 5. The shortness of the present proof might suggest the possibility of postponing Proposition 5 until the present section. However, Proposition 5 is needed to prove Proposition 6 and hence Theorem 2, and therefore is involved in the proof of the present corollary.

We also have the following partial converse to Corollary 1.

COROLLARY 2. *Let $\mathscr{G}$ be an extension of $\mathscr{F}$ that is finitely generated and constrained or is algebraic, and let $\mathscr{G}^\dagger$ be a constrained closure of $\mathscr{G}$. Then $\mathscr{G}^\dagger$ is a constrained closure of $\mathscr{F}$.*

Proof. Let $\mathscr{F}^\dagger$ be a constrained closure of $\mathscr{F}$. By Corollary 3 to Proposition 3 and the fact that $\mathscr{F}^\dagger$ is algebraically closed, there exists an embedding $\varphi\colon \mathscr{G} \to \mathscr{F}^\dagger$ over $\mathscr{F}$, and the extension $\varphi(\mathscr{G})$ of $\mathscr{F}$ is either finitely generated or algebraic. By Corollary 1, then $\mathscr{F}^\dagger$ is a constrained closure of $\varphi(G)$, and hence by Theorem 2(b), φ can be extended to an isomorphism $\mathscr{G}^\dagger \approx \mathscr{F}^\dagger$. Therefore, $\mathscr{G}^\dagger$ is a constrained closure of $\mathscr{F}$.

COROLLARY 3. *Let $\mathscr{F}^\dagger$ be a constrained closure of $\mathscr{F}$, and let $\mathscr{G}$ be an extension of $\mathscr{F}$ in $\mathscr{F}^\dagger$.*

(a) *If $\mathscr{G}$ is finitely generated or algebraic over $\mathscr{F}$, then every isomorphism $\mathscr{G} \approx \mathscr{G}'$ over $\mathscr{F}$ with $\mathscr{G}'$ an extension of $\mathscr{F}$ in $\mathscr{F}^\dagger$ can be extended to an automorphism of $\mathscr{F}^\dagger$.*

(b) *If $\mathscr{G}$ is normal over $\mathscr{F}$ in $\mathscr{F}^\dagger$, then every automorphism of $\mathscr{G}$ over $\mathscr{F}$ can be extended to an automorphism of $\mathscr{F}^\dagger$.*

Proof. This follows from Corollary 1 and part (b) of the theorem.

8. Nonuniqueness

The "uniqueness up to isomorphism" of a constrained closure of a differential field, as established in Theorem 2, is much weaker than that of an algebraic closure of a field. We shall see that if $\mathscr{U}$ is a universal extension of an ordinary differential field $\mathscr{C}$ of constants, then $\mathscr{U}$ contains infinitely many constrained closures of $\mathscr{C}$. Indeed, fix a solution $x \in \mathscr{U}$ of the ordinary differential equation $y' = 1$, and observe that for each constant $c \in \mathscr{U}$, $x + c$ is constrained over $\mathscr{C}$ (with constraint 1). Since $\mathscr{U}$ is constrainedly closed, $\mathscr{C}(x + c)$ has a constrained closure $\mathscr{F}_c$ in $\mathscr{U}$ and, by Corollary 2 to Theorem 2, $\mathscr{F}_c$ is a constrained closure of $\mathscr{C}$. Now, a constant that is constrained over $\mathscr{C}$ is algebraic over $\mathscr{C}$. Hence, if c runs over a set of constants no two of which have an algebraic difference (for example, a transcendence basis over $\mathscr{C}$ of the field of constants of $\mathscr{U}$), then the corresponding constrained closures $\mathscr{F}_c$ of $\mathscr{C}$ will be distinct.

Even more striking is the fact that in any constrainedly closed extension of $\mathscr{C}(x)$ there exists an infinite strictly decreasing sequence of constrained closures of $\mathscr{C}(x)$ (and hence of $\mathscr{C}$) or what is equivalent, any constrained closure $\mathscr{G}$ of $\mathscr{C}(x)$ has a proper differential subfield that also is a constrained closure of $\mathscr{C}(x)$.

To prove this, let Σ denote the set of all elements of $\mathscr{G}$ different from 0 and 1 that are solutions of the differential equation

$$y' = y^3 - y^2.$$

If n solutions $\eta_1, \ldots, \eta_n$ in some extension of $\mathscr{C}(x)$ are algebraically dependent over $\mathscr{C}(x)$ but no fewer of them are, then there is an irreducible polynomial $P \in \mathscr{C}[X, Y_1, \ldots, Y_n]$ such that $P(x, \eta_1, \ldots, \eta_n) = 0$, and the computation

$$\sum (\eta_j^3 - \eta_j^2) \frac{\partial P}{\partial Y_j}(x, \eta_1, \ldots, \eta_n) + \frac{\partial P}{\partial X}(x, \eta_1, \ldots, \eta_n) = P(x, \eta_1, \ldots, \eta_n)' = 0$$

shows that the polynomial $\sum (Y_j^3 - Y_j^2)\, \partial P/\partial Y_j + \partial P/\partial X$ is divisible by P; it follows, by the result proved in the appendix, that P is of the form cY_j or $c(Y_j - 1)$ or $c(Y_k - Y_j)$ and, hence, that either $n = 1$ and $\eta_1 = 0$ or 1 or else $n = 2$ and $\eta_2 = \eta_1$. This has several consequences:

(1°) Any family of n solutions different from 0 and 1 is algebraically independent over $\mathscr{C}(x)$ and is constrained over $\mathscr{C}(x)$, a constraint being $\prod y_j \cdot \prod (y_j - 1) \cdot \prod_{j<k} (y_k - y_j)$.

(2°) Σ is infinite.

(3°) Σ is an independent set of conjugates over $\mathscr{C}(x)$.

(4°) If Σ_1 is an infinite subset of Σ, then no element of $\Sigma - \Sigma_1$ is constrained over $\mathscr{C}(x)(\Sigma_1)$.

Now, $\mathscr{C}(x)(\Sigma)$ is normal over $\mathscr{C}(x)$ in $\mathscr{G}$, and hence, by Corollary 1 to Theorem 2, $\mathscr{G}$ is a constrained closure of $\mathscr{C}(x)(\Sigma)$. By (2°), there exists a bijection of Σ onto a proper subset Σ_1 of Σ, and because of (3°), this bijection extends to a differential field isomorphism $\mathscr{C}(x)(\Sigma) \approx \mathscr{C}(x)(\Sigma_1)$ over $\mathscr{C}(x)$. The constrainedly closed extension $\mathscr{G}$ of $\mathscr{C}(x)(\Sigma_1)$ contains a constrained closure $\mathscr{G}_1$ of $\mathscr{C}(x)(\Sigma_1)$, and $\mathscr{G} \neq \mathscr{G}_1$ by (4°). Finally, by Theorem 2(b), the isomorphism $\mathscr{C}(x)(\Sigma) \approx \mathscr{C}(x)(\Sigma_1)$ extends to an isomorphism $\mathscr{G} \approx \mathscr{G}_1$, and therefore $\mathscr{G}_1$ is a constrained closure of $\mathscr{C}(x)$.

9. Strongly Normal Extensions

In this section we consider strongly normal extensions in the sense of DAAG, Chap. VI, Section 3 (finitely generated extensions) and, more generally, in the sense of Kovacic [3] (not necessarily finitely generated ones). The purpose is to relate these extensions to the ideas considered above.

The following lemma will be helpful. Recall that the field of constants of a strongly normal extension of $\mathscr{F}$ coincides with that of $\mathscr{F}$.

Lemma 5. *Let $\mathscr{G}$ be a strongly normal extension of $\mathscr{F}$ and $\mathscr{H}$ be any extension of $\mathscr{G}$ having the same field of constants as $\mathscr{F}$ and $\mathscr{G}$. Then $\mathscr{G}$ is normal over $\mathscr{F}$ in $\mathscr{H}$* (in the sense of Section 6).

Proof. Let $\alpha \in \mathscr{G}$, $\alpha' \in \mathscr{H}$, and suppose that α' is a generic differential specialization of α over $\mathscr{F}$. We must show that $\alpha' \in \mathscr{G}$. To this end, fix a universal extension $\mathscr{U}$ of $\mathscr{H}$, denote the field of constants of $\mathscr{U}$ by $\mathscr{K}$, and set $\mathscr{C} = \mathscr{F} \cap \mathscr{K} = \mathscr{H} \cap \mathscr{K}$. Now, α is contained in some finitely generated strongly normal extension $\mathscr{G}'$ of $\mathscr{F}$ in $\mathscr{G}$, and the isomorphism $\mathscr{F}\langle\alpha\rangle \approx \mathscr{F}\langle\alpha'\rangle$ over $\mathscr{F}$ that maps α onto α' can be extended to an isomorphism σ of $\mathscr{G}'$ onto an extension $\sigma\mathscr{G}'$ of $\mathscr{F}$ in $\mathscr{U}$. Because $\mathscr{G}'$ is strongly normal over $\mathscr{F}$, σ is a strong isomorphism, so that $\sigma\mathscr{G}' \subset \mathscr{G}'\mathscr{K}$. However, $\mathscr{H}$ and $\mathscr{G}'\mathscr{K}$ are linearly disjoint over $\mathscr{G}'$, and hence

$$\alpha' \in (\mathscr{G}'\mathscr{K}) \cap \mathscr{H} = \mathscr{G}' \subset \mathscr{G}.$$

The following result shows that every strongly normal extension of $\mathcal{F}$ is contained in a constrained closure of $\mathcal{F}$ and hence, in particular, is constrained over $\mathcal{F}$.

THEOREM 3. *Let $\mathcal{G}$ be a strongly normal extension of $\mathcal{F}$ and $\mathcal{H}$ be an extension of $\mathcal{G}$. A necessary and sufficient condition that $\mathcal{H}$ be a constrained closure of $\mathcal{F}$ is that $\mathcal{H}$ be a constrained closure of $\mathcal{G}$.*

Proof. If $\mathcal{H}$ is a constrained closure of either $\mathcal{F}$ or $\mathcal{G}$, the field of constants of $\mathcal{H}$ is an algebraic closure $\mathcal{C}_a$ of the common field of constants $\mathcal{C}$ of $\mathcal{F}$ and $\mathcal{G}$. Now, $\mathcal{G}\mathcal{C}_a$ is a strongly normal extension of $\mathcal{F}\mathcal{C}_a$ with field of constants $\mathcal{C}_a$ (see DAAG, Chap. VI, Section 3, Theorem 2); also, by Corollaries 1 and 2 to Theorem 2, $\mathcal{H}$ is a constrained closure of $\mathcal{F}$ resp. $\mathcal{G}$ if and only if $\mathcal{H}$ is a constrained closure of $\mathcal{F}\mathcal{C}_a$ resp. $\mathcal{G}\mathcal{C}_a$. Thus, we may replace $\mathcal{F}$, $\mathcal{G}$ by $\mathcal{F}\mathcal{C}_a$, $\mathcal{G}\mathcal{C}_a$, that is, we may suppose that $\mathcal{F}$, $\mathcal{G}$, $\mathcal{H}$ have the same field of constants $\mathcal{C}$ which is algebraically closed.

By Lemma 5, $\mathcal{G}$ is normal over $\mathcal{F}$ in $\mathcal{H}$. It follows, by Corollary 1 to Theorem 2, that if $\mathcal{H}$ is a constrained closure of $\mathcal{F}$, then $\mathcal{H}$ is a constrained closure of $\mathcal{G}$. Henceforth, let $\mathcal{H}$ be a constrained closure of $\mathcal{G}$. We must prove that $\mathcal{H}$ is a constrained closure of $\mathcal{F}$.

Claim. It suffices to prove that there exists an embedding over $\mathcal{F}$ of $\mathcal{G}$ in a constrained closure of $\mathcal{F}$. Indeed, if $\varphi\colon \mathcal{G} \to \mathcal{F}^\dagger$ is such an embedding, then $\varphi(\mathcal{G})$ is a strongly normal extension of $\mathcal{F}$, the field of constants of $\mathcal{F}^\dagger$ is $\mathcal{C}$, and by Lemma 5, therefore $\varphi(\mathcal{G})$ is normal over $\mathcal{F}$ in $\mathcal{F}^\dagger$; by Corollary 1 to Theorem 2, then $\mathcal{F}^\dagger$ is a constrained closure of $\varphi(\mathcal{G})$, by Theorem 2(b), therefore φ can be extended to an isomorphism $\mathcal{H} \approx \mathcal{F}'$, and hence $\mathcal{H}$ is a constrained closure of $\mathcal{F}$.

Claim. It suffices to prove that every finitely generated strongly normal extension with algebraically closed field of constants is a constrained extension. Indeed, suppose that we have done this. By the definition of "strongly normal" (see Kovacic [3], Chap. II, Section 1), we have $\mathcal{G} = \mathcal{F}(\bigcup_{i\in I} \mathcal{G}_i)$ where, for each i, $\mathcal{G}_i$ is a finitely generated strongly normal extension of $\mathcal{F}$. Fix a constrained closure $\mathcal{F}^\dagger$ of $\mathcal{F}$, and consider the set $\mathfrak{M}$ of all pairs (J, ψ) with $J \subset I$ and ψ an embedding in $\mathcal{F}^\dagger$ over $\mathcal{F}$ of the differential field $\mathcal{G}_J = \mathcal{F}(\bigcup_{j\in J} \mathcal{G}_j)$. In $\mathfrak{M}$, define $(J_1, \psi_1) \leqslant (J_2, \psi_2)$ to mean that $J_1 \subset J_2$ and ψ_2 extends ψ_1. This makes $\mathfrak{M}$ an inductive ordered set. By Zorn's lemma, $\mathfrak{M}$ has a maximal element, say (J, ψ). Now, $\psi(\mathcal{G}_J)$ is a strongly normal extension of $\mathcal{F}$ and hence, by Lemma 5, is normal over

$\mathscr{F}$ in $\mathscr{F}^\dagger$; by Corollary 1 to Theorem 2, then $\mathscr{F}^\dagger$ is a constrained closure of $\psi(\mathscr{G}_J)$. If $J \neq I$, we can fix $i \in I - J$ and set $J' = J \cup \{i\}$; the extension $\mathscr{G}_{J'} = \mathscr{G}_J\mathscr{G}_i$ of $\mathscr{G}_J$ is a finitely generated strongly normal one (see DAAG, Chap. VI, Section 4, Theorem 5) with algebraically closed field of constants $\mathscr{C}$ and, hence, by what we are supposing, is constrained, so that, by Corollary 3 to Proposition 3, ψ can be extended to an embedding $\mathscr{G}_{J'} \rightarrow \mathscr{F}^\dagger$ contrary to the maximality of (J, ψ). Therefore, $J = I$ and ψ is an embedding over $\mathscr{F}$ of $\mathscr{G}_I = \mathscr{G}$ in $\mathscr{F}^\dagger$.

The claim established, let $\mathscr{G}$ now be any finitely generated strongly normal extension of $\mathscr{F}$ with algebraically closed field of constants $\mathscr{C}$. Fix a universal extension $\mathscr{U}$ of $\mathscr{G}$, and denote its field of constants by $\mathscr{K}$. Now, $\mathscr{U}$ contains an algebraic closure $\mathscr{F}_{\mathbf{a}}$ of $\mathscr{F}$, and by DAAG, Chap. VI, Section 4, Theorem 5, $\mathscr{G}\mathscr{F}_{\mathbf{a}}$ is a finitely generated strongly normal extension of $\mathscr{F}_{\mathbf{a}}$. Because $\mathscr{F}_{\mathbf{a}}$ is constrained over $\mathscr{F}$, Proposition 2(a) shows that it suffices to prove that $\mathscr{G}\mathscr{F}_{\mathbf{a}}$ is constrained over $\mathscr{F}_{\mathbf{a}}$. Replacing $\mathscr{F}$, $\mathscr{G}$ by $\mathscr{F}_{\mathbf{a}}$, $\mathscr{G}\mathscr{F}_{\mathbf{a}}$, we suppose henceforth that $\mathscr{F}$ is algebraically closed.

The Galois group $G(\mathscr{G}/\mathscr{F})$ is a connected $\mathscr{C}$-group relative to the universal field $\mathscr{K}$. There exist a $\mathscr{C}$-group G relative to the universal field $\mathscr{U}$ and a $\mathscr{C}$-isomorphism c: $G(\mathscr{G}/\mathscr{F}) \approx G_{\mathscr{K}}$, where $G_{\mathscr{K}}$ denotes the set of elements of G that are rational over $\mathscr{K}$ ($G_{\mathscr{K}}$ has a natural structure of $\mathscr{C}$-group relative to the universal field $\mathscr{K}$). Because $\mathscr{F}$ is algebraically closed, the differential Galois cohomology set $H^1(\mathscr{G}/\mathscr{F}, G)$ is trivial (see DAAG, Chap. VI, Section 8, Corollary 2 to Theorem 7). It follows (see DAAG, Chap. VI, Section 9, Theorem 8) that there exists a G-primitive α over $\mathscr{F}$ such that $\mathscr{G} = \mathscr{F}\langle\alpha\rangle$ and

$$\sigma\alpha = \alpha c(\sigma) \qquad (\sigma \in G(\mathscr{G}/\mathscr{F})).$$

Fix a $\mathscr{C}$-generic ($=$ isolated) element τ of $G(\mathscr{G}/\mathscr{F})$. Then α and $\tau\alpha$ are $\mathscr{F}$-generic elements of G, and the differential fields $\mathscr{G} = \mathscr{F}\langle\alpha\rangle$ and $\tau\mathscr{G} = \mathscr{F}\langle\tau\alpha\rangle$ are linearly disjoint over $\mathscr{F}$. There exists a generically invertible $\mathscr{C}$-mapping ξ of G into a $\mathscr{C}$-subset V of some affine space $\mathscr{U}^n$; the domains of bidefinition $\mathscr{O}$ and $\mathscr{O}'$ of ξ and its generic inverse are dense $\mathscr{C}$-open subsets of G and V, all respectively, and ξ maps $\mathscr{O}$ bijectively onto $\mathscr{O}'$. Since $V - \mathscr{O}'$ is a $\mathscr{C}$-closed proper subset of $\mathscr{U}^n$, there exists a polynomial $C \in \mathscr{C}[y_1, \ldots, y_n]$ that vanishes at every point of $V - \mathscr{O}'$ but not at every point of V. Setting $a = \xi(\alpha)$, we see that $\mathscr{G} = \mathscr{F}\langle a\rangle$, $\tau\mathscr{G} = \mathscr{F}\langle\tau a\rangle$, and $C(a) \neq 0$. We shall show that a is constrained over $\mathscr{F}$, with constraint C. By Proposition 1, this will show that $\mathscr{G}$ is a constrained extension of $\mathscr{F}$ and hence will complete the proof of the theorem.

Let a' be any differential specialization of a over $\mathscr{F}$ with $C(a') \neq 0$. We must show it generic. Evidently, $a' \in \mathcal{O}'$, so there is a unique element $\alpha' \in \mathcal{O}$ such that $\xi(\alpha') = a'$. Of course, a and a' are differential specializations over $\mathscr{F}$ of a and τa, respectively; it follows by the linear disjointness of $\mathscr{F}(a)$ and $\mathscr{F}(\tau a)$ that (a, a') is a differential specialization of $(a, \tau a)$ over $\mathscr{F}$.

Set $\gamma = \alpha^{-1}\alpha'$, and recall that $c(\tau) = \alpha^{-1}\tau\alpha$. Let χ be the everywhere defined $\mathscr{C}$-mapping $G^2 \rightarrow G$ given by the formula $\chi(\alpha_1, \alpha_2) = \alpha_1^{-1}\alpha_2$, and let ζ be the $\mathscr{C}$-mapping of V^2 into G^2 such that $\zeta(a, \tau a) = (\alpha, \tau\alpha)$. For any $\mathscr{C}$-function φ on G that is defined at γ,

$$\varphi(\gamma) = \varphi(\alpha^{-1}\alpha') = (\varphi \square \chi)(\alpha, \alpha') = ((\varphi \square \chi) \square \zeta)(a, a'),$$

$$\varphi(c(\tau)) = ((\varphi \square \chi) \square \zeta)(a, \tau a).$$

Now $(\varphi \square \chi) \square \zeta$ is a $\mathscr{C}$-function on V^2 defined at (a, a'). Therefore there exist polynomials

$$P, Q \in \mathscr{C}[y_1, \ldots, y_n, z_1, \ldots, z_n]$$

with $Q(a, a') \neq 0$ such that

$$\varphi(\gamma) = P(a, a')/Q(a, a'),$$

$$\varphi(c(\tau)) = P(a, \tau a)/Q(a, \tau a).$$

Since (a, a') is a differential specialization of $(a, \tau a)$ over $\mathscr{F}$, it follows that $\varphi(\gamma)$ is a differential specialization of $\varphi(c(\tau))$ over $\mathscr{F}$. Because $\mathscr{C}(\varphi(c(\tau))) \subset \mathscr{C}(c(\tau)) \subset \mathscr{K}$, that is, $\varphi(c(\tau))$ is a constant, we infer that $\varphi(\gamma)$ is a constant. Since this is the case for every $\mathscr{C}$-function φ on G defined at γ, it follows that $\gamma \in G_{\mathscr{K}}$. Therefore there exists an element $\sigma \in G(\mathscr{G}/\mathscr{F})$ with $\gamma = c(\sigma)$, and $\alpha' = \alpha\gamma = \alpha c(\sigma) = \sigma\alpha$, so that $a' = \xi(\alpha') = \xi(\sigma\alpha) = \sigma(\xi(\alpha)) = \sigma a$. This shows that a' is a generic differential specialization of a over $\mathscr{F}$, and completes the proof of the theorem.

Kovacic has proved (see [3, Chap. II, Section 4]) that when the field of constants of $\mathscr{F}$ is algebraically closed, then $\mathscr{F}$ has a *maximal* strongly normal extension (a strongly normal extension that is not contained in a larger one) and that it is unique up to isomorphism. It follows from Theorem 3 then that if $\mathscr{F}^\dagger$ is a constrained closure of $\mathscr{F}$, then $\mathscr{F}^\dagger$ contains a unique maximal strongly normal extension $\mathscr{S}$ of $\mathscr{F}$ and $\mathscr{F}^\dagger$ is a constrained closure of $\mathscr{S}$. Consequently, $\mathscr{F}^\dagger$ contains an infinite increasing sequence

$$\mathscr{F} \subset \mathscr{S} \subset \mathscr{S}' \subset \mathscr{S}'' \subset \cdots$$

in which each term after the first is a maximal strongly normal extension of the preceding term. The union $\mathscr{T} = \bigcup \mathscr{S}^{(n)}$ is a minimal extension of $\mathscr{F}$ that does not have a proper strongly normal extension, and is the unique such in $\mathscr{F}^{\dagger}$.

Appendix

Let K be a field of characteristic 0, let $n \in \mathbf{N}$, and let $(X, Y_1, \ldots, Y_n)$ be a family of indeterminates. The operator

$$\mathscr{D}_n = \sum_{1 \leqslant j \leqslant n} (Y_j^3 - Y_j^2) \frac{\partial}{\partial Y_j} + \frac{\partial}{\partial X}$$

is a derivation of $K(X, Y_1, \ldots, Y_n)$ over K that maps $K[X, Y_1, \ldots, Y_n]$ into itself. Trivial computations yield the following logarithmic derivatives:

$$\begin{aligned}
\mathscr{D}_n c \cdot c^{-1} &= 0 && (c \in K, c \neq 0),\\
\mathscr{D}_n Y_j \cdot Y_j^{-1} &= Y_j^2 - Y_j && (1 \leqslant j \leqslant n),\\
\mathscr{D}_n (Y_j - 1) \cdot (Y_j - 1)^{-1} &= Y_j^2 && (1 \leqslant j \leqslant n),\\
\mathscr{D}_n (Y_k - Y_j) \cdot (Y_k - Y_j)^{-1} &= Y_j^2 + Y_j Y_k + Y_k^2 - Y_j - Y_k &&\\
& && (1 \leqslant j \leqslant k \leqslant n).
\end{aligned}$$

It follows, for any nonzero polynomial of the form

$$P = c \prod_{1 \leqslant j \leqslant n} Y_j^{e_j} \cdot \prod_{i \leqslant j \leqslant n} (Y_j - 1)^{f_j} \cdot \prod_{1 \leqslant j < k \leqslant n} (Y_k - Y_j)^{g_{jk}} \tag{1}$$

with $c \in K$, that

$$\mathscr{D}_n P \cdot P^{-1} = A,$$

where

$$\begin{aligned}
A &= \sum_{1 \leqslant j \leqslant n} e_j (Y_j^2 - Y_j) + \sum_{1 \leqslant j \leqslant n} f_j Y_j^2\\
&\quad + \sum_{1 \leqslant j < k \leqslant n} g_{jk} (Y_j^2 + Y_j Y_k + Y_k^2 - Y_j - Y_k)\\
&= \sum_{1 \leqslant j \leqslant n} \Big(e_j + f_j + \sum_{1 \leqslant \nu < j} g_{\nu j} + \sum_{j < \nu \leqslant n} g_{j\nu} \Big) Y_j^2\\
&\quad + \sum_{i \leqslant j < k \leqslant n} g_{jk} Y_j Y_k - \sum_{1 \leqslant j \leqslant n} \Big(e_j + \sum_{1 \leqslant \nu < j} g_{\nu j} + \sum_{j < \nu \leqslant n} g_{j\nu} \Big) Y_j .
\end{aligned} \tag{2}$$

We shall prove the following converse of this remark.

Let $P \in K[X, Y_1, \ldots, Y_n]$, $P \neq 0$, $\mathscr{D}_n P \cdot P^{-1} \in K[X, Y_1, \ldots, Y_n]$. *Then* P *is of the form* (1).

It is an easy corollary of this result that *if* P_1, P_2 *are two nonzero polynomials with* $\mathscr{D}_n P_1 \cdot P_1^{-1} = \mathscr{D}_n P_2 \cdot P_2^{-1}$, *then* $P_2 = aP_1$ with $a \in K$.

The proof is by induction. When $n = 0$, the assertion is that a polynomial in $K[X]$ that divides its derivative must be in K, which is obvious. Let $n > 0$ and adopt the inductive hypothesis that the result is known for lower values of n.

If $P = P_1P_2$ with P_1, P_2 relatively prime, then evidently $\mathscr{D}_n P \cdot P^{-1} \in K[X, Y_1, \ldots, Y_n]$ if and only if $\mathscr{D}_n P_i \cdot P_i^{-1} \in K[X, Y_1, \ldots, Y_n]$ $(i = 1, 2)$. Since we may take P_2 to be the product of all the irreducible factors of P of the forms Y_j, $Y_j - 1$, $Y_k - Y_j$, this shows that we may add the hypothesis that P not have any such irreducible factors, the desired conclusion then being that $P \in K$. We assume, under the added hypothesis, that $P \notin K$, and seek a contradiction.

By the inductive hypothesis, then P is not free of any Y_j. Set $m_j = \deg_{Y_j} P$ $(1 \leqslant j \leqslant n)$ and write

$$\mathscr{D}_n P = AP, \tag{3}$$

where, by hypothesis, $A \in K[X, Y_1, \ldots, Y_n]$. We shall show that

$$A = \sum_{1 \leqslant j \leqslant n} m_j(Y_j^2 - Y_j). \tag{4}$$

Comparing degrees in X of the two members of (3), we see that A is free of X. Comparing degrees in Y_j, we see that $\deg_{Y_j} A = 2$. Comparing total degrees, we see that $\deg A = 2$. Comparing terms of degree $m_j + 2$ in Y_j, we see that the coefficient of Y_j^2 in A is m_j. Setting

$$Q = P(X, Y_1, \ldots, Y_{n-1}, 0), \qquad B = A(Y_1, \ldots, Y_{n-1}, 0),$$

we see from (3) that $\mathscr{D}_{n-1} Q = BQ$, and $Q \neq 0$ because P is not divisible by Y_n; by the inductive hypothesis, Q has the form of the second member of (1), with n replaced by $n - 1$, and hence B has the form of the third member of (2), also with n replaced by $n - 1$; therefore the term of degree 0 in B, which coincides with that in A, is 0. Thus, we may write

$$A = \sum_{1 \leqslant j \leqslant n} m_j Y_j^2 + \sum_{1 \leqslant j < k \leqslant n} c_{jk} Y_j Y_k + \sum_{1 \leqslant j \leqslant n} b_j Y_j .$$

Now setting

$$\begin{aligned} Q &= P(X, Y_1, \ldots, Y_{n-1}, 1), \\ B &= A(Y_1, \ldots, Y_{n-1}, 1) \\ &= \sum_{1 \leqslant j < n} m_j Y_j^2 + \sum_{1 \leqslant j < k < n} c_{jk} Y_j Y_k + \sum_{1 \leqslant j < n} (b_j + c_{jn}) Y_j + m_n + b_n, \end{aligned}$$

we again see that $\mathscr{D}_{n-1}Q = BQ$ and $Q \neq 0$ and, hence, that B has the form of the third member of (2), with n replaced by $n - 1$, whence $m_n + b_n = 0$. A similar argument, substituting 1 for Y_j instead of Y_n, shows that $b_j = -m_j$ $(1 \leqslant j \leqslant n)$. In completing the proof of (4), we may suppose that $n > 1$. Setting

$$\begin{aligned} Q &= P(X, Y_1, \ldots, Y_{n-1}, Y_{n-1}), \\ B &= A(Y_1, \ldots, Y_{n-1}, Y_{n-1}) \\ &= \sum_{1 < j < n-1} m_j Y_j^2 + (m_{n-1} + m_n + c_{n-1,n}) Y_{n-1}^2 + \sum_{1 \leqslant j < k < n-1} c_{jk} Y_j Y_k \\ &\quad + \sum_{1 \leqslant j < n-1} (c_{j,n-1} + c_{jn}) Y_j Y_{n-1} - \sum_{1 \leqslant j < n-1} m_j Y_j - (m_{n-1} + m_n) Y_{n-1}, \end{aligned}$$

we see once more that $\mathscr{D}_{n-1}Q = BQ$ and $Q \neq 0$ and, hence, that we may write

$$\begin{aligned} Q &= c \prod_{1 \leqslant j < n} Y_j^{e_j} \cdot \prod_{1 \leqslant j < n} (Y_j - 1)^{f_j} \cdot \prod_{1 \leqslant j < k < n} (Y_k - Y_j)^{g_{jk}}, \\ B &= \sum_{1 \leqslant j < n} \left(e_j + f_j + \sum_{i \leqslant \nu < j} g_{\nu j} + \sum_{j < \nu < n} g_{j\nu} \right) Y_j^2 \\ &\quad + \sum_{1 \leqslant j < k < n} g_{jk} Y_j Y_k - \sum_{1 \leqslant j < n} \left(e_j + \sum_{1 \leqslant \nu < j} g_{\nu j} + \sum_{j < \nu < n} g_{j\nu} \right) Y_j; \end{aligned}$$

comparing coefficients, here and in the preceding expression for B, of Y_{n-1}^2 and of Y_{n-1}, we see that

$$\begin{aligned} m_{n-1} + m_n + c_{n-1,n} &= e_{n-1} + f_{n-1} + \sum_{1 \leqslant \nu < n-1} g_{\nu,n-1}, \\ m_{n-1} + m_n &= e_{n-1} + \sum_{1 \leqslant \nu < n-1} g_{\nu,n-1}, \end{aligned}$$

whence $c_{n-1,n} = f_{n-1}$; but

$$\begin{aligned} e_{n-1} + f_{n-1} + \sum_{1 \leqslant \nu < n-1} g_{\nu,n-1} &= \deg_{Y_{n-1}} Q \leqslant \deg_{Y_{n-1}} P + \deg_{Y_n} P \\ &= m_{n-1} + m_n = e_{n-1} + \sum_{1 \leqslant \nu < n-1} g_{\nu,n-1}, \end{aligned}$$

so that $f_{n-1} = 0$, whence $c_{n-1,n} = 0$. A similar argument, replacing Y_k by Y_j instead of Y_n by Y_{n-1}, shows that $c_{jk} = 0$ $(1 \leqslant j < k \leqslant n)$, and completes the proof of (4).

Now set $m = m_n$ and $Y = Y_m$, and write

$$P = \sum P_i Y^i,$$

where $P_i \in K[X, Y_1, \ldots, Y_{n-1}]$ for each i. Also rewrite (4) in the form

$$A = mY^2 - mY + \sum_{1 \leqslant j < n} m_j(Y_j^2 - Y_j).$$

Because P is not divisible by Y and $m = \deg_Y P$,

$$P_0 P_m \neq 0, \qquad P_i = 0 \qquad (i > m \text{ or } i < 0).$$

Since $\mathscr{D}_n = (Y^3 - Y^2)\partial/\partial Y + \mathscr{D}_{n-1}$, when we compare terms in (3) of degree i in Y, we find that

$$\begin{aligned}(i-2)\, P_{i-2} - (i-1)\, P_{i-1} + \mathscr{D}_{n-1} P_i \\ = mP_{i-2} - mP_{i-1} + \sum_{1 \leqslant j < n} m_j(Y_j^2 - Y_j)\, P_i,\end{aligned}$$

whence

$$\mathscr{D}_{n-1} P_i - \sum_{1 \leqslant j < n} m_j(Y_j^2 - Y_j)\, P_i = (m - i + 2)\, P_{i-2} - (m - i + 1)\, P_{i-1}. \tag{5}$$

We claim that there exist $F_0, F_1, \ldots, F_m \in K[X]$ such that

$$F_i \neq 0, \qquad \deg F_i \leqslant i, \qquad P_i = F_i \prod_{1 \leqslant j < n} Y_j^{m_j}$$

for every index i with $0 \leqslant i \leqslant m$. Indeed, when $i = 0$, then (5) reduces to the equation $\mathscr{D}_{n-1} P_0 = \sum_{1 \leqslant j < n} m_j(Y_j^2 - Y_j)\, P_0$, and therefore the inductive hypothesis and Eq. (2) with n replaced by $n - 1$ show that $P_0 = c \prod_{1 \leqslant j < n} Y_j^{m_j}$ with $c \in K$, $c \neq 0$. This establishes the claim for $i = 0$. Now let $0 < i \leqslant n$, and suppose the claim established for lower values of i. Then by (5),

$$\begin{aligned}\mathscr{D}_{n-1} P_i - \sum_{1 \leqslant j < n} m_j(Y_j^2 - Y_j)\, P_i \\ = ((m - i + 2) F_{i-2} - (m - i + 1) F_{i-1}) \prod_{1 \leqslant j < n} Y_j^{m_j}\end{aligned}$$

(where $F_{i-2} = 0$ if $i = 1$). For any exponent $k \in \mathbf{N}$ and any polynomial $T \in K[X, Y_1, ..., Y_{n-1}]$ such that $\mathscr{D}_{n-1}T = CT$, where $C \in K[X, Y_1, ..., Y_{n-1}]$, a trivial computation shows that

$$\mathscr{D}_{n-1}(X^k T) - CX^k T = kX^{k-1}T,$$

and hence that

$$\mathscr{D}_{n-1}(FT) - CFT = (dF/dX)T$$

for every polynomial $F \in K[X]$. Taking $T = \prod_{1 \leqslant j < n} Y_j^{m}$, so that $C = \sum_{1 \leqslant j < n} m_j(Y_j^2 - Y_j)$, and choosing F so that

$$dF/dX = (m - i + 2) F_{i-2} - (m - i + 1) F_{i-1},$$

we see that $F \neq 0$ and $\deg F = i$, and that

$$\mathscr{D}_{n-1}(FT) - CFT = \mathscr{D}_{n-1}P_i - CP_i.$$

Thus, $\mathscr{D}_{n-1}(P_i - FT) = C(P_i - FT)$ and, hence, as we have already observed in the case of P_0, $P_i - FT = bT$ for some $b \in K$. Setting $F_i = F + b$, we see that $P_i = F_i \prod_{1 \leqslant j < n} Y_j^{m}$. This establishes the claim in general.

In particular, $P_{m-1} \neq 0$. But when $i = m + 1$, then (5) reduces to the equation $P_{m-1} = 0$. This provides the sought contradiction and completes the proof.

Note Added in Proof. Shelah has sent me a preprint of a paper in which he proves the result in Section 8 and the main result of Wood [11].

REFERENCES

1. L. BLUM, Generalized algebraic structures: A model theoretic approach. Ph.D. dissertation, Massachusetts Institute of Technology, Cambridge, MA, 1968.
2. E. R. KOLCHIN, "Differential Algebra and Algebraic Groups," Academic Press, New York, 1973. (Cited as DAAG.)
3. J. KOVACIC, Pro-algebraic groups and the Galois theory of differential fields, *Amer. J. Math.* to appear.
4. M. D. MORLEY, Categoricity in power, *Trans. Amer. Math. Soc.* **114** (1965), 514–538.
5. A. ROBINSON, On the concept of differentially closed field, *Bull. Res. Counc. Isr. Sect. F* 8 (1959), 113–118.
6. A. ROBINSON, "Introduction to Model Theory and the Metamathematics of Algebra," pp. 132–137 and 209–214, North Holland Publ., Amsterdam, 1963.

6a. M. Rosenlicht, The nonminimality of the differential closure, *Pacific J. Math.* to appear.
7. G. E. Sacks, "Saturated Model Theory," Benjamin, New York, 1972.
8. G. E. Sacks, The differential closure of a differential field, *Bull. Amer. Math. Soc.* **78** (1972), 629–635.
9. S. Shelah, Uniqueness and characterization of prime models over sets for totally transcendental first order theories, *J. Symbolic Logic* **37** (1972), 107–113.
10. C. Wood, The model theory of differential fields of characteristic $p \neq 0$, *Proc. Amer. Math. Soc.* to appear.
11. C. Wood, Prime model extensions for differential fields of characteristic $p \neq 0$, Dept. of Math. Report. Yale Univ., New Haven, CT, 1973.

Printed by the St Catherine Press Ltd., Tempelhof 37, Bruges, Belgium.

Reprinted from:
CONTRIBUTIONS TO ANALYSIS
A Collection of Papers Dedicated to Lipman Bers

ACADEMIC PRESS, INC.

Differential Equations in a Projective Space and Linear Dependence over a Projective Variety

E. R. KOLCHIN†
COLUMBIA UNIVERSITY

INTRODUCTION

Consider an irreducible algebraic variety V in complex projective space $\mathbf{P}_{\mathbf{C}}(n)$ and $n+1$ meromorphic functions $f_0, f_1, \ldots, f_n$ on some region of **C**. J. F. Ritt once remarked to me that there exists an irreducible ordinary differential polynomial $R \in \mathbf{C}\{y_0, y_1, \ldots, y_n\}$, dependent only on V and having order equal to the dimension of V, that enjoys the following property: A necessary and sufficient condition that there exist $c_0, c_1, \ldots, c_n \in \mathbf{C}$ not all zero such that $(c_0 : c_1 : \cdots : c_n)$ is a point of V and $\sum c_j f_j = 0$ is that $(f_0, f_1, \ldots, f_n)$ be in the general solution of the differential equation $R = 0$. When $V = \mathbf{P}_{\mathbf{C}}(n)$ this reduces to the classical theorem on the Wronskian determinant.

Ritt did not publish a proof of his result and to the best of my knowledge did not leave one when he died in 1951. The main purpose of this paper is to prove Ritt's result, generalized to include functions of several complex variables as well as one, and formulated in the setting of abstract differential fields of characteristic zero (see the theorem in Section 7 and its corollary). This provides an occasion to describe the beginnings of a theory of algebraic differential equations in a projective space. The second purpose is to describe these beginnings (see Sections 4–6).‡ The first part of the paper (Sections 1–3) describes some notions about differential polynomials and differential polynomial ideals that are used in the sequel.

The reader is expected to have some familiarity with the relevant parts of the Ritt theory of algebraic differential equations, as can be found in Chapters I–IV of my recent book [2]. Throughout the paper, $\mathcal{U}$ denotes a universal differential field of characteristic zero with derivation operators $\delta_1, \ldots, \delta_m$,§ and $\mathcal{K}$ denotes its field of constants. The set of all derivative operators $\delta_1^{i_1} \cdots \delta_m^{i_m}$ $(i_1, \ldots, i_m \in \mathbf{N})$ is denoted by $\mathbf{\Theta}$, and the set of all such derivative operators of order less than or equal to a given $s \in \mathbf{N}$

† This research was subsidized by a grant from the National Science Foundation.

‡ The work of Morikawa [3] is quite different. It deals with analytic aspects of a special class of systems of ordinary differential equations in a projective space.

§ When $\mathcal{U}$ is an ordinary differential field, that is, when $m = 1$, the successive derivatives $a, \delta_1 a, \delta_1{}^2 a, \ldots, \delta_1{}^r a, \ldots$, of an element $a \in \mathcal{U}$ are often denoted by $a, a', a'', \ldots, a^{(r)}, \ldots$.

is denoted by $\boldsymbol{\Theta}(s)$. $\mathscr{F}$ denotes a differential subfield of $\mathscr{U}$. $\mathscr{R}$ denotes a commutative differential ring that, except for a few brief moments, is an algebra over the field $\mathbf{Q}$ of rational numbers. The letters y and z, with and without various subscripts, denote differential indeterminates.

1. CANONICAL CHARACTERISTIC SETS

Let $\mathfrak{k}$ be a differential ideal if $\mathscr{F}\{y_1, \ldots, y_n\}$ and fix a ranking of $(y_1, \ldots, y_n)$. Then $\mathfrak{k}$ has a characteristic set A relative to the ranking, but A need not be unique, even when $\mathfrak{k}$ is prime and we impose, as we may, the condition that each element of A be irreducible over $\mathscr{F}$. For example, when $\mathscr{F}$ is the ordinary differential field $\mathbf{C}(x)$ of rational functions of a complex variable x, the set of differential polynomials in $\mathscr{F}\{y_1, y_2\}$ that vanish at $((x-c)^2, (x-c)^{-1})$ for every $c \in \mathbf{C}$ is a prime differential ideal; relative to the ranking of (y_1, y_2) for which every derivative of y_1 has lower rank than every derivative of y_2, the differential polynomials

$$y_1'^2 - 4y_1, \qquad 2y_1y_2 - y_1'$$

form a characteristic set, and so do

$$y_1'^2 - 4y_1, \qquad y_1'y_2 - 2.$$

In what follows, a condition will be put on A that ensures that A be unique.

For each element $P \in \mathscr{F}\{y_1, \ldots, y_n\}$ with $P \notin \mathscr{F}$ let u_P denote the leader of P and set $d_P = \deg_{u_P} P$. Even though A is not uniquely determined by $\mathfrak{k}$ and the ranking, the set of pairs (u_A, d_A) with $A \in$ A obviously is.

Let $\mathbf{Y}$ denote the set of all derivatives θy_j $(\theta \in \boldsymbol{\Theta}, 1 \leqslant j \leqslant n)$ and let $\mathbf{M}$ denote the set of all differential monomials in $(y_1, \ldots, y_n)$. Then $\mathbf{M}$ is the free commutative monoid generated by $\mathbf{Y}$, so that each $M \in \mathbf{M}$ has a unique expression of the form

$$M = \prod_{v \in \mathbf{Y}} v^{e(v)}$$

in which $e(v) \in \mathbf{N}$ for every $v \in \mathbf{Y}$ and $e(v) \neq 0$ for only finitely many $v \in \mathbf{Y}$. The ranking is a well ordering of $\mathbf{Y}$; extend it to $\mathbf{M}$ by defining $M < N$ (where M is as before and, similarly, $N = \prod v^{f(v)}$) to mean that $M \neq N$ and $e(v) < f(v)$ for the highest v with $e(v) \neq f(v)$. This extends the ranking to a well ordering of $\mathbf{M}$.

The set $\mathbf{M}$ is a basis of $\mathscr{F}\{y_1, \ldots, y_n\}$ when the latter is considered as a vector space over $\mathscr{F}$. For every nonzero $P \in \mathscr{F}\{y_1, \ldots, y_n\}$ there is a highest element of $\mathbf{M}$ that appears in P with a nonzero coefficient. Call it the *highest differential monomial* of P. Call P *unitary* when the coefficient of the highest differential monomial of P is 1.

Lemma 1. Let $\mathfrak{k}$ be a differential ideal of $\mathscr{F}\{y_1, \ldots, y_n\}$ and let a ranking of $(y_1, \ldots, y_n)$ be given. Then $\mathfrak{k}$ has a unique characteristic set A satisfying the following two conditions.

(i) For each element $A \in$ A and every nonzero element $B \in \mathfrak{k}$ that is reduced with respect to all the elements of A other than A, the highest differential monomial of A is lower than or equal to the highest differential monomial of B.

(ii) Each element of A is unitary.

Proof. Let A be any characteristic set of $\mathfrak{k}$. If an element $A \in$ A is replaced by any element $B \in \mathfrak{k}$ that is reduced with respect to all the elements of A other than A and that has the property that $(u_B, d_B) = (u_A, d_A)$, the set obtained will, like A, be a characteristic set of $\mathfrak{k}$. Take B so that its highest differential monomial is as low as possible and has coefficient 1, and do this for every $A \in$ A. The set obtained is a characteristic set of $\mathfrak{k}$ that satisfies the two conditions in the lemma.

Let A and A′ be two characteristic sets of $\mathfrak{k}$ that satisfy the two conditions. For any $A \in$ A there exists a unique $A' \in$ A′ with $(u_A, d_A) = (u_{A'}, d_{A'})$. Evidently A' (resp. A) is reduced with respect to the elements of A (resp. A′) other than A (resp. A'). Hence, by condition (i), A and A' have the same highest differential monomial. If $A - A'$ were not zero, then, by condition (ii), its highest differential monomial would be lower than that of A, contrary to condition (i). Therefore $A - A' = 0$. This implies the uniqueness of A, and completes the proof of the lemma.

Call the characteristic set of $\mathfrak{k}$ satisfying the two conditions in the lemma *canonical*.

Consider any extension $\mathscr{G}$ of $\mathscr{F}$. The ideal $\mathscr{G}\mathfrak{k}$ of $\mathscr{G}\{y_1, \ldots, y_n\}$ is a differential one. Any $P \in \mathscr{G}\{y_1, \ldots, y_n\}$ can be written in the form $P = \sum \gamma_i P_i$ with $P_i \in \mathscr{F}\{y_1, \ldots, y_n\}$, $\gamma_i \in \mathscr{G}$, and the γ_i linearly independent over $\mathscr{F}$; it is easy to see that $P \in \mathscr{G}\mathfrak{k}$ if and only if $P_i \in \mathfrak{k}$ for every i. It follows that every characteristic set of $\mathfrak{k}$ is a characteristic set of $\mathscr{G}\mathfrak{k}$, and that the canonical characteristic set of $\mathfrak{k}$ is the canonical characteristic set of $\mathscr{G}\mathfrak{k}$.

2. DIFFERENTIALLY HOMOGENEOUS AND DIFFERENTIALLY MULTIHOMOGENEOUS DIFFERENTIAL POLYNOMIALS

Let $\mathbf{y} = (y_0, y_1, \ldots, y_n)$ and consider the differential polynomial algebra $\mathscr{R}\{\mathbf{y}\} = \mathscr{R}\{y_0, y_1, \ldots, y_n\}$ over $\mathscr{R}$. For any element t of any differential overring of $\mathscr{R}\{\mathbf{y}\}$ set $t\mathbf{y} = (ty_0, ty_1, \ldots, ty_n)$.

Consider a differential polynomial $H \in \mathscr{R}\{\mathbf{y}\}$ and a number $d \in \mathbf{N}$. If the condition $H(t\mathbf{y}) = t^d H$ is satisfied when t is some differential indeterminate over $\mathscr{R}\{\mathbf{y}\}$, then the condition is satisfied when t is any element of any commutative differential overring of $\mathscr{R}\{\mathbf{y}\}$; H is then said to be *differentially homogeneous in* $\mathbf{y}$ *of degree* d. Since t can be an arbitrary constant, if H is differentially homogeneous in $\mathbf{y}$ of degree d, then H is homogeneous in $(\theta y_j)_{\theta \in \Theta, 0 \leqslant j \leqslant n}$ of degree d in the usual sense, so that when $H \neq 0$ then $d = \deg H$. The differential polynomial 0 is differentially homogeneous of every degree.

The differential ring $\mathscr{R}\{\mathbf{y}\}$ is canonically embedded in its differential ring of quotients over the multiplicatively stable set consisting of all the monomials $\prod_{0 \leqslant j \leqslant n} y_j^{k_j} (k_0, k_1, \ldots, k_n \in \mathbf{N})$, that is, in the differential ring

$$\mathscr{R}\{y_0, y_0^{-1}, y_1, y_1^{-1}, \ldots, y_n, y_n^{-1}\}.$$

For any $P \in \mathscr{R}\{y_1, \ldots, y_n\}$, when $d \in \mathbf{N}$ is sufficiently big then $y_0{}^d P(y_0^{-1}y_1, \ldots, y_0^{-1}y_n)$ is an element of $\mathscr{R}\{\mathbf{y}\}$ that is differentially homogeneous of degree d. Conversely, if $H \in \mathscr{R}\{\mathbf{y}\}$ is differentially homogeneous of degree d, then $H = y_0{}^d H(1, y_0^{-1}y_1, \ldots, y_0^{-1}y_n)$. Of course, in these two statements the role played by y_0 can be played by any y_j.

More generally, let $(y_{ij})_{1 \leqslant i \leqslant p,\, 0 \leqslant j \leqslant n_i}$ be a family of differential indeterminates, set $\mathbf{y}_i = (y_{i0}, y_{i1}, \ldots, y_{in_i})$ $(1 \leqslant i \leqslant p)$, and consider the differential polynomial algebra

$$\mathscr{R}\{\mathbf{y}_1, \ldots, \mathbf{y}_p\} = \mathscr{R}\{(y_{ij})_{1 \leqslant i \leqslant p,\, 0 \leqslant j \leqslant n_i}\}$$

over $\mathscr{R}$. Let $H \in \mathscr{R}\{\mathbf{y}_1, \ldots, \mathbf{y}_p\}$ and let $(d_1, \ldots, d_p) \in \mathbf{N}^p$. If, for each index i, H is differentially homogeneous in $\mathbf{y}_i$ of degree d_i, H is said to be *differentially p-homogeneous in* $(\mathbf{y}_1, \ldots, \mathbf{y}_p)$ *of degree* $(d_1, \ldots, d_p)$. This notion reduces to the preceding one when $p = 1$.

3. DIFFERENTIALLY HOMOGENEOUS AND DIFFERENTIALLY MULTIHOMOGENEOUS DIFFERENTIAL IDEALS

Continue the notation of Section 2. For any ideal ideal $\mathfrak{k}$ of $\mathscr{R}\{\mathbf{y}\}$, $\mathfrak{k}\colon \mathbf{y}^\infty$ denotes the set of elements $P \in \mathscr{R}\{\mathbf{y}\}$ such that $y_j{}^e P \in \mathfrak{k}$ $(0 \leqslant j \leqslant n)$ for some $e \in \mathbf{N}$. It is an ideal, and is a differential ideal whenever $\mathfrak{k}$ is. When the ideal $\mathfrak{k}$ is perfect then $\mathfrak{k}\colon \mathbf{y}^\infty$ is perfect and coincides with $\mathfrak{k}\colon \mathbf{y}$, that is, consists of all $P \in \mathscr{R}\{\mathbf{y}\}$ such that $y_j P \in \mathfrak{k}$ $(0 \leqslant j \leqslant n)$.

Similarly, when $\mathfrak{k}$ is an ideal of $\mathscr{R}\{\mathbf{y}_1, \ldots, \mathbf{y}_p\}$, then $\mathfrak{k}\colon (\mathbf{y}_1 \cdots \mathbf{y}_p)^\infty$ (or $\mathfrak{k}\colon \mathbf{y}_1{}^\infty \cdots \mathbf{y}_p{}^\infty$) denotes the set of all $P \in R\{y_1, \ldots, y_p\}$ such that $(y_{1j_1} \cdots y_{pj_p})^e P \in \mathfrak{k}$ $(0 \leqslant j_1 \leqslant n_1, \ldots, 0 \leqslant j_p \leqslant n_p)$ for some e. As before, $\mathfrak{k}\colon (\mathbf{y}_1 \cdots \mathbf{y}_p)^\infty$ is an ideal, is a differential ideal when $\mathfrak{k}$ is, and is a perfect ideal coinciding with $\mathfrak{k}\colon \mathbf{y}_1 \cdots \mathbf{y}_p$ when $\mathfrak{k}$ is perfect.

Proposition 1. Let $\mathscr{R}$ be a differential algebra over $\mathbf{Q}$. Let $\mathfrak{k}$ be a differential ideal of $\mathscr{R}\{\mathbf{y}_1, \ldots, \mathbf{y}_p\}$, and let X denote the set of elements of $\mathfrak{k}$ that are differentially p-homogeneous in $(\mathbf{y}_1, \ldots, \mathbf{y}_p)$. The following three conditions are equivalent.

(i) $\mathfrak{k} = (X)\colon (\mathbf{y}_1 \cdots \mathbf{y}_p)^\infty$.
(ii) $\mathfrak{k} = [X]\colon (\mathbf{y}_1 \cdots \mathbf{y}_p)^\infty$.
(iii) $\mathfrak{k}\colon \mathbf{y}_1 \cdots \mathbf{y}_p = \mathfrak{k}$ and, for every $P \in \mathfrak{k}$ and each index i and a differential indeterminate t over $\mathscr{R}\{\mathbf{y}_1, \ldots, \mathbf{y}_p\}$,

$$P(\mathbf{y}_1, \ldots, t\mathbf{y}_i, \ldots, \mathbf{y}_p) \in \mathscr{R}\{t\}\mathfrak{k}$$

in the differential ring $\mathscr{R}\{t, \mathbf{y}_1, \ldots, \mathbf{y}_p\}$.

Proof. It is obvious that (i) implies (ii). Let (ii) be satisfied. Then evidently $\mathfrak{k}\colon \mathbf{y}_1 \cdots \mathbf{y}_p = \mathfrak{k}$, and from this it is easy to prove that $(\mathscr{R}\{t\}\mathfrak{k})\colon \mathbf{y}_1 \cdots \mathbf{y}_p = \mathscr{R}\{t\}\mathfrak{k}$. For any $P \in \mathfrak{k}$ there exist an $e \in \mathbf{N}$ and finitely many elements $H_1, \ldots, H_r \in$ X, say of respective degrees $(d_{11}, \ldots, d_{p1}), \ldots, (d_{1r}, \ldots, d_{pr})$, such that

$$(y_{1j_1} \cdots y_{pj_p})^e P \in [H_1, \ldots, H_r] \qquad (0 \leqslant j_1 \leqslant n_1, \ldots, 0 \leqslant j_p \leqslant n_p).$$

When $t\mathbf{y}_i$ is substituted for $\mathbf{y}_i$ this shows that

$$t^e(y_{1j_1} \cdots y_{pj_p})^e\, P(\mathbf{y}_1, \ldots, t\mathbf{y}_i, \ldots, \mathbf{y}_p) \in [t^{d_{i1}}H_1, \ldots, t^{d_{ir}}H_r] \subset \mathscr{R}\{t\}\mathfrak{k}$$
$$(0 \leqslant j_1 \leqslant n_1, \ldots, 0 \leqslant j_p \leqslant n_p),$$

whence $t^e P(\mathbf{y}_1, \ldots, t\mathbf{y}_i, \ldots, \mathbf{y}_p) \in \mathscr{R}\{t\}\mathfrak{k}$, so that $P(\mathbf{y}_1, \ldots, t\mathbf{y}_i, \ldots, \mathbf{y}_p) \in \mathscr{R}\{t, t^{-1}\}\mathfrak{k}$. Since $P(\mathbf{y}_1, \ldots, t\mathbf{y}_i, \ldots, \mathbf{y}_p) \in \mathscr{R}\{t, \mathbf{y}_1, \ldots, \mathbf{y}_p\}$, and the $\mathscr{R}$ module $\mathscr{R}\{t, t^{-1}\}$ evidently has a basis that includes a basis of the submodule $\mathscr{R}\{t\}$, and $\mathscr{R}\{t, t^{-1}\}$ and $\mathscr{R}\{\mathbf{y}_1, \ldots, \mathbf{y}_p\}$

are linearly disjoint over $\mathscr{R}$, this implies that $P(\mathbf{y}_1, \ldots, t\mathbf{y}_i, \ldots, \mathbf{y}_p) \in \mathscr{R}\{t\}\mathfrak{k}$. Thus (ii) implies (iii).

Finally, let (iii) be satisfied, let $(t_1, \ldots, t_p, u_1, \ldots, u_p)$ be a family of differential indeterminates over $\mathscr{R}\{\mathbf{y}_1, \ldots, \mathbf{y}_p\}$, and consider any $P \in \mathfrak{k}$. By hypothesis

$$P(t_1\mathbf{y}_1, \ldots, t_p\mathbf{y}_p) = \sum_{a \leqslant \mu \leqslant g} C_\mu(t_1, \ldots, t_p)P_\mu, \tag{1}$$

where $P_\mu \in \mathfrak{k}$, $C_\mu(t_1, \ldots, t_p) \in \mathbf{Q}\{t_1, \ldots, t_p\}$ $(1 \leqslant \mu \leqslant g)$. Take g as small as possible; then $P_1, \ldots, P_g$ are linearly independent over $\mathbf{Q}$ and $C_1(t_1, \ldots, t_p), \ldots, C_g(t_1, \ldots, t_p)$ are, too. Similarly, for each index μ,

$$P_\mu(t_1\mathbf{y}_1, \ldots, t_p\mathbf{y}_p) = \sum_{0 \leqslant \nu \leqslant h} C_{\mu\nu}(t_1, \ldots, t_p)P_\nu,$$

where $h \geqslant g$ and $P_{g+1}, \ldots, P_h \in \mathfrak{k}$ and $P_1, \ldots, P_g, \ldots, P_h$ are linearly independent over $\mathbf{Q}$ (and hence over $\mathbf{Q}(t_1, \ldots, t_p, u_1, \ldots, u_p)$, too) and $C_{\mu\nu}(t_1, \ldots, t_p) \in \mathbf{Q}\{t_1, \ldots, t_p\}$ $(1 \leqslant \mu \leqslant g,\ 1 \leqslant \nu \leqslant h)$. Therefore the computation

$$\begin{aligned}\sum_{1 \leqslant \nu \leqslant g} C_\nu(t_1u_1, \ldots, t_pu_p)P_\nu &= P(t_1u_1\mathbf{y}_1, \ldots, t_pu_p\mathbf{y}_p)\\ &= \sum_{1 \leqslant \mu \leqslant g} C_\mu(t_1, \ldots, t_p)P_\mu(u_1\mathbf{y}_1, \ldots, u_p\mathbf{y}_p)\\ &= \sum_{1 \leqslant \mu \leqslant g}\ \sum_{1 \leqslant \nu \leqslant h} C_\mu(t_1, \ldots, t_p)C_{\mu\nu}(u_1, \ldots, u_p)P_\nu\end{aligned}$$

shows that $\sum_{1 \leqslant \mu \leqslant g} C_\mu(t_1, \ldots, t_p)C_{\mu\nu}(u_1, \ldots, u_p) = 0$ $(g < \nu \leqslant h)$, whence $C_{\mu\nu} = 0$ $(1 \leqslant \mu \leqslant g,\ g < \nu \leqslant h)$, so that

$$P_\mu(t_1\mathbf{y}_1, \ldots, t_p\mathbf{y}_p) = \sum_{0 \leqslant \nu \leqslant g} C_{\mu\nu}(t_1, \ldots, t_p)P_\nu \qquad (1 \leqslant \mu \leqslant g).$$

Now substitute $(y_{1j_1}^{-1}, \ldots, y_{pj_p}^{-1})$ for $(t_1, \ldots, t_p)$ here, and then multiply by $(y_{1j_1} \cdots y_{pj_p})^e$ for a big exponent $e \in \mathbf{N}$. The resulting equations shows that

$$(y_{1j_1} \cdots y_{pj_p})^e P_\mu(y_{1j_1}^{-1}\mathbf{y}_1, \ldots, y_{pj_p}^{-1}\mathbf{y}_p) \in \mathrm{X} \qquad (0 \leqslant j_1 \leqslant n_1, \ldots, 0 \leqslant j_p \leqslant n_p).$$

That being the case, substitute

$$(y_{1j_1}, y_{1j_1}^{-1}\mathbf{y}_1, \ldots, y_{pj_p}, y_{pj_p}^{-1}\mathbf{y}_p)$$

for $(t_1, \mathbf{y}_1, \ldots, t_p, \mathbf{y}_p)$ in Eq. (1) and then multiply by $(y_{1j_1} \cdots y_{pj_p})^e$. The result is that

$$(y_{1j_1} \cdots y_{pj_p})^e P \in (\mathrm{X}) \qquad (0 \leqslant j_1 \leqslant n_1, \ldots, 0 \leqslant j_p \leqslant n_p).$$

This shows that (iii) implies (i) and completes the proof of the proposition.

Define a *differentially p-homogeneous differential ideal* of $\mathscr{R}\{\mathbf{y}_1, \ldots, \mathbf{y}_p\}$ to be an differential ideal $\mathfrak{k}$ that satisfies the equivalent conditions in Proposition 1. In the special case in which $p = 1$ and $\mathbf{y}_1 = \mathbf{y}$, call $\mathfrak{k}$, simply, a *differentially homogeneous differential ideal* of $\mathscr{R}\{\mathbf{y}\}$.

It is easy to see that if $\mathfrak{k}$ is a differentially p-homogeneous differential ideal of $\mathscr{R}\{\mathbf{y}_1, \ldots, \mathbf{y}_p\}$, then $\mathfrak{k} \subset \mathscr{R} \cap \mathfrak{k} + [\mathbf{y}_1, \ldots, \mathbf{y}_p]$, and when $\mathfrak{k} \neq \mathscr{R}\{\mathbf{y}_1, \ldots, \mathbf{y}_p\}$ then $[\mathbf{y}_i] \not\subset \mathfrak{k}$ $(1 \leqslant i \leqslant p)$.

Corollary 1. Let $\mathscr{R}$ be a differential algebra over $\mathbf{Q}$. If $\mathfrak{k}$ is a differentially p-homogeneous differential ideal of $\mathscr{R}\{\mathbf{y}_1, \ldots, \mathbf{y}_p\}$, then so is the perfect differential ideal $\{\mathfrak{k}\}$.

Proof. If $P \in \{\mathfrak{k}\}: \mathbf{y}_1 \cdots \mathbf{y}_p$, then for some $e \in \mathbf{N}$

$$(y_{1j_1} \cdots y_{pj_p})^e P^e \in \mathfrak{k} \qquad (0 \leqslant j_1 \leqslant n_1, \ldots, 0 \leqslant j_p \leqslant n_p),$$

so that $P^e \in \mathfrak{k}:(\mathbf{y}_1 \cdots \mathbf{y}_p)^\infty = \mathfrak{k}$, whence $P \in \{\mathfrak{k}\}$. Thus $\{\mathfrak{k}\}:\mathbf{y}_1 \cdots \mathbf{y}_p = \{\mathfrak{k}\}$. Again, if $P \in \{\mathfrak{k}\}$, then $P^e \in \mathfrak{k}$ for some e, so that for each index i

$$P(\mathbf{y}_1, \ldots, t\mathbf{y}_i, \ldots, \mathbf{y}_p)^e \in \mathscr{R}\{t\}\mathfrak{k} \subset \mathscr{R}\{t\} \cdot \{\mathfrak{k}\};$$

but it is well known (and easy to prove) that if $\mathfrak{a}$ is a perfect ideal of a ring, then the ideal generated by $\mathfrak{a}$ in any polynomial algebra over the ring is perfect. Therefore

$$P(\mathbf{y}_1, \ldots, t\mathbf{y}_i, \ldots, \mathbf{y}_p) \in \mathscr{R}\{t\} \cdot \{\mathfrak{k}\}.$$

Corollary 2. Let $\mathscr{R}$ be a differential algebra over $\mathbf{Q}$, and let $\mathscr{R}'$ be a differential overring of $\mathscr{R}$ that is a free $\mathscr{R}$ module. If $\mathfrak{k}$ is a differentially p-homogeneous differential ideal of $\mathscr{R}\{\mathbf{y}_1, \ldots, \mathbf{y}_p\}$, then $\mathscr{R}'\mathfrak{k}$ is a differentially p-homogeneous differential ideal of $\mathscr{R}'\{\mathbf{y}_1, \ldots, \mathbf{y}_p\}$.

Proof. Fix a basis (a_ν) of $\mathscr{R}'$ over $\mathscr{R}$. For any $P \in \mathscr{R}'\{\mathbf{y}_1, \ldots, \mathbf{y}_p\}$ there are unique elements $P_\nu \in \mathscr{R}\{\mathbf{y}_1, \ldots, \mathbf{y}_p\}$ such that $P = \sum P_\nu a_\nu$, and $P \in \mathscr{R}'\mathfrak{k}$ if and only if $P_\nu \in \mathfrak{k}$ for every ν, whence also $P \in (\mathscr{R}'\mathfrak{k}): \mathbf{y}_1 \cdots \mathbf{y}_p$ if and only if $P_\nu \in \mathfrak{k}: \mathbf{y}_1 \cdots \mathbf{y}_p$ for every ν. Thus $(\mathscr{R}'\mathfrak{k}): \mathbf{y}_1 \cdots \mathbf{y}_p = \mathscr{R}'\mathfrak{k}$ if and only if $\mathfrak{k}: \mathbf{y}_1 \cdots \mathbf{y}_p = \mathfrak{k}$. Further, when t is a differential indeterminate over $\mathscr{R}\{\mathbf{y}_1, \ldots, \mathbf{y}_p\}$ then $P(\mathbf{y}_1, \ldots, t\mathbf{y}_i, \ldots, \mathbf{y}_p) \in \mathscr{R}'\{t\}\mathfrak{k}$ if and only if $P_\nu(\mathbf{y}_1, \ldots, t\mathbf{y}_i, \ldots, \mathbf{y}_p) \in \mathscr{R}\{t\}\mathfrak{k}$ for every ν. This proves the corollary.

Proposition 2. Let $\mathscr{R}$ be a differential algebra over $\mathbf{Q}$. Let $\mathfrak{a}$ be a perfect differential ideal of $\mathscr{R}\{\mathbf{y}_1, \ldots, \mathbf{y}_p\}$ and let X denote the set of elements of $\mathfrak{a}$ that are differentially p-homogeneous in $(\mathbf{y}_1, \ldots, \mathbf{y}_p)$. The following two conditions are equivalent.

(i) $\mathfrak{a}$ is differentially p-homogeneous.

(ii) $\mathfrak{a} = \{X\}: \mathbf{y}_1 \cdots \mathbf{y}_p$.

When $\mathscr{R}$ is the differential field $\mathscr{F}$, these conditions are equivalent to each of the following two conditions.

(iii) $\mathfrak{a}: \mathbf{y}_1 \cdots \mathbf{y}_p = \mathfrak{a}$ and, for every zero $(\boldsymbol{\eta}_1, \ldots, \boldsymbol{\eta}_p)$ of $\mathfrak{a}$ in $\mathscr{U}^{n_1+1} \times \cdots \times \mathscr{U}^{n_p+1}$ and each index i and every element $s \in \mathscr{U}^*$, $(\boldsymbol{\eta}_1, \ldots, s\boldsymbol{\eta}_i, \ldots, \boldsymbol{\eta}_p)$ is a zero of $\mathfrak{a}$.

(iv) each component of $\mathfrak{a}$ is differentially p-homogeneous.

Proof. If (i) is satisfied, then

$$\mathfrak{a} = [X]: (\mathbf{y}_1 \cdots \mathbf{y}_p)^\infty \subset \{X\}: \mathbf{y}_1 \cdots \mathbf{y}_p \subset \mathfrak{a}: \mathbf{y}_1 \cdots \mathbf{y}_p = \mathfrak{a},$$

so that (ii) is satisfied. Conversely, if (ii) is satisfied, then obviously $\mathfrak{a}: \mathbf{y}_1 \cdots \mathbf{y}_p = \mathfrak{a}$, and also for any $P \in \mathfrak{a}$ and index i and differential indeterminate t there is an $e \in \mathbf{N}$ such that P^e is in the differentially p-homogeneous differential ideal $[X]: (\mathbf{y}_1 \cdots \mathbf{y}_p)^\infty$, so that

$$P(\mathbf{y}_1, \ldots, t\mathbf{y}_i, \ldots, \mathbf{y}_p)^e \in \mathscr{R}\{t\} \cdot ([X]: (\mathbf{y}_1 \cdots \mathbf{y}_p)^\infty) \subset \mathscr{R}\{t\}\mathfrak{a};$$

since, as has already been observed in the proof of Corollary 1 to Proposition 1, $\mathscr{R}\{t\}\mathfrak{a}$ is perfect, this shows that $P(\mathbf{y}_1, \ldots, t\mathbf{y}_i, \ldots, \mathbf{y}_p) \in \mathscr{R}\{t\}\mathfrak{a}$, so that (i) is satisfied.

Thus (i) and (ii) are equivalent.

Now let $\mathscr{R} = \mathscr{F}$. If (ii) is satisfied, and hence also (i), then $\mathfrak{a} : \mathbf{y}_1 \cdots \mathbf{y}_p = \mathfrak{a}$ and for every $P \in \mathfrak{a}$ and any index i and any $s \in \mathscr{U}^*$

$$P(\mathbf{y}_1, \ldots, s\mathbf{y}_i, \ldots, \mathbf{y}_p) \in \mathscr{F}\{s\}\mathfrak{a},$$

so that if $(\boldsymbol{\eta}_1, \ldots, \boldsymbol{\eta}_p)$ is a zero of $\mathfrak{a}$, then so is $(\boldsymbol{\eta}_1, \ldots, s\boldsymbol{\eta}_i, \ldots, \boldsymbol{\eta}_p)$. Thus (ii) implies (iii). If (iii) is satisfied, and if P is any element of $\mathfrak{a}$, then for any index i the differential polynomial $P(\mathbf{y}_1, \ldots, t\mathbf{y}_i, \ldots, \mathbf{y}_p) \in \mathscr{F}\{t, \mathbf{y}_1, \ldots, \mathbf{y}_p\}$ vanishes at every zero of the perfect differential ideal $\mathscr{F}\{t\}\mathfrak{a}$ and hence is in this ideal. This shows that (iii) implies (i). Thus when $\mathscr{R} = \mathscr{F}$ then (i), (ii), and (iii) are equivalent.

Finally, still supposing that $\mathscr{R} = \mathscr{F}$, let $\mathfrak{p}_1, \ldots, \mathfrak{p}_r$ be the components of $\mathfrak{a}$. Each zero of $\mathfrak{a}$ is a zero of some $\mathfrak{p}_k$, and conversely. Therefore if $\mathfrak{a}$ satisfies (iv) and hence each $\mathfrak{p}_k$ satisfies (iii), then $\mathfrak{a}$ satisfies (iii). Conversely, let $\mathfrak{a}$ satisfy (iii) and hence also (i) and (ii). Then

$$\mathfrak{a} = \mathfrak{a} : \mathbf{y}_1 \cdots \mathbf{y}_p = \bigcap_{1 \leqslant k \leqslant r} \mathfrak{p}_k : \mathbf{y}_1 \cdots \mathbf{y}_p = \bigcap_{k \in \Lambda} \mathfrak{p}_k,$$

where Λ is the set of indices k such that $[\mathbf{y}_i] \not\subset \mathfrak{p}_k$ $(1 \leqslant i \leqslant p)$; by the uniqueness of the irredundant representation $\mathfrak{a} = \bigcap_{1 \leqslant k \leqslant r} \mathfrak{p}_k$, it follows that $k \in \Lambda$ for every k, so that $\mathfrak{p}_k : \mathbf{y}_1 \cdots \mathbf{y}_p = \mathfrak{p}_k$ $(1 \leqslant k \leqslant r)$. When the universal differential field $\mathscr{U}$ is replaced by a bigger one, condition (i) is unaffected and hence condition (iii) is preserved. Therefore it may be supposed that $\mathscr{U}$ is a universal extension of $\mathscr{F}$. Then for a given k there exist a generic zero $(\zeta_1, \ldots, \zeta_p)$ of $\mathfrak{p}_k$ in $\mathscr{U}^{n_1+1} \times \cdots \times \mathscr{U}^{n_p+1}$ and an element $u \in \mathscr{U}$ that is differentially transcendental over $\mathscr{F}\langle\zeta_1, \ldots, \zeta_p\rangle$. Of course, $(\zeta_1, \ldots, \zeta_p)$ is a zero of $\mathfrak{a}$, so that for any index i, $\{\zeta_1, \ldots, u\zeta_i, \ldots, \zeta_p)$ is a zero of $\mathfrak{a}$ and hence also of $\mathfrak{p}_l$ for some index l. Since $(\zeta_1, \ldots, \zeta_p)$ is obviously a differential specialization of $(\zeta_1, \ldots, u\zeta_i, \ldots, \zeta_p)$ over $\mathscr{F}$ and hence is a zero of $\mathfrak{p}_l$, it follows that $l = k$. For any zero $(\boldsymbol{\eta}_1, \ldots, \boldsymbol{\eta}_p)$ of $\mathfrak{p}_k$ and any $s \in \mathscr{U}^*$, $(\boldsymbol{\eta}_1, \ldots, s\boldsymbol{\eta}_i, \ldots, \boldsymbol{\eta}_p)$ is a differential specialization of $(\zeta_1, \ldots, u\zeta_i, \ldots, \zeta_p)$ over $\mathscr{F}$ and hence is a zero of $\mathfrak{p}_k$. Thus every $\mathfrak{p}_k$ satisfies (iii), so that $\mathfrak{a}$ satisfies (iv).

Proposition 3. Let $\mathfrak{p}$ be a prime differential ideal of $\mathscr{F}\{y_1, \ldots, y_p\}$ and let A denote the canonical characteristic set of $\mathfrak{p}$ relative to some ranking of $(y_{ij})_{1 \leqslant i \leqslant p,\, 0 \leqslant j \leqslant n_i}$. The following two conditions are equivalent.

(i) $\mathfrak{p}$ is differentially p-homogeneous.

(ii) $\mathfrak{p} : \mathbf{y}_1 \cdots \mathbf{y}_p = \mathfrak{p}$ and each element of A is differentially p homogeneous in $(\mathbf{y}_1, \ldots, \mathbf{y}_p)$.

Proof. As in the last part of the proof of Proposition 2, it may be supposed that $\mathscr{U}$ is universal over $\mathscr{F}$. Let (i) be satisfied, let $A \in$ A, and fix an element $u \in \mathscr{U}$ that is differentially transcendental over $\mathscr{F}$. Then $\mathscr{F}\langle u\rangle\mathfrak{p}$ is a prime differential ideal of $\mathscr{F}\langle u\rangle\{\mathbf{y}_1, \ldots, \mathbf{y}_p\}$ and A is its canonical characteristic set. For any index i the differential polynomial $A(\mathbf{y}_1, \ldots, u\mathbf{y}_i, \ldots, \mathbf{y}_p)$ of $\mathscr{F}\langle u\rangle\{\mathbf{y}_1, \ldots, \mathbf{y}_p\}$ is in $\mathscr{F}\langle u\rangle\mathfrak{p}$. Now, for any derivative $\delta_1^{e_1} \cdots \delta_m^{e_m} y_{ij}$

$$\delta_1^{e_1} \cdots \delta_m^{e_m}(u y_{ij}) = u\delta_1^{e_1} \cdots \delta_m^{e_m} y_{ij} + \cdots,$$

where the dots stand for terms of lower rank. It follows that if M denotes the highest differential monomial of A and $d = \deg M$, then M is also the highest differential monomial of $A(\mathbf{y}_1, \ldots, u\mathbf{y}_i, \ldots, \mathbf{y}_p)$, the difference

$$u^d A - A(\mathbf{y}_1, \ldots, u\mathbf{y}_i, \ldots, \mathbf{y}_p)$$

is reduced with respect to all the elements of A other than A, and every differential monomial appearing in this difference is lower than M. Because this difference is in $\mathscr{F}\langle u\rangle\mathfrak{p}$ and because A is the canonical characteristic set of $\mathscr{F}\langle u\rangle\mathfrak{p}$, this implies that the difference is zero. Thus A is differentially p-homogeneous.

Conversely, let (ii) be satisfied. Fix a generic zero $(\zeta_1, \ldots, \zeta_p)$ of $\mathfrak{p}$ and an element $u \in \mathscr{U}$ that is differentially transcendental over $\mathscr{F}\langle\zeta_1, \ldots, \zeta_p\rangle$. Then $(\zeta_1, \ldots, \zeta_p)$ is a zero of A and hence, because of (ii), so is $(\zeta_1, \ldots, u\zeta_i, \ldots, \zeta_p)$. But the latter is not a zero of the initial or separate of any element of A (because the former is not) and hence is a zero of $\mathfrak{p}$. For any zero $(\boldsymbol{\eta}_1, \ldots, \boldsymbol{\eta}_p)$ of $\mathfrak{p}$ and any element $s \in \mathscr{U}^*$, $(\boldsymbol{\eta}_1, \ldots, s\boldsymbol{\eta}_i, \ldots, \boldsymbol{\eta}_p)$ is a differential specialization of $(\zeta_1, \ldots, u\zeta_i, \ldots, \zeta_p)$ over $\mathscr{F}$ and hence is a zero of $\mathfrak{p}$. Therefore $\mathfrak{p}$ is differentially p-homogeneous.

Proposition 4. The set of all differentially p-homogeneous perfect differential ideals of $\mathscr{F}\{\mathbf{y}_1, \ldots, \mathbf{y}_p\}$ is a Noetherian perfect differential conservative system.

Proof. Let $\mathfrak{C}$ denote the set in question. If $\mathfrak{M}$ is any subset of $\mathfrak{C}$, then $\bigcap_{\mathfrak{c} \in \mathfrak{M}} \mathfrak{c}$ is a perfect differential ideal and (see Proposition 1, condition (iii)) is differentially p-homogeneous. When $\mathfrak{M}$ is nonempty and totally ordered by inclusion, $\bigcup_{\mathfrak{c} \in \mathfrak{M}} \mathfrak{c}$ is a perfect differential ideal and (again see Proposition 1, condition (iii)) is differentially homogeneous. If $\mathfrak{c} \in \mathfrak{C}$ and $S \in \mathscr{F}\{\mathbf{y}_1, \ldots, \mathbf{y}_p\}$, then, by Proposition 2, the components $\mathfrak{p}_1, \ldots, \mathfrak{p}_r$ of $\mathfrak{c}$ are differentially p-homogeneous and

$$\mathfrak{c}: S = (\mathfrak{p}_1 \cap \cdots \cap \mathfrak{p}_r): S = (\mathfrak{p}_1 : S) \cap \cdots \cap (\mathfrak{p}_r : S) = \mathfrak{p}_{i_1} \cap \cdots \cap \mathfrak{p}_{i_g},$$

where $\mathfrak{p}_{i_1}, \ldots, \mathfrak{p}_{i_g}$ are the components that do not contain S, and therefore $\mathfrak{c}: S$ is differentially p-homogeneous. This shows that $\mathfrak{C}$ is a perfect differential conservative system. Since the set of all perfect differential ideals of $\mathscr{F}\{\mathbf{y}_1, \ldots, \mathbf{y}_p\}$ is a Noetherian conservative system, $\mathfrak{C}$ is Noetherian, too.

Proposition 5. Let $\mathfrak{a}$ (resp. $\mathfrak{b}$) be a differentially p-homogeneous (resp. differentially q-homogeneous) perfect differential ideal of $\mathscr{F}\{\mathbf{y}_1, \ldots, \mathbf{y}_p\}$ (resp. $\mathscr{F}\{\mathbf{z}_1, \ldots, \mathbf{z}_q\}$), and let $\mathfrak{c}$ denote the ideal $(\mathfrak{a} \cup \mathfrak{b})$ generated by $\mathfrak{a} \cup \mathfrak{b}$ in $\mathscr{F}\{\mathbf{y}_1, \ldots, \mathbf{y}_p, \mathbf{z}_1, \ldots, \mathbf{z}_q\}$. Then $\mathfrak{c}$ is a differentially $(p+q)$-homogeneous perfect differential ideal.

Proof. First consider the special case in which $\mathfrak{a}$ and $\mathfrak{b}$ are both prime. It is obvious that $\mathfrak{c}$ is a differential ideal, and it is known that $\mathfrak{c}$ is perfect (this follows, for example, from [2, Chapter 0, Section 12, Lemma 11 and Corollary 1 to Proposition 7]). If $P \in \mathfrak{c}: \mathbf{y}_i$, that is, if $y_{ij}P \in (\mathfrak{a} \cup \mathfrak{b})$ $(0 \leqslant j \leqslant n_i)$, fix a generic zero $\zeta = (\zeta_{kl})$ of $\mathfrak{b}$. Then $y_{ij}P(\mathbf{y}_1, \ldots, \mathbf{y}_p, \zeta_1, \ldots, \zeta_q) \in \mathscr{F}\{\zeta\}\mathfrak{a}$, so that by Corollary 2 to Proposition 1, $P(\mathbf{y}_1, \ldots, \mathbf{y}_p, \zeta_1, \ldots, \zeta_q) \in \mathscr{F}\{\zeta\}\mathfrak{a}$. Therefore there exist differential polynomials $A_\lambda \in \mathfrak{a}$ and $D_\lambda \in \mathscr{F}\{\mathbf{z}_1, \ldots, \mathbf{z}_q\}$ such that $P(\mathbf{y}_1, \ldots, \mathbf{y}_p, \zeta_1, \ldots, \zeta_q) = \sum_\lambda D_\lambda(\zeta_1, \ldots, \zeta_q)A_\lambda$, so that $P - \sum_\lambda D_\lambda A_\lambda$ vanishes as a differential polynomial in $(\mathbf{z}_1, \ldots, \mathbf{z}_q)$ at ζ, whence $P - \sum_\lambda D_\lambda A_\lambda \in \mathscr{F}\{\mathbf{y}_1, \ldots, \mathbf{y}_p\}\mathfrak{b}$ and $P \in \mathfrak{c}$. Thus $\mathfrak{c}: \mathbf{y}_i = \mathfrak{c}$ $(1 \leqslant i \leqslant p)$. Similarly, $\mathfrak{c}: \mathbf{z}_k = \mathfrak{c}$ $(1 \leqslant k \leqslant q)$, so that $\mathfrak{c}: \mathbf{y}_1 \cdots \mathbf{y}_p \mathbf{z}_1 \cdots \mathbf{z}_q = \mathfrak{c}$. Finally, because $\mathfrak{a}$ satisfies condition (iii) in Proposition 1, for any $P \in \mathfrak{c} = (\mathfrak{a} \cup \mathfrak{b})$ and any index i

$$P(\mathbf{y}_1, \ldots, t\mathbf{y}_i, \ldots, \mathbf{y}_p, \mathbf{z}_1, \ldots, \mathbf{z}_q) \in \mathscr{F}\{t\}\mathfrak{c},$$

and similarly, for any index k

$$P(\mathbf{y}_1, \ldots, \mathbf{y}_p, \mathbf{z}_1, \ldots, t\mathbf{z}_k, \ldots, \mathbf{z}_k) \in \mathscr{F}\{y\}\mathfrak{c}.$$

Therefore $\mathfrak{c}$ is differentially $(p+q)$-homogeneous.

Now turn to the general case, and let $\mathfrak{p}_1, \ldots, \mathfrak{p}_g$ (resp. $\mathfrak{q}_1, \ldots, \mathfrak{q}_h$) be the components of $\mathfrak{a}$ (resp. $\mathfrak{b}$). By Proposition 2, these components are differentially p-homogeneous (resp. differentially q-homogeneous), and hence, by the special case already treated, each $(\mathfrak{p}_\alpha \cup \mathfrak{q}_\beta)$ is a differentially $(p+q)$-homogeneous perfect differential ideal, so that the intersection $\bigcap_{\alpha,\beta} (\mathfrak{p}_\alpha \cup \mathfrak{q}_\beta)$ is, too. This intersection obviously contains $\mathfrak{c}$. On the other hand, if (η, ζ) is any zero of $\mathfrak{c}$, then η is a zero of $\mathfrak{a}$ and hence of some $\mathfrak{p}_\alpha$, and, similarly, ζ is a zero of some $\mathfrak{q}_\beta$, so that (η, ζ) is a zero of some $(\mathfrak{p}_\alpha \cup \mathfrak{q}_\beta)$ and hence of the intersection. Therefore $\mathfrak{c}$ equals the intersection and hence is a differentially $(p+q)$-homogeneous perfect differential ideal.

4. ALGEBRAIC DIFFERENTIAL EQUATIONS IN PROJECTIVE AND MULTIPROJECTIVE SPACES

Let $n \in \mathbf{N}$ and consider the projective space $\mathbf{P}(n)$ over $\mathscr{U}$. The points of $\mathbf{P}(n)$ are the one-dimensional vector subspaces of the $(n+1)$-dimensional vector space $\mathscr{U}^{n+1}$ over $\mathscr{U}$. If α is some point of $\mathbf{P}(n)$, any element of α different from the origin is called a *representative* of α. Any element $(a_0, a_1, \ldots, a_n)$ of $\mathscr{U}^{n+1}$ different from the origin is a representative of a unique point of $\mathbf{P}(n)$, denoted by $(a_0 : a_1 : \cdots : a_n)$.

More generally, let $n_1, \ldots, n_p \in \mathbf{N}$ and consider the p-projective space

$$\mathbf{P}(n_1, \ldots, n_p) = \mathbf{P}(n_1) \times \cdots \times \mathbf{P}(n_p).$$

For any point $\alpha = (\alpha_1, \ldots, \alpha_p)$ of $\mathbf{P}(n_1, \ldots, n_p)$, if $\mathbf{a}_i = (a_{i0}, a_{i1}, \ldots, a_{in_i})$ is a representative of α_i $(1 \leqslant i \leqslant p)$, the element

$$(a_{ij})_{1 \leqslant i \leqslant p,\, 0 \leqslant j \leqslant n_i} = (\mathbf{a}_1, \ldots, \mathbf{a}_p)$$

of $\mathscr{U}^{n_1+1} \times \cdots \times \mathscr{U}^{n_p+1}$ is called a representative *of* α.

Consider a differential polynomial $P \in \mathscr{U}\{\mathbf{y}_1, \ldots, \mathbf{y}_p\}$ and a point $\alpha \in \mathbf{P}(n_1, \ldots, n_p)$. Say that P *vanishes at* α, and that α is a *zero of* P, if P vanishes at every representative of α. For a subset $\mathscr{M}$ of $\mathbf{P}(n_1, \ldots, n_p)$ say that P vanishes *on* $\mathscr{M}$ if P vanishes at every point of $\mathscr{M}$. For a subset Σ of $\mathscr{U}\{\mathbf{y}_1, \ldots, \mathbf{y}_p\}$ say that α is a zero *of* Σ if α is a zero of every differential polynomial in Σ.

Let $\mathfrak{A}_{\mathscr{F}}(\alpha)$ denote the set of differential polynomials in $\mathscr{F}\{\mathbf{y}_1, \ldots, \mathbf{y}_p\}$ that vanish at α. It obviously is a prime differential ideal that satisfies condition (iii) in Proposition 1, that is, that is differentially p-homogeneous. Call it the *defining differential ideal of* α *in* $\mathscr{F}\{\mathbf{y}_1, \ldots, \mathbf{y}_p\}$ (or *over* $\mathscr{F}$). Let $\mathfrak{A}_{\mathscr{F}}(\mathscr{M})$ denote the set of differential polynomials in $\mathscr{F}\{\mathbf{y}_1, \ldots, \mathbf{y}_p\}$ that vanish on $\mathscr{M}$, and write $\mathfrak{A}(\mathscr{M}) = \mathfrak{A}_{\mathscr{U}}(\mathscr{M})$. Evidently $\mathfrak{A}_{\mathscr{F}}(\mathscr{M}) = \bigcap_{\alpha \in M} \mathfrak{A}_{\mathscr{F}}(\alpha)$, so that $\mathfrak{A}_{\mathscr{F}}(\mathscr{M})$ is a differentially p-homogeneous perfect differential ideal of $\mathscr{F}\{\mathbf{y}_1, \ldots, \mathbf{y}_p\}$.

Let $\mathfrak{Z}_{n_1,\ldots,n_p}(\Sigma)$ denote the set of points of $\mathbf{P}(n_1, \ldots, n_p)$ that are zeros of the subset Σ of $\mathscr{U}\{\mathbf{y}_1, \ldots, \mathbf{y}_p\}$. When $(\Sigma_\lambda)_{\lambda \in \Lambda}$ is a family of subsets of $\mathscr{F}\{\mathbf{y}_1, \ldots, \mathbf{y}_p\}$ then $\bigcap_{\lambda \in \Lambda} \mathfrak{Z}_{n_1,\ldots,n_p}(\Sigma_\lambda) = \mathfrak{Z}_{n_1,\ldots,n_p}(\bigcup_{\lambda \in \Lambda} \Sigma_\lambda)$; when Λ is finite then $\bigcup_{\lambda \in \Lambda} \mathfrak{Z}_{n_1,\ldots,n_p}(\Sigma_\lambda) = \mathfrak{Z}(\prod_{\lambda \in \Lambda} \Sigma_\lambda)$. This shows that the subsets of $\mathbf{P}(n_1, \ldots, n_p)$ of the form $\mathfrak{Z}_{n_1,\ldots,n_p}(\Sigma)$ with

$\Sigma \subset \mathscr{F}\{\mathbf{y}_1, \ldots, \mathbf{y}_p\}$ are the closed sets for a topology on $\mathbf{P}(n_1, \ldots, n_p)$. Call it the *differential Zariski $\mathscr{F}$-topology* (or simply *differential $\mathscr{F}$-topology*) of $\mathbf{P}(n_1, \ldots, n_p)$, and when $\mathscr{F} = \mathscr{U}$ call it the *differential Zariski topology* of $\mathbf{P}(n_1, \ldots, n_p)$. Terms such as "differentially closed," "differentially continuous," etc., will mean "closed," "continuous," etc., relative to the differential Zariski topology; the terms "differentially $\mathscr{F}$-closed," "differentially $\mathscr{F}$-continuous," etc., will refer to the differential $\mathscr{F}$-topology.

Proposition 6. (a) For any subset $\mathscr{M}$ of $\mathbf{P}(n_1, \ldots, n_p)$, $\mathfrak{Z}_{n_1,\ldots,n_p}(\mathfrak{A}_{\mathscr{F}}(\mathscr{M}))$ is the differential $\mathscr{F}$-closure of $\mathscr{M}$ in $\mathbf{P}(n_1, \ldots, n_p)$.

(b) For any subset Σ of $\mathscr{F}\{\mathbf{y}_1, \ldots, \mathbf{y}_p\}$, $\mathfrak{A}_{\mathscr{F}}(\mathfrak{Z}_{n_1,\ldots,n_p}(\Sigma))$ is the smallest differentially p-homogeneous perfect differential ideal of $\mathscr{F}\{\mathbf{y}_1, \ldots, \mathbf{y}_p\}$ that contains Σ.

Proof. (a) Let $\mathscr{M}^+$ denote the differential $\mathscr{F}$-closure of $\mathscr{M}$. Since $\mathfrak{Z}_{n_1,\ldots,n_p}(\mathfrak{A}_{\mathscr{F}}(\mathscr{M}))$ is differentially $\mathscr{F}$-closed and contains $\mathscr{M}$, $\mathscr{M}^+ \subset \mathfrak{Z}_{n_1,\ldots,n_p}(\mathfrak{A}_{\mathscr{F}}(\mathscr{M}))$. Since $\mathscr{M}^+$ is differentially $\mathscr{F}$-closed, $\mathscr{M}^+ = \mathfrak{Z}_{n_1,\ldots,n_p}(\mathrm{T})$ for some $\mathrm{T} \subset \mathscr{F}\{\mathbf{y}_1, \ldots, \mathbf{y}_p\}$, and evidently $\mathrm{T} \subset \mathfrak{A}_{\mathscr{F}}(\mathscr{M})$, so that $\mathscr{M}^+ = \mathfrak{Z}_{n_1,\ldots,n_p}(\mathrm{T}) \supset \mathfrak{Z}_{n_1,\ldots,n_p}(\mathfrak{A}_{\mathscr{F}}(\mathscr{M}))$.

(b) Let $\mathfrak{a}$ denote the smallest differentially p-homogeneous perfect differential ideal containing Σ. Since $\mathfrak{A}_{\mathscr{F}}(\mathfrak{Z}_{n_1,\ldots,n_p}(\Sigma))$ is a differentially p-homogeneous perfect differential ideal containing Σ, $\mathfrak{a} \subset \mathfrak{A}_{\mathscr{F}}(\mathfrak{Z}_{n_1,\ldots,n_p}(\Sigma))$. Every zero of $\mathfrak{a}$ in $\mathbf{P}(n_1, \ldots, n_p)$ is a zero of Σ and hence also $\mathfrak{A}_{\mathscr{F}}(\mathfrak{Z}_{n_1,\ldots,n_p}(\Sigma))$; since the perfect differential ideals $\mathfrak{a}$ and $\mathfrak{A}_{\mathscr{F}}(\mathfrak{Z}_{n_1,\ldots,n_p}(\Sigma))$ are differentially p-homogeneous, it easily follows that the every zero of $\mathfrak{a}$ *in* $\mathscr{U}^{n_1+1} \times \cdots \times \mathscr{U}^{n_p+1}$ is a zero of $\mathfrak{A}_{\mathscr{F}}(\mathfrak{Z}_{n_1,\ldots,n_p}(\Sigma))$ and hence that $\mathfrak{A}_{\mathscr{F}}(\mathfrak{Z}_{n_1,\ldots,n_p}(\Sigma)) \subset \mathfrak{a}$.

Corollary 1. The mapping from the set of differentially $\mathscr{F}$-closed subsets of $P(n_1, \ldots, n_p)$ into the set of differentially p-homogeneous perfect differential ideals of $\mathscr{F}\{\mathbf{y}_1, \ldots, \mathbf{y}_p\}$, given by the formula $\mathscr{M} \mapsto \mathfrak{A}_{\mathscr{F}}(\mathscr{M})$, and the mapping in the opposite direction, given by the formula $\mathfrak{a} \mapsto \mathfrak{Z}_{n_1,\ldots,n_p}(\mathfrak{a})$, are bijective and inverse to each other.

Corollary 2. Relative to the differential $\mathscr{F}$ topology, $\mathbf{P}(n_1, \ldots, n_p)$ is a Noetherian topological space. A differential $\mathscr{F}$-closed subset $\mathscr{M}$ of $\mathbf{P}(n_1, \ldots, n_p)$ is $\mathscr{F}$-irreducible if and only if the differentially p-homogeneous perfect differential ideal $\mathfrak{A}_{\mathscr{F}}(\mathscr{M})$ is prime.

5. DIFFERENTIAL SPECIALIZATIONS

Consider a point $\alpha = (\alpha_1, \ldots, \alpha_p) \in \mathbf{P}(n_1, \ldots, n_p)$. Choose a representative $(\mathbf{a}_1, \ldots, \mathbf{a}_p)$ of α, where $\mathbf{a}_i = (a_{i0}, a_{i1}, \ldots, a_{in_i}) \in \mathscr{U}^{n_i+1}$ $(1 \leqslant i \leqslant p)$, and for each i choose an index j_i such that $a_{ij_i} \neq 0$. For any subfield K of $\mathscr{U}$ the field extension $K((a_{ij_i}^{-1}a_{ij})_{1 \leqslant i \leqslant p,\, 0 \leqslant j \leqslant n_i}$ is independent of the choice of representative $(a_1, \ldots, a_p)$ and indices j_i. This field extension is denoted by $K(\alpha)$. When $K(\alpha) = K$ the point α is said to be *rational* over K. The set of points of $\mathbf{P}(n_1, \ldots, n_p)$ that are rational over K is denoted by $\mathbf{P}_K(n_1, \ldots, n_p)$, and, more generally, for any subset $\mathscr{M}$ of $\mathbf{P}(n_1, \ldots, n_p)$ the set of points of $\mathscr{M}$ that are rational over K is denoted by $\mathscr{M}_K$.

The set $\mathbf{P}_K(n)$ is not quite the same as projective n-space over K, which consists of the one-dimensional vector subspaces of the vector space K^{n+1} over K, but each such subspace is the intersection with K^{n+1} of a unique one-dimensional subspace of the vector space $\mathscr{U}^{n+1}$ over $\mathscr{U}$, and the latter is, as a point of $\mathbf{P}(n)$, rational over K. This gives a canonical bijection of projective n-space over K onto $\mathbf{P}_K(n)$ that is used to identify the two. Similar remarks apply to $\mathbf{P}(n_1, \ldots, n_p)$ instead of $\mathbf{P}(n)$.

Consider again the point $\alpha \in \mathbf{P}(n_1, \ldots, n_p)$. Denote the differential field $\mathscr{F}\langle\mathscr{F}(\alpha)\rangle$ by $\mathscr{F}\langle\alpha\rangle$. Since the field $\mathscr{F}(\alpha) = \mathscr{F}((a_{ij_i}^{-1}a_{ij})_{1 \leqslant i \leqslant p,\, 0 \leqslant j \leqslant n_i}$ is independent of the choices made above, the differential transcendence polynomial of $(a_{ij_i}^{-1}a_{ij})_{1 \leqslant i \leqslant p,\, 0 \leqslant j \leqslant n_i}$ over $\mathscr{F}$ is independent of these choices, too [2, Chapter II, Section 12], and may therefore be called the differential transcendence polynomial *of* α *over* $\mathscr{F}$ and be denoted by $\omega_{\alpha/\mathscr{F}}$. It can be written in the form

$$\omega_{\alpha/\mathscr{F}} = \sum_{0 \leqslant i \leqslant m} a_i \binom{X+1}{i}.$$

where m is the number of derivation operators $\delta_1, \ldots, \delta_m$ and $a_i \in \mathbf{Z}$ $(0 \leqslant i \leqslant m)$; then

$$a_m = \text{dif tr deg}\, \mathscr{F}\langle\alpha\rangle/\mathscr{F}.$$

Consider a second point $\alpha' = (\alpha_1', \ldots, \alpha_p') \in \mathbf{P}(n_1, \ldots, n_p)$ and a representative $(\mathbf{a}_1', \ldots, \mathbf{a}_p')$ of α'; for each i write $\mathbf{a}_i' = (a_{i0}', a_{i1}', \ldots, a_{in_i}')$ and fix j_i' such that $a_{ij_i'}' \neq 0$. If $\mathfrak{A}_{\mathscr{F}}(\alpha) \subset \mathfrak{A}_{\mathscr{F}}(\alpha')$, then $a_{ij_i'} \neq 0$ $(1 \leqslant i \leqslant p)$, that is, the indices j_i can be chosen equal to the indices j_i', and evidently $(a_{ij_i'}'^{-1} a_{ij}')_{1 \leqslant i \leqslant p,\, 0 \leqslant j \leqslant n_i}$ is a differential specialization of $(a_{ij_i'}^{-1}a_{ij})_{1 \leqslant i \leqslant p,\, 0 \leqslant j \leqslant n_i}$ over $\mathscr{F}$. Conversely, if there exist indices $j_1, \ldots, j_p$ such that $a_{ij_i} \neq 0$, $a_{ij_i}' \neq 0$ $(1 \leqslant i \leqslant p)$, and $(a_{ij_i}'^{-1} a_{ij}')_{1 \leqslant i \leqslant p,\, 0 \leqslant j \leqslant n_i}$ is a differential specialization of $(a_{ij_i}^{-1}a_{ij})_{1 \leqslant i \leqslant p,\, 0 \leqslant j \leqslant n_i}$ over $\mathscr{F}$, then $\mathfrak{A}_{\mathscr{F}}(\alpha) \subset \mathfrak{A}_{\mathscr{F}}(\alpha')$. Under these conditions call α' a *differential specialization of* α *over* $\mathscr{F}$. When $\mathfrak{A}_{\mathscr{F}}(\alpha) = \mathfrak{A}_{\mathscr{F}}(\alpha')$, or equivalently when there exists an isomorphism $\mathscr{F}\langle\alpha\rangle \approx \mathscr{F}\langle\alpha'\rangle$ over $\mathscr{F}$ with $a_{ij_i}^{-1}a_{ij} \mapsto a_{ij_i}'^{-1} a_{ij}'$ $(1 \leqslant i \leqslant p$, $0 \leqslant j \leqslant n_i)$, call α' a *generic* differential specialization of α over $\mathscr{F}$.

It follows from the definitions and from the analogous considerations for points in affine space [12, Chapter III, Section 5] that if α' is a differential specialization of α over $\mathscr{F}$ then $\omega_{\alpha/\mathscr{F}} \geqslant \omega_{\alpha'/\mathscr{F}}$, and when the differential specialization is not generic then $\omega_{\alpha/\mathscr{F}} > \omega_{\alpha'/\mathscr{F}}$.

Let $\mathfrak{p}$ be a differentially p-homogeneous prime differential ideal of $\mathscr{F}\{\mathbf{y}_1, \ldots, \mathbf{y}_p\}$ and $\mathscr{V}$ be the corresponding differentially $\mathscr{F}$-closed subset of $\mathbf{P}(n_1, \ldots, n_p)$; thus $\mathscr{V}$ is $\mathscr{F}$-irreducible, $\mathscr{V} = \mathfrak{Z}_{n_1, \ldots, n_p}(\mathfrak{p})$, and $\mathfrak{p} = \mathfrak{A}_{\mathscr{F}}(\mathscr{V})$. It is easy to see that the following three conditions on a point $\alpha \in \mathbf{P}(n_1, \ldots, n_p)$ are equivalent.

(i) $\mathfrak{A}_{\mathscr{F}}(\alpha) = \mathfrak{p}$.
(ii) The smallest differentially $\mathscr{F}$-closed subset of $\mathbf{P}(n_1, \ldots, n_p)$ that contains α is $\mathscr{V}$.
(iii) The set of all differential specializations of α over $\mathscr{F}$ is $\mathscr{V}$.

Call such a point α a *generic zero of* $\mathfrak{p}$ *in* $\mathbf{P}(n_1, \ldots, n_p)$ or a *generic point of* $\mathscr{V}$ *over* $\mathscr{F}$. If $\mathscr{U}$ is a universal (or even semiuniversal) extension of $\mathscr{F}$, then such an α exists. In any case, $\mathscr{F}$ contains a differential field of definition of $\mathfrak{p}$, say $\mathscr{F}_0$, such that $\mathscr{U}$ is universal over $\mathscr{F}$; the intersection $\mathfrak{p}_0 = \mathfrak{p} \cap \mathscr{F}_0\{\mathbf{y}_1, \ldots, \mathbf{y}_p\}$ is a differentially p-homogeneous prime differential ideal of $\mathscr{F}_0\{\mathbf{y}_1, \ldots, \mathbf{y}_p\}$ and $\mathscr{V} = \mathfrak{Z}_{n_1, \ldots, n_p}(\mathfrak{p}_0)$, so that $\mathfrak{p}_0$ has a generic zero in $\mathbf{P}(n_1, \ldots, n_p)$, say α_0. Furthermore, $\omega_{\alpha_0/\mathscr{F}_0}$ is independent of the choice of $\mathscr{F}_0$ and of α_0, so that dif tr deg $\mathscr{F}_0\langle\alpha_0\rangle/\mathscr{F}_0$ is, too. Therefore $\omega_{\alpha_0/\mathscr{F}_0}$ may be called the *differential dimension polynomial of* $\mathscr{V}$ and be denoted by $\omega_{\mathscr{V}}$, and dif tr deg $\mathcal{F}_0\langle\alpha_0\rangle/\mathcal{F}_0$ may be called the *differential dimension of* $\mathcal{V}$ and be denoted by dif dim $\mathcal{V}$. It is apparent that

$$p\binom{X+m}{m} + \omega_{\mathcal{V}} = \omega_{\mathfrak{p}}, \tag{2}$$

and hence that

$$p + \text{dif dim } \mathcal{V} = \text{dif dim } \mathfrak{p}.$$

Also, if $\mathcal{V}'$ is an $\mathcal{F}$-irreducible differentially $\mathcal{F}$-closed proper subset of $\mathcal{V}$, then $\omega_{\mathcal{V}} > \omega_{\mathcal{V}'}$.

6. DIFFERENTIALLY COMPLETE DIFFERENTIALLY CLOSED SETS

Let $\mathcal{M}$ and $\mathcal{N}$ be differentially $\mathcal{F}$-closed subsets of $\mathbf{P}(n_1, \ldots, n_p)$ and $\mathbf{P}(r_1, \ldots, r_q)$, respectively, let

$$(\mathbf{y}_1, \ldots, \mathbf{y}_p, \mathbf{z}_1, \ldots, \mathbf{z}_q) = ((y_{ij})_{1 \leqslant i \leqslant p,\, 0 \leqslant j \leqslant n_i}, (z_{kl})_{1 \leqslant k \leqslant q,\, 0 \leqslant l \leqslant r_k})$$

be a family of differential indeterminates, and take

$$\mathfrak{A}_{\mathcal{F}}(\mathcal{M}) \subset \mathcal{F}\{\mathbf{y}_1, \ldots, \mathbf{y}_p\}, \qquad \mathfrak{A}_{\mathcal{F}}(\mathcal{N}) \subset \mathcal{F}\{\mathbf{z}_1, \ldots, \mathbf{z}_q\}.$$

The ideal

$$\mathfrak{c} = (\mathfrak{A}_{\mathcal{F}}(\mathcal{M}) \cup \mathfrak{A}_{\mathcal{F}}(\mathcal{N}))$$

of $\mathcal{F}\{\mathbf{y}_1, \ldots, \mathbf{y}_p, \mathbf{z}_1, \ldots, \mathbf{z}_q\}$ is, by Proposition 5, a differentially $(p+q)$-homogeneous perfect differential ideal and evidently

$$\mathfrak{Z}_{n_1, \ldots, n_p, r_1, \ldots, r_q}(\mathfrak{c}) = \mathcal{M} \times \mathcal{N}.$$

This shows that the Cartesian product of a differentially $\mathcal{F}$-closed subset of $\mathbf{P}(n_1, \ldots, n_p)$ with one of $\mathbf{P}(r_1, \ldots, r_q)$ is a differentially $\mathcal{F}$-closed subset of $\mathbf{P}(n_1, \ldots, n_p, r_1, \ldots, r_q)$. It implies, too, that when $\mathcal{M}$ and $\mathcal{N}$ are $\mathcal{F}$-irreducible and $\mathcal{P}$ is any $\mathcal{F}$-component of $\mathcal{M} \times \mathcal{N}$, then $\omega_{\mathcal{P}} = \omega_{\mathcal{M}} + \omega_{\mathcal{N}}$. It is easy to see that both projections

$$\text{pr}_1 : \mathcal{M} \times \mathcal{N} \to \mathcal{M}, \qquad \text{pr}_2 : \mathcal{M} \times \mathcal{N} \to \mathcal{N}$$

are differentially $\mathcal{F}$-continuous. Indeed, if $\mathcal{N}'$ is a differentially $\mathcal{F}$-closed subset of $\mathcal{N}$, then $\text{pr}_2^{-1}(\mathcal{N}') = \mathcal{M} \times \mathcal{N}'$, which is differentially $\mathcal{F}$-closed.

Call a differentially closed subset $\mathcal{M}$ of $\mathbf{P}(n_1, \ldots, n_p)$ *differentially complete* if for every differentially closed subset $\mathcal{N}$ of a multiprojective space the projection $\text{pr}_2 : \mathcal{M} \times \mathcal{N} \to \mathcal{N}$ is a differentially closed mapping (that is, maps each differentially closed subset of $\mathcal{M} \times \mathcal{N}$ onto a differentially closed subset of $\mathcal{N}$).

If $\mathcal{N}$ is a differentially closed subset of $\mathbf{P}(r_1, \ldots, r_q)$ then the diagram

$$\begin{array}{ccc} \mathcal{M} \times \mathbf{P}(r_1, \ldots, r_q) & \longrightarrow & \mathbf{P}(r_1, \ldots, r_q) \\ \uparrow & & \uparrow \\ \mathcal{M} \times \mathcal{N} & \longrightarrow & \mathcal{N} \end{array}$$

(in which the horizontal arrows are projections and the vertical arrows are inclusions) is commutative. Since the left inclusion is evidently a differentially closed mapping and the right inclusion is differentially continuous, if the upper projection is differentially closed, then so is the lower one. Thus $\mathscr{M}$ is differentially complete if and only if the projection $\mathscr{M} \times \mathbf{P}(r_1, \ldots, r_q) \to \mathbf{P}(r_1, \ldots, r_q)$ is differentially closed for every multiprojective space $\mathbf{P}(r_1, \ldots, r_q)$.

If $\mathscr{M}$ is differentially complete, then so is every differentially closed subset of $\mathscr{M}$. Indeed, if $\mathscr{M}' \subset \mathscr{M}$ and $\mathscr{M}'$ is differentially closed, then the diagram

$$\begin{array}{ccc} \mathscr{M} \times \mathscr{N} & & \\ \uparrow & \searrow & \\ & & \mathscr{N} \\ & \nearrow & \\ \mathscr{M}' \times \mathscr{N} & & \end{array}$$

is commutative.

If $\mathscr{M}_1$ and $\mathscr{M}_2$ are differentially complete, then so is $\mathscr{M}_1 \times \mathscr{M}_2$. Indeed, then each arrow in the sequence of projections $\mathscr{M}_1 \times \mathscr{M}_2 \times \mathscr{N} \to \mathscr{M}_2 \times \mathscr{N} \to \mathscr{N}$ is a closed mapping.

In view of the completeness of $\mathbf{P}(n)$ in the usual algebraico-geometric (Zariski topology) sense, the following result is at least mildy noteworthy.

Proposition 7. Let $n \in \mathbf{N}$, $n \neq 0$. Then $\mathbf{P}(n)$ is not differentially complete.

Proof. Let $L_1, \ldots, L_{2n(n+1)}$ be $2n(n+1)$ homogeneous linear polynomials in $\mathbf{Q}\{\mathbf{y}\} = \mathbf{Q}\{y_0, y_1, \ldots, y_n\}$ such that no $n+1$ of them are linearly dependent (over $\mathbf{Q}$ and hence over $\mathscr{U}$), and for each index v with $1 \leqslant v \leqslant 2n$ let W_v be one of the differential polynomials

$$\det(\delta_i{}^j y_{j'})_{0 \leqslant j \leqslant n,\, 0 \leqslant j' \leqslant n} \qquad (1 \leqslant i \leqslant m)$$

in $\mathbf{Q}\{\mathbf{y}\}$. Each differential ideal $[L_k]$ is prime, and every zero of it is a zero of every W_v; therefore

$$W_v \in [L_k] \qquad (1 \leqslant k \leqslant 2n(n+1), \quad 1 \leqslant v \leqslant 2n).$$

It is known and easy to prove (see Kolchin [1, Lemma 2]) that each W_v is differentially homogeneous in $\mathbf{y}$ of degree $n+1$; of course, each L_k is differentially homogeneous in $\mathbf{y}$ of degree 1. Let $\mathbf{z} = (z_0, z_1)$. Then the differential polynomial

$$H = z_0 \prod_{1 \leqslant k \leqslant 2n(n+1)} L_k + z_1 \prod_{1 \leqslant v \leqslant 2n} W_v$$

of $\mathbf{Q}\{\mathbf{y}, \mathbf{z}\}$ is irreducible (absolutely) and differentially bihomogeneous in $(\mathbf{y}, \mathbf{z})$ of degree $(2n(n+1), 1)$. For any subset Λ of the set of indices $1, 2, \ldots, 2n(n+1)$ such that $1 \leqslant \operatorname{Card} \Lambda \leqslant n$, H can be written in the form

$$H = \Big(z_0 \prod_{k \notin \Lambda} L_k\Big) \prod_{k \in \Lambda} L_k + \Big(z_1 \prod_{2l < v \leqslant 2n} W_v\Big) \prod_{k \in \Lambda} (W_{\mu(k)} W_{v(k)}),$$

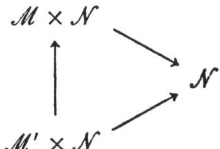

where $l = \text{Card}\,\Lambda$ and the indices $\mu(k)$, $\nu(k)$ $(k \in \Lambda)$ are the numbers $1, 2, \ldots, 2l$. It follows by the domination lemma [2, Chapter IV, Section 12, Lemma 6] that the perfect differential ideal $\{H\}$ contains a differential polynomial

$$\left(\prod_{k \in \Lambda} L_k\right)\left(\left(z_0 \prod_{k \notin \Lambda} L_k\right)^{e_\Lambda} + N_\Lambda\right),$$

where $e_\Lambda \geqslant 1$ and $N_\Lambda \in [(L_k)_{k \in \Lambda}] \cap [z_1]$, and hence that $(z_0 \prod_{k \notin \Lambda} L_k)^{e_\Lambda} + N_\Lambda$ is contained in the general component $\mathfrak{p}_{\mathbf{Q}}(H)$ of H. The set $\mathfrak{Z}_{n,1}(\mathfrak{p}_{\mathbf{Q}}(H))$ is a differentially $\mathbf{Q}$-closed subset of $\mathbf{P}(n) \times \mathbf{P}(1)$, and it is easily seen that $\text{pr}_2(\mathfrak{Z}_{n,1}(\mathfrak{p}_{\mathbf{Q}}(H)))$ is differentially dense in $\mathbf{P}(1)$. To prove the proposition, it therefore suffices to show that $\text{pr}_2(\mathfrak{Z}_{n,1}(\mathfrak{p}_{\mathbf{Q}}(H))) \neq \mathbf{P}(1)$, and for this it suffices to show that $(1:0) \notin \text{pr}_2(\mathfrak{Z}_{n,1}(\mathfrak{p}_{\mathbf{Q}}(H)))$.

Assume to the contrary that there exists an $\alpha = (a_0 : a_1 : \cdots : a_n) \in \mathbf{P}(n)$ such that $(\alpha, (1:0)) \in \mathfrak{Z}_{n,1}(\mathfrak{p}_{\mathbf{Q}}(H))$. Then the definition of H shows that α is a zero of L_k for some k, that is, the set Λ of indices k with $L_k(a_0, a_1, \ldots, a_n) = 0$ is not empty. If Card Λ were no bigger than n, then $\mathfrak{p}_{\mathbf{Q}}(H)$ would, by the preceding, contain a differential polynomial $(z_0 \prod_{k \notin \Lambda} L_k)^{e_\Lambda} + N_\Lambda$, and therefore α would be a zero of L_k for some $k \notin \Lambda$. Therefore Card $\Lambda \geqslant n + 1$. Since $n + 1$ elements of Λ are linearly independent, it follows that $\alpha_j = 0$ $(0 \leqslant j \leqslant n)$. This contradiction completes the proof.

The purpose of the rest of the present section is to prove that $\mathbf{P}_{\mathscr{K}}(n)$ is differentially complete. The main step is the following proposition.

Proposition 8. Suppose that $\mathscr{R}$ has no divisor of zero and that the field of constants of $\mathscr{F}$ is algebraically closed; let $f\colon \mathscr{R} \to \mathscr{F}$ be a homomorphism of differential rings, let γ be a constant in a differential field containing $\mathscr{R}$, and suppose that f can not be extended to a homomorphism $\mathscr{R}[\gamma] \to \mathscr{F}$. Then $\gamma \neq 0$ and f can be extended to a homomorphism $\mathscr{R}[\gamma^{-1}] \to \mathscr{F}$; the latter is unique and maps γ^{-1} onto 0.

Remark. Without the hypothesis that γ is a constant the proposition would be false (see [2, Chapter IV, Section 6, Exercise 6(c)]). In the theory of fields (non-differential) the analogous results (in which, of course, the notion of constant does not enter, and the characteristic can be arbitrary) is well known.

Proof. The proof closely follows Chevalley's proof of the well-known result just mentioned. Let $\mathfrak{p}$ denote the defining differential ideal of γ in the differential polynomial algebra $\mathscr{R}\{y\}$. Then $\delta_i y \in \mathfrak{p}$ $(1 \leqslant i \leqslant m)$ and f induces a surjective homomorphism $\varphi\colon \mathscr{R}\{y\} \to f(\mathscr{R})\{y\}$ that maps $\mathfrak{p}$ onto the differential ideal $\mathfrak{p}^f$ of $f(\mathscr{R})\{y\}$. The differential field of quotients $\mathscr{E}$ of $f(\mathscr{R})$ is contained in $\mathscr{F}$, and the ideal $\mathscr{E}\mathfrak{p}^f$ of $\mathscr{E}\{y\}$ generated by $\mathfrak{p}^f$ is a differential one.

Assume that $\mathfrak{p}^f \cap f(\mathscr{R}) = 0$. Then $1 \notin \mathscr{E}\mathfrak{p}^f$, so that $\mathfrak{p}^f$ has a zero β in some extension of $\mathscr{F}$; it may be supposed that β is constrained over $\mathscr{F}$ (see [2, Chapter III, Section 10, Proposition 6]), and since β is a constant (because $\delta_i y \in \mathfrak{p}^f$ for each i) it then follows that $\beta \in \mathscr{F}$ (see [2, Chapter III, Section 10, Proposition 7(d)]). The kernel of the composite homomorphism

$$\mathscr{R}\{y\} \xrightarrow{\varphi} f(\mathscr{R})\{y\} \subset \mathscr{F}\{y\} \to \mathscr{F}$$

(in which the final arrow denotes the substitution of β for y) obviously contains $\mathfrak{p}$, which is the kernel of the substitution homomorphism $\mathcal{R}\{y\} \to \mathcal{R}[\gamma]$. Therefore the composite homomorphism induces a homomorphism $\mathcal{R}[\gamma] \to \mathcal{F}$, which evidently extends f. This contradicts the hypothesis and shows that $\mathfrak{p}^f \cap f(\mathcal{R}) \neq 0$. Because $\mathfrak{p}$ contains each $\delta_i y$, this implies that $\mathfrak{p}$ contains a polynomial

$$P = a_0 + a_1 y + \cdots + a_n y^n$$

with $f(a_0) \neq 0, f(a_j) = 0$ $(1 \leqslant j \leqslant n)$. Fix such a P with minimal n.

Since $P(\gamma) = 0$, it follows that $\gamma \neq 0$. Let $\mathfrak{q}$ denote the defining differential ideal of γ^{-1} in $\mathcal{R}\{y\}$. It is obvious that $\mathfrak{q}$ contains each $\delta_i y$ and the polynomial

$$Q = a_0 y^n + a_1 y^{n-1} + \cdots + a_n.$$

Consider any polynomial $F \in \mathfrak{q}$. When $y^{n-1}F$ is divided by Q there results an equation

$$a_0^r y^{n-1}F = CQ + D,$$

where $r \in \mathbf{N}$ and $C, D \in \mathcal{R}[y]$ and $\deg D < n$, say

$$D = b_0 y^{n-1} + \cdots + b_{n-1}.$$

Obviously

$$f(a_0)^r y^{n-1} F^f = C^f f(a_0) y^n + D^f,$$

whence $f(b_j) = 0$ $(1 \leqslant j \leqslant n-1)$. Because $D \in \mathfrak{q}$ and hence $b_0 + \cdots + b_{n-1}y^{n-1} \in \mathfrak{p}$, it follows by the minimality of n that also $f(b_0) = 0$, that is, $D^f = 0$. Thus, F^f is divisible by y. This shows that $\mathfrak{q}^f \subset [y]$. In particular, $\mathfrak{q}^f \cap f(\mathcal{R}) = 0$. The argument used for $\mathfrak{p}$ now shows that f extends to a homomorphism $\mathcal{R}[\gamma^{-1}] \to \mathcal{F}$. The image of γ^{-1} is a zero of $\mathfrak{q}^f \subset [y]$ and hence is 0, so that the new homomorphism is unique.

Corollary 1. Let $f\colon \mathcal{R} \to \mathcal{F}$ be the same as in Proposition 6 and let $a_0, a_1, \ldots, a_n$ be constants, not all zero, in a differential field that contains $\mathcal{R}$. Then there exists an index j with $a_j \neq 0$ such that f can be extended to a homomorphism

$$\mathcal{R}\{a_j^{-1}a_0, a_j^{-1}a_1, \ldots, a_j^{-1}a_n\} \to \mathcal{F}.$$

Proof. When $n = 0$ the result is trivial. Let $n \geqslant 1$, suppose the result proved for lower values of n, and say $a_1 \neq 0$. By the proposition, either f can be extended to some $\mathcal{R}\{a_1^{-1}a_0\} \to \mathcal{F}$ or else $a_0 \neq 0$ and f can be extended to an $\mathcal{R}\{a_0^{-1}a_1\} \to \mathcal{F}$, say the former. By the case $n-1$, this in turn can, for some index j with $1 \leqslant j \leqslant n$, be extended to a homomorphism $\mathcal{R}\{a_1^{-1}a_0, a_j^{-1}a_1, \ldots, a_j^{-1}a_n\} \to \mathcal{F}$. Because $a_j^{-1}a_0 = a_j^{-1}a_1 \cdot a_1^{-1}a_0$, the desired homomorphism is obtained by restricting this one.

Corollary 2. $\mathbf{P}_{\mathcal{K}}(n)$ is a differentially complete irreducible differentially $\mathbf{Q}$-closed subset of $\mathbf{P}(n)$.

Proof. The differential ideal

$$\mathfrak{c} = [(y_j \delta_i y_{j'} - y_{j'} \delta_i y_j)_{1 \leqslant i \leqslant n,\, 0 \leqslant j < j' \leqslant n}] : \mathbf{y}^\infty$$

of $\mathbf{Q}\{\mathbf{y}\}$ is differentially homogeneous, and it is easy to see that if $\gamma_0, \gamma_1, \ldots, \gamma_n$ are elements of $\mathcal{K}$ algebraically independent over $\mathbf{Q}$ and if t is an element of $\mathcal{U}$ differentially transcendental over $\mathbf{Q}$ (and hence over $\mathscr{K}$), then $(t\gamma_0, t\gamma_1, \ldots, t\gamma_n)$ is a generic zero of $\mathfrak{c}$ in $\mathscr{U}^{n+1}$. Therefore $\mathfrak{c}$ is prime, $(\gamma_0:\gamma_1:\cdots:\gamma_n)$ is a generic zero of $\mathfrak{c}$ in $\mathbf{P}(n)$, and $\mathfrak{Z}_n(\mathfrak{c}) = \mathbf{P}_{\mathscr{K}}(n)$, so that $\mathbf{P}_{\mathscr{K}}(n)$ is a $\mathbf{Q}$-irreducible differentially $\mathbf{Q}$-closed subset of $\mathbf{P}(n)$. Because the entire discussion evidently remains valid when $\mathbf{Q}$ is replaced by its algebraic closure, $\mathbf{P}_{\mathscr{K}}(n)$ is irreducible.

Let $\mathscr{P}$ be any differentially closed subset of $\mathbf{P}_{\mathscr{K}}(n) \times \mathbf{P}(r_1, \ldots, r_q)$ and fix the differential field $\mathscr{F}$ so that $\mathscr{P}$ is differentially $\mathscr{F}$-closed and $\mathscr{U}$ is universal over $\mathscr{F}$. To complete the proof it suffices to show that $\mathrm{pr}_2(\mathscr{P})$ is differentially $\mathscr{F}$-closed in $\mathbf{P}(r_1, \ldots, r_q)$, that is, to show that for any point $\beta' = (\beta_1', \ldots, \beta_q')$ in the differential $\mathscr{F}$-closure of $\mathrm{pr}_2(\mathscr{P})$ there exists a point $\alpha' \in \mathbf{P}_{\mathscr{K}}(n)$ such that $(\alpha', \beta') \in \mathscr{P}$. Now, for some $\mathscr{F}$-irreducible component $\mathscr{W}$ of $\mathscr{P}$, β' is in the differential $\mathscr{F}$-closure of $\mathrm{pr}_2(\mathscr{W})$; this means that when $(\alpha, \beta) = (\alpha, (\beta_1, \ldots, \beta_q))$ is a generic point of $\mathscr{W}$ over $\mathscr{F}$ then β' is a differential specialization of β over $\mathscr{F}$. For each index l with $1 \leqslant l \leqslant q$ fix representatives $\mathbf{b}_l = (b_{l0}, b_{l1}, \ldots, b_{lr_l})$ of β_l and $\mathbf{b}_l' = (b_{l0}', b_{l1}', \ldots, b_{lr_l}')$ of β_l' and then fix an index k_l such that $b_{lk_l}' \neq 0$. Then $b_{lk_l} \neq 0$ $(1 \leqslant l \leqslant q)$ and there exists a homomorphism

$$\mathscr{F}\{(b_{lk_l}^{-1} b_{lk})_{1 \leqslant l \leqslant q,\, 0 \leqslant k \leqslant r_l}\} \to \mathscr{F}\{(y_{lk_l}'^{-1} b_{lk}')_{1 \leqslant l \leqslant q,\, 0 \leqslant k \leqslant r_l}\}$$

over $\mathscr{F}$ under which

$$b_{lk_l}^{-1} b_{lk} \mapsto b_{lk_l}'^{-1} b_{lk}' \qquad (1 \leqslant l \leqslant q, \quad 0 \leqslant k \leqslant r_l).$$

Take a representative $\mathbf{a} = (a_0, a_1, \ldots, a_n)$ of α with $a_j \in \mathscr{K}$ $(0 \leqslant j \leqslant n)$. By Corollary 1, there is an index j with $a_j \neq 0$ such that the above homomorphism can be extended to a homomorphism

$$\mathscr{F}\{a_j^{-1} a_0, a_j^{-1} a_1, \ldots, a_j^{-1} a_n, (b_{lk_l}^{-1} b_{lk})_{1 \leqslant l \leqslant q,\, 0 \leqslant k \leqslant r_l}\} \to \mathscr{U}.$$

Denote the image of $a_j^{-1} a_{j'}$ by $a_{j'}'$ $(0 \leqslant j' \leqslant n)$. Then $(a_0', a_1', \ldots, a_n')$ is a representative of a point $\alpha' \in \mathbf{P}_K(n)$ such that (α', β') is a differential specialization of (α, β) over $\mathscr{F}$, so that $(\alpha', \beta') \in \mathscr{W} \subset \mathscr{P}$.

7. LINEAR DEPENDENCE OVER PROJECTIVE VARIETIES

Let $\mathscr{C}$ be a subfield of the field of constants $\mathscr{K}$, and let V be an algebraic variety in $\mathbf{P}_{\mathscr{K}}(n)$ that is defined and irreducible over $\mathscr{C}$. Call an *element* $\mathbf{v} = (v_0, v_1, \ldots, v_n) \in \mathscr{U}^{n+1}$ *linearly dependent over* V if there exists a point $\gamma \in V$ such that for some (and hence every) representative $(c_0, c_1, \ldots, c_n)$ of γ, $\sum_{0 \leqslant j \leqslant n} c_j v_j = 0$. If some representative of a point $\eta \in \mathbf{P}(n)$ is linearly dependent over V, then every representative of η is. Say in this case that the *point* η is linearly dependent over V.

Theorem. Let $\mathscr{C}$ be a subfield of the field of constants $\mathscr{K}$ of $\mathscr{U}$, and let V be an algebraic variety in $\mathbf{P}_{\mathscr{K}}(n)$ that is defined and irreducible over $\mathscr{C}$. Let $\mathfrak{R}$ denote the set of points of $\mathbf{P}(n)$ that are linearly dependent over V. Then $\mathfrak{R}$ is a $\mathscr{C}$-irreducible differentially $\mathscr{C}$-closed subset of $\mathbf{P}(n)$ and

$$\omega_{\mathfrak{R}} = (n-1)\binom{X+m}{m} + \dim V.$$

Proof. Let $\mathfrak{p}_0$ denote the defining ideal of V in the polynomial algebra $\mathscr{C}[\mathbf{z}] = \mathscr{C}[z_0, z_1, \ldots, z_n]$. Since $\mathscr{C}$ may evidently be replaced by any smaller field of definition of $\mathfrak{p}_0$, it may be supposed that $\mathscr{U}$ is universal over $\mathscr{C}$. Consider the differential ideal

$$\mathfrak{s} = \left[\mathfrak{p}_0, (z_j\delta_i z_{j'} - z_{j'}\delta_i z_j)_{1 \leqslant i \leqslant m,\, 0 \leqslant j < j' \leqslant n}, \sum_{0 \leqslant j \leqslant n} y_j z_j\right] : (\mathbf{yz})^\infty$$

of $\mathscr{C}\{\mathbf{y}, \mathbf{z}\}$, and set $\mathscr{S} = \mathfrak{Z}_{n,n}(\mathfrak{s})$. Then $\mathfrak{s}$ is differentially bihomogeneous, and $\mathscr{S}$ is a differentially $\mathscr{C}$-closed subset of $\mathbf{P}(n, n)$. It is evident that $\mathfrak{R} = \mathrm{pr}_1(\mathscr{S})$, and hence, by Corollary 2 to Proposition 8, $\mathfrak{R}$ is differentially $\mathscr{C}$ closed in $\mathbf{P}(n)$.

The prime homogeneous polynomial ideal $\mathfrak{p}_0$ has a generic zero $\mathbf{c} = (c_0, c_1, \ldots, c_n) \in \mathscr{K}^{n+1}$. Fix an element $v \in \mathscr{U}$ that is differentially transcendental over $\mathscr{K}$ and set $\mathbf{v} = (vc_0, vc_1, \ldots, vc_n)$. Then $\mathbf{v}$ is a generic zero of $\mathfrak{p}_0$ in $\mathscr{U}^{n+1}$, and $\mathbf{c}$ and $\mathbf{v}$ are repre-, sentatives of one and the same point $\gamma \in \mathbf{P}_{\mathscr{K}}(n)$. Choose an index j_0 such that $c_{j_0} \neq 0$, then choose elements $u_j \in \mathscr{U}$ $(0 \leqslant j \leqslant n,\ j \neq j_0)$ that are differentially algebraically independent over $\mathscr{C}\langle\mathbf{v}\rangle$ (and hence over $\mathscr{K}\langle\mathbf{v}\rangle$, too), and set

$$u_{j_0} = -\sum_{j \neq j_0} u_j c_{j_0}^{-1} c_j, \qquad \mathbf{u} = (u_0, u_1, \ldots, u_n).$$

Then $(\mathbf{u}, \mathbf{v})$ is a zero of $\mathfrak{s}$ in $\mathscr{U}^{n+1} \times \mathscr{U}^{n+1}$, and the point $\eta = (u_0 : u_1 : \cdots : u_n) \in \mathbf{P}(n)$ has the property that $(\eta, \gamma) \in \mathscr{S}$, so that $\eta \in \mathfrak{R}$.

Consider any differential polynomial $P \in \mathscr{C}\{\mathbf{y}, \mathbf{z}\}$. There exists a congruence

$$z_{j_0}^h P \equiv P_0\left(\mathrm{mod}\left[\sum_{0 \leqslant j \leqslant n} y_j z_j\right]\right)$$

where $h \in \mathbf{N}$, $P_0 \in \mathscr{C}\{\mathbf{y}, \mathbf{z}\}$, and P_0 is differentially free of y_{j_0}. Because of the last property, P_0 can be written in the form $P = \sum P_M M$, where M runs over a finite set of differential monomials in $(y_j)_{0 \leqslant j \leqslant n,\, j \neq j_0}$ and $P_M \in \mathscr{C}\{\mathbf{z}\}$ for each M. Since $(u_j)_{0 \leqslant j \leqslant n,\, j \neq j_0}$ is differentially algebraically independent over $\mathscr{C}\langle\mathbf{v}\rangle$, it follows that if $P(\mathbf{u}, \mathbf{v}) = 0$, then $P_M(\mathbf{v}) = 0$ for every M. Now, the differential polynomials

$$z_{j_0}\delta_i z_j - z_j \delta_i z_{j_0} \qquad (1 \leqslant i \leqslant m,\quad 0 \leqslant j \leqslant n,\quad j \neq j_0)$$

in $\mathfrak{s}$ form an autoreduced set with respect to any ranking of $\mathbf{z}$ such that $z_{j_0} < z_j$ $(0 \leqslant j \leqslant n, j \neq j_0)$; the remainder P_M' of P_M with respect to this autoreduced set satisfies a congruence

$$z_{j_0}^k P_M \equiv P_M' (\mathrm{mod}\ \mathfrak{s})$$

and has the property that

$$P_M' \in \mathscr{C}\{z_{j_0}\}[(z_j)_{0 \leqslant j \leqslant n,\, j \neq j_0}].$$

Since the derivatives of vc_{j_0} of strictly positive order are obviously algebraically independent over $\mathscr{C}(\mathbf{v})$, it follows that if $P(\mathbf{u}, \mathbf{v}) = 0$, then $P_M' \in \mathscr{C}[z_0, z_1, \ldots, z_n]$ and hence $P_M' \in \mathfrak{p}_0$ for every M. This shows that $(\mathbf{u}, \mathbf{v})$ is a generic zero of $\mathfrak{s}$ in $\mathscr{U}^{n+1} \times \mathscr{U}^{n+1}$. In particular, $\mathfrak{s}$ is prime, so that $\mathscr{S}$ is $\mathscr{C}$-irreducible and hence $\mathfrak{R}$ is, too, and $\boldsymbol{\eta}$ is a generic point of $\mathfrak{R}$ over $\mathscr{C}$. Set $\mathfrak{r} = \mathfrak{A}_{\mathscr{C}}(\mathfrak{R})$.

There exist n derivative operators $\theta_j \in \mathbf{\Theta}(n-1)$ $(0 \leqslant j \leqslant n, j \neq j_0)$ such that

$$\det(\theta_j u_j)_{j \neq j_0,\, j' \neq j_0} \neq 0.$$

Since $u_{j_0} = \sum_{j' \neq j_0} u_{j'} c_{j_0}^{-1} c_{j'}$ and hence

$$\theta_j u_{j_0} = \sum_{\substack{0 \leqslant j' \leqslant n \\ j' \neq j_0}} \theta_j u_{j'} c_{j_0}^{-1} c_{j'} \qquad (0 \leqslant j \leqslant n, \quad j \neq j_0),$$

it follows that

$$\mathscr{C}(\gamma) = \mathscr{C}((c_{j_0}^{-1} c_j)_{0 \leqslant j \leqslant n}) \subset \mathscr{C}((\theta u_j)_{\theta \in \Theta(n-1),\, 0 \leqslant j \leqslant n}).$$

Therefore for every natural number $s \geqslant n - 1$

$$\mathscr{C}((\theta u_j)_{\theta \in \Theta(s),\, 0 \leqslant j \leqslant n}) = \mathscr{C}(\gamma)((\theta u_j)_{\theta \in \Theta(s),\, 0 \leqslant j \leqslant n,\, j \neq j_0})$$

and hence when s is sufficiently big

$$\omega_{\mathfrak{r}}(s) = \operatorname{tr deg} \mathscr{C}((\theta u_j)_{\theta \in \Theta(s),\, 0 \leqslant j \leqslant n})/\mathscr{C} = n\binom{s + m}{m} + \dim V,$$

whence (see Eq. (2) near the end of Section 5)

$$\omega_{\mathfrak{R}} = (n - 1)\binom{X + m}{m} + \dim V.$$

Corollary. Let $\mathscr{C}$ and V be the same as in the theorem, and suppose that the differential fields are ordinary (that is, that $m = 1$). Then there exists a differential polynomial $R \in \mathscr{C}\{\mathbf{y}\}$, irreducible over $\mathscr{C}$, such that an element of $\mathscr{U}^{n+1}$ is linearly dependent over V if and only if it is in the general solution of the differential equation $R = 0$. R is unique up to a nonzero factor in $\mathscr{C}$ and is differentially homogeneous. When V lies in the hyperplane $y_j = 0$ then R is differentially free of y_j; when V does not lie in the hyperplane $y_j = 0$ then the order of R in y_j equals the dimension of V.

Proof. As in the proof of the theorem, set $\mathfrak{r} = \mathfrak{A}_{\mathscr{C}}(\mathfrak{R})$. Then $\omega_{\mathfrak{r}} = n(X + 1) + \dim V$. It then follows by [2, Chapter VI, Section 7, Proposition 4] that there exists an irreducible differential polynomial $R \in \mathscr{C}\{\mathbf{y}\}$ such that $\mathfrak{p}_{\mathscr{C}}(R) = \mathfrak{r}$, and that ord $R =$ dim V. It is well known that if R_1 is any irreducible element of $\mathscr{C}\{\mathbf{y}\}$ with $\mathfrak{p}_{\mathscr{C}}(R) = \mathfrak{p}_{\mathscr{C}}(R_1)$, then $R_1 = cR$ for some $c \in \mathscr{C}$; indeed (see [2, Chapter IV, Section 6, Theorem 3(b)]), $R_1 \in \mathfrak{p}_{\mathscr{C}}(R)$, whence ord $R_1 \geqslant$ ord R, and similarly ord $R \geqslant$ ord R_1, so that ord $R_1 =$ ord R and therefore R and R_1 divide each other. This proves the required uniqueness. When V lies in the hyperplane $y_j = 0$, if an element $(u_0, u_1, \ldots, u_n)$ of $\mathscr{U}^{n+1}$ is linearly dependent over V, then so is $(u_0, \ldots, tu_j, \ldots, u_n\}$ for every $t \in \mathscr{U}$, that is, if the former element of $\mathscr{U}^{n+1}$ is a zero of $\mathfrak{p}_{\mathscr{C}}(R)$ then so is the latter for every t, and this implies that R is differentially free of y_j. This argument is reversible, and therefore when V does not lie in the hyperplane $y_j = 0$ then y_j is differentially present in R. Finally, let y_{j_0} be differentially present in R, and return to the generic zero $(\mathbf{u}, \mathbf{v})$ of $\mathfrak{s}$ used in the proof of the theorem; write $j(0) = j_0$ to simplify the typography. By Eq. (3) near the end of that proof,

$$\begin{aligned} \operatorname{ord}_{y_{j(0)}} R &= \operatorname{tr deg} \mathscr{C}\langle (u_j)_{0 \leqslant j \leqslant n} \rangle / \mathscr{C}\langle (u_j)_{0 \leqslant j \leqslant n,\, j \neq j(0)} \rangle \\ &= \operatorname{tr deg} \mathscr{C}\langle (u_j)_{0 \leqslant j \leqslant n,\, j \neq j(0)} \rangle(\gamma) / \mathscr{C}\langle (u_j)_{0 \leqslant j \leqslant n,\, j \neq j(0)} \rangle \\ &= \operatorname{tr deg} \mathscr{C}(\gamma)/\mathscr{C} = \dim V. \end{aligned}$$

8. EXAMPLES

Example 1. Let $V = \mathbf{P}_{\mathscr{K}}(n)$. Then $\mathscr{C}$ can be any subfield of $\mathscr{K}$ (for example, $\mathbf{Q}$}, and

$$\mathscr{R} = \mathfrak{Z}_n(\det(\theta_i y_j)_{0 \leqslant i \leqslant n,\, 0 \leqslant j \leqslant n} \quad (\theta_0 \in \Theta(0), \quad \theta_1 \in \Theta(1), \ldots, \quad \theta_n \in \Theta(n))),$$

as is well known (see [2, Chapter II, Section 1, Theorem 1]). When the differential fields are ordinary, R can be taken to be the Wronskian determinant $\det(y_j^{(i)})_{0 \leqslant i \leqslant n,\, 0 \leqslant j \leqslant n}$.

Example 2. More generally, let V be a linear subspace of $\mathbf{P}_{\mathscr{K}}(n)$ of dimension d, where $0 \leqslant d \leqslant n$. Then V is defined by a system of $n - d$ independent homogeneous linear equations

$$\sum_{0 \leqslant j \leqslant n} c_{ij} z_j = 0 \qquad (d + 1 \leqslant i \leqslant n)$$

with each coefficient an element of $\mathscr{K}$, $\mathscr{C}$ can be any subfield of $\mathscr{K}$ containing all the ratios of the coefficients, and $\mathfrak{R}$ is the set of zeros in $\mathbf{P}(n)$ of the set of differential polynomials

$$\det\begin{pmatrix} \theta_0 y_0 & \theta_0 y_1 & \cdots & \theta_0 y_n \\ \vdots & \vdots & & \vdots \\ \theta_d y_0 & \theta_d y_1 & \cdots & \theta_d y_n \\ c_{d+1,0} & c_{d+1,1} & \cdots & c_{d+1,n} \\ \vdots & \vdots & & \vdots \\ c_{n0} & c_{n1} & \cdots & c_{nn} \end{pmatrix} \qquad (\theta_0 \in \Theta(0), \quad \theta_1 \in \Theta(1), \ldots, \quad \theta_d \in \Theta(d)).$$

When the differential fields are ordinary, R can be taken to be

$$\det\begin{pmatrix} y_0 & y_1 & \cdots & y_n \\ y_0' & y_1' & \cdots & y_n' \\ \vdots & \vdots & & \vdots \\ y_0^{(d)} & y_1^{(d)} & \cdots & y_n^{(d)} \\ c_{d+1,0} & c_{d+1,1} & \cdots & c_{d+1,n} \\ \vdots & \vdots & & \vdots \\ c_{n0} & c_{n1} & \cdots & c_{nn} \end{pmatrix}.$$

Example 3. Let the differential fields be ordinary, let $n \geqslant 2$, and let V be the hypersurface defined by an irreducible homogeneous polynomial $P \in \mathscr{C}[z_0, z_1, \ldots, z_n]$. Then R may be taken as $P(W_0, -W_1, \ldots, (-1)^n W_n)$, where for each index k, W_k denotes the Wronskian determinant of $(y_0, \ldots, y_{k-1}, y_{k+1}, \ldots, y_n)$.

To prove this, consider any zero $\boldsymbol{\eta} = (\eta_0, \eta_1, \ldots, \eta_n)$ of $\mathfrak{p}_{\mathscr{C}}(R)$ in $\mathscr{U}^{n+1}$; by the corollary in Section 7, there exist constants $c_0, c_1, \ldots, c_n$ not all zero such that $P(c_1, c_0, \ldots, c_n) = 0$ and $\sum_{0 \leqslant j \leqslant n} c_j \eta_j = 0$, and hence such that also

$$\sum_{0 \leqslant j \leqslant n} c_j \eta_j^{(i)} = 0 \qquad (0 \leqslant i \leqslant n - 1).$$

When the rank of the matrix $(\eta_j^{(i)})_{0 \leqslant i \leqslant n-1,\, 0 \leqslant j \leqslant n}$ is $<n$ then each W_k vanishes at $\boldsymbol{\eta}$ and hence so does the polynomial $H = P(W_0, -W_1, \ldots, (-1)^n W_n)$; when the rank is equal to n then

$$(W_0(\eta): \; -W_1(\eta): \cdots : (-1)^n W_n(\eta)) = (c_0 : c_1 : \cdots : c_n)$$

and again H vanishes at η. Thus H vanishes at every zero of $\mathfrak{p}_{\mathscr{C}}(R)$ and hence is in $\mathfrak{p}_{\mathscr{C}}(R)$. Evidently ord $H = n - 1 = \dim V =$ ord R, so that H is divisible by R. The desired result therefore follows from the fact that H is irreducible over $\mathscr{C}$. (To establish this fact, observe for any fixed index k that as a polynomial in $(y_k^{(i)})_{0 \leqslant i \leqslant n-1}$, H is irreducible over the field $\mathscr{C}((y_j^{(i)})_{0 \leqslant i \leqslant n-1, 0 \leqslant j \leqslant n, j \neq k})$. Therefore if $H = AB$, where $A, B \in \mathscr{C}[(y_j^{(i)})_{0 \leqslant i \leqslant n-1, 0 \leqslant j \leqslant n}]$, then there exist subsets J, J' of $\{0, 1, \ldots, n\}$ with

$$J \cup J' = \{0, 1, \ldots, n\}, \qquad J \cap J' = \varnothing,$$

such that $A \in \mathscr{C}[(y_j^{(i)})_{0 \leqslant i \leqslant n-1, j \in J}]$ and $B \in \mathscr{C}[(y_j^{(i)})_{0 \leqslant i \leqslant n-1, j \in J'}]$. Assume that $J, J' \neq \varnothing$, fix $j \in J$ and $j' \in J'$, and suppose that, say $j < j'$. When $(y_{j'}^{(i)})_{0 \leqslant i \leqslant n-1}$ is replaced by $(y_j^{(i)})_{0 \leqslant i \leqslant n-1}$ then A remains unchanged, B becomes a nonzero polynomial C, W_j becomes $\pm W_{j'}$, $W_{j'}$ remains unchanged, and W_ν becomes zero $(0 \leqslant \nu \leqslant n,\ \nu \neq j,\ \nu \neq j')$, so that H becomes the differential polynomial

$$\begin{aligned} &P(0, \ldots, 0, \pm(-1)^j W_{j'}, 0, \ldots, 0, (-1)^{j'} W_{j'}, 0, \ldots, 0) \\ &\quad = W_{j'}^d P(0, \ldots, 0, \pm(-1)^j, 0, \ldots, 0, (-1)^{j'}, 0, \ldots, 0). \end{aligned}$$

Hence A divides $W_{j'}^d$, so that $J = \{0, 1, \ldots, n\} - \{j'\}$. Similarly, $J' = \{0, 1, \ldots, n\} - \{j\}$. Since $n \geqslant 2$, this forces the contradiction that $J \cap J' \neq \varnothing$. Therefore J and J' are not both nonempty, that is, H is irreducible.)

Added in proof. Professor Morikawa has pointed out to me that the ordinary differential polynomial R appearing in the Corollary in Section 7 equals the Cayley form of V computed at the signed minors of the matrix $(y_j^{(i)})_{0 \leqslant i \leqslant d, 0 \leqslant j \leqslant n}$, where $d = \dim V$. Equivalently, if $F((u_{ij})_{0 \leqslant i \leqslant d, 0 \leqslant j \leqslant n})$ denotes what van der Waerden calls the "zugeordnete Form" of V (see Section 36 of his "Einführung in die Algebraische Geometrie," Springer, Berlin, 1939), then $R = F((y_j^{(i)})_{0 \leqslant i \leqslant d, 0 \leqslant j \leqslant n})$.

REFERENCES

1. E. R. Kolchin, Rational approximation to solutions of algebraic differential equations, *Proc. Amer. Match. Soc.* **10** (1959), 238–244.
2. E. R. Kolchin, "Differential Algebra and Algebraic Groups." Academic Press, New York, 1973.
3. H. Morikawa, On osculating systems of differential equations of type (N, 1, 2), *Nagoya Math. J.* **31** (1968), 251–277.

DIFFERENTIAL ALGEBRAIC GROUPS

The last 30 years have seen the birth and growth to maturity of the theory of algebraic groups. This theory now plays an important role in many parts of algebra, number theory, geometry, and analysis. Basically, an algebraic group G is an algebraic set (that is, a set of points defined by a system of algebraic equations) that is also a group, such that the group law

$$G \times G \to G , \qquad (x,y) \mapsto xy$$

and the group symmetry

$$G \to G , \qquad x \mapsto x^{-1}$$

are rational mappings.

<u>Examples</u>. $\mathbb{G}_a$ = the additive group of all elements of a fixed (usually algebraically closed) field U . $\mathbb{G}_m$ = the multiplicative group of nonzero elements of U . $\mathbb{GL}(n)$ = the group of all invertible $n \times n$ matrices over U with the usual matrix multiplication. $\mathbb{SL}(n)$ = the subgroup of $\mathbb{GL}(n)$ consisting of the matrices x such that $\det x = 1$. $\mathbb{T}(n)$ = the subgroup of $\mathbb{GL}(n)$ consisting of the matrices (x_{ij}) such that $x_{ij} = 0 \ (i > j)$. The curve $\mathbb{W}(g_2,g_3)$ in the complex projective plane $\mathbb{P}_{\mathbb{C}}(2)$ obtained as the image of $\mathbb{C}$ under the mapping

$$\mathbf{C} \to W(g_2,g_3) , \qquad z \mapsto (1:\wp(z):\wp'(z)) ,$$

where $g_2,g_3 \in \mathbf{C}$, $g_2^3 - 27g_3^2 \neq 0$, and $\wp$ is the elliptic function of Weierstrass satisfying the differential equation $y'^2 = 4y^3 - g_2y - g_3$, the group law being given by the formula

$$\big((1:\wp(z_1):\wp'(z_1)), (1:\wp(z_2):\wp'(z_2))\big) \mapsto \big(1:\wp(z_1+z_2):\wp'(z_1+z_2)\big)$$

(which is a rational mapping because of the addition formulae for $\wp$ and $\wp'$).

Now, an algebraic equation is a very special case of an algebraic differential equation, and it is natural to try to generalize the notion of algebraic groups correspondingly. The first steps at such a generalization were taken by Cassidy [2,3], who defined the notion of affine differential algebraic group and proved many basic results for those that are linear (i.e., are differential algebraic subgroups of $\mathbf{GL}(n)$ for some n). That this notion will prove to be a fruitful one is indicated by the applications already made by W.Y. Sit [9,10]. While much remains to be done in this direction, I think it is time to launch a general theory of differential algebraic groups. Even for algebraic groups, linear ones do not tell the whole story. $W(g_2,g_3)$ and, more generally, Abelian varieties are not linear. These lectures describe the beginnings of such a general theory. The setting is abstract and the method axiomatic, more or less along the lines used in my book [5] to develop the theory of algebraic groups.

Let $\mathcal{U}$ be a fixed differential field of characteristic 0, that is, a field, of which $\mathbb{Q}$ is a subfield, equipped with a finite set Δ of derivation operators that commute with each other. Let $\mathcal{K}$ denote the field of constants of $\mathcal{U}$, that is, the set of elements $c \in \mathcal{U}$ such that $\delta c = 0$ $(\delta \in \Delta)$. We suppose that $\mathcal{U}$ is universal over $\mathbb{Q}$, i.e., that for every finitely generated extension $\mathcal{F}$ of $\mathbb{Q}$ and every finitely generated extension $\mathcal{G}$ of $\mathcal{F}$, there is an embedding of $\mathcal{G}$ in $\mathcal{U}$ that leaves the elements of $\mathcal{F}$ fixed. (Here, as elsewhere except when the contrary is stated, "extension" means "differential field extension," and words like "embedding" or "isomorphism" mean "differential embedding" of "differential isomorphism."). All differential fields, except those for which the contrary is obvious or explicitely stated, are differential subfields of $\mathcal{U}$ over which $\mathcal{U}$ is universal. Usually, the prefix "Δ-" is used instead of the adjective "differential."

Δ-algebraic groups (in the sense of Cassidy, or in any other concrete reasonable sense) that are defined over a Δ-field $\mathcal{F}$, i.e., that can be defined by a system of algebraic Δ-equations with coefficients in $\mathcal{F}$, have certain formal properties. Our approach is to take some of these formal properties as a system of axioms for what we call Δ-$\mathcal{F}$-group. These axioms, as well as axioms for the corresponding notion of <u>homogeneous</u> Δ-$\mathcal{F}$-<u>space</u>, will be discussed in Lecture 1, and elementary properties of these structures will be described. (Every Δ-$\mathcal{F}$-group is, of course, a homogeneous Δ-$\mathcal{F}$-space.) It will be shown, for a Δ-$\mathcal{F}$-group G and an extension $\mathcal{G}$ of $\mathcal{F}$, how to define on G a canonical structure of Δ-$\mathcal{G}$-group; likewise for a homogeneous Δ-$\mathcal{F}$-space. Certain subsets of a homogeneous Δ-$\mathcal{F}$-space M (those having a "generic element over $\mathcal{F}$") are said to be $\mathcal{F}$-*irreducible*, and a finite union of $\mathcal{F}$-irreducible subsets of M is called a Δ-$\mathcal{F}$-*subset* of M. It turns out that every Δ-$\mathcal{F}$-subset of M is a Δ-$\mathcal{G}$-subset of M for every extension $\mathcal{G}$ of $\mathcal{F}$. The subsets of A of M such that, for some extension $\mathcal{G}$ of $\mathcal{F}$, A is a Δ-$\mathcal{G}$-subset of M are the closed sets of a topology on M (the Δ-*Zariski topology* or, simply, the Δ-*topology*). This topology is Noetherian (i.e., every strictly decreasing sequence of closed sets is finite). The Δ-$\mathcal{F}$-subsets of M are the closed sets of another Noetherian topology (the Δ-$\mathcal{F}$-topology). Lecture 1 will end with a discussion of these topologies.

Lecture 2 will begin with a discussion of Δ-$\mathcal{F}$-subgroups of a Δ-$\mathcal{F}$-group and of Δ-$\mathcal{F}$-homomorphisms of Δ-$\mathcal{F}$-groups and homogeneous Δ-$\mathcal{F}$-spaces. It will be shown how to define direct products of Δ-$\mathcal{F}$-groups and of homogeneous Δ-$\mathcal{F}$-spaces, and how to define quotients of Δ-$\mathcal{F}$-groups by their Δ-$\mathcal{F}$-subgroups. The second part of Lecture 2 will be devoted to the definition and properties of Δ-rational mappings (of one Δ-$\mathcal{F}$-set into another) and of Δ-rational functions on a Δ-$\mathcal{F}$-set. Several distinctions between the theory of algebraic groups and the theory of Δ-algebraic groups will be pointed out.

The aim of Lecture 3 is to classify the principal homogeneous Δ-$\mathcal{F}$-spaces for a Δ-$\mathcal{F}$-group. As is well known, for a K-group G (i.e., for an algebraic group defined over a field K) the analogous problem is solved by means of the so called Galois cohomology set $H^1(K,G)$ of K in G : there exists a canonical bijection of the set of K-isomorphism classes of principal homogeneous K-spaces for G onto the set $H^1(K,G)$. (Thus, in particular, a necessary and sufficient condition that every principal homogeneous K-space for G have a point that is rational over K is that $H^1(K,G) = 1$.) However, when G is a Δ-$\mathcal{F}$-group, the Galois cohomology set does not, in general, solve the problem, because a principal homogeneous Δ-$\mathcal{F}$-space M for G need not have an element that is algebraic over $\mathcal{F}$. Even the Δ-Galois cohomology set of $\mathcal{F}$ in G, defined by Kovacic [7] in another connection, in which the elements of M that are algebraic over $\mathcal{F}$ are supplanted by those that are strongly normal over $\mathcal{F}$, is inadequate for our present purpose because, as shown by Kovacic, M need not have even an element that is strongly normal over $\mathcal{F}$. Fortunately, there does exist a suitable sufficiently ubiquitous class of elements, those that are constrained over $\mathcal{F}$. The notion of constrained extension of $\mathcal{F}$ and that of constrained closure of $\mathcal{F}$, which were developed (using a different terminology) by Blum [1] and Shelah [9] within the framework of model theory in

5

pursuing an idea of Robinson [8], and which received in [6] a unified exposition in the language of differential algebra, are in some ways analogous to the notion of algebraic extension and that of algebraic closure of a field. Using a fixed constrained closure $\mathcal{F}^{\dagger}$ of $\mathcal{F}$, it will be shown how to construct a constrained cohomology set $H^1_{\Delta}(\mathcal{F},G)$ of $\mathcal{F}$ in G, and, more generally, a cohomology set $H^1_{\Delta}(\mathcal{N}/\mathcal{F},G)$ where $\mathcal{N}$ is any extension of $\mathcal{F}$ in $\mathcal{F}^{\dagger}$ that is "normal over $\mathcal{F}$ in $\mathcal{F}^{\dagger}$." Relations between these cohomology sets, and the Galois cohomology set $H^1(\mathcal{F},G)$ and the Δ-Galois cohomology set of Kovacic, will be pointed out. There will then be defined, for any Δ-$\mathcal{F}$-set A, a Δ-$\mathcal{F}$-cohomology set $H^1_{\Delta,\mathcal{F}}(A,G)$ of A in G (analogous to the rational cohomology set of a K-set into a K-group). This notion will be used to compute the constrained cohomology $H^1_{\Delta}(\mathcal{F},G)$ in some special cases. To conclude Lecture 3, the constrained cohomology theory will be applied to the classification problem for principal homogeneous Δ-$\mathcal{F}$-spaces for G.

The purpose of Lecture 4 is to develop a suitable Lie theory for Δ-$\mathcal{F}$-groups. For linear Δ-$\mathcal{F}$-groups, this has been done by Cassidy [2]. If G is any connected Δ-$\mathcal{F}$-group and $\mathcal{F}(G)$ denotes the Δ-field of Δ-rational functions on G, a Δ-derivation on G is a derivation D of $\mathcal{F}(G)$ over $\mathcal{U}$ such that $D\delta\varphi = \delta D\varphi$ ($\delta \in \Delta$, $\varphi \in \mathcal{F}(G)$). G acts in a natural way (through right translations) on the set of all such Δ-derivations.

6

The set of Δ-derivations on G that are invariant under this action is a Lie algebra $\mathfrak{L}(G)$ over $\mathcal{K}$ called the <u>Lie algebra of</u> G. The usual argument that the Lie algebra of a connected algebraic group is as big as it should be, does not work in the present situation (essentially because a system of homogeneous linear equations with coefficients in the function field can always be solved in the function field, whereas for a system of homogeneous linear <u>differential</u> equations this is not the case). It will be shown how to overcome this difficulty by constructing in the expected way a vector space isomorphism between $\mathfrak{L}(G)$ and the space of Δ-<u>tangent vectors</u> to G at an element of G, and by noting that because of the homogeneity of G, this tangent space is as big as G itself. (In the language of Johnson [4] the local Δ-ring at each element of G is <u>regular</u>, but precisely because of the homogeneity, it is not necessary to introduce this concept in the present situation.) Even more, it will be shown how $\mathfrak{L}(G)$ has, in addition to its structure of Lie algebra over $\mathcal{K}$, a natural structure of Δ-algebraic group.

1. L. Blum, Generalized algebraic structures: A model theoretic approach, Ph.D. dissertation, Massachusetts Institute of Technology, 1968.
2. P.J. Cassidy, Differential algebraic groups, Amer. J. Math., 94(1972), 891-954.
3. ____________, The differential rational representation algebra on a linear differential algebraic group, Jour. Algebra, to appear.
4. J. Johnson, Regular local differential rings, in preparation.
5. E.R. Kolchin, Differential Algebra and Algebraic Groups, Academic Press, New York, 1973.
6. ____________, Constrained extensions of differential fields, Advances in Math., 12(1974), 141-170.
7. J. Kovacic, Pro-algebraic groups and the Galois theory of differential fields, Amer. J. Math., 95(1973), 507-536.
8. A. Robinson, On the concept of differentially closed field, Bull. Res. Council Israel, Sec. F, 8(1959), 113-118.
9. S. Shelah, Uniqueness and characterization of prime models over sets for totally transcendental first order theories, J. Symbolic Logic, 37(1972), 107-113.
10. W.Y. Sit, Differential algebraic subgroups of SL(2) and strong normality in simple extensions, Amer. J. Math., to appear.
11. ________, Typical differential dimension of the intersection of linear differential algebraic groups, Jour. Algebra, 32(1974).

8

E. R. KOLCHIN
(New York, USA)

DIFFERENTIAL ALGEBRAIC STRUCTURES

The present paper attaches to the theory of differential algebraic groups, a theory initiated by P. U. Cassidy [1—4] in analogy with the theory of algebraic groups, of which it is a generalization. Her work, which deals primarily with the affine (and, more particularly, linear) case, has been applied by W. Y. Sit [8, 9] in connection with the Galois theory of differential fields and with a problem on intersections. The present author has been generalizing the theory to the general (that is, not necessarily affine) case along the lines used in his exposition [5] of the theory of algebraic groups; the general theory, which will appear in a book in preparation [6] has already been enlarged upon and applied by J. Kovacic [7] to a «constrained» cohomology theory and the study of central differentially simple differential algebras. A different approach to such a generalization (via schemes) is currently being developed by J. L. Johnson. The purpose here is to indicate briefly how to introduce certain differential algebraic structures other than groups or their homogeneous spaces. That these concepts are not mere exercises in abstract definition, but arise naturally in the study of differential algebraic groups in connection with tangent spaces to a differential algebraic set and the Lie algebra of a differential algebraic group, is shown in outline in the final two sections. A complete treatment will appear in [6].

This work has been partially supported by a grant from the National Science Foundation.

Background

The terminology and notation are more or less like in [5]. We fix a universal differential field $\mathcal{U}$ of characteristic 0 with fundamental set Δ of derivation operators, denote the elements of Δ by $\delta_1, \ldots, \delta_m$, denote the field of constants of $\mathcal{U}$ by $\mathcal{K}$, and denote the monoid of derivative operators by Θ (Θ is commutative and consists of the operators $\delta_1^{e_1} \ldots \delta_m^{e_m}$ with $e_1, \ldots, e_m \in \in \mathbf{N}$). For brevity, we use the prefix «Δ» instead of the adjective «differential». All Δ-fields will be supposed to be Δ-subfields of $\mathcal{U}$, over which $\mathcal{U}$ is universal, except when the contrary is explicitly stated or obvious; $\mathfrak{F}$ will denote a Δ-field. The word «extension» will usually mean «Δ-field extension». We use a double prefix like «Δ — $\mathfrak{F}$»-instead of «differential rational ... (defined or rational) over $\mathfrak{F}$». For example, a Δ — $\mathfrak{F}$-polynomial is a differential polynomial over $\mathfrak{F}$, and a Δ — $\mathfrak{F}$-group is a differential algebraic group defined over $\mathfrak{F}$.

A Δ — $\mathfrak{F}$-group is, roughly speaking, a set G equipped with a structure of group and a structure of Δ — $\mathfrak{F}$-set, such that the group law $G \times G \to G$ and the group symmetry $G \to G$ are Δ — $\mathfrak{F}$-mappings. For example, a subgroup of $GL(n) = GL_U(n)$ that consists of all matrices $a = (a_{ij})_{1 \leqslant i \leqslant n, 1 \leqslant j \leqslant n} \in \in GL(n)$ that are zeros of a fixed subset of the Δ-polinomial algebra $\mathfrak{F}\{(y_{ij})_{1 \leqslant i \leqslant n, 1 \leqslant j \leqslant n}\}$ is a Δ — $\mathfrak{F}$-group; such a Δ — $\mathfrak{F}$-group is *linear*. A group that consists of all elements of $\mathcal{U}^n$ that are zeros of a fixed subset of $\mathfrak{F}\{y_1, \ldots \ldots, y_n\}$ that are not zeros of another fixed subset of $\mathfrak{F}\{y_1, \ldots, y_n\}$ and that has the property that the coordinates of any product ab (resp. any inverse a^{-1}) are expressible as Δ — $\mathfrak{F}$-functions of the coordinates of a and b (resp. of a), is a Δ — $\mathfrak{F}$-group; such a Δ — $\mathfrak{F}$-group is *affine*. A linear Δ — $\mathfrak{F}$-group is always affine, but not conversely (see [6]). The general notation of Δ — $\mathfrak{F}$-group is defined by a system of axioms that are easily verified in the affine case

We shall not give these axioms here. The reader who is troubled by this can think of the affine (or even linear) case.

If x, x' are elements of a $\Delta - \mathfrak{F}$-set, we use the notation $x \underset{\mathfrak{F}}{\rightarrow} x'$ (or even $x \rightarrow x'$ when the reference to $\mathfrak{F}$ is clear) to denote the relation «x' is a Δ-specialization of x over $\mathfrak{F}$», and use the notation $\mathfrak{F}\langle x\rangle$ to denote the extension of $\mathfrak{F}$ associated to x (in the affine case, this is the extension of $\mathfrak{F}$ generated by the coordinates of x).

1. Two lemmas

In this section G, H, I are $\Delta - \mathfrak{F}$-groups and f is a set-theoretic mapping of $G \times H$ into I. For each element $x \in G$, ${}_xf$ denotes the mapping of H into I defined by the formula ${}_xf\,(y) = f\,(x, y)$ $(y \in H)$, and for each element $y \in H$, f_y denotes the mapping of G into I defined by the formula $f_y\,(x) = f\,(x, y)$ $(x \in G)$.

L e m m a 1. *Suppose that for each* $x \in G$, ${}_xf$ *is a* $\Delta - \mathfrak{F}\langle x\rangle$*-homomorphism of* H *into* I, *and that for each* $y \in H$, f_y *is a* $\Delta - \mathfrak{F}\langle y\rangle$*-homomorphism of* G *into* I. *Suppose, too, that each of* G, H *has the property that all its* $\mathfrak{F}$*-components are irreducible (absolutely). Then* f *is a* $\Delta - \mathfrak{F}$*-mapping of* $G \times H$ *into* I.

P r o o f. For each $\mathfrak{F}$-component V of $G \times H$, fix an $\mathfrak{F}$-generic element (x_V, y_V) of V. Since ${}_{x_V}f$ is a $\Delta - \mathfrak{F}\langle x_V\rangle$-mapping, the element $f\,(x_V, y_V) = {}_{x_V}f\,(y_V)$ of I is rational over the Δ-field $\mathfrak{F}\langle x_V\rangle\langle y_V\rangle = \mathfrak{F}\langle x_V, y_V\rangle$; therefore there exists a $\Delta - \mathfrak{F}$-mapping φ of $G \times H$ into I such that $\varphi\,(x_V, y_V) = f\,(x_V, y_V)$ for every V. We claim that $\varphi\,(x, y) = f\,(x, y)$ for every $\mathfrak{F}$-generic element (x, y) of any $\mathfrak{F}$-component V of $G \times H$. Indeed, by the irreducibility of the $\mathfrak{F}$-components of G and of H, V is irreducible, and if we fix an $\mathfrak{F}\,(x, y, x_V, y_V)$-generic element (x_1, y_1) of V, then we have the following diagram of generic Δ-specializations over $\mathfrak{F}$:

$$(x, y) \leftrightarrow (x, y_1) \leftrightarrow (x_1, y_1) \leftrightarrow (x_V, y_1) \leftrightarrow (x_V, y_V).$$

Starting on the right, and letting $\tau : \mathfrak{F}\langle x_V, y_V\rangle \approx \mathfrak{F}\langle x_V, y_1\rangle$ denote the Δ-field isomorphism over $\mathfrak{F}$ associated with the generic Δ-specialization $(x_V, y_V) \leftrightarrow (x_V, y_1)$ over $\mathfrak{F}$, we see that

$$\varphi\,(x_V, y_1) = \varphi\,(\tau\,(x_V, y_V)) = \tau\varphi\,(x_V, y_V) = \tau f\,(x_V, y_V) = \tau\,({}_{x_V}f\,(y_V)) = {}_{x_V}f\,(\tau y_V) = f\,(x_V, y_1).$$

Similarly, we see, in succession, that

$$\varphi\,(x_1, y_1) = f\,(x_1, y_1), \quad \varphi\,(x, y_1) = f\,(x, y_1), \quad \varphi\,(x, y) = f\,(x, y).$$

This establishes our claim.

This being the case, consider any $(x', y') \in G \times H$ and fix an $\mathfrak{F}\langle x', y'\rangle$-generic element (x, y) of an $\mathfrak{F}\langle x', y'\rangle$-component of $G \times H$. When (x', y') is $\mathfrak{F}$-generic on some $\mathfrak{F}$-component of $G \times H$, then so too are (x, y'), (x', y), $(x'x, y'y)$, and hence by our claim we may write

$$\varphi\,(x'x, y'y) = f\,(x'x, y'y) = f\,(x', y')\,f\,(x, y')\,f\,(x', y)\,f\,(x, y) = \varphi\,(x', y')\,{}_xf\,(y')\,f_y\,(x')\,f\,(x, y),$$

whence

$$\varphi\,(x', y') = \varphi\,(x'x, y'y)\,({}_xf\,(y')\,f_y\,(x')\,f\,(x, y))^{-1}.$$

But when (x', y') is not $\mathfrak{F}$-generic, $(x'y, y'y)$ still is, and hence this equation shows that φ is defined at (x', y') and

$$\varphi\,(x', y') = f\,(x'x, y'y)\,(f\,(x, y')\,f\,(x', y)\,f\,(x, y))^{-1} = f\,(x', y').$$

This shows that $f = \varphi$ and hence f is a $\Delta - \mathfrak{F}$-mapping.

For any $\Delta - \mathfrak{F}$-group A and any normal $\Delta - \mathfrak{F}$-subgroup A' of A, we let $\pi_{A/A'} : A \to A/A'$ denote the canonical group homomorphism. This is, of course, a $\Delta - \mathfrak{F}$-homomorphism and is a quotient in the categorical sense.

Lemma 2. *Suppose that for each $x \in G$, ${}_xf$ is a group homomorphism of H into I, that for each $y \in H$, f_y is a group homomorphism of G into I, and that f is a $\Delta - \mathfrak{F}$-mapping of $G \times H$ into I. Let G', H', I' be normal $\Delta - \mathfrak{F}$-subgroups of G, H, I, respectively, with ${}_xf(H') \subset I'$ $(x \in G)$ and $f_y(G') \subset I'$ $(y \in H)$, and let f' denote the mapping of $(G/G') \times (H/H')$ into I/I' such that*

$$f'(\pi_{G/G'}x, \pi_{H/H'}y) = \pi_{I/I'}f(x, y) \qquad (x \in G, y \in H).$$

Then f' is a $\Delta - \mathfrak{F}$-mapping of $(G/G') \times (H/H')$ into I/I'.

Proof. For $\bar{x} \in G/G'$, let ${}_{\bar{x}}f'$ denote the mapping of H/H' into I/I' defined by the formula ${}_{\bar{x}}f'(\bar{y}) = f'(\bar{x}, \bar{y})$ $(\bar{y} \in H/H')$, and for each $\bar{y} \in H/H'$, let $f_{\bar{y}}'$ denote the mapping of G/G' into I/I' defined by the formula $f_{\bar{y}}'(\bar{x}) = f'(\bar{x}, \bar{y})$ $(\bar{x} \in G/G')$. Consider any $\bar{x} \in G/G'$ and any $x \in \bar{x}$. Since $\pi_{I/I'} \circ {}_xf$ evidently is a $\Delta - \mathfrak{F}\langle x\rangle$-homomorphism of H into I/I' with kernel, containing H' and ${}_{\bar{x}}f'$ evidently is the mapping of H/H' into I/I' such that ${}_{\bar{x}}f' \circ \pi_{H/H'} = \pi_{I/I'} \circ {}_xf$, ${}_{\bar{x}}f'$ must be a $\Delta - \mathfrak{F}\langle x\rangle$-homomorphism. If σ is any isomorphism of $\mathfrak{F}\langle x\rangle$ over $\mathfrak{F}\langle \bar{x}\rangle$, then, for any $y \in H$, the isomorphism σ^{-1} of $\mathfrak{F}\langle \sigma x\rangle$ can be extended to an isomorphism σ' of $\mathfrak{F}\langle \sigma x\rangle\langle y\rangle$, and

$$\begin{aligned}\sigma({}_{\bar{x}}f')(\pi_{H/H'}y) &= \sigma'^{-1}({}_{\bar{x}}f'(\sigma'\pi_{H/H'}y)) = \sigma'^{-1}({}_{\bar{x}}f'(\pi_{H/H'}\sigma'y)) = \\ &= \sigma'^{-1}(\pi_{I/I'}{}_xf(\sigma'y)) = \sigma'^{-1}(\pi_{I/I'}f(x, \sigma'y)) = \pi_{I/I'}f(\sigma x, y) = \\ &= f'(\pi_{G/G'}\sigma x, \pi_{H/H'}y) = f'(\sigma\pi_{G/G'}x, \pi_{H/H'}y) = f'(\bar{x}, \pi_{H/H'}y) = \\ &= {}_{\bar{x}}f'(\pi_{H/H'}y),\end{aligned}$$

whence $\sigma({}_{\bar{x}}f') = {}_{\bar{x}}f'$. This shows that ${}_{\bar{x}}f'$ is a $\Delta - \mathfrak{F}\langle \bar{x}\rangle$-homomorphism of H/H' into I/I'. Similarly, for any $\bar{y} \in H/H'$, $f_{\bar{y}}'$ is a $\Delta - \mathfrak{F}\langle \bar{y}\rangle$-homomorphism of G/G' into I/I'. Hence by lemma 1, f' is a $\Delta - \mathfrak{F}_a$-mapping of $(G/G') \times (H/H')$ into I/I' (where $\mathfrak{F}_a$ denotes the algebraic closure of $\mathfrak{F}$). But if τ is any automorphism of $\mathfrak{F}_a$ over $\mathfrak{F}$, then, for any $(x, y) \in G \times H$, τ^{-1} can be extended to an isomorphism τ' of $\mathfrak{F}_a\langle x, y\rangle$, and

$$\begin{aligned}\tau(f')(\pi_{G/G'}x, \pi_{H/H'}y) &= \tau'^{-1}f'(\tau'\pi_{G/G'}x, \tau'\pi_{H/H'}y) = \tau'^{-1}f'(\pi_{G/G'}\tau'x, \pi_{H/H'}\tau'y) = \\ &= \tau'^{-1}\pi_{I/I'}f(\tau'x, \tau'y) = \pi_{I/I'}f(x, y) = f'(\pi_{G/G'}x, \pi_{H/H'}y),\end{aligned}$$

whence $\tau(f') = f'$. Therefore, f' is a $\Delta - \mathfrak{F}$-mapping.

2. Differential algebraic rings

By a $\Delta - \mathfrak{F}$-*ring* we shall mean a set R equipped with a structure of ring and a structure of $\Delta - \mathfrak{F}$-group such that: i° the additive group structure of the ring and the group structure of the $\Delta - \mathfrak{F}$-group are identical, and ii° the ring multiplication is a $\Delta - \mathfrak{F}$-mapping of $R \times R$ into R.

A subset of a $\Delta - \mathfrak{F}$-ring R that is both a subring of the ring R and a $\Delta - \mathfrak{F}$-subgroup of the $\Delta - \mathfrak{F}$-group R has a natural structure of $\Delta - \mathfrak{F}$-ring; we shall call it a $\Delta - \mathfrak{F}$-*subring* of R. A subset of R that is both an ideal of the ring R (left, or right, or two-sided) and a $\Delta - \mathfrak{F}$-subgroup of the $\Delta - \mathfrak{F}$-group R, we shall call a $\Delta - \mathfrak{F}$-*ideal* of R (left, or right, or two-sided). (In general, when we use the term «ideal» unqualified by an adjective «left» or «right», we shall mean «two-sided ideal»).

By a $\Delta - \mathfrak{F}$-*homomorphism* of a $\Delta - \mathfrak{F}$-ring R into a $\Delta - \mathfrak{F}$-ring R' we shall mean a mapping of R into R' that is both a ring homomorphism of the ring R into the ring R' (and hence, by definition, maps the unity element of R into that of R') and a $\Delta - \mathfrak{F}$-homomorphism of the $\Delta - \mathfrak{F}$-group R into the $\Delta - \mathfrak{F}$-group R'. The kernel (resp. image) of such a $\Delta - \mathfrak{F}$-homomorphism is a $\Delta - \mathfrak{F}$-ideal of R (resp. $\Delta - \mathfrak{F}$-subring of R). Conversely, if $\mathfrak{k}$ is any $\Delta - \mathfrak{F}$-ideal of R, then the ring structure of the quotient ring $R/\mathfrak{k}$ and the $\Delta - \mathfrak{F}$-group structure of the quotient $\Delta - \mathfrak{F}$-group $R/\mathfrak{k}$ determine on the set $R/\mathfrak{k}$ a structure of $\Delta - \mathfrak{F}$-ring, and the canonical mapping $\pi_{R/\mathfrak{k}} : R \to R/\mathfrak{k}$ is a $\Delta - \mathfrak{F}$-homomorphism of $\Delta - \mathfrak{F}$-rings and is a quotient in the categorical sense. Indeed, the only point to establish that is not trivial is that the ring multiplication $(R/\mathfrak{k}) \times (R/\mathfrak{k}) \to R/\mathfrak{k}$ is a $\Delta - \mathfrak{F}$-mapping, and this is an immediate consequence of Lemma 2.

If R is a $\Delta - \mathfrak{F}$-ring, we denote the unity element of R by 1_R (or simply by 1 when there is no danger of confusion), and denote the multiplicative group of units of the ring R by R^*. For any isomorphism σ of $\mathfrak{F}\langle 1_R\rangle$ over σ^{-1} can be extended to an isomorphism σ' of $\mathfrak{F}\langle \sigma 1_R, 1_R\rangle$; since $\sigma 1_R = \sigma'^{-1} 1_R \cdot 1_R = \sigma'^{-1}(1_R \cdot \sigma' 1_R) = 1_R$, we infer that

$$\mathfrak{F}\langle 1_R\rangle = \mathfrak{F}$$

(and hence the $\mathfrak{F}$-component of R that contains 1_R is irreducible (absolutely)). A similar argument shows that if $x \in R^*$, then

$$\mathfrak{F}\langle x^{-1}\rangle = \mathfrak{F}\langle x\rangle$$

nd

$$\tau(x^{-1}) = (\tau x)^{-1}$$

for every isomorphism τ of $\mathfrak{F}\langle x\rangle$ over $\mathfrak{F}$. This suggests the following result.

P r o p o s i t i o n 1. *Let R be a $\Delta - \mathfrak{F}$-ring. The following two conditions are equivalent*:

(a) *whenever $x \in R$, $x' \in R^*$ and x' is a Δ-specialization of x over $\mathfrak{F}$, then $x \in R^*$;*

(b) *R^* is $\Delta - \mathfrak{F}$-open in R (that is, $R - R^*$ is a $\Delta - \mathfrak{F}$-subset of R).*

If these conditions are satisfied, then the group structure on R^* and the restriction to R^* of the $\Delta - \mathfrak{F}$-structure on R determine on R^* a structure of $\Delta - \mathfrak{F}$-group.

P r o o f. It is clear that (b) implies (a). Let (a) be satisfied.

We claim that if x, $x' \in R^*$ and $x \to x'$, then $x^{-1} \to x'^{-1}$. Indeed, fix an $\mathfrak{F}\langle x, x'\rangle$-generic element s of the component R^0 of R containing 0; then $1_R + s$ is an $\mathfrak{F}\langle x, x'\rangle$-generic element of the component R^1 of R containing 1_R, so that $1_R + s \in R^*$ by (a). Put $u = x(1_R + s)$, $u' = x'(1_R + s)$; then u resp. u' is an $\mathfrak{F}\langle x, x'\rangle$-generic element of the $\mathfrak{F}\langle x, x'\rangle$-component of R containing x resp. x' (because, in the case of u, if $v \xrightarrow[\mathfrak{F}\langle x,x'\rangle]{} u$, then $x^{-1}v \xrightarrow[\mathfrak{F}\langle x,x'\rangle]{} x^{-1}u = 1_R + s$, so that $x^{-1}v \underset{\mathfrak{F}\langle x,x'\rangle}{\leftrightarrow} x^{-1}u$ whence $v \underset{\mathfrak{F}\langle x,x'\rangle}{\leftrightarrow} u$). Also $(x, 1_R + s) \to (x', 1_R + s)$, whence $(u, 1_R + s) \to (u', 1_R + s)$, so that $u \leftrightarrow u'$, and the isomorphism $\tau : \mathfrak{F}\langle u\rangle = \mathfrak{F}\langle u'\rangle$ over $\mathfrak{F}$ with $\tau u = u'$ is compatible with $\mathrm{id}_{\mathfrak{F}(s)}$. Hence, $u^{-1} \leftrightarrow u'^{-1}$ and $\tau(u^{-1}) = u^{-1}$ so that $(u^{-1}, 1_R + s) \to (u'^{-1}, 1_R + s)$, whence $(1_R + s)u^{-1} \to (1_R + s)u'^{-1}$, that is, $x^{-1} \to x'^{-1}$ as claimed.

Let $V_1, \ldots, V_n$ be the $\mathfrak{F}$-components of R that intersect R^*, and fix an $\mathfrak{F}$-generic element x_j of V_j. Since $\mathfrak{F}\langle x_j{}^{-1}\rangle = \mathfrak{F}\langle x_j\rangle$, there exists a $\Delta - \mathfrak{F}$-mapping f of $V_1 \cup \ldots \cup V_n$ into itself such that $f(x_j) = x_j$ $(1 \leqslant j \leqslant n)$. For any $\mathfrak{F}$-generic element x of V_j, f is defined at x, there exists an isomorphism $\tau : \mathfrak{F}\langle x_j\rangle \approx \mathfrak{F}\langle x\rangle$ over $\mathfrak{F}$ with $\tau x_j = x$, and $f(x) = f(\tau x_j) = \tau f(x_j) = \tau(x^{-1}) = (\tau x_j)^{-1} = x^{-1}$. Consider any element $x \in V_j$. If f is defined at x, then $(x_j, f(x_j)) \to (x, f(x))$, so that $1_R = x_j f(x_j) \to x f(x)$ whence $x f(x) = 1_R$, and similarly $f(x)\, x = 1_R$; therefore $x \in R^*$ and $f(x) = x^{-1}$. Suppose, conversely, that $x \in R^*$. Fixing an $\mathfrak{F}\langle x\rangle$-generic element s of R^0, as before, we see that f is defined at $x\ (1_R + s)$ and $1_R + s$, and $f(x(1_R + s)) = (1_R + s)^{-1} x^{-1}$. The ring multiplication $R \times R \to R$ restricts to a $\Delta - \mathfrak{F}$-mapping g of $(V_1 \cup \dots \cup V_n)^2$ into $V_1 \cup \dots \cup V_n$, and for any $z \in V_1 \cup \dots \cup V_n$, the formula $y \mapsto zy$ resp. yz defines an everywhere defined $\Delta - \mathfrak{F}$-mapping of $V_1 \cup \dots \cup V_n$ into itself that we denote by ${}_z g$ resp. g_z. For any $\mathfrak{F}\langle s\rangle$-generic element y of any $\mathfrak{F}\langle s\rangle$-component of $V_1 \cup \dots \cup V_n$,

$$f(y) = (1_R + s)(y(1_R + s))^{-1} = {}_{1+s}g(f(g_{1+s}(y))).$$

Because g_{1+s} is defined at x and f is defined at $g_{1+s}(x) = x(1_R + s)$ and ${}_{1+s}g$ is defined at $f(g_{1+s}(x)) = (1_R + s)^{-1} x^{-1}$, it follows that f is defined at x.

We have shown that R^* is the domain of definition of the $\Delta - \mathfrak{F}$-mapping f, so that R^* is $\Delta - \mathfrak{F}$-open in $V_1 \cup \dots \cup V_n$ and hence in R, too, that is (b) is satisfied. Since the $\Delta - \mathfrak{F}$-mappings g and f are defined at every element of $R^* \times R^*$ and R^*, respectively, and map these sets into R^*, they determine on R^* a structure of $\Delta - \mathfrak{F}$-group.

By a $\Delta - \mathfrak{F}$-*division ring* we mean a $\Delta - \mathfrak{F}$-ring R such that the ring R is a division ring. By Proposition 1, then the set $R^* = R - \{0\}$ is a $\Delta - \mathfrak{F}$-group. By a $\Delta - \mathfrak{F}$-*field* we mean, of course, a commutative $\Delta - \mathfrak{F}$-division ring. The universe $\mathcal{U}$ has an obvious structure of $\Delta - \mathfrak{F}$-ring and, as such, is a $\Delta - \mathfrak{F}$-field. The field $\mathcal{K}$ of constants of $\mathcal{U}$ is a $\Delta - \mathfrak{F}$-subring of $\mathcal{U}$. It follows from a result of Sit [8, Prop. 3.2] that if R is any $\Delta - \mathfrak{F}$-subring of $\mathcal{U}$, then there exists a commuting linearly independent set Δ' of derivation operators of the form $a_1\delta_1 + \dots + a_m\delta_m$ with $a_1, \dots, a_m \in \mathfrak{F}$ such that R is the field of constants of $\mathcal{U}$ when $\mathcal{U}$ is considered as a Δ'-field.

For any extension $\mathcal{G}$ of $\mathfrak{F}$, a $\Delta - \mathfrak{F}$-ring has a natural structure of $\Delta - \mathcal{G}$-ring, and a $\Delta - \mathfrak{F}$-homomorphism of $\Delta - \mathfrak{F}$-rings is also a $\Delta - \mathcal{G}$-homomorphism.

3. Differential algebraic modules

Let R be a $\Delta - \mathfrak{F}$-ring. By a *left* $\Delta - \mathfrak{F}$-*module over* R we shall mean a set M equipped with a structure of left module over the ring R and a structure of $\Delta - \mathfrak{F}$-group such that: i⁰ the additive group structure of the module and the group structure of the $\Delta - \mathfrak{F}$-group are identical, and ii⁰ the external law of composition $R \times M \to M$ of the module is a $\Delta - \mathfrak{F}$-mapping of $R \times M$ into M. The definition of *right* $\Delta - \mathfrak{F}$-module over R is analogous. Of course, when R is commutative, we omit the adjective «left» or «right».

A subset of a left $\Delta - \mathfrak{F}$-module M over R that is both a submodule of the left module M over R and a $\Delta - \mathfrak{F}$-subgroup of the $\Delta - \mathfrak{F}$-group M has a natural structure of left $\Delta - \mathfrak{F}$-module over R; we shall call it a $\Delta - \mathfrak{F}$-*submodule* of M.

By a $\Delta - \mathfrak{F}$-*homomorphism* of a left $\Delta - \mathfrak{F}$-module M over R into a left $\Delta - \mathfrak{F}$-module M' over R we shall mean a mapping of M into M' that is both a homomorphism of left modules over R and a $\Delta - \mathfrak{F}$-homomorphism of $\Delta - \mathfrak{F}$-groups. The kernel resp. image of such a $\Delta - \mathfrak{F}$-homomorphism is a $\Delta - \mathfrak{F}$-submodule of M resp. M'. Conversely, if N is any $\Delta - \mathfrak{F}$-submodule of M, then the structure of left module over R of the quotient module M/N and the $\Delta - \mathfrak{F}$-group structure of the quotient $\Delta - \mathfrak{F}$-group M/N determine on the set M/N a structure of left $\Delta - \mathfrak{F}$-module over R, and the canonical mapping $\pi_{M/N} : M \to M/N$ is a $\Delta - \mathfrak{F}$-homomorphism of left $\Delta - \mathfrak{F}$-modules over R, and is a quotient in the categorical sense. (That the external law of composition $R \times (M/N) \to M/N$ is a $\Delta - \mathfrak{F}$-mapping follows from Lemma 2.)

When the $\Delta - \mathfrak{F}$-ring R is a $\Delta - \mathfrak{F}$-division ring, we call a left $\Delta - \mathfrak{F}$-module over R a *left* $\Delta - \mathfrak{F}$*-vector space over* R, and call a $\Delta - \mathfrak{F}$-submodule a $\Delta - \mathfrak{F}$-subspace.

For any natural number n, $\mathcal{U}^n$ has an obvious structure of $\Delta - \mathfrak{F}$-vector space over the $\Delta - \mathfrak{F}$-field $\mathcal{K}$ of constants of $\mathcal{U}$. If Σ is any set of homogeneous linear Δ-polinomials in $(y_1, \ldots, y_n)$ with coefficients in $\mathfrak{F}$, the set of zeros of Σ is a $\Delta - \mathfrak{F}$-subspace of the $\Delta - \mathfrak{F}$-vector space $\mathcal{U}^n$ over $\mathcal{K}$. It was shown by Cassidy [1, Prop. 11] that, conversely, every $\Delta - \mathfrak{F}$-subgroup of the $\Delta - \mathfrak{F}$-group $\mathcal{U}^n$ is the set of zeros of a set of homogeneous linear Δ-polinomials.

For any extension $\mathcal{G}$ or $\mathfrak{F}$, a left $\Delta - \mathfrak{F}$-module over a $\Delta - \mathfrak{F}$-ring has a natural structure of left $\Delta - \mathcal{G}$-module over R, and a $\Delta - \mathfrak{F}$-homomorphism of left $\Delta - \mathfrak{F}$-modules over R is also a $\Delta - \mathcal{G}$-homomorphism.

4. Differential algebraic Lie algebras

Let R be a commutative $\Delta - \mathfrak{F}$-ring. By a $\Delta - \mathfrak{F}$*-Lie algebra over* R we shall mean a set L equipped with a structure of Lie algebra over the ring R and a structure of $\Delta - \mathfrak{F}$-group such that: i⁰ the additive group structure of the Lie algebra and the group structure of the $\Delta - \mathfrak{F}$-group are identical, ii⁰ the external law of composition $R \times L \to L$ of the Lie algebra is a $\Delta - \mathfrak{F}$-mapping of $R \times L$ into L, and iii⁰ the bracket operation $L \times L \to L$ of the Lie algebra is a $\Delta - \mathfrak{F}$-mapping of $L \times L$ into L.

A subset of a $\Delta - \mathfrak{F}$-Lie algebra L over R that is both a subalgebra (resp. ideal) of the Lie algebra L over R and a $\Delta - \mathfrak{F}$-subgroup of the $\Delta - \mathfrak{F}$-group L has a natural structure of $\Delta - \mathfrak{F}$-Lie algebra over R; we shall call it a $\Delta - \mathfrak{F}$*-subalgebra* (resp. $\Delta - \mathfrak{F}$*-ideal*) of L.

By a $\Delta - \mathfrak{F}$*-homomorphism* of a $\Delta - \mathfrak{F}$-Lie algebra L over R into a $\Delta - \mathfrak{F}$-Lie algebra L' over R we shall mean a mapping of L into L' that is both a homomorphism of Lie algebras over R and a $\Delta - \mathfrak{F}$-homomorphism of $\Delta - \mathfrak{F}$-groups. The kernel resp. image od such a $\Delta - \mathfrak{F}$-homomorphism is a $\Delta - \mathfrak{F}$-ideal of L resp. $\Delta - \mathfrak{F}$-subalgebra of L'. Conversely, if $\mathfrak{l}$ is any $\Delta - \mathfrak{F}$-ideal of L, then the structure of Lie algebra over R of the quotient Lie algebra $L/\mathfrak{l}$ and the $\Delta - \mathfrak{F}$-group structure of the quotient $\Delta - \mathfrak{F}$-group $L/\mathfrak{l}$ determine on the set $L/\mathfrak{l}$ a structure of $\Delta - \mathfrak{F}$-Lie algebra over R, and the canonical mapping $\pi_{L/\mathfrak{l}}\colon L \to L/\mathfrak{l}$ is a $\Delta - \mathfrak{F}$-homomorphism of $\Delta - \mathfrak{F}$-Lie algebras over R, and is a quotient in the categorical sense. (In proving that the external law of composition $R \times (L/\mathfrak{l}) \to L/\mathfrak{l}$ and the bracket operation $(L/\mathfrak{l}) \times (L/\mathfrak{l}) \to L/\mathfrak{l}$ are $\Delta - \mathfrak{F}$-mappings, we use Lemma 2.)

For any Δ-field $\mathcal{G}$ that is an extension of $\mathfrak{F}$, a $\Delta - \mathfrak{F}$-Lie algebra over a commutative $\Delta - \mathfrak{F}$-ring R has a natural structure of $\Delta - \mathcal{G}$-Lie algebra over R, and a $\Delta - \mathfrak{F}$-homomorphism of $\Delta - \mathfrak{F}$-Lie algebras over R is also $\Delta - \mathcal{G}$-homomorphism.

5. Tangent spaces

Let V be an irreducible $\Delta - \mathfrak{F}$-set (that is, a subset of a homogeneous $\Delta - \mathfrak{F}$-space for a $\Delta - \mathfrak{F}$-group closed in the $\Delta - \mathfrak{F}$-topology).

The Δ-rational functions on V form a Δ-field that we denote by $\mathfrak{J}(V)$, and the set of elements of $\mathfrak{J}(V)$ that are defined over $\mathfrak{F}$ (the $\Delta - \mathfrak{F}$-functions on V) is a Δ-subfield of $\mathfrak{J}(V)$ that we denote by $\mathfrak{J}_{\mathfrak{F}}(V)$. For any $\alpha \in \mathcal{U}$, the everywhere defined function on V that has the value α at every element of V is an element of $\mathfrak{J}(V)$ that we identify with α; then $\mathcal{U}$ is a Δ-subfield of $\mathfrak{J}(V)$, $\mathfrak{J}_{\mathfrak{F}}(V)$ and $\mathcal{U}$ are linearly disjoint over $\mathfrak{F}$, and $\mathfrak{J}_{\mathfrak{F}}(V) \cdot \mathcal{U} = \mathfrak{J}(V)$.

Consider an element v of V. The set of elements of $\mathfrak{J}(V)$ that are defined at v is a Δ-subring of $\mathfrak{J}(V)$, and is a *local* Δ*-ring* (that is, is a Δ-ring and a local ring in which the maximal ideal is a Δ-ideal); we denote this local Δ-ring by

$\mathfrak{I}_v(V)$ and its maximal ideal by $m_v(V)$. The intersection $\mathfrak{I}_{\mathfrak{F},v}(V) = \mathfrak{I}_{\mathfrak{F}}(V) \cap \mathfrak{I}_v(V)$ is a local Δ-ring with maximal ideal

$$m_{\mathfrak{F},v}(V) = \mathfrak{I}_{\mathfrak{F},v}(V) \cap m_v(V).$$

Suppose given another irreducible $\Delta - \mathfrak{F}$-set W and a $\Delta - \mathfrak{F}$-mapping f of V into W that is defined at v. Then f induces a homomorphism

$$f_v^*: \mathfrak{I}_{f(v)}(W) \to \mathfrak{I}_v(V)$$

of Δ-algebras over $\mathcal{U}$ that maps $\mathfrak{I}_{\mathfrak{F},f(v)}(V)$ into $\mathfrak{I}_{\mathfrak{F},v}(V)$; for each $\psi \in \mathfrak{I}_{f(v)}(W)$, $f_v^*(\psi)$ is the Δ-rational function on V that assumes the value $\psi(f(v'))$ at v' whenever v' is an element of V such that f is defined at v' and ψ is defined at $f(v')$. When f is generically surjective, then f_v^* is injective; when f is generically invertible and its generic inverse is defined at $f(v)$, then f_v^* is an isomorphism.

Let $\eta_1, \ldots, \eta_n \in \mathfrak{I}_{\mathfrak{F},v}(V)$. There is a smallest subset W of $\mathcal{U}^n$ that is closed in the Δ-Zariski topology and contains the point $(\eta_1(v'), \ldots, \eta_n(v'))$ whenever v' is an element of V at which every η_j is defined; it is easy to see that W is an irreducible $\Delta - \mathfrak{F}$-subset of $\mathcal{U}^n$ and there is a generically surjective $\Delta - \mathfrak{F}$-mapping f of V into W that is defined at every such v' and has the value $(\eta_1(v'), \ldots, \eta_n(v'))$ there. If f is generically invertible and its generic inverse is defined at $f(v)$, we call $(\eta_1, \ldots, \eta_n)$ a *system of* $\Delta - \mathfrak{F}$*-affine coordinates* on V at v; then, for any extension $\mathcal{G}$ of $\mathfrak{F}$, $(\eta_1, \ldots, \eta_n)$ is also a system of $\Delta - \mathcal{G}$-affine coordinates on V at v. There exists a finitely generated extension $\mathfrak{F}'$ of $\mathfrak{F}$ such that, for every $v \in V$, there exists a system of $\Delta - \mathfrak{F}'$-affine coordinates on V at v. Also, for any finite family $\eta = (\eta_1, \ldots, \eta_n)$ of elements of $\mathfrak{I}_{\mathfrak{F}}(V)$ such that $\mathfrak{I}_{\mathfrak{F}}(V) = \mathfrak{I}\langle\eta\rangle$, the set of elements $v \in V$ at which η is a system of $\Delta - \mathfrak{F}$-affine coordinates on V is Δ-dense and $\Delta - \mathfrak{F}$-open in V; when this set in V, we call η a system of *global* $\Delta - \mathfrak{F}$-affine coordinates on V. When there exists a system of global $\Delta - \mathfrak{F}$-affine coordinates on V, we say that the $\Delta - \mathfrak{F}$-set V is $\Delta - \mathfrak{F}$*-affine.*

It is not hard to show that if $\eta = (\eta_1, \ldots, \eta_n)$ is a system of $\Delta - \mathfrak{F}$-affine coordinates on V at v, then $\mathfrak{F}\langle v\rangle = \mathfrak{F}\langle \eta_1(v), \ldots, \eta_n(v)\rangle$ and $\mathfrak{I}_{\mathfrak{F},v}(V)$ is the localization of the Δ-ring $\mathfrak{F}\{\eta\}$ at its prime Δ-ideal $m_v(V) \cap \mathfrak{F}\{\eta\}$. It then follows that in the local Δ-rings $\mathfrak{I}_v(V)$ and $\mathfrak{I}_{\mathfrak{F},v}(V)$ every perfect Δ-ideal is finitely generated (as a perfect Δ-ideal), and $\mathfrak{I}_v(V)$ is the localization of $\mathcal{U}[\mathfrak{I}_{\mathfrak{F},v}(V)]$ at its prime Δ-ideal $m_v(V) \cap \mathcal{U}[\mathfrak{I}_{\mathfrak{F},v}(V)]$.

Consider an element $v \in V$ and a Δ-subring $\mathcal{A}$ of the local Δ-ring $\mathfrak{I}_v(V)$. By a *local* Δ*-derivation* of $\mathcal{A}$ *at* v we shall mean any $(\mathcal{A} \cap \mathcal{U})$-linear mapping $T_0: \mathcal{A} \to \mathcal{U}$ such that

$$T_0(\varphi\psi) = T_0\varphi\cdot\psi(v) + \varphi(v)\,T_0\psi \quad (\varphi, \psi \in \mathcal{A}),$$

$$T_0(\delta\varphi) = \delta(T_0\varphi) \quad (\varphi \in \mathcal{A},\ \delta \in \Delta)$$

When Σ is a multiplicatively stable subset of $\mathcal{A}$ with $\Sigma \cap m_v(V) = \phi$, a local Δ-derivation of $\mathcal{A}$ at v can be extended to a unique local Δ-derivation at v of the Δ-ring of quotients $\Sigma^{-1}\mathcal{A}$.

A local Δ-derivation of $\mathfrak{I}_v(V)$ at v will be called a *tangent vector to* V *at* v. The set of all tangent vectors to V at v will be called the *tangent space to* V *at* v and will be denoted by $\mathfrak{T}_v(V)$.

The restriction to $\mathfrak{I}_{\mathfrak{F},v}(V)$ of a tangent vector to V at v is a local Δ-derivation of $\mathfrak{I}_{\mathfrak{F},v}(V)$. Conversely, *if there exists a system of* $\Delta - \mathfrak{F}$*-affine coordinates on* V *at* v, then any local Δ-derivation T_0 on $\mathfrak{I}_{\mathfrak{F},v}(V)$ can be extended to a unique tangent vector to V at v.

We are going to see that *if $\mathcal{G}$ is an extension of $\mathfrak{F}\langle v\rangle$ for which there exists a system of $\Delta - \mathcal{G}$-affine coordinates on V at v, then $\mathfrak{T}_v(V)$ has a natural structure of $\Delta - \mathcal{G}$-vector space over the $\Delta - \mathcal{G}$-field $\mathcal{K}$.*

To begin, fix a system $\eta = (\eta_1, \ldots, \eta_n)$ of $\Delta - \mathcal{G}$-affine coordinates on V at v, and let $\mathfrak{p}$ denote the defining Δ-ideal of η in the Δ-pollinomial algebra $\mathcal{G}\{y_1, \ldots, y_n\}$. For each $P \in \mathcal{G}\{y_1, \ldots, y_n\}$, put

$$P_v = \sum_{\theta\in\Theta, 1\leqslant j\leqslant n} \frac{\partial P}{\partial(\theta y_j)}(\eta_1(v), \ldots, \eta_n(v))\,\theta y_j.$$

Then P_v is a homogeneous linear Δ-polynomial in $\mathcal{G}\{y_1, \ldots, y_n\}$ and the ideal $\mathfrak{p}_v$ of $\mathcal{G}\{y_1, \ldots, y_n\}$ generated by the set of all P_v with $P \in \mathfrak{p}$ is a linear (hence prime) Δ-ideal. Hence (see Section 3), the set $\mathfrak{S}_v$ of zeros of $\mathfrak{p}_v$ is a $\Delta - \mathcal{G}$-vector space over $\mathcal{K}$. Now, $\mathfrak{T}_v(V)$ has a natural structure of vector space over $\mathcal{K}$. For any $T \in \mathfrak{T}_v(V)$ and $P \in \mathcal{G}\{y_1, \ldots, y_n\}$, we see (because T is a local derivation at v) that

$$TP(\eta_1, \ldots, \eta_n) = \sum_{\theta\in\Theta, 1\leqslant j\leqslant n} \frac{\partial P}{\partial(\theta y_j)}(\eta_1(v), \ldots, \eta_n(v))\,\theta(T\eta_j).$$

Therefore, the family $T\eta = (T\eta_1, \ldots, T\eta_n)$ is a zero of $\mathfrak{p}_v$. Thus, the formula $T \mapsto T\eta$ defines a mapping of $\mathfrak{T}_v(V)$ into $\mathfrak{S}_v$, and it is evident that this mapping is a homomorphism of vector spaces over $\mathcal{K}$. Since $\mathfrak{J}_v(V)$ is the localization of $\mathcal{U}\{\eta\}$ at $m_v(V) \cap \mathcal{U}\{\eta\}$, we see that if $T\eta = 0$, then $T = 0$, that is, the vector space homomorphism is injective. However, if $\alpha = (\alpha_1, \ldots, \alpha_n)$ is any element of $\mathfrak{S}_v$, then the mapping $\mathcal{G}\{y_1, \ldots, y_n\} \to \mathcal{U}$ defined by the formula $P \mapsto P_v(\alpha)$ is a homomorphism of Δ-vector spaces over $\mathcal{G}$ with kernel containing $\mathfrak{p}$; since tue mapping $\mathcal{G}\{y_1, \ldots, y_n\} \to \mathcal{G}\{\eta\}$ defined by the formula $P \mapsto P(\eta)$ is a homomorphism of Δ-algebras (and hence of Δ-vector spaces) over $\mathcal{G}$ with kernel $\mathfrak{p}$, we infer that there exists a homomorphism $T_0 : \mathcal{G}\{\eta\} \to \mathcal{U}$ of Δ-vector spaces over $\mathcal{G}$ such that $T_0 P(\eta) = p_v(\alpha)$. The computation

$$T_0(P(\eta)\,Q(\eta)) = (PQ)_v(\alpha) = P_v(\alpha)\,Q(\eta(v)) + P(\eta(v))\,Q_v(\alpha)$$

shows that T_0 is a local derivation of $\mathcal{G}\{\eta\}$ at v. Since evidently $T_0\eta = \alpha$, and since T_0 can be extended to a tangent vector $T \in \mathfrak{T}_v(V)$, we conclude that our mapping $\mathfrak{T}_v(V) \to \mathfrak{S}_v$ is surjective, and hence is an isomorphism of vector spaces over $\mathcal{K}$. This isomorphism permits us to identify $\mathfrak{T}_v(V)$ with $\mathfrak{S}_v$ and hence to equip $\mathfrak{T}_v(V)$ with a structure of $\Delta - \mathcal{G}$-vector space over $\mathcal{K}$. It is not hard to show that this structure is independent of the choice of the system η of $\Delta - \mathcal{G}$-affine coordinates on V at v.

For any $\varphi \in \mathfrak{J}_{\mathcal{G},v}$, the formula $T \mapsto T\varphi$ defines a mapping of $\mathfrak{T}_v(V)$ into $\mathcal{U}$ that is a $\Delta - \mathcal{G}$-homomorphism of $\Delta - \mathcal{G}$-vector spaces over $\mathcal{K}$; we shall denote it by

$$d_v\varphi : \mathfrak{T}_v(V) \to \mathcal{U}.$$

In particular $d_v\varphi \in \mathfrak{J}_{\mathcal{G},v}(\mathfrak{T}_v(V))$. We see from the above discussion that the family of $d_v\eta = (d_v\eta_1, \ldots, d_v\eta_n)$ is a system of global $\Delta - \mathcal{G}$-affine coordinates on $\mathfrak{T}_v(V)$. The defining Δ-ideal of $d_v\eta$ in $\mathcal{G}\{y_1, \ldots, y_n\}$ is $\mathfrak{p}_v$. We have proved the following theorem.

Theorem 1. *Let V be an irreducible $\Delta - \mathfrak{F}$-set, v be an element of V, and $\mathcal{G}$ be an extension of $\mathfrak{F}\langle v\rangle$ foz which there exists a system of $\Delta - \mathcal{G}$-affine coordinates on V at v. Then $\mathfrak{T}_v(V)$ has a structure of $\Delta - \mathcal{G}$-vector space over $\mathcal{K}$ with the following properties.*

(a) *For any $\varphi \in \mathfrak{J}_{\mathcal{G},v}(V)$, the mapping $d_v\varphi : \mathfrak{T}_v(V) \to \mathcal{U}$ given by the formula $T \mapsto T\varphi$ is a $\Delta - \mathcal{G}$-homomorphism of $\Delta - \mathcal{G}$-vector spaces over $\mathcal{K}$.*

(b) *For any system* $\eta = (\eta_1, \ldots, \eta_n)$ *of* $\Delta - \mathcal{G}$*-affine coordinates on* V *at* v, *the family* $d_v\eta = (d_v\eta_1, \ldots, d_v\eta_n)$ *is a system of global* $\Delta - \mathcal{G}$*-affine coordinates on* $\mathfrak{T}_v(V)$; *if* $\mathfrak{p}$ *denotes the defining* Δ*-ideal of* η *in* $\mathcal{G}\{y_1, \ldots, y_n\}$, *then* $\mathfrak{p}_v$ *is the defining* Δ*-ideal of* $d_v\eta$ *in* $\mathcal{G}\{y_1, \ldots, y_n\}$.

It is not hard to deduce the following corollary.

Corollary 1. *Let* V *be an irreducible* $\Delta - \mathfrak{F}$*-set and* $\eta = (\eta_1, \ldots, \eta_n)$ *be a finite family of* $\Delta - \mathfrak{F}$*-functions on* V *such that* $\mathfrak{F}\langle\eta\rangle = \mathfrak{J}_{\mathfrak{F}}(V)$. *Then there exists a nonempty* $\Delta - \mathfrak{F}$*-open subset* $\mathcal{O}$ *of* V *such that, for each* $v \in \mathcal{O}$, η *is a system of* $\Delta - \mathfrak{F}$*-affine coordinates on* V *at* v, $d_v\eta$ *is a system of global* $\Delta - \mathfrak{F}\langle v\rangle$*-affine coordinates on* $\mathfrak{T}_v(V)$, *and*

$$\omega_{\eta/\mathcal{U}} = \omega_{d_v\eta/\mathcal{U}}$$

In particular, for every $v \in \mathcal{O}$, $\mathfrak{T}_v(V)$ *and* V *have the same* Δ*-type and the same typical* Δ*-dimension.*

Now consider another irreducible $\Delta - \mathfrak{F}$-set W and a $\Delta - \mathfrak{F}$-mapping f of V into W that is defined at v. For any $T \in \mathfrak{T}_v(V)$, the composite mapping $T \circ f_v^*$ is evidently a tangent vector to W at $f(v)$. Therefore, we can define a mapping

$$f_v^{**} : \mathfrak{T}_v(V) \to \mathfrak{T}_{f(v)}(W)$$

by the formula

$$f_v^{**}(T) = T \circ f_v^*.$$

It is not hard to show that *if* $\mathcal{G}$ *is any extension of* $\mathfrak{F}$ *for which there exist systems of* $\Delta - \mathcal{G}$*-affine coordinates on* V *at* v *and on* W *at* $f(v)$, *then* f_v^{**} *is a* $\Delta - \mathcal{G}$*-homomorphism of* $\Delta - \mathcal{G}$*-vector spaces over* $\mathcal{K}$; *when* f *is generically invertible, then* f *is a* $\Delta - \mathcal{G}$*-isomorphism.*

This enables us, in the special case in which V is a homogeneous $\Delta - \mathfrak{F}$-space, to prove the following improvement of

Corollary 1. (In its statement, for each $x \in G$, ρ_x denotes the mapping $V \to V$ given by the formula $\rho_x v = vx$; ρ_x is an everywhere defined $\Delta - \mathfrak{F}\langle x\rangle$-mapping of V into V with inverse and generic inverse $\rho_{x^{-1}}$.)

Corollary 2. *Let* V, $\eta = (\eta_1, \ldots, \eta_n)$, *and* $\mathcal{O}$ *be as in Corollary* 1, *with* V *now a homogeneous* $\Delta - \mathfrak{F}$*-space for a connected* $\Delta - \mathfrak{F}$*-group* G. *For each* $v \in V$ *and each* $x \in G$ *with* $vx \in \mathcal{O}$, *the family* $(\rho_x)_v^*\eta = ((\rho_x)_v^*\eta_1, \ldots, (\rho_x)_v^*\eta_n)$ *is a system of* $\Delta - \mathfrak{F}\langle x\rangle$*-affine coordinates on* V *at* v, *the family* $d_v((\rho_x)_v^*\eta) = (d_v((\rho_x)_v^*\eta_1, \ldots, d_v((\rho_x)_v^*\eta_n))$ *is a system of global* $\Delta - \mathfrak{F}$*-affine coordinates on* $\mathfrak{T}_v(V)$, *and*

$$\omega_{\eta/\mathcal{U}} = \omega_{d_v((\rho_x)_v^*\eta)/\mathcal{U}}.$$

In particular, for every $v \in V$, $\mathfrak{T}_v(V)$ and V have the same Δ-type and the same typical Δ-dimension.

6. Lie algebra of a differential algebraic group

Let V be an irreducible $\Delta - \mathfrak{F}$-set.

It is easy to show that for every $\varphi\ \mathfrak{J}(V)$ there exists a smallest extension $\mathcal{G}$ of $\mathfrak{F}$ such that $\varphi \in \mathfrak{J}_{\mathcal{G}}(V)$, and $\mathcal{G}$ is finitely generated over $\mathfrak{F}$.

By a Δ-derivation on V me shall mean a $\mathcal{U}$-linear mapping D of $\mathfrak{J}(V)$ into itself such that

$$D(\varphi\psi) = D\varphi\cdot\psi + \varphi\cdot D\psi \quad (\varphi, \psi \in \mathfrak{J}(V)),$$

$$D(\delta\varphi) = \delta(D\varphi) \quad (\varphi \in \mathfrak{J}(V),\ \delta \in \Delta).$$

The set of Δ-derivations on V has a natural structure of Lie algebra over $\mathcal{K}$; we denote it by $\mathfrak{D}_\Delta(V)$.

By a $\Delta - \mathfrak{F}$-*derivation on* V we shall mean a $D \in \mathfrak{D}_\Delta(V)$ such that $D\mathfrak{J}_\mathfrak{F}(V) \in \in \mathfrak{J}_\mathfrak{F}(V)$. The set of all of these is a Lie algebra over the field $\mathfrak{C} = \mathfrak{F} \cap \mathcal{K}$ we denote it by $\mathfrak{D}_{\Delta,\mathfrak{F}}(V)$.

For each $D \in \mathfrak{D}_\Delta(V)$, there is a smallest extension $\mathcal{G}$ of $\mathfrak{F}$ such that $D \in \mathfrak{D}_{\Delta,\mathcal{G}}(V)$; this $\mathcal{G}$ is finitely generated over $\mathfrak{F}$.

Now let $\mathcal{G}$ be any extension of $\mathfrak{F}$. For any isomorphism σ of $\mathcal{G}$ over $\mathfrak{F}$, the formula $\varphi \mapsto \sigma(\varphi)$ defines an isomorfism $\mathfrak{J}_\mathcal{G}(V) \approx \mathfrak{J}_{\sigma\mathcal{G}}(V)$ of Δ-fields over $\mathfrak{J}_\mathfrak{F}(V)$. For any $D \in \mathfrak{D}_{\Delta,\mathcal{G}}(V)$, the formula $\psi \mapsto \sigma(D(\sigma^{-1}(\psi)))$ defines a Δ-derivation of $\mathfrak{J}_{\sigma\mathcal{G}}(V)$ over $\sigma\mathcal{G}$, which extends to a unique $\Delta - \sigma\mathcal{G}$-derivation $\sigma(D)$ on V. The formula $D \mapsto \sigma D$ defines an isomorphism $\mathfrak{D}_{\Delta,\mathcal{G}}(V) \approx \mathfrak{D}_{\Delta,\sigma\mathcal{G}}(V)$ of Lie algebras over $\mathfrak{C}$. It can be shown that if $\sigma(D) = D$ for every σ, then $D \in \mathfrak{D}_{\Delta,\mathfrak{F}}(V)$.

Let W be another irreducible $\Delta - \mathfrak{F}$-set and f be a generically invertible $\Delta - \mathfrak{F}$-mapping of V into W. Then f induces a Δ-field isomorphism $f^* : \mathfrak{J}(W) \approx \mathfrak{J}(V)$ over $\mathcal{U}$ that maps $\mathfrak{J}_\mathfrak{F}(W)$ onto $\mathfrak{J}_\mathfrak{F}(V)$, and the formula $D \mapsto f^{*-1} \circ D \circ f^*$ defines an isomorphism $f^{**} : \mathfrak{D}_\Delta(V) \approx \mathfrak{D}_\Delta(W)$, of Lie algebras over $\mathcal{K}$ that maps $\mathfrak{D}_{\Delta,\mathfrak{F}}(V) \approx \mathfrak{D}_{\Delta,\mathfrak{F}}(W)$. If g is a generically invertible $\Delta - \mathcal{G}$-mapping of V into W and σ is an isomorphism of $\mathcal{G}$ over $\mathfrak{F}$, then

$$\sigma(g^{**}(D)) = \sigma(g)^{**}(\sigma(D)) \quad (D \in \mathfrak{D}_{\Delta,\mathcal{G}}(V)).$$

If $D \in \mathfrak{D}_\Delta(V)$ and $v \in V$, we shall say that D is *localizable at* v, if $D\mathfrak{J}_v(V) \subset \mathfrak{J}_v(V)$. We denote the set of all $D \in \mathfrak{D}_\Delta(V)$ that are localizable at a given v by $\mathfrak{D}_{\Delta,v}(V)$; it is a Lie algebra over $\mathcal{K}$. The set of all v at which a given D is localizable is Δ-dense and Δ-open in V (and is $\Delta - \mathfrak{F}$-open when $D \in \mathfrak{D}_{\Delta,\mathfrak{F}}(V)$).

Let $v \in V$. For any $D \in \mathfrak{D}_{\Delta,v}(V)$, the formula

$$\varphi \mapsto (D\varphi)(v) \quad (\varphi \in \mathfrak{J}_v(V))$$

defines a tangent vector D_v to V at v. The formula

$$D \mapsto D_v \quad (D \in \mathfrak{D}_{\Delta,v}(V))$$

defines a $\mathcal{K}$-linear mapping

$$\mathfrak{D}_{\Delta,v}(V) \to \mathfrak{T}_v(V).$$

Its kernel consists of those D such that $D\mathfrak{J}_v(V) \subset \mathfrak{M}_v(V)$. If $D_v = 0$ for every v at which D is localizable, then $D = 0$.

When f is a generically invertible $\Delta - \mathfrak{J}$-mapping of V into W, as before, and $D \in \mathfrak{D}_{\Delta,v}(V)$, then $f^{**}(D) \in \mathfrak{D}_{\Delta,f(v)}(W)$ and we can see that $f^{**}(D)_{f(v)} = f_v^{**}(D_v)$.

We now suppose that V is a homogeneous $\Delta - \mathfrak{F}$-space for a connected $\Delta - \mathfrak{F}$-group G.

For any $x \in G$, the mapping $\rho_x : V \to V$ given by the formula $\rho_x(v) = vx$ is an invertible $\Delta - \mathfrak{F}\langle x\rangle$-mapping of V into V with inverse $\rho_{x^{-1}}$. We see from the above that

$$(\rho_x)^{**}(\mathfrak{D}_{\Delta,v}(V)) = \mathfrak{D}_{\Delta,vx}(V) \quad (v \in V,\ x \in G) \tag{1}$$

and

$$(\rho_x)_v^{**}(D_v) = (\rho_x)^{**}(D)_{vx} \quad (D \in \mathfrak{D}_{\Delta,v}(V),\ v \in V,\ x \in G). \tag{2}$$

If $\mathcal{G}$ is any extension of $\mathfrak{F}$ and σ is an isomorphism of $\mathcal{G}\langle x\rangle$ over $\mathfrak{F}$, then $\sigma(\rho_x) = \rho_{\sigma x}$. It follows that

$$\sigma((\rho_x)^*(\varphi)) = (\rho_{\sigma x})^*(\sigma(\varphi)) \quad (x \in G,\ \varphi \in \mathfrak{J}_{\mathcal{G}\langle x\rangle}(V)) \tag{3}$$

and also

$$\sigma\left((\rho_x)^{**}(D)\right)=(\rho_{\sigma x})^{**}(\sigma(D)) \quad (x\in G,\ D\in\mathfrak{D}_{\Delta,\mathcal{G}\langle x\rangle}(V)). \qquad (4)$$

A Δ-derivation D on V is said to be *invariant*, if $(\rho_x)^{**}(D)=D$ for every $x\in G$. (4) shows that when $D\in\mathfrak{D}_{\Delta,\mathcal{G}}(V)$ is invariant, then so is $\sigma(D)$ for every isomorphism σ of $\mathcal{G}$ over $\mathfrak{F}$. (1) shows that an invariant Δ-derivation is localizable at every $v\in V$. We shall denote the set of invariant Δ-derivations on V by $\mathfrak{C}_\Delta(V)$. We shall indicate below how, when V is a principal homogeneous $\Delta-\mathfrak{F}$-space for the connected $\Delta-\mathfrak{F}$-group G, then $\mathfrak{C}_\Delta(V)$ has a natural structure of $\Delta-\mathfrak{F}$-Lie algebra over the $\Delta-\mathfrak{F}$-field $\mathcal{K}$. We then shall call $\mathfrak{C}_\Delta(V)$, with this structure, the Lie algebra of V.

$\mathfrak{C}_\Delta(V)$ has an obvious structure of Lie algebra over $\mathcal{K}$ (subalgebra of $\mathfrak{D}_\Delta(V)$). To define on $\mathfrak{C}_\Delta(V)$ a structure of $\Delta-\mathfrak{F}$-group, we must define finitely generated extensions $\mathfrak{F}\langle D\rangle$ of $\mathfrak{F}$ $(D\in\mathfrak{C}_\Delta(V))$, define a pre-order $D\to D'$ on $\mathfrak{C}_\Delta(V)$ (Δ-specialization over $\mathfrak{F}$), and for all $D, D'\in\mathfrak{C}_\Delta(V)$ with $D\leftrightarrow D'$, define an isomorphism $S_{D',D}$: $\mathfrak{F}\langle D\rangle\approx\mathfrak{F}\langle D'\rangle$ over $\mathfrak{F}$, and then show that the appropriate axioms are satisfied. Since we have not described the axioms in this paper, we cannot give the details. They will apear in [6].

For any extension $\mathcal{G}$ of $\mathfrak{F}$, we put $\mathfrak{C}_{\Delta,\mathcal{G}}(V)=\mathfrak{C}_\Delta(V)\cap\mathfrak{D}_{\Delta,\mathcal{G}}(V)$; this is a Lie algebra over the field $\mathcal{K}$. Evidently $\sigma(\mathfrak{C}_{\Delta,\mathcal{G}}(V))=\mathfrak{C}_{\Delta,\sigma\mathcal{G}}(V)$ when σ is an isomorphism of $\mathcal{G}$ over $\mathfrak{F}$. It can be shown that, for $D\in\mathfrak{C}_\Delta(V)$, $\mathfrak{J}_\mathfrak{F}(V)(D\,\mathfrak{J}_\mathfrak{F}(V))\cap\mathcal{U}$ is the smallest extension $\mathcal{G}$ of $\mathfrak{F}$, such that $D\in\mathfrak{C}_{\Delta,\mathcal{G}}(V)$. We define $\mathfrak{F}\langle D\rangle$ to be this extension; it is finitely generated.

For $D, D'\in\mathfrak{C}_\Delta(V)$ we define $D\to D'$ to mean that there exists a homomorphism $\mathfrak{J}_\mathfrak{F}(V)[D\mathfrak{J}_\mathfrak{F}(V)]\to\mathfrak{J}_\mathfrak{F}(V)[D'\mathfrak{J}_\mathfrak{F}(V)]$ over $\mathfrak{J}_\mathfrak{F}(V)$ such that $D\varphi\mapsto D'\varphi$ $(\varphi\in\mathfrak{J}_\mathfrak{F}(V))$. The relation $D\to D'$ is a pre-order on $\mathfrak{C}_\Delta(V)$. When $D\leftrightarrow D'$, the homomorphism is an isomorphism; it extends to an isomorphism $\mathfrak{J}_\mathfrak{F}(V)(D\mathfrak{J}_\mathfrak{F}(V))\approx\mathfrak{J}_\mathfrak{F}(V)(D'\mathfrak{J}_\mathfrak{F}(V))$ of the Δ-fields of quotients, which, it can be shown, restricts to an isomorphism $\mathfrak{F}\langle D\rangle\approx\mathfrak{F}\langle D'\rangle$. We define $S_{D',D}$ to be this last isomorphism.

It turns out that there is not much difficulty in verifying the axioms, except for the following missing point: *There exists an element $D\in\mathfrak{C}_\Delta(V)$ such that $D\to D'$ for every element $D'\in\mathfrak{C}_\Delta(V)$ and $\mathfrak{F}$ is algebraically closed in $\mathfrak{F}\langle D\rangle$.* To establish this point, fix an element $v\in V$. The $\mathcal{K}$-linear mapping $\mathfrak{D}_{\Delta,v}(V)\to\mathfrak{T}_v(V)$ (given by the formula $D\mapsto D_v$) restricts to a $\mathcal{K}$-linear mapping $\mathfrak{C}_\Delta(V)\to\mathfrak{T}_v(V)$. It follows from eq. (2) that its kernel is 0. Much the same way as the analogous result is proved in the theory of algebraic groups, it can be shown that this mapping is surjective. Furthermore, if we fix an extension $\mathcal{G}$ of the algebraic closure of $\mathfrak{F}(v)$ such that there exists a system of $\Delta-\mathcal{G}$-affine coordinates on V at v, then

$$\mathcal{G}\cdot\mathfrak{F}\langle D\rangle=\mathcal{G}\langle D_v\rangle \qquad (D\in\mathfrak{C}_\Delta(V))$$

and (for $D, D'\in\mathfrak{C}_v(V)$) $D_v\underset{\mathcal{G}}{\to}D_v{}'$ if and only if there exists a homomorphism

$$\mathfrak{J}_\mathcal{G}(V)[D\mathfrak{J}_\mathcal{G}(V)]\to\mathfrak{J}_\mathcal{G}(V)[D'\mathfrak{J}_\mathcal{G}(V)]$$

over $\mathfrak{J}_\mathcal{G}(V)$ such that $D\varphi\mapsto D'\varphi$ $(\varphi\in\mathfrak{J}_\mathcal{G}(V))$. From the latter fact it follows that if $D_v\underset{\mathcal{G}}{\to}D_v{}'$, then $D\underset{\mathfrak{F}}{\to}D'$; since there exists an element $D_v\in\mathfrak{T}_v(V)$ such that $D_v\underset{\mathcal{G}}{\to}D_v{}'$ for every $D_v{}'\in\mathfrak{C}_\Delta(V)$, it follows that there exists an element $D\in\mathfrak{C}_\Delta(V)$ such that $D\to D'$ for every $D'\in\mathfrak{C}_v(V)$. Since $\mathfrak{T}_v(V)$ is a $\Delta-\mathcal{G}$-group, when we fix such a D, then $\mathcal{G}$ is algebraically closed in $\mathcal{G}\langle D_v\rangle$, and hence the former fact shows that $\mathcal{G}$ is algebraically closed in $\mathcal{G}\cdot\mathfrak{F}\langle D\rangle$. For any isomorphism σ of $\mathcal{G}\cdot\mathfrak{F}\langle D\rangle$ over $\mathfrak{F}$, $\sigma(D)\in\mathfrak{C}_\Delta(V)$

whence $\sigma(D)_v \in \mathfrak{T}_v(V)$, so that there exists a homomorphism

$$\mathcal{G}\mathfrak{J}_{\mathfrak{F}}(V)[D\mathfrak{J}_{\mathfrak{F}}(V)] = \mathfrak{J}_{\mathcal{G}}(V)[D\mathfrak{J}_{\mathfrak{F}}(V)] \to \mathfrak{J}_{\mathcal{G}}(V)[\sigma(D)\mathfrak{J}_{\mathfrak{F}}(V)] = \\ = \mathcal{G}\mathfrak{J}_{\mathfrak{F}}(V)[\sigma(D\mathfrak{J}_{\mathfrak{F}}(V))]$$

over $\mathcal{G}\,\mathfrak{J}_{\mathfrak{F}}(V) = \mathfrak{J}_{\mathfrak{F}}(V)$ with $D\varphi \mapsto \sigma(D)\,\varphi = \sigma(D\varphi)$ $(\varphi \in \mathfrak{J}_{\mathfrak{F}}(V))$. Letting $\mathfrak{F}^0$ denote the algebraic closure of $\mathfrak{F}$ in $\mathfrak{F}\langle D\rangle$, we see that this restricts to homomorphism

$$\mathfrak{F}^0\mathfrak{J}_{\mathfrak{F}}(V)[D\mathfrak{J}_{\mathfrak{F}}(V)] = \mathfrak{J}_{\mathfrak{F}^0}(V)[D\mathfrak{J}_{\mathfrak{F}^0}(V)] \to \mathfrak{J}_{\mathfrak{F}^0}(V)[\sigma(D)\mathfrak{J}_{\mathfrak{F}}(V)] = \\ = \mathfrak{F}^0\mathfrak{J}_{\mathfrak{F}}(V)[\sigma(D)\mathfrak{J}_{\mathfrak{F}}(V)]$$

and

$$\mathfrak{J}_{\mathfrak{F}}(V)[D\mathfrak{J}_{\mathfrak{F}}(V)] \to \mathfrak{J}_{\mathfrak{F}}(V)[\sigma(D)\mathfrak{J}_{\bar{\mathfrak{F}}}(V)].$$

But $D \leftrightarrow \sigma(D)$, and hence the last homomorphism is an isomorphism, and therefore so is the one before, which consequently must extend to an isomorphism

$$\mathfrak{J}_{\mathfrak{F}^0}(V)(D\mathfrak{J}_{\mathfrak{F}}(V)) \approx \mathfrak{J}_{\mathfrak{F}^0}(V)(\sigma(D)\mathfrak{J}_{\mathfrak{F}}(V)).$$

It coinsides with σ on $\mathfrak{F}\langle D\rangle$ and with the identity on $\mathfrak{F}^0$. This shows that σ leaves fixed each element of $\mathfrak{F}^0$, and hence that $\mathfrak{F}^0 = \mathfrak{F}$. Thus, $\mathfrak{F}$ is algebraically closed in $\mathfrak{F}\langle D\rangle$.

Thus, $\mathfrak{C}_\Delta(V)$ has a natural structure of $\Delta - \mathfrak{F}$-group, the group structure coinciding with the additive group structure of the natural Lie algebra structure of $\mathfrak{C}_\Delta(V)$. Two easy applications of Lemma 1 now show that the $\Delta - \mathfrak{F}$-group structure of $\mathfrak{C}_\Delta(V)$ and the Lie algebra structure of $\mathfrak{C}_\Delta(V)$ determine on $\mathfrak{C}_\Delta(V)$ a structure of $\Delta - \mathfrak{F}$-Lie algebra over $\mathcal{K}$.

Summarizing, we obtain the following result.

T h e o r e m 2. *Let V be a principal homogeneous $\Delta - \mathfrak{F}$-space for a connected $\Delta - \mathfrak{F}$-group. The set $\mathfrak{C}_\Delta(V)$ has a natural structure of $\Delta - \mathfrak{F}$-Lie algebra over the $\Delta - \mathfrak{F}$-field $\mathcal{K}$. For any $v \in V$ and any extension $\mathcal{G}$ of $\mathfrak{F}\langle v\rangle$ for which there exists a system of $\Delta - \mathcal{G}$-affine coordinates on V at v, the formula $D \mapsto D_v$ defines a $\Delta - \mathcal{G}$-isomorphism $\mathfrak{C}_\Delta(V) \approx \mathfrak{T}_v(V)$ of $\Delta - \mathcal{G}$-vector spaces over $\mathcal{K}$. The Δ-type and typical Δ-dimension of $\mathfrak{C}_\Delta(V)$ are equal to those of V.*

REFERENCES

1. *Cassidy P. J.* Differential algebraic groups.— Amer. J. Math., 1972, **94**, 891—954.
2. *Cassidy P. J.* The differential rational representation algebra on a linear differential group.— J. Algebra, 1975, **37**, 223—238.
3. *Cassidy P. J.* Unipotent differential algebraic groups.— In: Contriburions to Algebra. H. Bass, P. J. Cassidy, J. Kovacic (Eds). New York and London, Acad. Press (to appear).
4. *Cassidy P. J.* Differential algebraic Lie Algebras.— Trans. Amer. Math. Soc. (to appear).
5. *Kolchin E. R.* Differential algebra and algebraic groups. New York and London, Acad. Press, 1973.
6. *Kolchin E. R.* Differential algebraic groups. In preparation.
7. *Kovacic J.* Constrained cohomology.— In: Contributions to Algebra. H. Bass, P. J. Cassidy, J. Kovacic (Eds). New York and London, Acad. Press (to appear).
8. *Sit W. Y.* Differential algebraic subgroups of SL (2) and strong normality in simple extensions.— Amer. J. Math., 1975, **97**, 627—698.
9. *Sit W. Y.* Typical differential dimension of the intersection of linear differential algebraic groups.— J. Algebra, 1974, **32**, 476—487.

PACIFIC JOURNAL OF MATHEMATICS
Vol. 86, No. 1, 1980

ON UNIVERSAL EXTENSIONS OF DIFFERENTIAL FIELDS

E. R. KOLCHIN

Dedicated to Gerhard Hochschild on the occasion of his 65th birthday

The main result of this paper is the following:

THEOREM: **Let $\mathscr{U}$ be a universal extension of the differential field $\mathscr{F}$ of characteristic zero and let $\mathscr{G}$ be a strongly normal extension of $\mathscr{F}$ in $\mathscr{U}$. Then $\mathscr{U}$ is a universal extension of $\mathscr{G}$.**

Introduction. We deal with differential fields, always of characteristic zero, relative to a nonempty finite set of commuting derivation operators. By an *extension* of a differential field, we always mean a differential field extension. An extension $\mathscr{F}'$ of a differential field $\mathscr{F}$ is said to be *finitely generated* if $\mathscr{F}'$ has a finite subset Φ such that $\mathscr{F}' = \mathscr{F}\langle\Phi\rangle =$ the smallest extension of $\mathscr{F}$ in $\mathscr{F}'$ that contains Φ.

Let $\mathscr{F}$ be a differential field. Recall that an extension $\mathscr{U}$ of $\mathscr{F}$ is called *universal* if, for any finitely generated extension $\mathscr{F}_1$ of $\mathscr{F}$ in $\mathscr{U}$ and any finitely generated extension $\mathscr{G}$ of $\mathscr{F}_1$ not necessarily in $\mathscr{U}$, $\mathscr{G}$ can be embedded in $\mathscr{U}$ over $\mathscr{F}_1$, i.e., there exists an extension of $\mathscr{F}_1$ in $\mathscr{U}$ that is isomorphic (in the sense of differential fields) to $\mathscr{G}$ over $\mathscr{F}_1$. Such a universal extension of $\mathscr{F}$ always exists ([2] p. 132, Th. 2). It is not unique, but if $\mathscr{U}$ and $\mathscr{V}$ are two universal extensions of $\mathscr{F}$, then there exist universal extensions $\mathscr{U}'$ and $\mathscr{V}'$ of $\mathscr{F}$, lying in $\mathscr{U}$ and $\mathscr{V}$, respectively, such that $\mathscr{U}'$ is isomorphic to $\mathscr{V}'$ over $\mathscr{F}$ ([2] p. 135, Exerc. 7).

Let $\mathscr{U}$ be a universal extension of the differential field $\mathscr{F}$ and let $\mathscr{G}$ be an extension of $\mathscr{F}$ in $\mathscr{U}$. Under favorable conditions, $\mathscr{U}$ is then a universal extension of $\mathscr{G}$, too. For example, this is the case when $\mathscr{G}$ is finitely generated over $\mathscr{F}$ ([2] p. 133, Prop. 4), and also when $\mathscr{G}$ is algebraic over $\mathscr{F}$ ([2] p. 134, Exerc. 1). The main purpose of the present note is to point out another such favorable condition. We shall show (§1) that when $\mathscr{G}$ is a strongly normal extension of $\mathscr{F}$, in the general sense of Kovacic [4] (i.e., not necessarily finitely generated), then $\mathscr{U}$ is universal over $\mathscr{G}$. This result shows that, in the study of strongly normal extensions, it is not necessary to replace $\mathscr{U}$ by a larger universal extension of $\mathscr{F}$ (see Kovacic [4] p. 518).

Every strongly normal extension of $\mathscr{F}$ in $\mathscr{U}$ is embeddable over $\mathscr{F}$ in a constrained closure of $\mathscr{F}$ in $\mathscr{U}$ ([3] p. 162, Th. 3 or Blum [1] p. 42 (15)) and hence, in particular, is constrained over $\mathscr{F}$

([3] p. 148, Th. 1). It is tempting to conjecture that the above result generalizes to constrained extensions of $\mathscr{F}$ in $\mathscr{U}$. We shall show (§2) by a counterexample that $\mathscr{U}$ can fail to be universal over a constrained closure of $\mathscr{F}$ in $\mathscr{U}$.

1. Strongly normal extensions. Recall ([2] p. 393), for a finitely generated extension $\mathscr{G}$ of $\mathscr{F}$ in a given universal extension $\mathscr{U}$ of $\mathscr{F}$, that $\mathscr{G}$ is called strongly normal over $\mathscr{F}$ if every isomorphism σ over $\mathscr{F}$ of $\mathscr{G}$ onto an extension of $\mathscr{F}$ in $\mathscr{U}$ is strong, i.e., has the property that $\sigma c = c$ for every constant c in $\mathscr{G}$ and $\mathscr{G}\mathscr{K} = \sigma\mathscr{G} \cdot \mathscr{K}$, where $\mathscr{K}$ denotes the field of constants of $\mathscr{U}$. This definition is apparently a relative one, depending on the universal extension $\mathscr{U}$ of $\mathscr{F}$ in which $\mathscr{G}$ is embedded. It is eary to see, however, that if $\mathscr{G}$ is strongly normal over $\mathscr{F}$ relative to one $\mathscr{U}$, then $\mathscr{G}$ is strongly normal over $\mathscr{F}$ relative to every $\mathscr{U}$, so that the notion of strongly normal finitely generated extension is an absolute one. When $\mathscr{G}$ is not necessarily finitely generated over $\mathscr{F}$, $\mathscr{G}$ is said, following Kovacic [4] p. 518, to be strongly normal over $\mathscr{F}$ if $\mathscr{G}$ is the union of strongly normal finitely generated extensions. Hence, also this more general notion is absolute.

It follows from [2] pp. 402-403, Th. 5, and the definition that if $\mathscr{G}$ is any strongly normal extension of $\mathscr{F}$ and $\mathscr{E}$ is any extension of $\mathscr{F}$, both contained in an extension of $\mathscr{F}$ having the same field of constants as $\mathscr{F}$, then $\mathscr{G}\mathscr{E}$ is a strongly normal extension of $\mathscr{E}$, and $\mathscr{G}$ and $\mathscr{E}$ are linearly disjoint over $\mathscr{G} \cap \mathscr{E}$.

We now prove the main theorem of this paper which was stated in the opening paragraph.

Proof. (a) We must show that if $\mathscr{G}_1$ is a finitely generated extension of $\mathscr{G}$ in $\mathscr{U}$ and $\mathscr{H}$ is any finitely generated extension of $\mathscr{G}_1$ not necessarily in $\mathscr{U}$, then there exists an embedding $\mathscr{H} \to \mathscr{U}$ over $\mathscr{G}_1$. As before, denote the field of constants of $\mathscr{U}$ by $\mathscr{K}$, and put $\mathscr{C} = \mathscr{F} \cap \mathscr{K}$, $\mathscr{C}_1 = \mathscr{G}_1 \cap \mathscr{K}$. Then $\mathscr{C} = \mathscr{G} \cap \mathscr{K}$ ([2] p. 393, Prop. 9), $\mathscr{C}_1$ is a finitely generated field extension of $\mathscr{C}$ ([2] p. 113, Cor. 1 to Prop. 14), $\mathscr{U}$ is a universal extension of $\mathscr{F}\mathscr{C}_1$, and $\mathscr{G}\mathscr{C}_1$ is a strongly normal extension of $\mathscr{F}\mathscr{C}_1$ ([2] p. 396, Th. 2). Thus, we may replace $(\mathscr{F}, \mathscr{G}, \mathscr{G}_1, \mathscr{H})$ by $(\mathscr{F}\mathscr{C}_1, \mathscr{G}\mathscr{C}_1, \mathscr{G}_1, \mathscr{H})$, i.e., we may suppose that $\mathscr{F}$, $\mathscr{G}$, $\mathscr{G}_1$ have the same field of constants $\mathscr{C}$.

(b) That being the case, fix a finite family β of generators of $\mathscr{G}_1$ over $\mathscr{G}$. Then $\mathscr{U}$ is a universal extension of $\mathscr{F}\langle\beta\rangle$ and $\mathscr{G}_1 = \mathscr{G}\mathscr{F}\langle\beta\rangle$ is a strongly normal extension of $\mathscr{F}\langle\beta\rangle$. Thus, we may replace $(\mathscr{F}, \mathscr{G}, \mathscr{G}_1, \mathscr{H})$ by $(\mathscr{F}\langle\beta\rangle, \mathscr{G}_1, \mathscr{G}_1, \mathscr{H})$, i.e., we may suppose that $\mathscr{G}_1 = \mathscr{G}$.

(c) That being the case, let $\mathscr{D}$ denote the field of constants of $\mathscr{H}$. Then $\mathscr{D}$ is a finitely generated field extension of $\mathscr{C}$, so that there exists an isomorphism $\mathscr{D} \approx \mathscr{D}'$ over $\mathscr{C}$ with $\mathscr{D}'$ a field extension of $\mathscr{C}$ in $\mathscr{K}$. Because $\mathscr{G}$ and $\mathscr{D}$ are linearly disjoint over $\mathscr{C}$ ([2] p. 87, Cor. 1 to Th. 1), and likewise $\mathscr{G}$ and $\mathscr{D}'$, this can be extended to an isomorphism $\mathscr{G}\mathscr{D} \approx \mathscr{G}\mathscr{D}'$ over $\mathscr{G}$. This can in turn be extended to an isomorphism $\mathscr{H} \approx \mathscr{H}'$, where $\mathscr{H}'$ is a finitely generated extension of $\mathscr{G}\mathscr{D}'$ not necessarily in $\mathscr{U}$. Now, $\mathscr{U}$ is a universal extension of $\mathscr{F}\mathscr{D}'$, $\mathscr{G}\mathscr{D}'$ is a strongly normal extension of $\mathscr{F}\mathscr{D}'$ in $\mathscr{U}$, and $\mathscr{H}'$ is a finitely generated extension of $\mathscr{G}\mathscr{D}'$ with field of constants $\mathscr{D}'$. An embedding $\mathscr{H}' \to \mathscr{U}$ over $\mathscr{G}\mathscr{D}'$ would, when composed with the isomorphism $\mathscr{H} \approx \mathscr{H}'$ over $\mathscr{G}$, yield an embedding $\mathscr{H} \to \mathscr{U}$ over $\mathscr{G}$. Thus, we may replace $(\mathscr{F}, \mathscr{G}, \mathscr{H})$ by $(\mathscr{F}\mathscr{D}', \mathscr{G}\mathscr{D}', \mathscr{H}')$, i.e., we may suppose that the field of constants of $\mathscr{H}$ is $\mathscr{C}$.

(d) That being the case, fix a finite family α of generators of the extension $\mathscr{H}$ of $\mathscr{G}$, and put $\mathscr{E} = \mathscr{F}\langle\alpha\rangle$. Then $\mathscr{G} \cap \mathscr{E}$ is a finitely generated extension of $\mathscr{F}$ ([2] p. 112, Prop. 14), so that $\mathscr{U}$ is universal over $\mathscr{G} \cap \mathscr{E}$. Thus, we may replace $(\mathscr{F}, \mathscr{G}, \mathscr{H}, \mathscr{E})$ by $(\mathscr{G} \cap \mathscr{E}, \mathscr{G}, \mathscr{H}, \mathscr{E})$, i.e., we may suppose that $\mathscr{G} \cap \mathscr{E} = \mathscr{F}$. Since $\mathscr{G}$ is strongly normal over $\mathscr{F}$, then the differential field $\mathscr{H} = \mathscr{G}\mathscr{E}$ is strongly normal over $\mathscr{E}$ and $\mathscr{G}$ and $\mathscr{E}$ are linearly disjoint over $\mathscr{F}$.

(e) Because $\mathscr{U}$ is universal over $\mathscr{F}$, there exists an isomorphism $\mathscr{E} \approx \mathscr{E}_0$ over $\mathscr{F}$ with $\mathscr{E}_0$ an extension of $\mathscr{F}$ in $\mathscr{U}$, and this isomorphism can be extended to an isomorphism σ: $\mathscr{H} \approx \mathscr{H}_0$, where $\mathscr{H}_0$ is an extension of $\mathscr{F}$ (and of $\mathscr{E}_0$) not necessarily in $\mathscr{U}$. Put $\mathscr{G}_0 = \sigma\mathscr{G}$. Then $\mathscr{H}_0 = \mathscr{G}_0\mathscr{E}_0$, this differential field is a strongly normal extension of $\mathscr{E}_0$, and $\mathscr{G}_0$ and $\mathscr{E}_0$ are linearly disjoint over $\mathscr{F}$. Evidently $\mathscr{U}$ is universal over $\mathscr{E}_0$ (because $\mathscr{E}_0$ is finitely generated over $\mathscr{F}$), and hence the strongly normal extension $\mathscr{G}_0\mathscr{E}_0$ of $\mathscr{E}_0$ can be embedded in $\mathscr{U}$ over $\mathscr{E}_0$, i.e., there exists an isomorphism σ_0: $\mathscr{G}_0\mathscr{E}_0 \approx \mathscr{G}_2\mathscr{E}_0$ over $\mathscr{E}_0$ with $\sigma_0\mathscr{G}_0 = \mathscr{G}_2 \subset \mathscr{U}$. The field of constants of $\mathscr{G}_2\mathscr{E}_0$, like those of $\mathscr{H}_0 = \mathscr{G}_0\mathscr{E}_0$ and $\mathscr{H} = \mathscr{G}\mathscr{E}$, is $\mathscr{C}$, and hence $\mathscr{G}_2\mathscr{E}_0$ and $\mathscr{K}$ are linearly disjoint cover $\mathscr{C}$. Therefore $\mathscr{G}_2\mathscr{E}_0$ and $\mathscr{G}_2\mathscr{K}$ are linearly disjoint over $\mathscr{G}_2$. But by (d), $\mathscr{E}$ and $\mathscr{G}$ are linearly disjoint over $\mathscr{F}$, so that $\mathscr{E}_0$ and $\mathscr{G}_0$ are, too, and hence also $\mathscr{E}_0$ and $\mathscr{G}_2$. Therefore $\mathscr{E}_0$ and $\mathscr{G}_2\mathscr{K}$ are linearly disjoint over $\mathscr{F}$. But $\mathscr{G}$ is strongly normal over $\mathscr{F}$, so that $\mathscr{G} \subset \sigma_0\sigma\mathscr{G} \cdot \mathscr{K} = \mathscr{G}_2\mathscr{K}$. Hence $\mathscr{E}_0$ and $\mathscr{G}$ are linearly disjoint over $\mathscr{F}$. Therefore, $id_{\mathscr{E}_0}$ and the isomorphism $\mathscr{G}_2 \approx \mathscr{G}$ (restriction of $(\sigma_0 \circ \sigma)^{-1}$) extend to an isomorphism τ: $\mathscr{G}_2\mathscr{E}_0 \approx \mathscr{G}\mathscr{E}_0$. The composite isomorphism $\tau \circ \sigma_0 \circ \sigma$ is an embedding of $\mathscr{H}$ into $\mathscr{U}$ over $\mathscr{G}$.

2. A counterexample for constrained extensions. Recall that an extension $\mathscr{G}$ of a differential field is said to be *constrained* ([3] p. 144) if every finite family of elements of $\mathscr{G}$ is constrained over $\mathscr{F}$ in the sense of [2] p. 142, that a differential field is said to be *constrainedly closed* ([3] p. 145) if it has no constrained extension other than itself, and that $\mathscr{G}$ is said to be a *constrained closure* of $\mathscr{F}$ ([3] p. 147) if $\mathscr{G}$ is constrainedly closed and is embeddable over closed $\mathscr{F}$ in every constrainedly extension of $\mathscr{F}$. A constrained closure of $\mathscr{F}$ always exists, and it is a constrained extension of $\mathscr{F}$.

We are going to exhibit an ordinary differential field $\mathscr{F}$, a universal extension $\mathscr{U}$ of $\mathscr{F}$, and an extension $\mathscr{G}$ of $\mathscr{F}$ in $\mathscr{U}$ such that $\mathscr{G}$ is a constrained closure of $\mathscr{F}$ and $\mathscr{U}$ is not universal over $\mathscr{G}$.

Let $\mathscr{C}$ be any denumerable field of characteristic zero and put $\mathscr{F} = \mathscr{C}(x) =$ the field of rational fractions over $\mathscr{C}$ in an indeterminate x; $\mathscr{F}$ has a unique structure of ordinary differential field with field of constants $\mathscr{C}$ in which the derivative of x is 1. By [3] p. 149, Prop. 4, we may fix a denumerable universal extension $\mathscr{U}$ of $\mathscr{F}$. By [3] p. 146, Cor. 1 to Prop. 3, $\mathscr{U}$ is constrainedly closed.

The set of solutions in $\mathscr{U}$ different from 0 and 1 of the differential equation

$$y' = y^3 - y^2$$

is denumerable and hence can be arranged in a sequence

$$\eta_0, \eta_1, \eta_2, \cdots .$$

By [3] §8, this set is infinite and is an independent set of conjugates over $\mathscr{F}$, and $\mathscr{F}\langle\eta_0, \eta_1, \eta_2, \cdots\rangle$ is constrained over $\mathscr{F}$ (see [3] p. 144, Prop. 1). Because $\mathscr{U}$ is constrainedly closed, $\mathscr{F}\langle\eta_0, \eta_1, \eta_2, \cdots\rangle$ has a constrained closure $\mathscr{G}$ in $\mathscr{U}$. The differential ideal $[y' - y^3 + y^2]$ of the differential polynomial algebra $\mathscr{G}\{y\}$ is evidently prime and does not have a generic zero in $\mathscr{U}$ (because all its zeros in $\mathscr{U}$ are in $\mathscr{G}$). Therefore, $\mathscr{U}$ is not universal over $\mathscr{G}$. (The same argument shows that $\mathscr{U}$ is even not universal over $\mathscr{F}\langle\eta_0, \eta_1, \eta_2, \cdots\rangle$.) We are going to show that $\mathscr{G}$ is a constrained closure of $\mathscr{F}$.

By [3] p. 144, Prop. 2(a), $\mathscr{G}$ is constrained over $\mathscr{F}$. Let $\mathscr{H}$ be any denumerable constrained closure of $\mathscr{F}$ (e.g., any constrained closure of $\mathscr{F}$ in $\mathscr{U}$). The set of solutions in $\mathscr{H}$ of the above differential equation can be arranged in a sequence

$$\zeta_0, \zeta_1, \zeta_2, \cdots .$$

As before, this set is infinite and is an independent set of conjugates over $\mathscr{F}$. Therefore, there exists an isomorphism

$$\varphi: \mathscr{F}\langle \eta_0, \eta_1, \eta_2, \cdots \rangle \approx \mathscr{F}\langle \zeta_0, \zeta_1, \zeta_2, \cdots \rangle .$$

Now, $\mathscr{F}\langle \zeta_0, \zeta_1, \zeta_2, \cdots \rangle$ is normal over $\mathscr{F}$ in $\mathscr{H}$ (see [3] §6 p. 153). Hence, by [3] p. 159, Cor. 1 to Th. 2, $\mathscr{H}$ is a constrained closure of $\mathscr{F}\langle \zeta_0, \zeta_1, \zeta_2, \cdots \rangle$. Therefore, by [3] p. 158, Th. 2(b), φ can be extended to an isomorphism $\mathscr{G} \approx \mathscr{H}$, so that $\mathscr{G}$ is a constrained closure of $\mathscr{F}$.

REFERENCES

1. Lenore Blum, *Differentially closed fields: a model-theoretic tour, Contributions to Algebra,* Academic Press, New York, 1977, pp. 37–61.
2. E. R. Kolchin, *Differential Algebra and Algebraic Groups,* Academic Press, New York, 1973.
3. ———, *Constrained extensions of differential fields,* Advances in Math., **12** (1974), 141–170.
4. Jerald Kovacic, *Pro-algebraic groups and the Galois theory of differential fields,* Amer. J. Math., **95** (1973), 507–536.

Received March 13, 1978. Research supported by a grant from the National Science Foundation.

COLUMBIA UNIVERSITY
NEW YORK, NY 10027

DIFFERENTIAL ALGEBRAIC GROUPS

E.R. Kolchin
Columbia University

The lectures will attempt to describe the general theory of differential algebraic groups that has been developed in recent years in analogy with and as a generalization of the older theory of algebraic groups. Limitations of time make it necessary to omit a number of topics and to give broad descriptions instead of proofs.

I have given Professor Tuan at least one copy of each of several references, and I hope that these can be consulted by anyone interested in any of the details. The main reference is the manuscript of my forth-coming book [8]; two earlier works that may prove helpful on occasion are my 1973 book [6] and my paper [7]. Also included are reprints of five papers by P.J. Cassidy [1-5], one paper by J. Kovacic [9], and one paper by W.Y. Sit [10], all bearing on results in the theory that I shall not have time to describe.

These lectures are intended as a selective survey of the subject.

Algebraic groups

An algebraic group is a group, defined by a system of algebraic equations, such that the group law and group symmetry (inverse) are rational mappings. Thus, the additive group G_a, the multiplicative group G_m, the general linear group $GL(n)$, the special linear group $SL(n)$, and the upper triangular group $T(n)$ all are algebraic groups. So, too, is the elliptic curve $W(g_2, g_3)$ in the projective plane, given by the equation

$$y_0 y_2^2 - (y_1^3 - g_2 y_0^2 y_1) - g_3 y_0^3) = 0,$$

where g_2, g_3 are constants with $g_2^3 - 27g_3^2 \neq 0$. In general, when the coefficients in the algebraic equations and in the rational mappings are in a given field K, we say that the algebraic group is defined over K, or is a K-group.

Examples of differential algebraic groups.

Algebraic equations form a special case of algebraic differential equations. Therefore it is natural to try to generalize the idea of algebraic group to one of differential algebraic group. Of course, whatever defintion we use, every algebraic group should be a differential algebraic group. More generally, however, if G is one of the algebraic groups mentioned above, and Σ is a system of algebraic differential equations in the coordinates of the elements of G having the property

$$a,b \in G \text{ and } a,b \text{ are solutions of } \Sigma$$
$$\Longrightarrow ab,\ a^{-1} \text{ are solutions of } \Sigma,$$

then the set of elements of G that are solutions of Σ should be a differential algebraic group. Thus, for example, if Λ is a linear differential operator, the set L_Λ of solutions of the linear differential equation $\Lambda y = 0$ should be a differential algebraic subgroup of the additive group G_a.

The differential algebraic setting

Before we go further, the setting for our differential equations must be made clear. In what follows, all rings are commutative, with 1, and all fields are of characteristic 0.

Suppose given a ring R and a finite set Δ that operates on R as a commuting family of derivation operators:

$$\delta(a+b) = \delta a + \delta b,\ \delta(ab) = \delta a\cdot b + a\delta b,\ \delta\delta' a = \delta'\delta a$$
$$(a,b \in R,\quad \delta,\delta' \in \Delta).$$

Then we say that R is a differential ring (relative to Δ), or that R is a Δ-ring. When the ring is a field, we have a Δ-field. (In general, we use the prefix Δ- as a synonym for "differential" or "differentially".) We put $m = \mathrm{Card}\Delta$ and denote the elements of Δ by $\delta_1,\ldots,\delta_m$. The free commutative monoid Θ generated by Δ, the elements of which are the expressions $\delta_1^{e_1}\cdots\delta_m^{e_m}$ $(e_1 \in \mathbb{N},\ldots,e_m \in \mathbb{N})$, operates on any Δ-ring in an obvious way. The definitions of Δ-subring and homomorphism of Δ-rings are as expected.

If R is a Δ-subring of a Δ-ring R', and $s = (s_j)_{j \in J}$ is a family of elements of R', there is a unique smallest Δ-subring of R' containing the elements of R and every s_j; this is the Δ-ring generated by s over R, denoted by $R\{s\}$ (or $R\{s\}_\Delta$ if necessary). As a ring, $R\{s\}$ coincides with $R[(\theta s_j)_{\theta \in \Theta, j \in J}]$. If F is a Δ-subfield of a Δ-field F', and s is now a family of elements of F', the Δ-field generated by s over F is defined in a similar way; it is denoted by $F\langle s\rangle$ and is called, also, the extension of F generated by s. We say that s is a Δ-algebraically dependent over R if the family $\Theta s = (\theta s_j)_{\theta \in \Theta, j \in J}$ is algebraically dependent over R; in the contrary case, we say that s is Δ-algebraically independent over R or that s is a family of Δ-interdeterminates over R. For any set J, there exists a family $(y_j)_{j \in J}$ of Δ-indeterminates; the elements of $R\{(y_j)_{j \in J}\}$ are called Δ-polynomials in $(y_j)_{j \in J}$ over R.

Let G be an extension of the Δ-field F and $(y_1,\ldots,y_n)$ be a finite family of Δ-indeterminates. If $u = (u_1,\ldots,u_n)$ is a family of elements of G, there exists a unique homomorphism

$$F\{y_1,\ldots,y_n\} \longrightarrow G$$

over F with $y_j \to u_j (1 \leqslant j \leqslant n)$; for each Δ-polynomial $P \in F\{y_1,\ldots,y_n\}$, the image of P is denoted by $P(u)$. When $P(u) = 0$, we say that P vanishes at u; the kernel $\mathfrak{p}$ of the homomorphism, which is called the defining Δ-ideal of u over F, is a prime Δ-ideal of $F\{y_1,\ldots,y_n\}$. If Σ is a subset of $\mathfrak{p}$, we say that u is a solution of the system of Δ-equations

$$P = 0 \, (P \in \Sigma)$$

or that u is a zero of Σ. If also $u' = (u'_1,\ldots,u'_n)$ is a family of elements of an extension of F, and if $\mathfrak{p}'$ is its defining Δ-ideal over F, the following two conditions are equivalent to each other:

1) $\mathfrak{p} \subset \mathfrak{p}'$;

2) there exists a surjective homomorphism $\mathcal{F}\{u\} \to \mathcal{F}\{u'\}$ over $\mathcal{F}$ with $u_j \mapsto u'_j$ $(1 \leq j \leq n)$.

When these conditions are satisfied we say that u' is a Δ-specialization of u over $\mathcal{F}$ and we write $u \to u'$ (or $u \xrightarrow[\mathcal{F}]{} u'$ or $u \xrightarrow{\Delta} u'$ or $u \xrightarrow[\mathcal{F}]{\Delta} u'$, when necessary). When $\mathfrak{p} = \mathfrak{p}'$, the above homomorphism is an isomorphism and hence extends to an isomorphism $\mathcal{F}\langle u\rangle \longrightarrow \mathcal{F}\langle u'\rangle$, we then say that the Δ-specialization $u \longrightarrow u'$ is generic and write $u \longleftrightarrow u'$.

It will be convenient, in discussing Δ-equations over $\mathcal{F}$, to have an extension $\mathcal{U}$ of $\mathcal{F}$ so large that the solutions can always be taken in $\mathcal{U}$. More precisely, we require that for any finitely generated extension $\mathcal{F}'$ of $\mathcal{F}$ in $\mathcal{U}$ and any finitely generated extensions $\mathcal{G}$ of $\mathcal{F}'$ whatsoever, there exist an embedding of $\mathcal{G}$ into $\mathcal{U}$ over $\mathcal{F}'$. Such an extension $\mathcal{U}$ of $\mathcal{F}$ is said to be universal (over $\mathcal{F}$). That $\mathcal{F}$ always has a universal extension is a grand exercise in Zorn's lemma [6].

In what follows, we fix a universal extension $\mathcal{U}$ of $\mathcal{F}$. All Δ-fields that we discuss, except those for which the contrary is stated or is obvious, will be Δ-subfields of $\mathcal{U}$ over which $\mathcal{U}$ is universal. The field of constants of $\mathcal{U}$ (the set of elements $u \in \mathcal{U}$ such that $\delta_i u = 0 (1 \leq i \leq m)$) will be denoted by $\mathcal{K}$. It is obvious that $\mathcal{U}$ is not universal over $\mathcal{K}$.

Δ- $\mathcal{F}$-groups

If G is any one of the Δ-algebraic groups described above, or is any other "concrete" Δ-algebraic group defined over the Δ-field $\mathcal{F}$, we have the following "Δ- $\mathcal{F}$-data" on G:

for each $x \in G$, the extension $\mathcal{F}\langle x\rangle$;

the binary relation $x \to x'$ on G of Δ-specialization over $\mathcal{F}$;

for each $(x,x') \in G^2$ with $x \leftrightarrow x'$, the isomorphism

$$S_{x',x}: \mathcal{F}\langle x\rangle \longrightarrow \mathcal{F}\langle x'\rangle.$$

The group G, the extensions $\mathcal{F}\langle x\rangle$, the relation $x \to x'$, and the isomorphisms $S_{x',x}$ have certain formal properties. We use these formal properties as axioms for an abstract definition of Δ-$\mathcal{F}$-group.

This definition carries with it natural definitions of Δ-$\mathcal{F}$-subgroup and Δ-$\mathcal{F}$-homomorphism of Δ-$\mathcal{F}$-groups. The latter is a mapping $f: G \to G'$ between Δ-$\mathcal{F}$-groups that is a group homomorphism of their underlying groups and satisfies the conditions

$$\mathcal{F}\langle x\rangle \supset \mathcal{F}\langle f(x)\rangle,$$
$$x \to x' \implies f(x) \to f(x'),$$
$$x \leftrightarrow x' \implies S_{x',x} \text{ extends } S_{f(x'),f(x)}.$$

Homogeneous Δ-$\mathcal{F}$-spaces

For any group G, there is the notion of homogeneous space for G; this is a set M together with a transitive action of G on M, that is, a mapping $M \times G \to M$, for which we usually use the notation $(v,x) \mapsto vx$, that satisfies the identities

$$v(x_1x_2) = (vx_1)x_2, \quad v1 = v, \quad vG = M.$$

When G is a Δ-$\mathcal{F}$-group, we define, by axioms analogous to some of the axioms for Δ-$\mathcal{F}$-group, a companion notion of homogeneous Δ-$\mathcal{F}$-space for G and a corresponding notion of Δ-$\mathcal{F}$-homomorphism of homogeneous Δ-$\mathcal{F}$-spaces for G. When the action of G on the homogeneous Δ-$\mathcal{F}$-space M for G is simply transitive and some further axioms are satisfied, M is said to be a principal homogeneous Δ-$\mathcal{F}$-space for G. G itself is a principal homogeneous Δ-$\mathcal{F}$-space for G. An interesting problem, addressed later, is that of classifying, up to Δ-$\mathcal{F}$-isomorphism, the principal homogeneous Δ-$\mathcal{F}$-spaces for a given Δ-$\mathcal{F}$-group G.

$\mathcal{F}$-components

A subset V of a homogeneous Δ-$\mathcal{F}$-space M consisting of all the Δ-specialization over $\mathcal{F}$ of a fixed $v \in M$ is said to be $\mathcal{F}$-irreducible, and v is said to be an $\mathcal{F}$-generic element of V. The union of finitely many $\mathcal{F}$-irreducible subsets of M is said to be a Δ-$\mathcal{F}$-set; when none of these $\mathcal{F}$-irreducible sets is contained in another, they are unique, being the maximal $\mathcal{F}$-irreducible subsets of the Δ-$\mathcal{F}$-set (called its $\mathcal{F}$-components). One of the early results is that M itself is an $\mathcal{F}$-set of which the $\mathcal{F}$-components are pairwise disjoint. Since M can be G, it follows that G has a unique $\mathcal{F}$-component, which we denote by $G°$, that contains the neutral element 1 of G. This $G°$ is a normal Δ-$\mathcal{F}$-subgroup of G of finite index.

Varying $\mathcal{U}$ or $\mathcal{F}$ or Δ

In building the general theory there are a huge number of constructions to make and propositions to prove. Some of these are obvious for concrete Δ-$\mathcal{F}$-groups, but require substantial effort in general.

For example, consider an extension $\mathcal{U}_0$ of $\mathcal{F}$ in $\mathcal{U}$, over which $\mathcal{U}$ need not be universal, such that $\mathcal{U}_0$ is universal over $\mathcal{F}$. For any Δ-$\mathcal{F}$-group G relative to $\mathcal{U}$, abstract or concrete, the set $G_{\mathcal{U}_0}$, consisting of the elements $x \in G$ with $\mathcal{F}\langle x\rangle \subset \mathcal{U}_0$, has an easily described natural structure of Δ-$\mathcal{F}$-group relative to $\mathcal{U}_0$. Conversely, if G_0 is a Δ-$\mathcal{F}$-group relative to $\mathcal{U}_0$, we can associate to G_0, in a canonical way, a Δ-$\mathcal{F}$-group G relative to $\mathcal{U}$ such that $G_{\mathcal{U}_0} = G_0$. In the concrete case this is evident, but in general, for axiomatically defined Δ-$\mathcal{F}$-groups, this must be established.

Similarly, consider an extension $\mathcal{G}$ of $\mathcal{F}$. A concrete Δ-$\mathcal{F}$-group is obviously a Δ-$\mathcal{G}$-group (every algebraic Δ-equation that has coefficients in $\mathcal{F}$ has coefficients in $\mathcal{G}$). For an abstract Δ-$\mathcal{F}$-group this must be proved.

Finally, consider the set $\mathcal{F}\Delta$ consisting of the derivation operators $\Sigma a_i\delta_i$ with $a_1,\ldots,a_m \in \mathcal{F}$; this has obvious structures of vector space over $\mathcal{F}$ and of Lie ring. Let Δ' be a linearly independent subset of $\mathcal{F}\Delta$ the elements of which commute with each other. Every Δ-field may be regarded as a Δ'-field, and it is easy to see that, as Δ'-fields, $\mathcal{U}$ is universal over $\mathcal{F}$. A concrete Δ'-$\mathcal{F}$-group (relative to $\mathcal{U}$) is obviously a Δ-$\mathcal{F}$-group. For abstract Δ'-$\mathcal{F}$-groups the result is still true, but it is far from obvious. (In the extreme case in which Δ' is empty, the result states that every $\mathcal{F}$-group is a Δ-$\mathcal{F}$-group.)

The proof, for each of these results, consists in specifying the basic data (the extensions, the binary relation, the isomorphisms), verifying the axioms, and showing that the new structure bears the desired relation to the given structure. This can be a long sometimes boring process, but is always elementary. Of course, all these results extend to homogeneous Δ-$\mathcal{F}$-spaces.

Quotients and products

Some results that are difficult in the concrete case benefit from the axiomatic approach. For example, if H is a normal Δ-$\mathcal{F}$-subgroup of the Δ-$\mathcal{F}$-group G, how do we make G/H a Δ-$\mathcal{F}$-group? The method in broad terms is the same: specify the Δ-$\mathcal{F}$-data on the group G/H, verify the axioms, and show that the canonical group homomorphism $\pi: G \longrightarrow G/H$ is a Δ-$\mathcal{F}$-homomorphism with the correct universal mapping property. When H is not normal, the method exhibits G/H as a homogeneous Δ-$\mathcal{F}$-space for G. Similarly, if G_1 and G_2 are Δ-$\mathcal{F}$-groups, we can define on $G_1 \times G_2$ a structure of Δ-$\mathcal{F}$-group that deserves to be called the direct product of the Δ-$\mathcal{F}$-groups, and when M_i is a homogeneous Δ-$\mathcal{F}$-space for G_i $(i = 1,2)$, we can define on $M_1 \times M_2$ a structure of homogeneous Δ-$\mathcal{F}$-space for $G_1 \times G_2$.

The Δ-Zariski topologies

As in algebraic geometry, it is possible and useful to introduce into our subject the language of topology. Let M be a homogeneous Δ-$\mathcal{F}$-space for a Δ-$\mathcal{F}$-group. As stated above, for any extension $\mathcal{G}$ of $\mathcal{F}$, M is a homogeneous Δ-$\mathcal{G}$-space, too, and therefore it makes sense to speak of Δ-$\mathcal{G}$-subsets of M. If $\mathcal{H}$ is an extension of $\mathcal{F}$ in $\mathcal{U}$ over which $\mathcal{U}$ is not universal, we define a Δ-$\mathcal{H}$-subset of M to be a Δ-$\mathcal{G}$-subset of M for some extension $\mathcal{G}$ of $\mathcal{F}$ with $\mathcal{G} \subset \mathcal{H}$. It turns out that the Δ-$\mathcal{H}$-subsets of M are the closed sets of a topology on M; we call it the Δ-Zariski topology relative to $\mathcal{H}$ (or just the Δ-$\mathcal{H}$-topology) on M. The finest of these topologies is the Δ-$\mathcal{U}$-topology which we usually call the Δ-Zariski topology (or Δ-topology). As in algebraic geometry, the Δ-topology is Noetherian, that is, every strictly decreasing sequence of Δ-closed subsets of M is finite. This is, at bottom, a consequence of the Ritt-Raudenbush basis theorem, according to which the perfect Δ-ideals of a Δ-polynomial ring $\mathcal{H}\{y_1,\ldots,y_n\}$ satisfy the ascending chain condition.

Δ-rational mappings

Further development of the theory requires a good definition of Δ-$\mathcal{F}$-mapping of a Δ-$\mathcal{F}$-set A into a Δ-$\mathcal{F}$-set B. In the concrete case and with A $\mathcal{F}$-irreducible this means a mapping f of a nonempty Δ-$\mathcal{F}$-open subset of A into B such that, for any $\mathcal{F}$-generic element x of A, the coordinates of $f(x)$ are expressible rationally over $\mathcal{F}$ in terms of the coordinates of x and their derivatives of various orders, f being defined at an element x' of A if the rational expressions can be chosen with the denominators not vanishing at x'. In general, we use a fairly complicated definition in terms of the Δ-$\mathcal{F}_a$-data of the homogeneous Δ-$\mathcal{F}$-spaces containing A and B and of their respective Δ-$\mathcal{F}$-groups (where $\mathcal{F}_a$ denotes the algebraic closure of $\mathcal{F}$).

For any extension $\mathcal{G}$ of $\mathcal{F}$, it turns out that every Δ-$\mathcal{F}$-mapping of A into B is a Δ-$\mathcal{G}$-mapping. If $\mathcal{H}$ is an extension of $\mathcal{F}$ in $\mathcal{U}$ with $\mathcal{U}$ not universal over $\mathcal{H}$, we define a Δ-$\mathcal{H}$-mapping of A into B to be a Δ-$\mathcal{G}$-mapping for some extension $\mathcal{G}$ of $\mathcal{F}$ with $\mathcal{G} \subset \mathcal{H}$; such a $\mathcal{G}$ can always be taken finitely generated over $\mathcal{F}$. Every Δ-$\mathcal{F}$-homomorphism of Δ-$\mathcal{F}$-groups or of homogeneous Δ-$\mathcal{F}$-spaces is a Δ-$\mathcal{F}$-mapping, as are the group law $\mu_G\colon G \times G \to G$ of a Δ-$\mathcal{F}$-group G given by $(x,y) \mapsto xy$ and its group symmetry $\iota_G\colon G \to G$ given by $x \mapsto x^{-1}$; so are the action $\mu_M\colon M \times G \to M$ of a homogeneous Δ-$\mathcal{F}$-space M for G given by $(v,x) \mapsto vx$ and, when M is principal, the mapping $\psi_M\colon M \times M \to G$ given by $(v,w) \mapsto v^{-1}w$.

If f is a Δ-$\mathcal{F}$-mapping of A into B, then f is Δ-$\mathcal{F}$-continuous (that is, is continuous relative to the Δ-$\mathcal{F}$-topologies), and the domain of definition of f is a Δ-$\mathcal{F}$-open Δ-$\mathcal{F}$-dense subset of A. If also g is a Δ-$\mathcal{F}$-mapping of B into a Δ-$\mathcal{F}$-set C such that g is defined at $f(u)$ whenever u is an $\mathcal{F}$-generic element of A, then there exists a unique Δ-$\mathcal{F}$-mapping h of A into C such that $h(u) = g(f(u))$ for all such u. We call this h the generic composite of g and f, and denote it by $g \mathbin{\square} f$. When there exists a Δ-$\mathcal{F}$-mapping f' of B into A such that $f' \mathbin{\square} f$ exists and equals id_A and $f \mathbin{\square} f'$ exists and equals id_B, we say that f is generically invertible and that f' is its generic inverse; when this is the case, and $u \in A$ has the property that f is defined at u and f' is defined at $f(u)$, we say that f is bidefined at u; the set of all such u is a Δ-$\mathcal{F}$-open Δ-$\mathcal{F}$-dense subset of A called the domain of bidefinition of f.

Δ-rational functions

For Δ-$\mathcal{H}$-mappings of the Δ-$\mathcal{H}$-set A into G_a we have special terminology and notation. Such a Δ-$\mathcal{H}$-mapping is called a Δ-$\mathcal{H}$-function on A; the set of all Δ-$\mathcal{H}$-functions on A is denoted by $\mathfrak{F}_{\mathcal{H}}(A)$, and for $\mathfrak{F}_{\mathcal{U}}(A)$ we generally write just $\mathfrak{F}(A)$. The set of elements of $\mathfrak{F}_{\mathcal{H}}(A)$ resp. $\mathfrak{F}(A)$ that are defined at a given element $u \in A$ is denoted by $\mathfrak{F}_{\mathcal{H},u}(A)$ resp. $\mathfrak{F}_u(A)$. $\mathfrak{F}(A)$ has a natural structure of Δ-ring (indeed, of Δ-algebra over $\mathcal{U}$), of which $\mathfrak{F}_{\mathcal{F}}(A)$ and $\mathfrak{F}_u(A)$ are Δ-subrings. When A is $\mathcal{F}$-irreducible, then $\mathfrak{F}_{\mathcal{F}}(A)$ is a Δ-field (an extension of $\mathcal{F}$, not in $\mathcal{U}$, of course). When A is irreducible (that is, is $\mathcal{G}$-irreducible for every extension $\mathcal{G}$ of $\mathcal{F}$ or, what is equivalent, is $\mathcal{F}_a$-irreducible), then $\mathfrak{F}(A)$ is an extension of $\mathcal{U}$; in this case, $\mathcal{U}$ and $\mathfrak{F}_{\mathcal{F}}(A)$ are linearly disjoint over $\mathcal{F}$ and $\mathcal{U}.\mathfrak{F}_{\mathcal{F}}(A) = \mathfrak{F}(A)$, and for any $u \in A$, $\mathfrak{F}_u(A)$ is a local Δ-ring with field of quotients $\mathfrak{F}(A)$.

Now consider a Δ-$\mathcal{F}$-mapping of A into a Δ-$\mathcal{F}$-set B. If f is generically surjective (that is, f maps the set of $\mathcal{F}$-generic elements of A onto the set of $\mathcal{F}$-generic elements of B) then $\psi \circ f$ exists for every $\psi \in \mathfrak{F}(B)$ and hence f induces an injective mapping $f^*: \mathfrak{F}(B) \to \mathfrak{F}(A)$. This f^* is a homomorphism of Δ-rings over $\mathcal{U}$. If f is not generically surjective, this is no longer the case, but when $u \in A$ and f is defined at u, then f induces a homorphism $f_u^*: \mathfrak{F}_{f(u)}(B) \to \mathfrak{F}_u(A)$.

Principal homogeneous Δ-$\mathcal{F}$-spaces and constrained cohomology

One of the interesting problems in the theory of Δ-$\mathcal{F}$-groups is that of classifying, up to Δ-$\mathcal{F}$-isomorphism, the principal homogeneous Δ-$\mathcal{F}$-spaces for a given Δ-$\mathcal{F}$-group G.

Recall that G itself can be considered a principal homogeneous Δ-$\mathcal{F}$-space for G. Since G has an element rational over $\mathcal{F}$, namely the element $1 \in G$, every principal homogeneous Δ-$\mathcal{F}$-space for G that is Δ-$\mathcal{F}$-isomorphic to G has an element rational over $\mathcal{F}$. Conversely, if a principal homogeneous Δ-$\mathcal{F}$-space M for G has an element u with $\mathcal{F}\langle u\rangle = \mathcal{F}$, then the formula $x \mapsto ux\ (x \in G)$ defines a Δ-$\mathcal{F}$-isomorphism of G onto M.

Again, consider the set L_Λ of solutions of the Δ-equation

$$\Lambda y = 0$$

corresponding to a nonzero linear Δ-operator $\Lambda = \Sigma a_\theta \theta$ (where the sum extends over a finite set of derivative operators $\theta \in \Theta$ and $a_\theta \in F$ for each θ); L_Λ is a Δ-F-subgroup of G_a. For any element $a \in F$, the set $L_{\Lambda,a}$ of solutions of the Δ-equation

$$\Lambda y = a$$

has a natural structure of principal homogeneous Δ-F-space for L_Λ. We can ask two questions.

(1) If $a,b \in F$, when are $L_{\Lambda,a}$, $L_{\Lambda,b}$ Δ-F-isomorphic?

(2) Does there exist a principal homogeneous Δ-F-space for L_Λ that is not Δ-F-isomorphic to any $L_{\Lambda,a}$?

The answer to question (1) is easy: $L_{\Lambda,a}$ is Δ-F-isomorphic to $L_{\Lambda,b}$ if and only if the equation $\Lambda y = b - a$ has a solution in F, that is, $a \equiv b (\mathrm{mod}\ \Lambda F)$. In other words, the set of Δ-F-isomorphism classes of the $L_{\Lambda,a}$ with $a \in F$ is in bijective correspondence with $F/\Lambda F$. Thus these Δ-F-isomorphism classes have something to do with the possibility of solving certain Δ-equations <u>in</u> F. The answer to question 2 is less immediate.

We shall ultimately see, as a result of a general classification theorem, that the answer is negative. The theorem will classify the set $P(F,G)$ of Δ-F-isomorphism classes of principal homogeneous Δ-F-spaces for a Δ-F-group G in terms of a new kind of cohomology set $H^1_\Delta(F,G)$. This <u>constrained</u> cohomology set is somewhat analogous to the Galois cohomology set $H^1(K,G)$ of a field K in a K-group G (see e.g. [1]). The constrained cohomology is based on the notion of <u>constrained closure</u> of a Δ-field, a notion in differential algebra originated, curiously enough, by logicians, namely the late Abraham Robinson and his followers, Lenore Blum and Saharon Shelah(who used the term "differential closure"). A constrained closure of a Δ-field is somewhat analogous to an algebraic closure of a field, and has many of the same properties (see [7].) In particular, F has a constrained closure in U, unique up to a Δ-isomorphism over F; however,

U may contain more than one constrained closure of F, indeed, one can very well contain another. Also, for any Δ-F-set B, the set of elements of B that are rational over a given constrained closure of F is dense in B relative to the Δ-topology. In what follows, we fix a constrained closure F^+ of F in U (over which U need not be universal), and put $A = \mathrm{Aut}(F^+/F)$.

For any finitely generated extension G of F in F^+, the set $\mathrm{Iso}(G/F)$ of all isomorphisms of G over F onto an extension of F in U has a natural structure of Δ-F-set. If H is another such extension, with $G \subset H$, we have the obvious mappings

$$r : \mathrm{Iso}(H/F) \longrightarrow \mathrm{Iso}(G/F), \qquad i: \mathrm{Iso}(H/G) \longrightarrow \mathrm{Iso}(H/F)$$

of restriction and inclusion. It turns out that r is a Δ-F-mapping and i is a Δ-G-mapping. Similarly, we have the restriction mapping

$$A \longrightarrow \mathrm{Iso}(G/F),$$

the image of which is the set $\mathrm{Iso}_{F^+}(G/F)$ consisting of the elements of $\mathrm{Iso}(G/F)$ that are rational over F^+, and the diagram

$$\begin{array}{ccc} & A & \\ \swarrow & & \searrow \\ \mathrm{Iso}(H/F) & \xrightarrow{\ r\ } & (\mathrm{Iso}(G/F) \end{array}$$

is commutative. It follows that A may be regarded as a projective limit

$$A = \varprojlim_{G} \mathrm{Iso}_{F^+}(G/F)$$

A <u>constrained</u> <u>1-cocycle</u> of F in the Δ-F-group G is defined to be a mapping $\Phi: A \to G$ such that:

(a) There exist a finitely generated extension G of F and an everywhere defined Δ-G-mapping Φ_G: $\mathrm{Iso}(G/F) \to G$ such that $\Phi(\sigma) = \Phi(\sigma|G)$ for every $\sigma \in A$.

(b) $\Phi(\sigma\tau) = \Phi(\sigma)\cdot\sigma\Phi(\tau)$ for all $\sigma, \tau \in A$.

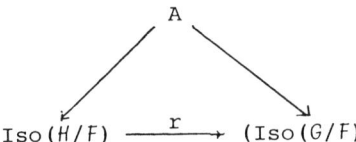

The subsequent definitions are as expected. Letting $Z^1_\Delta(F,G)$ denote the set of all constrained 1-cocycles of F in G, we say that two such cocycles Φ,Ψ are <u>cohomologous</u> if there exists an $x \in G_{F+}$ such that $\Psi(\sigma) = x^{-1}\Phi(\sigma)\sigma x \quad (\sigma \in A)$; this is an equivalence on $Z^1_\Delta(F,G)$. The <u>constrained cohomology set</u> of F in G, denoted by $H^1_\Delta(F,G)$, is the set of equivalence classes. It is a pointed set.

We can now state our classification theorem in the following form.

<u>Theorem</u>. <u>We have a canonical bijection</u> $P(F,G) \to H^1_\Delta(F,G)$.

To describe the bijection, consider any principal homogeneous Δ-F-space M for G. There exists an element $u \in M_{F+}$. The formula $\sigma \to u^{-1}\sigma u$ defines a mapping $\Phi_{M,u}: A \to G$. It is not very hard to show that when N is another principal homogeneous Δ-F-space for G and $v \in N_{F+}$, then N is Δ-F-isomorphic to M if and only if $\Phi_{N,v}$ is cohomologous to $\Phi_{M,u}$. Thus, we have an injection $P(F,G) \to H^1_\Delta(F,G)$. The final step is to prove that, for any $\Phi \in Z^1_\Delta(F,G)$, there exist M, u as above with $\Phi = \Phi_{M,v}$. This part of the proof is very long.

In order to compute $H^1_\Delta(F,G)$ in some cases, it is useful to put $H^0_\Delta(F,G) = G_F$. Every Δ-F-homomorphism $f\colon G \to G'$ of Δ-F-groups induces homomorphisms $f^0\colon H^0_\Delta(F,G) \to H^0_\Delta(F,G)$ of groups and $f^1 : H^1_\Delta(F,G) \to H^1_\Delta(F,G')$ of pointed sets. We have the following results.

<u>Theorem</u>. <u>Let</u>

$$1 \longrightarrow N \xrightarrow{j} G \xrightarrow{\pi} G' \longrightarrow 1$$

<u>be a short exact sequence of Δ-F-homomorphisms of Δ-F-groups. There exists a connecting homomorphism</u> $\delta\colon H^0_\Delta(F,G') \to H^1_\Delta(F,G)$ <u>such that the sequence</u>

$$1 \longrightarrow H^0_\Delta(F,N) \xrightarrow{j^0} H^0_\Delta(F,G) \xrightarrow{\pi^0} H^0_\Delta(F,G') \xrightarrow{\delta} H^1_\Delta(F,N) \xrightarrow{j^1} H^1_\Delta(F,G) \xrightarrow{\pi^1} H^1_\Delta(F,G')$$

<u>is exact</u>

<u>Theorem</u>. <u>When</u> G <u>is an</u> <u>F-group, then</u> $H^1_\Delta(F,G)$ <u>is canonically isomorphic to the Galois cohomology set</u> $H^1_\Delta(F,G)$.

The latter theorem shows, for example, that every principal homogeneous Δ-F-space for $GL(n)$ is Δ-F-isomorphic to $GL(n)$, because it is known that $H^1(F,GL(n)) = 1$. The former theorem shows, when applied to the exact sequence

$$0 \longrightarrow L_\Lambda \longrightarrow G_a \xrightarrow{\Lambda} G_a \longrightarrow 0,$$

that $H^1(F,L_\Lambda) \approx F/\Lambda F$ and every principal homogeneous Δ-F-space for L_Λ is Δ-F-isomorphic to some $L_{\Lambda,a}$.

Another application of constrained cohomology is to the classification of Δ-F-vector spaces over the universal Δ-field U. We are going to see that every such Δ-F-vecotor space is Δ-F-isomorphic to U^n for some n. More precisely, we have the following result.

__Theorem__. _Let_ V _be a_ Δ-F-_vector space over_ U. _Then the_ Δ-_dimension of_ V _is also its dimension as a vector space over_ U, _and_ V _has a basis that is rational over_ F.

Indeed, for any finite family $(u_1,\ldots,u_k)$ of elements of V_{F^+} that are linearly independent over U, the formula $(\alpha_1,\ldots,\alpha_k) \mapsto \Sigma \alpha_i u_i$ defines an injective Δ-F^+-homomorphism $U^k \to V$ of Δ-F^+-vector spaces over U. The image is a Δ-F^+-vector subspace of V of Δ-dimension k, so that $k \leqslant \Delta\text{-}\dim V$. If this image is not V, its complement in V, which is Δ-open, contains an element u_{k+1} of the Δ-dense set V_{F^+}. It follows that V has a basis $u = (u_1,\ldots,u_n)$ with $n = \Delta\text{-}\dim V$ that is rational over F^+.

For any automorphism $\sigma \in A$, the family $\sigma u = (\sigma u_1,\ldots,\sigma u_n)$ also is a basis of V, rational over F^+, so that there exists a matrix $a(\sigma) \in GL_{F^+}(n)$ such that $\sigma u = ua(\sigma)$. It can be shown that the mapping $a\colon A \to GL(n)$ given by the formula $\sigma \mapsto a(\sigma)$ is an element of $Z^1_\Delta(F,GL(n))$. Since $H^1_\Delta(F,GL(n)) = 1$ (see above), it follows that there exists an element $b \in GL_{F^+}(n)$ such that $a(\sigma) = b^{-1}\sigma b$ $(\sigma \in A)$. Hence, for the basis $v = ub^{-1}$ of V,

$$\sigma v = \sigma u \sigma b^{-1} = ua(\sigma)\sigma b^{-1} = ub^{-1} = v \quad (\sigma \in A).$$

This implies that the basis v of V is rational over F, and completes the proof.

The computation of $H^1_\Delta(F,G)$ in other cases would permit other applications.

Lie Theory

We conclude with a brief indication of how the mechanism of Lie theory can be introduced into our subject.

(a) Other Δ-$\mathcal{F}$-structures.

The first step is to define Δ-$\mathcal{F}$-structures other than groups and their homogeneous spaces. This is fairly straight-forward. A Δ-$\mathcal{F}$-*ring*, for example, is a ring R, together with Δ-$\mathcal{F}$-group structure on the additive group R, such that the ring multiplication $R \times R \to R$ is a Δ-$\mathcal{F}$-mapping. Thus, the field $\mathcal{K}$ of constants of $\mathcal{U}$ is a Δ-$\mathcal{F}$-ring (indeed, is a Δ-$\mathbf{Q}$-*field*). For any Δ-$\mathcal{F}$-ring R, a Δ-$\mathcal{F}$-*module* over R is a module V over the ring R, together with a Δ-$\mathcal{F}$-group structure on the additive group V, such that the external law of composition $V \times R \to V$ of the module is a Δ-$\mathcal{F}$-mapping. We shall be interested in the case in which R is $\mathcal{K}$, in which case V is called a Δ-$\mathcal{F}$-*vector space* over $\mathcal{K}$. Finally, we shall need the notion of Δ-$\mathcal{F}$-*Lie algebra* over $\mathcal{K}$, which is defined as expected.

(b) Tangent spaces.

Consider an irreducible Δ-$\mathcal{F}$-set V and an element $v \in V$. The definition of tangent space is analogous to the definition in algebraic geometry. The Δ-algebra $\mathfrak{F}_v(V)$ over $\mathcal{U}$, consisting of the Δ-rational functions on V that are defined at v, is now (because V is irreducible) a local Δ-ring. A *tangent vector* to V at v is a local Δ-derivation of $\mathfrak{F}_v(V)$ at v, that is, is a $\mathcal{U}$-linear mapping $T\colon \mathfrak{F}_v(V) \to \mathcal{U}$ such that

$$T(\varphi\psi) = T\varphi\cdot\psi(v) + \varphi(v)T\psi \qquad (\varphi,\psi \in \mathfrak{F}_v(V)),$$

$$T(\delta\varphi) = \delta(T\varphi) \qquad (\varphi \in \mathfrak{F}_v(V),\ \delta \in \Delta).$$

The set $\mathfrak{T}_v(V)$ of tangent vectors to V at v has an obvious structure of vector space over $\mathcal{K}$. Because it is usually infinite dimensional, we need more structure on it to keep it under control. It turns out that for a "large enough" extension $\mathcal{G}$ of $\mathcal{F}\langle v\rangle$, $\mathfrak{T}_v(V)$ has a natural structure of Δ-$\mathcal{G}$-vector space over $\mathcal{K}$. "Large enough" means that there must exist a generically invertible Δ-$\mathcal{G}$-mapping of V into a Δ-$\mathcal{G}$-subset of some affine space $\mathcal{U}^n$ that is bidefined at v, or as we shall say, that v *is* Δ-$\mathcal{G}$-*affine in* V. ("Most" elements of V

are Δ-$\mathcal{F}$-affine in V; there exists a finitely generated extension $\mathcal{G}$ of $\mathcal{F}$ such that every element of V is Δ-$\mathcal{G}$-affine in V.) We call $\mathfrak{T}_v(V)$, with its structure of Δ-$\mathcal{G}$-vector space over $\mathcal{K}$, the <u>tangent space</u> to V at v.

As remarked before, any Δ-$\mathcal{F}$-mapping f of V into an irreducible Δ-$\mathcal{F}$-set V', such that f is defined at v, induces a homomorphism $f_v^*: \mathfrak{F}_{f(v)}(V') \to \mathfrak{F}_v(V)$. For any $T \in \mathfrak{T}_v(V)$, the composite mapping $T \circ f_v^*$ is a tangent vector to V' at $f(v)$. Therefore the formula $T \mapsto T \circ f_v^*$ defines a mapping

$$f_v^{**} : \mathfrak{T}_v(V) \to \mathfrak{T}_{f(v)}(V').$$

When v and $f(v)$ are Δ-$\mathcal{G}$-affine in V and V', respectively, then f_v^{**} is a Δ-$\mathcal{G}$-homomorphism of Δ-$\mathcal{G}$-vector spaces over $\mathcal{K}$.

<u>(c) The Lie algebra</u>

The Lie algebra of a connected Δ-$\mathcal{F}$-group G, or more generally of a principal homogeneous Δ-$\mathcal{F}$-space V for G, is defined in terms of the invariant Δ-derivations on V. A Δ-<u>derivation</u> on V is a $\mathcal{U}$-linear mapping $D: \mathfrak{F}(V) \to \mathfrak{F}(V)$ such that

$$D(\varphi\psi) = D\varphi \cdot \psi + \varphi D\psi, \quad D(\delta\varphi) = \delta(D\varphi)$$

for all $\varphi, \psi \in \mathfrak{F}(V)$ and $\delta \in \Delta$. For any $x \in G$, the mapping $\rho_x: V \to V$, given by the formula $v \mapsto vx$, is a Δ-$\mathcal{F}\langle x\rangle$-mapping, and $\rho_{x^{-1}}$ is its inverse; hence ρ_x induces an automorphism $\rho_x{}^*$ of $\mathfrak{F}(V)$ over $\mathcal{U}$. Therefore the mapping $\rho_x{}^{**}(D) = \rho_{x^{-1}}{}^* \circ D \circ \rho_x{}^*$ is again a Δ-derivation on V. D is <u>invariant</u> when $\rho_x{}^{**}(D) = D\,(x \in G)$. The set $\mathcal{L}_\Delta(V)$ of invariant Δ-derivations on V has an obvious structure of Lie algebra over $\mathcal{K}$. This tends to be infinite dimensional, but there is a natural structure on $\mathcal{L}_\Delta(V)$ of Δ-$\mathcal{F}$-Lie algebra over $\mathcal{K}$. The difficulty here is not to specify the Δ-$\mathcal{F}$-data on $\mathcal{L}_\Delta(V)$, but rather to verify all the appropriate axioms. We call $\mathcal{L}_\Delta(V)$, with its structure of Δ-$\mathcal{F}$-Lie algebra over $\mathcal{K}$, the <u>Lie algebra</u> of V.

For any $D \in \mathcal{L}_\Delta(V)$ and $v \in V$, it turns out out that $D\mathfrak{F}_v(V) \subset \mathfrak{F}_v(V)$. Therefore the formula $\varphi \mapsto (D\varphi)(v)$ defines a mapping

$\mathfrak{F}_V(V) \to U$ which is easily seen to be a tangent vector to V at v and which we denote by D_v. The formula $D \mapsto D_v$ thus gives a canonical mapping $\mathcal{L}_\Delta(V) \to \mathcal{T}_v(V)$ which turns out to be an isomorphism of vector spaces over K. It is this isomorphism that allows us to complete the proof that $\mathcal{L}_\Delta(V)$ is a Δ-F-vector space over K, and then to conclude, for any extension G of $F\langle v\rangle$ such that v is Δ-G-affine in V, that the mapping is a Δ-G-isomorphism of Δ-G-vector spaces over K.

Consider a Δ-F-mapping f of V into a principal homogeneous Δ-F-space V', with f defined at v. If G is now large enough so that v is Δ-G-affine in V and $f(v)$ is Δ-G-affine in V', then we see that f induces a Δ-G-homomorphism $f_v^{\#} : \mathcal{L}_\Delta(V) \to \mathcal{L}_\Delta(V')$ of Δ-G-vector spaces over such that the diagram

$$\begin{array}{ccc} \mathcal{L}_\Delta(V) & \xrightarrow{\;f_v^{\#}\;} & \mathcal{L}_\Delta(V') \\ \downarrow & & \downarrow \\ \mathcal{T}_v(V) & \xrightarrow[\;f_v^{**}\;]{} & \mathcal{T}_{f(v)}(V') \end{array}$$

is commutative (the vertical arrows being the canonical ones). The functoriality of the Lie algebra $\mathcal{L}_\Delta(V)$ comes from the following corollary of a result about "crossed" Δ-F-homomorphisms:

<u>If</u> $f\colon V \to V'$ <u>is a "relative"</u> Δ-F-<u>homomorphism of principal homogeneous</u> Δ-F-<u>spaces, then</u> $f_v^{\#}$ <u>is independent of</u> v (and hence may be denoted by $f^{\#}$) <u>and is a</u> Δ-F-<u>homomorphism of</u> Δ-F-<u>Lie algebras over</u> K.

(d) <u>Logarithmic derivations</u>

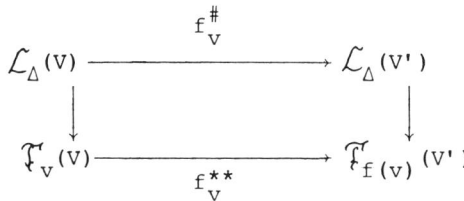

Let V be a principal homogeneous Δ-F-space for a connected Δ-F-group G. Let H be an extension of F with U not necessarily universal over H, and let ϵ be a Δ-derivation of H into U over F. Put $H_\epsilon = \operatorname{Ker} \epsilon$. For any $v \in V_H$, the formula $\varphi \mapsto \epsilon(\varphi(v))$ defines a mapping $\mathfrak{F}_{H_\epsilon}(V) \to U$; when v is Δ-H_ϵ-affine in V, this can be extended to a unique tangent vector to V at v which, by the above, is D_v for a unique $D \in \mathcal{L}_\Delta(V)$. We denote this D by $\ell\epsilon_V(v)$ or by $\ell\epsilon(v)$ and call it the <u>logarithmic derivative of</u> v <u>on</u> V <u>relative to</u> ϵ. The invariant derivation $\ell\epsilon(v)$ on V is characterized by the condition

$$\ell\epsilon(v)_V\varphi = \epsilon(\varphi(v)) \quad (\varphi \in \mathfrak{F}_{H_\epsilon, V}(V)).$$

Let V_ϵ denote the set of elements of V_H that are Δ- H_ϵ-affine in V. The formula $v \to \ell\epsilon(v)$ defines a mapping $\ell\epsilon = \ell\epsilon_V \colon V_\epsilon \to \mathcal{L}_\Delta(V)$ that we call the <u>logarithmic derivation on</u> V <u>relative to</u> ϵ. It is easy to see that $\ell\epsilon(v) = 0$ if and only if $v \in V_{H_\epsilon}$. The key property of logarithmic derivations is given by the following result.

<u>Theorem</u>. <u>Let the notation be as above, let</u> $v \in V_H$, $x \in G_H$ <u>and suppose that</u> v, x, vx <u>are</u> Δ- H_ϵ<u>-affine in</u> V, G, V, <u>respectively. Let</u> $\lambda_v \colon G \to V$ <u>denote the</u> Δ-$F\langle v\rangle$<u>-mapping defined by the formula</u> $y \mapsto vy (y \in G)$. <u>Then</u>

$$\ell\epsilon_V(vx) = \ell\epsilon_V(v) + \lambda_v^\#(\ell\epsilon_G(x))$$

When $V = G$ and G is commutative, this equation reduces to

$$\ell\epsilon_G(yx) = \ell\epsilon_G(y) + \ell\epsilon_G(x).$$

Logarithmic derivations relative to ϵ are most useful when $H = U$ and U is universal over F as a differential field relative to the set $\Delta(\epsilon) = \Delta \cup \{\epsilon\}$ of derivation operators. When that is the case, and ι_G denotes the symmetry mapping $x \mapsto x^{-1}$ of G, then $\ell\epsilon_G \circ \iota_G$ is a surjective everywhere defined crossed $\Delta(\epsilon)$-F-homomorphism of G into $\mathcal{L}_\Delta(G)$, and for every Δ-F-subgroup H of G, $\ell\epsilon_G(H) = in_{G,H}{}^\#(\mathcal{L}_\Delta(H^\circ))$, where $in_{G,H}$ denotes the inclusion mapping $H \to G$ (which is a Δ-F-homomorphism, of course). Also then ϵ induces a derivation $\epsilon^\#$ of the Lie algebra $\mathcal{L}_\Delta(G)$ for which

$$\lambda_x^\#(\epsilon^\# \lambda_{x^{-1}}{}^\#(D)) = \epsilon^\# D + [\ell\epsilon(x), D] \quad (x \in G,\ D \in \mathcal{L}_\Delta(G)),$$

and when W is a Δ-closed vector subspace of the Δ-F-vector space $\mathcal{L}_\Delta(G)$ over K, a necessary and sufficient condition that W be Δ-U_ϵ-closed in $\mathcal{L}_\Delta(G)$ is that $\epsilon^\# W \subset W$.

(e) The Lie-Cassidy-Kovacic method.

For the most effective application of logarithmic derivations, we need a Δ-derivation ε of $\mathcal{U}$ over $\mathcal{F}$ such that $\mathcal{U}$ is universal as a $\Delta(\varepsilon)$-field extension of $\mathcal{F}$. The Lie-Cassidy-Kovacic method satisfies this need and makes it possible to use logarithmic derivations to the same end that the logarithmic and exponential maps are used in classical Lie theory.

Fix an element ε of an overset of Δ with $\varepsilon \notin \Delta$ and put $\Delta(\varepsilon) = \Delta \cup \{\varepsilon\}$. The operation of Δ on $\mathcal{U}$ can be extended to one of $\Delta(\varepsilon)$ on $\mathcal{U}$ by putting $\varepsilon u = 0$ $(u \in \mathcal{U})$. This makes $\mathcal{U}$ (as well as every Δ-subfield of $\mathcal{U}$) a $\Delta(\varepsilon)$-field. Now fix a universal extension $\mathfrak{U}$ of the $\Delta(\varepsilon)$-field $\mathcal{U}$. Of course, $\mathfrak{U}$ is a universal extension of the Δ-field $\mathcal{U}$, too. Let $\mathfrak{K}$ denote the field of constants of $\mathfrak{U}$ as a Δ-field.

Now, to every Δ-$\mathcal{F}$-group G relative to the universal Δ-field $\mathcal{U}$ there is a canonically associated Δ-$\mathcal{F}$-group $\mathfrak{G}$ relative to the universal Δ-field $\mathfrak{U}$ such that $G = \mathfrak{G}_{\mathcal{U}}$. To a Δ-$\mathcal{F}$-homomorphism $f\colon G \to H$ of Δ-$\mathcal{F}$-groups there exists a unique Δ-$\mathcal{F}$-homomorphism $\mathfrak{f}\colon \mathfrak{G} \longrightarrow \mathfrak{H}$ of their canonically associated Δ-$\mathcal{F}$-groups such that $f = \mathfrak{f}|G$. When G and H are connected, then so too are $\mathfrak{G}$ and $\mathfrak{H}$; also $\mathcal{L}_\Delta(G)$ and $\mathcal{L}_\Delta(H)$ can be canonically identified with $\mathcal{L}_{\Delta,\mathcal{U}}(\mathfrak{G})$ and $\mathcal{L}_{\Delta,\mathcal{U}}(\mathfrak{H})$, and then $f^\# = \mathfrak{f}^\#|\mathcal{L}_\Delta(G)$.

This permits us to prove theorems about G and $\mathcal{L}_\Delta(G)$ by proving them instead for $\mathfrak{G}$ and $\mathcal{L}_\Delta(\mathfrak{G})$.

(f) Completing the theory.

There remains the task of forging the connection between properties of a connected Δ-$\mathcal{F}$-group and those of its Lie algebra. Using logarithmic derivations and the Lie-Cassidy-Kovacic method, we can obtain the following result.

Theorem. Let $H \xrightarrow{f} G \xrightarrow{g} G'$ be a sequence of Δ-$\mathcal{F}$-homomorphisms of connected Δ-$\mathcal{F}$-groups. A necessary and sufficient condition that $\operatorname{Im}(f) = \operatorname{Ker}(g)^\circ$ is that the induced sequence

$$\mathcal{L}_\Delta(H) \xrightarrow{f^\#} \mathcal{L}_\Delta(G) \xrightarrow{g^\#} \mathcal{L}_\Delta(G')$$ be exact.

The Lie algebras of the connected Δ-algebraic subgroups of F do not immediately reflect the inclusion relations among these subgroups; indeed, the way we have defined them, these Lie algebras are in general not even subsets of $\mathcal{L}_\Delta(G)$. However, for any connected say Δ-G-subgroup H of G, the inclusion mapping $in_{G,H}: H \longrightarrow G$, which is a Δ-G-homomorphism, induces an injective Δ-G-homomorphism $in_{G,H}{}^{\#} : \mathcal{L}_\Delta(H) \longrightarrow \mathcal{L}_\Delta(G)$. Its image is a Δ-G-Lie subalgebra of $\mathcal{L}_\Delta(G)$ that we denote by $\mathcal{L}^G_\Delta(H)$ and call the Lie algebra of H in G. Using logarithmic derivations and the Lie-Cassidy-Kovacic method, we can prove the following result.

Let G be a connected Δ-F-group.

(a) The formula $H \longrightarrow \mathcal{L}^G_\Delta(H)$ defines an injective mapping of the set of connected Δ-closed subgroups of G into the set of Δ-closed Lie subalgebras of $\mathcal{L}_\Delta(G)$.

(b) For any two connected Δ-closed subgroups H_1, H_2 of G,

$$H_1 \subset H_2 \iff \mathcal{L}^G_\Delta(H_1) \subset \mathcal{L}^G_\Delta(H_2).$$

(c) For any family $(H_i)_{i \in I}$ of connected Δ-closed subgroups of G,

$$\mathcal{L}^G_\Delta((\cap H_i)^\circ) = \cap \mathcal{L}^G_\Delta(H_i).$$

Other links between Δ-F-groups and their Lie algebras require the following partial analog of a result of Chevalley about linear algebraic groups ("Theorie des Groupes de Lie", vol. II, "Groups Algébriques". Hermann, Paris, 1951; see p. 172).

Theorem. Let G be a connected Δ-F-group, and V and V' be Δ-F-vector subspaces of $\mathcal{L}_\Delta(G)$ such that $V' \subset V$. Put

$$H = \{x \in G \mid \tau^{\#}_x(D) - D \in V' \quad (D \in V)\},$$

$$L = \{X \in \mathcal{L}_\Delta(G) \mid [X,D] \in V' \quad (D \in V)\}.$$

(a) H is a Δ-F-subgroup of G, L is a Δ-F-Lie subalgebra of $\mathcal{L}_\Delta(G)$, and $\mathcal{L}^G_\Delta(H^\circ) \subset L$.

(b) If V' is 0 or V, then $\mathcal{L}^G_\Delta(H^\circ) = L$.

The proof of this theorem uses logarithmic derivations and the Lie-Kovacic-Cassidy method. In the light of Chevalley's result, it is natural to ask whether the extra hypothesis in part (b) is needed. I have not been able to settle this question.

Using this theorem, we can prove a whole sequence of results of which we mention just three (in the statement of which G is a connected Δ-$\mathcal{F}$-group and N is a connected Δ-$\mathcal{F}$-subgroup of G.)

1. N _is normal in_ G _if and only if_ $\mathcal{L}_\Delta^G(N)$ _is an ideal of_ $\mathcal{L}_\Delta(G)$.

2. _If_ N _is normal in_ G _and_ $C_G(N)$ _denotes the commutant of_ N _in_ G, _then_ $\mathcal{L}_\Delta^G(C_G(N)^\circ)$ _is the commutant of_ $\mathcal{L}_\Delta^G(N)$ _in_ $\mathcal{L}_\Delta(G)$. _Hence, in particular, if_ $C(G)$ _denotes the center of_ G, _then_ $\mathcal{L}_\Delta^G(C(G)^\circ)$ _is the center of_ $\mathcal{L}_\Delta(G)$.

3. G _is solvable (resp. nilpotent) if and only if_ $\mathcal{L}_\Delta(G)$ _is solvable (resp. nilpotent)_.

References

1. P.J. Cassidy. Differential algebraic groups, _Amer. J. Math_. 94 (1972), 891-954.

2. P.J. Cassidy. The differential rational representation algebra on a linear differential algebraic group, _J. Algebra_ 37(1975), 222-238.

3. P.J. Cassidy. Unipotent differential algebraic groups, in "Contributions to Algebra" (H. Bass, P.J. Cassidy, and J. Kovacic, eds.), Academic Press, New York, 1977, pp. 83-115.

4. P.J. Cassidy. Differential algebraic Lie algebras, _Trans. Amer. Math. Soc_. 247(1979), 247-273.

5. P.J. Cassidy. Differential algebraic group structures on the plane, Proc. _Amer. Math. Soc_. 80 (1980), 210-214.

6. E.R. Kolchin. "Differential Algebra and Algebraic Groups." Academic Press, New York, 1973.

7. E.R. Kolchin. Constrained extensions of differential fields, _Adv. in Math_. 12(1974), 141-170.

8. E.R. Kolchin. "Differential Algebraic Groups." Academic Press, Orlando, 1985.

9. J. Kovacic. Constrained cohomology, in "Contributions to Algebra" (H. Bass, P.J. Cassidy, and J. Kovacic, eds.), Academic Press, New York, 1977, pp. 251-266.

10. W.Y. Sit. Differential algebraic subgroups of SL(2) and strong normality in simple extensions, _Amer. J. Math_. 97 (1975), 627-695.

Contemporary Mathematics
Volume **131**, 1992 (Part 2)

A PROBLEM ON DIFFERENTIAL POLYNOMIALS

By E. R. Kolchin
Columbia University, New York, NY 10027

§0. Denomination and the spaces $\mathcal{V}_n(d)$.

In what follows, $\mathcal{F}$ denotes an ordinary differential field of characteristic 0. The basic derivation operator of $\mathcal{F}$ is denoted by δ, and for any element $u \in \mathcal{F}$ we put $\delta u = u'$ and $\delta^k u = u^{(k)}$ $(k \in \mathbb{N})$. We use the prefix "δ-" as a synonym of "differential" or "differentially".

We deal with the δ-ring $\mathcal{F}\{y_1, \ldots, y_n\}$ of δ-polynomials over $\mathcal{F}$ in a finite family $(y_1, \ldots y_n)$ of δ-indeterminates; as a ring, $\mathcal{F}\{y_1, \ldots, y_n\}$ is just the polynomial ring over $\mathcal{F}$ in the infinite family $(y_j^{(k)})_{1 \le j \le n, k \in \mathbb{N}}$ of indeterminates.

Let y_0 be yet another δ-indeterminate. For any nonzero δ-polynomial $F = F(y_1, \ldots, y_n)$ in $\mathcal{F}\{y_1, \ldots, y_n\}$, $F(y_1/y_0, \ldots, y_n/y_0)$ can be written as a fraction in which the numerator is a δ-polynomial in $(y_0, y_1, \ldots, y_n)$ and the denominator is a power of y_0, that is, for sufficiently big $d \in \mathbb{N}$, we have $y_0{}^d F(y_1/y_0, \ldots, y_n/y_0) \in \mathcal{F}\{y_0, \ldots, y_n\}$; the smallest such d is called the *denomination* of F and is denoted by *den*F. The denomination of the δ-polynomial 0 is defined to be $-\infty$. For example, $y_1 y_1'' - 2y_1'^2$ has denomination 3 (although each of its two terms has denomination 4).

For any $d \in \mathbb{N}$, the set of elements of $\mathcal{F}\{y_1, \ldots y_n\}$ of denomination $\le d$ has a natural structure of vector space over $\mathcal{F}$; we shall denote this vector space by $\mathcal{V}_n(d)$. Our main concern is the following problem.

To find the dimension of $\mathcal{V}_n(d)$.

The analogous problem for a polynomial ring $K[X_1, \ldots, X_n]$ over a field K is trivial; the analog of denomination is just degree, and the polynomials of degree $\le d$ form a vector space over K of dimension $\binom{n+d}{n}$. The problem displayed above, however, is open and appears to be difficult.

In trying to clarify a paper by C. F. Osgood [2], Wolfgang Schmidt [3] proved that

1991 *Mathematics Subject Classification.* Primary 12H05.

$$\dim \mathcal{V}_n(d) \geq (n+1)^d,$$

and he asked whether perhaps this inequality is actually an equality. As far as I know, however, it has not been shown that $\dim \mathcal{V}_n(d)$ is always finite, even in the case n=1. In this special case, the inequality implies, as pointed out by Osgood, that an algebraic function $u = u(z)$ of degree k over $\mathbb{C}(z)$ is a zero of a nonzero δ-polynomial in $\mathbb{C}(z)\{y_1\}$ of denomination $\leq \frac{\log k}{\log 2} + 1$, and hence, by the main result in [1], u does not admit rational approximation at ∞ of order $> \left[\frac{\log k}{\log 2}\right] + 1$. Thus, the Schmidt-Osgood inequality yields a vast (and easy) improvement of the C. L. Siegel(-B. P. Gill) exponent, but of course not so good as the ultimate improvement achieved by K. F. Roth(-S. Uchiyama).

In what follows, we introduce some machinery that may be useful in attacking the problem. Much of the development can be extended from ordinary to partial differential polynomials, but for the sake of simplicity we do not do so here. The problem is hard enough in the ordinary case. Moreover, starting with §4 we shall consider only the case n=1. In §5 we conclude by showing, in that case, that the answer to Schmidt's question is affirmative when $d \leq 3$.

§1. Differential homogeneity and the spaces $\mathcal{B}_n(d)$.

A δ-polynomial $P \in \mathcal{F}\{y_0,\dots,y_n\}$ is called *δ-homogeneous* if, for a new δ-indeterminate t and some $d \in \mathbf{N}$, we have

$$P(ty_0,\dots,ty_n) = t^d P(y_0,\dots,y_n).$$

When $P \neq 0$, this d is unique, namely, $d = \deg P$. For example, $y_0y_1' - y_0'y_1$ is δ-homogeneous of degree 2 (whereas $y_0y_1' - 2y_0'y_0$ is not). Of course, 0 is δ-homogeneous of every degree. The set of elements of $\mathcal{F}\{y_0,\dots y_n\}$ that are δ-homogeneous of degree d has an obvious structure of vector space over $\mathcal{F}$; we shall denote this vector space by $\mathcal{B}_n(d)$.

It is clear that the formula $F \mapsto y_0^d F(y_1/y_0,\dots,y_n/y_0)$ defines a homomorphism $\mathcal{V}_n(d) \to \mathcal{B}_n(d)$, that the formula $P \mapsto P(1,y_1,\dots,y_n)$ defines a homomorphism $\mathcal{B}_n(d) \to \mathcal{V}_n(d)$, and that these are inverse to each other. In particular, therefore

$$\dim \mathcal{V}_n(d) = \dim \mathcal{B}_n(d).$$

In what follows, we find it advantageous to turn our attention to $\mathfrak{B}_n(d)$.

§2. Four easy lemmas.

Consider any $P \in \mathcal{F}\{y_0,\dots,y_n\}$.

If Φ is a vector space basis of $\mathcal{F}$ over a δ-subfield $\mathcal{F}_0$ of $\mathcal{F}$, there is a unique way of writing P in the form $P = \sum_{\varphi\in\Phi}\varphi P_\varphi$ where $P_\varphi \in \mathcal{F}_0\{y_0,\dots,y_n\}$ for every φ.

Lemma 1. *P is δ-homogeneous of degree d if and only if every P_φ is.*

Proof. We have the two equations $P(ty_0,\dots,ty_n) = \sum\varphi P_\varphi(ty_0,\dots,ty_n)$ and $t^d P(y_0,\dots,y_n) = \sum\varphi t^d P_\varphi(y_0,\dots,y_n)$. If P is δ-homogeneous of degree d, then the two left members are equal, and if every P_φ is δ-homogeneous of degree d, then the right members are.

Corollary. *The dimension of $\mathfrak{B}_n(d)$ is independent of the choice of the δ-field $\mathcal{F}$ of coefficients.*

Proof. By the lemma, there exists a basis of $\mathfrak{B}_n(d)$ consisting of elements of $\mathcal{F}_0\{y_0,\dots,y_n\}$.

If P is homogeneous of degree d, then there is a unique way of writing P in the form

$$P = \sum_{d_0+\cdots+d_n=d} P_{d_0\dots d_n}$$

where every $P_{d_0\dots d_n}$ is, for each index j, homogeneous in $(y_j, y_j', y_j'', \dots)$ of degree d_j.

Lemma 2. *Let P be homogeneous of degree d, as above. Then P is δ-homogeneous if and only if every $P_{d_0\dots d_n}$ is.*

Proof. We have the two equations $P(ty_0,\dots,ty_n) = \sum P_{d_0\dots d_n}(ty_0,\dots,ty_n)$

and $t^dP(y_0,\dots,y_n) = \Sigma t^d P_{d_0\dots d_n}(y_0,\dots,y_n)$. If P is δ-homogeneous, then the left members are equal, and if every $P_{d_0\dots d_n}$ is δ-homogeneous, then the right members are.

Recall that the *weight* of a δ-monomial $\prod_{0\le j\le n,\,k\in\mathbb{N}} (y_j^{(k)})^{d_{jk}}$ is the natural number $\sum_{0\le j\le n,\,k\in\mathbb{N}} kd_{jk}$, and that a δ-polynomial P is *isobaric* of weight w if every δ-monomial that appears in P with a nonzero coefficient has weight equal to w. Thus, 0 is isobaric of every weight, and the isobaric δ-polynomials of a given weight form a vector space over $\mathcal{F}$. Any δ-polynomial P can be written in a unique way in the form $P = \Sigma_{w\in\mathbb{N}} P_w$ where for each w, P_w is isobaric of weight w. When P is isobaric, we denote its weight by *wt*P.

<u>Lemma</u> 3. *P is δ-homogeneous of degree d if and only every P_w is.*

Proof. We have the two equations $P(ty_0,\dots,ty_n) = \Sigma P_w(ty_0,\dots,ty_n)$ and $t^dP(y_0,\dots,y_n) = \Sigma t^d P_w(y_0,\dots,y_n)$. If P is δ-homogeneous, the left members are equal and hence so are the right members; it is easy to see, for each w, that $P_w(ty_0,\dots,ty_n)$ and $t^dP(y_0,\dots,y_n)$ are isobaric of weight w as δ-polynomials in $(t,y_0,\dots,y_n)$, and hence are equal. If every P_w is δ-homogeneous, then the two right members are equal, and hence so are the left members.

<u>Lemma</u> 4. *Let P be δ-homogeneous of degree $d>0$, with $P\neq 0$, and let p denote the order of P. Then, for each index j, $\partial P/\partial y_j^{(p)}$ is δ-homogeneous of degree $d-1$.*

Proof. $t\dfrac{\partial P}{\partial y_j^{(p)}}(ty_0,\dots,ty_n) = \dfrac{\partial((ty_j)^{(p)})}{\partial y_j^{(p)}} \dfrac{\partial P}{\partial y_j^{(p)}}(ty_0,\dots,ty_n) = \dfrac{\partial}{\partial y_j^{(p)}}P(ty_0,\dots,ty_n) = \dfrac{\partial}{\partial y_j^{(p)}}t^dP(y_0,\dots,y_n) = t^d \dfrac{\partial P}{\partial y_j^{(p)}}(y_0,\dots,y_n)$, so that $\dfrac{\partial P}{\partial y_j^{(p)}}(ty_0,\dots,ty_n) = t^{d-1} \dfrac{\partial P}{\partial y_j^{(p)}}(y_0,\dots,y_n)$.

§3. The derivations $\mathfrak{D}_r$ and a criterion for differential homogeneity.

For each $r \in \mathbb{N}$ the formula

$$\mathfrak{D}_r = \sum_{k\in\mathbb{N}} \binom{r+k}{r} \sum_{0\leq j\leq n} y_j^{(k)} \frac{\partial}{\partial y_j^{(r+k)}}$$

defines a derivation $\mathfrak{D}_r$ of $\mathcal{F}\{y_0,\ldots,y_n\}$ over $\mathcal{F}$. If $P \in \mathcal{F}\{y_0,\ldots,y_n\}$ is homogeneous of degree d, then so is $\mathfrak{D}_r P$. If P is of order p, then $\mathfrak{D}_r P$ is of order $\leq p$ and $\mathfrak{D}_r P = 0$ when $r > p$. If P is isobaric of weight w, then $\mathfrak{D}_r P$ is isobaric of weight w-r.

We may regard δ, too, as a derivation of $\mathcal{F}\{y_0,\ldots,y_n\}$. If P is homogeneous of degree d, then so is $\delta P(=P')$. If $P \notin \mathcal{F}$ and $ord P = p$, then $ord\,\delta P = p+1$. If P is isobaric of weight w, then δP is isobaric of weight w+1

Lemma 5. $[\mathfrak{D}_r,\mathfrak{D}_s] = 0 \quad (r,s \in \mathbb{N})$.
$[\mathfrak{D}_0,\delta] = 0; \qquad [\mathfrak{D}_r,\delta] = \mathfrak{D}_{r-1} \quad (r \in \mathbb{N},\ r \neq 0)$.

Proof. Since the commutator of two derivations is a derivation, it suffices to verify the formulae at each $y_j^{(k)}$ and each element of $\mathcal{F}$.

The following theorem is the δ-analog of Euler's well known criterion for homogeneity of polynomials

Theorem. *Let* $P \in \mathcal{F}\{y_0,\ldots,y_n\}$. *A necessary and sufficient condition that* P *be* δ*-homogeneous of degree* d *is that*

$$\mathfrak{D}_0 P = dP, \qquad \mathfrak{D}_r P = 0 \quad (r \in \mathbb{N},\ r \neq 0).$$

Remark The condition $\mathfrak{D}_0 P = dP$ is, by Euler's criterion, just the condition that P be homogeneous of degree d.

Proof. Let $P(ty_0,\ldots,ty_n) = t^d P(y_0,\ldots,y_n)$. Then

$$\sum_{k\in\mathbb{N}}\binom{r+k}{r}\sum_{0\le j\le n} y_j^{(k)}\frac{\partial P}{\partial y_j^{(r+k)}}(ty_0,\ldots,ty_n)$$

$$= \sum_{k\in\mathbb{N}}\sum_{0\le j\le n}\frac{\partial P}{\partial y_j^{(r+k)}}(ty_0,\ldots,ty_n)\frac{\partial}{\partial t^{(r)}}(ty_j)^{(r+k)}$$

$$= \frac{\partial}{\partial t^{(r)}}P(ty_0,\ldots,ty_n) = \frac{\partial}{\partial t^{(r)}}t^dP(y_0,\ldots,y_n)$$

$$= \begin{cases} dt^{d-1}P(y_0,\ldots,y_n) & (r = 0) \\ 0 & (r \neq 0). \end{cases}$$

Putting $t = 1$, we see that the condition in the theorem is satisfied.

Conversely, let the condition be satisfied. For sufficiently big p (for example, for $p > \mathit{ord}P$) we have $\frac{\partial}{\partial t^{(p)}}P(ty_0,\ldots,ty_n) = 0$. Let $r \in \mathbb{N}$, $r \neq 0$, and suppose that $\frac{\partial}{\partial t^{(p)}}P(ty_0,\ldots,ty_n) = 0$ for all $p>r$. Then

$$t\frac{\partial}{\partial t^{(r)}}P(ty_0,\ldots,ty_n) = \sum_{0\le i<\infty}\binom{r+i}{r}t^{(i)}\frac{\partial}{\partial t^{(r+i)}}P(ty_0,\ldots,ty_n)$$

$$= \sum_{0\le i<\infty}\binom{r+i}{r}t^{(i)}\sum_{\substack{0\le j\le n\\ i\le k<\infty}}\frac{\partial P}{\partial y_j^{(r+k)}}(ty_0,\ldots,ty_n)\frac{\partial}{\partial t^{(r+i)}}(ty_j)^{(r+k)}$$

$$= \sum_{0\le k<\infty}\sum_{0\le i\le k}\sum_{0\le j\le n}\binom{r+i}{r}t^{(i)}\frac{\partial P}{\partial y_j^{(r+k)}}(ty_0,\ldots,ty_n)\binom{r+k}{r+i}y_j^{(k-i)}$$

(note that $\binom{r+i}{r}\binom{r+k}{r} = \binom{r+k}{r}\binom{k}{i}$)

$$= \sum_{k\in\mathbb{N}}\binom{r+k}{r}\sum_{0\le j\le n}\sum_{0\le i\le k}\binom{k}{i}t^{(i)}y_j^{(k-i)}\frac{\partial P}{\partial y_j^{(r+k)}}(ty_0,\ldots,ty_n)$$

$$= \sum_{k\in\mathbb{N}}\binom{r+k}{r}\sum_{0\le j\le n}(ty_j)^{(k)}\frac{\partial P}{\partial y_j^{(r+k)}}(ty_0,\ldots,ty_n)$$

$$= (\mathfrak{D}_rP)(ty_0,\ldots,ty_n) = 0.$$

Therefore $\frac{\partial}{\partial t^{(p)}}P(ty_0,\ldots,ty_n) = 0$ for every $p>0$, that is, $P(ty_0,\ldots,ty_n)$ is free of t', t'', t''',... . Now the same computation, this time with $r=0$, shows that $t\frac{\partial P}{\partial t}(ty_0,\ldots,ty_n) = dP(ty_0,\ldots,ty_n)$, so that $t^{-d}P(ty_0,\ldots,ty_n)$ is free also of t. Therefore $t^{-d}P(ty_0,\ldots,ty_n) = P(y_0,\ldots,y_n)$, and P is δ-homogeneous of degree d.

<u>Corollary</u>. *Let* $P \in \mathcal{F}\{y_0,\ldots,y_n\}$, $P \notin \mathcal{F}$. *Then* P' *is not* δ*-homogeneous.*

Proof. Assume that δP is δ-homogeneous. Then, by the theorem, $\mathfrak{D}_r\delta P = 0$ for all $r>0$. For sufficiently big p, we have $\mathfrak{D}_p P = 0$. let $r \in \mathbb{N}$ and suppose that $\mathfrak{D}_p P = 0$ for all $p>r$. By lemma 5 then $\mathfrak{D}_r P = \mathfrak{D}_{r+1}\delta P - \delta\mathfrak{D}_{r+1}P = 0$. This shows that $\mathfrak{D}_r P = 0$ for all $r\in\mathbb{N}$ and hence, by the theorem, that $P \in \mathcal{F}$, contradicting the hypothesis.

§4. The differential polynomials Q_p.

We suppose henceforth that $n = 1$, *and write* (y,z) *instead of* (y_0,y_1).

For any $p \in \mathbb{N}$, put $h = h(p) = \left[\frac{p-1}{2}\right]$, so that $p = \begin{cases} 2h+1 & (p \text{ odd}), \\ 2h+2 & (p \text{ even}), \end{cases}$ and define the δ-polynomial Q_p by the formula

$$Q_p = \sum_{0\leqq i\leqq h} \binom{p}{i} (y^{(i)}z^{(p-i)} - y^{(p-i)}z^{(i)}).$$

(Note that the condition $0\leqq i\leqq h$ for i here is just the condition $0\leqq i<\frac{p}{2}$.) Then Q_p is homogeneous of degree 2, isobaric of weight p, and of order p. It is clear, for any $k\in\mathbb{N}$ with $k<p$, that $Q_p,\ldots,Q_{p-k}$ are algebraically independent over the field $\mathcal{F}(y,z,\ldots,y^{(p-k-1)},z^{(p-k-1)})$.

We are going to compute $\mathfrak{D}_k Q_p$ for every $k\in\mathbb{N}$. of course, $\mathfrak{D}_0 Q_p = 2Q_p$, and if $k>p$, then $\mathfrak{D}_k Q_p = 0$. If $p-h\leqq k\leqq p$ (that is, $\frac{p}{2}<k\leqq p$), then

$$\begin{aligned}\mathfrak{D}_k Q_p &= \sum_{0\leq i\leq p-k} \binom{p}{i}\binom{p-i}{k} (y^{(i)}z^{(p-i-k)} - y^{(p-i-k)}z^{(i)}) \\ &= \binom{p}{k} \sum_{0\leq i\leq p-k} \binom{p-k}{i} (y^{(i)}z^{(p-k-i)} - y^{(p-k-i)}z^{(i)}) = 0.\end{aligned}$$

Now let $1\leq k\leq p-h-1$ (that is, $1\leq k\leq\frac{p}{2}$). Then

$$\begin{aligned}\mathfrak{D}_k Q_p = &\sum_{k\leq i\leq h} \binom{p}{i}\binom{i}{k} (y^{(i-k)}z^{(p-i)} - y^{(p-i)}z^{(i-k)}) \\ &+ \sum_{0\leq i\leq h} \binom{p}{i}\binom{p-i}{k} (y^{(i)}z^{(p-i-k)} - y^{(p-i-k)}z^{(i)})\end{aligned}$$

$$= \begin{bmatrix}p\\k\end{bmatrix}\sum_{k\le i\le h}\begin{bmatrix}p-k\\i-k\end{bmatrix}(y^{(i-k)}z^{(p-i)}-y^{(p-i)}z^{(i-k)})$$

$$+ \begin{bmatrix}p\\k\end{bmatrix}\sum_{0\le i\le h}\begin{bmatrix}p-k\\i\end{bmatrix}(y^{(i)}z^{(p-k-i)}-y^{(p-k-i)}z^{(i)})$$

$$= \begin{bmatrix}p\\k\end{bmatrix}\Big(\sum_{0\le i\le h-k}+\sum_{0\le i\le h}\Big)\begin{bmatrix}p-k\\i\end{bmatrix}(y^{(i)}z^{(p-k-i)}-y^{(p-k-i)}z^{(i)})$$

$$= \begin{bmatrix}p\\k\end{bmatrix}\Big(\sum_{0\le i\le h-k}+\sum_{0\le i\le p-k-h-1}+\sum_{p-k-h\le i\le h}\Big)\begin{bmatrix}p-k\\i\end{bmatrix}(y^{(i)}z^{(p-k-i)}-y^{(p-k-i)}z^{(i)}).$$

The last of these three sums Σ is evidently 0. Moreover, $\sum_{0\le i\le h-k}=\sum_{0\le i<\frac{p-k}{2}}-\sum_{h-k+1\le i<\frac{p-k}{2}}$ and $\sum_{0\le i\le p-k-h-1}=\sum_{0\le i<\frac{p-k}{2}}-\sum_{p-k-h\le i<\frac{p-k}{2}}$. Hence

$$\mathfrak{D}_kQ_p = 2\begin{bmatrix}p\\k\end{bmatrix}Q_{p-k}-\begin{bmatrix}p\\k\end{bmatrix}\Big(\sum_{h-k+1\le i<\frac{p-k}{2}}+\sum_{p-k-h\le i<\frac{p-k}{2}}\Big)\begin{bmatrix}p-k\\i\end{bmatrix}(y^{(i)}z^{(p-k-i)}-y^{(p-k-i)}z^{(i)}).$$

Since $p-k-h = h-k+1$ or $h-k+2$ according as p is odd or even, we may write

$$\mathfrak{D}_kQ_p = 2\begin{bmatrix}p\\k\end{bmatrix}Q_{p-k}-2\begin{bmatrix}p\\k\end{bmatrix}\sum_{p-k-h\le i<\frac{p-k}{2}}\begin{bmatrix}p-k\\i\end{bmatrix}(y^{(i)}z^{(p-k-i)}-y^{(p-k-i)}z^{(i)})$$

$$-f_p\begin{bmatrix}p\\k\end{bmatrix}\begin{bmatrix}p-k\\h+1\end{bmatrix}(y^{(p-k-h-1)}z^{(h+1)}-y^{(h+1)}z^{(p-k-h-1)}),$$

where we have put

$$f_p = \begin{cases}0 & (p \text{ odd}),\\ 1 & (p \text{ even}).\end{cases}$$

Every term in the sum Σ has order $< p-k$, and when $k < p-h-1$ then so does $y^{(p-k-h-1)}z^{(h+1)}-y^{(h+1)}z^{(p-k-h-1)}$. However, when p is even and $k = p-h-1 = h+1 = p/2$, then

$$y^{(p-k-h-1)}z^{(h+1)}-y^{(h+1+}z^{(p-k-h-1)} = yz^{(h+1)}-y^{(h+1)}z$$

$$= Q_{h+1}-\sum_{1\le i<\frac{h+1}{2}}\begin{bmatrix}h+1\\i\end{bmatrix}(y^{(i)}z^{(h+1-i)}-y^{(h+1-i)}z^{(i)}).$$

These results are summarized in the following lemma.

<u>Lemma</u> 6. *Let* $p,k \in \mathbb{N}$, $pk \ne 0$.

If $k > p/2$, *then* $\mathfrak{D}_kQ_p = 0$.

If $k = p/2$ *(so that* p *is even), then*

$$\mathfrak{D}_kQ_p = \begin{bmatrix}p\\k\end{bmatrix}(yz^{(k)}-y^{(k)}z) = \begin{bmatrix}p\\k\end{bmatrix}Q_{p-k}-\begin{bmatrix}p\\k\end{bmatrix}\sum_{1\le i<k/2}\begin{bmatrix}k\\i\end{bmatrix}(y^{(i)}z^{(k-i)}-y^{(k-i)}).$$

If $1 \le k < p/2$, *then*

$$\mathfrak{D}_k Q_p = 2\binom{p}{k}\sum_{0\le i<\frac{p}{2}-k}\binom{p-k}{i}(y^{(i)}z^{(p-k-i)}-y^{(p-k-i)}z^{(i)})$$
$$+ f_p\binom{p}{k}\binom{p-k}{\frac{p}{2}}(y^{(\frac{p}{2}-k)}z^{(\frac{p}{2})}-y^{(\frac{p}{2})}z^{(\frac{p}{2}-k)})$$
$$= 2\binom{p}{k}Q_{p-k} - 2\binom{p}{k}\sum_{\frac{p-2k}{2}<i<\frac{p-k}{2}}\binom{p-k}{i}(y^{(i)}z^{(p-k-i)}-y^{(p-k-i)}z^{(i)})$$
$$- f_p\binom{p}{k}\binom{p-k}{\frac{p}{2}}(y^{(\frac{p}{2}-k)}z^{(\frac{p}{2})}-y^{(\frac{p}{2})}z^{(\frac{p}{2}-k)}),$$

where $f_p = 0$ *or* $f_p = 1$ *according as* p *is odd or even.*

The usefulness of the δ-polynomials Q_p stems from the following lemma.

Lemma 7. *Let* $p\in\mathbb{N}$, $p\neq 0$, *write* $p=2h+1$ *or* $p=2h+2$ *according as* p *is odd or even, and let* $k\in\mathbb{N}$, $0\le k\le h$. *Let* $P \in \mathcal{F}\{y,z\}$, *ord*$P \le p$, *and suppose that* $\mathfrak{D}_{p-i}P = 0$ $(0\le i\le k)$. *Then*

$$P \in \mathcal{F}[(y^{(i)},z^{(i)})_{0\le i<p-k-1}, (Q_{p-i})_{0\le i\le k}].$$

Proof. We first let k=0. Let P be as stated, so that $y\partial P/\partial y^{(p)}+z\partial P/\partial z^{(p)} = \mathfrak{D}_pP = 0$. Therefore $\partial P/\partial y^{(p)}$ and $\partial P/\partial z^{(p)}$ are divisible by z and y, respectively, and there exists an $A \in \mathcal{F}\{y,z\}$ such that $\partial P/\partial y^{(p)} = -zA$ and $\partial P/\partial z^{(p)} = yA$. Evidently *ord*$A \le p$ and *deg*$A <$ *deg*P, and

$$\mathfrak{D}_pA = y\frac{\partial A}{\partial y^{(p)}} + z\frac{\partial A}{\partial z^{(p)}} = \frac{\partial^2 P}{\partial y^{(p)}\partial z^{(p)}} - \frac{\partial^2 P}{\partial z^{(p)}\partial y^{(p)}} = 0.$$

Arguing by induction on *deg*P, we may therefore suppose that $A \in \mathcal{F}[y,z,\ldots,y^{(p-1)},z^{(p-1)},Q_p]$, so that we may write $A = \sum C_iQ_p{}^i$, where $C_i \in \mathcal{F}[y,z,\ldots,y^{(p-1)},z^{(p-1)}]$ for each i. Then

$$\frac{\partial}{\partial y^{(p)}}(P - \sum\tfrac{1}{i+1}C_iQ_p{}^{i+1}) = \frac{\partial P}{\partial y^{(p)}} - \sum C_iQ_p{}^i\,\frac{\partial Q_p}{\partial y^{(p)}} = -Az + \sum C_iQ_p{}^iz = 0,$$

and similarly $\frac{\partial}{\partial z^{(p)}}(P - \sum\frac{1}{i+1}C_iQ_p{}^{i+1}) = 0$. Thus, $P - \sum\frac{1}{i+1}C_iQ_p{}^{i+1}$ is free of $y^{(p)}$ and $z^{(p)}$ and hence $P \in \mathcal{F}[y,z,\ldots,y^{(p-1)},z^{(p-1)},Q_p]$. This proves the lemma when $k=0$.

Now let $0<k\leq h$ and suppose the lemma proved for lower values of k. Since the hypothesis for k implies the hypothesis for $k-1$, we may write

$$P = \sum P_j M_j(Q_{p-k+1},\ldots,Q_p),$$

where the M_j are distinct monomials in k arguments, and the P_j are elements of $\mathcal{F}[y,z,\ldots,y^{(p-k)},z^{(p-k)}]$. Now, $p-k\geq p-h>\frac{p}{2}\geq\frac{p-i}{2}$ $(0\leq i\leq k-1)$, and hence by lemma 6, $\mathfrak{D}_{p-k}Q_{p-i}=0$ $(0\leq i\leq k-1)$. Therefore $\sum\mathfrak{D}_{p-k}P_j\cdot M_j(Q_{p-k+1},\ldots,Q_p) = \mathfrak{D}_{p-k}P = 0$. Since $Q_{p-k+1},\ldots,Q_p$ are algebraically independent over $\mathcal{F}(y,z,\ldots,y^{(p-k)},z^{(p-k)})$, this implies that $\mathfrak{D}_{p-k}P_j = 0$ for each j. Because $\mathit{ord}P_j \leq p-k$, what we have already proved shows that $P_j \in \mathcal{F}[y,z,\ldots y^{(p-k-1)},z^{(p-k-1)},Q_{p-k}]$ for each j, and hence that P satisfies the displayed condition in the statement of the lemma.

§5 The computation of $\mathcal{B}_1(d)$ for $d \leq 3$.

For any $d\in\mathbb{N}$, the elements of $\mathcal{B}_1(d)$ of weight 0, or what is the same thing, of order ≤ 0, are just the homogeneous polynomials in (y,z) of degree d, that is, are the linear combinations over $\mathcal{F}$ of the $d+1$ monomials $y^{d-i}z^i$ $(0\leq i\leq d)$. It remains to describe the elements of $\mathcal{B}_1(d)$ of order ≥ 1, and by lemma 3 we may confine ourselves to those that are isobaric. In this connection, the following lemma can be useful.

Lemma 8. *Let* $m,p \in \mathbb{N}$, $p\neq 0$, *let* $A,C \in \mathcal{F}\{y,z\}$, $\mathit{ord}A\leq p$, $\mathit{ord}C\leq p$, *and suppose that* C *is homogeneous in* $(y^{(p)},z^{(p)})$ *and is not in the ideal* (y, z). *Then* $\mathfrak{D}_p A \neq C(yz^{(p)}-y^{(p)}z)^m$.

Proof. First let $m=0$. If we had $C = \mathfrak{D}_p A = y\partial A/\partial y^{(p)}+z\partial A/\partial z^{(p)}$, then C would be in the ideal (y, z) contrary to hypothesis. Now let $m>0$ and suppose the lemma proved for lower values of m. Assume that $\mathfrak{D}_p A = C(yz^{(p)}-y^{(p)}z)^m$, and let e denote the degree of C in $(y^{(p)},z^{(p)})$. Then

$$0 = y\frac{\partial A}{\partial y^{(p)}}+z\frac{\partial A}{\partial z^{(p)}} - (yz^{(p)}-y^{(p)}z)C(yz^{(p)}-y^{(p)}z)^{m-1}$$
$$= y\left[\frac{\partial A}{\partial y^{(p)}} - z^{(p)}C(yz^{(p)}-y^{(p)}z)^{m-1}\right] + z\left[\frac{\partial A}{\partial z^{(p)}} + y^{(p)}C(yz^{(p)}-y^{(p)}z)^{m-1}\right].$$

Therefore the large parentheses are divisible by z and y, respectively, and

$$\frac{\partial A}{\partial y^{(p)}} - z^{(p)}C(yz^{(p)}-y^{(p)}z)^{m-1} = -zB, \quad \frac{\partial A}{\partial z^{(p)}} + y^{(p)}C(yz^{(p)}-y^{(p)}z)^{m-1} = yB,$$

where $B \in \mathfrak{F}\{y,z\}$ and $ord B \le p$. Applying $\partial/\partial z^{(p)}$ resp. $\partial/y^{(p)}$ to the first resp. second of these equations, and then subtracting, we find the equation $(m+1+e)C(yz^{(p)}-y^{(p)}z)^{m-1} = \mathfrak{D}_p B$, contradicting the lemma for m-1.

It is obvious that $\mathfrak{B}_1(0) = \mathfrak{F}$. Therefore 1 constitutes a basis of $\mathfrak{B}_1(0)$, and $dim\ \mathfrak{B}_1(0) = 1$.

If $P \in \mathfrak{B}_1(1)$, $P\neq 0$, $ord P\geq 1$, then we may write $P = ay^{(p)}+bz^{(p)}$ + (terms of order $\leq p-1$), with $a,b \in \mathfrak{F}$ not both 0; by our theorem then $0 = \mathfrak{D}_p P = ay+bz$, a contradiction. Therefore the nonzero elements of $\mathfrak{B}_1(1)$ all have order 0, so that y,z constitute a basis of $\mathfrak{B}_1(1)$, and $dim\ \mathfrak{B}_1(1) = 2$.

Let $P \in \mathfrak{B}_1(2)$, $P\neq 0$, $ord P=p\geq 1$, and suppose that P is isobaric. By our theorem, lemma 7, and the homogeneity of P of degree 2, we may write

$$P = \sum_{0\leq i\leq h} a_i Q_{p-i} + R,$$

where $a_i \in \mathfrak{F}$ $(0\leq i\leq h)$, and R is homogeneous of degree 2 and of order $\leq p-h-1$; since $ord P=p$, we have $a_0\neq 0$; since P is isobaric, then $wt P=p$, $a_i=0$ $(1\leq i\leq h)$, and $wt R=p$: $P = a_0 Q_p + R$. When p=1, then $ord R\leq 0$ whence R=0 (since $wt R=1$), so that $P=a_0Q_1$. It is clear, moreover, that every δ-polynomial of this form is in $\mathfrak{B}_1(2)$. When p=2, then h=0, $ord R\leq 1$, and $0 = \mathfrak{D}_1 P = a_0\cdot 2\ Q_1+ \mathfrak{D}_1 R$, which contradicts lemma 8. When $p\geq 3$, then h=1, $ord R\leq p-2$, and $0 = \mathfrak{D}_1 P = pa_0 Q_{p-1}$ + (terms of order $\leq p-2$), whence the contradiction that $a_0=0$. This shows that the δ-polynomials y^2, yz, z^2, Q_1 constitute a basis of $\mathfrak{B}_1(2)$ and hence that $dim\ \mathfrak{B}_1(2) = 4$

Let $P \in \mathcal{B}_1(3)$, $P\neq 0$, $ord P=p\geq 1$, and suppose that P is isobaric. By our theorem, lemma 7, and the homogeneity of P of degree 3, we may write

$$P = \sum_{0\leq i\leq h} A_i Q_{p-i} + R,$$

where, for each i, A_i is homogeneous of degree 1 and of order $\leq p-h-1$, and R is homogeneous of degree 3 and of order $\leq p-h-1$; since $ord P=p$, we have $A_0\neq 0$. By lemma 4, the δ-polynomial $\partial P/\partial z^{(p)} = A_0 y$ is δ-homogeneous, and hence so is A_0. By our results on $\mathcal{B}_1(1)$, then $A_0 = ay+bz$, with $a,b \in \mathcal{F}$ not both 0. Since P is isobaric, it follows that its weight is p, that $wt A_i=i$ $(0\leq i\leq h)$, and that $wt R=p$.

When $p=1$, then $h=0$, $wt R=1$, and $ord R\leq 0$, so that $R=0$ and the equation for P displayed above reduces to $P = ayQ_1+bzQ_1$. It is clear that every P of this form is in $\mathcal{B}_1(3)$.

When $p=2$, then $h=0$, $wt R=2$, and $ord R\leq 1$, so that the equation for P reduces to $P = A_0Q_2+R$. Therefore $0 = \mathcal{D}_1 P = A_0\cdot 2Q_1+\mathcal{D}_1 R = \mathcal{D}_1(2(ay'+bz')Q_1+R)$, whence by lemma 7, $2(ay'+bz')Q_1+R \in \mathcal{F}[y,z,Q_1]$; since this δ-polynomial is homogeneous of degree 3 and isobaric of weight 2, it must be 0, so that $R = -2(ay'+bz')Q_1$ and $P = a(yQ_2-2y'Q_1) + b(zQ_2-2z'Q_1)$. It is, morover, easy to see that every P of this form is in $\mathcal{B}_1(3)$.

Now assume that $p\geq 3$. Then $h\geq 1$ and, for every integer k with $1\leq k\leq h$, we see by our theorem and lemmas 6 and 7 that, for every $k\in\mathbb{N}$ with $1\leq k\leq h$,

$$\begin{aligned} 0 = \mathcal{D}_k P &= \sum_{k\leq i\leq h} \mathcal{D}_k A_i\cdot Q_{p-i} + \sum_{\substack{0\leq i\leq h\\ i\leq p-2k}} A_i \mathcal{D}_k Q_{p-i} + \mathcal{D}_k R \\ &= \mathcal{D}_k A_k\cdot Q_{p-k} + A_0\mathcal{D}_k Q_p + (\text{terms of order} < p-k) \\ &= \left(\mathcal{D}_k A_k+A_0\cdot 2\begin{bmatrix}p\\k\end{bmatrix}\right)(yz^{(p-k)}-y^{(p-k)}z) + (\text{terms of order} < p-k). \end{aligned}$$

Therefore $0 = \mathcal{D}_k A_k+2\begin{bmatrix}p\\k\end{bmatrix}A_0 = \mathcal{D}_k\left(A_k+2\begin{bmatrix}p\\k\end{bmatrix}A^{[k]}\right)$, where we have put

$$A^{[k]} = ay^{(k)}+bz^{(k)}.$$

Hence by lemma 7, $A_k+2\begin{bmatrix}p\\k\end{bmatrix}A^{[k]} \in \mathcal{F}[y,z,\dots,y^{(k-1)},z^{(k-1)},Q_k]$. Since this δ-polynomial is homogeneous of degree 1 and isobaric of weight k, it must be 0, that is

$$A_k = -2\begin{bmatrix}p\\k\end{bmatrix}A^{[k]} \quad (1\leq k\leq h).$$

It follows from the above and lemma 6 that

$$0 = \mathfrak{D}_h P = \mathfrak{D}_h A_h \cdot Q_{p-h} + \sum_{\substack{0\le i\le h\\ i\le p-2h}} A_i \mathfrak{D}_h Q_{p-i} + \mathfrak{D}_h R$$

$$= -2\binom{p}{h} A_0 Q_{p-h} + A_0 \mathfrak{D}_h Q_p - 2\binom{p}{1} A^{[1]} \mathfrak{D}_h Q_{p-1} - (1-\delta_{p,4}) f_p 2\binom{p}{2} A^{[2]} \mathfrak{D}_h Q_{p-2} + \mathfrak{D}_h R$$

$$= -2\binom{p}{h} A_0 \sum_{1\le i<\frac{p-h}{2}} \binom{p-h}{i}\left(y^{(i)}z^{(p-h-i)} - y^{(p-h-i)}z^{(i)}\right)$$

$$+ f_p\binom{p}{h}\binom{p-h}{1}A_0\left(y'z^{(p-h-1)} - y^{(p-h-1)}z'\right) - 2\binom{p}{1}(1+f_p)A^{[1]}\left(yz^{(p-h-1)} - y^{(p-h-1)}z\right)$$

$$-2\binom{p}{2}\binom{p-2}{h}(1-\delta_{p,4})f_pA^{[2]}\left(yz^{(p-h-2)}-y^{(p-h-2)}z\right) + \mathfrak{D}_hR.$$

When p=3, this reduces to the equation $0 = -6A^{[1]}(yz'-y'z) + \mathfrak{D}_1R$, contradicting lemma 8. When p=4, it reduces to

$$0 = -12A_0(y'z''-y''z') -16A^{[2]}(yz''-y''z) + \mathfrak{D}_1R$$

$$= y\left[-12a(y'z''-y''z')-16A^{[1]}z''+\frac{\partial R}{\partial y'}\right] + z\left[-12b(y'z''-y''z')+16A^{[1]}y'' + \frac{\partial R}{\partial z'}\right].$$

Hence the two large parentheses are divisible by z and y, respectively; as They are homogeneous of degree 2, isobaric of weight 3, and of order ≤ 3, they must be 0. Applying $\partial/\partial z'$ to the first one and $\partial/\partial y'$ to the second, and then subtracting, we find the contradiction $4(ay''+bz'')=0$.

Now assume that p≥5, so that h≥2. The above computations then yield

$$0 = -(2-f_p)\binom{p}{h}\binom{p-h}{1}A_0\left(y'z^{(p-h-i)}-y^{(p-h-i)}z'\right)$$

$$-2\binom{p}{h}A_0\sum_{2\le i<\frac{p-h}{2}}\binom{p-h}{i}\left(y^{(i)}z^{(p-h-i)}-y^{(p-h-i)}z^{(i)}\right)$$

$$-2\binom{p}{1}(1+f_p)A^{[1]}\left(yz^{(p-h-1)}-y^{(p-h-1}z\right)$$

$$-2\binom{p}{2}\binom{p-2}{h}f_pA^{[2]}\left(yz^{(p-h-2)}-y^{(p-h-2)}\right) + \mathfrak{D}_hR$$

$$= yY + zZ,$$

where we have put

$$Y = -(2-f_p)\binom{p}{h}\binom{p-h}{1}a(y'z^{(p-h-1)}-y^{(p-h-1)}z')$$
$$-2\binom{p}{h}a\sum_{2\le i<\frac{p-h}{2}}\binom{p-h}{i}(y^{(i)}z^{(p-h-i)}-y^{(p-h-i)}z^{(i)})$$
$$-2\binom{p}{1}(1+f_p)A^{[1]}z^{(p-h-1)}-2\binom{p}{2}\binom{p-2}{h}f_pA^{[2]}z^{(pih-2)}+\frac{\partial R}{\partial y^{(h)}},$$

$$Z = -(2-f_p)\binom{p}{h}\binom{p-h}{1}b(y'z^{(p-h-1)}-y^{(p-h-1)}z')$$
$$-2\binom{p}{h}b\sum_{2\le i<\frac{p-h}{2}}\binom{p-h}{i}(y^{(i)}z^{(p-h-i)}-y^{(p-h-i)}z^{(i)})$$
$$+2\binom{p}{1}(1+f_p)A^{[1]}y^{(p-h-1)}+2\binom{p}{2}\binom{p-2}{h}f_pA^{[2]}y^{(p-h-2)}+\frac{\partial R}{\partial z^{(h)}}.$$

Since $yY+zZ=0$, Y is divisible by z. As Y is homogeneous of degree 2, isobaric of weight $p-h$, and of order $\le p-h-1$, we have $Y=0$; similarly, $Z=0$. Computing $\partial Y/\partial z^{(h)}-\partial Z/\partial y^{(h)}$, we arrive, according as p is odd or even, at the equation

$$\left(2\binom{p}{h}\binom{p-h}{1}+4\binom{p}{1}+60\delta_{p,5}\right)A^{[1]}=0 \quad \text{or} \quad (3+\delta_{p,6})\binom{p}{h}\binom{p-h}{2}A^{[2]}=0,$$

contradicting the condition $A_0 \ne 0$. This shows that $p\le 2$.

Therefore the δ-polynomials

$$y^3,\ y^2z,\ yz^2,\ z^3,\ yQ_1,\ zQ_1,\ yQ_2-y'Q_1,\ zQ_2-z'Q_1$$

constitute a basis of $\mathfrak{B}_1(3)$, and *dim* $\mathfrak{B}_1(3) = 8$.

References.

1. E. R. Kolchin. Rational approximation to solutions of algebraic differential equations, *Proc. Amer. Math. Soc.* 10(1959), 238-244.

2. Charles F. Osgood. An effective lower bound on the "diophantine approximation" of algebraic functions by rational functions, *Mathematika* 20(1973), 4-15.

3. Wolfgang M. Schmidt. Contributions to diophantine approximation in fields of series, *Monatshefte f. Math.* 87(1979), 145-165.

PAINLEVÉ TRANSCENDENT

Let $\underline{F}$ be an ordinary differential field of characteristic zero which contains an element x such that $\delta x = x' = 1$. Let y be an element of a differential extension field of $\underline{F}$ such that $y'' = 6y^2 + x$ and such that the transcendence degree of $\underline{F}\langle y\rangle = \underline{F}(y,y')$ over $\underline{F}$ is 2. We claim that every constant in $\underline{F}\langle y\rangle$ is in $\underline{F}$. We prove several lemmas first.

Lemma 1. Suppose that $\underline{F}\langle y\rangle$ contains a constant not in $\underline{F}$, then there is a polynomial $P \in \underline{F}[y,y'] - \underline{F}$ and $\gamma \in \underline{F}$ such that

$$P^{\delta} + \frac{\partial P}{\partial y}y' + \frac{\partial P}{\partial y'}(6y^2 + x) = \gamma P \quad .$$

Proof. Let $P,Q \in \underline{F}[y,y']$, with $Q \neq 0$, be such that P/Q is a constant; we may assume that, say, $P \notin \underline{F}$ and that P and Q are relatively prime. Because P/Q is a constant, we find that

$$Q(P^{\delta} + \frac{\partial P}{\partial y}y' + \frac{\partial P}{\partial y'}(6y^2 + x)) = Q\delta P = P\delta Q = P(Q^{\delta} + \frac{\partial Q}{\partial y}y' + \frac{\partial Q}{\partial y'}(6y^2 + x)) \quad .$$

Because P and Q are relatively prime there must exist $L \in \underline{F}[y,y']$ such that

$$(*) \qquad P^{\delta} + \frac{\partial P}{\partial y}y' + \frac{\partial P}{\partial y'}(6y^2 + x) = L\cdot P \quad .$$

Since the total degree of the lhs of $(*)$ is at most one higher that that of P, L must be linear, say $L = \alpha y + \beta y' + \gamma$ $(\alpha,\beta,\gamma \in F)$. Let $\sum_{e\le i\le n} P_i y^i y'^{n-i}$ be the homogeneous part of P of highest degree $n > 0$ $(P_i \in F,\ P_e \neq 0)$. Then, from formula $(*)$ we obtain

$$6y^2\sum_{e\le i\le n}(n-i)P_i y^i y'^{n-i-1} = 6y^2\frac{\partial}{\partial y'}(\sum_{e\le i\le n}P_i y^i y'^{n-i}) =$$

$$= \alpha y\sum_{e\le i\le n}P_i y^i y'^{n-i} + \beta y'\sum_{e\le i\le n}P_i y^i y'^{n-i} \quad .$$

Comparing the coefficients of the monomial $y^e y'^{n-e+1}$ in the above equation, we find that $0 = \beta P_e$ so that $\beta = 0$.

Suppose that $e<n$, then, by comparing the coefficients of the monomial $y^{e+1}y'^{n-e}$ we find that $0 = \alpha P_e$ so that $\alpha = 0$. However if $e=n$ then the above equation is $0 = \alpha y P_e y^e y'^{n-e}$ so that $\alpha = 0$. We see therefore that $L = \gamma \in F$ which proves the lemma.

Lemma 2. *Let* $t = y'^2 - 4y^3$, $P \in \underline{F}[y,y']$, $A \in \underline{F}[y,y']$ *and* $h \in \underline{N}$. *Suppose that* $\frac{\partial P}{\partial y}y' + 6\frac{\partial P}{\partial y'}y^2 = A\cdot t^h$. *If* $2h > \deg_{y'}P$ *then* $A = 0$ *and* $P \in \underline{F}[t]$. *If* $2h = \deg_{y'}P$ *then there exists* $B \in \underline{F}[y]$ *such that* $A = y'\frac{dB}{dy}$ *and* $P - B\cdot t^h \in \underline{F}[t]$.

Proof. For any $C \in \underline{F}[y,y']$ let $L(C) = \frac{\partial C}{\partial y}y' + 6\frac{\partial C}{\partial y'}y^2$. We use induction on $\deg_{y'}P$ and may assume that $2h \geqslant \deg_{y'}P$.

Suppose that $\deg_{y'}P = 0$. Then $L(P) = \frac{dP}{dy}y' = A\cdot t^h$. If $2h > 0$ then $A = 0$ because the degree in y' of $L(P)$ is at most 1 . Evidently $P \in \underline{F}$. If $2h = 0$ then we set $B = P \in \underline{F}[y]$ and observe that $A = y'\frac{dB}{dy}$ and $P - B\cdot t^h \in \underline{F}$.

Suppose that $\deg_{y'}P = 1$. Then the condition $2h \geqslant \deg_{y'}P$ implies that $h \geqslant 1$ and $2h > \deg_{y'}P$. Since $\deg_{y'}L(P) \leq 2$ we must have that $h = 1$ and $A \in \underline{F}[y]$ or else $A = 0$, in which case we may replace h by 1 . Let $P = P_1y' + P_0$ where $P_1, P_0 \in \underline{F}[y]$. From the equation $L(P) = A\cdot t$ we obtain the three equations $\frac{dP_1}{dy} = A$, $\frac{dP_0}{dy} = 0$, and $6P_1y^2 = -4Ay^3$.

Thus $P_0 \in \underline{F}$ and $6A = 6\frac{dP_1}{dy} = \frac{d}{dy}(-4Ay) = -4\frac{dA}{dy}y - 4A$. If $A = A_ny^n + \cdots + A_0$ $(A_i \in \underline{F})$, then $5A_n = -2nA_n$ so that $A_n = 0$. Hence $A = 0$ and therefore $P_1 = 0$ so that $P = P_0 \in \underline{F}$.

Now suppose that $\deg_{y'}P \geq 2$ and that the lemma is proven for polynomials of smaller degree in y'. Let $Q, R \in \underline{F}[y,y']$ be such that $P = Q\cdot t + R$ and $\deg_{y'}R < \deg_{y'}t = 2$. Observe that $L(P) = L(Q)\cdot t + L(R)$, thus $L(R) = (A\cdot t^{h-1} - L(Q))t$. By the above, since $2 > \deg_{y'}R$, $A\cdot t^{h-1} = L(Q)$ and $R \in \underline{F}$. If $2h > \deg_{y'}P$ then $2(h-1) > \deg_{y'}Q$ so that, by the induction assumption, $A = 0$ and $Q \in \underline{F}[t]$, and therefore $P \in \underline{F}[t]$. If $2h = \deg_{y'}P$ then $2(h-1) = \deg_{y'}Q$ and hence there exists $B \in \underline{F}[y]$ such that $A = y'\frac{dB}{dy}$ and $Q - B\cdot t^{h-1} \in \underline{F}[t]$. Hence $P - B\cdot t^h \in \underline{F}[t]$. This completes the induction and proves the lemma.

<u>Theorem.</u> <u>Let $\underline{F}$ be an ordinary differential field of characteristic zero which contains an element x such that $\delta x = x' = 1$. Let y be an element of a differential extension field of $\underline{F}$ such that $y'' = 6y^2 + x$ and such that tr. deg. $\underline{F}\langle y\rangle/\underline{F}$ is 2. Then every constant in $\underline{F}\langle y\rangle$ is in $\underline{F}$.</u>

<u>Proof.</u> Suppose the contrary. Then, by Lemma 1, there exist $P \in \underline{F}[y,y'] - \underline{F}$ and $\gamma \in \underline{F}$ such that (*) $P^\delta + \frac{\partial P}{\partial y}y' + \frac{\partial P}{\partial y'}(6y^2 + x) = \gamma P$.

We define a grading on $\underline{F}[y,y']$ by assigning to the term $\alpha y^e y'^f$ the grade $2e + 3f$. We write $P = P_n + \cdots + P_0$ where every term in P_i is of grade i and $P_n \neq 0$. By considering the terms in (*) of grade $n + 1$ we find that $0 = \frac{\partial P_n}{\partial y}y' + 6\frac{\partial P_n}{\partial y'}y^2$ so that, by Lemma 2, $P_n \in \underline{F}[y'^2 - 4y^3]$.

Because every term in P_n is of grade n, $n = 6h$ for some $h \in \underline{\mathbb{N}}$ and $P_n = a(y'^2 - 4y^3)^h$ $(a \in \underline{F}, a \neq 0)$. By hypothesis $h > 0$ so that $n \geq 6$. By examining the terms in (*) of grade n we find that $\gamma P_n = P_n^\delta + {}$

$$+ \frac{\partial P_{n-1}}{\partial y} y' + \frac{\partial P_{n-1}}{\partial y'} 6y^2 \quad \text{i.e.} \quad a(y'^2 - 4y^3)^h = a'(y'^2 - 4y^3)^h + \frac{\partial P_{n-1}}{\partial y} y' + 6\frac{\partial P_{n-1}}{\partial y'} y^2 .$$

Now $\deg_{y'}P_{n-1} \leq (1/3)(n-1) = 2h - (1/3)$, hence $\deg_{y'}P_{n-1} \leq 2h-1 < 2h$. By Lemma 2 $\gamma a = a'$.

Now let $Q = P - a(y'^2 - 4y^3 - 2xy)^h$. By equation (*) we find that

$$Q^\delta + a'(y'^2 - 4y^3 - 2xy)^h + ha(y'^2 - 4y^3 - 2xy)^{h-1}(-2y) + \frac{\partial Q}{\partial y} y' +$$

$$+ ha(y'^2 - 4y^3 - 2xy)^{h-1}(-12y^2 - 2x)y' + \frac{\partial Q}{\partial y'}(6y^2 + x) +$$

$$+ ha(y'^2 - 4y^3 - 2xy)^{h-1}(2y')(6y^2 + x) = \gamma Q + \gamma a(y'^2 - 4y^3 - 2xy)^h .$$

Using the fact that $\gamma a = a'$ and simplifying, we have

$$Q^\delta + \frac{\partial Q}{\partial y} y' + \frac{\partial Q}{\partial y'}(6y^2 + x) = \gamma Q + 2hay(y'^2 - 4y^3 - 2xy)^{h-1} .$$

Because $2hay(y'^2 - 4y^3 - 2xy)^{h-1} \neq 0$, $Q \neq 0$.

We write $Q = Q_m + \cdots + Q_0$ where each term in Q_i is of grade i and $Q_m \neq 0$. Observe that $m < n = 6h$. By examining the terms in the last equation which are of grade $m + 1$ we find that

$$\frac{\partial Q_m}{\partial y} y' + 6\frac{\partial Q_m}{\partial y'} y^2 = 0 \qquad \text{if } m > 6h - 5$$

$$\frac{\partial Q_m}{\partial y} y' + 6\frac{\partial Q_m}{\partial y'} y^2 = 2hay(y'^2 - 4y^3)^{h-1} \qquad \text{if } m = 6h - 5$$

and

$$0 = 2hay(y'^2 - 4y^3)^{h-1} \qquad \text{if } m < 6h - 5 .$$

In the first case, Lemma 2 implies that $Q_m \in \underline{F}[y'^2 - 4y^3]$ so that m is divisible by 6 which is absurd. The third case is also absurd. In the second case $\deg_{y'} Q_m \leq (1/3)m = (1/3)(6h - 5) = 2h - (5/3)$. Thus $\deg_{y'} Q_m \leq 2(h-1)$. By Lemma 2 $\deg_{y'} Q_m = 2(h-1)$ since $ha \neq 0$. Then, by Lemma 2, there exists $B \in \underline{F}[y]$ such that $2hay = \frac{dB}{dy} y'$. This contradiction proves the theorem.

Painlevé transcendents

Let F be an ordinary differential field of characteristic zero, with derivation operator δ, which contains an element x such that $\delta x = x' = 1$. Let y be an element of a differential extension field of F such that $y'' = 6y^2 + x$ and such that the transcendence degree of $F\langle y\rangle = F(y, y')$ over F is 2. We claim that every constant in $F\langle y\rangle$ is in F. We prove several lemmas first.

LEMMA 1. *Suppose that $F\langle y\rangle$ contains a constant not in F; then there is a polynomial $P \in F[y, y'] - F$ and $\gamma \in F$ such that*

$$P^\delta + \frac{\partial P}{\partial y} y' + \frac{\partial P}{\partial y'}(6y^2 + x) = \gamma P,$$

[where P^δ is the polynomial obtained by differentiating the coefficients of P].[1]

PROOF. Let $P, Q \in F[y, y']$, with $Q \neq 0$, be such that P/Q is a constant; we may assume that, say, $P \notin F$ and that P and Q are relatively prime. Because P/Q is a constant, we find that

$$Q\Big(P^\delta + \frac{\partial P}{\partial y} y' + \frac{\partial P}{\partial y'}(6y^2 + x)\Big) = Q\delta P = P\delta Q = P\Big(Q^\delta + \frac{\partial Q}{\partial y} y' + \frac{\partial Q}{\partial y'}(6y^2 + x)\Big).$$

Because P and Q are relatively prime there must exist $L \in F[y, y']$ such that

$$(*) \qquad P^\delta + \frac{\partial P}{\partial y} y' + \frac{\partial P}{\partial y'}(6y^2 + x) = LP.$$

Since the total degree of the lhs of $(*)$ is at most one higher than that of P, L must be linear, say $L = \alpha y + \beta y' + \gamma$ $(\alpha, \beta, \gamma \in F)$. Let $\sum_{e \le i \le n} P_i y^i y'^{n-i}$ be the homogeneous part of P of highest degree $n > 0$ $(P_i \in F, P_e \neq 0)$. Then, from formula $(*)$ we obtain

$$\begin{aligned} 6y^2 \sum_{e \le i \le n} (n - i) P_i y^i y'^{n-i-1} &= 6y^2 \frac{\partial}{\partial y'}\Bigg(\sum_{e \le i \le n} P_i y^i y'^{n-i}\Bigg) \\ &= \alpha y \sum_{e \le i \le n} P_i y^i y'^{n-i} + \beta y' \sum_{e \le i \le n} P_i y^i y'^{n-i}. \end{aligned}$$

Comparing the coefficients of the monomial $y^e y'^{n-e+1}$ in the above equation, we find that $0 = \beta P_e$ so that $\beta = 0$.

Suppose that $e < n$, then, by comparing the coefficients of the monomial $y^{e+1} y'^{n-e}$, we find that $0 = \alpha P_0$ so that $\alpha = 0$. However if $e = n$ then the

This paper was found on Professor Kolchin's desk at home after his death.

[1] Editor's addition.

1

above equation is $0 = \alpha y P_e y^e y'^{n-e}$ so that $\alpha = 0$. We see therefore that $L = \gamma \in F$ which proves the lemma.

LEMMA 2. *Let $t = y'^2 - 4y^3$, $P \in F[y, y']$, $A \in F[y, y']$ and $h \in \mathbf{N}$. Suppose that $\frac{\partial P}{\partial y} y' + 6\frac{\partial P}{\partial y'} y^2 = A \cdot t^h$. If $2h > \deg_{y'} P$ then $A = 0$ and $P \in F[t]$. If $2h = \deg_{y'} P$ then there exists $B \in F[y]$ such that $A = y'\frac{dB}{dy}$ and $P - B \cdot t^h \in F[t]$.*

PROOF. For any $C \in F[y, y']$ let $L(C) = \frac{\partial C}{\partial y} y' + 6\frac{\partial C}{\partial y'} y^2$. We use induction on $\deg_{y'} P$ and may assume that $2h \geq \deg_{y'} P$.

Suppose that $\deg_{y'} P = 0$. Then $L(P) = \frac{dP}{dy} y' = A \cdot t^h$. If $2h > 0$ then $A = 0$ because the degree in y' of $L(P)$ is at most 1. Evidently $P \in F$. If $2h = 0$ then we set $B = P \in F[y]$ and observe that $A = y'\frac{dB}{dy}$ and $P - B \cdot t^h \in F$.

Suppose that $\deg_{y'} P = 1$. Then the condition $2h \geq \deg_{y'} P$ implies that $h \geq 1$ and $2h > \deg_{y'} P$. Since $\deg_{y'} L(P) \leq 2$ we must have that $h = 1$ and $A \in F[y]$ or else $A = 0$, in which case we may replace h by 1. Let $P = P_1 y' + P_0$ where $P_1, P_0 \in F[y]$. From the equation $L(P) = A \cdot t$ we obtain the three equations $\frac{dP_1}{dy} = A$, $\frac{dP_0}{dy} = 0$, and $6P_1 y^2 = -4Ay^3$. Thus $P_0 \in F$ and $6A = 6\frac{dP_1}{dy} = \frac{d}{dy}(-4Ay) = -4\frac{dA}{dy} y - 4A$. If $A = A_n y^n + \cdots + A_0$ ($A_i \in F$), then $5A_n = -2nA_n$ so that $A_n = 0$. Hence $A = 0$ and therefore $P_1 = 0$ so that $P = P_0 \in F$.

Now suppose that $\deg_{y'} P \geq 2$ and that the lemma is proven for polynomials of smaller degree in y'.[2] Let $Q, R \in F[y, y']$ be such that $P = Q \cdot t + R$ and $\deg_{y'} R < \deg_{y'} t = 2$. Observe that $L(P) = L(Q) \cdot t + L(R)$, thus $L(R) = (At^{h-1} - L(Q))t$. By the above, since $2 > \deg_{y'} R$, $A \cdot t^{h-1} = L(Q)$ and $R \in F$. [If $\deg_{y'} L(R) \leq 1$, the result follows immediately. The case $\deg_{y'} L(R) = 2$ is easily disposed of. We have $L(R) = ct$, $c \in F[y]$.

Suppose $c \neq 0$.

$$R = R_1(y)y' + R_0(y), R_i(y) \in F[y],\ i = 0, 1, R_1(y) \neq 0.$$

$$\begin{aligned} L(R) &= \frac{dR_1}{dy} y'^2 + \frac{dR_0}{dy} y' + 6R_1(y)y^2. \\ &= c(y)y'^2 - 4c(y)y^3. \end{aligned}$$

Therefore, $3R_1(y) = -2y\frac{dR_1}{dy}$. Set $R_1(y) = a_n y^n + \cdots + a_1 y$, where $a_i \in F$, $i = 1, \ldots, n$, $a_n \neq 0$. We have:

$$\begin{aligned} -2na_n y^n + \cdots + -2a_1 y &= 3a_n y^n + \cdots + 3a_1 y \\ -2n &= 3. \end{aligned}$$

This contradiction shows that $c = At^{h-1} - L(6) = 0$.][3] If $2h > \deg_{y'} P$ then $2(h-1) > \deg_{y'} Q$ so that, by the induction assumption, $A = 0$ and $Q \in F[t]$, and therefore $P \in F[t]$. If $2h = \deg_{y'} P$ then $2(h-1) = \deg_{y'} Q$ and hence there exists $B \in F[y]$ such that $A = y'\frac{dB}{dy}$ and $Q - B \cdot t^{h-1} \in F[t]$. Hence $P - B \cdot t^h \in F[t]$. This completes the induction and proves the lemma.

THEOREM. *Let F be an ordinary differential field of characteristic zero which contains an element x such that $\delta x = x' = 1$. Let y be an element of a differential extension field of F such that $y'' = 6y^2 + x$ and such that tr. deg. $F\langle y\rangle/F$ is 2. Then every constant in $F\langle y\rangle$ is in F.*

[2] Editor's addition: The case $h = 0$ is addressed above. So, we may assume $h \geq 1$.

[3] Editor's addition.

PROOF. Suppose the contrary. Then, by Lemma 1, there exist $P \in F[y, y'] - F$ and $\gamma \in F$ such that $(*)$ $P^\delta + \frac{\partial P}{\partial y} y' + \frac{\partial P}{\partial y'}(6y^2 + x) = \gamma P$.

We define a grading on $F[y, y']$ by assigning to the term $\alpha y^e y'^f$ the grade $2e + 3f$. We write $P = P_n + \cdots + P_0$ where every term in P_i is of grade i and $P_n \neq 0$. By considering the terms in $(*)$ of grade $n+1$ we find that $0 = \frac{\partial P_n}{\partial y} y' + 6\frac{\partial P_n}{\partial y'} y^2$ so that, by Lemma 2, $P_n \in F[y'^2 - 4y^3]$. Because every term in P_n is of grade n, $n = 6h$ for some $h \in \mathbf{N}$ and $P_n = a(y'^2 - 4y^3)^h$ $(a \in F, a \neq 0)$. By hypothesis $h > 0$ so that $n \geq 6$. By examining the terms in $(*)$ of grade n we find that $\gamma P_n = P_n^\delta + \frac{\partial P_{n-1}}{\partial y} y' + \frac{\partial P_{n-1}}{\partial y'} 6y^2$ i.e. $\gamma a(y'^2 - 4y^3)^h = a'(y'^2 - 4y^3)^h + \frac{\partial P_{n-1}}{\partial y} y' + 6\frac{\partial P_{n-1}}{\partial y^1} y^2$. Now $\deg_{y'} P_{n-1} \leq (1/3)(n-1) = 2h - (1/3)$, hence $\deg_{y'} P_{n-1} \leq 2h - 1 < 2h$. By Lemma 2 $\gamma a = a'$.

Now let $Q = P - a(y'^2 - 4y^3 - 2xy)^h$. By equation $(*)$ we find that

$$
\begin{aligned}
&Q^\delta + a'(y'^2 - 4y^3 - 2xy)^h + ha(y'^2 - 4y^3 - 2xy)^{h-1}(-2y) + \frac{\partial Q}{\partial y} y' \\
&\quad + ha(y'^2 - 4y^3 - 2xy)^{h-1}(-12y^2 - 2x)y' + \frac{\partial Q}{\partial y'}(6y^2 + x) \\
&\quad + ha(y'^2 - 4y^3 - 2xy)^{h-1}(2y')(6y^2 + x) = \gamma Q + \gamma a(y'^2 - 4y^3 - 2xy)^h.
\end{aligned}
$$

Using the fact that $\gamma a = a'$ and simplifying, we have

$$Q^\delta + \frac{\partial Q}{\partial y} y' + \frac{\partial Q}{\partial y'}(6y^2 + x) = \gamma Q + 2hay(y'^2 - 4y^3 - 2xy)^{h-1}.$$

Because $2hay(y'^2 - 4y^3 - 2xy)^{h-1} \neq 0$, $Q \neq 0$.

We write $Q = Q_m + \cdots + Q_0$ where each term in Q_i is of grade i and $Q_m \neq 0$. Observe that $m < n = 6h$. By examining the terms in the last equation which are of grade $m+1$ we find that

$$
\begin{aligned}
&\frac{\partial Q_m}{\partial y} y' + 6\frac{\partial Q_m}{\partial y'} y^2 = 0 && \text{if } m > 6h - 5 \\
&\frac{\partial Q_m}{\partial y} y' + 6\frac{\partial Q_m}{\partial y'} y^2 = 2hay(y'^2 - 4y^3)^{h-1} && \text{if } m = 6h - 5
\end{aligned}
$$

and

$$0 = 2hay(y'^2 - 4y^3)^{h-1} \quad \text{if } m < 6h - 5.$$

In the first case, Lemma 2 implies that $Q_m \in F[y'^2 - 4y^3]$ so that m is divisible by 6 which is absurd. The third case is also absurd. In the second case $\deg_{y'} Q_m \leq (1/3)m = (1/3)(6h - 5) = 2h - (5/3)$. Thus $\deg_{y'} Q_m \leq 2(h-1)$. By Lemma 2 $\deg_{y'} Q_m = 2(h-1)$ since $ha \neq 0$. Then, by Lemma 2, there exists $B \in F[t]$ such that $2hay = \frac{dB}{dy} y'$. This contradiction proves the theorem.

Part II

Commentary

Algebraic Groups and Galois Theory in the Work of Ellis R. Kolchin *

Armand Borel

The Galois theory of differential fields, a generalization of the Picard-Vessiot theory, was a major concern of E. Kolchin's during the first thirty years or so of his scientific life. From the beginning, it appeared that these Galois groups would be algebraic groups, or rather be naturally isomorphic to such. However, the theory of algebraic groups was not suitably developed in Kolchin's view for his purpose when he first needed it, so that a minor, but persistent and essential, theme in his work is the theory of algebraic groups.

As implied by the title, this lecture will emphasize both themes, but I still do not claim to provide an even-handed discussion of all of Kolchin's work in this area. More specifically, I shall take for granted whatever is needed from the theory of differential fields, or more generally from differential algebra, to which Kolchin had to add in no small measure, in order to arrive as directly as possible to the two main topics and their relationships.

My starting point in describing Kolchin's work will be his Annals paper on algebraic matric groups and Picard-Vessiot theory [Kolchin 1948a]. It already contains important results, but also a program which was to occupy him for about 25 years. In order to put this paper in context, I would like first to make some historical remarks, to indicate how the theory presented itself to him then.

*First Kolchin Memorial Lecture, given at Columbia University on April 16, 1993

1. The idea of using Lie groups to develop some sort of Galois theory for differential equations is as old as the Lie theory itself. It was in the mind of S. Lie when he first developed it, as can already be seen from an 1874 letter to A. Mayer.[1] He had noticed that several classical methods to find the solutions of certain differential equations could be given a unified treatment by using a one-parameter subgroup leaving the equation invariant, whence the idea that if a Lie group leaves invariant a given system of ordinary differential equations this should help one to reduce the search of solutions to the study of a system of lower degree. The existence of such a group would imply certain properties of the solutions, reminding one of the role of the Galois group of an algebraic equation. But the analogy is rather weak: every algebraic equation has a Galois group, the structure of which gives essential insight into the properties of the roots, whereas most differential equations do not admit a non-trivial Lie group of transformations.

2. A point of view closer to the model of the Galois theory was proposed by Emile Picard for homogeneous linear differential equations, in a series of C.R. Notes and papers between 1883 and 1898, see [P2], and in the third volume of his "Traité d'Analyse" [P1]. Consider the ordinary homogeneous differential equation

$$(\mathcal{E}) \qquad \frac{d^n y}{dx^n} + p_1(x)\frac{d^{n-1}y}{dx^{n-1}} + \cdots + p_n(x)y = O$$

where the p_i's belong to some field F of functions meromorphic in a given region of the plane, stable under the taking of derivatives and containing $\mathbf{C}$. It has a "fundamental system of solutions" $y_1, \cdots, y_n$, i.e. a set of solutions linearly independent over $\mathbf{C}$. Then every other solution is a linear combination of the y_i's with constant coefficients. As was well-known, the condition for linear independence is that the Wronskian

$$W(y_1, \cdots, y_n) = \begin{vmatrix} y_1 & \cdots & y_x \\ y_1^{(1)} & \cdots & y_n^{(1)} \\ & \vdots & \\ y_1^{(n-1)} & \cdots & y_n^{(n-1)} \end{vmatrix}$$

[1]Cf the letter of Febr.3, 1874 in vol.V of S. Lie's Collected Papers, p.586.

2

be $\neq 0$. Here and below, $y_i^{(j)}$ stands for $\frac{d^j y_i}{dx^j}$. Let now

$$A = (a_i^j) \epsilon GL_n(\mathbf{C}).$$

Then

$$\overline{y}_j = \sum_i a_j^i y_i$$

is another fundamental set of solutions. We want to impose on A certain conditions analogous to those defining the Galois group of an algebraic equation. In that case it can be required that A is an automorphism of the splitting field of the equation or, equivalently, without using that notion, that it respects all algebraic relations between the roots. Here we use an analogue of that second formulation. Let

$$Y_i^{(j)} \qquad (i = 1, \cdots, n; j = 0, 1, 2, \cdots)$$

be indeterminates. Let $\mathcal{P}(\mathcal{E})$ be the set of polynomials $P(Y_i^{(j)})$ in the $Y_i^{(j)}$, with complex coefficients, such that $P(y_i^{(j)})$, viewed as a function of x, belongs to F. The condition imposed on A is then that $P(\overline{y}_j^{(j)})$ is the same function of x as $P(y_i^{(j)})$ for all $P \epsilon \mathcal{P}(\mathcal{E})$. These A's form the "transformation group of $(\mathcal{E})$", in E. Picard's terminology, to be denoted $G(\mathcal{E})$, or the "rationality group of $(\mathcal{E})$", in that of F. Klein. The latter also pointed out that if $n = 2$ and the equation has regular singular points, then $G(\mathcal{E})$ is the smallest algebraic group containing the monodromy group of the equation.[2]

To justify the use of "group", Picard pointed out only that $G(\mathcal{E})$ contains the product of any two of its elements. That $A \,\epsilon\, G(\mathcal{E})$ implies $A^{-1} \,\epsilon\, G(\mathcal{E})$ was shown later by A. Loewy [L].[3]

The group $G(\mathcal{E})$ is first of all a Lie group but Picard pointed out that it is moreover "algebraic" in the sense that the matrix coefficients are algebraic functions of suitably chosen parameters (rather that merely analytic), which led him to study some properties of such groups. In particular, he determined

[2]This is also true of a linear homogeneous equation with regular singular points of any degree, cf. F. Beukers, *Differential Galois theory*, in From number theory to physics, Waldschmidt et al eds. Springer, 1992, 413-439, Thm. 2.5.1. Here it is understood that $F = \mathbf{C}(z)$.

[3]There were precedents in Lie theory where this condition had been forgotten. However, Picard includes it when it comes to defining finite groups of substitutions (cf [P1], p. 452).

the structure of those of dimensions 1 or 2 (cf (96) in [P2]) and stated that the previous parameters can be chosen so that the matrix coefficients are rational functions of them ([P1], XVII, n^0 12) (i.e. that the group variety is unirational in present-day language).

The theory was further developed by E. Vessiot (1892 - 1904), see his account in [V], where earlier references are also given. His main result is in substance that if $G(\mathcal{E})$ is connected solvable ("integrable" in Lie's sense) if and only if $(\mathcal{E})$ is solvable by "quadratures". That notion is not precisely or consistently defined, as pointed out by Kolchin in [Kolchin 1948a], and we shall come back to it later (see §6). Of course, it means adjoining a primitive but, apparently, the exponential of a primitive was also allowed.

3. The whole theory was somewhat obscure and the necessity of various clarifications or reformulations was felt already at the time. In fact, in giving the definition of $G(\mathcal{E})$ above, I have followed [L] and [V] rather than Picard himself. There was little further development until [Kolchin 1948a]. There after summarizing the history of the subject and pointing out various ways in which it was lacking in precision or rigor, Kolchin states:

> *the purposes of the present paper are, first to develop a set of theorems on algebraic matric groups at least adequate to meet the demands of the Picard-Vessiot theory and second, to algebraize, rigorize, round out, and augment that theory*

The framework for the algebraization needed to discuss differential equations was already available, namely the differential algebra, developed by his teacher J. F. Ritt. On the group side, Kolchin felt the need to build a theory of algebraic matric groups, rather than rely on the theory of Lie groups, which he viewed as far deeper than the one to be developed. His paper then has three main parts.

A) algebraic matric groups (Chap. I)

B) Galois theory of differential fields (Chap. III)

C) The Picard-Vessiot theory (Chap. IV, V)

(Chap.II being devoted to preliminaries on differential algebra.)

Before coming to (A), let me mention that there was more at the time to the theory of algebraic matric groups than he implied. In particular L. Maurer had published several interesting papers (1888-94) in which he notably

4

characterized the Lie algebras of algebraic groups and essentially proved that a group variety is rational (rather than unirational). His theory had been revived and generalized by Chevalley and Tuan, and then by Chevalley.

Since B) and C) are in characteristic zero, this might have filled Kolchin's needs. Whether he was aware of it or not I do not know, but the fact that it was not taken into account was ultimately a "good thing" because Kolchin decided that:

> *To emphasize the purely algebraic nature of the subject matter the proofs are carried out in manner valid for fields of non-zero as well as zero characteristic.*

so that Part (A) and the companion paper [Kolchin 1948b] constitute in fact the birth certificate of the theory of linear algebraic groups over algebraically closed fields of *arbitrary* characteristic. In the following summary, I shall discuss both together.

4. Let C be an algebraically closed commutative field, p its characteristic. The notion of algebraic matrix group over C introduced by Kolchin is the nowadays usual one: a subgroup G of $GL_n(C)$ is algebraic if it is the set of all invertible matrices the coefficients of which annihilate a given set of polynomials in n^2 indeterminates with coefficients in C, i.e. G is the intersection of $GL_n(C)$ with a closed subvariety of $M_n(C)$, called by Kolchin the underlying manifold of G.

4.1. A first main result is that if G is connected and solvable, it can be put in triangular form, since then known as the Lie-Kolchin theorem. He also shows that G is solvable, as an abstract group, if and only it is solvable as an algebraic group: the series of the smallest algebraic groups containing the successive derived groups stops at $\{1\}$. (In fact, it was proved later that the abstract derived groups are automatically algebraic subgroups, cf. e.g. [B]).

4.2. If G is commutative, then it is the direct product of two subgroups G_s and G_u consisting respectively of the semisimple and unipotent elements in G. [An immediate consequence, not drawn by Kolchin, is that any linear algebraic group is stable under the (multiplicative) Jordan decomposition.]

5

4.3. Kolchin also introduces the notions of "anticompact" and "quasicompact" groups of matrices, i.e consisting of unipotent and semisimple matrices respectively. He shows that any anticompact group can be put in triangular form. If G is quasicompact, algebraic, connected, then it can be put in diagonal form (i.e. is a torus in the usual terminology). Also he shows that if G is connected, the k-th power map $x \mapsto x^k$ is a dominant morphism, if $(k, p) = 1$ in case $p \neq 0$.

4.4. These papers also contain some general comments about connected components, Jordan-Hölder series and algebraic subgroups. Altogether, they present the first body of theorems on linear algebraic groups with proofs insensitive to the characteristic of the groundfield, carried out in the framework of algebraic geometry, without any recourse to Lie algebras and the usual mechanism of Lie theory. They were influential on my own work, as can be seen from my first paper on this topic (Annals Math. **64** (1956), 20-82), in which these results are all proved again. They are also incorporated in [B] and belong to any systematic exposition of the theory.

5.1. For (B) and (C) the basic notion is that of a *differential field.* A (commutative) field F is differential if it is endowed with a derivation,[4] i.e. self-map δ satisfying

$$\delta(a+b) = \delta(a) + \delta(b), \ \ \delta(a.b) = \delta(a).b + a.\delta(b) \ \ (a, b \epsilon F).$$

Then

$$C_F = \{a \epsilon F, \delta(a) = 0\}$$

is a subfield, the *field of constants* of F. The nth derivative of a is $\delta^n.a$ and is written $a^{(n)}$. An *automorphism* of F is an automorphism of the underlying field commuting with the derivation.

In fact, this is an *ordinary* differential field. Kolchin develops the theory more generally for *partial* differential fields, i.e. fields endowed with a finite set of commuting derivations.

[4]Called in [Kolchin 1948a] a derivative. From [Kolchin 1953] on, Kolchin switched to *derivation operator* and let *derivative operators* be the operators on F defined by the powers of δ (similarly for partial differential fields). I shall use derivation and derivatives.

In the sequel, unless otherwise stated, F is a differential field of characteristic zero with algebraically closed field of constants C_F or C. The field F is partial from § 7 on, ordinary before, unless otherwise stated.

5.2. Let $E \supset F$ be a field containing F as a subfield. It is a *differential extension* of F it is a differential field with derivation leaving F stable and inducing on F its given derivation. It is a *finitely generated differential extension* if there exist moreover finitely many elements $\eta_1 \cdots, \eta_n$ in E such that E is generated over F by the η_i's and their derivatives of all orders, in which case we write

$$E = F < \eta_1, \cdots, \eta_n > .$$

It is not necessarily a finitely generated field extension of F, i.e. its transcendence degree $\partial^0(E/F)$ over F may be infinite.

In the present context, the equation $(\mathcal{E})$ takes the form

$$(\mathcal{E}) \qquad L(y) = y^{(n)} + p_1.y^{(n-1)} + \cdots + p_n.y = 0, \ (p_i \epsilon F, i = 1, \cdots, n)$$

and we are looking for solutions y in some differential extension of F. The standard facts about the solutions of $(\mathcal{E})$ in the classical case recalled earlier generalize to the following theorem (see [Kolchin 1948a], §§ 14, 15).

Theorem. *There exists a finitely generated differential extension E of F containing n solutions $\eta_1 \cdots, \eta_n$ of $(\mathcal{E})$ such that*

$$W(\eta_1, \cdots, \eta_n) = \begin{vmatrix} \eta_1, & \cdots & , \eta_n \\ \eta_1^{(1)}, & \cdots & , \eta_n^{(1)} \\ \eta_1^{(n-1)}, & \cdots & , \eta_n^{(n-1)} \end{vmatrix} \neq 0.$$

In any differential extension E' of E, the η_i are linearly independent over $C_{E'}$ and span over $C_{E'}$, the space of solutions of $(\mathcal{E})$ in E'. No differential extension F' of F contains more than n solutions of $(\mathcal{E})$ linearly independent over the constants.

A set of n linearly independent solutions of $(\mathcal{E})$ is called a *fundamental set of solutions* of $(\mathcal{E})$.

7

5.3. A differential extension E of F is a *Picard-Vessiot extension* (a P.V. extension for short) of F if it is of the form

$$E = F < \eta_1, \cdots, \eta_n >$$

where the η_i form a fundamental set of solutions of an equation $(\mathcal{E})$ *and* if $C_E = C_F$.

It is clear from $(\mathcal{E})$ that the derivatives of all orders of the η_i's are linear combinations with coefficients in F of the $(n-1)$ first ones. Therefore $\partial^0(E/F)$ is finite.

Kolchin proved later that given $(\mathcal{E})$, one can find a fundamental set of solutions $\eta_1, \cdots, \eta_n$ in some extension such that $F < \eta_1, \cdots, \eta_n >$ has the same field of constants as F (cf [Kolchin 1948c] or [Kolchin 1973], Prop. 13, p.412). This is under our standing assumption that C_F is algebraically closed (of char. 0). If not, then Seidenberg has produced an equation $(\mathcal{E})$ such that $C_E \neq C_F$ for all differential field extensions E generated over F by a fundamental set of solutions of that equation (cf. [S] or [Kolchin 1973], Exercise 1, p.413).

5.4. Let E be a P.V. extension of F. Then, by definition, its *Galois group over* F is the group $G = G(E/F) = Aut(E/F)$ of automorphisms of E (as a differential field) leaving fixed each element of F. Let $g \epsilon G$. Then

$$g.\eta_i = \sum_j c_i^j(g).\eta_j$$

where the $c_i^j(g)$ belong to C_E at first, hence to C_F since both are assumed to be equal. The map

$$g \mapsto c(g) = (c_i^j(g))$$

is a faithful linear representation of G into $GL_n(C_F)$.

Before stating the main theorem concerning $G(E/F)$, we note that if E_1 is a differential field intermediate between E and F then its field of constants is equal to C_F, since $C_E = C_F$. Therefore, it is clear from the definition that E is a P.V. extension of E_1 and $G(E/E_1)$ is defined as before.

8

5.5. Theorem. *Let E be a P.V. extension of F. We keep the previous notation.*

(i) The group $c(G)$ is the group of rational points over C_F of an algebraic subgroup of $GL_n(C_F)$ defined over C_F, of dimension equal to $\partial^0(E/F)$.

(ii) The map which assigns to an algebraic subgroup H of $c(G)$ the subfield E^H of elements fixed under H is a bijection between algebraic subgroups and intermediate differential subfields. The inverse bijection assigns to a differential subfield E_1 the group $G(E/E_1)$ of automorphisms of E over E_1.

(iii) H is a normal subgroup if and only if $\sigma(E^H) = E^H$ for all $\sigma \epsilon G(E/F)$. In that case, the restriction of an automorphism to E^H defines an isomorphism of G/H onto $Aut(E^H/F)$.

See §§ 17 to 20 in [Kolchin 1948a]. There F is ordinary. The generalization to partial differential fields is carried out in [Kolchin 1952]. We indicate briefly how Kolchin define P.V. extensions in the general case. To this effect he first remarks that $p_i = \pm W_i/W_0$, where

$$W_i = \det\ (\eta_k^{(j)}), (1 \leq i, k \leq n; 0 \leq j \leq n, j \neq n-i)$$

by Cramer's rule. Given now n elements $y_1, \cdots, y_n$ in a partial differential field E and partial derivatives $\theta_1, \cdots, \theta_n$, let

$$W(\theta_1, \cdots, \theta_n, y_1, \cdots, y_n) = \det\ (\theta_j y_i).$$

If the y_i's are linearly independent over constants, there is always a choice of partial derivatives of orders $< n$, say $\sigma_1, \cdots, \sigma_n$ such that

$$W(\sigma_1, \cdots, \sigma_n; y_1, \cdots, y_n) \neq 0. \tag{1}$$

([Kolchin 1952]), lemma 1). Assume now F to be a partial differential field. A differentiable extension of E of F is a P.V. extension if

$$C_E = C_F, E = F < \eta_1, \cdots, \eta_n >,$$

where the η_i are linearly independent over constants and

$$W(\theta_1, \cdots, \theta_n; \eta_1, \cdots, \eta_n)/W(\sigma_1, \cdots, \sigma_n; \eta_1, \cdots, \eta_n) \epsilon F$$

for all choices $\theta_1, \cdots, \theta_n$ of partial derivatives of orders $\leq n$, the σ_i's being a set of partial derivatives of orders $< n$ such that (1) is satisfied by the η_i's.

The initial remark above shows readily that this definition is equivalent to the original one in the ordinary case. For other characterizations and further discussion of P.V. extensions, see [Kolchin 1973], VI, $n^0 6$, and the exercises following it, pp.409-418.

In [Kolchin 1948a], Kolchin does not raise the question as to whether E^H in (*iii*) is a P.V. extension of F. Later on, he provided a positive answer in [Kolchin 1955], III, n^0 3, Corollary to Theorem 2 (see also 9.2 below).

6. The last sections of the paper (§§ 23 to 27) are devoted to a precise version and generalization of Vessiot's theorem about equations solvable by "quadratures". Consider first two special cases:

6.1. The primitives of $a \epsilon F$, i.e. the solutions of $y' = a$, differ by a constant. The equation $y' = a$ is not homogeneous but any solution η_2 and the constant $\eta_1 = 1$ span the space of solutions of

$$(1) \qquad y'' - (a'/a).y' = 0.$$

The extension $E = F < \eta_2 >$ is P.V. If $g \epsilon G(E/F)$ then

$$g.\eta_1 = \eta_1, \qquad g.\eta_2 = \eta_2 + c = \eta_2 + c.\eta_1,$$

so that G can be identified to the group of upper triangular unipotent matrices in $\mathbf{GL}_2(C_F)$, i.e. to the additive group $\mathbf{G}_a$ of C_F.

6.2. Let now $E = F < \eta >$, where η is a solution of

$$y' - a.y = 0.$$

Then $\eta = \exp b$, where b is a primitive of a. Any other fundamental solution is a non-zero multiple of η, whence an isomorphism of $G(E/F)$ onto a subgroup of $\mathbf{GL}_1$. There are then two cases:

a) η is transcendental, which is equivalent to $G(E/F) = \mathbf{GL}_1$,

b) η is algebraic. Then $G(E/F)$ is finite cyclic, of some order h, and η is an h-th root of an element of F.

The results of §27 imply conversely that if E is a P.V. extension with a Galois group of one of the three previous types, then we are in the corresponding case.

10

6.3 From this it follows first that a P.V. extension has a solvable Galois group if and only it can be obtained from F by a succession of steps 6.1, 6.2. In fact, Kolchin carries out the discussion a bit more generally so as to allow for more general algebraic extensions.

An extension $E = F < \alpha_1, \cdots, \alpha_q >$ of F is said to be *Liouvillian* if $C_E = C_F$ and α_{i+1} is either algebraic over $F < \alpha_1, \cdots, \alpha_i >$ or a primitive or an exponential of a primitive of an element in that field ($i = 0, \cdots, q - 1$). Let now E be a P.V. extension of F. Then (§25, Theorem), if E is contained in a Liouvillian extension, the group $G(E/F)$ has a solvable identity component. Conversely, the latter property implies that E is Liouvillian. More precisely, Kolchin distinguishes ten types of Liouvillian extensions and ten types of algebraic groups and show that they match in case of P.V. extensions contained in a Liouvillian extension.

7. The above sections 5,6 summarize part C in my initial division of [Kolchin 1948a]. Part B is Kolchin's first attempt at a Galois theory of differential fields, not necessarily associated to differential equations and is much more tentative. To this effect, he has to introduce a notion of normal extension of F, but can prove only part of the sought for Galois correspondence. See the introduction of [Kolchin 1953] for a discussion of the "blemishes" of that theory. There he defines the concept of "strongly normal extension", which was to be the basis of all his further work on Galois theory, so I shall directly go over to it. The main results are proved first in [Kolchin 1953], [Kolchin 1955] and in the paper [Kolchin 1958], written jointly with S. Lang. The whole theory was given a comprehensive exposition in [Kolchin 1973].

8.1. Kolchin first introduces a *universal field* in analogy with the universal field in Weil's Foundations of Algebraic Geometry [W].

A differential extension U of F is a *universal extension* if given a finitely generated differential extension F_1 of F in U and a finitely generated differential extension E of F_1, there exists an isomorphism of E into U over F_1. It is algebraically closed and its field of constants C_U is a universal field for C_F in Weil's sense, (i.e. algebraically closed, of infinite transcendence degree over C_F). Then nothing is lost by restricting oneself to finitely generated extensions of F contained in U and we shall do so. (Cf [Kolchin 1953 : I, no.5], and [Kolchin 1973: III, No.7] for an extension to arbitrary characteristics.)

8.2. Let now E be a finitely generated differential extension of F. Wanted is a condition of normality. The most obvious one is that E be stable under any automorphism of U over F. However, Kolchin noticed that this forces E to be algebraic over F, hence is too strong, so that isomorphisms of E into U over F have to be allowed. In order to avoid endless repetitions of "into U", I shall follow Kolchin's convention and call this simply an isomorphism of E over F, (whereas an "automorphism" of E over F is of course an isomorphism of E onto itself fixing F pointwise). However, they must be restricted in some fashion. On the other hand, whatever the definition of normal extension is, a P.V. extension has to be one. Assume then that $E = F < \eta_1, \cdots, \eta_n >$ is a P.V. extension, where the η_i are a fundamental set of solutions of the underlying equation $L(y) = 0$. Let σ be an isomorphism of E over F. Then the $\sigma.\eta_i$ form a fundamental set of solutions in $\sigma(E)$, therefore

$$\sigma.\eta_i = \sum c_i^j \eta_j, (i = 1, \cdots, n)$$

where the c_i^j belong to the constant field of $\sigma(E)$, hence also to C_U. Therefore

$$\sigma(E) \subset E.C_U.$$

Kolchin saw that this was the decisive restriction and he introduced the following definitions:

Definitions.

(i) An isomorphism σ of E over F is *strong* it satisfies the two conditions

(1) $\sigma \mid C_E = Id.$

(2) $\sigma(E) \subset E.C_U$ and $E \subset \sigma(E).C_U$ i.e. $E.C_U = \sigma(E).C_U$.

(ii) E is a *strongly normal extension* of F if it is finitely generated over F (as a differentiable field) and every isomorphism of E over F is strong.

See [Kolchin 1973], VI, n° 2,3. These definitions appear first in [Kolchin 1953] except that (1) is not needed since $C_E = C_F$ by a standing assumption there (on p.772).

If E is strongly normal over F, then it is clearly so over every differential subfield containing F.

12

In the sequel, $C(\sigma)$ denotes the field of constants of the field $E.\sigma(E)$. The condition (2) may also be written

$$E.C(\sigma) = E.\sigma(E) = \sigma(E).C(\sigma). \tag{3}$$

If E is strongly normal, $C_E = C_F$ ([Kolchin 1973], V, Prop. 9, p.393) and E has finite transcendence degree over F ([Kolchin 1973] VI, Prop.11, p.394; the latter statement is also contained in Theorem 2, III, $n^0 2$ of [Kolchin 1953]).

8.3. *The Galois group.* If E is any finitely generated differential extension of F, then it is linearly disjoint from C_U over its own field of constants C_E of E (see [Kolchin 1953], I, $n^0 4$, Prop.3 and the comment following it, or [Kolchin 1973], II, n^0 1, Cor.1, p.87). As a consequence, a strong isomorphism of E over F extends uniquely to an automorphism of $E.C_U$ fixing C_U pointwise, i.e. to an automorphism of $E.C_U$ over $F.C_U$. Conversely, if τ is an automorphism of $E.C_U$ over $F.C_U$, then its restriction to E is a strong isomorphism of E over F which determines τ completely. Assume now E to be strongly normal. Then every isomorphism of E over F is strong, by definition, whence a canonical bijection

$$\{\mathit{Isom}\ E/F\} = \text{Aut }(E.C_U/F.C_U),$$

which allows one to view the left-hand side as a *group*. By definition, this is the Galois group $G(E/F)$ of E over F.

8.4. In the sequel we write C for C_F and C_U is viewed as a universal field for C, i.e. algebraic varieties defined over C are assumed to have their points in C_U. If M is an irreducible one, then the residue field $C(x)$ of $x \epsilon M$ at x is the subfield of C_U generated over C by the values at x of the rational functions on M which are defined at x.

We now give a first, provisory as far as (i) is concerned, formulation of the fundamental theorem of Kolchin's Galois theory of strongly normal extensions:

Theorem. *Let E be a strongly normal differential extension of F.*

(i) There exists an algebraic group $\mathbf{G}$ *defined over* C *such that* $G(E/F)$ *may be identified with* $\mathbf{G}(C_U)$ *so that* $C(\sigma)$ *is the residue field of* $\sigma \epsilon G(E/F)$. *For a finitely generated extension* K *of* C *in* C_U, *this isomorphism identifies* $\mathbf{G}(K)$ *with* $\{\sigma \epsilon G(E/F) \mid \sigma(E) \subset E.K\}$. *In particular,* $\mathbf{G}(C_F) = Aut(E/F)$. *The dimension of* $\mathbf{G}$ *is equal to* $\partial^0(E/F)$.

(ii) The map which assigns to an algebraic subgroup H *of* $G(E/F)$ *defined over* C *the field* E^H *of invariants of* H *establishes a bijection between algebraic* C*-subgroups of* $G(E/F)$ *and intermediate differential fields between* F *and* E. *The inverse bijection assigns to such a field* F_1 *the subgroup of* $G(E/F)$ *of elements of* $G(E/F)$ *fixing* F_1 *pointwise.*

(iii) The C*-subgroup* H *is normal if and only if* E^H *is strongly normal over* F. *In that case, the restriction of* $\sigma \epsilon G(E/F)$ *to* E^H *defines an isomorphism of algebraic groups of* $G(E/F)/H$ *onto* $G(E^H/F)$.

For *(ii)* and *(iii)*, see [Kolchin 1973] VI, $n^0 4$, Theorem 3,4 respectively. It is also shown in the latter that the conditions in *(iii)* are equivalent to either of:

(a) For each element α of E^H not contained in F, there exists a strong isomorphism of E^H over F not leaving α fixed.

(b) $\sigma(E^H) \subset E^H C_U$ for every $\sigma \epsilon G(E/F)$.

We shall soon come to *(i)* and to what is already proved in [Kolchin 1953] and [Kolchin 1955]. Before doing so, we list some complements to the fundamental theorem. We write G for $G(E/F)$ and view it as an algebraic group over C.

9.1. The group $G(E/F)$ is not necessarily connected. The fixed point set of the identity component G^0 of G is the algebraic closure F^0 of F in E and we have

$$[F^0 : F] = [G : G^0].$$

9.2. In this theorem, the algebraic groups are not necessarily linear. In fact, Kolchin proves that E is P.V. if and only if $G(E/F)$ is linear. The new implication is of course that $G(E/F)$ linear implies E to be P.V. (see [Kolchin 1955], III, n^0 3, Theorem 2, p. 891 or Cor.2 to Theorem 8, p.427 in

14

[Kolchin 1973]). Now, given a linear algebraic group G and a closed normal subgroup N there always exists a rational representation of G with kernel N (a by now standard fact, proved first in [Kolchin 1951]); it follows from this characterization that if E is P.V., any strongly normal extension of F in E is also P.V. ([Kolchin 1955], Cor. to Theorem 2, p.891). A slightly more direct way to derive the corollary from the theorem is also hinted at in exercise 2, p.427 of [Kolchin 1973].

9.3. Assume now that G is connected of dimension one. The two cases where it is linear were discussed in 6.1, 6.2. There remains the one where G is an elliptic curve. This corresponds to the case where E is *Weierstrassian* over F, first discussed in [Kolchin 1953] III $n^0 6$ then also in [Kolchin 1955] III. $n^0 3$ and in [Kolchin 1973], VI $n^0 5$. Briefly, $E = F < \alpha >$, with α transcendental and there exist $g_2, g_3 \epsilon C_F, a \epsilon F$ such that

$$g_2^3 - 27g_3^2 \neq 0, (\alpha')^2 = a^2(4\alpha^3 - g_2\alpha - g_3).$$

9.4. By the Theorem, every Galois group is a group variety over C. As a step towards a converse, Kolchin shows that given a connected group variety over C, there exists a differential field with field of constants C and a strongly normal extension of that field with Galois group G. This is still far from a solution to the "inverse problem" of Galois theory, namely, given F, to determine the possible Galois groups of strongly normal extensions of F.[5]

9.5. Let G be connected. Then by the Chevalley-Barsotti structure theorem G has a biggest normal linear subgroup N such that G/N is an abelian variety (see e.g. [C], a proof is also given in [Kolchin 1973], V, n^0 24). Let $E' = E^N$. Then, in view of 9.2, E' is the smallest strongly normal extension of F in E such that E is P.V. over E', and $G(E'/F)$ is an abelian variety.

9.6. In [Kolchin 1958] and in the last section of [Kolchin 1973] the realization of strongly normal extensions by function fields of principal homogeneous spaces is studied. Let G be a connected group variety over C. A variety V over F is a principal homogeneous space (over F) for G if G acts simply

[5]Kolchin has not contributed further to the inverse problem, which puts it outside the purview of this talk. For a historical survey, see the paper by M. Singer in this volume.

transitively on it. Given $v \epsilon V$, the map $g \mapsto v.g$ is an isomorphism of varieties over $F(v)$. There is always such a v which is algebraic over F. Then G operates over $F(V)$ by right translations. It is shown that a strongly normal extension of F with Galois group G may be so realized.

10.1. We now come back to (i) in the fundamental theorem (8.4) and to the reason why its formulation had been called provisory. It is correct as stated and references will be provided. Still, it misrepresents an important aspect of Kolchin's approach. The assertion "is isomorphic to an algebraic group over C" refers implicitly to the standard notion of algebraic group, namely, an algebraic variety over C endowed with a group structure such that inverse and product are defined by C-morphisms of varieties. In the sequel, I shall refer to those as group varieties over C. But Kolchin proceeds differently: he first defines a structure on $G(E/F)$ analogous to that of an algebraic group, in which the fields $C(\sigma)$ play the role of residue fields, so that (i) is in fact the combination of two statements:

(a) The Galois group $G(E/F)$, endowed with the fields $C(\sigma)$, is an algebraic group in Kolchin's sense.

(b) $G(E/F)$ may be identified to a group variety over C_F so that $C(\sigma)$ is the residue field at σ $(\sigma \epsilon G(E/F))$.

10.2. To describe this, let me first recall, however briefly, some concepts familiar in [W], which play a great role in Kolchin's approach.

Let K be an algebraically closed groundfield. All extensions of K are to be contained in a fixed universal field. Let V be a variety over K. Then, as already recalled, every point $x \epsilon V$ has a residue field $K(x)$, a finitely generated extension of K. The points of V are partially ordered by a notion of specialization over K, noted $x \underset{K}{\rightarrow} x'$ or simply $x \rightarrow x'$. [If x, x' are in an open affine subset, every regular function vanishing at x also vanishes at x']. It implies that $\partial^0(K(x)/K) \geq \partial^0(K(x')/K)$. The specialization is *generic* if there is equality, in which case $x' \rightarrow x$, *non-generic* otherwise. The locus Z_x of x over K is the set of specializations (over K) of x. It is the smallest K-irreducible subset of V containing x, and x is a generic point of Z_x over K.

16

If W is another variety over K, a map $f : V \to W$ is a morphism if $f^0(K(f(x)) \subset K(x)$ for all $x \epsilon V$ and if it is compatible with specializations.

10.3. We now return to [Kolchin 1953] and the strongly normal extension E of F, let G^* be the group defined by the strong isomorphisms of E over F (see 8.3) and G the subgroup of automorphisms of E over F. By definition, G^* is the Galois group of E over F, but the main results of [Kolchin 1953] pertain to G.

Kolchin introduces the notion of specialization (over C) for isomorphisms of E over F and shows that the specialization of a strong one σ is strong (II, n^0 5, Prop. 5, p.774). The strong isomorphism σ is isolated if it not a non-generic specialization of another strong isomorphism. G^* has finitely many isolated elements, up to generic specialization. This, and some further properties of strong isomorphisms allow Kolchin to define on G^* a structure of algebraic group and morphisms of such groups with the usual properties. Then G is endowed with the induced structure of algebraic group, in which the algebraic subgroups are the intersections of G with algebraic subgroups of G^* defined over C, i.e. in fact the groups of rational C-points of such subgroups. It is this notion of algebraic subgroup which underlies the statement of the fundamental theorem in [Kolchin 1953], which deals only with G. (See n^0s 2,3 in III). The question as to whether G^* may be naturally identified with a group variety over C, in Weil's sense, is raised in [Kolchin 1953: p.759] and positively answered in [Kolchin 1955], II, n^0 1, Thm. 1, p.878. More precisely, the fields $C(\sigma)$ are the residue fields in the group variety structure. This then realizes (a) and (b) in 10.1 and yields a proof of (i) in 8.4.

In [Kolchin 1958] this identification is used to put the discussion of realization via principal homogeneous spaces (see 9.6) in the framework of the usual algebraic group theory. But Kolchin comes back in [Kolchin 1973] to his original point of view and develops it much more systematically.

10.4. Kolchin's treatment of "algebraic sets and groups" is the subject matter of Chapter V in [Kolchin 1973]. The references in this section and the two following ones are to that Chapter.

Let K be as in 10.2. The starting notion is that of pre-K set ($n^0 2$). The set V is a pre$-K$ set if to each point $x \epsilon V$ is assigned a finitely generated

17

extension $K(x)$ of K, subject to relations of specializations, satisfying certain natural conditions. V is a K-group if moreover it is a group and we have

$$K(x.y) \subset K(x).K(y) \qquad K(x^{-1}.y) \subset K(x).K(y), \; (x, y \epsilon V)$$

and some further conditions relative to specializations. V is a homogeneous K-space if it is a pre-K-set acted upon transitively by a K-group G, again with certain natural conditions (p.220). A K-set is a pre-K-subset of a homogeneous K-space. Thus, in this algebraic geometry, the algebraic sets (the K-sets) are all contained in homogeneous spaces. Kolchin then develops his theory along lines familiar in other contexts : effect of extensions of the groundfield, Zariski topology, regular extensions and, for K-groups: connected components, products, morphisms, quotients (to be discussed below) linear groups, Galois cohomology, ending up with a proof of the Chevalley-Barsotti theorem mentioned in 9.5.

10.5. We now want to relate those concepts to the corresponding ones in algebraic geometry. There, in any treatment familiar to me, one starts with a notion of affine variety (or scheme) and the varieties (or schemes) are by definition locally affine. Kolchin's approach is *a priori* very different as we saw. However, a product of $\mathbf{G}_a$'s, i.e. the additive group of a vector space over K, is homogeneous and its K-subsets are included in Kolchin's set up. They are essentially affine varieties in disguise and provide the link between the two points of view.

A K-affine set is by definition a K-set which is isomorphic to a K-subset of some $\mathbf{G}_a^n$. The main point is then to show that in a K-set A, any finite set of points is contained in an open affine K-subset, from which it follows that A has a finite open cover by affine K-subsets (n^0 16, Thm. 4 and Cor., p.311). This then implies that Kolchin's K-groups or homogeneous K-spaces may be naturally identified with group or homogeneous varieties (*loc. cit.* Remark p.311-312). [In view of this identification, Thm.4 quoted above is not surprising since it is known that any homogeneous variety is quasi-projective (Chow's theorem).]

10.6. Kolchin achieves in this way autonomy, at the cost of a huge (his word) chapter on algebraic groups. Whether it is worth the effort or whether

18

a shorter treatment of the Galois theory using the standard theory of algebraic groups could be given, I will not try to assess, but I would like to mention two points, one minor, one major, in which Kolchin's approach has paid off. The former is the construction of the quotient space G/H of a K-group G by a closed K-subgroup H. Wanted is a structure of K-variety on G/H with a universal property, such that $\pi : G \to G/H$ is a separable K-morphism commuting with the action of G. For linear groups, there is a simple trick to achieve this (see [B]). For non-linear groups the original approach of Weil [W1] is rather awkward. The usual method nowadays, which also proceeds from the local to the global, is to start from the existence of the quotient of some invariant Zariski-open set, proved by M. Rosenlicht, and to use translations (see [Ss]). From Kolchin's point of view, it is straightforward and directly global. The starting point is Theorem 4, p.240 according to which, given a closed subset A of a homogeneous K-space, there is a smallest field $L \supset K$ such that A is a L-subset. It is finitely generated over K and denoted $K(A)$. Then, Kolchin assigns to $x \epsilon G/H$ the field $K(\pi^{-1}(x))$. If $x' \epsilon G/H$, then $x \to x'$ if there exist $y \epsilon \pi^{-1}(x), y' \epsilon \pi^{-1}(x')$ such that $y \to y'$. It is then rather easily seen that his axioms for a homogeneous K-space are satisfied and that π is separable (n^0 11).

10.7. The second point reaches further, to the last major project of Kolchin: the foundation of a theory of differential algebraic groups i.e., roughly speaking, groups locally defined by algebraic differential equations [Kolchin 1985]. In the seventies, one of affine or linear differential algebraic groups had been developed by P.J. Cassidy, but Kolchin decided to supply the foundations for a general theory. In spite of many technical complications, he could use Chapter V of [Kolchin 1973] almost as a blueprint: his axiomatic treatment there could be adapted to this more general situation and in fact, a number of sections could be taken almost verbatim, with minor changes in notation. It initiates more broadly a "differential algebraic geometry", at any rate for subsets of homogeneous spaces, as in [Kolchin 1973].

In the linear case, the theory was pursued by P.J. Cassidy and led to a classification of semisimple differential groups and Lie algebras [Ca]. On the other hand, A. Buium applied similar ideas to the study of diophantine problems over function fields, thus establishing completely new connections with other topics in mathematics. This whole area is in full development and well worth a report, which I have unfortunately neither the time nor the competence to give.[6] Still, I wanted to mention it as a counterpart to the main topic of my talk: the latter is a well-rounded theory, with statements of a considerable aesthetic beauty, seemingly in final form. The former opens up new directions and points out to work for the future.

[6]but for which I can now refer to the paper by A. Buium and P. Cassidy in this volume.

REFERENCES

The symbol [Kolchin x] refers to item x in the bibliography of E. Kolchin's work, at the end of this volume.

[B] A. Borel, Linear algebraic groups, 2nd enlarged edition, GTM **126**, Springer 1991.

[Ca] P.J. Cassidy, *The classification of the semisimple differential algebraic groups and the linear semisimple differential algebraic Lie algebras*, Jour. Algebra **121** (1988), 169-238.

[C] C. Chevalley, *Une démonstration d'un théorème sur les groupes algébriques*, Jour. Math.pur. appl. (9) **39** (1960), 307-317.

[L] A. Loewy, *Die Rationalitätsgruppe einer linearen homogenen Differentialgleichung*, Math. Annalen **65** (1908), 129-160.

[P1] E. Picard, Traité d'Analyse, t. III, Gauthier-Villars, Paris 1898 or 1908.

[P2] E. Picard, Oeuvres, t.2, éditions du CNRS, Paris, 1979.

[Se] A. Seidenberg, *Contribution to the Picard-Vessiot theory of homogeneous linear differential equations*, Amer. J. Math. **78** (1956), 808-817.

[Ss] C. S. Seshadri, *Some results on the quotient space by an algebraic group of automorphisms*, Math. Annalen **149** (1963), 286-301.

[V] E. Vessiot, *Méthodes d' intégration élémentaires*, Encycl. des sci. math. pures et appliquées, tome II, Vol. 3 (1910), pp.58-170.

[W] A. Weil, Foundations of algebraic geometry, A.M.S. Colloquium Publ. **29**, New-York, 1946.

[W1] A. Weil, *On algebraic groups and homogeneous spaces,* Amer. J. Math. **77** (1955), 493-512.

School of Mathematics, Institute for Advanced Study, Olden Lane, Princeton, NJ 08540

Direct and Inverse Problems in Differential Galois Theory

Michael F. Singer

ABSTRACT. This paper surveys recent work on the problems of calculating Galois groups of differential equations and of constructing differential equations with given groups as their Galois groups.

1. Introduction

In his address to the 1966 International Congress of Mathematicians [**89**], Ellis Kolchin described five problems with the hope that "my description of a few of the more challenging unsolved problems will lure some into accepting the challenge." The fourth problem, concerning the Picard-Vessiot theory, was presented in the following way:

> Now let $\mathfrak{F}$ be an ordinary differential field and denote its field of constants by $\mathfrak{C}$; denote the field of constants of $\mathfrak{U}$[1] by $\mathfrak{K}$. Consider a homogeneous linear differential polynomial $L = Y^{(n)} + a_1 Y^{(n-1)} + \ldots + a_n Y$ with coefficients in $\mathfrak{F}$. If $\mathfrak{C}$ is algebraically closed then L has a fundamental system of zeros $(\eta_1, \ldots, \eta_n)$ such that the extension $\mathfrak{G} = \mathfrak{F}\langle \eta_1, \ldots \eta_n \rangle$ of $\mathfrak{F}$ has the same field of constants $\mathfrak{C}$, and then the set of all the isomorphisms over $\mathfrak{F}$ of $\mathfrak{G}$ onto an extension of $\mathfrak{F}$ in $\mathfrak{U}$ has a natural group structure; by means of the fundamental system of zeros $(\eta_1, \ldots, \eta_n)$ this group can be identified with an algebraic subgroup G of $\mathrm{GL}_{\mathfrak{K}}(n)$ defined over $\mathfrak{C}$. This G can be called the *Galois group of* L over $\mathfrak{F}$. Of course, a different choice of fundamental systems of zeros in general gives a different Galois group; G is determined up to conjugation in $\mathrm{GL}(n)$ by a matrix that is rational over $\mathfrak{C}$. (This ambiguity can be removed by taking for G an algebraic subgroup of the group $\mathrm{GL}(V)$ of all automorphisms of the vector space V of zeros of L).
>
> As in the ordinary Galois theory, there are two general problems:
>
> 1° *Given L, to determine G.*
>
> 2° *Given G, to find an L of which G is the Galois group.*

1991 *Mathematics Subject Classification.* Primary 12H05, 14E20; Secondary 12F12, 12Y05.
The author was supported in part by NSF Grant #CCR-93222422.

[1]Kolchin previously defined $\mathfrak{U}$ to be a universal differential field, that is, a field "so large that it contains enough elements for every purpose we can have." Kolchin also assumes that this field is of characteristic zero and we shall make this assumption for all fields mentioned in this paper.

> Of course, the nature of these problems depends on the differential field $\mathfrak{F}$ which is given. For example, when $\mathfrak{F} = \mathfrak{C}$ then it is easy to see that, for every L, G is reducible to triangular form (and hence is solvable). A. Bialynicki-Birula [**25**] has shown that if $\mathfrak{F}$ has finite transcendence degree over $\mathfrak{C}$ and G is a connected and nilpotent algebraic subgroup of $\mathrm{GL}(n)$ defined over $\mathfrak{C}$, then there exists a homogeneous linear differential polynomial L with coefficients in $\mathfrak{F}$ such that the Galois group of L over $\mathfrak{F}$ is isomorphic to G. Aside from this and some work in a slightly different direction by Lawrence Goldman [**58**], even when $\mathfrak{F} = \mathbb{C}(x)$ very little of a general nature is known. For the second problem this is not unexpected, in view of the history of the analogous problem in classical Galois theory. The obstinacy of the first problem is somewhat of a surprise, and it now appears that we should be grateful for almost any kind of information about G; the difficulty stems in part from the fact that a zero of L that is not a zero of any linear differential polynomial over $\mathbb{C}(x)$ of lower order may very well be a zero of a nonlinear one.

We shall refer to problem 1° as the *direct problem* and problem 2° as the *inverse problem*. The aim of this paper is to describe some of the work that has been done on these problems since Kolchin's talk and to give a guide to the literature. Excellent guides to the 19th Century literature are [**70**], [**134**], [**135**] and [**163**] as well as [**59**] and [**60**] which give historical insight into thi s era. A good recent survey of work on linear meromorphic differential equations is [**161**] (see also [**162**]). We refer to Borel's article [**26**] or any of [**22**], [**54**], [**55**], [**79**], [**90**], [**97**], [**102**], [**121**] for the basic facts concerning differential Galois theory and [**27**] for the basic facts concerning linear algebraic groups.

2. The Direct Problem

At present, we do not know a general algorithm that will compute the Galois group of a linear differential equation with coefficients in a differential field k, even when $k = \bar{\mathbb{Q}}(x)$, where $\bar{\mathbb{Q}}$ is the algebraic closure of the rational numbers. In contrast, algorithms for calculating the Galois group of a polynomial with coefficients in $\mathbb{Q}$ or $\bar{\mathbb{Q}}(x)$ have been known for a long time [**39**], [**119**], [**167**]. The key idea behind these methods is to represent the splitting field of a polynomial in terms of generators and relations. The Galois group is then the set of permutations of the generators that preserve the relations. In the differential case, the Picard-Vessiot extension is the analogue of the splitting field. The obstruction to mimicking the ideas from the Galois theory of polynomials is that, at present, we do not know how to effectively present a general Picard-Vessiot extension in terms of generators and relations. Nonetheless, there are many *properties* of the Galois group that can be computed and, in some cases, one can even determine this group (see [**143**] for an introduction to differential Galois theory with an emphasis on computational aspects).

Let k be a differential field with algebraically closed field of constants C. Let $L(y)$ be a homogeneous linear differential polynomial of order n with coefficients in k, $K = k\langle y_1, \ldots, y_n\rangle$ the associated Picard-Vessiot extension, and G its Galois group. We will denote by V the solution space of $L(y) = 0$ in K. Note that V

is a finite dimensional C-vector space and a G-module. We shall now consider algorithms to determine various properties of G, V and L. Throughout we shall assume that the field of constants C is a computable field with a factorization algorithm, that is, a field whose field theoretic operations $+, -, \times, ^{-1}$ are recursive functions and over which we can effectively factor polynomials. For example, the algebraic closure of a finitely generated extension of $\mathbb{Q}$ is such a field [**167**].

2.1. Triviality of G. The Galois group is trivial if and only if $K = k$, so we wish to consider algorithms to decide if $L(y) = 0$ has a full set of solutions in k. In fact algorithms exist to find the space $W \subset V$ of solutions of $L(y) = 0$ in k for large classes of fields k. Consider first the simpler problem of finding solutions $a \in \mathbb{Q}$ of $p(x) = a_n x^n + \ldots a_0 = 0$, $p(x) \in \mathbb{Z}[x]$. One way to proceed is to note that factors of the denominator of a must divide a_n and the factors of the numerator must divide a_0 (assuming that a has relatively prime numerator and denominator). One could then try all possibilities. For differential equations $L(y) = a_n y^{(n)} + \ldots + a_0 y = 0$ with $a_i \in C(x)$, $x' = 1$ one can attempt a similar approach. One can see that any *irreducible* factor q of the denominator of a solution $y \in C(x)$ must divide a_n, but the largest power q^m of q dividing the denominator of y may be greater than the largest power of q dividing a_n. For example, $y = x^m$ is a solution of $xy' - my = 0$. To bound m in general, one expands q-adically each $a_i = a_{i,n_i} q^{n_i} + a_{i,n_1+1} q^{n_i+1} + \ldots$, $deg(a_{i,j}) < deg(q)$, substitutes $y = A_m q^m + A_{m+1} q^{m+1} + \ldots$ into $L(y) = 0$ and notices that in order for equality to hold the exponent m must satisfy a polynomial equation over C (the *indicial equation at q*). The smallest integer root m_q of this equation will bound the possible m. Substituting $y = z \prod q^{m_q}$ into $L(y) = 0$, reduces the problem to finding polynomial solutions of $\tilde{L}(z) = L(z \prod q^{m_q}) = 0$. In order for equality to hold in $\tilde{L}(z) = 0$ after one substitutes $z = \alpha_N x^N + \ldots + \alpha_0$, one again sees that N must satisfy a polynomial equation (the indicial equation of $\tilde{L}$ at infinity). If N is the largest integer root of this polynomial, we can substitute $z = \alpha_N x^N + \ldots + \alpha_0$ into $\tilde{L}(z) = 0$ to yield a system of linear equations for the α_i. A basis of the solution space of this system will yield a basis of the solution space of $L(y) = 0$ in $C(x)$.

This naive algorithm can be improved in several ways. For example, one would like to work in the smallest field $C_0(x) \subset C(x)$ containing the coefficients of L and one would like to find efficient algorithms to find polynomial solutions of linear differential equations. These and related issues are discussed in [**1**], [**3**], [**4**], [**35**]. Before turning to other fields, we note that in many situations one is given a system $Y' = AY$ of differential equations where A is an $n \times n$ matrix with coefficients in $C(x)$ and asked to determine a basis for all solutions in $(C(x))^n$. In theory one can reduce this problem to finding all solutions of an associated scalar equation $L(y) = 0$ in $C(x)$ (this is described in Section 2.4.1) but finding this associated equation can be costly. An algorithm to find rational solutions of the system $Y' = AY$ directly has been given by M. A. Barkatou [**14**].

Algorithms to find solutions of $L(y) = 0$ in a given differential field k when k is an algebraic extension of $C(x)$, or more generally, certain types of liouvillian extensions (this term will be defined below) of $C(x)$ are discussed in [**144**]. Finally algorithms to find solutions of inhomogeneous linear differential equations in various fields are discussed in [**30**], [**31**], [**32**], [**33**], [**45**].

2.2. Reducibility of V. The G-module V contains a proper nontrivial G-module W of dimension r if and only if one can write the differential polynomial $L(y) = L_{n-r}(L_r(y))$ as a *composite* of homogeneous linear differential polynomials L_{n-r} and L_r of orders $n-r$ and r respectively with coefficients in k. To see this, let $\{y_1, \dots, y_r\}$ be a basis of W and let $L_r(y) = \mathrm{Wr}(y, y_1, \dots, y_r)/\mathrm{Wr}(y_1, \dots y_r)$, where Wr(...) denotes the wronskian determinant. Assume that W is a G-invariant subspace of V. The coefficients of L_r are left invariant by G since applying $g \in G$ to these wronskians multiplies each of these determinants by the determinant of the linear map induced by g on W. Writing $L(y) = L_{n-r}(L_r(y)) + R(y)$ where the order of $R(y)$ is less than r (c.f., [**120**], Ch. III) we see that $R(y) = 0$ has r independent solutions in K and therefore must be 0. Conversely, if $L(y) = L_{n-r}(L_r(y))$, the solution space W of $L_r(y) = 0$ is a G-invariant subspace of V.

This characterization of reducibility leads naturally to considering the ring $\mathcal{D} = k[D]$ of linear operators. This is the ring of noncommutative polynomials in D with coefficients in k where $D \cdot a = a \cdot D + a'$. To each homogeneous linear differential polynomial $L(y) = a_n y^{(n)} + \dots + a_0 y$ we can associate the operator $L = a_n D^n + \dots + a_0$ in $\mathcal{D}$. We then have that V is reducible as a G-module if and only if L factors in $\mathcal{D}$ as a product of operators of smaller positive order. Furthermore, to any operator $L \in \mathcal{D}$ one can associate the left $\mathcal{D}$-module $\mathcal{D}/\mathcal{D} \cdot L$. The operator L factors as a product of operators of smaller positive order if and only if the $\mathcal{D}$-module $\mathcal{D}/\mathcal{D} \cdot L$ contains a proper nontrivial $\mathcal{D}$-submodule (see [**146**] for a discussion of this ring and historical comments on the work of Frobenius, Jordan, Beke, Loewy, Ore and Jacobson concerning $\mathcal{D}$). Applying the Jordan-Hölder Theorem one sees that every operator $L \in \mathcal{D}$ can be written as a product of irreducible operators $L = L_1 \cdot \dots \cdot L_r$. Furthermore, if $L = \tilde{L}_1 \cdot \dots \cdot \tilde{L}_s$ with the $\tilde{L}_i$ irreducible, then $r = s$ and (after a possible renumbering) the $\mathcal{D}$-modules $\mathcal{D}/\mathcal{D}L_i$ and $\mathcal{D}/\mathcal{D}\tilde{L}_i$ are isomorphic or, in more pedestrian terms, the orders of L_i and $\tilde{L}_j$ are equal and there exist nonzero operators R and S whose orders are less than the order of L_i satisfying $L_i R = S\tilde{L}_i$.

Before we discuss algorithms to determine if V is reducible we investigate the more restrictive question of whether V has a one dimensional G-submodule. As we have seen this is equivalent to L having a right factor of order one. If we write $L = L_{n-1}L_1$ where $L_1 = D - r$, then one sees that L has a first order right factor if and only if $L(y) = 0$ has a solution of the form $y = \exp(\int r)$ with $r \in k$ (y is then called an *exponential over* k). Algorithms to find such solutions when $k = \overline{\mathbb{Q}}(x)$ are classical ([**134**], Section 177) and rest on the fact that at any point $x_0 \in \mathbb{C}$, one can formally find the solution space of $L(y) = 0$ (one can also consider the point at infinity by letting $x = 1/z$ and $d/dx = -z^2 d/dz$). If x_0 is a nonsingular point (i.e., a point where $a_n(x_0) \neq 0$), then all solutions are given by convergent power series. In general, if a linear differential equation has coefficients that are meromorphic at x_0, there exist n linearly independent solutions of the form

$$y_i = (x - x_0)^{\lambda_i} e^{Q_i(t)} (\phi_{i,0} + \phi_{i,1}(t) \log(x - x_0) + \dots \phi_{i,s_i}(t)(\log(x - x_0))^{s_i}) \tag{2.1}$$

where $\lambda_i \in \mathbb{C}$, $t = (x - x_0)^{-1/q_i}$ for some positive integer q_i, Q_i is a polynomial in t^{-1} without constant term, and $\phi_{i,j} \in \mathbb{C}[[t]]$. At each point x_0 there are only a finite number of λ_i (called *exponents*) and Q_i (called *determining factors*) that can occur and these can be calculated directly from the operator L (or a system $Y' = AY$) ([**134**], Section 110, [**11**], [**13**], [**49**], [**69**], [**73**], [**96**], [**103**], [**151**], [**153**]). If $\exp(\int r)$ is a solution of $L(y) = 0$ and $r = \alpha_m (x - x_0)^{-m} + \dots + \alpha_1 (x - x_0)^{-1} + \dots$ is the

Laurent expansion of r at x_0, then $Q(x) = \frac{\alpha_m}{-m+1}(x-x_0)^{-m+1} - \ldots - \alpha_2(x-x_0)^{-1}$ is a determining factor and $\lambda = \alpha_1$ is an exponent at x_0. In particular, if x_0 is a nonsingular point, then $m = 1$ and α_1 is an integer between 0 and $n-1$. Therefore to find the possible $r \in \overline{\mathbb{Q}}(x)$ with $L(\exp(\int r)) = 0$, we write

$$r = p(x) + \sum_{\alpha \in \mathcal{S}} \left(\sum_{i=1}^{n(\alpha)} \frac{c_{\alpha,i}}{(x-\alpha)^i}\right) + \sum_{\beta \in \mathcal{N}} \frac{m_\beta}{(x-\beta)}$$

where $\mathcal{S}$ is the set of finite singular points, $\mathcal{N}$ is the set of finite non-singular points, the $c_{\alpha,i}$ are constants, the m_β are integers between 0 and $n-1$, and $p(x)$ is a polynomial. For $\alpha \in \mathcal{S}$, the possible terms $\sum_{i=1}^{n(\alpha)} \frac{c_{\alpha,i}}{(x-\alpha)^i}$ can be found by examining the exponents and determining factors at α. The possibilities for the polynomial $p(x)$ c an be found by examining the determining factors at infinity. For each choice of these possibilities, we consider $z = y\exp(-\int \tilde{r})$ where $r = p(x) + \sum_{\alpha \in \mathcal{S}} (\sum_{i=1}^{n(\alpha)} \frac{c_{\alpha,i}}{(x-\alpha)^i})$. The element $z = \exp(\int \sum_{\beta \in \mathcal{N}} \frac{m_\beta}{(x-\beta)}) = \prod (x-\beta)^{m_\beta}$ satisfies the linear differential equation $\tilde{L}(z) = \exp(-\int \tilde{r}) L(z\ exp(\int \tilde{r})) = 0$. The original equation $L(y) = 0$ has an exponential solution if and only for some choice possibilities the equation $\tilde{L}(y) = 0$ has a polynomial solution. In the previous section we discussed how to decide this latter question. One can furthermore use this approach to find all such solutions. There are many places where this algorithm can be improved. For example, one would like to work in the coefficient field for as long as possible. One would like to efficiently find the exponents and determining factors. Also, it is not always necessary to try all possible combinations of the determining factors and exponents. These and other implementation issues as well as generalizations to other fields are discussed in [**34**], [**45**], [**72**], [**144**] . There are also algorithms to find exponential solutions of systems $Y' = AY$, that is solutions $Y = (y_1, \ldots, y_n)^T$ where each y_i is exponential over $\overline{\mathbb{Q}}(x)$ [**117**].

We now turn to the problem of deciding if V has invariant subspaces W of arbitrary dimension r. As noted in the beginning of this section, this is equivalent to finding a right factor of L of order r. Such an operator will be of the form $L_r = \mathrm{Wr}(y, y_1, \ldots, y_r)/\mathrm{Wr}(y_1, \ldots y_r)$ where $y_1, \ldots y_r$ forms a basis of W. The coefficients of L_r are quotients whose numerators are $r \times r$ subdeterminants of

$$\begin{pmatrix} y_1 & \cdots & y_r \\ y_1' & \cdots & y_r' \\ \cdots & \cdots & \cdots \\ y_1^{(r)} & \cdots & y_r^{(r)} \end{pmatrix} \tag{2.2}$$

and whose denominators are $\mathrm{Wr}(y_1, \ldots, y_r)$. For example the coefficient of $y^{(r-1)}$ in L_r is $\mathrm{Wr}(y_1, \ldots, y_r)'/\mathrm{Wr}(y_1, \ldots, y_r)$. Since each of these quotients is in k, one can show that each of these subdeterminants is an exponential over k. Furthermore, one can show ([**134**], Section 167) that there exists an operator $L^{\wedge r}$ of order at most $\binom{n}{r}$ with coefficients in k, such that for any basis $y_1, \ldots, y_n$ of V, all $r \times r$ subdeterminants of $\mathrm{Wr}(y_1, \ldots, y_n)$ satisfy $L^{\wedge r}(y) = 0$. We again restrict ourselves to operators whose coefficients lie in the field $k = \overline{\mathbb{Q}}(x)$ and outline an algorithm originally due to Beke [**16**]. To find a right factor of order r of L, one first finds a maximal set $\{z_i\}$ of linearly independent solutions of $L^{\wedge r}(y) = 0$ that are exponential over k. One can then give a bound N on the degrees of the numerators and denominators of all elements of $\overline{\mathbb{Q}}(x)$ that equal a quotient $(\sum c_i z_i)/(\sum d_i z_i)$. One forms an operator $\tilde{L}_r$ of order r whose coefficients are quotients of polynomials of degree N in x with indeterminate coefficients. Formally dividing $\tilde{L}_r$ into L yields a remainder that is an operator R of order less than r. The coefficients of this remainder will be quotients of polynomials in x whose coefficients are polynomials in the indeterminates introduced above. The condition $R = 0$ is equivalent to systems of polynomial equalities and inequalities that the indeterminates must satisfy in order that $\tilde{L}_r$ divides L. One then decides if these equalities and inequalities can be satisfied in C (details and refinements of this basic algorithm are given in [**16**], [**36**], [**42**] Appendix, [**134**] Section 176, [**136**]). One can refine this algorithm to give a factorization into irreducible factors. Furthermore, one can enumerate all possible factorizations [**157**].

There have been other improvements in this algorithm. In [**64**] Grigoriev gives several improvements in the algorithm together with a careful analysis of the running time and the size of the resulting factors. He also gives a polynomial time algorithm to find the greatest common right divisor of a family of operators.

Another improvement comes from the observation that not all exponential solutions of $L^{\wedge r}(y) = 0$ yield information about factors. Generically, the solution space of $L^{\wedge r}(y) = 0$ is G-isomorphic to $\bigwedge^r V$ (although, there are situations when this solution space is isomorphic to a quotient of $\bigwedge^r V$). Not all G-invariant lines in $\bigwedge^r V$ correspond to decomposable vectors, i.e., vectors of the form $v_1 \wedge \ldots \wedge v_r$; the Plücker relations define those that do. Therefore not all exponential solutions of $L^{\wedge r}(y) = 0$ must correspond to subdeterminants of 2.2. Tsarev [**156**] has shown how one can use the Plücker relations to narrow down the possible exponential solutions of $L^{\wedge r}(y) = 0$ that enter into the factorization problem.

Recent work of van Hoeij [**72, 73**] has led to good implementations of algorithms to factor linear operators. In [**73**] van Hoeij takes the methods of [**103**] and turns them into efficient algorithms for factoring in $C((x))[D]$. In [**72**] he uses the local methods of [**73**] to give good ways to find exponential solutions of linear differential operators with the aim of efficiently determining the coefficients in a factorization. In addition he minimizes the need to go beyond the smallest field containing the coefficients of L.

When one considers restricted classes of differential operators, the problem of factorization becomes more tractable. We say that a differential operator L is *completely reducible* if it is the least common left multiple of irreducible operators. This is equivalent to any of the following: a) $\mathcal{D}/\mathcal{D}L$ is the sum of irreducible $\mathcal{D}$-modules; b) V is the sum of irreducible G-modules; c) G is a reductive group. In [**146**] algorithms are given to decide if an operator in $C(x)[D]$ is completely reducible (this is generalized to algebraic extensions of $C(x)$ in the appendix of [**42**]). Furthermore, once it is known that the operator is completely reducible, it can be factored by finding solutions *in* k of certain auxiliary systems; one does not need to search for elements that are exponential over k. Although this class seems to be restrictive, we shall see below that in other decision problems one is frequently given an operator that is known in advance to be completely reducible and for which one needs to determine a factorization. In [**71**], van Hoeij shows how his methods may be applied to this class of differential operators (as well as some more general classes).

For differential systems $Y' = AY$, the solution space has an invariant G-submodule of dimension s if and only if there is an invertible matrix $B \in \mathrm{GL}(n,k)$ such that

$$B'B^{-1} + BAB^{-1} = \begin{pmatrix} A_{1,1} & 0 \\ A_{2,1} & A_{2,2} \end{pmatrix}$$

where $A_{2,2}$ is in $M(s,k)$, [**63**]. In this latter paper, Grigoriev gives a procedure (with complexity analysis) for determining this property when $k = C(x)$.

Finally, we note that although Gauss's Lemma does not hold in $\mathcal{D}$, an Eisenstein-like criterion for irreducibility was developed for this ring in [**93**].

2.3. Solvability of the Identity Component. One consequence of Kolchin's work on the Galois theory of differential equations was to give a precise version of Vessiot's theorem concerning equations solvable by quadratures (see [**87**] for a discussion of Vessiot's original work). One says that an extension $E = k(t_1, \dots t_r)$ of k is a *liouvillian extension* of k if for each i either:

1. t_i is primitive over $k(t_1, \dots t_{i-1})$, i.e. $t_i' \in k(t_1, \dots t_{i-1})$, or
2. t_i is exponential over $k(t_1, \dots t_{i-1})$, i.e. $t_i'/t_i \in k(t_1, \dots t_{i-1})$, or
3. t_i is algebraic over $k(t_1, \dots t_{i-1})$.

We say that an element u of an extension of k is liouvillian over k if it belongs to a liouvillian extension of k. Kolchin showed [**87**] that a linear homogeneous differential equation is solvable in terms of liouvillian functions (i.e., its Picard-Vessiot extension lies in a liouvillian extension of k) if and only if the component G^0 of the identity of its Galois group G is solvable.

An algorithm to find a basis for the space of liouvillian solutions of a homogeneous linear differential equation $L(y) = 0$ with coefficients in an algebraic extension k of $C(x)$ was presented in [**139**]. To get a taste of the ideas involved, we will describe an algorithm to decide if all solutions of $L(y) = 0$ are liouvillian. An effective version of the Lie-Kolchin Theorem asserts that in this case G will have a subgroup H such that the elements of H can be simultaneously put in upper triangular form and such that the index of H in G is bounded by a computable function $I(n)$ of n, the order of L. If y is a common eigenvector of H, then $\sigma(y'/y) = cy'/cy = y'/y$ so y'/y is left fixed by all elements σ of H. This implies that y'/y is algebraic over k of degree bounded by $I(n)$ and so will be algebraic over $C(x)$ of degree bounded by $M = I(n)[k : C(x)]$. Therefore, if $L(y) = 0$ is solvable in terms of liouvillian functions over k, $L(y) = 0$ has a solution y such that y'/y is algebraic over $C(x)$ of degree bounded by M. To decide if $L(y) = 0$ has such a solution we will look for candidates for the minimal polynomial of $u = y'/y$. If $p(u) = u^N + b_{N-1}(x)u^{N-1} + \dots + b_0(x)$, $N \leq M$, is the minimal polynomial of such a u, then one can show that there exist solutions $z_1, \dots, z_N$ of $L(y) = 0$ such that each b_i is the ith symmetric function of the z_i'/z_i. From the exponents and determining factors at the singular points of L, we can bound the degrees of the numerators and denominators of the b_i. Therefore, if $L(y) = 0$ has only liouvillian solutions, it will have a solution y such that $u = y'/y$ satisfies a polynomial over $C(x)$ of degree at most M whose coefficients have numerators and denominators of effectively bounded degrees. Elimination theory then allows us to decide if such a solution exists and produces u. We then use the change of variables $y = z\exp(\int u)$ to get a new equation $\tilde{L}(z) = 0$ of lower order and proceed by induction.

This algorithm is not a practical algorithm. For example when $n = 2$ and the equation has coefficients in $\mathbb{Q}(x)$, the bound for $I(n)$ given in [**139**] is 384,064 while it is known that a second order linear differential equation with a liouvillian solution will have a solution whose logarithmic derivative is algebraic of degree at most 12 (see the description of Kovacic's algorithm below). In [**158**], Ulmer modified the algorithm in [**139**] by refining the group theoretic techniques involved. For second order equations, Ulmer's algorithm yields the correct bound. He was not only able to get significantly improved bounds in all cases but he developed conditions to further narrow down the set of possible degrees of the logarithmic derivative of an exponential solution of $L(y)$. Similar algorithms have been developed to find liouvillian solutions of inhomogeneous linear differential equations [**45**] and linear differential equations over more general fields [**144**]. A different approach to the general algorithm based on the fact that a linear differential equation has a liouvillian solution if and only if the Galois group has a semi-invariant linear form that factors into linear factors, is given in [**150**]. Finally, the substitution $y = z\exp(\int u)$ and other order reducing substitutions are discussed in [**2**], [**5**], and [**6**].

In 1979, previous to [**139**], Kovacic [**94**] discovered an algorithm to decide if a second order linear differential equation $L(y) = y'' + ay' + by = 0$ with coefficients in $C(x)$ has liouvillian solutions (see [**51**] or [**122**] for other expositions of this algorithm). With hindsight, this algorithm is also based on the idea that when such solutions exist, the equation will have a solution that is exponential over an algebraic extension of the base field. Yet Kovacic was able to produce an algorithm that is much more efficient than any general algorithm produced since because he also took advantage of knowledge of all the groups that can occur as Galois groups for second order equations. Kovacic's first observation is that after the substitution $y = \exp(-\frac{1}{2}\int a)z$ one has a differential equation of the form $z'' - rz = 0$ whose Galois group G is unimodular and has liouvillian solutions if and only if the original equation has solutions of this type. We will describe the rest of the algorithm using a trichotomy used by Ulmer in [**158**] to improve the results of [**139**]. If G is a group acting on a vector space V we say that G is *reducible* if V is reducible as a G-module. The group G is said to be *imprimitive* if G is not reducible and if there exist nontrivial, proper subspaces $V_1, \ldots, V_r$ of V such that $V = V_1 \oplus \ldots \oplus V_r$ and such that G permutes the V_i. We say G is *primitive* if it is not reducible and not imprimitive. If $G \subset \mathrm{SL}(2, C)$ is reducible then it must have an invariant one-dimensional subspace (so the linear differential equation will have a solution that is exponential over $C(x)$) and Kovacic gives a very explicit algorithm to decide this. If $G \subset \mathrm{SL}(2, C)$ is imprimitive then G^o is diagonalizable and of index 2 in G. Kovacic shows that this implies that the equation will have linearly independent solutions y_1, y_2 such that $y_1 y_2$ is exponential over a quadratic extension of $C(x)$ and shows how one can decide this. Finally, if $G \subset \mathrm{SL}(2, C)$ is primitive, Kovacic shows that G is either $\mathrm{SL}(2)$ or a specific finite group of order 24, 48, or 120. Furthermore, in each of these latter three cases he shows that there are linearly independent solutions y_1, y_2 and a homogeneous form f in y_1, y_2 with constant coefficients, of degree 4, 6, and 12 respectively, such that $f^n \in C(x)$ with $n = 3, 2$, or 1 respectively. In particular f will be exponential over $C(x)$. Kovacic shows how one can decide each of these latter three cases. Since these are the only possibilities, we have that the equation has liouvillian solutions if and only if the Galois group is reducible, imprimitive or one of these three cases hold Furthermore, a decision procedure results from these considerations. Kovacic is able to construct a liouvillian solution when one exists and is also able to give very strong necessary conditions by looking at the possible exponents that can occur at singular points (these conditions are generalized to third order equations in [**149**]). We note that Kovacic's algorithm finds a solution of the form $\exp(\int u)$ where u is algebraic over $C(x)$ when the equation has liouvillian solutions. When the equation has only algebraic solutions, the algorithm does not find the minimal polynomials of such solutions, even when the Galois group is one of the finite primitive groups. For these groups this task was begun in [**56**] and [**57**] and completed and generalized to third order equations in [**147**] and [**148**].

Several improvements have been made on Kovacic's original algorithm. Duval and Loday-Richaud [**51**] have given a more uniform treatment of the considerations concerning singular points and have as well applied the algorithm to decide which parameters in the hypergeometric equations (as well as several other classes of second order equations) lead to liouvillian solutions. In [**122**], van der Put shows how the algorithm can be simplified if the equation has only one or two singular points. In [**160**], Ulmer and Weil improve the algorithm by using the observation that any homogeneous polynomial with constant coefficients of degree m in any number of solutions of the equation must satisfy an auxiliary equation $L^{Ⓢm}(y) = 0$ called the mth symmetric power of L (see also [**37**]). They then show that except in the reducible case, one can decide if there is a liouvillian solution (and find one) by looking for solutions of appropriate symmetric powers *that lie in* $C(x)$. This eliminates some of the nonlinear considerations of Kovacic's algorithm. A general necessary and sufficient condition for the existence of liouvillian solutions in terms of the existence of G-invariant homogeneous forms is given in [**150**] with improvements in [**74**]. If the equation has coefficients in $C_0(x)$ where C_0 is not algebraically closed, it is important to know in advance how large an algebraic extension of C_0 is required. In [**67**] and [**168**] sharp results are given for Kovacic's algorithm as well as a general framework for higher order equations. In [**159**], sharp results are given concerning what constant fields are needed for equations of all prime orders. Applications of the algorithm to questions concerning integrability of Hamiltonians are given in [**9**] and the references cited there. The structure of the space of second order equations with one singular point and having liouvillian solutions is studied in [**164**]. We finally note that it was recently discovered that an algorithm (with some mistakes) to find the minimal polynomial of an element u algebraic over $C(x)$ with $\exp(\int u)$ satisfying a given second order linear differential equation was found by Pepin [**115**] in 1881. The problem of deciding when an nth order linear differential equation has a solution y with y'/y algebraic is also discussed by Marotte in [**104**] and concrete results are given when $n = 2, 3, 4$.

The question of deciding if a linear differential equation has only algebraic solutions (or even one nonzero algebraic solution) has a long history. In 1872, Schwarz [**137**] gave a list of those parameters for which the hypergeometric equation has only algebraic solutions (for higher hypergeometric functions this was done by Beukers and Heckman [**24**]). Using his algorithm, Pépin [**115**] was able to reproduce this list as well. Using invariant theory, Fuchs [**56**], [**57**] was able to find the minimal polynomial of an algebraic solution of a second order linear differential equation assuming that the Galois group is the finite imprimitive group of order 48 or 120 (this method is generalized in [**148**]). In [**85**], [**86**], Klein shows that any second order linear differential equation with only algebraic solutions comes from some hypergeometric equation via a rational change in the independent variable $x = r(z)$. This approach was turned into an algorithm by Baldassarri and Dwork [**10**]. Jordan [**78**] considered the problem of deciding if a linear differential equation of order n has only algebraic solutions. He showed that a finite subgroup of $\mathrm{GL}(n)$ has an abelian normal subgroup of index bounded by a computable function $f(n)$ of n (this fact and its generalizations are at the heart of the algorithms for finding liouvillian solutions in [**139**] and [**158**] described above). This implies that such an equation has a solution whose logarithmic derivative is algebraic of degree at most $f(n)$. Jordan's approach was made algorithmic in [**138**] (see also [**28**] and [**114**] for similar but incomplete algorithms due to Boulanger and Painlevé). It should be noted that the algorithms of Boulanger, Klein, Painlevé, and Pépin, are all incomplete in at least one regard. Each of these algorithms, at one point or another, is confronted with the following problem: Given an element u, algebraic over $C(x)$, decide if $\exp(\int u)$ is also algebraic over $C(x)$. Boulanger refers to this as *Abel's Problem* ([**28**], p. 93) and none of these authors gives an algorithm to answer this question. In 1970, Risch [**133**] showed that this problem could be solved if one could decide if a given divisor on a given algebraic curve is of finite order. Risch showed how one could solve this latter problem by reducing the jacobian variety of the curve modulo two distinct primes and bounding the torsion of the resulting finite groups. For other work concerning Abel's Problem, see [**10**], [**31**] [**44**], [**154**], [**165**], [**166**]. The introduction to [**104**], the book [**60**] and the article [**59**] give historical accounts of work concerning algebraic solutions of linear differential equations.

Over the field $C((x))$ of formal power series, all linear differential equations are solvable in terms of liouvillian functions (see section 2.4.2). An algorithm to solve the direct problem for second order equations whose Galois group is a su bgroup of $\mathrm{SL}(2)$ is given in [**62**].

The question of when a linear differential equation has liouvillian or algebraic solutions leads to the general question of when such an equation can be solved in terms of other special functions. An equivalent definition of liouvillian extension is the following; an extension $E = k(t_1, \dots t_r)$ of k is a *liouvillian extension* of k if each t_i is either algebraic over $k(t_1, \dots t_{i-1})$ or satisfies a first order linear differential equation $t_i' + at_i = b$ with $a, b \in k(t_1, \dots t_{i-1})$. We can generalize this to define a linear differential equation of order n to be solvable in terms of lower order equations if its Picard-Vessiot field lies in a field of the form $k(t_1, \dots t_r)$ where each t_i is either algebraic over $k(t_1, \dots t_{i-1})$ or satisfies a linear differential equation of order *less than* n over $k(t_1, \dots t_{i-1})$. In [**142**] it is shown that a linear differential equation of order n is *not* solvable in terms of lower order linear differential equations if and only if the Lie algebra $\mathfrak{g}$ of G has the following two properties:

- $\mathfrak{g}$ is simple, and
- $\mathfrak{g}$ has no nontrivial representation of degree less than n.

An algorithm for third order equations is described in [**140**] (see also [**141**] for related material). An algorithm to solve linear differential equations in terms of hypergeometric functions is given in [**116**].

2.4. Calculating Galois Groups.

2.4.1. *Algebraic Methods.* The key to understanding modern algebraic techniques for calculating Galois groups comes from the formulation of the Galois theory in terms of Tannakian Categories [**18**], [**19**], [**21**], [**46**], [**47**], [**48**], [**80**]. Instead of a linear differential equation we consider a *connection*, that is, a finite dimensional

k-space $\mathcal{M}$ with an operator $\nabla : \mathcal{M} \to \mathcal{M}$ satisfying

$$\begin{aligned} \nabla(u+v) &= \nabla(u) + \nabla(v) \\ \nabla(fu) &= f'u + f\nabla(u) \end{aligned}$$

for all $u, v \in \mathcal{M}$ and $f \in k$ (c.f., [**65**]). If $e_1, \ldots, e_n$ is a k-basis of $\mathcal{M}$, we may write

$$\nabla e_i = -\sum_j a_{j,i} e_j \tag{2.3}$$

where $A = (a_{i,j}) \in M(n,k)$. If $u = \sum_i u_i e_i$, then $\nabla(u) = \sum_i (u_i' - \sum_j a_{i,j} u_j) e_i$. Therefore, once a basis of $\mathcal{M}$ has been selected and the identification $\mathcal{M} \simeq k^n$ has been made, we have that $u \in k^n$ satisfies $u' = Au$ if and only if $\nabla u = 0$. Conversely, given a system $Y' = AY$, $A \in M(n,k)$ one can use equation 2.3 to define a connection ∇_A on k^n.

All of the usual constructions of linear algebra can be carried out for connections. A connection $(\mathcal{N}, \nabla_N)$ is a subconnection of $(\mathcal{M}, \nabla)$ if $\mathcal{N} \subset \mathcal{M}$ and $\nabla_N = \nabla|_N$. Given a connection and a subconnection one can define a quotient connection and if $(\mathcal{M}_1, \nabla_1)$ and $(\mathcal{M}_2, \nabla_2)$ are two connections one can form the direct sum $(\mathcal{M}_1 \oplus \mathcal{M}_2, \nabla_1 \oplus \nabla_2)$ and the tensor product $(\mathcal{M}_1 \otimes \mathcal{M}_2, \nabla_1 \otimes 1 + 1 \otimes \nabla_2)$ in the obvious ways. A morphism $\phi : (\mathcal{M}_1, \nabla_1) \to (\mathcal{M}_2, \nabla_2)$ is a k-linear map $\phi : \mathcal{M}_1 \to \mathcal{M}_2$ such that $\phi \circ \nabla_1 = \nabla_2 \circ \phi$. If $\{e_1, \ldots, e_n\}$ is a basis of $\mathcal{M}_1$ (resp. $\{f_1, \ldots, f_m\}$ is a basis of $\mathcal{M}_2$) and $Y' = A_1 Y$ (resp. $Y' = A_2 Y$) is the equation associated with $(\mathcal{M}_1, \nabla_1)$ (resp. $(\mathcal{M}_2, \nabla_2)$) then $U \in \mathrm{Hom}_k(\mathcal{M}_1, \mathcal{M}_2)$ defines a morphism if and only if $U' = A_2 U - U A_1$. One can define a connection $(\mathrm{Hom}_k(\mathcal{M}_1, \mathcal{M}_2), \nabla_{\mathrm{Hom}})$ by the equation $\nabla_{\mathrm{Hom}} \phi(u) = \nabla_2(\phi(u)) - \phi(\nabla_1 u)$. One sees that $\phi \in \mathrm{Hom}_k(\mathcal{M}_1, \mathcal{M}_2)$ defines a morphism if and only if $\nabla_{\mathrm{Hom}} \phi = 0$. When $\mathcal{M}_2 = k$ and ∇_2 is the trivial connection, then we say that $(\mathrm{Hom}_k(\mathcal{M}_1, \mathcal{M}_2), \nabla_{\mathrm{Hom}})$ is the dual connection $(\mathcal{M}_1^*, \nabla_1^*)$. The differential equation associated with $(\mathcal{M}_1^*, \nabla_1^*)$ is $Y' = -A^T Y$. If $\mathcal{M}_1 = \mathcal{M}_2$ and U defines an isomorphism, we then have

$$A_2 = U'U^{-1} + UA_1U^{-1} \tag{2.4}$$

We therefore define the systems $Y' = A_1 Y$ and $Y' = A_2 Y$ to be *equivalent* if there exists a matrix $U \in \mathrm{GL}(n,k)$ such that equation 2.4 holds. Finally we note that on any finite dimensional $\mathcal{D}$-module $\mathcal{M}$ one can define a natural connection via the equation $\nabla u = Du$.

The relationship with linear differential equations or, more specifically linear operators, is the following. Given an operator $L = D^n + a_{n-1}D^{n-1} + \ldots + a_0 \in \mathcal{D}$ one can associate to it the system $Y' = A_L Y$ where

$$A_L = \begin{pmatrix} 0 & 1 & 0 & \ldots & 0 \\ 0 & 0 & 1 & \ldots & 0 \\ \ldots & \ldots & \ldots & \ldots & \ldots \\ -a_0 & -a_1 & \ldots & \ldots & -a_{n-1} \end{pmatrix} \tag{2.5}$$

We denote the associated connection (k^n, ∇_L). One can easily check that this $\mathcal{D}$-module is isomorphic to $(\mathcal{D}/\mathcal{D} \cdot L)^*$. It is well known [**7**], [**17**], [**20**], [**46**], [**77**], [**82**], [**126**], that if k contains a non-constant element then any connection $(\mathcal{M}, \nabla)$ is cyclic, that is, there exists an element $u \in \mathcal{M}$ such that the elements $u, \nabla u, \nabla^2 u, \ldots, \nabla^{n-1} u$ form a k-basis of $\mathcal{M}$. Applying this fact to the dual $(\mathcal{M}^*, \nabla^*)$, we see that with respect to a basis of the form $v, \nabla^* v, \ldots, (\nabla^*)^{n-1} v$, the connection will have a matrix of the form

$$\begin{pmatrix} 0 & 0 & 0 & \dots & -a_0 \\ 1 & 0 & 0 & \dots & -a_1 \\ 0 & 1 & 0 & \dots & -a_2 \\ \dots & \dots & \dots & \dots & \dots \\ 0 & 0 & \dots & 1 & -a_{n-1} \end{pmatrix}$$

Therefore $(\mathcal{M}^{**}, \nabla^{**}) \simeq (\mathcal{M}, \nabla)$ is associated with the equation $Y' = -B^T Y$ where $-B^T = A_L$ for the operator $L = D^n + a_{n-1}D^{n-1} + \ldots + a_0$. Let $Y' = AY$ be a differential equation and (k^n, ∇_A) the associated connection. We shall refer to an operator L so that $(k^n, \nabla_A) \simeq (k^n, \nabla_L)$ as an operator *equivalent to the system* $Y' = AY$ or *equivalent to the connection* ∇_A. Finding an equivalent linear operator allows one to reduce questions concerning systems (e.g., existence of rational or liouvillian solutions) to the corresponding questions for linear operators.

The Picard-Vessiot extension K associated to a system $Y' = AY$ is a field of the form $k(z_{1,1}, \ldots z_{n,n})$ having no new constants where $Z = (z_{i,j})$ is an invertible matrix satisfying $Z' = AZ$. The columns of Z span the solution space V of this equation. Equivalent equations have isomorphic Picard-Vessiot extensions and so we can define the Picard-Vessiot extension K corresponding to a connection. If $(\tilde{\mathcal{M}}, \tilde{\nabla})$ is a connection gotten from $(\mathcal{M}, \nabla)$ via iteration of the operations of direct sums, tensor products and duals, we say that $(\tilde{\mathcal{M}}, \tilde{\nabla})$ is a *construction* of $(\mathcal{M}, \nabla)$. If V is the solution space in K^n of the connection $(\mathcal{M}, \nabla)$ we can also consider all iterations of the operations of direct sums, tensor products and duals on V and call a resulting vector space $\tilde{V}$ a *construction* of V. One can show that if the same sequence of operations are applied to both $(\mathcal{M}, \nabla)$ and V, resulting in $(\tilde{\mathcal{M}}, \tilde{\nabla})$ and $\tilde{V}$, then the solution space of $(\tilde{\mathcal{M}}, \tilde{\nabla})$ is $\tilde{V}$. Let $\mathbf{M}$ be the collection of all subconnections of all constructions of $(\mathcal{M}, \nabla)$ and let $\mathbf{V}$ be the collection of all solutions spaces of elements of $\mathbf{M}$. Clearly, $\mathbf{V}$ is a collection of certain subspaces of constructions of V. For any construction $\tilde{V}$ of V there is a natural representation $\mathrm{GL}(V) \hookrightarrow \mathrm{GL}(\tilde{V})$. One can show [**47**], [**48**], [**80**], that the group

$$G = \{g \in \mathrm{GL}(V) \mid g(W) = W \text{ for all } W \in \mathbf{V}\}$$

is the Galois group of K over k. In more pedestrian terms, reducibility properties of the solutions spaces of associated differential equations determine the Galois group of a given differential equation.

This idea is at the heart of many results determining the Galois groups of linear differential equations. One distinguishes a certain group from other groups by the manner in which constructions decompose into invariant subspaces. One then gives criteria (or uses the tests described in Section 2.2) for the associated connections to decompose into the subconnections in the same way. For example, Beukers, Brownawell and Heckman show in [**23**] that for a linear algebraic group $G \subset \mathrm{GL}(V)$, G modulo its center is infinite and the second symmetric power S^2V is irreducible under G if and only if G is an extension by scalars of $\mathrm{SL}(V)$ or $\mathrm{Sp}(V)$. Using this they were able to show that large classes of hypergeometric equations have Galois groups of this type (for related work see also [**66**]). Katz and Pink [**84**] give a sufficient condition for the Galois group of a linear differential equation to contain $\mathrm{SL}(n), \mathrm{Sp}(n)$ or $\mathrm{SO}(n)$. The differential Galois groups of all irreducible confluent hypergeometric differential equations were determined by Katz and Gabber,[**81**], [**83**] using representation theoretic characterizations of semisimple

Lie algebras. An algebraic study of the Galois groups of reducible hypergeometric equations was made by Boussel in [**29**]. We note that many of the calculations were motivated by questions in the theory of transcendental numbers where one needs to know if certain functions satisfying differential equations and their derivatives are algebraically independent. This can be translated into questions concerning the dimension of certain Galois groups. Kolchin also wrote on this question. In [**88**], Kolchin calculated the Galois groups of Picard-Vessiot extensions generated by sets of Bessel functions but he relied on simple group theoretic considerations rather than the representation theoretic approach described above.

Algorithms based on this philosophy for computing Galois groups and their properties are given in [**147, 148, 149**]. For example, it is shown that $L(y) = y'' + ry = 0$ has liouvillian solutions if and only if $L^{\circledS 6}(y)$ is reducible, where $L^{\circledS 6}(y) = 0$ is the seventh order equation whose solution space is the sixth symmetric power of the solution space of $L(y) = 0$. We note that if L is irreducible its Galois group is reductive. This implies that $L^{\circledS 6}$ will be completely reducible and so its reducibility can be tested using the results of [**146**]. Similar criteria are given for properties of third order equations as well as criteria to determine if a particular group is the Galois group of such an equation. Fourth order equations have been treated by Hessinger in [**68**].

If one knows in advance that a linear differential equation has a reductive Galois group (that is, if the equation is completely reducible), then one can calculate a presentation of the associated Picard-Vessiot extension and its Galois group [**43**]. In [**41, 40**], Compoint showed that if a Picard-Vessiot extension has a reductive unimodular Galois group then the relations defining the Picard-Vessiot extension come from the invariants of the Galois group. To be more specific, let k be a differential field of characteristic zero with algebraically closed field C of constants and let $Y' = AY$ be a differential equation where A is an $n \times n$ matrix with entries in k. Let $G \subset \mathrm{SL}(n)$ be the Galois group and let its action on the polynomial ring $C[Y_{1,1}, \dots, Y_{n,n}]$ be defined by letting each element of G act on the $n \times n$ matrix $[Y_{i,j}]$ by multiplication on the left. Since G is reductive, the ring $C[Y_{1,1}, \dots, Y_{n,n}]^G$ of invariants is finitely generated. Compoint showed that if this ring is generated by polynomials of degree at most m, then the Picard-Vessiot extension is the quotient field of a ring $k[Y_{1,1}, \dots, Y_{n,n}]/I$, where I is an ideal generated by polynomials of degree at most m as well. It furthermore is known that given such an m, one can calculate the generators of I directly from the equation $Y' = AY$, without a priori knowledge of the Galois group [**75**]. Therefore the question of determining the Galois group of an equation $Y' = AY$ with reductive unimodular group is reduced to the question of finding a bound on the generators of the ring of invariants. The main result of [**43**] is that there is an effective method to find such a bound. It is also shown how this result can be extended to equations with arbitrary (not necessarily unimodular) reductive Galois groups as well as be used to calculate G/G^o, that is, the group of connected components of the Galois group of a linear differential equation with arbitrary Galois group.

Reduction modulo a prime can also be used to get information concerning the Galois group. Let $L(y) = 0$ be a linear differential equation with coefficients in $k_0(x)$ where k_0 is a number field. For almost all primes $\mathfrak{p}$ of k_0, it makes sense to reduce the coefficients of L modulo $\mathfrak{p}$ and get an operator $L_\mathfrak{p}$ with coefficients in $k_\mathfrak{p}(x)$, where $k_\mathfrak{p}$ is the residue field at $\mathfrak{p}$. A conjecture of Grothendieck states: *All solutions of $L(y) = 0$ are algebraic over $k_0(x)$ if and only if, for almost all $\mathfrak{p}$,* *the equation* $L_{\mathfrak{p}}(y) = 0$ *has a full set of solutions in* $k_{\mathfrak{p}}(x)$, [**19**], [**38**],[**53**], [**76**], [**80**], [**123**]. In [**125**], van der Put uses the classification of differential equations in positive characteristic and ideas around the Grothendieck conjecture to factor differential operators and find symbolic solutions over $\mathbb{Q}(x)$.

2.4.2. *Analytic Methods.* Classically [**60**], analytic/topological techniques were used to associate a group with a differential equation $Y' = AY$ having coefficients in $\mathbb{C}(x)$. This group is defined as follows [**95**]. Let $\{\alpha_1, \dots, \alpha_m\}$ be the singular points of the equation on the Riemann sphere S^2, let $\alpha_0 \in S^2 - \{\alpha_1, \dots, \alpha_m\}$ and let $\{y_1, \dots, y_n\}$ be a fundamental set of solutions, analytic in a neighborhood of α_0. If $[\gamma] \in \pi_1(S^2 - \{\alpha_1, \dots, \alpha_m\}, \alpha_0)$, one can analytically continue each solution y_i along γ. The resulting function y_i^γ is independent of the homotopy class representative γ used and is again a solution of $Y' = AY$. From this one sees that there is an invertible matrix A_γ such that $(y_1^\gamma, \dots, y_m^\gamma) = (y_1, \dots y_n)A_\gamma$. Furthermore, the map $[\gamma] \mapsto A_\gamma$ is a group homomorphism of $\pi_1(S^2 - \{\alpha_1, \dots, \alpha_m\}, \alpha_0)$ into $\mathrm{GL}(n, \mathbb{C})$. Selecting a different fundamental set of solutions at α_0 will give a conjugate homomorphism. The conjugacy class of this homomorphism is called the *monodromy representation* of the differential equation and the conjugacy class of the image is called the *monodromy group* [**120**]. The monodromy group is contained in the Galois group, but the Galois group can be strictly larger. For example the equation $y' - y = 0$ has trivial monodromy group since $\pi_1(S^2 - \infty, 0) = 0$ while its Galois group is $\mathbb{C}^*$. We say that a singular point α of the operator L is a *regular* singular point if for any open sector Ω at α, all solutions y, analytic in Ω, satisfy $\lim_{x \to \alpha} x^N y = 0$ for some $N \geq 0$. An equation is called *regular* if all of its singular points are regular. Fuchs showed that for a differential operator, $\alpha \in \mathbb{C}$ is a regular singular point if and only if the coefficient of D^{n-i} has a pole of order at most i. For a system $Y' = AY$, $\alpha \in \mathbb{C}$ is a regular singular pount if and only if there is a $U \in \mathrm{GL}(n, C((x - \alpha)))$ such that $U'U^{-1} + UAU^{-1} = (x - \alpha)^{-1}A_0$ for some constant matrix A_0 (this latter property is decidable [**12**]). It was known classically ([**134**], Section 160) that if all singular points are regular singular points then the monodromy group is Zariski dense in the Galois group. In general this is not true as the above example shows. For arbitrary singular points Ramis [**105**], [**127**], [**129**], [**128**] was able to define analytically a subgroup of each local Galois group (defined below) that is dense in this latter group.

To understand Ramis's results, we begin with some definitions (we shall follow the presentation in [**109**]; see also [**124**]). Let $\alpha \in S^2$ and let t_α be the local coordinate of S^2 at α. Let $\mathbb{C}(\{t_\alpha\})$ be the field of fractions of the ring of convergent power series at α and let $\mathbb{C}((t_\alpha))$ be the field of fractions of the ring of formal power series. Any differential equation $Y' = AY$ with $A \in M_n(\mathbb{C}(x))$ can be thought of as a differential equation over $\mathbb{C}(\{t_\alpha\})$ or $\mathbb{C}((t_\alpha))$ as well. The Galois group of $Y' = AY$ over $\mathbb{C}(\{t_\alpha\})$ is called the *local Galois group at* α and the Galois group over $\mathbb{C}((t_\alpha))$ is called the *formal Galois group at* α. It is known [**109**] that the Zariski closure of the group generated by all local Galois groups is the Galois group of the equation over $\mathbb{C}(x)$. For simplicity of notation, we shall restrict ourselves to the situation where $\alpha = 0$.

A differential equation $Y' = AY$ with $A \in \mathbb{C}(\{x\})$ admits a formal solution

$$\hat{Y} = \hat{H}(x)x^L e^{Q(t)} \tag{2.6}$$

where $t = x^{1/r}$, r a positive integer , $L \in M_n(\mathbb{C})$, $\hat{H} \in \mathrm{GL}(n, \mathbb{C}((x)))$ and $Q = diag(q_1, \dots q_n)$ where $q_i \in t^{-1}\mathbb{C}[t^{-1}], i = 1, \dots, n$ (this is the matrix version of formula 2.1). Furthermore, one may assume that the polynomials q_i are permuted by the map $t \mapsto te^{2i\pi/r}$.

We now define two groups. The first is the *formal monodromy group* G_M. This is the group generated by the matrix $\hat{M}$ defined $\hat{Y}(te^{2i\pi/r}) = \hat{Y}\hat{M}$. The second is the *exponential torus* $\mathcal{T}$. This is the Galois group of $\mathbb{C}((t))\langle e^{q_1}, \ldots, e^{q_n}\rangle$ over $\mathbb{C}((t))$. Both of these groups may be identified with subgroups of the local Galois group of $Y' = AY$ at 0 and both of these groups may be calculated once one has the formal solution 2.6. It can be shown that the Zariski closure of the group generated by G_M and $\mathcal{T}$ is precisely the formal Galois group at 0 [**19**], [**109**]. Furthermore, when 0 is a regular singular point, the local and formal Galois groups coincide. For general singular points, this is not true and one needs additional elements to generate G topologically.

To define these elements, Ramis shows that one can canonically select open sectors $\mathcal{S}_1, \ldots, \mathcal{S}_m$ at 0 that fill out a neighborhood of 0 and canonically (in fact, functorially) select fundamental solution matrices $\mathcal{Y}_i$ of $Y' = AY$, holomorphic in $\mathcal{S}_i$ (this latter process is called *multisummation* [**99**], [**101**], [**124**]). On the intersection $\mathcal{S}_i \cap \mathcal{S}_{i+1}$ the fundamental solution matrices must differ multiplicatively by a constant matrix S_i, $\mathcal{Y}_i = \mathcal{Y}_{i+1}S_i$. The matrices S_i are called the *Stokes matrices.* The fundamental result of Ramis is that the Zariski closure of the group generated by the formal monodromy, the exponential torus and the Stokes matrices is the local Galois group ([**105**] or [**100**] for a proof based on Tannakian considerations). When the differential equation has at most two singular points, at worst a regular one at 0 and a possibly irregular one at ∞, then one also knows that the Zariski closure of the local Galois group at infinity is the entire Galois group of the equation over $\mathbb{C}(x)$.

In general, there is no effective procedure to compute the Stokes matrices. There are numerical schemes to approximate the canonically selected holomorphic functions on the sectors (and so approximate the Stokes matrices) [**113**] as well as to directly approximate the Stokes matrices for systems of dimension 2 [**98**]. For large classes of differential equations, one has integral representations of the solutions and this does allow one to calculate the Stokes matrices. This has been done by Duval and Mitschi [**50**], [**52**] and applied to calculate Galois groups of some confluent hypergeometric differential equations. This work has been extended and generalized by Mitschi [**108, 109**] to calculate the Galois groups (over $\mathbb{C}(x)$) of large classes of confluent hypergeometric differential equations. Toulouse [**152**] uses this approach to calculate Galois groups of classes of equations of the form $Y' = (A_0 + xA_1)Y$ where $A_i \in M_n(\mathbb{C})$ with $n = 2$ or 3.

3. The Inverse Problem

The inverse problem is: *Given a differential field k, what groups are Galois groups of Picard-Vessiot extensions of k?* As Kolchin already noted in the extract quoted at the beginning of this article, the answer depends both on k and G. Expanding on Kolchin's comments, let $k = \mathbb{C}$ and let $Y' = AY$ be a differential equation with $A \in M_n(\mathbb{C})$. Classically, it is known that by reducing the matrix A to rational normal form one can reduce the problem of finding solutions of this equation to the problem of finding solutions of constant coefficient linear differential equations $L(y) = 0$ of order possibly greater than one. All solutions of such equations are of the form $y = \sum p_i(x)e^{\alpha_i x}$ where $p_i \in \mathbb{C}(x)$ and the $\alpha_i \in \mathbb{C}$.

Therefore the associated Picard-Vessiot extension K lies in a field of the form $F = \mathbb{C}(x, e^{\alpha_1 x}, \dots, e^{\alpha_t x})$. One sees that F is a Picard-Vessiot extension of $\mathbb{C}$ with Galois group $G = \mathbb{C} \times \mathbb{C}^* \times \dots \times \mathbb{C}^*$. Therefore the Galois group of K over $\mathbb{C}$ is a quotient of such a group, which can be shown to be of the form $\mathbb{C}^s \times \mathbb{C}^{*t}$ where s and t are nonnegative integers and $0 \leq s \leq 1$. Conversely, one can adjoin appropriate elements to $\mathbb{C}$ and show that any such group is a Galois group. Group theoretically, one can characterize such groups as linear algebraic groups that are (i) connected, (ii) abelian and (iii) whose largest unipotent subgroup has dimension at most one (a linear group is unipotent if all of its elements are unipotent). These three properties therefore characterize the groups that can appear as Galois groups of Picard-Vessiot extensions of $\mathbb{C}$. Once again, one can approach the general problem using algebraic and analytic techniques.

3.1. Algebraic Methods. Preceding Kolchin's 1966 talk, Bialynicki-Birula [**25**] had already shown that, for any differential field k of characteristic zero with algebraically closed field of constants C, if the transcendence degree of k over C is finite and nonzero then any connected nilpotent group is a Galois group of a Picard-Vessiot extension of k. Goldman [**58**] developed the notion of a generic differential equation with group $G \subset \mathrm{GL}(n)$, that is, a linear differential operator L with parameters such that any linear differential operator with Galois group contained in G is a specialization of L (this is analogous to the notion E. Noether developed for algebraic equations). In his thesis [**107**], Miller developed the notion of a differentially hilbertian differential field and gave a sufficient condition for the generic equation of a group to specialize over such a field to an equation having this group as a Galois group.

A breakthrough was made by Kovacic [**91**], [**92**] who proposed an ultimately successful program to solve the inverse problem for connected groups (over fields of finite, nonzero transcendence degree over their algebraically closed field of constants) and solved the problem for connected solvable groups. A key element of Kovacic's program is an emphasis on the role of the Lie algebra of the Galois group. Let G be a connected linear algebraic group defined over the algebraically closed field of constants C of a differential field k and let $\mathfrak{g}$ be its Lie algebra. We denote by $\mathfrak{g}_k$ the k-points in the Lie algebra, that is, $\mathfrak{g}_k = \mathfrak{g}_C \otimes_C k$. Kovacic shows that if $A \in \mathfrak{g}_k$ then the Picard-Vessiot extension of k corresponding to $Y' = AY$ has Galois group H that is a subgroup of G. Furthermore, for $k = C(x)$ (or any cohomologically trivial field), if the Galois group $H \subset G$ of $Y' = AY$ with $A \in \mathfrak{g}_k$ is connected, then there exists an invertible $g \in G(k)$ such that $g'g^{-1} + gAg^{-1} \in \mathfrak{h}_k$, where $\mathfrak{h}$ is the Lie algebra of H. In particular, to construct a differential equation $Y' = AY$ with coefficients in $C(x)$ having a given connected Galois group G one must find a matrix $A \in \mathfrak{g}_k$ such that:

1. the Galois group of $Y' = AY$ is connected and
2. for any $g \in G(k)$, $g'g^{-1} + gAg^{-1}$ does not lie in the Lie algebra of a smaller connected group.

Let us consider an example. Let

$$G = \{\begin{pmatrix} 1 & c \\ 0 & 1 \end{pmatrix} \mid c \in C\}$$

G is isomorphic to the additive group C. The Lie algebra of this group is

$$\mathfrak{g} = \{\begin{pmatrix} 0 & c \\ 0 & 0 \end{pmatrix} \mid c \in C\}$$

If we start with a matrix

$$A = \begin{pmatrix} 0 & f \\ 0 & 0 \end{pmatrix}$$

with $f \in C(x)$ then the equation $Y' = AY$ will have a Galois group that is an algebraic subgroup of the additive group C. The only such groups are the trivial group and C itself, so the Galois group will be connected. If the Galois group is trivial then there is an element

$$g = \begin{pmatrix} 1 & h \\ 0 & 1 \end{pmatrix} \in G(C(x))$$

such that $g'g^{-1} + gAg^{-1}$ is the zero matrix. This is equivalent to the existence of an element $h \in C(x)$ such that $h' + f = 0$, that is, f has an antiderivative in $C(x)$. On can easily select an f so that this is not the case (e.g., $f = 1/x$) and so find a differential equation whose Galois group is the additive group C.

This criterion can be thought of in more geometric terms. Kolchin showed ([**90**] or the paper of Borel [**26**] in this volume) that a Picard-Vessiot extension of a differential field k with Galois group G is the function field of a principal homogeneous G-space. If $k = C(x)$ or any other cohomologically trivial field, all principal homogeneous spaces for a connected group G are isomorphic to G. Therefore, a Picard-Vessiot extension K of $C(x)$ can be thought of as the function field of a principal G-bundle over the projective line. The derivation on K gives a connection on this bundle and the condition that for any $g \in \mathrm{GL}(n, k)$, $g'g^{-1} + gAg^{-1}$ does not lie in the Lie algebra of a smaller connected group is equivalent to the differential geometric condition that the connection is not reducible to a connection on an H-bundle for some connected subgroup H of G.

Another key element of Kovacic's program was to give a procedure to inductively apply the above criterion to lift a solution of the inverse problem for G/R_u to a solution for the group G (R_u is the unipotent radical of G, that is, the largest normal subgroup of G all of whose elements are unipotent). Using this, Kovacic showed that to give a complete solution of the inverse problem for connected groups, one needed only to solve this problem for reductive groups (since G/R_u is reductive). He was able to solve the problem for tori and so could give a solution when G/R_u is a torus, that is, when G is a solvable group. He furthermore showed that one could reduce the problem to the case when G is the power of a simple connected linear algebraic group (see also [**61**] for extensions and modifications of Kovacic's methods to partial differential fields).

The solution of the inverse problem for such groups (or more generally semisimple groups) is one of the main contributions of [**110**] (see [**111**] for a simplified exposition). In this paper, the authors use an idea of Kovacic to show that to solve the inverse problem for connected groups over a differential field k of finite transcendence degree over its algebraically closed field of constants C it is enough to solve this problem for connected groups over $C(x)$. For this field and connected linear algebraic groups G, the authors show there are integers $d(G)$ and $e(G)$ depending only on G such that G is the Galois group of an equation of the form

$$Y' = (\frac{A_1}{x-\alpha_1} + \ldots + \frac{A_{d(G)}}{x-\alpha_{d(G)}} + A_\infty)Y$$

where A_i, $i = 1, \ldots, d(G)$ are constant matrices and A_∞ is a matrix with polynomial entries of degree at most $e(G)$. In particular, the only possible singularities of this system are $d(G)$ regular singular points in the finite plane and a (possibly irregular) singular point at infinity. The integers $d(G)$ and $e(G)$ are defined as follows. Let G be a connected linear algebraic group and let R_u be its unipotent radical and P a Levi factor, that is, a reductive group P such that $G = R_u \rtimes P$. The group $R_u/(R_u, R_u)$, where (R_u, R_u) is the commutator subgroup, is a commutative unipotent group and so isomorphic to a vector group C^n. The group P acts on R_u by conjugation and this action factors to an action on $R_u/(R_u, R_u)$. Therefore we may write $R_u/(R_u, R_u) = U_1^{n_1} \oplus \ldots \oplus U_s^{n_s}$, where each U_i is an irreducible P-module. We shall assume that U_1 is the trivial one dimensional P-module (and so allow the possibility that $n_1 = 0$). We may write $P = T \cdot H$, where T is a torus and H is a semisimple group. Let $m_i = n_i$ if the action of H on U_i is trivial and let $m_i = n_i + 1$ if the action of H on U_i is nontrivial. Let $N = 0$ if H is trivial and $N = 1$ if H is nontrivial. We define the *defect* $d(G)$ of G to be the number n_1 and the *excess* $e(G)$ of G to be $\max\{N, m_2, \ldots, m_s\}$. We note that any two Levi factors are conjugate so that these numbers are independent of the choice of P. Furthermore, one can show that $d(G)$ is the dimension of $R_u/(G, R_u)$

For example, for any connected reductive group G, we have that $R_u(G) = (0)$. Therefore $d(G) = 0$ and $e(G) = 1$ so there exists constant matrices A and B such that G is the Galois group of an equation of the form $Y' = (A + xB)Y$. It is also shown in [**110**] that the bound on the number of singular points given by $d(G)$ is sharp: there are examples of groups G that cannot be Galois groups of differential equations with fewer than $d(G) + 1$ singular points. Finally, the proof in [**110**] is constructive.

Kovacic also considered fields that are not finitely generated over their constant subfields. He showed that a solvable connected linear algebraic group G is a Galois over $C((x))$ if and only if it is commutative and its unipotent radical R_u is of dimension at most one. Equation 2.1 implies that the Picard-Vessiot extension of a linear differential equation over $C((x))$ lies in liouvillian extension of $C((x))$. Therefore, the identity component G^o of the Galois group G is solvable (in fact, G/G^o is cyclic so G is solvable). Therefore Kovacic's result implies that a connected group is a Galois group over $C((x))$ if and only if it is commutative and its unipotent radical is of dimension at most 1. Kovacic also considers the field $C(\{x\})$ of convergent power series and shows that a connected solvable linear algebraic group is a Galois group over this field if and only if the unipotent radical of the center of $G/(R_u, R_u)$ has dimension at most one. Using the notation of the previous paragraphs, this is equivalent to the condition that $d(G) \leq 1$. The complete characterization of Galois groups over $C(\{x\})$ when $C = \mathbb{C}$ rests on analytic techniques due to Ramis and will be described in the next section.

It should be noted also that Kovacic [**91**], [**92**] considered the inverse problem for arbitrary connected algebraic groups (not just linear groups), that is, classifying those groups that arise as Galois groups of strongly normal extensions. He showed that one can reduce this problem to the inverse problem for connected linear algebraic groups and for abelian varieties. In [**91**] he solved the problem for abelian varieties over fields of finite nonzero transcendence degree over their algebraically closed constant subfields. In [**118**], Pillay develops a Galois theory of differential fields where the Galois groups are differential algebraic groups. He also considers inverse problems in this context.

In the soon to be corrected edition of [**102**], Magid uses geometric techniques to solve the inverse problem for several classes of groups. The proof given in the first edition of [**102**] to demonstrate that all connected groups having no subgroups of codimension one are Galois groups (Theorem 7.13) is flawed.

It should be finally noted that algebraic techniques have not yet been successful in attacking the inverse problem for groups that are not connected. At present, one must rely on analytic techniques.

3.2. Analytic Methods. Analogous to the inverse problem in differential Galois theory, there is an inverse problem in the analytic theory of linear differential equations. This is the *Riemann-Hilbert Problem.* In its weakest form the problem asks if, for any representation $\rho : \pi_1(S^2 - \Sigma) \to \mathrm{GL}(n, \mathbb{C})$ of the Riemann Sphere S^2 minus a finite set Σ of points, does there exist a differential equation $Y' = AY, A \in M_n(\mathbb{C}(x))$ with regular singular points at the points of Σ whose monodromy representation is given by ρ. In this form, the answer is always yes. This positive answer can be used to show that there is an equivalence of categories between the category of connections having only regular singular points in some set $\Sigma \subset S^2$ and the category of finite dimensional representations of $pi_1(S^2 - \Sigma)$. This equivalence is called the *Riemann-Hilbert Correspondence* and, in this form, was established and generalized by Deligne [**46**]. A more restrictive form of the Riemann-Hilbert problem is to ask if there is a differential equation of the form

$$Y' = \sum_{\alpha \in \Sigma} \frac{A_\alpha}{x - \alpha} Y$$

(with $A_\alpha \in M(n, \mathbb{C})$ and $\sum A_\alpha = 0$) having ρ as its monodromy representation. Bolibruch has shown that the answer to this question is no for certain representations (for a history of this problem and its solution, see [**15**] or [**8**]).

C. and M. Tretkoff [**155**] used the solution of the weak Riemann-Hilbert Problem to solve the inverse problem for all linear algebraic groups G over the field $\mathbb{C}(x)$. They showed that any such G contains a Zariski - dense finitely generated subgroup H. One can select generators $h_1, \dots h_m$ of H such that $\prod_{i=1}^{m} h_i = 1$ and therefore there exists a representation ρ of the fundamental group of the Riemann Sphere minus m points onto H. Using the solution of the weak Riemann-Hilbert Problem, there is a differential equation $Y' = AY$ with regular singular points whose monodromy representation is given by ρ. Since, for equations with regular singular points, the monodromy group is Zariski dense in the Galois group, the equation has Galois group G.

Ramis [**105**], [**106**] has shown how one can use analytic methods to completely solve the inverse problem over $\mathbb{C}(\{x\})$. To do this he defined a *wild monodromy group* $\pi_{1,S}$ and showed that it has the following two properties:

1. If $Y' = AY$ is a differential equation with $A \in \mathrm{GL}(n, \mathbb{C}(\{x\})$, there is a representation $\rho : \pi_{1,S} \to \mathrm{GL}(n, \mathbb{C})$ whose image is the group generated by the formal monodromy, exponential torus and Stokes matrices of this equation.
2. Any representation $\rho : \pi_{1,S} \to \mathrm{GL}(n, \mathbb{C})$ can be realized as in 1.

These two facts yield a Riemann-Hilbert Correspondence between the categories of local connections (i.e., connections over the field $\mathbb{C}(\{x\})$) and finite dimensional representations of $\pi_{1,S}$. To define the wild monodromy group, one notes that the possible formal monodromy groups are all homomorphic images of $\mathbb{Z}$. Furthermore, the possible exponential tori form an inverse system and we let $\mathcal{T}$ denote their inverse limit. There is a natural action of $\mathbb{Z}$ on $\mathcal{T}$ and so we can form their semi-direct product $\pi_{1,Sf} = \mathbb{Z} \ltimes \mathcal{T}$. The group $\pi_{1,Sf}$ is called the wild formal fundamental group and any formal Galois group is a quotient of this group. One can furthermore define a group $\mathcal{R}$ such that any group generated by Stokes matrices is a homomorphic image of $\mathcal{R}$ and an action of $\pi_{1,Sf}$ on $\mathcal{R}$ [**106**]. The group $\pi_{1,S}$ is defined to be $\pi_{1,Sf} \ltimes \mathcal{R}$.

We have already noted that the group generated by the formal monodromy, exponential torus and Stokes matrices is Zariski dense in the Galois group of the equation over $\mathbb{C}(\{x\})$. Therefore to solve the inverse problem over this field it is enough to identify those linear algebraic groups G containing a Zariski dense homomorphic image of $\pi_{1,S}$. Ramis gave necessary and sufficient conditions on the Lie algebra of G for this to be true and equivalent group theoretic conditions were given in [**112**]. The final result is the following [**131**] (see also [**130**], [**132**]): a linear algebraic group G with identity component G^o and unipotent radical R_u is a Galois group of a Picard-Vessiot extension of $\mathbb{C}(\{x\})$ if and only if

1. the finite group G/G^o is cyclic,
2. the dimension of $R_u/(R_u, G^o)$ is at most one, and
3. the group G/G^o acts trivially on $R_u/(R_u, G^o)$

Ramis is furthermore able to show that these three conditions are equivalent to the quotient $G/L(G)$ being the Zariski closure of a cyclic group, where $L(G)$ is the subgroup of G generated by all the tori of G. Ramis took this local result and combined it with a glueing technique to show that any linear algebraic group is the Galois group of a differential equation $Y' = AY$ over $\mathbb{C}(x)$ (or even over an algebraic extension of $\mathbb{C}(x)$) . Furthermore he can bound the number and type of singular points in terms of the number of generators of any Zarisk-dense subgroup of $G/L(G)$.

Analytic methods only work when the field of constants is $\mathbb{C}$. Nonetheless, some transfer techniques have been developed to allow one to draw conclusions about $C(x)$ where C is any algebraically closed field. Let $\mathcal{L}(n,m)$ be the vector space of differential operators

$$L = \sum_{i=0}^{n} \sum_{j=0}^{m} a_{i,j} x^j D^i$$

of order at most n whose coefficients are polynomials of degree at most m with coefficients in $\mathbb{C}$. Let G be a linear algebraic group, V a finite dimensional G-module and let $\mathcal{L}(n,m,G,V)$ be the set of operators in $\mathcal{L}(m,n)$ with Galois group G and solution space G-isomorphic to V. In general, $\mathcal{L}(n,m,G,V)$ is not a Zariski-closed subset (or even a constructible subset) of $\mathcal{L}(n,m)$. For example, an operator $L_c = xD - c$, $c \in \mathbb{C}$, has its solution space spanned by x^c. Therefore it has Galois group $\mathbb{C}^*$ if and only if c is not rational. This implies that $\mathcal{L}(n,m,\mathbb{C}^*,\mathbb{C})$ where $\mathbb{C}^*$ acts on $\mathbb{C}$ by multiplication, cannot be constructible. Note that c is an exponent at the singular point 0. This suggests that one fix a certain amount of information at the singular points. Let $\mathcal{S}$ be a finite subset of $k_0[x], k_0 \subset \mathbb{C}$ and let $\mathcal{L}(n,m,\mathcal{S},G,V)$ be the set of $L \in \mathcal{L}(n,m,G,V)$ such that at any singular point the exponents and determining factors at that point lie in $\mathcal{S}$ (c.f., the discussion following equation 2.1). In [**145**], large classes of groups are determined for which this set is constructible. To be more precise, let G be a linear algebraic group with identity component G^o. Let $\mathrm{Ker}\chi(G)$ be the intersection of the kernels of all characters $\chi : G^o \to \mathbb{C}^*$. The group $G^o/\mathrm{Ker}\chi(G^o)$ is abelian (in fact it is a torus) and so one can use the sequence

$$1 \to G^o/\mathrm{Ker}\chi(G^o) \to G/\mathrm{Ker}\chi(G^o) \to G/G^o \to 1$$

to define an action of G/G^o on $G^o/\mathrm{Ker}\chi(G^o)$. The main result of [**145**] is: if the action of G/G^o on $G^o/\mathrm{Ker}\chi(G^o)$ is trivial, then for any fixed G-module V and fixed n, m and $\mathcal{S}$, the set $\mathcal{L}(n,m,\mathcal{S},G,V)$ is constructible and defined over k_0. For example, connected groups, finite groups, and groups where G^o is semisimple all satisfy these hypotheses. It is also shown in [**145**] that the assumption of triviality of the action is crucial; there is a family of second order differential operators with regular singular points and fixed exponents such that their Galois groups are all subgroups of $\mathbb{C}^* \rtimes \mathbb{Z}/2\mathbb{Z}$ and such that the set of parameters for which the Galois group is all of $\mathbb{C}^* \rtimes \mathbb{Z}/2\mathbb{Z}$ is not constructible.

If G is a group satisfying the hypothesis of this result, one can use the results of the Tretkoffs or Ramis to show that for any faithful G-module V and for some fixed n, m and $\mathcal{S}$, the set $\mathcal{L}(n,m,\mathcal{S},G,V)$ contains a $\mathbb{C}$-point. The Hilbert Nullstellensatz then implies that it will contain a point in any algebraically closed field containing k_0. This allows one to solve the inverse problem for many classes of groups over $C(x)$ where C is an arbitrary algebraically closed field of characteristic zero.

References

1. S. A. Abramov, *Rational solutions of linear differential and difference equations with polynomial coefficients (in Russian)*, Journal of Computational Mathematics and Mathematical Physics **29** (1989), no. 11, 1611–1620.
2. S.A. Abramov, *On D'Alembert substitution*, Proceedings of the 1993 International Symposium on Symbolic and Algebraic Computation (ISSAC'93) (M. Bronstein, ed.), ACM Press, 1993, pp. 20–26.
3. S.A. Abramov, M. Bronstein, and M. Petkovšek, *On polynomial solutions of linear operator equations*, Proceedings of the 1995 International Symposium on Symbolic and Algebraic Computation (ISSAC'95) (A. H. M. Levelt, ed.), ACM Press, 1995, pp. 290–296.
4. S.A. Abramov and K. Yu. Kvansenko, *Fast algorithms for rational solutions of linear differential equations with polynomial coefficients*, Proceedings of the 1991 International Symposium on Symbolic and Algebraic Computation (ISSAC'91) (S. M. Watt, ed.), ACM Press, 1991, pp. 267–270.
5. S.A. Abramov and M. Petkovšek, *D'Alembert solutions of linear differential and difference equations*, Proceedings of the 1994 International Symposium on Symbolic and Algebraic Computation (ISSAC'94) (J. von zur Gathen, ed.), ACM Press, 1994, pp. 169–180.
6. S.A. Abramov and E. V. Zima, *D'Alembert solutions of inhomogeneous linear equations (differential, difference and otherwise)*, Proceedings of the 1996 International Symposium on Symbolic and Algebraic Computation, ACM Press, 1996, pp. 232–239.
7. K. Adjamagbo, *Sur l'effectivité du lemme du vecteur cyclique*, Comptes Rendus de l'Académie des Sciences, Paris **306** (1988), 543 – 546.
8. D. V. Anosov and A. A. Bolibruch, *The Riemann-Hilbert Problem*, Vieweg, Braunschweig, Wiesbaden, 1994.
9. A. Baider, R. C. Churchill, D. L. Rod, and M. F. Singer, *On the infinitesimal geometry of integrable systems*, Mechanics Day (W. F. Shadwick, ed.), American Mathematical Society, 1996, pp. 5–56.

10. F. Baldassarri and B. Dwork, *On second order linear differential equations with algebraic solutions*, American Journal of Mathematics **101** (1979), 42–76.
11. M. A. Barkatou, *Rational Newton algorithm for computing formal solutions of linear differential equations*, Symbolic and Algebraic Computation - ISSAC'88 (P. Gianni, ed.), ACM Press, 1988, pp. 183–195.
12. ———, *A rational version of Moser's algorithm*, Proceedings of the 1995 International Symposium on Symbolic and Algebraic Computation (ISSAC'95) (A. H. M. Levelt, ed.), ACM Press, 1995, pp. 290–296.
13. ———, *An algorithm to compute the exponential part of a formal fundamental matrix solution of a linear differential system*, Applicable Algebra in Engineering, Communication and Computing **8** (1997), no. 9, 1 – 24.
14. ———, *On rational solutions of systems of linear differential equations*, Tech. report, Université de Grenoble I, IMAG-LMC, 1997.
15. A. Beauville, *Monodromie des systèmes différentieles à pôles simples sur la sphère de Riemann (d'après A. Bolibruch)*, Astérisque **216** (1993), 103–119, Séminaire Bourbaki, No. 765.
16. E. Beke, *Die Irreducibilität der homogenen Differentialgleichungen*, Mathematische Annalen **45** (1894), 278–294.
17. D. Bertrand, *Constructions effectives de vecteurs cycliques pour un D-module*, Publication du Groupe d'Étude d'Analyse Ultramétrique, 12e année, 1984/85, no. 11.
18. ———, *Groupes algébriques linéaires et théorie de Galois différentielle*, Cours de 3e cycle à l'Université Pierre et Marie Curie (Paris 6); notes de cours rédigées par René Lardon, 1985-86.
19. ———, *Groupes algébriques et équations différentielles linéaires*, Astérisque **206** (1992), 183–204, Séminaire Bourbaki, Vol. 1991/92.
20. ———, *Un analogue différentiel de la théorie de kummer*, Approximations Diophantiennes et Nombres Transcendents, Luminy 1990 (P. Philippon, ed.), Walter de Gruyter and Co., Berlin, 1992, pp. 39–49.
21. ———, *Review of Lectures on Differential Galois Theory by A. Magid*, Bulletin (New Series) of the American Mathematical Society **33** (1996), no. 2, 289–294.
22. F. Beukers, *Differential Galois Theory*, From Number Theory to Physics (Les Houches, 1989) (Berline) (M. Waldschmidt et al, ed.), Springer-Verlag, 1992, pp. 413–439.
23. F. Beukers, D. Brownawell, and G. Heckman, *Siegel normality*, Annals of Mathematics **127** (1988), 279 – 308.
24. F. Beukers and G. Heckman, *Monodromy for the hypergeometric function* ${}_nF_{n-1}$, Inventiones Mathematicae **95** (1989), 325–354.
25. A. Bialynicki-Birula, *On the inverse problem of Galois theory of differential fields*, Bulletin of the American Mathematical Society **69** (1963), 960–964.
26. A. Borel, *Algebraic groups in the work of Ellis R. Kolchin*, this volume.
27. ———, *Linear Algebraic Groups, Second Enlarged Edition*, Springer-Verlag, New York, 1991.
28. A. Boulanger, *Contribution à l'étude des équations linéaires homogènes intégrables algébriquement*, Journal de l'École Polytechnique, Paris **4** (1898), 1 – 122.
29. K. Boussel, *Groupes de Galois des équations hypergéométriques*, Comptes Rendus de l'Académie des Sciences, Paris **309** (1989), 587–589.
30. M. Bronstein, *Fast reduction of the Risch differential equation*, Symbolic and Algebraic Computation - ISSAC'88 (P. Gianni, ed.), ACM Press, 1988, pp. 64–72.
31. ———, *The Risch differential equation on an algebraic curve*, Symbolic and Algebraic Computation - ISSAC'88 (P. Gianni, ed.), ACM Press, 1988, pp. 64–72.
32. ———, *Integration of elementary functions*, Journal of Symbolic Computation **9** (1990), no. 3, 117–174.
33. ———, *The transcendental Risch equation*, Journal of Symbolic Computation **9** (1990), no. 1, 49 – 60.
34. ———, *Linear ordinary differential equations: breaking through the order 2 barrier*, Proceedings of the International Symposium on Symbolic and Algebraic Computation- ISSAC'92 (P. Wang, ed.), ACM Press, 1992, pp. 42–48.
35. ———, *On solutions of linear ordinary differential equations in their coefficient field*, Journal of Symbolic Computation **13** (1992), no. 4, 413 – 440.

36. ———, *An improved algorithm for factoring linear ordinary differential operators*, Proceedings of the 1994 International Symposium on Symbolic and Algebraic Computation (ISSAC'94) (J. von zur Gathen, ed.), ACM Press, 1994, pp. 336–347.
37. M. Bronstein, T. Mulders, and J.-A. Weil, *On symmetric powers of differential operators*, Proceedings of the 1997 International Symposium on Symbolic and Algebraic Computation (ISSAC'97), ACM Press, 1997, pp. 156 – 163.
38. D. Chudnovsky and G. Chudnovsky, *Application of Padé approximation to the Grothendieck conjecture on linear differential equations*, Number Theory (New York, 1983-84), Springer-Verlag, 1985, Lecture Notes In Mathematics, v. 1135, pp. 52–100.
39. H. Cohen, *A Course in Computational Algebraic Number Theory*, Graduate Texts in Mathematics, Springer-Verlag, New York, 1993.
40. E. Compoint, *Équations différentielles, relations algébriques et invariants*, Thèse, Université de Paris VI, 1996.
41. ———, *Differential equations and algebraic relations*, Tech. report, Université de Paris VI, 1996.
42. E. Compoint and M. F. Singer, *Calculating Galois groups of completely reducible linear operators*, Manuscript, North Carolina State University, 1997.
43. ———, *Relations linéaires entre solutions d'une equation différentielle*, Manuscript, North Carolina State University, September 1996.
44. J. Davenport, *On the Integration of Algebraic Functions*, Lecture Notes in Computer Science, vol. 102, Springer-Verlag, Heidelberg, 1981.
45. J. Davenport and M. F. Singer, *Elementary and liouvillian solutions of linear differential equations*, Journal of Symbolic Computation **2** (1986), no. 3, 237–260.
46. P. Deligne, *Equations Différentielles à Points Singuliers Réguliers*, Lecture Notes in Mathematics, vol. 163, Springer-Verlag, Heidelberg, 1970.
47. ———, *Catégories tannakiennes*, Grothendieck Festschrift, Vol. 2 (P. Cartier et al., ed.), Birkhäuser, 1990, Progress in Mathematics, Vol. 87, pp. 111–195.
48. P. Deligne and J. Milne, *Tannakian categories*, Hodge Cycles, Motives and Shimura Varieties (P. Deligne et al., ed.), 1982, Springer Lecture Notes in Mathematics, Vol. 900, pp. 101–228.
49. J. Della Dora, C. di Crescenzo, and E. Tournier, *An algorithm to obtain formal solutions of a linear differential equation at an irregular singular point*, Computer Algebra - EUROCAM '82 (*Lecture Notes in Computer Science, 144* (J. Calmet, ed.), 1982, pp. 273–280.
50. A. Duval, *Biconfluence et groupe de Galois*, Publication IRMA, Lille, Vol 18, No. 1, 1989.
51. A. Duval and M. Loday-Richaud, *Kovacic's algorithm and its application to some families of special functions*, Applicable Algebra in Engineering, Communication and Computing **3** (1992), no. 3, 211–246.
52. A. Duval and C. Mitschi, *Matrices de Stokes et groupe de Galois des équations hypergéométriques confluentes généralisées*, Pacific Journal of Mathematics **138** (1989), no. 1, 25–56.
53. B. Dwork, *Differential operators with nilpotent p-curvature*, American Journal of Mathematics **112** (1990), 749–786.
54. A. Fahim, *Extensions galoisiennes d'algèbres différentielles*, Publication IRMA, Lille, 1993; to appear in the *Pacific Journal of Mathematics.*
55. ———, *Extensions galoisiennes d'algèbres différentielles*, Comptes Rendus de l'Académie des Sciences, Paris **314** (1992), 1–4.
56. L. Fuchs, *Über die linearen Differentialgleichungen zweiter Ordnung, welche algebraische Integrale besitzen, und eine neue Anwendung der Invariantentheorie*, Journal für die reine und angewandte Mathematik **81** (1875), 97–147.
57. ———, *Über die linearen Differentialgleichungen zweiter Ordnung, welche algebraische Integrale besitzen, zweite Abhandlung*, Journal für die reine und angewandte Mathematik **85** (1878), 1–25.
58. L. Goldman, *Specialization and Picard-Vessiot theory*, Transactions of the American Mathematical Society **85** (1957), 327–356.
59. J. J. Gray, *Fuchs and the theory of differential equations*, Bulletin of the American Matehmatical Society **10** (1984), no. 1, 1 – 26.
60. ———, *Linear Differential Equations and Group Theory from Riemann to Poincaré*, Birkhäuser, Boston, Basel, Stuttgart, 1986.

61. N. V. Grigorenko, *G-Primitive extensions of differential fields*, Mathematical Notes **23** (1978), no. 3-4, 231–235, Originally published in Akademiya Nauk SSR. Matematicheskie Zametki, V. 23, No. 3, 1978.
62. ———, *Two problems in the Galois theory of differential fields for the field of formal power series*, Mathematics of the USSR- Sbornik **37** (1980), no. 3, 327–335, Originally published in Matematicheskiu i Sbonik. Novaya Seriya, V. 109(151) No. 3, 1979.
63. D. Yu. Grigoriev, *Complexity for irreducibility testing for a system of linear ordinary differential equations*, Proceedings of the International Symposium on Symbolic and Algebraic Computation- ISSAC'90 (M. Nagata and S. Watanabe, eds.), ACM Press, 1990, pp. 225–230.
64. ———, *Complexity of factoring and calculating the gcd of linear ordinary differential operators*, Journal of Symbolic Computation **10** (1990), no. 1, 7–38.
65. A. Haefliger, *Local theory of meromorphic connections in dimension one (Fuchs theory)*, Algebraic D-Modules, *Borel et al*, Academic Press, 1987, pp. 129–149.
66. P. Hendriks, *Shidlovskii irreducibility*, Indagationes Mathematicae **5** (1994), no. 4, 439–456.
67. P. A. Hendriks and M. van der Put, *Galois action on solutions of a differential equation*, Journal of Symbolic Computation **19** (1995), no. 6, 559 – 576.
68. S. Hessinger, *Computing Galois groups of fourth order linear differential equations*, Ph.D. thesis, North Carolina State University, 1997.
69. A. Hilali, *On the algebraic and differential Newton-Puiseux polygons*, Journal of Symbolic Computation **4** (1987), no. 3, 335–349.
70. E. Hilb, *Lineare Differentialgleichungen im komplexen Gebiet*, Encyclopädie der mathematischen Wissenschaften, Vol. II, B. 5, Teubner, Leipzig, 1915, pp. 471–562.
71. M. van Hoeij, *Rational solutions of the mixed differential equation and its application to factorization of differential operators*, Proceedings of the 1996 International Symposium on Symbolic and Algebraic Computation, ACM Press, 1996, pp. 219 – 225.
72. ———, *Factorization of differential operators with rational function coefficients*, to appear in the Journal of Symbolic Computation, 1997.
73. ———, *Formal solutions and factorization of differential operators with power series coefficients*, Journal of Symbolic Computation **24** (1997), no. 1, 1–31.
74. M. van Hoeij, J.F Rgot, F. Ulmer, and J-A Weil, *Liouvillian solutions of linear differential equations of order three and higher*, Tech. report, Université de Rennes, 1997.
75. M. van Hoeij and J.-A. Weil, *An Algorithm for Computing Invariants of Differential Galois Groups*, to appear in Journal of Pure and Applied Algebra, 1996.
76. T. Honda, *Algebraic differential equations*, Symposia Mathematica **24** (1981), 169–204.
77. N. Jacobson, *Pseudo-linear transformations*, Annals of Mathematics **38** (1937), 484–507.
78. C. Jordan, *Mémoire sur les équations différentielles linéaires à intégrale algébrique*, Journal für die reine und angewandte Mathematik **84** (1878), 89 – 215.
79. I. Kaplansky, *An Introduction to Differential Algebra*, second ed., Hermann, Paris, 1976.
80. N. Katz, *A conjecture in the arithmetic theory of differential equations*, Bulletin de la Société Mathématique de France **110** (1982), 203–239.
81. ———, *On the calculation of some differential Galois groups*, Inventiones Mathematicae **87** (1987), 13–61.
82. ———, *A simple algorithm for cyclic vectors*, American Journal of Mathematics **109** (1987), 65–70.
83. ———, *Exponential Sums and Differential Equations*, Annals of Mathematics Studies, vol. 124, Princeton University Press, Princeton, 1990.
84. N. Katz and R. Pink, *A note on pseudo-CM representatives and differential Galois groups*, Duke Mathematics Journal **54** (1987), no. 1, 57–65.
85. F. Klein, *Über lineare Differentialgleichungen, I*, Mathematische Annalen **11** (1877), 115–118.
86. ———, *Über lineare Differentialgleichungen, II*, Mathematische Annalen **12** (1878), 167–179.
87. E. R. Kolchin, *Algebraic matric groups and the Picard-Vessiot theory of homogeneous linear ordinary differential equations*, Annals of Mathematics **49** (1948), 1–42.
88. ———, *Algebraic groups and algebraic dependence*, American Journal of Mathematics **90** (1968), 1151–1164.
89. ———, *Some problems in differential algebra*, Proceedings of the International Congress of Mathematicians, Moscow, 1966 (Moscow), 1968, pp. 269–276.

90. ———, *Differential Algebra and Algebraic Groups*, Academic Press, New York, 1976.
91. J. Kovacic, *The inverse problem in the Galois theory of differential fields*, Annals of Mathematics **89** (1969), 583–608.
92. ———, *On the inverse problem in the Galois theory of differential fields*, Annals of Mathematics **93** (1971), 269–284.
93. ———, *An Eisenstein criterion for noncommutative polynomials*, Prroceedings of the American Mathematical Society **34** (1972), no. 1, 25–29.
94. ———, *An algorithm for solving second order linear homogeneous differential equations*, Journal of Symbolic Computation **2** (1986), 3–43.
95. M. Kuga, *Galois' Dream*, Birkhäuser, Boston, 1993.
96. A. H. M. Levelt, *Jordan decomposition for a class of singular differential operators*, Archiv für Mathematik **13** (1975), no. 1, 1–27.
97. ———, *Differential Galois theory and tensor products*, Indagationes Mathematicae, New Series **1** (1990), no. 4, 439 – 450.
98. M. Loday-Richaud, *Calcul des invariants de Birkhoff des systèmes d'ordre deux*, Funkcialaj Ekvacioj **33** (1990), 161–225.
99. ———, *Introduction à la multisommabilité*, Gazette des Mathématiciens, SMF **44** (1990), 41–63.
100. ———, *Stokes phenomenon, multisummability and differential Galois groups*, Annales de l'Institut Fourier **44** (1994), no. 3, 849–906.
101. ———, *Solutions formelles des systèmes différentiels linéaires méromorphes et sommation*, Expositiones Mathematicae **13** (1995), 116–162.
102. A. Magid, *Lectures on Differential Galois Theory*, University Lecture Series, American Mathematical Society, 1994, Reprinted with corrections 1997.
103. B. Malgrange, *Sur la réduction formelle des équations différentielles à singularités irrégulières*, manuscript, 1979.
104. M. F. Marotte, *Les équations différentielles linéaires et la théorie des groupes*, Annales de la Faculté des Sciences de Toulouse **12** (1887), no. 1, H1 – H92.
105. J. Martinet and J.-P. Ramis, *Théorie de Galois différentielle et resommation*, Computer Algebra and Differential Equations (E. Tournier, ed.), Academic Press, 1989, pp. 115–214.
106. ———, *Elementary acceleration and multisummability*, Annales de l'Institut Henri Poincaré, Physique Théorique **54** (1991), no. 4, 331–401.
107. J. Miller, *On differentially hilbertian differential fields*, Ph.D. thesis, Columbia University, 1970.
108. C. Mitschi, *Differential Galois groups and G-functions*, Computer Algebra and Differential Equations (M. F. Singer, ed.), Academic Press, 1991, pp. 149–180.
109. ———, *Differential Galois groups of confluent generalized hypergeometric equations: An approach using Stokes multipliers*, Pacific Journal of Mathematics **176** (1996), no. 2, 365–405.
110. C. Mitschi and M. F. Singer, *Connected linear groups as differential Galois groups*, Journal of Algebra **184** (1996), 333–361.
111. ———, *The inverse problem in differential Galois theory*, The Stokes Phenomenon and Hilbert's 16th Problem (B. L. J. Braaksma et al., ed.), World Scientific, River Edge, NJ, 1996, pp. 185–197.
112. ———, *On Ramis's solution of the local inverse problem of differential Galois theory*, Journal of Pure and Applied Algebra **110** (1996), 185–194.
113. F. Naegele and J. Thomann, *Algorithmic approach of the multisummation of formal power series solutions of linear ODE applied to Stokes phenomena*, The Stokes Phenomenon and Hilbert's 16th Problem (Braaksma et. al., ed.), World Scientific, 1996, pp. 197–214.
114. P. Painlevé, *Sur les équations différentielles linéaires*, Comptes Rendus de l'Académie des Sciences, Paris **105** (1887), 165– 168.
115. P. Th. Pépin, *Méthodes pour obtenir les intégrales algébriques des équations différentielles linéaires du second ordre*, Atti dell'Accad. Pont. de Nuovi Lincei **36** (1881), 243–388.
116. M. Petkovšek and B. Salvy, *Finding all hypergeometric solutions of linear differential equations*, Proceedings of the 1993 International Symposium on Symbolic and Algebraic Computation (ISSAC'93) (M. Bronstein, ed.), ACM Press, 1993, pp. 27–33.

117. E. Pflügel, *An algorithm for computing exponential solutions of first order linear differential systems*, Proceedings of the 1997 International Symposium on Symbolic and Algebraic Computation (ISSAC'97), ACM Press, 1997, pp. 225–230.
118. A. Pillay, *Differential Galois theory, I, II*, Preprint, University of Notre Dame, 1996.
119. M. Pohst and H. Zassenhaus, *Algorithmic Algebraic Number Theory*, Encyclopedia of Mathematics and its Applications, Cambridge University Press, Cambridge, 1989.
120. E. G. C. Poole, *Introduction to the Theory of Linear Differential Equations*, Dover Publications, Inc, New York, 1960.
121. M. van der Put, *Galois theory of differential equations, algebraic groups and Lie algebras*, To appear in the *Journal of Symbolic Computation.*
122. ———, *Symbolic analysis of differential equations*, To appear in *Some Tapas of Computer Algebra*, A.M.Cohen, H.Cuypers, H. Sterk, eds., .
123. ———, *Differential equations in characteristic p*, Compositio Mathematica **97** (1995), 227–251.
124. ———, *Singular complex differential equations: An introduction*, Nieuw Archief voor Wiskunde, vierde serie **13** (1995), no. 3, 451–470.
125. ———, *Reduction modulo p of differential equations*, Indagationes Mathematicae **7** (1996), 367–387.
126. J.-P. Ramis, *Théorèmes d'indices Gevrey pour les équations différentielles ordinaires*, Memoirs of the American Mathematical Society, vol. 296, American Mathematical Society, 1984.
127. ———, *Filtration Gevrey sur le groupe de Picard-Vessiot d'une équation différentielle irrégulière*, Informes de Matematica, Preprint IMPA, Series A-045/85, 1985.
128. ———, *Phénomène de Stokes et filtration Gevrey sur le groupe de Picard-Vessiot*, Comptes Rendus de l'Académie des Sciences, Paris **301** (1985), 165–167.
129. ———, *Phénomène de Stokes et resommation*, Comptes Rendus de l'Académie des Sciences, Paris **301** (1985), 99–102.
130. ———, *About the solution of some inverse problems in differential Galois theory by Hamburger equations*, Differential Equations, Dynamical Systems and Control Science, Lecture Notes in Pure and Applied Mathematics, 152 (K.P. Elsworthy et. al., ed.), Marcel Dekker, 1994, pp. 277–300.
131. ———, *About the inverse problem in differential Galois theory: Solutions of the local inverse problem and of the differential Abhyankar conjecture*, Tech. report, Université Paul Sabatier, Toulouse, 1996.
132. ———, *About the inverse problem in differential Galois theory: The differential Abhyankar conjecture*, The Stokes Phenomenon and Hilbert's 16th Problem (Braaksma et. al., ed.), World Scientific, 1996, pp. 261 – 278.
133. R. H. Risch, *The solution of the problem of integration in finite terms*, Bulletin of the American Mathematical Society **76** (1970), 605–608.
134. L. Schlesinger, *Handbuch der Theorie der Linearen Differentialgleichungen*, Teubner, Leipzig, 1887.
135. L. Schlesinger, *Bericht über die Entwicklung der Theorie der linearen Differentialgleichungen seit 1865*, Jahresbericht den Deutschen Vereinigung, XVIII, G. Reimer, Berlin, 1909, pp. 133–266.
136. F. Schwarz, *A factorization algorithm for linear ordinary differential operators*, Proceedings of the ACM-SIGSAM 1989 International Symposium on Symbolic and Algebraic Computation (G. Gonnet, ed.), ACM Press, 1989, pp. 17–25.
137. H. A. Schwarz, *Ueber diejenigen Fälle, in welchen die Gaussische hypergeometrische Reihe eine algebraische Funktion ihres vierten Elements darstellt*, Journal für die reine und angewandte Mathematik **75** (1872), 292–335.
138. M. F. Singer, *Algebraic solutions of nth order linear differential equations*, Proceedings of the 1979 Queen's Conference on Number Theory, Queen's Papers in Pure and Applied Mathematics, **59**, 1979, pp. 379–420.
139. ———, *Liouvillian solutions of nth order homogeneous linear differential equations*, American Journal of Mathematics **103** (1981), 661–681.
140. ———, *Solving homogeneous linear differential equations in terms of second order linear differential equations*, American Journal of Mathematics **107** (1985), 663–696.
141. ———, *Algebraic relations among solutions of linear differential equations*, Transactions of the American Mathematical Society **295** (1986), 753–763.

142. ———, *Algebraic relations among solutions of linear differential equations: Fano's theorem*, American Journal of Mathematics **110** (1988), 115–144.
143. ———, *An outline of differential Galois theory*, Computer Algebra and Differential Equations (E. Tournier, ed.), Academic Press, 1989, pp. 3–57.
144. ———, *Liouvillian solutions of linear differential equations with liouvillian coefficients*, Journal of Symbolic Computation **11** (1991), 251–273.
145. ———, *Moduli of linear differential equations on the Riemann sphere with fixed Galois group*, Pacific Journal of Mathematics **106** (1993), no. 2, 343–395.
146. ———, *Testing reducibility of linear differential operators: a group theoretic perspective*, Applicable Algebra in Engineering, Communication and Computing **7** (1996), 77–104.
147. M. F. Singer and F. Ulmer, *Galois groups of second and third order linear differential equations*, Journal of Symbolic Computation **16** (1993), no. 3, 9–36.
148. ———, *Liouvillian and algebraic solutions of second and third order linear differential equations*, Journal of Symbolic Computation **16** (1993), no. 3, 37–73.
149. ———, *Necessary conditions for liouvillian solutions of (third order) linear differential equations*, Applied Algebra in Engineering, Communication, and Computing **6** (1995), no. 1, 1–22.
150. ———, *Linear differential equations and products of linear forms*, to appear in the Journal of Pure and Applied Algebra, 1996.
151. R. Sommeling, *Characteristic classes for irregular singularities*, Ph.D. thesis, Catholic University of Nijmegen, 1993, An extended abstract appeared as *Characteristic Classes for Irregular Singularities*, Proceedings of ISSAC'94, ACM Press, 1994.
152. M. Toulouse, *Étude de systèmes différentieles linéaires d'ordre deux et trois: Algorithmes de calcul du groupe de Galois différentiel*, Thèse, Université Louis Pasteur, Strasbourg, 1996.
153. E. Tournier, *Solutions formelles d'équations différentielles*, Thèse, Faculté des Sciences de Grenoble, 1987.
154. B. M. Trager, *On the integration of algebraic functions*, Ph.D. thesis, MIT, 1984.
155. C. Tretkoff and M. Tretkoff, *Solution of the inverse problem in differential Galois theory in the classical case*, American Journal of Mathematics **101** (1979), 1327–1332.
156. S. P. Tsarev, *Problems that appear during factorization of ordinary linear differential operators*, Programming and Computer Software **20** (1994), no. 1, 27–29.
157. ———, *An algorithm for complete enumeration of all factorizations of a linear ordinary differential operator*, Proceedings of the 1996 International Symposium on Symbolic and Algebraic Computation (Lakshman Y. N., ed.), ACM Press, 1996, pp. 226–231.
158. F. Ulmer, *On liouvillian solutions of differential equations*, Applicable Algebra in Engineering, Communication and Computing **2** (1992), 171–193.
159. ———, *Irreducible linear differential equations of prime order*, Journal of Symbolic Computation **18** (1994), no. 4, 385–401.
160. F. Ulmer and J.-A. Weil, *A note on Kovacic's algorithm*, Journal of Symbolic Computation **22** (1996), no. 2, 179 – 200.
161. V. S. Varadarajan, *Meromorphic differential equations*, Expositiones Mathematicae **9** (1991), no. 2, 97–188.
162. ———, *Linear meromorphic differential equations: A modern point of view*, Bulletin (New Series) of the American Mathematical Society **33** (1996), no. 1, 1 – 42.
163. E. Vessiot, *Méthodes d'intégration élémentaires*, Encyclopédie des Sciences Mathématiques Pures et Appliquées, Tome II, Vol. 3, Fasc. 1, Gauthier-Villars, Paris, 1910, pp. 58–170.
164. R. Vidunas, *Differential equations of order two with one singular point*, Tech. report, University of Groningen, 1997.
165. E. Volcheck, *Resolving singularities and computing in the jacobian of a plane algebraic curve*, Ph.D. thesis, UCLA, 1994.
166. ———, *Testing torsion divisors for symbolic integration*, The ISSAC'96 Poster Session Abstracts (Zurich, Switzerland) (Wolfgang W. Küchlin, ed.), ETH, July 24–26 1996.
167. B. L. van der Waerden, *Modern Algebra*, second ed., Frederick Ungar Publishing Co., New York, 1953.
168. A. Zharkov, *Coefficient fields of solutions in Kovacic's algorithm*, Journal of Symbolic Computation **19** (1995), no. 5, 403–408.

Department of Mathematics, North Carolina State University, Raleigh, NC 27698-8205
E-mail address: singer@math.ncsu.edu

Math. Japonica **42**, No. 3(1995), 575–585

LES CORPS DIFFÉRENTIELLEMENT CLOS, COMPAGNONS DE ROUTE DE LA THÉORIE DES MODÈLES.

BRUNO POIZAT

Received May 16, 1995

Resumo. Tiu ĉi artikolo estas dediĉata a la historio de la paralelaj progresoj de la Teorio de la modeloj kaj de la Diferenciala algebro. Tre ofte en la kvardek lastaj jaroj, la Teorio de la modeloj usis la konon de diferenciale ŝlosata korpo pri elpensi aŭ ne triviale ilustri la siajn novajn konseptojn : modelplenigeco, Morleya ranko, primmodelo, DOP, elĵeteco de la imagatoj, perpendikulareco kaj interneco, Zilbera tritranĉeco, ktp.

Introduction.

Cet article est consacré à l'histoire d'un cheminement parallèle, celui de la Théorie des modèles et de l'Algèbre différentielle. La Théorie des modèles est la partie de la Logique mathématique qui traite de la sémantique, c'est-à-dire de l'interprétation du langage, et non pas de ses développements formels ; elle en mesure le pouvoir d'expression, ainsi que ses limites : son activité favorite est de décrire la classe des structures satisfaisant les mêmes énoncés du premier ordre (i.e. qui ne quantifient que les éléments des structures considérées ; on a d'excellentes raisons de se limiter au premier ordre) qu'une structure donnée, ou bien - ce qui correspond à un point-de-vue plus récent - à décrire la famille formée par les objets définissables, chacun par une formule du premier ordre, dans une structure donnée.

En tant que discipline mathématique dûment identifiée, la Théorie des modèles est une science neuve : elle est née des oeuvres de TARSKI, VAUGHT, ROBINSON, et quelques autres, en même temps que le Rock'n roll (HALLEY, BERRY, VINCENT, LEWIS, etc...) ; l'expression "theory of models" apparaît dans [TARSKI 1954], et son premier résultat significatif a été obtenu par MORLEY en 1962. Ce qui est remarquable, et fait l'objet de cet article, c'est qu'au cours des quarante années passées elle a illustré chacune des étapes cruciales de son développement en faisant appel aux corps différentiellement clos, soit comme mine d'exemples élaborés, soit comme cource d'inspiration pour les techniques nouvelles qu'elle créait : on verra que bien des outils de la Théorie des modèles ont été à l'origine forgés spécifiquement pour ces corps différentiellement clos.
Disons dès à présent de quoi il s'agit ; nous appelons corps différentiel un corps K *de caractéristique nulle* (sauf mention expresse du contraire) muni d'une dérivation ; dans ce contexte, un polynôme différentiel en n variables est un polynôme ordinaire, à coefficients dans K , en ces variables et leurs dérivées successives ; la notion de solution d'une équation différentielle polynomiale $P(x_1, \dots x_n) = 0$ est alors claire. Les constantes de K sont par définition ses éléments de dérivé nul ; elles forment un sous-corps C de K .

Le corps différentiel K est dit différentiellement clos s'il satisfait la version différentielle du Nullstellensatz, à savoir que tout système formé d'un nombre fini d'équations et d'inéquations ($\neq 0$) différentielles polynomiales, en un nombre n de variables, à coefficients dans K , et qui a une solution dans une extension de K , en a une dans K ; cela implique

que le corps des constantes C de K est algébriquement clos, ainsi d'ailleurs que K lui-même. Dans le cas des corps ordinaires, sans dérivation, et des équations et inéquations polynômes, cette condition caractérise les corps algébriquement clos ; mais vous en connaissez certainement une définition plus simple, qui ne fait intervenir que les polynômes en une variable ! Une chose analogue se produit dans le cas différentiel : en 1968, Lenore BLUM a montré que K était différentiellement clos pourvu que tout système formé d'une équation $P(x) = 0$ et d'une inéquation $Q(x) \neq 0$ polynomiales différentielles, à coefficients dans K , en une seule variable, l'ordre de Q (= indice de la plus haute dérivée de x intervenant dans Q) étant strictement inférieur à celui de P, ait une solution dans K .

1. La préhistoire.

En 1968 les corps différentiellement clos, à défaut d'une longue histoire, avaient déjà une longue préhistoire modèle-théorique. Quant à l'Algèbre différentielle, elle était bien plus ancienne encore.

L'Algèbre différentielle commence quand on oublie la nature des objets à qui on applique la dérivation (ce ne sont plus des fonctions qu'on différencie) pour ne s'intéresser qu'aux propriétés formelles du calcul différentiel. Les premiers travaux de ce genre sont dus à LIOUVILLE ; ils aboutiront à la définition du groupe de Galois d'une équation différentielle linéaire, et au critère pour qu'elle soit résoluble par primitives et primitives exponentielles : il faut que son groupe de Galois soit résoluble !

A dire vrai, pas plus que GALOIS n'envisage, pour ses équations, de solutions autres que des nombres complexes, LIOUVILLE ne considère, comme coefficients ou comme solutions des équations différentielles, que des fonctions analytiques dans un domaine complexe. On ne raisonne pas en termes plus abstraits avant la fin du siècle, et l'algébrisation du contexte différentiel sera faite par PICARD, VESSIOT, DRACH et surtout RITT, qui, bien qu'analyste dans l'âme, jettera les bases de la théorie des idéaux différentiels (voir [RITT 1950] et [KAPLANSKY 1957]). J'en dis deux mots ; si K est un corps différentiel, $K[X_1, \ldots X_n]_d$ est l'anneau des polynômes en les variables $X_1, \ldots X_n$, $X_1', \ldots X_n'$, $\ldots X_1^{(k)}$, $\ldots X_n^{(k)}, \ldots$, muni de la dérivation prolongeant celle de K et pour laquelle la dérivée de $X_i^{(k)}$ est $X_i^{(k+1)}$; un idéal de cet anneau est dit différentiel s'il est clos par dérivation : la dérivation passe au quotient par un tel idéal, si bien que les idéaux différentiels premiers décrivent les extensions finiment engendrées de K . RITT et RAUDENBUSH ont obtenu l'analogue du Théorème de la base, de HILBERT, mais avec une restriction (et valable seulement en caractéristique nulle) : dans $K[X_1, \ldots X_n]_d$ il n'y a pas de suite infinie strictement croissante d'idéaux différentiels *radicaux*.

Quant à la Théorie de Galois, elle prend sa forme finale chez KOLCHIN : certaines équations, qui ne sont pas toutes linéaires, ont leurs solutions qui se laissent bien paramétrer par les constantes. Cela permet de leur associer un groupe de Galois, qui est un groupe algébrique sur le corps C des constantes de K , qui agit sur leurs solutions. La somme des oeuvres de KOLCHIN est reprise dans [KOLCHIN 1973].

Bien des notions introduites par KOLCHIN et ses prédécesseurs trouveront plus tard une interprétation modèle-théorique, mais il en est une où ce sont les logiciens qui ont la priorité : c'est cette notion de clôture différentielle, de domaine universel où les équations différentielles ont suffisamment de solutions, qui, assez curieusement, a échappé aux algébristes. Voici comment les choses se sont passées.

Le Théorème des zéros pour les corps est lié à la théorie de l'élimination : dans un système d'équations et d'inéquations polynomiales, on peut éliminer l'une des variables grâce des conditions s'exprimant par des combinaisons d'équations et d'inéquations portant sur les autres variables. C'est un algorithme aussi vieux que les mathématiques ; les logiciens le qualifient d'"élimination des quanteurs dans la théorie des corps algébriquement clos" ; pour les géomètres algébriques, c'est le "Théorème de Chevalley", affirmant que la projection d'un constructible est un constructible. Eh bien, cette théorie de l'élimination a été étendue par [SEIDENBERG 1956] au cadre différentiel ; Abraham ROBINSON s'en est alors immédiatement emparé pour illustrer un concept qu'il venait d'inventer : puisqu'il y avait élimination, les corps différentiels de caractéristique nulle devaient avoir une modèle-complétion, caractérisable par des axiomes du premier ordre, qu'il baptisa "Théorie des corps différentiellement clos" (voir [ROBINSON 1959]).

On voit donc que les corps différentiellement clos, dès l'origine de la Théorie des modèles, ont constitué la première illustration non-triviale de la notion de modèle-complétion ; il faut dire qu'ils sont restés bien mystérieux tant qu'on ignorait les axiomes de BLUM.

2. Dans la foulée de MORLEY.

En 1968, on sait déjà beaucoup de choses : dans ses travaux sur les théories aleph-un catégoriques, Michael MORLEY avait introduit la notion de structures totalement transcendantes, et Lenore BLUM a eu le mérite de reconnaître que les corps différentiellement clos entraient dans ce cadre. Sa thèse est d'ailleurs résolument orientée vers ces développements nouveaux de la Théorie des modèles : bien que la postérité en retienne surtout cette axiomatisation des corps différentiellement clos, elle n'est pas spécifiquement consacrée à l'algèbre différentielle ; bien au contraire, son axiomatisatisation repose sur une idée modèle-théorique de portée générale, qui est de forcer l'existence d'un type de rang de Morley oméga, l'idéologie étant de considérer "de rang de Morley fini" comme un analogue d'une liaison algébrique. Pour des raisons discutées plus en détail dans [POIZAT 199?], cette tentative ne pouvait aboutir avec le rang de Morley, mais, sans le savoir, BLUM contemplait le premier exemple concret d'un type régulier de rang supérieur à un ; c'était aussi le premier exemple algébrique d'une théorie totalement transcendente de rang de Morley infini, qui échappait au cadre aleph-un-catégorique où restaient enfermés MORLEY et ses successeurs immédiats (ce rang de Morley généralise la notion de dimension des variétés en Géométrie algébrique ; son existence, dans le cas présent, est intimement liée au Théorème de Ritt-Raudenbush, l'une se déduisant à partir de l'autre, et réciproquement).

Une conséquence immédiate de la totale transcendance, c'est la densité des types isolés (qui sont la même chose que les "idéaux contraints" de KOLCHIN), d'où l'existence de modèles premiers : en conséquence, tout corps différentiel K a une clôture différentielle K_{dc}, c'est-à-dire une plus petite extension différentiellement close : tout plongement de K dans un corps différentiellement clos s'étend à K_{dc}. En 1968, hormis le cas où K est dénombrable, on ne sait pas montrer l'unicité de cette clôture différentielle ; il faudra pour cela attendre [SHELAH 1972], qui est une preuve générale d'unicité du modèle premier pour les théories totalement transcendentes : la situation est de ce point de vue bien plus compliquée que pour la clôture algébrique, et le cadre différentiel n'apporte aucune simplification à la démonstration générale. L'existence et l'unicité de la clôture différentielle constituent certainement une contribution importante de la Théorie des modèles à l'Algèbre.

Je signale au passage qu'il y a une notion de corps différentiellement clos en caractéristique p, qui a été explorée par Carol WOOD [WOOD 1973] ; leur traitement est plus complexe, la difficulté principale étant que toute puissance p° est une constante : cela interdit l'élimination des quanteurs, puisqu'il faut bien distinguer les constantes dont la racine p° est constante des autres ! La théorie correspondante est modèle-compagne, et pas modèle-complétion, de celle des corps différentiellement clos de caractéristique p ; elle est stable non superstable, mais se laisse quand même apprivoiser ; par exemple elle admet des modèles premiers (le cadre superstable est le plus général où on peut définir l'analogue de la dimension ; au delà, il y a les structures stables, où on ne dispose que d'une version "locale" de cette dimension : si vous désirez combler des lacunes dans vos connaissances sur les fondements de la Théorie des modèles, consultez [POIZAT 1985]). Ces corps différentiels sont les protagonistes d'une curieuse anecdote : [MACINTYRE 1971] a montré que tout corps infini aleph-un catégorique est algébriquement clos, démonstration étendue plus tard au cas superstable par [CHERLIN-SHELAH 1980] ; on s'est alors demandé s'il existait d'autres corps stables, sans se rendre compte qu'il y en avait déjà en évidence : prendre un corps K différentiellement clos de caractéristique p, qui, en tant que corps, est séparablement clos non algébriquement clos, et oublier la dérivation ! En fait, tout corps séparablement clos est stable ([WOOD 1979]) ; on ne sait pas s'il existe d'autres corps stables.

3. La course aux types triviaux.

Au début de la décennie suivante, [KOLCHIN 1974], [ROSENLICHT 1974] et [SHELAH 1973] mettent simultanément en évidence une propriété un peu déroutante de la clôture différentielle : si C est un corps de constantes, sa clôture différentielle C_{dc} n'est pas minimale ; il y a des corps différentiellement clos qui sont strictement compris entre C et C_{dc} ! Cela n'affecte pas l'unicité de la clôture différentielle : ils sont naturellement C-isomorphes à C_{dc}.

Cette propriété vient d'un phénomène qui contredit une intuition venant de l'Analyse : on vit dans l'idée que la solution générale d'une équation différentielle du premier ordre dépend d'un paramètre constant. Cela peut bien se produire dans notre cadre, par exemple pour les équations différentielles linéaires : dès qu'on a une solution a de l'équation $x' = 1$ - et tout corps différentiellement clos contient une primitive de 1 - les autres se mettent sous la forme $x = a + c$, où c est une constante. Autrement dit, si K est différentiellement clos, ajouter à K une solution de $x' = 1$, ou bien lui ajouter une constante, c'est pareil (on dit alors que l'équation $x' = 1$ est C-interne ; voit la section suivante).

De même, si b est une solution non nulle de l'équation $x' = x$, toutes les autres se mettent sous la forme $x = c.b$. Vous avez peut-être envie d'intégrer cette équation $x'/x = 1$ en $\text{Log}\,|x| = a + c$, mais ça n'a pas de sens dans le contexte des corps différentiels : il n'y a pas de fonction logarithme, ni d'exponentielle. Toutes les primitives de 1, ou de x'/x se ressemblent, il n'y en n'a pas de privilégiées ! C'est d'ailleurs la source de la Théorie de Galois : le groupe de Galois est le groupe additif C^+ pour la première équation, et le groupe multiplicatif C^* pour la seconde. Pour paramétrer les solutions de $x' = 1$ et de $x' = x$ par les constantes, il faut fixer une solution a de la première et une solution b de la seconde (il n'y en a pas de privilégiées, comme en Analyse la fonction identité pour la première, et la fonction exponentielle pour la seconde) ; de plus, il n'y a pas moyen d'établir un lien direct entre a et b, du genre $b = e^a$.

Si maintenant nous considérons l'équation $E(1)$: $x.x' + x' - x = 0$, qui s'écrit aussi, si on néglige sa solution nulle, $x' + x'/x = 1$, on ne peut plus faire apparaître de lien entre les solutions de $E(1)$ et les constantes, même en fixant des paramètres. La raison en est que si on intègre, ça donne $x+ \text{Log}\,|x| = a + c$, c'est-à-dire rien du tout ; dans notre contexte, le théorème d'existence et d'unicité des solutions des équations différentielles, qui repose sur l'existence de fonctions explicites ou implicites, ne peut s'exprimer. Ici, nous n'avons à notre disposition que les fonctions rationnelles au niveau explicite, et les fonctions algébriques au niveau implicite.

Le paragraphe précédent est une explication, pas une démonstration. Ce qu'il faut démontrer, c'est que $E(1)$ est *orthogonale* aux constantes, ce qui signifie que, si K est un corps différentiellement clos, et qu'on lui ajoute une nouvelle solution a de $E(1)$, alors la clôture différentielle $K(a)_{dc}$ de l'extension obtenue ne contient pas de constantes autres que celles de K . En répétant l'opération, on fabrique des corps différentiellement clos contenant autant de solutions de $E(1)$ qu'on veut, sans ajouter de constantes. Symétriquement, on peut ajouter des constantes sans ajouter de solutions de $E(1)$. On voit que l'orthogonalité à C est la notion exactement opposée à la C-internalité.

Cette orthogonalité est conséquence facile de la propriété de *trivialité* que j'expose maintenant. Si x est une solution de $E(1)$, sa dérivée s'écrit $x' = x/1 + x$; en différenciant, on obtient des expressions de x'', x''', etc ... en fonction de x. Donc la seule chose qui peut arriver à des solutions non nulles x_1, ... x_n de cette équation, c'est d'être algébriquement dépendantes ou bien d'être algébriquement indépendantes sur C . Eh bien, la propriété remarquable de $E(1)$, c'est que ses solutions sont indépendantes dès qu'elles sont distinctes ! Leur relation de dépendance est *triviale*, ce qui signifie que l'indépendance par paires implique l'indépendance globale ; on comprend pourquoi cette propriété implique l'orthogonalité à C , car l'indépendance algébrique dans les corps n'est pas triviale du tout !

On observe que cette relation de dépendance (sur C) est *atomique*, ce qui signifie qu'on peut exprimer finiment que x_1, ... x_n sont indépendants (on imagine fort bien une relation de dépendance triviale, où l'expression de l'indépendance nécessite une infinité d'inégalités polynomiales du type $P_i(x_1) \neq P_i(x_2)$).

Un corps différentiellement clos K a une solution pour chaque système formé de $E(1)$ et de l'équation d'ordre zéro $(x - a_1)$. ... .$(x - a_n) \neq 0$; il contient donc une infinité de solutions de $E(1)$. Si C est un corps de constantes, sa clôture différentielle C_{dc} en contient une famille dénombrable (même si C est plus que dénombrable : c'est dû à l'orthogonalité) x_0, x_1, ... x_n, ... ; elles forment ce qu'on appelle un ensemble *indiscernable*, c'est-à-dire que toute permutation de cet ensemble s'étend en un C-automorphisme de C_{dc}. Il est alors facile de voir que tout plongement dans C_{dc} de la clôture différentielle de $C(x_1, \ldots\, x_n, \ldots)$ évite x_0 : on obtient ainsi des plongements non surjectifs de C_{dc} dans lui-même.

Bien. Donc tout se ramène à prouver la trivialité et l'atomicité de la relation de dépendance de certaines équations. Par un calcul élémentaire, KOLCHIN traite de l'équation $x' = x^3 - x^2$, soit encore $x'/x-1 \;-\; x'/x \;-\; x'/x^2 \;=\; 1$, ce qui donnerait $\mathrm{Log}\,|x-1/x|+1/x = a+c$ si on pouvait l'intégrer. SHELAH considère notre équation $E(1)$, et utilise une approche analytique : il montre que certains corps différentiels se plongent dans le corps des fonctions méromorphes sur un domaine complexe judicieusement choisi, et que certaines fonctions sont algébriquement indépendantes en vérifiant que sont algébriquement indépendantes leurs valeurs en un point qui fait également l'objet d'un choix judicieux ; quand on fait ça, il faut être sûr que les fonctions considérées, qui ont été obtenues par prolongement analytique, sont bien définies en ce point : SHELAH n'y est pas maniaquement attentif, mais [GRAMAIN 1983] a vérifié que sa méthode était politiquement correcte. Quant à ROSENLICHT, il traite de toutes les équations, à coefficients constants, de la forme $x'.P(x) - Q(x) = 0$: si elles sont orthogonales aux constantes, se qui se produit quand la forme différentielle $(Q(x)/P(x)).dx$ a à la fois une partie exacte et une partie logarithmique, la relation de dépendance des solutions de l'équation est à la fois triviale et atomique. ROSENLICHT fait des calculs dans l'espace des différentielles de Kähler des extensions $K(a)/K$ obtenues en ajoutant une solution a de l'équation considérée ; comme l'observe GRAMAIN, sa méthode, qui est purement algébrique, est la seule qui ait une portée générale.

L'article de SHELAH contient un lemme 6, pour la démonstration duquel l'auteur se contente d'indiquer qu'elle peut être obtenue par les méthodes employées auparavant, qui déclare que si c_1 et c_2 sont différentiellement indépendants (i.e. si $P(c_1, c_2) = 0$, pour un polynôme différentiel P à coefficients rationnels, ne se produit que s'il est identiquement nul) alors les équations $E(c_1)$ $x' + x'/x = c_1$ et $E(c_2)$ $x' + x'/x = c_2$ sont (triviales et) orthogonales ; à ma connaissance ce résultat reste encore aujourd'hui du domaine de la conjecture ! Voici le verdict de GRAMAIN : la méthode de SHELAH permet de conclure à l'orthogonalité quand c_1 et c_2 sont deux *constantes* algébriquement indépendantes, et celles de ROSENLICHT quand ce sont deux *constantes* distinctes.

SHELAH en déduit rapidement qu'en tout cardinal κ non dénombrable il y a le maximum possible de corps différentiellement clos de ce cardinal, deux-à-deux non isomorphes, soit 2^κ. Maintenant que nous sommes familier de la pensée shelahienne, l'évidence de cette conclusion nous aveugle : les corps différentiellement clos ont la DOP ! On sait l'importance que cette Dimensionnal Order Property, qui n'est pas nommée dans [SHELAH 1973], a pris dans les travaux de SHELAH sur la classification ([SHELAH 1978/90]) : c'est pour les corps différentiellement clos qu'elle a été inventée.

A l'attention des non initiés, je dis quelques mots sur la façon de construire beaucoup de structures dans ce cas particulièrement simple de DOP. A toute relation binaire R, définie sur κ éléments, nous associons un corps différentiellement clos $K(R)$ de ce cardinal, de sorte que tout isomorphisme entre $K(R_1)$ et $K(R_2)$ induise un isomorphisme entre R_1 et R_2. Pour cela, on considère un corps de constantes C de cardinal κ, dont on choisit une base ..... c_i .... en tant que $\mathbf{Q}$-espace vectoriel, indexée par κ. Nous lui ajoutons κ solutions de chaque $E(c_i)$, chaque $E(2.c_i)$, ainsi que de $E(c_i + 2.c_j)$ chaque fois que le couple (i, j) satisfait la relation R ; $K(R)$ est la clôture différentielle de tout ça ; on y retouve R car les autres $E(c)$ n'y ont qu'une famille dénombrable de solutions.

On voit que notre grande famille d'équations deux-à-deux orthogonales nous permet de construire beaucoup de corps différentiellement clos en tout cardinal non dénombrable, mais que leur atomicité les prive de toute utilité pour distinguer des corps différentiellement clos dénombrables. La question du nombre de corps différentiellement clos dénombrables est posée dans [SHELAH 1973] ; elle restera ouverte plus de vingt ans ; ses liens avec la nature des relations de dépendance associées aux types de rang un sont discutés dans [LASCAR 1983].

Ce nombre est maintenant connu : il vous sera révélé à la fin de cet article. Comme conclusion de cette section, sur les types triviaux, je rappelle quelques mystères toujours hermétiques de la théorie des corps différentiellement clos ; ils sont de nature un peu technique : je m'en excuse auprès des non-spécialistes.

Le premier problème est lié aux différents rangs modèle-théoriques définis dans les corps différentiellement clos ; par exemple on ne sait pas caractériser le rang de Morley en termes algébriques, ni s'il est toujours égal au rang U de Lascar. Il serait intéressant d'avoir un type de rang U un et de rang de Morley supérieur ; il devrait alors être trivial d'après un théorème buechlerien dans le genre de [BUECHLER 1985]. La seule chose connue, c'est que ce rang, pour un type en une variable, peut être strictement inférieur à l'ordre de son équation différentielle minimale, ainsi qu'à sa hauteur de Krull comme idéal différentiel ; un exemple est donné dans [POIZAT 1978] ; la relation de dépendance de ce dernier type, et de quelques autres, est étudiée par [BRESTOVSKI 1982 & 1989] : elles sont triviales et atomiques ; BRESTOVSKI suit la voie tracée par ROSENLICHT, mais comme il doit travailler dans un espace de différentielles de dimension deux, c'est beaucoup plus compliqué ; il est de plus le seul à avoir étudié des équations différentielles à coefficients non constants. Comme on le verra à la fin de cet article, les types triviaux sont ceux des types de rang un qui restent les moins connus ; il convient de rappeler une question de [POIZAT 1983a] : est-il exact qu'une équation différentielle non-linéaire "générale" (i.e. à coefficients différentiellement indépendants) définisse un type de rang un trivial ?

4. La mauvaise influence des séries américaines.

Lors d'un congrès de Logique mathématique et de Linguistique, tenu à Lyon en décembre 1994, l'un des co-organisateurs (Fabrizio PENNACCHIETTI) a demandé si l'élimination des imaginaires était conséquence de la médiocrité des programmes de télévision américaine importés sur nos chaînes nationales. Eh bien non : son origine est à rechercher dans la Théorie de Galois pour les équations différentielles.

Les éléments imaginaires ont été définis par SHELAH (voir [SHELAH 1978/90]) ; étant données une structure M arbitraire, et une relation d'équivalence définissable E entre n-uples d'éléments de M, SHELAH associe à chaque $\underline{a}$ de M^n un "élément imaginaire" α qui représente la classe de $\underline{a}$ modulo E. On dit que structure M élimine les imaginaires si justement on n'a pas besoin de les ajouter, si tout imaginaire est co-définissable avec un uple réel (i.e. formé d'éléments de M).

Prenons un petit exemple : soit E l'équivalence entre couples déclarant qu'ils ont même ensemble sous-jacent, c'est-à-dire $(x,y)\ E\ (u,v)$ ssi $(x = u \wedge y = v) \vee (x = v \wedge y = u)$. Quand la structure M est un corps, les imaginaires correspondants sont éliminés, car deux n-uples représentent le même ensemble à permutation près ss'ils donnent mêmes valeurs aux fonctions symétriques élémentaires. Donc $(x,y)\ E\ (u,v)$ ssi $x + y = u + v \wedge x.y = u.v$; l'imaginaire $(x,y)/E$ peut être remplacé par le couple réel $(x + y, x.y)$ car, pour un automorphisme de corps, conserver globalement l'ensemble $\{x,y\}$, c'est-à-dire fixer x et y ou bien les échanger, c'est la même chose que de fixer $x + y$ et $x.y$.

Cette élimination a été introduite pour expliquer un versant de la Théorie de Galois ; soit K un corps différentiellement clos, dont on suppose que le corps de constantes C est algébriquement clos, et soit X une partie définissable de K_{dc} qui se laisse paramétrer par les constantes (dans le vocabulaire de KOLCHIN, l'extension $K(X)_d/K$ est *fortement normale* ; dans celui de HRUSHOVSKI, X est C-interne ; comme exemple pour X, on peut prendre les solutions d'une équation différentiellle linéaire, qui se mettent toutes sous la forme $c_1.a_1 + ... + c_n.a_n$, une fois qu'on a choisi un système fondamental $(a_1, ...a_n)$ de solutions ; mais il y en a d'autres !). KOLCHIN a montré que le groupe des automorphismes de $K(X)_d/K$ était un groupe G algébrique sur C , et que les extensions intermédiaires entre $K(X)_d$ et K correspondaient aux sous-groupes algébriques de G. Il est observé dans [POIZAT 1983] d'une part que le fait que G soit un groupe définissable dans C n'est qu'un cas particulier d'une construction modèle-théorique générale (employée par [ZIL'BER 1980] dans un cadre oméga-catégorique, sous le nom de "groupe de liaison", et reprise par [HRUSHOVSKI 1986] dans le cadre de sa notion d'internalité) ; d'autre part que tout sous-groupe H définissable de G apparaît naturellement comme le groupe de préservation d'une relation définissable r, soit encore du paramètre canonique permettant de la définir, lequel est imaginaire. Si on peut remplacer cet imaginaire par quelques réels, c'est que les corps différentiellement clos éliminent les imaginaires !

Les corps algébriquement clos aussi, d'ailleurs ; si on y regarde bien, la démonstration originelle de GALOIS consiste à remplacer une relation (l'orbite de H sur le uple des racines du polynôme considéré) par un élément : c'est une élimination d'imaginaire !

Quand au premier aspect de la chose, on se demande pourquoi KOLCHIN obtient-il pour G une structure de groupe algébrique sur C , alors que le Général Nonsense, commandant la Théorie des modèles, n'en fait qu'un groupe définissable dans le corps algébriquement clos C , c'est-à-dire un groupe constructible ? Y-a-t'il identité entre groupes algébriques et groupes constructibles ?

Une question âprement débattue : dès qu'elle fut posée, LPD Van den DRIES fit remarquer que la réponse se trouvait dans un théorème de WEIL, qui reconstituait un groupe algébrique à partir d'un "group chunk", un tel fragment de groupe étant de toute évidence fourni par un groupe constructible ; il faudra attendre l'intervention de HRUSHOVSKI pour avoir une démonstration complète, claire et généralisable de cette identité, reposant sur la manufacture d'un groupe définissable à partir de données génériques, dans un contexte stable (voir [POIZAT 1987, ch. 4]).

Les corps différentiellement clos ont cette fois encore bien mérité de la Science ; ils en ont été récompensés, puisque [PILLAY 1990] a montré que tout groupe définissable dans un corps différentiellement clos devait être isomorphe à un "groupe algébrique différentiel", généralisation obvie au contexte différentiel de la notion de groupe algébrique, définie dans [KOLCHIN 1985].

5. Les types modulaires.

Nous en venons à un succès plus contemporain, et tout-à-fait inattendu, des corps différentiellement clos, dont l'histoire est intéressante. Elle commence par la question parallèle à celle résolue dans [MACINTYRE 1971] et [CHERLIN-SHELAH 1980] : existe-t'il des corps différentiels superstables non constants et non différentiellement clos ?

Cette question est toujours ouverte ; on sait seulement qu'un tel corps ne peut pas avoir d'extensions d'un certain type. En reprenant l'héritage de [MICHAUX 1986] à propos des équations linéaires, valable pour les corps différentiels éliminant les quanteurs (une condition sans signification structurelle en général, mais qui ici implique la superstabilité), [PILLAY-SOKOLOVIC 1992] ont montré que ces corps n'avaient pas d'extensions fortement normales.

C'est en fait un problème d'imaginaires ; considérons une extension fortement normale de K , dont le groupe de Galois soit le groupe $G(C)$, algébrique sur C ; nous notons $G(K)$ l'extension naturelle de $G(C)$ à K , c'est-à-dire l'ensemble des points K-rationnels du groupe algébrique G. D'après un théorème de KOLCHIN, dont PILLAY et SOKOLOVIC donnent une démonstration purement modèle-théorique, l'extension est engendrée par un élément G-primitif, qui correspond à un point du quotient à droite $G(K)/G(C)$, et cet imaginaire peut se remplacer par un uple réel dans la clôture différentielle K_{dc} de K , mais qui ne doit pas être dans K puisqu'il engendre l'extension.

Si par exemple G est le groupe multiplicatif du corps, la dérivée logarithmique $x \rightarrow x'/x$ est un homomorphisme de K^* sur K^+, dont le noyau est C^* ; autrement dit, l'espace imaginaire K^*/C^* s'identifie à l'espace réel K^+. Dans un corps différentiel superstable non-constant, la dérivée logarithmique est surjective (un simple calcul de rang - version modèle-théorique de la "dimension" des géomètres - le prouve), ce qui est une façon de dire que tout élément de K y a une primitive exponentielle.

Plus généralement, deux matrices X et Y de $GL_n(K)$ sont congrues à droite modulo $GL_n(C)$ ssi $X'.X^{-1} = Y'.Y^{-1}$. L'application $X \rightarrow X'.X^{-1}$ permet d'identifier le quotient droit $GL_n(K)/GL_n(C)$ à une partie de $M_n(K)$; elle est surjective si K est non-constant superstable, elle ne l'est pas si K a des extensions fortement normales dont le groupe de Galois est $GL_n(C)$.

PILLAY et SOKOLOVIC cherchent, et trouvent, un argument de ce genre valable pour tous les groupes algébriques, lesquels ne sont pas tous affines. A l'opposé des groupes linéaires, il y a les variétés abéliennes, et en particulier les courbes elliptiques : si A est la cubique d'équation $y^2 = x.(x-1).(x-c)$, où c est une constante ($c' = 0$, $c \neq 0, 1$), par quoi est représenté le groupe quotient $A(K)/A(C)$? PILLAY et SOKOLOVIC découvrent que la réponse à cette question est connue depuis [MANIN 1958], car l'application $(x, y) \rightarrow$

x'/y est un homomorphisme de groupe de $A(K)$ dans K^+, qui est surjectif quand K est différentiellement clos, ou même seulement superstable.

Mais ils constatent aussi que Yuri MANIN, chaque fois qu'il dispose d'une dérivation, définit un homomorphisme μ de n'importe quelle variété abélienne dans un groupe vectoriel : il obtient donc un homomorphisme μ de A dans K^+, surjectif si K est différentiellement clos, même si c n'est pas une constante (l'expression de μ fait alors intervenir la dérivée seconde de x). D'où question troublante : comme il n'est plus question de points C-rationnels pour A, en quoi consiste le noyau de l'homomorphisme μ de MANIN ?

C'est alors que HRUSHOVSKI intervient dans le débat, en plaçant ce problème, d'apparence modeste, au coeur de la grande question qui a agité la Théorie des modèles depuis plus de dix ans : il s'agit de la trichotomie de ZIL'BER, démolie dans [HRUSHOVSKI 1993], et reconstruite, dans un cadre beaucoup plus contraignant, dans [HRUSHOVSKI-ZIL'BER 1993]. Cadre vraiment contraignant, mais dans lequel entrent les corps différentiellement clos : [HRUSHOVSKI-SOKOLOVIC 199?] ont vérifié qu'ils satisfaisaient les conditions des "structures de type Zariski", donc que la trichotomie y est valide. Cela signifie qu'un type de rang un y donne lieu à une relation de dépendance soit triviale, soit modulaire, soit intimement liée à la dépendance algébrique d'un corps, qui ne peut être ici que le corps des constantes C .

Nous avons vu des types C-internes, ainsi que des types triviaux, mais y-en a t'il de modulaires ? Et d'abord, de quoi s'agit-il ? Cela signifie que la relation de dépendance du type est semblable à la dépendance linéaire d'un espace vectoriel, que le type est fortement correllé à un groupe (commutatif) H de rang un, qui a la propriété que toute partie définissable de ses puissances cartésiennes H^n soit combinaison booléenne d'un nombre fini de cossettes modulo des sous-groupes de H^n. C'est une structure essentiellement linéaire, bien moins riche que celle qu'apporte un corps.

HRUSHOVKI a montré que les noyaux de MANIN d'une variété abélienne, sans sous-groupes algébriques isomorphes à une variété abélienne ($\neq 0$!) définie sur C , étaient modulaires. Donc les types modulaires existent, ce qui n'a pas manqué de surprendre l'auteur de ces lignes ; [PILLAY 199?] a montré qu'ils provenaient tous d'un noyau de MANIN.

Inspiré par [BUIUM 1992], HRUSHOVSKI a vu comment cette modularité pouvait être exploitée pour montrer une conjecture de LANG de type mordellien. Il emploie les corps différentiellement clos en caractéristique nulle, cas où la réponse à la conjecture était connue, mais il a donné une démonstration semblable, utilisant une autre structure de type Zariski, pour la caractéristique p : c'est la première démonstration complète pour la caractéristique p. Cette autre saga est racontée dans [GOODE 199?] ; on citera seulement cette appréciation de l'observateur aussi pénétrant de la Théorie des modèles contemporaine qu'est le Professeur GOODE " ... c'est la première fois que la Logique sert vraiment à montrer un résultat impliquant des objets mathématiques classiques".

Redescendant sur terre, HRUSHOVSKI montre aussi que le problème posé par SHELAH en 1973 est essentiellement résolu par MANIN en 1958. Si $c' \neq 0$, la relation de dépendance du noyau de MANIN de la cubique A est celle d'un espace vectoriel sur $\mathbf{Q}$, ce qui lui interdit d'être atomique ; par conséquent, on peut fabriquer des corps différentiellement clos, contenant c, où cet espace vectoriel est d'une dimension fixée arbitrairement, finie ou infinie. De plus, si les cubiques A_1 et A_2 ne sont pas isomorphes, ce qui ne signifie pas tout-à-fait que leurs paramètres c_1 et c_2 soient différents, mais presque, alors leurs noyaux de MANIN sont orthogonaux, si bien que les dimensions correspondantes peuvent être choisies en toute indépendance.

Avec tout ça, en jouant sur le motif dessiné par les dimensions des noyaux de MANIN des cubiques, on fabrique une famille continupotente de corps différentiellement clos dénombrables, ce qui contredit la conjecture perversement attribuée à l'auteur par [LASCAR 1983].

Pour les détails, ainsi que sur l'intérêt portée par la Théorie des modèles à certains travaux récents en Algèbre différentielle, voir [PILLAY 199?].

Conclusion.

Il demeure quelque opacité autour des corps différentiellement clos, surtout, comme il a été dit, autour des types triviaux. La Théorie des modèles, elle, conserve beaucoup plus de mystère : cela signifie-t'il que ces braves corps différentiels ne pourront plus lui servir, comme ils l'ont fait par le passé ? Nous verrons bien ; il semble que la notion d'équation différentielle soit importante dans certains domaines des mathématiques ; pourquoi pas en Logique ?

RÉFÉRENCES.

[BLUM 1968] L. Blum, *Generalized Algebraic Structures*, PhD Dissertation, MIT { voir aussi : G. Sacks, *The differential closure of a differential field*, Bull. Amer. Math. Soc., **52** (1974), 629–634 }.

[BRESTOVSKI 1982] M. Brestovski, *Déviation et indépendance algébrique de solutions génériques d'équations différentielles du second ordre*, C.R. Acad. Sc. Paris, **294** no. 1, 609–612.

[BRESTOVSKI 1989] M. Brestovski, *Algebraic independence of solutions of differential equations of the second order*, Pacific Journal of Math., **140**, 1–19.

[BUECHLER 1985] S. Buechler, *The geometry of weakly minimal types*, Journ. Symb. Logic, **50**, 1044–1053.

[BUIUM 1992] A. Buium, *Intersections in jet spaces and a conjecture of S. Lang*, Annals of Mathematics, **136**, 557–567.

[CHERLIN-SHELAH 1980] G. Cherlin and S. Shelah, *Superstable fields and groups*, Ann. Math. Logic, **18**, 227–270.

[GOODE 199?] J.B. Goode, *H.L.M. (Hrushovski-Lang-Mordell)*, à paraître.

[GRAMAIN 1983] F. Gramain, *Non-minimalité de la clôture différentielle, I & II*, Groupe d'étude de Théories Stables, **3**, Institut Henri Poincaré.

[HRUSHOVSKI 1986] E. Hrushovski, *Contributions to stable model theory*, PhD Dissertation, Berkeley.

[HRUSHOVSKI 1993] E. Hrushovski, *A new strongly minimal set*, Ann. Pure Appl. Logic, **62**, 147–166.

[HRUSHOVSKI 199?] E. Hrushovski, *The Mordell-Lang conjecture for function fields*, preprint.

[HRUSHOVSKI-SOKOLOVIC 199?] E. Hrushovski and Z. Sokolovic, *Minimal subsets of differentially closed fields*, preprint.

[HRUSHOVSKI-ZIL'BER 1993] E. Hrushovski and B. Zil'ber, *Zariski's geometries*, Bull. Amer. Math. Soc., **28**, 315–324.

[KAPLANSKY 1957] I. Kaplansky, *An Introduction to Differential Algebra*, Hermann.

[KOLCHIN 1973] E.R. Kolchin, *Differential Algebra and Algebraic Groups*, Academic Press.

[KOLCHIN 1974] E.R. Kolchin, *Constrained extensions of differential fields*, Advances in Math., **12**, 141–170.

[KOLCHIN 1985] E.R. Kolchin, *Differential Algebraic Groups*, Academic Press.

[LASCAR 1983] D. Lascar, *Les corps différentiellement clos dénombrables*, Groupe d'étude de Théories Stables, **3**, Institut Henri Poincaré.

[MANIN 1958] Y. Manin, *Courbes algébriques sur un corps avec dérivation (en russe)*, Izvestia Akad. Nauk SSSR, **22**, 737–756.

[MACINTYRE 1971] A. Macintyre, *On omega-one categorical theories of fields*, Fund. Math.,, 1–25.

[MICHAUX 1986] C. Michaux, *Sur l'élimination des quantificateurs dans les anneaux différentiel*, C. R. Acad. Sc. Paris, **302** no. 1, 287–290.

[PILLAY 1990] A. Pillay, *Differentially algebraic group chunks*, Journ. Symb. Logic, **55**, 1138–1142.

[PILLAY 199?] A. Pillay, *Differentially algebraic groups and the number of countable differentially closed fields*, preprint.

[PILLAY 199?a] A. Pillay, *Some foundational questions concerning differential algebraic groups*, preprint.

[PILLAY-SOKOLOVIC 1992] A. Pillay and Z. Sokolovic, *Superstable differential fields*, Journ. Symb. Logic, **57**, 97–108.

[PILLAY-SOKOLOVIC 1992a] A. Pillay and Z. Sokolovic, *A remark on differential algebraic groups*, Communications in Algebra, **20**, 3015–3026.

[POIZAT 1978] B. Poizat, *Rangs des types dans les corps différentiels*, Groupe d'étude de Théories Stables, **1**, Institut Henri Poincaré.

[POIZAT 1983] B. Poizat, *Une théorie de Galois imaginaire*, Journ. of Symb. Logic, **48**, 1151–1170.

[POIZAT 1983a] B. Poizat, *C'est beau et chaud !*, Groupe d'étude de Théories Stables, **3**, Institut Henri Poincaré.

[POIZAT 1985] B. Poizat, *Cours de Théorie des Modèles*, Nur al-Mantiq wal-Ma'rifah (diffusé par OFFILIB).

[POIZAT 1987] B. Poizat, *Groupes Stables*, Nur al-Mantiq wal-Ma'rifah (diffusé par OFFILIB).

[POIZAT 199?] B. Poizat, *Autour du Théorème de Morley : la Théorie des modèles entre 1960 et 1970*, à paraître.

[RITT 1950] J.R. Ritt, *Differential Algebra*, AMS publ., **33**.

[ROBINSON 1959] A. Robinson, *On the concept of a differentially closed field*, Bull. Res. Council Israel, **F8**, 113–128, { aussi dans : Selected Papers of Abraham Robinson, 1979, North Holland}.

[ROSENLICHT 1974] M. Rosenlicht, *The non minimality of the differential closure*, Pacific Journ. of Math., **12**, 140–170.

[SEIDENBERG 1956] A. Seidenberg, *An elimination theory for Differential Algebra*, Univ. Calif. Math. Publ., **3**, 31–35.

[SHELAH 1972] S. Shelah, *Uniqueness and characterization of prime models over sets for totally transcendental first-order theories*, Journ. Symb. Logic, **37**, 107–113.

[SHELAH 1973] S. Shelah, *Differentially closed fields*, Israel Journ. Math., **16**, 314–328.

[SHELAH 1978/90] S. Shelah, *Classification Theory and the number of non-isomorphic models*, North Holland ; 2° édition.

[SOKOLOVIC 1992] Z. Sokolovic, *Model Theory of Differential Fields*, PhD Dissertation, Notre Dame.

[TARSKI 1954] A. Tarski, *Contributions to the theory of model, I & II*, Indag. Math., **16**, 572–581 & 582–588.

[WOOD 1973] C. Wood, *The model theory of differential fields of characterisic* $p \neq 0$, Proc. Amer. Math. Soc., **40**, 577–584.

[WOOD 1974] C. Wood, *Prime model extensions for differential fields of characterisic* $p \neq 0$, Journ. Symb. Logic, **39**, 469–477.

[WOOD 1976] C. Wood, *The model theory of differential fields revisited*, Israel Journ. Math., **27**, 331–352.

[WOOD 1979] C. Wood, *Notes on the stability of separably closed fields*, Journ. Symb. Logic, **44**, 412–416.

[ZIL'BER 1980] B.Y. Zil'ber, *Totally categorical theories : structural properties and the non-finite axiomatizability*, Lecture Notes in Math., **834**, 381–410.

Institut Girard Desargues, Université de Lyon-1, Mathématiques (bâtiment 101), 43, boulevard du 11 novembre 1981, 69622 Villeurbanne cedex, France

E-mail: poizat@lan1.univ-lyon1.fr

DIFFERENTIAL ALGEBRAIC GEOMETRY AND DIFFERENTIAL ALGEBRAIC GROUPS: FROM ALGEBRAIC DIFFERENTIAL EQUATIONS TO DIOPHANTINE GEOMETRY

ALEXANDRU BUIUM AND PHYLLIS J. CASSIDY

CONTENTS

1991 *Mathematics Subject Classification.* 01, 12H, 03, 11G

Alexandru Buium is partially suppported by NSF Grant #DMS 9500331 .

1

0. Introduction

The object of algebraic geometry is the geometric study of solution sets of algebraic equations. Just as algebraic equations are special cases of algebraic differential equations, algebraic geometry may be viewed as a special case of the more general geometry, called differential algebraic geometry, of solutions sets of algebraic differential equations. This "new geometry" turns out to be related in deep ways to a great variety of mathematical topics, ranging from analytic function theory, simple group theory, and mathematical physics, to mathematical logic, and, more recently, Diophantine geometry.

The history of differential algebraic geometry can be said to have started in 1932, when Joseph Fels Ritt [1893-1951], published, in the Colloqui ım series of the American Mathematical Society, *Differential Equations from the Algebraic Standpoint.* A complex analyst, who competed strenuously with Julia and Fatou for the prize offered by the French Academy for work on the iteration of rational functions [1918], [1920], [1923a], [1923b], Ritt had a life-long interest in the properties of complex functions. He immersed himself in the literature of the eighteenth and nineteenth centuries that was concerned with the "transcendents" defined by algebraic differential equations with rational function coefficients. This concentration on what would be called "rationality questions" in algebraic geometry, which drew him to the work of Lagrange, Laplace, Liouville, Picard, Painlevé, and Drach, led Ritt to the "new algebra" of Noether and van der Waerden. He patterned his approach to differential equations theory, which Kolchin named "differential algebra," on algebraic geometry, eschewing the transcendental methods of Lie. In his paper commemorating the semicentennial year of the American Mathematical Society, Ritt wrote:

> For want of proper algebraic viewpoints, questions which date from the beginnings of analysis remained without systematic treatment. The problem of the number of arbitrary constants in the solution of a system failed to receive even a sound formulation. The most striking of all is perhaps that in the literature on singular solutions. The greatest source of light on the nature of the singular solutions of an algebraic differential equation is probably the fact that the solutions separate, in a unique manner, into collections analogous to irreducible manifolds. [1938]

In his second Colloquium publication, *Differential Algebra*, Ritt, looking back, commented on the influences on his work:

> When I began to work on algebraic differential equations, early in 1930, van der Waerden's excellent *Moderne Algebra* had not yet appeared.

> However, Emmy Noether's work of the twenties was available. The form in which the results of differential algebra are presented has been deeply influenced by her teachings. She was a prime mover of our period, who, in continuing Julius König's development of Kronecker's ideas, brought mathematics to know algebra as it was never known before. [1950a]

Although Ritt, "fascinated by the elemental idea from which springs all of contemporary algebra - - the notion of structure" [LORCH 1951], devoted the greater part of his life to algebraicizing differential equations theory, his life blood was classical analysis. "The 'modern' exploratory spirit in algebra, topology, or linear spaces was not for him" [LORCH 1951]. A beginning of the complete algebraization of differential algebra was made by Ritt's students Raudenbush and Levi. However, it was his student Ellis Kolchin, deeply influenced by the " modern exploratory spirit" of Weil and Chevalley, who completed the task. Without deviating from the central philosophy of his teacher, whom he called differential algebra's "principal prophet and practitioner" [1973], Kolchin deepened and modernized differential algebra, and developed differential algebraic geometry and differential algebraic groups.

> It has been said that every mathematician (with the possible exception of Gauss) has stood on the shoulders of his predecessors. In my case, the shoulders were those of three giants: J. F. Ritt, C. Chevalley, and A. Weil. [KOLCHIN 1985]

It was Kolchin's work that led to the approach to diophantine geometry through differential algebraic geometry; a phenomenon, unanticipated by Ritt, that went beyond and even reversed the application of algebro-geometric methods to differential equations theory. The papers by Armand Borel and Michael Singer in this volume describe Kolchin's contributions to differential Galois theory, and the important recent work, part of his legacy, on the inverse problem and the calculation of Galois groups. Here, in this paper, we present Kolchin's "new geometry" in the context of the evolution of differential algebra and explore its applications to diophantine geometry. We leave untouched, reluctantly, the applications of differential algebra to nonlinear control theory, and refer the reader to the excellent expository paper [FLIESS, GLAD 1993].

In 1885, Poincaré described the work of Painlevé in these words:

> Les mathématiques constituent un continent solidement agencé, dont les pays sont bien relié les uns aux autres; l'oeuvre de Paul Painlevé est une île originale et splendide dans l'océan voisin.

This observation had its counterpart some 70 years later, in the oft-quoted sentence of Kaplansky's beautiful, and concise introduction to differential algebra [1957]: "Differential algebra is easily described: it is (99 percent or more) the work of Ritt and Kolchin." But, to paraphrase Hadamard's answer to Poincaré: recently, bridges and roads have been constructed that lead from the island of differential algebra to the continent of contemporary mathematics—both to neighboring territories and to more distant lands. In this paper, we will explore both the island and some of these roads.

1. Differential Algebraic Geometry

1.1. **The Language of Differential Algebraic Geometry.** The fields of rational, real and complex numbers will be denoted by $\mathbb{Q}, \mathbb{R}$ and $\mathbb{C}$, respectively. All rings will be commutative, associative, with unity 1, and will have characteristic 0. For details, see [KOLCHIN 1973].

The central notion in differential algebra is that of *differential ring.* Kolchin posits the existence of a distinguished set $\Delta = \{\delta_1, ..., \delta_m\}$ of derivation operators and forms the commutative monoid Θ generated by it. A ring R on which Δ acts is called a *differential ring* (or $\Delta - ring$). Thus,

$$\delta(\eta + \zeta) = \delta\eta + \delta\zeta, \text{and } \delta(\eta\zeta) = \eta\delta\zeta + \zeta\delta\eta \,\text{for all } \delta \in \Delta, \eta, \zeta \in R.$$

$$\delta_i \circ \delta_j = \delta_j \circ \delta_i, \; i, j = 1, ..., m.$$

This defines on R the distinguished Δ-structure. Unless otherwise indicated, the set Δ is kept fixed during any discussion, and the adjective *differential* indicates the Δ-structure. For example, an ideal I in R is a *differential ideal* if I is stable under the operation of Δ, i.e., if η is in I and δ is in Δ, then $\delta\eta$ is in I. The quotient ring R/I has defined on it a natural structure of differential ring. If I is any ideal of R, and S is any subset of R, an ideal central in the discussion of singular solutions of algebraic differential equations, is the ideal $I : S$, called a *colon ideal,* of all elements r in R such that for all $s \in S$, sr is in I. If S consists of a single element s, we simply write $I : s$. The ideal $\bigcup_{n \in N} I : s^n$ is denoted by $I : s^\infty$. I is clearly contained in $I : S$ and in $I : s^\infty$, and if I is a differential ideal, then, so are $I : s$ and $I : s^\infty$. The set of all elements c in a differential field F such that $\delta c = 0$ for all $\delta \in \Delta$ is a differential subfield of F called the *field of constants* of F.

We use Kolchin's notation $\Theta(s)$ to denote the elements of Θ of order less than or equal to s.

If R is a differential subring of a differential ring S, and $(\eta_i)_{i \in I}$ are elements of S, the *differential algebra* $R\{(\eta_i)_{i \in I}\}$ *differentially generated* over R by the family $(\eta_i)_{i \in I}$ is the algebra $R\left[(\theta\eta_i), \theta = \delta_1^{e_1} \cdots \delta_m^{e_m}, i \in I\right]$, generated over R by the members of the family and their higher derivatives. Note that if I is finite, this differential algebra is finitely generated over R as a differential algebra, but can be infinitely generated as an algebra. If R and S are differential fields, the quotient field of $R\{(\eta_i)_{i \in I}\}$, denoted by $R\left\langle(\eta_i)_{i \in I}\right\rangle$, is called the *differential field extension of R differentially generated by $(\eta_i)_{i \in I}$*. A homomorphism φ from a differential ring R to a differential ring S is a *differential homomorphism if* $\varphi \circ \delta = \delta \circ \varphi$. If R_0 is a common differential subring of R and S and the homomorphism φ leaves every element of R_0 invariant, it is said to be *over* R_0. If, in addition R is an integral domain and S is a field, φ is called a *differential specialization* of R into S.

Early work on differential fields was done by Vessiot in 1892, who treated base fields in differential Galois theory that were not restricted to the field of rational functions. Although Vessiot's base fields always contained $\mathbb{C}(x)$, he allowed them to be arbitrary fields of functions, closed under differentiation. Later, Reinhold Baer [1927], studied abstract differential fields equipped with a single derivation operator, and obtained necessary and sufficient conditions for the existence of a differential field with a given field of constants. He also showed that any differential field can be extended to a differential field that is closed under integration, a first step toward the concept of a differential closure. For Ritt, differential fields were subfields of the field of functions meromorphic on a region of complex $m-$ space, and he tended to fall back on function-theoretic techniques. The replacement of differential fields of functions, equipped with the standard commuting derivations of analysis, by abstract differential fields whose elements need not be functions, was a crucial one for the development of differential algebra. It is interesting to note, though, that the fact that Kolchin, and earlier, Raudenbush and Levi were able to give direct algebraic proofs of theorems proved by Ritt by analytic methods, is partly explained by Seidenberg's analogue of the Lefschetz Embedding Principle:

> A differential field equipped with m commuting derivation operators and finitely differentially generated over the field $\mathbb{Q}$ of rational numbers is differentially isomorphic to a differential field of meromorphic functions of m complex variables. [SEIDENBERG 1958] and [1969]

We begin differential algebraic geometry with a base Δ - field F, which, if F is finitely differentially generated over $\mathbb{Q}$, can be taken to be a field of meromorphic functions of m complex variables. In particular, if F is the field of all functions meromorphic in a region of complex m-space, the field of functions meromorphic in a subregion is an extension differential field of F; the field of algebroid functions on $\mathbb{C}^m$ is a differential extension field of the differential field of all meromorphic functions on $\mathbb{C}^m$ (for the case $m = 1$, see, for example, [LAINE 1993]). Kolchin creates *differential affine n-space* $A_n(E)$, E a differential extension field of F, by defining on affine $n-$ space of algebraic geometry what we will call the *Kolchin topology.* In the analytic case, the $n-$ tuples in $A_n(E)$ are meromorphic parametrizations of a portion of complex n - space. The closed sets in the Kolchin topology on differential affine n - space are interpretations of the solution sets of algebraic differential equations (whose left hand sides are somewhat arbitrary functions of the independent variables and polynomial functions of the dependent variables and their derivatives). We must make this more precise.

The left hand sides of algebraic differential equations are *differential polynomials* with coefficients that lie in the base differential field F. The analogue in differential algebra of algebraic independence in commutative algebra and analytic independence in analysis is that of *differential algebraic independence*, which was first formulated by Ernest Vessiot. A family $(\eta_i)_{i\in I}$ of elements of a differential extension field E of F is called *differentially algebraically independent* over F (or a family of *differential indeterminates*) if the family $(\theta\eta_i)_{\theta\in\Theta,i\in I}$ is algebraically independent over F. An element η of a differential extension field of F that is differentially algebraically independent over F is called *differentially transcendental* over F (otherwise, *differentially algebraic* over F). For $F = \mathbb{C}(x)$, x a complex variable, η is called transcendentally transcendental in the classical literature. The Γ function is an example of a transcendentally transcendental function. As we will see later, when we distinguish the ordinary from the partial case in the section on differential closures, this definition differs from that of A. Ostrowski [1920], which implies that the algebraic transcendence of $F\langle\eta\rangle$ over F is finite. A differential extension field E of F is called *differentially algebraic over* F if all its elements are differentially algebraic over F. By Zorn's Lemma, there exists a family $(\eta_i)_{i\in I}$ of elements of E that are differentially algebraically independent over F and which satisfies the property that E is differentially algebraic over $E\left\langle(\eta_i)_{i\in I}\right\rangle$. The family is called a *differential transcendence basis of* E over F. All differential transcendences bases of E over F have the same cardinality, called the *differential transcendence degree* of E over F. If $y_1, ..., y_n$ are differential indeterminates, the differential ring $F\{y_1, ..., y_n\}$ is called the *differential polynomial ring (*over F) in n differential indeterminates. Its quotient field is called the field of *differential rational functions over* F. The differential transcendence degree over F of the field of differential rational functions over F is n. Let E be a differential extension field of F and let $\eta = (\eta_1, ..., \eta_n)$ be in E^n. The substitution homorphism, which maps θy_i to $\theta \eta_i$, $\theta \in \Theta, i = 1, ..., n$, is a differential homomorphism over F from $F\{y_1, ..., y_n\}$ into E. We write $f(\eta_1, ..., \eta_n)$, or, simply, $f(\eta)$ for the value of f at η. (This notation, first used by Vessiot, differs from the classical notation in differential equations theory; it assumes, implicitly, that $\theta \eta_i$ is substituted for θy_i whenever the latter appears in f.) Thus, if E is any differential extension field of F, we identify the differential polynomial f with a polynomial function on the space of infinite-tuples $(\theta \eta_i)_{\theta \in \Theta, i \in I}, \eta_i \in E$ or, equivalently, from the point of view of differential algebra, as a differential polynomial function on E^n. If $f(\eta) = 0$, we call η a zero of f. A subset V of E^n is Kolchin F - closed if V is the set of zeros in E^n of a family $(f_i)_{i \in I}$ (which may be infinite) of differential polynomials in $F\{y_1, ..., y_n\}$. Clearly, every Zariski F-closed set is also Kolchin F-closed. It is clear that a zero of the family of differential polynomials is also a zero of all elements of the differential ideal generated by the family, and so, we speak somewhat loosely of the set of zeros of a differential ideal in the differential polynomial ring.

In his desire to create an elimination theory for algebraic differential equations, one of Ritts' first tasks was to decide how he would interpret the classical terminology in the particular case when the differential equations were algebraic. In 1894, Tresse, for example, notes that if a manifold satisfies a system of partial differential equations between the (independent and dependent variables), it satisfies also all the equations obtained from them by differentiation. One gets, of course, an infinity of equations. He asks whether every system is equivalent to (has the same solutions as) a finite system. He goes on to state and attempt to prove two theorems.

1. *A system of partial differential equations, defined in any manner whatsoever, is finite, i.e., there exists a finite order s such that all the equations of order higher than s in the system can be deduced by simple differentiations from equations of order less than or equal to s.*
2. *Given an arbitrary system of partial differential equations, one can, with a finite number of differentiations and eliminations either show that they are incompatible or put them in the form of a completely integrable system, whose general solution depends on either arbitrary functions or constants.*

1.2. **The Basis Theorem.** If $(\eta_i)_{i \in I}$ is a family of elements of a differential ring R, the differential ideal $[(\eta_i)_{i \in I}]$ generated by the family is the ideal generated by the family $(\theta \eta_i)_{\theta \in \Theta, i \in I}$, consisting of the η_i and their derivatives. If the indexing set is finite, we say that the *differential* ideal is *finitely generated,* or has a *finite basis.* Tresse's observation is captured in differential algebra by the statement that if $(f_i)_{i \in I}$ is a family of differential polynomials, then the zero set of the family equals the zero set of the differential ideal generated by the family. The key to an elimination theory for algebraic differential equations is to prove an analogue of Hilbert's basis theorem for algebraic equations. Tresse's first theorem translates into an analogue of Hilbert's basis theorem for differential ideals. Unfortunately, Ritt gives a counterexample which shows that Tresse's theorem was too optimistic, and there is no exact analogue of Hilbert's theorem for differential ideals.

Let F be an ordinary differential field with derivation δ, and let R be the differential polynomial ring $F\{y\}$. Write $\delta y = y'$, $\delta^2 y = y''$, ..., $\delta^k y = y^{(k)}$. Ritt shows that the differential ideal generated by $y'y'', y''y''', ..., y^{(k)}y^{(k+1)}, ...$ has no finite basis. [1932, p12].

The Ritt-Raudenbush Basis Theorem, which substitutes radical differential ideals for differential ideals, was first proved by Ritt by analytic methods in [RITT 1932]; later, Raudenbush gave a direct algebraic proof [RAUDENBUSH 1934]. The basis theorem answers Tresse's question affirmatively. The idea of proving a basis theorem came to Ritt from the 1898 dissertation of Jules Drach, who quotes Tresse's second theorem.

A differential ideal J in a differential ring R is called a *radical differential ideal* (Ritt and Kolchin used *perfect differential ideal*) if whenever a positive integer power of an element of R is in the differential ideal J, then the element itself is in J, *i.e.*, the differential ideal J is radical as an ideal. If $(\eta_i)_{i\in I}$ is a family of elements of R, the radical differential ideal generated by the family is the radical of the differential ideal generated by the family, and is denoted by $\{(\eta_i)_{i\in I}\}$; it contains all those ring elements ζ such that some positive integral power of ζ can be written as a linear combination with coefficients from R of a finite subfamily of the family $(\theta\eta_i)_{\theta\in\Theta, i\in I}$ consisting of the η_i and their derivatives. (It is easy (in characteristic 0) to see that the radical of a differential ideal is itself a differential ideal.) If J is a radical ideal of R and s is an element of R, the colon ideal $J : s$ is also a radical ideal. If J is finite, we say that the *radical differential* ideal J is *finitely generated* or has a *finite basis.* (Note that this is not a finite basis of the differential ideal J.)

It is easy to see that any zero of a family $(f_i)_{i\in I}$ of differential polynomials is also a zero of the radical differential ideal it generates. If E is a differential extension field of a differential field F, and V is a subset of differential affine space $A_n(E)$, the set of differential polynomials in the differential polynomial ring $F\{y_1, ..., y_n\}$ vanishing at every element of V is a radical differential ideal and is called the *defining differential ideal of V in $F\{y_1, ..., y_n\}$.*

The Ritt-Raudenbush Basis Theorem. *Let F be a differential field and let $y_1, ..., y_n$ be differential indeterminates. Then every radical differential ideal in $R = F\{y_1, ..., y_n\}$ has a finite basis.* [RITT 1932], [RAUDENBUSH 1934]

Corollary. *The set of radical differential ideals in $F\{y_1, ..., y_n\}$ satisfies the ascending chain condition.*

This theorem, which states that if J is a radical differential ideal in the differential polynomial ring R, then there is a finite family of differential polynomials in J generating it as a radical differential ideal, makes it evident that every system of algebraic differential equations has the same solutions as a finite system, *i.e.*, every Kolchin $F-$ closed set is defined by a finite system of differential polynomial equations. However, this weakened analogue of the Hilbert Basis Theorem for differential ideals is further weakened by the fact that, as Kolchin discussed in his dissertation [1941], an analogue of the notion of the *exponent* of an ideal is lacking.

> In the theory of polynomial ideals, in algebra, there are methods, stemming from the theorem of M. Noether, and associated with the names of E. Bertini, E. Lasker, F.S. Macauley, K. Hentzelt, H. Kapferer and P. Dubreil, for finding the exponent of an ideal, or at least a bound for the exponent.
>
> When one seeks to create a notion of exponent for ideals of differential polynomials, one is forced, because of a situation revealed by H. W.

> Raudenbush [1936], to admit infinite exponents as well as finite ones. [KOLCHIN 1941]

If J is a radical differential ideal in the differential polynomial ring $F\{y_1, ..., y_n\}$, then there is a finite subset $f_1, ..., f_k$ of J such that J is the radical of the differential ideal $[f_1, ..., f_k]$. If there is a positive integer e such that for all elements $f \in J$, $f^e \in [f_1, ..., f_k]$, then e is called the *exponent* of the differential ideal $[f_1, ..., f_k]$. If no such integer exists, we say that $[f_1, ..., f_k]$ has infinite exponent. Raudenbush shows that even in the case where the cardinality of Δ is one and the differential ideal in $F\{y\}$ is generated by a single differential polynomial of order zero, it can have infinite exponent; he gives the very simple example of the differential ideal generated by y^3. Raudenbush remarks, "This result shows, in particular, why the ideal theory of differential polynomials as far as it has been developed is marked by the absence of primary ideals." The situation is markedly different in differential rings that are finitely generated as *rings* over a base differential field. Abraham Seidenberg proved that if R is a noetherian ring containing the field of rational numbers, equipped with an arbitrary set Δ of derivation operators (not necessarily either commuting or finite), each differential ideal has an irredundant finite primary decomposition, where the primary ideals are differential ideals [SEIDENBERG 1967]. He also proved that the prime ideals associated with the given ideal are differential ideals. In his dissertation, Kolchin proves that if the cardinality of Δ is one and a differential polynomial f in $F\{y\}$ has order 0 (is in $F[y]$), then the differential ideal generated by f has exponent 1 or ∞ according to whether f is algebraically irreducible or not. If f in $F\{y\}$ has order 1, the exponent of the differential ideal generated by f is infinity if f has a singular zero of multiplicity greater than 1 (even if f is algebraically irreducible). If the number of differential indeterminates is greater than one, nothing is known. In particular, the nature of the exponent of the differential ideal generated by an algebraically irreducible order zero differential polynomial in $F\{y_1, ..., y_n\}$ has not yet been explored. The problem of infinite exponents of differential ideals is intriguingly linked to the nature of the singular solutions of f.

1.3. **Prime Differential Ideals of Differential Polynomials: The Concept of Irreducibility and the Finite Decomposition of Radical Differential Ideals.** It is interesting that Kolchin began thinking about the "new geometry" in his papers on the differential Galois theory. Some of the first efforts at creating differential algebraic geometry were made at the close of the nineteenth century by Picard, Vessiot, and Drach in attempts to realize Sophus Lie's dream of a Galois theory of differential equations. Picard, in his beautiful chapter XVII of [PICARD 1896] on the Galois theory of a single ordinary homogeneous linear differential equation, in essence, constructs what we would now call a principal homogeneous space for a differential algebraic group. Searching for a definition of irreducibility stronger than Frobenius' irreducibility of linear differential operators, he modifies slightly a definition of irreducibility due to Koenigsberger [KOENIGSBERGER 1889], p. 155. Later, in his thesis of 1898, Drach formulated a definition of irreducibility that was central in his highly original, but flawed attempts to extend the Picard-Vessiot theory to arbitrary systems of ordinary algebraic differential equations. (See the preface of [RITT 1932].) It is not difficult to see that the definitions of Koenigsberger and Drach are equivalent.

Drach-Koenigsberger irreducibility takes its inspiration from the Galois theory of a polynomial equation in one indeterminate. A differential polynomial f is Drach-Koenigsberger irreducible if and only if it is algebraically irreducible and the radical differential ideal it generates is a maximal differential ideal, *i.e.*, every differential polynomial equation satisfied by one solution is satisfied by all solutions. Using Tresse's first theorem, which provides a finite basis for any system of differential equations, Drach shows that by adjoining differential equations that are compatible with the basis, he can reach an irreducible system. In the language of differential algebra, a radical differential ideal in a differential polynomial ring is contained in a maximal differential ideal. So, the "atomic" ideals for Drach were maximal differential ideals. In the polynomial ring $F[y]$ in one indeterminate, maximality and primality are the same. A radical ideal in $F[y]$ decomposes into a *finite* intersection of maximal ideals. This is no longer true in the polynomial ring in more than one indeterminate, and certainly fails in all differential polynomial rings.

It is easy to prove that maximal differential ideals in a differential ring containing the field of rational numbers are prime (be careful, they are not in general maximal *ideals*). For Ritt, following the lead both of Noether and van der Waerden in algebraic geometry and Lagrange and Poisson in their analysis of the singular solutions of a single ordinary algebraic differential equation, and differing from Koenigsberger and Drach, the atomic differential ideals are prime differential ideals. The zero sets of prime differential ideals are called *irreducible.*

Theorem. *Every radical differential ideal in the differential polynomial ring $F\{y_1, ..., y_n\}$ is a finite intersection of prime differential ideals, none of which contains any other (an irredundant decomposition).* [RITT 1932], [RAUDENBUSH 1934]

The set of prime differential ideals in the decomposition is unique, and they are the distinct minimal prime differential ideals containing the given radical differential ideal. Ritt called the prime differential ideals in the decomposition the $F-$ *components* of the radical differential ideal.

A differential polynomial f in the differential polynomial ring $R = F\{y_1, ..., y_n\}$ is *algebraically irreducible* if it is irreducible as an element of the integral domain R. The following example shows that the radical differential ideals generated by algebraically irreducible differential polynomials are not necessarily prime.

Example. Clairaut's Equation [1734].

Let F be an ordinary differential field, with derivation operator δ, containing an element x such that δx is 1. Denote δy by y', and $\delta^2 y$ by y''.

$$y = xy' - \frac{1}{4}y'^2$$

$$f(y) = y - xy' + \frac{1}{4}y'^2.$$

The differential polynomial f is algebraically irreducible in $F\{y\}$, for any base field F, yet the radical differential ideal I it generates is not prime. For, it contains the derivative $\delta f = -(x - \frac{1}{2}y')y''$. Neither factor is in I, and the radical differential ideals $\{f, x - \frac{1}{2}y'\}$ and $\{f, y''\}$ are prime. It is not difficult to see that I is the intersection of these prime differential ideals.

1.4. The Differential Nullstellensatz. The Ritt-Raudenbush basis theorem does not answer the question, implicit in Tresse's second theorem, of whether, given a system of algebraic differential equations, one can complete it in such a way that all algebraic differential equations that are satisfied by the solutions of the first system are "algebraic consequences of the complete system", and whether this complete system is equivalent to a finite system. This question is answered by the *Differential Nullstellensatz,* the analogue for differential algebraic geometry of Hilbert's Theorem of Zeros. It was first proved in [RITT 1932] for differential polynomials with meromorphic coefficients, and later in [RAUDENBUSH 1934] for arbitrary coefficient differential fields.

The Differential Nullstellensatz. *Let F be a differential field and let $y_1, ..., y_n$ be differential indeterminates. Let I be a radical differential ideal in $R = F\{y_1, ..., y_n\}$ different from the unit ideal. Then I has a zero in some differential extension field of F, and if f is a differential polynomial in R that vanishes on the set of all zeros of I, then f is in I.*

Corollary. *Let F be a differential field and let $y_1, ..., y_n$ be differential indeterminates. Let I be a radical differential ideal in $R = F\{y_1, ..., y_n\}$. Then I has no zero in a differential extension field of F if and only if 1 is in I.*

This important corollary of the Differential Nullstellensatz states that a system of differential polynomial equations

$$f_1(y_1, ..., y_n) = 0$$
$$\dots$$
$$f_k(y_1, ..., y_n) = 0$$

with coefficients in F has no zero in a differential extension field of F if and only if 1 can be written as a linear combination

$$1 = \sum_{\theta \in \Theta(s), 1 \leq i \leq k} A_{\theta,i} \theta f_i, \; A_{\theta,i} \in R.$$

Of course, there is no bound on s.

It follows from the differential Nullstellensatz that, given any system of differential polynomials, there is a finite system of differential polynomials with the same zeros, such that any differential polynomial that vanishes on the zeros has some positive integral power in the differential ideal generated by the finite system. (This is weakened somewhat by the lack of finite exponents for differential ideals.) In particular, the original system is a subset of the radical differential ideal generated by the finite system. The differential polynomials in the radical differential ideal generated by a system of differential polynomials are Ritt's interpretations of the vague classical terminology "algebraic consequences of the given system." In Ritt's language, the radical differential ideal generated by $y - xy' + \frac{1}{4}y'^2$ is the defining differential ideal of the manifold consisting of all solutions of the Clairaut equation in all finitely generated differential extension fields of F. The radical differential ideal is the largest set of differential polynomials equivalent to the given system.

Ritt's proofs of the basis theorem and the Nullstellensatz are inextricably intertwined. The basis theorem and application of the so-called "Trick of Rabinowitsch", which greatly simplified the algebraic Nullstellensatz, reduced the differential Nullstellensatz to the problem of determining the insolvability of a finite system of differential polynomials by writing the unity 1 as a linear combination of the differential polynomials in the system and a finite number of their derivatives. Ritt, in Chapter VII of [RITT 1932], and Richard Cohn, in 1941, gave proofs that are constructive if factorization over the coefficient field can be effectively carried out. In 1956, Seidenberg, in his influential paper "An Elimination Theory for Differential Algebra" [1956], used elimination theory to give the first constructive proof of the Nullstellensatz.

1.5. **Universal Differential Fields.** The "new geometry" created by Ritt and Kolchin requires the coordinates of the zeros of differential polynomials to lie in differential extension fields of a base coefficient field. As we mentioned earlier, for Ritt the base field was a differential subfield F of the field of functions meromorphic on a region Ω of complex $m-$ space. This is still the case today for the complex analysts who combine the techniques of Nevanlinna theory and differential algebra [LAINE 1993]. Ritt defined an *analytic zero* of a differential polynomial with coefficients in F to be a pair, consisting of a tuple of functions analytic in a subregion Ω' of Ω, together with the region Ω', with the tuple a zero of the given differential polynomial. Such a tuple generates a finitely differentially generated differential extension field E of F. Ritt then proves that every radical differential ideal I in $F\{y_1, ..., y_n\}$ has an analytic zero, and if a differential polynomial f in $F\{y_1, ..., y_n\}$ vanishes at every analytic zero of I, it must lie in I. This global sheaf-theoretic approach is in the spirit of Painlevé, who, according to [GARNIER 1972], initiated the global study of differential equations in the complex domain. Garnier writes "Jusqu'alors les disciples de Cauchy, comme Briot et Bouquet, se bornaient à une étude locale des solutions des équations différentielles, pour en déduire-abusivement-une vue d'ensemble sur ces solutions." As Ritt remarks, although each analytic zero in his sense generates a finitely differentially generated differential extension field of the base differential field, there is no "universal" differential field containing all of them; there is no set of all finitely differentially generated differential field extensions of a differential field F. Since, then, the manifold of zeros of a differential ideal is not a set, this leads to awkwardness. It is very useful to have a universal differential field for differential algebraic geometry, a differential field with "enough solutions.." This is what Kolchin set out to create. He first constructed a universal differential field extension of a differential field in connection with differential Galois theory [1953]. He writes, "The use of a universal extension, which follows the now well-known procedure of modern algebraic geometry, makes it possible to avoid certain logical difficulties connected with phrases like 'the set of all extensions'."

The essential requirement of a universal differential extension field U of F is that if E is any finitely differentially generated differential field extension of F in U, and E_1 is any finitely differentially generated differential field extension of E, then it can be embedded over E in U. This will ensure that every prime differential ideal P in $F\{y_1, ..., y_n\}$ has a generic zero in $A_n(U)$, *i.e.*, a zero $\eta \in A_n(U)$ such that a differential polynomial in $F\{y_1, ..., y_n\}$ that vanishes at η is in the prime differential ideal; the substitution homomorphism induces a differential isomorphism over F from the differential ring $F\{y_1, ..., y_n\}/P$ to $F\{\eta_1, ..., \eta_n\}$. It follows that the differential Nullstellensatz holds in U, *i.e.*, in its hypothesis (and that of the corollary) we may replace "zero in some differential extension field of F" by zero in U.

Until recently, the underlying differential field of differential affine space was U^n, although every radical differential ideal in $U\{y_1, ..., y_n\}$ has a basis in a differential field F that is finitely differentially generated over $\mathbb{Q}$ [KOLCHIN 1973]. If F has the property that the radical differential ideal has a basis with coefficients in F, then F is called a *field of definition* of the radical differential ideal. The smallest field of definition is finitely differentially generated over $\mathbb{Q}$. For a complete discussion of universal differential fields, see [KOLCHIN 1973].

1.6. **Differentially Closed Differential Fields.** Universal differential fields are large. Even though Kolchin proved in [KOLCHIN 1974a] that if F is any differential field, then F has a universal differential extension field U with the same cardinality, it is nonetheless true that U has infinite differential transcendence degree over F, and the field of constants of U has infinite transcendence degree over the field of constants of F. Because of the size of U relative to F, and the number of transcendental constants of U, it is natural to wonder if there is an analogue in differential algebra of algebraically closed fields and algebraic closures in algebra. If so, it seems reasonable to expect that the differential Nullstellensatz would hold in such fields. Surprisingly, the affirmative answer came not from differential algebra, but, instead, from logic, in the 1959 paper, "On the Concept of a Differentially Closed Field," by Abraham Robinson. In this paper, which initiated the intriguing theory of differentially closed fields, Robinson demonstrated that the theory of ordinary differential fields has a model completion. We will concentrate here on the theory as inspiration for new insights in differential algebraic geometry. For the history of the remarkable and exciting confluence of model theory and differential algebra initiated by Robinson's work, we direct the reader to Poizat's 1995 paper, reprinted in this volume.

In 1956, Seidenberg gave an algorithm for deciding whether the system

$$f_1(y_1, ..., y_n) = 0, ..., f_s(y_1, ..., y_n) = 0, g(y_1, ..., y_n) \neq 0,$$

where $f_1, ..., f_s, g$ are differential polynomials with coefficients in F, has a solution in some differential extension field of F. Although he divides the proof into separate parts for the ordinary and partial cases, Seidenberg allows the cardinality of Δ to be an arbitrary positive integer.

The Seidenberg Elimination Theorem. *Given a finite system*

1 $f_1(a_1, ..., a_N, y_1, ..., y_n) = 0, ..., f_s(a_1, ..., a_N, y_1, ..., y_n) = 0,$ $g(a_1, ..., a_N, y_1, ..., y_n) \neq 0,$ *where* $f_1, ..., f_s, g$ *are in the differential polynomial ring* $\mathbb{Z}\{a_1, ..., a_N, y_1, ..., y_n\}$ *with integer coefficients, there are a finite number of finite systems*

1′ $f_{i1}(a_1, ..., a_N) = 0, ..., f_{is_i}(a_1, ..., a_N) = 0, g_i(a_1, ..., a_N) \neq 0,$*where* $f_{i1}, ..., f_{is_i}$, g *are in* $\mathbb{Z}\{a_1, ..., a_N\}$, *such that for any differential field* F *and for any* $\bar{a}_1, ..., \bar{a}_N$ *in* F, *the system*

$\bar{1}$ $f_1(\bar{a}_1, ..., \bar{a}_N, y_1, ..., y_n) = 0, ..., f_s(\bar{a}_1, ..., \bar{a}_N, y_1, ...y_n) = 0, g(\bar{a}_1, ..., \bar{a}_N,$ $y_1, ..., y_n) \neq 0$ *has a solution in some differential extension field of* F *if and only if there is an* i *such that* $(a_1, ..., a_N)$ *is a solution of* $(1')$. *Moreover, there is an algorithm for producing the finite systems* $(1')$. [SEIDENBERG 1956]

We should note that under the assumption that F is closed under the extraction of pth roots, Seidenberg proved an elimination theorem in characteristic p. In 1973, Carol Wood developed a theory of differentially closed fields in characteristic p.

See [WOOD 1973], [WOOD 1976]. This theory is complicated by the fact that pth powers of elements of differential fields are constants.

The Seidenberg Elimination Theorem implies the following analogue in differential algebra of the Chevalley lifting theorem in commutative algebra. It was proved independently by Azriel Rosenfeld in 1959.

The Lifting Theorem. *Let S be a differential integral domain, let R be a differential subring over which S is finitely differentially generated, and let u be a nonzero element of S. Then there exists a nonzero element a in R such that for every prime differential ideal P of R, with a not in P, there is a prime differential ideal Q of S such that*

$$Q \cap R = P, \text{ and } u \notin Q.$$

[RITT 1940], [KOLCHIN 1973], p.140

Seidenberg's Elimination Theorem implies the differential Nullstellensatz; moreover, it shows that if h vanishes at every zero of $f_1, ..., f_s$, the smallest positive integer p such that h^p is in the differential ideal generated by $f_1, ..., f_s$ is effectively computable. However, as we remarked earlier, the fact that differential ideals can have infinite exponents means that p depends on h as well as on $f_1, ..., f_s$.

Robinson, following Seidenberg's lead, defines a differentially closed field to be a differential field F such that if system (1) with coefficients in F has a solution rational over a differential extension field of F, then it has a solution rational over F. It follows immediately that the field of constants of a differentially closed field is algebraically closed. Using the language and techniques of model theory, he also gives a constructive proof of the differential Nullstellensatz. He proves, also, the fundamental theorem that every ordinary differential field can be extended to a differentially closed field. In this way, Robinson invents for differential algebra (at least for the theory of ordinary differential fields) the analogue of algebraically closed fields [DAUBEN 1995], p.273. The analogy between differentially closed fields and algebraically closed fields is not perfect. Differentially closed fields can have proper differentially algebraic extensions. That this is true was the reason why Kolchin objected to the name, and called such fields "constrainedly closed." Our second example illustrates the complexity introduced when algebraic equations are replaced by differential equations. Algebraically closed fields are characterized up to isomorphism by the characteristic and the transcendence degree (over the prime field). This implies, in particular, that for every uncountable cardinal κ there is a unique algebraically closed field up to isomorphism. In contrast, if κ is uncountable, there are 2^κ non-isomorphic differentially closed differential fields of characteristic zero that are of cardinality κ [SHELAH 1973]. Recently, Ehud Hrushovski and Z. Sokolovic [1994], proved that there are continuum many non-isomorphic countable differentially closed fields of characteristic zero, confuting a conjecture of Lascar [1980]. For a new proof, see [PILLAY 1996]. Using the theory of definable groups (differential algebraic groups) and elliptic curves, Pillay is able to avoid both deep Zariski geometry and the properties of jet bundles of algebraic groups.

What is missing from Robinson's paper is the proof that every differential field F has a differential closure $\hat{F}$ - - a differential extension field that can be embedded in every differentially closed differential extension field of F - - the analogue of the algebraic closure of a field. The existence of such a "prime" differentially closed field was proved by Lenore Blum in her dissertation of 1968. Blum also simplified

Robinson's definition of a differentially closed ordinary differential field. In her clear and lively exposition in [BLUM 1977], Blum describes Robinson's definition as being like *defining* a field F to be algebraically closed if and only if every finite consistent system of equations and inequations with coefficients in F has a solution rational over F. Blum gives a simple axiom defining a differently closed ordinary differential field, and shows that Robinson's definition is equivalent to it. Her axiom, which closely resembles the axiom defining an algebraically closed field, is the definition of choice in the literature today. Unfortunately, there is as yet no such simple, elegant axiom characterizing differentially closed partial differential fields; for the moment, Seidenberg's differential elimination theorem will have to suffice.

Blum's Axiom. *Let F be a differential field equipped with a single derivation operator δ. Let f and g be differential polynomials in a single differential indeterminate, with coefficients in F, and with the order of g greater than or equal to -1 and less than the order of f. Then there exists in F a zero of f that is not a zero of g.*

The beauty of the axiom is that it reduces the question of the solvability of differential equations in several unknowns to the case of a single unknown. In [SACKS 1972], Gerald Sack writes, "The reduction of several variables to one is a general phenomenon discovered by Blum, and applies to all model completions of universal theories." Since the theory of partial differential fields is a universal theory, reduction of the solvability question to one variable should be possible.

Theorem. *Let F be a differential field equipped with a single derivation operator δ. Then F is differentially closed (in the sense of Robinson) if and only if F satisfies Blum's axiom.* [BLUM 1977]

1.7. **Differential Closures.** Although Robinson proved the existence of differentially closed fields, he seemed to question the existence of "prime models of differentially closed fields"- - the analogues of algebraic closures. A differentially closed differential extension field $\hat{F}$ of a differential field F is called a *differential closure* of F if every differential isomorphism from F into a differentially closed field extends to $\hat{F}$. In her thesis, Blum proved the existence and conjectured the uniqueness up to differential isomorphism of a differential closure of an ordinary differential field F. In 1972, Saharon Shelah proved the crucial property of uniqueness by establishing a remarkable criterion that holds for any "complete first-order totally transcendental theory", and which informed Kolchin's characterization of differential closures in the language of differential algebra. (Blum had earlier proved that the theory of differentially closed fields is totally transcendental.) But, there remained unproved for differential closures the minimality property of algebraic closures, which holds also for real closures and Henselizations, and which seems an almost obvious consequence of the definition. The differential closure $\hat{F}$ of F is *minimal* if there is no differentially closed differential extension field of F properly contained in $\hat{F}$. Sacks conjectured the minimality of differential closures [SACKS 1972]. In 1973-1974, Shelah [1973], Rosenlicht [1974], and Kolchin [1974a] independently found counterexamples. Indeed, if C is a field of constants, its differential closure has an infinite strictly decreasing sequence of constrained closures of C.

Lenore Blum, in an extended correspondence with Ellis Kolchin in 1972, took him on what he referred to in his 1974 paper on constrained extensions as a private guided tour of the work of Robinson, Blum, and Shelah. This inspired him to write the paper, which became in differential algebra the definitive paper on differentially closed fields, and one that he very much enjoyed writing. In this paper, he was in his element, as he re-cast the proofs of the existence of differentially closed fields and the existence and uniqueness of differential closures, in the language of differential algebra, extending these theorems to partial differential fields, and presenting a counterexample to the minimality of differential closures in the ordinary case. (This extension to the case of partial differential fields was typical of his work. Kolchin always insisted that the proofs in differential algebra should, whenever possible, not distinguish between ordinary and partial differential fields.) In addition, he proved a theorem that indicates that the proper context for any differential Galois theory over a differential field F, either for algebraic or differential algebraic groups, is the differential closure of the ground field F. If E is a strongly normal extension of F (for a definition of strongly normal extensions of F, see Borel's paper in this volume), whether finitely differentially generated or not, then E can be embedded over F in the differential closure of F. Moreover, a differential extension field of E is a differential closure of E if and only if it is a differential closure of F.

In characterizing differential closures, Kolchin, in the spirit of Shelah, focused his attention on a certain property shared by elements of differential closures, called *isolated* by the logicians. Let F be a differential field equipped with an arbitrary finite set Δ of commuting derivation operators. Let $\eta = (\eta_1, ..., \eta_n)$ and $\zeta = (\zeta_1, ..., \zeta_n)$ be finite families of elements of a differential extension field of F. We say that η *specializes differentially* over F to ζ if every differential polynomial in $F\{y_1, ..., y_n\}$ that vanishes at η also vanishes at ζ. In other words, the defining differential ideal of η in $F\{y_1, ..., y_n\}$ is contained in that of ζ. This implies that there is a differential homomorphism over F from the differential algebra $F\{\eta\}$ onto $F\{\zeta\}$, mapping η_i onto $\zeta_i, i = 1, .., n$. The differential specialization is *generic* if η and ζ have the same defining differential ideals in $F\{y_1, ..., y_n\}$, *i.e.*, the above homomorphism is an isomorphism. η is *constrained* over F if there is a differential polynomial c in $F\{y_1, ..., y_n\}$, with $c(\eta) \neq 0$, such that $c(\zeta) = 0$ whenever ζ is a non-generic differential specialization of η over F. It is not difficult to see that η is constrained over F if and only if $\eta_1, ..., \eta_n$ are constrained over F. A differential extension field E of F is *constrained* over F if all its elements are constrained over F. It follows immediately from the definition that E is differential algebraic over F, and that its field of constants is algebraic over the field of constants of F. In particular, if the field of constants of F is algebraically closed, then E introduces no new constants, a fact crucial to the differential Galois theory.

It is interesting to note that if P is a prime differential ideal in $F\{y_1, ..., y_n\}$, and η is a zero of P in $A_n(E)$, E a differential extension field of F that is constrained over F, and if η has constraint c, then the defining differential ideal of η over F is maximal among differential ideals not containing c. Tying constrained extensions of F to differential Galois theory in yet another way, we observe that this is precisely Picard's modification of Drach-Koenigsberger irreducibility.

We remarked earlier that Kolchin did not like Robinson's terminology "differentially closed" since differentially closed differential fields have proper differential algebraic extensions. He defines a differential field to be *constrainedly closed* if it has no proper constrained differential extension fields, and proves that his definition is equivalent to the statement that given a prime differential ideal in $F\{y_1, ..., y_n\}$ and a differential polynomial c in $F\{y_1, ..., y_n\}$ not in the prime differential ideal, there exists a zero η of the prime differential ideal, but not a zero of c, that is rational over F. This shows that for F a constrainedly closed differential field, if V is a Kolchin $F-$ closed subset of differential affine space over a differential field U universal over F, the points of V that are rational over F are dense in V in the Kolchin topology. It also shows that if F is an ordinary field, Kolchin's definition and Robinson's are equivalent. So, following the model theorists, we will call constrainedly closed fields differentially closed, even though Kolchin's terminology reflects better the fact that although a differentially closed field may have proper differential algebraic extensions, it does not have proper constrained extensions.

Kolchin uses the model theory definition of differential closure. So, if F is a differential field equipped with an arbitrary finite set Δ of commuting derivation operators, the differential closure $\hat{F}$ of F satisfies the following properties, all verified by Kolchin in the language of differential algebra:

1. Every finite set of differential equations and inequations with coefficients in $\hat{F}$ that has a zero rational over a differential extension field of $\hat{F}$ has a zero rational over $\hat{F}$ (the differential Nullstellensatz holds in $\hat{F}$).
2. $\hat{F}$ can be embedded over F in any differentially closed differential extension field of F.
3. $\hat{F}$ is constrained over F (all the elements of $\hat{F}$ are differential algebraic over F, with a constraint).
4. The field of constants of $\hat{F}$ is the algebraic closure of the field of constants of F.
5. The cardinality of $\hat{F}$ equals that of F.
6. $\hat{F}$ contains the strongly normal closure of F.
7. If E is a differentially finitely generated differential extension field of F, or is algebraic over F, then every differential isomorphism of E over F into $\hat{F}$ can be extended to a differential automorphism of $\hat{F}$.

We add an eighth property that follows from Seidenberg's embedding theorem and appears in David Marker's article on differentially closed fields [MARKER 1996].

8. Let F be a countable ordinary differential field (in particular, if F is finitely differentially generated over the algebraic closure of the field of rational numbers). Then the differential closure of F is differentially isomorphic over F to a differential subfield of the field of germs of meromorphic functions at the origin.

It seems evident that an analogous theorem holds in the partial case.

If F is an *ordinary* differential field, and if E is a finitely differentially generated constrained differential extension field of F, then E has finite transcendence degree over F. This follows from the fact that every generator satisfies an ordinary polynomial differential equation with coefficients in F. Now, suppose Δ has arbitrary finite cardinality and E has n generators, and let $\eta = (\eta_1, ..., \eta_n)$ be a point in affine $n-$ space over some differentially closed differential extension field of E such that $F\{\eta_1, ..., \eta_n\}$ is differentially isomorphic over F to E. We will see in the section on dimension that the finiteness of the transcendence degree of E translates into the language of differential algebra the statement in analysis that the differential equations defining η can be reduced to ordinary differential equations, *i.e.*, that η "depends on" a finite family of "arbitrary constants", and can be generated by an analog computer [LIPSCHITZ and RUBEL 1987], [POUR-EL and RICHARDS 1979]. Suppose η is a constrained solution of a finite system of partial differential polynomial equations with coefficients in F. Then, finite transcendence degree would mean that we have found a solution of the system that is obtained by solving ordinary differential equations. This is what, for example, the separation of variables technique is designed to do. Now, by the minimality of the part of the locus of the constrained point η that avoids the constraint, it is a natural conjecture that finite transcendence degree over F characterizes finitely differentially generated constrained differential extension fields of the base field F. This would, in particular, indicate the impossibility of realizing infinite-dimensional differential algebraic groups as differential Galois groups. Also, since points constrained over F are dense in the Kolchin $F-$ topology, this would realize the 19th century dream of solving partial differential equations by "reducing" them to ordinary differential equations! But, alas, it is not so.

In 1995, Joseph Johnson, George Reinhart, and Lee Rubel gave an algebraic argument that the first order non-homogeneous linear partial differential equation in one differential indeterminate u :

$$\frac{\partial u}{\partial y} + \left(\frac{x}{y} - 1\right) u = 1,$$

with coefficients in $F = \mathbb{C}(x, y)$, $\mathbb{C}$ the field of complex numbers, x, y complex variables, has no solution η in a differential extension field of F such that $F < \eta >$ has finite transcendence degree over F (this partial differential equation has no analog-computable solutions). We thus have many counterexamples to the conjecture. For more details about functions that generate differential fields of finite transcendence degree over a base differential field of functions, see [OSTROWSKI 1920].

Suppose that F is a differentially closed differential field. The maps

$$\begin{array}{ccl} I & \longmapsto & \text{the set of zeros of } I \\ V & \longmapsto & \text{the defining differential ideal of } V \end{array}$$

between the set of radical differential ideals in the differential polynomial ring $F\{y_1, ..., y_n\}$ and the set of all Kolchin $F-$ closed subsets of $A_n(F)$ are inclusion reversing, and inverse to one another. If F is differentially closed, we will say, simply, Kolchin closed, and we will often call a Kolchin closed set a *differential algebraic variety.* If V is a differential algebraic variety and in addition is Kolchin F_0-closed, where F_0 is a differential subfield of F, we will also say that V is *defined over* F_0.

By definition, irreducible differential algebraic varieties are associated in the above 1-1 correspondence with prime differential ideals. But, we can also describe them topologically, as in algebraic geometry: Irreducible differential algebraic varieties are precisely those that cannot be written as a union of two proper differential algebraic subvarieties. The finite decomposition of a radical differential ideal into prime differential ideals then give us the fundamental theorem:

> Let V be a differential algebraic variety. Then V can be written $V = V_1 \cup ... \cup V_s$, where $V_1, ..., V_s$ are distinct maximal irreducible differential algebraic subvarieties of V, called the components of V. If V is defined over a differential subfield F_0 of F, then so are its components.

We should note that since every strictly increasing chain of radical differential ideals of the differential polynomial ring is finite, it follows that every strictly decreasing chain of differential algebraic subvarieties of a differential algebraic variety is finite, *i.e.*, the Kolchin topology is Noetherian.

1.8. **Ritt's Theory of a Single Algebraic Differential Equation: The Low Power Theorem.** Beginning with the early work of Lagrange, Laplace, and Poisson on singular and general solutions of algebraic differential equations, the later discovery by Painleve of "new transcendents" defined by second order nonlinear algebraic differential equations whose solutions have no movable critical points, and continuing to contemporary work on the Korteweg-de Vries equation, the theory of a single algebraic differential equation has had considerable importance in the classical literature and its applications to physics. The first precise definition of the general solution of an ordinary differential equation was given by Ritt, and the analysis of the distribution of the singular solutions is the most profound work of his early years in differential algebra. We will call the problems surrounding the distribution of singular solutions the *Ritt problem*. A clear exposition of the work on the Ritt problem before 1973, covering both the ordinary and partial cases in an integrated way, is given in [KOLCHIN 1973].

We assume throughout this section that F is a differentially closed field with finite set Δ of commuting derivations operators and set Θ of derivative operators. In his *Colloquium* publication of 1932, Ritt defined on the set of all derivatives θy_i of the differential indeterminates $y_1, ..., y_n$ a total ordering, called a *ranking* , which, if the cardinality of Δ is > 1, is based on the "markings" assigned to the derivatives by C. H. Riquier [RIQUIER 1910], and, which is similar in spirit to a term ordering in computational algebraic geometry. (See, for example, [BUCHBERGER 1985], [GEBAUER and MÖLLER 1988], [ROBBIANO 1985], [BAYER and STILLMAN 1987], [SCHWARTZ 1991].) Giuseppa Carrá - Ferro and William Y. Sit recently studied rankings and term-orderings, using a generalized concept of Dedekind cut, in an article that provides a good historical survey [CARRÁ - FERRO and SIT 1994].

The total ordering must satisfy the additional two conditions

$$u \leq \theta u, \text{ and } u \leq v \Longrightarrow \theta u \leq \theta v.$$

Let $f \in F\{y_1, ..., y_n\}$, $f \notin F$. The highest ranking derivative θy_i appearing in f is called the *leader* of f and is denoted by u_f. If f is an ordinary differential polynomial in one differential indeterminate y, there is a unique ranking, by order, and the leader is the derivative of y of highest order appearing in f. A differential polynomial g is *partially reduced* with respect to f if g is free of all proper derivatives of the leader of f. A partially reduced differential polynomial g is *reduced* with respect to f if $\deg_{u_f}(g) < \deg_{u_f}(f)$. In the expression for f as a polynomial in its leader, the leading coefficient is called the *initial* of f, denoted by I_f, and $\frac{\partial f}{\partial u_f}$ is called the *separant* of f, denoted by S_f. A zero η of f is called *nonsingular* if $S_f(\eta) \neq 0$.

Theorem. *Let f be an algebraically irreducible differential polynomial in $F\{y_1, ..., y_n\}$.*

1. *Among the components of the radical differential ideal $\{f\}$ generated by f, there is one, called the general component of f, that does not contain the separant of f. Each of the other components of $\{f\}$, called the singular components, contains the separant of f.*
2. *The general component of f is the colon ideal $[f] : S_f^\infty$.*
3. *A differential polynomial in the general component of f that is partially reduced with respect to f is divisible by f.*

4. *If P is a singular component of f, there is an algebraically irreducible differential polynomial g in $F\{y_1, ..., y_n\}$ such that P is the general component of g. f involves a proper derivative of the leader of g, and the order of g is less than the order of f.*

[RITT 1932], for ordinary differential polynomials with meromorphic coefficients; [RITT 1945a], for the partial case; [KOLCHIN 1973], for an integrated and much modernized approach.

If V is the set of zeros in $A_n(F)$ of the differential polynomial f, then the components of V are defined by the components of f. Kolchin called the component defined by the general component of f the *general component* of V. Ritt often called it the *general solution* of f. In this, Ritt deviated from much of the classical literature, where the general solution of an ordinary differential equation of order n is an $n-$ parameter family of solutions. We will follow Ritt's usage in this section instead of Kolchin's. We will also use his terminology "singular *solution* of f," in place of singular zero of f.

Part 4 of the theorem tells us that the singular components of the solution set of a first order differential polynomial are *algebraic* varieties, since they are defined by differential polynomials of order 0, and in the case of a single differential indeterminate, consist of isolated points, algebraic over the differential field of definition of f, which is startling. In the case of a single differential indeterminate y, Ritt points out in [RITT 1938] that the functions constituting the singular components of the solution set of f were called "particular solutions without differences" by Laplace in 1772. In the analytic setting, if the order of f is 1, they are envelopes of the general solution. We saw this earlier for the Clairaut equation. Let $f = 0$ be the Clairaut equation on p. 9. The differential polynomial $\frac{1}{2}y' - x$ is the separant of f. So, $\{f, x - \frac{1}{2}y'\}$ defines the singular zero $y = x^2$ of f. $\{f, y''\}$ defines the "general solution" $y = cx - \frac{1}{4}c^2$ of Clairaut's equation. The singular zero is an envelope of the general solution. (For a more rigorous theory of first order ordinary algebraic differential equations in one unknown function, see [HAMBURGER 1893].) We should note that in [1945b], Ritt considers first order partial differential equations in two independent variables and one unknown function, and analyzes Darboux's 1883 treatment of the behavior of the characteristics in the vicinity of a singular solution, from the viewpoint of differential algebra. The theory of characteristics has not been explored in differential algebra.

In [RITT 1938], Ritt writes:

> Laplace sets the following two problems: Being given a differential equation of any order
>
> 1. to determine whether an equation of lower order which satisfies it is contained or not contained in the general integral;.
> 2. to determine all of the particular solutions of the given equation. [LAPLACE 1772]

By "particular solution," Laplace meant a singular solution not contained in the general solution. "General integral" is a synonym for "general solution."

Part 1 of the theorem tells us that every nonsingular solution of f lies in the general solution of f. Indeed, it is evident that the set of nonsingular solutions is a Kolchin open subset of the general solution. However, the general solution can also contain singular solutions. For example, let $f(y) = y'^2 - 4y^3$. The complex functions that satisfy this differential equations are the 1-parameter family $y = \frac{1}{(x+c)^2}$ and

$y = 0$. Letting $|c|$ increase, we see that a differential polynomial that vanishes for every $\frac{1}{(x+c)^2}$ vanishes at $y = 0$. So, $y = 0$ is in the general solution, and is not an envelope of the family.

1.8.1. *The Component Problem: Laplace's first problem.* As we noted in the Introduction, for Ritt, the fact that the differential algebraic variety defined by a single differential polynomial equation separates into irreducible components, one of which is the general solution, was the compelling reason to turn to algebraic geometry as the context for his theory of algebraic differential equations.It was this phenomenon, appearing already in the fragmentary but inspired work of Lagrange, Laplace, and, most of all Poisson, but largely ignored by "the era of great progress which followed the appearance of the fundamental existence theorems of Cauchy" [RITT 1938], that led Ritt and later Kolchin to the new geometry. And so, it is not surprising that the classification of singular solutions preoccupied Ritt for so many years and produced his deep and beautiful *low power theorem.*

Let f be an algebraically irreducible differential polynomial in $F\{y_1, ..., y_n\}$. Part 4 of the last theorem tells us that there exist algebraically irreducible differential polynomials $g_1, ..., g_s$, of order lower than the order of f, such that the general solutions of $g_1, ..., g_s$ are the singular components of the solution set of f. Let g be an algebraically irreducible differential polynomial such that the solution set of f contains the general solution of g. The low power theorem gives a necessary and sufficient condition for the general solution of g to be a component of the solution set of f.

The first step in the proof is to write $S_g f$ as a differential polynomial in g with coefficients that are differential polynomials in $F\{y_1, ..., y_n\}$ not contained in the general component of g. This is called the *preparation congruence* for f with respect to g.

The Low Power Theorem. *The general solution of g is a component of the solution set of f if and only if the preparation congruence of f with respect to g contains a term cg^k, free of proper derivatives of g, which, considered as a differential polynomial in g, has lower degree than every other term.* [RITT 1936a]

An example due to Chazy [1909], quoted in [INCE 1927], p. 355, illustrates the low power theorem, and, in addition, demonstrates that care must be taken when defining a differential equation whose solutions are without movable (parametric) critical points. (For a discussion of differential equations without movable critical points, see [INCE 1927], and for a modern treatment, [IWASAKI, *et alia,* 1991].) In this example, we take F to be the differential closure of the field $\mathbb{C}(x)$ in the field of germs of functions meromorphic at the origin, $\mathbb{C}$ the field of complex numbers, x a complex variable, $\delta = \frac{d}{dx}$.

$$f(y) = (y'' + y^3 y')^2 - y^2 y'^2 (4y' + y^4).$$

It is not hard to see that $f(y)$ is algebraically irreducible. The separant of f is $2(y'' + y^3 y')$. Thus, η is a singular zero of f if and only if $4\eta' + \eta^4 = 0$, or $\eta' = 0$, or $\eta = 0$. $\{f\} = \{f\} : S_f \cap \{f, S_f\}$. $\{f\} : S_f$ is the general component of f, and the singular zeros of f are zeros of the radical ideal $\{f, S_f\} = [y] \cap [y'] \cap [4y' + y^4] = [y'] \cap [4y' + y^4]$. Therefore, $\{f\} = \{f\} : S_f \cap [y'] \cap [4y' + y^4]$. The differential polynomials y' and $4y' + y^4$ are easily seen to be algebraically irreducible, and have separants that are free of y and its derivatives. Thus, $[y']$ and $[4y' + y^4]$ are the general components of their generators, and clearly neither ideal is contained in the other. They are the components of $\{f, S_f\}$. They are, therefore, candidates for the components of f. However, and this is the crux of the matter, either one could contain the general component of f. Let $g = y'$.

$$f = -4y^2g^3 + 2y^3gg' + g'^2.$$

The term $-4y^2g^3$ in the preparation equation does not have lower degree than the other terms. Therefore, the general component of y' is not a component of f. The last possibility is $g = 4y' + y^4$, which is algebraically irreducible since y' has degree 1. Its separant is 4, and the general component of g is the differential ideal generated by g.

$$4f = -4y^2y'^2g + \frac{1}{4}g'^2.$$

As we observed above, the coefficient $c = -4y^2y'$ of g in the preparation equation is not in the general component of g. Thus, we see from the low power theorem, that the general solution of g is a component of the solution set of f. Now, we are left with the puzzle: Does the solution set of y' lie in the general solution of f? Luckily, since the order of y' is one less than the order of f, we can call on a theorem, first proved by Hamburger [1899]: *Suppose the cardinality of Δ is 1. If f and g are algebraically irreducible differential polynomials in $F\{y\}$, of order n and $n-1$, respectively, then if the solution set of f contains the general solution of g, but the general solution of g is not a singular component of f, then the general solution of g is contained in the general solution of f. So, the solution set of y' is contained in the general solution of f.* The subset of the set of singular solutions, consisting of the constant functions $y = c$, is contained in the general solution of f. It follows immediately that

$$\{f\} = [f] : S_f^\infty \cap \left[4y' + y^4\right],$$

and that $\left[4y' + y^4\right]$ is the only singular component of f.

It is not difficult to solve the differential equations. The general solution consists of the 2-parameter family

$$y = c_1 \tan(c_1^3 x + c_2), c_1, c_2 \,\text{constants},\; c_1 \neq 0,$$

together with the constant functions. The general solution has no movable critical points (poles are allowed). But, the singular component of the solution set is the 1-parameter family.

$$y = \sqrt[3]{\frac{4}{3(x-c)}},$$

which has movable branch points.

It is instructive to re-phrase the low power theorem in the special case of $g = y$:

> Let f vanish for $y = 0$. For $y = 0$ to be a component of the solution set of f, it is necessary and sufficient that the differential polynomial f in y contain a term in y alone, *i.e.*, a term free of all proper derivatives of y, that is of lower degree than every other term of f.

For example, $y = 0$ is a component of the solution set of $y'y'' - y$, but not for $yy''' - y''$ or $y''y''' - y^2$.

Ritt, in his work during the 1930's on the Ritt problem, assumed that the differential polynomial f has meromorphic coefficients, and his proofs combined the techniques of function theory and differential algebra. It was not at all evident how to algebraicize the proof of, for example, the low power theorem. A crucial first step was taken by Ritt's student Howard Levi [1942] and [1945], who proved a central theorem in the algebraicization, called *Levi's lemma.* Levi's work enabled Ritt [1945a], and [1950], to remove the remaining function theory by "adapting his Newton polygon method to the abstract case" [KOLCHIN 1973], p. 165. Ritt's polygon process, which is difficult to follow, was simplified by Kolchin's student Abraham P. Hillman [1952], and by Hillman and D. G. Mead [1962]. Hillman and Mead unified all the existing algebraic techniques used in work on the Ritt problem. They proved the *leading coefficient theorem,* which together with Levi's lemma, forms the cornerstone of Kolchin's discussion in [KOLCHIN 1973].

The Ritt problem (Laplace's second problem) has proved daunting. As Kolchin remarks in [1973], p. 191, "a solution...appears to be remote at present." Suppose we assume, even, that $f \in F\{y\}$, where F is equipped with a single derivation operator δ, and y is a differential indeterminate, *i.e.,* we consider only ordinary differential fields and a single unknown function. The first order case was solved by Hamburger [1893], who proved in addition that if a singular solution of a first order differential equation is not contained in the general solution, then it is an envelope of nonsingular solutions. If it is contained in the general solution, it is analytically embedded among nonsingular solutions. Ritt re-proved Hamburger's theorem in the combined languages of differential algebra and function theory [1932]. See also [MATSUDA 1977]. (These results do not hold for higher order equations. For such equations, Ritt replaced the result about singular solutions in the general component by a weaker analytic condition.) Except for sporadic results (see footnote 2 in [RITT 1936a]), nothing was known about the problem until Ritt's own work in the 1930's. In his remarkable, definitive, but complicated 1936 paper, in which he first presented the low power theorem, Ritt solved the problem completely for second order algebraically irreducible differential polynomials, in the analytic context, combining function theory and differential algebra in ingenious ways. Nothing much is known about the distribution of the singular zeros of higher order ordinary differential polynomials in one differential indeterminate. The case of ordinary differential equations in several differential indeterminates or partial differential polynomials in one or several differential indeterminates, of order exceeding 1, has not been explored. It is interesting to note, though, as Ritt did in [1936a], the link between the Ritt problem and the theory of differential equations whose solutions have no movable critical points. Indeed, Ritt mentions that the transformations of the independent and dependent variable, which he uses in the proof of the sufficiency of the condition in the low power theorem, were used independently by Painlevé in his work on the order 2 case [RITT 1936a]. He points out also that "Hamburger's methods were suggested by those which Fuchs employed for the study of the critical points of solutions of equations of the first order" [RITT 1936a], p. 554. Interestingly, the important results on algebraic differential equations with no movable critical points, including the deep work of Painlevé, whose "new transcendents" have proved to be so interesting in the modern work on soliton theory and the isomonodromy theory and isospectral theory of ordinary differential operators, are limited to differential equations of order less than or equal to 2. For a modern treatment of the theory of first order differential equations free from parametric singularities, developed originally by Fuchs [1884], and Poincaré [1885], and recast in the language of differential algebra, see [MATSUDA 1978] and [MATSUDA 1980]. [BUIUM 1986], combining differential algebra and algebraic geometry, generalizes Matsuda's results.

Let F be an ordinary differential field and let $f \in F\{y\}$, y a differential indeterminate. Suppose f has order 2. The components of the radical differential ideal I generated by f and S_f are the general components of differential polynomials of order ≤ 1. Because of this, Ritt is able to reduce the problem of the distribution of its singular zeros to the problem: Determine if a given singular solution is in the general solution of f. Then, since F is differentially closed, and, therefore, all our solutions are rational over F, we can translate a given singular solution of f to 0, in which case the problem becomes: *Given that 0 is a solution of f, determine if 0 is in the general solution of f.*

Ritt's solution of this problem in his 1936 paper on the order 2 case, rested on his construction of what he called *y-solutions* of the differential polynomial f. They are formal fractional power series in y for its derivative y' that formally annul f, and are variants of Puiseux expansions for solutions of the first order differential polynomials defining certain of the singular components of the radical differential ideal generated by f. As Richard Cohn remarks [1986a], Ritt's paper "received little attention, perhaps because a formidable apparatus of formal power series constructed by means of generalized Newton polygons must be worked through in order to obtain a very special result" (but consider our remarks above about a similar lack of general results in the study of differential equations whose solutions have no movable critical points).

Kolchin synthesized, modernized, and greatly deepened the work on singular solutions, introducing both the above preparation process and the notion of the domination of one differential monomial over another. [KOLCHIN 1965] and [KOLCHIN 1973]. The resulting *domination* lemma, which drew upon an unpublished combinatorial result of Arnold Shapiro, enabled Kolchin to generalize Levi's lemma enough to provide a sufficient condition that (0,0,...,0) not be in the general solution of f.

In 1976, Cohn gave the first algebraic proof of Ritt's theorem [RITT 1950] that gives a bound on the number of differentiations needed to find a basis for the general component, as a radical differential ideal, of a first order ordinary algebraically irreducible differential polynomial in one differential indeterminate. In a series of papers [1977], [1986a], and [1986b], Cohn, using the elegant theory of differential places, initiated by Peter Blum [BLUM 1969], and developed by Blum's student Sally Morrison, algebraicized Ritt's "order 2" 1936 paper, and replaced Ritt's fractional power series arguments by differential valuations.

1.8.2. *Extending Differential Specializations: Differential Places.* In commutative algebra, an integral domain is a valuation ring if and only if given η in its quotient field, either η or $\frac{1}{\eta}$ is in the valuation ring. In Exercise 6, p. 159 of [1973], Kolchin gives an example of the following anomaly in differential algebra:

> Let R_0 be a differential integral domain, equipped with a single derivation operator, and containing the field of rational numbers, and let $\varphi_0 : R_0 \to F_0$ be a differential specialization. Let L be a differential extension field of the quotient field of R_0. Let $\eta \in L$. Then φ_0 need

not extend to a differential specialization from either $R_0\{\eta\}$ or $R_0\left\{\frac{1}{\eta}\right\}$ into a differential extension field of F_0.

Therefore, "valuation rings" in differential algebra must be approached somewhat differently. In 1978, Morrison, a student of Blum, abandoned this property that φ_0 extend either to η or $\frac{1}{\eta}$, and asked whether φ_0 even *extends* to a differential domain with quotient field L [MORRISON 1978].

So, let R_0 and R be differential integral domains with $R_0 \subset R$, and let $\varphi_0 : R_0 \to F_0$ be a differential specialization. Then there is a differential extension field F of F_0 such that φ_0 extends to a differential specialization $\varphi : R \to F$ if and only if there is a prime differential ideal P of R such that $P \cap R_0 = \ker \varphi_0$. Of course, $P = \ker \varphi$. We say, also that φ_0 extends to R. Suppose R_0 is contained in a differential field L. By Zorn's lemma, there are maximal differential subrings R of L to which φ_0 extends. A differential specialization $\varphi : R \to F$ extending φ_0 is called a *maximal differential specialization of* L. If, in addition, the quotient field of R is L, the maximal differential specializationφ is called a *differential place* of L, with *differential valuation ring* R.

A differential integral domain R is a *local differential ring* if R has a unique maximal ideal, and the unique maximal ideal M of R is a differential ideal. Suppose R_0, with maximal ideal M_0, and R with maximal ideal M, are local differential rings, with $R_0 \subset R$. R *dominates* R_0 if $M \cap R_0 = M_0$. This is the case if and only if the differential specialization with kernel M extends the differential specialization with kernel M_0. It is easy to see that a differential valuation ring must be a local differential ring.

There are valuation rings that are also differential valuation rings. The polynomial ring $C[x]$, for example, can be made into a differential ring by regarding the elements of C as constants and defining $x' = x$. Then the localization $(C[x])_{(x)}$ is both a differential valuation ring and a valuation ring. [MORRISON 1978]

Morrison proves the following:

> Let R be a differential valuation ring with a single nonzero prime differential ideal. Then R is a rank one valuation ring. [MORRISON 1978]
>
> Let L be an algebraic extension of a differential field L_0, and R_0 a differential valuation ring of L_0. Then there is a differential valuation ring R of L dominating R_0. Therefore, every differential place of L_0 extends to a differential place of L. [MORRISON 1978]
>
> Let L_0 be a differential field, let η be an element of a differential extension field of L_0 that is a solution of a linear differential equation with coefficients in L_0. Let R_0 be a differential valuation ring of L_0. Then there is a differential valuation ring of $L_0(\eta)$ dominating R_0. [MORRISON 1979]

It follows from the last result that a differential place of L_0 extends to a differential place of any Picard-Vessiot extension of L_0. See also [BRZEZINSKI 1962].

Let R_0 be a differential integral domain containing the field of rational numbers, and let L be a differential extension field of the quotient field of R_0. (L may equal the quotient field.) Then, as Seidenberg pointed out in [1966], the integral closure R of R_0 in L need not be a *differential* extension ring of R_0. He gives an example when

L is the quotient field of R_0. Indeed, Seidenberg proves, under this assumption, that the derivation can be extended to R if and only if the formal power series ring $R[[t]]$ is integrally closed. If R_0 is Noetherian, the derivation always extends to its integral closure in the quotient field. The problem in the non-Noetherian case arises because a place of L finite at η need not be finite at the derivative of η. Morrison proves that if L is any differential extension field of the quotient field of R_0, and η is an element of L integral over R_0, then a differential specialization of R_0 into F extends to $R_0\{\eta\}$. [MORRISON 1978] Indeed, there is a differential ring R containing the integral closure of R_0 in L that satisfies properties analogous (for differential algebra) to those of an integral closure. It follows that a *maximal differential ring of a differential field is integrally closed.*

Cohn's key observation was that the proof by function-theoretic methods, of Ritt's central *Theorem A* in [1936a] could be recast in the language of differential valuation theory. He remarks in [1986a], p.509 that the remainder of Ritt's paper is algorithmic in nature, and seems to be free of its context in analysis.

Theorem A. *Let $f \in F(y\}$ be algebraically irreducible and of second order. Let 0 be a solution of f. Then 0 is in the general solution of f if and only if there is a y-solution in the general solution of f.*

Cohn proved, using [MORRISON 1978], that since the order of f is ≤ 2, the existence of a y-solution in the general solution of f is equivalent to the existence of a discrete valuation of rank 1 at a differential place [1986a]. This replacement of Ritt's function-theoretic and fractional power series methods by the language of differential places was an important step in differential algebra.

1.9. Ritt's Characteristic Sets: Constructive Differential Algebra.

1.9.1. *The Wu-Ritt Method of Mechanical Theorem Proving.* It was Ritt's theory of characteristic sets of differential polynomials that made rigorous Tresse's second theorem. Characteristic sets are the analogues in differential algebra of Elie Cartan's involutive systems and the passive orthonomic systems of C. H. Riquier, M. Janet, and J. M. Thomas. See [CARTAN 1899], [RIQUIER 1910], [JANET 1920], [THOMAS 1929], [THOMAS 1934], [POMMARET 1978]. In the early 1930's, about the same time that Gröbner started thinking about symbolic computing in algebraic elimination theory. Ritt developed constructive methods in the elimination theory of algebraic differential equations.

In the preface of [RITT 1932], Ritt alludes once again to Kronecker's *Festschrift* of 1882, and writes, "The contributions of Mertens, Hilbert, Koenig, Lasker, Macaulay, Henzelt, Emmy Noether, van der Waerden...have brought to the theory of algebraic elimination and the general theory of algebraic manifolds a high degree of perfection." In the wider mathematical community, elimination theory, hence the theory of characteristic sets, languished during Ritt's lifetime. It took the development of high-speed digital computers, followed by powerful symbolic computational systems, to revive interest in algorithmic algebra, and, therefore, in elimination theory. In differential algebra, Ellis Kolchin, in his definitive treatment of the Ritt theory in [KOLCHIN 1973], and in his graduate course and ongoing seminar at Columbia University, synthesized the work of Ritt and Azriel Rosenfeld, modernized the language, and emphasized the role of characteristic sets in simplifying the formidable complexity that marks computation in differential polynomial rings. Because of his interest in differential Galois theory and algebraic groups, Kolchin brought to the theory of characteristic sets his love of structure. Characteristic sets of prime differential polynomial ideals are especially effective computational tools in the presence of structure and symmetries, since they tend to be in canonical form. This is especially true in the theory of differential algebraic groups and Lie algebras.

In his monograph, *Differential Algebra* [1950a], Ritt gives a new exposition of the theory of characteristic sets, including what is now called in computer algebra the pseudo-reduction process, and applies it first to the *polynomial* ring in n indeterminates (the order zero case). He writes "...the theory of algebraic equations can be developed from the algorithmic standpoint, so that every entity whose existence is established is constructed with a finite number of operations." In this, he echoes Hensel's words in the preface to Kronecker's lectures on number theory:

> [Kroenecker] believed that one could, and that one must, in these parts of mathematics, frame each definition in such a way that one can test in a finite number of steps whether it applies to any given quantity. [MISHRA 1993]

Recently, G. Gallo and B. Mishra proved the constructivity of Ritt's characteristic sets for polynomial ideals [GALLO and MISHRA 1991].

In the late 1970's, Wu Wen-tsün discovered an unexpected application of Ritt's characteristic sets for *polynomial* ideals to the problem, dating back to Descartes and Leibniz in Europe and to the Song-Yuan dynasties (960-1368) in China, of mechanical theorem proving in elementary geometry. (See [WU 1978], [1984], [1994].) The Wu-Ritt process (known as the *China prover*) applies to geometric theorems that can be translated into a set of polynomial equations and inequations. Using characteristic sets, Wu used a desktop HP 9835A computer with a RAM of 2.56 megabytes, not only to prove theorems in elementary geometry, but, also, to rediscover such 19th century favorites as the Pascal Conic Theorem. An elegant and deservedly famous theorem of elementary geometry, tricky to prove synthetically, but with an easy proof by Wu's methods, is Karl Wilhelm Feuerbach's (1800-1834) *nine-point circle* theorem, which states,

> *The nine-point circle of a triangle is tangent to the inscribed and three escribed circles.*

Wu's first proof (repeated in [WU 1994],) by exhibiting the system of polynomial equations and inequations associated with a modified hypothesis of the theorem, emphasizes the generic nature of Feuerbach's Theorem, which holds only in the non-degenerate case. A later proof decomposes into its components the solution set of the system of polynomial equations, on 4 of which the conclusion of the theorem holds and on 4 of which it doesn't. We encountered a similar phenomenon in Chazy's example of section 1.8, which we decomposed into its components by the low power theorem.

The Wu-Ritt characteristic set method in constructive algebraic geometry has been recently implemented in *Maple* by Dongming Wang [1992], [1995]. We should mention that significant work on the Wu-Ritt method of mechanical theorem proving has been done by Shang-Ching Chou [1988], and H. Wang [1981].

1.9.2. *Characteristic Sets.* As Wu points out in his preface to [WU 1994], the algebraic relations representing the geometric relations in a theorem in geometry are usually disorderly and unsystematic. In order to implement the proofs on a computer, it is necessary to order these algebraic relations. So, to define characteristic sets of differential polynomial ideals, we return to Ritt's concept of ranking the differential indeterminates (section 1.8). We follow Kolchin's exposition in [KOLCHIN 1973], Chapter I. Unless otherwise noted, the cardinality of the set Δ of derivation operators is an arbitrary positive integer m, $R = F\{y_1, ..., y_n\}$, F an arbitrary differential field.

A ranking always well orders the set of derivatives of the differential indeterminates. It is *orderly* if the rank of θy_j is less than the rank of $\theta' y_{j'}$ if $\theta < \theta'$. An example of an orderly ranking is the ordering of the set of derivatives $\delta_1^{i_1} \cdots \delta_m^{i_m}$ lexicographically with respect to $(\sum i_j, j, i_1, ..., i_m)$;

Fix a ranking of the differential indeterminates. We must now try to extend the notion of comparative rank to whole of R. We obtain a *pre-order* relation on R.

(1) Every element of F has lower rank than every element of R not in F.
(2) If f and g are in R but not in F, and if either $u_f < u_g$ or $u_f = u_g$ and $\deg_{u_f} f < \deg_{u_g} g$, then f has lower rank that g.
(3) Two elements of R that are either both in F or have the same leader and the same degree in that leader, have the same rank.

It follows that if $f \in R$, $f \notin F$, then f has higher rank than both its initial and its separant.

Clearly, in every nonempty subset of R there is an element that has rank lower than or equal to the rank of every element of the subset. Any such element is called an *element of lowest rank* in the subset.

Let f be an element of R not in F. Recall that a differential polynomial g is *partially reduced* with respect to f if g is free of every proper derivative of f. If g is partially reduced with respect to f and, in addition, the degree of g as a polynomial in the leader of f is less than the degree of f in its leader, then g is said to be *reduced* with respect to f. If S is any subset of R not intersecting F, g is *partially reduced (*or *reduced)* with respect to S if it is partially reduced (or reduced) with respect to every element in S. S is *autoreduced (ascending set* or *chain* in Ritt's terminology) if each element is reduced with respect to all the other elements of S. Clearly, every autoreduced set is finite.

We begin with an autoreduced set A in a given ranking, and an element g of R. There exist several algorithms for reducing g with respect to A. (See, for example, [RITT 1950a], p. 5, [KOLCHIN 1973], p. 77, [BOULIER 1995].) These algorithms all produce non-negative integers $i_f, s_f, f \in A$, and a differential polynomial g_0, reduced with respect to A, such that

$$\prod_{f \in A} I_f^{i_f} S_f^{s_f} g \equiv g_0([A]).$$

g_o is called the *remainder* of g with respect to A, and has lower rank than that of g. It is far from unique. The unavoidable presence of the factor consisting of products of the initials and separants of the elements of A is the reason why the Ritt reduction process is called pseudoreduction in computer algebra, and is responsible for the fact that Wu's China prover will demonstrate the truth of the conclusion of a theorem in geometry *generically*. We should note that $\prod I_f^{i_f} S_f^{s_f} g - g_0$ can be written as a linear combination over R of derivatives θf, $f \in A$, $\theta u_f \leq \theta u_g$.

The notion of comparative rank in the set of all autoreduced sets in R is analogous to a similar concept in Gröbner basis theory. Let A, with elements $f_1, ..., f_r$ and

B, with elements $g_1, ..., g_s$, be autoreduced subsets of R, in each case arranged in order of increasing rank.

(1) If there exists a positive integer k with $k \leq r$ and $k \leq s$ such that

$$\text{rank } f_i = \text{rank } g_i (1 \leq i < k), \quad \text{rank } f_k < \text{rank } g_k,$$

or, if $r > s$ and

$$\text{rank } f_i = \text{rank } g_i \ (1 \leq i \leq s),$$

then A is said to have *lower rank* than B.

(2) If $r = s$ and rank f_i = rank $g_i(1 \leq i \leq r)$, then A is said to have the same rank as B.

A fundamental property of rankings implies that *in every nonempty set of autoreduced subsets of R there is an autoreduced subset of lowest rank.* For any differential ideal I of R there is an autoreduced subset A of I such that no separant of an element of A is in I. Such an autoreduced set of lowest rank is called a *characteristic set* of I. It follows easily that for characteristic sets of I, no initial of an element of the characteristic set is in I.

Until recent work in computer algebra, cited below, the theory of characteristic sets was limited to prime differential ideals. The *Ritt-Kolchin algorithm* (KONDRATEVA, *et alia,* 1995], refining the Ritt-Raudenbush decomposition theorem, computes, for a radical differential ideal with given basis $f_1, ..., f_r$, a set of autoreduced sets $A_1, ..., A_k$ such that A_i is a characteristic set of P_i, and the radical differential ideal

$$\{f_1, ..., f_r\} = P_1 \cap ... \cap P_k.$$

Unfortunately, we cannot conclude that $P_1, ..., P_r$ are the prime components of the radical differential ideal, since the following problem is wide open:

> Given characteristic sets A and B for prime differential ideals P and Q, respectively, determine if $P \subset Q$.

1.9.3. *Coherence of Autoreduced Sets: Rosenfeld's Lemma.* It was Kolchin and his student Rosenfeld who provided coherence conditions (integrability conditions) for autoreduced sets that single out the completely integrable systems of Tresse's Theorem 2. In his dissertation, [ROSENFELD 1959], Rosenfeld proved a lemma that became an important tool in elimination theory. Seidenberg proved a slightly weaker version in [SEIDENBERG 1956]. In [BOULIER 1995], François Boulier points out that Rosenfeld's lemma is the manifestation in differential algebra of the famous Knuth-Bendix theorem [KNUTH and BENDIX 1967]; its manifestation in algebra is Buchberger's Gröbner basis algorithm. Rosenfeld introduced the lemma in order to prove both the lifting theorem for differential specializations (cited earlier, and proved independently by Seidenberg), and a catenary theorem for affine differential algebras. Rosenfeld's coherence conditions are the analogues in differential algebra both of the S - polynomial conditions in the Buchberger algorithm and of E. Cartan's compatibility conditions for involutive systems of partial differential equations (see the article [CARTAN 1931], written for Albert Einstein in the language of partial differential equations, rather than his customary Pfaffian forms.) The resemblance both to Cartan's compatibility conditions and Riquier's conditions for a system to be passive orthonomic becomes more evident when the initials and separants of the differential polynomials in the given autoreduced set are in the ground differential field, the case most generally studied in classical differential equations theory.

Rosenfeld's lemma and Seidenberg's elimination theorem have been combined recently with Buchberger's algorithm to given an algorithm, called *Rosenfeld-Gröbner*, that solves the radical differential ideal membership problem [BOULIER, *et alia*, 1997]. The algorithm was implemented by Boulier, using the big number library of PARI and the software GB[FGLM], and, then, in MAPLE. The algorithm was improved by Morrison in [MORRISON 1997]. She also gives a proof of Lazard's lemma, an important step in the Rosenfeld-Gröbner algorithm, and removes from the coherence conditions the hypothesis that the given set be autoreduced.

Let A be an autoreduced subset of R relative to some fixed ranking, and set $I = \prod_{f \in A} I_f$ and $S = \prod_{f \in A} S_f$. A is *coherent* if whenever $f, g \in A$ and v is a common derivative of u_f, u_g, say $v = \theta u_f = \theta' u_g$, then the *S-polynomial* $S_g \theta f - S_f \theta' g \in (A_v) : (IS)^\infty$ where A_v is the set of all differential polynomials $\tau h, h \in A, \tau \in \Theta$, $\tau u_h < v$. It is enough to verify the condition for the lowest common derivative of the leaders of f and g.

The coherence conditions require that after multiplying by a suitable power of the product of the initials and separants you can represent an S-polynomial as a linear combination over R of derivatives of the elements of the autoreduced set that have leaders of lower rank than the common derivative v.

Rosenfeld's Lemma. *Let A be a subset of R. If A is a characteristic set of a prime differential ideal P, then $P = [A] : (IS)^\infty$, A is coherent, and $(A) : (IS)^\infty$ is a prime ideal not containing a nonzero element of P that is reduced with respect to A. Conversely, if A is a coherent autoreduced subset of R such that $(A) : (IS)^\infty$ is prime and does not contain a nonzero element reduced with respect to A, then A is a characteristic set of a prime differential ideal of R, namely $[A] : (IS)^\infty$.* [ROSENFELD 1959]

The following are corollaries: Let A be a coherent autoreduced subset of R.

1. *Let g be an element of R that is partially reduced with respect to A. Then A has a zero that does not annihilate ISg if and only if A, considered as a set of polynomials in the polynomial ring $F[(y_i), ..., (\theta y_i), ...]$ in infinitely many algebraic indeterminates has such a zero, called an algebraic zero of A.*
2. $[A] : (IS)^\infty$ *is a prime differential ideal if and only if* $(A) : (IS)^\infty$ *is a prime ideal.*
3. $[A] : (IS)^\infty$ *is a radical differential ideal if and only if* $(A) : (IS)^\infty$ *is a radical ideal.*
4. *If every differential polynomial in A has degree 1 in its leader (A is quasi-linear) then $P = [A] : (IS)^\infty$ is a prime differential ideal, and no nonzero differential polynomial in R that is reduced with respect to A can be in P.*

Riquier's notion of passive orthonomic set is also captured by Rosenfeld's lemma:

1. *If $IS = 1$ (i.e., A is an orthonomic autoreduced set), then A is coherent if and only if A is a characteristic set for the prime differential ideal $[A]$.*

One of the most important corollaries of Rosenfeld's lemma is that it solves the ideal membership problem for prime differential ideals.

Let P be a prime differential ideal of R, let A be a characteristic set of P, and let $g \in R$. Then $g \in P$ if and only if the remainder of g with respect to A is 0.

Ritt proved in [RITT 1950a] that it is possible to remove the initials from the characterization of the prime differential ideal associated with a coherent autoreduced set:

Let P be a prime differential ideal and let A be a characteristic set of P. Then $P = [A] : S^{\infty}$.

Examples. (1) In this example, we illustrate both the first corollary, which relates a coherent autoreduced set to the algebraic ideal that it generates, and the Ritt-Kolchin algorithm for determining a characteristic set for a prime differential ideal. Set F equal to the differential closure of $C(x,y)$, where $\Delta = \left\{\frac{\partial}{\partial x}, \frac{\partial}{\partial y}\right\}$. $R = F\{z\}$, where z is a differential indeterminate. We use the Mongean notation $p = \frac{\partial z}{\partial x}$, $q = \frac{\partial z}{\partial y}$, $r = \frac{\partial p}{\partial x}$, $s = \frac{\partial p}{\partial y} = \frac{\partial q}{\partial x}$, $t = \frac{\partial q}{\partial y}$. We consider the system of differential polynomials

$$\begin{aligned} f &= p^2 - (x+y+1)p + x + y \\ g &= q^2 - (x+1)q + x. \end{aligned}$$

We shall show that f and g are compatible, *i.e.*, define a consistent system of differential equations. We choose an orderly ranking that ranks q higher than p. Then the leader of f is p and of g is q, and the set consisting of f and g is autoreduced.

$$\begin{aligned} S_f &= 2p - (x+y+1). \\ S_g &= 2q - (x+1). \end{aligned}$$

The initials of f and g are both equal to 1. We check Rosenfeld's coherence condition. The lowest common derivative of the leaders of f and g is $v = s$. So, we set $h = S_g \frac{\partial f}{\partial y} - S_f \frac{\partial g}{\partial x} = (1-x-y)q + (x-1)p + y$, which has lower rank than g. However, h does not satisfy the coherence condition. We must adjoin it to the system consisting of f and g. A straightforward, but tedious, computation shows that the differential ideal generated by f and g equals the differential ideal generated by f and h, whence the radical differential ideal generated by f and g equals the radical differential ideal generated by f and h. So, if we are interested only in the existence of a zero of f and g, hence only in the radical differential ideal they generate, we may replace the system consisting of f and g by the system consisting of f and h. The new system has the same solutions as the original system.

Let A be the autoreduced set consisting of f and h. Since the leader of h is q, A is autoreduced.

$$S_h \frac{\partial f}{\partial y} - S_f \frac{\partial h}{\partial x} = (1-x)\frac{\partial f}{\partial x} - \frac{2y}{x+y-1} f + \frac{2p-x-y}{x+y-1} h.$$

Since $r = \frac{\partial p}{\partial x} < \frac{\partial p}{\partial y} = s$, the autoreduced set of A is coherent. Now, $f = (p - (x+y))(p-1)$. We can easily find an algebraic zero of A, considered as a subset of $F[z,p,q,r,s,t,...]$, that does not annihilate the separant of f, hence does not annihilate the product $S_f S_h$. Take, for example, (0,1,1,0,...,0,...) *i.e.*, all coordinates following $q = 1$ are 0. So, in particular, the radical differential ideal generated by f and h has a zero. It follows that the original system of differential polynomials defines a consistent system of differential equations.

(2) Cayley establishes the following result. Let $L = \delta^2 + 2b\delta + c$ be in the noncommutative ring $F[\delta]$ of linear differential operators with coefficients in a differentially closed field F equipped with the derivation operator δ. If $a \in F$, $\delta a = a\delta + a'$, where

$a' = \delta(a)$. Cayley proves that L factors into a product of two first order operators. $L = (\delta + a_1)(\delta + a_2)$. To determine a_1 and a_2, we must solve the nonlinear system

$$a_1 + a_2 = 2b$$

$$a_1' + a_1 a_2 = c.$$

Set $f = a_2 + a_1 - 2b$, $g = a_1' + a_1 a_2 - c$. Choose an orderly ranking with $a_1 < a_2$. Then the leader of f is a_2 and the leader of g is a_1'. The set consisting of f and g is not autoreduced. So, we reduce g with respect to f. We see that

$$g = a_1 f + h, \text{ where } h = a_1' - a_1^2 + 2ba_1 - c.$$

So, the ideal (differential ideal, radical differential ideal) generated by f and g equals the ideal (differential ideal, radical differential ideal) generated by f and h. Let A be the set whose elements are f and h. It is autoreduced. The leader of h is a_1'. Since f and h are led by derivatives of different differential indeterminates, their leaders cannot have a common derivative. So, A is a coherent autoreduced set. Since it is quasilinear, it follows from corollary (4) that $[A] : (S_f S_g)^\infty$ is a prime differential ideal. It therefore has a zero, which is a zero of the radical differential ideal that does not annul the separants of f. [CAYLEY 1886]

1.10. **Differential Rational Maps in Affine Differential Algebraic Geometry.** In his 1979 article on nonlinear differential equations, Yu Manin describes three possible languages for the variational formalism: the classical language, differential algebra, and the geometric language. Following his comments on the classical language, he writes:

> The language of differential algebra is better suited for expressing such properties [invariant properties of the differential equations] and puts at the disposal of the investigator the extensive apparatus of commutative algebra, differential algebra, and algebraic geometry....The numerous "explicit formulas" for the solutions of the classical and newest differential equations have good interpretations in this language; the same may be said for conservation laws. However, the language of differential algebra which has been traditional since the work of Ritt does not contain the means for describing changes of the functions (dependent variables) and the variables x_i (independent variables) and for clarifying properties which are invariant under such changes. This is one of the main reasons for the embryonic state of the theory of so-called "Bäcklund transformations" in which there has been a recent surge of interest. [MANIN 1979]

It was Kolchin's work on differential algebraic geometry and differential algebraic groups that made differential algebra less static. The concept of differential rational map (changes in the "dependent variables") was central to its development, beginning with affine differential algebraic geometry [CASSIDY 1972], and culminating in Kolchin's powerful axiomatic formulation of the theory of differential algebraic groups and their homogeneous spaces, which was the focus of his 1985 book. (See Borel's article in this volume.) Differential rational maps are the analogues in differential algebra of Bäcklund transformations. Furthermore, in the classification of semisimple differential algebraic groups, it became necessary to allow transformations of the set of derivation operators by arbitrary Jacobian matrices; in his 1973 book, Kolchin used only changes by constant matrices.

Let F be a differentially closed field with finite commuting set Δ of derivation operators. Throughout this section, unless otherwise indicated, all differential algebraic varieties will be irreducible. So, *Kolchin closed* will mean irreducible Kolchin closed.

Let V be a Kolchin closed subset of $A_n(F)$, and let P be its defining differential ideal in the differential polynomial ring $R = F\{y_1, ..., y_n\}$. Then P is a prime differential ideal. The quotient ring R/P is a differential integral domain, which we can write $F\{\bar{y}_1, ..., \bar{y}_n\}$, where $\bar{y}_j$, is the residue class of y_j, and which is, therefore, finitely differentially generated over the ground field. It is called the *differential coordinate ring* of V, and we denote it by $F\{V\}$. We can identify its elements with $F-$ valued functions on V and so we call them *differential polynomial functions* on V. The field of fractions of the differential coordinate ring is called the *field of differential rational functions on* V, and is denoted by $F\langle V\rangle$. If $\eta \in V$, a differential rational function on V is *defined* at η if it can be represented as a quotient of differential polynomial functions whose denominator does not vanish at η. The subring of $F\langle V\rangle$ consisting of all differential rational functions on V that are defined at η is a local differential ring denoted by $F_\eta\langle V\rangle$. The intersection of the local differential rings $F_\eta\langle V\rangle, \eta \in V$, is called the *ring of everywhere defined differential rational functions* on V. In contrast with affine algebraic geometry, the ring of everywhere defined differential rational functions on V need not equal the differential coordinate ring, and need not be finitely differentially generated over F [CASSIDY 1972].

Let $V \subset A_n(F)$ and $W \subset A_m(F)$ be closed Kolchin sets. A *differential rational map* $\varphi : V \cdots \to W$ is a family $(f_1, ..., f_m)$ of differential rational functions on V such that $\varphi(\eta) = (f_1(\eta), ..., f_m(\eta)) \in W$ whenever the *coordinate functions* $f_1, ..., f_m$ are defined at η. The set of all $\eta \in V$ such that coordinate functions of f are defined at η is called the *domain* of f and is a nonempty Kolchin open subset of V. It follows that the domain of f is dense in V in the Kolchin topology. The Kolchin closure of the image of its domain is irreducible in the Kolchin topology. The differential rational map is called *dominant* if the Kolchin closure of the image of its domain is the differential algebraic variety W. φ is called a *differential birational map* if it is dominant and there is a dominant differential rational map $\psi : W \cdot\cdot \to V$, called the *generic inverse* of φ, such that

- if φ is defined at η and ψ is defined at $\varphi(\eta)$, then $\psi(\varphi(\eta)) = \eta$;
- if ψ is defined at ζ and φ is defined at $\psi(\zeta)$, then $\varphi(\psi(\zeta)) = \zeta$.

A differential rational map from V to W is called a *differential polynomial* map if its coordinate functions lie in the differential coordinate ring of V. Of course, everywhere defined differential polynomial maps need not be differential polynomial maps.

Examples. (1) Assume that the cardinality of Δ is 1. Let $R = F\{y\}$, y a differential indeterminate. Set $f = y'^2 - y^3$. It is not difficult to show that f is algebraically irreducible. $S_f = 2y'$. By the low power theorem, the only singular solution, 0, is in the general solution of f. Therefore, the differential ideal P generated by f is prime and equal to its general component. Let $V \subset A_1(F)$ be the differential algebraic variety defined by P. $F\{V\} = F\{\bar{y}\} = F[\bar{y}, \bar{y}']$, since $\bar{y}'' = \frac{3}{2}\bar{y}^2$. Now, set $g = y' - \frac{1}{2}y^2$.

Since g is quasilinear, it is algebraically irreducible, and has no singular solutions. Therefore, the differential ideal Q generated by g is prime and equal to its general component. Let $W \subset A_1(F)$ be the differential algebraic variety defined by Q. We'll call it a *Riccati variety* since its defining differential polynomial equation is a Riccati equation. We define a dominant differential rational map $\varphi : V \cdots \to W$, with domain the complement of 0 in V, that maps its domain surjectively onto W. $\varphi = \frac{\bar{y}'}{\bar{y}}$. (A short calculation shows that $\left(\frac{\bar{y}'}{\bar{y}}\right)' = \frac{1}{2}\left(\frac{\bar{y}'}{\bar{y}}\right)^2$.) Now, it is not hard to prove that every prime differential ideal in $F\{y\}$ is the general component of an algebraically irreducible differential polynomial h, and that if it contains P properly, h must have order 0. Therefore, every proper Kolchin closed subset of the Riccati variety must be finite. Since the Kolchin closure of the image of the domain of a differential rational map is irreducible, the image of the domain of φ is either a point or the whole Riccati variety. Since φ is clearly non-constant on its domain, it is dominant. We now show that φ is a differential birational map by defining a generic inverse. Set $\psi(\zeta) = \zeta^2$, $\zeta \in W$. ψ is everywhere defined on W. A short calculation shows that $\left((\zeta^2)'\right)^2 = (\zeta^2)^3$. So, the image of ψ is contained in V. Indeed, ψ is a bijective everywhere defined differential map from W onto V. Let $\zeta \in W$. $\varphi(\zeta^2) = 2\frac{\zeta'}{\zeta} = 2\left(\frac{1}{2}\frac{\zeta^2}{\zeta}\right) = \zeta$. Let $\eta \in V$, $\eta \neq 0$. $\psi\left(\frac{\eta'}{\eta}\right) = \left(\frac{\eta'}{\eta}\right)^2 = \frac{\eta^3}{\eta^2} = \eta$. So, φ is a differential birational map from V onto the Riccati variety W whose generic inverse is the squaring function, and maps the Riccati variety bijectively onto V. It is easy to see that φ maps the complement of 0 bijectively onto the Riccati variety.

(2) [KAUP 1980] The Cole-Hopf transformation from the solution set of the heat equation to the solution set of Burgers' equation is an example of a differential rational map, and the proof of its surjectivity illustrates the use of Rosenfeld's lemma. Here, our differentially closed field F is the differential closure of $\mathbb{R}(t, x)$, $\mathbb{R}$ the field of real numbers, t, x, real variables, and $\Delta = \left\{\frac{\partial}{\partial t}, \frac{\partial}{\partial x}\right\}$. Let y be a differential indeterminate. Let $V(h)$ be the solution set in $A_1(F)$ of the heat equation,

$$h(y) = \frac{\partial y}{\partial t} + \frac{\partial^2 y}{\partial x^2} = 0.$$

Let $V(b)$ be the solution set in affine 1-space of Burgers' equation,

$$b(y) = \frac{\partial y}{\partial t} + \frac{\partial^2 y}{\partial x^2} + 2\frac{\partial y}{\partial x} y = 0,$$

which attempts to model turbulent flows in a channel. Let η be a nonzero element of $V(h)$. A short calculation shows that $\lambda(\eta) = \frac{\frac{\partial \eta}{\partial x}}{\eta} \in V(b)$. So, we have the differential rational map $\lambda : V(h) \cdots \to V(b)$, with domain $V(h)\backslash\{0\}$, called the Cole-Hopf transformation. We wish to show that the Cole-Hopf transformation maps its domain surjectively onto the solution set of Burgers' equation. This enables physicists to solve Burgers' equation with given initial conditions, if they can solve the heat equation, thus "linearizing" Burgers' equation. Interestingly, the differential rational map that does this is the same one that linearizes Riccati's equation.

We set an orderly ranking that ranks $\frac{\partial}{\partial t}$ lower than $\frac{\partial}{\partial x}$.

Let ζ be an arbitrary element of $A_1(F)$. We want to show that if $\zeta \in V(b)$, then there exists $\eta \in V(h)$ such that $\lambda(\eta) = \zeta$. So, we want to demonstrate the compatibility of the differential equations

$$\begin{aligned} \frac{\partial y}{\partial t} + \frac{\partial^2 y}{\partial x^2} &= 0 \\ \frac{\partial y}{\partial x} - \zeta y &= 0. \end{aligned}$$

Set $f_1 = \frac{\partial y}{\partial x} - \zeta y$. We first reduce h with respect to f_1, whose leader is $\frac{\partial y}{\partial x}$, and whose separant and initial both equal 1.

$$h = \frac{\partial f_1}{\partial x} + \zeta f_1 + \left(\frac{\partial y}{\partial t} + \left(\frac{\partial \zeta}{\partial x} + \zeta^2\right) y\right).$$

Set $f_2 = \frac{\partial y}{\partial t} + \left(\frac{\partial \zeta}{\partial x} + \zeta^2\right) y$. Then, the differential ideal generated by f_1 and h equal the differential ideal generated by f_1 and f_2. Now, we set A equal to the set consisting of the differential polynomials f_1 and f_2. The leader of f_2 is $\frac{\partial y}{\partial t}$, and its separant and initial both equal to 1 (the equations of physics tend to be orthonomic). A is clearly autoreduced. We turn now to the Rosenfeld coherence condition:

$$\frac{\partial f_1}{\partial t} - \frac{\partial f_2}{\partial x} = -\left(\zeta f_2 + \left(\frac{\partial \zeta}{\partial x} + \zeta^2\right) f_1 + b(\zeta) y\right),$$

where b is the differential polynomial defining Burgers' equation. So, the vanishing of Burgers' equation at ζ is a sufficient condition for coherence. Therefore, if $\zeta \in V(b)$, the set A is coherent as well as autoreduced. It follows from the fifth corollary of Rosenfeld's lemma that the differential ideal P generated by f_1 and f_2 is prime, and A is a characteristic set of P. Since the differential polynomial y is reduced with respect to A, it is not in P. Therefore, λ is a surjective differential rational map from its domain onto the solution set of Burgers' equation.

1.11. **Dimension Theory in Differential Algebraic Geometry.** In this section, we will again assume that, unless otherwise stated, all differential algebraic varieties are irreducible. We work in differential affine n−space over a differentially closed field F, and set m equal to the cardinality of Δ.

1.11.1. *Kolchin's Dimension Polynomial.*

> Of all the theorems of analysis situs, the most important is that which we express by saying that space has three dimensions. It is this proposition that we are about to consider, and we shall put the question in these terms: When we say that space has three dimensions, what do we mean?
> HENRI POINCARÉ, quoted in [EISENBUD 1995]

Eisenbud begins his chapter on dimension theory in algebraic geometry with this quote. As he points out,

> In nineteenth-century geometry, the idea of dimension was used intuitively. Euclidean n-space was, by agreement, of dimension n; and in general an object was said to be n-dimensional if the least number of parameters needed to describe its points, in some unspecified way, was n....
>
> From these beginnings, and from the axiom, like that of the topologists, that affine d-space has dimension d, came the algebraic definition that was used early in the century: The dimension of an irreducible variety

in affine r−space over a field k (initially $\mathbb{C}$) is the transcendence degree of the field of rational functions on X. [EISENBUD 1995]

The first attempt at formulating a measure of the "size" of the set of solutions of a system of algebraic differential equations was made by Ritt (see, for example, [RITT 1938].) The *differential dimension* $d = d(V)$ of a differential algebraic variety V is the differential transcendence degree of the differential field $F\langle V\rangle$ of differential rational functions on V. The "arbitrary unknowns" or "parametric indeterminates" in Ritt's work form a differential transcendence basis over F of $F\langle V\rangle$, differential affine n-space has differential dimension n, and differential dimension is a differential birational invariant. Furthermore, in the analytic setting, it captures the rather vague concept, usually defined by way of examples, of the number of "arbitrary functions of m variables in the solution." So, if the differential dimension is 0, *i.e.,* $F\langle V\rangle$ is differentially algebraic over F, then there are 0 arbitrary functions of m variables in the solution, but a certain number of functions of m-1 variables. If F is an *ordinary* differential field, this means that the "solution depends on a finite number of arbitrary constants."

Although it is true that if a differential algebraic variety V is contained in a differential algebraic variety W, then $d(V) \leq d(W)$, the inequality need not be strict if the inclusion is strict; differential dimension is not a fine enough measure of size. A more discriminating measure was needed.

For model theorists, the answer lay in the notion of *Morley rank* [POIZAT, this volume], [MARKER 1996]. Kolchin found the solution in algebraic geometry. The idea of the Hilbert polynomial associated with a homogeneous polynomial ideal suggested to him that the differential coordinate ring of a differential algebraic variety might be viewed as an inductive limit of algebras that are finitely generated over F, obtained by truncating the coordinate ring by order, and that the transcendence degrees over F of their quotient fields might be measured by a numerical polynomial. The *dimension polynomial* was the numerical polynomial he sought [KOLCHIN 1964], and [1973]. In 1969, Kolchin's student Joseph L. Johnson made the metaphor real, by using homological methods to prove that Kolchin's dimension polynomial is equal to the Hilbert polynomial of the graded module over the polynomial ring $F[X_1, ..., X_m]$ associated with the left $F[\Delta]$− module of Kähler differentials of $F\langle V\rangle$ [JOHNSON 1969 a, b], and [1974]. A general theory of dimension polynomials is being developed by the Russian mathematicians Marina V. Kondrateva, Alexandr B. Levin, Alexandr V. Mikhalev, and Evgeny V. Pankratev [1995]. They also devise algorithms, which have been implemented on the computer, for calculating dimension polynomials. See also [PANKRATEV 1989], [KONDRATEVA *et al* 1992], [LEVIN 1997].

Let $\mathbb{R}$ be the field of real numbers. A polynomial $\omega \in \mathbb{R}[x]$ is *numerical* if $\omega(s)$ is an integer for all sufficiently large natural numbers s. Any numerical polynomial ω can be written in the form

$$\omega = \sum_k a_k \binom{X+k}{k},$$

where a_k is an integer, and

$$\binom{X+k}{k} = (X+1)(X+2)\cdots(X+k)/k!.$$

We define a total ordering on the set of numerical polynomials: $\omega_1 \leq \omega_2$ if $\omega_1(s) \leq \omega_2(s)$ for all sufficiently large s.

Kolchin defines the dimension polynomial of a Kolchin closed subset V of $A_n(F)$ by defining the dimension polynomial of its defining prime differential ideal. So, let P be a prime differential ideal of the differential polynomial ring $R = F\{y_1, ..., y_n\}$. Regard R as a polynomial ring in countably many indeterminates: Define $R_s = F[(\delta_1^{i_1} \cdots \delta_m^{i_m} y_j), i_1 + \cdots + i_s \leq s, j = 1, ..., n]$. Then R_s is a polynomial ring in $n \binom{s+m}{m}$ indeterminates, and $P \cap R_s$ is a prime ideal in R_s. It therefore is the defining ideal of an irreducible algebraic variety, whose dimension we call $\dim(P \cap R_s)$. Kolchin proved the following theorems:

Theorem. *There is a unique numerical polynomial ω_p such that* $\dim(P \cap R_s) = \omega_p(s)$ *for all sufficiently big natural numbers s.*

Theorem.

(1) $0 \leq \omega_p \leq n \binom{s+m}{m}$, *so that* $\deg \omega_p \leq m$.
(2) $\omega_p = 0$ *if and only if the set of zeros of P is finite.*
(3) $\omega_p = n \binom{s+m}{m}$ *if and only if the set of zeros of P is affine $n-$space.*
(4) *If we write* $\omega_p = \sum_{i=0}^{m} a_k \binom{s+m}{m}$, *then a_m is the differential dimension of the set of zeros of P.*
(5) *If Q is a prime differential ideal of R, with $P \subset Q$, then $\omega_p \geq \omega_Q$ and equality holds if and only if $P = Q$.*

$\omega_p = \omega_V$ is called the *dimension polynomial* of both the prime differential ideal P and of its set V of zeros, and is, by property (5) a fine enough measure of size. Unfortunately, if W is a differential algebraic variety differentially birationally isomorphic to V, in which case we can think of $F\langle W\rangle$ as being obtained from $F\langle V\rangle$ by a change of generators, it does not follow that their dimension polynomials are the same. The dimension polynomial, in contrast to the differential dimension, is not a differential birational invariant, although William Sit proved that associated with the differential field $F\langle V\rangle$ is a minimal dimension polynomial, which is, of course a differential birational invariant [SIT 1978]. See, also, [KONDRATEVA 1989]. The dimension polynomial is invariant under birational changes (free of proper derivatives) of the dependent variables, and "arbitrary changes" of the independent variables, *i.e.*, transformations by Jacobian matrices of the set of distinguished derivation operators.

But, what quantities associated with the dimension polynomial of a differential algebraic variety *are* invariant under differential birational transformations? Kolchin's next theorem specifies two important numerical invariants. Suppose $\omega_V \neq 0$. Set $\tau_V = \deg \omega_V$. Then a_τ is the last nonzero coefficient in the representation of the dimension polynomial, in part (4) of the above theorem, as a linear combination of binomial coefficients. Denote it by a_V.

Theorem. *Suppose $\omega_V \neq 0$; τ_V and a_V are differential birational invariants.* [KOLCHIN 1964]

τ_V is called the *differential type* of V (and of its defining differential ideal P), and a_V is the *typical differential dimension* of V (and of P). If $\omega_V = 0$, then τ_V is defined to be -1, and a_V to be 0. In this case, V is a finite set. If $\tau_V = 0$, then $\omega_V = a_V$, and the differential field of differential rational functions on V is a finitely generated field extension of F of transcendence degree a_V. If this is the case, we will, with a slight abuse of language, call V *finite-dimensional.*

What is the significance of the differential type and typical differential dimension? Kolchin proves the following theorem:

Theorem. *There exists a set Δ' of commuting derivation operators on F, of cardinality τ_V, obtained from Δ by a linear transformation with constant coefficients, such that if we regard F and $F\langle V\rangle$ as Δ'-fields, the differential transcendence degree is a_V.*

By changing the distinguished derivation operators, we change the differential structure on the ground field and the field of rational functions on the differential algebraic variety V. This is tantamount to regarding the solutions as functions of τ_V new independent variables. The Δ'-field $F\langle V\rangle$ has a differential transcendence basis over the ground field of cardinality a_V. The differential type τ_V and the typical differential dimension a_V are the interpretations in differential algebra of the statements in analysis that the solution depends on a_V arbitrary functions of $\tau_V < m$ independent variables, but does not depend on any arbitrary functions of more than τ_V independent variables. The typical differential dimension measures the degree of arbitrariness of the data that determine the general solution. For example, consider the heat equation

$$h = \frac{\partial^2 y}{\partial x^2} - \frac{\partial y}{\partial t} = 0.$$

Assume that the differentially closed field F is contained in the differential field of germs of meromorphic functions at the origin of complex 2-space. Since the heat equation is linear, the differential ideal it generates is prime, and there are no singular solutions. One can say that the general solution depends on two arbitrary functions of t, *viz.*, the functions $f(t)$ and $g(t)$, which are restrictions of y and $\frac{\partial y}{\partial x}$, respectively, to $x = 0$. These functions being given, one has

$$y(t,x) = y(t,0) + \frac{\partial y}{\partial x}(t,0)x + \frac{\partial^2 y}{\partial x^2}(t,0)\frac{x^2}{2!} + \frac{\partial^3 y}{\partial x^3}(t,0)\frac{x^3}{3!} + \ldots, \text{ and, so}$$

$$y(t,x) = f(t) + g(t)x + f'(t)\frac{x^2}{2!} + g'(t)\frac{x^3}{3!} + \ldots$$

in a neighborhood of the origin [DEBEVER 1979], p. 217.

The fundamental tool used to construct all algorithms for computing the differential dimension polynomial of a prime differential ideal can be found in section 17, chapter 0 on lattice points, and in Theorem 6, chapter II, of [KOLCHIN 1973]. Kolchin's instructions in Theorem 6 for computing the differential dimension polynomial start with the choice of a characteristic set for the ideal, with respect to an *orderly* ranking, and then refers the reader to chapter 0 to count lattice points. It is essential that the ranking be orderly. Thus, the leader of the heat equation must be $\frac{\partial^2 y}{\partial x^2}$; the dimension polynomial cannot be computed if $\frac{\partial y}{\partial t}$ is chosen as leader. If we followed the algorithm, we would see that the differential type of the heat equation is 1, and its typical dimension is 2. The general solution of the heat equation depends on 2 arbitrary functions of t. This is in accord with the above power series representation of the general solution. There is a unique solution of the heat equation, analytic at (0,0), when the temperature and temperature gradient are functions of time that are analytic at $t = 0$. Elie Cartan and Albert Einstein carried on a long correspondence (from 1929 to 1932) on the subject of Einstein's model of a unified field theory based on absolute parallelism [DEBEVER 1979]. One of the topics they debated throughout their correspondence was the problem of determining what Cartan called the *index of generality (degree of indeterminacy, etc.)* and Einstein called the *strength* of a system of differential equations in involution. The smaller the index of generality the more deterministic the system. The analogue in Kolchin's dimension theory of Cartan's "index of generality" is "typical differential dimension." Cartan remarks in his letter to Einstein of April 29, 1932 (p. 217), that for systems of partial differential equations, in contrast to ordinary differential equations, the specification of the index of generality is, to a great extent, a matter of convention. For physicists, who do not insist on analytic solutions and analytic Cauchy data, the solutions of the heat equation depend on *one* arbitrary function of x, the initial temperature distribution $y(0, x)$, corresponding to the choice of $\frac{\partial y}{\partial t}$ as leader of the heat equation. The fact that there need not exist an analytic solution at $(0, 0)$ with prescribed initial temperature distribution analytic at $x = 0$, was first shown by Sonya Kovalevaskaya in her dissertation [von KOVALESVSKY 1875]. Her elegant counterexample has Cauchy datum $y(0, x) = \frac{1}{x-1}$. There is no solution of the heat equation analytic at $(0, 0)$ with this prescribed initial temperature distribution. Cauchy-Kovalevsky theory is an analytic theory, and Cartan's choice of index of generality was influenced by his preference for analytic solutions and Cauchy data. Furthermore, according to his letters to Einstein, it was also influenced by his wish that the index of generality be invariant under transformations of the independent variables, and under transformations of the dependent variables whose coordinate functions are analytic functions of the dependent variables and their derivatives. As we noted above, the differential type and the typical differential dimension are differential birational invariants (the coordinate functions are rational functions of the dependent variables and their derivatives), as well as invariants under transformations of the distinguished derivation operators by Jacobian matrices (arbitrary transformations of the independent variables). We should also note that the underlying context of differential algebra chosen by Ritt and Kolchin is the theory of analytic functions.

In his address to the 1966 International Congress of Mathematicians, one of the challenging problems Kolchin described concerned a conjecture made by the physicist Maurice Janet [1921], which he formulated as follows:

> Assume the cardinality of Δ is m, let S be a subset of cardinality n of the differential polynomial ring $R = F\{y_1, ..., y_n\}$, and suppose P is a component of the radical differential ideal generated by S of differential type less than $m - 1$. Then $\omega_p = 0$. [KOLCHIN 1968a]

As originally formulated, the Janet conjecture was restricted to homogeneous linear differential equations that satisfy a certain independence condition. A family $L_1, ..., L_n$ of homogeneous linear differential polynomials in the differential polynomial ring $R = F\{y_1, ..., y_n\}$ is *differentially linearly independent* over F if the family of derivatives of $L_1, ..., L_n$ is linearly independent over F. The differential ideal generated be a family of homogeneous linear differential polynomials is easily seen to be prime. In the language of differential algebra, the original Janet conjecture for homogeneous linear differential ideals reads:

> Assume the cardinality of Δ is m, let $L_1, ..., L_n$ be a family of n homogenous linear differential polynomials that is differentially linearly independent over F. Then the differential type of the differential ideal generated by $L_1, ..., L_n$ is either $m-1$ or -1.

So, the Janet conjecture states that for an independent system of n homogeneous linear differential equations in n unknown functions, either 0 is the only solution, or the general solution depends on at least one arbitrary function of $m-1$ independent variables. Using a theorem about differential operator rings [GOODEARL 1975], Johnson proved the Janet conjecture in the homogeneous linear case [JOHNSON 1978].

1.11.2. *Estimating the Typical Differential Dimension: the Ritt-Jacobi Bound.* Computing (or even estimating) the differential birational invariant a_V, when the differential dimension is 0, is a difficult problem. In the case of ordinary differential equations ($m = 1$) of differential dimension 0, the typical differential dimension is the transcendence degree over F of the differential field of differential rational functions on V (the number of arbitrary constants in the "general solution"). Ritt considered this special case $m = 1$ for a system of n differential polynomials $f_1, ..., f_n$ in $F\{y_1, ..., y_n\}$. Set e_{ij} equal to the smallest natural number such that f_i does not involve a derivative of y_j of order greater than e_{ij}. Set $h = \max_\sigma(e_{1,\sigma(1)}+\cdots+e_{n,\sigma(n)})$, $\sigma \in S_n$, where S_n is the symmetric group. He proved that if $f_1, ..., f_n$ are linear, or if $n = 2$, then every component of the radical differential they generate that has differential dimension 0 has typical dimension $a \leq h$ [RITT 1935] and [1950a]. This bound h was achieved heuristically, without precise definitions, but with an ingenious combinatorial argument, by C. G. J. Jacobi. (Richard Cohn has worked through the Latin.) It is known as the Jacobi-Ritt bound, and is an improvement in the special case of ordinary differential equations of the bound achieved by Kolchin in the general case:

Theorem. *Let $e_1, ..., e_n$ be natural numbers, let S be a subset of the differential polynomial ring $F\{y_1, ..., y_n\}$ such that the order of each differential polynomial in S in y_j is $\leq e_j$. Let P be a component of the radical differential ideal generated by S that has differential type $m-1$. Then the typical differential dimension of P is $\leq e_1 + \cdots + e_n$.* [KOLCHIN 1973], p. 199

It is surprising that a natural conjecture of Ritt's, called the *dimension conjecture,* is still unproved:

> Let S be a system of $k < n$ ordinary differential polynomials in n differential indeterminates such that the set of zeros of S is not empty. Then every component of the radical differential ideal generated by S has positive differential dimension. [RITT 1950a]

Attempts have been made to prove the Ritt-Jacobi bound in more generality, to generalize it to the partial case, and to improve it. See [LANDO 1970], [TOMASOVIC 1976], [COHN 1983]. However, Cohn showed that a much improved bound, called the *strong Jacobi bound,* true in the linear case, fails for certain systems of ordinary differential polynomials if the dimension conjecture is false.

1.11.3. *Cartan's example, letter to Einstein of December 3, 1929.* The system that models the theory of matter and graviation. [DEBEVER 1979]. The differential indeterminates are written in Cartan's notation.

The 7 physical quantities in the equations are the components X, Y, Z of the gravitational acceleration, the matter density ρ, and the components u, v, w of an element of matter. There are 4 independent variables, x, y, z, t. $\Delta = \left\{ \frac{\partial}{\partial t}, \frac{\partial}{\partial x}, \frac{\partial}{\partial y}, \frac{\partial}{\partial z} \right\}$.

$$\begin{aligned}
f_1 &= \frac{\partial Y}{\partial z} - \frac{\partial Z}{\partial y} = 0 \\
f_2 &= \frac{\partial Z}{\partial x} - \frac{\partial X}{\partial z} = 0 \\
f_3 &= \frac{\partial X}{\partial y} - \frac{\partial Y}{\partial x} = 0 \\
f_4 &= \frac{\partial X}{\partial x} + \frac{\partial Y}{\partial y} + \frac{\partial Z}{\partial z} = 4\pi a \rho \\
f_5 &= \frac{\partial \rho}{\partial t} + \frac{\partial}{\partial x}(\rho u) + \frac{\partial}{\partial y}(\rho v) + \frac{\partial}{\partial z}(\rho w) = 0 \\
f_6 &= \frac{\partial u}{\partial t} + u\frac{\partial u}{\partial x} + v\frac{\partial u}{\partial y} + w\frac{\partial u}{\partial z} = X \\
f_7 &= \frac{\partial v}{\partial t} + u\frac{\partial v}{\partial x} + v\frac{\partial v}{\partial y} + w\frac{\partial v}{\partial z} = Y \\
f_8 &= \frac{\partial w}{\partial t} + u\frac{\partial w}{\partial x} + v\frac{\partial w}{\partial y} + w\frac{\partial w}{\partial z} = Z, \text{ where } a \in F.
\end{aligned}$$

To describe our choice of ranking, we set $\delta_1 = \frac{\partial}{\partial t}, \delta_2 = \frac{\partial}{\partial x}, \delta_3 = \frac{\partial}{\partial y}, \delta_4 = \frac{\partial}{\partial z}$, $y_1 = X, y_2 = Y, y_3 = Z, y_4 = \rho, y_5 = u, y_6 = v, y_7 = w$. We then rank the derivatives $\delta_1^{e_1}\delta_2^{e_2}\delta_3^{e_3}\delta_4^{e_4}y_j$ of the differential indeterminates lexicographically with respect to (order, e_1, e_2, e_3, e_4, j). With this ranking, the leaders of the differential polynomials are:

$$u_{f1} = \frac{\partial Z}{\partial y}, u_{f_2} = \frac{\partial Z}{\partial x}, u_{f_3} = \frac{\partial Y}{\partial x}, u_{f_4} = \frac{\partial X}{\partial x}, u_{f_5} = \frac{\partial \rho}{\partial t}, u_{f_6} = \frac{\partial u}{\partial t}, u_{f_7} = \frac{\partial v}{\partial t}, u_{f_8} = \frac{\partial w}{\partial t}.$$

Set A equal to the set whose elements are the f_i's. A is clearly autoreduced. The only coherence condition we have to verify is the $S-$polynomial associated with f_1 and f_2.

$$\frac{\partial f_1}{\partial x} - \frac{\partial f_2}{\partial y} = \frac{\partial}{\partial z}\left(\frac{\partial Y}{\partial x} - \frac{\partial X}{\partial y}\right) = -\frac{\partial f_3}{\partial z}.$$

The common derivative of $\frac{\partial f_1}{\partial x}$ and $\frac{\partial f_2}{\partial y}$ is $v = \frac{\partial^2 Z}{\partial z \partial y}$. But, $\frac{\partial}{\partial z}u_{A_3} = \frac{\partial^2 Y}{\partial x \partial z} < v$. So, the autoreduced set A is coherent. For Cartan, the set is in involution. Since the differential polynomials are linear in their leaders and have initials and separants equal to 1, the differential ideal generated by A is prime. In particular, the differential equations are compatible, and their solution set is irreducible in the Kolchin topology. Using Kolchin's algorithm, it is not difficult to compute the dimension polynomial of the differential algebraic variety defined by the given system of differential polynomials:

$$6\binom{X+3}{3} + \binom{X+2}{2}.$$

Therefore, the type is 3, and the typical dimension is 6. The "general solution" depends on 6 arbitrary functions of 3 variables. The type and typical differential dimension agree with Cartan's computation of the "index of generality" of the system.

1.11.4. *The Anomaly of Differential Dimension.* To simplify the exposition, we will assume in this section and the next that F is a δ−field. Ritt begins his article [RITT 1939] with the sentence, "This note on algebraic differential equations deals with a peculiar situation, in which three or more functions are uniquely determined by what may very reasonably be called two conditions." In algebraic geometry, every component of the nonempty intersection of two algebraic varieties of dimensions r and s in affine n-space has dimension greater than or equal to $r+s-n$. In this paper, Ritt gives an example of an irreducible Kolchin closed subset V of differential dimension 2 in differential affine 3-space that intersects the hyperplane $y_3 = 0$ at the single point $(0,0,0)$. V is the general component of the differential polynomial

$$y_1^5 - y_2^5 + y_3\left(y_1 y_2' - y_1' y_2\right)^2.$$

This example also shows, in contrast to algebraic geometry, that the maximal ideal of the local differential ring on V of the origin is generated as a radical differential ideal by the residue class of y_3. However, the situation is better for differential algebraic groups (Section 2 of this paper). Sit establishes the following formulas:

If V and W are Kolchin closed subsets of $A_n(F)$ defined by homogeneous linear differential polynomials, then diff $\dim(V \cap W) \geq$diff $\dim(V)$+diff $\dim(W) - n$.

If G and H are connected differential groups of nonsingular $n \times n$ matrices, then diff $\dim(G \cap H) \geq$ diff $\dim(G)$+ diff $\dim(H) - n^2$. [SIT 1974]

1.11.5. *Differential Krull Dimension.* Let R be an integral domain, and a finitely generated algebra over a field k. The length of a chain $P_r \supset P_{r-1} \supset \cdots \supset P_0$ of prime ideals of R (strict containments) is r. The *Krull dimension* of R is the supremum of the lengths of chains of prime ideals in R. It is the topological dimension of an affine algebraic variety V with coordinate ring R. The Krull dimension equals the dimension of V, *i.e.*, the transcendence degree over k of the quotient field of R. Moreover, if η is a point of V, the Krull dimension tells us the least number of parameters needed to describe η, *i.e.*, the minimum number of generators of the maximal ideal of the local ring of η on V, as a radical ideal. The above example shows us that the number of parameters needed to describe a point of an irreducible differential algebraic variety of V of positive differential dimension may be smaller than the differential dimension. Kolchin asked what natural number plays the role of Krull dimension for differential affine rings? In order to answer this question, Rosenfeld and Johnson independently investigated the concept of "large gap chains" [ROSENFELD 1959], [JOHNSON 1969a] in affine differential rings.

Let R be an integral domain, and a finitely differentially generated differential algebra over a δ−field of F, and set d equal to the differential transcendence degree over F of the quotient field of R, *i.e.*, the differential dimension of an affine differential algebraic variety with coordinate ring R. Although R satisfies the ascending chain condition on prime differential ideals, it can have infinite strictly descending chains of prime differential ideals. Let $P \supset Q$ be prime differential ideals in R. The gap between P and Q is *large* if there is an infinite strictly descending chain of prime differential ideals between P and Q. A chain $P_r \supset P_{r-1} \supset \cdots \supset P_0$ of prime differential ideals of R is a *large gap chain* of length r if the gap between any two successive links in the chain is a large gap. Johnson defined the *differential Krull dimension* of R to be the supremum of the lengths of large gap chains of prime differential ideals of R, proved that there is a large gap chain of length d, and that the Krull dimension equals d. So, the "'topological dimension" of an affine differential algebraic variety equals its differential dimension. But, the Krull dimension does not pin down the minimum number of parameters need to describe points of differential algebraic varieties. And, there is not as yet a catenary theorem. Let us state the catenary theorem geometrically - - for an irreducible differential algebraic variety V - - and start with an arbitrary point $\eta \in V$. Does there exist a large gap chain of irreducible differential algebraic subvarieties, of length d, beginning at η and ending at V? Rosenfeld gave an affirmative answer under the condition that η is nonsingular with respect to a certain ranking of the differential indeterminates. But, he expressed doubt that his nonsingularity condition is needed. It is interesting that this nonetheless gives a catenary theory on a Kolchin open subset of V. It is also interesting that the Ritt example illustrating the anomaly of differential dimension provides evidence supporting Rosenfeld's doubts about the necessity of a nonsingularity condition. The origin is a singular point of the variety, in Rosenfeld's sense, and yet is not difficult to show that there is a large gap chain of length 2, beginning at the origin and ending at V. Indeed, such a chain exists beginning at any point of V. So, the question remains: Does every irreducible differential algebraic variety of positive differential dimension d satisfy the property that if $U_1 \subset U_2$ are irreducible Kolchin closed subsets of V, with d_i the differential dimension of U_i, and $d_1 < d_2$, does there exist a large gap chain of irreducible Kolchin closed subsets of V, of length $d_2 - d_1$,beginning at U_1 and ending at U_2 ?

1.12. **Differential Projective Space and a Conjecture of Schmidt-Kolchin.** It was because of what may be described as an idle remark of Ritt's made years earlier that Kolchin began to think about non-affine differential algebraic varieties. He repeats the remark in [KOLCHIN 1974b]:

> Consider an irreducible algebraic variety V in complex projective space $P_n(C)$ and $n+1$ meromorphic functions $f_0, ..., f_n$ on some region of $\mathbb{C}$. J.F. Ritt once remarked to me that there exists an irreducible ordinary differential polynomial $h \in \mathbb{C}\{y_0, ..., y_n\}$, dependent only on V and having order equal to the dimension of V, that enjoys the following property: A necessary and sufficient condition that there exist $c_0, ...c_n \in \mathbb{C}$ not all zero such that $(c_0 : c_1 : \cdots : c_n)$ is a point of V and $\sum c_j f_j = 0$ is that $(f_0, ..., f_n)$ be in the general solution of h.

The remark of Ritt's generalizes the Wronskian condition for linear independence over constants. Then $n+1$ -tuple $(f_0, ..., f_n)$ is said to be *linearly dependent over* V.

Kolchin writes, "This provides an occasion to describe the beginnings of a theory of algebraic differential equations in a projective space." So, let F be a differentially closed field, with set Δ of commuting derivation operators. Let $R = F\{y_0, ..., y_n\}$ be the differential polynomial ring in $n+1$ differential indeterminates. Let $f \in R$ and let d be a non-negative integer. Then F is *differentially homogeneous of degree* d if $f(ty) = t^d f(y)$, where $y = (y_0, ..., y_n)$, and t is a differential indeterminate over R. When $f \neq 0$, d is unique, namely, the degree of f. It is evident that every differentially homogeneous differential polynomial is homogeneous. The property of being differentially homogeneous depends very much on the coefficients of the differential polynomial. If F is a δ–field, and $n = 1$, the Wronskian $y_0 y_1' - y_0' y_1$ is differentially homogeneous, but $y_0 y_1' - 2y_0' y_1$ is not.

Kolchin devoted two papers to differential projective space $P_n(F)$: [KOLCHIN 1974b], and [KOLCHIN 1992], published posthumously. In the former, he develops the foundations for a theory of differentially homogeneous differential ideals and their zero sets in $P_n(F)$, which define the Kolchin topology. Since the affine open subsets of projective n-space are patched together birationally, the dimension polynomial of an irreducible Kolchin closed subset of projective space is well-defined. Kolchin proves the following theorem:

Theorem. *Let C_0 be a subfield of the field of constants C of F, and let V be an irreducible algebraic C_0- variety in $P_n(C)$. Let W be the set of points of $P_n(F)$ that are linearly dependent over V. Then W is a C-irreducible Kolchin C-closed subset of $P_n(F)$ and* $\omega_W = (n-1)\left(\begin{array}{c} X+m \\ m \end{array}\right) + \dim V$.

Corollary. *Let C and V be the same as in the theorem, and suppose that the differential fields are ordinary (that is, $m = 1$). Then there exists a differential polynomial $h \in C\{y_0, ..., y_n\}$, algebraically irreducible over C, such that an element of F^{n+1} is linearly dependent over V if and only if it is in the general solution of h. h is unique up to a nonzero factor in C and is differentially homogeneous. When V lies in the hyperplane $y_j = 0$ then h is differentially free of y_j; when V does not lie in the hyperplane $y_j = 0$ then the order of h in y_j equals the dimension of V.*

This establishes Ritt's claim. At the end of the paper, Kolchin adds:

> Professor Morikawa has pointed out to me that the ordinary differential polynomial h appearing in the Corollary...equals the Cayley form of V computed at the signed minors of the matrix $\left[y_j^{(i)}\right]_{0\leq i\leq d, 0\leq j\leq n}$, where d equals $\dim V$. Equivalently, if $g\left((u_{ij})_{0\leq i\leq d, 0\leq j\leq n}\right)$ denotes what van-der Waerden calls the "zugeordnete Form" of V (see section 36 of his "Einführing in die Algebraische Geometrie," Springer, Berlin, 1939), then $h = g\left(y_j^{(i)}\right)_{0\leq i\leq d, 0\leq j\leq n}$. [KOLCHIN 1974b]

Projective differential algebraic varieties are intriguing because of their startling "incompleteness." There are so many *affine* projective differential algebraic varieties. For example, the reason for the anomaly of differential dimension provided by Ritt's example discussed in section 1.11.4 is that the affine differential algebraic variety in the example is equal to its projective Kolchin closure. Another example is provided (in the ordinary differential field case) by $P_1(C)$, which is δ-isomorphic to a Riccati variety! The phenomenon, which is even more surprising, of the existence of non-constant $\delta-$regular maps from projective algebraic varieties to affine spaces was explored by Buium, and will be discussed in sections 2.5 ff. of this paper.

Kolchin's second paper [1992] touching on differential projective space was also his last paper. In this note, he begins work on a conjecture of Wolfgang Schmidt. In a paper [SCHMIDT 1979], in which he clarifies a paper by Charles Osgood [OSGOOD 1973], Schmidt proved that the dimension of the vector space $H_d(n)$ over F of differentially homogeneous differential polynomials of degree d in $F\{y_0, ..., y_n\}$ is $\geq (n+1)^d$. This is known as the Schmidt-Osgood inequality. These papers by Osgood and Schmidt were very interesting to Kolchin, because they are devoted to a subject that was close to his heart, namely, his desire to prove an analogue for differentially algebraic functions (*i.e.*, differentially algebraic over $C(x)$) of the

Thue-Siegel-Roth improvement on Liouville's theorem for algebraic numbers. In [KOLCHIN 1959], he proved an analogue for differentially algebraic functions of Liouville's theorem, using the concept of a valued differential field. He also expressed the wish that his analogue of Liouville's theorem could be refined to give an analogue of the Thue-Siegel-Roth theorem. This problem is now known as Kolchin's "$2+\varepsilon$" problem for differentially algebraic functions.

The case $n=1$ of the Schmidt-Osgood inequality implies, as is shown in Osgood's paper, that an algebraic function $u=u(x)$ of degree k over $C(x)$ satisfies the main result of [KOLCHIN 1959], and does not admit rational approximation at ∞ of order $> \left[\frac{\log k}{\log 2}\right]+1$. Kolchin goes on to point out that "the Schmidt-Osgood inequality yields a vast (and easy) improvement of the C. L. Seigel (-B. P. Gill) exponent, but of course not so good as the ultimate improvement achieved by K. F. Roth (-S. Uchiyama) [KOLCHIN 1992], p. 450. We should mention that in 1984 D. V. and G. V. Chudnovsky, using Padé approximations, solved Kolchin's "$2+\varepsilon$" problem in the case in which the differentially algebraic function to be approximated is a solution of a *linear* differential equation with coefficients in $C(x)$. [CHUDNOVSKY and CHUDNOVSKY 1984].

The Schmidt conjecture replaces the Schmidt-Osgood inequality with an equality. As Schmidt remarks in [SCHMIDT 1979], it is not even clear that $\dim H_d(n)$ is finite. In [KOLCHIN 1992], Kolchin is able to prove only that in the case $n=1$ (projective 1-space), the Schmidt conjecture is true for $d \leq 3$. Recently, Georg Reinhart proved that $\dim H_d(1)$ is 2^d, thus establishing the truth of the Schmidt conjecture for $n=1$ [REINHART 1997].

2. Differential Algebraic Groups and Diophantine Geometry

We now set F equal to a $\Delta-$closed field with field C of constants. Recall that C is algebraically closed. We will often indicate the differential structure by replacing "differential", "differentially", "differential algebraic", *etc.*, by the prefix "$\Delta -$ " (or "$\delta -$ " if the cardinality of Δ is 1). If F_0 is a $\Delta-$subfield of F, we will write "$\Delta - F_0-$set" or "$\delta - F_0-$set" in place of "differential algebraic variety defined over F_0." An algebraic group defined over F_0 will often be called an F_0 -group. We will also use "$\Delta-$regular" in place of "everywhere defined."

In 1985, Kolchin completed what Armand Borel, in a paper in this volume, calls his last major project. It was the subject both of his Colloquium Lectures given at the seventy-ninth summer meeting of the American Mathematical Society [1975], and of his second book. In this last work, which occupied him for a decade, Kolchin established the foundations for a general theory of differential algebraic groups and differential algebraic geometry for subsets of homogeneous spaces. A theory of $\Delta-$varieties in differential algebra had earlier been developed by Buium [1982] and William Keigher [1982a], [1982b], and [1983]. However, the anomalies following from the fact that the ring of $\Delta-$regular functions on an affine differential algebraic variety need not be the coordinate ring, as well as technical difficulties arising from rationality problems, influenced Kolchin to base the theory of differential algebraic groups on a set of axioms, as he had done earlier for algebraic groups. As Borel points out in section 10.7 of his paper, Kolchin was able to use Chapter V of his first book [1973] "almost as a blueprint." See Borel's paper, sections 10.1-10.6, for a discussion of Kolchin's axiomatic treatment of algebraic groups. Recently, Anand Pillay gave a model-theoretic proof that the category of differential algebraic groups defined as group objects in the category of $\Delta-$varieties, the category of differential algebraic groups as defined axiomatically by Kolchin, and the category of groups definable in differentially closed fields, are equivalent [PILLAY 1997]. In the same paper, Pillay answers affirmatively the three questions posed by Kolchin in the preface of [1985]:

> Let F be a Δ-field, and let G be a $\Delta - F-$group. Let x be an element of G. Is G locally $F-$affine at x?
> Is there an analogue for differential algebraic groups of the "Chevalley-Barsotti theorem" for algebraic groups?
> Can every differential algebraic group be embedded in an algebraic group?

There are as yet no differential algebraic proofs.

Kolchin's meticulous axiomatic treatment lends itself readily to the development of his "constrained cohomology" that classifies the principal homogenous spaces of differential algebraic groups, treats rationality questions delicately, and provides a natural context for a Galois theory of "infinite-dimensional" differential algebraic groups. The beautiful Galois theory of *finite-dimensional* differential algebraic groups developed by Pillay and David Marker [1996] includes Kolchin's differential Galois theory as a special case, but is by no means limited to this case, due to the presence of the "exotic" differential algebraic groups that do not descend to constants [BUIUM 1993a]. Finite-dimensional differential algebraic groups will be discussed in section 2.5 ff. of this paper.

2.1. **Basic Notions.** The theory of differential algebraic groups, initiated in [CASSIDY 1972], was first developed in the affine case. Following "Groupes linéaires algébriques" as a model [BOREL 1956; see also BOREL 1991], a $\Delta - F_0-$group G was defined to be a $\Delta - F_0-$ subset of $A_n(F)$, equipped with a group structure defined by multiplication and inversion maps that are $\Delta-$regular maps whose coordinate functions have coefficients in F_0. A morphism in the category of $\Delta-$groups (*resp.*Δ–F_0-groups) will be called a $\Delta-$homomorphism (*resp.*Δ–F_0-homomorphism). We call an arbitrary $\Delta - F_0-$group *affine* (or $\Delta - F_0-$affine, if we want to emphasize rationality over F) if it is embeddable in some differential affine space by a $\Delta-$isomorphism (or $\Delta - F_0-$isomorphism). The theory of differential algebraic groups departs here from the theory of algebraic groups. Assume that F is a $\delta-$field and that x is an element of F whose derivative is 1. The $\delta-$group of $C-$rational points of an elliptic curve defined over C is a $\delta - C(x)-$affine differential algebraic group, although it is not, of course affine as an algebraic $C-$group.

Every algebraic group G can be given a canonical structure of differential algebraic group [KOLCHIN 1985], Chapter II, section 3. If G is a Zariski closed subset of affine $n-$space, we can just regard it as a differential algebraic group whose defining differential ideal is generated by differential polynomials or order 0, and whose multiplication and inversion operations have order 0 (are rational). If $F\{G\} = F\{\bar{y}_1, ..., \bar{y}_n\}$ is the differential coordinate ring of G, viewed as a differential algebraic group, then $F[\bar{y}_1, ..., \bar{y}_n]$ can be identified with the coordinate ring of the algebraic group G. By Kolchin's Irreducibility Theorem [1973], p. 200, G is connected as an algebraic group if and only if it is connected as a differential algebraic group, and if $F\langle G\rangle = F\langle\bar{y}_1, ...\bar{y}_n\rangle$, then $F(\bar{y}_1, ..., \bar{y}_n)$ can be identified with the field $F(G)$ of rational functions on the algebraic group G. The interplay between the algebraic and differential algebraic group structures is ingeniously treated as a special case of the "change of derivation operator" process in [KOLCHIN 1985]. It is not surprising that the dimension of G as an algebraic group equals the differential dimension of the differential algebraic group G. In the study of linear differential algebraic groups, we place on the algebraic group $GL_n(F)$, closed up in the usual way, the canonical structure of differential algebraic group. Although if G is a $\Delta-$subgroup of $GL_n(F)$, it follows that the ring of $\Delta-$regular functions equals the coordinate ring [CASSIDY 1975], this is not necessarily true for all affine differential algebraic groups. As a result, there are affine differential algebraic groups that are not linear, the group of C- rational points of an elliptic curve defined over C providing an example. In contrast, every Δ-regular function on a connected F -group is a differential polynomial function, a remarkable fact about all smooth F - varieties [BUIUM 1994a].

The most important tools in the classification of differential algebraic groups that can be embedded in a C-group have been the Lie algebra of the C-group and the "logarithmic derivative" map. Kolchin's student Jerald J. Kovacic used these tools in a significant way in his papers on the inverse problem in differential Galois theory [1969], and [1971]. Michael Singer writes, "A breakthrough was made by Kovacic, who proposed an ultimately successful program to solve the inverse problem for connected groups (over fields of finite, nonzero transcendence degree over the algebraically closed field of constants of a differential field k) and solved the problem for connected solvable groups. A key element of Kovacic's program is an emphasis on the role of the Lie algebra of the Galois group G" [SINGER, this volume].

2.2. **The Lie Algebra of a Differential Algebraic Group.** For any $\Delta-$subgroup G of $GL_n(F)$, the *Lie algebra* $\ell(G)$ *of matrices* of the group is easy to define, following [CHEVALLEY 1951]. If G is an algebraic group, $\ell(G)$ is a Lie subalgebra of $g\ell_n(F)$, if we regard $g\ell_n(F)$ as a Lie algebra over the differential field F. The dimension of $\ell(G)$ over F equals the dimension of G. Since F is a vector space over the field C of constants of F, and the Lie bracket is defined over Q, $\ell(G)$ is also a Lie subalgebra (infinite-dimensional) of $g\ell_n(F)$, regarded as a Lie algebra over C. The Lie algebra of matrices of the *algebraic* group G is equal to the Lie algebra of matrices of the *differential algebraic group* G. For a linear *differential* algebraic group, the most we can say is that $\ell(G)$ is defined by homogeneous linear differential polynomials in the differential polynomial ring $F\{(y_{ij}, i, j = 1, ..., n\}$, and, so, is a Lie subalgebra over C of $g\ell_n(F)$, and, in addition, is closed in the Kolchin topology on $g\ell_n(F)$. It thus satisfies Sophus Lie's requirement that the "infinitesimal transformations" of an "infinite (dimensional) continuous transformation group" satisfy homogeneous linear differential equations. The dimension polynomial of $\ell(G)$ is equal to that of the identity component of G (all the components of G have the same dimension polynomial), and G has the same Lie algebra of matrices as its identity component. Differential algebraic groups resemble their Lie algebras in the expected ways: $\ell(G)$ is abelian, solvable, simple, or semi-simple if and only if the identity component of G has the same property.

Suppose G is a connected $\Delta-$subgroup of $GL_n(F)$. Then, we can represent the Lie algebra of matrices of G as the Lie algebra of right invariant derivations over F of $F\langle G\rangle$ that commute with the distinguished derivations $\delta \in \Delta$. We call them *invariant* Δ *- derivations* on G. They preserve order, and leave stable the coordinate ring of G. If G is an algebraic group, an invariant $\Delta-$derivation over F

on G is obtained by prolonging in a natural way an invariant derivation over F of the field $F(G)$ of rational functions on G. Thus, if F is a δ-field, the right invariant $\delta-$derivations on the multiplicative group $G_m = GL_1(F)$ can be written

$$D = uy\frac{\partial}{\partial y} + (uy' + u'y)\frac{\partial}{\partial y'} + (uy'' + 2u'y' + u''y)\frac{\partial}{\partial y''} + ...,$$

where $u \in g\ell_1(F) = G_a$, the additive group of F.

In [1985], Kolchin defined a $\Delta-$*Lie algebra over* C to be a Lie algebra over C whose additive group is a $\Delta-$group, and whose operations of scalar multiplication and Lie bracket are Δ-regular maps [KOLCHIN 1985], p. 152. It is not known whether every $\Delta-$Lie algebra over C is affine. Kolchin devotes Chapter VIII of [1985] to the Lie theory of differential algebraic groups. The Lie algebra of a connected differential algebraic group is the set of right invariant $\Delta-$derivations over F of $F\langle G\rangle$, and has the structure of $\Delta-$Lie algebra over C. Although it is known that the Lie algebra of G is affine [KOLCHIN 1985], Chapter VIII, it is not known whether it is embeddable in $g\ell_n(F)$ (is linear); there are many classical examples of affine, non-linear $\Delta-$Lie algebras over C. Although Kolchin's dimension polynomial does not extend to differential algebraic groups that are not embedded in affine space, since it is not a differential birational invariant, the differential type and typical differential dimension can be defined, and are the same for the group and its Lie algebra. Kolchin also defined the Lie algebra of a principal homogeneous space for a $\Delta-$group.

2.3. **The Logarithmic Derivative Map.** Let $A = [a_{ij}]$ be an $n \times n$ matrix with entries in F, and let $\delta \in \Delta$. Then $\delta A = [\delta a_{ij}]$. If $g \in GL_n(F)$, then $\ell\delta(g) = \delta(g)g^{-1}$. If the cardinality of Δ is 1, this formula defines the *logarithmic derivative map* from $GL_n(F)$ into $g\ell_n(F)$. For $\Delta = \{\delta_1, ..., \delta_m\}$, $m > 1$, we define the logarithmic derivative map from $GL_n(F)$ into $g\ell_n(F) \times \cdots \times g\ell_n(F)$ by the following formula:

$$\ell\Delta(g) = (\ell\delta_1(g), ..., \ell\delta_m(g)).$$

Suppose G is a connected $C-$subgroup of $GL_n(F)$. First, assume that F is a δ-field.

Theorem. *$\ell\delta$ maps G surjectively onto $\ell(G)$.* [Kovacic 1969]

If $m > 1$, integrability conditions come into play:

Define the $\Delta-$subgroup I of the additive group of $\ell(G) \times \cdots \times \ell(G)$ by the "integrability conditions"

$$\delta_i A_j - \delta_j A_i = [A_i, A_j],$$

where $[A_i, A_j] = A_iA_j - A_jA_i$, the Lie bracket of A_i, A_j. Then $\ell\Delta$ maps G surjectively onto I.

G acts on $\ell(G) \times \cdots \times \ell(G)$ by the adjoint action on each factor. This induces an action of G on the Δ-group I. $\ell\Delta$ is a crossed homomorphism under this action, with kernel the group $G(C)$ consisting of the matrices in G with entries in C. If G is commutative, $\ell\Delta$ is a homomorphism from G onto I.

In his second paper on the inverse problem [1971], Kovacic defined the logarithmic derivative map $\ell\delta$ for arbitrary connected C-groups and proved that it is a crossed homomorphism for the adjoint action. Later, Kolchin studied logarithmic derivative maps $\ell\varepsilon$ for connected $\Delta - F-$groups, where ε is a $\Delta-$derivation over F of a $\Delta-$extension field of F, [KOLCHIN 1985], Chapter VIII, Sections 8-10, but, if $m > 1$, the concept of the logarithmic derivative map $\ell\Delta$, with its integrability conditions, has not yet been extended to non-linear groups.

2.4. **The Classification of Differential Algebraic Groups.** In section 3 of his paper in this volume, Borel mentions that in his 1948 Picard-Vessiot paper, Kolchin chose to develop a "theory of linear algebraic groups over algebraically closed fields of *arbitrary* characteristic." We mentioned in section 1.10 that, as Manin pointed out, Ritt and Kolchin were cautious about transforming the differential indeterminates or the distinguished derivation operators. The twin desires: (1) to remove the theory of algebraic matrix groups from Lie theory, and replace "infinitesimal methods" by global algebraic geometry; (2) to take care not to blur the particular analytic nature of the solutions of a system of differential equations by trying to reduce the system to "canonical form," influenced the evolution of differential algebra and gave it an arithmetic rather than a Lie-theoretic flavor. It was important to recent developments in differential algebra geometry that, although (1) and (2) were modified both in differential Galois theory and differential algebraic geometry, the arithmetic emphasis remained.

2.4.1. *Unipotent Differential Algebraic Groups.* A $\Delta-$group is *unipotent* if it has a descending normal sequence whose factor groups are $\Delta-$subgroups of G_a. A unipotent $\Delta-$group is connected, and a unipotent $\Delta-$subgroup of $GL_n(F)$ is conjugate to a subgroup of the upper triangular group with 1's on the diagonal. It is probably true that every unipotent $\Delta-$group is linear. On the one hand, it is easy to describe the $\Delta-$subgroups of the additive group G_a^n of the vector space F^n; they are defined by homogeneous linear differential equations. On the other hand, the cardinality of Δ influences their structure as $\Delta-$groups. The finite-dimensional $\Delta-$group $G_a(C)$ has no proper non-trivial $\Delta-$subgroups, hence satisfies the property that every non-trivial $\Delta-$homomorphism is a $\Delta-$isomorphism. G_a has an abundance of $\Delta-$subgroups, since every homogeneous linear differential ideal in $F\{y\}$ defines one. However, if F is a δ-field, every non-trivial homomorphic image of G_a is $\delta-$isomorphic. This is no longer true if the cardinality of Δ is greater than 1. Furthermore, if F is a $\delta-$field, the theory of modules over left-and-right-Euclidean domains [van der WAERDEN 1931], Zweiter Teil, pp. 120-126[1], gives us a direct product decomposition of commutative linear unipotent $\delta-$groups G. G is $\delta-$isomorphic to $G_a^d \times G_a(C)^e$, where d is the differential dimension of G. So, if F is an ordinary differential field, a unipotent $\delta-$group has a composition series, and the factor groups are $\delta-$isomorphic either to G_a or $G_a(C)$.

The underlying $\Delta-$variety of a linear unipotent $\Delta-$group is $\Delta-$birationally isomorphic to the underlying $\Delta-$variety of a $\Delta-$subgroup of G_a^e for some e. An open question is whether every $\Delta-$group whose underlying $\Delta-$variety is differential affine n-space is unipotent, as is the case for algebraic groups [LAZARD 1955]. The problem is untouched if the cardinality of Δ is greater than 1. If F is a $\delta-$field, the problem is solved only for $n \leq 2$. The affirmative answer for $n \leq 2$ rests on what Lorch calls "the last phase of Ritt's writing" [LORCH 1951], a series of papers on a generalization of formal Lie groups, which Ritt called "differential groups" [RITT 1950a], [1950b], [1950c], [1951a], [1951b], [1952]. See also [NICHOLS and WEISFEILER 1982]. In [CASSIDY 1979], Ritt's differential groups, which are essentially $n-$tuples of differential power series, were replaced by $\delta-$Lie algebras

[1] In the English edition of 1950, the Euclidean domains were assumed to be commutative.

obtained by antisymmetrizing the terms of degree 2 in the power series (see Lazard's 1975 monograph on commutative formal Lie groups).

The only δ−group structure on the differential affine line is G_a, but there are infinitely many δ−group structures on the differential affine plane $A_2(F)$. Here, differential algebraic group theory deviates from the theory of algebraic groups over fields of characteristic 0, but parallels algebraic group theory if the ground field has characteristic $p > 0$. This is not surprising, since unipotent δ-groups bear a striking resemblance to unipotent algebraic groups defined over perfect fields of positive characteristic. If k is a perfect field of positive characteristic p, and ϕ is the Frobenius endomorphism, the ring $k[\phi]$ is also a left- and right- Euclidean domain.

Let G be a δ−group with underlying δ−set the plane. It follows that G is linear and we can represent the Lie algebra of matrices of G as a δ−Lie algebra structure on the plane. But, these Lie algebras were already classified in [RITT 1951b], and it was not difficult to see that only the solvable Lie algebras can possibly be the Lie algebras of δ−groups. Therefore, G is solvable, and a weak form of Lazard's theorem shows that G is unipotent, and, indeed, is a central extension of G_a by G_a.

In Chapter VII of *Differential Algebraic Groups,* Kolchin develops two cohomology theories, called *constrained cohomology* and *differential rational cohomology theory*, the latter paralleling rational cohomology. He proves the following important theorems, where F_0 is a Δ-subfield of F, and G is a $\Delta - F_0-$ group, card $\Delta = m$.

Theorem. *The first constrained cohomology set $H^1_\Delta(F_0, G)$ is canonically isomorphic to the first Galois cohomology set $H^1(F_0, G)$.*

Corollary. *$H^1_\Delta(F_0,G)$ is trivial if G is one of the $\Delta-\mathbb{Q}-$groups G_a, $GL_n(F)$, $SL_n(F)$, or if G is an F_0- group and F_0 is algebraically closed.*

Theorem. *There is a bijective correspondence between the set of $\Delta - F_0-$isomorphism classes of principal homogeneous $\Delta - F_0-$spaces for the $\Delta - F_0-$group G and the first constrained cohomology set $H^1_\Delta(F_0, G)$.*

Corollary. *Every central extension of G_a by G_a has a $\Delta-$rational cross section.*

The following results follow [CASSIDY 1977]:

Theorem. *Every central extension of G_a by G_a has a $\Delta-$polynomial cross section.*

Theorem. *Every commutative $\Delta-$group whose underlying $\Delta-$set is the plane is $\Delta-$isomorphic to $G_a \times G_a$.*

Theorem. *The second differential rational cohomology group $H^2_{\delta-rat}(G_a, G_a)$ is a left module over the ring $F[\delta]$ of linear differential operators, generated by the cohomology classes of the differential polynomials $y_1 y_2^{(j)}$, j an odd positive integer.*

The last theorem, as well as the following description of the δ−group structures on the plane, follows from Ritt's remarkable computation of a complete set of representatives of the 2-cohomology classes. A well-thumbed guide on the road to these theorems was provided by [SERRE 1959], Chapter VII. If k is a field of characteristic $p > 0$, the left $k[\phi]$−module $H^2_{rat}(G_a, G_a)$ of cohomology classes of rational 2-cocycles from G_a into G_a is generated by the cohomology classes of the polynomial 2-cocycles $y_1 y_2^{p^r}$, r a positive integer, and the image modulo p of the polynomial $\frac{1}{p}((y_1 + y_2)^p - y_1^p - y_2^p)$ [DEMAZURE and GABRIEL 1970],

[ROSENLICHT 1957]. The latter generates the classes belonging to commutative extensions. So, the analogies are again striking.

Since a δ-group G whose underlying δ-set is the plane is a central extension of G_a by G_a, it follows that there exists a 2-cocycle of the form $f(y_1, y_2) = \sum_{i<j} a_{ij} y_1^{(i)} y_2^{(j)}$ such that

$$(\eta_1, \eta_2) + (\zeta_1, \zeta_2) = (\eta_1 + \zeta_1, \eta_2 + \zeta_2 + f(\eta_1, \zeta_1)).$$

2.4.2. *The Differential Algebraic Subgroups of G_m^n.* A Δ-homomorphism from a Δ-group G into G_a is called a Δ-*character.* There are no rational characters on the algebraic group G_m^n, where G_m is the multiplicative group of F. On the other hand, the group of Δ-characters is large. They are all of the form $L \circ \ell\Delta$, where L is a homogeneous linear differential polynomial function on G_a^{mn} [CASSIDY 1972].

Theorem. *Every proper Zariski dense Δ-subgroup G of G_m^n is connected, contains $G_m(C)^n = \ker \ell\Delta$, and is the intersection of the kernels of Δ-characters.*

If F is a δ-field, the structure of the lattice of connected δ-subgroups of G_m^n greatly resembles that of an abelian variety. The δ-subgroups of an abelian variety were classified in [BUIUM 1993a]. This is most apparent when $n = 1$, and we have an exact sequence

$$0 \to G_m(C) \to G_m \xrightarrow{\ell\delta} G_a \to 0.$$

$G_m(C)$ is the Kolchin closure of the torsion group of G_m, and is the smallest Zariski dense δ-subgroup of G_m. If G is a Zariski dense δ-subgroup of G_m, G is connected and contains $G_m(C)$. If $G \neq G_m$, there exists a homogeneous linear differential polynomial $L \in F\{y\}$ such that G is the solution set in G_m of the differential equation $L\left(\frac{y'}{y}\right) = 0$. G is the kernel of the δ-character $L \circ \ell\delta$. Multiplying by an appropriate power of y, we obtain the defining differential polynomial of the group G. It is a differential semi-invariant of G. For example, the subgroup of G_m defined by the differential equation $\left(\frac{y'}{y}\right)' = 0$ has defining differential polynomial $yy'' - y'^2$. Let E be an elliptic curve on the projective F-plane. If E is defined over C, we have the exact sequence

$$0 \to E(C) \to E \xrightarrow{\ell\delta} G_a \to 0.$$

$E(C)$ is the Kolchin closure of the torsion group of E. If G is a Zariski dense δ-subgroup of E, G is connected and contains $E(C)$. If $G \neq G_m$, there exists a homogeneous linear differential polynomial $L \in F\{y\}$ such that G is the kernel of the δ-character $L \circ \ell\delta$. Note that the order of the δ-polynomial map $\ell\delta$ is 1. If the elliptic curve E does not descent to constants (its j-invariant is not in C), the role of $\ell\delta$ is played by an order 2 δ-polynomial map, defined by the Picard-Fuchs linear differential operator, whose kernel is the closure of the torsion group of E.

2.4.3. *Semisimple Differential Algebraic Groups and Lie Algebras.* Recall that if F_0 is a Δ-subfield of F, an F_0-group is an algebraic F-group defined over F_0.

An ideal of a Δ-Lie algebra over C is a Δ-ideal if it is closed in the Kolchin topology. A Δ-Lie algebra is *simple* it is non-abelian and has no proper non-trivial Δ-ideals. It is *semisimple* if it has no non-trivial abelian Δ-ideals. A connected Δ- group G is *simple* if G is not commutative and has no proper nontrivial connected normal Δ- subgroups. G is *semisimple* if it has no nontrivial connected normal commutative $\Delta-$ subgroups. If G is a connected F - group, then G is simple (*resp.* semisimple) as an algebraic group if and only if it is simple (*resp.* semisimple) as a differential algebraic group. It follows from Pillay's work [1996] that every semisimple $\Delta-$ group G is linear. Although the Zariski closure of a semisimple $\Delta-$ subgroup of an algebraic group is not always semisimple, it is the case that a connected $\Delta-$ group G is semisimple (*resp.* simple) if and only if it can be embedded in a connected semisimple (*resp.* simple) $F-$ subgroup A as a Zariski dense subgroup. Indeed, A can be taken to be an $F-$ group defined over $\mathbb{Q}$- - a $\mathbb{Q}$- group - - and, if G is simple, A can even be taken to be a Chevalley group (a connected simple $\mathbb{Q}$ - group containing a maximal torus diagonalizable over $\mathbb{Q}$).

First, suppose F is a $\delta-$ field. Then we have the following result:

Theorem. *Let G be a connected simple $\delta-$ group. There exists a Chevalley group A such that G is $\delta-$ isomorphic either to A or to $A(C)$.* [CASSIDY 1989], [BUIUM 1993a], [PILLAY 1997].

So, if a connected simple $\delta-$ group has positive $\delta-$ dimension, it is $\delta-$ isomorphic to an algebraic simple group relative to the universe F, equipped with the canonical $\delta-$ structure. It it has zero $\delta-$ dimension, *i.e.*, is finite-dimensional, it is isomorphic to the group of $C-$ rational points of a simple $C-$ group.

Cassidy's proof, which does not separate ordinary from partial differential fields, reduces most of the classification of the semisimple $\Delta-$groups to the parallel classification of the semisimple $\Delta-$ Lie algebras, and combines the methods of differential algebra with the classical theory of semisimple Lie algebras, as presented in [BOURBAKI 1968] with its useful *épinglages*. Buium and Pillay give greatly simplified proofs for ordinary differential fields, Buium's is a geometric proof that makes use of "jet groups," Pillay's assuming finite-dimensionality (finite Morley rank), is model-theoretic. The theorem implies the result, in model theory, that a definable field of finite Morley rank is definably isomorphic to the field C of constants of F.

The above representation theorem does not address the nature of the embedding of a simple $\delta-$ group in a simple $F-$ group, except to say that it is Zariski dense in the $F-$ group. From the point of view of the theory of algebraic differential equations, the nature of the embedding is as important as the Chevalley group representation. The study of the Zariski dense $\delta-$ subgroups of $SL_n(F)$ in [1972] began with two questions inspired by the arithmetic metaphor:

> Is $SL_n(C)$ a maximal $\delta-$ subgroup of $SL_n(F)$?
> Is every proper Zariski dense $\delta-$ subgroup of $SL_n(F)$ conjugate to $SL_n(C)$?

The affirmative answer to the second question implies an affirmative answer to the first. The proof reduced the problem to the parallel one for the Lie algebra $sl_n(F)$. Out of the symbolic computation, using Ritt-Rosenfeld characteristic sets, came a matrix differential equation, now called a Lax equation [LAX 1968]:

$$\delta Y = [A, Y],$$

where $[A, Y] = AY - YA$ is the Lie bracket of matrices. The set of Y in $sl_n(F)$ satisfying the Lax equation is a proper Zariski dense $\delta-$ Lie subalgebra over C (*i.e.*, it contains a basis over F of $sl_n(F)$) and is conjugate by a matrix in $SL_n(F)$ to $sl_n(C)$ In particular, the eigenvalues of each solution of the Lax equation are in C. So, if we interpret the solution Y as a flow on $sl_n(C)$, the eigenvalues of each matrix in the flow are the same as those of the matrix at time $t = 0$. The converse is also true: if $\mathfrak{g}$ is a proper Zariski dense $\delta-$ Lie subalgebra over C of $sl_n(F)$,there here exists a matrix $A \in sl_n(F)$ such that $\mathfrak{g}$ is the set of solutions in $sl_n(F)$ of the corresponding Lax equation.

Sit [1975] used the "conjugacy theorem" to classify the $\delta-$ subgroups of $SL_2(F)$, and applied his classification to solve the following problem, partially solved in [KOLCHIN and SOUNDARAJAN 1972]:

> Let F_0 be a $\delta-$ subfield of F and t a differential indeterminate. Determine all $\delta-$ subfields intermediate to F_0 and $F_0 \langle t \rangle$ over which $F_0 \langle t \rangle$ is strongly normal.

(Ritt had earlier proved the "differential Lüroth theorem," which states that every intermediate $\delta-$ field is generated differentially over F_0 by an element of F_0 differentially transcendental over F_0 [RITT 1950a].)

Let F be a $\Delta-$ field. Set $\mathbf{D}$ equal to the vector space over F with basis Δ. $\mathbf{D}$ has defined on it two useful structures: that of *Lie space* [NELSON 1967], and that of $\Delta-$ Lie algebra over C. The Lie bracket is given by the bracket of operators, and is a quadratic differential polynomial map. Determining canonical forms of simple $\Delta-$ groups exploits the structure on $\mathbf{D}$ of Lie space, and requires the flexibility of being able to change the distinguished $\Delta-$ structures, both by replacing Δ by an arbitrary commuting familyΔ' of derivation operators in $\mathbf{D}$ that are linearly independent over F, and by defining flat connections on finite dimensional Lie algebras over F. The interplay among the differential structures is a central focus of [KOLCHIN 1985] and [CASSIDY 1989]. See Singer's article, section 2.4.1, in this volume, for a discussion of connections and their use in the formulation of differential Galois theory in the language of Tannakian categories. We will put them aside here by limiting the discussion to $C-$ groups and their Lie algebras.

As is the case for ordinary differential fields, because of the surjectivity onto I of the logarithmic derivative map $\ell\Delta$, the problem of deciding the conjugacy classes of the proper Zariski dense $\Delta-$ subgroups of a connected simple $C-$ subgroup of $GL_n(F)$ reduces to the parallel problem for its Lie algebra of matrices. So, let G be a connected simple $C-$ subgroup of $GL_n(F)$, and let $\mathfrak{g} = \ell(G)$. Let Δ' be a commuting subset of cardinality m' of $\mathbf{D}$, linearly independent over F, and let $C(\Delta')$ be the field of $\Delta'-$ constants of F. Every $\Delta'-$ group is canonically a $\Delta-$ group [KOLCHIN 1985], p. 56. Let $\ell = \mathfrak{g} \times \cdots \times \mathfrak{g}$. Then, the additive group of ℓ is a $\Delta'-$ group and a $\Delta-$ group and the subgroup ℓ of the additive group of ℓ, consisting of the solutions of the integrability conditions of section 2.3, with Δ' replacing Δ, is both a $\Delta'-$ subgroup and a $\Delta-$ subgroup of ℓ.

Theorem. *A $\Delta-$ Lie subalgebra $\mathfrak{h}$ of $\mathfrak{g}$ over C is Zariski dense in $\mathfrak{g}$ if and only if there exist a commuting subset Δ' of $\mathbf{D}$, linearly independent over F, and an $m'-$ tuple $(A_1, ..., A_{m'})$ of matrices in I such that $\mathfrak{h}$ is the solution set in $\mathfrak{g}$ of the system of Lax equations*

$$\delta_i' Y = [A_i, Y], \quad i = 1, ..., m'.$$

Corollary. *A $\Delta-$ Lie subalgebra $\mathfrak{h}$ of $\mathfrak{g}$ is Zariski dense in $\mathfrak{g}$ if and only if there exist a commuting subset Δ' of $\mathbf{D}$, linearly independent over F, and a matrix $\eta \in G$ such that $\eta \mathfrak{h} \, \eta^{-1} = \mathfrak{g} \cap gl_n(C(\Delta'))$.*

So, every proper Zariski dense $\Delta-$ Lie subalgebra over C of a simple Lie subalgebra $\mathfrak{g}$ over F of $gl_n(F)$ is conjugate to the Lie subalgebra over C consisting of matrices with entries that are constants of a commuting linearly independent subset of $\mathbf{D}$. The parallel result describes the proper Zariski dense $\Delta-$ subgroups of the simple C - group G. The description of the connected simple $\Delta-$ groups is:

Theorem. *Let G be a connected simple $\Delta-$ group. There exist a commuting subset Δ' of $\mathbf{D}$, linearly independent over F, and a Chevalley group H such that G is $\Delta-$ isomorphic to the group of $C(\Delta')-$ points of H.*[2]

Theorem. *Let $\mathfrak{g}$ be a semisimple linear $\Delta-$ Lie algebra over C.*[3]

(1) $\mathfrak{g} = \mathfrak{g}_1 \oplus \cdots \oplus \mathfrak{g}_k$, *where $\mathfrak{g}_i$ is a $\Delta-$ ideal of $\mathfrak{g}$.*
(2) *There exist k commuting subsets $\Delta_1, ..., \Delta_k$ of $\mathbf{D}$, linearly independent over F, such that $\mathfrak{g}_i$ is a finite dimensional linear simple Lie algebra over $C(\Delta_i)$.*

Theorem. *Let G be a connected semisimple $\Delta-$ group. Then there exist commuting subsets $\Delta_1, ..., \Delta_k$ of $\mathbf{D}$, linearly independent over F, Chevalley groups $H_1, ..., H_k$, and a $\Delta-$ isogeny $\sigma : H_1(C(\Delta_1)) \times \cdots \times H_k(C(\Delta_k)) \to G$.*

Example. Let $\Delta = \left\{\frac{\partial}{\partial x}, \frac{\partial}{\partial y}\right\}$, and let F be the differential closure of $\mathbb{C}(x,y), x, y$ complex variables. Define the $\Delta-$ polynomial map $\lambda : G_a \times G_a \to sl_3(F) \times sl_3(F)$ by the formula $(u,v) \mapsto (A,B)$, where

$$A = \begin{bmatrix} 0 & 1 & 0 \\ 0 & 0 & 1 \\ -3\frac{\partial u}{\partial x} - v & -3u & 0 \end{bmatrix}, \quad B = \begin{bmatrix} 2u & 0 & 1 \\ -\frac{\partial u}{\partial x} - v & -u & 0 \\ -6\frac{\partial^2 u}{\partial x^2} & -2\frac{\partial u}{\partial x} - v & -u \end{bmatrix}$$

We will find a sufficient condition on (u,v) for (A,B) to satisfy the integrability condition

$$\frac{\partial B}{\partial x} - \frac{\partial A}{\partial y} = [A,B].$$

It will follow that the Lax equations

$$\begin{aligned} \frac{\partial Y}{\partial x} &= [A,Y] \\ \frac{\partial Y}{\partial y} &= [B,Y] \end{aligned}$$

are compatible, and define a proper Zariski dense $\Delta-$ Lie subalgebra over C of $sl_3(F)$.

---

[2]In [CARTAN 1909], E. Cartan describes those of his "infinite simple groups not isomorphic to a transitive group," as being obtained by taking a "simple transitive group and making its elements depend in the most general way possible on p variables invariant under the group." He goes on to say that a "finite simple transitive group of order r becomes an infinite intransitive group depending on r arbitrary functions of p arguments." Although it is not quite the same, the fact that every simple $\Delta-$ Lie algebra over the field C of contants of F can be obtained from one of the classical simple $\mathbb{Q}-$ Lie algebras relative to the universe C by extension of scalars to the field of constants of a commuting, linearly independent set of operators in $\mathbf{D}$ is strikingly similar.

[3]Not all semisimple $\Delta-$ Lie algebras over C are linear, *e.g.,* $\mathbf{D}$, which is simple, is non-linear. Since all connected simple $\Delta-$ groups are linear, $\mathbf{D}$ cannot be the Lie algebra of a $\Delta-$ group.

The integrability condition is equivalent to the system of differential equations

$$f = 2\frac{\partial v}{\partial x} - 3\frac{\partial u}{\partial y} + 3\frac{\partial^2 u}{\partial x^2} = 0$$

$$g = \frac{\partial v}{\partial y} - \frac{\partial^2 v}{\partial x^2} + 3\frac{\partial^2 u}{\partial x \partial y} + 6u\frac{\partial u}{\partial x} - \frac{\partial^3 u}{\partial x^3} = 0.$$

Choose a ranking that ranks any derivative of v higher than every derivative of u. Then the autoreduced set

$$f_1 = 2\frac{\partial v}{\partial x} - 3\frac{\partial u}{\partial y} + 3\frac{\partial^2 u}{\partial x^2}$$

$$f_2 = 2\frac{\partial v}{\partial y} + \frac{\partial^3 u}{\partial x^3} + 12u\frac{\partial u}{\partial x} + 3\frac{\partial^2 u}{\partial x \partial y}$$

generates the same differential ideal as f, g. The leader of f_1 is $\frac{\partial v}{\partial x}$, and the leader of f_2 is $\frac{\partial v}{\partial y}$. It is easy to see that the differential ideal $[f_1, f_2]$ generated by f_1, and f_2 is prime if and only if the autoreduced set is coherent.

$$2\left(\frac{\partial f_2}{\partial x} - \frac{\partial f_1}{\partial y}\right) = 2\left(\frac{\partial^4 u}{\partial x^4} + 12\frac{\partial}{\partial x}\left(u\frac{\partial u}{\partial x}\right) + 3\frac{\partial^2 u}{\partial y^2}\right).$$

Let $b(u) = \frac{\partial^4 u}{\partial x^4} + 12\frac{\partial}{\partial x}\left(u\frac{\partial u}{\partial x}\right) + 3\frac{\partial^2 u}{\partial x^2}$. $b(u) = 0$ is called the *Boussinesq equation*, and since $b(u)$ is in the differential ideal generated by f and g, it can be quite properly called a consequence of the Lax equations (or at least of the integrability conditions). If the Boussinesq equation is satisfied by the pair (u, v), then the system of Lax equations is compatible. For each solution u of the Boussinesq equation, the pair (u, v) gives us a proper Zariski dense $\Delta-$ Lie subalgebra over C of $sl_2(F)$. All of them are conjugate to $sl_2(C)$ by matrices of determinant 1. (This example is adapted from example 1, of section 2.2.10 of [PREVIATO 1993].)

There is vast literature concerned with Lax pairs, isospectral theory and spectral curves, isomonodromy theory, Hamiltonian systems, Painlevé analysis, and other related topics. For an approach to Hamiltonian systems, especially Ziglin theory, that uses differential Galois theory and Kovacic's algorithm [KOVACIC 1986] instead of the more traditional monodromy theory, see [CHURCHILL, ROD, SINGER 1995].

2.5. **Differential Algebraic Groups of Finite Dimension.** In the following sections, we make the simplifying assumption that we are in the ordinary case (one derivation operator δ), and use the language of the category of $\delta-$ varieties, as defined in [BUIUM 1982]. We fix a differentially closed $\delta-$ field F, with field C of constants. The geometry on a Kolchin closed subset V of affine $n-$ space $A_n(F)$ can be sheafified in a straightforward way, as follows. Equip V with the Kolchin topology. A $\delta-$ rational function on a Kolchin open subset Ω of V is $\delta-$ *regular on* Ω if it is everywhere defined on Ω. The set of $\delta-$ regular functions on Ω is a $\delta-$ algebra over F, and we can form the associated sheaf O_V on V. Now, let us consider ringed spaces in the "naive sense," *i.e.*, pairs, consisting of a topological space plus a sheaf of rings of $F-$ valued functions on the open sets of the space. A ringed space (V, O_V), as above, will be called an *affine $\delta-$ variety*. A ringed space locally isomorphic to an affine $\delta-$ variety will be called a $\delta-$ variety. If V is an irreducible $\delta-$ variety, the field $F\langle V\rangle$ of $\delta-$ rational functions on V is equal to the field $F\langle\Omega\rangle$, Ω a non-empty affine Kolchin open subset of V. The $\delta-$ dimension,

$\delta-$ type and typical $\delta-$ dimension are well-defined for V, since they are differential birational invariants, and, therefore, are the same for all non-empty Kolchin open subsets of V. So, the $\delta-$ dimension of V is the differential transcendence degree over F of $F\langle V\rangle$, and, if it is zero, V is *finite dimensional* (type 0), and the typical $\delta-$ dimension is the transcendence degree over F of $F\langle V\rangle$. The typical $\delta-$ dimension was called the "absolute dimension" in Buium's papers; here we will call it simply the *dimension* of V. A morphism of ringed spaces between two $\delta-$ varieties will be called a *$\delta-$ regular map*. $\delta-$ varieties and $\delta-$ regular maps define a category $\{\delta-$ varieties$\}$, which is probably the most natural frame for "global $\delta-$ algebraic geometry." The ideal of this geometry would be to attain the same level of depth and sophistication as the "usual " algebraic geometry We are far from this ideal today; however, a lot can be done in two special cases, which we now describe.

The first case is that of $\delta-$ groups defined to be group objects in the category of $\delta-$ varieties. As we noted, at the beginning of section 2, Pillay proved that the category $\{\delta$ - groups$\}$ is equivalent to the category of $\delta-$ groups as defined axiomatically by Kolchin. So, the terminology used in section 2.1 will carry over here. In particular,we have $\delta-$ homomorphisms, $\delta-$ characters, and $\delta-$ linear representations.

The second case in which one can develop a quite satisfactory theory is that of the category $\{$classical $\delta-$ varieties$\}$, defined as follows. Just as is the case for $F-$ groups in Kolchin's axiomatic treatment of $\delta-$ groups, an $F-$ variety (algebraic variety) can be equipped with a canonical structure of $\delta-$ variety (the Kolchin topology instead of the Zariski topology, and the sheaf of $\delta-$ regular functions instead of the sheaf of regular functions). A $\delta-$ variety obtained in this way will be called a *classical $\delta-$ variety*. Then $\{$classical $\delta-$ varieties$\}$ is taken to be the full subcategory of $\{\delta-$ varieties$\}$ whose objects are the classical $\delta-$ varieties. In other words, the objects of $\{$classical $\delta-$ varieties) are the "old" varieties of algebraic geometry, while the morphisms in $\{$classical $\delta-$ varieties$\}$ are the "new" ones, the $\delta-$ regular maps.

This section will be devoted to explaining the results in [BUIUM 1992a] on finite-dimensional $\delta-$ groups. The next section will be devoted to an exposition of the results in [BUIUM 1993a], [1994b], [1995a]. Finally, in section 2.7, we explain how $\delta-$ groups can be applied to solving Diophantine problems over functions fields. See [BUIUM 1992b] and subsequent developments, originating in [BUIUM 1992b], especially those in [HRUSHOVSKI 1996a], [HRUSHOVSKI and SOKOLOVIC 1994], [BUIUM and PILLAY 1997].

In order to state some of the results in [BUIUM 1992a], we need a definition. By a $D-$ group, we will understand an $F-$ group G (algebraic group over F), equipped with a derivation δ on its structure sheaf O_G that extends the derivation on F, and for which the multiplication, inversion, and unit morphisms are "horizontal" (in the obvious sense). So, note that $D-$ groups belong entirely to algebraic geometry, rather than to differential algebraic geometry. They form a category in the obvious way. We have the following:

Theorem. *The category of $\delta-$ groups of finite dimension is equivalent to the category of $D-$ groups.* [BUIUM 1992a]

For any $\delta-$ group Γ of finite dimension, we shall denote by $G(\Gamma)$ the corresponding $D-$ group; note that Γ will appear as a $\delta-$ subgroup of $G(\Gamma)$, and, for irreducible Γ, $F\langle\Gamma\rangle$ identifies with the field $F(G(\Gamma))$ of rational functions on $G(\Gamma)$. The above theorem reduces the study of $\delta-$ groups of finite dimension to two problems in algebraic geometry:

1. Characterize those $F-$ groups G which admit a structure of $D-$ group.
2. Describe, for any G that admits a structure of $D-$ group, the set of its $D-$ group structures.

Both problems are deformation-theoretic: the first is related to the deformation theory of the group itself, while the second is related to the deformation theory of its automorphisms. Problem 1 is solved in [BUIUM 1992a] in case G is either linear or commutative. It is, however, open in general. We shall only quote the result in the linear case:

Theorem. *A linear $F-$ group admits a structure of $D-$ group if and only if G descends to C as an algebraic group.* [BUIUM 1992a]

For the solution of problem 1 in the commutative case, as well as for results on problem 2, we refer the reader to [BUIUM 1992a]. It suffices to say that there are lots of commutative, non-affine $F-$ groups, not descending to C, that admit $D-$ group structures. These structures are related to the Grothendieck-Mazur-Messing cyrstalline theory [MAZUR and MESSING 1974].

Among $\delta-$ groups of finite dimension there is a "trivial" class of groups that we call *split*: they are the $C-$ groups (algebraic groups with respect to the universe C). We saw such groups in section 2.4. In view of the discussion above, we can see that they correspond to $F-$ groups G that descend to C, equipped with the "trivial" $D-$ group structure. There exist interesting, very simple examples of non-split $\delta-$ groups, for example, the $\delta-$ subgroup of G_m mentioned in section 2.2.

An $F-$ group is *semi-abelian* if it is an extension of an abelian variety by a torus. If G is a connected $F-$ group, the subfield of $F(G)$ identified with the rational function field of the maximum semi-abelian quotient of G is called the *maximum semi-abelian subfield* of $F(G)$. The following general splitting criterion was proved in [BUIUM 1992a]:

Theorem. *A $\delta-$ group Γ is split if and only if δ preserves the maximum semiabelian subfield of $F\langle\Gamma\rangle = F(G(\Gamma))$. Moreover, a $\delta-$ subgroup of $GL_n(F)$ is split if and only if all group-like elements of the Hopf $\delta-$ coordinate algebra $F\{\Gamma\}$ are $\delta-$ constants.* [BUIUM 1992a]

One of the most interesting outcomes of the theory in [BUIUM 1992a] is the analysis of abelian varieties, viewed as classical $\delta-$ varieties.

Theorem. *Let A be an abelian $F-$ variety. Let $A^\#$ be the Kolchin closure of the torsion group A_{tors} in A. Then $A^\#$ is a $\delta-$ group of finite dimension, it is Zariski dense in A (being the smallest $\delta-$ subgroup with this property), and has no non-trivial $\delta-$ linear representation. Moreover, $A^\#$ has dimension between g and $2g$. It has dimension g if and only if A descends to C. Finally, there is a $\delta-$ open subset of the moduli space of principally polarized abelian varieties over F such that, for any A belonging to that set, $A^\#$ has dimension $2g$.* [BUIUM 1992a], [1993a]

The $\delta-$ groups $A^\#$ are the incarnation, in this theory, of the "Kernel of the Manin map," [MANIN 1966] and [1990]; see, also, the second theorem of the next section. Note, however, that Manin never left the realm of function fields with derivation, and, so, the *geometric* object $A^\#$ remains hidden in his theory. In other words, in Manin's work we see groups, but never $\delta-$ groups. On the contrary, the *geometry* of $A^\#$ is what enabled differential algebraic geometry to become relevant in its applications to the Lang conjecture [BUIUM 1992b], [1993b], [1994c].

Finally, let us note that [BUIUM 1992a] contains a complete classification of $\delta-$ groups with no nontrivial $\delta-$ linear representations.

2.6. **Classical $\delta-$ varieties.** Projective $F-$ varieties have, of course, no nonconstant regular maps into affine space. It turns out that they often have, however, interesting $\delta-$ regular maps into affine space. It also is true that the "number" of such maps is related to the Kodaira dimension of our projective varieties. We illustrate this by stating three theorems. These theorems cover, in particular, the case of curves of genus $0, 1$, and ≥ 2, respectively.

Theorem. *If X is a smooth projective rational $F-$ variety, then any $\delta-$ regular map $X \longrightarrow A_n(F)$ must be constant. In particular, $\delta-$ regular maps from projective varieties into affine space must contract all rational curves.* [BUIUM 1993a]

Theorem. *If X is an abelian $F-$ variety, then there is a $\delta-$ homomorphism $\psi : X \longrightarrow G_a^n$ whose kernel is $A^\#$, and with the property that any $\delta-$ regular map $X \longrightarrow A_m(F)$ factors through ψ.* [BUIUM 1993a]

Theorem. *Let X be a smooth projective curve over F of genus $g \geq 2$. If X does not descend to C, then there exists an injective $\delta-$ regular map $\varphi : X \longrightarrow A_n(F)$ such that any $\delta-$ regular map $X \longrightarrow A_m(F)$ factors through φ. If X descends to C, then any $\delta-$ regular map from X into affine space must contract the set of $C-$ points X_C to a point, and there exists a $\delta-$ regular map $\varphi : X \longrightarrow A_n(F)$ that is injective on $X \backslash X_C$ (so that φ looks like "blowing down" X_C).* [BUIUM 1994b]

For a detailed analysis of these maps in case X has "general moduli," we refer to [BUIUM 1994b]. Here is one more striking result about the relation between $\delta-$ regular functions on curves and $\delta-$ regular functions on their Jacobians.

Theorem. *Let X be a smooth projective curve over F of genus ≥ 2 that does not descend to C. Fix any point on X, and embed X into its Jacobian $J(X)$ via the Abel-Jacobi map. Then, the field $F\langle X\rangle$ of $\delta-$ rational functions on X is generated by restrictions to X of $\delta-$ characters on $J(X)$.* [BUIUM 1995a]

2.7. **Applications to Diophantine Geometry.** The theory of differential algebraic groups can be applied to Diophantine questions over functions fields. This was first done in [BUIUM 1992b]. Indeed, we have the following result on differential algebraic subgroups of an abelian variety:

Theorem. *Let A be an abelian $F-$ variety that has F/C trace zero, and let Γ be a $\delta-$ subgroup of finite dimension. Then the Zariski closure in A of any Kolchin closed subset of Γ is a finite union of translates of abelian subvarieties.* [BUIUM 1992b]

This result can be used to prove the geometric analogue of the following celebrated conjecture:

The Lang Conjecture. *Let A be a complex abelian variety, let $X \subset A$ be a proper Zariski closed subvariety, and let $S \subset A$ be a finite rank subgroup. Then there exist in X a finite number of translates of abelian subvarieties whose union covers $X \cap S$.* [LANG 1965].

Here, the rank of a group is the vector space dimension over the field $\mathbb{Q}$ of rational numbers of the group tensored with $\mathbb{Q}$. If S is finitely generated, and X is a curve, the conjecture reduces to the Mordell Conjecture [MORDELL 1922]. If $S = A_{tors}$, and X is a curve, the conjecture reduces to the Manin-Mumford Conjecture (see [LANG 1997]). Lang's conjecture is now a theorem due to the work of Faltings [FALTINGS 1994] and Hindry [HINDRY 1988]. However, the following was previously proved in [BUIUM 1992b]:

Theorem. *Lang's conjecture is true in case A has trace zero over the field of algebraic numbers.*

This should be viewed as the "geometric case" of Lang's conjecture, and follows formally from the preceding theorem, the second theorem in section 2.6, and the fourth theorem in section 2.5. Here is the argument. We may replace the field of complex numbers in the preceding theorem by a differentially closed field F such that A has F/C trace zero. Let ψ be as in section 2.6, Theorem 2. Since S has finite rank, it follows that $\psi(S)$ is contained in a finite-dimensional vector space over C. By section 2.6, Theorem 2 and section 2.5, Theorem 5, $\psi^{-1}(\psi(S))$ is contained in a $\delta-$ subgroup Γ of A of finite dimension. By the first theorem of this section, the Zariski closure in A of $X \cap \Gamma$ is a finite union of translates of abelian subvarieties of A, which are obviously contained in X. This proves the theorem.

A careful analysis of the proof of the theorem leads to an effective version of it:

Theorem. *We assume the trace condition in Lang's conjecture. The number of translates of abelian varieties can be made smaller than an explicit constant $C(g,d,e,r)$ depending only on the dimension g of A, the degree d^2 of a fixed polarization on A, the degree e of X with respect to this polarization, and the rank r of S. If, in addition, X is a curve, then, the cardinality of $X \cap S$ can be bounded by a constant $C(q,r)$ depending only on the genus q of X, and the rank r of S.* [BUIUM 1993b], [BUIUM 1994c]

This theorem answers, in the geometric case, a question of Mazur [MAZUR 1986]. In obtaining this kind of uniformity, differential algebraic geometry shows its full strength; no such bounds (for $r > 0$) are known without the trace condition.

An interesting development originating in the first theorem of this section occurred in logic. The proof in [BUIUM 1992b] involved an analytic argument based on the "Big Picard Theorem" (see [KOBAYASHI and OCHIAI 1975]). Hrushovski was able to replace this analytic argument by a remarkable model - theoretic development [HRUSHOVSKI 1996], based, in its turn, on the difficult model theory developed in [HRUSHVOSKI and ZILBER 1996]. This was immediately recognized as a spectacular success of logic (specifically, of model theory). Hrushovski's methods based on model theory subsequently led to new, striking results on differential algebraic groups [HRUSHOVSKI and SOKOLOVIC]; see the last theorem of this section. Of course, it was a challenge to find proofs of these results independent of model theory. A substantial part of this task is achieved in [BUIUM and PILLAY 1997] as follows. Note that the theorem [BUIUM 1992b] gives no information in the case in which the Kolchin closed subset of Γ under consideration is Zariski dense A. It turns out that this is exactly the case that needs to be analyzed in order to recapture the results of the logicians. One first proves:

Theorem. *Let A be an abelian $F-$ variety of dimension g that has F/C trace zero, and let $\Gamma \subset A$ be a $\delta-$ subgroup of finite dimension. Then any Kolchin closed* *subset of Γ that is Zariski dense in A has dimension $\geq g+1$.* [BUIUM and PILLAY 1997]

Here is the remarkable result obtained by the logicians, and for which an analytic-differential algebraic proof was found in [BUIUM and PILLAY 1997]. Note that, in some sense, we have come full circle, since analytic-differential algebraic proofs are very much "in the spirit of Ritt."

Theorem. *Let A be a simple abelian $F-$ variety that does not descend to C. Then the following properties hold:*

1. *Any proper Kolchin closed subset of $A^{\#}$ is finite.*
2. *Any Kolchin closed subset of $A^{\#} \times A^{\#}$ is a finite union of translates of $\delta-$ subgroups.*
3. *If B is another simple abelian $F-$ variety that is not isogenous to A, and does not descend to C, then, any proper Kolchin closed subset of $A^{\#} \times B^{\#}$ is a finite union of points or points times one of the factors.* [HRUSHOVSKI and SOKOLOVIC], [BUIUM and PILLAY 1997]

In the language of model theory [HRUSHOVSKI and ZILBER 1996], assertion (1) above says that $A^{\#}$ is "strongly minimal," assertion (2) implies that $A^{\#}$ is "locally modular," and assertion (3) says that $A^{\#}$ and $B^{\#}$ are "orthogonal." The theorem also implies the following "dichotomy theorem":

> Any strongly minimal group definable in a differentially closed field ($\delta - F-$ group, F differentially closed) is either locally modular or nonorthogonal to the field of constants.

It also follows that:

> The non-orthogonality classes of locally modular strongly minimal groups are in 1-1 correspondence with the isogeny classes of simple abelian varieties that do not descend to constants.

We should note, however, that the analytic-differential algebraic methods in [BUIUM and PILLAY 1997] do not recapture *all* the differential algebraic results obtained by the logicians in their full strength.

2.8. **Arithmetic Analogue of Differential Algebraic Geometry.** There are no non-zero derivations on number fields, and no continuous non-zero derivations on local fields. This was generally believed to lead to a total failure of differential algebraic methods in arithmetic. However, an arithmetic analogue of differential algebraic geometry was initiated in [BUIUM 1995b], [1996a], [1996b], and [1997] that has led to several arithmetic applications. We will not explain this development in any detail here, but, instead, content ourselves with giving a few hints of the beginnings of the theory. The differentially closed field F of differential algebra must be replaced, in the arithmetic theory, by an absolutely unramified complete discrete valuation ring R of characteristic zero with algebraically closed residue field of characteristic $p > 0$. The derivation δ on F must be replaced by the operator δ on R, defined by the formula

$$\delta\eta = \frac{\phi(\eta) - \eta^p}{p},$$

where $\phi : R \to R$ is the unique lifting of the Frobenius endomorphism of the residue field. This is, of course, not a derivation, but, morally, plays the role of one. The ring $F\{y_1, ..., y_n\}$ of differential polynomials must be replaced by the $p-$ adic completion $R\{y_1, ..., y_n\}\hat{}$ of the ring $R[y, y', y'', ...]$. The derivation δ on $F\{y_1, ..., y_n\}$ must be replaced by the following operator on $R\{y_1, ..., y_n\}\hat{}$:

$$f(y, y', y'', ...) \mapsto \frac{f^{\phi}(y^p + py', y'^p + py'', ...) - (f(y, y', ...))^p}{p}.$$

The theory obtained in this way is strikingly similar to the differential algebraic geometry discussed in this paper. In particular, appropriate analogues of the first three theorems of section 2.6, can be proved. One can also obtain, using this theory, a simple, effective proof of the Manin-Mumford conjecture [BUIUM 1996a] (first proved, but not effectively, by Raynaud [1983]), an arithmetic analogue [BUIUM 1995b] of the Manin-Chai Theorem of the Kernel [MANIN 1966], [1990], [CHAI 1991], and a special, but significant case [BUIUM 1996b] of a conjecture of Tate and Voloch [1996].

The arithmetic analogue of differential algebraic geometry described above is very different from the "difference algebra" initiated by Ritt, Raudenbush, and Cohn ([RITT and DOOB 1933], [RITT 1934], [RITT and RAUDENBUSH 1939], [RITT 1941], [COHN 1948]).[4] The idea in difference algebra is to replace fields with derivations with fields with automorphisms. Recently, the logicians were able to apply difference algebra to Diophantine problems over number fields [HRUSHOVSKI 1996], [HRUSHOVSKI and CHATZIDIKIS].

REFERENCES

[BAER 1927] Baer, R., Algebraische Theorie der differentierbaren Funktionen körper, S.-B. *Heidelberger Akad. Wiss. Math-Natur. Kl.*, 8 Abh, 15-32.

[BAIDER et al 1996] Baider, A. R., R.C. Churchill, D. L. Rod, M. F. Singer, On the infinitesimal geometry of integrable systems, *Fields Inst. Comm.* 7 (1996), 5-56.

[BAYER and STILLMAN 1987] Bayer, D., and M. Stillman, A theorem on refining division orders by the reverse lexicographical order, *Duke Math. J.* 55 (1987), 321-328.

[BIALYNICKI-BIRULA 1962] Bialynicki-Birula, A., On Galois theory of fields with operators, *Amer. J. Math.* 84 (1962), 89-109.

[BLUM, L. 1968] Blum, L., Generalized algebraic structures: a model theoretic approach, Ph.D. dissertation, Massachusetts Institute of Technology, Cambridge, MA (1968).

[BLUM, L. 1977] Blum, L., Differentially closed fields: a model theoretic tour, in *Contributions to Algebra: A Collection of Papers Dedicated to Ellis Kolchin* (H. Bass, P. J. Cassidy, J. Kovacic, eds.), Academic Press, New York (1977), 37-61.

[BLUM, P. 1969] Blum, P., Complete model of differential fields, *Trans. Amer. Math Soc.* 137 (1969), 309-325.

[BOULIER 1995] Boulier, F., Some improvements on a lemma of Rosenfeld, preprint (1995).

[BOULIER 1997] Boulier, F., D. Lazard, F. Ollivier, M. Petitot, Computing representations for radicals of finitely generated differential ideals, preprint (1997).

[4]See Cohn's book [1965] for an extensive bibliography. For a Galois theory of difference equations, see [BIALYNICKI-BIRULA 1962], [FRANCKE 1963], [KREIMER 1965a], [1965b], and, especially, [van der PUT and SINGER 1997].

[BOREL 1956] Borel, A., Groupes linéaires algébriques, *Ann. of Math.* (2), 64 (1956), 20-82.

[BOREL 1991] Borel, A., *Linear Algebraic Groups*, 2nd enl. ed., Springer-Verlag, New York, Berlin, Heidelberg (1991).

[BOREL] Borel, A., Algebraic Groups and Galois theory in the work of Ellis R. Kolchin, this volume.

[BOURBAKI 1968,1975] Bourbaki, N., Groupes et algèbres de Lie, Chaps. 1, 7, 8 in *Élements de Mathématiques*, Hermann, Paris (1968), (1975).

[BRZEZINSKI 1962] Brzezinski, J., On differentially integral elements, *Bull. Acad. Polon. Sci. Sér. Sci. Math. Astron. Phys.* 10 (1962), 325-328.

[BUCHBERGER 1985] Buchberger, Bruno, Gröbner bases: an algorithmic method in polynomial ideal theory, in *Multidimensional Systems Theory* (N. K. Bose, ed.), D. Reidel, Dordrecht (1985), 184-232.

[BUIUM 1982] Buium, A., Ritt schemes and torsion theory, *Pacific J. Math.* 98, 281-293.

[BUIUM 1986] Buium, A., *Differential Function Fields and Moduli of Algebraic Varieties,* Lecture Notes in Math. 1226 (1986), Springer-Verlag, Berlin, Heidelberg, New York.

[BUIUM 1992a] Buium, A., *Differential Algebraic Groups of Finite Dimension,* Lecture Notes in Math. 1506 (1992), Springer-Verlag, Berlin, Heidelberg, New York.

[BUIUM 1992b] Buium, A., Intersections in jet spaces and a conjecture of S. Lang, *Ann. of Math.* 136 (1992), 583-593.

[BUIUM 1993a] Buium, A., Geometry of differential polynomial functions I: Algebraic groups, *Amer. J. Math.* 115 (1993), 1385-1444.

[BUIUM 1993b] Buium, A., Effective bound for the geometric Lang conjecture, *Duke Math. J.* 71 (1993), 475-499.

[BUIUM 1994a] Buium, A., *Differential Algebra and Diophantine Geometry,* Hermann, Paris (1994).

[BUIUM 1994b] Buium, A., Geometry of differential polynomial functions II: Algebraic curves, *Amer. J. Math.* 116 (1994), 785-819.

[BUIUM 1994c] Buium, A., On a question of B. Mazur, *Duke Math J.* 75 (1994), 639-644.

[BUIUM 1994d] Buium, A., The **abc** theorem for abelian varieties, *Internat. Math. Res. Notices* 5 (1994), 219-233.

[BUIUM 1995a] Buium, A., Geometry of differential polynomial functions III: Moduli spaces, *Amer. J. Math.* 117 (1995), 1-73.

[BUIUM 1995b] Buium, A., Differential characters of abelian varieties over $p-$adic fields, *Invent. Math.* 122 (1995), 309-340.

[BUIUM 1996a] Buium, A., Geometry of $p-$ adic jets, *Duke Math J.* 82 (1996), 349-367.

[BUIUM 1996b] Buium, A., An approximation property for Teichmüller points, *Math. Res. Lett.* 3 (1996), 453-457.

[BUIUM 1997] Buium, A., Differential characters and the characteristic polynomial of Frobenius, *J. Reine Angew. Math.* 485 (1997), 209-219.

[BUIUM and PILLAY 1997] Buium, A., and A. Pillay, A gap theorem for abelian varieties over differential fields, *Math. Res. Lett.* 4 (1997), 211-219.

[CARRÀ-FERRO and SIT 1994] Carrà-Ferro, G. and W. Sit, On term-orderings and rankings, in *Lecture Notes in Pure and Appl. Math.* 151, Dekker, New York (1994), 31-77.

[CARTAN 1899/1984] Cartan, E., 1899/1984, Sur certaines expressions différentielles et le problème de Pfaff, *Ann. Sci. École Norm. Sup.* 16 (1899), 239-332, reprinted in *Oeuvres Complètes,* partie II, 2nd ed., Springer-Verlag, New York, Heidelberg, Berlin (1984), 303-396.

[CARTAN 1909/1984] Cartan, E., 1909/1984, Les groupes de transformations continus, infinis, simples, *Ann. Sci. École Norm. Sup.* 26 (1909),

93-161, reprinted in *Oeuvres Complétes*, partie II, 2nd ed., Springer-Verlag, New York, Heidelberg, Berlin (1984), 857-925.

[CARTAN 1931/1984] Cartan, E., 1931/1984, Sur la théorie des systèmes en involution et ses applications a la rélativité, *Bull. Soc. Math. France* 59 (1931), 88-118, reprinted in *Oeuvres Complètes*, Partie II, 2nd ed., Springer-Verlag, New York, Heidelberg, Berlin (1984), 1199-1229.

[CASSIDY 1972] Cassidy, P. J., Differential algebraic groups, *Amer. J. Math* 94 (1972), 891-954.

[CASSIDY 1975] Cassidy, P. J., The differential rational representation algebra on a linear differential algebraic group, *J. Algebra* 37 (1975), 223-238.

[CASSIDY 1977] Cassidy, P. J., Unipotent differential algebraic groups, in *Contributions to Algebra: A Collection of Papers Dedicated to Ellis Kolchin* (H. Bass, P. J. Cassidy, J. Kovacic eds.), Academic Press, New York (1977), 83-115.

[CASSIDY 1979] Cassidy, P. J., Differential algebraic Lie algebras, *Trans. Amer. Math. Soc.* 247 (1979), 247-273.

[CASSIDY 1980] Cassidy, P. J., Differential algebraic group structures on the plane, *Proc. Amer. Math. Soc.* 80 (1980), 210-214.

[CASSIDY 1989] Cassidy, P. J., The classification of the semisimple differential algebraic groups and the linear semisimple differential algebraic Lie algebras, *J. Algebra* (1989), 169-238.

[CAYLEY 1886] Cayley, Arthur, *Quart. J. Math.* 21 (1886), p. 21.

[CHAI 1991] Chai, Ching-Li, A note on Manin's theorem of the kernel, *Amer. J. Math.* 113 (1991), 387-389.

[CHAZY 1909/1927] Chazy, J., *C. R. Acad. Sci. Paris* 148 (1909), 157, reprinted in [INCE 1927], 355.

[CHEVALLEY 1951] Chevalley, Claude, *Théorie des groupes de Lie 2: Groupes algébriques*, Hermann, Paris (1951).

[CHOU 1988] Chou, S. C., *Mechanical Theorem Proving*, D. Reidel, Dordrecht (1988).

[CHUDNOVSKY and CHUDNOVSKY 1984] Chudnovsky, D. V. and G. V. Chudnovsky, Padé approximations to solutions of linear differential equations and applications to diophantine analysis, Number theory Seminar, New York 1982, Lecture Notes in Math. 1052, Springer-Verlag, Berlin, Heidelberg, New York (1984), 85-167.

[CHURCHILL *et al. 1995*] Churchill, R. C., D. L. Rod, M. F. Singer,1995, Group-theoretic obstructions to integrability, *Ergod. Theory Dynam. Systems* 15, 15-48.

[COHN 1941] Cohn, R. M., On the analog for differential equations of the Hilbert-Netto theorem, *Bull. Amer. Math. Soc.* 47 (1941), 268-270.

[COHN 1948] Cohn, R. M., Manifolds of difference polynomials, *Trans. Amer. Math. Soc.* 64 (1948), 133-172.

[COHN 1965] Cohn, R. M., *Difference Algebra,* Interscience, New York. 1965.

[COHN 1976] Cohn, R. M., The general solution of a first order differential polynomial, *Proc. Amer. Math. Soc.* 55 (1976), 14-16.

[COHN 1977] Cohn, R. M., Solutions in the general solution, in *Contributions to Algebra: A Collection of Papers dedicated to Ellis Kolchin* (H. Bass, P. J. Cassidy, J. Kovacic, eds.), Academic Press, New York (1977), 117-128.

[COHN 1983] Cohn, R. M., Order and Dimension, *Proc. Amer. Math. Soc.* 87 (1983), 1-6.

[COHN 1986a] Cohn, R. M., Solutions in the general solution of second order algebraic differential equations, *Amer. J. Math* 108 (1986), 505-522.

[COHN 1986b] Cohn, R. M.,Valuations and the Ritt problem, *J. Algebra* 101 (1986), 1-15.

[DARBOUX 1883] Darboux, G., Solutions singulières des équations aux dérivées partielles du premier ordre, Mémoires présentés par divers savants étrangers a l'Académie des Sciences 27 (1883), 1-243.

[DAUBEN 1995] Dauben, J. W., *Abraham Robinson: The Creation of Nonstandard Analysis: A Personal and Mathematical Odyssey*, Princeton University Press, Princeton (1995).

[DEBEVER 1979] Debever, R., ed., *Elie Cartan-Albert Einstein: Letters on Absolute Parallelism, 1929-1932,* tr. by Jules Leroy and Jim Ritter, Princeton University Press and Académie Royale de Belgique, Princeton (1979).

[DEMAZURE and GABRIEL 1970] Demazure, M. and P. Gabriel, *Groupes algébriques,* Vol. I, Masson, Paris (1970).

[DRACH 1898] Drach, J., Essai sur une théorie générale de l'intégration et sur la classification des transcendantes, *Ann. Sci. École Norm. Sup.* (3), 15 (1898), 243-384.

[EISENBUD 1995] Eisenbud, D., *Commutative Algebra with a View Toward Algebraic Geometry,* Springer-Verlag, New York, Berlin, Heidelberg (1995).

[FALTINGS 1994] Faltings, G., The general case of S. Lang's conjecture, in *Barsotti Symposium in Algebraic Geometry,* Academic Press, San Diego (1994), 175-182.

[FLIESS and GLAD 1993] Fliess, M., and S. T. Glad, An algebraic approach to linear and nonlinear control, in *Essays on Control: Perspectives in the Theory and its Applications* (H. Trentelman and J. Willems, ed.), Birkhäuser, Boston (1993), 223-267.

[FRANKE 1963] Franke, C. H., Picard-Vessiot theory of linear homogeneous difference equations, *Trans. Amer. Math. Soc.* 103 (1963), 491-515.

[FUCHS 1884] Fuchs, L. Über Differentialgleichungen, deren Integrale feste Verzweigungspunkte besitzen, *Berl. Ber.* 32 (1884), 699-710.

[GALLO and MISHRA 1991] Gallo, G., and B. Mishra, Wu-Ritt characteristic sets and their complexity, *Discrete Mathematics and Computational Geometry; Paper from the DIMACS Special Year*, vol. 6, (J.E. Goodman, R. Pollack, and W. Steiger eds.), Amer. Math. Soc. and Assoc. of Comput. Mach, Providence (1991), 111-136.

[GARNIER 1972] Garnier, R., préface, *Oeuvres de Paul Painlevé,* t. 1, Centre Nat. Rech. Sci., Paris (1972-1975).

[GEBAUER and MÖLLER 1988] Gebauer, R. and H. M. Möller, On an installation of Buchberger's algorithm, *J. Symbolic Comput.* 6, special issue on *Computational Aspects of Commutative Algebra* (1988), 275-286.

[GOODEARL 1975] Goodearl, K. R., Global dimension of differential operator rings II, *Trans. Amer. Math. Soc.* 209 (1975), 65-85.

[HAMBURGER 1893] Hamburger, M., Über die singulären Lösungen der algebraischen Differenzialgleichungen erster Ordnung, *J. Reine und Angew. Math.* 112 (1893), 205-246.

[HAMBURGER 1899] Hamburger, M., Über die singulären Lösungen der algebraischen Differenzialgleichungen höher Ordnung, *Berl. Ber.* 8 (1899), 140-145.

[HILLMAN 1952] Hillman, A. P., On the differential algebra of a single differential polynomial, *Ann. of Math.* 56 (1952), 157-168.

[HILLMAN and MEAD 1962] Hillman, A. P. and D. G. Mead, On the Ritt polygon process, *Amer. J. Math.* 84 (1962), 629-634.

[HINDRY 1988] Hindry, M., Autour d'une conjecture de Serge Lang, *Invent. Math.* 94 (1988), 575-605.

[HRUSHOVSKI 1996a] Hrushovski, E., The Mordell-Lang conjecture over function fields, *J. Amer. Math. Soc.* 9 (1996), 667-690.

[HRUSHOVSKI 1996b] Hrushovski, E., The Manin-Mumford conjecture and the model theory of difference fields, preprint (1996).

[HRUSHOVSKI and CHATZIDIKIS 1996] Hrushovski, E. and Z. Chatzidikis, Model theory of difference fields, preprint (1996).

[HRUSHOVSKI and SOKOLOVIC] Hrushovski, E. and Z. Sokolovic, Minimal subsets of differentially closed fields, preprint (1994), to appear in *Trans. Amer. Math. Soc.*

[HRUSHOVSKI and ZILBER 1993] Hrushovski, E. and B. Zilber, Zariski geometries, *Bull. Amer. Math. Soc.* 28 (1993), 315-323.

[HRUSHOVSKI and ZILBER 1996] Hrushovski, E. and B. Zilber, Zariski geometries, *J. Amer. Math. Soc.* 9 (1996), 1-56.

[INCE 1927] Ince, E. L., *Ordinary Differential Equations,* Longmans (1927), reprinted by Dover, New York (1944).

[IWASAKI *et al.* 1991] Iwasaki, K., H. Kimura, S. Shimomura, Y. Masaaki, *From Gauss to Painlevé: A Modern Theory of Special Functions, dedicated to Professor Tosihusa Kimura, Aspects of Mathematics: E;* 16 (1991), Vieweg, Braunschweig.

[JACOBI] Jacobi, C. G. J., De investigando ordine systematis aequationem differentialium vulgarium cujusqunque, *Gesammelte Werke,* vol. 5, 191-216, D. Reimer, Berlin (1881), 2nd ed. Chelsea, New York (1969).

[JANET 1920] Janet, M., Sur les systèmes d'équations aux dérivées partielles, *J. Math. Pures Appl.* 8 (3), 1920, 65-151.

[JANET 1921] Janet, M., Sur les systèmes aux dérivées partielles comprenant autant d'équations que de fonctions inconnues, *C. R. Acad. Sci. Paris* 172 (1921), 1637-1639.

[JOHNSON 1969a] Johnson, J., A notion of Krull dimension for differential rings, *Comm. Math. Helv.* 44 (1969), 207-216.

[JOHNSON 1969b] Johnson, J., Differential dimension polynomials and a fundamental theorem on differential modules, *Amer. J. Math* 91 (1969), 239-248.

[JOHNSON 1969c] Johnson, J., Kähler differentials and differential algebra, *Ann. of Math.* 89 (1969), 92-98.

[JOHNSON 1974] Johnson, J., Kähler differentials and differential algebra in arbitrary characteristic, *Trans. Amer. Math. Soc.* 192 (1974), 201-208.

[JOHNSON 1978] Johnson, J., Systems of n partial differential equations in n unknown functions: The conjecture of M. Janet, *Trans. Amer. Math. Soc.* 242 (1978), 329-334.

[JOHNSON *et al.* 1995] Johnson, J., G. M. Reinhart, and L. A. Rubel, Some counterexamples to separation of variables, *J. Differential Equations* 121 (1995), 42-66.

[KAPLANSKY 1957] Kaplansky, I., *An Introduction to Differential Algebra*, Hermann, Paris (1957).

[KAUP 1980] Kaup, D. J., The Estabrook-Wahlquist method with examples of application, *Physical 1D* (1980), 391-411.

[KEIGHER 1982a] Keigher, W. F., Spectra of differential rings, *Cahiers Topologie Géom. Différentielle Categ.* 23 (1982), 37-50.

[KEIGHER 1982b] Keigher, W. F., Differential Schemes and premodels of differential fields, *J. Algebra,* 79 (1982), 37-50.

[KEIGHER 1983] Keigher, W. F., On the structure presheaf of a differential ring, *J. Pure Appl. Algebra* 27 (1983), 163-172.

[KNUTH and BENDIX 1967] Knuth, D. and P. B. Bendix, Simple word problems in universal algebras, in *Computational Problems in Abstract Algebra,* (J. Leech ed.), Pergamon Press (1967), 263-297.

[KOBAYASHI and OCHIAI 1975] Kobayashi, S. and T. Ochiai, Meromorphic mappings into compact complex spaces of general type, *Invent. Math.* 31 (1975), 7-16.

[KOENIGSBERGER 1889] Koenigsberger, L., *Lehrbuch zur Theorie der Differentialgleichungen mit einer unhabhängiger Variabeln,* Teubner, Leipzig (1889), 155.

[KONDRATEVA 1989] Kondrateva, M. V., A minimal dimension polynomial of a field extension that is given by a system of linear differential equations, *Math. Notes* 45 (1989), 80-86.

[KONDRATEVA *et al.* 1992] Kondrateva, M. V., A. B. Levin, A. V. Mikhalev, E. V. Pankratev, Computation on dimension polynomials, *J. Algebra Comput.* 2 (1992), 117-137.

[KONDRATEVA *et al.* 1995] Kondrateva, M. V., A. B. Levin, A. V. Mikhalev, E. V. Pankratev, *Differential and Difference Dimension Polynomials,* 3 vols, preprint (1995).

[KOVACIC 1969] Kovacic, J., The inverse problem in the Galois theory of differential fields, *Ann. of Math.* 89 (1969), 583-608.

[KOVACIC 1971] Kovacic, J., On the inverse problem in the Galois theory of differential fields, *Ann. of Math.* 93 (1971), 269-284.

[KOVACIC 1986] Kovacic, J., An algorithm for solving second order linear homogeneous differential equations, *J. Symbolic Comput.* 2 (1986), 3-43.

[von KOVALEVSKY 1875] von Kovalevsky, S., Zur Theorie der partiellen Differentialgleichungen, *J. Reine Angew. Math.* 80 (1875), 1-32.

[KREIMER 1965a] Kreimer, H. F., An extension of differential Galois theory, *Trans. Amer. Math. Soc.* 118 (1965), 247-256.

[KREIMER 1965b] Kreimer, H. F., An extension of the Picard-Vessiot Theory, *Pacific J. Math.* 15 (1965), 191-205.

[KRONECKER 1845/1882] Kronecker, L., *Grundzuge einer arithmetischen Theorie der algebraischen Grössen* (1845), 2nd edition (1882); Notes: *Festschrift zu Herrn Ernst Edouard Kummer's fünfzigjährigem Doctor-Jubiläum, 1881,* G. Reimer, Berlin (1882).

[LAGRANGE 1772] Lagrange, J. L. Sur les solutions particulières des équations différentielles, *Oeuvres de Lagrange,* Vol. 4 (J. A. Serret, ed.), Gauthier - Villars, Paris (1867-1892), reprinted by Georg Ohms Verlag, Hildesheim and New York (1973), 5-108.

[LAINE 1993] Laine, I., *Nevanlinna Theory and Complex Differential Equations,* de Gruyter Studies in Math. 15, Walter de Gruyter, Berlin, New York (1993).

[LANDO 1970] Lando, B. A., Jacobi's bound for the order of systems of first order differential equations, *Trans. Amer. Math. Soc.* 152 (1970), 119-135.

[LANG 1965] Lang, S., Division points on curves, *Ann. Mat. Pura Appl.* (4) 70 (1965), 229-234.

[LANG 1997] Lang, Serge, *Survey of Diophantine Geometry,* tr. fr. the Russ., 2nd pr., Springer-Verlag, Berlin, Heidelberg, New York (1997).

[LAPLACE 1772] Laplace, P. S., Mémoire sur les solutions particulières des équations différentielles, *Oeuvres Complètes,* Vol. 8 (1878), Gauthier-Villars, Paris, 326-365.

[LASCAR 1980] Lascar, D., Les corps différentiellement clos denombrables, in *Théorie stables,* Paris Sem. Notes (1980-1983).

[LAX 1968] Lax, P. D., Integrals of nonlinear equations of evolution and solitary waves, *Comm. Pure Appl. Math.* 21 (1968), 467-490.

[LAZARD 1955] Lazard, M., Sur nilpotence de certains groupes algébriques, *C. R. Acad. Sci. Paris* 241 (1955), 1687-1689.

[LAZARD 1975] Lazard, M., *Commutative Formal Groups,* Lecture Notes in Math. 443, Springer-Verlag, Berlin, Heidelberg, New York (1975).

[LEVI 1942] Levi, H., On the structure of differential polynomials and on their theory of ideals, *Trans. Amer. Math. Soc.* 51 (1942), 532-568.

[LEVI 1945] Levi, H., The low power theorem for partial differential polynomials, *Ann. of Math.* 46 (1945), 113-119.

[LEVIN 1997] Levin, A. B., Computation of Hilbert polynomials in two variables, preprint (1997).

[LIPSCHITZ and RUBEL 1987] Lipschitz, L. and L. A. Rubel, A differentially algebraic replacement theorem, and analog computability, *Proc. Amer. Math. Soc.* 99 (1987), 367-372.

[LORCH 1951] Joseph Fels Ritt, *Bull. Amer. Math Soc.* 57 (1951), 307-318.

[MANIN 1966] Manin, Yu. I., Rational points on an algebraic curve over a function field, *Amer. Math. Soc. Transl.* Ser. 2, 50 (1966), 189-234.

[MANIN 1979] Manin, Yu. I., Algebraic aspects of nonlinear differential equations, *J. Sov. Math.* 11 (1979), 1-122.

[MANIN 1990] Manin, Yu. I., Letter to the Editor, *Izv. Akad. Nauk USSR* 34 (1990), 465-466.

[MARKER 1996] Marker, D., Model theory of differential fields, in *Model Theory of Fields* (D. Marker, M. Messmer, A. Pillay, eds.), Lecture Notes Logic 5, Springer-Verlag, Berlin, Heidelberg, New York (1996), 38-113.

[MATSUDA 1977] Matsuda, M., An application of Ritt's low power theorem, *Nagoya Math. J.* 68 (1977), 17-19.

[MATSUDA 1978] Matsuda, M., Algebraic differential equations of the first order free from parametric singularities from the differential-algebraic standpoint, *J. Math. Soc. Japan* 30 (1978), 447-455.

[MATSUDA 1980] Matsuda, M., *First Order Algebraic Differential Equations,* Lecture Notes in Math 804, Springer-Verlag, Berlin, Heidelberg, New York (1980).

[MAZUR 1986] Mazur, B., Arithmetic on curves, *Bull. Amer. Math. Soc.* 14 (1986), 207-259.

[MAZUR and MESSING 1974] Mazur, B. and W. Messing, *Universal Extensions and One-Dimensional Crystalline Cohomology,* Lecture Notes in Math. 370, Springer-Verlag, Berlin, Heidelberg, New York (1974).

[MISHRA 1993] Mishra, B., *Algorithmic Algebra*, Texts Monog. Comput. Sci., Springer-Verlag, New York, Berlin, Heidelberg (1993).

[MORDELL 1922] Mordell, L. J., On the rational solutions of the indeterminate equations of the third and fourth degrees, *Math. Proc. Camb. Philos. Soc.* 21 (1922), 179-192.

[MORRISON 1978] Morrison, S. D., Extensions of differential places, *Amer. J. Math.* 100 (1978), 245-261.

[MORRISON 1979] Morrison, S. D., Differential specializations in integral and algebraic extensions, *Amer. J. Math.* 101 (1979), 1381-1399.

[MORRISON 1997] Morrison, S. D., The differential ideal $[P] : M^{\infty}$, preprint (1997).

[NELSON 1967] Nelson, E., *Tensor Analysis,* Princeton University Press, Princeton (1967).

[NICHOLS and WEISFEILER 1982] Nichols, W., and B. Weisfeiler, Differential formal groups of J. F. Ritt, *Amer. J. Math.* 104 (1982), 943-1003.

[OSGOOD 1973] Osgood, C. F., An effective lower bound on the "diophantine approximation" of algebraic functions by rational functions, *Mathematika* 20 (1973), 4-15.

[OSTROWSKI 1920] Ostrowski, A., Über Dirichletschen Reihen und algebraische Differentialgleichungen, *Math. Z.* 8 (1920), 241-298.

[PANKRATEV 1989] Pankratev, E. V., Computations in differential and difference modules, *Acta. Appl. Math.* (1989), 167-189.

[PICARD 1896] Picard, E., *Traité d'analyse,* Vol. 3, Ch. 18, Gauthiers - Villars, Paris (1898 or 1908 or 1928); reprinted as *Analogies entre la théorie des équations différentielles linéaires et la théorie des équations algébriques,* Gauthiers - Villars, Paris (1936).

[PILLAY 1995] Pillay, A., Differential Galois theory I, preprint, Univ. Notre Dame (1995), to appear in *Illinois J. Math.*

[PILLAY 1996] Pillay, A., Differential algebraic groups and the number of countable differentially closed fields, in *Model Theory of Fields* (D. Marker, M. Messmer, A. Pillay, eds.), Lecture Notes Logic 5, Springer-Verlag, Berlin, Heidelberg, New York (1996), 114-134.

[PILLAY 1997] Pillay, A., Some foundational questions concerning differential algebraic groups, *Pacific J. Math.* 179 (1997), 179-200.

[PILLAY 1998] Pillay, A., Differential Galois theory II, *Ann. Pure Appl. Logic 88* (1998), 181-192.

[PILLAY and MARKER 1997] Pillay, A. and D. Marker, Differential Galois theory III: Some inverse problems, *Illinois J. Math.* 41 (1997), 453-461.

[POINCARÉ 1885] Poincaré, H., Sur un théorème de M. Fuchs, *Acta Math.* 7(1885), 1-32.

[POISSON 1806] Poisson, S. -D., Sur les solutions particulières des équations différentielles et des équations aux différences, *J. l'École Poly.,* Vol. 6 (1806), 60-125.

[POIZAT 1995] Poizat, B., Les corps différentiellement clos, compagnons de route de la théorie des modèles, *Math. Japon.* 42 (1995), 575-585, reprinted in this volume.

[POMMARET 1978] Pommaret, J. M., *Systems of Partial Differential Equations and Lie Pseudogroups,* Gordon & Breach, New York (1978).

[POUR-EL and RICHARDS 1979] Pour-El, M., and I. Richards, A computable ordinary differential equation which possesses no computable solution, *Ann. of Math. Logic* 17 (1979), 61-90.

[PREVIATO 1993] Previato, E., Seventy years of spectral curves: 1923-1993, *Proc. CIME* (1993), 553-569.

[VAN DER PUT and SINGER 1997] van der Put, M., and M. Singer, *Galois Theory of Difference Fields,* Lecture Notes in Math. 1666, Springer-Verlag, Berlin, Heidelberg, New York (1997).

[RAUDENBUSH 1934] Raudenbush, H. W., Ideal theory and algebraic differential equations, *Trans. Amer. Math. Soc.* 36 (1934), 361-368.

[RAUDENBUSH 1936] Raudenbush, H. W., On the analog for differential equations of the Hilbert-Netto theorem, *Bull. Amer. Math. Soc.* 42 (1936), 371-373.

[RAYNAUD 1983] Raynaud, M., Courbes sur une variété abélienne et points de torsion, *Invent. Math.* 71 (1983), 207-235.

[REINHART 1997] Reinhart, G. M., The Schmidt-Kolchin conjecture, preprint (1997).

[RIQUIER 1910] Riquier, C. H., *Les systèmes d'équations aux dérivées partielles,* Gauthier-Villars, Paris (1910).

[RITT 1918] Ritt, J. F., Sur l'itération des fonctions rationnelles, *C.R. Acad. Sci. Paris* 166 (1918), 380-381.

[RITT 1920] Ritt, J. F., On the iteration of rational functions, *Trans. Amer. Math. Soc.* 21 (1920), 313-320.

[RITT 1923a] Ritt, J. F., Sur les fonctions rationnelles permutable, *C.R. Acad. Sci. Paris* 176 (1923), 60-68.

[RITT 1923b] Ritt, J. F, Permutable rational functions, *Trans. Amer. Math. Soc.* 25 (1923), 399-448.

[RITT 1932] Ritt, J. F., *Differential Equations from the Algebraic Standpoint,* Amer. Math. Soc. Colloq. Pub. 14, Amer. Math. Soc., New York (1934).

[RITT 1934] Ritt, J. F., Algebraic difference equations, *Bull. Amer. Math. Soc.* 40 (1934), 303-308.

[RITT 1935] Ritt, J. F., Jacobi's problem on the order of a system of differential equations, *Ann. of Math.* (2) 36 (1935), 303-312.

[RITT 1936a] Ritt, J. F., On the singular solutions of algebraic differential equations, *Ann. of Math.* (2) 37 (1936), 552-617.

[RITT 1936b] Ritt, J. F., On certain points in the theory of algebraic differential equations, *Amer. J. Math.* 60 (1936), 1-43.

[RITT 1938] Ritt, J. F., Algebraic aspects of the theory of differential equations, *Amer. Math. Soc. Semicentennial Pubs., Vol. 2, Semicentennial Addresses,* Amer. Math. Soc., New York (1938), 35-55.

[RITT 1939] Ritt, J. F., On the intersections of algebraic differential manifolds, *Proc. Nat. Acad. Sci., U.S.A.* 25 (1939), 214-215.

[RITT 1940] Ritt, J. F., On a type of algebraic differential manifold, *Trans. Amer. Math. Soc.* 48 (1940), 542-552.

[RITT 1941] Ritt, J. F., Complete difference ideals, *Amer. J. Math.* 63 (1941), 681-690.

[RITT 1945a] Ritt, J. F., On the Manifolds of Partial Differential Polynomials, *Ann. of Math.* 46 (1945), 102-112.

[RITT 1945b] Ritt, J. F., Analytical theory of singular solutions of partial differential equations of the first order, *Ann. of Math.* 46 (1945), 120-143.

[RITT 1950a] Ritt, J. F., *Differential Algebra*, Amer. Math. Soc. Colloq. Pub. 33, Amer. Math. Soc., New York (1950).

[RITT 1950b] Ritt, J. F., Associative differential operations, *Ann. of Math. 51* (1950), 756-765.

[RITT 1950c] Ritt, J. F., Differential groups and formal Lie theory for an infinite number of parameters, *Ann. of Math.* 52 (1950), 708-726.

[RITT 1951a] Ritt, J. F., Differential groups of order two, *Ann. of Math.* 53 (1951), 491-519.

[RITT 1951b] Ritt, J. F., Subgroups of differential groups, *Ann. of Math.* 54 (1951), 110-146.

[RITT 1952] Ritt, J. F. Differential groups, *Proc. 1950 Int. Congress of Mathematicians,* Vol. 1 (1952).

[RITT and DOOB 1933] Ritt, J. F. and J. L. Doob, Systems of algebraic difference equations, *Amer. J. Math.* 55 (1933), 505-514.

[RITT and RAUDENBUSH 1939] Ritt, J. F. and Raudenbush, H. W. , Ideal theory and algebraic difference equations, *Trans. Amer. Math. Soc.* 46 (1939), 445-452.

[ROBBIANO 1985] Robbiano, L., Term orderings on the polynomial ring, *Proc. EUROCAL* 1985, *LNCS* 204 (1985), 513-517.

[ROBINSON 1959] Robinson, A., On the concept of a differentially closed field, *Bull. Res. Council Israel* 8F (1959), 113-128.

[ROSENFELD 1959] Rosenfeld, A., Specializations in differential algebra, *Trans. Amer. Math. Soc.* 90 (1959), 394-407.

[ROSENLICHT 1957] Rosenlicht, M., Commutative algebraic group varieties, in *Algebraic Geometry and Topology: A Symposium in Honor of S. Lefschetz* (R. H. Fox, D. C. Spencer, A. W. Tucker, eds.), Princeton University Press, Princeton (1957), 153-156.

[ROSENLICHT 1969] Rosenlicht, M., On the explicit solvability of certain transcendental equations, *Instit. Hautes Etudes Sci. Publ. Math.,* no. 36 (1969), 15-22.

[ROSENLICHT 1974] Rosenlicht, M., The nonminimality of the differential closure, *Pacific J. Math.* 52 (1974), 529-537.

[SACKS 1972] Sacks, G. E., The differential closure of a differential field, *Bull. Amer. Math. Soc.* (1972), 629-634.

[SCANLON 1997] Scanlon, T., $p-$ Adic distance from torsion points of semi-abelian varieties, preprint, 1997.

[SCHMIDT 1979] Schmidt, W. M., Contributions to diophantine approximations in fields of series, *Monatsh. Math.* 87 (1979), 145-165.

[SCHWARTZ 1991] Schwartz, F., Monomial orderings and Gröbner bases, *ACM SIGSAM Bull.* (1991).

[SEIDENBERG 1956] Seidenberg, A., An elimination theory for differential algebra, *Univ. of Calif. Pub. in Math* (N. S.) 3 (1956), 31-66.

[SEIDENBERG 1958] Seidenberg, A., Abstract differential algebra and the analytic case, *Proc. Amer. Math. Soc.* 9 (1958), 159-164.

[SEIDENBERG 1966] Seidenberg, A., Derivations and integral closure, *Pacific J. Math.* 16 (1966), 167-173.

[SEIDENBERG 1967] Seidenberg, A., Differential ideals in rings of finitely generated type, *Amer. J. Math.* 89 (1967), 22-42.

[SEIDENBERG 1969] Seidenberg, A., Abstract differential algebra and the analytic case II, *Proc. Amer.Math. Soc.* 23 (1969) 689-691.

[SERRE 1959] Serre, J.-P., *Groupes algébriques et corps de classes,* Hermann, Paris (1959).

[SHELAH 1972] Shelah, S., Uniqueness and characterization of prime models over sets for totally transcendental first order theories, *J. Symbolic Logic* 37 (1972), 107-113.

[SHELAH 1973] Shelah, S., Differentially closed fields, *Israel J. Math.* 16 (1973), 314-328.

[SINGER] Singer, M. F., Direct and inverse problems in differential Galois theory, this volume.

[SIT 1974] Sit, W. Y., Typical differential dimension of the intersection of linear differential algebraic groups, *J. Algebra* 32 (1974), 476-487.

[SIT 1975] Sit, W. Y., Differential algebraic subgroups of $SL(2)$ and strong normality in simple extensions, *Amer. J. Math.* 97 (1975), 627-695.

[SIT 1978] Sit, W. Y., Differential dimension polynomials of finitely generated extensions, *Proc. Amer. Math. Soc.* 68 (1978), 251-257.

[SIT 1989] Sit, W. Y., Some comments on term-ordering in Gröbner basis computations, *ACM SIGSAM Bull.* 23 (2), 34-38.

[TATE and VOLOCH 1996)] Tate, J. and F. Voloch, Linear forms in $p-$ adic roots of unity, *Internat. Math. Res. Notices* 12 (1996), 589-601.

[THOMAS 1929] Thomas, J. M., Riquier's Theory, *Ann. of Math* 30 (1929).

[THOMAS 1934] Thomas, J. M., *Differential Systems*, Amer. Math. Soc. Colloq, Publ. 21, Amer. Math. Soc., New York (1934).

[TOMASOVIC 1976] Tomasovic, J. J., A generalized Jacobi conjecture for arbitrary systems of algebraic differential equations, Dissertation, Columbia University (1976).

[TRESSE 1894] Tresse, A., Sur les invariants différentiels des groupes continus de transformations, *Acta Math.* 18 (1894), 1-88.

[VESSIOT 1892] Vessiot, E., Sur les intégrations des équations differéntiellles linéaires, *Ann. Sci. École Norm. Sup.* (3) 9 (1892), 192-280.

[van der WAERDEN 1931] van der Waerden, B. L., *Moderne Algebra,* Vol. 2, Springer-Verlag, Berlin (1931).

[van der WAERDEN 1939] van der Waerden, B. L., *Einfuhrung in die algebraische Geometrie,* Springer, Berlin (1939).

[WANG, D. 1992] Wang, D., Some improvements on Wu's method for solving systems of algebraic equations, *Proc. Int. Workshop Math. Mech.,* Beijing, July 16-18 (1992), 89-100.

[WANG, D. 1995] Wand, D., An implementation of the characteristic set method in Maple, Research Institute for Symbolic Computation, Johannes Kepler University, Linz, Austria, preprint (1995).

[WANG, H. 1981] Wang, H. *Popular Lectures on Mathematical Logic,* Science Press, Beijing (1981).

[WOOD 1973] Wood, C., The model theory of differential fields of characteristic $p \neq 0$, *Proc. Amer. Math. Soc.* 40 (1973), 469-477.

[WOOD 1976] Wood, C., The model theory of differential fields revisited, *Israel J. Math.* 25 (1976), 331-352.

[WU 1978] Wu, W.-t., On the decision problem and the mechanization of theorem proving in elementary geometry, *Sci. Sinica* 21 (1978). 157-179.

[WU 1984] Wu, W.-t., Some recent advances in mechanical theorem proving of geometries, *Automated Theorem Proving: After 25 Years, Contemporary Mathematics,* Vol.29 (1984), 235-242.

[WU 1994] Wu, W.-t., *Mechanical Theorem Proving in Geometries: Basic Principles,* tr from the Chinese by Xiaofan Jin and Dongming Wang, Texts Monog. Comput. Sci., Springer-Verlag, New York, Berlin, Heidelberg (1994).

DEPARTMENT OF MATHEMATICS AND STATISTICS, UNIVERSITY OF NEW MEXICO, ALBUQUERQUE, NM 87131

E-mail address: buium@math.unm.edu

DEPARTMENT OF MATHEMATICS, SMITH COLLEGE, NORTHAMPTON, MA 01063

E-mail address: pcassidy@sophia.smith.edu

Acknowledgments

The American Mathematical Society gratefully acknowledges the kindness of these institutions and individuals in granting the following permissions:

Academic Press, Inc.

Constrained extensions of differential fields, Adv. Math. **12**, 141–170.

Differential equations in a projective space and linear dependence over a projective variety, in Contributions to Analysis: A Collection of Papers Dedicated to Lipman Bers (L. Ahlfors, I. Kra, B. Maskit, L. Nirenberg, eds.), Academic Press, New York and London, 195–214.

Annals of Mathematics

On the exponents of differential ideals, Ann. of Math. **42**, 740–777.

Extensions of differential fields, I, Ann. of Math. **43**, 724–729.

Extensions of differential fields, II, Ann. of Math. **45**, 358–361.

Algebraic matric groups and the Picard–Vessiot theory of homogeneous linear ordinary differential equations, Ann. of Math. **49**, 1–42.

On certain concepts in the theory of algebraic matric groups, Ann. of Math. **49**, 774–789.

Catherine Chevalley

Two proofs of a theorem on algebraic groups (with C. Chevalley), Proc. Amer. Math. Soc. **2**, 126–134.

Elsevier Science Ltd.

Reprinted with permission from *Singular solutions of algebraic differential equations and a lemma of Arnold Shapiro*, Topology **3**, Suppl. **2**, 309–318, 1965, Elsevier Science Ltd., Pergamon Imprint, Oxford, England.

Institut Henri Poincaré

Differential fields and group varieties, First lecture, (mimeographed) Lecture Notes prepared in connection with the Colloque Henri Poincaré, Paris.

Differential fields and group varieties, Second lecture, (mimeographed) Lecture Notes prepared in connection with the Colloque Henri Poincaré, Paris.

Le théorème de la base finie pour les polynomes différentiels, (mimeographed) Lecture Notes prepared in connection with Sém. Dubreil-Pisot **14** (1960/1961), No. 7, Secrétariat Mathématique, Paris.

Japanese Association of Mathematical Sciences

Les corps différentiellement clos, compagnons de route de la théorie des modèles by Bruno Poizat, Math. Japonica **42**, No. 3 (1995), pp. 575–585.

Johns Hopkins University Press

Galois theory of differential fields, Amer. J. Math. **75**, 753–824. Reprinted by permission of the Johns Hopkins University Press.

On the Galois theory of differential fields, Amer. J. Math. **77**, 868–894. Reprinted by permission of the Johns Hopkins University Press.

Algebraic groups and the Galois theory of differential fields (with S. Lang), Amer. J. Math. **80**, 103–110. Reprinted by permission of the Johns Hopkins University Press.

Abelian extensions of differential fields, Amer. J. Math. **82**, 779–790. Reprinted by permission of the Johns Hopkins University Press.

Algebraic groups and algebraic dependence, Amer. J. Math. **90**, 1151–1164. Reprinted by permission of the Johns Hopkins University Press.

Differential polynomials and strongly normal extensions (with T. Soundararajan), Amer. J. Math. **94**, 467–472. Reprinted by permission of the Johns Hopkins University Press.

Kate Kolchin

Algebraic matric groups, Proc. Nat. Acad. Sci. U.S.A. **32**, 306–308.

The Picard–Vessiot theory of homogeneous linear ordinary differential equations, Proc. Nat. Acad. Sci. U.S.A. **32**, 308–311.

Algebraic groups and differential equations, (mimeographed) Lecture Notes prepared in connection with the Conference on Algebraic Geometry and Algebraic Number Theory, University of Chicago, Jan. 24–27, 1949.

Painlevé transcendents, unpublished manuscript.

Pacific Journal of Mathematics

On universal extensions of differential fields, Pacific J. Math. **86**, 139–143.

Serg Lang

Existence of invariant bases (with S. Lang), Proc. Amer. Math. Soc. **11**, 140–148.

Springer-Verlag, New York, Inc.

Differential algebraic groups, in Group Theory (Beijing 1984), Lect. Notes in Math. **1185**, Springer-Verlag, Berlin, Heidelberg, New York, 155–174.

ISBN 0-8218-0542-8

9 780821 805428